T0139365

Genetics

Genetics

Genes, Genomes, and Evolution

Philip Meneely
Haverford College

Rachel Dawes Hoang
Haverford College

Iruka N. Okeke
University of Ibadan, Nigeria

Katherine Heston
Haverford College

OXFORD
UNIVERSITY PRESS

OXFORD
UNIVERSITY PRESS

Great Clarendon Street, Oxford, OX2 6DP,
United Kingdom

Oxford University Press is a department of the University of Oxford.
It furthers the University's objective of excellence in research, scholarship,
and education by publishing worldwide. Oxford is a registered trade mark of
Oxford University Press in the UK and in certain other countries

Impression: 10

Published in the United States of America by Oxford University Press
198 Madison Avenue, New York, NY 10016, United States of America

British Library Cataloguing in Publication Data
Data available

Library of Congress Control Number: 2016953997

ISBN 978–0–19–871255–8 (pbk)
978–0–19–879536–0 (hbk)

Printed in Great Britain by
CPI Group (UK) Ltd, Croydon, CR0 4YY

BRIEF CONTENTS

FULL CONTENTS

PREFACE

Recent advances that allow scientists to quickly and accurately sequence a genome have revolutionized our view of the structure and function of genes as well as our understanding of evolution. A new era of genetics is under way—one that allows us to fully embrace Dobzhansky's famous statement that "Nothing in biology makes sense except in the light of evolution." *Genetics: Genes, Genomes, and Evolution* reflects the excitement of this new era, presenting the fundamental principles of genetics and molecular biology from an evolutionary perspective informed by genome analysis.

By using what has been learned from the analyses of bacterial and eukaryotic genomes as the basis of our book, we are able to unite evolution, genomics, and genetics in one narrative approach. Genome analysis is inherently both molecular and evolutionary, and we approach every chapter from this unified perspective. Thus, rather than relying on separate chapters on genome analysis or evolutionary principles and expecting the student to synthesize them with the principles of classical genetics, we include these as part of each topic. Similarly, the study of genomes has provided a deeper appreciation of the profound relationships between all organisms; we reflect this in our decision not to separate bacterial from eukaryotic evolution, genetics, and genomics. There are chapters in which bacterial genetics, molecular genetics, or evolutionary principles are more prominent, but all chapters include and integrate these concepts.

Audience and approach

This book is written to be an introductory genetics text that emphasizes the connections within and between topics. For several years, we have used drafts of this book in the first semester of our introductory genetics and molecular biology course—a second-year class—at Haverford College, a small US liberal arts institution; the book includes the topics that we are typically able to cover in one semester. The students in this class have usually taken college-level chemistry classes but are just beginning the biology curriculum.

The four of us have co-taught this course numerous times. The resulting synergy has energized our efforts and allowed us to combine our broad, but overlapping, areas of expertise to deliver a student-focused, coherent approach to teaching modern genetics. We have found that students who have completed our introductory genetics class have a highly integrated view of biology and a strong conceptual framework that allows them and us to fill in more detailed information in later upper-level courses. We feel that this integrated approach provides them with a uniquely flexible and contemporary view of genetics, genomics, and evolution.

Instructors and students can choose from numerous introductory genetics or molecular biology books, many of which attempt to cover all aspects of genetics and molecular biology. So what makes ours different? We have tried to maintain an accessible narrative voice and provide analogies and references that engage students; in addition, we have attempted to include the amount of detail appropriate for students at this level and that can be covered in a single term. We have integrated the topics across many fields of biology, including both eukaryotes and bacteria, drawing parallels and comparisons between them. Perhaps most importantly, however, our text uses genomes and the information gained from genome analysis as its foundation, providing a truly contemporary approach to understanding genetics and evolution.

The book does not assume any particular scientific background in biology or other sciences, but most students will likely have a broad background either from a prior college biology or chemistry course or from advanced courses in high school.

Organization

The core of the book covers the topics found in most introductory genetics courses, with a strong molecular biology component needed to understand genome structure and function. These topics are introduced and developed after a discussion of evolutionary history as recorded in the genome, and evolutionary perspectives are emphasized throughout. Thus, you will find chapters that cover traditional topics in introductory genetics such as Mendelian genetics, single and two-factor crosses, X-linkage, pedigree analysis, mapping, meiosis, and linkage, but this coverage has been integrated with information about genomes from a wide range of organisms and viewed through the lens of evolution.

A chapter-by-chapter guide

We begin the Prologue of this book with the Five Great Ideas of Biology, as outlined by Sir Paul Nurse. We use these ideas throughout the text to interconnect concepts about the chemistry of biological molecules, the idea of a gene, cells as basic building blocks of life, the organization of living systems, and natural selection and evolution. As such, these ideas provide a framework for the information presented in the text. We have found that many students are adept at acquiring biological information on specific topics but are challenged when asked to make connections between concepts and topics. However, once these integration skills have been emphasized and developed, the depth of understanding increases dramatically. We therefore model this approach explicitly throughout the text.

In **Chapter 1**, we introduce a recent study of the genomes of Darwin's finches that identified genes that contribute to the differences in beak shape among these species. In this way, we link one of the most important and familiar examples of natural selection among Darwin's finches with the underlying genetic and genomic basis for the differences observed among these birds. Using this example in our opening chapter prepares students to think across scales, from DNA to molecules to phenotypes to species and then to evolution in a community of organisms. This perspective has only become possible because of our ability to sequence and compare genomes.

We then move on to describe in **Chapter 2** the structure of DNA and the Central Dogma, highlighting features of the DNA molecule that have an impact on its function. We point out modifications to the traditional view of the Central Dogma and introduce students to the basic structure of a gene. **Chapter 3** covers the structure of the genome and the variation in genome organization found in different species, which are both the outcome of, and the ingredients for, natural selection. This chapter discusses the structure of chromosomes, extrachromosomal DNA, and changes in the genome, and is one of the most important for integrating genomic findings with evolutionary and genetic principles.

In **Chapter 4**, we discuss DNA replication and repair, tying these topics to Darwin's concept of "descent with modification," a fundamental principle of natural selection and evolution. Several challenges to DNA replication that arise from its length and anti-parallel structure are presented, along with the processes that have evolved to address these challenges. We introduce the many types of mutation that occur and how the occurrence of such mutations and the effects of selection can be revealed by comparing the genomes of different organisms. These comparisons provide us with the ability to reconstruct evolutionary history and develop phylogenetic trees based on DNA sequence changes.

The next section of the text, comprising **Chapters 5 to 9**, discusses Mendelian genetics, meiosis, the inheritance of two genes, sex-linked traits, and linkage and mapping. These chapters present all the material found in a traditional genetics course, framed in the context of genomics and evolution. For example, the relationship of the process of meiosis to Mendel's Laws of Segregation and independent assortment is explored in a progressive approach that encourages students to think between topics and across scales.

We then bring these fundamental concepts of genetics and genomes together in **Chapter 10** on complex traits and genome-wide association studies, which focuses primarily on human traits and diseases. We explore how these studies build upon the basic principles outlined in earlier chapters, to identify contributing genes and causative mutations. While chapters on complex traits are found in many books, an approach that shows how genome-wide associations integrate genomic variation, complex phenotypes, and evolutionary history is novel.

Chapter 11 introduces the process of horizontal gene transfer, originally found among bacteria, but now known to occur in other types of organisms too. **Chapters 12 and 13** focus on the essential processes of transcription and translation that link genotype to phenotype.

In **Chapter 14**, operons in bacteria are introduced as an example of gene regulatory networks and are followed by a discussion of transcriptional regulatory networks in eukaryotic organisms, using recent information from many genome annotation projects such as ENCODE. The tools of genetic analysis are covered in **Chapter 15**, beginning with Beadle and Tatum's experiments in *Neurospora* and Jacob and Monod's work with *lac* mutants, and moving into a discussion of more recent genetic screens for the identification of genes essential to embryonic development in *Drosophila*.

The final section of the text moves beyond individuals to populations and communities of organisms. Population genetics with a human focus is the subject of **Chapter 16** where we explore the assumptions of the Hardy–Weinberg equilibrium model. We review the many different types of evolutionary change that can operate, in addition to natural selection, to shape the genetic structure of a population and the imprints these leave at the level of the genome. Long-term studies of bacterial populations are featured as a method to explore evolution experimentally. **Chapter 17** concludes the text by introducing the relatively new field of metagenomics, in which genomic information is extracted directly from communities of organisms living in their natural environments, revealing evidence for their interdependence and co-evolution.

Pedagogical features

We have used many pedagogical features in the text to improve the accessibility for students and allow instructors to tailor each chapter for their own courses. Each chapter begins with an introductory paragraph—**IN A NUTSHELL**—which provides a brief preview of the upcoming chapter content, while **KEY POINTS** highlighting important concepts are found throughout the text. A summary paragraph concludes each chapter, followed by a **CHAPTER CAPSULE**, which presents a bulleted list of the important ideas we have discussed.

Boxes that expand on specific topics are found in every chapter and are grouped into several categories. **COMMUNICATING GENETICS** boxes present the terminology, notation, and diagrammatic conventions used to communicate genetics concepts. **TOOL BOXES** illustrate how the biological processes described in the chapter form the bases for essential experimental techniques. Other boxes feature ***A Human Angle, An Historical Perspective,*** or ***Quantitative Toolkits*** to a topic presented in the chapter, each giving a different flavor to the study of genetics. The ***Going Deeper*** boxes provide more advanced content about a process or system than is included in the main text of each chapter to allow instructors to include more details without losing the connections between the topics. We regard all of these boxes as an essential feature of the book, allowing each instructor to adjust the content for the particular audience and emphasis in different classes.

Every chapter also includes **study questions and problems**, for which all of the solutions are provided online. One goal of these questions is to have students work through examples that integrate more than one concept introduced in the chapter or from more than one chapter. We use realistic examples in the problems; since few experimental problems map neatly onto the concepts from only one chapter, we ask the students to do a similar integration of ideas with the result that they will gain a more unified view of the subject. For example, our chapter on meiosis has questions that draw on natural variation, Mendelian principles, and sex determination and sex linkage—concepts that were introduced in previous chapters or that will be covered in the next chapter. In addition, problems on the comparative genomics of Darwin's finches are contained in several chapters of the book to emphasize the connections to what has been covered before.

Supplements

Accompanying online resources include the solutions for all of the study problems, instructional videos for each chapter, and Journal Clubs that lead the class through a structured discussion of a publication from the scientific literature.

The **videos** have one of us walking students through the solution of a problem from the text, highlighting a research technique or providing additional analysis of an original research figure in a 5-minute summary.

The **Journal Clubs** offer the possibility of an active learning environment that reverses the typical pedagogical process in most large introductory courses. Class time could be spent discussing the Journal Club, reinforcing students' understanding of the chapter by exploring a more detailed example from the original literature, for instance.

We describe the learning features and supplements package in more detail in our guide to Teaching and Learning with *Genetics* on **p. xviii**.

Acknowledgements

We would like to sincerely acknowledge all the helpful feedback we have received from our students in the years since we first began purposefully reconfiguring our introductory genetics class to integrate genomics and evolution more fully. The assistance of the students in the class, the teaching assistants, and the peer tutors has been indispensable. They have encouraged us to improve and clarify our explanations, and have provided input on figures, study questions, and solutions. The students' responses have shaped every part of this book, and we regard them as contributors.

We would also like to thank our colleagues in the Biology Department at Haverford College for their suggestions and their patience as the four of us worked to make this idea a reality. Others at Haverford College have provided assistance as well, including those who provided technical support in creating videos and helped with the administrative tasks associated with this project.

We would like to thank our editor, Jonathan Crowe, for his insightful suggestions, his patience, and his unflagging support of this endeavor.

Finally, we want to acknowledge the steadfast support of our friends and families during the development of this text.

Haverford, PA
March 2017

TEACHING AND LEARNING WITH
GENETICS: GENES, GENOMES, AND EVOLUTION

Genetics offers a range of features and resources—in print and online—to make your study of genetics as effective as possible. We outline the key features in this section.

Using *Genetics* to remember the key ideas

The sheer amount of information we now know about the field of genetics can often be overwhelming; you may find yourself wondering 'what do I really need to remember to make sense of the subject?' *Genetics* offers several features to help you identify and remember the key ideas.

In a nutshell

Each chapter opens with a succinct overview of what the chapter covers to give you a mental map of the topics that follow.

IN A NUTSHELL

The DNA sequence of an organism, know
information that underpins and directs m

Key points

Key point panels throughout each chapter highlight important concepts right after they are introduced.

KEY POINT DNA replication and repair provide a balance between opposing evolutionary pressures for accurate transmission of the genome to the next generation and the opportunity for production of new genomic variants.

Chapter capsule

Each chapter ends with a chapter capsule that recapitulates the key themes of the chapter in any easy-to-review format—perfect for revision and exam preparation.

CHAPTER CAPSULE

- The molecular processes of replication d
  with modification occurs.

Using *Genetics* to master challenging concepts

As you study genetics, you may find some concepts easy to understand, even when encountering them for the first time. You may find others more challenging, and be left wishing for a little more support to really make sense of them. *Genetics* is on hand to help.

Video tutorials

Available free online, the authors have produced a series of videos that provide deeper, step-by-step explanations of a range of topics featured in the text. Think of these videos as one-on-one tutorials. Perfect when you're after that little extra guidance!

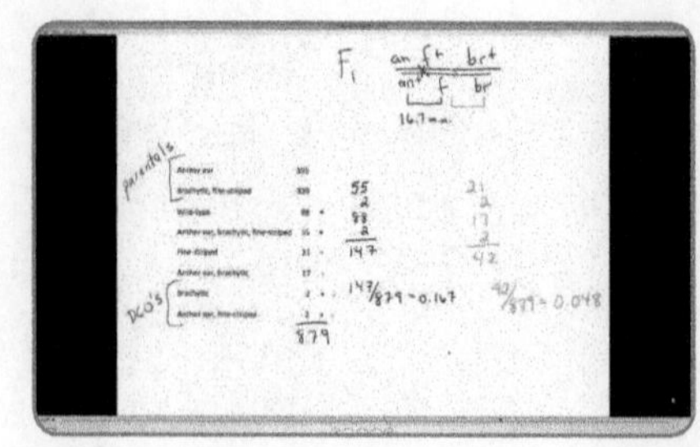

Communicating genetics panels

The language of genetics can often appear bewildering when you encounter it for the first time. *Communicating genetics* panels are written to help you avoid confusion, and master the terminology and notation that is used to communicate genetic ideas and concepts to others.

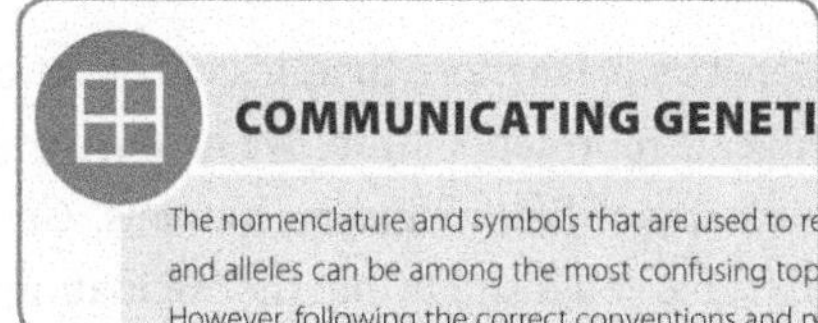
COMMUNICATING GENETI

The nomenclature and symbols that are used to re
and alleles can be among the most confusing top
However, following the correct conventions and n

Using *Genetics* to develop problem solving and critical thinking skills

Our current understanding of genetics is the outcome of the work of many individuals over many years, all of whom have been united by a spirit of exploration and enquiry. *Genetics* offers a range of features to encourage you to explore and enquire for yourself.

Study questions

Every chapter has an extensive set of study questions and problems, many of which stimulate you to make connections between different parts of the subject.

The study questions are grouped into three categories:

Concepts and definitions—these questions prompt you to recall information discussed in the chapter in order to reinforce your learning of key principles and terminology.

Beyond the concepts—these are more challenging questions that encourage you to think more critically about the topic under discussion. These questions may require you to draw from material presented in previous chapters.

Applying the concepts—these questions require you to draw on and apply concepts you have studied in order to understand and interpret potentially unfamiliar biological problems or scenarios, often involving more than a single concept or topic. These questions may also prompt you to look ahead to concepts in upcoming chapters.

STUDY QUESTIONS

Concepts and Definitions

8.1 Which of the rules of probability is important genotypic ratios from crosses involving more

8.2 What are the similarities and differences amo polymorphism? (A more precise technical de Chapter 10, so this is asking for a conceptual

8.3 Why are dominant lethal traits not passed on reproduction?

8.4 What is a complementation test, and why is a

8.5 What is epistasis, and how does it differ from

8.6 What are the molecular or cellular origins of p common?

Beyond the Concepts

8.7 Diagram how the orientation of two different metaphase I of meiosis leads to the expected unlinked genes.

8.8 In a cross involving parents who are heterozy many boxes will be found in a Punnett square

8.9 Geneticists working with model organisms in

Journal Clubs

Journal Clubs, provided online, feature a series of questions that guide you through the reading and interpretation of a research paper that relates to the subject matter of a given chapter. The Journal Clubs help to reinforce your understanding of a given topic by working through a more detailed example—and help you to read and interpret the primary literature with more confidence.

Chapter 10 Journal Club

Paper: Sutter et al., 2007. A single IGF1 allele is a major determinant of small size in dogs. *Science* **316**: 112-115.

Background information:

This paper uses the principles of positional cloning and genome association studies to identify a gene that appears to play a significant role in the size of different dog breeds. In addition to working with a familiar and easily measured phenotype, the paper illustrates many of the important techniques and strategies that are important in genome-wide association studies (GWAS) and other mapping and association-based approaches to find the causative genes for phenotypes. Once you have read the paper and supporting information, be prepared to answer the following questions. Solutions to the questions are available to registered instructors.

Using *Genetics* to go beyond the basics

Genetics is far more than just a collection of abstract theories and knowledge: it represents a tool we can use to further our understanding of the diversity of life, including humans. *Genetics* includes a range of features that give you insights into the fascinating world of genetics beyond the basic concepts.

Tool Boxes

Our growing understanding of the behavior of genes and genomes at a molecular level has enabled us to develop new tools for exploring how life operates at the level of genetics. *Tool Boxes* throughout the book introduce many of the key experimental tools and techniques that are used to reveal more about the structure and function of genetic systems, and the way they evolve over time.

TOOL BOX 2.4 cDNA

DNA molecules are usually very stable and no
work with in the lab. While many enzymes kn
ases (or, more formally, deoxyribonucleases sinc
DNA) are capable of cutting or digesting DNA r
activity of most nucleases that work on DNA can

Boxes

Boxes throughout are written to enrich the main text with additional perspectives and insights. These boxes explore four themes:

A Human Angle

What is the genetic basis of color blindness? How can genetics be used to analyze the Black Death? Genetic concepts influence the lives of every organism on the planet, and humans are no exception. The *A Human Angle* panels explore a range of intriguing questions about the impact of genetics on humans.

BOX 9.2 *A Human Angle* Red

Mutations that arise spontaneously usually occu
unrepaired errors during replication, as discussed i
as transposable element movement, as discusse
However, a few mutant alleles arise during reco
most familiar example is red–green color blindn
The same process that produces red–green colo
also been an important evolutionary force for sha
and gene families.

The primary photoreceptors in our retina are a
teins known as the opsin proteins; the opsins in
absorb light of different wavelengths and allow
ors. Opsins that absorb short wavelengths are th
the blue opsin gene is autosomal and is not in
green color blindness. Opsins that absorb lon
are called the red opsins, while those that ab
wavelengths are called the green opsins; stri
the maximum absorbance for the red opsins is

Quantitative Toolkits

Quantitative tools are an essential element of scientific enquiry, yet many students lack the confidence to use these tools in practice. The *Quantitative Toolkit* panels are designed to remove this 'fear factor', explaining in a clear, straightforward way a number of quantitative topics that help us understand genetics.

Going Deeper

It is impossible to distil all we know about genetics into a single textbook. But your studies may have left you wanting to know that little bit more. The *Going Deeper* panels are designed to give you a taste of genetics beyond an introductory course by exploring some more challenging—but intellectually stimulating—topics.

An Historical Perspective

Our current understanding of genetics has been shaped by experimentation, enquiry, and the refinement of ideas over the course of several hundred years. The *An Historical Perspective* panels reveal some of the historical events that have influenced genetics as we know it today.

BOX 5.2 *An Historical Perspec*

In 1936, the statistician and geneticist Ronald A. Fi
Mendel's results from his experiments with peas
to be true" in how well they fit his hypothesis ab
tance of traits, that is, any experiment that involv
samples also requires some margin of error in the
flip a coin, we expect that it will come up heads ha
tails half the time, but we are not concerned if, af
got 11 heads and nine tails, rather than ten of ea
always exactly conform to our expectations. Fis
that, in effect, the variations in Mendel's data wer
that Mendel must have biased his results—or wors
quite a controversy then and is still debated today,
is no evidence that Mendel deliberately altered his
data are statistically closer than expected to the 3:
anticipated from a cross of two heterozygotes.

At a distance of 150 years and with no access
experiments and methodology other than the da
presented in his paper, it is impossible to know

SUPPORT PACKAGE

Oxford University Press offers a comprehensive ancillary package for instructors and students using *Genetics: Genes, Genomes, and Evolution*. Go to www.oup.com/uk/meneely to find out more.

For students

- *Video tutorials:* A series of videos that provide deeper, step-by-step explanations of a range of topics featured in the text.
- *Flashcards:* Electronic flashcards covering the key terms from the text.
- *Further reading:* Suggested research and review articles that build on topics discussed in the text, to take your learning one step further.

For registered adopters of the text

- *Digital image library:* Includes electronic files in PowerPoint format of every illustration, photo, graph and table from the text
- *Lecture notes:* Editable lecture notes in PowerPoint format for each chapter help make preparing lectures faster and easier than ever. Each chapter's presentation includes a succinct outline of key concepts, and incorporates the graphics from the chapter
- *Library of exam-style questions:* A suite of questions from which you can pick class assignments and exams.
- *Solutions to all questions featured in the book:* Solutions written by the authors help make the grading of homework assignments easier.
- *Journal Clubs:* A series of questions that guide your students through the reading and interpretation of a research paper that relates to the subject matter of a given chapter. Each Journal Club includes model answers for instructors.
- *Instructor's guide:* The instructor's guide discusses the educational approach taken by *Genetics: Genes, Genomes, and Evolution* in more detail, why this approach has been taken, what benefits it offers, and how it can be adopted in your class.

Oxford University Press's **Dashboard learning management system** features a streamlined interface that connects instructors and students with the functions that they perform most often, simplifying the learning experience in order to save instructors' time and put students' progress first. Dashboard's pre-built assessments were created specifically to accompany *Genetics*, and are automatically graded so that instructors can see student progress instantly. Section-level progress reporting in the gradebook gives instructors a quick indication of which areas of each chapter students are having the most difficulty with, making office hours more effective and guiding lecture and class time planning.

Dashboard includes a quiz for each chapter of the textbook. These quizzes are designed to be used by the instructor as an assessment of student mastery of the important facts and concepts introduced in the chapter, after the student has read the chapter and attended the relevant lecture/class period/discussion section.

ABOUT THE AUTHORS OF *GENETICS: GENES, GENOMES, AND EVOLUTION*

Philip Meneely (PhD, U of Minnesota) is a Professor of Biology at Haverford College where he has taught both introductory and advanced genetics for more than 20 years, as well as courses in genomics and bioinformatics. He previously was on the faculty of the Fred Hutchinson Cancer Research Center. His research with *C. elegans* has included publications on chromosome rearrangements, polyploidy, meiosis, sex determination, dosage compensation, and gene interactions. He is also the author of *Genetic Analysis: Genes, Genomes, and Networks in Eukaryotes* (Oxford University Press), now in its second edition, which was shortlisted by the Royal Society of Biology (London) in 2015 for Undergraduate Biology Textbook of the Year.

Rachel Dawes Hoang (Ph.D. Cambridge University, UK) is an Associate Professor in the Biology Department at Haverford College. She has published research and review articles in the fields of developmental biology and evolutionary developmental biology. Her current research investigates the evolution of genes controlling cell shape changes as well as the interactions between endosymbiotic bacteria and host cells during embryonic development of insects. She regularly teaches courses in genetics, evolution, and development. She is currently the chair of the Biology Department. She is a former Helen Hay Whitney fellow.

Iruka N. Okeke (PhD, Obafemi Awolowo University, Nigeria) taught biology at Haverford College, PA, USA from 2002 until 2014. She is presently Professor of Pharmaceutical Microbiology at the University of Ibadan, Nigeria and has also taught in other African and UK applied health programs. Her research on bacterial genetics and microbiology focuses on intestinal pathogens and on antimicrobial resistance. She is co-author of two books and approximately one hundred articles and chapters. She has been the recipient of Fulbright, Branco Weiss and Institute for Advanced Study (Berlin) Fellowships. Okeke serves on editorial, higher education, health policy and science policy advisory panels and boards in the US, Europe and Africa.

Katherine Heston (M.S. University of Wisconsin, A.B. Princeton University) is an Instructor in Biology at Haverford College. She has been involved in teaching undergraduates for thirty years at Lake Forest College (IL), Northwestern University, Villanova University, and Haverford College. Her teaching background includes botany, ecology, genetics, cell and molecular biology. Working closely with students in the teaching lab has developed her sense of the student perspective, which informs her development of effective teaching materials for her classes.

MANUSCRIPT REVIEWERS

The authors and publisher gratefully thank the academic reviewers whose invaluable feedback during the preparation of *Genetics* helped to shape the final manuscript. In addition to a number of anonymous reviewers, we thank:

William Bradley Barbazuk, University of Florida

Claire Cronmiller, University of Virginia

Wayne Forrester, Indiana University

Michelle F. Gaudette, Tufts University

Michael A. D. Goodisman, Georgia Institute of Technology

J. L. Henriksen, Bellevue University

Margaret Hollingsworth, University at Buffalo, The State University of New York

Barbara C. Hoopes, Colgate University

Eric Liebl, Denison University

Jorge Mena-Ali, Franklin and Marshall College

Mary Montgomery, Macalester College

Lewis E. Obermiller, Arizona State University

Aaron Schrey, Armstrong State University

Len Seligman, Pomona College

Clarissa Shearer, Bellevue University

Edwin Stephenson, University of Alabama

James N. Thompson, Jr., University of Oklahoma

Prologue

Genetics, genomes, and the Great Ideas of Biology

On 26 June 2000, then President of the United States Bill Clinton and then Prime Minister of the United Kingdom Tony Blair held a joint press conference to announce the completion of the first draft of the human genome. This announcement heralded the dawn of the post-genome era, bringing with it the promise of a greater understanding of human diseases and their effective diagnosis and treatment.

Nearly two decades later, genome sequence information is now widely available for thousands of organisms. This recent development has resulted in an explosion in our knowledge and understanding of many areas of biology. One practical outcome of all of this biological information is that it can be easy to feel overwhelmed by the amount of data and the pace of new discoveries. At a time when there is the most to be learned and the excitement is the greatest, the sheer volume of information makes learning biology challenging; some of the most interesting and important insights seem inaccessible. How we can hope to comprehend it all?

The short answer is that no one can learn and grasp all of it. But the good news is that accurate and up-to-date information is also readily accessible from our computers, phones, and tablets. We can find biological information as easily as we can find a review for a nearby restaurant. However, in order to navigate and make sense of all of this online information, we need a solid foundation in the important principles of biology, especially genetics. Part of our goal in this book is to help you understand some of the principles that unify biology and their interconnectedness.

Great Ideas and unifying themes

This text focuses on inheritance and genetic information. But what you learn about genetics and genomics doesn't operate in some abstract setting. It happens in a particular biological context—often a cell, a tissue, an organism, or a population—and it happens over evolutionary time. To properly understand genetics and genomics, we need to appreciate how it fits into the bigger picture of biology.

The science of biology is very broad, but the concepts that will unfold in subsequent chapters are integral to an understanding of living organisms and their function. We will make connections between genetics and other fields of biology, referring to a few basic themes as a framework for our discussions throughout this book.

Many scientists have identified principles that they see as unifying biology. We will use the Nobel Laureate Paul Nurse's unifying themes of biology as our starting point in this book. You can see a summary of Nurse's ideas in Figure P.1.

According to Professor Nurse, the Five Great Ideas of Biology are:

- The cell as the basic unit of living systems
- The gene as the unit of heredity
- Life as chemistry (most, if not all, of the fundamental biological processes that occur in living organisms arise from the chemical components that comprise the organism and the reactions that occur among them)
- Biology as an organized system or as sets of organized systems
- Evolution by natural selection.

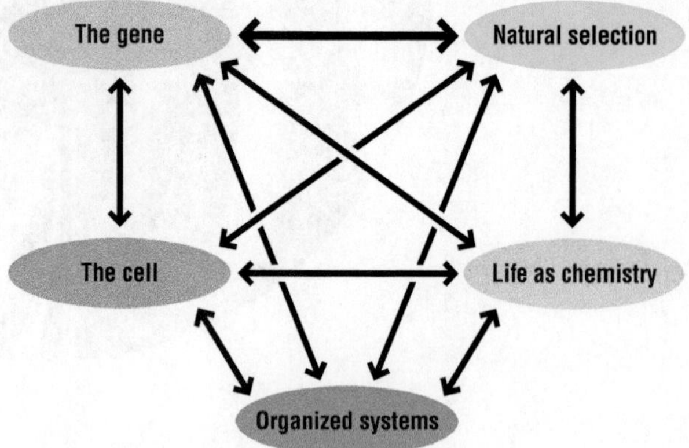

Figure P.1 The Five Great Ideas that unify biology, according to the Nobel Laureate Paul Nurse. These ideas have many connections, as indicated by the arrows, which provide the actual unifying framework.

WEBLINK: You can find a video of Nurse's lecture on the Five Great Ideas of Biology and a written version of the talk at **http://online.itp.ucsb.edu/online/plecture/nurse/**.

It is important that we do not treat these as separate and unrelated ideas but instead consider the interactions among them. Part of their greatness lies in how they are connected to one another, indicated by the arrows in Figure P.1. Although not all of the connections among these ideas are important for every topic, we will keep each of these themes and the connections in mind as we continue our exploration of genetics.

Let's now consider each of Nurse's Great Ideas in turn.

The cell as the basic unit of living systems

All organisms are composed of cells. Some organisms are capable of life as a single cell—they exhibit the principles of inheritance, chemistry, organization, and natural selection with one cell. These organisms are microbial, meaning that microscopes are required to see them. (Not all microbes are single-celled, but virtually all single-celled organisms are microbial.) Other organisms may consist of hundreds, thousands, tens of thousands, or millions of cells. The same processes of life occur in multicellular and unicellular organisms, and they occur in the organism as a whole as well as in individual cells. All of the processes that are described in this book occur in the context of cells.

The structure of a cell

Cells are both the structural and functional units of organisms, so it is important to recognize their internal organization. The component parts of a cell do not slosh around in a gelatinous cytoplasm, bumping into other components as they move, but are instead highly organized, with extensive intracellular communication and interactions. A diagram of an animal cell is shown in Figure P.2.

The key part of the cell is the information content contained in its DNA, primarily in the sequence of its nucleotide bases. DNA and the flow of information in the cell are the main topic of Chapter 2. From a cellular perspective, a key difference among cells is whether or not the DNA is physically separated from the other components or organelles in the cell. This makes the distinction between bacteria and archaea on the one hand (sometimes referred to as **prokaryotes**) and **eukaryotes** on the other. In bacteria, as well as archaea, there is no physical barrier between the DNA and most other cellular components. As a consequence, DNA is housed in the same compartment in which it is expressed. All multicellular organisms and the eukaryotic microbes have their DNA physically separated from the other components of the cell by a specialized structure called the nucleus. The nucleus is surrounded by a membrane consisting of two different layers; because the membrane is used to enclose the nucleus away from other cellular components, it is also called the nuclear envelope.

DNA is the repository for all of the information in the cell, but it is not itself the molecule that carries out cellular functions. Cellular activities require proteins as the principal (although by no means the only) molecules by which the information in DNA is turned into function. A mammal, such as a human, is estimated to make more than 70,000 proteins, although only a fraction of these are made in one type of cell at one particular time. Proteins are not made directly from the DNA sequence. The intermediate between the information content of DNA and the functional activity of proteins is RNA, as we describe in Chapter 2. In other words, a gene exerts its function via an RNA copy, which can be deployed and later degraded while the original DNA sequence is left intact.

There are many subcellular structures within a cell. Within the cytoplasm of both bacteria and eukaryotes,

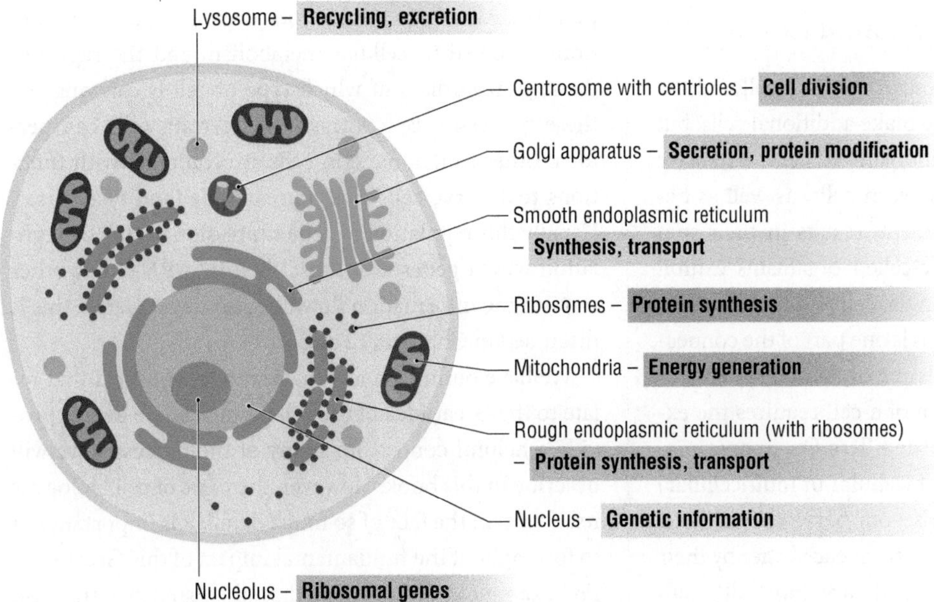

Figure P.2 The basic features of an animal cell. The cell membrane that separates the cell from its environment is a double layer of lipids. Within the cytoplasm are various organelles and structures, some of whose functions are summarized here. The nucleus, in pale orange, has its own double-layered membrane that separates the DNA of the genome from the cytoplasm. This diagram summarizes only the most general features of animal cells, and most cells have other specialized structures.

the translation from RNA to protein happens in the ribosome, a complex intracellular structure made up of proteins and RNA molecules. This process is described in more detail in Chapters 12 and 13.

Cell division

Living organisms can reproduce. Since living organisms consist of cells, the cell must also be capable of reproduction or cell division and must transmit its information content to the next generation of daughter cells, so that no information is lost. Thus, cell division begins with the duplication or replication of the DNA, a process described in Chapter 4. Cell division involves dividing both the DNA content and the cell itself; most of our attention in this book will be on the DNA content.

In bacteria, cell division occurs by binary fission, that is, simply by a cell splitting in half. The DNA replicates itself and attaches the replicated molecules to the cell membrane; as it is doing so, the cell splits into two, and the DNA molecules are segregated to each daughter cell. Thus, cell division generally involves a series of doubling steps, from one cell to two, two cells to four, and so on.

Virtually all cellular organisms have a cell cycle, part of which is the organized and specialized process of cell division. The organization is probably most apparent with eukaryotic cell division, known as mitosis. In a single-celled eukaryote, mitosis creates new cells that have the same DNA content; these daughter cells can form a loosely organized colony of cells, each of which could divide and grow on its own. Mitotic growth is also called **vegetative growth**, particularly in fungi and plants, or **asexual growth**. Bacteria and archaea also reproduce vegetatively by a process that is quite different from mitosis, although conceptually similar to it.

In animals, particular groups of cells form a reproductive organ—an ovary or a testis—which is set aside from the rest of the body. Within the reproductive organs, some cells divide by mitosis, while other cells become specialized into gametes to form the next generation; this requires a specialized form of cell division known as meiosis. The meiotic cells involved in reproduction are also known as the germ line or germ cells. All other cells that divide by mitosis are known as the somatic cells. Both mitosis and meiosis are discussed in more detail in Chapter 6. Eukaryotes begin as a single cell arising from the union of two other cells, such as a sperm and an ovum, or a pollen grain and an egg sac (or the female gametophyte). Growth from this single-celled zygote into a differentiated metazoan involves many mitotic divisions.

Cell specialization and communication

We have briefly described the structure of a cell and the process by which cells divide to make additional cells, but no organism is just a ball of unorganized cells. Extensive communication within and between cells, as well as between cells and their environment, results in their specialized functions. Even single-celled organisms exhibit functional specialization within the cell and in their communication with other cells. This is one part of the connection between the themes of cells and organized systems.

The functional specialization of a cell requires the expression of its genes to be regulated. To a first approximation, all cells in an organism (unicellular or multicellular) have the same DNA sequence. Yet our nerve cells and our skin cells are easily distinguished from each other by their shapes, positions in the body, and functions, although they contain nearly the same DNA sequence; in fact, any somatic cell in a multicellular organism carries the entire genome of that organism. How then do cells bearing the same genetic information become specialized in form and function?

The answer is that any cell expresses only a fraction of its total DNA sequence and thus is carrying out just some of the potential functions it has the information to perform. All cells in a eukaryotic organism perform functions involved in cellular metabolism and the repair of damage, regardless of which type of cell is carrying out these processes. By contrast, only certain cells can perform other functions; skin cells are equipped with functions that nerve cells lack, for example (and vice versa). Usually this regulation of gene expression occurs by regulating which genes are transcribed into RNA, but other steps in the information flow can also be regulated; this is discussed in Chapters 12 and 14.

We have outlined a few foundational concepts that relate to the Great Idea of the cell because these details provide a helpful context for many of the processes we will describe in this book. However, the topic of cell biology is itself vast. In the face of so much detail, it is important not to lose sight of the fundamental impact of this Great Idea. For example, the very fact that all organisms are built from the same basic building block—the cell—is a consequence of our shared evolutionary history, that is, that all life on earth arose from a common ancestor. Similarly, all of the cellular components we have described are themselves encoded by genes. Relationships between genes for basic cellular components can, in fact, be traced across the entire tree of life, reflecting the intimate connections between genes, genomes, and evolution.

The gene as the unit of heredity

Living things reproduce, and the offspring are similar (although not identical) to their parents. This is heredity, and the gene as the unit of heredity is another Great Idea. Genes are our major topic in this text, and we will discuss them from many different perspectives. For Mendelian genetics, the gene is the unit of inheritance and is usually described in terms of the phenotype or appearance of the organism. It now comes as no surprise that the molecule of heredity is DNA. In the 1920s, Griffith showed that a chemical substance referred to as the "transforming principle" could transfer heritable characteristics between different strains of bacteria; about 15 years later, Avery and colleagues demonstrated that this transforming principle was DNA.

The fact that genes are made up of DNA allows us, and even requires us, to explain the activities of genes in terms of biochemical processes. Genes have to be copied and transmitted to the next generation, and genes have to be actively expressed to produce phenotypes. Both of these activities are inherent in the biochemical properties of DNA. Heredity—and thus genes—also connects directly to the processes that happen during evolutionary change, illustrating another set of connections among the Great Ideas.

Life as chemistry

We can't think fully about biological processes, unless we can also think about some of the biochemical and molecular principles that make them possible. We will introduce a few chemical principles here, and others will be discussed throughout the book as needed.

The structures of molecules affect their functions, and vice versa

One chemical principle that arises repeatedly is that structure and function are inextricably connected to each another. The structure of a molecule—such details as precisely where a hydroxyl group (represented by-OH) is attached—profoundly affects its function. As discussed more fully in Chapter 2, the nucleotide bases in DNA can pair with one another. Adenines (As) pair with thymines (Ts), and cytosines (Cs) pair with guanines (Gs). This pairing is crucial to the structure and function of DNA. Figure P.3 shows the chemical structures of base pairs of DNA, with the interactions between the bases (the hydrogen bonds) illustrated with pink dashed lines.

Note that the chemical structures of the nucleotides are central to the way specific base pairs are formed; in turn, this specific base pairing—A:T and C:G—is central to the way DNA is both inherited and expressed. Adenine does not interact with cytosine as stably as it does with thymine, and a base pair between adenine and guanine would require a distortion of the entire DNA structure. In other words, if the structures of the molecules or macromolecules are different, the function may also be affected.

Conversely, the functions of a molecule or a larger macromolecule also affect its structure. This illustrates an important connection between biochemistry and evolution.

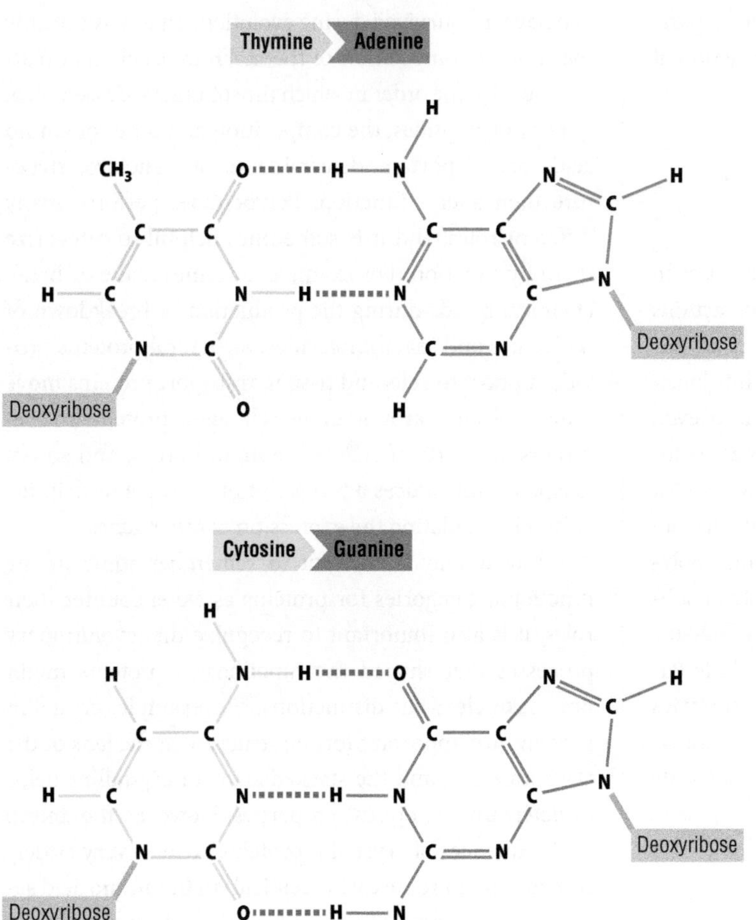

Figure P.3 The chemical structures of the nucleotide bases of DNA and their base pairing. Note that adenine and guanine have two rings in their structures, while thymine and cytosine have one ring. In DNA, adenine pairs with thymine and guanine pairs with cytosine because of the chemical interactions between the atoms on the rings; these chemical interactions are known as hydrogen bonds and are represented by the dashed pink lines. The specificity of the base pairing underlies both the expression of the gene and the inheritance of the gene, and comes from the chemical structures of the rings.

All biological processes are the outcome of evolution and thus are subject to change. The functions of a macromolecule or a cellular structure are what matters most for the cell's or organism's survival and reproduction; the functions are recognized and acted on by evolutionary forces. Although evolutionary forces may not directly affect the structure—any structure that functions well enough could survive—they can have an indirect impact on the structure. The functions of a molecule, including its associations with other molecules into macromolecules, can allow or constrain the amount of change in its structure. As a result, evolutionary comparisons between molecules can often allow us to recognize the parts of a structure that are essential to its function because those parts are the least likely to change; in other words, they are often the most **evolutionarily conserved**. While evolutionary conservation does not always indicate which parts of a molecule's structure are important for its function, it makes a useful starting point.

Macromolecular polymers are made from monomer subunits

A second important principle of chemistry that arises in biological processes relates to the assembly or interactions of chemical moieties with one another. Atoms are assembled into molecules, molecules are assembled into macromolecules, macromolecules are assembled into even larger macromolecules (sometimes referred to as nanomachines), and so on. For biological processes, we can list four major types of monomer molecular subunits that are assembled into different types of macromolecular polymers. These are listed in Table P.1. The assembly of subunits into larger macromolecules is known as **anabolism**.

While there are four types of monomers in Table P.1, each of those monomer subunits has different subtypes as well. Polymers are assembled from different combinations of subunit types, and the sequence of the subunits is usually (if not always) essential for the function of the macromolecule. For example, the monomer subunit for DNA is the deoxynucleotide base. Four different deoxynucleotide bases make up DNA, and a different, but related, four make up RNA. Those four bases can generate 16 (4^2) different dinucleotides and 64 (4^3) different trinucleotides. Both the specificity and the differences arising from DNA stem from the fact that sequence is important; although TGA and GAT use the same three bases, they do not convey the same information.

Table P.1 The building blocks of macromolecules

Monomer subunit	Polymer macromolecule
Nucleotide bases	Nucleic acids—DNA and RNA
Amino acids	Polypeptides
Sugars	Carbohydrates
Fatty acids	Lipids

Similarly, 20 different amino acids are used to make a polypeptide, and an average polypeptide comprises about 550 amino acids. That means that 20^{550} amino acid sequences are possible; in principle, each of these sequences could have slightly or dramatically different functions; the vast, vast majority of them have no biochemical function and have not survived during evolution. Thus, it is not only the subunit composition of the macromolecule that matters, but also the order in which the subunits are assembled.

For polypeptides, the composition and order of amino acids are important determinants of structure; structure then affects function. Polypeptides perform many different roles, and it is sometimes helpful to categorize them by function. For example, enzymes make or break chemical bonds during the production or breakdown of molecules and macromolecules. Structural proteins provide support to cells and tissues, transport proteins move other molecules around in the cell, signal proteins and receptors are parts of cellular communication, and so on. Chapter 1 introduces a transcription factor, a protein involved in regulating the expression of other genes.

While it may be helpful to remember some of the functional categories for proteins as we encounter their roles, it is also important to recognize that evolutionary processes that shaped the functions of proteins might not make clear-cut distinctions. For example, crystallin proteins are important for the structure of the lens of the vertebrate eye, and the stacked array of crystallins helps to determine its optical properties. However, the amino acid sequences of crystallin proteins are also very closely related (and in some cases identical) to the amino acid sequences of important enzymes in the cell. In mammals, for instance, the sequence of one of the crystallin proteins that allows you to focus on, and read, these words is very similar to, and probably derived from, the enzyme alcohol dehydrogenase, which helps our cells to break down alcohol. So this amino acid sequence can work as both a structural protein and an enzyme. Of course, we also use everyday objects for different purposes as well; the same book can be read for information or used to level an uneven table leg.

While we have focused here on the assembly of monomer subunits into polymeric macromolecules, we also need to point out that these macromolecules have to be broken down again. The breakdown of macromolecules into its smaller parts is called **catabolism**. Catabolism allows the monomer subunits (or their components) to be taken up, recycled, or excreted, as needed. We will discuss some specific examples of catabolism and how they are regulated in Chapter 14.

Because structure and function are so closely connected to one another, and thus to the chemical properties of the components and the processes, the dominant view of biology during the past 60 years has been based on the principle of knowing structures. A typical set of experimental questions in genetics and molecular biology is to ask, "How do we determine what the component parts are—the proteins, nucleic acids, lipids, sugars, and other metabolites? How can we use their chemical and molecular properties to explain biological processes?"

Molecular biology and biochemistry have been incredibly successful at identifying the component parts and relying on these to forge links to the other Great Ideas. This has often been a very early step for understanding how a biological process occurs. But the biochemistry of a process is not free-standing. In order to understand a process, it is often helpful to analyze it in terms of the other ideas as well.

Biology as a set of organized and regulated systems

In our discussion of the principles of biochemistry, we noted that monomers are assembled into macromolecular polymers, and macromolecules work together as parts of an organized system. What do we mean by a system? The key to this idea lies in being "organized," that is, the various parts interact with each other and affect one another's functions and roles in a coherent way. They stimulate or repress each other, enhance or suppress each other, and synthesize or degrade each other. They maintain stable equilibria (homeostasis) through complicated feedbacks and connections. Because biology is a series of organized systems, it is not enough to think only about the parts as separate pieces; we must also think about their interactions.

One key to thinking about biology as organized systems is an issue of scale. We can try to understand biological organization on the scale of atoms and molecules. That would fall into the broad field of biochemistry. We can approach it from the scale of individual cells and their molecular components interacting to make a tissue or an organ or a physiological process. This would fall into the broad field of cell biology. We can think about biological organization when we refer to the interactions among the tissues, organs, and molecules that make up an organism. This would fall into the broad fields of physiology, developmental biology, neurobiology, and immunology. We can also talk about interactions among organisms to make communities of species, or among communities to make ecosystems. This is what we would think of as ecology and environmental biology. All of these are organized biological systems, but they differ dramatically in scales of size and time.

Since we now have the sequence of many complete genomes, we may be able to think about biology in a different way, as systems of selected, organized parts from the full catalog of available parts. Of course, physiologists and ecologists (among others) have used systems-based approaches like this successfully for many decades. Based on our knowledge of the sequence of the genome, it is now feasible to know all of the parts—the genes and gene products—in a living organism. We can fairly accurately predict the functions of most of them or, if not the precise function, at least the overall type of function. Therefore, we can begin to analyze the system as the outcome of those parts and their interactions. The goal of the approach based on the properties of the system is then synthetic (learning by putting things together), rather than reductionist (learning by taking things apart)—we are attempting to assemble the parts intellectually in order to understand the function of the whole. Because these interactions form a network, this approach is also sometimes referred to as **network biology**.

By introducing the genome into our discussion, we have also introduced the last of the Great Ideas, that of evolution by natural selection. Evolution by natural selection and other evolutionary processes is one of the dominant themes of this book, so an accurate perspective is essential.

Evolution by natural selection

Biology is a historical science. One of the best expressions of this concept was made by the Nobel Laureate geneticist Max Delbrück, who said, "Any living cell carries with it the experiences of a billion years of experimentation by its ancestors." This statement summarizes in a nutshell the impact of the Great Idea on evolution by natural selection, that is, when we examine genes or cells or molecules or the interactions among them or the genome that encodes them, we are getting glimpses of the processes of evolution that occurred in the past.

Evolution and tinkering

One of the most profoundly important, and often misunderstood, topics in the discussion of evolution arises from this historical perspective. We are always looking back from what exists now to what has happened in the past, and attempting to re-create the path that led from there to here. We can begin to think that this path was inevitable, or even that this represents the best possible solution to a biological problem. This type of reasoning can be very misleading about what has actually occurred during evolution; an analysis of the genomes can help to better inform us.

The molecular geneticist François Jacob (whose research appears in Chapters 14 and 15) referred to evolution as a "tinkerer," rather than an engineer; his influential and provocative essay has been posted on this book's website. A tinkerer takes whatever pieces are available and uses them to make something that performs a particular function, sometimes without regard to the previous functions of that piece. In modern jargon, these pieces are re-purposed. Similarly, organisms are constrained by their evolutionary history to use mostly pre-existing parts to survive and reproduce in their environment at the moment. They can't plan for what might happen in the future, and relatively few completely new parts (that is, new genes and gene products) originate at one time.

According to Jacob, living organisms do not design a process from scratch, as an engineer often does. As a result, a biological process may not be the most efficient or simplest way we can imagine to accomplish something. The use of certain proteins as both enzymes and structural components of the lens of an eye is an example of such tinkering, or re-purposing of an existing enzyme for a completely different function. Many other proteins might have had the necessary optical properties, but organisms re-purposed the ones that were available, and those work well enough.

Inherent, but sometimes overlooked, in this description is that nearly all biological processes include some level of errors or variability in their outcomes. This variability results in "evolvability," that is, the ability of a system to change as the environment changes. The precision of a particular biological process needed to sustain reproduction often arises not only because the process is fairly precise biochemically, but also because there are additional steps to correct the imprecisions or errors that always occur. Thus, precision and reproducibility in biological processes usually arise from a balance of forces; on the one hand, there is a process that is as precise as feasible based on its biochemical nature but which still produces errors, and, on the other hand, proofreading steps fix many of the errors and checkpoints exist to halt the process until the errors can be repaired. All aspects of this system influence how the individual is subject to selection—the genes, RNA, and protein molecules that carry out the original process and those that do the proofreading and form the checkpoints. We will see this especially in Chapter 4 with the discussion of DNA replication, in which a highly precise process is made even less error-prone by a series of proofreading steps and checkpoints.

We see evidence for natural selection in every field of biology. More than 40 years ago, the *Drosophila* geneticist Theodosius Dobzhansky wrote, "Nothing in biology makes sense except in the light of evolution." Almost all biologists know this quotation and embrace it. With the advent of genome sequences, we can apply it even more fully than Dobzhansky could. Because we now have the DNA sequence that an organism uses to live and reproduce and carry out all of its functions, in principle, we have access to all of the parts from which it is made; by comparing it to other species, we also know some of the evolutionary history of the origin of those parts. Similar DNA sequences point to a common ancestor, and the degree of similarity can be a rough measure of the time since the species diverged.

Genomes record the billion years of experimentation lived out by ancestors, reaching back to the origin of life. Evolution may not "make sense" in a way that seems logical or simple to us, but the genome does provide part of the record of what happened. As we learn to read genomes, we can see connections and relationships in a new way; we can also find places that are simply head-scratching. Nonetheless, to quote Darwin, from the conclusion of *The Origin of Species*:

> There is grandeur in this view of life, with its several powers, having been originally breathed into a few forms or into one; and that, whilst this planet has gone cycling on according to the fixed law of gravity, from so simple a beginning endless forms most beautiful and most wonderful have been, and are being, evolved.

One of the most important practical implications of evolution is that biologists can focus their attention on understanding relatively few organisms but can extrapolate what they learn in order to understand the key biological processes found in many species. These **model organisms** serve as stand-ins for understanding the functions of genes and gene products, and their interactions with each other into pathways, networks, and systems in other organisms, including humans. We can use model organisms because fundamental biological properties are shared by all living things. The use of model organisms is discussed more fully in Tool Box P.1 where some key model organisms are introduced.

TOOL BOX P.1 Model organisms

Anyone who has gone for a stroll in the park or spent a day at a zoo or an arboretum recognizes the many thousands of different species to be found in nature. Each species has its own unique DNA sequence, so even if we are not aware of it, we are seeing the consequences that different DNA sequences can bring to bear on the nature and diversity of life. The leaves of a red oak tree are easily distinguished from the leaves of a London plane tree (if a person knows what to look for) because, in part, these different species have somewhat different DNA sequences, which are expressed in the form of different morphologies and physiologies.

There are millions of different species of bacteria, fungi, protozoa, flowering plants, animals, and so on in the world. In response to this amazing diversity, biologists have chosen to study a few organisms in detail as models and then extrapolate our knowledge of them to other species. These model organisms cannot possibly represent all of the diversity of life, but they provide a starting point for us to think about other species. In this box, we introduce a few model organisms, discuss the intellectual basis for studying model organisms, and describe a few properties that are important for model organisms.

Introducing some model organisms

Many different organisms are used in research laboratories or studied in the field, but much of what we learn has been drawn from just a few. Like a list of characters appearing at the beginning of a novel to help the reader, we will briefly introduce these model organisms. Also like the characters in a novel, only a few comments are made here because their true qualities are revealed in the rest of the book. Some other properties of these and other widely used organisms are found in Table 3.1, and these organisms are shown in Figure A.

Escherichia coli

E. coli is the best studied bacterium and quite possibly the best studied species on earth. It is a facultative anaerobe, that is, it can grow in the presence or absence of oxygen. This allows it to inhabit the largely anaerobic mammalian intestinal tract and also to be easily cultured in laboratory broths and on plates in the presence of air. *E. coli* reproduces by binary fission, or splitting of one cell into two. Laboratory strains of *E. coli* are not hazardous, and most natural strains are harmless as well. A few can cause disease, and these are responsible for the "*E. coli* outbreaks" we hear about in the news.

Saccharomyces cerevisiae

S. cerevisiae is commonly referred to as budding yeast or simply "yeast;" because of its other uses, it is also called bakers' yeast or brewers' yeast. It is a single-celled eukaryote and exhibits most of the biological properties common to eukaryotic cells. In the laboratory, it can be grown in liquid cultures or on agar plates where it forms colonies by mitotic growth. It can also undergo meiosis, known as sporulation, like other eukaryotes.

Caenorhabditis elegans

C. elegans is a free-living nematode worm that normally lives in the soil; its natural habitat is relatively little studied. It is usually referred to as "the worm" by geneticists. It is typically grown on agar plates spread with *E. coli*, which it eats. It is about 1 mm long and is barely visible with the unaided eye, so most research is done using microscopes. *C. elegans* can reproduce by self-fertilization, since the "female" is actually a hermaphrodite that first produces some sperm before producing ova and can use those sperm to fertilize ova internally; males can also mate with hermaphrodites for cross-fertilization.

Drosophila melanogaster

D. melanogaster, or more commonly *Drosophila*, the fruit fly, or "the fly," has the longest and most intensive history of genetics research of any laboratory animal. It is typically grown in plastic or glass vials on corn mash media where it eats yeast. Some of its morphological features are visible to the unaided eye, but many structures and mutant phenotypes are more easily observed with a microscope.

TOOL BOX P.1 Continued

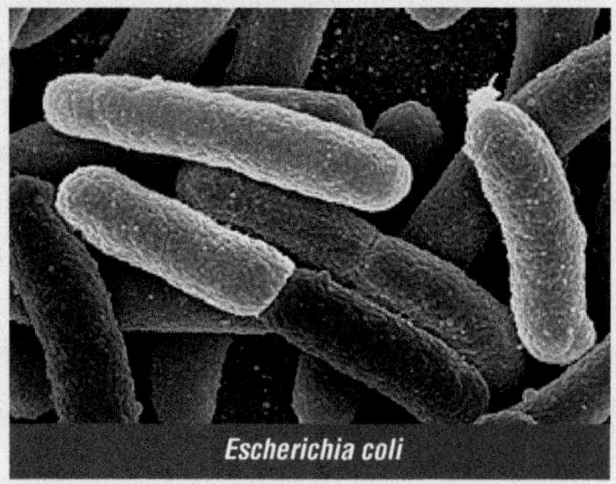

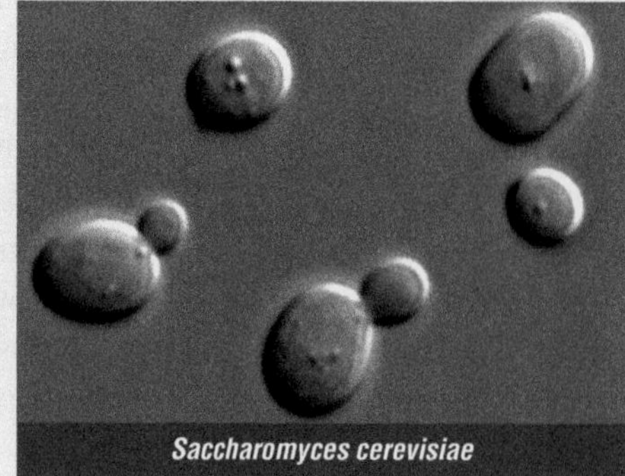

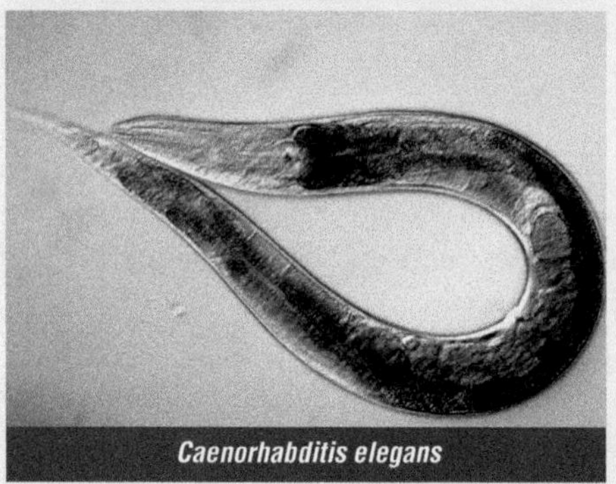

Figure A Some commonly used model organisms.

Source: Image of *E. Coli* courtesy of NIAID/ CC BY 2.0. Image of *S. cerevisiae* courtesy of Masur. Image of *C. elegans* courtesy of National Human Genome Research Institute. Image of *D. melanogaster* courtesy of André Karwath/ CC BY-SA 2.5. Image of *M. musculus* courtesy of Rama/ CC BY-SA 2.0 FR. Image of *A. thaliana* courtesy of Alberto Salguero/ CC BY-SA 3.0.

Mus musculus

The laboratory mouse is the most commonly used mammal for research purposes. Since mice have also been domesticated and kept as pets, many coat color variations and other phenotypes were originally found by mouse fanciers and hobbyists, rather than by geneticists. Most of the genes and biological properties in mice are also found in other placental mammals, including humans.

Arabidopsis thaliana

Arabidopsis is the most widely studied flowering plant in genetics. The plant grows naturally worldwide but has little or no agricultural or horticultural significance; like *C. elegans*, it was intentionally chosen as a model organism because of its rapid life cycle and ease of growth in the laboratory or greenhouse. It is about 20–25 cm tall and can reproduce by either self-pollination or cross-pollination.

Model organisms can be used because there are universal biological properties shared among living things

The intellectual basis for using model organisms lies in Darwin's theory of evolution. All living things arose by a process of "descent with modification," to use Darwin's phrase, from other living things. All mammals, to pick one related group of organisms, are descended with modifications from a common ancestral mammal.

TOOL BOX P.1 Continued

This common ancestor no longer exists as a distinct species, but its heritage is found in the genomes of all species of present-day mammals. By making comparisons among the current species, we can infer the properties of their last common ancestor.

There is a common misperception here that needs to be corrected with an example. No biologist claims that humans are descended from chimpanzees or gorillas. Rather humans, chimps, and gorillas had a common ancestor that no longer exists because it has evolved into these other species. Many of the species that it evolved into—the so-called "missing links"—have themselves been extinct for a long time, including many species in the genus *Homo*. If we wait long enough—say, another 10 million years—the species that we see now may also be unrecognizable because they have evolved into something different. Yet we have undeniable similarities to chimps and gorillas, because all three current species shared a common ancestor roughly 7–10 million years ago. The properties of this common ancestor, in morphology, behavior, physiology, and biochemistry, live on in its current representatives, although some of these properties and the genes that encode them have been lost in the process (at least in some species) and a few new ones have been gained. This is especially evident when we look at the genomes of these current species. Genomics has not only "proved" the theory of evolution (as if it needed more proof), but the use of genomics as an experimental tool is only feasible because evolution has occurred.

Model organisms are so valuable in biological research because they provide insights into biological properties exhibited by—and shared with—their common ancestors. The French molecular biologist and philosopher Jacques Monod wrote, "If it is true for *E. coli*, it is true for elephants," a comment that we discuss and illustrate in Chapter 15. While this is clearly not fully accurate, this quotation does describe the many fundamental biological properties that are shared by all living things. *E. coli* and elephants are very different, but they use the same genetic code and transcribe and translate their DNA in remarkably similar ways.

It may be worth pointing out that the emphasis on model organisms makes geneticists subject to some ridicule in the eyes of the public, however good-natured and well-intentioned, but misinformed, this may be. People joke about studying fruit flies, for example, as if this is a frivolous activity. Geneticists proudly embrace fruit flies, not because we think fruit flies are especially valuable or interesting in themselves—although they do have their own form of beauty—but because they serve as the model organism for understanding key biological properties in other species of animals and many properties in plants as well.

To take one specific example that we use in Chapter 9, the linear arrangement of genes on chromosomes was initially worked out in fruit flies and then found to be true across eukaryotic organisms. Fruit flies share basic cellular structures, as well as biochemical components and processes, found in all living things. Thus, results obtained from studying a gene and its function in the fruit fly are likely to be applicable to many other species.

What makes a good model organism?

With that background about why we use model organisms, it is worth asking about the properties that make a good model organism. After all, the intellectual basis for using model organisms applies equally well to any species, so why did we study these particular species as models, rather than some others?

The reality is that we probably could have studied some other organisms as models, rather than the ones we do. Nonetheless, the widely used model organisms share the following properties.

1. They are easy and safe to work with in the laboratory. This may seem so obvious that it is overlooked. If an organism is going to be used as a model, it must be able to be grown and analyzed without health hazards by scientists at all stages of expertise using many different types of facilities.
2. They have simple nutritional requirements, so they can grow and reproduce easily. Many otherwise interesting species fail on this point, since they have specialized nutritional requirements. This particularly includes some parasites or organisms that live in unusual environments.
3. Their life cycles are simple and relatively short, so many generations can be observed in a short period of time.
4. They have a large number of offspring, so that statistically significant results can be obtained and rare events can be observed.
5. They are relatively small. Large organisms require more space and food, so smaller organisms are preferred if there are methods to observe them such as magnifying glasses or microscopes. Mice have been a more widely used model organism than dogs, for example.
6. There is a community of workers who are willing to share their findings and their knowledge. This is an important factor that can be lost if one just focuses on the science. Research on most model organisms can be traced back to a small group of people working together, who then attracted more people to work with them. The people, rather than the organism itself, may have been the main attraction. Other potential model organisms lost out because there were too few individuals in the field or the members of a small community began to feud among themselves.

TOOL BOX P.1 Continued

7. The organism is relatively easy to save and disseminate to others. Geneticists create varieties and lineages of model organisms that could be of use to other current or future scientists. It is much easier to answer a question with an existing strain than to begin or return to a project by having to construct or reconstruct the strain in the laboratory. Thus, many microbial and invertebrate model organisms are species that can be cheaply dried or frozen and are often mailed.
8. They were lucky. Many other species had these properties but lost out in the model organism lottery. It seems likely that *Caenorhabditis briggsae*, *Drosophila pseudobscura*, and *Arabidopsis lyrata* could probably have been as successful as *C. elegans*, *D. melanogaster*, and *A. thaliana* if they had been chosen instead.

What do model organisms model?

All model organisms are representatives for other species. It is appropriate to ask ourselves, "What species are they representing?" We might believe that the role of a model organism is to tell us about humans and human disease; indeed, many of them have that property. (Some model organisms tell us about humans in other ways. *E. coli* initially became of interest because it was one of the first few bacteria found in the intestines of newborn babies. The yeast *S. cerevisiae* was developed as a model organism because of its utility in brewing beer and making bread. Mice developed as a model organism, at least in part, because it was fashionable to keep "fancy" mice as pets a century ago.) This human-centric view of biology is understandable, albeit short-sighted. Biology is much richer than just us.

With that in mind, here is a short list of what we hope to learn from model organisms.

1. Causes and therapies for human diseases and conditions. The simple reality is that most model organisms are studied because they provide information about those things that are important to us, whether it is crop improvement, aging and other conditions, cancer and other diseases, infections, and so on. Scientific research is expensive, and it is hard to justify spending a lot of money (much of it tax money) to sequence the genome of a nematode worm simply because we love biology. Instead, we expect that the biology of *C. elegans* can provide information about human health issues or about parasitic nematodes that affect our crops or our farm animals. There is ample justification for this expectation, and research on model organisms, such as fruit flies and nematodes, has made many fundamental contributions to human health.
2. Universal biological properties shared among all living things. This is the basis for the Monod statement about *E. coli* and elephants. DNA replication is not identical in *E. coli* and elephants, but the fundamental principles of base pairing and directionality (as explained in Chapter 2) are the same, and many of the important molecules in DNA replication are highly similar. One of the most compelling examples, which we may overlook, is that the genetic code used for translation is identical in bacteria and mammals, and essentially every other living organism, as we discuss in Chapter 13.
3. Evolutionary comparisons with related species. For many contemporary biologists, evolutionary change is among the most interesting reasons to use model organisms. For example, we can use our knowledge of *D. melanogaster*, which is not an especially economically important species or the most beautiful representative among the Class Insecta, as a starting point to learn about the diversity in other species such as flour beetles, mosquitoes, butterflies, and dragonflies, which all have highly similar genomes. We can use *A. thaliana*, an unprepossessing flowering plant, to understand far more lovely flowers. This is the most important underlying justification for the use of mice as the model mammal for understanding human biology.

Going beyond the Great Ideas

Many of the topics introduced in this Prologue deserve to be explained in more detail because remarkable, breath-taking, peculiar, and sometimes puzzling information lies behind our paragraphs. It is a great time to be a biologist, with so much being learned. Any attempt to describe it all or even summarize all of the connections among the Great Ideas is almost certainly doomed to failure because it will be out-of-date before it is published. The Great Ideas provide a starting point and an organizing theme. As we describe genetic principles, these other ideas will crop up repeatedly. Biology is evolution, carried out by genes, cells, and molecules acting alone and in organized systems, as recorded in the genomes.

CHAPTER 1

Evolution, Genomes, and Genetics

1

IN A NUTSHELL

Genetics is the study of inheritance. The total genetic information in an individual organism consists of DNA and constitutes its genome. In the middle of the twentieth century, the geneticist Theodosius Dobzhansky wrote that "Nothing in biology makes sense except in the light of evolution". The recent availability of genomic information from many species of bacteria, plants, and animals allows us to more closely follow inheritance and to see anew how evolution has shaped DNA sequences, genes, and genomes throughout biology. This can be illustrated by examining the genes and genomes of Darwin's finches.

1.1 Darwin's finches: evolution, a story written in the genome and performed by the genes

The Galapagos Islands are 19 volcanic islands strung over a distance of about 140 miles, straddling the equator, 500 or more miles west of Ecuador off the South American coast; their location is seen in the map in Figure 1.1. The environment varies from island to island. While the Galapagos Islands are the permanent home to more than 25,000 people, they will always be closely associated with one person who visited them for 5 weeks in September and October of 1835. Charles Darwin, a 26-year-old naturalist aboard HMS *Beagle*, immortalized the islands in *The Voyage of the Beagle* and *The Origin of Species*, the books that described the wildlife he had encountered and that provided evidence for his theory of evolution by **natural selection**. In return, the theory immortalized Darwin.

Darwin recognized that the wildlife on the islands was related to, but distinct from, what he had recently observed and collected on the South American mainland. He recognized that each island had its own distinctive set of animals and plants, and observed differences in many physical characteristics in comparison to related species on different islands. Some of these changes appeared to make those species better suited to survival in the environment of that particular island than in the environment of another island. In modern language, some of the changes were **adaptive** and allowed the species to thrive in its environment.

A well-known example was found with the shape of the beaks among certain birds from different islands, the birds now known as Darwin's (or Galapagos) finches.

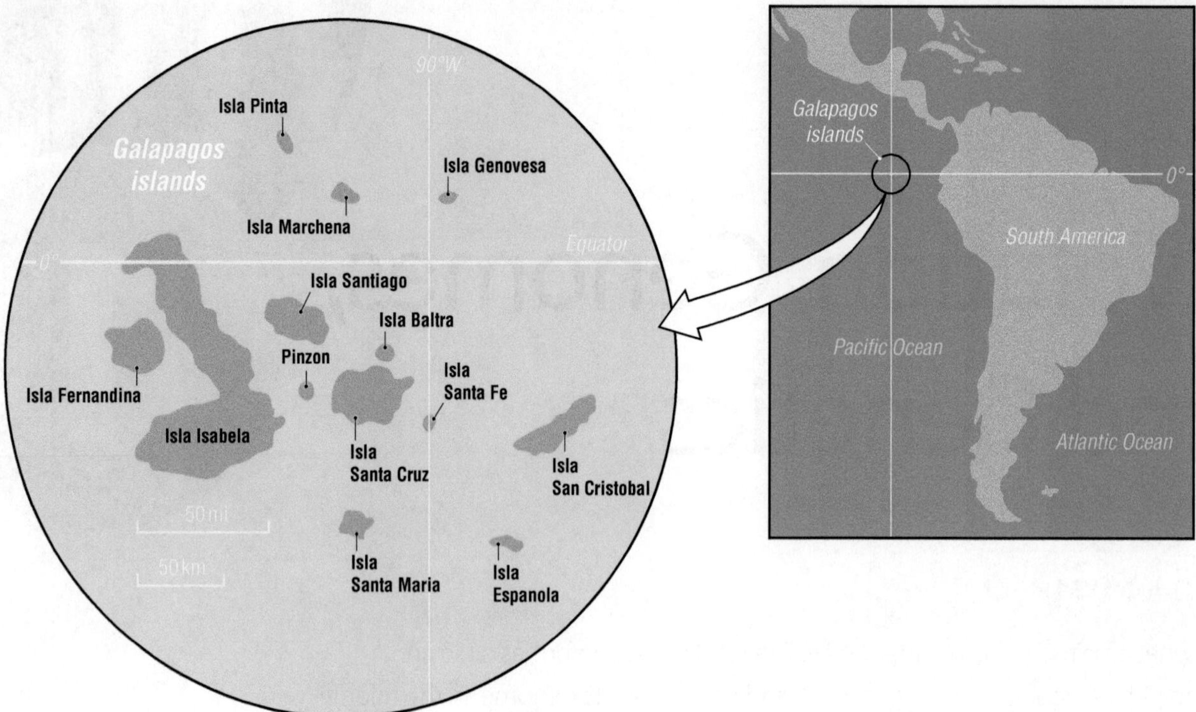

Figure 1.1 A map showing the locations of the Galapagos Islands where Darwin found his finches. This volcanic archipelago is west of Ecuador in the Pacific Ocean.

Darwin's finches comprise at least 15 different species, which was brought to his attention by the ornithologist John Gould when Darwin brought specimens back to England. What Darwin recognized is recorded clearly in the second edition of *The Voyage of the Beagle* from 1845; the boldface highlight has been added to the original text, and Figure 1.2(a) is reproduced from his figure in that book.

> The remaining land-birds form a most singular group of finches, related to each other in the structure of their beaks, short tails, form of body and plumage: there are thirteen species, which Mr. Gould has divided into four subgroups. All these species are peculiar to this archipelago; and so is the whole group . . . The most curious fact is the perfect gradation in the size of the beaks in the different species of *Geospiza*, from one as large as that of a hawfinch to that of a chaffinch, and (if Mr. Gould is right in including his sub-group, *Certhidea*, in the main group) even to that of a warbler. The largest beak in the genus *Geospiza* is shown in Fig. 1, and the smallest in Fig. 3; but instead of there being only one intermediate species, with a beak of the size shown in Fig. 2, there are no less than six species with insensibly graduated beaks. The beak of the sub-group *Certhidea*, is shown in Fig. 4. The beak of *Cactornis* is somewhat like that of a starling, and that of the fourth subgroup, *Camarhynchus*, is slightly parrot-shaped. **Seeing this gradation and diversity of structure in one small, intimately related group of birds, one might really fancy that from an original paucity of birds in this archipelago, one species had been taken and modified for different ends.**

Darwin would continue thinking and writing about these finches, and in using the shapes of their beaks to support his theory. Here is a reference to them in *The Origin of Species*, published 14 years later. Again, we have added the boldface highlights.

> The most striking and important fact for us in regard to the inhabitants of islands, is their affinity to those of the nearest mainland, without being actually the same species. [In] the Galapagos Archipelago . . . almost every product of the land and water bears the unmistakable stamp of the American continent. There are twenty-six land birds, and twenty-five of these are ranked by Mr. Gould as distinct species, supposed to have been created here; yet the close affinity of most of these birds to American species in every character, in their habits, gestures, and tones of voice, was manifest . . . The naturalist, looking at the inhabitants of these volcanic islands in the Pacific, distant several hundred miles

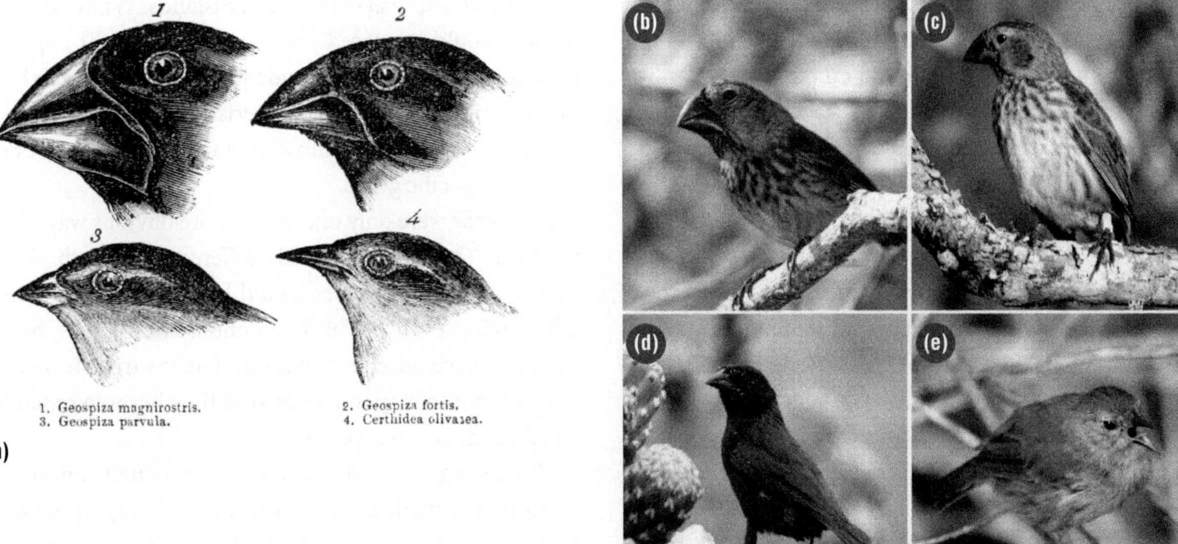

Figure 1.2 (a) The beaks of different species of finches, as reproduced from Darwin's *The Voyage of the Beagle*. The numbers shown are from the original. (b–e) A selection of finches photographed on the Galapagos Islands. (b) Female *Geospiza magnirostris*, the large ground finch, on Isla Genovesa. (c) Female *G. fortis*, the medium ground finch, on Isla Santa Cruz. (d) Male *G. difficilis*, the sharp-beaked finch, on Isla Genovesa. (e) Male *Certhidea fusca*, the warbler finch.

Source for photos: Abzhanov A. Darwin's Galapagos finches in modern biology. Philos Trans R Soc Lond B Biol Sci. 2010 Apr 12;365(1543):1001–7. doi: 10.1098/rstb.2009.0321. PubMed PMID: 20194163; PubMed Central PMCID: PMC2830240. Adapted from Fig 1.

from the continent, yet feels that he is standing on American land. Why should this be so? Why should the species which are supposed to have been created in the Galapagos Archipelago, and nowhere else, bear so plain a stamp of affinity to those created in America? There is nothing in the conditions of life, in the geological nature of the islands, in their height or climate, or in the proportions in which the several classes are associated together, which resembles closely the conditions of the South American coast: in fact there is a considerable dissimilarity in all these respects. On the other hand, there is a considerable degree of resemblance in the volcanic nature of the soil, in climate, height, and size of the islands, between the Galapagos and Cape de Verde Archipelagos: but what an entire and absolute difference in their inhabitants! The inhabitants of the Cape de Verde Islands are related to those of Africa, like those of the Galapagos to America. I believe this grand fact can receive no sort of explanation on the ordinary view of independent creation; whereas on the view here maintained, **it is obvious that the Galapagos Islands would be likely to receive colonists, whether by occasional means of transport or by formerly continuous land, from America; and the Cape de Verde Islands from Africa; and that such colonists would be liable to modification;—the principle of inheritance still betraying their original birthplace.**

KEY POINT The shape of the beaks in finches in the Galapagos Islands provided some important insights for Darwin in developing his theory of evolution by natural selection.

Evolution: adaptive radiation

The beaks of Darwin's finches, each of them different but appropriate for the habitat on its island, are an example of **adaptive radiation**. Adaptive radiation is the divergence of a single group or species into a series of distinct, but related, groups or species, each of them suited to survive in its specific niche. It occurs frequently in archipelagos like the Galapagos Islands, strings of islands separated from one another and providing different types of habitat, some of them newly emerged and lacking predators and competitor species. Adaptive radiation can result in rapid morphological divergence and speciation, which is what caught Darwin's eye.

Knowledge of this and other radiation events allows us to offer a more thorough description of what is likely to have occurred. Finch species found on the mainland migrated to one or more of the islands where they encountered different environments. Birds from one island could eventually come to colonize another island nearby. There was **genetic diversity** among the birds, that is, not all of the birds in the species were exactly the same, and their specific features could be passed on to their offspring. The diversity included the shapes and structures of their beaks. Some of the diversity was present in the population that originally colonized each island, and more changes arose once the population was established on the island, so the birds on different islands became dissimilar from one another, yet sharing common ancestors. As Darwin proposed, those individuals whose beaks were well suited to obtaining food on that particular island were more fit, that is, being better fed, they reproduced at a higher rate than birds whose beaks were less suitable for obtaining food. Thus, the number of birds with a particular beak shape increased relative to those with other shapes. Eventually, the shape of the beak became **fixed**, that is, all birds in that habitat had beaks of the same shape.

Meanwhile, the genetically diverse birds on one island were also migrating to nearby islands where similar events occurred. However, since the environments on the islands were different, a beak that might be well suited for feeding on one island might not be well suited on another; likewise, a beak shape that might be poorly suited on one island might be more favorable on another. Finches from the mainland in South America generally have pointed beaks, so that is considered the **ancestral characteristic**; blunt beaks are a **derived characteristic** that favored survival on particular islands. While characteristics could be traced back to their ancestral population, the populations descended from that ancestor had modifications in those characteristics; today, we would recognize that many of those characteristics are the outcomes of specific genes.

The beaks were only one of the more obvious ways that the birds on different islands differed from each other because similar processes would be occurring for many other traits. Thus, the birds on one island became more similar to each other but diverged in many traits from the birds on another island, such that they became separate and distinguishable species.

This divergence and fixation of a particular trait often occur more rapidly on islands than in other geographical settings for a few reasons. In the case of Darwin's finches, the birds from one island were relatively isolated from those on another island, and the number of birds of that species on each island was relatively small; the "original paucity of birds" that colonized these islands has recently been estimated to have been about 10,000. In addition, the trait was fundamentally important for survival and reproduction. As Darwin realized, "one species had been taken and modified for different ends." The habitats didn't induce the changes in the shapes of the beak; these underlying changes occurred naturally. However, the changes that occurred had different impacts on survival and reproduction, that is, on the **fitness** of the species, because of the different habitats.

KEY POINT Differing environments on the Galapagos Islands and genetic variation among the finches resulted in species with beaks of different shapes.

1.2 Genome analysis of Darwin's finches

Adaptive radiation is a key principle in evolution and is readily observable from the morphological differences such as the shapes of beaks. But what are the underlying causes of these morphological differences? What biological processes can produce beaks of such different shapes and ensure that this shape will be passed on to succeeding generations?

Answers to such questions are written in the genetic information of organisms, which contains a plan for the structure and function of the organism that can be transmitted from one generation to another. Early geneticists inferred the content of the genetic information, or **genotype**, by making deductions from traits, or **phenotypes**, they saw in parents and their offspring. This is still an important methodology in genetics, as discussed extensively in this book. In addition, because many of the principles connecting the genotype to the phenotype are now known, it is also possible to infer many phenotypes of an organism from its genotype, without examining individuals from successive generations.

Genetic information in living organisms is contained in a very large linear molecule known as **deoxyribonucleic** acid or DNA. The information content of DNA resides in the order of its four nucleotide bases, usually abbreviated A, T, G, and C. We know how to interpret much of the information in DNA sequences, and cataloging the complete complement of genetic information of organisms, that is, their genome, is now within our reach. We are living at a great time to be a biologist, quite possibly the most interesting time in scientific history. New species of microbes, plants, and animals are being discovered more rapidly than ever before. New species have always been discovered—Darwin himself discovered 15 species of finches in just one group of islands. The excitement of today is not merely that we are discovering new species, but that we also have tools to analyze these species in greater detail than ever before. These new tools were recently used to analyze the genomes of Darwin's finches. As we describe this analysis, you may not be familiar yet with the tools or some of the terms, but all of these will be described in more detail in later chapters, so we will return to this example frequently.

Finding evolutionary variation in the genome

A team of biologists investigated the genetic and biological origins of the finches' beaks in a paper in *Nature* in 2015, almost 180 years after Darwin's visit. They used modern techniques that allowed them to determine the DNA sequences of the genomes of 120 different birds, including representatives of each of the species Darwin found. The genome of each of these species of birds is a sequence slightly longer than 10^9 base pairs of DNA. The scientists determined the sequence of the A, T, G, and C bases in the DNA of each genome, and compared the DNA sequences from different species to one another to find similarities and differences.

Let's pause to describe what they could expect to see; we discuss the types of variation found in genomes more fully in Chapter 3. The genomes of these species of birds are different in many ways, but the primary difference is that the DNA has different bases at specific sites in the genome sequences; one species might have the base A at one site where another species has the base G at the corresponding site, for example. Such changes occur randomly (by processes that we discuss in Chapter 4). Many of these changes are inconsequential or neutral for the fitness of the organism, while others might be deleterious, and still others may be advantageous. Some of these changes in the base sequence could result in particular genes with slightly different functions, and a few of these genes with changes may be responsible for the shape of the beak. By looking at many millions of base pairs throughout the genomes and comparing them to the ancestral species, we can recognize those species that have changed at each site in the genome and those species that have not.

Birds of different species on different islands will have many hundreds of thousands of differences in the base sequences of their genomes; even birds from the same species with the same shaped beak have many thousands of differences, just as your genome has many thousands of sequence differences from that of another person. With contemporary DNA sequencing technology, finding differences in the DNA sequences of the genomes of the finches is not that difficult. But all of these differences introduce some other questions, which might not be so easy to answer. Of these many hundreds of thousands of changes, which are the ones that are the most important for the phenotype they are studying? Could they find the changes in the genes that explain the differences in the shape of the beak? Of the many DNA differences they find, how could they be confident that these are the ones that help to explain the changes in the shape of the beak? Finally, how exactly do these differences in the DNA sequence and the genes explain the differences in the shape of the beak? The flow chart of their experimental analysis is shown in Figure 1.3.

KEY POINT The genomes of the species of finches in the Galapagos Islands have been sequenced and analyzed to find the changes in the DNA that led to the changes in the shape of the beak.

Finding the regions of the genome that affect the shape of the beak

Let's think for a moment about what happens to the genomes of individuals and species over time by using a hypothetical, but realistic, example. There is an ancestral species, and individual members of that species had slightly different DNA sequences and thus different phenotypes (such as slightly different shaped beaks). The genome comprises 10^9 base pairs, and most of the DNA sequence is the same in two different closely related species; the total number of sequence differences could be about a million, which seems like a lot, but, in a genome of 10^9 base pairs, differences in 10^6 of them is still only 0.1% of the total. The rest of the sequence is the same

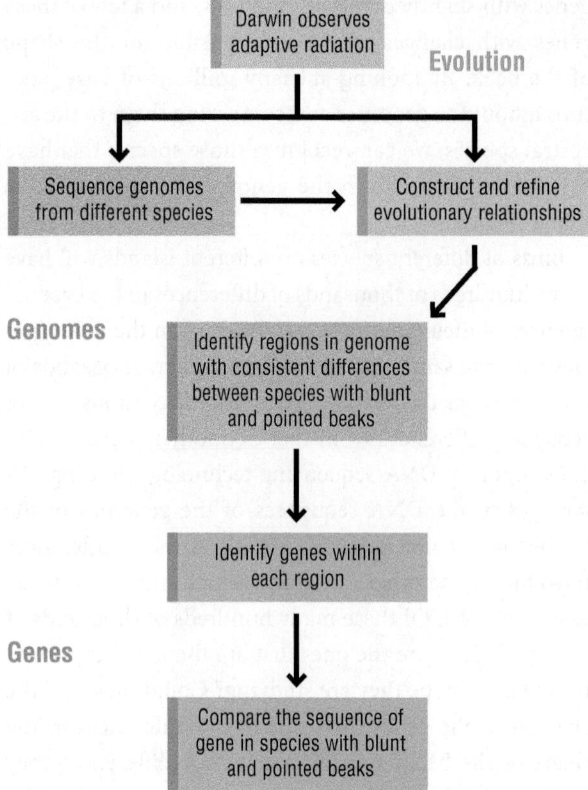

Figure 1.3 The genomic analysis of the evolutionary changes associated with beak shape. This flow chart summarizes the strategy used to identify genes involved in the evolution of beak shape. The colors represent different intellectual and experimental approaches to the analysis; green boxes are based on evolution; orange boxes are based on genomes, while blue boxes are based on genes and molecules. These approaches overlap, so the distinctions between them are somewhat arbitrary. Darwin (and others) used the shape of the beak and other characteristics to infer the evolutionary and ancestral relationships among the different species. More recently, the DNA sequences of genomes from different species were determined, and those sequences were used to refine the evolutionary relationships. The genomes were then scanned for regions that showed consistent differences between closely related species with different shaped beaks. Within each of those regions of the genome, the genes were identified, based on the sequence and known properties of the genes. Genes that are likely to be involved in affecting the shape of the beak were identified among the other genes that happen to be located in the same region of the genome. The sequences of these candidate genes were determined and analyzed in species with blunt and pointed beaks to find specific changes associated with beak shape. Thus, the analysis went progressively to smaller scales, from birds to regions of their genome to sequence changes in individual genes.

between the two species, so it cannot be responsible for changes in the shape of the beaks.

By comparing one species to an ancestral species, the G:C base pair that is usually found at (say) position 15,426,249 on chromosome 3 has mutated to become an A:T base pair. That species and the ancestral species continued to change, with more **mutations** arising over time. Among some of the individuals in that species, another mutation arises at position 7,231,397 on chromosome 6 that changes a T:A base pair to a C:G base pair, while other individuals do not have this change. This process of accumulating sequence changes or mutations continues for every individual and at many locations in the genome.

Because many of the changes in the genome that occur during evolution occur stepwise in the ways that we have just described, species that are more closely related to one another have DNA sequences that are more similar throughout their genomes than species that are less closely related. To borrow Darwin's phrase, the principle of inheritance (that is, the DNA sequence) betrays their original birthplace (that is, their ancestry). Since the species are all derived from a common ancestor on the mainland, the DNA sequences of their genomes will be highly similar to one another. Thus, the first step in identifying the specific differences in DNA that account for beak shape variations across the islands was to use the total genome sequences to construct and revisit the evolutionary history and relatedness of these birds as predicted from morphological differences, before DNA sequence information was available.

The team of biologists in 2015 identified species whose beaks are of different shapes—either blunt or pointed—but whose genome sequences demonstrated that they are closely related overall. In other words, each individual in the species consistently had one beak shape, so the shape of the beak was fixed for that species. When the researchers found two species that were closely related overall but differed in the shapes of their beaks, they could predict that the DNA sequences would be similar overall but differ in regions that affect the shape of the beak.

Fixation for the shape of the beak implies that the underlying DNA sequences that contribute to the shape of the beak should also be fixed, or at least much more common in one species than in another, as summarized in Figure 1.4. Because different traits (blunt or pointed beaks) are fixed in each species, a species with a blunt beak is expected to have one sequence in a certain region of the genome, while a related species with a pointed beak would have a somewhat different DNA sequence. The researchers computationally scanned the DNA sequences of these species to find regions in which most of the birds with blunt beaks had one long sequence (several hundreds of thousand base pairs in length, as we shall see in

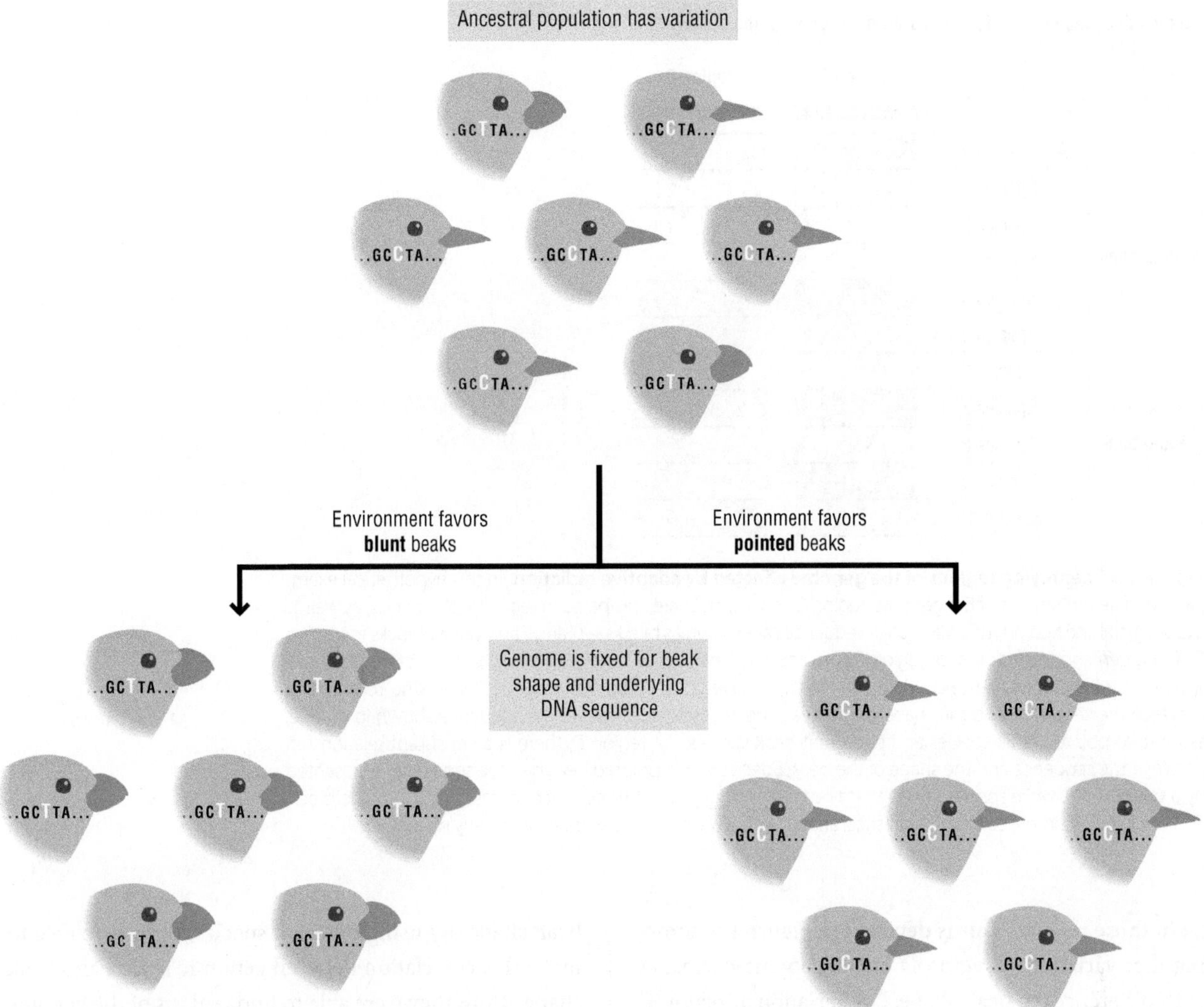

Figure 1.4 Fixation of the beak shape results in fixation of the genome sequence. The ancestral founding population of birds had beaks of different shapes, shown here as blunt and pointed. Each of these phenotypes has an underlying genotype, shown as the DNA sequence. The environments in different locations confer an advantage on beaks of different shapes, so the beaks in each group or species become fixed for that shape. Correspondingly, the underlying DNA sequence is also fixed. While this diagram shows only two beak shapes and a single base difference between them for simplicity, the same principle holds for other shapes and other DNA sequence differences.

Section 1.3), whereas most of the birds with the pointed beaks had a **different** long sequence.

Many regions of the genome are expected to vary in sequence, but this variation is only predicted to correlate with beak shape in regions that include genes affecting the beak. Let's think a bit about how they did this experiment, and imagine another way they might have approached this. They did not begin by asking, "What genes are likely to affect beak shape?" and then testing those candidate genes for variation, although that is also a reasonable approach that is often used in such experiments. The advantage of an approach based on candidate genes is that it focuses on regions or genes that are thought to be important for the trait being considered. The disadvantage of that approach is that it depends on knowing what genes are good candidates for affecting beaks, so it could easily miss genes whose role is important but was previously unknown.

The strategy shown in Figure 1.3 was different. They began with the evolutionary principle of adaptive radiation and its expected impact on the genome. They used this information to find the regions of the genome where differences are present, regardless of what genes

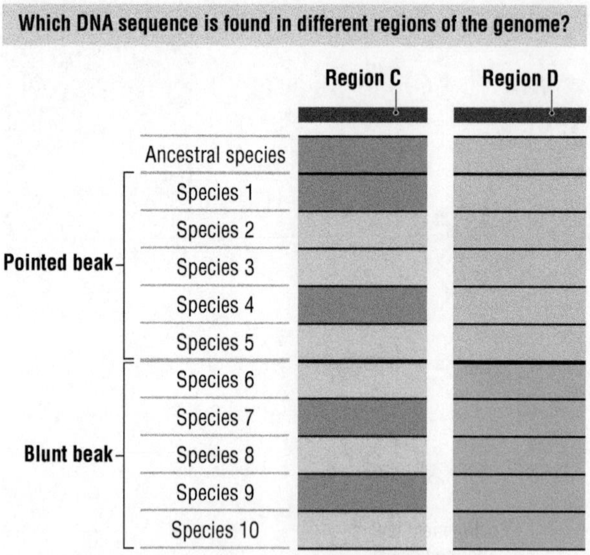

Figure 1.5 Identifying regions of the genome affected by adaptive radiation. In this hypothetical example, two different regions of the genome, region C and region D, are compared in ten different species, as well as the ancestral species. The two alternative sequences are shown as blocks—orange and yellow blocks for region C, and green and blue blocks for region D. The ancestral species has a pointed beak, and five of our example species also have pointed beaks, while the other five have blunt beaks. For region C, there is no relationship between the sequence and the shape of the beak; the ancestral species had the sequence shown in orange, as did two pointed beak species and two blunt beak species. For region D, there is a consistent relationship between the sequence and the shape of the beak. Species with a pointed beak had the sequence represented by a green block, while species with a blunt beak consistently had the sequence represented by a blue block. This diagram is a simplification of a statistical analysis of each sequence to show the underlying logic.

lie in those regions. This is depicted in Figure 1.5, showing that variation at region D of the genome correlates with differences in beak shape, but variation at region C does not. The advantage of this approach is that it can find genes and regions whose functions are not known, even though we know them to be important. The disadvantage is that many changes will be found, so it is essential to have good methods to filter the meaningful variation from differences that have nothing to do with beak shape. By using different species, they were able to make this correlation between genomic region and beak shape. Thus, they were able to find regions of the genome that contain candidate genes responsible for the evolutionary change.

KEY POINT Inferences about evolutionary changes in the sequences of the genomes were used to identify the regions containing genes involved in the evolution of the beak shape.

1.3 Connecting genome variation to beak variation

Fifteen regions of the finch genomes were found to fit their criterion—the same long sequence (or very similar sequences) is found in birds with blunt beaks, but a different long sequence is found in the genomes of birds with pointed beaks. Any or all of these 15 regions could contain crucial genes affecting beak shape, while others might not have anything to do with beak shape; after all, these are different species, so they vary in more than just one characteristic. It then became important to determine which genes were located in each region—something the investigators could achieve by studying the DNA sequences in more detail.

Deciding which changes are the most important candidates

Not all of the DNA comprising the genomes of vertebrates is part of a gene. As we will discuss in more detail in

Chapter 2, the DNA sequences of genes have characteristic structures that signal their starting and end points, for example. Genes can also be identified from the RNA that is transcribed from them. Thus, the investigators located the genes in these 15 regions of the genome and asked what is known about functions of these genes. This would help them determine at least one gene that is important for changes in the shape of the beak.

The DNA sequence of the gene can be used to predict the amino acid sequence of its protein product (as we will discuss in Chapter 2), and the amino acid sequence of the protein can be compared to other proteins to make inferences about the function of the protein. At this point, they turned to previous knowledge from birds and other vertebrates about the functions of genes and the gene product in each of the regions. Six of the regions contained genes that are known from other studies in birds or other vertebrates to affect craniofacial or beak development, including one that included a gene that had previously been found to affect beak shape in finches. This does not mean that the other nine regions do not contain genes affecting beak shape, but rather that these six regions are the ones that might be the most productive to analyze initially.

Among these six, the region of the genome exhibiting the strongest correlation with the beak shape is about 240,000 base pairs (referred to as 240 kilobase pairs and abbreviated as 240 kb) in length and includes a gene called *ALX1*, diagrammed in Figure 1.6 and referred to as region D. The *ALX1* gene itself is about 5000 base pairs (5 kb) in length, well within the normal range for the size of a gene and the 355 amino acid protein it encodes, if a bit smaller than average. The known biological role of *ALX1* also indicates that it is a good candidate for affecting beak development. In birds and other vertebrates, *ALX1* is actively expressed during the normal development of the embryonic tissues and processes that give rise to craniofacial structures. In fact, a few children who have been born with malformations of craniofacial structures, including a severely cleft palate, have changes or mutations in *ALX1*. Furthermore, mice in which the *ALX1* gene has been deleted suffer from severe craniofacial abnormalities and die *in utero*.

Let's summarize this evidence so far. *ALX1* lies in a region of the genome that correlates strongly with a difference in the shape of the beak, and is the only gene in that region known to affect craniofacial development. It is expressed at the time and location that suggest it might be involved in the normal development of tissues that give rise to the beak and other craniofacial structures

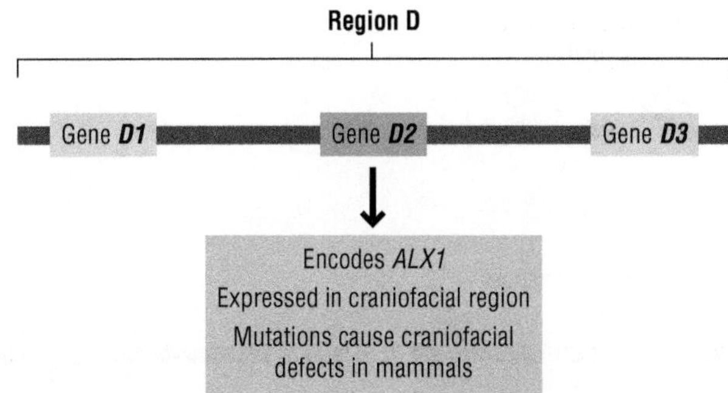

Figure 1.6 Identifying candidate genes within region D. In the example in Figure 1.5, changes in region D were associated with the differences between blunt and pointed beaks. Thus, the sequence of region D was analyzed to identify the genes and to identify which of these are the best candidates to affect beak shape. In our hypothetical region D, three genes are found: *D1*, *D2*, and *D3*. Of these three, gene *D2* is known to be expressed in the craniofacial region from research in other vertebrates, and changes in gene *D2* result in craniofacial abnormalities in mammals. Thus, gene *D2*, whose actual name is *ALX1*, is a strong candidate to be a gene associated with changes in the shape of the beak.

in birds and other vertebrates. Mammals (humans and mice) with changes that reduce or eliminate the function of *ALX1* have craniofacial abnormalities, indicating that it is needed for normal craniofacial development. Taken together, this evidence makes *ALX1* a very good candidate for being one of the genes that cause differences in the finches' beaks.

WEBLINK: You can find out more about the genomic analysis of the Galapagos finches in Video 1.1. Find it at **www.oup.com/uk/meneely**

So, we see how the investigators used the changes in the sequence and structure of the genome, as predicted by evolutionary changes, in order to find a gene whose role is known to be important in craniofacial development. But what is the molecular function of *ALX1*, and how do changes in *ALX1* contribute to changes in the beaks?

KEY POINT An important gene associated with beak shape was identified, based on its expected role during development and its effects in other vertebrates.

Gene: the role of *ALX1* in beak development

A gene is a sequence of DNA, which (in the case of *ALX1* and many other genes) includes the information necessary to make a particular protein. The protein carries out specific functions within the cells and tissues, but

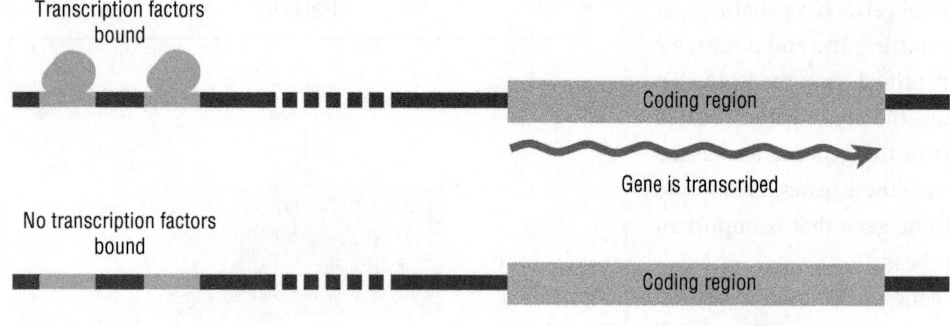

Figure 1.7 Transcription factors control gene expression. While all cells in a multicellular organism have the same DNA sequence, and thus the same genes, they express or transcribe different genes. The binding sites for a class of proteins known as transcription factors are shown as green line segments; the gene has that sequence, regardless of whether the transcription factor is present or not. The differences in transcription pattern depend on the activity of the transcription factors. When a transcription factor protein, shown here in green, is bound to the DNA near the gene, the gene is transcribed, represented by the wavy purple line. When no transcription factor is bound, the gene is not transcribed.

not all proteins are made at the same time or in the same cells. Genes are actively expressed—**transcribed** into ribonucleic acid (RNA) and then often **translated** into proteins—in particular tissues at particular times. Some genes are expressed broadly over time and space, while others are expressed in a more limited pattern in only certain cells at specific times, with many different possible expression patterns, depending on the gene being studied. The specific expression pattern of any gene arises from the functions of a type of protein known as a **transcription factor**, as summarized in Figure 1.7 and discussed in Chapters 2 and 12. Transcription factor proteins, which are themselves encoded by genes, direct the expression of their gene targets to make sure that they are expressed in the proper pattern in time and space.

The *ALX1* gene encodes a transcription factor protein (also called ALX1), and thus it regulates the expression of other genes. (Communicating Genetics 1.1 provides some guidance on conventions used to name genes and proteins.) Based on when the *ALX1* gene and the ALX1 protein are expressed and what happens when *ALX1* is altered or missing, the target genes it regulates must include

COMMUNICATING GENETICS 1.1 How do we represent gene and protein names?

It is probably safe to say that no other topic in genetics has more potential for confusion than the names of genes and proteins and how they are written. The potential for confusion arises from many sources. Different organisms have different systems of nomenclature; gene names differ among *Escherichia coli* bacteria, nematode worms, fruit flies, *Arabidopsis* plants, and humans in terms of how many letters are used and whether these names include numbers. Sometimes, but certainly not always, the product of a gene—usually a protein—has the same name as the gene. In addition, similar genes in different organisms may have been assigned quite different names before their resemblance was noted.

As you read this book, you may find it difficult at first to know when we are referring to a gene and when we are discussing the resulting protein, but we will use a convention that will help to distinguish them. This convention is found in the nomenclature system for most, but not all, organisms, but we will use it for every organism:

- Gene names are italicized
- The first letter or whole name of the protein is capitalized but protein names are not italicized.

For example, in this chapter, we describe a gene identified in the analysis of the genomes of Darwin's finches. The gene itself is known as *ALX1* (with italics), while the protein encoded by it is written as ALX1 (with no italics).

Remembering this simple convention will help you to distinguish between genes and proteins throughout this book.

Communicating Genetics 5.1 adds more information about gene names, which will be helpful once we have provided more examples of genes and the proteins they encode.

the ones that are needed for normal craniofacial development in vertebrates. We don't know what all of the targets of ALX1 are; we only know the targets of a relatively few transcription factors in a relatively few species, and ALX1 in finches is not one of those few cases. Some of the target genes of ALX1 regulation could encode other transcription factors, while other target genes might encode enzymes and structural proteins necessary to make a beak and other craniofacial structures. We don't know if ALX1 directly regulates one target gene or 100 target genes, although if we were to guess, based on what we know about other transcription factors like ALX1, we might expect that it directly regulates between ten and 20 other genes; we might further guess that some of those direct targets encode transcription factors that regulate even more genes. In Chapter 2, when we introduce transcription factors, we will consider the DNA sequence changes that were found in the *ALX1* gene that might explain its contribution to changes in the shape of the beak.

KEY POINT ALX1 regulates the expression of other genes, some of which may have roles in the development of beaks.

Evolution, genomes, and genetics

We have started this book with an analysis of the genetics of the beaks of Darwin's finches not only because it is such an important historical example, but also because it exemplifies our approach in this book. The changes that Darwin observed in the shape of the beak of the finches on the Galapagos Islands can be traced back to changes in the DNA sequence of at least one specific gene, a transcription factor controlling the expression of other genes. We introduced many terms, experimental approaches, and ideas that will be explained more thoroughly in other chapters. The analysis of Darwin's finches

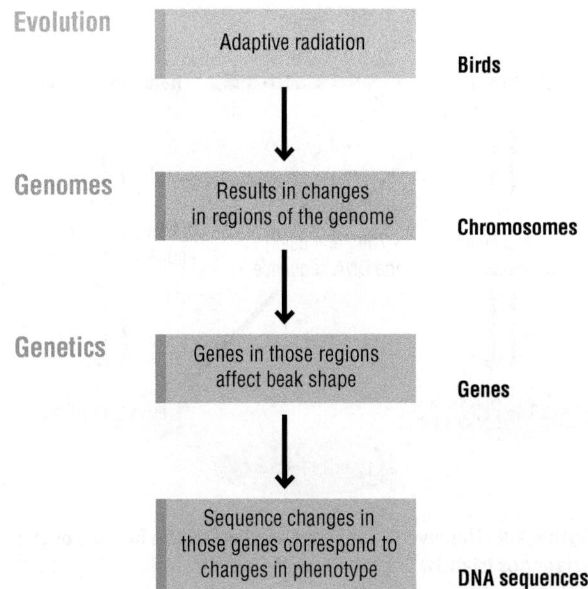

Figure 1.8 Evolution, genomes, and genetics summarized. Evolutionary change, observed as differences in beak shape in finches, affects their genomes. This, in turn, indicates possible genes that might be involved, and DNA sequence changes in those genes can confirm and help to explain the nature of the changes. The analysis is carried out at an increasingly smaller scale, from organisms like birds to genomic regions and chromosomes, to individual genes, and finally to nucleotide base changes in those genes.

combines the principles of evolution with an analysis of the genome and an identification of the functions of specific genes, as summarized in Figure 1.8. The beaks of these finches illustrate relationships among evolution, genomics, and genetics in a way that could not have been determined even a decade ago. It is not that evolution, genomics, and genetics are three different tributaries that come together from different sources, but rather that these are all parts of the same stream. It is only now that we can see them this way. This is an exciting time to be a student of biology.

1.4 Summary: Darwin's finches and the Great Ideas of Biology

To summarize the chapter, we return to where we began—in fact, to where significant parts of evolutionary biology began—with Darwin's finches. In the Prologue, we discussed how five Great Ideas (or major themes) can help us to see the connections between different topics in biology, and summarized them in Figure P.1. Most of the great ideas and connections between them were present in our discussion of Darwin's finches. The differences in the shape of the beak led Darwin to recognize and propose the principle of natural selection, so the Great Idea about evolution and natural selection was the starting point for understanding this process, as summarized in Figure 1.9.

Recent investigators then exploited the effect that evolution, particularly adaptive radiation, has on the genome, which allowed them to select *ALX1* as a strong candidate gene affecting beak shape. The genome connected the

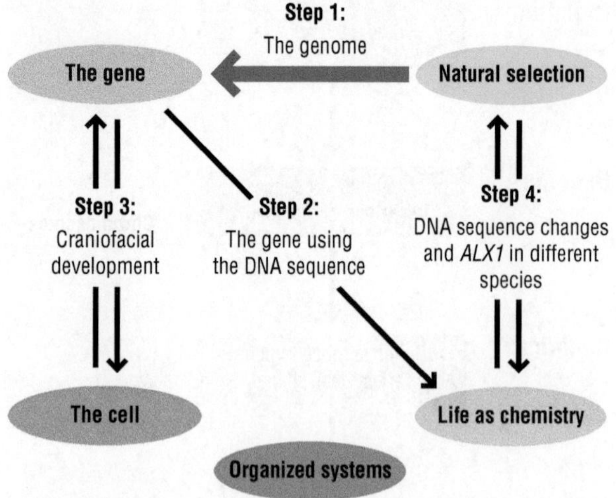

Figure 1.9 The Five Great Ideas applied to the finches of the Galapagos Islands. Since the Great Ideas unify biology, an analysis could begin with any one of them and work to any of the others. For the analysis of the finches, the analysis began with evolution by natural selection and then moved from the sequences of the genomes of different species to individual genes, cellular processes, and the underlying biochemical and molecular parts. The particular experiments we discussed did not include a further analysis of the system, although that was implied.

Great Idea on natural selection and evolution to the Great Idea on the gene theory, shown in Figure 1.9 as Step 1.

From the *ALX1* gene and its corresponding protein, the investigation used the well-established tools of molecular biology to infer a function for the gene and its effect on beak shape; these tools are based on the Great Idea about the chemical principles that underlie biological processes and are shown in Figure 1.9 as Step 2. This connection has been made so strongly by molecular biologists in the past 70 years that we may not recognize that we have moved from genes to molecules to chemical principles, but we did. This allowed them to return to the Great Idea about evolution, shown in Figure 1.9 as Step 4.

The Great Idea about cells was used in the analysis as well, in the discussion of where the gene is expressed during normal development, shown in Figure 1.9 as Step 3. Because all of biology occurs in the context of cells, this may be taken for granted, but we could not have interpreted the biological connection between *ALX1* and beak shape without also thinking about the cells and cellular processes that give rise to the beak.

Finally, the Great Idea that biology consists of organized and regulated systems was important in knowing how a transcription factor protein like ALX1 can organize the expression of other (unidentified) genes that function in the development of the beak; the study of genes affecting the shape of the beak did not identify the system involving *ALX1*, so that arrow has been omitted from Figure 1.9, but it certainly played a role in their thinking, and further studies can provide more detail.

It may seem artificial to interpret the analysis of beak shape according to the Five Great Ideas in this way, and it may not be a useful method for other examples; in every case, some of the Great Ideas will be more prominent than others. But an exercise like this illustrates how closely the biological principles are connected to one another. Any student of biology who wants to think more deeply about processes in living organisms will need to think about the connections between these ideas.

KEY POINT Any biological process can be thought of using the connections among two or more of these five unifying ideas.

During the remaining chapters of this book, we will develop these ideas and explain many of the principles we introduced as we discussed the genomic analysis of Darwin's finches. As Darwin wrote, there is grandeur in a view of life—evolution, as told by the genomes and performed by the genes.

CHAPTER CAPSULE

- The analysis of the genomes of Darwin's finches allowed connections to be made between their evolutionary history and the genetic, cellular, and molecular processes affecting beak shape.
- Genetics is an inextricable part of a web of principles that come together to form the science of biology. The unifying connections between these principles, as exemplified by the Five Great Ideas of Biology, are an essential foundation for the

study of genetics and genomics. Much more information can be learned about each of the topics represented in the Great Ideas.

- For the investigation of any particular biological process, one or more of the Great Ideas could hold center stage, but all of them will have played a role.

STUDY QUESTIONS FOR THE PROLOGUE AND CHAPTER 1

Concepts and Definitions

1.1 Define the following terms:

a. Phenotype

b. Fitness

c. Genome

d. Ancestral and derived conditions

e. Haplotype

f. Anabolism and catabolism

1.2 Briefly explain the five unifying ideas of biology, as described in the Prologue. Which of these ideas is the easiest for you to understand, and which is the most difficult?

1.3 What are the main differences (based on the Prologue) between bacterial cells and eukaryotic cells? Which of the organelles and cellular structures discussed in this chapter are common to both bacteria and eukaryotes?

1.4 What are the monomer subunits that become assembled into biologically important macromolecules, and what macromolecules do they make?

1.5 Explain, in your own words, Darwin's statement about how the principle of inheritance betrays the original birthplace of a species or a group of species. What are some other examples of this idea?

Beyond the Concepts

1.6 Outline the process by which adaptive radiation occurs. What makes a string of islands or lakes an especially good opportunity to observe adaptive radiation?

Looking ahead **1.7** Based on what you know from this chapter or information from other courses, what are the basic steps by which a gene becomes expressed into a phenotype?

1.8 Look at the arrows connecting the five unifying ideas in Figure P.1 in the Prologue. What course in your college experience might emphasize each of these connections? Would any of them not be important in a particular course?

1.9 Explain the concept that the function of a macromolecule is affected by its structure (that is, the shape and sequence of monomer subunits). How does this important concept play out in the evolution of biological systems?

1.10 It is now possible to know the entire DNA sequence of an organism, and it is feasible to predict all of its genes. What are some of the ways that this has changed the ways that we can analyze biological questions?

1.11 Geneticist Max Delbrück said, "Any living cell carries with it the experiences of a billion years of experimentation by its ancestors."

a. Explain what this means.

b. How have cells carried out this experimentation?

c. How have technological and scientific advances in the past 20 years allowed us to explore this statement more fully?

Applying the Concepts

1.12 The investigators used a genome-based analysis to find genes that affected the shape of the beak, rather than an approach based on candidate genes.

a. Outline how they did this.

b. Compare and contrast the advantages of each approach.

Challenging **1.13** Other studies had implicated the *BMP4* gene as being important in the shape of birds' beaks, but *BMP4* was not among the candidate genes identified here. Speculate about why *BMP4* was not found in these studies.

Challenging **1.14** The changes in the *ALX1* gene alter its function but do not eliminate its function. Hypothesize why changes that would eliminate the function of *ALX1* were not found.

Looking ahead **1.15** A phenotype is the outcome of both genes and the environment in which the organism is found, as well as the interaction between genes and the environment.

a. What is an example of a phenotype (from humans or other organisms) that is primarily, if not entirely, affected by the genes?

b. What is an example of a phenotype (from humans or other organisms) that is primarily, if not entirely, affected by the environment?

Challenging **c.** What might be some of the ways that you could determine the relative contributions of genes and the environment to a particular phenotype?

1.16 All the statements below are taken from Wikipedia or an introductory college biology/AP Biology book. For each of the following statements, discuss which of the Great Ideas are being used. Feel free to use online resources to look up words or concepts that are not yet familiar to you. The goal of this question is not to determine the "one right answer" but to engage you in thinking about how the Great Ideas relate to one another.

- "Glycolysis produces pyruvate molecules in the cytosol and an active transport mechanism moves them into the mitochondrial matrix."
- "Eventually, most of the solar energy absorbed by green plants is converted into heat energy as the activities of life take place."
- "This model [referring to the Watson–Crick model for double-stranded DNA] fit all of the known data about the arrangement of atoms, and made it apparent how genetic information is stored and how it could be replicated faithfully."
- "Embryological and molecular evidence suggests that bilaterally symmetrical animals are divided into two lineages, the protostomes and the deuterostomes."
- "Immunological memory forms the basis for vaccinations, in which antigens in the form of living or dead pathogens are introduced into the body."
- "In nature, prokaryotes often live in communities where they interact in a variety of ways."

- "Occasionally, differentiated cells of complex multicellular organisms deviate from their normal genetic program and begin to divide and grow, giving rise to tissue masses called tumors."
- "The generation of genetic variability is a prime evolutionary advantage of sexual reproduction."

1.17 Jacob described evolutionary processes as "tinkering."

a. What does this tinkering mean?

b. How does this affect your view of how biological processes occur?

c. What does this imply about the common perception that evolution by natural selection produces a "better" organism or system?

d. Give an example of such evolutionary tinkering, either one that you see in the chapter or one from other information or courses.

1.18 Much of contemporary molecular biology and genetics is based on the use of model organisms such as the bacteria *Escherichia coli*, the budding yeast *Saccharomyces cerevisiae*, the fruit fly *Drosophila melanogaster*, the nematode *Caenorhabditis elegans*, or the flowering plant *Arabidopsis thaliana*, among a few others.

a. In light of the Great Ideas, why do we use model organisms?

b. What are some of the properties that make an organism a good model organism?

c. What are some of the disadvantages of using model organisms to understand biological questions?

d. Were Mendel's peas (introduced in Chapter 5) an example of a model organism? Explain your reasoning.

e. Would additional model organisms be helpful? Why or why not?

CHAPTER 2

2 The Central Dogma of Molecular Biology

IN A NUTSHELL

The Central Dogma of molecular biology is that DNA provides the template to make RNA, that RNA provides the template to make a polypeptide, and that polypeptides contribute to phenotypes. The information content that flows from DNA to RNA to polypeptide is inherent in the structures of the molecules, particularly in the base sequence of the nucleic acids, which encodes the amino acid sequence of the polypeptide. With the ability to determine the complete DNA sequence of genomes, it is now possible to learn detailed information about where and when this information flow occurs. In fact, as genome analysis has shown, a surprising number of RNA molecules do not serve as templates for making polypeptides but are functional themselves.

2.1 Overview: the Central Dogma and the nature of the gene

The concept of a "gene" as the unit of biological inheritance has a very long intellectual history, although the word "gene" was not used by either Mendel or Darwin. The Dutch botanist Hugo de Vries referred to a "pangen" as the "smallest particle representing one hereditary characteristic" about a decade before he rediscovered Mendel's research in 1900. The Danish botanist Wilhelm Johannsen shortened this to "gen" or "gene", and the British geneticist William Bateson referred to "genetics" in the early years of the twentieth century, so genes have been part of our vocabulary for more than a century. These men are shown in Figure 2.1.

Although we use the term "gene" and have an idea of what we mean when we do, a comprehensive definition of a gene is surprisingly elusive. Genes have many different structures and roles, and it is hard to include all of these in one definition. Nonetheless, we understand that all genes share certain properties. For example, they are **inherited**—this was the original definition used more than a century ago. Even this definition has some subtlety to it, however. When an individual passes on his genes to his offspring, he has not lost his own genes; you inherited genes from your parents, but your parents still have them too. Thus, genes have to be copied or replicated, so that

Hugo de Vries   Wilhelm Johannsen   William Bateson

Figure 2.1 The gene as the unit of inheritance. The use of the name "gene" to describe the particulate unit of inheritance was largely due to the influence of these three men working in the early years of the twentieth century—Hugo de Vries from The Netherlands, Wilhelm Johannsen from Denmark, and William Bateson from England.

Source: Image of Hugo de Vries from *The Popular Science Monthly*, Vol. 67, 1905. Image of Wilhelm Johannsen from *The History of Biology de Erik Nordenskiöld*, Ed. Knopf, 1928.

each generation is passing on copies of its genes. The replication process also produces some variation, which we will introduce in Chapter 3 and describe in more detail in Chapter 4.

Furthermore, genes cause a **phenotype** or a morphological characteristic. We use the term phenotype frequently in this book, so we should provide a formal definition. A phenotype is any quality that can be measured by some assay, whether or not this characteristic is inherited. For phenotypes affected by genes, we can thus make an additional statement about the roles or definitions of a gene. We can talk about a gene affecting phenotypes like eye color or blood type, for instance. So genes have to be **expressed.** There are other properties of genes that we will introduce in subsequent chapters, but these two properties are the ones that will be discussed for now. Whatever biochemical or molecular substance comprises a gene, it must have these characteristics.

- A gene has to be replicated before it can be transmitted to the next generation, and the replication process includes the potential for variation.
- A gene has to be expressed to produce a phenotype or some component of a phenotype.

Of course, when we write above "Whatever biochemical or molecular substance comprises a gene . . . ," we are being somewhat disingenuous. It is well known that genes, as introduced in Chapter 1, are composed of DNA. The beauty of DNA as the genetic material is that these two properties of genes could be envisioned as soon as Watson and Crick published the structure of DNA, that is, some of the properties of genes as the units of inheritance could be immediately recognized from the structure of the DNA molecule. In this chapter, we discuss the structure of DNA and briefly introduce how its structure allows its two functions of replication and expression.

2.2 The structure of DNA

DNA is a very long polymer made by bringing together subunits known as deoxyribonucleotides, often abbreviated to dNTPs. The biochemical structures of dNTPs are shown in Figure 2.2. A dNTP has three component parts: the "d" for "deoxyribo," the "N" for "nucleotide base," and the "TP" for "triphosphate." The base varies with each nucleotide, while the deoxyribose and the phosphate are constant features of all the nucleotide subunits. Let's discuss these three components in turn.

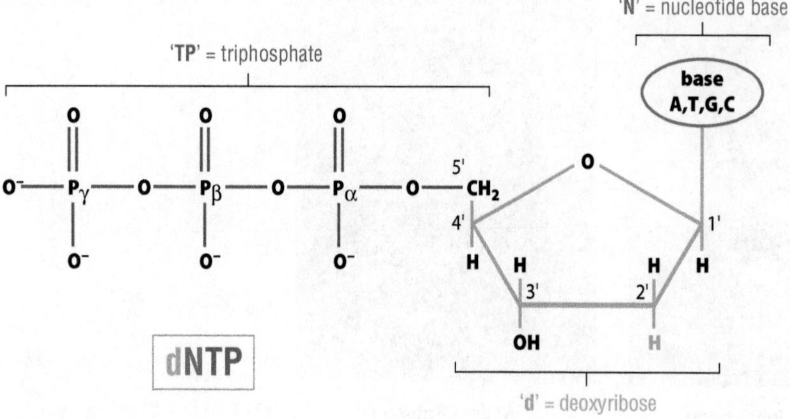

Figure 2.2 The structure of a nucleotide triphosphate, abbreviated dNTP. Note the five-member deoxyribose ring structure, with the –OH at the 3′ position. Three phosphate groups are attached at the 5′ position, while the nucleotide bases (such as adenine, guanine, cytosine, and thymine) are attached at the 1′ position. The numbered positions indicate the directionality of the molecule.

The sugar phosphate backbone

Deoxyribose is a sugar with five carbon atoms, four of which form a ring with an oxygen atom, as shown in Figure 2.2. The five carbon atoms in the sugar ring are numbered 1′ to 5′ (that is, one prime to five prime), based on their positions with respect to the oxygen atom, with the 5′ being the carbon that is not part of the ring. Notice that there is a hydroxyl or –OH group at the 3′ position on the sugar ring, but there is no hydroxyl group or –OH at the 2′ position in the ring; the absence of an –OH at the 2′ position is why this sugar is called "deoxyribose." (We will discuss the ribose sugar, with an –OH in the 2′ position, when we consider RNA.) For our current discussion, the most important aspect is that each carbon atom in the molecule can be numbered, and the ring has a specific orientation. The orientation of the carbons, and the groups that are attached to them, mean that dNTPs and the DNA molecule that is assembled from them also have an inherent directionality. The directionality or the polarity of DNA is one of its most important characteristics.

As seen in Figure 2.2, phosphate groups are found attached at the 5′ position. When DNA is made, the phosphate group closest to the ring (labeled α in Figure 2.2) will be attached to the 3′ position of the ring above it. Thus, each nucleotide that makes up DNA is hooked to the one above it by the attachment of its 5′ phosphate group to the 3′ OH of the preceding base, as shown in Figure 2.3. The building block for DNA is actually dNTP, with the "T" standing for "triphosphate," as shown in Figure 2.2. Notice that the diagram of DNA in Figure 2.3 has only a single phosphate, while the dNTP has three phosphates. When DNA is assembled from the dNTP subunits, two of the phosphate groups are released and produce the energy for the reaction, so the subunit in the DNA molecule is a monophosphate.

The connection of the 5′ phosphate of one dNTP to the 3′ OH of the one before it is what produces the polarity of DNA, as indicated by the arrows in Figure 2.3. This is the backbone of one strand of the iconic double helix. There are two distinct ends to each strand of DNA—a 5′ end with a phosphate group and a 3′ end with an –OH group.

KEY POINT DNA is made up of subunits known as deoxyribonucleotides, each comprising a nitrogenous base, a deoxyribose sugar, and a triphosphate group. As a result of its structure, DNA has a polarity with distinct 5′ and 3′ ends.

So far, we have described one of the two strands of DNA that comprise the double helix. In Figure 2.4, we expand our view, in diagrammatic form, to represent both strands of DNA. The phosphate backbone is represented by the heavy black line, so its chemical structure has been removed for simplicity, and the arrows have been added to show the polarity or orientation of the molecule.

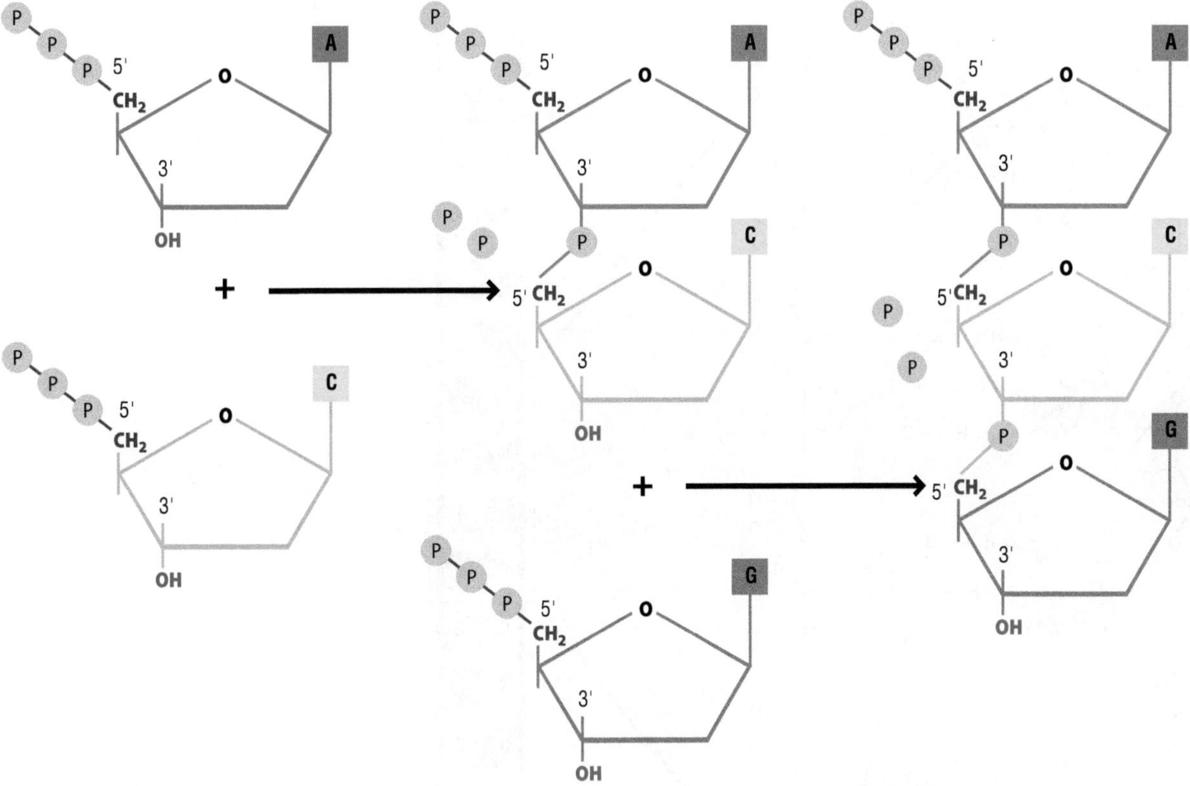

Figure 2.3 The direction of synthesis for nucleic acids. The processes that require the synthesis of nucleic acid polymers, that is, DNA replication and RNA transcription, occur with the same polarity. The phosphate groups at the 5′ position of the incoming nucleotide triphosphate are added to the –OH group at the 3′ position of the preceding nucleotide. In this synthesis, two inorganic phosphates are generated. Synthesis always occurs by adding the 5′ phosphate of the next nucleotide to the existing 3′ OH.

Notice in this figure that the two DNA strands run in opposite directions, as indicated by the arrows, that is, they are said to be **anti-parallel**. Notice also that the phosphate group has a negative charge, which is why DNA is an acid. That negative charge plays an important role in the interaction of DNA with water molecules and in the structure of chromosomes. The net negative charge also provides the basis for separating DNA molecules by size, using gel electrophoresis; this technique is discussed in Tool Box 2.1. In order to detect the presence of DNA during gel electrophoresis or other molecular biology procedures, there have to be methods to visualize it. These are discussed in Tool Box 2.2.

The base pairs

We have discussed the "d" for deoxyribose and the "TP" for triphosphate in the dNTP, so now we turn our attention to the "N" in Figure 2.2. The "N" refers to the nitrogenous base that is attached to carbon at position 1′ on the sugar ring. There are four different bases, and we use "N" to stand for any of them. The four bases fall into two categories: cytosine (abbreviated C) and thymine (T) are **pyrimidines** and have a single ring structure, while adenine (A) and guanine (G) are **purines** and have a double ring structure. These are shown in Figure 2.5(a). So we refer to an unspecified deoxyribonucleotide as a dNTP; in reality, they are dCTP, dTTP, dATP, or dGTP, with the structures shown in Figure 2.5.

The four bases project into the center of the double helix, with the phosphate backbone on the outside, as shown in Figure 2.4. Each strand of the DNA molecule has a specific sequence of the four bases. That sequence of the four bases encodes all of the information in a DNA molecule—a sequence of 5′-TGATCTG-3′ means something different from a sequence of 5′-TTATCTG-3′ because the second base is not the same. The sequence

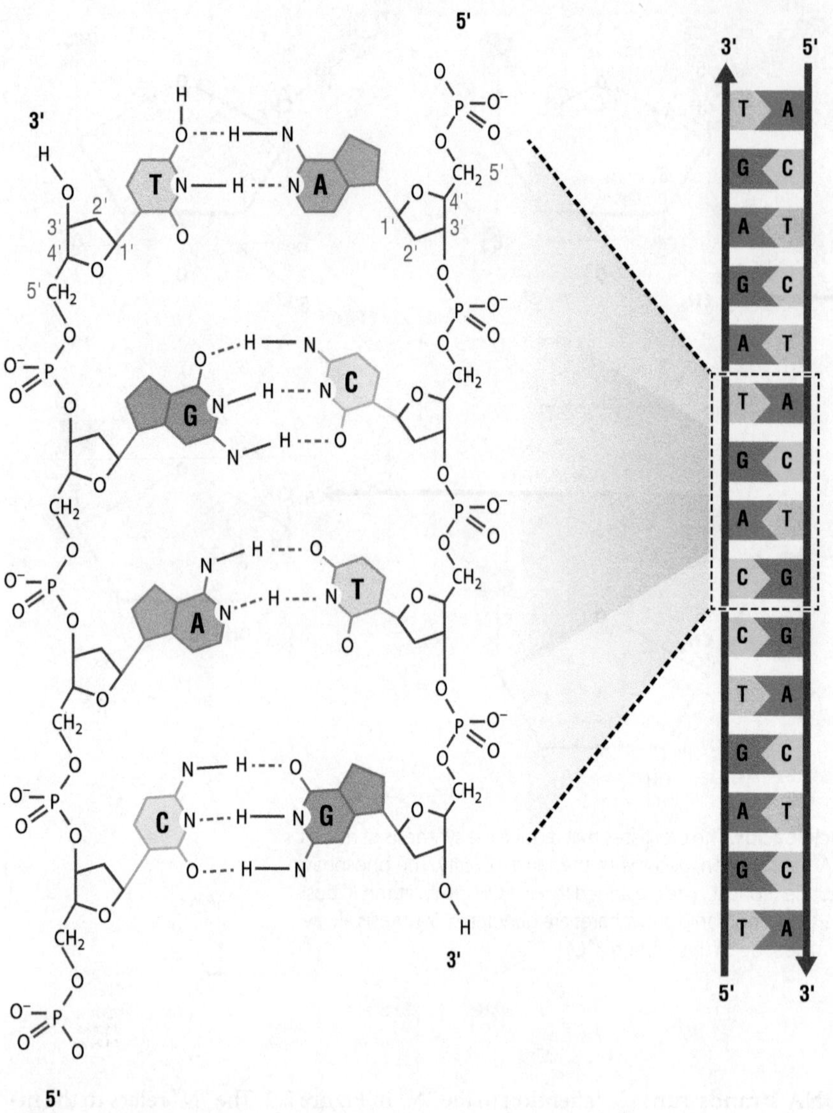

Figure 2.4 The anti-parallel structure of DNA. DNA is double-stranded, but the two strands are oriented in opposite directions, so that the 5′ end of one strand is matched with the 3′ end of the other. The colors used for the bases in this diagram will be used for drawings of DNA throughout the book.

TOOL BOX 2.1 The importance of the DNA backbone: separating DNA molecules by size

The DNA molecule has a long backbone of phosphate molecules on each strand; the backbone can be compared to the uprights on a ladder. A number of different important laboratory techniques have been developed, based on the physicochemical properties of the phosphate backbone; even more techniques are designed around the base pairs, which we discuss in other boxes. In this box, we describe one of the most important techniques based on the phosphate backbone—gel electrophoresis.

Consider a solution in which DNA molecules of different lengths or sizes are present. This is a common laboratory situation, either because there has been a reaction to produce more DNA and the *in vitro* synthesis has resulted in molecules of different sizes or because a much larger DNA molecule has been sheared or broken down into smaller pieces, for example. What method can be used to separate the molecules of different sizes from each other?

Notice from Figure 2.4 that each DNA molecule is negatively charged because of the phosphate backbone. In addition, notice that the negative charges are evenly spaced, with one negative charge in the backbone between each base pair.

The negative charge on the phosphate backbone is the property underlying the technique of electrophoresis to separate DNA molecules. If we put DNA into an ionic solution, such as a salt solution, and apply an electric current to the solution, DNA will migrate to different positions, based on its size and electric charge.

TOOL BOX 2.1 Continued

In fact, DNA is an anion—it has a net negative charge—so it will always run towards the positive pole.

We could put DNA into a salt solution and apply a current, but this will not provide a readily visible method to separate DNA by size. To achieve that objective, we first make a colloidal substance or a gel for the procedure of gel electrophoresis. We then put the DNA in the gel, immerse that in a salt solution, and apply an electric current for electrophoresis. The gel anchors the DNA in place before the current is applied.

However, the gel introduces another property that affects how the DNA will migrate in an electric field, and that property is actually more significant than the charge. If we put DNA molecules of different sizes into a gel and apply an electric current to them, you may think, based on what we have said, that the longer molecules will migrate further (or faster) since they have more negative charges. That is not true because the gel offers resistance to macromolecular movement and acts like a sieve. Shorter molecules migrate faster than long ones because shorter ones can move through the pores of the gel more easily. Think of the gel as being a forest, and the DNA molecules being animals running through the forest. Small animals, like mice and chipmunks, can run through a forest more quickly than large animals like bears and elk because they get to move through small openings between the trees more quickly. In addition, the more densely packed the trees in the forest, the greater the difference between the pace of migration between small and large animals. In the same way, small molecules of DNA run faster than large molecules in the same gel, and both the length and the charge on the backbone determine its position on the gel. In fact, there is a logarithmic relationship between the size of a linear DNA molecule and the distance it migrates on a gel.

Figure A(i) shows a gel in an electrophoresis apparatus (or gel box), with the positive and negative poles indicated. Since DNA is negatively charged and migrates towards the positive pole, we make wells in the gel next to the negative pole for loading the DNA, before we apply the current. The gel is usually made of agarose, a synthetic version of the natural polymer agar, although molecules of DNA less than 150 base pairs long can be separated using a different polymer called polyacrylamide. Polyacrylamide is routinely used to separate proteins by a similar procedure of electrophoresis. Agarose gels are made by dissolving agarose in a buffered ionic solution; the most commonly used solution has the buffer Tris to regulate the pH, acetate ions to provide the ionic environment, and a chemical called ethylenediaminetetraacetic acid or EDTA, so this buffer is called TAE. The role of EDTA is to reduce the degradation or breakdown of the DNA during the procedure. Enzymes that degrade DNA, known as nucleases, require magnesium ions for their activity. EDTA chelates or attaches to magnesium ions in an inactive form, so that nucleases are also inactive. A typical gel is about 1.5% agarose, although lower percentages (such as 0.8%) can be used for long DNA molecules and higher percentages (such as 2%) can be used for shorter DNA molecules.

While the gel is still a liquid, a plastic comb is inserted at one end; when the gel is solid, the comb is removed. The teeth of the comb form the wells into which DNA can be added, as shown in

(i) Gel is prepared

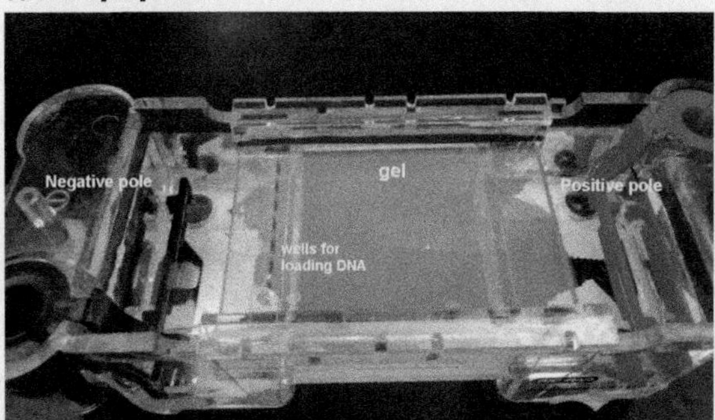

(ii) DNA samples loaded

(iii) Samples begin to run

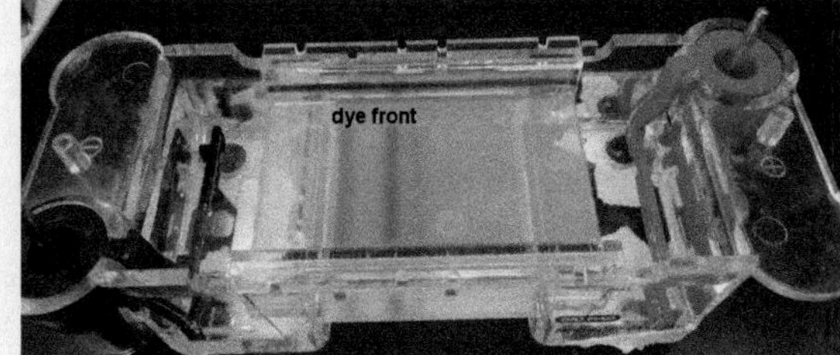

Figure A Gel electrophoresis. One of the most widely used methods to separate DNA molecules of different sizes is agarose gel electrophoresis. As shown in Panel (i), a gel is prepared from agarose and placed in a gel box with electrodes at each end. Note that a comb was placed in the gel before it set, to create wells at one end. These wells will hold the DNA samples. In Panel (ii), the DNA samples have been loaded into the wells. A dark blue loading dye solution is included with the samples. The loading dye includes glycerol, so that the samples sink into the wells, and different dyes that allow visualization of the DNA samples as they are being loaded and the gel is run. Note that the samples are loaded at the negative pole, and electrophoresis will cause them to run towards the positive (red) pole. In Panel (iii), an electrical current has been applied, and the samples have begun to run into the agarose gel. The different dyes included in the loading solution separate, based on size, so the dye fronts provide a rough indication of how far the samples have run. The cover of the gel box has been removed to show the gel as it runs.

TOOL BOX 2.1 Continued

Figure A(ii). The gel is immersed in TAE; the comb is removed, and the DNA is pipetted or loaded into the wells. The wells are placed at the negative terminus, and an electric current is applied. DNA molecules migrate towards the positive pole, which separates them by size. In order to have a standard basis for comparisons, one lane of the gel includes commercially available size markers—that is, a solution that has DNA molecules of defined sizes. A commercially available loading dye is also typically included in each lane, so that the progress of the DNA through the gel can be monitored visually; the loading dye is suspended in a dense liquid, such as glycerol, so that the DNA solution will sink to the bottom of the well when it is being loaded. This is seen in Figure A(iii).

Once DNA molecules have been separated, it is necessary to see where they have migrated. The most common way to detect DNA molecules in the gel is to use a molecule such as ethidium bromide, as we discuss in Tool Box 2.2. Ethidium bromide is typically added to the agarose solution before it has solidified, and the gel is examined under ultraviolet (UV) light.

As described here, agarose gel electrophoresis separates linear, double-stranded DNA molecules, based on the charge on the phosphate backbone and the length of the molecule. Separation by gel electrophoresis depends on the size of the DNA molecule, rather than on its base sequence, so two molecules of about the same length, but of different sequences, will migrate similarly. Related methods can be used to separate RNA and single-stranded DNA molecules, but the process is a bit more complicated since a single-stranded molecule can make intra-strand base pairs. It can also be used for circular DNA molecules, such as plasmids, but the size estimates are dependent upon the conformation of the circular molecules, which can be hard to predict.

TOOL BOX 2.2 Making DNA visible: DNA stains and labels

What does the chemical DNA look like, and how does one see it? DNA can be precipitated out of solution using alcohol and salt. The DNA, which can be spooled out like long pasta if it is unbroken, is colorless. Thus, when DNA molecules are separated by electrophoresis or hybridized with one another, we need to be able to see the location of the DNA. Two general methods are used to visualize or label DNA, so that investigators can track it.

Fluorescently staining DNA

Many dyes that fluoresce under UV light can be used to visualize or stain DNA; these are particularly useful during gel electrophoresis. The most commonly used stain is ethidium bromide (often given the abbreviation EtBr). During gel electrophoresis, EtBr is typically added to the gel before it solidifies. EtBr is a flat molecule with a ring structure, as shown in Figure A(i), that has a natural orange fluorescence under UV light. When added to a

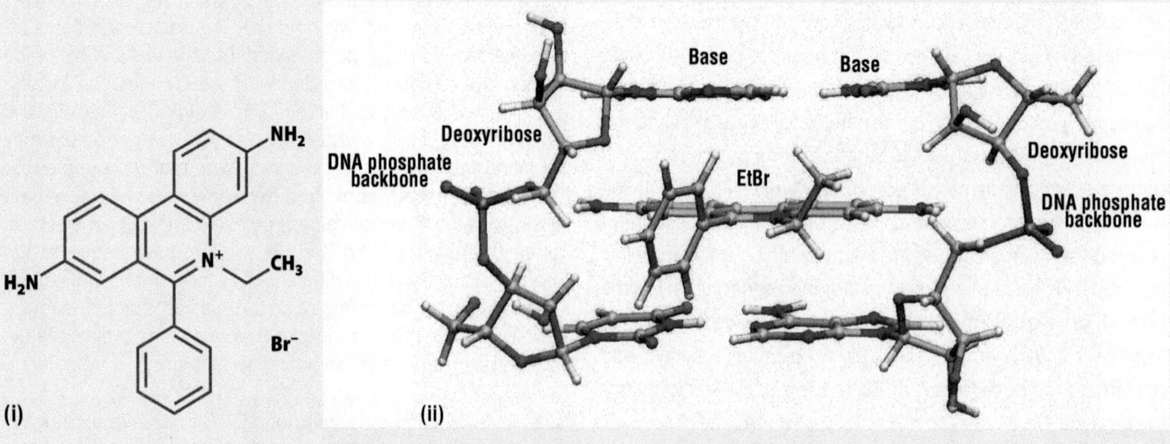

Figure A The structure and action of ethidium bromide. As shown in Panel (i), ethidium bromide (often abbreviated EtBr) is a flat, planar molecule with multiple rings. In Panel (ii), EtBr is shown stacked between the bases of the DNA molecule. The two strands of the DNA molecule are shown on the left and right, with the phosphate backbone, the deoxyribose sugar, and the bases labeled.

TOOL BOX 2.2 Continued

solution with double-stranded DNA, EtBr stacks or intercalates itself between the base pairs, as shown in Figure A(ii). Upon its intercalation between the bases, the natural fluorescence of EtBr increases dramatically. This increased fluorescence is probably because the internal environment of DNA excludes water, and water quenches the fluorescence. Thus, when it intercalates in a reduced water environment, EtBr is unquenched and fluoresces more brightly. This bright fluorescence can be readily detected by UV and photographed, as shown in Figure B.

Because the change in fluorescence requires stacking between the base pairs, EtBr is not widely used for visualizing single-stranded molecules, including RNA. RNA and single-stranded DNA can form intra-strand base pairs, so EtBr can be used for this, but it is not very sensitive. EtBr staining is probably the one molecular biology tool that depends on the most familiar property of DNA, namely the double helix.

While EtBr is widely used and very inexpensive, the potential hazards associated with its use have led to the development of other dyes. These include SYBR dyes that are more sensitive than EtBr and so can detect lower concentrations of DNA. Their toxicity is reported by their manufacturers to be less hazardous than EtBr. SYBR dyes are considerably more expensive than EtBr, so most laboratories use EtBr for routine staining of DNA during gel electrophoresis.

Labeling DNA by sequence

EtBr and related molecules stain DNA non-specifically, so any double-stranded nucleic acid can be seen. This is useful for detecting the presence of DNA, such as for molecules that have been resolved by size on a gel, but it does not detect specific sequences. Thus, methods have been developed to make it possible to label DNA molecules of specific sequences, and a variety of different labels and methods of detection have been developed. Nearly all

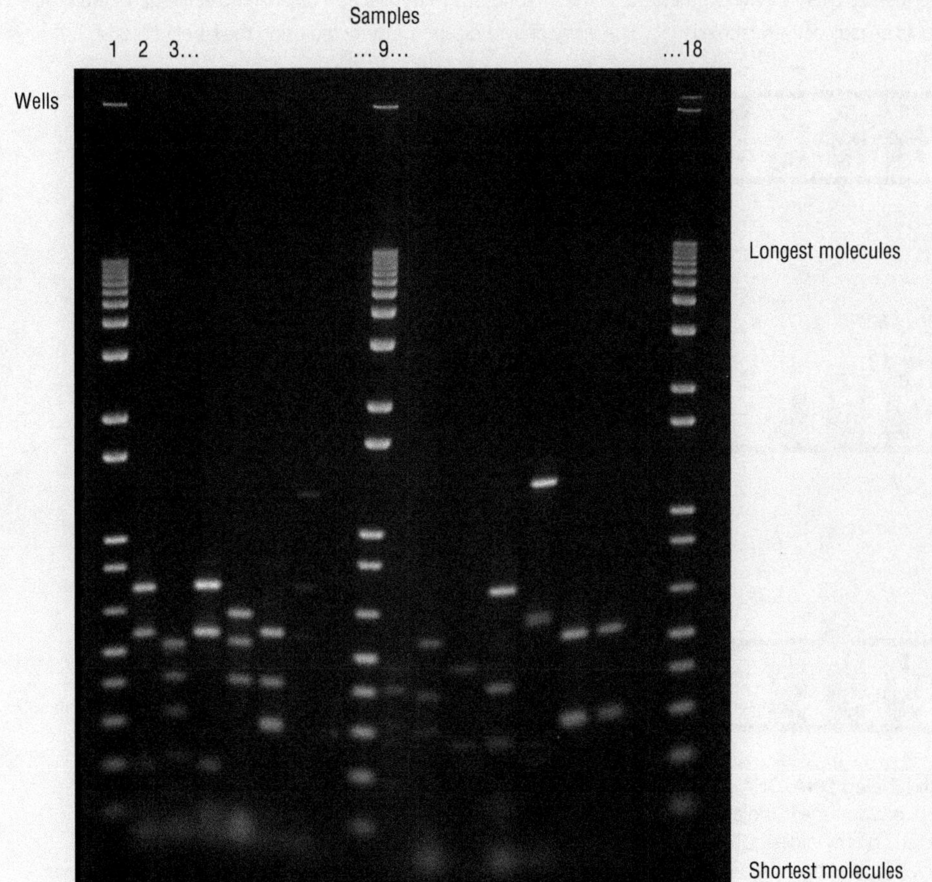

Figure B An ethidium bromide-stained gel. The gel has been stained with ethidium bromide and exposed to ultraviolet light. The image shows the bands for DNA molecules of different sizes, with each lane of the gel corresponding to the sample loaded in each well. Approximately 1 μg of DNA was loaded into each lane of the gel. Smaller molecules are at the bottom, and longer molecules are at the top. The wells are visible at the very top of the gel.

TOOL BOX 2.2 Continued

of these methods begin with a process for synthesizing or replicating DNA *in vitro*. DNA replication, which will be discussed in Chapter 4, requires the enzyme DNA polymerase, a short nucleotide or deoxynucleotide sequence to act as a primer for the reaction, appropriate salts, and the four deoxyribonucleotides (the dNTPs). Since the specificity of DNA functions depend on its base sequence, methods for labeling specific sequences depend on labeling these dNTPs.

In an *in vitro* labeling reaction, one of the four dNTPs has a label attached to it, as illustrated in Figure C. For example, in older experiments, the dTTP solution used for the reaction included some dTTP molecules radioactively labeled with ^{32}P at the α-phosphate position. dTTP was used to prevent the accidental labeling of RNA molecules; RNA has uracil, rather than thymidine. Thus, whenever the copy of the DNA molecule was synthesized *in vitro*, radioactively labeled thymidine could be inserted, rather than regular thymidine. Not every dTTP nucleotide had the radioactive label, so some thymidines are not labeled. Nevertheless, every newly synthesized DNA molecule made *in vitro* would be radioactively labeled.

Radioactive labels have been largely replaced by a variety of non-radioactive labels that are safer to use, but the same basic procedure is used. One of the dNTPs is labeled with some molecule that does not affect its interactions and functions, so that the synthesized DNA can be recognized. Among the other labels that are in widespread use are the fluorescent dyes Cy3 and Cy5, and the small molecule biotin. Cy3 fluoresces in the yellow–green range, while Cy5 fluoresces in the red range. Biotin (also known as vitamin B7) is a small molecule that is "avidly" and specifically bound by the bacterial protein streptavidin. Streptavidin can be purchased with many different tag molecules on it, so the interaction of the labeled streptavidin with the biotinylated DNA molecule indicates the presence of the DNA.

In Tool Box 2.3, we combine the methods for labeling DNA with information about hybridization. In a typical hybridization reaction, a probe is synthesized *in vitro* with a particular label. This is then hybridized to a collection of DNA molecules, and the specificity of hybridization is detected using the labeled probe.

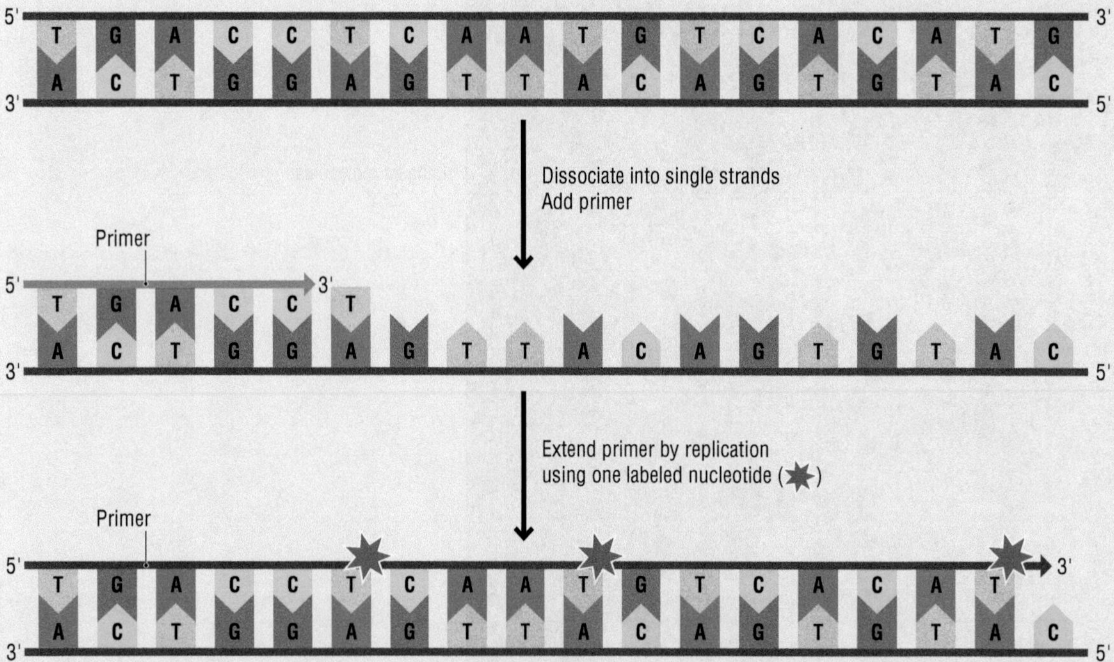

Figure C Hybridization with labeled DNA. DNA can be labeled when it is synthesized or copied by attaching a radioactive, chemical, or fluorescent molecule to one of the deoxyribonucleotides. In the figure, the label is depicted by the red star on thymidine. DNA replication is described in Chapter 4. For labeling, the double-stranded DNA is dissociated into single strands, usually by heating the solution, and a short sequence, known as a primer, is added. The primer, shown here with a blue backbone, has been made *in vitro*, often commercially, with a sequence that is complementary to a sequence in the original DNA. The primer here is five bases long, but this is much shorter than actual primers, which are usually more than 20 bases in length. The primer serves as the sequence to initiate replication *in vitro*, which proceeds by adding new bases to the 3′ end. Although not all thymidines are labeled, there are enough that the entire newly synthesized double-stranded DNA sequence can be detected.

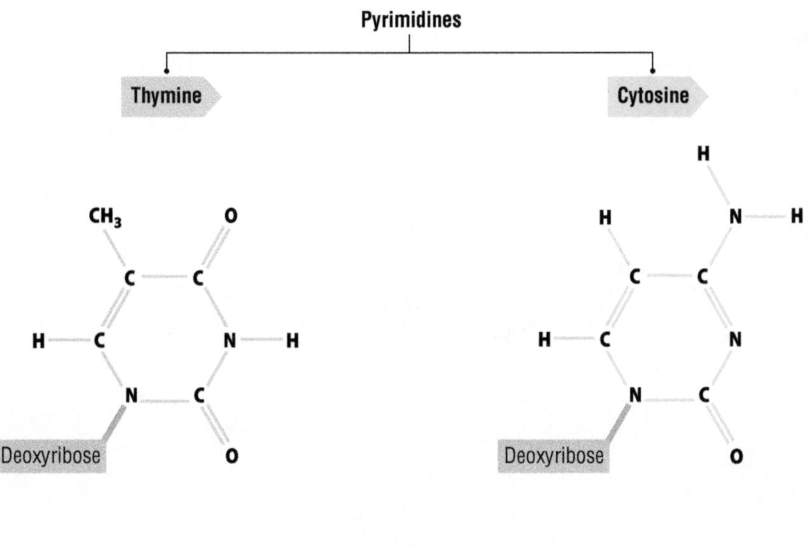

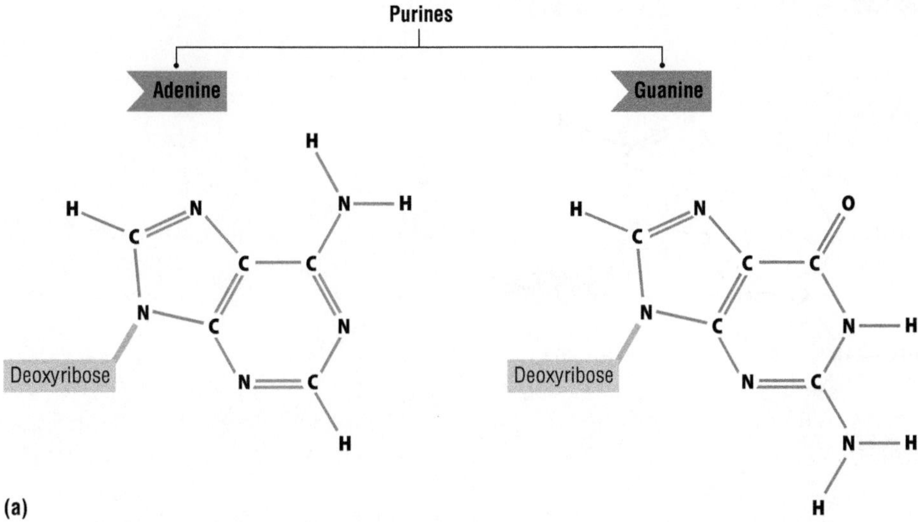

Figure 2.5 Hydrogen bonds formed by the base pairs. The four nucleotide bases fall into two structure categories, as shown in (a). The pyrimidines cytosine and thymine have a single ring, whereas the purines adenine and guanine have double ring structures. (As a memory aid, note the Y in "pYrimidine," "cYtosine," and "thYmine.") The two DNA strands in Figure 2.4 are held together by hydrogen bonds between the corresponding nucleotide bases, shown in (b) as dotted lines with the H. Adenine can form two hydrogen bonds with thymine, while guanine can form three hydrogen bonds with cytosine. While it is structurally feasible to have thymine paired with guanine or adenine paired with cytosine, without distorting the double helix, these pairs are not as stable.

of bases on one strand of the DNA molecule has to be determined experimentally by one of several techniques known as DNA sequencing. However, once we know the sequence of bases on one DNA strand, we automatically know the sequence of bases on the other strand because of base pairing. As spelled out in Communicating Genetics 2.1, the sequence of a section or molecule of DNA is conventionally written from 5′ to 3′ using the first letter of each base.

The bases form pairs with each other, held together by hydrogen bonds. If we use the common metaphor that DNA is like a ladder, the base pairs form the rungs, as is diagrammed in Figure 2.5(b). Since purines and pyrimidines are molecules of different sizes, the base pair consists of one purine with one pyrimidine, so that the distance between the two strands is constant at each base pair. Based on the sizes of the molecules and the positions of the double bonds, there can be weak

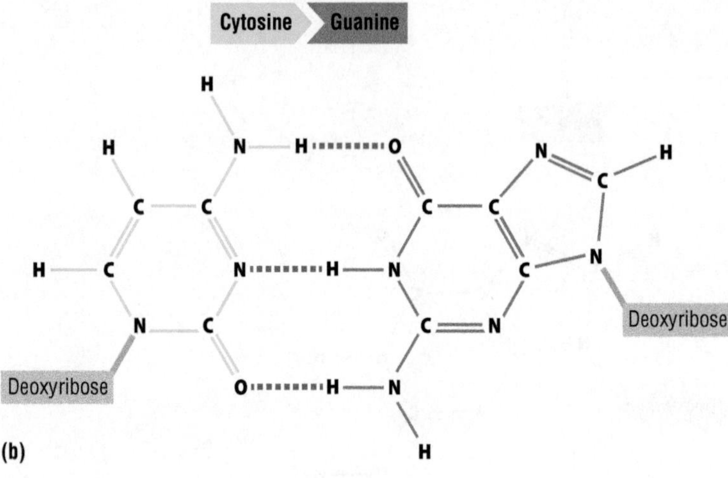

Figure 2.5 Continued

hydrogen bonding between any purine and any pyrimidine, including the ones that are not naturally found in DNA. However, the base pairs that are the most stable are the ones that form between adenine and thymidine as one base pair, and guanine and cytosine as the other base pair. As expressed in the common shorthand, A pairs with T, and G pairs with C, as shown in Figures 2.4 and 2.5(b).

We need to look more closely at these base pairs since base pairing is among the most important principles of molecular biology. While the double helical structure of DNA is very well known to non-biologists, it really doesn't have much impact on how we think about, and work with, DNA in the laboratory. The base pair is much more significant, if not as familiar to the general public. Note that the base pairs are held together with hydrogen bonds (H bonds) between the purine on one strand and the pyrimidine on the other strand. The G:C base pair has three H bonds, while the A:T base pair has two H bonds, as shown in Figures 2.4 and 2.5(b).

KEY POINT Base pairs form between the nitrogenous bases adenine and thymine, and between cytosine and guanine, which are held together by hydrogen bonds.

The difference in the number of hydrogen bonds has both practical and biological implications. Suppose that we have two different fragments of double-stranded DNA in solution in different tubes, and we want to separate these into single strands. In the jargon of molecular biology, we want to **denature** the double-stranded DNA. Both of the fragments are 1000 base pairs long (known as 1 kilobase pair or 1 kbp), but one of the DNA fragments has 653 G:C base pairs and 347 A:T base pairs, and the other fragment has 479 G:C base pairs and 521 A:T base pairs. By convention, this is always expressed in terms of the GC

COMMUNICATING GENETICS 2.1 Sequence shorthand

The nucleotide base sequence of a portion of DNA that a scientist might be studying can range from a few bases for a very specific function to several thousands of bases (or kilobases, abbreviated kb) for a gene. If the topic of interest is an entire chromosome, the sequence may be several millions of bases (or megabases or Mb) in length. The most important information in the DNA is not its chemical formula but the sequence in which the bases are connected together. The easiest way to document these sequences is to write the bases in the order that they occur from 5′ to 3′ as single letters, in which C stands for cytosine, T for thymine, A for adenine, and G for guanine. In RNA, uracil is denoted by U and replaces T in the sequence. Researchers and computers can make inferences from sequences presented in this way, and that is the expected format when one orders DNA of a specific sequence from a commercial vendor.

The abbreviations or symbols A, T, G, C, and U are so widely used that we often do not have to describe them further. A few other notations are helpful as well. If a base's identity is uncertain or unspecified, it is conventionally written as N. We use that notation in the chapter when we refer to dNTPs. If the base is a pyrimidine, but it makes no difference which pyrimidine (or if this is not known), a Y is used; similarly, R stands for a purine.

Just as the information content of a nucleic acid like DNA or RNA lies in the sequence of bases, the information content of a polypeptide lies in its sequence of amino acids, so these are also written out in order. The 20 naturally occurring amino acids can be designated by a three-letter abbreviation—usually the first three letters of the amino acid, except that glutamic acid (Glu) has to be differentiated from glutamine (Gln), and aspartic acid (Asp) has to be distinguished from asparagine (Asn). There is also a single-letter abbreviation for each of the 20 amino acids. This is often the first letter of the name of the amino acid. However, many amino acids begin with the same letters, so many exceptions exist; for example, T stands for threonine, W for tryptophan, and Y for tyrosine. A person who works with polypeptide sequences often usually has these single-letter abbreviations memorized. Both the single- and three-letter abbreviations for the 20 naturally occurring amino acids are shown in Figure 2.15.

When an amino acid residue is unspecified, it is denoted X.

(or G + C) content, so the first of these is 65.3% GC, and the other is 47.9% GC. We gradually raise the temperature of these two tubes and monitor what percentage of the DNA in each tube is still double-stranded at each temperature. At the beginning of our experiment, both solutions consist of 100% double-stranded DNA, with the same salt concentration and the same pH. Which of the two tubes will denature into single strands first?

The answer is that the one with the lower GC content will denature first, that is, the one that is 47.9% GC. As the temperature rises, the H bonds become less stable and come apart randomly. The fragment with 47.9% GC has fewer hydrogen bonds than the one with 65.3% GC, so it requires less energy to separate the strands and will therefore denature at a lower temperature. As a practical matter for research, we don't usually focus on when the entire double-stranded molecule comes apart. We usually focus instead on the **melting temperature**, abbreviated T_m, the temperature at which half of the hydrogen bonds have broken. As a very rough approximation, the melting temperature of the DNA molecule with 47.9% GC content will be about 7°C lower than the one with the GC content of 65.3%.

The GC content (or more precisely the AT content) of a stretch of DNA is also important inside the cell. Other than some viruses in which the DNA is naturally single-stranded, cellular DNA rarely, if ever, denatures into single strands along its entire length. However, there are frequently regions of the genome inside the cell in which the DNA is single-stranded, that is, in which the base pairs separate transiently. We will encounter this when we discuss DNA replication in Chapter 4 and transcription from DNA into RNA in Chapter 12. In both cases, the DNA molecule becomes locally single-stranded for a time to allow one or both strands to be copied. In both cases, the process begins at regions on the DNA molecule that are rich in A:T base pairs with two hydrogen bonds. A:T base pairs can be separated more easily than G:C base pairs—in one helpful expression, the DNA can "breathe" at these locations more easily. These regions in which the base pair composition varies are functionally more significant than the overall GC content of a genome. So while we might write that the average GC content of the human genome is 41%, there are 100 kb regions in which the GC content is nearly 60% and other 100 kb regions in which the GC content is about 35%.

While the nucleotide base sequence of one strand of DNA needs to be determined experimentally, the sequence of the other strand can be immediately and accurately inferred because of the specificity of base pairing. The two strands are said to be **complementary**. If one strand has the sequence 5′-ATCC-3′ (notice that the polarity is being indicated), the other strand will have the sequence 3′-TAGG-5′ or, using the same polarity indicators for both strands, 5′-GGAT-3′. The strands are complementary and anti-parallel. They read as base pair complements, but they read in the opposite directions.

The complementary base pair is quite possibly **the single most important principle** in molecular biology, both in the laboratory and in nature. In nature, complementary base pairs form the basis for DNA replication, transcription, RNA splicing, translation, and many more processes in the cell that will be discussed in future chapters.

Complementary base pairs are also fundamentally important to many laboratory techniques with nucleic acids. Some of the techniques that use base pair complementarity are described in Tool Box 2.3.

Base pairing allows biologists to seek or measure a specific portion of DNA or RNA if some of the sequence of the target nucleic acid is known. The DNA or RNA molecule used as starting material is often called a **probe**, since it is being used to probe or search a complex mixture of nucleic acid molecules. As described in Tool Box 2.3, the probe could be a DNA fragment, several hundred bases long, or simply a short string of nucleotides, that is, an **oligonucleotide**, which is typically under 50 bases long. The process to pair the complementary sequence of the probe with its target is called **hybridization**, as shown in Figure 2.6. Nucleic acid hybridization is based on the principle of base pairing.

TOOL BOX 2.3 The importance of the base pair: hybridization

The DNA molecule has a long backbone of phosphate molecules on each strand, similar to the uprights on a ladder, with the base pairs projecting across the middle, similar to the steps on the ladder. In this box, we describe some important techniques based on the base pair or, more generally, the process of **hybridization**. Hybridization with nucleic acid sequences refers to a process in which one single-stranded molecule forms a double-stranded molecule with another single-stranded molecule with the complementary sequence. The complementary sequence is usually found on another molecule of DNA or RNA, although intra-strand base pairing, in which a single strand forms complementary base pairs with another part of itself, is also significant, particularly for RNA molecules.

Recall that the bases on one strand pair specifically with the bases on the other strand, adenine with thymine and guanine with cytosine. The base pair is due to hydrogen bonds, so the stability of the base pair depends on the temperature and the ionic strength of the solution, which affect hydrogen bonds. The key is the specificity of the base pair. This specificity is central to the fidelity of DNA replication, transcription of RNA from a DNA template, translation from mRNA to amino acid sequences, and many other biological processes. The functional beauty of the DNA molecule lies not in its iconic double helix but in the base pairing—guanine pairs with cytosine, and adenine pairs with thymine (or uracil). Other base pairs can occur, but these are much less stable and are known as mismatches; most of these are transient, rather than stable.

The specificity of the base pairs has also been the foundation for many techniques in molecular biology. Let's imagine an experiment that is conceptually similar to how several laboratory techniques are actually done. We have DNA molecules in solution, with random lengths of 1000 base pairs or shorter. The solution is heated up (to 94°C, for example), so all of the hydrogen bonds come apart and the two strands of each molecule separate; the solution now has single-stranded DNA. This process of separating the two strands is known as **denaturation**. The solution is slowly cooled (to about 53°C, for example). Hydrogen bonds begin to form between the complementary bases, but the strengths of the individual hydrogen bonds for a single base pair, or even for a few base pairs, is not enough to hold the two strands together. On the other hand, as the single-stranded molecules form many base pairs with another strand, the double-stranded structure starts to reform, a process known as **renaturation**. The ability to form stable base pairs between two strands depends on the base sequence of the strands, and the most stable double-stranded molecule will be the one in which the base sequence on one strand is completely complementary to the base strand on the other strand. In other words, given the proper conditions and adequate time, a single-stranded sequence will "find" and pair with its specific complementary sequence to make a double-stranded sequence.

This describes the process of hybridization, and it forms the basis for almost every technique that relies on the specific sequences of DNA and RNA. One very common variation is that hybridization is not done with two full, or equal-length, sequences, as described above. Rather, the investigator has in his possession a short sequence that corresponds precisely to a sequence within a much longer gene or region of the genome. "Short" in this case

TOOL BOX 2.3 Continued

refers to a sequence that is roughly 20–50 nucleotides in length, but this can vary; specific sequences like this are called oligonucleotides, often shortened to "oligos," and can be purchased from many companies at a cost of less than 50 cents a base. Most molecular biology laboratories have boxes with dozens or hundreds of different specific oligonucleotides in the freezer. These are likely to be single-stranded sequences, although double-stranded oligos are not uncommon for some applications. Because denaturation is an early step in many of these experimental techniques, double-stranded sequences can be readily made into single-stranded sequences. The oligo is identical to the sequence of one strand but complementary to the sequence of the other strand.

For many experiments, these specific oligonucleotides are used to locate the particular complementary sequence in a collection or library of longer and more complex DNA sequences. In the jargon of molecular biology, these oligos are used as a **probe** to find a specific location in the genome that has that sequence. For example, suppose that the laboratory is studying a particular gene in an organism and knows the DNA sequence of this gene and its flanking location in the genome. They might have had oligos that are complementary to particular regions of the gene, which allow them to identify and analyze their gene of interest. Depending on the application, the probe may need to have a label that allows the hybridization to be detected; methods for labeling and visualizing DNA are described in Tool Box 2.2.

The concepts of oligos, probes, and hybridization will be used, as other techniques are described. One of the most important aspects of these techniques is that the oligonucleotide needs to be specific to a particular region and does not cross-hybridize to another region in the genome. The two parameters that affect specificity are the base sequence of the oligonucleotides and the conditions (including temperature) under which the hybridization is done. Specificity of the sequence typically is done by increasing the length of the oligo; a ten-base oligo is much more likely to bind non-specifically and have cross-reacting sequences than a 50-base oligo. On the other hand, a 50-base oligo may be longer than is necessary for most applications, which affects how long it takes to find it its complementary sequence. However, even an oligo that has exactly one complementary sequence will hybridize to multiple targets under certain conditions; for example, if the temperature is too low, mismatches with related, but non-identical, sequences will occur. Sometimes the investigator will want to allow mismatches—for example, in searching for related, but not identical, genes—but commonly the frequency of mismatches is minimized by raising the temperature of hybridization.

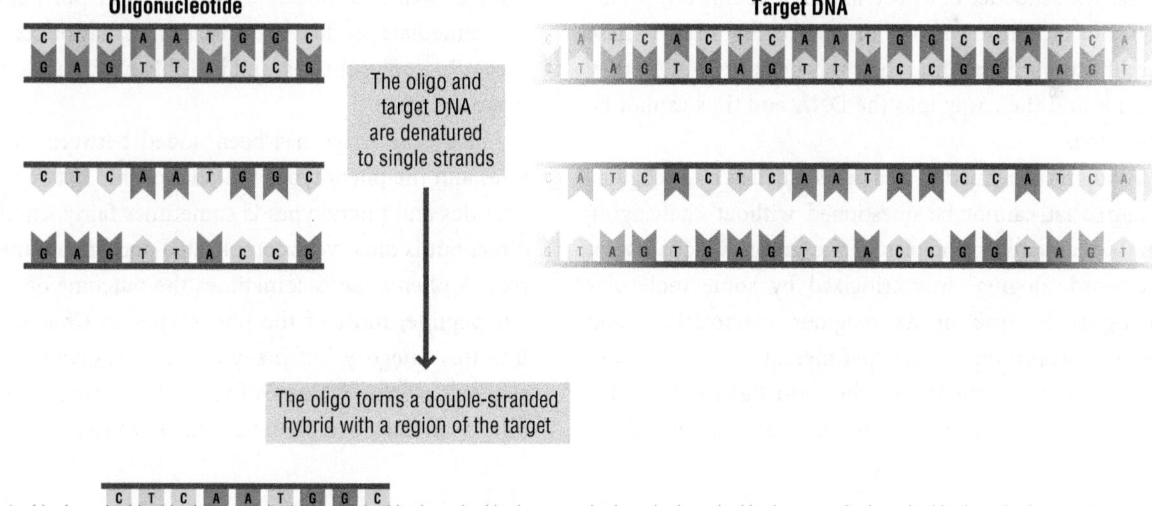

Figure 2.6 Hybridization of short DNA sequences to longer target sequences. The specificity of the base pairs provides an important experimental tool that is used repeatedly in molecular biology techniques. A short sequence of DNA (or RNA) with a defined sequence can be synthesized *in vitro*; these are usually known as oligonucleotides or "oligos." Oligonucleotides can often be labeled, as discussed in Tool Box 2.2. The oligonucleotide and the much longer target sequence, which can also be either DNA or RNA, are mixed under conditions that allow base pairs to form. The oligonucleotide forms base pairs with its complementary sequence to form a double-stranded hybrid molecule. Tool Box 2.3 provides additional information on hybridization methods.

KEY POINT Complementary base pairing is essential to DNA replication, transcription, and other important processes in the cell, as well as to techniques used frequently in the laboratory.

Having looked at the structure of DNA, let's see how that structure is involved in the two fundamental properties of a gene: gene replication and gene expression. We will provide an overview of gene replication and variation in Chapter 3 and a more detailed account in Chapter 4. We will provide an overview of gene expression in this chapter and will fill in the details in Chapters 12 and 13. We begin our overview of gene expression with the Central Dogma of molecular biology—DNA makes RNA, and RNA makes protein.

2.3 DNA and the Central Dogma

The Central Dogma of molecular biology was first stated directly by Francis Crick in 1958, and, with some modifications, it has formed the centerpiece of molecular biology since then. The Central Dogma is shown in diagrammatic form in Figure 2.7.

DNA forms the **template** that is used to make RNA. This process is called **transcription**. RNA forms the template that is used to make protein in a process called **translation**. Notice that, in Figure 2.7, the process of gene expression runs in just one direction, as indicated by the arrows. (There is an important and useful exception to these arrows that is found in some viruses as part of their infection process, as described in Tool Box 2.4.) This simple diagram has some very important implications. The sequence of a protein can be chemically modified, but, according to the arrows in Figure 2.7, these modifications do not find their way back to RNA, so they cannot find their way into the DNA and thus cannot be inherited.

A dogma is defined to be a belief (such as by a religious group) that cannot be questioned without challenging the fundamental principles of the group. Crick's use of the word "dogma" was criticized by some molecular biologists because of its religious connotations and because everything in an experimental science needs to be challenged. Nonetheless, the term has stuck. (Crick said, in his autobiography, that he was thinking of the adjective "dogmatic," meaning something that a person insists on during a discussion or an argument.) But even a dogma requires clarification and elaboration by later scholars. So it is with the Central Dogma.

Figure 2.8 shows a more elaborated version of the Central Dogma of molecular biology, with a few additional arrows and terms.

First, notice that an arrow has been added that shows DNA being used as a template to make more DNA; this is the process of DNA replication.

Second, the word "protein" has been replaced by the more accurate term "polypeptide." In brief, a protein can be composed of more than one polypeptide. (For example, the protein hemoglobin consists of four polypeptides, as well as a molecule of heme.) The polypeptide is the immediate product of translation; it may be combined with other polypeptides, and possibly other molecules, to make a protein.

Finally, an arrow has been added between polypeptides and the phenotype. The relationship between polypeptides and phenotypes is sometimes fairly simple and direct but is often very complicated and often quite indirect. A phenotype is sometimes the outcome of a single polypeptide; most of the phenotypes in Chapter 5 fall into this category, but many of the phenotypes in other chapters are the outcome of more than a single polypeptide. As indicated in Figure 2.8, the RNA that is translated

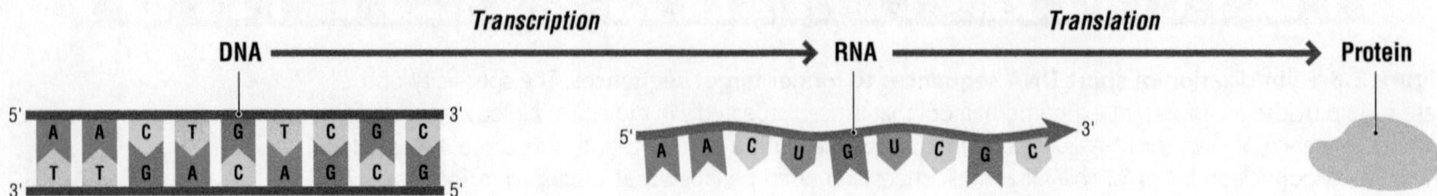

Figure 2.7 The Central Dogma of molecular biology. The arrows indicate the direction of information flow. DNA provides the template to make RNA via transcription, while RNA provides the template to make a protein or a polypeptide via translation.

TOOL BOX 2.4 cDNA

DNA molecules are usually very stable and not very hard to work with in the laboratory. While many enzymes known as nucleases (or, more formally, deoxyribonucleases since they work on DNA) are capable of cutting or digesting DNA molecules, the activity of most nucleases that work on DNA can be inactivated readily. We noted in passing in Tool Box 2.1 that agarose gels are run in solutions with the chemical EDTA since EDTA inactivates most deoxyribonucleases. With a bit of care and the appropriate ionic conditions, DNA can be isolated and manipulated in the laboratory.

RNA, on the other hand, is not very stable and requires much more care to prevent its degradation. Ribonucleases, the enzymes designed to degrade RNA in the cell, are notoriously resistant to inactivation, so it is much less simple to isolate and manipulate RNA in the lab. This is particularly true for mRNA, the transcripts that will be translated into amino acid sequences; we often want to isolate and study the transcripts, either for a gene or for many genes, so it helps to have methods to make mRNA easier to work with. (Other types of RNA molecules that we will encounter in Chapters 12 and 13, namely tRNA and rRNA, are covalently modified in the cell after transcription, which makes them resistant to ribonucleases and easier to isolate.) The best of the methods to study mRNA require making a DNA copy of it first. A DNA copy of an RNA molecule is known as a cDNA (for complementary DNA) and is made by a class of enzymes known as reverse transcriptases. In this box, we provide a brief overview of cDNA and reverse transcriptases.

What is cDNA?

Generally speaking, cDNA is not a naturally occurring molecule but is instead a DNA molecule that has been synthesized in vitro using mRNA as the template. Because it is made from an mRNA template, the cDNA for a gene lacks the regulatory regions of genomic DNA that control expression of the gene, and it lacks the introns present in most eukaryotic genes. The cDNA therefore retains all the protein-coding information of the original gene but is more compact and less unwieldy than the full genomic gene sequence. Since DNA is more stable than RNA, it is also a more stable version of the protein-coding sequence than the original mRNA was. Furthermore, because a cDNA contains the protein-coding sequence in DNA form (unlike the mRNA), it can be integrated into other DNA contexts, such as bacterial plasmids, or into other genomes.

How is cDNA made?

The Central Dogma dictates that information flows in one direction, from DNA to RNA to protein. However, in the 1970s, it was shown that certain viruses defy the Central Dogma and are, in fact, capable of "going backwards" by converting RNA into DNA. Such viruses are termed retroviruses; these include the human immunodeficiency virus (HIV) and Rous sarcoma virus, as well as many more. These retroviruses carry their genetic information in the form of an RNA genome. Upon infecting a cell, they then convert their RNA genome into DNA, which integrates into the host genome. Once integrated into the host genome, the virally derived DNA then proceeds through the usual information flow, as dictated by the Central Dogma, from DNA to RNA and proteins, to produce the progeny viruses.

The key enzyme in the initial conversion of the RNA viral genome into a DNA copy is called reverse transcriptase; this enzyme is the one that scientists use to make cDNA in the laboratory. The discovery of the reverse transcriptase enzyme was an important part of the work that led to the award of the 1975 Nobel Prize for Physiology and Medicine to Howard Temin and David Baltimore. These scientists shared the Nobel Prize with Renato Dulbecco for their combined discoveries "concerning the interaction between tumour viruses and the genetic material of the cell."

To make cDNA, we use the reverse transcriptase enzyme to convert an mRNA template into a DNA copy, just as the retroviruses do, as shown in Figure 2.10 in the text. The reverse transcription of the mRNA results in a single-stranded DNA molecule that is complementary to the mRNA sequence. This single-stranded cDNA molecule is then usually converted to double-stranded cDNA. A cDNA can be labeled *in vitro* by the same methods described to label DNA in Tool Box 2.2, by having one of the deoxyribonucleotides tagged by fluorescence, for example.

To make the double-stranded cDNA, the original mRNA template is removed chemically or by ribonuclease, and DNA polymerase is then used to synthesize the complementary second strand of DNA from the first cDNA template. Once DNA polymerase has completed its job, the final product is a double-stranded DNA version of the original mRNA sequence, one strand identical in sequence to the mRNA and the other strand complementary to it. Although the term cDNA originally referred only to the initial single-stranded DNA copy of the RNA (the initial "complementary" copy of the RNA), the term cDNA is now frequently used to also refer to the double-stranded DNA copy of the mRNA.

Why is cDNA made?

A cDNA copy of a gene includes the protein-coding sequence from the mRNA but is more stable than mRNA; this is useful for a variety of experiments. For example, cDNA is used widely in gene cloning and in studies of the function of the encoded protein product. A cDNA can be cloned into a bacterial plasmid, as a

TOOL BOX 2.4 Continued

means of maintaining and producing more of the protein-coding sequence of the gene of interest. Most cDNAs are kept in the lab as inserts into bacterial plasmids.

The cDNA can also be placed under the control of other regulatory sequences to express the protein in a variety of cell culture systems, to examine the protein product. When bacteria are used to express a eukaryotic gene, they would usually be unable to perform the necessary splicing to remove the introns present in the original gene's RNA (see Section 2.4), but since the cDNA lacks introns, it can be expressed successfully in bacterial systems. A cDNA can also be placed into a genome under a specific regulatory region and promoter of the scientist's choosing, to express the gene at a particular time and place in an organism. This can be used to test the function of the encoded protein in a different context to its usual expression or to provide the protein in situations that would otherwise lack the protein (as forms the basis for many medical gene therapy approaches). Since the cDNA is more compact than the original genomic gene sequence, it is more easily manipulated and will usually be expressed more efficiently in these scenarios. When cDNAs are produced from a mixed pool of mRNAs, they can also form a useful library of all the mRNA sequences that were being expressed in the cells and tissues that the mRNA was collected from. Such cDNA libraries can be used to understand which genes are important in the development and function of particular cell types and how genes get spliced. Overall, the use of cDNA is widespread and varied, and the ability to make cDNA has enabled many experiments that would otherwise have been impossible to perform. All thanks to retroviruses and the scientists who discovered their Central Dogma-defying enzyme—the reverse transcriptase!

FIND OUT MORE

DNA From The Beginning. Concept 25. *Some viruses store genetic information in RNA*. Available at: www.dnaftb.org/25/ [accessed 2 August 2016]

Nobelprize.org. *The Nobel Prize in Physiology or Medicine 1975*. Available at: www.nobelprize.org/nobel_prizes/medicine/laureates/1975/ [accessed 2 August 2016]

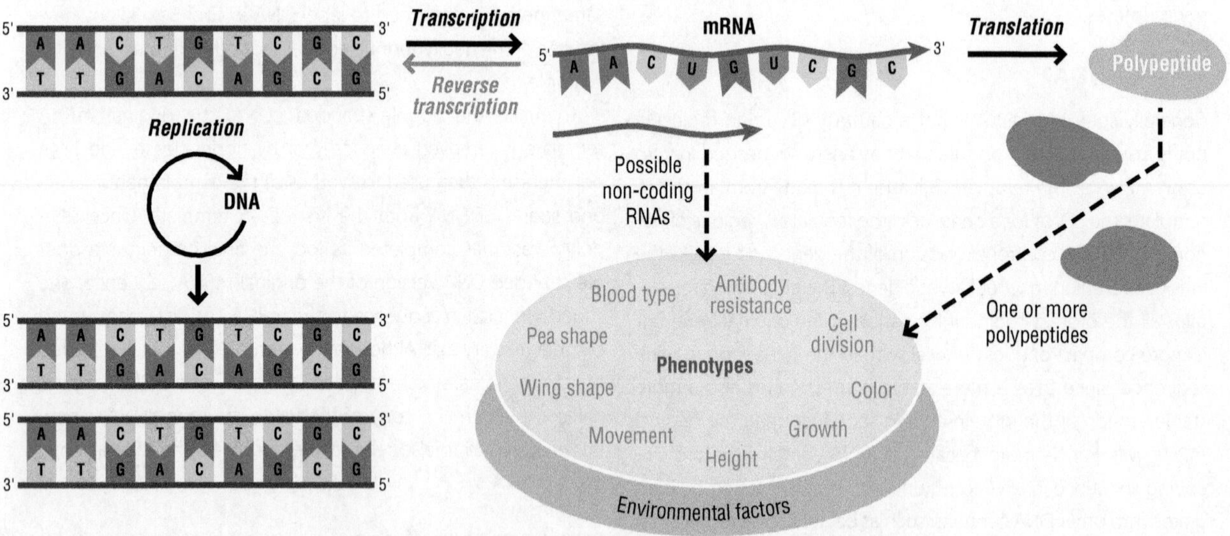

Figure 2.8 An expanded view of the Central Dogma. Arrows have been added to indicate that the DNA also provides the template for its own synthesis, a process known as replication. Some viruses encode enzymes that allow RNA to serve as a template for the synthesis of DNA, a process known as reverse transcription. The RNA molecules that become translated are the mRNAs. The polypeptide provides the basis for a phenotype, either by itself, in combination with the polypeptide products of other genes, in combination with the functional RNA product of other genes, or in combination with environmental factors. Depending on what phenotype is being considered, any or all of these may make significant contributions.

into proteins is known as the messenger RNA or mRNA. However, some phenotypes arise from RNA molecules that are not translated into proteins. A phenotype may be the outcome of many different proteins, some RNAs, and the environment.

The overview of the Central Dogma that is provided in the next few pages sets a foundation that is important throughout the book. Even if the concepts seem familiar, they may not be simple, and the connections between the concepts can provide many nuances.

Gene expression begins with transcription

A common metaphor is to describe the DNA sequence in a genome as a book. But a book sitting untouched on a night table does not convey much information; the information is expressed when the book is read. Similarly, the DNA sequences of genes carry out their functions by being expressed. What do we mean by gene expression? A gene, which is made up of a sequence of DNA, is transcribed into a sequence of RNA. **Transcription**, the first step of gene expression, results in an RNA copy of one of the strands of the DNA molecule.

We need to describe how a transcript, comprising RNA, is similar to, and different from, the original gene comprising DNA. The chemical differences between a sequence of DNA and a sequence of RNA may seem small, but they are significant.

First, RNA is usually single-stranded in the cell, while DNA is usually double-stranded. There are many processes in which a section of DNA is single-stranded locally, but overall DNA has two strands. RNA may have regions that are double-stranded, most often when base pairs form between different regions of the same strand but occasionally from two different complementary RNA sequences. This can be quite important, but, generally speaking, RNA is single-stranded and DNA is double-stranded.

Second, the core sugars that are used to make the nucleotide building blocks are slightly different between them; DNA uses deoxyribose as its core, while RNA uses ribose as its core, as shown in Figure 2.9. Notice

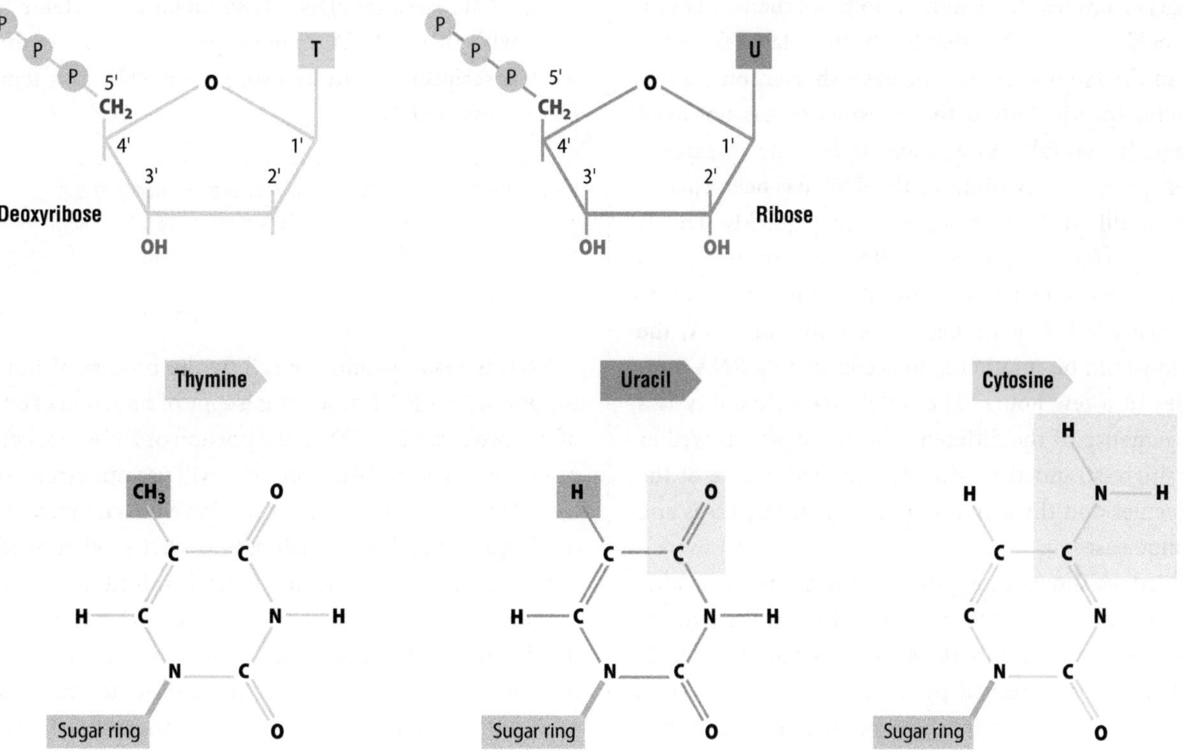

Figure 2.9 DNA and RNA. The product of transcription is ribonucleic acid or RNA. DNA and RNA are chemically similar, but the differences are critically important for their functions. First, RNA is built on a ribose ring, rather than a deoxyribose ring. The difference is the presence of the –OH at the 2′ position in the ribose ring, as shown at the top of the figure. Second, RNA uses the base uracil (U), rather than thymine which differs by a CH_3 group, as highlighted in purple. Uracil and cytosine differ by the presence of a double-bonded oxygen (U) or an NH_2 group (C).

that ribose has a hydroxyl (–OH) group on the carbon at position 2′. This seemingly small difference between the sugars results in an even more significant difference in the chemical properties and thus the functions of DNA and RNA. DNA is extremely stable. DNA has been isolated from specimens that are tens of thousands of years old, and, while the DNA molecule is not fully intact, its base sequence can still be determined. We have DNA sequence data from Neanderthal bones that are 45,000 years old, for example. Less dramatically than these ancient DNA extractions, a researcher only needs to take a few simple precautions to store a DNA sample in the refrigerator for a decade or more and still have it be undamaged. The inherent stability of DNA confers an evolutionary advantage, since it means that the molecule of inheritance is stable and intact from one generation to the next.

On the other hand, RNA is unstable, both because of its inherent chemical structure and because there are enzymes in the cell known as ribonucleases that degrade it. The median half-life of mRNA molecules in a mammalian cell is about 9 hours, although the range is from less than an hour to more than 15 hours. The half-life of mRNA that has been isolated from the cell in the laboratory may be even shorter, only a few minutes (particularly if the investigator has not been especially careful). As a result, unless the researcher takes specific precautions or the RNA has been chemically modified, RNA is degraded very quickly. This is also important evolutionarily. RNA is a transcript, so its function is to be synthesized (transcribed), used, and degraded. If RNA had the stability of DNA, the world would be smothered in accumulating RNA molecules in a few hours. The difference in stability is a consequence of the difference in the sugars, as well as the single-stranded vs. double-stranded nature of the molecules and the activities of different nucleases and ribonucleases.

Third, as shown in Figure 2.9, one of the nucleotide bases is different between DNA and RNA; the pyrimidine base thymine (abbreviated T) is found in DNA, while the closely related pyrimidine base uracil (U) is found in RNA. The change of bases actually may be relatively minor; DNA can exist with uracil, rather than thymine, temporarily, although repair mechanisms identify and replace the Us with Ts at or before the time of DNA replication. Compared to uracil, thymine appears to offer DNA a number of selective advantages in its function as the repository of genetic information. Thymine is more limited in its pairing and thus more specific in its information content, whereas uracil pairs more readily with bases other than adenine. Also, when DNA is degraded, cytosines are deaminated to uracil. (Notice from the structures of the bases in Figure 2.9 that if NH_2 of cytosine is replaced by oxygen with a double bond, cytosine becomes uracil.) The cell's repair systems recognize that a U in the DNA sequence is a consequence of chemical damage and fix the sequence; this could not happen if U occurred naturally in DNA.

Because mRNA is unstable, it is more difficult to work with in the laboratory than DNA. Thus, in the laboratory, mRNA is often isolated carefully and immediately "reverse transcribed" into a DNA molecule with deoxyribose as the sugar and thymine, rather than uracil, as the base. We discuss this technique in Tool Box 2.4. This DNA copy of an RNA molecule is referred to as the **complementary DNA** or **cDNA**; it is the copy of a transcript only in DNA form, rather than RNA form. The process of making cDNA is summarized in Figure 2.10. Because cDNA is so much more stable to work with than mRNA, many experiments that examine transcription begin by using the mRNA as a template to make cDNA.

KEY POINT RNA differs from DNA in its sugar ribose and its incorporation of uracil, instead of thymine, bases. RNA is short-lived and unstable, while DNA is very stable and remains intact for long periods of time.

RNA is made within the cell by the process of transcription. An RNA transcript is a copy of a portion of one of the two strands of DNA, the portion of DNA that corresponds to a gene. The transcript will use one strand of the DNA as its template, so it will have the complementary sequence to this template strand. On the other hand, it will have the same sequence (with U substituted for T) as the other strand of the DNA molecule. This terminology is easy to misunderstand. As shown in Figure 2.11, the **coding strand** is the one with the **same sequence** as the RNA sequence (with the replacement of T by U). The coding strand is sometimes called the non-template strand or the sense strand. The other strand is called the **template strand**, since that is the one that forms the template for the RNA transcript. It will be the complement of the RNA transcript. The template strand is also called the

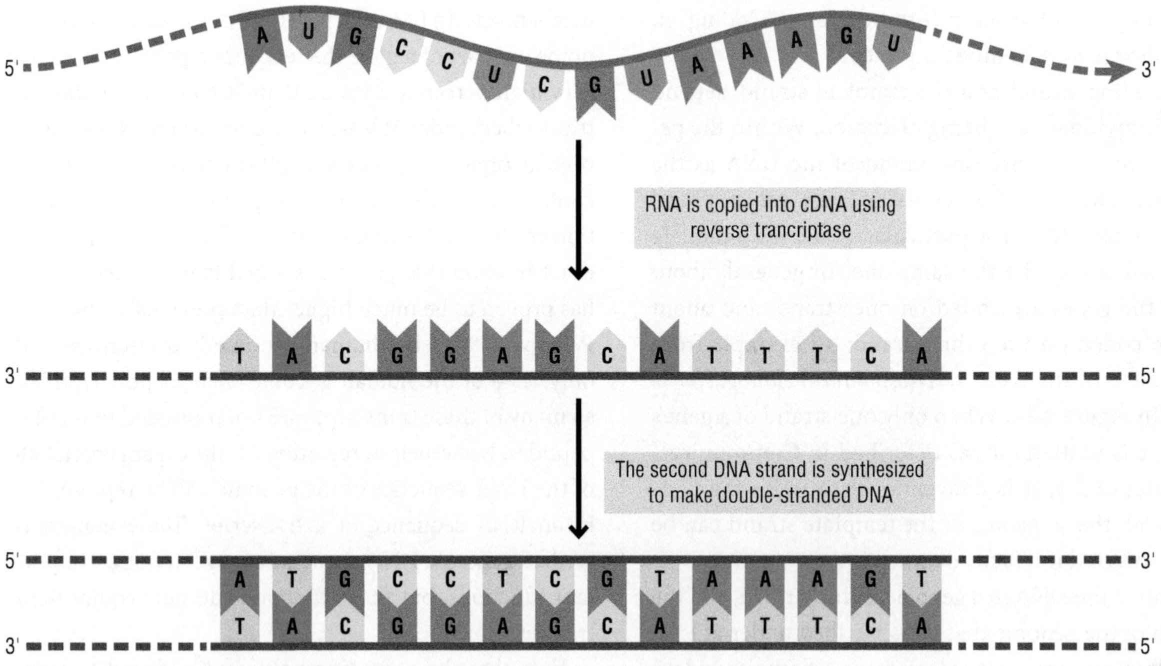

Figure 2.10 Making complementary DNA or cDNA. Because RNA is chemically unstable, while DNA is quite stable, many techniques for working with RNA in the laboratory include a step to make the corresponding DNA molecule. The RNA to be copied is combined with an enzyme known as reverse transcriptase, a class of enzyme encoded in the genomes of certain viruses. Reverse transcriptase is able to use RNA as a template to make DNA; this is the backwards "reverse transcription" arrow in Figure 2.8. The DNA made from such an RNA is called complementary DNA or cDNA. There is usually a second step that uses the first cDNA strand to make the second DNA strand, since double-stranded DNA is much more stable than single-stranded DNA. Note that the sequence is written with the 5′ ends of the RNA on the left, but the bases at the 5′ end of the cDNA are the reverse complement of the ones at the 3′ end of the RNA; similarly, the bases at the 3′ end of the cDNA are the reverse complement of the 5′ end of the RNA. This illustrates the anti-parallel nature of the nucleic acids.

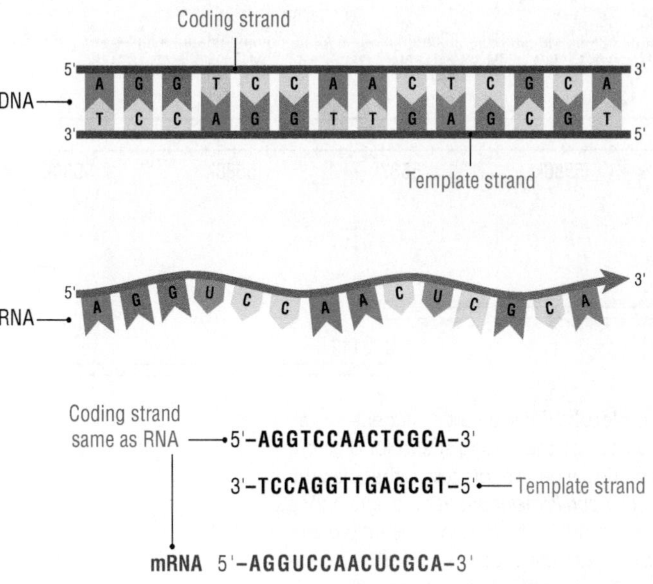

Figure 2.11 Coding strands and template strands. Since one strand of DNA forms the template for the synthesis of the other strand, the two strands are named differently. Although other names are sometimes used, we will refer to the DNA strand with the same sequence (with T, rather than U) running in the same direction as the RNA transcript as the coding strand of DNA. The reverse complement of the RNA transcript is the template strand. Usually, only the coding strand is given when the base sequence of a gene is written.

non-coding strand or the anti-sense strand. "Coding" in this context means "coding for protein."

The coding strand and the template strand depend on the individual gene being discussed. Within the genome, some genes use one strand of the DNA as the template, whereas other genes use the other strand as their template; for a particular gene, the template strand will always be the same one. In general, about half of the genes are coded on one strand, and about half are coded on the other strand; an example from the genome of the fruit fly *Drosophila melanogaster* is shown in Figure 2.12. When only one strand of a gene's sequence is written out, as described in Communicating Genetics 2.1, it is conventional to write the coding strand; the sequence of the template strand can be readily inferred if it is needed.

Not all of the DNA in a genome is transcribed, and the fraction of the genome that is transcribed under at least some circumstances or in at least some cells varies widely. In addition, the fraction of the transcripts that are translated into polypeptide also varies. As a result, every step in the information flow of the Central Dogma represents only some of the information available from the preceding step, and thus every step offers potential regulatory mechanisms. In bacteria, nearly all of the bases in the genome are part of a gene that encodes a polypeptide, so a very high percentage (more than 90%) of the genome is transcribed under at least some circumstances. In multicellular organisms, only a small fraction of the genome contains genes that encode polypeptides, so only a fraction of the total genome is transcribed. In humans, the exact fraction that gets transcribed is under debate and has proven to be much higher than previously expected. Perhaps 70% of the human genome is transcribed, but only 1.5% of the human genome encodes polypeptides, so many of these transcripts are not translated into polypeptides. Nonetheless, regardless of the organism, not all of the DNA sequence of the genome will be represented in an RNA sequence of a transcript. These non-transcribed genomic DNA sequences may have other important functions, but those functions do not require their transcription.

Note that the preceding paragraph referred to transcription "under at least some circumstances," indicating that not all cells are making the same transcripts all the time. With rare exceptions, all of the somatic (or non-reproductive) cells in a multicellular organism have all of the genes in a genome; DNA extracted from

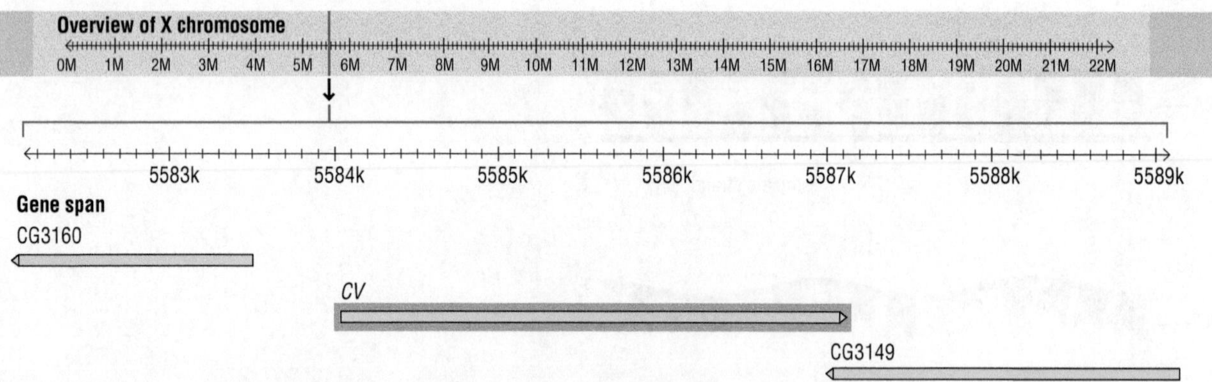

Figure 2.12 Genes are encoded on both strands of the DNA molecule. While a particular gene will always use the same strand of DNA as its coding strand, both strands can be the coding strand for different genes. Thus, the terms coding strand and template strand refer to a particular gene or process, rather than the entire molecule. In this example of a small region of the genome in *Drosophila melanogaster*, a diagram for a gene known as *crossveinless* is shown (labeled *cv* and highlighted in the center), with its two adjacent protein-coding genes (CG3160 and CG3149). The point at the end of the line indicates the direction in which the gene is transcribed, and thus which strand is the coding strand for that particular gene. The *crossveinless* gene is encoded on the opposite strand from the neighboring genes. While neighboring genes are encoded on opposite strands in this particular example, they may be on the same strand or opposite strands in other cases. Note also that the end of the *crossveinless* gene overlaps slightly with the CG3149 gene to its right. This figure has been redrawn slightly from the *Drosophila* database Flybase. The top of the figure indicates the base pair location about 5,584,000 bases from the left end of the X chromosome.

a nerve cell has the same nucleotide sequence as DNA extracted from a skin cell of the same individual. But these cells do not have the same functions or the same morphology. Why not? The answer is that cells do not transcribe the same genes; nerve cells transcribe some genes that skin cells do not transcribe, and vice versa. Some genes, commonly called "housekeeping genes," are transcribed in nearly all cells under nearly all conditions. These genes encode proteins whose functions are necessary for all cells. Other genes are transcribed only in certain cells under certain conditions. The catalog of transcripts that are made by a cell type, called the **transcription profile** or **expression profile**, is characteristic of that cell type and that stage of the life cycle. The concept of an expression profile is developed in more depth in Section 2.5.

Transcription is followed by translation

As noted earlier, transcription of a section of DNA into an RNA sequence does not explain how all genes carry out their functions. The next step in the gene expression process for some genes is that the nucleotide sequence of the RNA transcript is **translated** into the amino acid sequence of a polypeptide (or protein). The RNA transcripts that will be translated into amino acid sequences are called **messenger RNAs** or **mRNAs**. The amino acid sequence of the polypeptide determines the functional activity of the protein. Or, to be a little more precise, the amino acid sequences that comprise a polypeptide will fold and assemble into a particular structure, and the structure of that folded protein determines the function. The three-dimensional structure of a polypeptide is inherent in its amino acid sequence, although folding the polypeptide into the correct functional shape (and predicting the shape from looking at the amino acid sequence) is an extremely difficult problem. **Translation** is the process by which the nucleotide sequence of the nucleic acids (DNA and RNA) becomes the amino acid sequence of a polypeptide.

A rule of thumb that prevailed for many decades in molecular biology is that a gene encodes the information to make a single polypeptide. The experiments that led to this conclusion, done by George Beadle and Edward Tatum, are described in more detail in Chapter 15. This correspondence between a gene and a polypeptide was originally determined using genes encoding enzymes in eye pigmentation in *Drosophila* and metabolic pathways in the fungus *Neurospora*, and was expressed as "one gene, one enzyme" or, more generally, as "one gene, one polypeptide," as summarized in Figure 2.13. This principle is a useful starting point in understanding the functions of genes, but it is only the starting point. As will be discussed in more detail in Chapter 12, the same gene often has the information to make numerous related, but distinct, polypeptides because the RNA transcript is processed in different ways. Furthermore, many genes with important functions do not encode polypeptides at all—the functional product is the RNA molecule itself. The exceptions to the "one gene, one polypeptide" rule of thumb cannot be ignored, but it is still the first principle that most biologists use in thinking about the functions of genes.

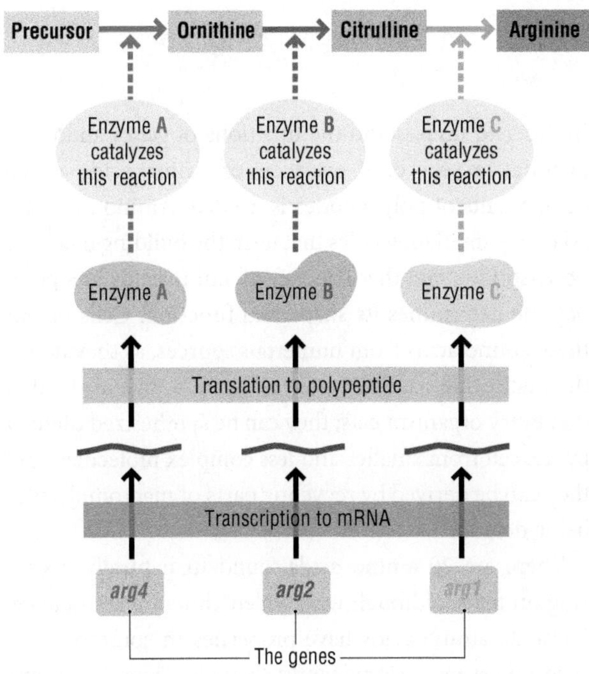

Figure 2.13 One gene, one polypeptide. For many decades, the rule of thumb for molecular biology has been that a gene encodes the information to make a single polypeptide. Part of the evidence is summarized here, based on the experiments of Nobel Laureates George Beadle and Edward Tatum in the fungus *Neurospora*. *Neurospora* synthesizes the amino acid arginine in three steps from a precursor molecule. As described in more detail in Chapter 15, Beadle and Tatum identified mutant strains of *Neurospora* that could not synthesize arginine. The three genes that they found, termed *arg4*, *arg2*, and *arg1*, each encodes one of the enzymes that catalyzes a step in this biosynthetic pathway. While "one gene, one polypeptide" remains a good starting principle for understanding gene function, many different important exceptions are now known.

Amino acids are the building blocks of polypeptides

In order to understand the functions of genes and polypeptides, a very brief overview of amino acids as the components of polypeptides is needed. Amino acids are relatively small molecules that form the building blocks of polypeptides, and the sequence of amino acids in a polypeptide determines its shape and function. Cells obtain these amino acids from numerous sources, as they do for the nucleotide bases. Amino acids are found in the diet that every organism eats; they can be synthesized directly by the cell from smaller and less complex molecules, and they can be derived by recycling parts of macromolecules being degraded.

There are 20 amino acids found in naturally occurring proteins, although with widely different abundance. These 20 amino acids have properties in common with each other; most significantly, they also have a polarity, with an amino ($-NH_2$) end and a carboxyl ($-COOH$) end, as shown in Figure 2.14. The amino acids are connected into a chain by hooking the amino-terminus of one onto the carboxyl-terminus of the preceding one, releasing H_2O in the process. Thus, a fully assembled polypeptide is a string of amino acids; a typical polypeptide might have 500 or so amino acids strung together, but there is a very wide range of sizes. Since each amino acid has a polarity, the polypeptide chain also has a distinct polarity or orientation. It has an amino (abbreviated N)-terminus and a carboxyl (abbreviated C)-terminus. Just as the nucleotide sequence of DNA or RNA is read from 5′ to 3′, the amino acid sequence of a polypeptide is read from N-terminus to C-terminus.

In addition to the common properties shared by all 20 amino acids, each amino acid has its own distinct chemical properties, represented as the R group in Figure 2.14; each R group is diagrammed separately in Figure 2.15. Some R groups are positively charged, some are negatively charged, and many have no net charge. Some are relatively bulky and inflexible; others are small and quite flexible. Some amino acids are hydrophobic and are usually not exposed on the surface of the polypeptide in aqueous environments, while others are hydrophilic. All of these properties affect the shape and function of the protein, so that replacing one amino acid with another can change or eliminate the function of a protein. Thus, just as the function of a DNA or RNA molecule is contained in the sequence of its nucleotide bases, the function of a polypeptide is contained in the sequence of its amino acids.

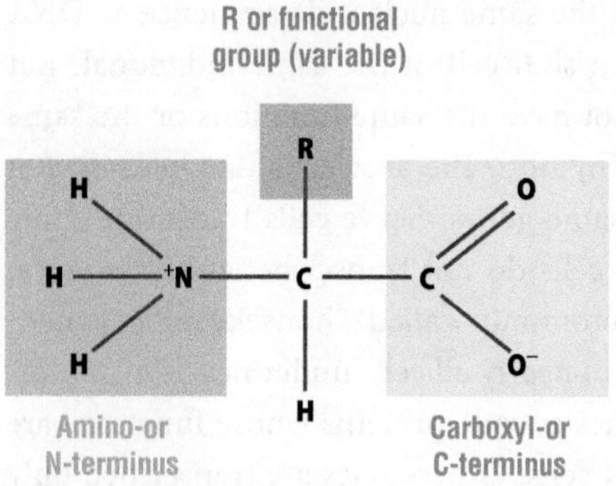

Figure 2.14 The overall structure of amino acids. Just as nucleotide triphosphates comprise the structural subunits of nucleic acids, amino acids form the structural subunits of polypeptides. The overall structure of an amino acid is shown. Two features stand out. First, the N (amino) group and the C (carboxyl) group provide a distinct directionality to the structure with their positive and negative charges, respectively. Second, the chemical and functional differences between amino acids reside in their R groups.

During translation from an mRNA sequence to an amino acid sequence, the C-terminus of one amino acid is joined to the N-terminus of the following amino acid, as shown in Figure 2.16. This is known as a peptide bond. The next amino acid sequence in this sequence will be added at the C-terminus, and so on. Thus, both nucleic acid sequences and peptide sequences have an inherent directionality, from 5′ to 3′ for nucleic acids and from N-terminus to C-terminus for polypeptides. The nature of translation ensures that the 5′ end of the mRNA corresponds to the N-terminus of the polypeptide, and the 3′ end of the mRNA corresponds to the C-terminus. (There are regions at the ends of the transcript that are not translated, so the amino acid at the N-terminus is not encoded by the first bases in the mRNA, but the direction is the same.)

A few additional points about the process of translation need to be introduced here and will be further developed in Chapter 13. While the mRNA transcript is simply a sequence of bases, it is a sequence of bases packed with information. The bases are translated or read in groups of three, so that three bases have the information for one amino acid. Those three bases are called a **codon**. The dictionary that lists which codon corresponds to which amino acid is called the **genetic code**, and will be discussed in Chapter 3 and more fully in Chapter 13. For example, the mRNA sequence

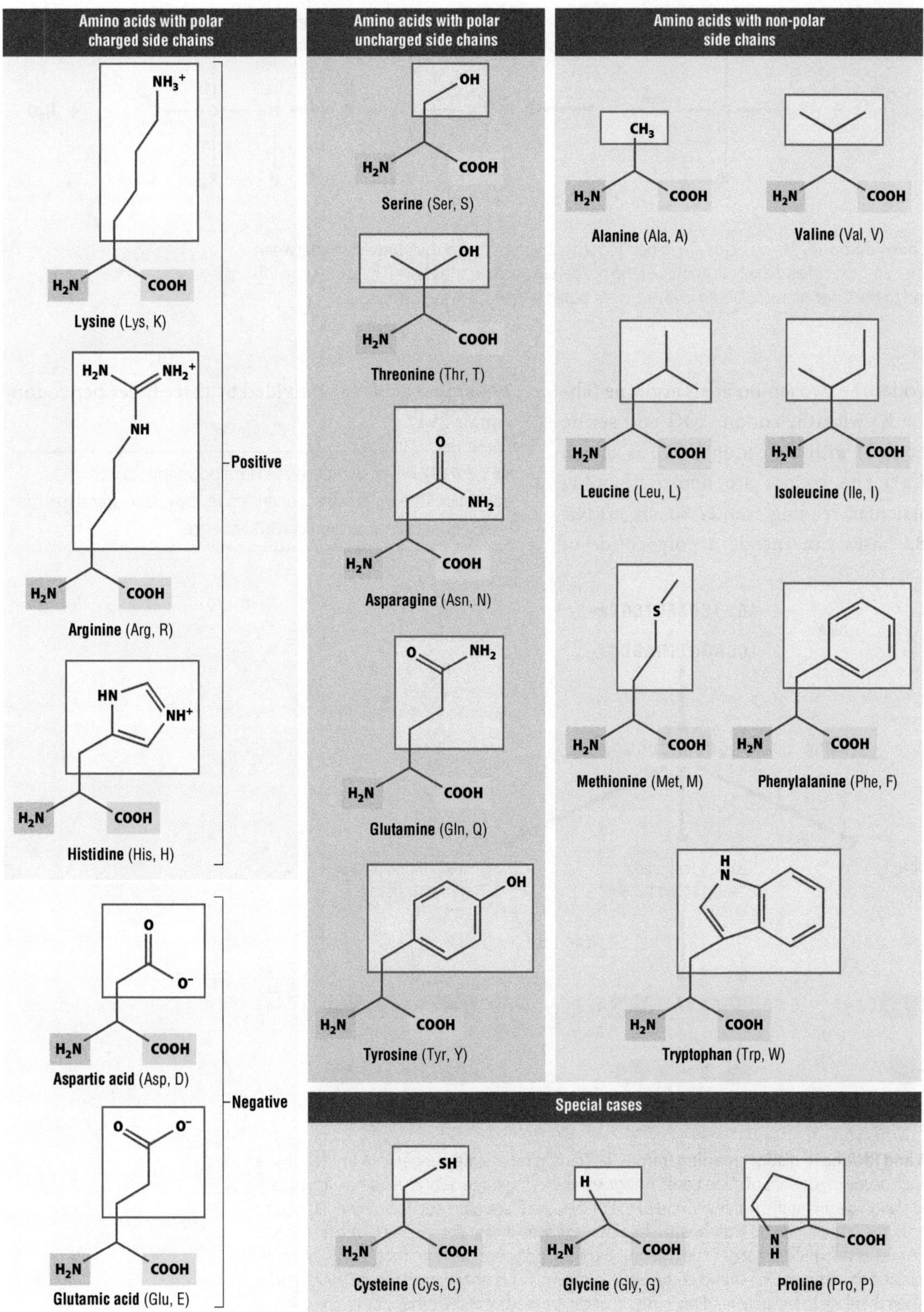

Figure 2.15 The structures of the 20 amino acids. The chemical structures of the 20 amino acids are shown, illustrating the variety of their R groups. For example, the R group for alanine is CH_3 and the R group for serine is O. By biochemical convention, not all carbon and hydrogen atoms are labeled.

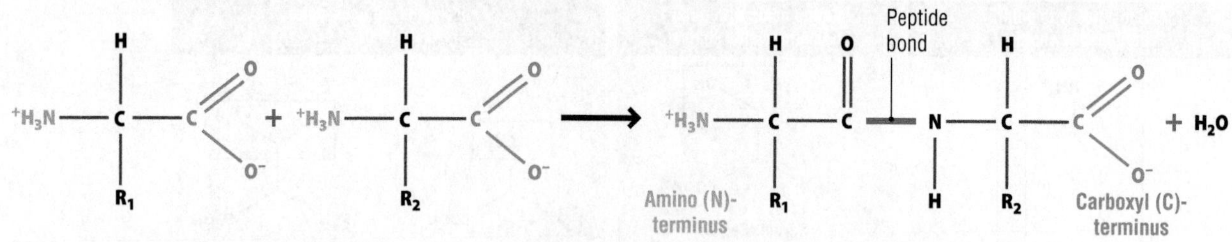

Figure 2.16 The peptide bond. Amino acids are joined together by peptide bonds. These form between the C-terminus of one amino acid and the N-terminus of the next amino acid in the chain. Thus, the growing chain will always add to the C-terminus, resulting in an inherent polarity of the polypeptide chain.

AGG–UCC encodes the two amino acids arginine (abbreviated Arg or R) with the codon AGG and serine (abbreviated Ser or S) with the codon UCC, as shown in Figure 2.17(a). The codons are non-overlapping and have a particular reading frame. So an mRNA sequence of 882 bases can encode a polypeptide of 294 amino acids (882 divided by three bases per codon equals 294).

KEY POINTS mRNA is translated into polypeptides during translation. The three-letter genetic code, the codon, translates the RNA sequence into an amino acid sequence.

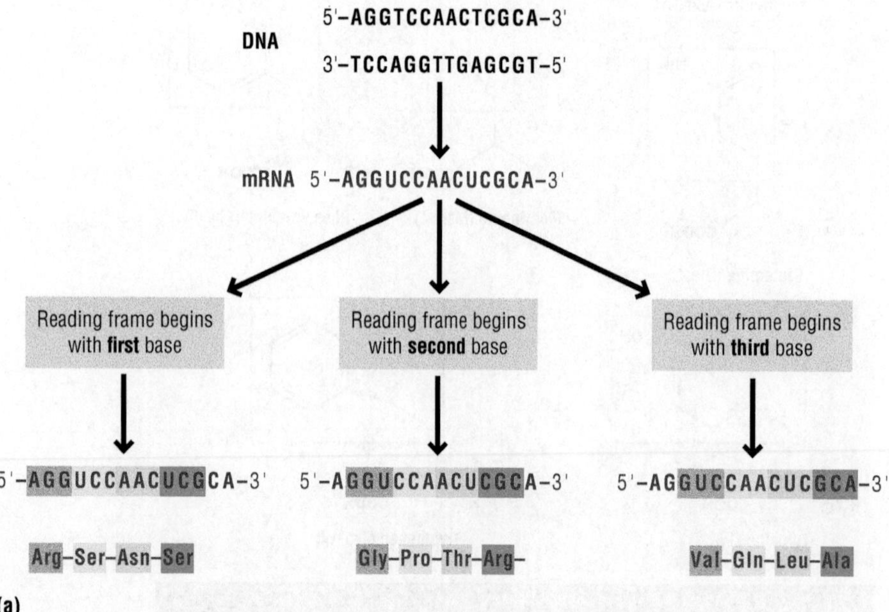

Figure 2.17 DNA and RNA have distinct reading frames. (a) Translation of a messenger RNA (mRNA) into a polypeptide sequence occurs by codons of three bases. Any given mRNA has three distinct reading frames, as illustrated here. If translation begins at the A at the 5′ end of the mRNA, the first amino acid is arginine. If it begins one base later at the G, the first amino acid is glycine. If it begins two bases later at the second G, the first amino acid is valine. The 5′ end of the mRNA encodes the N-terminus of the polypeptide. Note that, in this example, the mRNA sequence, and thus the coding strand of the DNA, is known; if it is not known which DNA strand is the coding strand, then there are three more reading frames that can be read from the other strand. The genetic code used to translate this base sequence into an amino acid sequence is discussed in Chapters 3 and 12. (b) The distinct reading frames are shown on a larger scale with these examples of a portion of the genomes from the bacterium *Staphylococcus aureus* and the unicellular eukaryote *Plasmodium knowlesi*. Genes are indicated by the blue arrows, and the reading frames are shown by the distance from the dark gray numbered genome line in the middle. For example, for *Plasmodium*, the genes *Phat14* and *Phat16* have the same reading frame, while the reading frame for *Phat15* is offset by a base. Reading frames often shift at the borders of introns and exons in eukaryotes. Note also that both strands serve as the coding strand, depending on the gene.

Source: Images generated from Artemis, courtesy of Pathogen Informatics, Wellcome Trust Sanger Institute

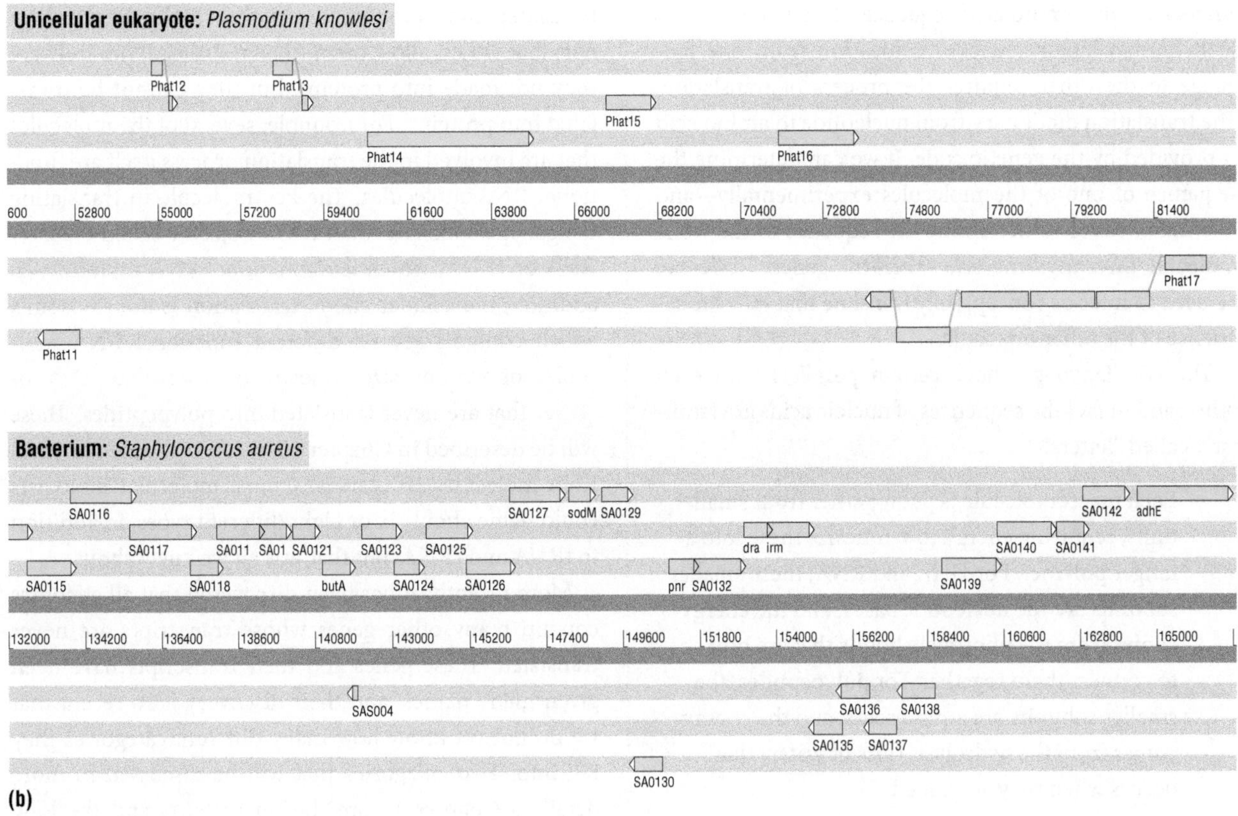

Figure 2.17 Continued

Figure 2.17(a) illustrates how this reading frame matters. If the sequence of the mRNA fragment is 5′-AGGUCCAA-3′ and the reading frame is not known, we could read the code for the first two amino acids as AGG, which codes for arginine (Arg), and UCC, which codes for serine (Ser). But the code could also be read as aGGU and CCA by assuming that the first A is the last base of a preceding codon; this would give the amino acid glycine (Gly) from GGU, followed by proline (Pro) from CCA. It is also possible that the first two bases of this sequence are the last two bases of the preceding codon and read as agGUC and CAA which would translate as amino acids valine (Val) and glutamine (Gln).

Clearly then, the reading frame makes a difference to what polypeptide is produced. Therefore, given a stretch of DNA with no other information, there are three possible reading frames that would produce different peptides. The example in Figure 2.17(a) gave the mRNA sequence, so it specified the coding strand of the DNA. If this had been expressed as a double-stranded DNA sequence, and the coding and template strands were not known, there would have been three possible reading frames from each strand of DNA, giving six possible reading frames in total. Indeed genomes make use of all six possible reading frames of the double-stranded DNA, as illustrated in Figure 2.17(b). As we will describe in Chapter 12, genes and mRNA molecules begin at certain locations, and the first codon in the protein-coding portion of a gene is AUG (ATG in the DNA), which encodes methionine; this AUG sets the reading frame, the coding strand, and the start position of the amino acid sequence.

WEBLINK: Video 2.1 illustrates the importance of reading frames. Find them at **www.oup.com/uk/meneely**

The languages of nucleic acids and polypeptides

To summarize this section, there are two great languages that communicate information in molecular biology. The first of these languages is the nucleotide sequence of DNA and RNA. The process of "reading the DNA" into an RNA copy is called transcription, because DNA and RNA use the same language of nucleotide sequences. The second language is the amino acid sequence of polypeptides. As with human languages, communication from one language to the other requires the process of translation. The translation dictionary from nucleotide to amino acid is provided by the genetic code. If we can determine the sequence of one of the molecules experimentally—and it is much easier to determine the sequence of the DNA molecule than anything else—then the genetic code can be used to infer or conceptually translate that into the sequence of the other molecule.

The two languages have certain parallels with each other, and in fact the sequences of nucleic acids are familiarly called "letters."

- Each macromolecule is put together from smaller units, requiring energy to synthesize these into a longer polymer. For DNA and RNA, these smaller subunits are the nucleotide bases, and the energy comes from the phosphate bonds that are broken to connect them together. For polypeptides, the smaller subunits are amino acids, and the energy comes from the hydrolysis (loss of water) that occurs when they are joined.
- Each macromolecule is read in only one direction, and the direction of the one is also the direction of the other, that is, the 5′ end of a coding strand of DNA or the 5′ end of the RNA molecule will contain the information for the N-terminus of the polypeptide, and the 3′ end will correspond to the C-terminus.
- In each case, the informational content is embedded in the sequence of the subunits, so that knowing the sequence allows us to make hypotheses about the function of the overall macromolecule.

All of these topics will be explored in more depth in later chapters, particularly Chapters 4, 12, and 13.

Functional RNA molecules are transcripts that are not translated

The Central Dogma expressed that RNA is translated into polypeptides, and our extension was that polypeptides produce or contribute to the phenotype. While many RNA molecules are translated into a polypeptide, research in the past decade or so has led to a significant modification of this principle for other RNA molecules.

It has been known for decades that some RNA transcripts are functional in and of themselves and are never translated into a protein. In fact, these RNAs lack the signals that would allow them to be translated; not only are they not made into proteins, but they **cannot** be translated into proteins. For example, several of the molecules that are involved in the translation process itself are functional RNA molecules. The key molecule in translating the genetic code is a small RNA molecule called **transfer RNA** or **tRNA**, which is a functional RNA molecule. In addition, the cellular site of translation is the ribosome, which contains several different functional RNA molecules of various sizes known as **ribosomal RNA** or **rRNA** that are never translated into polypeptides. These will be described in Chapter 13. If asked about functional RNA molecules 10 years ago, most molecular biologists could name about six or eight different types, in addition to tRNA and rRNA, but that was all we knew about.

More recently, it has been discovered that all genomes contain many other genes whose transcripts are never translated. These genes and their transcripts have been given many names, and their discovery is so recent that we do not yet know how many different categories they fall into. Two categories that will be discussed in more detail in Chapter 12 are the microRNAs and the long non-coding RNAs. In each case, the names refer to their lengths, rather their functions.

A **microRNA** is made as an initial transcript of about 120 nucleotides in length and processed to a functional molecule of only 22 nucleotides in length. A **long non-coding RNA** or **long ncRNA** is defined to be longer than 200 bases in length and ranges from 200 nucleotides to about 1000 nucleotides, but longer ones are also known. By contrast, a typical transcript for a protein-coding gene in humans is initially about 10,000 nucleotides in length, although this varies widely, depending on the gene. Thus, long ncRNAs and microRNAs are shorter than an mRNA molecule; in fact, a microRNA molecule is much much shorter. The short length is one reason that these non-coding RNAs went undetected for so long.

The functions of microRNAs are to reduce the expression of their target genes, with an individual microRNA having one to many specific target genes; we discuss the mechanism by which they work in Chapter 12. The functions of long ncRNAs are diverse and largely unknown, with more functions to be elucidated in the near future. Both microRNAs and long ncRNAs will be discussed more fully in Chapter 12.

KEY POINT Some RNA transcripts are functional without being translated into a polypeptide.

2.4 The structure and function of genes: gene regulation

In this overview section of gene expression and regulation, we are going to focus on genes that encode the information to make a protein. Gene expression can be regulated at many different steps in the process. This is summarized in Figure 2.18. Most of these processes will be described in more detail in Chapters 12 and 13, so this is only a very quick summary to lay the foundation for some topics in the subsequent chapters.

As summarized in Figure 2.18(a), most of the regulation of gene expression in bacteria occurs at the initiation of transcription. In eukaryotes, as shown in Figure 2.18(b), there are additional steps at which regulation occurs.

- A gene has to be transcribed at the correct time and in the proper cells, so transcription is regulated. Transcriptional regulation occurs in both bacteria and eukaryotes and is probably the most significant regulatory step overall in both groups. In eukaryotes, most of the regulation occurs before transcription is initiated when all of the components of the transcriptional machinery are assembled on the chromatin in the right position; this is usually known as the pre-initiation step.
- In most eukaryotes, the original transcript is much longer than the final mRNA, because significant and defined portions of it are spliced out. The regulation of splicing is another key regulatory step.
- Most (if not all) transcripts are modified in other ways after transcription; furthermore, all transcripts in eukaryotes that are going to be translated into protein have to be transported from the nucleus to the cytoplasm. Editing and transport of spliced transcripts can be regulated, although this seems to be a fairly minor method for gene regulation.
- In eukaryotes, microRNAs can block an mRNA or target it for degradation, which occurs in the cytoplasm. In addition, long ncRNAs have largely unknown roles in the regulation of gene expression, some of which appear to occur in the nucleus and others in the cytoplasm, so the expression of a gene may somehow be regulated by a long ncRNA.
- Once the edited and spliced mRNA is in the cytoplasm in eukaryotes, it is translated into a polypeptide, so the regulation of the time of translation can be used to control gene expression. In bacteria, the DNA is not in a membrane-bound nucleus. Transcription and translation are coupled in location and time, so most mRNAs are translated as they are being transcribed.
- If the mRNA is stable enough, it will probably get translated again and again. Thus, RNA stability and the regulation of RNA degradation provide another step to regulate gene expression.
- After translation, the polypeptide can be chemically modified in many different ways to render it active or inactive; post-translational modifications are another important step in regulation. By this point, it is not gene regulation but gene product or protein regulation that is occurring, but these still affect phenotypes.

All of these many ways to regulate gene expression are important in some process in the cell or the organism. Almost every gene will be regulated by more than one of these mechanisms. The expression of most genes includes regulation of the time and location of transcription, that is, with transcription pre-initiation and initiation, but this is not the only step. Examples of regulation at some of these post-transcriptional steps will be developed in Chapters 12, 13, and 14. Nonetheless, despite these notable counter-examples, when most geneticists speak of **gene regulation**, they are usually referring to the **regulation of transcription**. Even more specifically, gene regulation usually refers to the regulation of the **initiation** of transcription. The process of transcription and its control will be covered in Chapter 12, but here is a summary of some key concepts.

Genes have a modular structure

A gene that encodes the information for a polypeptide has two main functional parts, as shown in Figure 2.19. The **coding region** of the gene, also known as the structural portion of the gene, is the part that will be represented in the final mRNA transcript and will be translated into the amino acid sequence. In Figure 2.19, this would be the region beginning with

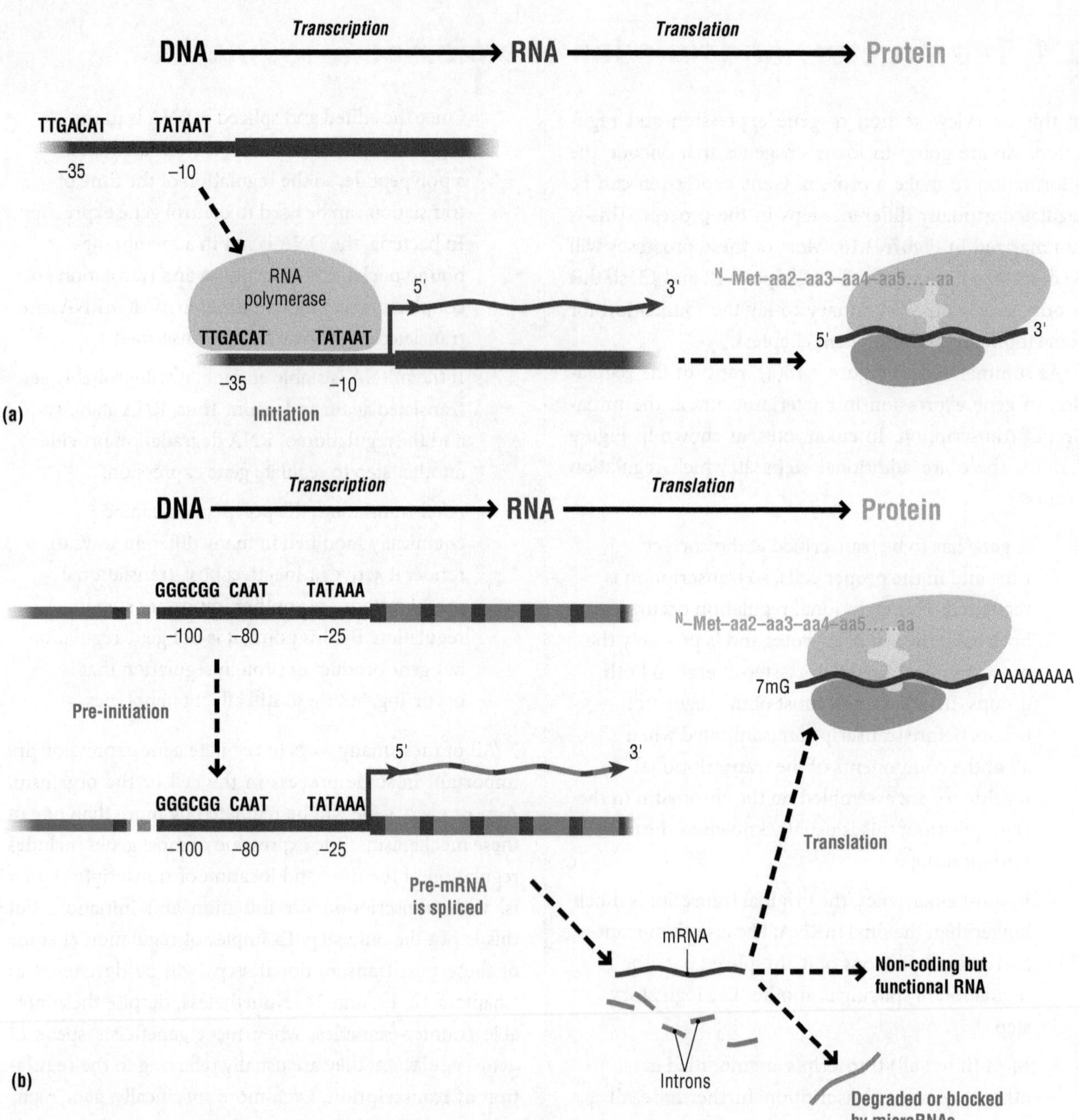

Figure 2.18 The basis for gene regulation in bacteria and eukaryotes. The Central Dogma is used to show the most common steps in which gene expression is regulated indicated in red text. (a) In bacteria, most, but not all, of the regulation of gene expression occurs by regulating the initiation of transcription. (b) In eukaryotes, regulation before the initiation of transcription, or pre-initiation, is probably the most important step in regulating gene expression. However, additional regulation events also occur during the splicing and processing of the transcript, at translation itself, and mediated by microRNAs and other functional, but non-coding, RNAs.

ATG and ending with TGA. It begins upstream of the gene and ends at a region downstream of the gene. For genes that are not spliced after transcription, which includes all genes in bacteria and some genes in eukaryotes, the gene sequence consists of one continuous and uninterrupted open reading frame. An open reading frame (abbreviated ORF and pronounced "oarf") is the length of the DNA or RNA sequence that has no stop signals in it. An ORF generally indicates a section that probably will be translated into an amino acid sequence.

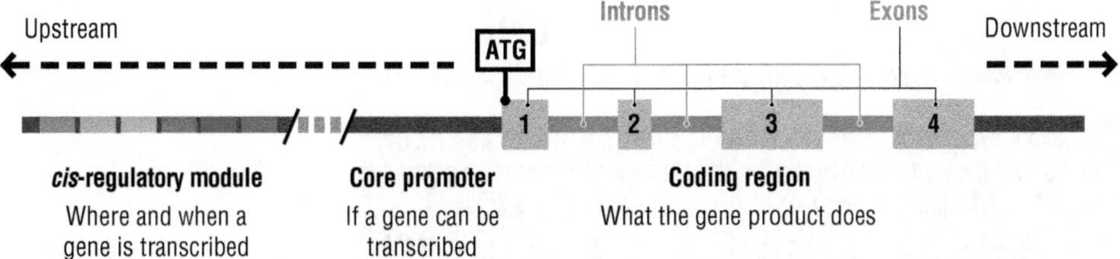

Figure 2.19 Genes have a modular structure. An example of a typical eukaryotic gene is shown. The coding region of the gene is the portion that is transcribed into RNA and, for protein-coding genes, translated into a polypeptide sequence. This part of the gene controls what the gene does. Immediately upstream of the coding region is the core promoter, a region that is not transcribed itself but controls if the gene can be transcribed. Further upstream in this diagram is the *cis*-regulatory module, or CRM, which controls when and where the gene is transcribed. The CRM is usually upstream of the gene, but examples are known in which important regulatory sequences are located downstream of, or within, the gene. A bacterial gene lacks introns—in other words, it consists of a single uninterrupted open reading frame—and regulates transcription at the core promoter, rather than with a CRM.

The coding region of the gene has the information about the amino acid, and thus the function of the protein, that is, on **what** it does.

As noted earlier, eukaryotic genes typically have internal sections of the transcript that are removed (or "spliced out") to make the final mRNA; splicing introduces an important level of regulation of gene expression and evolutionary flexibility to genes. Thus, the sequence of a coding region of a eukaryotic gene will have sections with an ORF, followed by sections that have stop codons in them and no long ORF. This is illustrated in the gene sequence in Figure 2.20.

The regions with a sustained ORF will be retained in the final mRNA and expressed into an amino acid sequence when the mRNA is translated; these regions are known as **exons**. The regions that are removed from the initial transcript and have no sustained ORF (that is, that intervene in the coding region) are called intervening sequences or **introns**. The length and number of introns vary among eukaryotes. Genes in mammals usually have dozens of introns whose length greatly exceeds the combined length of the exons; in mammals, the exons comprise only about 1–2% of the genome, or less than 5% of the total amount of RNA that is transcribed. Invertebrates like *Drosophila* and nematodes generally have fewer introns with shorter lengths than mammals, and flowering plants seem to be somewhat intermediate, although with much variation.

As shown in Figure 2.19, genes also have **regulatory regions**. The regulatory region of the gene has the information on **when**, **where**, and **how much** of the protein product is made. This information is important to ensure that each cell type is transcribing the appropriate genes in the appropriate amount at the appropriate time. It is also important for evolutionary change, which, as will be discussed in Chapter 3, can occur not only through changes in the sequences of the coding regions, but also through changes in the sequences of regulatory regions. Immediately upstream of a coding region of the gene is the **core promoter** or **transcriptional start site** or **TSS**. This regulatory region is needed for transcription to occur at all, and it is generally similar for most genes. (However, recent work in the human genome has shown that the core promoters are not nearly as similar to one another as previously thought.) For most bacterial genes, this core promoter region has the information to regulate the specificity of expression; we will elaborate further on this point in Chapters 12 and 14.

For eukaryotes, the core promoter is necessary for a gene to be transcribed, but it does not regulate the specificity of transcription. For most genes, a regulatory region, sometimes known as a ***cis*-regulatory module** or **CRM**, is usually located further upstream of the core promoter, although examples are known in eukaryotes in which important regulatory signals are found downstream of the coding region or even embedded within the gene in an intron, for example. Figure 2.19 shows a typical case in which the CRM is upstream.

The CRM can be relatively simple in structure and compact in length, or very complex and spread out over a long distance. The overall structure of a CRM is diagrammed in Figure 2.21; a more detailed example will be explained in Chapter 12. Within the CRM are very short sequences of base pairs, usually in the order of 8–12 base pairs, that are sites for other proteins to bind to and control transcription. A protein that binds to DNA and

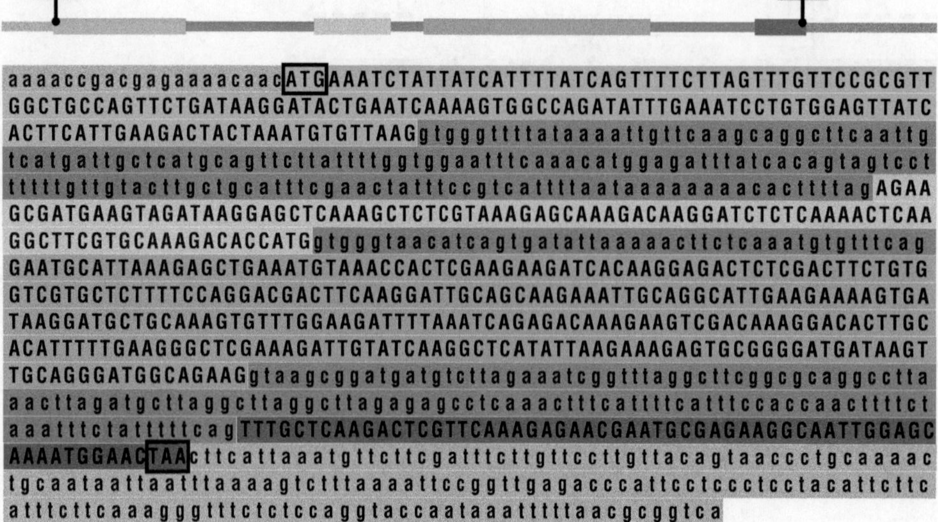

Figure 2.20 The structure and sequence of the coding region of a gene from *Caenorhabditis elegans*. This is a gene of unknown function, referred to as D1086.3, and is used here as an example of a typical gene in a metazoan. The colors in the diagram at the top correspond to the shaded regions of the sequence of the coding strand below. The gray lines upstream and downstream of the coding region are transcribed but are not translated. The introns are orange, while the exons are blue, green, turquoise, and pink. Note that the first exon begins with an ATG codon and the final exon ends with TAA, one of the three stop codons. By convention, the nucleotide sequences of exons are shown in uppercase letters, while the sequences of introns are in lowercase letters.

regulates the transcription of other genes (or itself) is called a **transcription factor**. In most eukaryotic organisms, between 5 and 10% of the genes in the genome encode proteins that are transcription factors, so there are hundreds of different transcription factors.

Many transcription factors regulate the transcription of one or two genes, but some of them directly regulate the transcription of hundreds of genes. Even a transcription factor that directly regulates only one or two genes might indirectly regulate dozens or hundreds of genes; one of its targets could be another transcription factor, which then regulates other genes, including other transcription factors. This can set up a cascade of regulatory factors, as discussed more fully in Chapter 14.

KEY POINT A typical gene includes a coding region and a regulatory region, which includes the core promoter.

ALX1 encodes a transcription factor

We can return to Darwin's finches for an example to illustrate many of the points we have made about gene structures, transcription, and the regulation of transcription. You may recall from Chapter 1 that one gene implicated in changes in the shape of the beak in Darwin's finches is a gene known as *ALX1*. *ALX1* encodes a transcription factor, so it must regulate the transcription of other genes. (As we noted in Communicating Genetics 1.1, gene names like *ALX1* are italicized, while the polypeptide product made from the gene is shown without italics like ALX1.)

If you look again at Figure 2.21, you might imagine that one of the transcription factors is ALX1, located in the CRM of this gene. The genes whose transcription is regulated by ALX1 have not been identified. However, based on when the *ALX1* gene and protein are expressed and what happens when *ALX1* is altered or missing in other vertebrates, its target genes must include the ones that are needed for normal craniofacial development. ALX1 may well regulate genes that affect other cellular functions, but our focus is on the beak and craniofacial structures. Some of the target genes of ALX1 regulation could encode enzymes and structural proteins necessary to make a beak and other craniofacial structures, while other target genes might encode other transcription

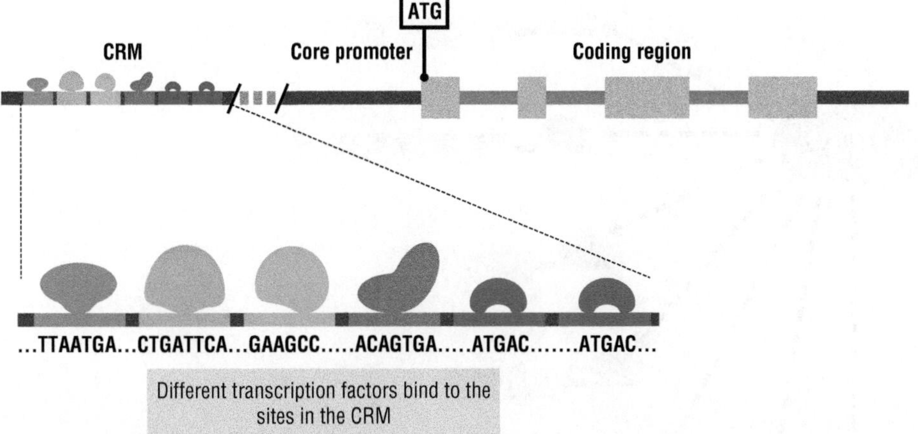

Figure 2.21 The *cis*-regulatory module (CRM) of a gene. The CRM is a region of variable length, position, and sequence that controls which cells express the gene and when the gene is transcribed. This occurs when transcription factor proteins bind to specific sequences in the CRM. The binding sites and the corresponding transcription factors are shown in different colors. A particular cell or developmental stage might express different combinations of these transcription factors, rather than all of them at the same time.

factors, but this is not yet known. In other words, *ALX1* is probably part of a gene regulatory network that directs craniofacial development, as summarized in Figure 2.22. Its expression is regulated, and it directly regulates the expression of other genes, some of which regulate additional genes.

While the molecular function of ALX1 appears to be the regulation of the transcription of genes involved in craniofacial or beak development, how are changes in the DNA sequence of the *ALX1* gene itself responsible for the changes in the beak shape? In order to address this question, the investigators turned again to the analysis of the DNA sequences of birds with blunt or pointed beaks, looking specifically at the changes in the *ALX1* gene itself, as well as in the surrounding region of the chromosome. The entire region that was found to differ in species with different beak shapes is 240,000 bp (240 kb) long, of which *ALX1* itself occupies only about 5 kb or about 2% of the sequence. In this 240 kb region, there were 335 changes in the DNA sequence found in the blunt-beaked birds (called the B region or B haplotype, a term we define more fully in Chapter 9), compared to the DNA sequence of the pointed-beaked birds (called the P region or P haplotype); in other words, at a certain site in the sequence of this region, the B haplotype has an A while the P haplotype has a G (for example). This is summarized in Figure 2.23. The ancestral species on the mainland has a pointed beak and the P haplotype, so the changes that apparently affected beak shape will be those that occurred to make the B haplotype or the derived version.

The diagram in Figure 2.23 summarizes the analysis of the changes that are the most likely to affect the sequence and function of the *ALX1* gene. Many of the 335 derived changes on the B haplotype are located in or near the *ALX1* gene, which could mean that they affect its function. In addition, by comparing different species of finches that all had blunt beaks and the B haplotype, the investigators found which of these changes of the B haplotype were most consistently associated with the blunt phenotype; this focused their attention on eight sequence changes within and near the *ALX1* gene, as shown by the arrows in Figure 2.23.

These changes are expected to affect the function of the *ALX1* gene in two different ways. Remember that ALX1 is a transcription factor and directly regulates the expression of other genes. Transcription factors work via the specific interactions between certain amino acids in the transcription factor protein and particular sequences in the DNA near the gene they regulate; the transcription factor binds to its DNA target site and allows the nearby gene to be transcribed. If either interacting component is changed, the transcription pattern changes. A change in the binding site on the DNA can eliminate the ability of a particular transcription factor to bind there, or decrease or increase its binding; it may even make the sequence into the binding site for a different transcription factor. A change in the amino acid

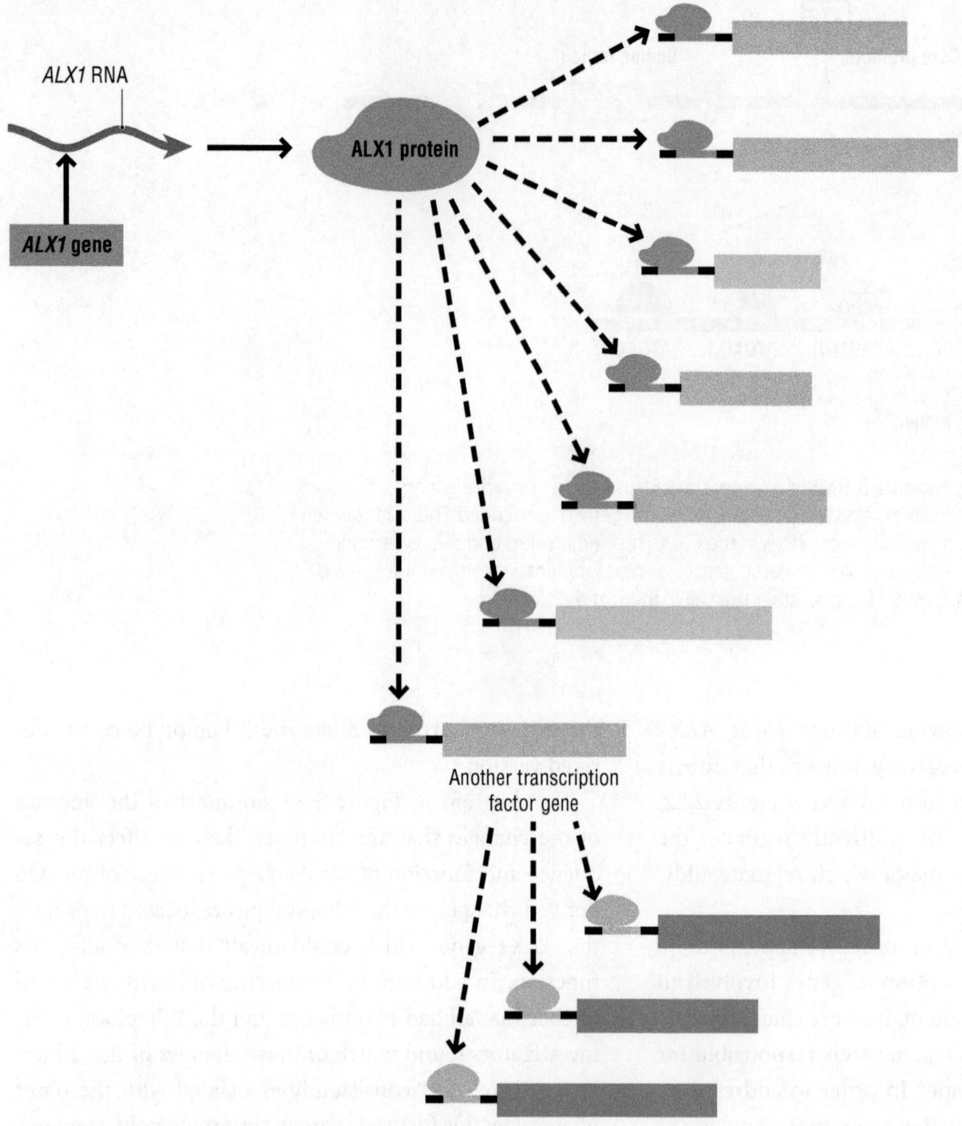

Figure 2.22 ***ALX1*** **encodes a transcription factor.** The protein encoded by the *ALX1* gene is predicted to be a transcription factor, also called ALX1. The gene is shown as an orange box; the RNA transcript from the gene is shown as a wavy purple line, while the protein produced from that is the orange shape. ALX1 is likely to affect craniofacial development, and thus beak shape, because it regulates the expression of other genes by binding to DNA sequences near each gene. Seven gene targets of ALX1 are shown, but the actual number and identity are not known, so this is an illustrative, but hypothetical, network. Genes whose functions are thought to be involved in craniofacial development are represented by orange boxes, whereas those with other roles are represented by green boxes; this diagram suggests that ALX1 has other functions, in addition to craniofacial development, which would be true for most transcription factors. Note that one of its hypothetical targets is itself a gene that also encodes a transcription factor, and that gene regulates the transcription of three other genes involved in craniofacial development, as shown in dark orange. While this is a completely hypothetical regulatory network for ALX1, its characteristics are similar to the known regulatory networks for genes similar to *ALX1*.

sequence of the transcription factor protein can change which sequence it binds to and how well it binds to that sequence. Both components are important and specific, and both were found with *ALX1*.

Three of the changes related to blunt beaks lie within the coding region of the *ALX1* gene and would change the amino acid sequence of the ALX1 protein. Thus, as a result of these changes, the version of the ALX1 protein

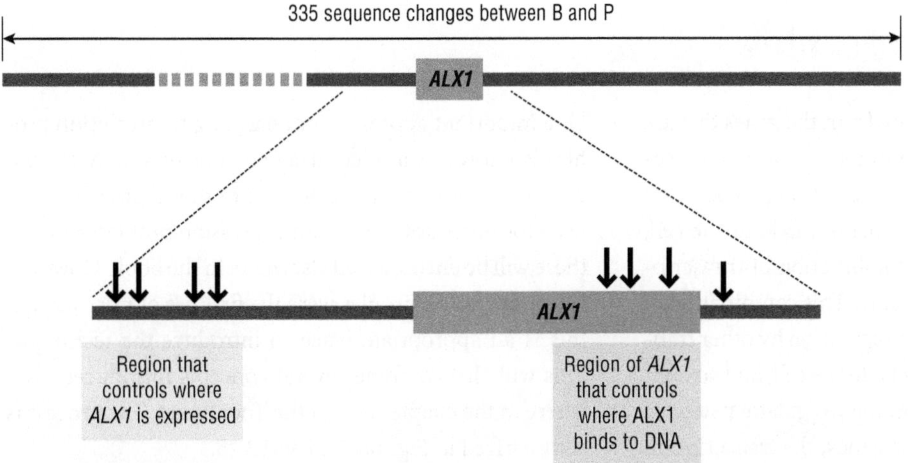

Figure 2.23 The sequence changes in *ALX1* are likely to be associated with beak shape. The *ALX1* region is 240 kb in length, with 335 sequence changes between species with blunt beaks (B) and species with pointed beaks (P), as shown by the double-headed arrow. Eight of these changes are near or within the *ALX1* gene itself, shown in the expanded diagram at the bottom. The locations of these changes with respect to the *ALX1* gene are indicated by the black arrows. Four changes fall within the upstream region that is thought to regulate *ALX1* transcription, so *ALX1* could be expressed in different tissues or at different times in species with different beak shapes. Three changes are within the *ALX1* gene sequence itself and could change the ALX1 protein, so that it binds to a slightly different target sequence. Thus, the ALX1 protein could potentially regulate some genes in the B species that it does not in the P species, or vice versa. One change is found downstream of the *ALX1* gene and is not predicted to have any effect, although this is not known. The drawing is not to scale, although the relative positions are correct.

made by the B haplotype and the version of the ALX1 protein made by the P haplotype probably would not bind to precisely the same DNA sequence or would bind with different affinities. In other words, some of the target genes regulated by the one version of the ALX1 protein may be different from the target genes regulated by the other version of the ALX1 protein. Perhaps some of the changes in the shape of the beak are due to changes in exactly which target genes ALX1 regulates.

The other changes on the B and P haplotypes lie near the *ALX1* gene, but not within it, and so would not change its amino acid sequence or its binding specificity. However, four of these changes lie in the region where the transcription factors regulating *ALX1* are expected to bind. These sequence changes could change where and when *ALX1* is transcribed or how much of the protein is made. Thus, the *ALX1* genes in these species differ in both of the modular structures of a typical eukaryotic gene. First, the upstream changes could result in the gene being transcribed in different cells or at different times during development, or in different amounts. Second, the changes within the gene could alter which genes are targets for ALX1 regulation. Either or both of these could be crucial changes that resulted in different beak shapes.

The DNA changes that occurred during beak evolution raise intriguing possibilities about changes in the pattern of *ALX1* gene expression and the function of its protein. Studies of both changes in target genes and changes in expression pattern involve tools and experimental approaches that are widely used in genetic studies in many organisms, so we can expect that these could be done in Darwin's finches as well. Evolutionary changes among finch species affected both the proposed regulatory region for the *ALX*1 gene—that is, when and where the gene is expressed—and the coding region of the *ALX1* gene that likely resulted in slight changes in the function of the protein.

KEY POINT Evolutionary changes in the *ALX1* gene apparently affected both its expression pattern and its predicted targets for regulation. These appear to have been involved in changes in beak shape in Darwin's finches.

2.5 Transcription profiles

The unique functions of a cell arise from the genes that are being transcribed at a given time, or its **transcription profile** or **expression profile**. The transcription profile is the collection of the RNA molecules being made in the cell, which arises, in turn, from the combination of transcription factors that the cell is expressing. This combination of transcription factors comes from regulation by other transcription factors, signals from outside the cell, and so on.

Because transcription is the primary regulatory step for regulating the expression of most genes, the transcription profile of a cell can help reveal which specific genes and associated molecular pathways are functioning in that cell. Many different and sensitive techniques have therefore been developed in the laboratory to study transcription. In fact, the application of these techniques demonstrated that transcription was far more pervasive than previously expected and that many transcripts are not translated.

One important approach for analyzing transcription profiles is known as a microarray analysis or simply **microarrays**. The underlying technique of this approach is also used for purposes other than expression profiles; some of these will be encountered elsewhere in the book. However, since expression profiles were the first use of microarrays, this is an appropriate place to introduce the technique. This will also combine several concepts introduced elsewhere in the chapter and in the Tool Boxes. The process is summarized in Figures 2.24 and 2.25.

Recall that every gene has a unique DNA sequence. When it became possible to sequence entire genomes, it was also possible to identify sequences that are uniquely found in one gene, as shown for three genes in Figure 2.24. Defined sequences of DNA (or RNA) that uniquely identify each gene can be synthesized in the laboratory as a series of oligonucleotides, as noted in Tool Box 2.3. In

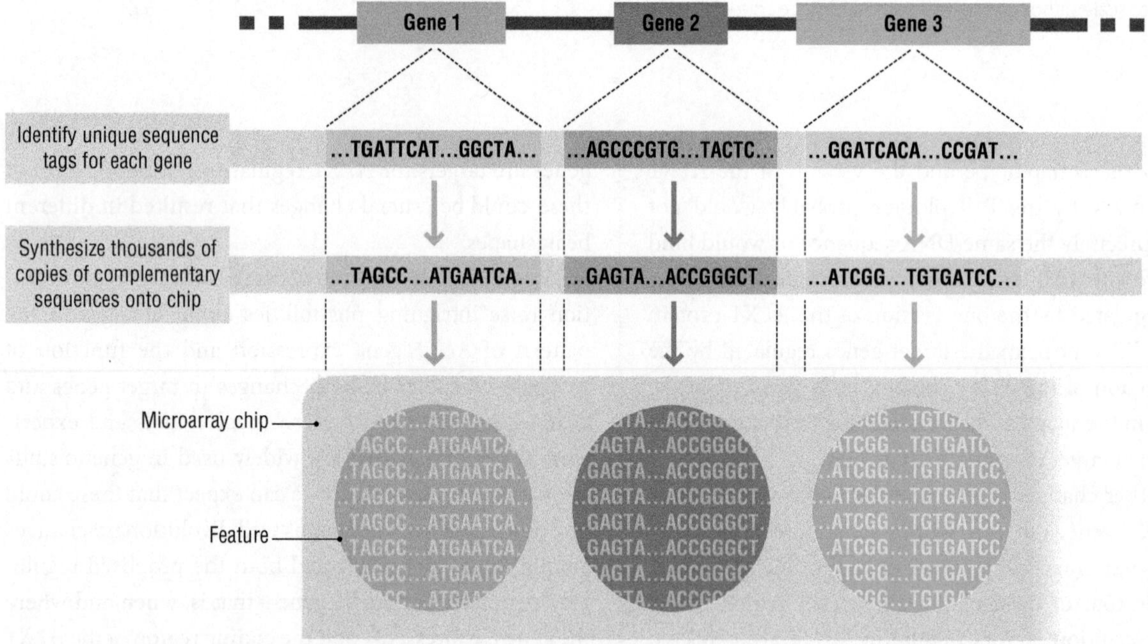

Figure 2.24 Preparing a microarray chip. A microarray allows the investigator to determine all of the genes being transcribed in a cell type or tissue. The first step is to prepare the chip. In this example, a unique sequence from each gene is found by examining the DNA sequence of the genome. The complementary single-stranded sequence from the unique tag is synthesized *in vitro* and printed onto the microarray chip in a defined arrangement. This creates a feature on the chip. A feature has tens of thousands of copies of the sequence at the same location on the chip, vastly more copies than the possible number of copies of any mRNA in the sample. This diagram shows one sequence tag from three genes in different colors, which are printed as three adjacent features. In reality, adjacent features on a chip do not correspond to neighboring genes, and, since a million features can be printed onto a chip, there may be several features that correspond to different parts of the same gene or to other regions of the genome. Microarray chips for well-studied organisms can be purchased from commercial vendors, so the individual investigator usually does not need to carry out these steps.

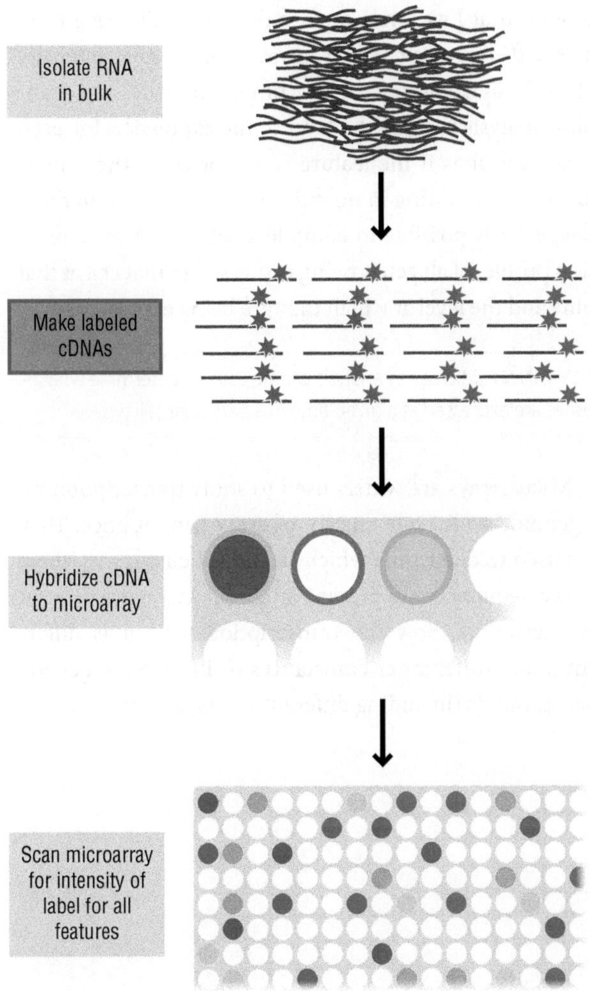
Isolate RNA in bulk
Make labeled cDNAs
Hybridize cDNA to microarray
Scan microarray for intensity of label for all features

Figure 2.25 Hybridization to a microarray chip. In order to find all of the genes being transcribed in a sample, the investigator isolates total mRNA in bulk from the sample. This RNA is reverse transcribed to make labeled cDNAs corresponding to every transcript in the sample. The labeled cDNAs are hybridized in bulk to the microarray chip, and the intensity of the signal is scanned computationally and recorded. The same three features corresponding to the three genes in Figure 2.24 are shown for illustration. The feature for gene 1 (in green) shows a strong signal from hybridization with the cDNA, indicating that this gene is transcribed in the sample at a high level. Gene 2 (in purple) is not transcribed in this sample, and gene 3 (in blue) is transcribed at a lower level.

fact, it is not too complicated for commercial vendors to make specific oligonucleotides about 50 bases long corresponding to every gene in a genome; these are pre-made robotically for organisms that are widely studied such as humans. Each of these oligonucleotides has a sequence that is unique to a given gene, and no other gene contains exactly the same sequence.

In a microarray, the oligonucleotides are fixed or synthesized onto a silicon or glass slide in a specified and defined pattern, in the same way that microcircuits are embedded or synthesized onto silicon chips. In fact, these silicon or glass slides with the oligonucleotides embedded on them are called microarray chips. Imagine a microarray chip as a giant grid, like a game of *Battleship*™. Oligonucleotides fixed in this way are referred to as **features**, as shown in Figure 2.24. Each feature represents a specific gene and has tens of thousands of copies of the same oligonucleotide tag sequence from that particular gene. A microarray chip can hold many such features, as many as a million, on a support no larger than a microscope slide, roughly 3 cm × 8 cm. Since the human genome has about 22,000 protein-coding genes and a microarray chip can hold a million features, it is possible to "download the genome onto a chip" with a unique feature for each gene; in fact, there are often multiple features with different sequences from the same gene at different places on the chip. In addition, since the oligonucleotide is usually made as a single-stranded sequence, other features can be made with the complementary sequences, so that each DNA strand is represented.

This chip with its defined pattern of specific sequences is used for hybridization with nucleic acids isolated from the cell or the organism; hybridization of nucleic acids is described in Tool Box 2.3. The original use of the microarray chip was to examine patterns of transcription, so the nucleic acid being hybridized to the microarray was mRNA. This is summarized in Figure 2.25. Since many of the unique characteristics of a cell lie in its transcription profile, microarrays can be used to identify exactly which genes are being transcribed in any given cell type; because the number of features on the chip is so large, the experiment is testing the transcription of every gene in the genome.

There are two types of microarrays used to study transcription, which serve slightly different purposes. If the goal is to determine **which** sequences are transcribed under a particular set of conditions, a one-channel or one-color array can be used. If the goal is to determine the **changes** in the transcription pattern under different conditions, a two-channel or two-color array is used. We will describe how a one-channel array is done and will save the description of two-channel arrays for Chapter 14.

In the experiment summarized in Figure 2.25, mRNAs are isolated from a sample of tissue; ideally, this will represent all of the mRNAs being made in that tissue at that time. The mRNA is used to make cDNA, a process described in Tool Box 2.4. When the cDNA is made, one of the nucleotides is labeled, as described in Tool Box 2.2, so the cDNA for each mRNA is also labeled. Thus, if the procedure worked well, every mRNA present in the tissue at the time the sample was taken has been copied into cDNA, and every cDNA is labeled.

A solution with all the labeled cDNAs is hybridized to the microarray chip. Let's think about what happens, as shown in Figure 2.25. A cDNA will form base pairs with the feature on the microarray chip that has the complementary sequence; every gene in the genome is represented by features on the microarray chip, so every cDNA will find complementary sequences among the features. The cDNA is labeled, so the double-stranded hybrid between the cDNA and the feature will now also be labeled, that is, by scanning the microarray chip, every feature that is labeled with cDNA can be identified. Many features on the chip will not be labeled, because no cDNA in this sample has that sequence, but every cDNA will find a feature with a complementary sequence and label it as well. Because the features were embedded onto the microarray chip in a known pattern, scanning for labeled features results in a catalog of every gene that is being transcribed in that tissue.

The signal from the label is very sensitive, so very low amounts of signal can be recognized; many transcripts produced at low levels or in only a few cells have been detected by one-channel arrays. Since the intensity of each signal is measured computationally, the scanning analysis records the level of the expression for each gene as well as if the feature is unlabeled or the gene is turned off, resulting in no mRNA expression. From these images, it is possible to compile a catalog, or an expression profile, of all genes being expressed in that cell at that time and the level at which they are being expressed.

KEY POINT Microarray analysis can be used to determine what genes are expressed at a given time in a particular cell type.

Microarrays are widely used to study transcription on a genome-wide scale, that is, of every gene at once. They are used to determine which strand is being transcribed, which sequences are present or absent in a particular set of transcripts, how the transcription pattern is different under different circumstances or in different genetic backgrounds (including different sexes), and so on.

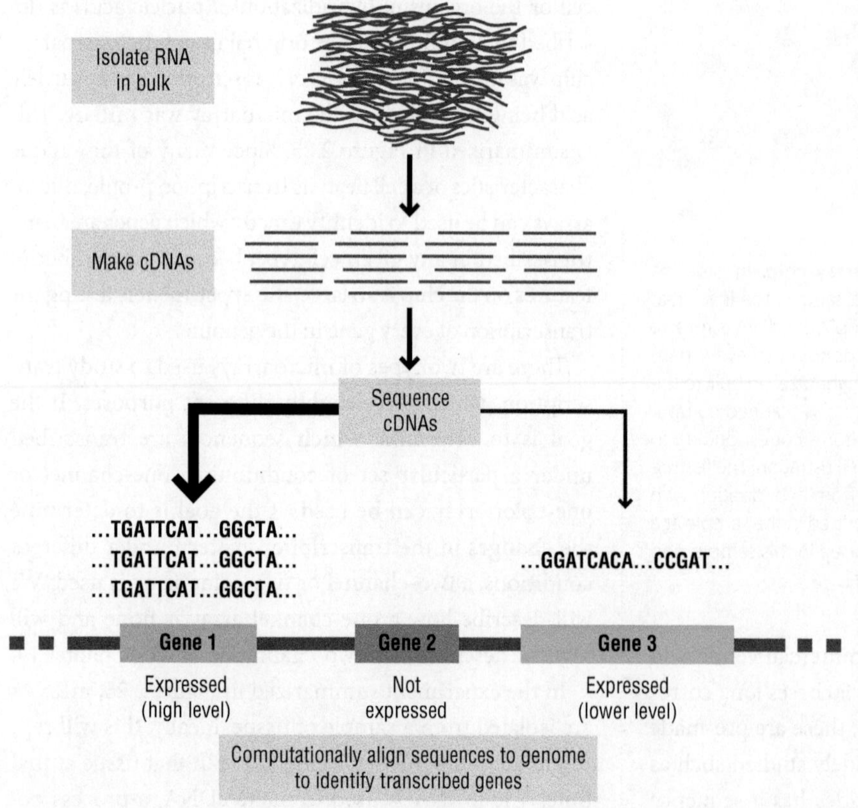

Figure 2.26 Using RNA-seq to compile an expression profile. In the method known as RNA-seq (for sequencing), mRNA is isolated from a sample. The mRNA is reverse transcribed into cDNA, but no label is added to the cDNA. The cDNAs are then sequenced directly, and the sequences are aligned to the genome to determine what genes are being expressed. The frequency at which a particular cDNA is found indicates the level of expression of its mRNA in the sample. While the method is called RNA-seq, it is usually cDNA that is sequenced, rather than RNA.

As technologies have made it cheaper and faster to sequence DNA, an expression profile can also be compiled from direct sequencing of cDNA, as shown in Figure 2.26. The method is called RNA sequencing or RNA-seq, although the sequences are actually cDNAs made using the mRNA as a template. (Methods are being developed to sequence RNA molecules directly, without the need for making them into cDNA, but the most common current methods rely on making cDNA before sequencing.) Microarray technology and expression profiles in general have revolutionized our ability to look at transcriptional patterns in cells and to know precisely the level of expression of every gene in a tissue sample. With all of these data, most of the Central Dogma holds—DNA provides the template to make RNA, which provides the template to make polypeptides. But with tools based on genome sequence, the scale at which this can be understood has greatly increased, and many transcripts that are not translated have been detected.

2.6 Summary: the Central Dogma

The Central Dogma of molecular biology describes how information flows from the DNA molecule, that is, the genes, to RNA and then to polypeptides, as summarized in Figure 2.27. This provides a foundation for almost every other topic that we consider, although it may be more apparent for some topics than for others. Heritable differences in phenotypes, whether between generations or over evolutionary time, occur because of differences in this information and its transmission. The differences, seen as changes in DNA sequence, can produce changes in the information content (that is, the sequence) of the RNA and polypeptide, or changes in its pattern of expression. As noted in Chapter 1, the gene concept is one of the Great Ideas of Biology. The

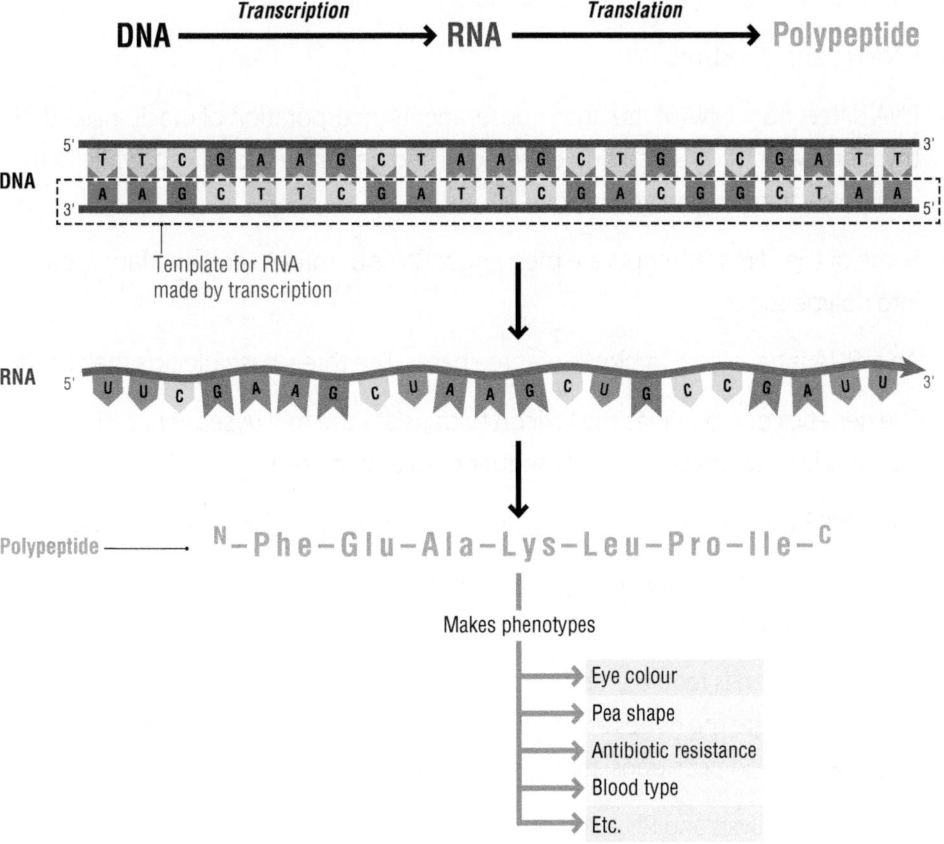

Figure 2.27 The Central Dogma summarized.

gene concept is intimately connected to the biochemistry of the DNA molecule, another of the Great Ideas of Biology.

While this chapter has focused on DNA, RNA, and proteins as macromolecules, it should be recognized that these are all found and function within the context of cells. Within cells, DNA forms complexes with proteins, based, in part, on the electrostatic interactions between the negative charges of the phosphate backbone of DNA and the positive charges of particular proteins. In eukaryotes, these complexes form chromosomes, which reside in the nucleus.

In addition, the DNA sequence of each gene is part of the DNA sequence of the genome, which is many thousands of times longer. Chapter 3 places the gene into the context of genomes and changes in genomes, while subsequent chapters provide many other elaborations of the principles introduced in this chapter. But it is the Central Dogma that lays the foundation, an informational mechanism, by which the rest of the principles of life must occur.

CHAPTER CAPSULE

- DNA is a polymer of four deoxyribonucleotides: adenine, thymine, cytosine, and guanine.
- Base pairs held together by hydrogen bonds form between the nitrogenous bases adenine and thymine, and between cytosine and guanine. A pairs with T, while G pairs with C.
- Base pairing is essential to DNA replication, transcription, and other important processes in the cell, as well as techniques in the laboratory.
- DNA is transcribed into RNA.
- RNA differs from DNA in its sugar, ribose, and its incorporation of uracil, instead of thymine, bases. RNA is also single-stranded and generally short-lived, while DNA is double-stranded and very stable.
- Some of the RNA transcripts are messenger RNAs or mRNAs and are translated into polypeptides.
- An mRNA is translated in blocks of three bases. The three-base block is the codon.
- The genetic code provides the lexicon to translate the mRNA sequence of nucleotides into the polypeptide sequence of amino acids.
- Some RNA transcripts are not translated and are functional in and of themselves.
- A typical gene includes a coding region and a regulatory region, which includes the core promoter.
- Gene expression is regulated by numerous processes, including control of when and where the genes are transcribed, mRNA splicing, regulation of RNA degradation, and post-translational modification of polypeptides.
- Microarray analysis and RNA-seq can be used to determine which genes are being transcribed at any given time in a cell type.

STUDY QUESTIONS

Concepts and Definitions

2.1 The Central Dogma is a unifying principle in biology.

a. Why is the Central Dogma so "central" to biology?

b. In what sense can this be considered to be a "dogma", and how is it not like a dogma?

c. The Central Dogma is sometimes expressed as "DNA makes RNA, which makes polypeptides." How is this way of expressing the Central Dogma helpful? How might it be confusing or misleading?

Challenging

d. Since its original description, the Central Dogma has been modified somewhat, and some small, but significant, exceptions to it have been found that required an arrow in the opposite direction. Other arrows have been added.

i. What are some of these changes that modified an arrow or added an arrow?

ii. One of the "arrows" has remained unchanged. Which step in the Central Dogma process has not been changed, and why is this so important?

2.2 What are the three categories of chemical components that make up a DNA molecule, and what is one functional importance of each component?

2.3 The structure of the DNA molecule is defined by its phosphate backbone and its base pairs. Each of these is important for one or more laboratory techniques.

a. Briefly describe how the phosphate backbone is used to separate DNA molecules based on their size.

b. Briefly describe how the specificity of the base pairs has been used for different laboratory techniques.

c. Many of the techniques that use the specificity of the base pairs in part (b) for hybridization can be performed at different temperatures, which affect the results. What happens if the temperature for hybridization is raised slightly? What happens if the temperature for hybridization is lowered?

2.4 What is cDNA, and what makes cDNA so important to molecular biology?

2.5 DNA sequence changes can occur in any part of the genome, including within genes.

a. Describe how a DNA sequence change in the coding region of a gene might affect the function of the gene.

b. Describe how a DNA sequence change in the *cis*-regulatory region or CRM for a gene might affect the function of a gene.

2.6 What is an expression profile?

Beyond the Concepts

2.7 Here is the sequence of a portion of one strand of DNA: 5′-GTCCTAACGACTGATCGT-3′

a. What is the sequence of the other DNA strand?

b. Is it possible to determine if this sequence is from the template strand or from the coding strand? Why or why not?

c. Suppose that this is the sequence from the **template** strand of DNA. What is the corresponding RNA sequence that would be transcribed, using this sequence as the template?

d. The RNA sequence in part (c) is made into cDNA. What is the sequence of the first strand of the cDNA that would be made from this gene?

Challenging
Looking ahead

e. Suppose that this sequence is from an **exon** in the **middle** of a gene. What is the amino acid sequence that would be translated from this gene? (This question requires a little more insight than it might seem. The codon table needed for the translation is found in Figure 13.13(a).)

2.8 While all bacteria and eukaryotes use double-stranded DNA as their genetic material, not all viruses have double-stranded DNA (dsDNA). Some viruses have single-stranded DNA (ssDNA), and some have single-stranded RNA (ssRNA). Before DNA sequencing was a routine method, biochemists would analyze nucleic acids for their base composition by determining the concentration of different components. Suppose that a biological sample was isolated from a deep sea vent, with the following results. Fill in the rest of this table. If some part cannot be answered, explain why.

Organism	% A	% C	% G	% T	% U	Type of genome
Bacteria		23				dsDNA
Virus	29		21			
Virus					26	

2.9 When the DNA sequences of genes from two closely related species are compared, sequence changes can be observed in the coding region of the gene and in the CRM for the gene. However, few changes are seen in the core promoter.

a. What might be the evolutionary effect of a change in the coding region of a gene?

b. What might be the evolutionary effect of a change in the CRM of a gene?

Challenging

c. Why are relatively few changes observed in the core promoter region of a gene?

Applying the Concepts

2.10 Nucleic acids are isolated and analyzed from different cells in a mammal.

a. How would the DNA sequence of a skin cell be different from the DNA of a nerve cell?

b. How would the RNA isolated from a skin cell and a nerve cell be different?

c. A stem cell is a cell that is not currently differentiated into a specific cell type but that retains the capacity to differentiate into different cell types (such as nerve and skin). How does the existence of stem cells help verify your answer to the preceding questions?

2.11 Design an experimental strategy that would allow an investigator to determine if the codons in mRNA are overlapping or non-overlapping in sequence. You do not need to describe how the experiment would be done, but you should imagine a simple approach and the expected outcomes that could distinguish between the alternative possibilities.

2.12 Two sample tubes each have about 100 nanograms (ng) of a 25-base single-stranded oligonucleotide and about 10 picograms (pg) of double-stranded genomic DNA from the same species in 100 μL of buffer. (Recall that there are 1000 picograms in a nanogram, so that the oligo is in excess). The sequences of each oligonucleotide are shown below. Each sample tube is heated to 95°C and then cooled to about 56°C. The percentage of double-stranded DNA is determined at various times and plotted (Figure Q2.1).

a. Which oligonucleotide yields reannealing curve A, and which one gives reannealing curve B? Explain your reasoning.

Oligo 1 5′-CGCTAGGATCGAACCATACTCGGAC-3′

Oligo 2 5′-CATAGAATTTACGCATACCTAAGAT-3′

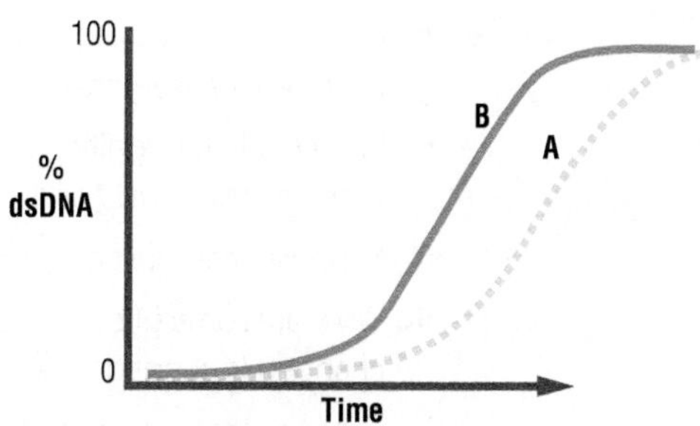

Figure Q2.1

Challenging

b. Why is it important to have a vast excess of the oligonucleotide over the genomic DNA in these experiments?

c. A third oligonucleotide is used in a different tube in a similar experiment, with the same concentration and the same amount of genomic DNA. The sequence of this oligonucleotide is shown below. However, its reannealing curve is completely different, since it **immediately** forms double-stranded DNA as soon as the cooling step begins. Explain what is occurring with this oligonucleotide and why it could not be used for further experiments. (Hint: Look closely at the oligonucleotide sequence.)

Oligo 3 `5′-TACGTACGTACGCGTACGTACGTA-3′`

Challenging
Looking ahead

d. A fourth oligonucleotide is used in a different tube in a similar experiment, with the same concentration and the same amount of genomic DNA. The results are shown as curve D in Figure Q2.2, with the results for curves A and B included for comparison. How can you explain this result? (Hint: Knowing the sequence of this oligonucleotide would not be useful in answering this question, unless you also knew the sequence of the genomic DNA.)

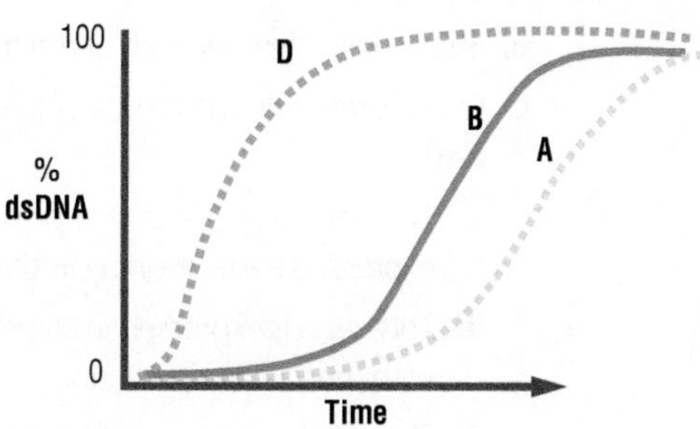

Figure Q2.2

2.13 Figure 2.17(b) is a screen capture of a portion of the genome of the unicellular eukaryote *Plasmodium knowlesi*. This figure contains many of the key points introduced in this chapter and to be developed in subsequent chapters. Assume that the coding portion of each gene is represented by the blue arrows. Each of the light gray bars indicates a different potential reading frame.

a. Indicate the location of the 5′ and 3′ ends of both DNA strands and the likely mRNA of the *Phat14* gene.

b. Show how this diagram supports the key points that:

i. Genes are encoded on each strand.

ii. DNA can be read with different reading frames.

c. Assume that genes in *Plasmodium* have essentially the same modular structure as genes in other eukaryotes.

i. What could be significant about the region from 52,800 to 54,000 which is located between the *Phat11* and *Phat12* genes?

ii. Suppose that most of this region were deleted. What is the possible result?

iii. What are some of the possible outcomes if the sequence of this region were changed (but not deleted)?

Looking ahead

iv. Suppose that you examined the corresponding region of the genome from a distinct, but closely related, species of *Plasmodium*. Would you predict that the sequence at 52,800–54,000 would be more similar or less similar than the region between 68,200 and 70,000 (that is, between *Phat15* and *Phat 16*)? Explain your answer.

Challenging Looking ahead

v. How can the structure of Phat17 be explained? Diagram the structures of the initial RNA transcript and the final mRNA product.

2.14 A one-channel microarray was used to investigate the expression of genes involved in muscle cell differentiation in chickens. The results from three cell types are shown in Figure Q2.3, each represented by its own microarray. A myoblast is a cell that will differentiate into a muscle cell but has not yet done so; the muscle cells and skin cells are fully differentiated cells from adults. Each of the four features on the microarray represents a different gene.

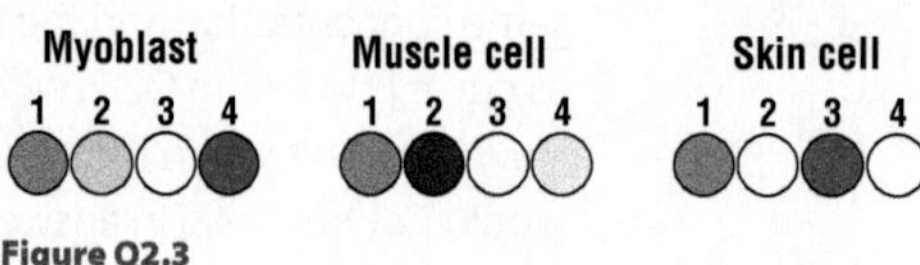

Figure Q2.3

a. What precisely is represented by each of the features shown as the circles?

b. Briefly describe how each of the three experiments was done.

c. From comparing the results of three microarrays, which gene is described by these results?

i. This gene is specifically or preferentially expressed in the musculature and is expressed more strongly as the myoblasts differentiate into mature muscle cells.

ii. This gene is expressed in all cell types shown and may have nothing specific to do with muscle differentiation.

iii. This gene is not involved in muscle differentiation but may be involved with the specification differentiation of skin cells.

iv. This gene is specifically or preferentially expressed in the musculature but is expressed at lower levels as the cells differentiate.

Looking ahead

d. A transcription factor known as Myf-5 is expressed in myoblasts and appears to be important for muscle cell differentiation. The experiment was done by the same methods as before, with four features representing different genes from those in the experiment above. In the results shown in Figure Q2.4, expression profiles were prepared from myoblasts from normal embryos in which Myf-5 is functional and from myoblasts from embryos in which myf-5 is not functional.

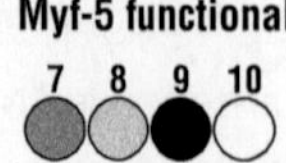

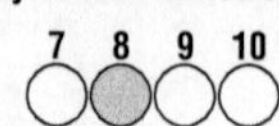

Figure Q2.4

i. Which of the features or features shown may correspond to genes whose transcription is regulated by Myf-5?

ii. Which of the feature or features correspond to genes expressed in myoblasts but not regulated by Myf-5?

Challenging

iii. Will all of the genes whose expression is found to be regulated by Myf-5 in this experiment have binding sites for the Myf-5 protein in their *cis*-regulatory modules? Very briefly, explain your answer.

CHAPTER 3
Genome Structure, Organization, and Variation

IN A NUTSHELL

The DNA sequence of an organism, known as its genome, encodes the information that underpins and directs most of the biological processes taking place within that organism. Genomes of different species differ in their DNA sequences and in the organization of their genes and other genetic elements. These variations are the product of natural selection and other evolutionary processes, but are also the means by which these same processes shape the genome—and hence the species itself.

3.1 Overview: genomes, variation, and evolution

In this chapter, we start the process of connecting genomic information with well-defined evolutionary and genetic concepts. This unifying perspective provides a framework to weave together the details of genetics, genomes, and evolution, as we describe them in other chapters. We start with natural selection and move to the concepts of the gene and the genome. We discuss the size and organization of genomes, and then introduce the notion of genetic variation and its relationship to gene functions. In Chapter 2, we learned that the expression of genes yields phenotypes, and that genes must be replicated or copied to be passed to the next generation. The replication process, introduced in this chapter and explored in more detail in Chapter 4, produces variation in the DNA sequence. Because of this variation, genes are altered and so may give different phenotypes when they are expressed. Natural selection acts on this phenotypic diversity.

Natural selection

Many great ideas appear to be simple, or even obvious, in retrospect. The adage that hindsight has 20–20 vision certainly applies when we think about the Great Ideas of Biology. Sometimes this retrospective simplicity occurs because the original Great Idea used observations and concepts that were widely known at the time and connected them in a new and creative way. Sometimes the retrospective clarity occurs because the Great Idea has changed our perspective so much that it is difficult to imagine a time when this was not known. Evolution by

Figure 3.1 Charles Darwin and Alfred Russel Wallace. The British naturalists who independently developed the fundamental ideas of evolutionary change by natural selection.

Source: Portrait of Charles Darwin by George Richmond; Photo of Alfred Russel Wallace taken by E. O. Hoppé.

natural selection fits both of these categories. Charles Darwin and his contemporary, Alfred Wallace, shown in Figure 3.1, took ideas about population growth, variation, and fertility that were widely known to most naturalists of the time and drew them together into a new explanatory framework. Furthermore, the power of the idea of natural selection changed how many people, including non-biologists, viewed nature, history, and the world.

Darwin's Theory of Evolution by Natural Selection, which unifies biology, boils down to two somewhat obvious points and a logical inference that can be drawn from them. Darwin accumulated years of data to support his ideas and the conclusions he drew, and a vast array of supporting information has also been added since then, but we still come back to the same two facts and reasoning from them. Darwinian evolution by natural selection rests on these simple principles.

1. Among the members of a population of a species, individuals vary in the number of offspring that they leave and in how many of these offspring go on to reproduce for themselves. Put another way, individuals vary in their **fitness**—the probability that they will survive and reproduce in that particular environment.
2. Some of the differences in fitness are heritable. There are individuals within a species or population who leave more offspring, and their offspring inherit the same ability to leave more offspring.

Based on these two points, here is the logical inference that Darwin drew—certain genetic constitutions (now referred to as **genotypes**) will increase in frequency over time, while others will decrease in frequency and may even become extinct because of differences in their fitness. This presumes that the population has genetic variation, and many different methods have been developed to recognize and study genetic variation. Some of these are found in Tool Box 3.1.

Herbert Spencer referred to this as "survival of the fittest," a term that quickly became a popular shorthand

TOOL BOX 3.1 Identifying natural variation

Not all individuals of a species are identical to each other. Natural variation is a foundation for genetics and evolutionary biology and is detected and studied using many different methods. The oldest, most familiar, and still among the best methods are simple, but detailed, observations of the morphology, structure, and behavior of organisms. The eminent corn geneticist Barbara McClintock was often described as having a "feeling for the organism" in relation to her plants, a quality that many biologists share whether the organism is corn, *Escherichia coli*, fruit flies, mice, or any other species that is studied closely. This detailed, and sometimes intuitive, observation is a key tool in identifying and understanding biological variation.

Other methods supplement such direct observations. For example, agglutination or clotting assays are widely used for studying natural variation in blood types, and various transplantation or grafting assays are used as well in both plants and animals. To a geneticist, all of these methods reveal different phenotypes—any quality that can be measured by some assay. The phenotype may have an underlying basis in the genotype or the genome, or it may be due to non-genetic factors. For example, the length of our hair is a common phenotype with almost no underlying genetic basis; the color and texture of our hair are phenotypes that have both an underlying genetic basis and significant non-genetic components (which can often disguise the genetic basis).

A key to thinking about natural variation is to distinguish between the genetic variation, which is due to the underlying variation in the genome, and non-genetic or environmental variation. Genetic variation can be transmitted to future generations and provides the foundation for evolutionary change. The importance and some of the challenges of distinguishing genetic from environmental variation will be discussed in more detail in Chapters 10 and 16. In the current era, genomes can be sequenced readily and genetic variation observed directly from the DNA sequence. However, this is a very recent development in the history of

TOOL BOX 3.1 Continued

studying biological variation. In this Tool Box, we discuss a few of the tools that were used before genome sequencing was common and that are still valuable methods for observing variation. These tools were originally developed for other experimental purposes, and their application to looking at natural biological variation came shortly later.

Protein variation using polyacrylamide gel electrophoresis

Gel electrophoresis was described in detail in Tool Box 2.1. In that Tool Box, the focus was on separating nucleic acids by size using the charges on their phosphate backbones. With nucleic acids, the gel matrix is typically made of agarose, and the gels are often run horizontally. Similar principles of electrophoresis can be used to separate polypeptides. For separating polypeptides, the gel matrix is usually made of polyacrylamide, and the gels are often run vertically; the method is called polyacrylamide gel electrophoresis, abbreviated PAGE. In addition, the charges on a polypeptide chain do not come from the backbone, but from the side chains on the amino acids, as shown in Figures 2.14 and 2.15. Two polypeptides, each of 500 amino acids, may have different electric charges (known as isoelectric points) because they do not have the same amino acid sequences. Thus, PAGE can separate polypeptides based not only on their sizes, but also on their electric charges, and is very widely used for separating and analyzing polypeptides, with many variations for different applications.

PAGE has also been widely used to look at natural variation. The proteins of different individuals can be isolated and separated, based on their sizes and/or charges, and then stained and compared. If a protein is found at different locations on the gels from two individuals, there must be a change in the amino acid sequence. From that, it is reasonable to infer that there has been an underlying change in the nucleic acid sequence of the gene encoding that protein. PAGE alone usually cannot determine what the amino acid change has been, but it can reveal that the organisms differ. PAGE was the method used to show that sickle-cell anemia arises from a single amino acid substitution in the β-globin chain of hemoglobin, for example, which we depict in Figure 16.16. The phenotype of the disease state (sickle-cell anemia) could be correlated with the phenotype using electrophoretic mobility.

While PAGE has been important for recognizing the extent of natural variation and can be readily applied to all organisms, it has some limitations. First, it can underestimate the extent of variation in polypeptides, since not all amino acid changes result in changes in electrophoretic mobility on PAGE. Second, since it examines only changes in polypeptides, it detects only changes in the protein-coding regions for genes, and not in any other part of the genome. As discussed in Section 3.6, changes in protein-coding regions are only one source of the genomic variation important in evolution.

DNA sequence changes using restriction endonucleases

A more direct method to look at the DNA sequence variation in all parts of the genome is provided by classes of enzymes known as **restriction endonucleases**. As we will discuss in Chapter 11, restriction endonucleases are enzymes that are important in protecting bacteria from invading DNA. While that is their natural function, they are very widely used for many experimental applications for molecular biology without considering their origins; most genetics and molecular biology laboratories have racks of different restriction endonucleases (or restriction enzymes) in their freezers, for which no one knows their origins, except as coming from one of the commercial vendors. They are a standard tool for many experiments in molecular biology.

Restriction enzymes bind to a DNA sequence at a specific sequence and make a precise cut in the backbones of the two strands of DNA. The recognition sequences for different restriction enzymes range from four bases to as many as 12 bases; most of the commonly used ones have recognition sites of five or six bases. For example, as shown in Figure A, the widely used restriction endonucleases *Eco*RI recognizes and cuts the sequence 5′-GAATTC-3′, *Bam*HI recognizes and cuts 5′-GGATCC-3′, *Sma*I recognizes and cuts 5′-GGGCCC-3′, and so on; hundreds of different enzymes are commercially available. The sites are usually palindromes (that is, they read the same on the other anti-parallel strand) because the enzymes function as a dimer with two subunits that bind and cut the two strands.

As tools for manipulating DNA sequences in the laboratory, restriction enzymes are invaluable simply because they precisely cut a specific sequence. The process is shown in Figure A. *Sma*I cuts between the G and the C in its recognition site and leaves blunt ends on the molecules that it cuts. Other restriction enzymes, such as *Eco*RI and *Bam*HI, make asymmetric cuts on the two strands, between the G and the A for *Eco*RI and between the two Gs for *Bam*HI. Thus, they leave an overhang of five bases on each strand, or a "sticky end." Suppose one wants to insert a segment of DNA from one source (a plant, for example) into a bacterial plasmid, so that the bacteria can produce many copies of the sequence or express the specific gene. The process is shown in Figure B. DNA from both the plant and the plasmid are purified separately and cut with *Bam*HI, so that both have complementary sticky ends. The DNA molecules are combined in solution; the sticky ends from the plant DNA form base pairs with the sticky ends from the plasmid DNA, and the plant DNA is inserted into the plasmid; the backbones are sealed up by DNA ligase, an enzyme involved in DNA replication, as discussed in Chapter 4.

TOOL BOX 3.1 Continued

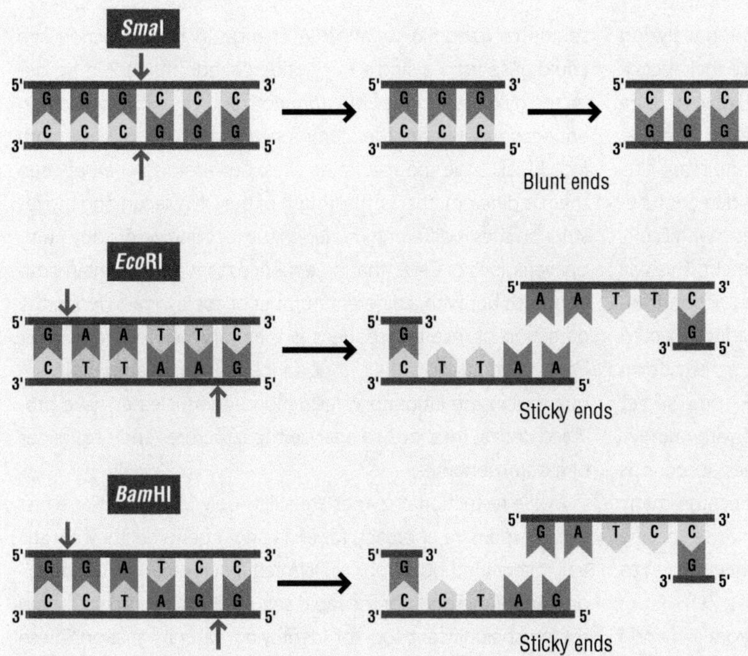

Figure A Restriction enzyme recognition sites. Restriction endonucleases or restriction enzymes bind to, and cut, specific sequences in DNA. The recognition sites for three commonly used restriction enzymes are shown, with the red arrows showing the location of the cut. Note that cutting with *Sma*I leaves a blunt end, whereas cutting with *Eco*RI and *Bam*HI leaves sticky ends.

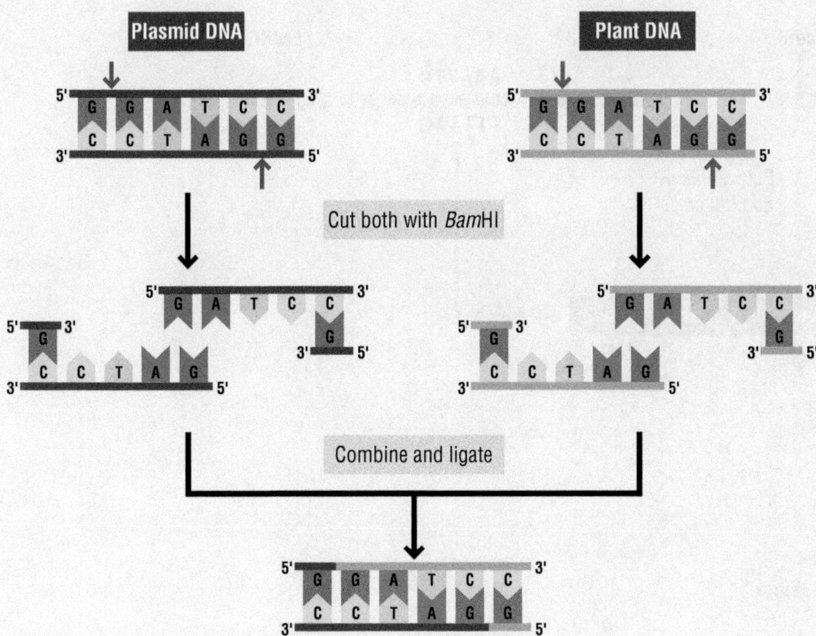

Figure B Inserting a restriction fragment into a plasmid vector. The plasmid and the DNA of interest (here from a plant) are cut with the same restriction enzyme, *Bam*HI, which leaves sticky ends. The digested DNA is combined and the sticky ends of one fragment pair with the complementary bases of the other, inserting the restriction fragment from the plant into the plasmid. The nicks at the insertion sites are repaired by DNA ligase.

TOOL BOX 3.1 Continued

Restriction enzymes can also be used as tools for analyzing natural variation in DNA sequences simply because their recognition sites are specific. An example is shown in Figure C. One individual may have the sequence GAATTC at a specific site, while a different individual may have AAATTC at the same site. The change in the nucleotide sequence can be detected because the site in the first individual can be cut *in vitro* with *Eco*RI, while the site in the second individual cannot be cut. This will result in a size difference in the length of the restriction fragment between the two individuals that can be readily detected by agarose gel electrophoresis. This is known as a **restriction fragment length polymorphism** or **RFLP**. RFLPs provide yet another phenotype, one that could be correlated with one observed from another assay but that does not need to be; it is a direct look at the underlying DNA sequence. Because many hundreds of restriction enzymes are known, each with a specific recognition site, tens of thousands of RFLPs have been found as natural variation among genomes.

While we have discussed the variation arising from a change in a recognition site, RFLPs also revealed another important source of genomic variation. A change in the restriction site could arise from a single nucleotide change that could be detected by digesting the DNA from both sources with many different enzymes. But another common variation occurs when both individuals have the restriction sites, but the distance between them is different, that is, the length of the DNA separating the restriction sites is different. This led to the recognition of copy number variations or CNVs, that is, sequences that are present in both genomes but with a different number of copies. It also led to the detection of inserted sequences that vary between genomes. Thus, in addition to their importance in manipulating DNA sequences in the laboratory, restriction endonucleases have provided critical methods to analyze the structures and sequences of natural genomes.

While restriction enzymes are still widely used, often as a first and inexpensive approach for analyzing a genome, they are still an indirect method to look at DNA sequence variation. The development of methods for the rapid sequencing of genomes is now the most significant tool for identifying natural variation. These methods are considered in Tool Box 3.2.

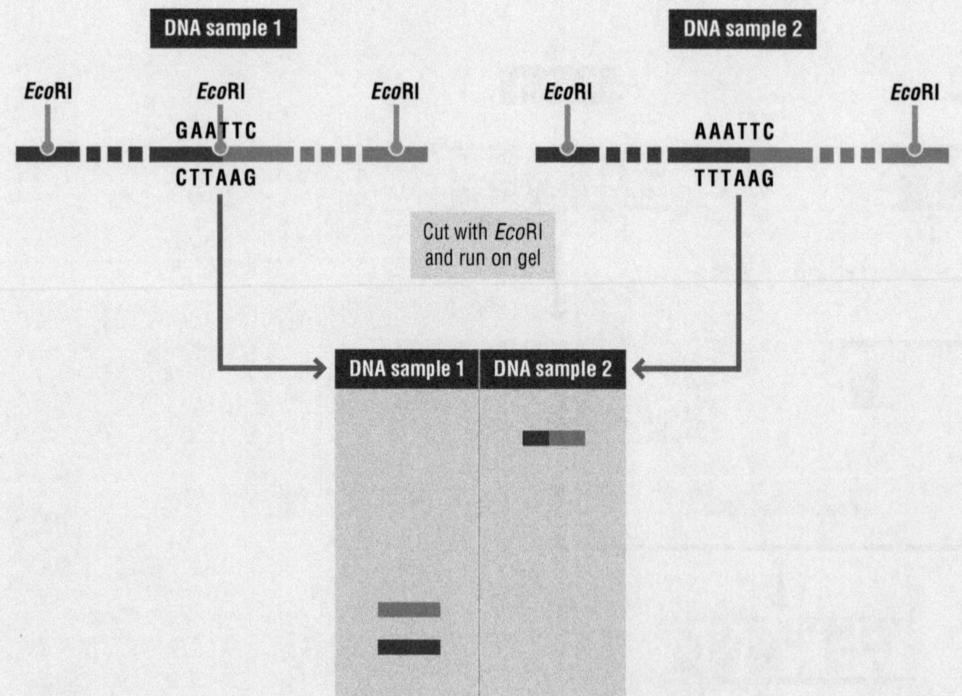

Figure C Restriction fragment length polymorphisms. The two DNA samples differ by a single base at a particular location. This base is in the recognition sequence for a particular restriction enzyme, here *Eco*RI. Thus, one sample is cut with *Eco*RI and gives two fragments when analyzed on a gel, while the other sample is not cut and leaves only one large fragment. This type of phenotype is a restriction fragment length polymorphism (RFLP) and has been widely used to analyze genomic variation.

(and remains so today), but which Darwin did not adopt until *The Origin of Species* went into its fifth edition. Darwin referred to this inference as "natural selection" and, more commonly, as "descent with modification."

KEY POINT Natural selection depends on the fact that different genotypes give rise to different numbers of offspring.

An enormous body of literature, both biological and non-biological, has grown up around the principles of Darwinian evolution, but these two facts and this inference remain its core components. Selection, in its simplest form, implies that some genetic characteristics are conferring a fitness advantage on some individuals, relative to other individuals, in the population. Since those individuals have a fitness advantage, they will leave more offspring in the next generation, and so those characteristics will begin to predominate in the population. The differences in which genotypes are increasing in frequency are referred to as the direction of selection. The rate at which the changes in frequency occur is called the strength of selection. These two ideas are summarized in Figure 3.2.

Darwin recognized that natural selection depends on the population, the environment, and the time. Characteristics that confer a selective advantage in one population at one time might not confer a selective advantage in all populations or at other times; in fact, they might even be disadvantageous to some populations in some environments or at some times. Chapter 16 will discuss some examples known from human populations, but consider our lives now, compared to the lives of our ancestors of hundreds of generations ago. In a population living in an environment in which food supplies are irregular—for example, a fishing community that occasionally makes a large catch but then can experience times when fish are scarce—genetic variants that increase the ability to store fat might confer an advantage to survive the times when food supplies run low. But the descendants of that population no longer live in an environment where food supplies are irregular; the grocery store is a few blocks away and is open at all hours. These same alleles that were once advantageous for increasing the ability to store fat might now predispose the descendants of this population to obesity, hypertension, and diabetes.

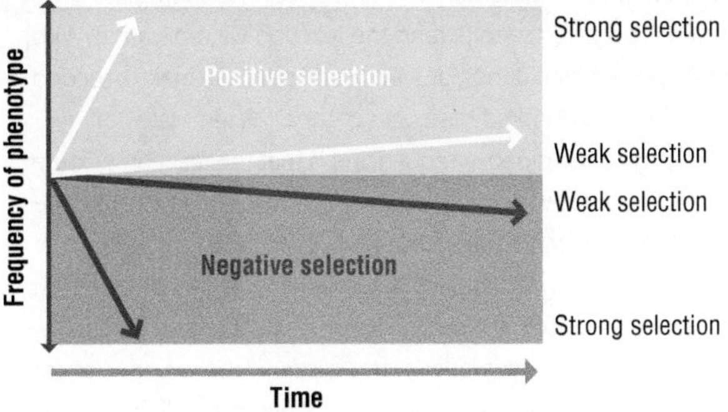

Figure 3.2 The strength and direction of selection. Natural selection results in the increase of the frequency of certain phenotypic traits (on the Y-axis) at the expense of some others. Positive selection results in an increase in the frequency of a trait over time (on the X-axis), while negative selection results in a decreased frequency. The rate at which the frequency of the trait changes is an indication of the strength of selection.

While our attention is drawn here to the effects of natural selection, we want to remind you that other evolutionary processes are also at work to bring about changes in genotype frequencies in populations. The ways that selection, population size, mating structure, and population movements affect the genetic structures of a population are explored in more detail in Chapter 16. The question is usually about the relative importance of these various factors.

Let's pause to make a connection back to Chapter 2. The genotype is the DNA sequence for the organism. Some of this DNA sequence includes genes, as discussed in Chapter 2, and the concepts that we elaborate in this chapter are more readily explained by discussing genes that encode polypeptides. Natural selection is based upon differences among individuals (or sometimes, as discussed in Chapter 17, on the interrelationships among the population), which arise, in part, from these differences in the base sequences of their genomes. Natural selection favors the alleles and DNA sequences that produce greater reproductive fitness. The effect of selection is to change the frequency at which particular DNA sequences are found within a population as a consequence of its action on the individuals that have those DNA sequences. As a result, evolution of the population and the genome occurs.

Fitness relates to DNA sequences

It is worth thinking about the effect of selection and other evolutionary processes on the genome a bit more. While the fitness advantage involves individuals and their DNA sequences, these genomic DNA changes can impact processes acting at many levels. Most functional DNA sequences are transcribed into RNA, which, for messenger RNA (mRNA), is translated into the amino acid sequence of proteins. Selective differences can arise based on the time at which transcription occurs, on the amount of transcript that is made, and on the altered amino acid sequence.

But we can continue and consider the effects of selection that occur at every step of processes that produce a phenotype discussed in Chapter 2. Selection can also happen because a particular combination of proteins work together better than another combination of proteins to increase the fitness of an individual, but selection that arises from an underlying protein interaction (or systems, if you like) also entails selection for particular amino acid sequences, and thus for certain underlying DNA sequences. Particular DNA sequences can even confer a fitness advantage to an individual, even when the sequences are not themselves expressed into RNA or protein, such as sequences that play a role in the structure or maintenance of the chromosome.

In other words, a heritable fitness advantage that occurs at any level—a population, a system, an individual, or a protein or another macromolecule—will inevitably result in changes in the frequency of particular DNA sequences in a population. This has to be true because of the second observation that Darwin made—the changes in fitness that drive evolution are heritable from one generation to the next.

KEY POINT As a result of natural selection, the frequency of specific DNA sequences in a population might increase or decrease, leading to evolutionary changes in the genome.

Genome sequencing

As we wrote at the beginning of the book, biology is in an exciting era. New information is pouring in because **genome sequencing** can be done very rapidly; a sequencing robot can determine the sequence of tens of millions of base pairs of DNA in less than a week. An introduction to some of the methods involved in DNA sequencing is found in Tool Box 3.2.

TOOL BOX 3.2 DNA sequencing

No technologies have been as important to genome analysis as those used to determine the base sequence of DNA. The earliest chemical methods to analyze DNA were able to look only at the overall base composition such as the fraction of the bases that are G, the fraction that are A, and so on. These analyses provided the basis for the insight into the nature of base pairing that led Crick and Watson to the double helix models and gave rise to Chargaff's rules of base pairing. But the information in a DNA molecule is conveyed by virtue of the sequence of its bases, so the overall composition of bases was less helpful than knowing the order of the bases.

Methods to determine the base sequence of DNA were developed in the late 1970s, with two methods being widely used. Maxam–Gilbert sequencing combined radioactive labeling with four different chemical cleavage reactions, which chemically modified bases and resulted in cleavage at different locations; although it was the preferred method by which individual scientists sequenced DNA for more than a decade, it is little used nowadays because it could not be readily automated. By contrast, Sanger (or "dideoxy") sequencing, which was less commonly used by individuals, became the preferred method once sequencing machines were developed.

Maxam–Gilbert and **dideoxy sequencing** are sometimes referred to as the first-generation sequencing methods; Nobel Prizes for these methods were awarded to Walter Gilbert and Frederick Sanger in 1980. (Since a Nobel Prize can only be given to, at most, three people, their students and co-workers Allan Maxam and Alan Coulson were not recognized by the prizes but are still widely known and highly respected for this work.)

Dideoxy sequencing

Unlike any of the more recent methods we will discuss later in this box, dideoxy sequencing was originally designed to be done by hand by an individual investigator. Dideoxy sequencing uses a single double-stranded DNA template, typically an isolated purified fragment of a few thousand base pairs in length, such as the insert from a plasmid or a polymerase chain reaction (PCR) product. In addition to the template DNA, the reaction mix also includes the four deoxyribonucleotides (dNTPs) described in Chapter 2, a short DNA primer complementary to a sequence on the template DNA, and DNA polymerase to extend the reaction via replication. With the proper reaction conditions, such a mixture can make a second copy of several hundred base pairs of the DNA template.

The key to the dideoxy sequencing is that the reaction mixture also includes a low concentration of the four modified dideoxyribonucleotides, which lack the –OH group on the 3′ carbon of the sugar. The ring structure of a dideoxyribonucleotide is shown in Figure A. Each of the dideoxy bases used in the sequencing reaction has a fluorescent label, with a distinct label being used for each base. DNA polymerization always works by adding the next base to the 3′ OH in the ring of the preceding base. Because the dideoxyribonucleotide does not have a free –OH group at that site, the reaction stops as soon as one of the dideoxynucleotides is incorporated. This chain termination is the basis for the dideoxy sequencing, since the dideoxynucleotide will identify the last base in the DNA polymer that was made.

TOOL BOX 3.2 Continued

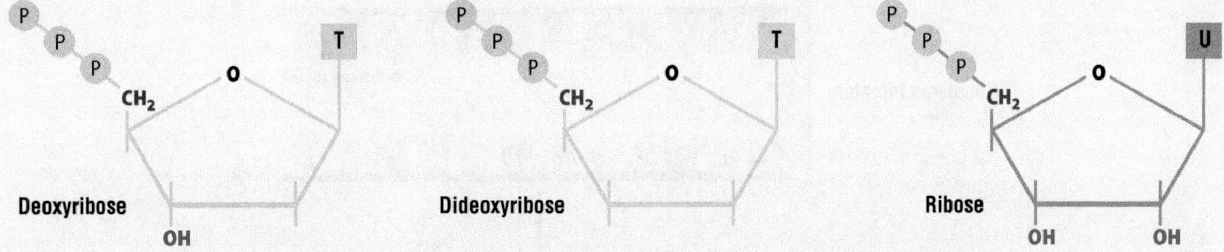

Figure A The structure of dideoxynucleotides. The key to dideoxy sequencing is the structure of the dideoxynucleotides, shown here compared to the structure of deoxynucleotides and ribonucleotides. As shown here, dideoxynucleotides have no –OH group at the 3′ position on the ring, which stops the polymerization process.

Dideoxy sequencing is diagrammed in Figure B. For clarity, we depict four separate reaction vials, each with its own dideoxy base, which was often the way it was done by hand; within a sequencing machine, these reactions are done in the same vial.

Consider what happens in the vial with dideoxy C. Polymerase begins the reaction where the primer hybridized to the template and extends the reaction until the incorporation of a dideoxy C terminates the reaction. This may happen within the first few bases, or it may happen after several hundreds of bases have been incorporated; the length will depend, in part, upon the concentration of the dideoxy base, relative to the unmodified base. The reaction products in each vial are then of different lengths, depending on when the dideoxy nucleotide was incorporated, with a distinct label on the last nucleotide in each sequencing product. In our diagram, the four vials are then run in separate lanes on a gel. The sequencing products are separated by size, and the sequence is read up from the bottom (that is, from the smallest fragments). If, for example, a fragment of 53 nucleotides in length ends in a C and one of 54 nucleotides in length ends in an A, the sequence at bases 53 and 54 is C–A.

As summarized in Table A, the typical extension reaction—a **sequencing read**—is about 700 base pairs or more. When dideoxy sequencing is automated, the fragments are labeled with a fluorescent dye, analyzed by capillary electrophoresis, instead of on a gel, and the data presented as a chromatogram with different colors for each base. An example of a sequencing chromatogram is shown in Figure C. Although the reactions are now sent out to be done by machines at separate sequencing facilities, rather than by an individual working in her own laboratory, dideoxy sequencing is still routinely used to obtain long and accurate sequence reads from a single template and a single primer.

Massively parallel sequencing

It is helpful to keep dideoxy sequencing in mind when we think about more recent sequencing methods. However, these newer methods are fundamentally different from dideoxy sequencing and are also different from one another. The new sequencing technologies are frequently called **next-generation sequencing (NGS)**, but calling these 'next-generation' presents a problem in the vocabulary, since new technologies are still being developed to replace these; there will undoubtedly be another 'next generation' of methods appearing regularly for the foreseeable future. These technologies are also called **massively parallel sequencing** methods because many thousands of DNA templates are sequenced simultaneously. Those who use the sequences typically refer to the methods by the name of the instrument or the manufacturer—the most common are **454 sequencing** and **Illumina sequencing**, although each of these are being improved and additional new methods are being developed.

The process of obtaining sequence data from any of these methods is often referred to as **deep sequencing**, because every nucleotide is sequenced many times in the course of a reaction, unlike dideoxy sequencing, which sequences the template once. It is not uncommon to report an average coverage of 40-fold, meaning that each base was sequenced an average of 40 times in the course of the experiments.

Both of the primary manufacturers—Roche Applied Science and Illumina—have excellent websites, with instructional videos that describe these procedures and more recent advancements in more detail. Some of the relative information for comparative purposes is shown in Table A.

Once the DNA is isolated from the sample, it is sheared by sonication into random fragments of about 1 kb. The fragments are then anchored to a support, and PCR is done to amplify each fragment template. The amplified products are then sequenced. The various methods differ in how the fragments are anchored, the details of the PCR amplification, and the sequencing method. As a result of these differences, other differences arise in the methods and cost. For example, a typical run for 454 sequencing generates about 400–500 megabases (Mb) of sequence in 8 or 9 hours, or as much as 1 gigabase (Gb) in 3 days with different sample preparations.

Individual sequencing runs for 454 are relatively long, as much as 700 bp or nearly as long as a dideoxy sequence run; this can be

TOOL BOX 3.2 Continued

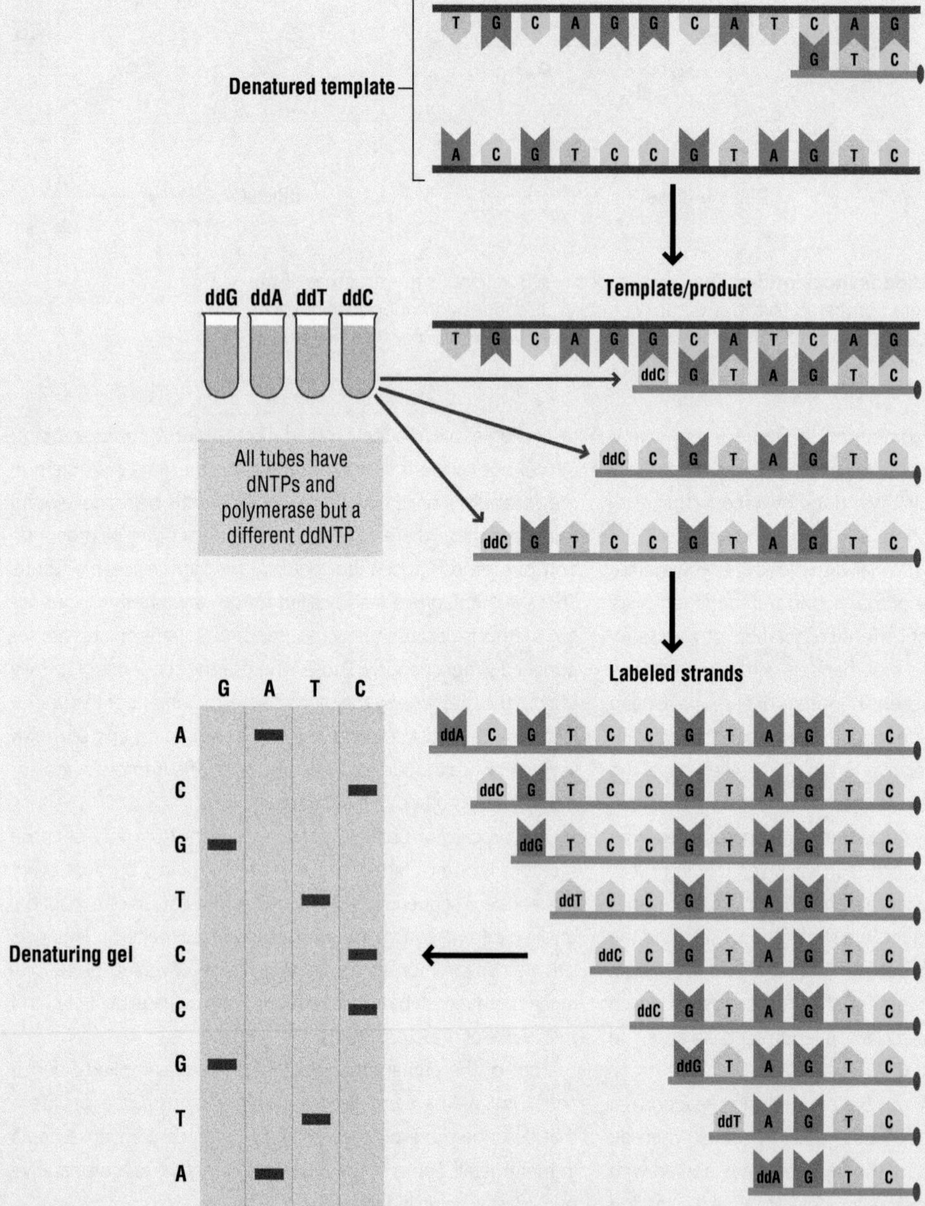

Figure B The process of dideoxy sequencing. A primer that is complementary to part of the DNA template to be sequenced is added and extended by polymerase. The reaction mix has the four deoxynucleotides and a small percentage of each of the four dideoxynucleotides with a different fluorescent label. Thus, when a dideoxynucleotide is incorporated, the reaction is stopped, with the terminal nucleotide having a label. The reaction fragments are separated, and the fluorescent labels are read in order to determine the sequence of the nucleotide at each position.

Source: Reproduced from Moran, P. *Overview of commonly used DNA techniques*. NOAA.

useful for assembling the sequence of an unknown genome for the first time. Imagine that assembling a complete genome from randomly generated sequence fragments, some of which overlap with each other, is analogous to assembling a jigsaw puzzle. A puzzle with fewer large pieces is typically easier to assemble than one with many small pieces.

Sequence runs from the Illumina machine are very short, usually 70–100 bp—but this short length also makes them very rapid. In addition, depending on how the template DNA has been prepared, roughly 20 Gb are generated per run. Because the Illumina runs are so short, they are more difficult to assemble into a longer genome sequence and thus are usually compared against a reference

TOOL BOX 3.2 Continued

Table A

Method	Template, primer, and reaction	Sample preparation	Sequencing read	Approximate rate	Comments
Sanger dideoxy	Single DNA template, single primer per reaction	Purified product	Approximately 700 bp	One run in 20–150 minutes	Useful for single applications, but impractical for large genomes
Roche 454	Thousands of DNA templates anchored on beads in an oil emulsion for PCR	Fragmentation	Up to 700 bp	1 Gb in 3 days, or 200–300 Mb in 9 hours	Long reads quickly, but individual sequence runs more expensive
Illumina	Thousands of DNA templates anchored on a solid support for PCR	Fragmentation	50–300 bp	Up to 20 Gb in 1–10 days	Very high yields with a low cost per sequence run, but can be the most expensive to purchase and might require more DNA

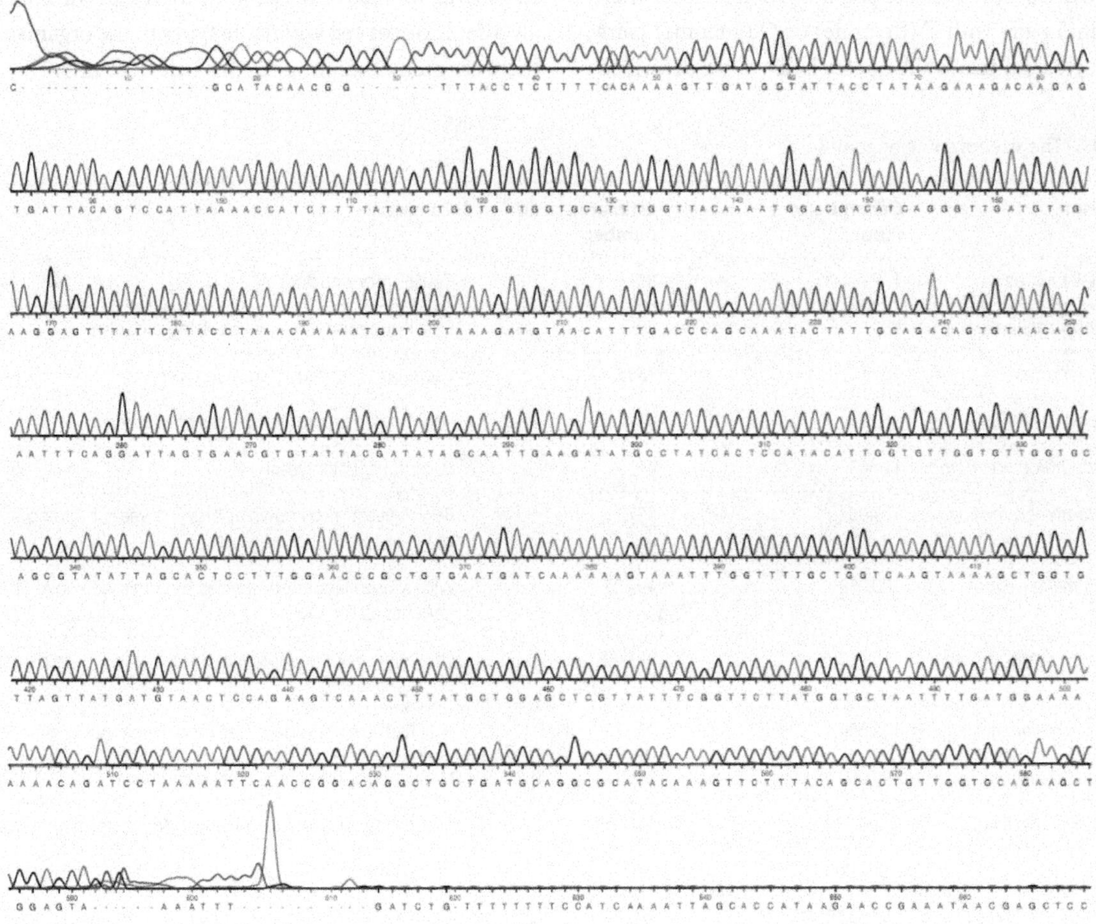

Figure C A scan from a run done by dideoxy sequencing. Each peak represents the fragments that terminate at that base, each colored with fluorescent dye. The height and separation of the peaks are indicators of the reliability of the sequence at that location. The sequences closest to the start and end of the read are usually not very reliable.

TOOL BOX 3.2 Continued

genome as a scaffold. (In our analogy, the reference genome scaffold is a highly detailed picture on the top of the box of the puzzle.) Both methods are very accurate—more than 99.9% accuracy for 454 sequencing and about 99% accuracy for Illumina. Many genome sequencing facilities have both types of machines, and often other massively parallel methods as well, running simultaneously.

RNA sequencing

The original chemical sequencing was done in 1976 with the RNA virus MS2 whose genome is about 4000 bases. However, it is currently uncommon for RNA to be sequenced directly, and no sequencing machines currently on the market are designed to do so. Rather, the RNA is first converted to cDNA, as described in Tool Box 2.4, and the cDNA copy is then sequenced by one of the massively parallel sequencing methods used for DNA sequencing. This approach is known as **RNA sequencing or RNA-seq**, although it is cDNA, and not RNA, that is actually being sequenced, and these methods are very widely used to analyze transcription patterns.

RNA-seq is extremely sensitive, so much so that RNA that is present at only a few copies per cell can be detected and analyzed. The vast number of previously unknown transcripts detected by RNA-seq has changed our views of how much of a eukaryotic genome is transcribed, and is stimulating us to formulate new hypotheses about functional non-coding RNAs.

The size of a genome is typically measured in base pairs (bp), kilobase pairs (kb or 10^3 base pairs), or megabase pairs (Mb or 10^6 bp). The base pair is the unit formed when A (adenine) pairs with T (thymine) or G (guanine) pairs with C (cytosine) in a DNA double helix, as we described in Chapter 2. The sizes of some genomes are shown in Table 3.1; some of these species are depicted in Figure 3.3. In general, the size of the genome increases with the overall complexity of cell and tissue types in the organism, but there are many exceptions to this rule of thumb.

Table 3.1 The sizes of some genomes

Organism	Genome size (kbp)	Estimated gene number	Comments
Epstein–Barr virus	172	80	Causes mononucleosis
Mycoplasma genitalium	580	485	Smallest genome of any free-living organism
Heliobacter pylori	1667	1589	Bacterium that causes stomach ulcers
Escherichia coli K-12	4639	4377	Widely used laboratory species of bacteria
Schizosaccharomyces pombe	12,463	4929	Eukaryotic fission yeast
Saccharomyces cerevisiae	12,496	5770	Baker's yeast, also known as brewer's yeast, budding yeast, or, in genetics, as Yeast
Caenorhabditis elegans	100,258	19,427	Widely used laboratory species, a nematode known in genetics as The Worm
Arabidopsis thaliana	115,410	28,000	Flowering plant (angiosperm). Most widely used plant in genetics research
Drosophila melanogaster	122,654	13,379	Fruit fly, the most widely used model organism in genetics
Rice	390,000	28,236	Most commonly eaten cereal by humans
Zebrafish (*Danio rerio*)	1,200,000	15,761	Common aquarium fish, widely used model in genetics
Humans	3,300,000	22,000	Nearly all mammals have genomes of about this size
Norway Spruce (*Picea abies*)	19,600,000	28,358	Conifer pine (gymnosperm). Conifer genomes are several gigabases large.
Xenopus laevis	Approximately 3,100,000	~17,000	African clawed frog, tetraploid genome
Whisk ferns	Approximately 350,000,000		Huge genome, but has no flowers or fruit

Figure 3.3 Genome sizes are as diverse as the organisms they encode. These are some of the organisms whose genomes are included in Table 3.1. The organisms that are illustrated elsewhere in the book, such as *Escherichia coli*, *Caenorhabditis elegans*, and *Drosophila melanogaster*, are not included.

Source: *Mycoplasma genitalium* courtesy of THOMAS DEERINCK, NCMIR/SCIENCE PHOTO LIBRARY; *Helicobacter pylori* courtesy of Professor Yutaka Tsutsumi, M.D. Department of Pathology, Fujita Health University School of Medicine; *Schizosaccharomyces pombe* courtesy of David O. Morgan; *Saccharomyces cerevisiae* courtesy of Masur; Norway spruce courtesy of JLPC / Wikimedia Commons / CC-BY-SA-3.0; *Arabidopsis thaliana* courtesy of NASA; Whisk fern courtesy of Eric Guinther/ CC BY-SA 3.0; *Danio rerio* courtesy of Azul; *Xenopus laevis* courtesy of H. Krisp/ CC BY 3.0.

The first DNA genome sequenced was that of bacteriophage φX174, a virus that infects bacteria, published in 1977. The genome of *Haemophilus influenzae*, a bacterial species, was sequenced in 1995, and the first eukaryotic genome to be sequenced was the budding yeast *Saccharomyces cerevisiae* in 1996. The first genome from a multicellular organism was from the nematode worm *Caenorhabditis elegans*, completed in 1998. The sequence of the human genome was published soon after, in February 2001, utilizing the technology developed both to sequence prior genomes and to share and annotate their data. The strategies by which DNA sequences are assembled into a sequenced genome are described in Tool Box 3.3.

A genome that previously took years to sequence now takes only days or weeks; in fact, a bacterial genome can be sequenced in a few hours. Within the last 10 years, biologists now have obtained the complete or nearly complete genome sequence for thousands of different species. This includes nearly all of the ones that are used as experimental organisms in laboratories, many domesticated

TOOL BOX 3.3 Genome assembly and annotation

In Tool Box 3.2, we described how a genome is sequenced and noted that a sequencing run with dideoxy sequencing generates sequences of about 700 bases. The *Escherichia coli* genome is about 4.5 million base pairs (4.5×10^6 bp), so a sequence run of 700 bp is approximately 0.02% of the entire genome; the difference in scale between an individual sequencing run and the size of a genome is even greater for a eukaryotic genome. Thus, each sequence run of a few hundred bases has to be assembled with all of the other sequence runs (also of a few hundred bases) to produce a completely assembled and sequenced genome. If we continue our analogy with a jigsaw puzzle, each sequence run is an individual piece, while the puzzle (the genome) might have 10,000 pieces for a bacterial genome and more than 10 million pieces for a mammalian genome. Assembly of the individual sequence runs into a genome sequence presents a formidable problem. On the other hand, once a genome has been assembled, it does not have to be assembled repeatedly. Instead, the completed or reference genome can be used as a scaffold to locate and assemble new sequence runs that are done.

Then, even when the genome sequence has been assembled, it is necessary to annotate the genome, that is, to assign functional characteristics, such as the locations of genes and transcripts, to the sequences. Annotation of a genome, like annotating any other text, is open-ended; even when the text is completely known, new information can be found in it. Think of the many volumes that continue to annotate the works of Shakespeare whose work has been completely assembled for five centuries.

In this box, we present a brief introduction to the assembly and annotation of a genome once the sequences are generated by the methods described in Tool Box 3.2. Assembly is now largely done computationally, with little human input, and the technical challenges are primarily the ones that interest computer scientists interested in large data sets. Our introduction to assembly is more general and historical. Annotation will be covered only briefly with some introduction to what kinds of information are sought.

Genome assembly

There are two general approaches to genome sequencing and assembly, both of which were used for the human genome. Directed sequencing and assembly (although other similar terms were used) was the only method available 20–25 years ago when genome sequencing began. The other approach, called **shotgun sequencing** and assembly, was first used about 20 years ago in bacteria and about 15 years ago in eukaryotes, and is almost exclusively the only method used for assembly nowadays.

Directed sequencing

In directed sequencing, the genome is first subdivided into many randomly sized (but large) fragments, and the fragments are cloned into different cloning vectors for growth in a host cell, as shown in Figure A(i). We can illustrate this by describing the process used to sequence and assemble the *Caenorhabditis elegans* genome of about 100 Mbp, the first metazoan genome to be completed. Genomic DNA was isolated from a large population of worms and was cut into fragments of about 40 kb using different restriction enzymes. Each of these fragments was cloned into a vector known as a **cosmid**, derived from bacteriophage λ (described in Section 11.4), which could be grown in *E. coli* to obtain large amounts of the worm DNA fragments; each *E coli* strain had one cosmid with a single insert of worm DNA. A collection of cloned fragments like this is known as a cosmid library, and several different cosmid libraries were constructed of the *C. elegans* genome. The cosmid, whose position in the genome was usually unknown, was then sequenced, and about 700 bp of sequence was obtained. The ends of this sequence were then used to make primers to sequence the adjoining sequence in the cosmid (as shown in Figure A(i)), and the entire cosmid was thus sequenced. This was done repeatedly for thousands of different cosmids representing many thousands of different random fragments, each of about 40 kb.

At the outset of the process, all of the inserts that were sequenced were unlike any other sequence; the worm genome is 10^8 bp, and a cosmid has about 4×10^4 bp, so it would be unlikely to find the same insert in two different cosmids. But as more sequences were completed, archived, and compared to one another (by computer alignment using the relatively limited computational power of 30 years ago), the sequence obtained from an end of the insert of one cosmid would sometimes be found to be the same as the sequence from an end of the insert of another cosmid. Thus, the inserts of these two cosmids must be sequences that are contiguous in the worm genome; the overlapping cosmids were said to form a **contig** (as shown in Figure A(i)). As more cosmids were sequenced, the number of contigs grew. Eventually, two different contigs would be joined together to make one longer contig, because a cosmid would be found that overlapped both of them; the number of contigs began to shrink, but the lengths of each contig grew. Each contig represented a section of the genome, and the cosmids formed a tiling path through that section of the genome.

This basic approach, with a few modifications, was how most of the *C. elegans* genome was sequenced and assembled and how the directed sequencing of the human genome began. The modifications were important. First, the adjacent cosmids were not always found simply by sequence comparisons and, in the early days of the sequencing project, were only rarely found by sequence comparisons; the software and computers to perform such comparisons did not exist. Instead, the ends of a cosmid were isolated and used as a probe (described in Tool Box 2.3) to

TOOL BOX 3.3 Continued

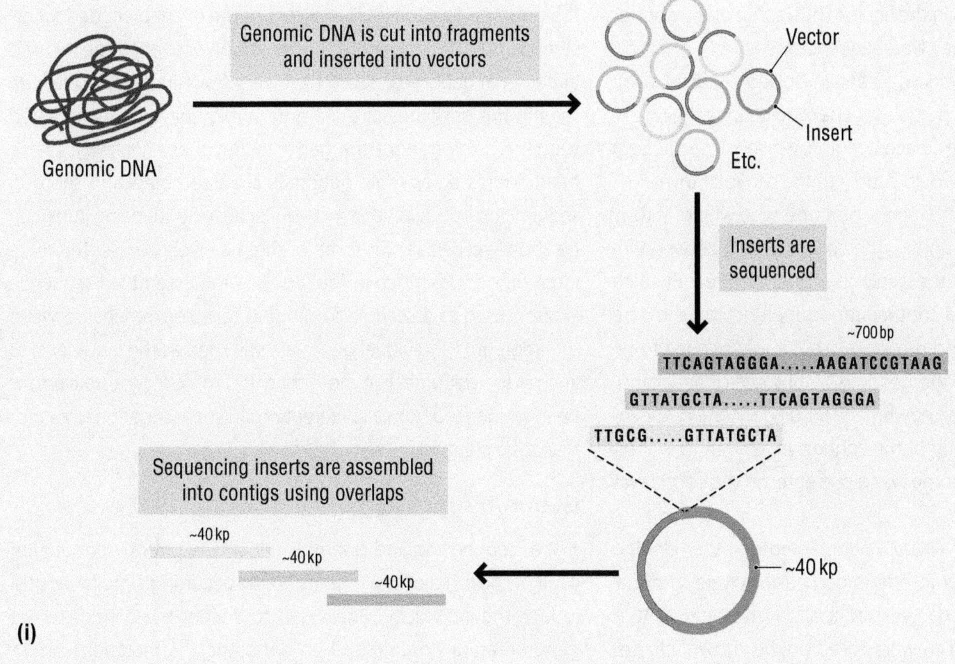

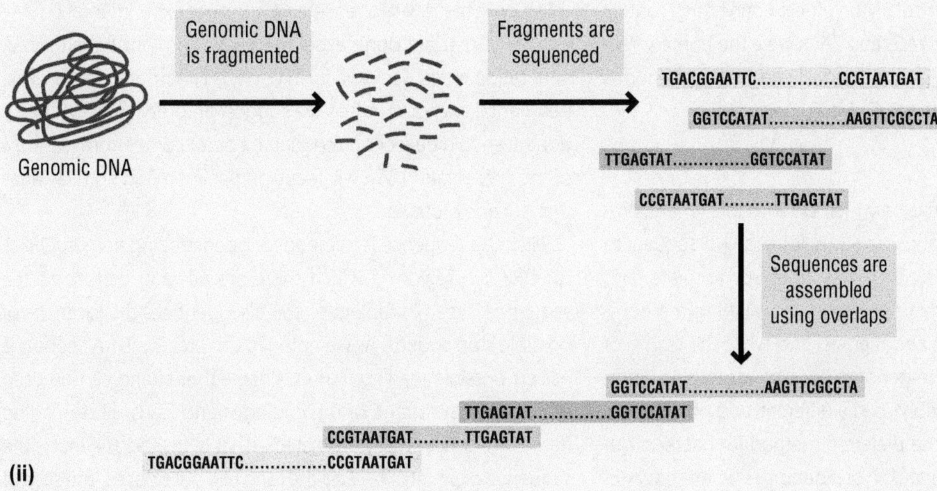

Figure A Sequencing and assembly strategies. (i) Directed sequencing. In directed sequencing and assembly, genomic DNA is cut into fragments with restriction enzymes and inserted into cloning vectors. In this figure, the average size of the insert is about 40 kb, which is typical for a cosmid. The vectors with the inserts make up a library, as shown in the collection of circles at the top right. An actual library has hundreds of thousands of vectors with inserts. A clone is isolated, and the insert is sequenced. A sequencing run generates about 700 bp from the 40-kb insert. The end of that sequence is used as a primer for the next sequencing run, and so on, until the entire insert is sequenced. The sequences from different cosmids are aligned with each other, or used as a probe, to find overlapping cosmids to produce a contig. (ii) Shotgun sequencing. In shotgun sequencing and assembly, genomic DNA is cut into small fragments, which are then sequenced without cloning into a vector. The sequences are aligned computationally and assembled into regions of overlap or contigs.

TOOL BOX 3.3 Continued

find the adjacent cosmids in the library. Thus, some of the contigs were assembled before they were sequenced.

Second, none of this work was done in isolation. Individual investigators interested in a particular gene would clone their gene of interest and some of the adjacent region, which formed its own small contig, which they would supply to the group of investigators sequencing and assembling the genome, who would, in turn, supply other parts of the contig. The collaborations between individual investigators and the genome sequencing team (which was extensive) also served another important function. The contig could be assembled and sequenced, but it also needed to be anchored to a location in the genome. Until some genetic landmark was found, such as a known gene studied by one of the individual investigators, many of the contigs were floating islands, and which part of the genome was represented in the contig was not known.

Third, cosmids were eventually supplanted by other cloning vectors that allowed larger inserts, in particular by yeast artificial chromosomes (YACs) and bacterial artificial chromosomes (BACs). These cloning vectors, which were based on the known features of yeast and bacterial chromosomes, rather than viruses, could have inserts as long as 1000 kb (1 Mb) or as much as 25 cosmids. The larger inserts were particularly helpful in connecting contigs to make one longer contig. YACs and BACs were the primary vectors used for the directed sequencing of the human genome, which is 30 times larger than the *C. elegans* genome.

Shotgun sequencing

Directed sequencing was the only method available when computer power was limited and programs for aligning sequences were primitive or non-existent. It was slow but accurate, and there was a certain anticipatory pleasure in watching the number of contigs first increase and then decline, while the length of the contigs increased as the genome became assembled. Shotgun sequencing and assembly used a different, and possibly less elegant, approach. It was the preferred method for bacterial genomes and was possible for eukaryotic genomes when sequencing became easier and computers became more powerful.

The method is shown in Figure A(ii). Rather than isolate each fragment, insert it into a vector, and complete a tiling path, the fragments are sequenced directly without assembling them first. The sequences of each fragment are read into a computer—with sequencing robots, this happens directly—and the sequences are immediately compared and aligned with each other for overlaps.

Shotgun sequencing and assembly is very highly efficient, but it presents some challenges. For example, errors arise during sequencing, so the overlapping sequences are probably not identical; one must decide what mismatch rate is acceptable, and what sequence differences represent truly different parts of the genome. In addition, the extent of the overlap needed for assembly must be specified; a shorter region of overlap can be used in bacterial genomes, which have very few repeated sequences, than in the genomes of a mammal, which have many repeated sequences throughout the genome. But these challenges have been addressed as more genomes are assembled, and shotgun sequencing is now the standard. Because sequencing runs can be short—especially for some of the massively parallel sequencing methods described in Tool Box 3.2—an assembled sequence is often used as a scaffold for finding reliable overlaps between the fragments. Individual genes are still used as the landmarks to anchor the genome, but more than 100 million genes have now been sequenced from all different organisms, so identifiable landmarks are plentiful.

Genome annotation

Just as sequencing and assembly occur concurrently, annotation of all of the functional elements is also occurring simultaneously. In fact, the individual genes used to anchor the contigs to the genome are an example of an annotation. Annotation can include any of the many functional features found in a genome. It includes both steps that can be done computationally—that is, using only the raw sequence and no other experimental data—and steps that are done experimentally. As might be expected, annotation works back and forth between computational and experimental analysis. Even as we suggest an order to which this is done, it should be recognized that the order comes from the logic of the presentation. We will focus primarily on finding the genes that encode proteins.

First, the sequence is scanned for open reading frames (ORFs). An ORF is a region in which there are no stop codons as the sequence is read in triplets, so this also identifies the locations of possible stop codons. As we noted in Chapter 2, a DNA sequence has six possible reading frames, since either strand can be used as the template strand or the coding strand. Thus, all six reading frames are conceptually translated—that is, treated as if that is the sequence of an mRNA—and the long ORFs are noted. The sites of possible start codons (ATG) are also recorded. An ATG at the beginning of an ORF is an indicator that this could be the start of the gene (for a bacterial gene) or of the first exon (for a eukaryotic gene). For a eukaryotic gene, ORFs that do not begin with an ATG might be internal exons of the gene, and these can be used with surrounding ORFs to predict a possible splicing pattern of exons and introns. Of course, ATG is also found internally, so it is the ones that might be start codons that are the most helpful.

Second, the region upstream of potential ATG start codons is scanned for possible transcriptional start sites (TSSs). As noted in Chapter 2 and again in Chapter 12, the TSS is usually about 25–45 bases upstream of the ATG start codon and consists of a series of

TOOL BOX 3.3 Continued

T and A bases. The presence of a likely TSS in the appropriate location is an indication that the gene is being correctly identified.

Third, the possible amino acid sequence that might be encoded by the ORFs is compared to the amino acid sequences of known proteins, using an alignment program such as **BLAST**. Since most genes have **orthologs** in other species, a significant alignment "hit" is an indication that this is a genuine protein-coding gene. Not all genes have clear orthologs, or the species used for comparison might be too distantly related to provide convincing alignments, so the failure to find a hit is not an indication that the predicted gene is a false positive. Alignment with other proteins in the databases also provides important information about the likely functions of the gene.

This has described annotation of a gene—or what is often nicknamed "gene finding"—as a computational exercise; certainly, very good predictions about the locations and functions of genes can be obtained from an assiduous use of these methods. But these are always being supplemented by experimental data. RNA is isolated and sequenced to provide expressed sequence tags (ESTs), which can be aligned against the genome to identify transcribed regions. For genes whose products are functional RNAs, rather than proteins, the presence of a transcript is often the only evidence that the gene exists; scanning for ORFs misses all of these genes, since they do not encode proteins. Transcript predictions, such as from ESTs and full-length cDNAs, refine the predictions about TSSs, exons and introns, untranslated regions at the ends of the transcript, and so on. As sequencing has become easier and very fast, identification of transcripts is done very early in the process of annotation. Transcripts are also the best evidence that can be used for a gene, and the isolation of a transcript is far more convincing than any prediction program.

Finding genes and transcripts, that is, gene annotation, is only part of the annotation process, but it is an essential part. Many of the functional elements in the genome are not genes but include such important elements as origins of replication, *cis*-regulatory modules for regulating transcription, sites of recombination, and epigenetic modifications. Information compiled from such an annotation project are discussed in more detail in Chapter 14.

and agricultural species, many of the ones that are medically important, and hundreds more.

As impressive as these efforts are, the ongoing genome projects capture only a tiny fraction of the roughly 2 million named species currently alive on earth. Even these 2 million named species are only a fraction of the species that probably exist, and a very tiny fraction of the ones that once lived. Fortunately, because species are descended from common ancestors, we do not need to study every individual species to be able to understand most biological processes.

The evolutionary history of an organism is both revealed and affected by what is found in its genome. Since, as Max Delbrück said, "Any living cell carries with it the experiences of a billion years of experimentation by its ancestors," exploring the DNA sequences of genomes is like reading the laboratory notebooks of evolution. We are only beginning to learn how to read genomes in detail, but already we see evidence for innovations, unexpected leaps, false starts, failed experiments, and unsuspected connections between processes and organisms.

KEY POINT The complete nucleotide sequences of many genomes are increasingly available. Understanding these sequences is essential to uncovering how life works now and how it worked in the past.

3.2 Genome size and organization

It probably comes as no particular surprise that genomes became larger as more complex organisms with novel biological processes arose. But there is a corollary to this observation that may not be so intuitive. This increased genome size means that mechanisms must exist to **add** genetic information (that is, DNA sequences) to what is already present. Some of the mechanisms that add DNA sequences to genomes are summarized in Box 3.1. Many of these mechanisms for adding DNA sequences are discussed again in Section 3.6 and in later chapters. There are also mechanisms by which sequences are lost from genomes, but these must contribute less than the mechanisms that add to the size of genomes.

BOX 3.1 *Going Deeper* How are sequences added to genomes?

We note in this chapter that nuclear genomes have generally increased in size as organisms have become more complex, although there are many exceptions. This raises questions about the evolutionary origins of this additional DNA and about the processes by which these additions have occurred during evolution. We summarize some of the most important processes here that increase the sizes of genomes and which have been found to have occurred during evolution. Many of these are discussed again in other chapters.

Polyploidization

Animals are typically diploid, although a haploid phase—such as gametogenesis—occurs at some stage of the life cycle of all eukaryotes. Many plants, however, have more than two copies of each chromosome, a constitution known as **polyploidy**. More than half of all flowering plant species are polyploids, and nearly all commercially grown grains and ornamental flowers show evidence for polyploidy. Polyploids have additional copies of a complete genome, and not just copies of some genes or some chromosomes. The terms for the different polyploid constitutions follow the Greek prefixes—triploids have three copies of each chromosome; tetraploids have four copies, and so on. Polyploidy will be discussed in more detail in Box 6.3 and is illustrated in Figure 3.20(a).

Meiosis results in a reduction of the chromosome number during gametogenesis, a process described in Chapter 6. Polyploidization occurs in nature by any of several different mechanisms. These include the production of diploid gametes when the chromosome number is not reduced during meiosis, a round of DNA replication in premeiotic cells that is not followed by meiosis, or the fusion of diploid cells. The exact process by which polyploidy has occurred in a particular species is usually not known.

If the duplicated genomes are duplicates of the same genome, as would occur by a failure of meiosis, the organism is said to be **autopolyploid**. Potatoes are an autotetraploid. If the duplicated genomes are derived from different species, the organism is said to be an **allopolyploid**. Durum wheat (used in pasta) is an allotetraploid, while bread wheat is an allohexaploid with the genomes of three different species.

By whatever mechanisms they arise, polyploids are sometimes referred to as whole-genome duplications, since they originally have entire sets of additional chromosomes. Duplication of the genome, like duplication and divergence that occur within gene families, can be followed by increasing diversification in expression patterns and functions among the duplicated genes. During subsequent evolution, the whole-genome duplication has often been followed by the loss of some parts of the genome, again much like the process of duplication and divergence in gene families.

The evidence for whole-genome duplication during evolution, also referred to as **paleopolyploidy**, is that entire gene families are duplicated in a taxonomic group. Such a process of whole-genome duplication, followed by divergence and loss, is thought to have occurred twice as the vertebrates evolved; this is referred to as the 2R hypothesis (for two rounds of duplication). Thus, fish and all other vertebrates have extra copies of many gene families, compared to invertebrates; the budding yeast *Saccharomyces cerevisiae* has extra copies of genes and gene families, compared to its relative *Kluyveromyces waltii*, and flowering plants have extra copies of gene families, compared to non-flowering plants. Polyploidization is the mechanism that results in the greatest increase in the size of genomes, and it has clearly occurred many times in many different taxonomic groups.

Horizontal gene transfer

Horizontal gene transfer is the lateral transfer of part of the genome of one species into another species and will be described in detail in Chapter 11. Horizontal gene transfer is particularly common in bacteria, but examples are known in plants and animals as well, and may be quite common but undetected. Horizontal gene transfer increases the size of the genome by a few percent, so it does not greatly increase the size of the genome; it does, however, increase the coding capacity of the genome. Horizontal gene transfer is often mediated by mobile genetic elements or by viruses, as described in Chapter 11.

Expansion of gene families

As noted in the chapter, the number of members of a gene family can also increase. For example, the β-globin gene family in humans has five functional members and two non-functional members known as pseudogenes, two of which are the result of a recent gene duplication in primates. Most other mammals have only four members, and chickens and many other vertebrates have three members.

A number of different cellular mechanisms result in such expansion of gene family members, and it is often difficult to know which one has occurred. As noted in Section 3.6 and illustrated in Figure 3.20(c), the additional members of the family can result in a diversification in expression pattern and function. While such expansion of family members is very important for diversification of functions, and thus adding copy capacity, it adds a few genes to the genome, rather than large-scale changes arising from polyploidization and an increase in transposable elements.

Even considering all of these processes that add sequences to the genome, it is somewhat hard to understand how genomes increase dramatically in size. The haploid genomes of *Caenorhabditis elegans* and *Drosophila* are about 3% of the size of the genome of mammals, although the number of protein-coding genes is observed to be about the same. Thus, most of the genome expansion has occurred without the expansion of the number of protein-coding regions, but hypotheses that can account for all of this difference remain speculative.

An analogy between sweaters and genes

The observation that genome size tends to increase and only rarely decreases is worth thinking about, so we can use an analogy to illustrate it. Most of us accumulate possessions over time—clothes, books, appliances, furniture, and other things. Even if we make a personal rule that every time we buy a new item (say, a new sweater), we have to give one of our old sweaters to charity, most of us are not disciplined enough to keep the rule consistently. As a result, we end up with a closet full of sweaters, only some of which ever get worn. For the most part, there is not much harm in having too many sweaters so long as our closets and budgets are big enough. After all, it takes work and foresight to sort through the sweaters that no longer fit, that are damaged or out of style, or that we simply no longer wear. Getting rid of them becomes more involved than letting them accumulate. In addition, it is better to have sweaters that we no longer wear than it is to lack a sweater on a cold morning. There is only a small penalty associated with having too many sweaters, but there is a much larger penalty (or disadvantage) associated with not having a sweater when we need it.

The genome appears to be similar to our closets—it accumulates DNA sequences. Although some energetic harm (that is, a selective disadvantage) arises from a genome getting too large, which may keep genomes from growing ever larger, there is far more of a disadvantage in lacking functions that are needed. The disadvantage to having the genome increase in size varies among species, depending, in part, upon the selective pressures on the species. This again has a parallel with our sweater analogy. A person who lives in a small apartment in the city is under greater pressure to monitor the sweater collection than a person who lives in a country estate with large closets. Thus, there is greater disadvantage to increasing genome sizes in bacteria and viruses whose genomes are relatively small than in most eukaryotes with relatively large genomes.

Furthermore, as we noted in the Prologue, a genome has no awareness of what is needed or what might be needed in the future. We know that the warm days of summer will be replaced by the chilly days of autumn, so we keep a supply of sweaters on hand. A genome does not have that foresight. Thus, the small selective disadvantage in having DNA sequences that are never used is offset by the huge disadvantage that would arise from deleting something that is necessary. Randomly deleting sequences is far more deleterious than randomly adding them. Thus, genomes get bigger.

We can take the sweater analogy a little further. The sweaters that we wear regularly are kept clean and in good condition. If we don't wear a sweater for a while, we will not know if it still fits, if moths have eaten holes in it, or if it is no longer in style. It is remotely possible that we will find a sweater in the back of the closet that perfectly matches the new outfit that we bought. (We will encounter an apparent example of this in the human genome with a gene known as *CCR5*, described in Section 16.5.) But it is far more likely that the unused sweater has become damaged or useless when we pull it out. So it is with genes and DNA sequences in the genome. If the genes are used, natural selection keeps them in good working order. If a gene is not used for long periods of time—that is, if the selective pressure on its function is relaxed—it may accumulate so much damage that it can't be used again. But since the mechanisms to get rid of these sequences are relatively ineffective, genomes accumulate damaged and non-functional sequences.

At the risk of going one step too far, we can make one final analogy between sweaters in our closet and genomes. Sometimes sweaters can be reused for purposes other than what they were originally intended for. For example, it is possible that someone with a closet full of disused sweaters will one day require a pair of leg warmers at short notice. She could make do with the sleeves of a disused sweater. These won't be as good as a tailor-made pair of leg warmers she could get at the store, but they might do for a start and could be improved over time. This concept of evolutionary tinkering with gene functions was also introduced in the Prologue. While the genome cannot anticipate its needs or explicitly tailor one part to suit another function because selection acts only on genetic material that is available, disused sequences occasionally do serve new functions. This is one source of novel genes and functions.

KEY POINT Genome size is loosely correlated with biological complexity. A genome is as much a record of evolutionary history as it is of current function and phenotype.

Genome reduction

Let us briefly consider the related, but opposite, phenomenon of gene loss. Variants that lose essential genes become less fit; on the other hand, loss of unused genetic material may not result in an evolutionary disadvantage.

Indeed, reducing the size of the genome that must be replicated through the loss of non-functional DNA can actually increase fitness.

Genomes can lose functions by accumulating inactivating mutations such as certain nucleotide changes or deletions. The loss of several genes simultaneously is expected to be severely incapacitating or lethal for an organism, so large deletions usually result in inviable organisms. However, not all genes are necessary in all environments at all times. Therefore, a gene, or a series of genes, can be lost from a population if the DNA does not encode any functions that are essential for competition and survival in a specific habitat. Once a gene that is not under selection is lost from the genome, the organism may become confined to a specific habitat, even though its recent ancestor may have been capable of inhabiting many. Multiple examples of gene loss are known, from many nutritional functions in parasitic organisms to genes for eye development and pigmentation in cave-dwelling fish, as shown in Figure 3.4. Gene loss also occurs in organisms that spend all of their lives within other organisms, either as obligate parasites or causing no harm, in which case they are referred to as obligate endosymbionts.

One example is *Wolbachia*, bacteria that are obligate endosymbionts of insects and nematodes. *Wolbachia* are closely related to the free-living bacterium *E. coli*, but their genomes are only about 20% of the size of the standard *E. coli* genome. The common ancestor of *Wolbachia* and *E. coli* most likely had a larger genome, similar in size to that of *E. coli*, which implies that *Wolbachia*

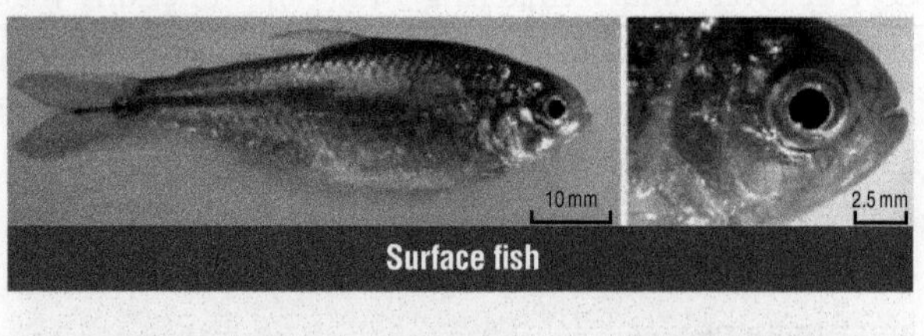

Figure 3.4 Surface and cave-dwelling fish of the species *Astyanax mexicanus*. The cave-dwelling fish belong to the same species but have lost the function of genetic material encoding eyes and pigmentation.

Source: Reproduced from McGaugh, S.E. et al. (2014). The cavefish genome reveals candidate genes for eye loss. *Nat Commun*. 5: 5307. Image courtesy of Bethany Stahl, University of Cincinnati.

have undergone extensive **genome reduction**. In contrast to *E. coli*, *Wolbachia* spend their entire life cycle inside host cells and cannot survive extracellularly. They are passed from mother to offspring, so that they never have to exit their insect or nematode host. Genes required for many anabolic pathways are inactivated or missing from the *Wolbachia* genome, which resulted in its reduced size.

KEY POINT Just as genomes can acquire blocks of DNA, they can also lose them. Genome reduction is a common feature of endosymbionts and parasites.

Genomes of bacteria and eukaryotes

Where is the genome found in an organism, and how is it arranged? The key subcellular structure that is crucial in thinking about genomes and in classifying organisms is the **nucleus**. The nucleus is an organelle that contains DNA and is set apart from the rest of the cell by a double-layered membrane; organisms with a nucleus are termed **eukaryotes**. A nucleus has its own substructures, and the DNA sequences within the nucleus are organized in discrete structures known as chromosomes, which we will discuss in depth in further text.

Bacteria do not have a nucleus bounded by a nuclear membrane; the absence of a nucleus means that these organisms are called prokaryotic, or **prokaryotes**; several examples of different types of bacteria are shown in Figure 3.5(a). As we will discuss in Chapter 4, two very different kingdoms of organisms—the true bacteria and the archaea—were originally grouped together as prokaryotes, so we will typically recognize this distinction and refer to bacteria. As shown in Figure 3.5(b), the genome of a bacterium is, for the most part, found in a single large double-stranded DNA molecule located in the cytoplasm. This structure is sometimes referred to as a **nucleoid** but more familiarly called the bacterial chromosome because it is analogous to a eukaryotic chromosome, although with an entirely different structure. Typically, such as in the most widely studied bacterium *E. coli*, the bacterial chromosome is a closed circle that is highly compacted and supercoiled. The chromosome for the *E. coli* strain K-12 (a widely used laboratory strain) is about 4.6×10^6 base pairs long.

In addition to the bacterial chromosome, bacteria may have other smaller circular molecules of DNA known as **plasmids**, as illustrated in Figure 3.5(b) and discussed in more detail in Chapter 11. Plasmids range in size from

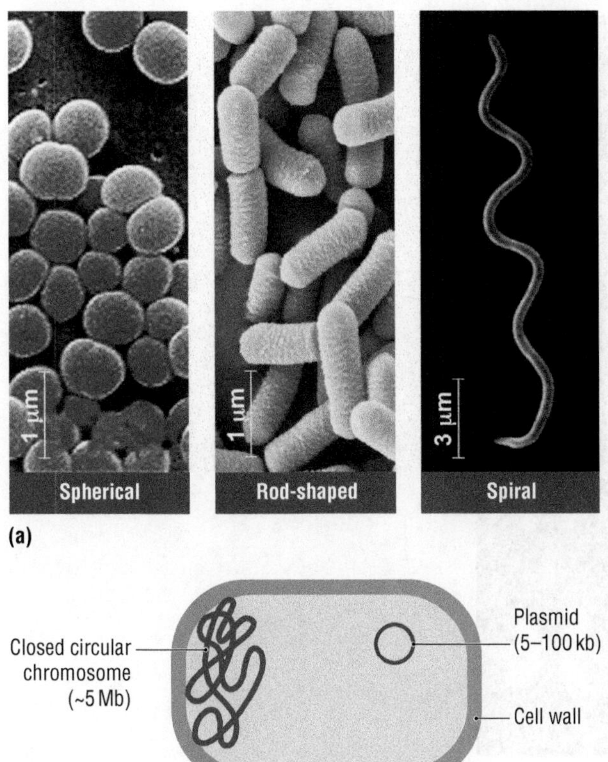

Figure 3.5 Bacterial genomes. (a) One characteristic used to distinguish different types of bacteria is the shape of the cells. Three different shapes are illustrated in these micrographs. (b) Nearly all bacteria have a closed circular DNA molecule referred to as the chromosome that typically encodes all of the necessary functions for that cell. Bacteria can have a variable number of additional and smaller circular DNA molecules known as plasmids. Plasmids encode functions important in specific processes or environments.

less than 10 kb to about 200 kb. They also have different genetic content than the chromosomes, with genes that are dispensable for the essential activities of the bacterial cell but which might encode functions important under some conditions. For example, genes on naturally occurring plasmids in *E. coli* encode their ability to metabolize unusual substrates, their ability to exchange genes with other bacteria, their pathogenicity, and their resistance to antibiotics.

KEY POINT Most bacteria have a single circular chromosome, and many bacteria have additional extrachromosomal circles of DNA known as plasmids.

Eukaryotic chromosomes

The structure and arrangement of eukaryotic genomes are completely different from those found in bacteria. Within the nucleus—the feature by which eukaryotes are named—DNA is organized into **chromosomes**. Chromosomes are not naked DNA molecules. Instead, the DNA molecule forms a complex with many different **chromosomal proteins**. The complex of DNA with the chromosomal proteins is known as **chromatin** and is summarized in Figure 3.6. There are hundreds of different chromosomal proteins, some of which are common to all chromosomes at all times, and others of which are associated only with particular chromosomes or at particular times.

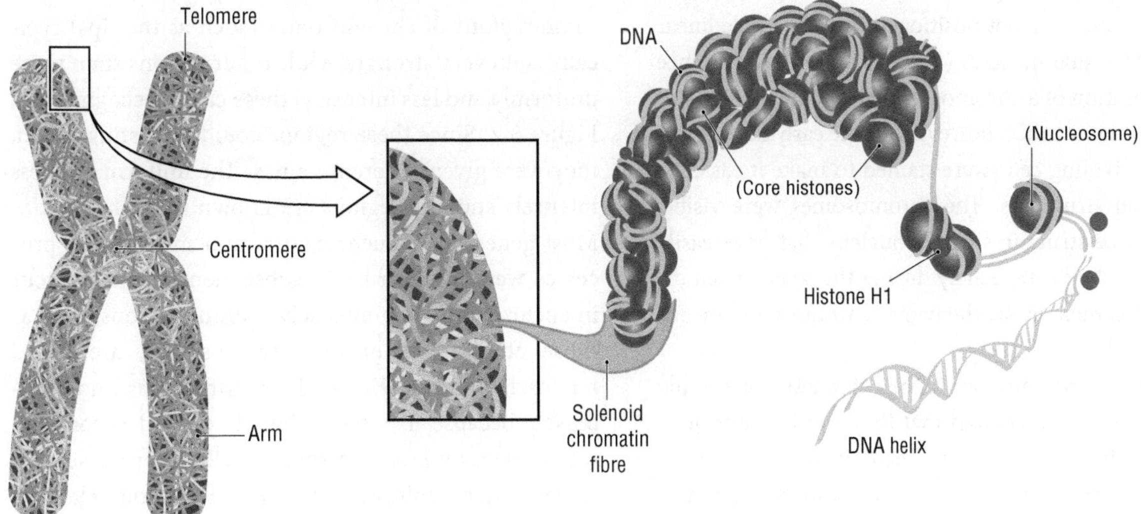

Figure 3.6 An overview of chromatin structure. DNA in a eukaryote is packaged with chromosomal proteins to form the much more compact chromatin. Chromatin is then folded into higher-order structures and eventually into the chromosome, as diagrammed here.

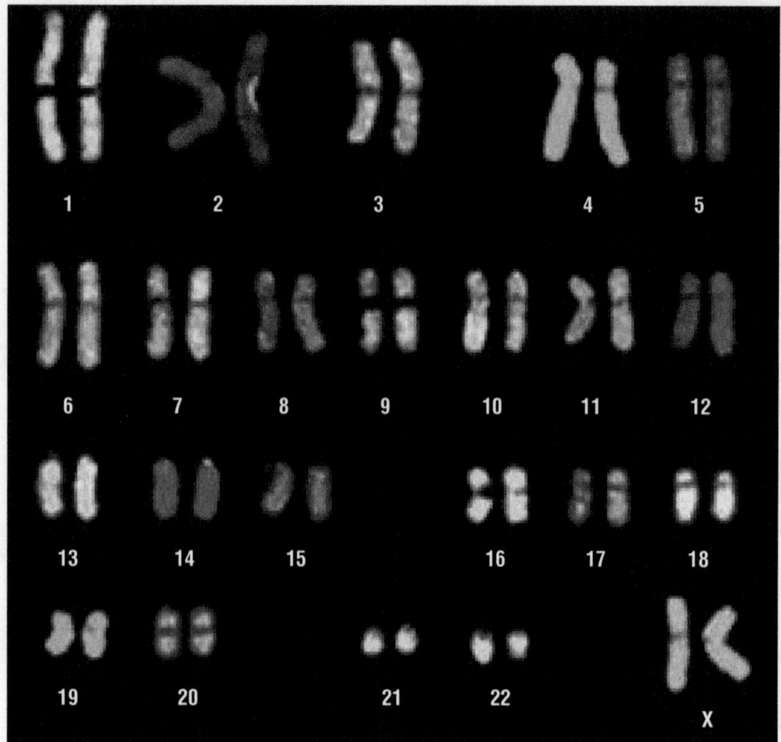

Figure 3.7 The human karyotype. The complete set of 23 pairs of chromosomes in a human cell, treated with a dye that stains DNA. Note the particularly intense staining near the tips where the DNA is particularly compacted.

Source: Courtesy of the National Institutes of Health/ genome.gov.

The complete set of human chromosomes, known as the **karyotype**, is shown in Figure 3.7. The pairs of chromosomes in the figure have been arranged in order by size, but they are not positioned in such a nice linear order in an actual nucleus. (There is increasing evidence that the location of a chromosome within the nucleus has functional significance, however.) In the early days of microscopy, dividing cells were stained to make it easier to see internal structures. The chromosomes were visible as thread-like structures in the nucleus that were easily stained or colored, especially during the stage of cell division called metaphase; the name "chromosome" means "colored body."

While chromosomes vary in their size and shape, many of them are similar enough that they can be difficult to distinguish from one another. The development of a variety of dyes and improved microscopy made it possible to observe differences in the size and shape of individual chromosomes. Some dyes, such as one called Giemsa, generated banding patterns that made it possible to tell one chromosome from another, so that the chromosomes could be distinguished and numbered more consistently; Giemsa staining or the **G band** pattern of human chromosomes is shown in Figure 3.8.

In addition to the banding patterns with specific dyes, some regions of chromosomes (such as the tips) typically stain very strongly, while other regions stain more uniformly and less intensely; these can be recognized in Figure 3.7. Since these regions could be distinguished, they were given different names. The uniform and less intensely staining regions are known as **euchromatin**. Most genes lie in euchromatin, and most of the processes we will describe in subsequent chapters occur in euchromatin. The intensely staining regions, such as those at the tips of the chromosomes, are called **heterochromatin**. Heterochromatin stains more intensely because it is more densely packed chromatin, which makes it less accessible to cellular processes like transcription, replication, and recombination; relatively few genes lie in heterochromatin. Heterochromatin and euchromatin were initially distinguished from each other, based on their DNA staining intensity, but now that more is known about the molecular structure of chromosomes, these regions also differ in the presence

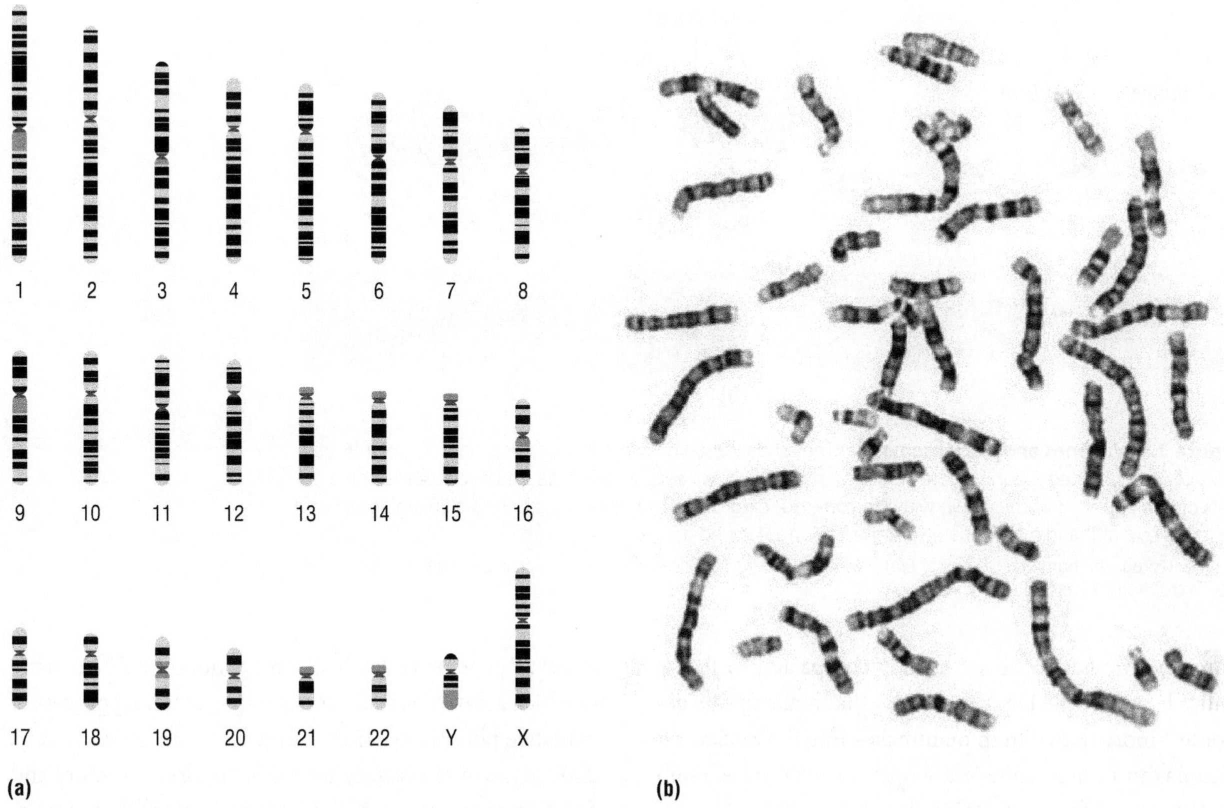

Figure 3.8 G-banded human chromosomes. Chromosomes can be stained with a dye known as Giemsa, which results in a reproducible pattern of bands. These G bands, along with the size and shape of each chromosome, are important for distinguishing one human chromosome from another. (a) Diagram of a human male karyotype. (b) Photomicrograph of Giemsa-stained chromosomes from a diploid cell from a human male.

Source: Reproduced from Craig, N.L. et al. (2014) *Molecular Biology: Principles of Genome Function* 2nd edition, with permission from Oxford University Press.

or absence of certain chromosomal proteins and in other aspects of their chromatin structure. It is also clear that having only two categories is a simplification. Regions of heterochromatin that stain similarly with DNA dyes may differ from one another in chromatin composition.

The most important structural proteins of the chromosome are the **histones**. The histones are a group of five different proteins, known as H1, H2A, H2B, H3, and H4. At the most fundamental structural level of the chromosome, 146 base pairs of DNA are wound around a core particle (or "bead") of histone proteins: two molecules of H2A, two molecules of H2B, two molecules of H3, and two molecules of H4, as shown in Figure 3.9. The histones are very positively charged proteins, so there are electrostatic interactions between the core histones and the negative charge on the phosphate backbone of DNA. Taken together, the 146 base pairs of DNA and the core histone particle around which they are wound are referred to as a **nucleosome**.

The four core histone proteins are among the most highly conserved proteins known. For example, the amino acid sequence of histone H4 is almost the same in humans as it is in corn. A change in the amino acid sequence of any of these core histones likely decreases the fitness of the organism. As a result, their function and sequence have been highly conserved over evolutionary time. (This is not the only explanation for the nearly identical histone sequences, but it is an important reason.) Chromosome structure is an important topic that is explored in more detail in Chapter 12.

The individual nucleosome beads (referred to as the 10-nm fiber or, more commonly, "beads on a string") are connected by H1 and other proteins into a 30-nm chromatin fiber. The 30-nm fiber is further compacted by other chromosomal proteins into a 100-nm fiber, and this is further compacted into the chromosomes that are seen in the cell; the sizes refer to the diameter of the fiber as the length of DNA is compacted. Unlike the 10-nm fiber, this higher-order structure of chromatin is dynamic. The

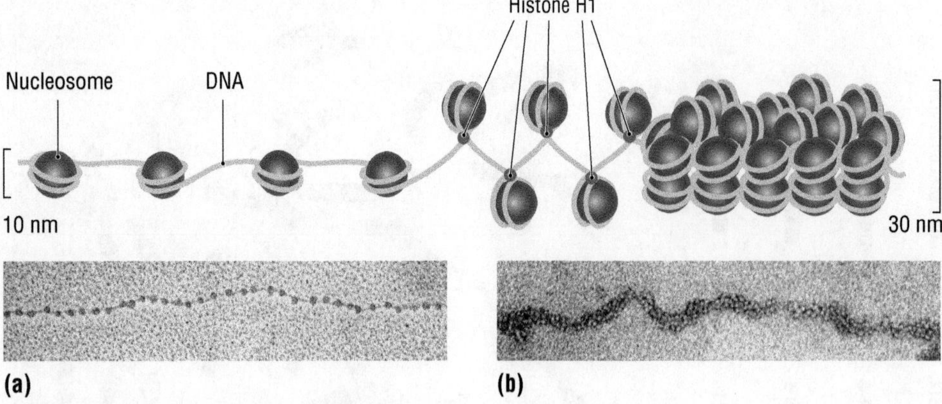
Histone H1
Nucleosome
DNA
10 nm
30 nm
(a)
(b)

Figure 3.9 Histones and nucleosomes. (a) The fundamental unit of chromatin is a complex of 146 base pairs of DNA wrapped around a core particle of proteins known as a nucleosome. The nucleosome core consists of two copies of each protein with the corresponding DNA. (b) Adjacent histones are linked and folded together by a fifth and more variable histone known as H1 (red dots).

Source: Reproduced from Craig, N.L. et al. (2014). *Molecular Biology: Principles of Genome Function* 2nd edition, with permission from Oxford University Press.

chromosome has to be sufficiently compacted to fit the entire length of the DNA molecules that make up the genome—more than 2 m in mammals—into the nucleus. It also has to remain sufficiently open or exposed enough that many different proteins can access the DNA sequence, so that the normal cellular processes, such as replication, transcription, and repair, can occur.

KEY POINT Nuclear DNA is separated into chromosomes where it is associated with chromosomal proteins. Histones are the most important structural proteins found in chromosomes. DNA wraps around core histone particles to create a more compact structural unit known as a nucleosome.

DNA outside of the nucleus

Most multicellular eukaryotes do not have plasmids like bacteria (although some unicellular ones do). In general, plasmids among eukaryotes are uncommon enough that they are ignored when discussing the genome. However, the chromosomes are not the only DNA found inside a eukaryotic cell—eukaryotes have copies of one or possibly two additional genomes outside of the nucleus.

In the cytoplasm of all cells, eukaryotes have many copies—in the order of 1000 or more in some cells—of an organelle known as the mitochondrion, shown in Figure 3.10. Mitochondria are known as the "power plants of the cell" because the biochemical reactions that generate energy occur primarily here. Mitochondria have their own genome, a circular DNA molecule consisting of about 16,500 base pairs in mammals. Like the nuclear genome, the mitochondrial genome encodes genes—in mammals, there are about 37 genes encoded by the mitochondrial DNA, many of which are necessary for the energy-generating process of oxidative phosphorylation. It is estimated that as many as 3000 genes are necessary for mitochondrial structure and function in mammals, but the vast majority of these are encoded by genes on the chromosomes in the nucleus, and only a few are present in the mitochondrial genome.

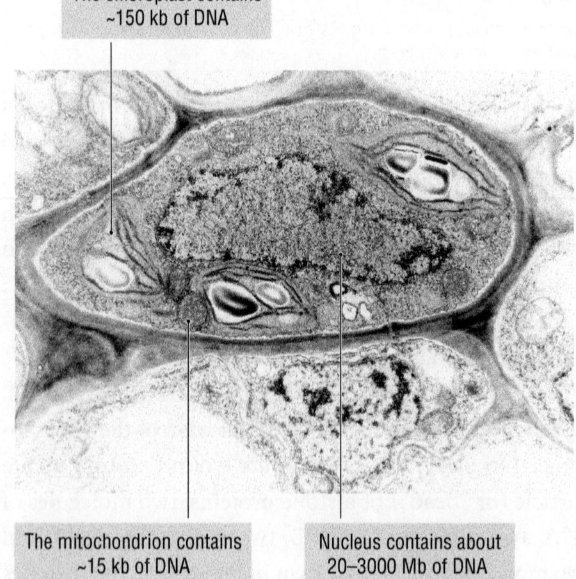
The chloroplast contains ~150 kb of DNA
The mitochondrion contains ~15 kb of DNA
Nucleus contains about 20–3000 Mb of DNA

Figure 3.10 Genomes in eukaryotes. In addition to its nucleus (shown here in orange) and the chromosomes, a eukaryotic cell such as the plant cell shown here, has multiple mitochondria in the cytoplasm (shown here in red), which have their own genome. A plant cell also has chloroplasts (dark green), which also have their own genome. Both mitochondria and chloroplasts originated as bacteria that became intracellular symbionts in an ancient eukaryotic cell.

Source: DR. MARTHA POWELL, VISUALS UNLIMITED /SCIENCE PHOTO LIBRARY

In addition to mitochondria, the cytoplasm of the cells of green plants contains chloroplasts, which also have their own circular genome. The size of the chloroplast genome varies by plant species but consists of about 150,000 base pairs encoding 120 or so genes, with somewhat more genes than are found in algae. These genes are largely involved in photosynthesis.

Mitochondria and chloroplasts are almost certainly derived from prokaryotes that were engulfed by an ancient eukaryotic cell more than 1 billion years ago, as summarized in Figure 3.11. This type of intracellular relationship is known as **endosymbiosis**. After engulfment, the mitochondrial or chloroplast predecessor developed such a mutually beneficial relationship with the host cell that neither could survive without the other. Chloroplasts in green plant cells probably originated from engulfment of a blue–green cyanobacterium. The closest living relative of the mitochondria are bacteria of the genus *Rickettsia*, but because both the mitochondria and the bacteria have continued to evolve, the precise origin is not certain. Bacterial genes that duplicated functions present in the eukaryotic cell were subsequently lost, so that the mitochondrial genome is a greatly reduced version of the original symbiont. Some other fraction of the ancestral mitochondrial genome was not lost entirely but migrated to the nucleus in horizontal transfer events, becoming part of the nuclear genome.

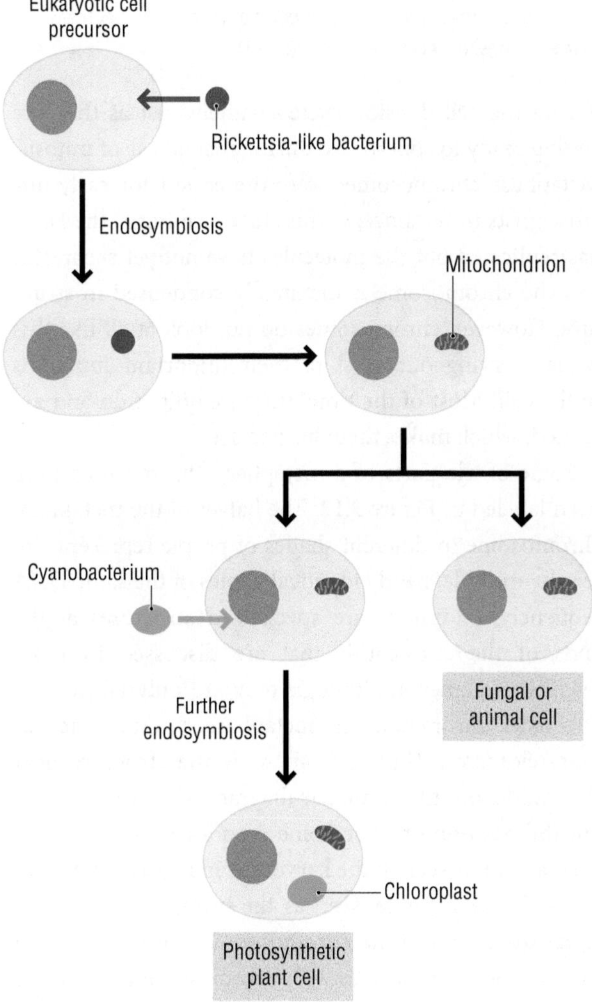

Figure 3.11 The symbiogenesis hypothesis. Both mitochondria and chloroplasts apparently arose from bacteria that were once intracellular parasites in a primitive eukaryotic cell. These became organelles with their own genomes, providing essential functions to the cell.

Although a handful of examples are known, endosymbiotic relationships, in which the symbiont became an organelle, probably occurred very rarely. In fact, all mitochondria seem to be descended from a single endosymbiotic ancestor, suggesting that the series of events that resulted in mitochondria occurred only once. Mitochondria have continued to evolve and accumulate changes alongside their host genomes, so that mitochondrial genomes in different eukaryotic species are recognizably similar in sequence to one another, but also have many differences as well. The use of mitochondrial genomes to reconstruct evolutionary history will be discussed in Box 16.3.

KEY POINT Eukaryotic organisms have extranuclear DNA contained within mitochondria and chloroplasts, organelles that are descended from bacteria that lived within eukaryotic cells.

Back to the nucleus

Most of the genome of eukaryotes is found in the nucleus in the form of chromosomes. The number of chromosomes varies widely among eukaryotic species, from as few as two or four to many hundreds. The chromosome number does not correlate with anything other than evolutionary history, and even that can be difficult to reconstruct. Humans have 46 chromosomes, dogs have 78, *D. melanogaster* has eight, and some species of ferns have hundreds.

Each chromosome consists of one double-stranded molecule of DNA that extends from tip to tip. The identity of each chromosome, initially detected by microscopy, is reflected in its DNA sequence—chromosome 1 and chromosome 2 have a completely different base sequence and encode completely different genes, for example. Chromosomes also differ in the length of the

DNA molecule they contain. In humans, chromosome 1 is the largest, both cytologically and molecularly, comprising a DNA molecule that is 249 million base pairs (Mb) in length. Chromosome 2 is 243 Mb in length, and the smallest chromosome—chromosome 21—is 50 Mb. Human chromosomes are numbered from largest to smallest, except that chromosome 21 proved to be slightly smaller than chromosome 22 once the DNA sequences were determined, and it was too late to change the numbering system. The X and Y chromosomes, involved in sex determination, are the exceptions to this numbering scheme. The X chromosome consists of 164 Mb, whereas the Y chromosome has about 59 Mb. Sex chromosomes are discussed further in Chapter 7.

Look again at the karyotype of human chromosomes in Figure 3.7. The chromosomes have been individually numbered to distinguish them from one another, but you will quickly see that there are two copies of each chromosome. Our genome is organized in 46 chromosomes, found in 23 pairs. Thus, humans have a pair of chromosome 1 and a pair of chromosome 14. One member of each pair is inherited from each parent. A pair of chromosomes has the same size, shape, and staining pattern. They also have similar (but not identical) DNA sequences and encode the same genes, albeit possibly with slight differences in their sequences. The slightly different DNA sequences for the two copies of the same gene are the basis for **alleles**, or different forms of the gene. All multicellular organisms and most unicellular eukaryotes have two copies of every chromosome during most stages of their life cycle.

The number of sets of chromosomes found in the nucleus of an organism is referred to as its **ploidy**. In other words, the **diploid** number of chromosomes in humans is 46; the **haploid** number of different chromosomes is 23. In females, the two X chromosomes constitute a pair and have similar DNA sequences. For a variety of reasons, the X and Y chromosomes are considered a pair, even though they differ in size, shape, staining pattern, and DNA sequence. For many multicellular organisms, the somatic cells are diploid and have two copies of each chromosome; the only cells that are haploid are eggs and sperm. As discussed in Box 3.1 and in Chapter 6, many plants and a few animals are polyploid and have more than two copies of each chromosome.

If asked to draw a chromosome, you might draw something that looks like Figure 3.12, which features a diagram of a generalized chromosome on the left and an electron micrograph of a human chromosome on the right. In fact, a chromosome resembles the letter X only at certain times

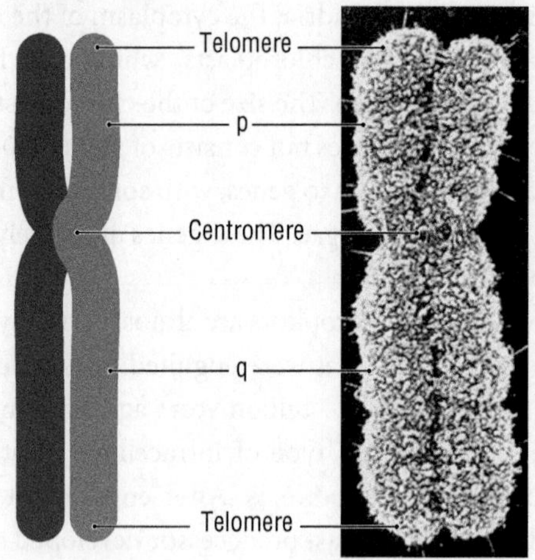

Figure 3.12 A chromosome at metaphase. This familiar image of a chromosome, shown as both a diagram and in an electron micrograph, is only present at metaphase at mitosis and meiosis. The chromatids, in different shades of purple, have duplicated as sister chromatids held together at the centromere. The short and long arms of the chromosome are referred to as p and q, respectively.
Source: Courtesy of Abogomazova/CC BY-SA 3.0.

during the cell division cycle—namely, just as they are getting ready to split in two during metaphase of mitosis. Metaphase chromosomes were the easiest for early microscopists to see since, at this stage of mitosis, the DNA has replicated but the molecules have not yet separated, and the chromosome is maximally condensed in structure. However, chromosomes do not look at all like this while carrying out most of their important functions in the cell. Most of the time, they are quite thin and extended, which makes them hard to see.

Some of the parts of a metaphase chromosome have been labeled in Figure 3.12. The halves of the metaphase chromosome in different shades of purple represent **sister chromatids**, nearly identical copies of the same DNA sequence. **Telomeres** are specialized structures at the ends of the chromatids that are discussed in more detail in Chapter 4; although they are only labeled for one sister chromatid, this metaphase chromosome has four telomeres. The **centromere** is the structure near the middle on this particular diagram of a chromosome, but the position of a centromere on a chromosome can vary, as can be seen in the karyotype in Figure 3.7. Sometimes the centromere, seen as the primary constriction along the length of the chromosome, is near the end of the chromosome (such as with chromosome 14 in Figure 3.7), while other chromosomes have their centromere close to, or at, the middle (such as seen for chromosome 1 in Figure 3.7). The role of the centromere will be explored in Chapter 6. In humans and many other organisms, the longer of the two arms of the chromosome is called "q," while the shorter of the two arms is "p;" thus, 8q refers to the long arm of chromosome 8, the one below the centromere.

In order to re-enforce some important concepts about the DNA sequences, chromosomes, and genetic variation, Figure 3.13 represents chromosome 1 in humans, which is 249,240,621 base pairs or 249 Mb long. The chromatid on the left (shown in darker purple) extends from the telomere at the top through the centromere and down to the telomere at the bottom, a total of 249 Mb. Its sister chromatid (lighter purple) is also 249 Mb long and is nearly identical in sequence to the sister chromatid.

The DNA molecules in the sister chromatids are the two products of DNA **replication** from a single DNA molecule, a copying process described in more detail in Chapter 4. If the two DNA sequences of 249 Mb in the sister chromatids were compared, there might be about five to ten base pair differences that arose from copying errors during this replication process. This is a very approximate number, which we use only to show the scale of the differences, but sister chromatids have nearly identical DNA sequences.

DNA replication was completed a few hours before the electron micrograph of the metaphase chromosome in Figure 3.12 was taken or before the drawings shown in Figure 3.13. If the chromosomes hadn't been prepared for microscopy, the sister chromatids would have gone on to separate from each other along their length, splitting apart at the centromere, resulting in two nearly identical copies of this chromosome, each consisting of 249 Mb. Thus, during most cellular processes, this chromosome will be a double-stranded molecule of 249 Mb that is much longer in physical length and much thinner than the structure shown in Figure 3.12, but with its centromere, chromatids, and telomeres in the same places.

Homologous chromosomes

While only one copy of chromosome 1 has been shown, a typical somatic cell in diploid organisms like humans actually has two copies, as evident from the karyotype in Figure 3.7. This other copy of the same chromosome is referred to as the **homolog** or **homologous chromosome**. At metaphase, it will also look like that shown in Figure 3.13.

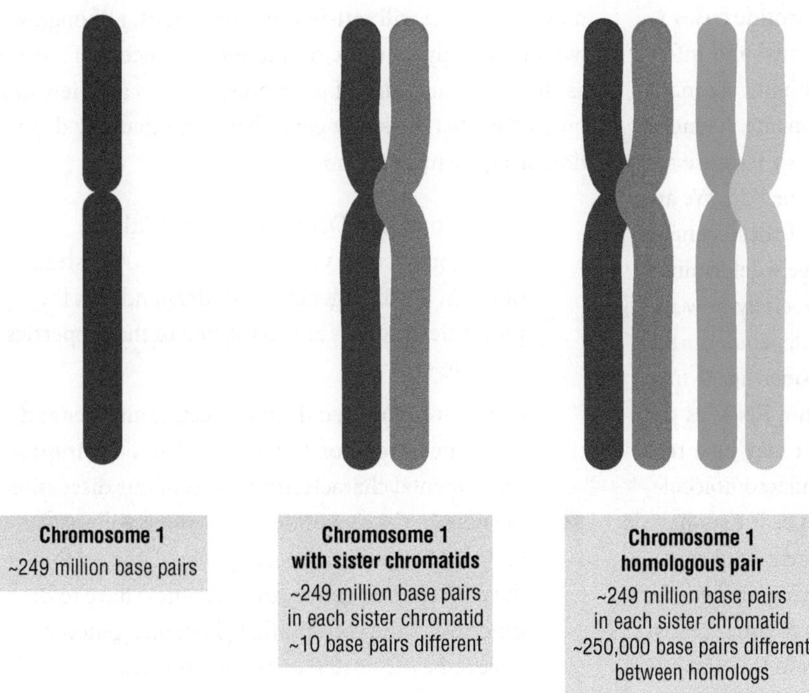

Figure 3.13 Sister chromatids and homologous chromosomes. The diagram represents chromosome 1 in humans, which is a double-stranded DNA molecule of 249 million base pairs. Estimates based on the measured mutation rate are that sister chromatids will differ by about 10 base changes as a result of replication errors. Homologous chromosomes will have the same overall genetic information but will have slightly different sequences. Estimates based on human variation are that homologs will differ by about 250,000 base changes, but this is very approximate.

Although the sequences of a particular chromosome's two sister chromatids will be nearly identical to each other, the DNA sequences of two homologous chromosomes are similar but slightly different from each other. While sister chromatids are expected to have only five to ten differences among the 249 million base pairs, two homologous chromosomes will have about 250,000 sequence differences between them—about 0.1% of the total number of bases in the chromosomes, as summarized in Figure 3.13. This is an approximation to give you a sense of scale for what might be typical for human populations, and the number of differences might be greater or less, depending on the population being considered.

The differences in DNA sequence between the two sister chromatids arose during the most recent round of DNA replication. By contrast, the differences between the homologues are the consequence of being inherited from different parents and might have originated hundreds or thousands of years earlier. The reason that the differences between homologs was presented as an approximation is that the homologs come from different origins; if the parents were from two very different populations, the amount of difference might be greater, whereas the difference is less if the parents are, even distantly, related to one another, as occurs in small populations. Even though the difference in the DNA sequence of homologous chromosomes is slight, it is enough to have major ramifications. In fact, these differences are the foundation of evolution and genetics.

KEY POINT Chromosome copy number is referred to as ploidy. Diploid organisms have pairs of homologous chromosomes, which differ slightly in their sequence. However, this difference between the sequences of homologous chromosomes is the basis for genetic variation.

3.3 Genes and composition of the genome

At its most fundamental, the DNA sequence of the genome is a very long string of the four nucleotide bases A, T, G, and C. Yet that sequence has functional and informational content embedded within it. Genomes are made up of several different types of genetic elements. "Genetic elements" is a vague term, intentionally so because we don't always know what they are or what they do. We are reasonably confident that the functions of these genetic elements, whatever they might be, are largely determined by the base sequence of the DNA. But precisely how the base sequence determines the function depends on the type of genetic element that we are considering. It may encode information that is transcribed into RNA, as the Central Dogma in Chapter 2 describes. It may be a recognition site that interacts with another macromolecule, or it may fold into a particular structure, or it may have some other function. Many genetic elements are transposable elements that are capable of moving around the genome or were capable of doing so at one time. Transposable elements and virus integration into the genome are discussed in Chapter 11. Irrespective of the specific function, nearly all of the informational functions of DNA somehow rely on its base sequence.

For the most part, our focus is on genes—that is, on the transcribed functional elements in the genomes. We have to first ask ourselves, "What do we mean by a gene?" and immediately we run into complications. No single definition of a gene will satisfy every geneticist, although we would probably all agree on the broad concept of a gene as the functional unit of a genome. Let's recap a few key properties that we know genes have and that any definition of a gene must include.

1. **Genes consist of DNA sequences.** With the exception of some viruses that use RNA, instead of DNA, for their genes, we understand that the properties of genes are connected to the properties of DNA.
2. **Genes are inherited** from one cell to another and from one generation to the next. This is their most fundamental characteristic. Most of our discussion of genes in the remainder of the book will focus on the inheritance aspect of genes. Coupled to the concept that genes (and genomes) have to be inherited is the concept that genes and genomes have to be replicated or copied, something we will discuss in some detail in Chapter 4. Thus, every generation of a cell, an organism, or a population makes a copy of its genes and genomes and passes that on. So we could rewrite this subheading to be a bit more accurate to say that "Genes are first copied and then the copies are inherited."

3. **Genes are expressed.** "Gene expression" is a very broad term, which we introduced in Chapter 2 and will develop in more depth in Chapters 12 and 13. For now, we will simply say that genes have functional activities in particular cells at particular times and in particular places. In order to carry out those functions, the gene has to be expressed.

We can capture these last two concepts in a somewhat simplified version. "Inheritance" involves how your offspring will use your genes some day in the future. "Expression" describes how you are using them now.

KEY POINT Genes are composed of DNA. They are copied, inherited, and expressed in living cells. Genes and other genetic elements make up the complete genome.

Where are genes located in the genome?

A fundamental concept in genetics is that a gene occupies a particular and characteristic location in the genome and that this location is on a particular chromosome; it comprises a particular set of base pairs on the DNA sequence of that chromosome. Every individual in the species (with rare exceptions) has the same gene at the same location on the same chromosome. This is true in both bacteria and eukaryotes, regardless of the way that the genome is organized into chromosomes. For this reason, a gene is often associated with a **locus**—an address for a gene, which is consistent for all members of the species. We use the same concept in everyday speech. For example, we refer to "1600 Pennsylvania Avenue" as shorthand for the White House or for the President of the United States, regardless of who is actually serving as the President at the time.

Because we can talk about a locus for a gene, we can also use other terms related to position. For example, we can talk about a **genetic map** of a species—the atlas of the positions of the genes in a genome. Because we have a map, we can refer to the distance between genes. And, just as we do for maps and distance in everyday life, we can define maps and distances in genetics in different ways but still refer to the same locus. How far is it from your house to the nearest grocery store? We could say that it is 15 minutes on a bicycle, a 30-minute walk, or 1.25 miles. Although these are different ways of expressing the distances, the locations of your house and the grocery store have not changed. The same is true for genetics—we have different ways of describing the distance between two genes, but the genes themselves have not changed position. The types of maps used in genetics are discussed in more detail in Chapter 9.

What fraction of the genome consists of genes, and what fraction consists of other types of genetic elements?

Although we do not have a very good definition of a gene or of other genetic elements, it is reasonable to ask how much of the genome falls into each category. The answer depends on what organism is being studied, as well as how exactly we try to define these categories. As a result, the answers will be very approximate. For instance, our numbers will change if we consider a gene to be a sequence that is transcribed or if we define it to be the information that encodes a polypeptide (that is, both transcribed and translated). For this illustration, we will discuss genes that encode polypeptides.

A short answer for now is that most of the genome in bacteria—probably more than 95%—consists of genes. In eukaryotes, the genomes from more complicated organisms, such as humans, have a **lower** percentage of the genome that seems to be genes; generally speaking, the more complex the organism, the lower the percentage of its genome that is devoted to genes. This may seem counterintuitive, but it is important. Consider these numbers—the worm *C. elegans* has about 22,000 protein-coding genes in a haploid genome of 10^8 base pairs; humans have about 21,000 protein-coding genes in a genome of 3×10^9 base pairs, or about 30 times larger than that of worms.

Another way to ask the same question is to carry out the following thought experiment. Let's select a specific nucleotide at random from some point in the genome. What is the probability that the selected base is transcribed or that it encodes some other functional information such as a binding site or recognition sequence? Again, the answer depends on the organism, as illustrated in Figure 3.14. If the genome in which the selected nucleotide is found is the *E. coli* chromosome or one from another species of bacteria, there is an approximately 90% probability that it will be expressed or will encode functional information since nearly every base in the bacterial chromosome is expressed or has functional information. (If the genome is from a virus, the answer could be considered to be greater than 100%, if that were possible. Not only is every base used, but many bases are used for more than one function. Viruses

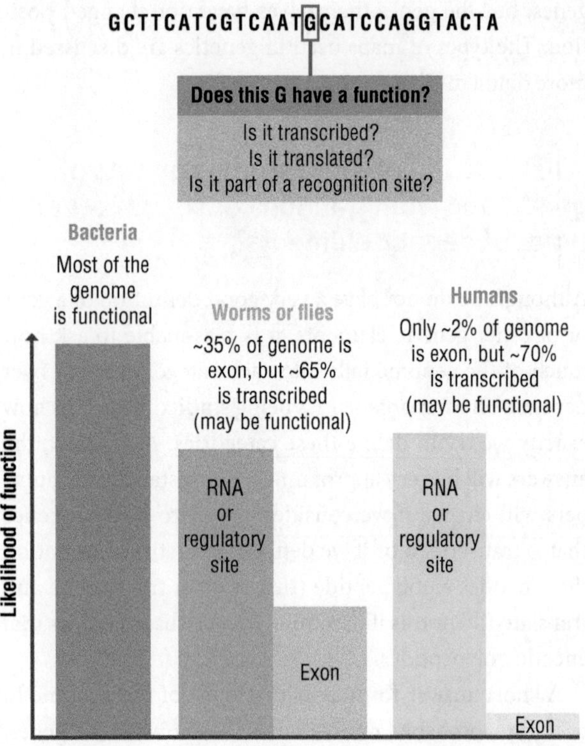

Figure 3.14 Functions in the genome. A short sequence from a genome is shown, with one particular nucleotide base (a G) highlighted. In order to determine if this G is functionally important, we may want to ask if it is transcribed or translated or serves as a recognition site. Estimated results are shown in the bar graph. In bacteria, most bases have some function, so a particular base is very likely to encode a function. In invertebrates, such as fruit flies and nematodes, about a third of the genome consists of exons (which are transcribed and translated), and about two-thirds are transcribed or serve as a recognition site, so a specific base is less likely to be functional in the organism. In humans, exons comprise less than 2% of the genome, but more than 70% are transcribed or serve as a recognition site. These are approximations, based on current information, so this should be considered illustrative only.

encode the maximal amount of information in the smallest genomes.)

On the other hand, if the chosen nucleotide is from the genome of an invertebrate, such as the fruit fly *D. melanogaster* or the nematode worm *C. elegans*, the probability that the nucleotide is functional is much lower. For *C. elegans*, about two-thirds of the genome consists of genes or has functional information; for *D. melanogaster*, slightly less than half of the genome falls into these categories, but these numbers are very approximate. So we would say that the probability that a randomly selected nucleotide has functional information is perhaps 60%.

For humans or other mammals, the probability that the selected nucleotide will be functional is much lower yet. Only about 1–2% of the human genome consists of genes that encode proteins. Even if we take the most generous definition of known functional information associated with a gene, the percentage would probably increase to only about 3–5%. Nearly all of our genome consists of other genetic elements, rather than genes that encode proteins. However, we now know that most of the sequence of the human genome—more than two-thirds by the most conservative estimates—is transcribed, and so it could be functional. The distinction between "gene" and "other genetic element" is increasingly blurred.

KEY POINT All genes are composed of DNA, but not all DNA constitutes genes. The exact percentage of the genome that encodes protein-coding genes varies among organisms. In general, more complex organisms contain more protein-coding genes, but these make up a lower overall percentage of their DNA.

3.4 Genetic change, redundancy, and robustness

Let's consider the question about what fraction of the genome consists of genes and other functional elements in yet one more way. Suppose that we were to pick out a base in the genome at random and change it to another base. Would this have functional consequences for the organism? This, after all, is what happens during errors in the copying of DNA and during many genetics experiments in the laboratory or the field, and is what has occurred routinely during evolution. In effect, evolution has repeatedly done this experiment for us. A base is changed to some other base (or is deleted or duplicated), and the consequences for the organism or the cell are observed.

These consequences are summarized in Figure 3.15. Note that there is no scale on the Y-axis in Figure 3.15 because many different assumptions are being made; however, all of the bars in Figure 3.14 are higher than the corresponding bars in Figure 3.15. Many changes in the nucleotide sequence will not have functional consequences on the organism because the information content in all organisms has multiple types of redundancies, which serve to buffer the genome against functional changes.

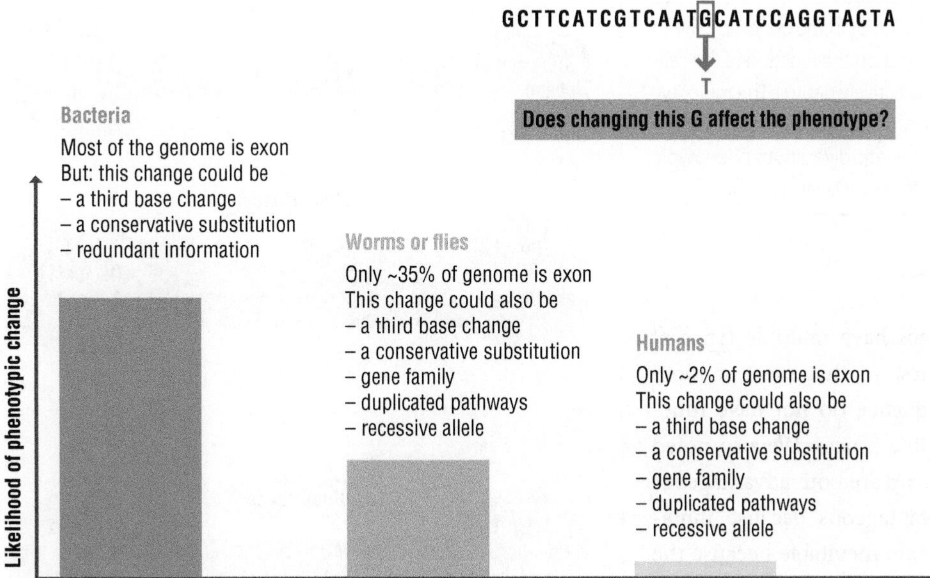

Figure 3.15 Mutations, redundancy, and robustness. One way to determine if a particular nucleotide has a function is to change it to another one. Some of these changes will result in a functional difference, but many do not. In bacteria, most of these changes will affect a base in a gene, whereas this is less likely in flies or worms and unlikely in humans. Even if the mutation does affect a gene, all organisms have different mechanisms to buffer against changes in function. The change might affect the third base in a codon, result in a neutral substitution, or have no effect because there is redundant information in the genome. In addition, in flies, worms, and humans, the change could affect a member of a gene family, might affect genes in duplicate pathways, or might be recessive; none of these changes would have an effect on the function. Thus, changes to a nucleotide often do not result in a change in function or phenotype; the likelihood of observing a phenotypic change following a single base change is less than the likelihood of each base exhibiting a function (as shown in Figure 3.14). Since this is illustrating a general concept, the relative size of the bars is more important than the actual scale, which is usually not known.

These redundancies are important evolutionarily because mutations of different types occur constantly.

Let's think about this question specifically for humans, for which most random changes in a single base are expected to have very little or no effect on the function or the phenotype. A random change is very unlikely to affect a protein-coding sequence, since only about 1–2% of the genome consists of genes that encode proteins. It may affect the sequence of an RNA that is not translated, but we know very little about the effects of single base changes in RNA molecules. Thus, most random changes like this in humans will be selectively neutral; they will not confer a reproductive advantage or disadvantage on the individual.

This situation presents a conundrum. Mutations are occurring at all sites in the genome in all cells at all times. They are the source of genetic diversity, and genetic diversity is the most essential ingredient for evolutionary change. On the other hand, many mutations do not have significant functional consequences and do not affect the phenotype. Thus, while they are creating changes in the DNA sequence, these mutations are not generating phenotypic changes. This presents a problem—how can natural selection occur if mutations rarely affect the phenotype of the organism?

The solution to our conundrum is that mutations are occurring all the time. While the probability that changing any **individual** base in the genome at random will make a functional change in the organism is low, the accumulated effect of many mutations occurring at many different locations over a long period of time provides the genetic diversity for evolution. Most individual changes in the human genome will have no effect at all because they will not affect a gene. Another large set of mutations will affect a gene but will not have much, if any, effect on its function because of the many sources of functional redundancy discussed below. Only a few mutations will actually affect the function of a gene, but those few mutations provide the diversity needed for evolutionary change.

KEY POINT Mutations can accrue at all sites in the genome in all cells, but most mutations have little effect on the function of the cell or organism and are expected to be selectively neutral. The minority of mutations that have significant functional consequences are the ones that contribute most directly to genetic diversity, to phenotypic differences between individuals, and to evolutionary change.

Redundancy and robustness

We wrote above that organisms have multiple types of redundancies in their genomes, so that most random changes in the nucleotide sequence do not have functional consequences or affect the fitness. We also noted that such systems of redundancies are both advantageous and inevitable. They are advantageous because mutations happen constantly. They are inevitable because the changes that we observe in most natural populations are the ones that are capable of surviving to reproduce; most changes that affected the function (and these do occur) probably did not leave many offspring.

This brings us to the distinct, but related, ideas of redundancy and robustness. These are concepts used by people who think about complex systems of many types, and they apply to biological systems as well. **Redundancy** means having more than one version of the same information. Thus, a change in one set of the information makes very little or no difference to the overall function of the organism or the system because there is another version of the same information that is still in place. Think of the pathways and networks in a cell as a system of interconnected highways, with each protein analogous to a section of the road and the functions analogous to towns, as shown in Figure 3.16. The top figure shows a single two-lane road connecting the two towns. The middle diagram is an example of redundancy; the same road has been widened to four lanes, rather than two.

Here are some examples of redundancy in biology, in which highly similar duplicate copies reduce the likelihood that a change in the sequence of one copy will change the overall function.

1. The genetic code uses 64 codons to specify 20 amino acids and a stop signal; there is more than a single codon for most amino acids. This degeneracy, introduced in Chapter 2, is an important example of redundancy in a biological system. This topic will be discussed in greater detail in Chapter 13.
2. Many genes are very closely related in sequence and function to other genes in the genome; these groups are known as **gene families**, and the individual members of the gene family present in an individual are known as **paralogs**. In some cases, the genes in a family are functional duplicates, so that removing one member of the family has no functional consequence because another family member is carrying out the same role. Histone genes are

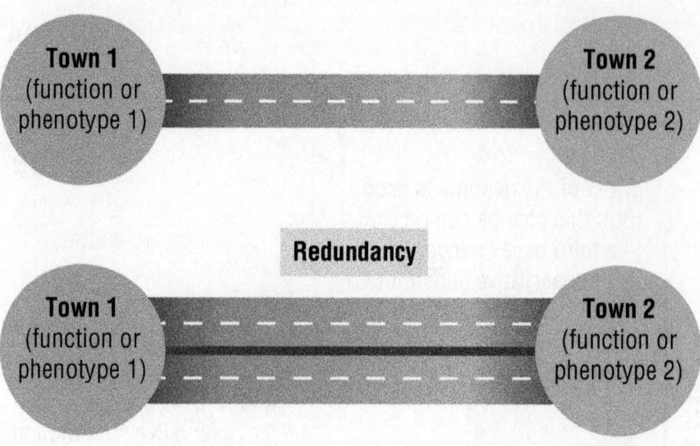

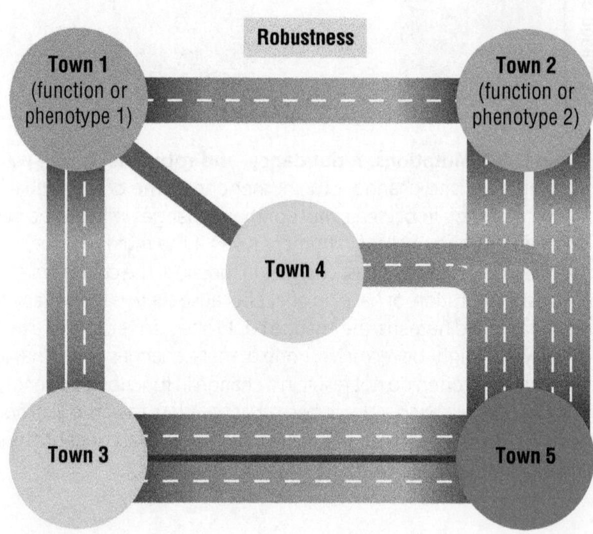

Figure 3.16 Redundancy and robustness. Changes in a gene or DNA sequence may not have a phenotypic effect because the genome has redundancy and robustness. The concepts are illustrated using the analogy of towns and roads. Towns 1 and 2 are connected by a highway in the top panel; in an organism, these "towns" could be different functions, processes, or molecules, and the highway between them might correspond to a gene that affects the conversion from one to another. Redundancy is found when duplicate functions occur, as shown in the middle panel. The road has widened to have additional lanes that provide two equivalent (duplicate) routes between the two towns. In the organism, these may be duplicate genes of a gene family with highly similar functions. Robustness is found when there are multiple pathways connecting the same processes. In the bottom panel, there are other roads, other towns, and many routes connecting Towns 1 and 2. In the organism, these may be multiple biological pathways with some shared genes, metabolites, and molecules.

encoded by a family with functional redundancy. For example, the histone H3 gene family has 15 members in the human genome. There are many more examples, and at least half of the genes in our genome are thought to have a functional duplicate or be a member of a gene family. In Section 3.6, we discuss how the presence of a gene family can also lead to the evolution of new or slightly altered functions.

3. A more subtle, but possibly more familiar, example of redundancy foreshadows Chapter 5. A diploid organism has two copies of every chromosome, which implies two alleles of every gene. We will call these ***A*** and ***a***. Suppose that the ***a*** allele arises from a change that eliminates the function of that gene, and the ***A*** allele makes a normal gene product. As noted by Mendel and discussed more fully in Chapter 5, an organism with the ***Aa*** genotype may look exactly like one that has the ***AA*** genotype, that is, one with one functional copy is indistinguishable from the one that has two functional copies. Because the presence of one functional allele is all that is necessary for the normal activity of the gene, the second allele is redundant. Or in a language that you might have encountered before, the ***a*** allele is recessive. Recessive alleles often arise because of genetic redundancy.

KEY POINT Redundancy means that there is more than one version of the same information present in the cell's DNA. The genetic code, the existence of gene families, and the presence of recessive alleles are examples of redundancy in genetics.

Another concept that helps to explain why random changes in the genome may not have functional consequences is robustness. **Robustness** is a broader term than redundancy and has a very similar meaning in many different fields, including engineering and computer science, as well as evolutionary biology. A system is considered to be **robust** when random changes in one component do not change the overall function of the system. In our analogy in Figure 3.16, if one road is closed or jammed, traffic can be diverted onto another road, often with minimal or no effect on the overall traffic flow. This is an example of robustness. If one protein is altered, the function can be diverted into another molecular pathway that accomplishes the same end or allows the organism to achieve the same biological function.

Robustness is a property of most organized systems and is a common property of biological systems as well. For example, there are many different paths for blood to travel between any two parts of the body because our circulatory system is robust. One way that robustness arises in molecular or genetic systems is from the many interactions within a cell or an organism. Proteins interact to form pathways or networks that lead to particular functions. Often, eliminating a gene or its protein product in one pathway has no effect because other pathways are available to carry out the same function. There is another path to the same end.

KEY POINT Robustness refers to situations where a change to one component of a system does not lead to failure, as there are alternative systems that can be utilized to reach the same result.

3.5 Which genes are found in the genome?

We noted earlier that the number of protein-coding genes has not changed very much, even as genomes have grown larger. This may seem rather surprising, since it seems like it should take many more parts (in the form of genes and gene products) to make a human being than it does to make a worm. Let's consider this concept a bit more carefully, using the genomes of *C. elegans* and humans for our comparison.

Humans really do have more gene products than worms

Humans and worms have almost the same number of protein-coding genes, according to the best current estimates, which seems unexpected. Perhaps it is our ego, but most of us would have assumed that the difference between a worm and a human should be more genetically obvious than this. Has any worm ever composed a symphony or played in the World Cup, or even passed a genetics course? Did nematodes sequence the human genome, or even the nematode genome for that matter? Surely we must be more richly endowed genetically than a mere worm. Yet humans and worms have about the same number of genes.

While humans and worms have about the same number of genes, humans do appear to have more **gene products** than invertebrates. The rule of thumb from the Central Dogma in Chapter 2 is "one gene, one polypeptide"—that each gene has the information to make a single polypeptide. That is a good rule of thumb, but, like most rules of thumb, it is not completely accurate. In Chapter 2, we noted that the initial RNA transcript for most eukaryotic genes is much longer than the final mRNA product that is translated and that the initial transcript is spliced to make the final mRNA. This is summarized in Figure 3.17. We will discuss splicing in more detail in Chapter 12.

Not only does splicing occur for most genes in multicellular organisms, but it is also highly regulated and is subject to evolutionary pressures. A single gene might be spliced in more than one way to produce more than one final mRNA transcript, with each transcript being related but distinct. This is known as **alternative splicing**.

Alternative splicing is exceptionally common in mammals. Most genes—perhaps more than 75% of genes in mammals—produce more than one related, but distinct, mRNA from the same initial transcript because of alternative splicing. Put another way, one gene in mammals can produce more than one polypeptide, sometimes dozens of different, but related, polypeptides. As such, the "one gene, one polypeptide" rule of thumb really doesn't hold true. Conversely, most genes in a yeast cell are not spliced at all—the initial transcript is very similar to the final transcript, so the "one gene, one polypeptide" rule of thumb works well for yeast.

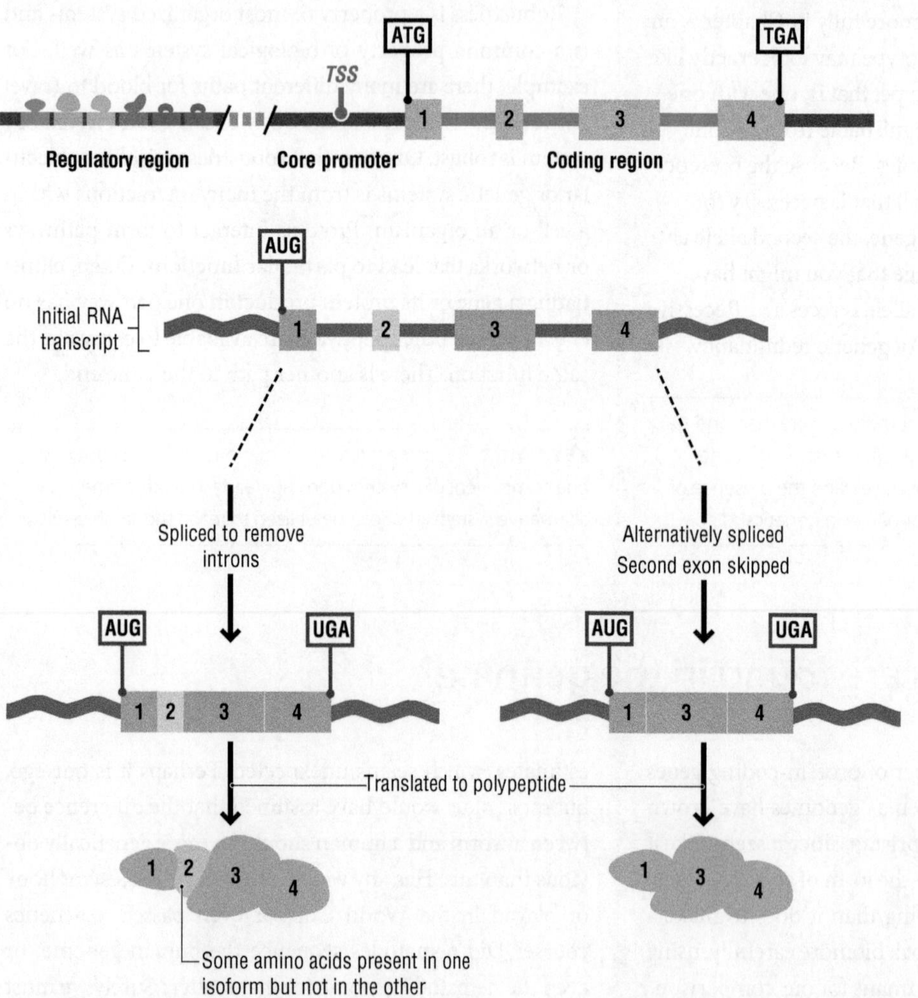

Figure 3.17 Splicing and alternative splicing. A typical gene from a multicellular organism is shown, with four exons downstream of the transcriptional start site (TSS). The entire region is transcribed into RNA and then spliced, so that the exons are spliced together and a long reading frame with no stop codons is produced. The gene may be spliced in multiple ways, known as alternative splicing. In isoform 1, all four exons are present in the mature mRNA. In isoform 2, the second exon is skipped, so that exon 1 is directly spliced to exon 3. This difference in splicing results in two polypeptides with different amino acid sequences, since one has the amino acids encoded in exon 2 and the other does not.

Alternative splicing has generally increased with the complexity of organisms. The 22,000 genes in worms are estimated to yield about 35,000 proteins, whereas the 21,000 genes in humans produce about 100,000 proteins. So humans probably do make more gene products; we just don't have that many more genes to encode them. This is an important difference between humans and worms.

We are all worms (as well as yeast and flies)

Humans have about the same number of genes as a worm, and we apparently use those genes to make more proteins due to alternative splicing. But what do we know about the genes and gene products that we make? How many of them are unique to humans, and how many are common to all mammals? How many of them are common to all vertebrates or to all multicellular animals or to all living things? As many genomes have now been analyzed, we realize that nearly all of our genes have equivalent copies, related by sequence and function, in other organisms. We have more of the same (evolutionarily) old parts, but not that many new parts.

The exact numbers that would help us make a compelling comparison between organisms depend on precisely what is meant by the "same" parts. Let's express it in this way. Imagine that we isolate a particular protein found in humans. If we take the amino acid sequence of this protein, can we find a similar sequence in other organisms? For technical reasons, it is easier to sequence DNA than proteins, but we use the inferred amino acid sequences of the proteins to make the comparison. We need to define "similar" to make this comparison precisely, but we can make some realistic statistical assumptions and come up with some illustrative estimates.

In many cases, directly equivalent genes can be found in two different species. These equivalent genes, present in different species as a result of their shared ancestry, are known as **orthologs**. An example from the electron transport chain of a protein involved in cellular respiration is shown in Figure 3.18. These are genes that are identified because they encode very similar proteins, which likely have similar functions.

If we make estimates of the similarity of genes across species, about 20–25% of human genes have a related gene in *E. coli* that is probably carrying out the same basic molecular function, that is, about a fourth of the genes in our genome were present in the last common ancestor between humans and bacteria, billions of years ago. More than a third of human genes (and, by some estimates, more than half) have identifiable relatives in a yeast cell that are probably carrying out similar functions in the two organisms; certainly more than half of our genes have identifiable relatives in a fly or a worm. Thus, many of our genes are present in all eukaryotes, and most are found in all other animals.

Furthermore, more than 96% of our genes have one or more functionally similar relatives in the mouse genome, and more than 99% have one or more functionally similar relatives in chimpanzees and gorillas. Looked at in another way, species-specific genes definitely do exist, but the number of these is not very large. Of the roughly 21,000 protein-coding genes found in humans and chimps, 200 or fewer are found in a human but not in a chimp, and another set of about 150 genes are found in a chimp but not in a human.

Does it surprise you that we have so many of our genes in common with other species, including all other living species? We share common ancestors with all living things, so it should not be all that unexpected that we have common genes. In addition, many of the fundamental and essential functions being performed in our cells are the same or slightly modified versions of activities that occur in all cells, or at least in all eukaryotic cells, and

Human	MLWNLLALHQIGQRTISTASHRHFKNKVPEKQKLFQEDDGIPLYLKGGIADALLHRATM ILT
Dog	MLRNLLALRQIAQRTISTASRRQFENKVPEKQKLFQEDNGIPVHLKGGVADALLYRATMMLT

Human	VGGTAYA I YQLAVASFPNKGVTSIIPAITWFTFIQLSMDQKSDK
Dog	VGGTAYAMYQLAVASFPKKQD

Figure 3.18 Orthologous proteins from different species. The amino acid sequences of cytochrome C oxidase subunit VIIA2 from humans and dogs are shown. Cytochrome C oxidase is the final enzyme in the electron transport chain in cellular respiration; VIIA2 is one of 14 protein subunits in this enzyme. It is a very highly conserved protein, with only 13 substitutions (shown in black) among the first 80 amino acids between humans and dogs, although the human protein is about 20 amino acids longer. Many orthologs are not as similar as these two proteins and so are recognized by statistical analysis, rather than direct observation of aligned sequences.

the underlying biochemical and molecular processes are about the same. These functions include the processes of transcription and translation, DNA replication, cell division, cellular respiration, and so on. This is why we can use model organisms so effectively to understand molecular processes in humans, as discussed in Tool Box P.1 in the Prologue. Furthermore, the similarities in sequence that arise from common descent allow us to construct phylogenetic trees that show relatedness and evolutionary history, as discussed in Chapter 4.

3.6 Genetic changes and morphological differences

If different species, particularly closely related species like humans and chimps, have about the same number of genes, and the genes are carrying out the same functions, why aren't species even more alike? How are we able to tell humans from chimps? Or, more generally, what changes in our genomes make us different?

When we compare individuals within a species or two different species, we are seeing the functional and morphological changes that have occurred to give rise to different phenotypes. The phenotypic changes arise from underlying changes in the sequence and structure of the genome. Until the genomes have been examined, it may not be possible to know the origins of phenotypic differences, and the connections between genomic changes and phenotypic changes can be complex. We will begin by discussing the changes that are observed when genomes are compared and then discuss how these changes in the sequence and overall organization of the genome can lead to phenotypic changes. We also need to consider the differences between individuals of the same species, between two closely related species, and between two more distantly related species.

Changes in the sequence and organization of the genome

The most apparent and common changes between two genomes will be in the sequence of nucleotide bases in the DNA. These are summarized in Figure 3.19. There are substitutions of one base for another, insertions or deletions of one or more bases, and changes in the number of copies of a sequence. Because comparisons of two genomes make it difficult to determine if bases have been inserted in one or deleted in the other, insertions and deletions are collectively referred to as **indels**. The mechanisms by which base substitutions arise are discussed in Chapter 4; some of these same mechanisms probably are involved with the formation of indels, while changes in the overall content occur by other processes.

Base substitutions and indels occur throughout the genome, with different phenotypic consequences depending on their locations. Sequence changes in the protein-coding region of a gene can change the amino acid sequence of the protein it encodes; sequence changes at the splice junction sites can alter the splicing pattern

Normal or wild-type	TTGATCTCGGATTCTAATGGGCATGCTA
Base substitution	TTGATCTCGGATTCCAATGGGCATGCTA
Base deletion	TTGATCTCGGATTC-AATGGGCATGCTA
Base insertion	TTGATCTCGGATTCTAAATGGGCATGCTA
Copy number variation	TTGATCTCGGTATCTAATTCTAATTCTAATGGGCATGCTA

Figure 3.19 Types of changes in the DNA sequence. The "normal" or wild-type sequence is shown in the top line, with the base that will change shown in red. A base substitution is the replacement of one base by another, here the T in the original sequence with a C. An addition or deletion changes the number of bases at the site; since it is difficult to know if the event was an insertion or a deletion, these are often grouped as "indels." Copy number variation does not change the individual bases but changes the number of times a particular sequence is found. Although this figure shows the change in a six-base block, copy number variation is usually considered with blocks of 1000 bases or longer.

for mRNAs, and thus the amino acid sequences, and sequence changes in a regulatory region can alter the transcription pattern of the gene. Any of these changes may have functional or phenotypic consequences. On the other hand, sequence changes that occur in regions of the genome that do not encode any function probably have no phenotypic consequences. Since such changes may have no effect on the fitness of the organism, they are neutral changes or neutral mutations.

Base substitutions and indels are the most common changes when two members of a species are compared. Because these sequence changes usually occur during DNA replication, as discussed in Chapter 4, their numbers increase as different species are compared. In fact, the number of sequence changes is one measure of the distance between two species. As we will discuss in Chapter 16, neutral changes are especially useful to indicate some of the evolutionary histories of the populations or species.

Changes in the number of copies of genes and other sequences

The number of copies of some genes or genetic elements also varies, both within and between species, that is, changes are not only in the base sequence of DNA, but also in the amount of DNA, the size of the genome, or the organization of the genome. Changes in copy number are primarily found in eukaryotes. In bacteria, copy number effects are largely limited to genes located on multi-copy extrachromosomal elements such as plasmids. Different sources of gene and copy number changes are at work among eukaryotes, operating on different scales and occurring by different mechanisms. Changes in the copy number of genes and sequences include polyploidization of entire genomes, copy number variation (CNV), and the expansion and divergence of gene families, as summarized in Figure 3.20. We need to be clear and careful about our language here. The general topic is changes in the number of copies of genes and sequences in a genome, of which the specific phenomenon, known as CNV, is one example.

In a broad sense, changes in the number of copies of genes and sequences among members of the same species are less common and less significant than base substitutions and indels; however, there are many specific exceptions, so this broad rule should be applied with caution. On the other hand, changes in the number of copies of genes and sequences are quite commonly observed between species. For any given species or pair of species being compared, one or more of these changes in the number of copies may be more significant than the others.

Polyploidization

Let's consider polyploidization as an example of how the number of copies of a gene in the genome changes, as introduced in Box 3.1 and summarized in Figure 3.20(a). We wrote in Section 3.2 that most eukaryotes are diploid, with two copies of each set of chromosomes. Among animals, this is true. Individuals within an animal species do not differ in their ploidy. Even when current animal species are compared, there are very few examples of naturally occurring ploidy differences, other than a few amphibians. When the evolutionary history of animal species is considered, polyploidization has been enormously important; ancestral polyploidization is thought to have been one of the key drivers of evolutionary change among animals, as we explore in Box 3.1. But we don't see that in the species present today, since most animal species are stable diploids, without naturally occurring differences in chromosome or ploidy number.

Plants are a completely different story. Most flowering plants are polyploids, and polyploidization happens even among plants in the same field. Indeed, when different species of flowering plants are compared, polyploidization is the rule, rather than the exception. Examples of diploids, triploids, tetraploids, hexaploids, and octaploids are seen among different varieties or specimens within a species, particularly among ornamental or agricultural plants; double flowers in marigolds and chrysanthemums are examples of individuals within a species that differ in their ploidy. These differences in ploidy occur naturally, but they may also have been selected by breeders and agronomists. Polyploids are larger than diploids, which may make them desirable for many ornamental plants. In addition, triploids are seedless and more convenient to eat; bananas and seedless watermelons are among the triploid plants we regularly eat.

Polyploidization can be followed or accompanied by other changes that result in even greater morphological differences. Changes in the number of genes produce quantitative alterations in the levels of gene expression. This also changes the selective pressure, since the functions that were being carried out by one gene can now be

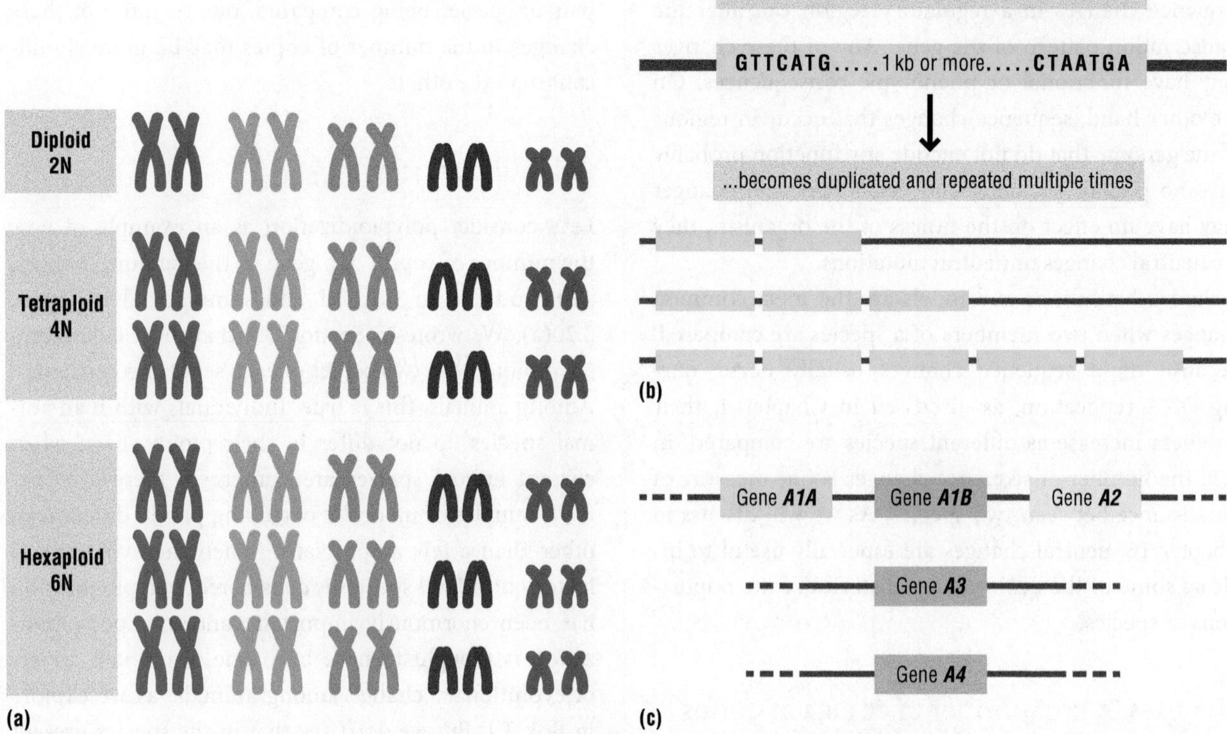

Figure 3.20 Types of variation in the copy number of genes and sequences. One source of variation in the genome comes from differences in the number of copies of the genomes, genes, and other sequences. These arise from different mechanisms and have somewhat different consequences but are considered together here for simplicity. (a) Polyploidy arises from differences in the number of copies of the genome itself and is especially common among plants. The hypothetical diploid (2N) species has five pairs of chromosomes. A related, but distinct, tetraploid (4N) has two complete sets of the diploid genome, albeit with variations in some of the genes. A related, but distinct, hexaploid (6N) has three complete sets of the diploid chromosomes, again with some sequence and functional variations. (b) Copy number variation (CNV) arises from multiple copies of a discrete sequence of 1 kb or larger in the genome. CNV is extensive in primate genomes (including the human genome) and has been associated with some disease and cognitive traits in humans, but its role in functional variation among individuals is not clear. (c) Gene families. Many genes in eukaryotic genomes are members of a gene family, with several copies of genes with closely related sequences and functions. In this hypothetical example, there are five copies of the related gene *A* in the haploid genome located at three distinct locations; two nearly identical copies of gene *A* (*A1A* and *A1B*) are found together, with the related, but distinct, copies *A2*, *A3*, and *A4* found nearby or elsewhere. The different shades of blue indicate the sequence and functional similarity.

diversified among two or more genes. Modified functions and changes in expression patterns occur, similar to what we discuss with gene families later. We will consider polyploidy again in Box 6.3.

Copy number variation

Analysis of the genomes of multicellular organisms uncovered a second type of change in copy number in which blocks of DNA are duplicated, resulting in multiple copies of the same sequence, as illustrated in Figure 3.20(b). This is known as **copy number variation** or **CNV**. CNV refers to the observation that a specific sequence or genetic element—ranging in size from 1 kb to more than 25 kb—is present in different numbers of copies when species or individuals are compared. CNV has been investigated most thoroughly in humans and other primates.

Different investigators and sources use somewhat different definitions of CNV—some include sequences as short as an individual codon—so we may be conservative in measuring the impact of CNV by limiting our discussion to sequences of at least 1 kb in length. CNV of a particular sequence is found both between species and between individuals within a species; in fact, identical twins may differ in their copy number of particular elements, suggesting that some CNV arises after fertilization by mechanisms occurring during mitotic cell division.

Roughly 0.4% of the genomes of two unrelated people (that is, as much as 2 million base pairs) is estimated to differ because of CNV.

Interestingly, the regions that vary in copy number seem to be discrete segments, rather than random blocks of sequence. More than 1400 different regions in the human genome have been found to have CNVs, indicating that these are widespread but that not every sequence in the genome is involved. Their precise role in phenotypic variation is not entirely clear, although CNVs have been associated with some genetic disorders and particularly some cognitive and personality differences, including schizophrenia. It seems likely that the impact of the CNV on phenotypic variation depends on what region is affected and the amount of CNV in that region. While CNV has been most extensively considered in the human genome, individual CNV has also been observed in other eukaryotes.

Between species, one of the most extensive comparisons of CNV involves humans and chimps, which suggests that CNV has been important in primate evolution. The protein-coding sequences of the genomes of humans and chimps differ by about 1.2%, that is, the protein-coding region (or exons) of the human genome comprises about 30 million base pairs, of which about 400,000 base pairs are different between humans and chimps. CNV between the two genomes is more extensive than this and is thought to amount to a difference of about 800,000 to 1 million base pairs, so that the genomes are not the same size, even though they have highly similar sequences. This result can be extended to other primates. Many of the genome regions that experience CNV are shared among humans and the great apes; about 40% of the variable regions in one genome are variable among humans, chimps, gorillas, bonobos, and orangutans. However, about 35% of the variable regions are specific to an individual species, and many of these vary in copy number among individuals **within** a species. Clearly, CNV is very common, with the extent and impact both within and between species being actively investigated.

Gene families

The third change in the number of copies of a gene involves gene families, summarized in Figure 3.20(c) and introduced briefly in Section 3.4. A **gene family** is composed of genes within a genome that have readily identifiable sequence similarity, such that they carry out similar or highly related functions; individual members of a gene family in the same genome are known as **paralogs**. Gene families (or protein families if the gene products are being considered) are very widespread. As noted in our earlier discussion of redundancy, at least half of the genes in our genome are members of some gene family.

In general, members of a species do not differ in the number of paralogs, but there are exceptions, and the distinction between CNV and gene families can become somewhat arbitrary. One of the exceptions involves amylase, the principal enzyme in our saliva that allows us to digest starch. Humans have 2–16 amylase paralogs in our diploid genomes, which affects the amount of amylase produced. Thus, people vary in their ability to digest starch, according to the number of paralogs they possess. Interestingly, people from populations that traditionally ate high-starch diets (such as rice and potatoes) have more amylase paralogs on average than people from populations that ate low-starch diets, as summarized in Figure 3.21.

Variation in the number of paralogs among members of the same species is somewhat unusual, but variation in the size of gene families between different species is extremely widespread and of tremendous importance when it comes to determining the differences between species. In fact, one of the primary differences between species is that gene families can be expanded or contracted to have more or fewer paralogs.

Duplication and divergence among gene family members

Gene families evolve through the process of duplication and divergence, as summarized in Figure 3.22(a). In the simplest version of this process, an ancestral species has one functional copy of a gene in its haploid genome. By one of many different cellular processes, this gene becomes duplicated, so the haploid genome now has two copies. The original member of the gene family is still capable of carrying out the original function, so mutations that occur in the newly arisen paralog could slightly change its function without affecting aspects of fitness that are associated with the original member of the gene family. For example, paralogs may come to be expressed at different times or in different tissues, which leads to greater diversification in their functions. Alternatively, the functions of the original gene could now be subdivided between the paralogs, which also allows greater diversification.

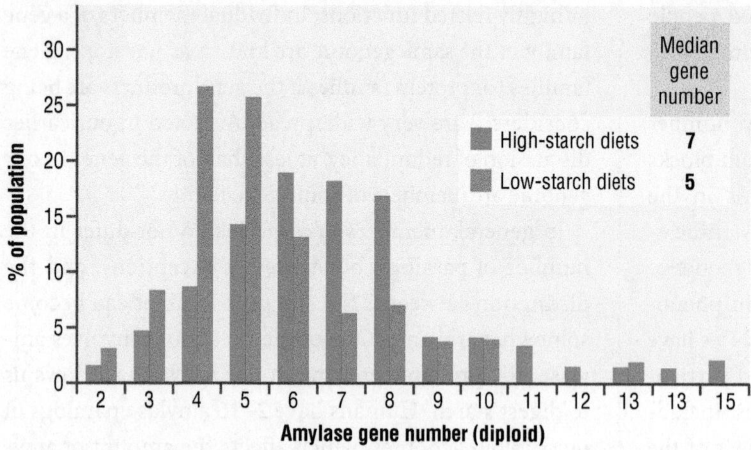

Figure 3.21 Amylase copy number and diet. One example of a gene family with an apparently adaptive effect among individuals comes from the gene amylase, involved in the digestion of starch. Modern humans who are descended from populations that ate a diet high in starch in the past, such as rice or potatoes, tend to have more paralogs of the amylase gene, and thus more amylase in their saliva. The increase in the number of paralogs may have provided a nutritional advantage, and therefore increased fitness, among populations that ate high-starch diets, whereas the high copy number would not confer an advantage in populations with a low-starch diet.

Many different examples could be used to show how changes in a gene family create functional diversity between species because this is very common. We will describe just a few gene families that affect familiar biological properties to illustrate what is often observed.

A well-known example in mammals is the globin gene family, responsible for oxygen and carbon dioxide transport, summarized in Figure 3.22(b). A gene that encodes a protein known as myoglobin is found in all vertebrates and is required for oxygen and carbon dioxide transport in muscle cells. (More distantly related proteins are found in invertebrates as well.) This gene family has undergone a series of duplications and divergence during vertebrate, and specifically mammalian, evolution, while myoglobin has remained an essential gene. Other members of the gene family have become specialized for expression in the hematopoietic system in mammals. These, in turn, have diversified to create two distinct subfamilies responsible for transport of oxygen from the lung cells to the rest of the body; these subfamilies are known as the α-globin and β-globin families. Each of these subfamilies has undergone further duplication and divergence to generate paralogs that are expressed at different times during embryonic and fetal development. An inevitable part of this duplication and divergence process is that some members of the family can acquire a mutation that renders them non-functional. These are examples of **pseudogenes**, the gene family equivalent of sweaters stored in the back of the closet that have become damaged beyond repair. In humans, both the α-globin and β-globin gene families have two pseudogenes in addition to the functional family members.

There are hundreds of other examples of this type of gene duplication and divergence in eukaryotic genomes. Two interesting examples of duplication and divergence with gene families are seen in the genome of the coffee plant. This divergence may have originally occurred naturally, but centuries of selection by coffee growers have increased the diversification and altered the properties of the plant. Coffee is known for two qualities found in its beans—its aroma and its caffeine. Both are due to the expansion of gene families.

The primary chemical that contributes to the distinctive aroma of roasted coffee beans is the fatty acid linoleic acid; the synthesis of linoleic acid is catalyzed by an enzyme that is the product of the *FAD2* gene. The model plant *Arabidopsis* has a single *FAD2* gene and makes low levels of linoleic acid, as seen in Figure 3.23. The coffee genome has six paralogs of this gene, and linoleic acid is also expressed at much higher levels in coffee than in *Arabidopsis*. In addition to an increased level of expression, two paralogs are expressed at high levels in the perisperm and endosperm (that is, the coffee bean), while *Arabidopsis* does not express the gene at all in these tissues; no one roasts *Arabidopsis* beans for their aroma.

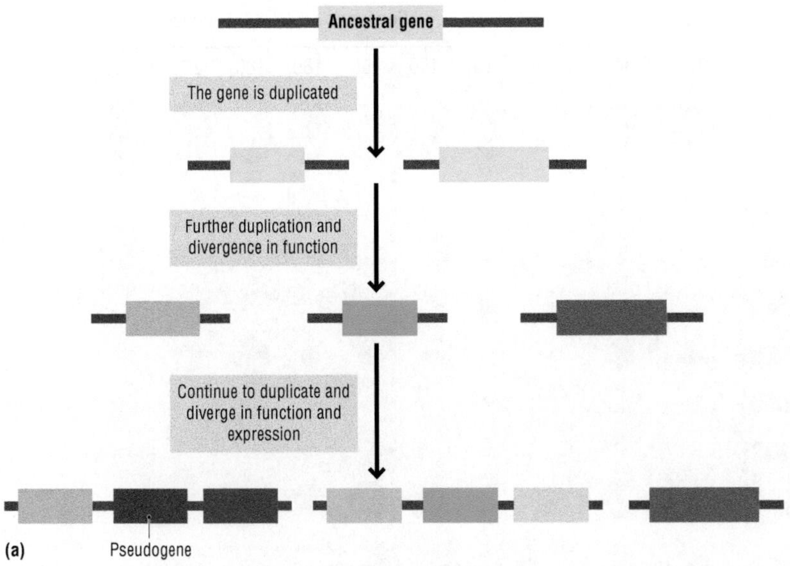

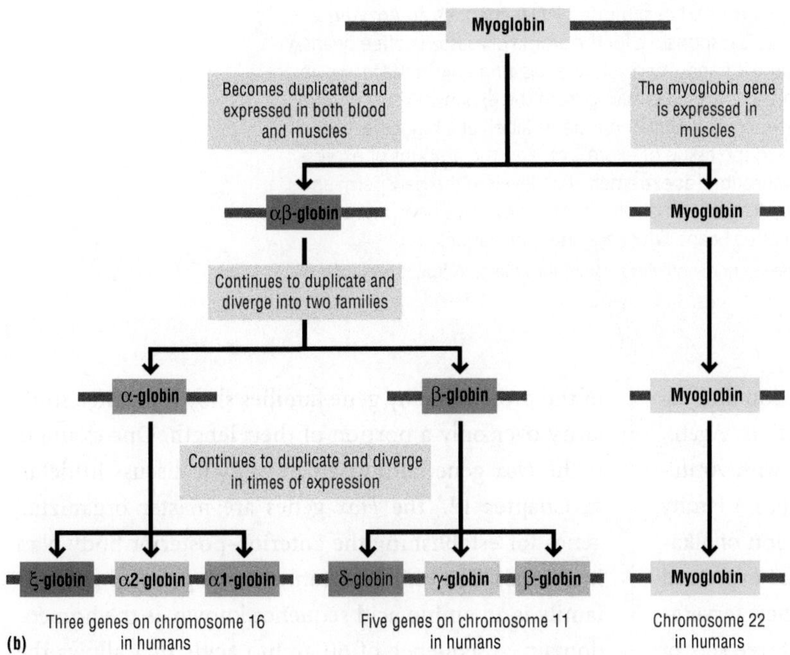

Figure 3.22 Gene duplication and divergence in a gene family. (a) An overview of duplication and divergence. A gene family often arises from the process of duplication and divergence. An ancestral gene is duplicated, so that two genes are now fulfilling the function previously performed by a single gene. This changes the selective pressure on the duplicate genes, so they can diverge in function and in times of expression, as illustrated by changes in the colors. It is possible that one or more of the duplicated genes retains the original function or that the original function becomes subdivided among the gene family and that new functions arise. Mutations will also result in non-functional paralogs, known as pseudogenes. (b) The globin gene family. Myoglobin is an oxygen transport protein in vertebrate muscle cells. The additional α- and β-globin gene families that make up the protein components of hemoglobin were generated by a series of gene duplications, while myoglobin has retained its original function in muscles. The α- and β-globin families are expressed in different tissues from myoglobin and have varying oxygen transport capacities. In mammals, the α- and β-globin families have been further diversified by changes in the times of expression, with the genes on the left of each cluster expressed at embryonic stages and the genes on the right of the cluster expressed at birth. While three functional α-globin genes and five functional β-globin genes are present in humans, each cluster also includes two pseudogenes, not shown in this drawing, which have mutations that prevent their function. The generation of pseudogenes is an inevitable consequence of the duplication and divergence process.

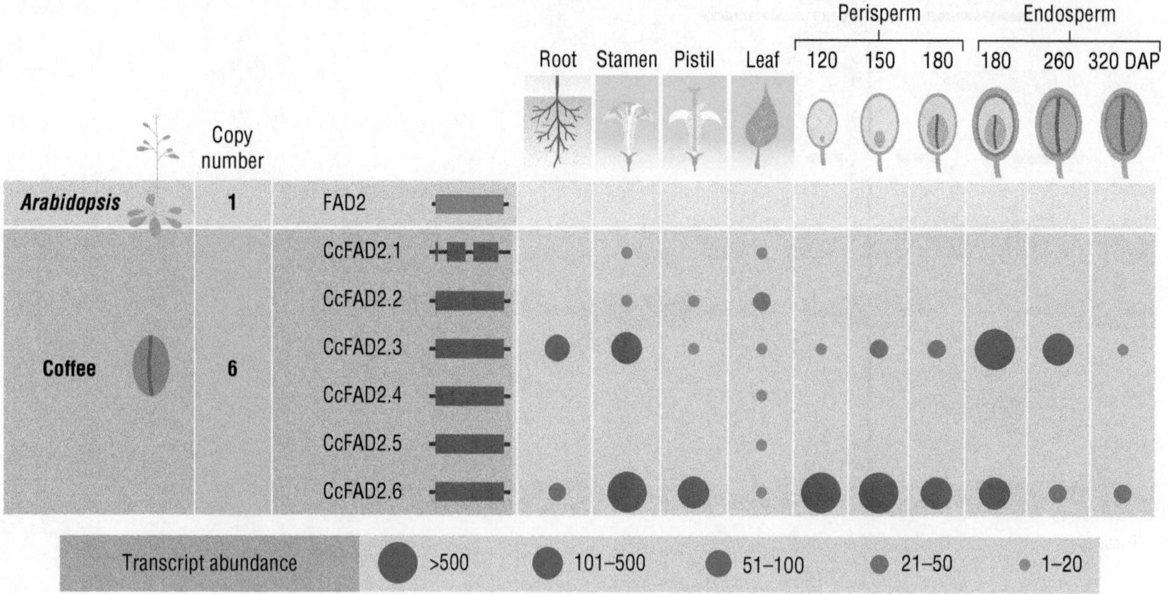

Figure 3.23 ***FAD2* gene family in coffee.** The gene fatty acid desaturase (*FAD2*) encodes an enzyme involved in the production of linoleic acid, a fatty acid that is responsible for the aroma of roasted coffee beans. In *Arabidopsis*, a single *FAD2* gene in the haploid genome is transcribed at low levels, primarily in the leaves. In the coffee genome, there are six *FAD2* genes in the haploid genome arising from the expansion of the gene family. While *FAD2.1* retains a similar transcription pattern to the *Arabidopsis* gene, albeit at a higher level, the other five members of the family have diverged both in the tissue of expression and the amount of expression. Note particularly *FAD2.3* and *FAD2.6*, which are transcribed at extremely high levels in the perisperm and endosperm, which produce the coffee bean. Thus, duplication and divergence of *FAD2* have been important in increasing the strength of the distinctive aroma of coffee beans. DAP, days after pollination.

Source: Reproduced from Denoeud, F. et al. (2014). The coffee genome provides insight into the convergent evolution of caffeine biosynthesis. *Science* Vol. 345, Iss. 6201.

A second example of gene family duplication and divergence in coffee involves caffeine production. Again, it is helpful to compare the coffee genome with *Arabidopsis*. The haploid genome of *Arabidopsis* has a family of six genes that are involved in the production of alkaloids, such as nicotine and caffeine, which are expressed principally in the leaves and roots where they serve as insecticides. By what appears to be at least three sets of duplication and divergence events, this family has been expanded to include 23 paralogs in coffee. Some genes continue to be expressed in the leaves and roots, while most of them are now expressed at very high levels in the beans. Thus, both the aroma and caffeine jolt from roasted coffee beans come from the duplication and divergence of gene families.

Gene families based on protein domain structures

While members of the gene families, such as globin and *FAD2*, show sequence similarity over the entire length of the protein, many gene families show sequence similarity over only a portion of their length. One example is the *Hox* gene family, which we will discuss in detail in Chapter 12. The *Hox* genes are master organizing genes for establishing the anterior–posterior body plan in animals. The unifying feature of the gene (or protein) family is an amino acid sequence known as the homeodomain, a sequence of 60 amino acids that allows the proteins to bind to DNA where they regulate the transcription of other genes. All proteins encoded by members of the family have a homeodomain, but amino acid sequence differences within the homeodomain allow the classification of different, but related, families of the *Hox* genes. You can explore this and other important features of the *Hox* genes in Box 3.2; this gene family is one of the most important for the evolution of the body plan of animals.

WEBLINK: VIDEO 3.1 provides more detail about the *Hox* gene family and its importance in evolution. Find it at **www.oup.com/uk/meneely**

BOX 3.2 *Going Deeper* *Hox* genes as an example of genome and morphological differences

While no single gene, genome, or gene family can be considered to be representative of all of the sequence changes that arise in genomes during evolution, the *Hox* genes include nearly all of the examples of sequence changes that we have discussed in the text and are certainly one of the best to choose. The *Hox* genes were discovered in *Drosophila melanogaster* by the striking mutant phenotypes that some of them exhibit. Many mutations in the *Hox* genes result in **homeotic** transformations along the anterior–posterior axis of the fly, whereby one normal body part is transformed into the likeness of another body part. These transformations are reflected in the names of the original mutants; two of these mutants are shown in Figure A.

Antennapedia mutants have a particularly unforgettable phenotype—they have a foreleg growing on the head where the antennae normally form. Bithorax mutants, which result from the mis-expression of a gene known as *Ubx*, have an extra pair of wings. The extra wings are formed because the third thoracic segment is transformed into the likeness of the second thoracic segment, and the balancing structures normally formed on the third thoracic segment now develop into wings typical of the second thoracic segment.

In general, the mutant phenotypes arise because the normal functions of the genes are missing or are mis-expressed in a different body part; as a consequence, one body part is transformed to another. Thus, the *Hox* genes are considered the master regulators of the anterior–posterior axis. As discussed in the chapter, they encode transcription factors with a particular DNA-binding domain, so the expression of a particular combination of *Hox* genes determines which structures will form on which segment in *Drosophila* by the target genes that they regulate. The genes are expressed during larval development, but their effects are seen most clearly in adult flies. The genes are discussed in Chapter 12, and gene regulatory networks are a topic in Chapter 14. The *Hox* genes are found in all animals, and changes in these genes have been associated with morphological differences in body plans and structures among animals.

Drosophila has a split cluster comprising eight *Hox* genes in total, which are all expressed in different, but overlapping, regions of the fly, as summarized in Figure B. The gene *lab* is located at the left end of the cluster in Figure B and affects the most anterior parts of the fly; strikingly, the *Abd-B* gene at the other end of the cluster affects the most posterior parts of the fly. As such, the order of genes in the cluster parallels the anterior–posterior organization of the body plan that they regulate. The *Hox* genes in other animals are also found in one or more clusters, and, in general, this organization has also been retained throughout animal development, so that the *Hox* genes at one end of the cluster regulate structures at the most anterior part of the body, the adjacent *Hox* genes in the cluster regulate structures that are slightly posterior to that, and so on. The combinations of the *Hox* genes being expressed in a particular region of the body specify what that region will become by activating the transcription of specific target genes, which then trigger or repress the formation of particular structures.

The *Hox* genes have proved to be a rich source of information to help us understand how molecular changes in one set of genes found in all animals can produce such obvious morphological differences. Let's consider amino acid sequence differences, regulatory differences, and changes in the number of genes in the family, the three specific examples of how the genetic changes

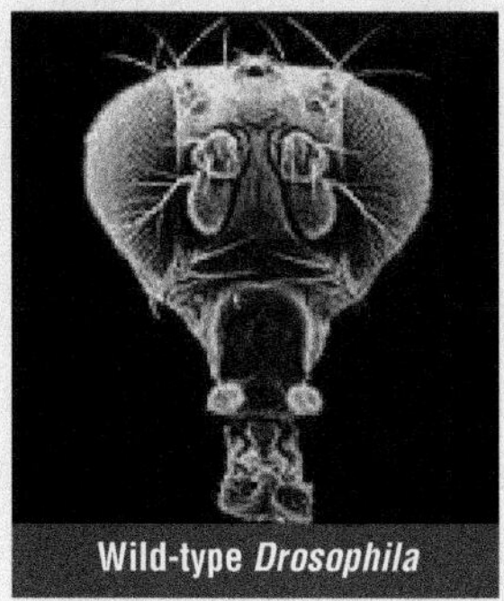

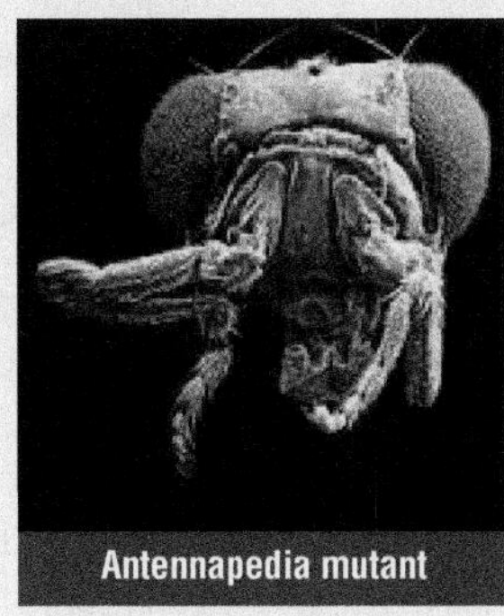

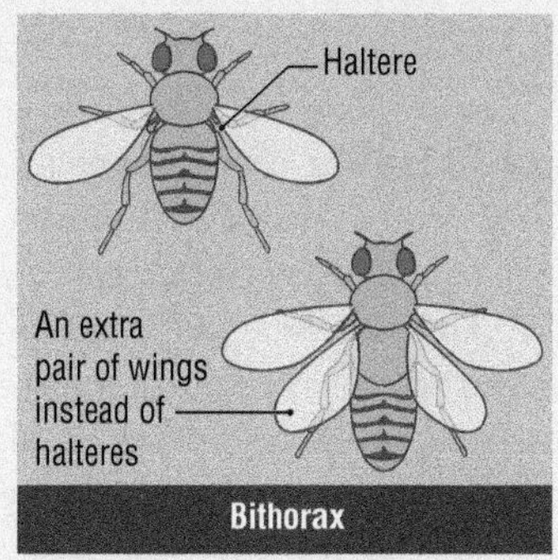

Figure A The homeotic phenotypes of *Drosophila Hox* genes. The *Hox* genes were recognized and named for the homeotic phenotypes arising from mutations in the genes. In Antennapedia mutants, arising from mutations and mis-expression of the *Antp* gene, forelegs form where antennae should be. In Bithorax mutants, arising from mis-expression of the *Ubx* gene, the balancing organs, known as halters, found on the third thoracic segment are replaced by a second set of wings, so the fly appears to have four wings.

Source: Wild-type *Drosophila* courtesy of EYE OF SCIENCE/SCIENCE PHOTO LIBRARY. Antennapedia mutant courtesy of Science VU/Dr. F. Rudolph Turner, VISUALS UNLIMITED /SCIENCE PHOTO LIBRARY.

BOX 3.2 Continued

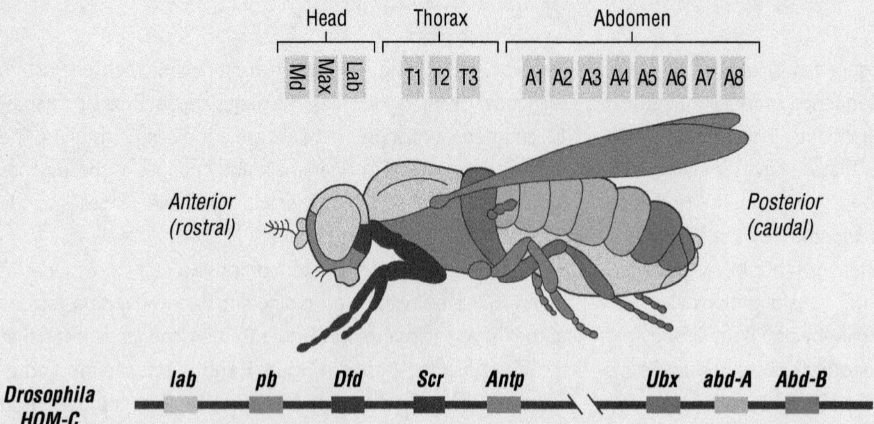

Figure B ***Hox*** **genes and the organization of the** ***Drosophila*** **body axis.** *Drosophila melanogaster* has the eight *Hox* genes in the split cluster shown here. They can be grouped in three broad subfamilies, based on similarities and differences in the homeodomain. These subfamilies are shown by the related colors: *lab* and *pb*; *Dfd*, *Scr*, *Antp*, *Ubx*, and *abd-A*; and *Abd-B*. The regions and time of expression of various *Hox* genes establish the anterior–posterior body axis of the embryo and the structures in the fly, as seen in the corresponding colors between the genes and the body regions that they affect..

listed in Section 3.6 have been observed among the *Hox* genes, resulting in evolutionary change.

Amino acid sequence differences

The homeodomain is the portion of the Hox protein that binds to specific DNA sequences to regulate transcription of their target genes. Changes in the amino acid sequence of the homeodomain, of the type shown in Figure 3.24, result in binding to different DNA sequences and thus allow different Hox proteins to regulate different target genes. It is the regulation of these target genes that produces the different structures along the body axis, and changes in the target genes that determine many of the local differences.

Regulatory differences

There are certainly quantitative differences in the levels of transcription of different *Hox* genes, when different species are compared. However, the most striking changes are differences in the spatial patterns of expression of different genes in different species. An example is shown in Figure C. In primitive crustaceans, like the brine shrimp *Artemia*, the *Hox* gene *Ubx* is expressed throughout a domain that includes all the thoracic segments, and these segments make highly similar appendages. In other species of crustaceans, the first thoracic segment no longer expresses *Ubx*, so that the domain of *Ubx* gene expression has shifted more posteriorly. The appendages that form in this anterior thoracic segment that lacks *Ubx* expression look distinctly different from the appendages that form in the domain of *Ubx* expression. The anterior thoracic appendages, called maxillipeds, are reduced in size and bear similarities to the feeding appendages of the head segments. Changes in the regulation of *Ubx* gene expression therefore correlate with changes in the patterning of the body plan.

Such changes are not limited to arthropods, however. Remarkably, changes in *Hox* gene expression patterns also contribute to the morphological differences seen in other animals. Among vertebrates, changes in the dynamics of *Hox* gene expression are evident when many different body structures, including limbs and vertebrae, are compared. We will discuss some of these examples in Chapter 12. In fact, changes in the expression pattern of the various *Hox* genes are one of the most important differences when comparing animal body plans. If the pattern of the expression of the *Hox* gene itself has not changed, changes in the amino acid sequence of the protein it encodes have probably resulted in changes in the expression pattern of some of its target genes.

Changes in the numbers of *Hox* genes

The *Hox* gene family has been subject to duplication, divergence, and gene loss throughout evolution. This expansion in the size of the gene family has allowed even greater diversity in the patterns of expression of various genes. This, in turn, has allowed even more diversity in the body plans of animals. The most likely inferred history is summarized in Figure D.

The last common ancestor of flies and humans probably had four types of *Hox* genes, although multiple copies of some family members were likely. *Drosophila* has the eight *Hox* genes shown in the split cluster in Figure D. (*D. melanogaster* has this cluster split into two halves at different chromosomal locations, but the overall anterior–posterior structure is similar to other organisms.)

BOX 3.2 Continued

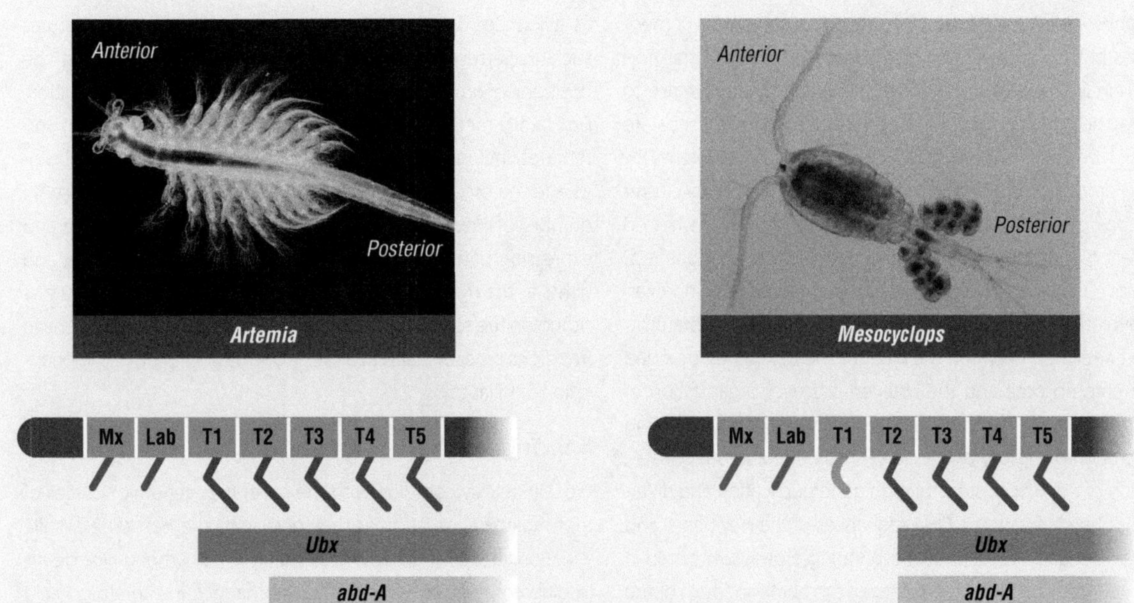

Figure C Changes in *Hox* gene expression lead to morphological diversity between species. A change in the domain of *Ubx* gene expression between the crustaceans *Artemia* and *Mesocyclops* corresponds with a difference in the types of appendages formed. In *Artemia*, the head segments (Mx, Lab) bear small feeding appendages, while the thoracic segments, expressing the *Hox* genes, *Ubx*, and *abd-A*, form larger locomotory appendages. In *Mesocyclops*, a difference in the domain of *Ubx* gene expression corresponds with a difference in the types of appendages formed. The first thoracic segment (T1) of *Mesocyclops* no longer expresses *Ubx*, and the appendage that forms takes on a new identity as an additional feeding appendage referred to as a maxilliped. Not all of the segments are diagrammed.

Source: *Artemia* courtesy of djpmapleferryman/ CC BY 2.0. *Mesocyclops* courtesy of Simon Kutcher/AFAP/ CC BY 2.0.

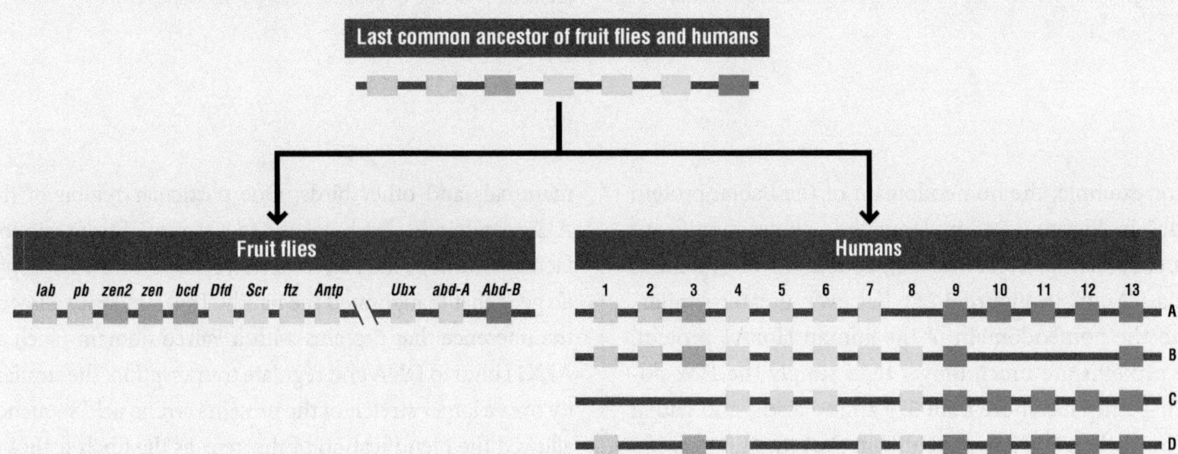

Figure D The *Hox* gene family. The *Hox* genes can be divided into four main types (colored yellow, green, blue, and orange), based on the sequence of the homeodomains that they encode. The ancestral *Hox* cluster common to all animals is thought to have comprised at least seven genes, which were independently expanded in the lineages giving rise to *Drosophila melanogaster*, humans, and other animals. The names of the genes in *Drosophila* are shown; the gene cluster in *D. melanogaster* is also broken into two clusters. In the human lineage, the ancestral gene cluster was duplicated to give four clusters, called A, B, C, and D, from which various individual genes have diversified in function and expression and some members have subsequently been lost.

BOX 3.2 Continued

Four genes—*zen2*, *zen*, *bcd*, and *ftz*—encode proteins with homeodomains but do not seem to be involved in homeotic transformations in *Drosophila*. In the vertebrate lineage that gave rise to humans, the *Hox* gene family has been expanded to four separate clusters, labeled A to D in Figure D, each with 9–11 different *Hox* genes; humans have 39 *Hox* genes, for example. Not every cluster has a copy of every *Hox* gene type—cluster C in humans has no members of the *labial* family, for example—but at least one representative of all of the *Hox* genes found in *Drosophila*, and even more evolutionarily ancient animals, is also found in vertebrates.

The correlation between the order of the *Hox* genes arranged on the chromosome and the anterior–posterior organization of expression patterns has also been generally preserved, reflecting the evolutionary history of the clusters from the ancestral cluster. Evolution of the *Hox* clusters through gene duplication and divergence is also shown in the DNA sequences of the *Hox* genes and amino acid sequences of the corresponding proteins. As noted in Figure 3.24, the amino acid sequence of the homeodomain of the *Drosophila* Lab protein is more similar to the human Hox protein A1 than it is to the *Drosophila* Antp protein, since *Drosophila* Lab and human A1 share a common ancestor with one another more recently than they do with *Drosophila* Antp. Expansion of the *Hox* gene number has allowed a tremendous diversification in gene expression patterns in both time and location, which has, in turn, resulted in changes in far more target genes regulated by the Hox transcription factors.

The *Hox* genes in *Drosophila* regulate segmental identity. The *Hox* genes in vertebrates continue to regulate segment diversity, as evidenced by the distinctions among segmental structures such as vertebra and spinal nerves in our backbone and the positions of limbs, and the types of limbs that form. In addition, *Hox* family members also regulate many other structures in vertebrates, including the structure and spatial organization of digits and the organization of the ear. Without the expansion in the number of *Hox* genes and their diversification of sequence and expression patterns, we would not have shoulders, elbows, and wrists at the right positions, and we would not have our thumbs sitting on the space bar when we type; in fact, we would not have five digits on each hand at all, let alone having them in the orientation familiar to us.

Splicing and novel genes

We will quickly continue with the other two genomic sources of morphological variation as they pertain to the *Hox* genes. While variations in splicing patterns certainly occur among *Hox* genes in different species, which result in changes in the amino acid sequences of the proteins, splicing changes are not currently thought to be a major source of diversification among these genes or their activities. Conversely, some novel genes have clearly arisen. For example, the *Hox* gene *ftz* in *D. melanogaster* (shown in blue in Figure D) is involved in the regulation of segment number; the gene appears to be novel, since few other organisms have a functionally equivalent *Hox* gene at this position. Based on its nucleotide sequence and the amino acid sequence of the protein, *ftz* appears to have arisen from the duplication and modification of one of the neighboring *Hox* genes.

For example, the homeodomain of the Labial protein ("Lab" in Figure 3.24) in *Drosophila* shows significant sequence similarity to the homeodomain of Antennapedia ("Antp" in Figure 3.24), but even greater similarity to the homeodomain of the human HoxA1 protein. The proteins are much longer than simply the Hox domain—Antp has more than 370 amino acids, and Labial has more than 600 amino acids, but the remaining amino acid sequences are not particularly similar. The family similarity among the proteins is found almost entirely in the homeodomain region.

We previously introduced the concepts of protein domain structures and gene families in Chapter 1, when we discussed the biological role of the *ALX1* gene in finches. The role of the ALX1 protein in finches has not been tested experimentally, but inferences about its function were based on its amino acid similarity to the ALX1 protein in mammals and other birds. One particular region of the ALX1 protein is clearly related to a family of transcription factors known as the Paired family, named for the *Drosophila* gene that inaugurated the family. That similarity allowed the inference that proteins with a Paired domain (such as ALX1) bind to DNA and regulate transcription; the similarity over a larger stretch of the protein's amino acid sequence allowed the identification of the gene as the finch ortholog of ALX1, rather than another member of the Paired family.

A helpful analogy can be made between gene or protein families and human families. Sometimes, when we discuss a human family, we are referring to only the first-degree relatives—parents and siblings. Sometimes we can extend this to include a broader collection of more distant relatives such as second cousins, great uncles, and so on. Similarly, when discussing gene and protein families, we are sometimes referring only to those most closely related

LAB	NNSGRTNFTNKQLTELEKEFHFNRYLTRARRIEIANTLQLNETQVKIWFQNRRMKQKKRV
HOXA1	PNAVRTNFTTKDLTELEKEFHFNKYLTRARRVEIAASLQLNETQVKIWFQNRRMKQKKRE
ANTP	RKRGRQTYTRYQTLELEKEFHFNRYLTRRRRIEIAHALCLTERQIKIWFQNRRMKWKKEN

Figure 3.24 The homeodomain sequences of three Hox proteins. The *Hox* gene family all encode a 60-amino acid sequence known as the homeodomain; the domain is responsible for the ability of Hox proteins to bind to DNA sequences to regulate the transcription of specific target genes. The amino acid sequences of the homeodomain from three Hox proteins are shown, with the Labial protein (Lab) from *Drosophila* being used as the standard. Differences between Labial and the human HoxA1 protein are shown in red, while differences between Labial and the *Drosophila* Antp protein are shown in blue. Notice how the amino acid sequences of the homeodomain from these three proteins are highly similar but not identical; the differences in amino acid sequences give these proteins distinct transcriptional targets. Because there are nine amino acid differences between Lab and HoxA1 and 19 differences between Lab and Antp, HoxA1 is considered to be more closely related to Lab than in Antp.

family members and other times to more distant members with less sequence similarity. Most of these differences are seen when different species are compared. When comparing members of the same species, it is somewhat unusual to see variation in the number of members of a gene family. By contrast, when comparing separate species, it is unusual not to see variation in copy number, sequences, and expression patterns of gene families.

KEY POINT The major sources of genome differences are base substitutions, insertions–deletions, and changes in the number of copies of genes and other sequences. Base changes are the primary source of variation between individuals of the same species, while all of these are significant when different species are compared. When comparing the genomes of different species, expansion and diversification of gene families is especially common.

Phenotypic and morphological effects of genomic variation

Having discussed the types of changes that occur in genome sequence and structure, we can now consider the phenotypic consequences of these changes. Changes that affect the functions of genes are different amino acid sequences, different transcription patterns, and variation in splicing. We will also briefly consider newly arisen or novel genes with their functions and changes involving RNA-encoding genes. The phenotypic consequences of these various changes are summarized in Figure 3.25, which compares their relative effects between individuals in the same species, between closely related species, and between distantly related species.

Variation in amino acid sequences

Within a species, individuals differ in the nucleotide sequences of their genes, which may change the amino acid sequences of the same protein as it exists in each individual. The differences in the nucleotide sequence could arise from a combination of base substitutions and indels. These differences in amino acid sequence may be enough to produce different phenotypes. Most of the examples of the genetic differences we discuss in this book fall into this category of change. For example, all humans have the gene for the ABO blood type, but DNA sequence changes in that gene result in proteins with different amino acid sequences, which result in different individuals having different blood types. The alleles that produce round and wrinkled peas differ by the presence of an indel.

Amino acid sequence differences are also common between species, with the number of changes increasing as the evolutionary distance between the species is greater. While the equivalent genes and gene products in two different species are similar or very highly similar, they usually do not have exactly the same amino acid sequence, as illustrated previously with cytochrome C oxidase in Figure 3.18 or Lab and HoxA1 in Figure 3.24. Because genes that have slightly different nucleotide sequences often encode slightly different amino acid sequences, the proteins from different species usually do not have exactly the same biochemical characteristics.

This source of genetic change is found when we compare both very similar species and very distant species to an extent that depends on the gene. For example, while humans and *E. coli* recognizably share many of the enzymes that repair damage to DNA, there are enough amino acid differences between them that the human version does not work well in *E. coli* and the *E. coli* version would not work well in human cells. Equally, if we consider more closely related species, certain mouse genes can work in human cells (grown in culture), and the equivalent human gene can work in mice. For other

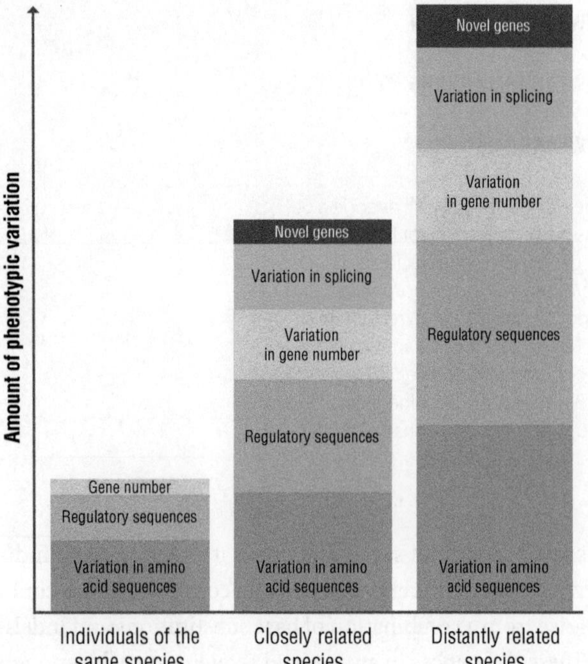

Figure 3.25 The sources of phenotypic and genomic variation. Five different sources of genomic variation are considered—changes in amino acid sequences, changes in regulatory sequences, changes in splicing patterns, changes in gene and sequence number, and novel genes. The bar graphs summarize the relative importance of these, when comparing two individuals of the same species, individuals of closely related species, and individuals of distantly related species. The "amount of phenotypic variation" on the Y-axis is an estimate of what fraction of the total variation will come from what source. For two individuals of the same species, most of the phenotypic variation can be explained by changes in the amino acid sequences and changes in regulatory sequences, particularly changes that alter the amount of a gene product. There are relatively few changes in gene and sequence number, although this depends on the species. Changes in splicing patterns and the existence of novel genes are generally not found. For closely related species, however, the sources of variation between individuals are greater; some of the changes in regulatory sequences probably also change the time and location of expression, as well as the amount. In addition, changes in splicing patterns are common, and novel genes unique to each species are found. When distantly related species are compared, all of the genomic sources of variation are greater. This is a summary of the sources of variation, but many deviations from this summary are known, and the relative importance depends on the species, as discussed in the text.

genes, the mouse and human are different enough that they cannot replace one another in function in cell cultures or *in vivo*.

Variation in gene regulation

The other principal source of genetic difference between individuals in the same species comes from the regulation of gene expression. Genes are expressed in particular cells, at particular times, and in certain amounts. Two individuals of the same species probably do not differ much in which cells or at what times the genes are expressed, but the **level** of expression may differ quite a bit. Thus, there are quantitative differences in the amount of the gene product, which results in observable differences in the phenotype.

An example in humans is found with hair pigmentation, shown in Figure 3.26. One difference between blond hair found among northern Europeans and brown hair is due to the level of transcription of the gene that encodes a protein known as the kit ligand. This protein is involved in the differentiation of the pigment-producing cells or melanocytes. A change in the regulatory region of the gene for the kit ligand means that blond-haired people make less of the transcript and the protein than brown-haired people. The change seems to be only in the level of expression, since the protein sequence and its function are not different. While this is only one of five or more genes involved in hair pigmentation in humans, a quantitative change in just this one can explain a very familiar phenotypic difference in some northern European populations.

The number and types of changes in the regulation of gene expression are greater when different species are compared. These changes are now thought to be among the most important genetic differences giving rise to morphological distinctions, particularly between closely related species. Just as is seen among individuals within a species, there are many quantitative differences in the level of expression of related genes in different species. Furthermore, when species are compared, it is common to discover that the same genes carry out highly similar functions, but they are not expressed at equivalent times or locations. These changes in the time, place, and amount of the expression of a gene can result in profound changes in morphology and development. Such changes in gene regulation can occur more rapidly than alterations in gene function, at least in evolutionary time, because the function of the protein itself is not being changed.

These were the two sources of changes noted for the *ALX1* gene in the genomes of Darwin's finches in Chapter 1 and discussed in Chapter 2; these changes were summarized in Figure 2.23. Three of the eight changes among the *ALX1* genes of different finch species were base substitutions that changed the amino acid sequence of the ALX1 protein; these are predicted to affect its function, particularly the specific DNA sequences to which it binds to regulate transcription in different species. Four of the eight changes are predicted to change the transcription pattern of the *ALX1* gene, suggesting that the gene is expressed in

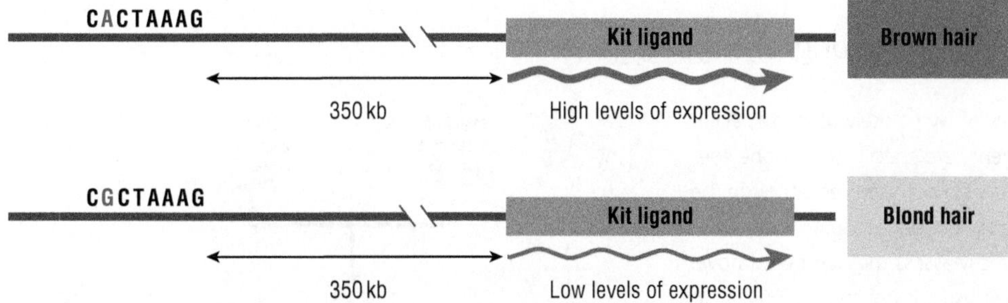

Figure 3.26 Quantitative changes in gene expression and phenotypic variation. Within a species, one of the main sources of phenotypic variation is a quantitative difference in the level of transcription seen in different individuals. One familiar difference comes from hair color differences among northern Europeans, including Icelanders, Scandinavians, and the Dutch. Many of the blond individuals from these populations have a single base difference (shown in red) in the *cis*-regulatory module upstream of a gene called the kit ligand. This base change results in reduced transcription of the kit ligand gene, compared to the amount from brown-haired people from the same population. The overall function of the kit ligand is not affected, simply its level of expression. The kit ligand is one of at least five genes that affect pigmentation in humans and other mammals; the kit ligand does not affect pigmentation in the eye, for example, and this change appears not to be associated with blue eyes.

different tissues or at different times in these species. (The eighth change is immediately downstream of the gene; its predicted phenotypic effect, if any, is less clear.)

Variation in splicing

As discussed in Chapter 2, the initial RNA transcript from a gene in most eukaryotes is spliced, with some parts of the sequence retained (the exons) and other parts of the RNA sequence removed (the introns). While the pattern of splicing can be different in an individual at different times and in different cells, the transcripts from a gene are nearly always spliced at the same locations in the sequence and in the same pattern among different individuals of the same species. Thus, very little of the morphological variation we observe within a species can be explained by splicing differences, although there are exceptions. One of the most notable exceptions is the splicing of a gene known as *doublesex* (*dsx*) in *D. melanogaster*, which regulates sexual differentiation. The alternative splicing of *dsx* in males and females will be discussed in Box 7.5.

However, when different species are compared, splicing differences become much more common and are a significant source of morphological variation. These splicing differences take two forms. First, some species exhibit much more splicing than others—only about 5% of the genes in yeast have more than one exon; more than 70% of the genes in flies and worms consist of more than one exon, and nearly all genes in vertebrates have multiple exons. As we move from yeast to flies and worms, and to vertebrates, we see the number of exons increase, and the lengths of the introns that separate the exons also increase. It is questionable, however, how much of the morphological differences between related species can be explained by an overall increase in splicing.

Second, the initial transcript from a gene can be spliced in more than one way, to give more than one related mRNA and be translated into more than one related amino acid sequence. We noted above that alternative splicing is a significant source of variation when the transcription profiles of species are compared. Thus, alternative splicing clearly serves as a mechanism that increases the morphological diversity between species.

Novel genes and functions

The final source of morphological diversity between species, summarized in Figure 3.25, is the presence of novel genes unique to one species. Although most genes are the same when closely related species are compared, there are a few genes that are unique to that species; these genes may have a significant impact on defining one species from another. Of the roughly 200 genes found in humans and not in chimps, many are expressed in the brain, while others are expressed in the skin or the immune system. Thus, novel genes and functions do arise during evolution, as described in Box 3.3.

Fewer species-specific genes exist than we once thought, but the fact that they exist is important. Species-specific genes can result in species–species differences; different species do not have exactly the same genes,

BOX 3.3 *Going Deeper* The origins of novel genes

When comparing the genomes of two closely related species, relatively few protein-coding genes are specific to just one species. Rather, most genes are present in both species, although the exact DNA sequences of the equivalent genes usually vary somewhat. It can be difficult to correctly assess the number of novel genes in a particular species because it is not straightforward to establish what would be considered a truly novel gene, rather than a highly derived version of an ancestral gene. A conservative definition would only include genes for which there is no related DNA sequence found in the other lineage that is being used for the comparison. But a broader definition would include genes for which there are generally related sequences in the other species, but not a directly equivalent gene.

Comparisons of genomes at different evolutionary distances are helpful. For example, estimates for the number of genes specific to humans and not found in chimpanzees range from approximately ten to 300, depending on the exact definition of what constitutes a novel gene. Some of these are not found in the Neanderthal genome, so they arose in the few hundreds of thousands of years since humans and Neanderthals shared a common ancestor. Others are found in both Neanderthals and modern humans, but not in chimps, so these probably arose when the hominid lineage diverged from the chimpanzee lineage roughly 5 million years ago.

Even when we compare over larger evolutionary distances—say a fruit fly to a non-insect genome—we still find only a few genes that lack at least a detectably related gene in the other genome. In the fruit fly example, about 20% of protein-coding genes do not have a clear match with non-insect genes. It has been proposed that these lineage-specific genes are more likely to be involved in the evolution of adaptive traits that are specific to that lineage. But where do these novel genes come from?

The origin of any individual novel gene tells a unique story of change, chance, and evolution. However, as more genomes are analyzed and compared, it is clear that several basic mechanisms are at play when it comes to the origination of new protein-coding genes, as summarized in Figure A. These mechanisms fall into two broad categories: (1) tinkering with existing protein-coding genes to create new ones and (2) changes to the DNA sequence of non-coding regions to create new protein-coding genes.

Rearranging and combining existing genes

There are many ways that existing protein-coding genes can be rearranged and combined to produce new ones. The process of DNA replication (discussed in Chapter 4) can "stutter" and copy the same piece of DNA more than once, duplicating a stretch of DNA and providing an opportunity for new sequence changes and combinations. Similarly, during the process of recombination (discussed in Chapter 6), errors can result in the bringing together

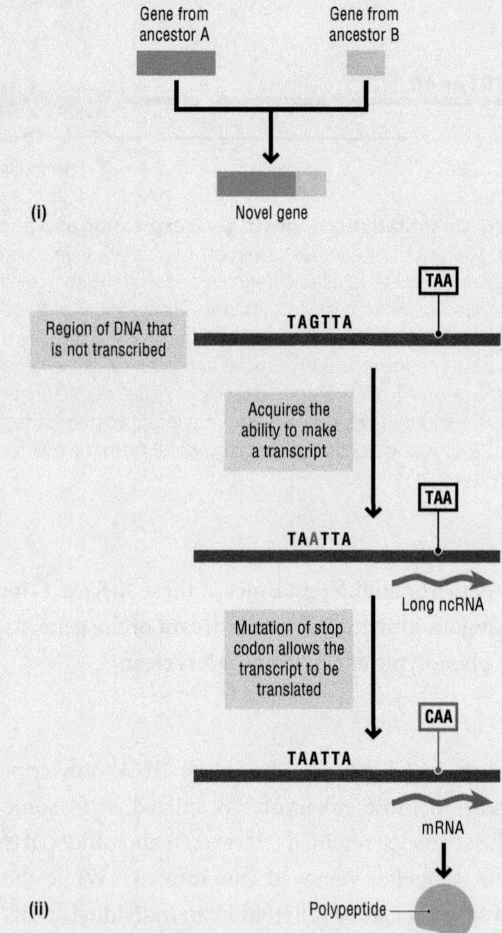

Figure A The origins of novel genes. Novel genes, that is, genes that are specific to that species or to a particular phylogenetic group of related species, arise from many sources. Two of them are diagrammed here. (i) Some novel genes are composites of other genes found in other species, including unrelated species from which the gene may have been acquired by horizontal gene transfer. This type of novel gene can be recognized because parts of the amino acid sequence will be similar to other genes, although the occurrence of these amino acid modules together is novel. (ii) Some novel genes appear to arise from a process involving non-coding RNAs. Since transcriptional start sites in eukaryotes do not have a specifically defined sequence, a transcriptional start can arise within a region of the genome that has not been previously transcribed. This is shown in the figure as a mutation that generates a TAATTA or TATA box to initiate transcription, but many processes are possible. Transcription generates an RNA that lacks a long open reading frame or other features associated with a protein-coding gene. However, it may subsequently acquire some function as a non-coding RNA. If it makes a peptide at all, it is likely to be short and non-functional. A short open reading frame in the non-coding RNA can be lengthened by a mutation that eliminates a stop codon (as shown here), and a novel polypeptide can result. This can sometimes be recognized because the DNA sequence is similar to a sequence found in a non-coding region in a related organism.

BOX 3.3 Continued

of stretches of DNA that were not previously adjacent to one another. Furthermore, all the mechanisms capable of adding DNA to a genome, discussed in Box 3.1, also provide an opportunity for the subsequent formation of new genes from the newly introduced DNA. These mechanisms may involve entire protein-coding genes, individual exons, or even smaller regions that may encode a structural motif of a protein, bringing that motif into the protein-coding sequence of a gene that did not previously encode it. This is shown in Figure A(i).

Let's explore this mechanism of creating new genes from old genes with an example. The oskar protein from the fruit fly *Drosophila melanogaster* plays a role in assembling the specialized "germ plasm" needed for the formation of the germ cells; it also plays a role in anchoring the mRNA for specific proteins needed to specify the development of the posterior region of the developing embryo.

Interestingly, the N-terminus of the encoded oskar protein shows sequence similarity to *Tdrd-7* genes found in all metazoans; like one of the roles of oskar, *Tdrd-7* orthologs are also involved in a variety of processes that involve the binding of the protein to RNA to regulate its position in the cell, its transport, or its translation. It is likely that evolution of the novel insect *oskar* gene involved duplication and divergence of an ancestral *Tdrd-7* gene. However, the C-terminus of the oskar protein shows sequence similarity to entirely different genes, a class of enzymes called hydrolases. Furthermore, the hydrolases that the oskar sequence shows the most similarity to are bacterial genes. It has therefore been proposed that the insect *oskar* gene may have evolved through the fusion of an insect *Tdrd-7*-like gene with a bacterial hydrolase gene that had been horizontally transferred into the insect genome.

An ortholog of the *oskar* gene has recently been identified in crickets. This gene does not appear to play a role in development of the germ line, but it has a presumably more ancestral function in the nervous system. However, the cricket does show low levels of expression of *oskar* in the early embryo in the region where the germ cells develop. This early expression appears to be of no functional consequence to the cricket and is likely just a result of the particular "wiring" of the transcriptional control regions of the *oskar* gene. It is then easy to see how the gene product might have subsequently been co-opted into a role in germ line development in other more derived insect lineages, such as *D. melanogaster*, since it was likely already being expressed at the right place and at the right time.

New coding regions

The other way that new protein-coding genes can arise is through changes to stretches of DNA sequence that previously did not encode proteins. One version of how this might occur is shown in Figure A(ii). This might be recognized because a closely related species has a similar DNA sequence at the corresponding location in the genome, but no expression. At least 11 protein-coding genes in humans that appear to have arisen by this mechanism have recently been characterized. It appears that these genes first acquired DNA sequence changes that impart regulated transcription of that stretch of DNA. A subsequent step appears to be the acquisition of changes to the DNA that impart protein-coding potential to the transcript.

With this view of the evolution of new protein-coding genes from non-coding transcripts in mind, scientists are starting to look at the large numbers of non-coding transcripts present in many eukaryotic genomes in a new light; if the first step in the origin of a novel gene is its transcription, these non-coding RNAs might be a "nursery" of potential new protein-coding genes. In support of this hypothesis that novel protein-coding genes arise from some non-coding RNAs, novel protein-coding genes tend to have shorter transcripts than the average-sized protein-coding genes in the genome.

while individuals within a species do. Species-specific genes arise from a variety of mechanisms, many of which are not known. One known mechanism is the introduction of novel genes by horizontal gene transfer, as will be discussed in Chapter 11. Specific novel genes often have arisen from different processes, so every origin story is a bit different.

Changes involving RNA-encoding genes

All of our previous points discussed evolutionary and morphological change from the perspective of changes in genes that encode polypeptides. We have discussed these at length because these are very important and well-understood examples. However, it has become clear that much of the genome of a multicellular animal or plant is transcribed into RNAs that are not translated, known as non-coding RNAs. Thus, the roles that these non-coding RNAs play in morphological change must also be considered. At this time, however, there are few comparative data to use.

As we will discuss in more detail in Chapter 12, non-coding RNAs fall into different categories, based on size, referred to as microRNAs and long non-coding RNAs. The overall function of both microRNAs and long non-coding RNAs can be summarized to be the regulation of the expression of other genes, albeit by different

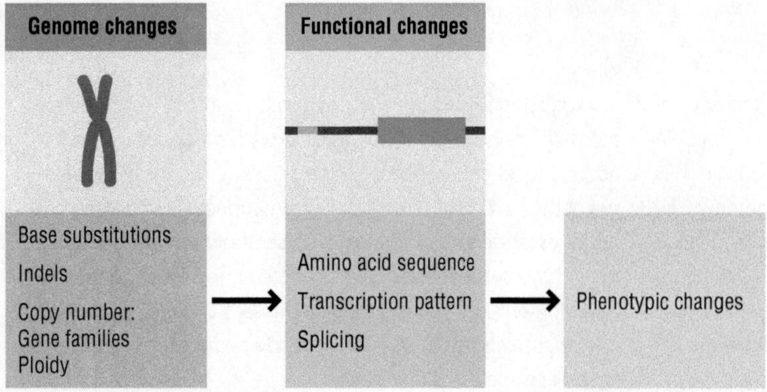

Figure 3.27 Genomic changes, functional changes, and phenotypic variation. Changes in the genome, such as base substitutions, indels, and copy number changes, can produce changes in functions of genes. These functional changes include changes in the amino acid sequence, in the pattern and level of expression, and in splicing. Functional changes, in turn, can lead to phenotypic variation.

mechanisms. Both categories of non-coding RNA genes are less highly conserved between species than are protein-coding genes. Some, but not all, microRNAs are evolutionarily conserved when related species are compared, while others are found in a single, or a few closely related, species. On the other hand, long non-coding RNAs are quite diverse and much less conserved between even closely related species. This lack of conservation suggests that the non-coding RNAs may provide rapidly evolving systems of gene regulation. As we learn more about them, they could prove to be among the most important factors in species-specific (or possibly individual) differences, but very little is known about them so far.

We have considered changes in the genome and their effect on phenotypes from a broad perspective, with only a few examples. These are summarized in Figure 3.27. Any of the types of genomic change can produce any of the types of change in the function of genes, which can, in turn, produce phenotypic differences among individuals and between species. Every genome provides thousands of examples that illustrate these points in more detail. In Box 3.2, we have gone into more detail with the specific example of genome changes and morphological changes related to the *Hox* genes. We will consider *Hox* genes again in Chapter 12, but this example is one to contemplate.

KEY POINT Genome changes produce phenotypic variation due to changes in amino acid sequences, changes in the expression pattern, and changes in RNA splicing. In addition, when the genomes of two species are compared, a few novel genes have typically arisen. Changes in genes encoding non-coding RNAs may allow very rapid changes in gene regulation.

3.7 Summary: genome structure and variation

With current sequencing technology, it is not as expensive, difficult, or time-consuming to sequence the genome of a species as it was even a decade ago. The technology is improving so rapidly that we will undoubtedly regard the current sequencing methods as being old-fashioned in a few years. The widespread availability of genome sequences has reshaped biological research. Many biological questions related to evolution, domestication, phenotypic variation, basic cellular functions, diseases, and more can be approached differently because of the widespread availability of the sequences of genomes of many species.

But it is a biological simplification to speak of the genome of a species as if it were a single sequence. Individual members of a species each have their own genomes, and the sequences of these genomes vary. Among members of the same species of animals, the variation is primarily in the base sequence, rather than in the number of copies of genes or genetic elements, although that occurs as well. For plants and bacteria, variation in the size and content of genomes is common, in addition to the variation in sequences. For plants, much of this variation in size comes from changes in ploidy. For bacteria, the size and content vary due to the presence of plasmids and horizontal gene transfer.

Variation between species arises from the same sources as variation within a species, with some additional origins as well, and increases as more distantly related species are compared. For eukaryotes, changes in the sizes and sequences of gene families are particularly important. Some novel genes also arise, although not as many as some might have expected; most differences among species come not from novel genes, but from differences in genes that the species have in common.

Although we have discussed variation in terms of analyzing genomes, that is not the way that most of us encounter biological variation. Unless we are actually looking at the sequence of the genome, we are seeing its outputs. We spot faces in a crowd, walk around a zoo or an arboretum, and see cultures growing on surfaces without thinking about the underlying genomic changes that have produced these phenotypic differences. Genomic variation becomes phenotypic variation through changes in the amino acid sequences of proteins, changes in the patterns of gene expression, and (for many eukaryotes) changes in RNA splicing, as reviewed in Chapter 2.

Genetic and phenotypic diversity are necessary for evolutionary change to occur. We can see phenotypic variation, while, at the same time, the genome is keeping a record of the history of these genetic changes. Some of the origins of the genetic continuity and variation in a genome will be described in detail in Chapter 4.

CHAPTER CAPSULE

- Individuals in a population vary in how many offspring they have that also reproduce, which is known as fitness. Some of the differences in fitness are heritable.
- Evolution by natural selection, or descent with modification, results in certain genotypes becoming more frequent in a population, with others decreasing or disappearing entirely.
- The complete DNA sequences of many genomes are known. A genome records both the evolutionary history and the molecular functions for an organism.
- Genes are made up of DNA. They are copied, inherited, and expressed in living cells. While genes are composed of DNA, not all DNA constitutes genes, and some DNA has no known function.
- Most bacteria have a single circular chromosome, and many bacteria have extrachromosomal circles of DNA known as plasmids.
- Eukaryotic organisms have their nuclear DNA organized into a number of linear chromosomes. Multicellular organisms, and some stages of the life cycle in unicellular eukaryotes, have two copies of each chromosome.
- Some organelles in eukaryotic organisms, such as mitochondria and chloroplasts, contain DNA sequences descended from prokaryotic organisms that lived endosymbiotically within eukaryotic cells in the past.
- Mutations occur in all cells all the time and at all sites in the genome, but most mutations have very little effect on the function of the cell or organism, particularly in higher eukaryotes. This is due to redundancy,

robustness, and the relatively small fraction of a eukaryotic genome that encodes functional genes.

- Although only a small fraction of a mammalian genome consists of protein-coding genes, much of the genome is transcribed into RNA.
- Most of the genes that are found in the genome of one species are related to genes found in the genomes of other species, since organisms arise from common ancestors.
- Genomic changes include changes in the base sequences, changes in the copy number of genes and other elements in the genome, and the introduction of novel genes. These genomic changes affect both morphological differences between individuals of the same species, as well as morphological differences between related species, by changing the amino acid sequences of proteins, the transcription patterns of genes, and the splicing of RNAs.

STUDY QUESTIONS

Concepts and Definitions

3.1 Describe the process of evolution by natural selection. What are the two key facts (highlighted in this chapter) and the inference that Darwin drew from these facts?

3.2 What is meant by the strength of selection, and how does this affect a population? What is meant by the direction of selection, and how does this affect a population?

Looking ahead **3.3** In addition to natural selection, what other processes have an impact on the genetic structures of populations?

3.4 Define the term genome.

3.5 In which organelles is DNA found in eukaryotes?

3.6 Define and compare the terms chromatin, heterochromatin, and euchromatin.

3.7 Define a gene or protein family.

3.8 Define redundancy and robustness, and give an example of each.

Beyond the Concepts

3.9 We write that "A heritable fitness advantage that occurs at any level—a population, a system, an individual, or a protein or another macromolecule—will inevitably result in changes in the frequency of particular DNA sequences." Describe how this fitness advantage (or disadvantage) might be seen for a sequence variation that affects each of the following, and where (in a gene or genome) such as variation might be located.

a. The same gene is transcribed at different levels and at different times and locations in two different individuals of a species.

b. A protein carries out similar functions in two individuals, but not equally well in each individual.

c. Two different species have similar genes, but these are spliced to make slightly different mRNAs.

d. In one species, protein A interacts with proteins B and C, while, in another species, protein A interacts with proteins C and D.

3.10 The genomes for many different species, including dozens of different insects, have now been sequenced. The genome of a praying mantis has apparently not yet been sequenced (as far as we can determine). What information can be learned from sequencing a genome from another insect like a praying mantis that might not already be known? You should answer this generally about the rationale for additional genome sequences for any organism, and more specifically about some particular property of the praying mantis that interests you. Feel free to research the biology of the praying mantis; there are about 2400 different species of the praying mantis, any of which could be used to answer this question.

3.11 Discuss how it has been important to the evolution of species that biological processes that reduce the size of a genome are less efficient than those that increase the size of the genome. What parts of the sweater analogy are most helpful in thinking about genome evolution, and which parts are less helpful or least applicable?

3.12 What are the differences in the organization of the genome within a bacterial and eukaryotic cell? How might the differences in cellular organization of the genome affect some of the basic processes of gene expression and transmission?

3.13 Both bacteria and eukaryotes have parts of their genomes, in addition to the main chromosome; bacteria have plasmids, while eukaryotes have mitochondria and chloroplasts. Compare and contrast these types of "non-chromosomal DNA" with the following questions.

a. Approximately what fraction of the total genome is not found in the chromosome in bacteria, compared to eukaryotes?

b. What are the functions that are encoded by non-chromosomal genes?

c. Is the non-chromosomal genome essential or dispensable?

3.14 Draw a metaphase chromosome, and label the sister chromatids, centromere, and telomeres.

3.15 You have a purebred cocker spaniel. Your neighbor has a purebred greyhound.

a. If the genomes of these two dogs were compared, what types of differences are likely to be encountered?

b. Suppose these dog genomes were compared to the genome of a wolf. What types of differences might be encountered, and how might this be different from what is seen by comparing two breeds of dogs?

3.16 α-actinin is a gene family in humans, with four paralogs. Using online resources, try to determine what is known about some of the fundamental properties of each of the four members of this gene family—that is, how have the functions of the four paralogs diverged from each other (or have they)? What is unusual about α-actinin-3 in humans?

Applying the Concepts

Looking ahead **3.17** Suppose that an individual has a mutation in a mitochondrial gene.

a. What would you predict might be a phenotypic effect of a mutation that affects a mitochondrial gene?

b. What would be distinctive about the inheritance pattern of such a mutation?

3.18 A beginning graduate student decides to try to reconstitute chromatin from *Arabidopsis in vitro*. He has a tube of purified *Arabidopsis* DNA, but he does not have purified histones from *Arabidopsis*. He does have tubes of purified histones from other species, however. He has H1 from humans, H2A and H2B from *Drosophila*, H3 from *C. elegans*, and H4 from maize. He knows the exact salt conditions and the proper pH for reconstitution, so he decides to mix them together. The experiment works, at least somewhat. (This type of reconstruction experiment has been done, although not with these components from these organisms.)

a. What part of the structure of a chromosome do you think he was able to reconstitute? Describe this structure in as much detail as you can.

b. He learns that he can reconstitute the same chromatin structure, even if he leaves out one component. Which component could he leave out? (In other words, which of these components is missing when the structure in (a) is assembled?)

c. The fact that he is able to reconstitute this structure provides some important information about the underlying structure of the chromosome. What are some of the inferences that can be drawn from this successful experiment?

3.19 Chromosome III in *Caenorhabditis elegans* is approximately 13 Mb. At position 10,743,645, one worm has a G:C base pair on both of its chromosomes, while another worm has an A:T base pair.

a. Draw out the chromosomes that would be found in a nerve cell, and show the difference between these two worms.

b. The nerve cell enters mitosis and is preparing to divide. Draw out the chromosomes that would be found in the nerve cell in these two worms.

Looking ahead

c. Suppose that the worm with the G:C base pair mates with the worm that has an A:T base pair. Draw out the chromosomes that would be found in a nerve cell in one of their offspring.

Challenging

d. In a model eukaryotic organism like *C. elegans*, it is entirely possible that this single base change on chromosome III is the only difference in the genomes of these two worms.

i. Briefly explain why that might be true.

ii. Suppose that chromosome III from a worm from nature was sequenced and compared to the sequence of the laboratory strain. Approximately how many base pair changes would you expect to see, and why? What other types of genetic changes in the sequences might you see when comparing a natural isolate of *C. elegans* with the laboratory standard strain?

3.20 A small sample of seawater is taken from deep sea vents, which yields one particular type of bacteria. The genome from this bacterial isolate is shotgun-sequenced, and the sequences are assembled computationally.

a. The sequencing robot used has an average sequence length of 300 bp. The bacterium is estimated to have a genome of 5 Mb. What is the minimum number of sequences that must be assembled to have the complete genome sequence for this bacterium? Assume that an overlap of 30 base pairs is needed at each end to make the assembly.

b. By hand, try to assemble the following three sequences, and determine if these represent one, two, or three separate contigs.

i. GCTAGTCAG..........CAAGTTTCAG

ii. GTACTAGCAT.........CCTATAGA

iii. CCTATAGA............CTGACTAGC

c. Once the genome has been sequenced and assembled, what are the next steps in determining the locations and functions of genes?

d. Shotgun assembly is more complicated for a metazoan (that is, a multicellular organism) genome than in bacteria. List some of the factors that make genome assembly more difficult in metazoans.

e. Annotation of metazoan genomes is also more complicated that annotation of bacterial genomes. List some of the challenges encountered in annotation in metazoans that are not encountered with bacteria.

Looking back and ahead **3.21** The investigators identified the *ALX1* as one candidate gene for affecting beak shape in finches, as described in each of these first three chapters. Let's review this example in more detail, using the concepts you have learned so far.

a. Explain how searching for regions of the finch genomes that were fixed in different species was an appropriate way to begin the analysis.

b. One of these regions contains the *ALX1* gene, but the others have not been investigated. What might be the reasons that sequences in these other regions are also fixed?

c. What important principle, based, in part, upon model organisms, allowed them to focus on *ALX1* as a candidate gene?

d. ALX1 is a transcription factor in the Paired family. Explain what this means.

e. How do the observed changes in the *ALX1* gene and surrounding region of the genome affect the functions of *ALX1*, and how are these changes in functions postulated to affect the shape of beaks?

Challenging and speculative

f. One change was located immediately downstream of the *ALX1* gene. Speculate about possible reasons that this change might have been found. We will return to this question in Chapter 12.

3.22 You make a mutation in a DNA sequence from a mouse and reintroduce that mutated piece of DNA back into the mouse, replacing the DNA that was already present. (The experimental procedures by which these techniques are done in mice go beyond the scope of this book, but this kind of experiment is feasible.) The mouse with the mutated DNA sequence looks exactly the same as a mouse with the original DNA sequence. List all of the explanations for why this particular mutation may not have had a functional effect on the mouse.

3.23 Since the genome of a species of the praying mantis has been sequenced and assembled (see Study Question 3.10) and at least partially annotated, you are interested in how changes related to the *Hox* genes might have contributed to the unique morphology of a praying mantis. You decide to compare its *Hox* gene cluster to those of other insects.

a. What characteristic features of the sequences of the genes (and their predicted protein products) will help you find the *Hox* gene cluster in the praying mantis genome?

b. Describe overall what you expect to find when you examine the *Hox* gene cluster. In other words, what is the approximate number of genes, and how do you expect that they will be arranged?

c. List the types of changes that you can expect for the *Hox* gene cluster of a praying mantis compared to that of another insect, such as *D. melanogaster*. Is there a particular type of change in the *Hox* genes that you do not expect to see when two insects are compared like this?

Biologically challenging question

d. Look carefully at Box 3.2 Figure B, and look at a picture of a praying mantis, if you are not familiar with their distinctive body types. Ideally, you want to study all of the *Hox* genes and their targets to understand the distinctive mantis body type. Practically, you might want to start with the gene or genes that you think might be responsible for the most obvious changes. Although there are many possibilities and evolutionary change is sometimes unpredictable, which *Hox* genes will be your starting point, and why?

Speculative and challenging

e. When comparing the sequences of the genes of *Drosophila* and the praying mantis, you notice that the proteins encoded by the *Labial* genes are almost identical, at least in the relevant part of the characteristic feature from (a) above. However, when looking at the head and mouth parts of *Drosophila* and the praying mantis (which are presumably regulated by the *Lab* gene), the insects are not that similar. Speculate about how changes in the head and mouth parts of the two species might have arisen, even if the regulatory gene itself is not that different.

CHAPTER 4

Descent with Modification: Continuity and Variation in the Genome

IN A NUTSHELL

The survival and evolution of a species depends on its ability to replicate its genome faithfully. Replication occurs accurately, because a new copy of the genome is made using the old version as a template and because repair mechanisms survey the genome for errors during and after replication. However, survival and evolution also depend on variation. Even if it were possible, perfectly accurate DNA replication would not necessarily be advantageous to the organism, since the imperfections in these replication and repair mechanisms are an important source of genetic variation. As a result, every generation has DNA sequence changes not found in the preceding generation, a few of which are favorable, some of which are unfavorable, and many of which are neutral. Because changes occur in every generation at a more or less predictable frequency, the number of changes present at a particular point in time can be used to estimate the time since two genomes last shared a common ancestor.

4.1 Overview: evolution, replication, and repair

The Great Idea of natural selection in evolution depends on a fundamental principle that Darwin himself referred to as "descent with modification," that is, organisms reproduce and their descendants are very similar to the parents but are not exactly the same—they have modifications. As we discussed in Chapter 1, Darwin himself recognized that the finches on the Galapagos Islands were similar to the ancestral species on the mainland but also showed significant differences; this is depicted in Figure 4.1. This principle of descent with modification

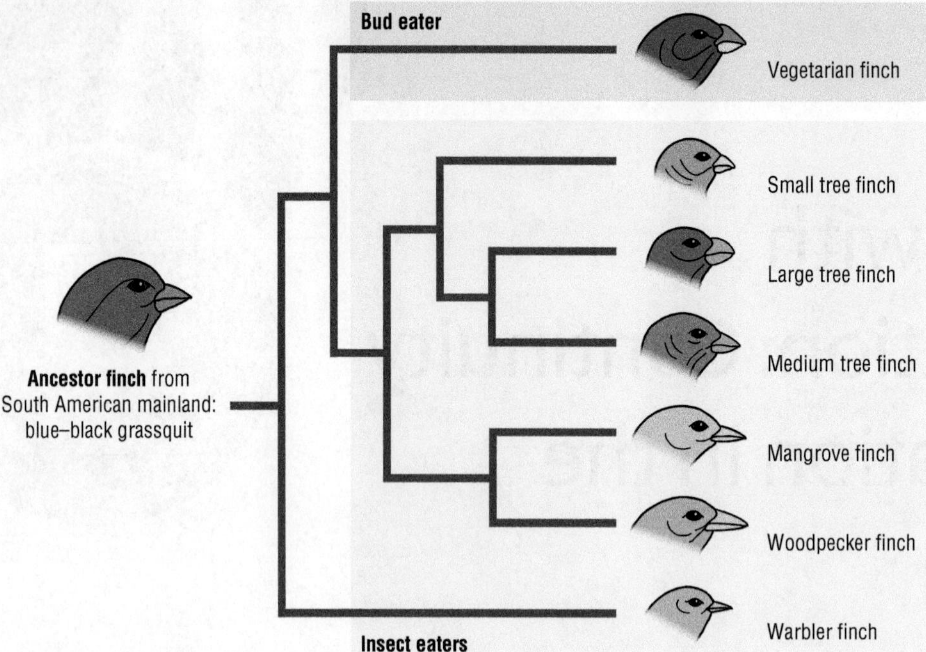

Figure 4.1 Descent with modification. The finches Darwin observed on the Galapagos Islands led him to propose evolution by natural selection, or descent with modification. The ancestral finch on the South American mainland had a blunt beak and dark coloration. When populations migrated to different islands, new species arose from the genetic variation among the finch populations by the effects of selection and small population sizes.
Source: Reproduced from Sadava et al. *Life: The Science of Biology* 8th edition. With permission from Sinauer Associates, Inc.

holds true whether we are looking at families of humans, bacteria growing on our skin, or any other organism. Although we can see this principle by examining the phenotypic similarities and differences of organisms, as Darwin did, it is also true when we look at the underlying DNA sequences of their genomes. The focus of this chapter will be on the principle of descent with modification as it plays out in DNA sequences and genomes that, as we know from Chapters 2 and 3, affect the organism as a whole.

The principle of descent with modification demonstrates the balance between two competing evolutionary pressures. On the one hand, individuals that are well adapted to their environment and have a higher level of **fitness** (that is, an increased rate of survival and reproduction) will be favored by natural selection. Barring other evolutionary pressures or changes to the environment, the genomes of these individuals will then be present at an increased frequency in the next generation. Remember that natural selection has no ability to look ahead and anticipate what properties might be useful in the future. But it provides an outstanding record of what has been successful in the past. Thus, there is a pressure for continuity of the genome from one generation to the next.

On the other hand, if all of the organisms in the next generation are identical to one another because of the need to maintain the continuity of the genome, they will continue to compete directly for resources and be susceptible to all of the same pathogens and predators. Thus, there is also a benefit to producing diversity, so that all of the individuals do not have exactly the same genome sequence.

The balance between the need for both continuity and diversity means that, as depicted in Figure 4.2, successful populations contain variation and continue to diversify within constraints. Diversity is essential for the survival of a species. How then can the genome, which must copy itself correctly, also produce diversity? The answer is that the copying process, or DNA replication, is imperfect—errors do arise, and some of them persist. These errors produce much of the variation on which evolutionary change depends—genotypes that vary in fitness for a given environment.

DNA replication represents an intricately selected balance between the need for continuity and the possible advantages offered by variation. We were careful to write "possible advantages," because most of the effects of this new variation are deleterious. Only a few are the

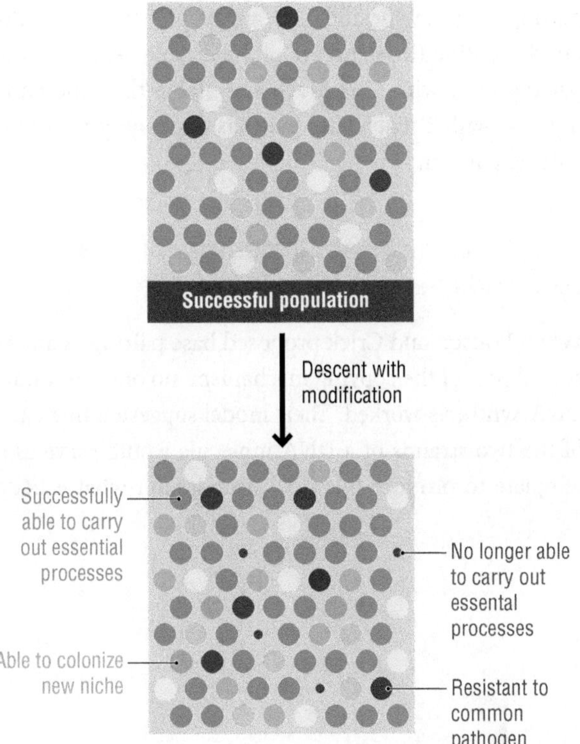

Figure 4.2 Continuity and diversity. Any population has a balance between selection for the continuation of a successful genome and the occurrence of variants that may be beneficial or harmful. Each of the circles represents an individual, with the colors representing variants. As variation occurs for all genes, most individuals will continue to reproduce successfully (in shades of turquoise); some individuals will die out because they can no longer carry out some essential process (in black), and some will acquire variants that confer an advantage if a particular pathogen arises (in purple) or in a certain niche (in green).

advantageous changes that allow an organism to adapt better to its environment, and other changes are neutral in their effects and neither improve nor decrease fitness.

The steps of variation and selection are separate processes. Desirable variants are not anticipated or produced on demand. Rather, random variation occurs first (good, bad, and neutral), and natural selection follows, favoring the variants with the greatest fitness for that particular time and place. By calling this principle "descent with modification," Darwin was unknowingly describing the processes of DNA replication and repair.

KEY POINT DNA replication and repair provide a balance between opposing evolutionary pressures for accurate transmission of the genome to the next generation and the opportunity for production of new genomic variants.

The general molecular mechanisms of DNA replication and repair are highly conserved over evolutionary time, with the same fundamental properties and involving many very similar molecules in bacteria and eukaryotes. However, these processes do differ between organisms in a few ways. As we noted in Chapter 3, the genomes of bacteria and eukaryotes have different structures; most of the differences in DNA replication and repair between bacteria and eukaryotes are therefore adaptations to the different sizes and structures of their genomes. The basic process of DNA synthesis can be readily carried out *in vitro* with relatively few components, at least for small molecules. However, DNA replication in living organisms is more complicated—more elements are needed for genome replication to proceed successfully in the cell than to copy a few thousand bases in a test tube. We will describe the fundamental process as it might be carried out *in vitro*, with further explanations of the processes occurring in the cell and the differences between bacterial and eukaryotic genomes.

KEY POINT DNA replication and repair are highly conserved among organisms. Variations in the processes of DNA replication and repair that do exist relate primarily to differences in genome size and structure.

4.2 Continuity from templates

One of the Great Ideas in the Prologue identifies the gene as the unit of heredity. We will discuss what genes are and what they do, together with how they are transmitted to progeny, throughout the rest of this book. But you can imagine that DNA must be copied if the information it contains is to physically pass from parents to offspring. In eukaryotes, this copying takes place during the S phase (DNA synthesis phase) of the cell cycle, which we will discuss further in Section 6.2. Replication and cell division are also connected processes in bacteria, but the bacterial cell cycle is very different and less well understood.

The key feature underpinning DNA replication is the existence of the base pair—one of the clearest connections between structure and function in all of biology. As we noted in Chapter 2, Watson and Crick made this connection when they concluded their manuscript with the statement:

> It has not escaped our notice that the specific [base] pairing we have postulated immediately suggests a possible copying mechanism for the genetic material.
>
> (Watson and Crick, 1953)

Watson and Crick's conclusion was referring to the hydrogen-bonding interactions between adenine and thymine and between guanine and cytosine that keep the complementary strands of DNA together—the base pairs shown in Figure 4.3. Remember that this base pairing is quite specific—A cannot pair with G without disrupting the structure of the double helix, and the pairing of A with C is much less stable than the pairing of A with T. Each base pair has one highly preferred pairing partner.

The Meselson–Stahl experiment and semi-conservative replication

When Watson and Crick proposed base pairing as an essential part of the copying mechanism, no one knew how DNA synthesis worked. Their model suggested that each of the two strands of a DNA molecule would serve as a template to produce two new strands. But what exactly

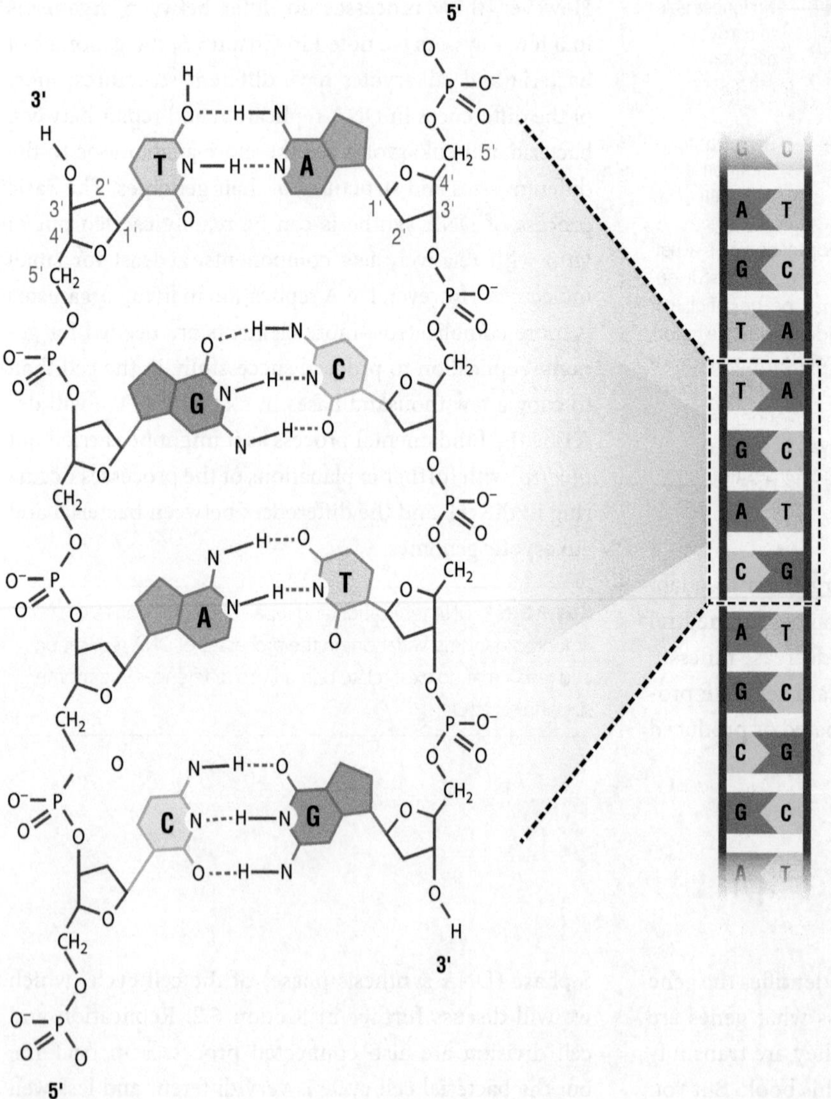

Figure 4.3 DNA structure and base pairing. Adenine pairs with thymine via two hydrogen bonds, and cytosine pairs with guanine with three hydrogen bonds. Guanine and adenine are purines, while cytosine and thymine are pyrimidines.

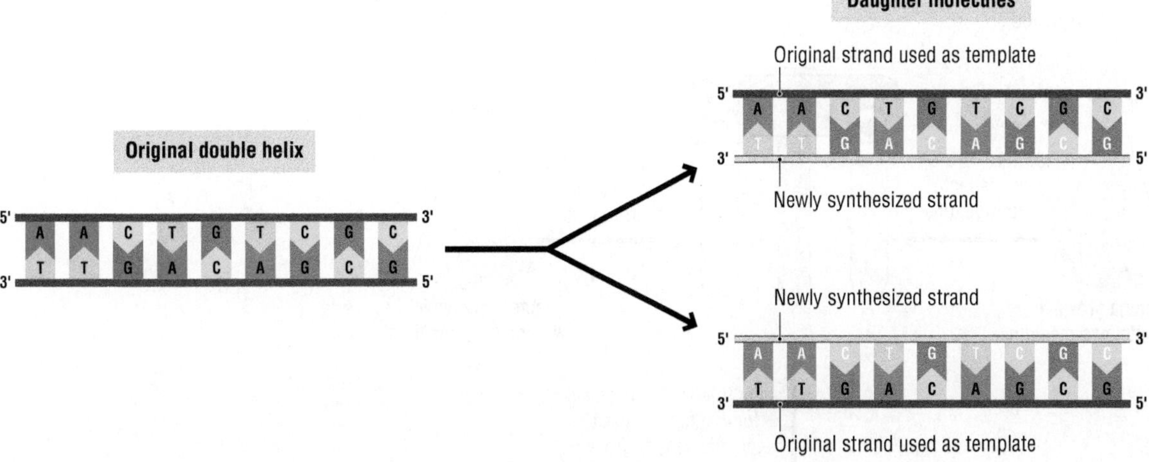

Figure 4.4 Semi-conservative replication. The Meselson–Stahl experiment demonstrated that replication resulted in daughter DNA double helices comprising one parental strand and one newly synthesized strand. The newly synthesized strand is shown with a gray backbone. Note that the two daughter molecules are identical because each strand is used as a template.

would be the composition of the newly replicated DNA double helix? In principle, at least two models for a template-driven process are possible, one in which the new DNA double helix consists of two newly synthesized strands and the other in which the new DNA double helix consists of one new strand and one of the parental template strands. (Other models were also envisioned but are of interest primarily for historical reasons.)

One of the most beautiful experiments in molecular biology, done by Matthew Meselson and Franklin Stahl, demonstrated that DNA replication is semi-conservative, that is, the outcome of replication is two "daughter" double helices, each of which would comprise an old strand from the parent and a newly synthesized strand, as shown in Figure 4.4. These experiments are described in Box 4.1. Meselson and Stahl's work revealed that each strand of the original molecule is used as a template and ends up in one of the two replication products, paired with a newly synthesized strand. Consider this for a minute. If no other process occurred that shuffled or rearranged the DNA strands, then one of the strands of DNA in one of your cells in generation **n** came from a DNA double helix in generation **n−1**, which, in turn, came from one in generation **n−2**, then from generation **n−3**, from generation **n−4**, and so on, back to the original DNA helix. (As we will see in subsequent chapters, processes that break,

BOX 4.1 *An Historical Perspective* The Meselson–Stahl experiment and semi-conservative replication

Semi-conservative replication yields two DNA molecules, each comprising one of the original "old" strands and one newly replicated strand. This may seem so fundamental that we overlook the experiment that proved that this was the mechanism. The experiment, done by Matthew Meselson and Franklin Stahl in 1958, was among the most conceptually simple and beautiful in the annals of molecular biology research. Few experiments produce such clear results that anyone can simply look at the data and accurately conclude, "Now I understand how that works." This is one such experiment.

The experiment took advantage of a newly developed technique called analytical ultracentrifugation, in which molecules could be separated based on their density. A gradient of the salt cesium chloride (CsCl) was prepared in a tube; the macromolecule, such as DNA, was added to the salt gradient, and the tube was spun at very high speeds (around 70,000 rpm) in the ultracentrifuge until the molecule reached the equilibrium point—the point at which the density of the salt gradient corresponded to the density of the molecule. Meselson and Stahl used the ability of analytical ultracentrifugation to

BOX 4.1 Continued

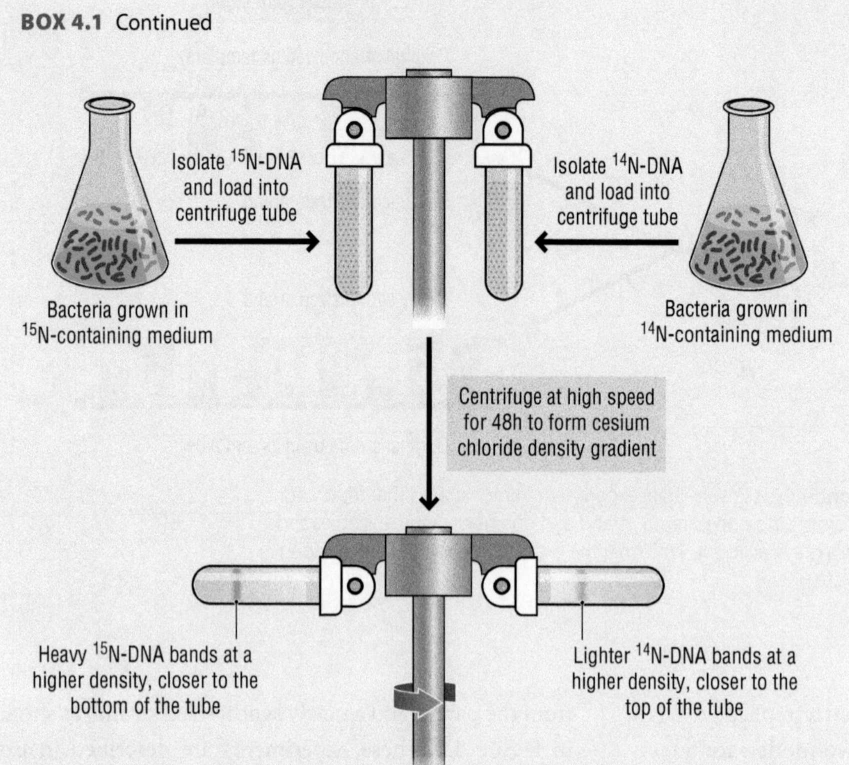

Figure A The technique used to demonstrate semi-conservative replication. Bacteria are grown in medium containing the heavy isotope ^{15}N, rather than the normal isotope ^{14}N. The DNA is isolated and separated by density, using analytical centrifugation.

Source: *Essential Cell Biology* (© Garland Science 2010).

separate DNA molecules to demonstrate that replication was semi-conservative.

The bases in a DNA molecule include several nitrogen atoms; in nature, nearly all of these are the naturally occurring nitrogen ^{14}N. Meselson and Stahl grew *Escherichia coli* cells in medium containing only the heavier isotope ^{15}N for multiple generations, so that all of the cells had DNA containing ^{15}N, rather than ^{14}N, as shown in Figure A. The cells grew well, and their DNA showed a higher density in the salt gradient than the "standard" DNA containing ^{14}N.

In order to investigate the mechanism of replication, cells were grown in medium with ^{15}N until all of their DNA was labeled, at which point the cells were switched to a standard medium with ^{14}N for enough time for one round of replication to occur. After this single round of replication, some of the cells were removed, and their DNA was extracted and analyzed for its density on CsCl. The results are shown in Figure B. The newly replicated DNA formed a single band on the gradient, which was intermediate in density between the ^{14}N-DNA and the ^{15}N-DNA. Thus, after one round of replication, all the DNA consisted of one ^{15}N ("old") strand and one ^{14}N ("new") strand.

We return to these results in a study problem at the end of the chapter.

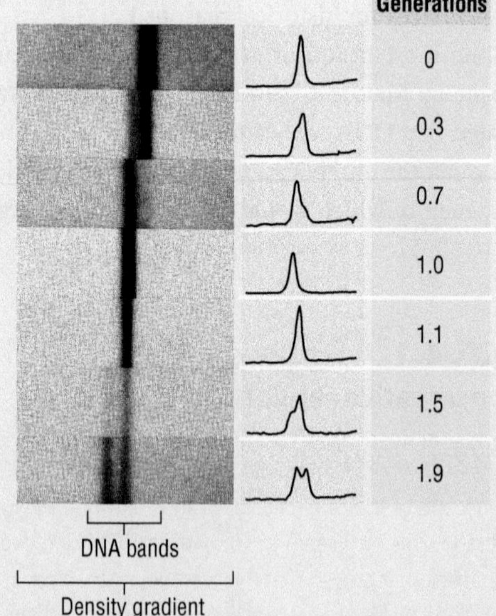

Figure B The evidence for semi-conservative replication. The DNA is isolated and separated by centrifugation at different times, with the results shown. At time 0, a single band is observed at the high ^{15}N density. By the end of the first generation, a single band is seen at a slightly lower density, demonstrating that the DNA has one strand labeled with ^{14}N and one strand labeled with ^{15}N.

Source: Meselson and Stahl (1958) *PNAS*, 44, 671.

shuffle, and reunite DNA strands occur during each generation.) When it comes to DNA strands, we always carry the past with us; in fact, it is not even the past since we are still using it.

KEY POINT DNA is replicated semi-conservatively—every DNA molecule generated during replication includes one old strand and one newly synthesized strand.

Base pairing drives template-directed synthesis of new strands of DNA

Hydrogen bonds between the nucleotide bases of the two separate DNA strands are strong enough to maintain the double-stranded structure of DNA in a cell but are much weaker than the covalent bonds that hold the nucleotides together within a strand. This balance between the hydrogen bonds being strong enough to maintain the structure but not so strong that DNA strands are locked together is important. Strong covalent bonds are essential to maintain the primary structure of each strand of DNA, but weaker interactions between the paired strands allow the strands to separate for copying under appropriate conditions. Much less energy is needed to separate strands in a double helix than to disrupt the covalent bonds that keep the nucleotides strung together. As a result, if DNA is heated to near boiling *in vitro*, the double helix falls apart, or **denatures**, into two single strands, but each strand and its nucleotide sequence remains intact and stable, as shown in Figure 4.5. As we shall see in Section 4.3, changes in temperature are not the mechanism used to separate the strands within a cell, and the DNA strands do not separate along their entire length during DNA replication in the cell, but the strands do come apart in local regions, known as a replication bubble, to allow replication to occur.

Once the two strands are locally separated from each other, the nucleotides along the single-stranded DNA can **anneal** to their complementary bases; this association is stabilized by the formation of hydrogen bonds between them. These bonds will form according to the chemistry of the bases—A will bind with T, and G with C. Consequently, the base pairs will exactly mirror those in the original molecule, as shown in Figure 4.4. In other words, the two single strands have acted as templates for the synthesis and re-assembly of the double helical structure, forming two copies of that double helical structure in the process—the original DNA helix has been replicated (if we are describing the process within the cell), or new DNA molecules have been synthesized (if we are describing the process *in vitro* or more generally).

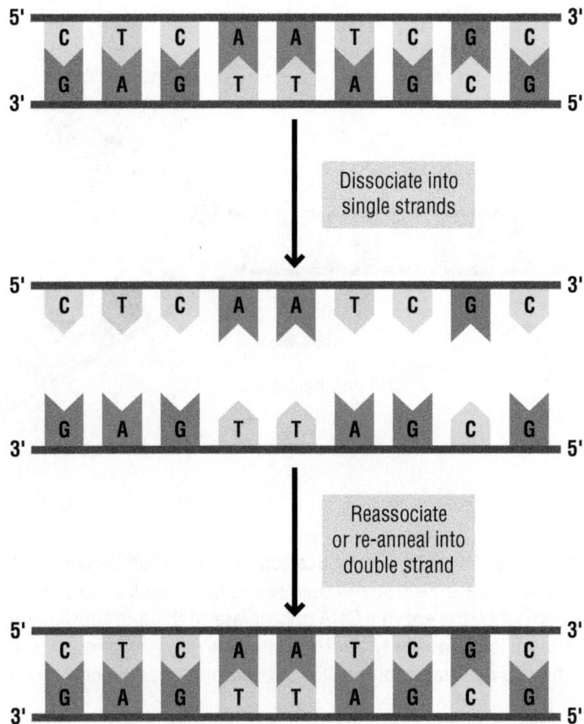

Figure 4.5 Annealing of DNA strands. DNA strands can be separated, becoming single-stranded along their entire length (which happens by heating the molecule *in vitro*) or in local regions (which happens in the cell). These single strands will re-associate or anneal into double-stranded DNA, as bases pair again once the solution is cooled, for example.

It is the chemistry of DNA, specifically the hydrogen bonding of purine and pyrimidine bases described in Chapter 2, that allows the genetic material to serve as a **template** for its own synthesis, as Watson and Crick recognized. The template-directed synthesis of DNA may be the most fundamental point of connection between the Great Idea of the gene as the unit of heredity and the Great Idea of life as depending on chemical processes.

DNA polymerase synthesizes the new strands

Base pairing accounts for how the fidelity of the copying of a DNA sequence is achieved, whether *in vitro* or *in vivo*. During DNA replication in the cell, however, the strands are not heated to near boiling to separate them. Instead, replication is initiated at regions known as replication origins by the action of proteins. As part of the initiation process, the strands are separated by an enzyme known

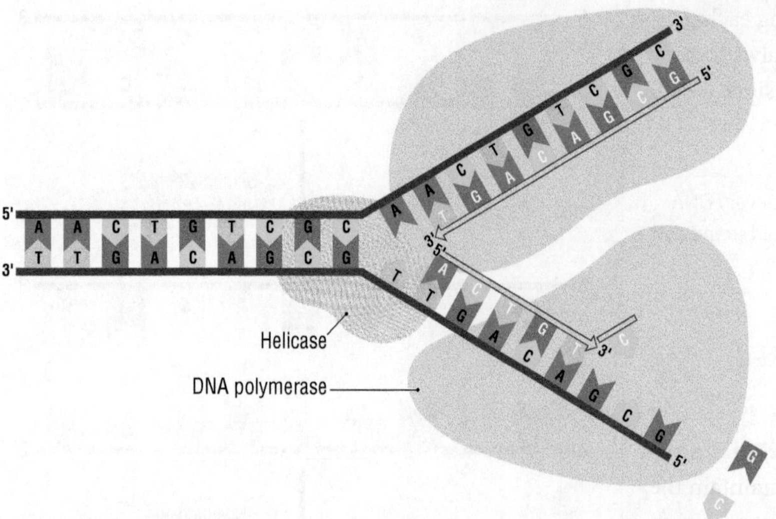

Figure 4.6 Overview of replication. As replication begins, the two strands of DNA are separated by the enzyme helicase. Each strand then serves as a template for the synthesis of a new strand of DNA, a process carried out by the enzyme DNA polymerase. In this and subsequent drawings, the original strand is shown in black, while the newly synthesized strand is gray. The arrows indicate the direction of polymerization. Both of the original strands of the DNA are templates for the new strands, which preserves the continuity of the information in the sequence.

as DNA helicase, as shown in Figure 4.6. Each resulting piece of single-stranded DNA then acts as a template for the **semi-conservative** synthesis of a new strand, resulting in two new double helices, each comprising one old strand and one new strand, as was shown in Figure 4.4. The key enzyme for the synthesis of the new strands is **DNA polymerase**.

KEY POINT During DNA replication, the strands of the double helix are unwound by DNA helicase. The chemistry of DNA base pairing allows each strand of DNA to serve as a template for the recruitment of complementary bases. These bases are then joined together into a newly synthesized strand of DNA by the enzyme DNA polymerase.

In one sense, DNA polymerase has one job to perform—to catalyze the joining of each incoming nucleotide to a second nucleotide located at the end of a growing complementary chain. As the name suggests, polymerases convert single nucleotide residues into polymeric chains of many nucleotide bases. Specifically, they catalyze the formation of a covalent bond between the OH group on the 3′ carbon of the deoxyribose sugar of the nucleotide at the end of a DNA chain to the phosphate group on the 5′ carbon of an incoming nucleotide triphosphate, as shown in Figure 4.7. This covalent bond is called a **phosphodiester** bond. Once the reaction is complete, the phosphodiester bond attaches the new base to the nucleotide chain, and a pyrophosphate molecule ($2PO_4$) is released.

Thus, the building blocks for DNA synthesis are nucleotide triphosphates, with DNA being synthesized in the 5′ to 3′ direction; the phosphate on the 5′ carbon of the sugar of the incoming nucleotide base is attached to the -OH on the 3′ carbon of the sugar on the preceding base.

Although we have referred to DNA polymerase as if it were a single enzyme, there are many different, but highly related, versions of DNA polymerase; these perform

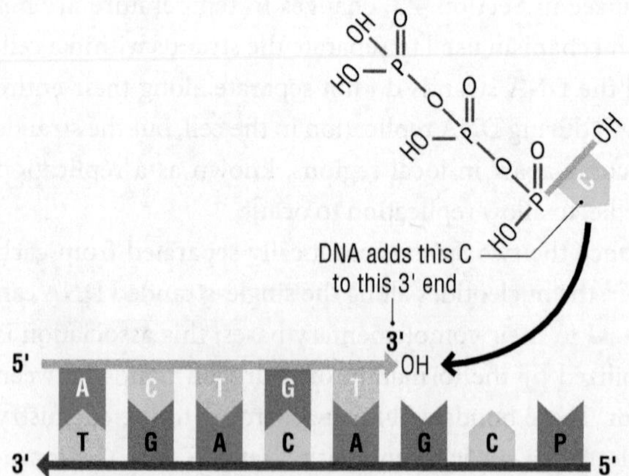

Figure 4.7 DNA polymerase adds new bases only at the 3′ end. The addition of bases occurs at the 3′ end of the DNA strand, with DNA polymerase catalyzing the formation of a covalent bond between the 3′ OH group on the deoxyribose sugar of the base at the end of a DNA chain and the phosphate group closest to the sugar on the 5′ OH of an incoming nucleotide triphosphate. Although the polarity of replication can be confusing, the key principle is that new bases are only added to the 3′ OH of the preceding base.

Table 4.1 Some DNA polymerases

Activity	Bacterial polymerases	Eukaryotic polymerases	Description
Most of DNA replication	Pol III	Pol α, δ, and ε	In eukaryotes, a complex of primase and Pol α is responsible for the synthesis of the RNA primer and the first 20 or so bases on each strand before polymerase switching occurs. Pol ε synthesizes the leading strand, while Pol δ synthesizes the lagging strand. In bacteria, primase is a separate enzyme, and two molecules of Pol III are responsible for the synthesis of both strands
Processing Okazaki fragments	Pol I	Pol δ and ε	Pol I removes the RNA primer and fills in the gap in bacteria. In eukaryotes, Pol ε and δ synthesize past the primer at the next replication fork, and other proteins remove the primer and the overlapping region
Repair	Pol II, IV, and V	Pol β, λ, μ, σ, η, ι, and κ	All of these are error-prone polymerases. Pol II, IV, and V are part of the SOS response. Pol β is used in base excision repair, while pol λ and μ are used in double-stranded break repair. Pol η is used in nucleotide excision repair, following UV damage. The roles of Pol σ, ι, and κ in DNA repair are not fully understood. Pol σ may be involved in double-stranded break repair. Pol κ and ι may be involved when replication forks stall at the sites of DNA damage
Mitochondrial DNA synthesis		Pol γ	Pol γ has similarity to Pol I from bacteria

similar functions under different circumstances, as summarized in Table 4.1. DNA synthesis is important not only for the overall replication of the genome but also for repairing damaged DNA. The DNA polymerases in bacteria and eukaryotes are also not precisely the same; the DNA polymerases in the cellular organelles with their own genomes, that is, mitochondria and chloroplasts, are also distinct from the other polymerases. DNA polymerase III (Pol III) is the principal enzyme involved in general replication in bacteria, while DNA polymerases α, δ, and ε play a similar role in eukaryotes.

KEY POINT A variety of DNA polymerases exist, but they all catalyze the formation of a phosphodiester bond between adjacent nucleotides in the newly forming strands of DNA. The phosphodiester bond is formed between the 3′ OH end of the sugar group of the growing DNA chain and the 5′ phosphate end of the sugar group of the next nucleotide. DNA synthesis therefore proceeds in a 5′ to 3′ direction.

The process we have just described essentially captures how DNA is copied in the cell. The entire reaction is so simple that it can be done in a test tube, using heat to separate double-stranded DNA. This forms the basis of a laboratory technique known as the polymerase chain reaction (PCR), described in Tool Box 4.1. Replication works exceptionally well and is remarkably fast. In eukaryotic cells, DNA replication occurs at an average rate of 2000–3000 base pairs per minute; in bacteria, replication can occur even faster, at a rate of about 800 base pairs per second or more than 45,000 base pairs per minute. Many factors affect these rates, and the rates of synthesis achieved experimentally may not be the same as the rates for replication *in vivo*, but the process is rapid. Of course, it has to occur rapidly in order for all the DNA in a cell to be replicated within every cell division cycle. The rapid rate of replication has some important implications, as we will discuss in Section 4.4.

Replication origins

Replication begins at specialized sites in the genome known as origins of replication. Most bacteria have a single origin of replication, as depicted in Figure 4.8, and replication progresses in both directions from this site around the circular chromosome. Eukaryotes have multiple origins of replication, even at the level of individual chromosomes. In fact, there are probably nearly 10,000 origins of replication in the human genome. Some simple arithmetic will illustrate one reason why so many

TOOL BOX 4.1 The polymerase chain reaction

The premise of the technique known as the polymerase chain reaction (PCR) is that very low concentrations of DNA are specifically amplified or synthesized *in vitro*. PCR is used to make enough DNA from a region of interest to yield detectable and useful amounts of a specific sequence. The process requires that DNA sequences flanking the region of interest are known. The steps involved in this process are best understood by thinking through an example.

Figure A shows a stretch of DNA that will be used as the template for our hypothetical PCR. The first step in PCR is to design two **primers**. These are short stretches of single-stranded DNA (usually 15–25 bases in length) that can be synthesized in the laboratory or by a commercial vendor. The first primer matches the sequence at the 5′ end of the region of interest on the upper strand. Of course, which of the two strands of DNA is designated as upper or lower depends on which way the double-stranded DNA is drawn. This will be arbitrary, unless the DNA corresponds to a region that codes for a protein, in which case the convention is to write out the DNA sequence so that the code can be read directly from the upper strand. This first primer is often referred to as the "forward primer." The second primer is taken from the other end of the region of interest, matching the sequence on the other strand and is often referred to as the "reverse primer." These primers are shown in blue in the figure. The primers provide the 3′ OH needed to begin polymerization, and they target DNA amplification to a specific region.

The reaction itself involves three steps. In step 1, the primers and template DNA are mixed together and heated to near boiling. As described in the chapter, this heat will break the hydrogen bonds that hold the two strands of template DNA together, so that all of the DNA in the PCR will now become single-stranded. This is the **denaturation** step.

In step 2, this DNA mixture is cooled, and the hydrogen bonds between complementary bases naturally start to reform. Since the concentration of the two primers is in vast excess, compared to the concentration of the template DNA, the base pairing most likely to occur will be between a primer and the template DNA (rather than between the two template strands), as shown in Figure A. The binding of the primers to the template is the **annealing** step.

The reaction mix also includes a DNA polymerase enzyme and all of the free nucleotide triphosphates dATP, dGTP, dCTP, and dTTP, collectively called the dNTPs. In step 3, the DNA polymerase carries out the normal function of a polymerase—to extend the primer sequences by combining bases in a 5′ to 3′ direction, following the sequence of the template strand. This extension step will proceed until the end of the template strand is reached or until the reaction is heated again.

The whole process of denaturing, annealing, and extending is then repeated (represented by step 4 in Figure A), except this time there will be twice as many long stretches of DNA to serve as template than there were in the previous round. Thus, the amplification is an exponential process, doubling the concentration of amplified DNA in each round of the PCR. The polymerization becomes a chain reaction, since it uses the products of the previous round of synthesis as the templates for the next round of synthesis. The newly synthesized strands will also always begin with a primer sequence, so before long these newly synthesized strands far outnumber the copies of the original template DNA. The end product of PCR is therefore a high concentration of DNA of specific length (sometimes called an amplicon), corresponding to the distance between the two primer sequences.

One of the keys to the success of PCR is the use of a thermostable DNA polymerase taken from thermophilic bacteria that live in high-temperature thermal vents. Thermostable DNA polymerases are able to withstand the 20–30 cycles of heating and cooling involved in a typical PCR. The first and most widely used polymerase was taken from *Thermus aquaticus* and called *Taq* polymerase. As in nature, the synthesis of the new strands of DNA in PCR is not an error-free process, and an incorrect nucleotide can sometimes become incorporated. *Taq* polymerase lacks the "proofreading" activity that can help detect and correct such errors as they occur. Therefore, other thermostable DNA polymerases that possess some proofreading activity, such as *Pfu* and *Pfx*, are now widely used, especially when the fidelity of the amplified sequence is more critical. It is a trade-off. *Taq* is error-prone but fast; *Pfu* and *Pfx* make fewer errors but are slower.

How can we optimize the PCR?

Although the logic of PCR is fairly straightforward, obtaining a successful amplification can sometimes be tricky. Parameters that are adjusted to optimize a PCR include the selected primer sequences (trying to minimize the chance that they may base pair to themselves, to each other, or to regions of the template DNA other than the desired target sequence) and the temperature at which the reaction is held for the primers and template to anneal to one another. If the annealing temperature is too high, it is harder for the primers and target DNA to maintain their hydrogen bonding; if it is too low, incorrect hydrogen bonding to other regions of the template DNA that are a close, but not perfect, match to the primer sequence might be tolerated. The precise salt and buffer composition of the reaction is also sometimes adjusted to maximize correct annealing and extension. The extension time for the reaction must also be set to allow enough time to fully extend the newly synthesized strand along the entire region of interest but should not be so long that additional or inappropriate extension and synthesis might occur.

PCR works best for shorter stretches of DNA; the amplification of regions longer than about 5 kb can be more challenging. The

TOOL BOX 4.1 Continued

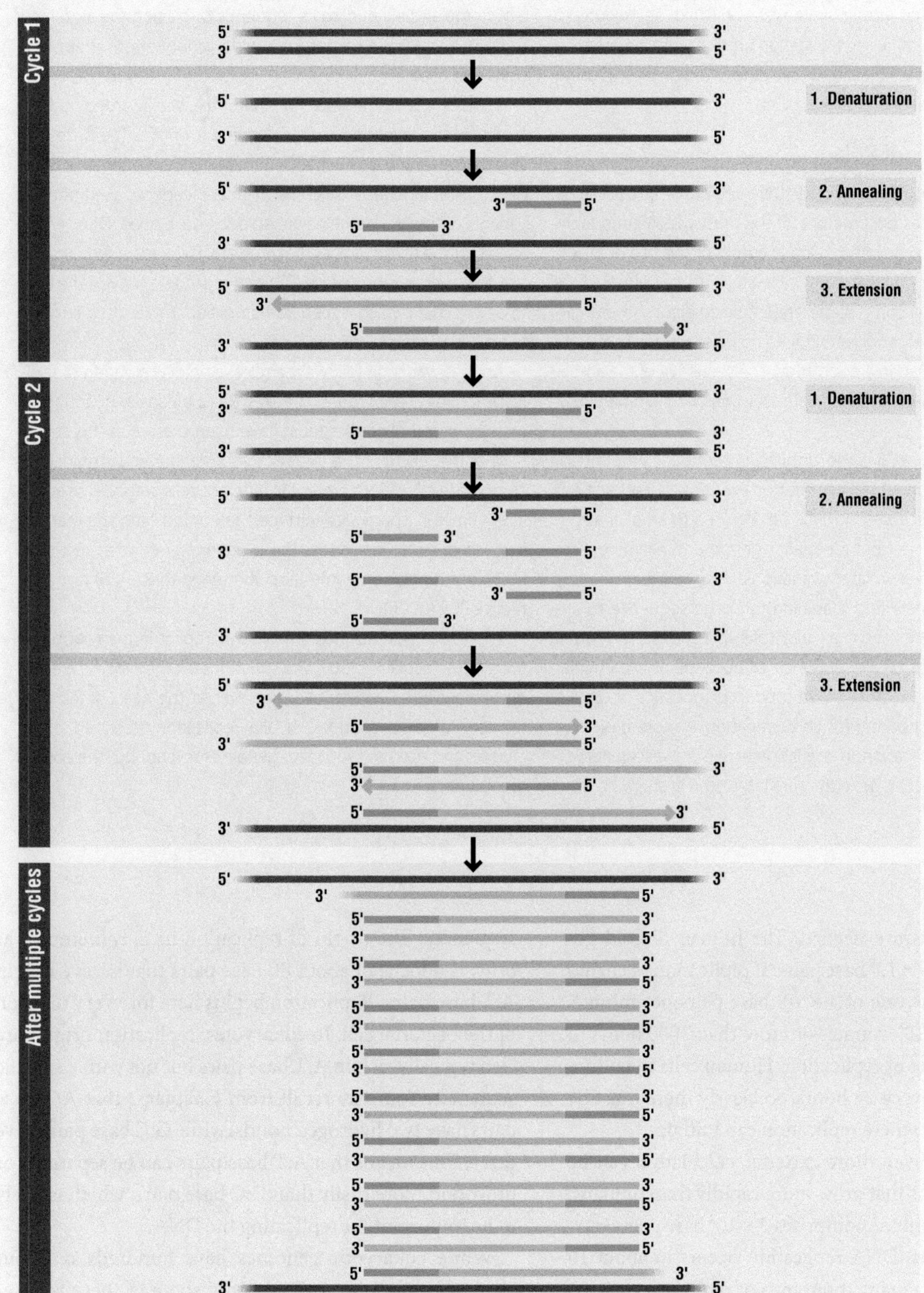

Figure A An overview of the polymerase chain reaction or PCR. Double-stranded DNA to be amplified is denatured (1). Primers complementary to the sequences flanking the region to be copied are added in excess. These primers anneal to their targets (2), and *Taq* (or another) polymerase is used to amplify the region (3). The product is again denatured by heat and then cooled so that the primers can anneal to their targets for the second round. With each subsequent round, more target sequences for the primers are created, so a chain reaction results, synthesizing much more of the DNA product.

TOOL BOX 4.1 Continued

rule of thumb is that *Taq* can synthesize about 1 kb per minute, so a region of 3 kb requires an elongation time of slightly more than 3 minutes. An entire PCR with 30 cycles for a region of 3 kb using *Taq* polymerase takes only a few hours.

The many uses of PCR

PCR amplification of a particular DNA region of interest is so useful that PCR has become one of the most widely used techniques of molecular biology. One example of the use of PCR is in DNA fingerprinting, whereby trace amounts of DNA, such as may be found at a crime scene, are amplified across several target regions of the genome that are known to vary in sequence between individuals in a population. The sequence of the amplified template DNA can then be precisely matched to a particular individual.

Another use of PCR is the amplification of trace amounts of DNA from environmental samples or even fossils and ancient remains. In a clinical setting, PCR can be used to amplify a specific stretch of DNA from a patient's genome to examine it for possible disease-associated variations. Also, since PCR only works if the primers find matching template sequence to anneal to, PCR can be used in a variety of applications that require a simple read-out of the presence or absence of a particular DNA in a sample, such as those used in diagnostics of certain infections or in checking for contamination. A more basic and widely used application in the laboratory is for researchers to amplify the particular gene they are interested in studying from a genome and examining its sequence or placing the amplified DNA into a plasmid vector for a wide variety of other applications and uses.

In 1993, the Nobel Prize for Chemistry was awarded to Kary Mullis for the invention and development of PCR. The prize was shared that year with Michael Smith for his development of site-directed mutagenesis. Since its original inception, a number of variants of the technique have also been developed. One variant allows for the use of RNA as the template. This variant requires the use of a reverse transcriptase (RT) enzyme to first convert the RNA to a DNA copy, which is then PCR-amplified. This is often referred to as reverse transcriptase PCR or RT-PCR.

Another variation follows the release or incorporation of fluorescent dyes in each round of the PCR as a means to quantify the amount of template that must have been present at the beginning of the reaction. The two techniques are often combined to compare levels of RNA present in different cells or tissues or at different times in a particular cell type, thereby indicating differences in levels of gene expression. This is often referred to as real-time PCR (confusingly sometimes also shortened to RT-PCR) or quantitative PCR (qPCR).

The use of PCR has had a profound impact in many applications of molecular biology, and variations on the process are still being developed and refined. However, at the heart of PCR lies the very same basic process of DNA replication described in this chapter and used to propagate genomes throughout the course of evolution.

---

replication origins are needed. The human diploid genome comprises 6×10^9 base pairs. If replication occurred at the approximate rate of 3×10^3 base pairs per minute, it would take 2×10^6 minutes or more than 3000 hours to complete one cycle of replication. Human cells in culture can divide once every 24 hours, so clearly there must be thousands of sites where replication can initiate.

A similar, but even more extreme, calculation can be used for organisms that grow more rapidly than humans. The *Drosophila* genome comprises 2×10^8 base pairs; during embryogenesis, DNA replication occurs in about 10 minutes, so clearly many thousands of replication origins are used. Cell cycles vary widely in length, but the rate of replication itself is relatively constant. The variation in the time to complete the S phase arises generally from the number of replication origins that are used, rather than speeding up or slowing down the rate of replication.

Replication origins represent specific sites in the genome, but these sites don't necessarily feature specific sequences. The origin of replication in *E. coli*, known as *oriC*, is a locus of about 80 base pairs that is very rich in A:T base pairs. Replication begins here for every division of the bacterial cell. In eukaryotes, replication origins are also typically rich in A:T base pairs but not with a specific sequence. You may recall from Chapter 2 that A:T base pairs have two hydrogen bonds, while G:C base pairs have three. This means that A:T base pairs can be separated or unwound more easily than G:C base pairs, which is likely to be important for replicating the DNA.

While eukaryotic genomes have hundreds or thousands of replication origins, only some of these replication origins are used for each cell cycle. We don't yet know what determines which replication origins are used and how these are coordinated with each other. In addition to the A:T sequence, it is clear that the structure of the chromosome at the origin is also important; this structure depends, in part, on the underlying nucleotide sequence but more significantly on the organization and composition

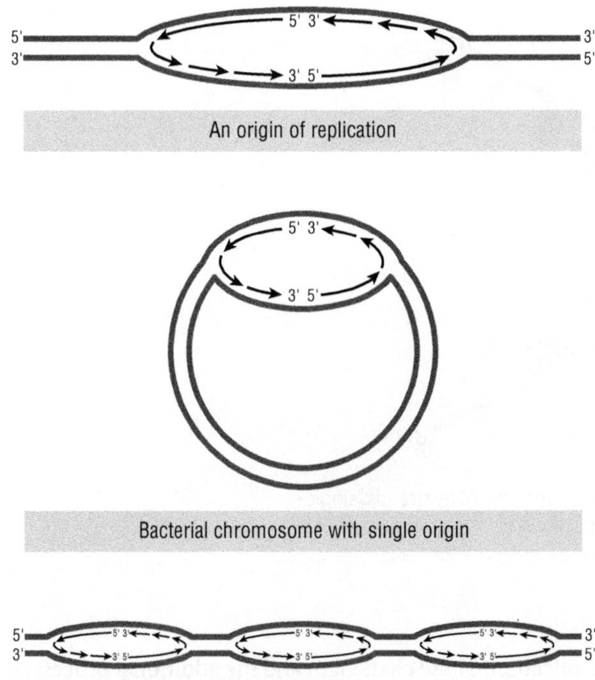

Figure 4.8 Replication begins at specific sites in the genome. The diagram at the top shows DNA replication at a single origin, known as a replication bubble. The *Escherichia. coli* chromosome has a single origin of replication, known as *oriC*, as shown in the middle. Eukaryotic chromosomes have multiple origins of replication on each chromosome. Note that both strands are synthesized at each origin, so replication is bidirectional.

of the histones and other proteins found there. (As mentioned in Chapter 3, DNA in eukaryotes exists as a complex with histones and other proteins, collectively called chromatin.)

While DNA synthesis *in vitro* under perfect conditions requires only a few components, DNA replication inside cells requires a number of other proteins, as summarized in Table 4.2. For example, once the origin is exposed, helicases unwind the DNA and separate the two strands from each other. This forms a replication bubble in which the two DNA strands are separated locally, and single-strand regions of DNA occur. Proteins known as single-strand binding proteins bind to the exposed single strands to keep them separate long enough for copies to be made, as shown in Figure 4.9. Together, these and other proteins constitute a **replication machine**, summarized in Table 4.2, that moves along each **replication fork** as the DNA is copied. The fork "handle" is a double helix awaiting replication, and its two prongs are the two double helices that result from copying each of the original strands.

The replicating machine's key components in bacteria are two molecules of DNA polymerase III, which work together, each replicating one strand of a separated double helix. In eukaryotes, different polymerases are used at different steps in the process; polymerase α is responsible for the replication of the first 20 or so bases, before polymerases δ and ε take over, each of which synthesizes one of

Table 4.2 Some key accessory proteins for DNA replication

Protein(s)	Function and description
DNA helicase	Unwind double-stranded DNA to make single strands accessible
Topoisomerases	Relax supercoiling in the double-stranded DNA ahead of the replication fork to relieve torsional stress. At least five different categories are known, with slightly different activities
Single-strand binding protein (SSB)	Bind unwound single strands at the replication fork to prevent their reannealing. *Escherichia coli* has a single SSB, while eukaryotes have a complex of three subunits known as replication origin A
Origin-binding proteins in *E. coli*	At least ten different proteins are involved. DnaA binds to the origin (*oriC*) site on the DNA, and local unwinding occurs. DnaC binds to facilitate binding of helicases, including DnaB. DnaB progressively unwinds the strands, while other replication proteins bind to DnaB
Origin-binding proteins in eukaryotes (yeast)	The replication origin is bound by the origin recognition complex (ORC) of six proteins. In conjunction with the binding of Cdc6 and Cdt1, the ORC forms the pre-replication complex (PRC), which loads the minichromosome maintenance complex (MCM) consisting of six proteins that include helicase activity. While this is best studied in the yeast *Saccharomyces cerevisiae*, the process is likely to be highly similar in other eukaryotes
Clamp loader and sliding clamp	Bind at the 3′ OH after the RNA primer has been made. The sliding clamp is a ring-shaped protein complex that keeps DNA polymerase attached to the DNA during replication. Replicating DNA passes through the center of the ring. The clamp loader is a protein complex of five subunits that opens the sliding clamp, uses topoisomerase activity to thread the 3′ OH of the primer through the opening in the clamp, and then closes the clamp
Ligase	Connects the Okazaki fragments on the lagging strand

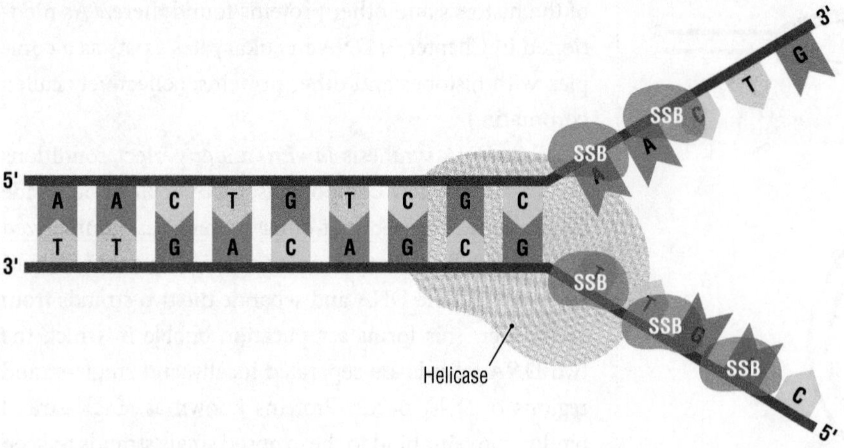

Figure 4.9 Beginning replication. Once helicase has unwound and separated the DNA strands, single-strand binding proteins (SSBs) keeps the strands apart long enough for replication to occur. Additional proteins are added to make a multi-enzyme replication machine that moves along the replication fork.

the two strands. Despite the differences in polymerases, the overall process is highly similar in bacteria and eukaryotes.

Although DNA replication in the cell is a complex and coordinated process, it is important to remember that the sequence of the DNA being copied is determined by the strand providing the template for the reaction. DNA polymerase simply catalyzes the formation of phosphodiester bonds, and the numerous other required proteins aid DNA polymerase in its task.

DNA replication involves more than base pairing and polymerase

The Meselson–Stahl experiments provided compelling evidence for how DNA replication uses complementary base pairing with each strand as the template for the synthesis of new DNA molecules. We know that replication of both strands occurs at a fork, because these structures have been visualized by electron microscopy; such an image is shown in Figure 4.10.

Within a few years of the structure of DNA being elucidated, most biologists were convinced that DNA was copied by strand separation at a replication fork, base pairing, and then phosphodiester bonding, as we have described so far. However, before this very simple mechanism for DNA replication was universally accepted, additional experimental evidence was necessary to supply a more complete picture of the process of replication and the events at the replication fork. This additional evidence is important because most DNA molecules are very long; the scale and peculiarities of the copying process present challenges to replication. These challenges and the additional processes that address them are summarized in Figure 4.11 and described in the succeeding sections.

1. Challenge 1: How does DNA polymerase get started? **Priming**: DNA polymerase can only add nucleotides to an existing chain; there is no way for the enzyme to begin replication on its own. This problem is solved by beginning replication with short RNA primers, synthesized by the RNA polymerase primase, at each point along a chromosome where DNA polymerase will begin synthesis.

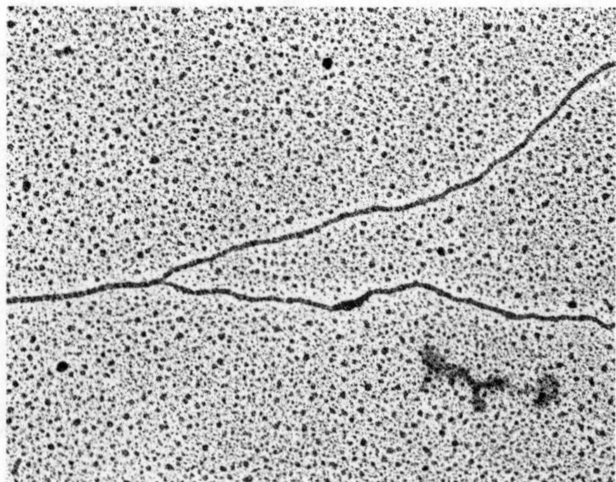

Figure 4.10 Electron micrograph of a replication fork. This image, taken with an electron microscope, shows the replication fork of a DNA molecule. The unreplicated region is the single line on the left, with the two newly synthesized double-stranded molecules branching off on the right.

Source: DR GOPAL MURTI/SCIENCE PHOTO LIBRARY.

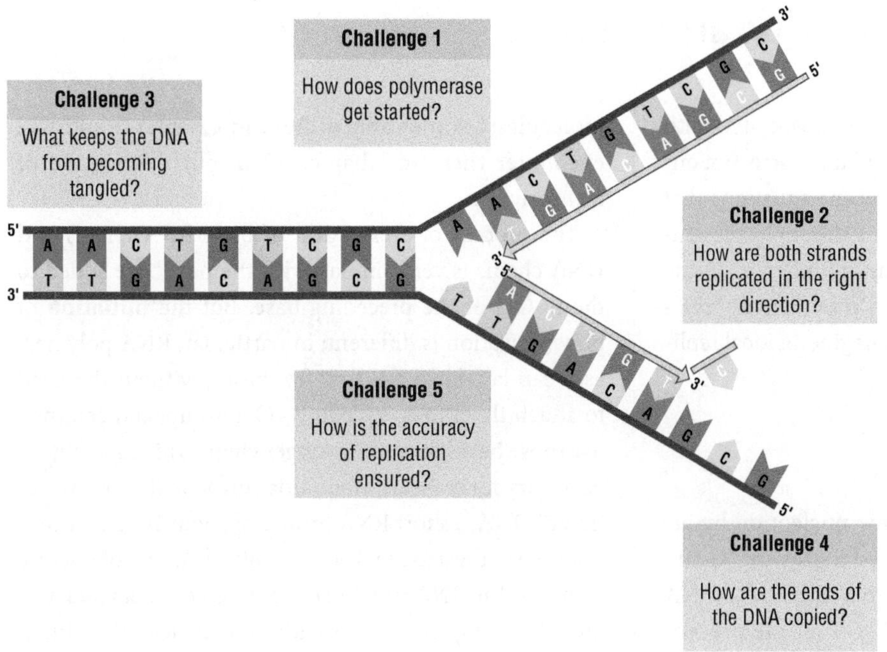

Figure 4.11 Challenges encountered in replication. Some of the challenges encountered during the process of replication are summarized here.

2. Challenge 2: How are both strands replicated in the right direction? **Simultaneous continuous and discontinuous replication**: each replication fork moves along double-stranded DNA, synthesizing two identical double strands of DNA, with both parent strands being used as a template. DNA is always synthesized in the 5′ to 3′ direction, yet the parent DNA strands are arranged in an anti-parallel fashion, with one running 5′ to 3′ and the other running 3′ to 5′. Thus, it is not possible for one DNA polymerase molecule to catalyze the synthesis of both strands continuously. This problem is solved by synthesizing only one strand continuously (the "leading strand"), with the simultaneous, but discontinuous, synthesis of a second "lagging strand." In bacteria, this is done with two different molecules of DNA polymerase III, while the leading and lagging strands are synthesized by different DNA polymerases in eukaryotes.
3. Challenge 3: What keeps the DNA from being tangled? **Unwinding by topoisomerases**: if the two strands in a double helix are separated by pulling apart, the rest of the double helix will coil around itself, eventually producing a mass of knots that would stall replication. This "winding" problem (sometimes referred to as the "untwiddling" problem) is solved by "unwinding" enzymes known as topoisomerases.
4. Challenge 4: How are the ends of DNA copied? **End replication by telomerase**: the removal of the RNA primers used to initiate replication creates another problem in eukaryotes; there is no primer for DNA polymerase to complete the replication to the ends of the strand. This end-replication problem is solved in eukaryotic cells by the enzyme telomerase. Bacteria do not have this problem because most of them have circular chromosomes and do not make telomerase. The few bacteria that have linear chromosomes solve the end-replication problem in other ways that are not fully understood.
5. Challenge 5: How is the accuracy of replication ensured? **Proofreading**: DNA is synthesized very rapidly. A single attempt at hydrogen bond-dependent base pairing by DNA polymerase is insufficient to guarantee fidelity. If DNA replication depended on only a single base-pairing opportunity, mistakes would arise too frequently to ensure the fidelity of the information of the base sequence. Improved accuracy in the copying process is solved by proofreading.

Let's discuss each of these challenges—and their solutions—in a bit more detail.

4.3 The initiation, polarity, and direction of replication

We have said that there are "challenges" associated with DNA replication. But the use of such language is not entirely accurate. Instead, we are describing situations that arise because of the biochemistry of the molecules involved in DNA replication and thus are inherent in the structure of DNA and chromosomes. They are only "challenges" if we are trying to create or imagine an ideal replication machine from scratch.

Priming

Since DNA polymerase adds a new nucleotide by attaching it to the free 3′ OH end of a growing chain, DNA must be replicated in the 5′ to 3′ direction. DNA polymerase can only add to the 3′ OH of the preceding nucleotide—so where does the first 3′ OH come from? The solution lies with a different type of polymerase, an RNA polymerase known as **primase**, as shown in Figure 4.12. As the name implies, RNA polymerases are enzymes that make chains of RNA molecules; we mentioned these in Chapter 2 and will encounter these in Chapter 12 in our description of transcription.

The process of **elongation** of both RNA chains and DNA chains is very similar, with the next base added to the 3′ OH of the preceding base. But the **initiation** of polymerization is different; in particular, RNA polymerases can lay down the first few bases, without the need to attach them to a preceding -OH group, and can then use those bases to create a longer chain. This capability is necessary for transcription initiation as well. For replication of DNA, a short RNA primer of about 10–12 nucleotides is synthesized by the specialized RNA polymerase primase. The RNA primer sequence is complementary to the DNA template and provides the critical 3′ end that DNA polymerase requires; two different RNA primers are made at each replication fork, one for each DNA strand.

As DNA synthesis continues, the RNA primer is removed and replaced with DNA through a filling-in

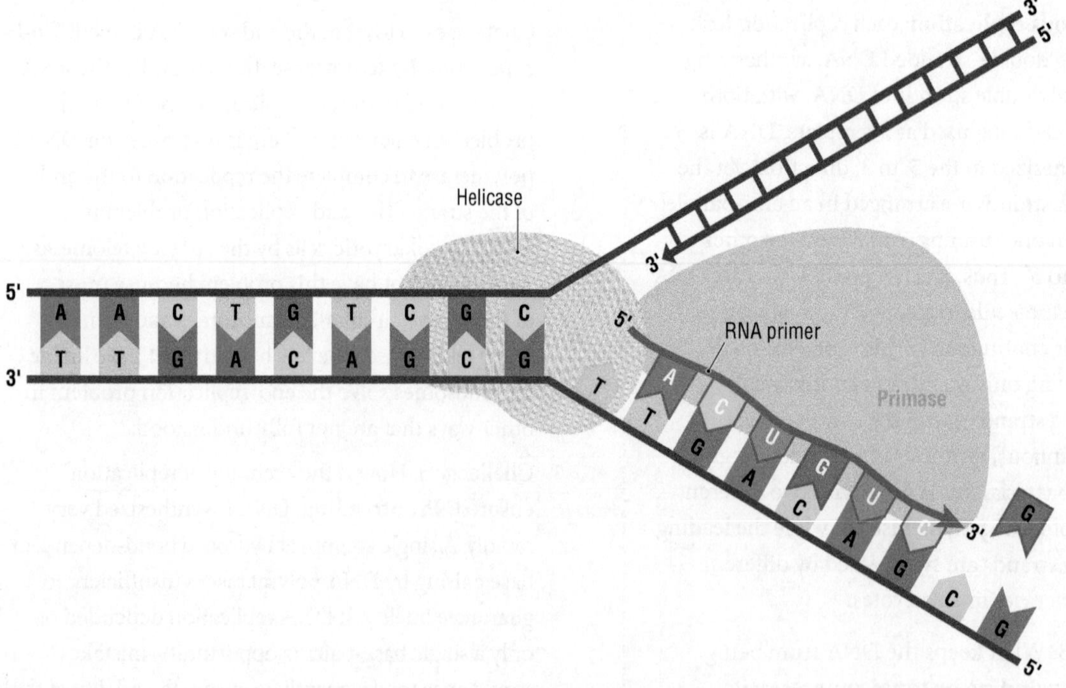

Figure 4.12 The priming challenge—how does polymerase get started? A type of RNA polymerase known as primase can initiate replication because it does not require the presence of a 3′ OH end. Primase creates a short RNA primer (the wavy purple line, with ribonucleotide bases outlined in purple) complementary to the template DNA, which serves as the 3′ end required for DNA polymerase activity. The length of the RNA primer in this and subsequent drawings is not to scale; typically, the primer is 10–12 bases in length.

process carried out by DNA polymerase I in bacteria and several different DNA polymerases in eukaryotes, all of which consist of multiple functional subunits. In eukaryotes, primase is part of DNA polymerase α; once primase makes the short RNA primer, the DNA polymerase activity of DNA polymerase α extends this with deoxyribonucleotides.

Polarity

Another formidable challenge arises once replication is primed and elongation begins. Each strand of DNA in a double helix serves as a template for new strand synthesis. Because the strands are anti-parallel and DNA is only synthesized in the 5′ to 3′ direction, one of the strand pairs must be synthesized towards the replication fork, while the other strand is synthesized away from the replication fork. This has been shown in Figures 4.6, 4.11, and 4.12 but might be best demonstrated by drawing a replication fork yourself and attempting to polymerize DNA synthesis in each direction. The simultaneous replication of both strands is perhaps the most orderly way to copy very long molecules of DNA. The strands only have to be separated once, allowing hydrogen bonding to free nucleotides to occur on both single strands. The cell prevents reannealing of the exposed nucleotides on separated strands by coating one or both of them with single-strand binding proteins; these proteins stay in place until they are displaced by DNA polymerase.

Because DNA strands in a helix are anti-parallel, the DNA polymerases are moving in opposite directions from each other as the two strands are replicated from 5′ to 3′. Electron micrographs demonstrate that DNA uncoiling and re-coiling happen only once at each replication fork and that both strands are copied simultaneously as the replication fork migrates along the double helix. This behavior reveals something that is central to the nature of DNA replication—one strand is replicated continuously or as one long DNA strand, while the other is replicated discontinuously or as a series of fragments. But what does this actually mean in practice?

We diagram this process in Figure 4.13(a) to (d). In Figure 4.13(a), the replication fork from the previous figures has progressed to the left in our diagram, as indicated by the arrow, which exposes single-stranded regions on each template strand and leaves some newly synthesized double-stranded DNA. Note that the situation on the two strands is different, as shown in Figure 4.13(b). The strand that is elongated continuously, as its template becomes exposed upon unwinding at the replication fork, is the **leading strand**; in eukaryotes, the leading strand is replicated by DNA polymerase ε, which takes over from polymerase α after about 20 nucleotides. The DNA polymerase moves along the template strand in a 3′ to 5′ direction, synthesizing the leading strand in a 5′ to 3′ direction as it goes. This point bears some emphasis; while the polymerase is moving along the template strand in a 3′ to 5′ direction, the strand that it is synthesizing, that is, the leading strand, assembles in a 5′ to 3′ direction—the incoming nucleotide is added to the 3′ end of the growing strand.

But, in order for the other DNA strand to be synthesized by addition of nucleotides to its 3′ end, its synthesis has to happen behind the movement of the replication fork, that is, it must be synthesized **discontinuously** in small sections. The discontinuously replicated strand is known as the **lagging strand**, since its synthesis takes slightly longer than the synthesis of the leading strand. The lagging strand is synthesized by DNA polymerase δ in small sections of about 100 nucleotides; these small sections are known as Okazaki fragments, after Tsuneko and Reiji Okazaki, the couple who discovered them. We show two Okazaki fragments in Figure 4.13(b).

As illustrated in Figure 4.13(b) and (c), each Okazaki fragment begins with a short RNA primer, to which a DNA polymerase adds nucleotides to extend the chain for a short distance. As a result, the lagging strand consists of blocks of DNA interspersed with RNA primers, as shown in Figure 4.13(c). The portion of DNA being processed in this manner loops out on the lagging strand, making the replication fork look a bit like a trombone. Once DNA is synthesized, the RNA primers are removed (by polymerase I in bacteria or polymerase δ in eukaryotes) and replaced with DNA, as shown in Figure 4.13(d). The last base of each segment of DNA is attached to the first base of the next segment by a DNA-joining enzyme known as DNA **ligase**. If the gene encoding ligase is mutated, such that the ligase enzyme is non-functional, the lagging strand consists only of short fragments, and the cell cannot survive.

In addition to its essential role in DNA replication, DNA ligase is used in different experimental procedures

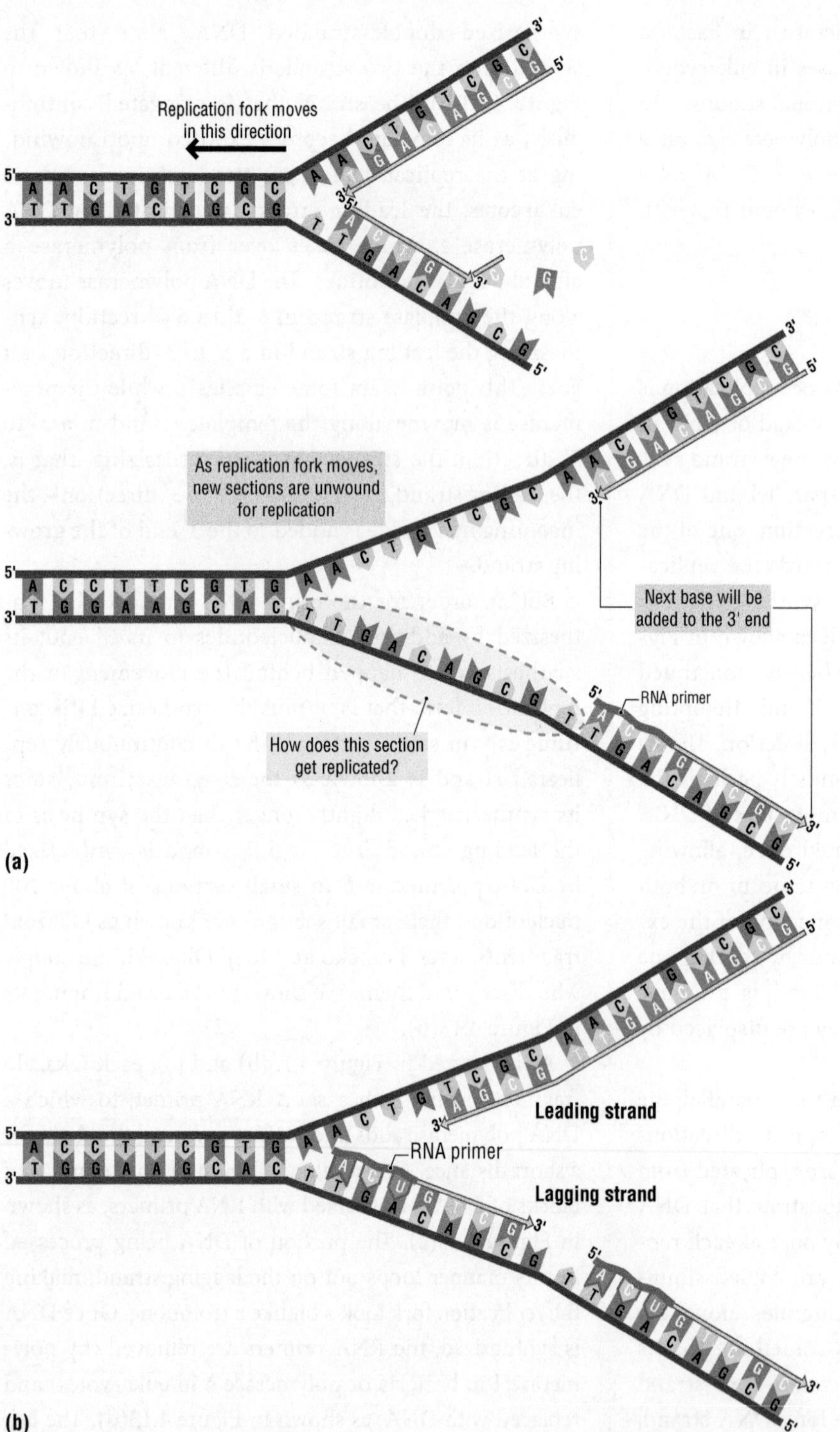

Figure 4.13 The challenge of DNA polarity—how are both strands replicated in the right direction? These four panels illustrate how both strands are replicated. (a) DNA strands in a helix are anti-parallel, so DNA polymerases must move in opposite directions as the two strands are replicated. Note the places where the next base will be added on each strand. How is the section in the dotted blue oval replicated? (b) Leading and lagging strands. The solution to the question posed in (a) is that one strand is replicated continuously (the leading strand), while the other strand is made in short, discontinuous fragments (the lagging strand, since it is expected to take longer to synthesize). Note that each fragment on the lagging strand begins with its own RNA primer. Neither the length of the primer nor the size of the fragments on the lagging strand is drawn to scale. (c) Okazaki fragments. The short sequences on the lagging strand, of approximately 70 nucleotides, are known as Okazaki fragments, after their discoverers. Note the purple RNA primers and the direction of synthesis. (d) RNA primers are removed and replaced, with DNA synthesized by DNA polymerase (polymerase I in bacteria, polymerase δ in eukaryotes). DNA ligase joins these segments together.

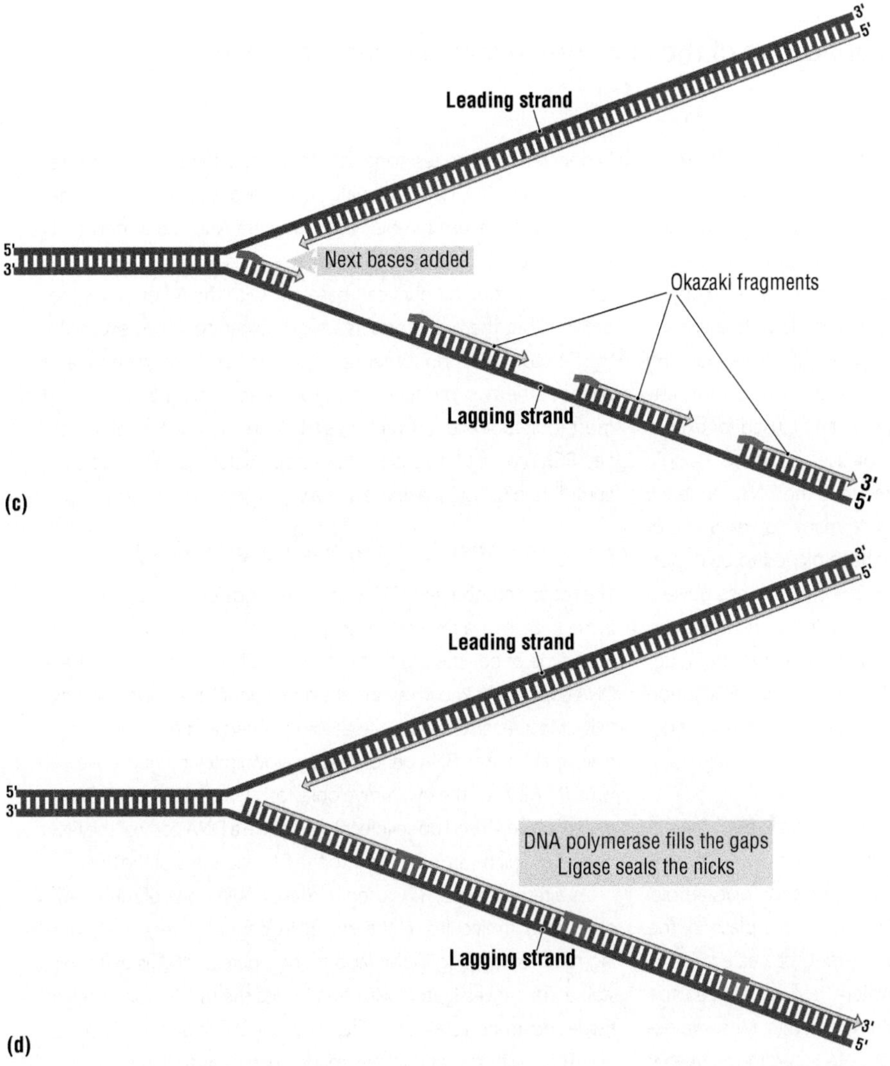

Figure 4.13 continued

that involve joining two DNA molecules *in vitro*, as described in Tool Box 4.2.

KEY POINT DNA replication begins with an RNA primer. One strand is made continuously, while the other is made discontinuously. The discontinuous synthesis of the lagging strand arises because DNA must be assembled in a 5′ to 3′ direction, yet the replication fork is moving in the opposite direction.

The unwinding problem

The third of our challenges to DNA replication faced by the cell arises from fact that the DNA double helix, itself a twisted structure, has to be tightly wound, so that it fits into the cell. In addition to its association with packing proteins (such as nucleosomes in eukaryotes), the DNA is coiled in on itself, generating **supercoils** and overwinding of the unreplicated double-stranded portion of the DNA ahead of the replication fork. Unravelling the long, twisted double helix for replication introduces even more supercoiling and overwinding. This can be difficult to visualize or describe, but if you have ever tried to unwind a long string of lights or strands in a lengthy piece of wool that has been stored as a coil without getting it knotted, you are familiar with some of the problems introduced by supercoiling and overwinding; the long, twisted object easily becomes more highly twisted around itself. In order to prevent the formation of knots as the helix is unwound during replication, cells have categories of enzymes called **topoisomerases** that relieve supercoiling and prevent overwinding of the DNA, as shown in Figure 4.14.

TOOL BOX 4.2 The application of the enzymes involved in replication in molecular cloning

Many of the experiments described throughout this book require that the DNA sequence corresponding to a gene has been isolated, so it can be manipulated *in vitro*. The various processes involved in these *in vitro* experiments are collectively referred to as **molecular cloning**. The goal of molecular cloning is to obtain enough identical copies of a DNA sequence to allow its analysis.

A gene is considered to be cloned when its corresponding DNA sequence is readily available as a reagent, much like any other chemical or molecular reagent that can be used for experiments. Because it was rarely possible or useful to have a single molecule of the cloned gene as the reagent, the DNA sequence was often synthesized *in vitro* to produce many copies. Some of the enzymes described in this chapter have proved to be important tools for molecular cloning and *in vitro* DNA synthesis. Several of these tools are still in widespread use, while others have been largely supplanted by the polymerase chain reaction (PCR), described in Tool Box 4.1. In this box, we provide a brief description of some other uses for these enzymes associated with replication.

DNA ligase to connect molecules

DNA ligase is among the most versatile enzymatic tools in molecular biology. There are many experiments in which two different DNA molecules need to be covalently connected; for example, when a cloned fragment of DNA is inserted into a plasmid, the sites of insertion have small gaps that need to be sealed. This is the role of DNA ligase—to make covalent bonds between the backbones of DNA molecules to "tie" them together. Most molecular biology laboratories have tubes of DNA ligase in the freezer that can be used whenever two different molecules need to be covalently connected to each other.

Topoisomerase and DNA cloning

The biological role of topoisomerases is to unwind supercoiled double-stranded DNA molecules. In order to do this unwinding, topoisomerases cut the DNA, pass the molecule through the cut or nick, and then reseal the cut or nicked DNA. It is this cutting and resealing activity (rather than the unwinding activity) that provides the basis for a technique known as topoisomerase cloning; the molecular biology company Invitrogen markets a widely used TOPO-TA cloning kit based on topoisomerase. Topoisomerase cloning, in general, uses the resealing activity of topoisomerase in a similar way to ligase. TOPO®-TA cloning takes advantage of the fact that *Taq* polymerase used in PCR adds an extra A to the 3′ end of its products; thus, if another DNA molecule ends with an overhanging T, the T from one product can base pair with the A from the other product, but the two products are not joined together, except by the T:A base pair. Topoisomerase is then added to seal the ends and connect the two products. Topoisomerase cloning is one of the most foolproof and fast methods to clone and connect an amplified PCR product into a plasmid vector. TA cloning can also be accomplished using DNA ligase and a vector with A overhangs.

Reverse transcriptase is used to make cDNA

The protein component TERT of telomerase belongs to a class of enzyme known as a reverse transcriptase (RT). Reverse transcriptases are a type of polymerase that can use an RNA template to make a DNA copy; this is known as an RNA-dependent DNA polymerase activity. Most reverse transcriptases are encoded in the genome of viruses that have an RNA genome such as Moloney murine leukemia virus (M-MLV) or the avian myeloblastosis virus (AMV); the reverse transcriptase is used upon infection to make a DNA copy of the RNA genome, which can integrate into the DNA genome of the host.

As a research tool, RT is used to make DNA copies of mRNA and other RNA molecules in the cell. RNA degrades very easily during routine handling in the laboratory, whereas DNA is extremely stable. Thus, investigators will often isolate the mRNA from a cell or tissue and immediately use RT to make a single-stranded DNA copy of it. This DNA copy of the RNA, made *in vitro* using RT, is referred to as complementary DNA or cDNA. (We discussed cDNA in more detail in Tool Box 2.4.) Since the single-stranded cDNA copied from the mRNA would have the sequence of the non-coding or template strand, rather than the coding strand, the isolation of cDNA usually includes a second round of replication, this time using the single-stranded cDNA to make the second strand and thus the more stable double-stranded cDNA. Most RTs also include the DNA-dependent DNA polymerase activity that makes the second strand.

Using RTs to make cDNA is standard practice in most molecular biology laboratories. For a eukaryotic gene, it is important to note that the cDNA copy will be a copy of the mRNA, so it will include the protein-coding region, as well as the ends, but will not include the introns.

Replication forks occasionally stall—that is, the DNA is damaged or entangled in such a way that continued replication cannot progress easily. This can be analogous to having a small knot or kink in the long, twisted object you are trying to unwinding. In eukaryotes, some specialized polymerases are involved in replication at stalled forks, as summarized in Table 4.1.

Replication to the ends

Bacterial chromosomes are usually circular, so replication can progress around the circle until the entire molecule is duplicated. However, eukaryotic chromosomes are linear, which presents a particular challenge when the end of a DNA molecule is reached. As shown in Figure 4.15(a),

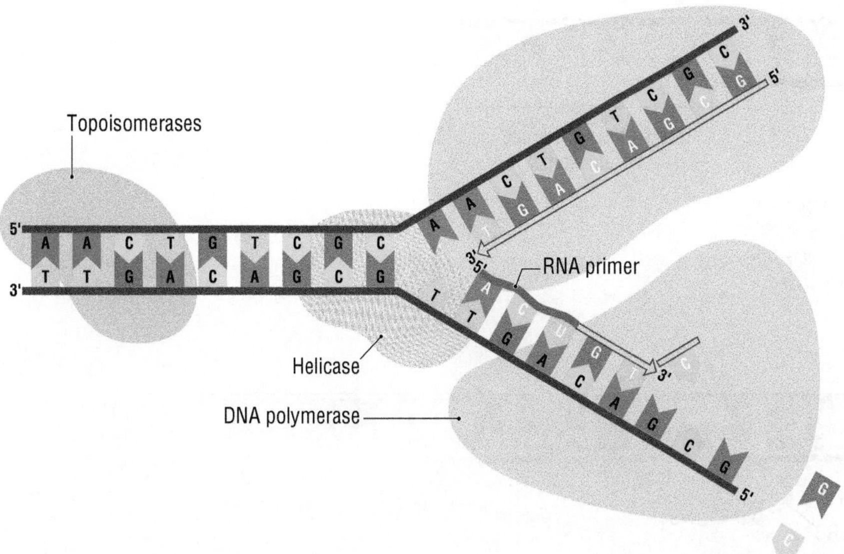

Figure 4.14 The challenge of unwinding DNA—what keeps DNA from becoming tangled? Unwinding the DNA helix introduces additional coiling. Topoisomerases relieve the supercoiling of the DNA by cutting one or both strands and resealing after the torsion of the supercoil has been resolved.

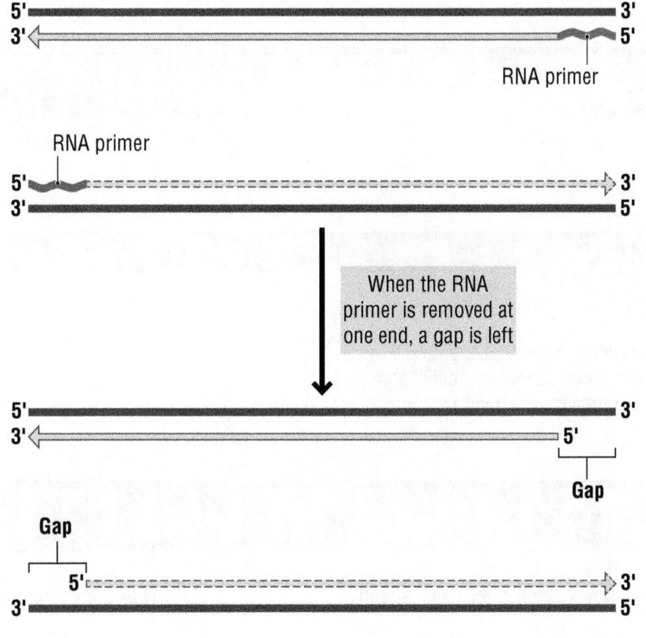

Figure 4.15 The challenge of replicating the ends of the molecule—how are the ends of DNA copied? (a) On the extreme 5′ end of the newly synthesized strands, removal of the RNA primer leaves a gap. If this gap is not replaced, it will grow larger with each replication cycle. (b) This end-replication problem is solved by telomerase extending the ends of the chromosomes. Telomerase is a riboprotein, with both a protein subunit (known as TERT in mammals) and an RNA molecule (known as TERC in mammals and shown in purple). In humans, TERC is more than 500 bases in length, with multiple copies of the repeat sequence CCCUAA. This is the reverse complement of the sequence found at telomeres—TTAGGG. TERT is a reverse transcriptase, an enzyme that can use an RNA template to make a DNA copy. TERT uses the sequence of TERC to add dozens or hundreds of bases to the 3′ end of the opposite strand. (c) Completing the other strand. The extended region of the lower strand provides the template needed for replication of the upper strand, which is made in the same way as any other replication fork. Thus, when the RNA primer is degraded, a gap remains. However, this gap is dozens or hundreds of nucleotides further downstream of the previous end of the chromosome. As a result of telomerase activity, the ends of eukaryotic chromosomes are variable in length.

Source: Part (a) partially adapted from Shay, J.W. and Wright, W.E. (2000). Hayflick, his limit, and cellular ageing. *Nature Reviews Molecular Cell Biology* 1, 72–76.

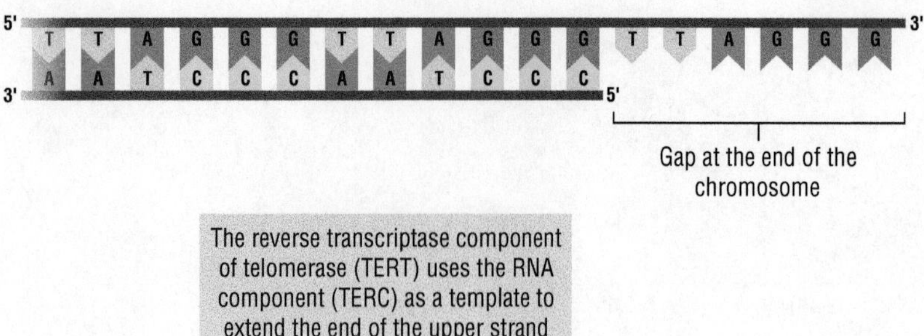

The reverse transcriptase component of telomerase (TERT) uses the RNA component (TERC) as a template to extend the end of the upper strand

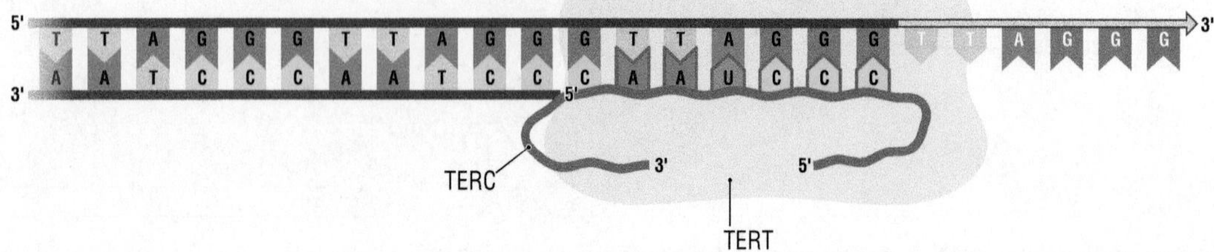

The end continues to be extended

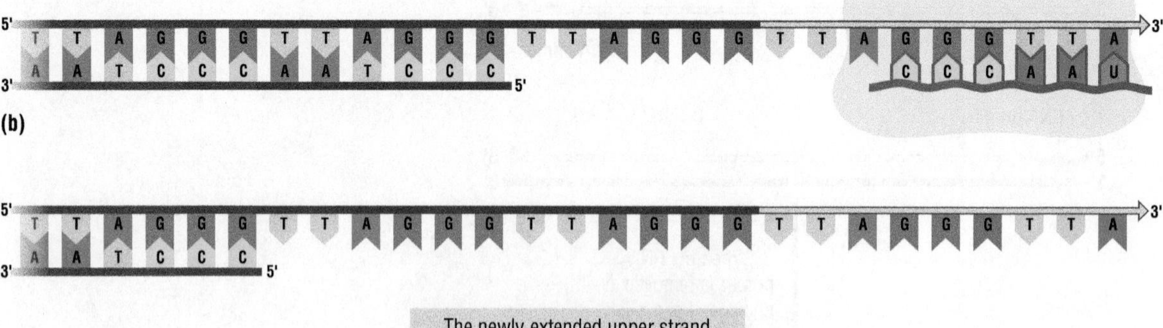

(b)

The newly extended upper strand provides the template for the replication of the lower strand

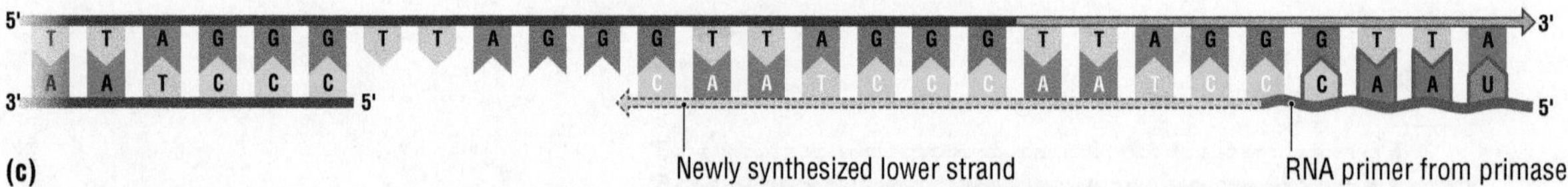

(c)

Figure 4.15 continued

the 3′ end of the newly made strands can be synthesized all the way to the end of the chromosome simply by completing the extension of the newly synthesized DNA in the 5′ to 3′ direction. However, the 5′ end of the newly synthesized DNA strand presents a problem. Remember that DNA polymerase adds to an existing 3′ end, and the initial 3′ OH is provided by an RNA primer. Removal of the primer at the extreme 5′ end of the newly made DNA results in a gap that no DNA polymerase can fill. The 5′ end of the newly synthesized DNA will therefore always be shorter than the template that has been copied. If left unaddressed, these gaps would lead to a slight shortening of each DNA molecule with every round of replication, and potentially the loss of DNA sequence information.

A partial solution to this problem is that the ends of eukaryotic chromosomes have a specialized structure known as a **telomere**, illustrated in Figure 4.15(b). We refer to this as a partial solution, since replication still does not extend to the very end. Telomeres are extended during replication, so that the DNA sequences lost when replication is completed are not missed. While replication to the ends of the chromosomes is the primary activity of telomerase, telomerase activity has been implicated in other cellular functions as well, as described in Box 4.2.

The DNA sequence at telomeres comprises thousands of copies of a very short repeated sequence. In vertebrates, the sequence is 5′-TTAGGG-3′, and most other eukaryotes have this sequence or a slight variant of it. This sequence does not code for anything, so some variation in the number of copies of this repeat can be tolerated by the cell without loss of critical coding or structural information. However, loss from the ends of chromosomes cannot be tolerated indefinitely, not over the hundreds of mitotic cell divisions that occur during growth and differentiation in multicellular organisms, and certainly not over the many hundreds of thousands of generations during eukaryotic evolution. Consequently, the sequence at telomeres needs to be replenished, so that the ends of chromosomes are not constantly eroded.

Eukaryotic cells rely on an unusual enzyme called **telomerase** to solve this end-replication problem and to resynthesize the sequence at the telomeres of chromosomes. Telomerase is an example of a ribonucleoprotein, a macromolecule with both a functional RNA and a protein component. The protein component of telomerase is a type of enzyme known as a **reverse transcriptase**; the reverse transcriptase in telomerase is called TERT. Reverse transcriptases were first identified in the genomes of viruses that have RNA, rather than DNA, as their genetic material. In order to multiply in host cells, these RNA viruses use reverse transcriptase to make a DNA copy of their genome. This is the source of the arrow going back from RNA to DNA in the Central Dogma diagram in Figure 2.9, and the tool for making cDNA as described in Tool Box 2.4.

The reverse transcriptase component of telomerase is apparently derived from a viral invader in an ancient extinct eukaryotic genome. The gene from this virus moved into the ancient eukaryotic genome by some form of horizontal gene transfer, a topic we will discuss in Chapter 11. The reverse transcriptase from this (unknown) viral invader has apparently been re-purposed by the eukaryotic cell for making the ends of linear chromosomes; this appears to be an example of evolutionary "tinkering," or taking advantage of an existing component and using it for a different function, as discussed in Chapter 1.

This ability of reverse transcriptases, like TERT, to use an RNA template to synthesize a DNA molecule allows

BOX 4.2 *A Human Angle* Telomeres

Telomeres are the specialized structures at the ends of eukaryotic chromosomes. Like many other fundamental properties of chromosomes, the existence of telomeres was first recognized by genetic experiments using *Drosophila*, performed by the "Fly Group" led by Thomas H. Morgan in the first decades of the twentieth century. The key experiments that recognized telomeres were done by Hermann J. Muller. Muller was a pioneer of the use of X-rays to produce mutations. He plotted dose–response curves showing the relationship between radiation dosage and number of lethal events, including cytologically visible internal deletions within chromosomes. A deletion that removes an internal region of a chromosome requires two radiation "hits," one on each side. Muller noted that "terminal deletions"—deletions that apparently removed the ends of chromosomes—also followed a two-hit response. This was enigmatic, since it seemed that terminal deletions should require only a single radiation hit that removed the ends of chromosomes. He postulated that, although these events were not truly terminal deletions, they appeared that way as viewed under the microscope; rather some structure at the very ends of chromosomes was essential and thus was being retained in these deletion events. These essential structures were termed telomeres ("end pieces"), without any knowledge of their structure or other functions.

What do telomeres do?

Muller's radiation experiments identified one of the important functions of telomeres—to seal the ends of chromosomes. The end of a linear molecule, such as the DNA in a chromosome, is structurally equivalent to a double-stranded break in the molecule. Double-stranded breaks in DNA occur often, either as part of the normal activities of the cell (such as recombination of DNA sequences, discussed in Chapters 6 and 9) or as damage to the DNA. Most such breaks are rapidly repaired to protect the integrity of the DNA. Breaks that are not repaired at all, or that are not repaired rapidly, are subject to a process known as non-homologous end joining, in which the broken ends of DNA molecules are joined together without regard to the sequence or origin; chromosome rearrangements like translocations, mentioned in Section 4.5, probably arise through non-homologous end joining of double-stranded breaks on two different chromosomes.

BOX 4.2 Continued

If we can infer a premise underlying a biological function for a moment, the cell connects a broken DNA molecule to something, even if it is not the same DNA molecule, rather than leave it broken. In fact, if telomeres are removed by a true terminal deletion, the ends of chromosomes do fuse with one another. Thus, telomeres protect the ends of chromosomes from being treated like another double-stranded break.

The second and more widely recognized function of telomeres is their role in DNA replication, as discussed in Section 4.3. The story goes that this role was recognized by Watson when he was drawing diagrams of DNA replication for his seminal undergraduate textbook *The Molecular Biology of the Gene* in the early 1970s. Watson noticed that replication cannot extend to the very ends of the chromosome, what is often called "the end-replication problem." He postulated in his book that telomeres were responsible for the end replication of chromosomes. Even if this version of the story can't be easily verified, it is true that his textbook drew attention to the end-replication problem. (The Russian geneticist Alexey Olovnikov had reached this same conclusion in 1971, but his work was published in Russian and is not widely recognized.) In the mid-1970s, Elizabeth Blackburn, working in the laboratory of Joseph Gall, one of the most influential cytogeneticists of the past half century, began working on the structure and replication functions of telomeres. Most of what is described in Section 4.3 concerning telomere biology is work done by Blackburn, her then student Carol Greider, and Jack Szostak in the 1980s and early 1990s, all of whom shared a Nobel Prize for this work in 2009.

The role of telomeres in disease

While solving the end-replication problem and protecting the ends of chromosomes from unwanted fusions are the key known roles of telomeres, they have also been associated with cellular immortality in cancer cells and germ cells, as well as with aging at the cellular, and even organismal, levels known as senescence. The exact nature of this role is not clear. Telomerase is not active in most cells of an adult, but it becomes reactivated in tumor cells and is active in germline and rapidly dividing embryonic cells. Thus, in the somatic cells of an adult, telomeres are getting shorter as the cells divide, resulting in an association between the age of a cell (that is, the number of times it has divided) and the length of the telomeres. As such, it has been widely postulated that the length of the telomere affects or regulates the number of times a cell can divide.

Much work by many scientists over the past two decades has been devoted to understanding the connection between telomere length and cellular immortality and senescence. After all, if telomere length is a key component for cancer and aging, drugs or therapies that alter telomerase function and telomere length might hold significant medical promise. To date, no therapies based on telomerase and telomere length have been developed, so the connection remains strong but elusive. It may be that telomerase function and telomere length are not key for cancer and senescence but rather that both of these are consequences or symptoms of another unknown cellular function or program.

Ironically, the one well-studied organism that seems to lack telomerase and a conventional telomere structure is *Drosophila. melanogaster*, the organism in which Muller proposed the existence of telomeres. In *Drosophila*, the end-replication problem is solved not by telomerase but rather by the addition of certain classes of transposable elements to the ends of chromosomes. These additional transposable elements have a role analogous to telomerase in extending the normal end of the chromosome to allow replication of the lagging strand, but the mechanism is quite distinct.

eukaryotic cells to replenish the sequence at telomeres. The RNA component of telomerase is the complementary sequence of the telomeric repeat. In humans, the telomerase RNA component (called TERC) is a sequence of 451 nucleotides featuring multiple copies of 5′-CCCUAA-3′. When telomerase binds to the end of a chromosome, the TERT protein component of telomerase uses the RNA component TERC as a template to synthesize DNA sequences that are complementary to TERC and identical in sequence to the telomerase repeat, as shown in Figure 4.15(b). Telomerase extends the free 3′ end of the linear DNA in the chromosome by adding more copies of the telomeric repeat. Telomerase continues to add these repeats by moving along the telomere sequence on the linear DNA molecule, using its RNA sequence as a template for elongation, as shown in Figure 4.15(c).

The number of copies of the repeat that are added is variable; the important thing is that the length of this DNA strand has been extended and can now serve as a template to synthesize the complementary strand of DNA, thereby extending that strand too. Of course, the extreme 5′ end of this newly synthesized DNA strand is still lost, but this loss is occurring after more copies of the telomeric repeat at the end have been added. This effectively solves the problem of replicating the ends of the chromosome, by making a new end with additional bases.

KEY POINT A specialized enzyme known as telomerase extends the 3′ ends of linear eukaryotic chromosomes.

As might be inferred from this description, the length of the ends of chromosomes is dynamic, with cycles of lengthening and shortening. Not all cells produce telomerase—most cells in an adult human do not, for example—so when these cells replicate their DNA, the extreme ends of the chromosome are, in fact, lost. Although the extended repeat structure at telomeres ensures that no essential protein-coding regions are lost, the shortening process itself has biological importance, as described in Box 4.2.

In the absence of telomerase to maintain the ends of the chromosomes, there is a built-in limit to how many times a cell can divide without encountering problems from shortened chromosome ends. It has also been found that cancer cells often reactivate the expression of telomerase as they continue to divide. This has led to suggestions that reducing telomerase function could be a therapeutic option to block the proliferation of the cancer cell. However, it is not clear if reactivation of telomerase is an event that allows the cells to avoid the limits on cell division, and thus become immortal, or if reactivation of telomerase occurs in response to the loss of the limit on cell division.

4.4 Errors, repair, and variation

We now turn to the fifth challenge faced by the cell when carrying out DNA replication—the propensity for error during the replication process. How do errors arise? How does the cell correct for them, and what benefits might errors actually bring?

In order to ensure continuity of genetic information from one generation to the next, DNA must be copied accurately. Hydrogen bonding of base pairs to the template is sufficient to confer a high level of accuracy, but mispairing of the bases still can occur during replication. If you have some background in organic chemistry, you may recognize that the nucleotide bases have a chemical structure that allows them to exist as tautomers—that is, alternative structures that form when a double bond and proton shift position, as shown in Figure 4.16. Sometimes mispairing occurs when the tautomeric form of a nucleotide base preferentially forms hydrogen bonds with the "wrong" nucleotide—that is, a non-Watson–Crick base pair forms. For example, thymine with the double bond between the ring and the oxygen (the keto form) pairs with adenine, as we have shown before; thymine with a single bond to the oxygen (and a double bond within the ring—the enol form) can base pair with guanine. Thus, a G may be incorporated, instead of an A, if thymine on the template strand is in its less common enol form.

In addition, since DNA synthesis occurs rapidly, the wrong base may be targeted to a position, even though hydrogen bonding is less than optimal. *In vitro* analysis of DNA replication indicates that mispairing is typically present at rates of 0.0001–0.001%, or one base in 100,000 to one base in 1,000,000. This rate might seem insignificant, but a DNA sequence is very long; one mispaired base among 100,000 (that is, a rate of 10^{-5} or 0.001%) would produce scores of replication errors in a bacterial genome each generation, and tens of thousands in a mammalian genome. As a result, all organisms have mechanisms to increase the fidelity of replication above that which is possible, simply as a consequence of the stability of the base pair. Thus, while the rate at which errors occur during replication is in the order of 10^{-5} to 10^{-6} per base per replication cycle, the rate of errors that are passed on without being corrected is 1000-fold lower, in the order of 10^{-8} to 10^{-9} per base per replication cycle.

Proofreading and checkpoints

The process by which a polymerase removes mispaired bases during replication is known as **proofreading** and is outlined in Figure 4.17. Proofreading for errors takes numerous forms, and several different DNA polymerases include proofreading functions. The polymerases used for routine DNA replication in most cellular organisms are able to "check" the distance between paired nucleotides, remove mispaired bases, and create a new opportunity for base pairing. The mispaired base is often removed from the 3′ end of the growing DNA chain by DNA polymerase itself, so the polymerase is said to have 3′ exonuclease activity. The 3′ exonuclease activity can be thought of as an eraser that immediately catches the error (the mispaired bases) and removes it. Some DNA polymerases responsible for repair processes do not have a 3′ exonuclease activity; these are known as error-prone polymerases because they do not proofread.

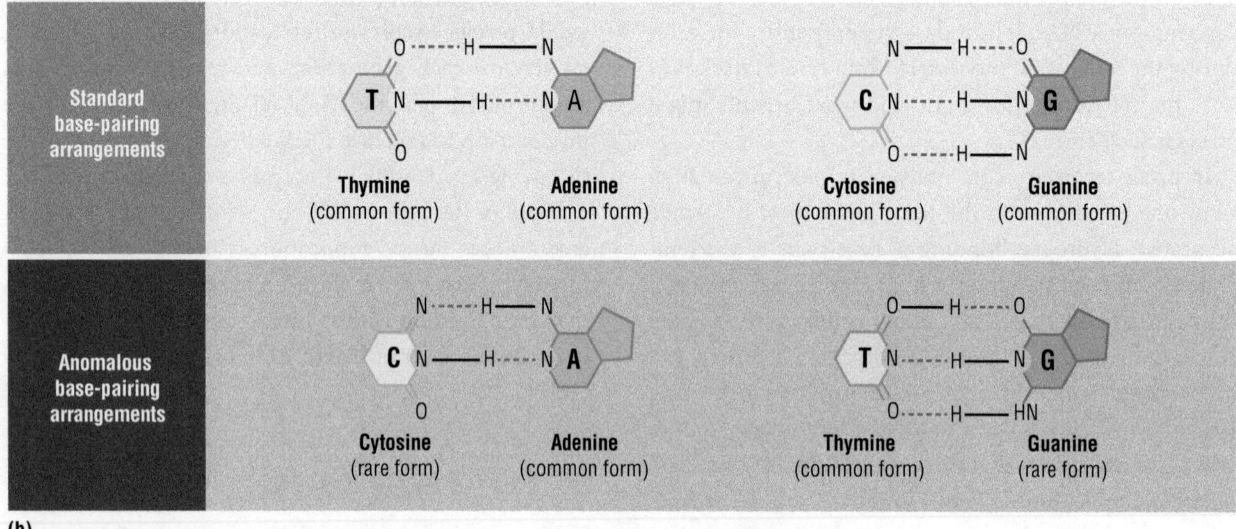

Figure 4.16 Tautomers of nucleotide bases. All four of the bases in DNA have chemical structures that enable them to adopt alternative forms, called tautomers. While the more energetically stable form of cysteine pairs with guanine, the less common tautomeric form can pair with adenine. This produces a change in the sequence if it is not recognized and repaired.

Proofreading DNA, like proofreading a written page, takes longer than leaving it unedited, and these error-prone polymerases are required for repair functions that need to occur rapidly.

In order to carry out the repair processes, eukaryotic cells have different surveillance systems that detect damage and prevent the cell from further division until repair is completed. These are known as **checkpoints**. The cell has to complete the process successfully before the checkpoint is released and the cell can move on to the next step, as outlined in Figure 4.18. Checkpoints are found for many cellular processes. We will encounter the ones important in both mitosis and meiosis in Chapter 6.

The checkpoints important for replication work by pausing the cell division cycle until the damage can be repaired or the replication fork can move forward. The proteins involved in the checkpoint make covalent alterations to proteins involved in signal transduction, notably adding phosphate groups to them, which halts the cell division cycle until the damage can be repaired. If the damage is too extensive to be repaired, the cell triggers a suicide process known as **apoptosis**, which results in programmed cell death.

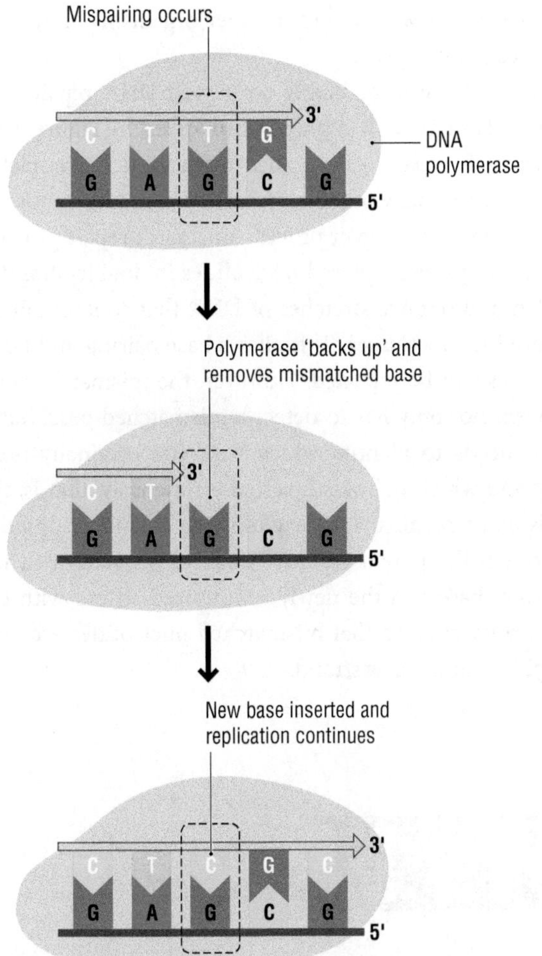
Mispairing occurs
3'
C T T G
G A G C G
5'
DNA polymerase
Polymerase 'backs up' and removes mismatched base
3'
C T
G A G C G
5'
New base inserted and replication continues
3'
C T C G C
G A G C G
5'

Figure 4.17 The challenge of proofreading—how is the accuracy of replication ensured? During replication, DNA polymerase identifies mispaired bases and corrects errors. A mispaired base is removed from the 3′ end of the DNA chain, and thus the polymerase is said to have 3′ exonuclease activity. A new base is inserted to correct the mispairing.

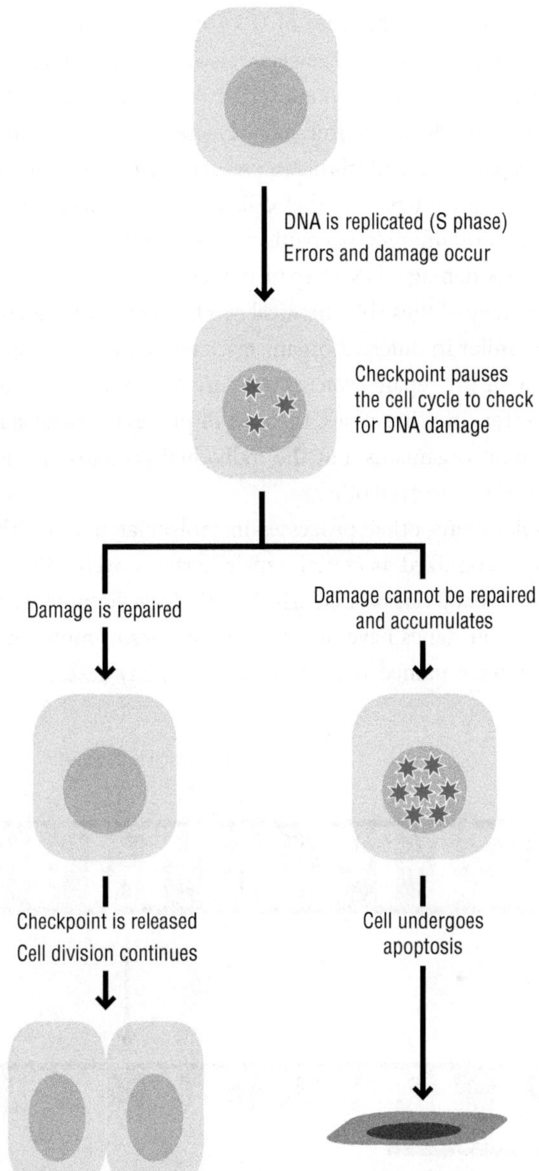
DNA is replicated (S phase)
Errors and damage occur
Checkpoint pauses the cell cycle to check for DNA damage
Damage is repaired
Damage cannot be repaired and accumulates
Checkpoint is released
Cell division continues
Cell undergoes apoptosis

Figure 4.18 Checkpoints in cell division. DNA damage and errors arise during replication, as shown here as the red stars. Cells can "survey" the DNA to detect damage and activate processes that prevent further cell divisions until this damage is repaired. If the DNA cannot be repaired, more damage can accumulate, and the cell may undergo programmed cell death (apoptosis).

Bacteria do not have the same type of checkpoints as eukaryotic cells. However, they have biochemically distinct, but conceptually similar, processes. One of these is known as the SOS response. Genes involved in DNA repair are transcribed as needed. When DNA damage or stalled replication forks lead to accumulation of single-stranded DNA in a bacterial cell, an "SOS response" becomes activated. This response causes the normally repressed genes involved in DNA repair and cell cycle inhibition to be transcribed. The de-repressed repair genes and error-prone polymerases (which work rapidly) repair the damage, while cell division is inhibited. Once repair is complete, all genes under SOS control are repressed, and the cell can replicate its DNA and divide normally.

Cells have different processes to repair different types of damage

Proofreading during replication catches many, but not all, errors due to mispaired bases; other mispaired or mismatched bases are corrected when replication is completed. We will focus on the repairs made once DNA replication has been completed, that is, the post-replication repair. However, DNA damage can occur at any time, and many different types of DNA damage arise in cells. Other types of DNA damage are also repaired, and the repair process happens whenever DNA damage is detected, rather than during or immediately after replication. Since DNA plays the central role in preserving genetic continuity, it is not surprising that cells in all kingdoms have a number of different mechanisms to repair the many types of DNA damage. DNA repair processes are well studied, and many of these biochemical mechanisms and proteins are similar in different organisms, suggesting that at least some of the repair proteins are derived from a common ancestor. In other cases, the overall process is similar in different organisms, but the individual proteins are not that related to each other.

Like many other processes in molecular biology, the first recognized bacterial repair systems were discovered in bacteria, particularly *E. coli*. Because mutants in the repair genes have unusually high rates of mutations, these were named the *mut* genes. In many cases, when similar genes were found in other organisms, they were also called *mut* genes.

One of the most highly conserved DNA repair processes is **mismatch repair**, summarized in Figure 4.19. Mismatch repair occurs once replication is completed and catches many of the errors that proofreading missed. In the mismatch repair process, mismatch repair proteins detect improperly paired nucleotides in double-stranded DNA and remove stretches of DNA that contain one or more incorrect bases. This allows base pairing and DNA synthesis to be repeated. However, the mismatch repair system not only has to detect a mismatched base, but it also needs to identify which base was originally present and which one was inserted erroneously, that is, the mismatch repair machinery must be able to distinguish between the parent (or "old") strand of DNA with the correct base and the newly synthesized strand with the incorrect base, so that mismatched nucleotides are only replaced in the new strand.

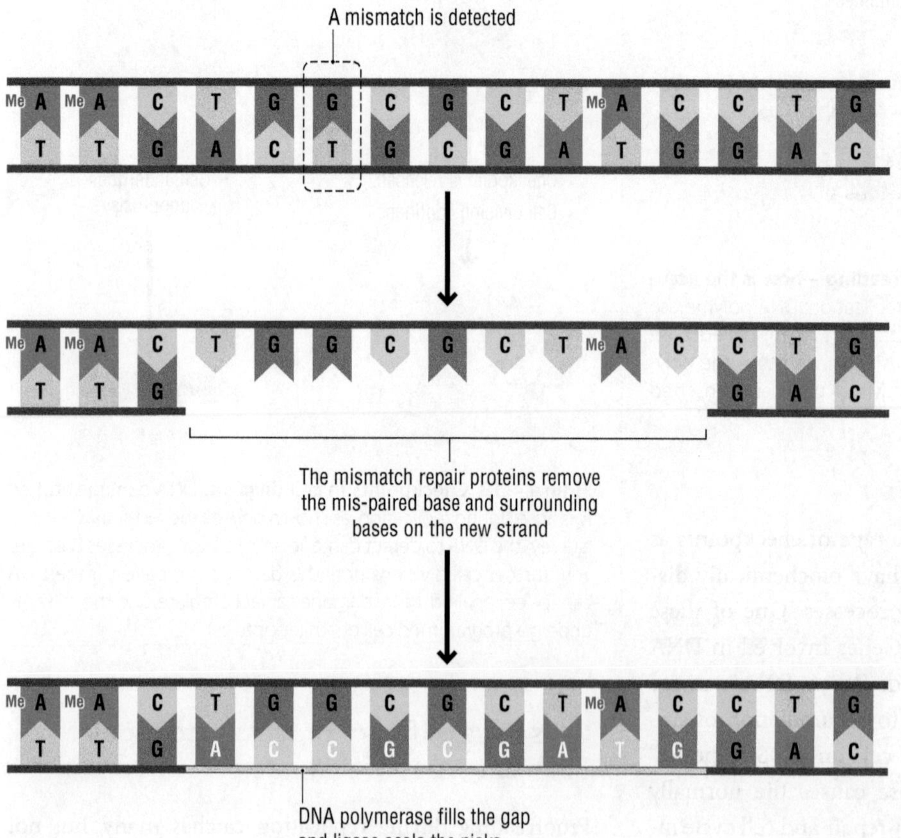

Figure 4.19 Mismatch repair. Once replication and proofreading have been completed, some errors still remain. Mismatch repair proteins detect improperly paired nucleotides in double-stranded DNA and remove stretches of newly synthesized DNA that contain these replication errors. DNA polymerase can then enter, and DNA synthesis will repair the mismatch. Note that the adenines on the original strand are methylated. Bacteria use methylation of adenines on the old strand to indicate which strand had the original sequence and which strand has the error.

It is not altogether clear how this distinction between old and new strands happens in eukaryotes, but the process in *E. coli* has been well studied. After replication in bacteria, DNA is often chemically modified, for example, by the addition of methyl groups to the 6′ carbon of some adenines. As a result, the old strand after replication will feature methylation and other chemical modifications that the new strand lacks. The mismatched bases are recognized by a protein known as MutS, which then recruits two additional proteins, MutH and MutL, to repair the damage. MutH binds to the methylated strand, thereby distinguishing between the old and new DNA strands for making the repair.

While proteins with significant amino sequence similarity to MutS and MutL are found in most organisms, including humans, proteins related to MutH are only found in some bacteria, so a different mechanism for distinguishing the old and new strands for mismatch repair must exist in these other organisms.

Mismatch repair is only one of several mechanisms by which organisms correct errors and repair damaged DNA. Many different types of DNA damage occur, including chemical changes to the nucleotide bases, the loss of the attachment between the base and the sugar–phosphate backbone, and breaks in the backbone itself. We have summarized some of these in Table 4.3. While we have focused in this section on the repair of errors that arise during DNA replication, DNA damage occurs throughout the cell cycle.

One of the best understood of the repair systems is one that detects damage due to ultraviolet (UV) light exposure. UV light, which has high energy, produces a distinctive type of damage to the DNA molecule; adjacent pyrimidines on the same strand, particularly adjacent thymines, become covalently bound to each other to form a structure known as a **thymine (or pyrimidine) dimer**, as shown in Figure 4.20. These dimers result an intra-strand bulge in the DNA molecule, which triggers the repair process. In most organisms other than mammals, an enzyme known as photolyase recognizes and breaks the bond in the covalently bound pyrimidine dimer. Photolyase requires light in the visible spectrum for its activity, so this process is known as **photoreactivation**. Because photoreactivation is a very effective repair process, DNA damage from UV light is minimized, although it still occurs.

UV repair in mammals does not involve photoreactivation and instead uses nucleotide excision repair, summarized in Table 4.3. UV light primarily affects cells on the surface of an organism and has little ability to penetrate into the deeper layers of cells. For most mammals other than humans, these surface cells (that is, skin cells) are protected from sunlight by hair. UV damage is also reduced when skin cells are more highly pigmented or when UV rays are absorbed or reflected by the mixture of chemicals in different commercial sunscreens. Some manufacturers advertise the inclusion of photolyase in their sunscreens, but this is probably more effective for marketing purposes than it is for repairing UV damage.

The enzymes involved in nucleotide excision repair in humans were identified because of a family of diseases known as xeroderma pigmentosum (XP). People with XP have an extreme sensitivity to UV light (including sunlight) and form many basal cell carcinomas and other

Table 4.3 Repair processes for DNA damage

DNA damage	Repair type	Description
Methylation of bases	Direct repair	Many environmental agents attach methyl groups to bases, especially guanine. This can cause guanine to mispair with thymine. Chemically modified bases can be directly repaired by the removal of methyl groups by methyltransferases
Breakage of bond connecting the base to the sugar–phosphate backbone	Base excision repair	No enzyme can directly reconnect the base to the sugar–phosphate backbone. The loss of the base creates an AP (apurinic or apyrimidinic) site. The AP site is recognized by AP endonuclease, which cleaves the backbone on each side of the base. DNA polymerases can fill in this gap
Changes to multiple adjacent bases or more extensive damage	Nucleotide excision repair	When damage to the nucleotides is greater than a single base, the damaged region is recognized; the strands are separated, and one strand is degraded to create a gap of 10–20 nucleotides. This gap is filled in by DNA polymerases. While the overall process is similar in bacteria and eukaryotes, the proteins are not highly conserved
Double-stranded DNA breaks	Break repair	Double-stranded breaks to DNA are repaired by reattaching the broken end to another DNA molecule. Two different processes, with many variations, are used. Commonly, the DNA molecules are not similar in sequence, so the breaks are repaired by non-homologous end joining. Less commonly, the DNA molecules are highly similar in sequence, and homologous recombination occurs. Homologous recombination is discussed in Chapter 6, and the process is described in Box 6.3

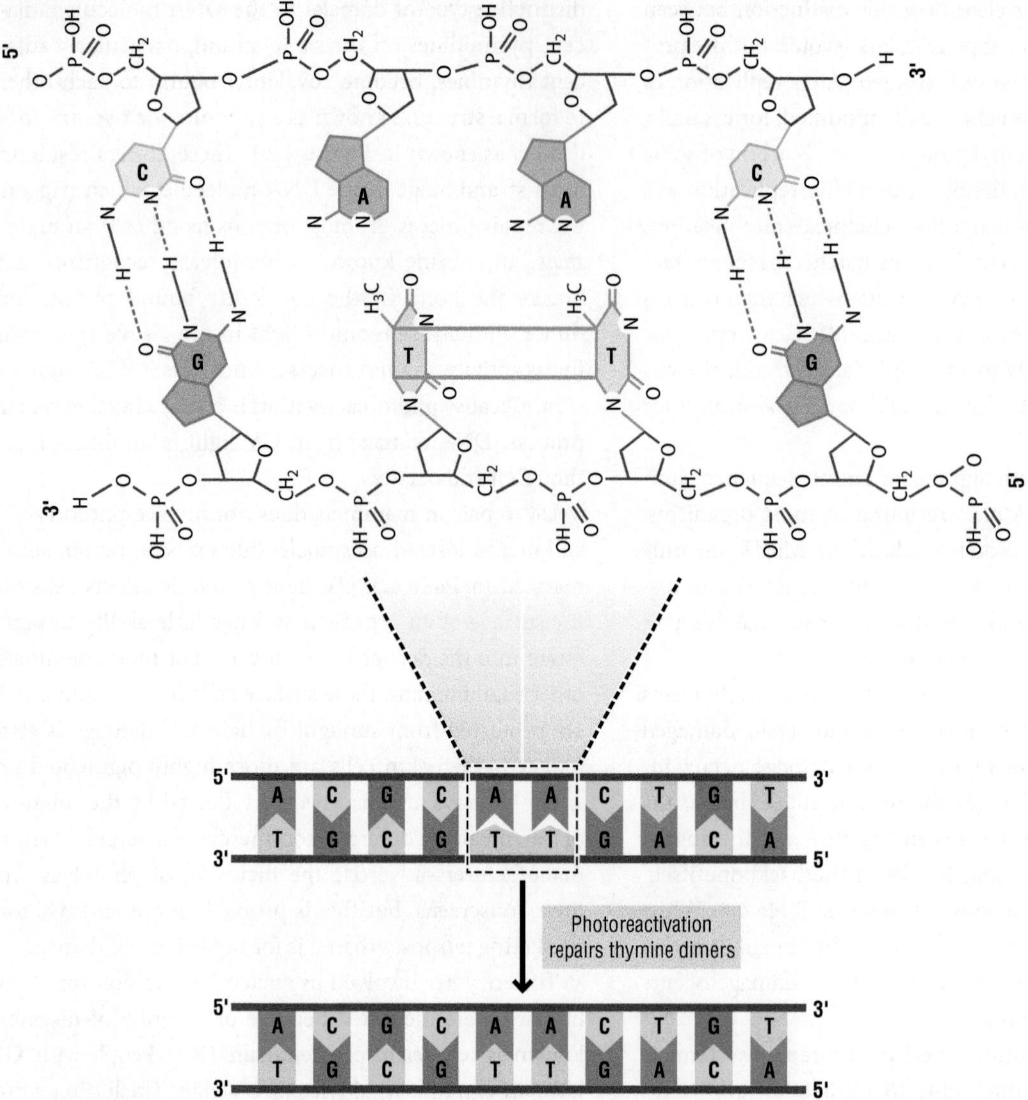

Figure 4.20 UV damage repair—photoreactivation. UV light can damage DNA due to its high energy, causing adjacent pyrimidines on the same strand, particularly adjacent thymines, to become covalently bound to each other. This forms a structure known as a pyrimidine (or thymine) dimer, which causes a bulge in the DNA. In many organisms, photolyase recognizes and breaks apart this dimer. Because photolyase requires light for its activity, this repair process is called photoreactivation. In mammals, pyrimidine dimers are repaired by nucleotide excision repair.

skin disorders; most affected people do not live beyond age 20. At least seven distinct genes give rise to XP in humans, each of them encoding a protein involved in nucleotide excision repair.

KEY POINT The fidelity of replication is greatly improved by the proofreading function of DNA polymerase and mismatch repair after replication. Other processes repair DNA damage that arises separately from replication.

Since DNA replication occurs in all cells, DNA damage and repair also occur in all cells. In cells in the germline, the unrepaired errors are mutations—sequence variations in the genome that are passed on to subsequent generations. From a genetic and evolutionary perspective, these are the sequence changes of greatest interest. As we noted in Chapter 3, many different types of sequence variation are found in genomes, and much of this variation is not directly connected to errors during replication. Variations affecting single bases, that is, the single-nucleotide polymorphisms, or SNPs, are probably the most directly connected to uncorrected replication errors; SNPs will be discussed in subsequent chapters, most significantly in Chapter 10.

Although there have been relatively few direct sequence comparisons of the parents' genomes with the child's genome in humans, these comparisons indicate that each individual has about 70 newly arisen single base changes. Errors that arise in somatic cells are far more common, simply because we have many more somatic cells with more cell divisions; they are less important from an evolutionary standpoint than germline mutations but are still significant for medical reasons in humans, as we discuss in Section 4.5.

4.5 Types of mutations

Changes in the DNA sequence, that is, mutations, are an inevitable consequence of DNA replication. The altered DNA sequence will be passed to the next generation, so mutations are heritable. Mutations can be classified according to their effect on the DNA sequence itself or on the protein sequence that it would encode. As shown in Figure 4.21, when considering the DNA sequence, changes in which one pyrimidine replaces the other pyrimidine or one purine replaces the other purine are referred to as **transitions**. Changes in which a purine replaces a pyrimidine are referred to as **transversions**. As we noted in Chapter 3, insertion/deletion mutations, which may involve one base or many bases, are referred to as **indels**.

Mutations can arise for any base, whether it is part of the coding region of a gene, part of the *cis*-regulatory module for the gene, or not part of a gene at all. Both the type of change, such as a transition, a transversion, or an indel, and the location of the mutation with respect to the gene affect the consequences of the change. We described the consequences of mutations or changes in both the coding region and the regulatory regions of the *ALX1* gene in finches at the beginning of Chapter 1. We noted, in passing, that other mutations were found in the same region of the chromosome that did not appear to affect the *ALX1* gene. This example is typical of many other genes in many other organisms, and we will consider many examples of different kinds of mutations in later chapters.

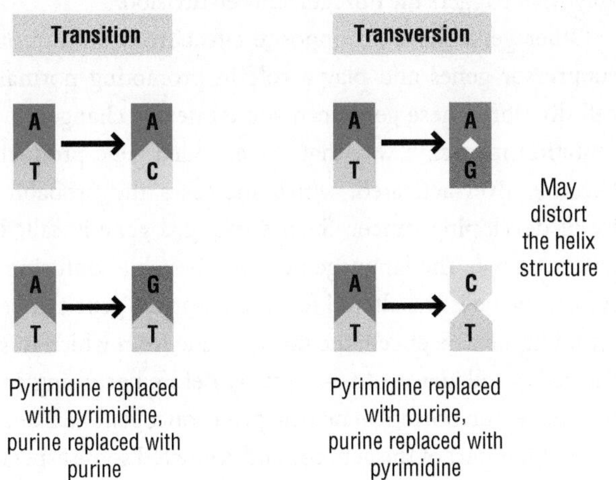

Figure 4.21 Transitions and transversions. Mutations in which one pyrimidine replaces the other pyrimidine or one purine replaces the other purine are referred to as transitions. Mutations in which a purine replaces a pyrimidine, or vice versa, are referred to as transversions.

For mutations that arise in the protein-coding region of the gene, classification can be based on their effects on the amino acid sequence, as shown in Figure 4.22. Some mutations will result in the replacement of one amino

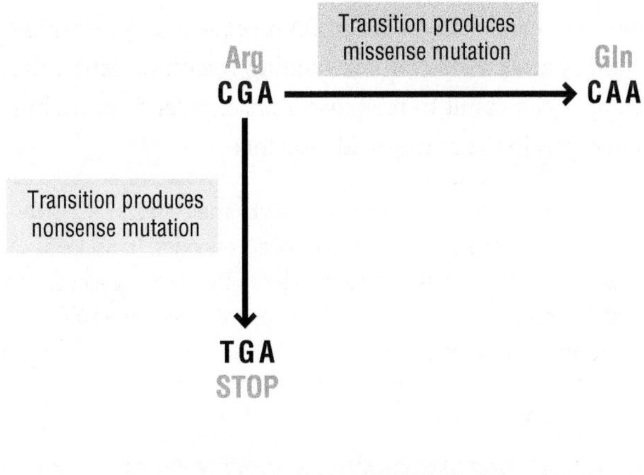

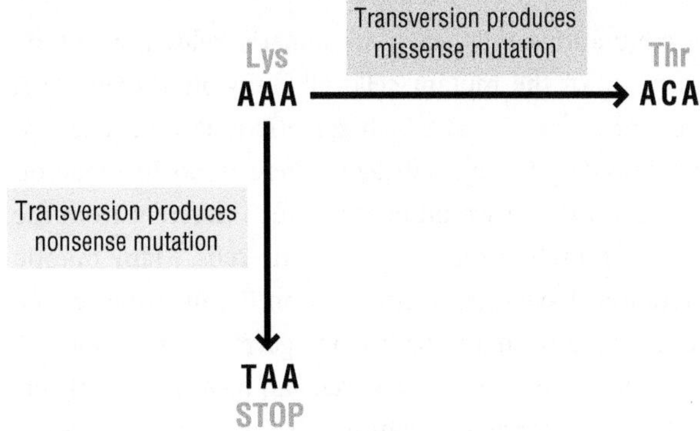

Figure 4.22 Missense and nonsense mutations—mutations at the level of the codon. Changes in which one amino acid is replaced by a different amino acid (regardless of what event occurred in the DNA sequence) are referred to as missense mutations. Changes in which a codon encoding an amino acid is replaced by a stop codon are called nonsense mutations.

acid by another amino acid; these are known as **missense mutations**. The protein is still produced, but the missense changes in its amino acid sequence may have changed its function. Other mutations will change the DNA sequence, such that the mRNA transcript has a stop codon (UAA, UAG, or UGA), rather than a codon for an amino acid, that is, these mutations generate a premature stop codon. These mutations are known as **nonsense mutations** because an amino acid sequence is usually not produced. As a general rule, with many exceptions, nonsense mutations have a more severe effect on the function of the gene than do missense mutations. Indels have a different effect on the protein. The insertion or deletion of a base or bases changes the reading frame (unless, by change, the base change involves a multiple of three bases). These are known as **frameshift mutations**. By changing the reading frame, indels generate both missense and nonsense mutations at the same time.

While it may seem a bit confusing to think of mutations in these different terms, the key is to think about the molecule being affected. Transitions, transversions, and indels are changes in the DNA sequence. When these changes arise in the protein-coding region of genes, the changes can result in missense, nonsense, and frameshift mutations in the amino acid sequence.

KEY POINT Mutations can be classified by their effects on a DNA sequence or on the amino acid sequence it encodes. Transitions, transversions, and indels refer to changes in the DNA sequence. Missense, nonsense, and frameshift mutations refer to the effects on the amino acid sequence.

Somatic mutations can give rise to cancer

For mutations that occur in somatic cells, the mitotic offspring of the mutant cell will show an altered DNA sequence. The rate at which somatic mutations occur is not known precisely, although there is no information to suggest that it would be very much different from the level of mutation seen in premeiotic cells. Many mitotic mutations have little or no effect on the functions of the cells, a feature shared with many germline mutations. A few mitotic mutations, however, can have a dramatic effect, as summarized in Figure 4.23. Our genome encodes many genes whose function is to regulate the growth and division of the cell. Mutations that affect these genes can affect the ability of the cell to divide or to regulate its ability to divide. If the cell cannot divide as a result of a mutation or other damage, it is simply eliminated as part of the normal process of growth and development. If, on the other hand, the mutation affects a gene such that the cell divides without the normal regulatory controls, the cell may continue to divide and become a tumor, as depicted in Figure 4.24(a). Two broad categories of genes have been found to be associated with cancers.

There are genes in our genomes that encode proteins that function to keep cell division in check. These are often referred to as **tumor suppressor** genes. A mutation in such a gene that renders it non-functional can remove the check on cell division, leading to over-proliferation of cells and the development of cancer. Many of the genes involved in the cell cycle checkpoints are considered tumor suppressor genes. Since we have two copies of each of our genes, both copies of the gene usually must be inactivated to see an effect on cancer formation. These two inactivated copies can occur in different ways, but one apparently common way involves both a germline (inherited) mutant allele and a somatic mutant allele. Suppose that a person has an inherited inactivation of one copy of a tumor suppressor gene. If a mutation arises in the somatic cells that happens to inactivate the same gene, neither allele is functional and cancer can develop. Examples include the inherited inactivation of the *APC* gene that is associated with a high risk of developing colon cancer, and inherited inactivation of *Rb*, the first identified tumor suppressor gene, which is associated with the formation of retinoblastoma. It is usually the "second hit," the one that inactivates the remaining functional copy of the gene in somatic cells, that leads to cancer. The inherited germline mutation is a risk factor for the cancer, but the second somatic mutation in the same gene is probably what triggers the unregulated cell division.

Other genes have an opposite effect from the tumor suppressor genes and play a role in promoting normal cell division. These genes can sometimes be changed by mutation in such a way that the encoded gene product becomes over-activated, which increases the probability of developing cancer. Such a changed gene is called an **oncogene**. The language here can be a bit confusing. An oncogene is the altered form of a normal gene that is found in the cancer cell; the normal gene from which it is derived is called a **proto-oncogene**. Before becoming an oncogene, the normal gene (the proto-oncogene) existed as a regular part of the genome and regulated some aspect of normal cell growth and differentiation. Some sort of error, such as those that can arise during DNA replication, caused the normal gene to be altered. This mutation created an oncogene that now misregulates cell division and thereby has the potential to produce cancer.

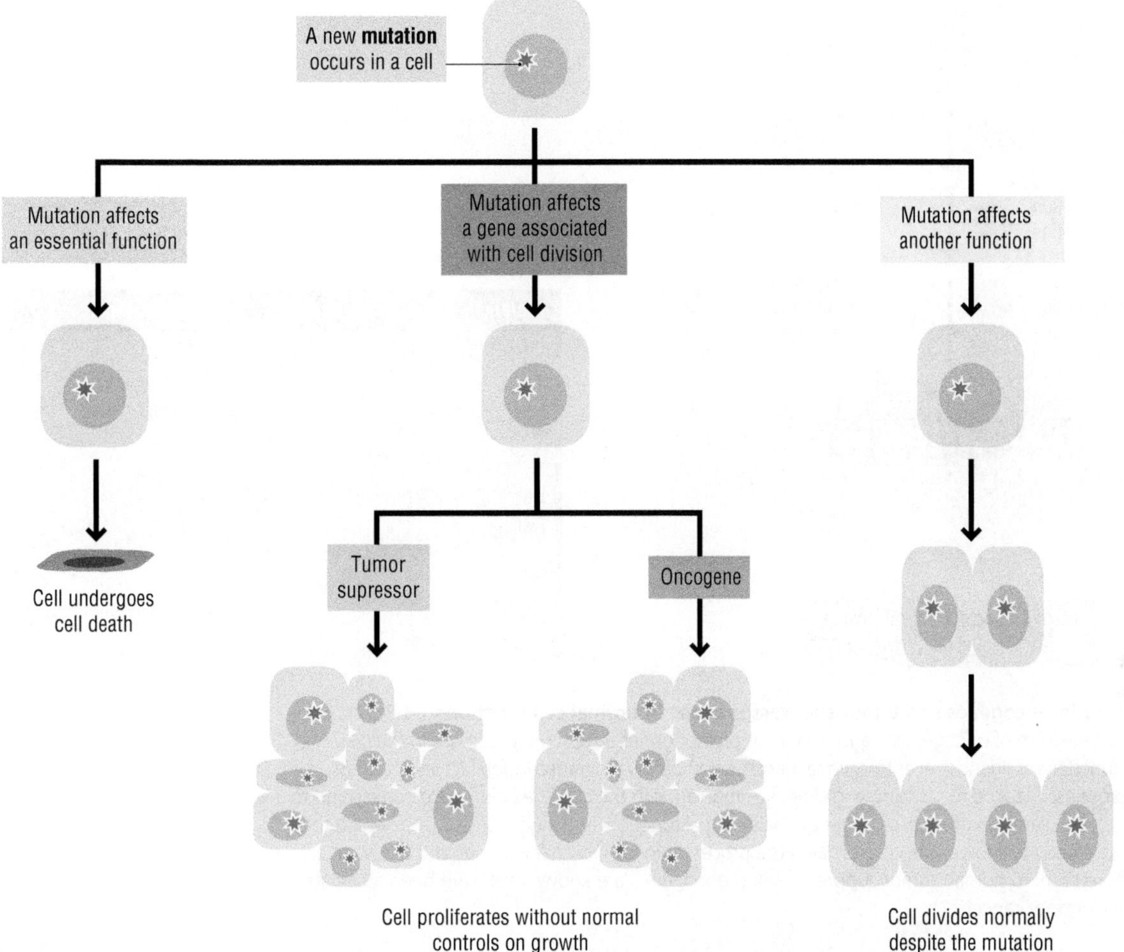

Figure 4.23 Somatic mutations and cancer. Mutations arise during mitosis in somatic cells, as well as in germline cells. Many mitotic mutations have little or no effect on the functions of the cells, shown in blue on the far right. Many other mutations trigger apoptosis or necrosis, so that the cell cannot divide and subsequently dies, shown in black on the far left. A few mitotic mutations can have a dramatic effect, particularly if they affect genes that regulate the growth and division of the cell, shown in purple in the center. Two types of genes can be affected. Some genes have the normal function of keeping cell division in check; these are known as tumor suppressor genes. Mutations in these genes may allow unregulated growth of the cell and produce a tumor. Other genes, known as proto-oncogenes, normally function to promote cell division. Some mutations that affect a proto-oncogene result in an altered version of the gene, called an oncogene. An oncogene allows the cell to grow and divide without the normal regulatory controls, so that the cell may continue to divide and become a tumor.

Several dozen proto-oncogenes and tumor suppressor genes have been identified in mammalian genomes, with functions that affect nearly every part of normal cellular life and gene expression that we mentioned in Chapters 2 and 3. Some of these genes encode transcription factors; some affect chromatin structure; some are involved in intracellular signaling, others in signaling between cells, and more. The names and roles of some proto-oncogenes and tumor suppressor genes are summarized in Figure 4.24(b). The *ras* gene, a proto-oncogene involved in intracellular signaling, will be discussed again in Chapter 15 in the context of its normal cellular function. The oncogenic and over-activated form of *ras* has been implicated in as many as two-thirds of human cancers. Almost all of the genes involved in the DNA repair process are proto-oncogenes. A mutation in one of these genes predisposes the person carrying them to recurrent cancers; as a corollary, many cancers have been associated with a failure to repair DNA damage.

By the time the tumorous growth has been detected, the initial cancer cell and its descendants have accumulated hundreds of mutations affecting dozens of different

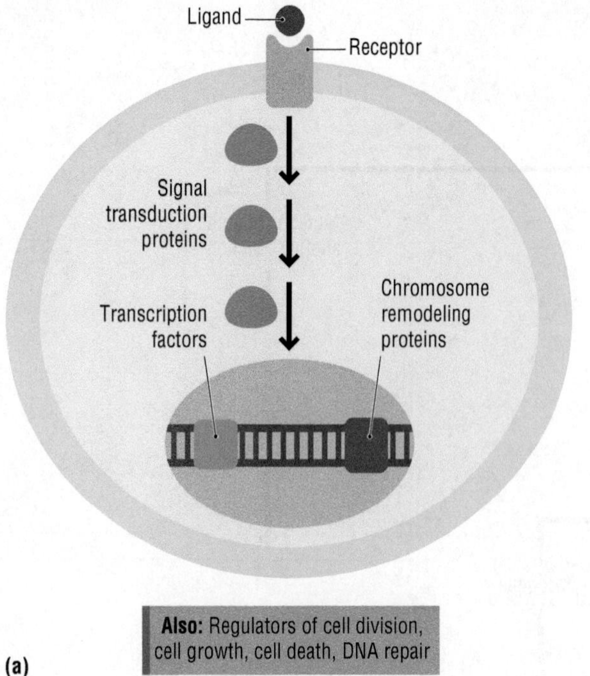

(b)

Protein	Biological role	Type of cancer
Int-2	Ligand	Breast
erb-B	Receptor	Glioblastoma
Ras	Signal transduction	Many
c-myc	Transcription factor	Many
Rb	Chromatin remodeling	Retinoblastoma
CDKN2A	Cell division	Melanoma
Abl	Cell growth	CML
Bcl2	Cell death	Follicular lymphoma
MSH2	DNA repair	Many sporadic

Figure 4.24 Proto-oncogenes and tumor suppressor genes in normal cell functions. (a) Proto-oncogenes—the normal form of oncogenes—and tumor suppressor genes affect core processes in the cell. They encode such functions as ligands, receptors, signal transduction pathways, transcription factors, chromosome remodeling proteins, checkpoint proteins, and other key regulators in the cell. The colors here correspond to the colors in (b). (b) Examples of a few oncogenes, their normal function, and a type of cancer in which they have been implicated are shown. The colors match the processes in (a). Only a few genes and a few cancers are shown; more than 20 human tumor suppressors and oncogenes are known and have been associated with additional forms of cancer.

genes, as illustrated in Figure 4.25. Direct sequence comparisons show that some of the mutated genes—the oncogenes or tumor suppressor genes—are common to different types of cancer, while others are particularly common in one type of cancer. These are called **driver mutations**, implying that these are the ones responsible for the transformation from normal to unregulated growth. Other mutations, known as **passenger mutations**, arise as the cell divides. Passenger mutations tend to affect a more variable set of genes when different types of cancers are compared, but they can affect how quickly a cancer grows at the original site or metastasizes to a new location, and how well it responds to particular types of therapies. In practice, it can be difficult to distinguish the driver mutations from the passenger mutations because cancer cells have many mutations affecting many different genes.

KEY POINT Mutations can occur in somatic cells, as well as germline cells, and some somatic mutations in mammals are important contributors to the onset of cancer.

Other changes to the DNA

Our focus in the first part of this chapter has been on the process of DNA replication and many of the types of mutation that can arise as errors in this process. However, it is important to note that, while errors during replication are an important source of mutations, they are not the only source. When DNA is exposed to harmful conditions, it can become damaged and altered, even in non-dividing cells, and this can produce many alterations to the DNA, including some of the same types of changes as we have described here. It is also important to remember from our discussion of genomic variation in Chapter 3 that many different kinds of alterations in DNA sequences can occur, in addition to single base changes. Some of these other alterations include duplications, insertions, deletions, and larger chromosomal rearrangements, as well as changes involving the insertion or movement of transposable elements. In Chapter 9, we will also discuss the effects of recombination between pairs of chromosomes. The many mechanisms that alter DNA sequences, including errors

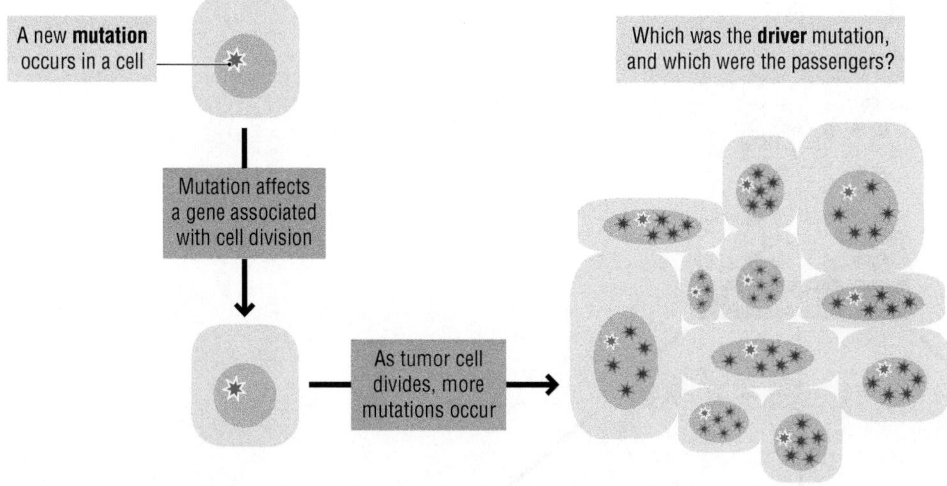

Figure 4.25 Passenger and driver mutations. Some mutated genes are common to different types of cancer or particularly common in one type of cancer and are called driver mutations, implying that these are the ones responsible for the transformation from normal growth to unregulated growth. Other mutations, known as passenger mutations, arise and accumulate as the cell divides. Passenger mutations are expected to affect a wide range of genes, many of which are not themselves involved with the cancer. Genomic analysis of different tumors has allowed the distinction between driver and passenger mutations to become clearer.

in DNA replication, all contribute to the differences that accumulate in genomes over time, and they also provide important variation for evolution to act upon.

KEY POINT Mutations can also change the copy number of sequences, insert or delete sequences, and invert or move sequences. These changes may or may not occur as part of replication.

4.6 The fate and future of sequence changes

We saw in the preceding sections how different types of sequence changes arise, particularly during replication. Mutations arise constantly throughout the genomes. Any base in any gene affecting any phenotype in any cell in any organism can be altered at any time. While the process of mutation itself is generally random, although particular sites in the genome are hot spots or cold spots for mutations, the impact of mutations on the organism can be decidedly non-random. While not all mutations arise during DNA replication, all of them figure into "descent with modification" for an organism or a species. Mutations are the modification, and, for many mutations, DNA replication forms the means of descent. We want to reunite these ideas and consider the evolutionary impact of mutations.

Neutral and selective effects

Many mutations have no observable impact on the survival and fitness of the organism. Mutations with no impact on fitness are called **neutral mutations** or neutral substitutions. The exact frequency at which neutral mutations arise is not clear, in part because it is difficult to be certain that a mutation really has no impact. Defining a mutation as being neutral also implies that it has no effect on fitness in that particular environment. In another environment, the same mutation may not be neutral. Nonetheless, sequence changes at some locations are very likely to have a neutral effect on fitness. For example, not all parts of the genome have functions, so changes in those locations are likely to have no effect. (Of course, we often do not know if a location is important for a function until we see the effects of a change in the sequence at that location, so the reasoning here is a bit circular.) Most tellingly, as we noted in Chapter 2 in our discussion of the genetic code, the third base of a codon often has no effect on the coding capacity of the codon. For example, CU_ is leucine, GU_ is valine, UU-pyrimidine is phenylalanine, CA-purine is glutamine, and so on. Thus, changes in the third position of a codon that do not change the amino acid sequence can almost certainly be considered to be neutral mutations.

On the other hand, many mutations **do** have an observable effect on the survival and fitness of the organism, as illustrated in Figure 4.26. This figure shows these effects

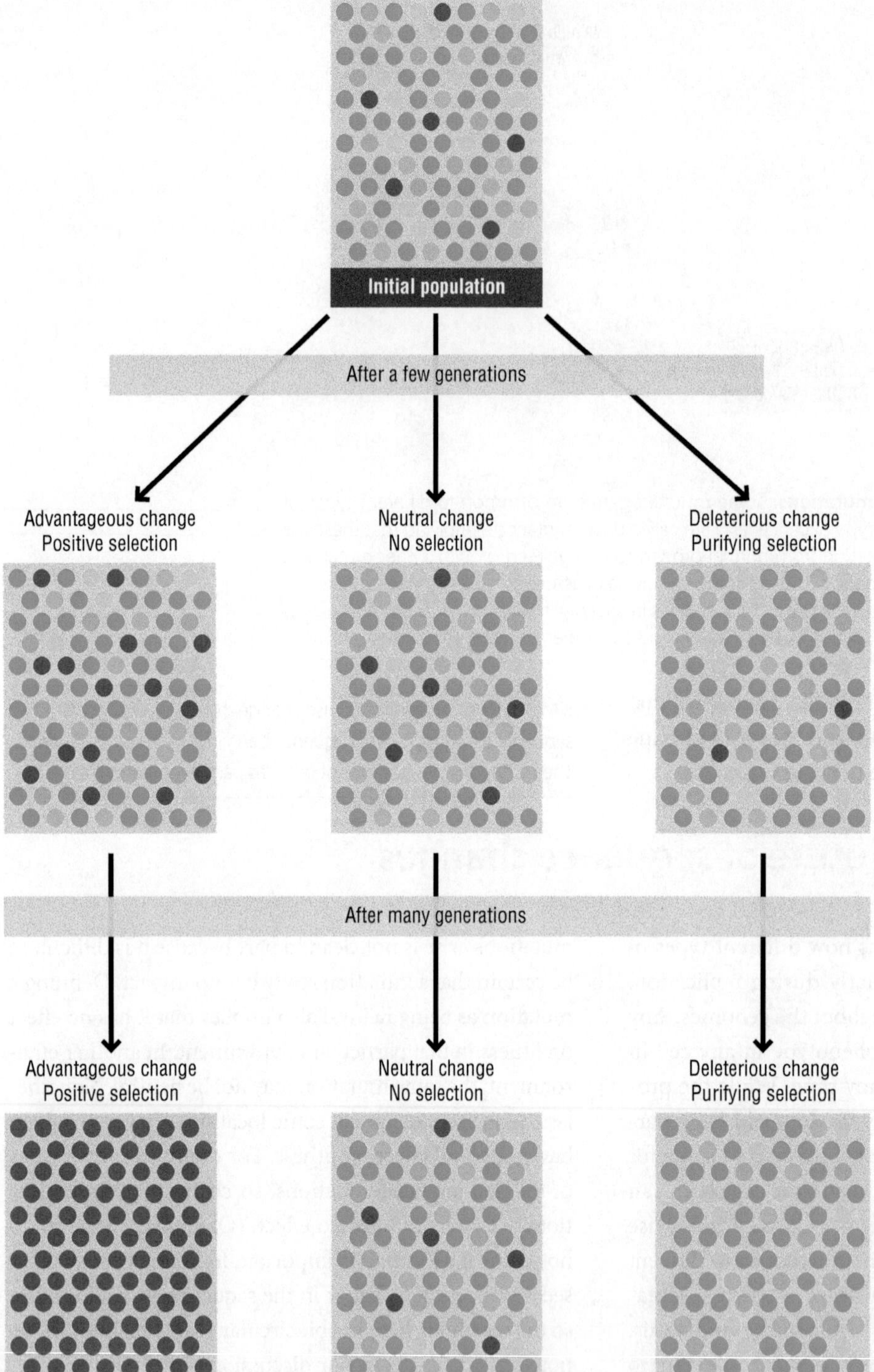

Figure 4.26 Purifying selection and positive selection. A genetically variable population is shown, with each shade of turquoise representing a genotype. The newly arisen mutation is shown in pink. In the original population, the mutation is at a relatively low frequency of about 5%. When a mutation occurs that is deleterious to the fitness of an organism, purifying selection acts to remove this mutation from the population, as seen on the right. If a mutation is advantageous, positive selection will increase the frequency of this mutation in the population, as shown on the left. If the mutation is neutral, its frequency will remain about the same, as seen in the middle panel. After time has elapsed and the population has come to equilibrium, the effects of these selective pressures are visible in the population, such that the deleterious change is absent, an adaptive change becomes fixed, and a neutral change remains at about the same frequency. As described in Chapter 16, other factors also affect the frequency of mutations, so this should be understood as a generalization.

during population change and once the population is stable. Most of these mutations are deleterious and greatly reduce the fitness and survival of the individual. This might be the most common consequence of a mutation, particularly those that occur within the coding region of a gene. These mutations are expected to be selected against, so the number of individuals in the population who carry them is predicted to be low. In fact, in the absence of other effects, such deleterious mutations are likely to be eliminated from the population over time by natural selection. Thus, this type of selection is sometimes called **purifying selection**, with the connotation that selection is keeping deleterious mutations out of the gene pool. Note that we wrote "in the absence of other effects." As will be discussed in Chapter 16, these other effects are important and cannot be ignored; some can result in a deleterious mutation being maintained in the population at an appreciable frequency. Nonetheless, in general, many mutations are deleterious and will be eventually eliminated from the population by purifying selection.

A few mutations are beneficial to the organism and result in higher fitness. Perhaps a change in an amino acid sequence results in a protein that functions better in that particular environment. Perhaps a change in those upstream sequences involved in regulation of the gene alters the time or place at which a gene is transcribed, and this change in expression is advantageous. Changes that affect the level of transcription or the pattern of splicing for a gene could also be advantageous to the organism under some, often unknown, circumstances. Mutations that are beneficial to the organism, at least those that confer a **fitness** advantage to the organism in a certain environment at a certain time, are called **adaptive**.

KEY POINT Mutations that have no observable effect on the fitness of the organism in a particular environment are called neutral mutations. Most mutations that affect a phenotype will have a negative effect on fitness, while a few may have a beneficial or adaptive effect.

It is important to recognize that a mutation arises without regard to its effect on fitness or any other phenotype. Natural selection in a given environment determines if the change is adaptive or deleterious. Mutations do not arise because they are adaptive. This may seem obvious, but even experienced molecular biologists can sometimes fall prey to the seductive way of thinking that a mutation arose **because** it conferred a fitness advantage. The important experiment that demonstrated that mutations arise without regard to fitness was Luria and Delbrück's fluctuation test, which we describe in Box 4.3.

When a mutation or an allele confers an adaptive advantage, its frequency in the population is predicted to

BOX 4.3 *Quantitative Toolkit* The Luria–Delbrück fluctuation test

Do mutations arise randomly before selection acts on them, or does exposure to a selective agent produce mutations? This is a profoundly important question that underlies much of how we think evolution by natural selection works; as we described this in Chapter 3, there is genetic variation in a population, and some of that variation confers a selective advantage. But what are the experiments that show that genetic variation and mutations arise before selection is applied? Salvador Luria and Max Delbrück relied on the rapid generation times and large size of bacterial populations to address this question experimentally and quantitatively in 1943. While the experiments were done in bacteria, the results are applicable to other organisms and are among the most intellectually influential experiments in genetics.

Like other organisms, bacteria can be infected by viruses. Luria and Delbrück used cultures of the bacterium *Escherichia coli* B and looked at resistance to viral infection in these cells. The experiment is summarized in Figure A. In the control experiment, shown in Figure A(i), they started a culture of *E. coli* B by adding a few cells to liquid medium and allowing them to divide until a high cell density was obtained. They then spread samples of this culture on multiple agar plates that had been covered with a bacterial virus. Most bacterial cells were sensitive to the virus and died. After an incubation period, Luria and Delbrück counted the number of bacterial colonies on each plate that were resistant to the virus. The variation among the number of colonies on each plate was small, as expected for replicate samples taken from a single population of cells. To put this in more formal statistical terms, the variance among the samples was similar to the mean number of resistant colonies on the plates.

To test whether resistance mutations arose because of exposure to the virus, a hypothesis Luria and Delbrück called "acquired hereditary immunity," they grew *E. coli* cells in many separate individual liquid cultures until the populations had all increased to a similar density, as shown in Figure A(ii). The cells were then plated on agar plates containing the bacterial virus. Again, the number of *E. coli* B colonies on each plate that were resistant to the virus was counted.

Two possible outcomes were envisioned. If selection—in this case, the presence of a virus—induces a mutation in the bacteria,

BOX 4.3 Continued

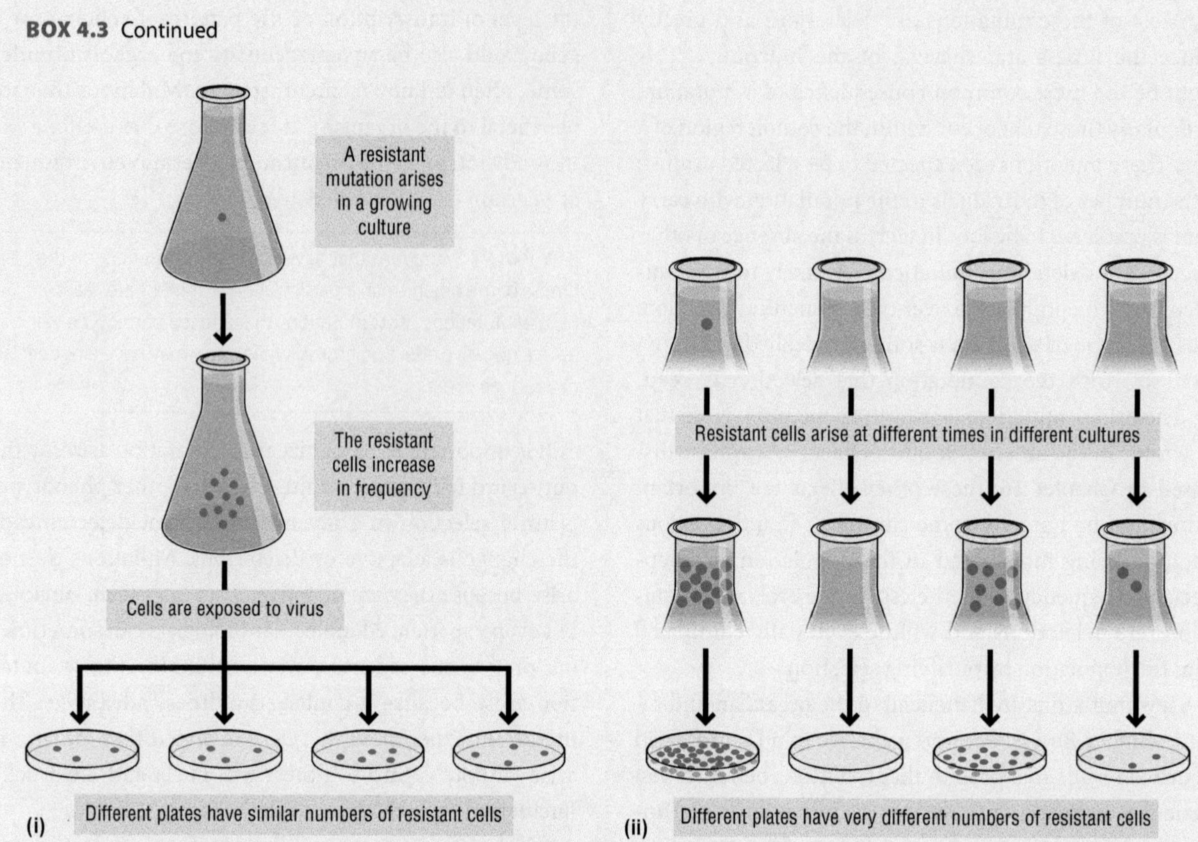

Figure A The fluctuation test demonstrated the randomness of mutations. Luria and Delbrück used resistance to viral infection, a single gene trait in *E. coli*, to demonstrate that mutations occur at random, rather than in response to selection. The resistant mutant cells are shown in pink. In the control experiment in Panel (i), bacteria grown together in culture are plated separately on Petri dishes containing the virus. The plates will have similar numbers of resistant cells, from which they could calculate the mean and variance for their experiment. In the experimental case in Panel (ii), bacteria are grown in separate small cultures and then plated on Petri dishes containing the virus. If mutations to viral resistance arise at random, different cultures will have very different numbers of resistant colonies. On the other hand, if exposure to the virus induced resistance, different cultures were expected to have similar numbers, since they were exposed to the selective agent on the plate at the same time. Again, Luria and Delbrück could calculate the mean number of the resistant cells and the variance by combining data from different cultures.

we would expect that each plate would have a similar number of resistant colonies, even if these samples came from different original cultures, with results similar to those in Figure A(i), that is, since the cultures were exposed to the selective agent on the plate at the same time and the cultures grew at about the same rate, all of the independent cultures should have about the same number of resistant colonies; these would be equivalent to replicate samples from the same large population as before. On the other hand, if mutations arise spontaneously in the separate populations, without regard to the presence of the virus as a selective agent, then the number of resistant colonies would vary widely or fluctuate among the different plates, depending on when the mutation arose during the growth of the liquid culture, as shown in Figure A(ii). If a mutation occurred spontaneously soon after the cultures were started, many of the offspring would carry this mutation. In other cultures, a mutation might occur in a much later generation, and fewer offspring would show the trait.

What did these experiments find?

Some data from Luria and Delbrück's experiments are shown in their original table in Figure B. Note that their data are presented in the opposite order from our discussion. In the samples all taken from one bulk culture, shown as Panel (ii) in their table, the number of resistant colonies on each plate is fairly similar, ranging from

BOX 4.3 Continued

Culture number	# of resistant colonies
1	1
2	0
3	3
4	0
5	0
6	5
7	0
8	5
9	0
10	6
11	107
12	0
13	0
14	0
15	1
16	0
17	0
18	64
19	0
20	35
Mean	11.3
Variance	694

(i) Individual cultures

Sample number	# of resistant colonies
1	14
2	15
3	14
4	21
5	15
6	14
7	26
8	16
9	20
10	13
Mean	16.7
Variance	15

(ii) Samples from bulk cultures

Figure B Data from Luria and Delbrück's fluctuation test. Note that the data in Panel (ii) correspond to Panel (i) of our drawing in Figure A. The data obtained by Luria and Delbrück clearly show that samples from separate individual cultures vary widely in the number of resistant colonies, while replicate samples from a bulk culture had similar numbers of resistant colonies. Although the means for both experimental set-ups are similar, the variance for the individual culture approach is very high.

13 to 26, with a mean of 16.7 resistant colonies and a variance of 15, which is close to the mean. (If you are familiar with probability, you may recognize this as being a Poisson distribution.)

If the introduction of the virus induced resistance, a similar result is expected for the individual cultures as well. However, this is not what they observed, as shown in Panel (i) of their table in Figure B. The mean for the ten cultures was 11.3 resistant colonies.

But the key result lies in the variance, which is equal to 694, and can also be observed in the range of resistant colonies. Eleven of the 20 cultures had no resistant colonies, indicating that no mutation to resistance had occurred in these cultures. Others had a few resistant colonies, but three cultures had 35 or more resistant colonies, with one culture having 107; in a statistical sense, the variance was many times larger than the mean number of colonies. (In fact, while the mean of resistant colonies was 11.3, none of the cultures was particularly close to the mean.) In other words, the resistance mutation arose in different cultures at different times, even though the virus was introduced at the same time.

The data from this experiment clearly rule out the hypothesis that the induction of the virus as a selective agent resulted in beneficial mutations. They are consistent with the hypothesis that mutations to resistance occur randomly, without regard to the presence of selection by the virus. In 11 of the 20 cultures, the mutation did not occur at all, while in six other cultures, the mutation arose late in the growth phase, resulting in one, three, five, or

BOX 4.3 Continued

six resistant colonies. But in three cultures, the mutation occurred early—earlier than the introduction of the virus, in fact, since all three cultures had many more resistant colonies that expected from the control. Thus, the cultures were not "responding" to the presence of the virus by producing beneficial mutations; the beneficial mutations were present already and were revealed when the selective pressure by the virus was applied.

FIND OUT MORE

Luria, S. E. and Delbruck, M. (1943) Mutations of bacteria from virus sensitivity to virus resistance. *Genetics* **28**: 491–511

Meneely, P. M. (2016) Pick your Poisson: an educational primer for Luria and Delbruck's classic paper. *Genetics* **202**: 371–4

Murray, A. (2016) Salvador Luria and Max Delbrück on random mutation and fluctuation tests. *Genetics* **202**: 367–8

increase over time; it might eventually become **fixed** in the population, as shown for the adaptive mutation in Figure 4.26, that is, if a mutation is adaptive and the population itself and its overall environment remain stable, every individual in the population is eventually expected to become homozygous for it and no other allele will be common. Not all adaptive mutations do become fixed, however, for reasons that we will discuss in Chapter 16.

4.7 Inferring evolutionary history from sequence changes

The occurrence of mutations and the effects of selection or other evolutionary forces can sometimes be revealed by comparisons between genomes. Finding evidence for evolutionary change in the genome is a bit like snooping in your grandparents' attics—we find relics of their lives, such as photographs and clothing, that help us understand what they must have been like in the past. These relics also remind us that times and fashions change, and what now seems current, or even avant-garde, will soon seem dated. By comparing the clothing with the hairstyles in the photographs, for example, we can infer temporal relationships between different items in the past. In fact, we can often assign a date to a photograph or an article of clothing because of the knowledge of when such things were in fashion.

Genome sequence data can play a similar role in giving us not only a glimpse of the past but also a framework with which to date events in the past and to determine their relationships to one another. Darwin recognized this ability to see past relationships and reconstruct evolutionary history, or **phylogeny**, based on current similarities. We encountered this with our example of the finches that began Chapter 1. In fact, *The Origin of Species* has the single figure shown in Figure 4.27(a), a branching tree to interpret relationships among morphological phenotypes; the field notebook from 1837 leading to his conclusions features a tree diagram with the stunning handwritten notation, "I think," as reproduced in Figure 4.27(b). With genome sequences, we can turn Darwin's contemplations into quantitative examinations of evolutionary history.

The key principle to descent with modification is that two species (or organisms or even sequences) had a common ancestor, that is, there was a species in the past which, through the effects of mutations that produce variation and evolutionary forces that act on those variations, gave rise to two species. Neither species is exactly like the common ancestral species, since there have been modifications, but both species have recognizable similarities to it and to each other, since there has been descent. These similarities can be observed in many characters or phenotypes, but, since genome sequences are so widely available, the similarities can be quantified through DNA sequence changes. The phylogeny can be represented as a branching diagram or **phylogenetic tree**.

KEY POINT A description of the evolutionary history of a group of organisms is referred to as a phylogeny. As a result of common descent and descent with modification, evolutionary histories are branching in nature. A diagrammatic representation of these relationships is referred to as a phylogenetic tree.

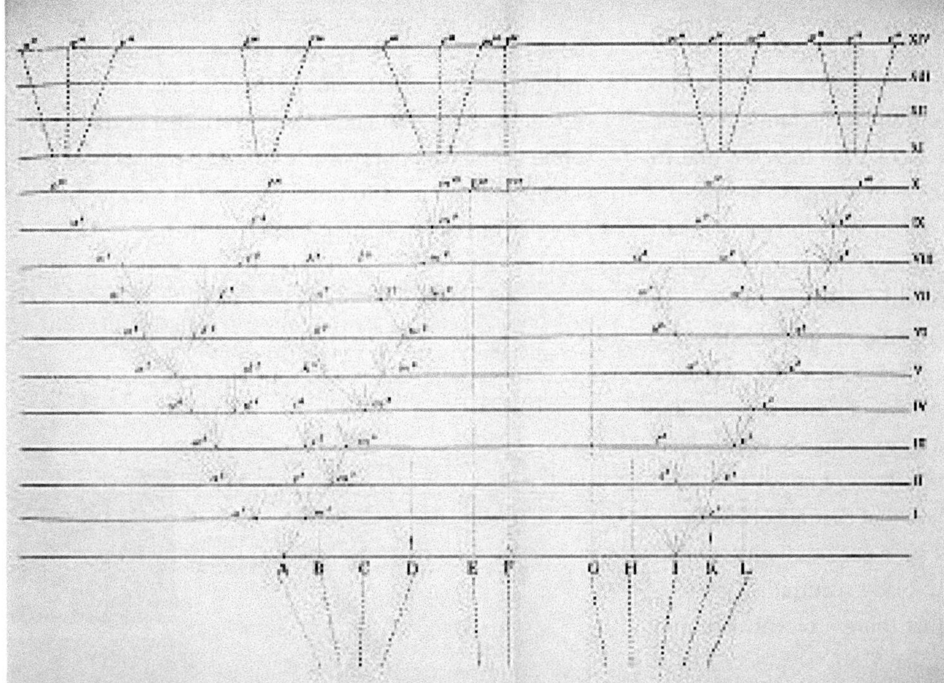

(a) Illustration of tree from *the Origin of Species*

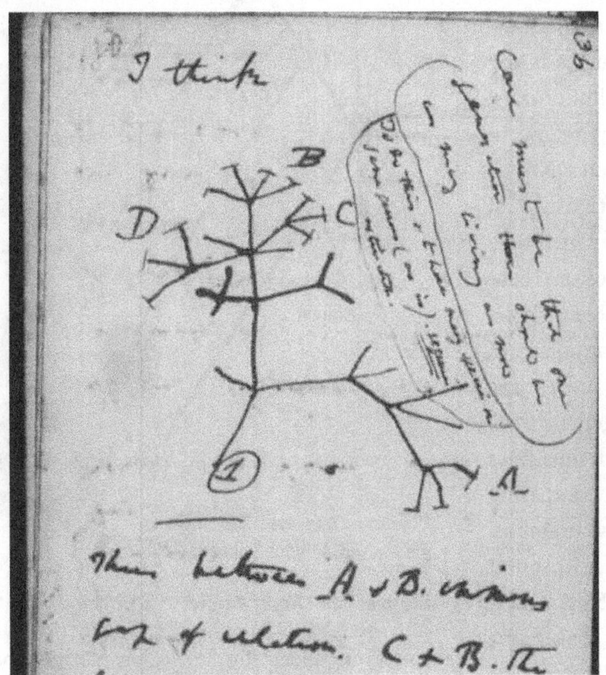

(b) Tree diagram from Darwin's notes

Figure 4.27 Darwin's trees. (a) A phylogenetic tree that appeared in Darwin's *The Origin of Species* as the only figure. (b) A figure from Darwin's field notebook B from 1837, in which he drew a phylogenetic tree and the words "I think." A branching pattern of descent with modification lies at the heart of Darwin's theories of evolution.

Navigating DNA-based phylogenetic trees

Genetic change typically occurs slowly over time. The rate of change is not always constant, and, at certain junctures, the rate at which mutations occur may increase due to environmental conditions such as exposure to **mutagens**. Changes in the environment also affect which changes in the genome survive better. While we may not recognize all of these changes in the genome as they occur, we can summarize them as a substitution rate. The substitution rate refers to the number of differences in the sequence among current species when compared to a particular standard. Variations in substitution rates have implications for building phylogenetic trees, but we will start with the simplifying assumption that the rate of change is more or less constant over the time periods being considered. As a consequence of this steady accumulation of sequence changes, organisms that share a recent common ancestor will have more similar DNA sequences than those whose common ancestor is more distant.

Thus, if two DNA sequences are closely related, we can infer that they must have shared a recent common ancestor. (This is especially useful if neutral changes in the DNA sequence are used to estimate the substitution rate, so that the effects of selection can be treated separately.) In this way, the similarities and differences between DNA sequences can be used to infer evolutionary relationships of populations or species, including the order in which they may have each diverged from a common ancestor. These sequence relationships can be presented as a table with the number of changes between each sequence being analyzed, but it is often easier and more informative to present the relationships pictorially in the form of a phylogenetic tree.

One such tree is shown in Figure 4.28, which illustrates the evolutionary relationships inferred by comparing DNA sequences from humans, chimpanzees, bonobos, gorillas, and orangutans. We can use this tree to point out some basic features and terminology. Notice first that the tree consists of branches and nodes, that is, the points from which two or more branches diverge. The **branches** represent evolutionary lineages. A **node** within the tree—an internal node—represents the hypothetical last common ancestor of the lineages descended from it. The internal nodes are therefore typically inferred extinct populations for which no direct DNA data exist, although, in some cases, such as rapidly reproducing microbes, the ancestor at the node might not be extinct.

In our example in Figure 4.28, only two branches diverge from any one node, so this tree is binary. This is the most common way to represent the divergence, although there may be unusual situations when more than two branches arise from the node. There are also nodes at the very ends of the branches. These "terminal nodes" correspond to the data points such as DNA sequences or individuals that are used to build the tree. These are the data we collect in order to infer the rest of the tree.

KEY POINT Branches on a phylogenetic tree represent evolutionary lineages. The nodes where the branches intersect represent the last common ancestor to those lineages.

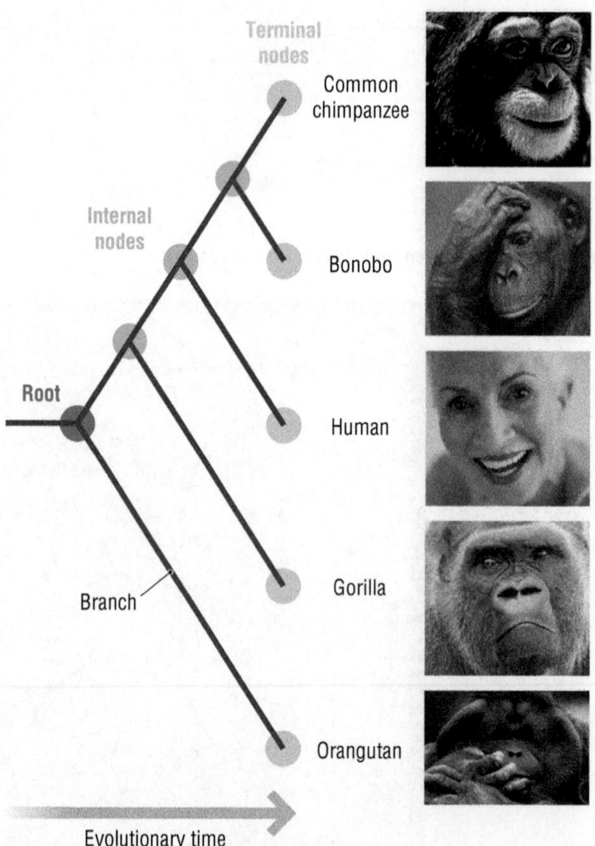

Figure 4.28 An example of a phylogenetic tree. The tree depicts evolutionary relationships based on DNA sequence data among higher primates. Key features of phylogenetic trees are highlighted: the branches of the tree represent evolutionary lineages; internal nodes from which two branches diverge represent the presumed last common ancestor of the lineages descended from it; the terminal nodes point to the individuals/data used to build the tree; the node at the root of the tree represents the last common ancestor of all the lineages in the tree; and evolutionary time proceeds from the root to the terminal nodes.

 Tree layout reproduced from Olson, M.V. and Varki, A. (2003). Sequencing the chimpanzee genome: insights into human evolution and disease. *Nature Reviews Genetics* 4, 20–28.

In a DNA-based phylogenetic tree, the terminal node may be listed as a species. This may require some clarification. For example, when a terminal node of a DNA tree is labeled as "human," it is typically a specific DNA sequence from one person, rather than the entire human genome from many people. So strictly speaking, the arrangement of nodes in the tree really represents the relationships between the DNA sequences of particular individuals in different species. This is usually, although not always, the same as the relationships between the species if many sequences were used.

Phylogenetic trees can be drawn vertically or horizontally. Although it is more usual to draw them horizontally, as in Figure 4.28, the actual orientation has no significance. Overall, the arrangement of the nodes and branches in the tree lays out the deduced evolutionary relationship and pattern of shared ancestry between the lineages in the study. It would be the same pattern, even if the entire diagram were rotated into a different orientation.

At the base of the tree is the **root**, the oldest point on the tree. A node at the root therefore represents the hypothetical last common ancestor of all the individuals in the tree. Knowing which point on the tree is the oldest enables us to determine the order in which the other branches occurred over the course of evolution. Taking Figure 4.28 as an example, knowing the root of the tree allows us to determine that the branch in the phylogeny leading to the gorilla lineage occurred before the branch that separated the chimpanzees and humans into separate lineages. Rooted trees therefore have directionality, proceeding from the root as the oldest point in time to the terminal nodes as the most recent points in time.

Trees are rooted using an "outgroup"—that is, an organism known to have shared a common ancestor further in the past than all the individuals whose evolutionary relationship is being studied. For example, the phylogenetic tree of humans and apes shown in Figure 4.28 could have been rooted by including DNA data from an Old World monkey since the monkey lineage is known to have diverged from a common ancestor prior to the lineage that gave rise to the humans and apes in this tree.

While rooted trees reflect the relative times since the divergence from a common ancestor, an unrooted tree shows relationships between the organisms of interest without necessarily providing information about the last common ancestor for the group or the temporal order of the branching. Unrooted trees can be used when data from an outgroup are not available. As an alternative to an outgroup, other data, such as evidence from the fossil record, may be used to estimate the position of the root of the tree instead.

KEY POINT Trees can be rooted by using an outgroup that shared a common ancestor to all of the other species on the trees. Rooted trees reflect the times of divergence from a shared common ancestor. Unrooted trees show the relationships among the species without indicating the relative times since they shared a common ancestor.

We pause here in our description of evolutionary trees to emphasize a point that is often missed, at least in the way many people think about evolutionary change. Note that the branches in the tree come from a common ancestor, but neither branch is itself the common ancestor. In the tree in Figure 4.28, the human and chimpanzee lineages can be traced back to a shared internal node (highlighted in blue). This node corresponds to the last common ancestor of chimps and humans, which lived roughly 5 million years ago. That common ancestor is extinct, and neither humans nor chimps are that common ancestor, although both are similar to it. We often hear people talk about humans being descended from chimps, but this is not what evolutionary trees are depicting or what biologists believe happened. The trees show that there was a species in the past that diverged into two separate species roughly 5 million years ago, neither of which is that original species.

The node itself is not the point at which the changes occurred but rather the last common ancestor from which those changes occurred. With many changes in sequence and phenotype (and probably many other branches) since that time of divergence, one species ultimately became what we see as the modern chimpanzee, while the other became what we now see as humans, or more properly as anatomically modern humans. Similarly, as illustrated in Box 4.4, although modern-day plague bacteria descended from the Black Death strain, none of them is the Black Death strain. The Black Death strain is located at a node, but it is extinct, and the data were obtained from fossilized bacteria.

While it might seem obvious upon reflection, it is important to understand that a phylogenetic tree of DNA sequences only depicts the evolutionary relationship between individuals whose DNA sequence was included in the analysis. A DNA-based phylogenetic tree does not show the many related lineages that likely arose and produced individuals not included in the analysis. In reality,

BOX 4.4 *A Human Angle* Analyzing the Black Death by phylogenetics

Yersinia pestis is a bacterial species that causes a deadly disease known as the plague. Currently, outbreaks of the plague occur in very few parts of the world, but ancient outbreaks had a major role in shaping human history. In the fourteenth century, a plague outbreak, referred to as the Black Death, swept through Europe, as shown on the map in Figure A, killing 30 million people—over a third of the population—and causing perhaps as many to flee, further spreading the disease. In addition to its impact on the genetic structure of modern European populations, the Black Death affected European art and culture; stories and songs about the Black Death are common, and plague statues to commemorate survivors are found in many European cities.

A few years ago, modern genomic technology examined the Black Death. Victims of a medieval plague outbreak were exhumed from East Smithfield in London; DNA from the infecting bacteria was extracted from their teeth, and the genomes were sequenced. The sequences of those strains were compared with sequences from present-day plague outbreaks; sources of these modern outbreaks are represented by green dots on the map in Figure B. A phylogenetic tree representing the genetic relatedness of the strains was constructed, based on 1694 informative positions across the genome.

Present-day strains that cause the plague differed from the Black Death strain by fewer than 100 nucleotide sites. As shown in Figure B, the fourteenth-century East Smithfield strain shared an ancestral node with present-day strains, with the exception of lineages from Asia, strongly suggesting that a single lineage was responsible for the medieval strain and

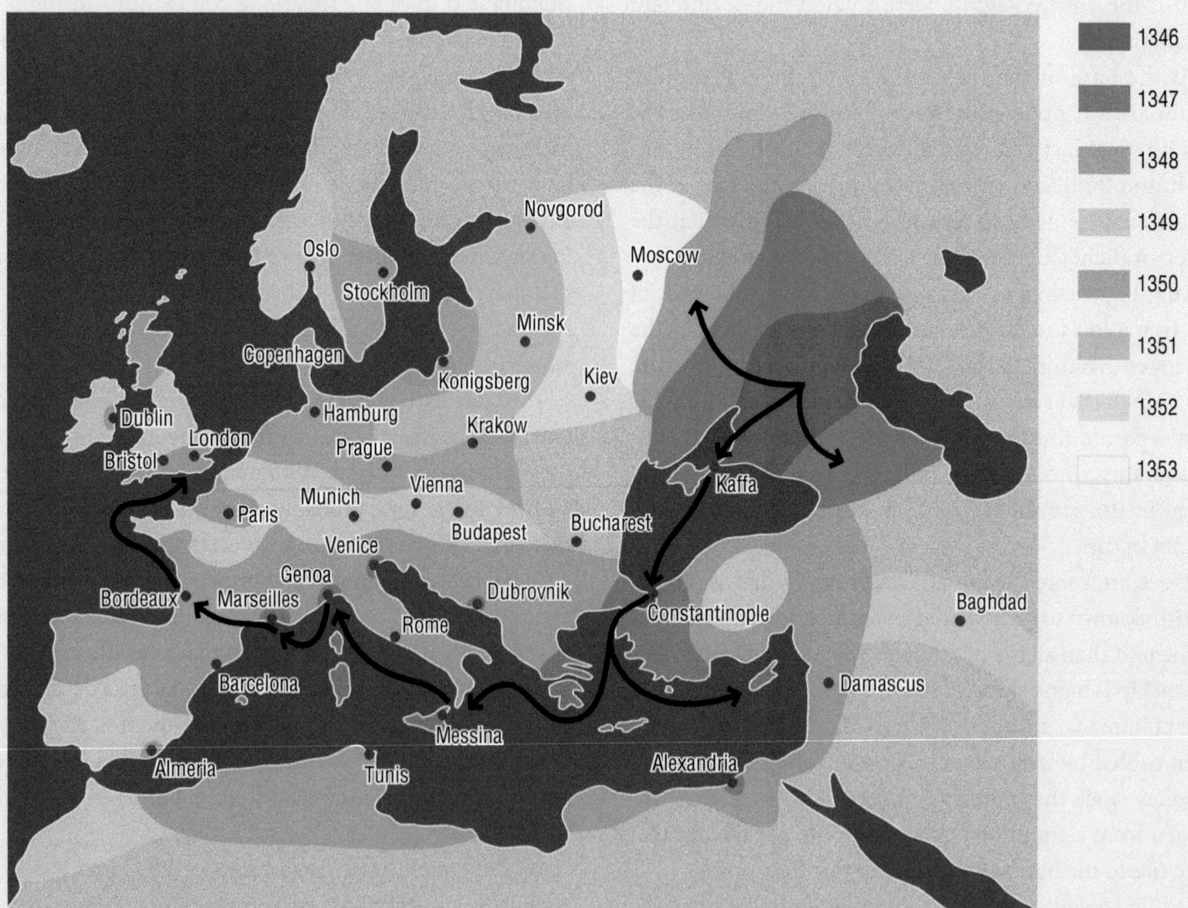

Figure A The spread of the plague. The plague, also known as the Black Death, arose in ports around the Black Sea in 1346 and spread rapidly across Europe in the next decade. The years when its appearance were recorded are shown by the different colors.

Source: Reprinted by permission from Macmillan Publishers Ltd: Bos KI, et al. Draft genome of *Yersinia pestis* from victims of the Black Death. *Nature*. (2011) Oct 12. Vol. 478, Iss. 7370.

BOX 4.4 Continued

all non-Asian outbreaks since. As the tree calculations placed the East Smithfield strain at an ancestral node, the Black Death strain is the probable ancestor of modern disease-causing *Yersinia pestis.*

FIND OUT MORE

Bos, K. I., Schuenemann, V. J., Golding, G. B., *et al.* (2011) Draft genome of *Yersinia pestis* from victims of the Black Death. *Nature* **478**: 506–10

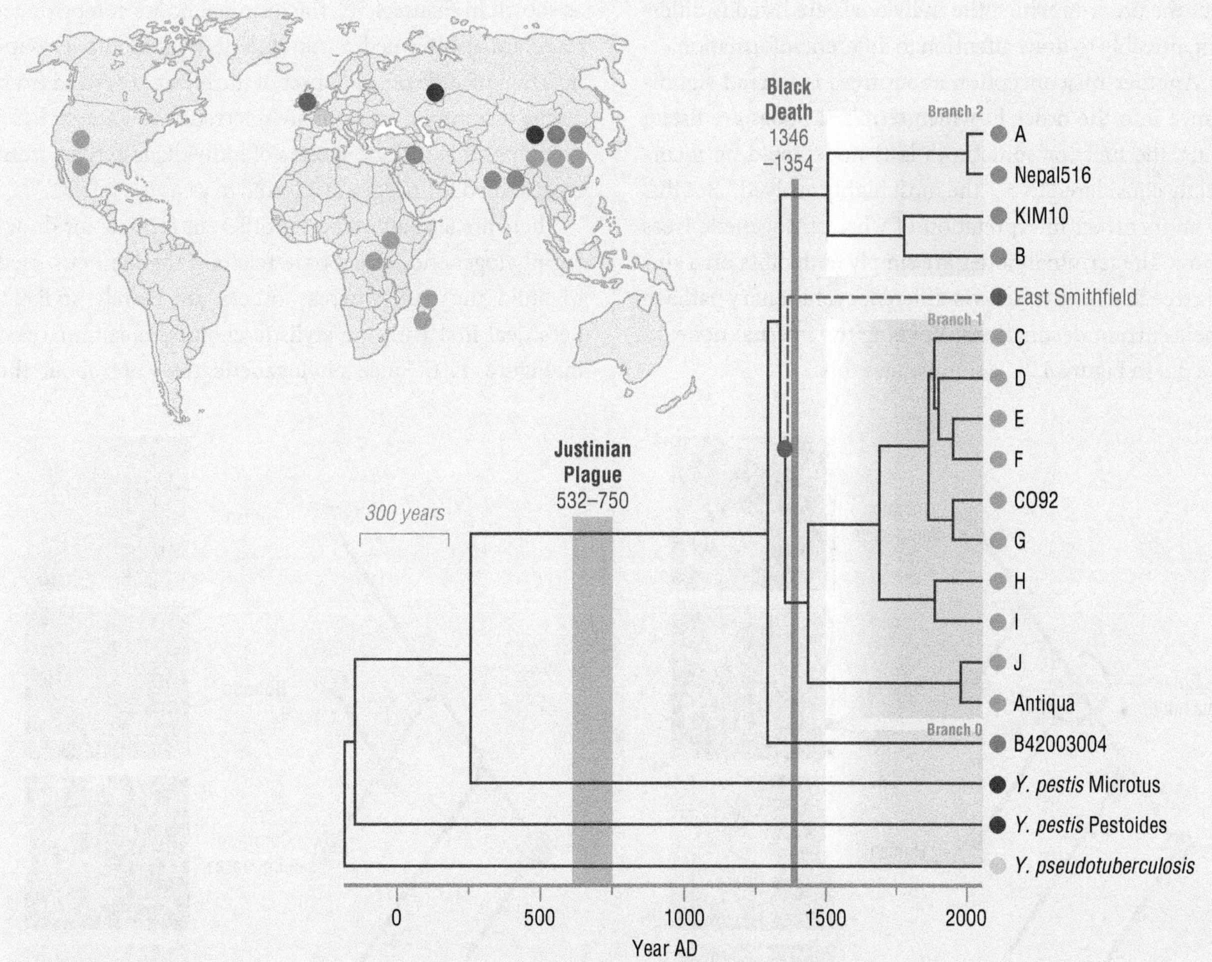

Figure B Phylogeny of *Yersinia pestis*, the bacterium responsible for the plague. By comparing sequences from victims of the plague who died at different times and in different locations, a tree showing the relationship among the isolates can be constructed. Modern samples are represented by green dots, while the sample from approximately 1350 in Smithfield, England, is shown in red. The tree was rooted using samples from *Yersinia pseudotuberculosis*, which does not cause the plague. Note the close similarity between the clusters shown by the sequences and the spread of the bacteria in Figure A.

Source: Reprinted by permission from Macmillan Publishers Ltd: Bos KI, et al. Draft genome of *Yersinia pestis* from victims of the Black Death. *Nature.* (2011). Vol. 478, 7370.

many related lineages could have resulted in individuals that are now extinct and for which DNA data are not generally available. So while a larger number of terminal nodes than internal nodes might reflect greater species diversity at the end point of the tree, this may also simply reflect bias in the data available to construct the tree.

Different ways trees are drawn

Many nodes on a tree can be rotated with respect to one another without altering the overall topology or branching pattern of the tree. As a result, the proximity of terminal nodes to one another is not necessarily indicative of relationships and may simply reflect convenience or emphasis. For example, we could have drawn the tree in Figure 4.28 just as accurately by rotating the central internal node highlighted in red in Figure 4.29(a), with humans at the top followed by bonobos and then chimpanzee; this revised tree is shown in Figure 4.29(b). The evolutionary relationships between the individuals in both versions of the tree are identical—the tree topology has not changed. But the order in which the individuals are listed is different, possibly to draw attention to different information.

Another misconception about trees is to read significance into the order in which terminal nodes are listed; thus, the first (or sometimes last) node could be incorrectly considered to be "the most highly evolved." But this is an incorrect interpretation of what phylogenetic trees show. The terminal nodes are simply endpoints on a single tree that have arisen via different evolutionary paths to their current destinations. Rotating the internal node, as we did in Figure 4.29, demonstrates this.

The groupings that are found in a tree are more informative than the order in which the terminal nodes are listed. Organisms are more closely related if they share more common ancestors, that is, if they branch from the same node. The term **clade** refers to all the lineages that are grouped together, such that they would all be removed from the tree if an internal node was conceptually "clipped" from the tree, as shown in Figure 4.30. The terminal nodes comprising a clade are also termed a **monophyletic** grouping. Groupings that include only a subset of the lineages derived from a shared common ancestor are referred to as **paraphyletic**; groupings that include subsets of individuals derived from more than one lineage are referred to as **polyphyletic**.

There are also a number of different formats for drawing phylogenetic trees. Some relate to the methods used to build the tree, whereas others are merely stylistic. Let's deal first with the stylistic choices, as summarized in Figure 4.31. Since phylogenetic trees are about the

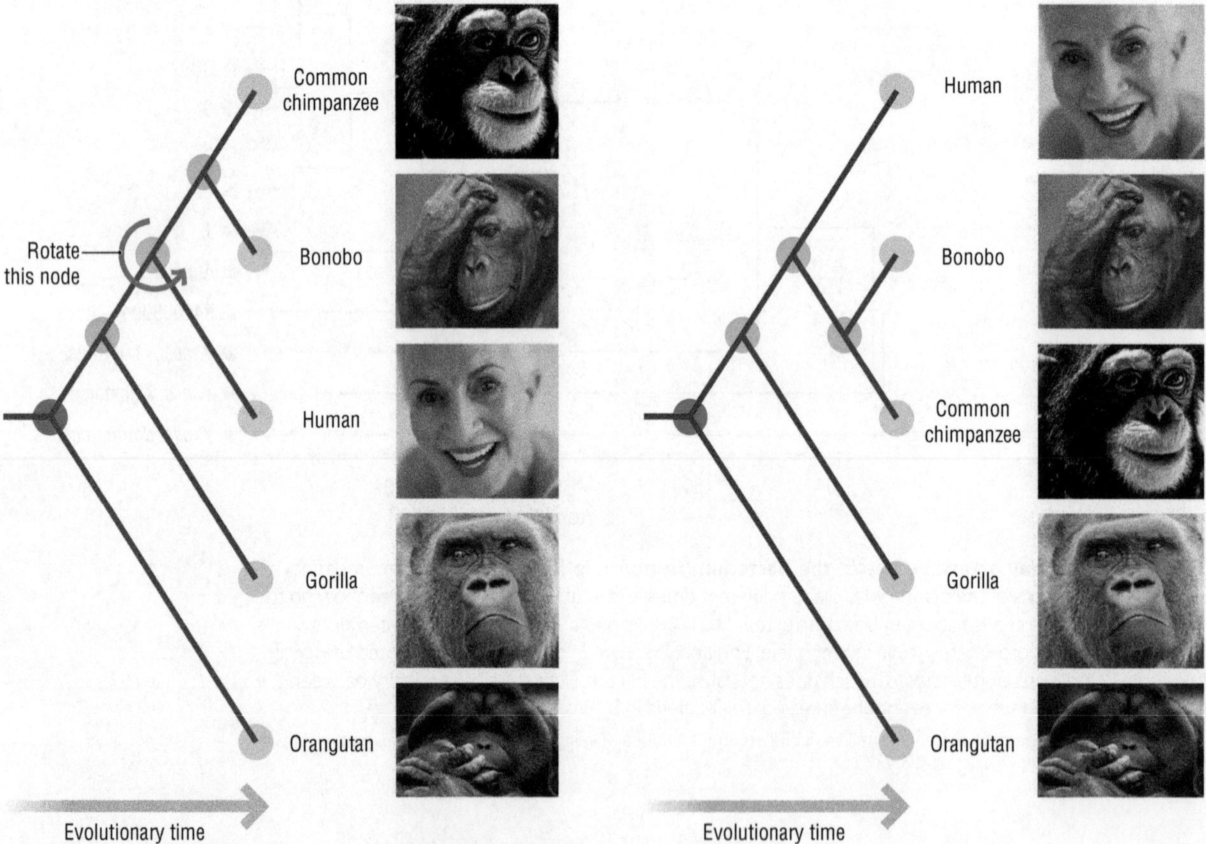

Figure 4.29 The internal nodes in a tree can be rotated. The trees shown are equivalent to one another. The tree on the left is the same as the one in Figure 4.28. The tree on the right differs in that the node from which the human, bonobo, and chimpanzee lineages split has been rotated, but the branching pattern and relationship between the lineages are conserved. When a node is rotated, it will change the order in which the terminal nodes are listed without changing the relationships between them.

Source: Image of common chimpanzee courtesy of Thomas Lersch/ CC BY-SA 3.0. Image of bonobo courtesy of Evanmaclean. Image of gorilla courtesy of Kabir Bakie/ CC BY-SA 2.5. Image of orangutan courtesy of Zyance/ CC BY-SA 2.5. Tree layout reproduced from Olson, M.V. and Varki, A. (2003). Sequencing the chimpanzee genome: insights into human evolution and disease. *Nature Reviews Genetics* 4, 20–28.

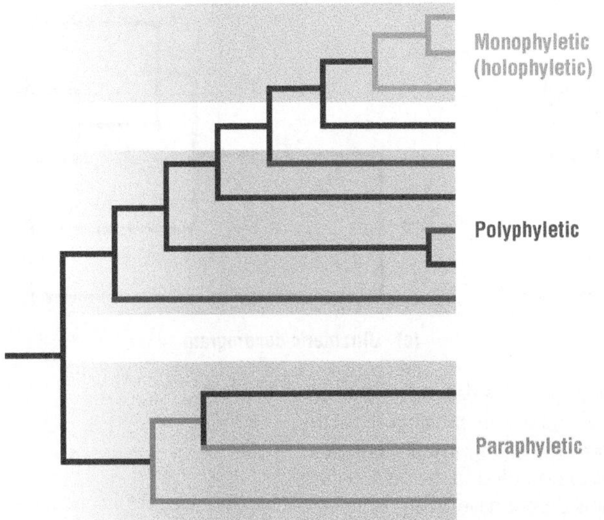

Figure 4.30 Nodes on a tree can be grouped in different ways. A grouping that comprises an internal node (an ancestor) and all lineages derived from that node is termed "monophyletic." Everything in a monophyletic group is descended from a single common ancestor and comprises a clade. A grouping comprising an internal node but leaving out some of the lineages derived from that node is referred to as "paraphyletic." A grouping that does not contain a common ancestor for the members in the group is referred to as "polyphyletic."

Figure adapted from article by Sandra L. Baldauf, *Trends in Genetics* Vol.19 No.6 June 2003.

relationships between the nodes and branches, it doesn't really matter how these are drawn so long as the relationships are depicted correctly. There are three basic style choices that are most commonly used: a V pattern of branching as in Figure 4.31(a), a pattern that connects the lineages through a series of brackets as in Figure 4.31(b), or a radial pattern. Any phylogenetic tree can be drawn with any of these styles. Trees drawn in a radial style can help to avoid some of the misinterpretations associated with the order in which terminal nodes are listed in other styles of trees, but all three styles are widely used.

The next difference in the ways trees may be drawn relates to the lengths of the branches, as summarized in Figure 4.32. Trees can be drawn in which the lengths of the branches have no meaning. These types of trees, referred to as **cladograms** and depicted in Figure 4.32(a), are often used when drawing trees inferred from morphological comparisons but are not used as commonly for phylogenies built using DNA sequence data for which the amount of change can be calculated. In phylogenetic trees built using DNA sequence data, the lengths of the branches are usually proportional in length to the amount of change that has occurred since they last shared a common ancestor. Trees with branch lengths proportional to the degree of sequence change are referred to as **phylograms**; an example is shown in Figure 4.32(b). Some trees are built with meaningful branch lengths, but the terminal nodes are constrained to all be equidistant from the root, as shown in Figure 4.32(c). These are referred to as ultrameric trees or **dendrograms**. These tend to work well when the distances are being measured in units of time, since all the current species date to the same point in time (the present) since the existence of the root of the tree.

KEY POINT Cladograms reflect the relatedness of the organisms but not the amount of divergence among them. Phylograms include information on the amount of divergence by having the terminal nodes in different places, while dendrograms align the terminal nodes at the same place and show the divergence through the placement of the internal nodes and branches.

Trees vary in the type of data used to build them. Originally, evolutionary trees were constructed using phenotypic and morphological data, whose connection

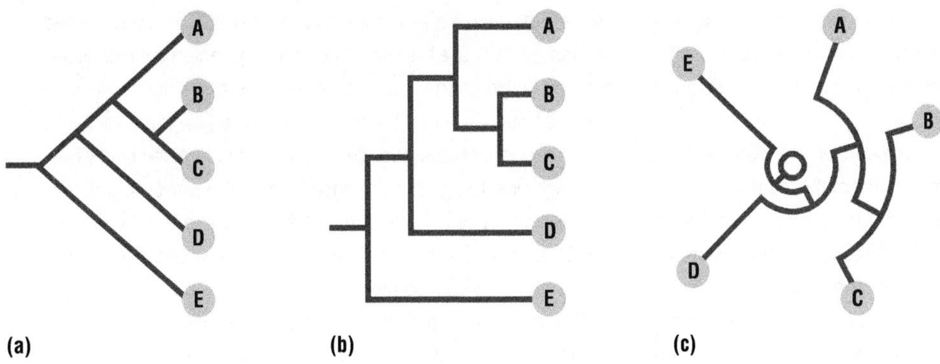

Figure 4.31 Different styles of representing the same phylogenetic tree data. The same phylogenetic tree can be drawn in various different styles such as a V-shaped branching tree in (a), the bracketed branching version in (b), and radial branching in (c). The style in which the tree is drawn does not affect the relationships.

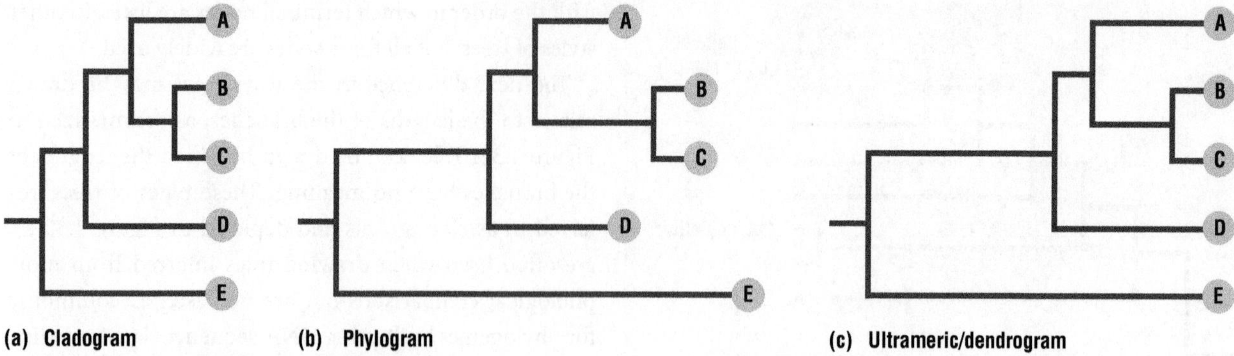

Figure 4.32 Different types of phylogenetic trees. Different types of phylogenetic trees convey different information about the lineages on the tree. A cladogram, as shown in (a), conveys only information about the relationships between lineages, and the branch lengths have no significance or meaning. In a phylogram, as shown in (b), the branch lengths are proportional to the number of changes occurring to the DNA in each lineage. Thus, phylograms include quantitative information that cladograms do not have. In an ultrametric tree or dendrogram, as shown in (c), the branch lengths are proportional as in a phylogram, but the distance from root to all terminal nodes is constrained to be equal. This aligns the terminal nodes at the same place in the diagram.

to the genotype was unclear and possibly even indirect. DNA sequence information is now widely used to build phylogenetic trees, as the data have become easier to procure and theoretical models for how sequences change over time have been developed. Other types of molecular data, such as amino acid sequences, can also be used by analogous methods. Even with sequence data becoming available, morphological data from the fossil record still provide an important component to understanding evolutionary relationships and histories.

In Tool Box 4.3, we use some simple and short DNA sequences to show some principles involved in constructing phylogenetic trees.

TOOL BOX 4.3 Building phylogenetic trees using DNA sequence data: some examples

Many different methods are used to construct phylogenetic trees using DNA sequences. As we noted in the chapter, none of these can guarantee the best possible tree, even when only a few sequences of modest length are being compared. While many methods are used, all of them begin by tabulating the number of differences between the two sequences being compared. A distance matrix is a table that records the number of differences between each pair of sequences.

Different methods can be used to convert the data from the distance matrix into a tree diagram; many of these same methods are used for clustering data sets in fields such as computer science, economics, and linguistics. In the following two examples, we will use small stretches of DNA sequence to illustrate how a simple phylogenetic tree is built. In actual research applications, much longer DNA sequences would be used to create these trees. Similarly, we write out each method in full here to illustrate the logic of the approach, but, in reality, a number of algorithms and computer software packages are available to process the sequence data.

Distance methods are the easiest approaches, in terms of the required computation, and work very well when the divergence between the sequences is low, that is, for closely related sequences. Because the similarities and differences are calculated over the length of the sequence, the simplest versions of distance methods, such as we describe here, are based on the underlying assumption that the sequence can be treated as a unit of change, and each position in the sequence is given the same weight, that is, a change at position 35 is treated the same as a change at position 36, even if one of them is the third base in a codon or part of some functional component.

This simplifying assumption that every position in the sequence is of equal likelihood to change allows distance methods to be quick and computationally simple but generally does not reflect the complex realities of how biological molecules change

TOOL BOX 4.3 Continued

over time. A variety of models have therefore been developed that attempt to account for some of these complexities, which go beyond our discussion here. For example, there may be hidden differences—a position has changed more than once or changed back to its original base—such that similarities in two sequences may have arisen independently.

Unweighted pair-group method with arithmetic mean

A distance method of tree building, the unweighted pair-group method with arithmetic mean (UPGMA) identifies DNA sequences with the greatest similarity and groups these together. The series of panels in Figure A walks through the process step by step. The

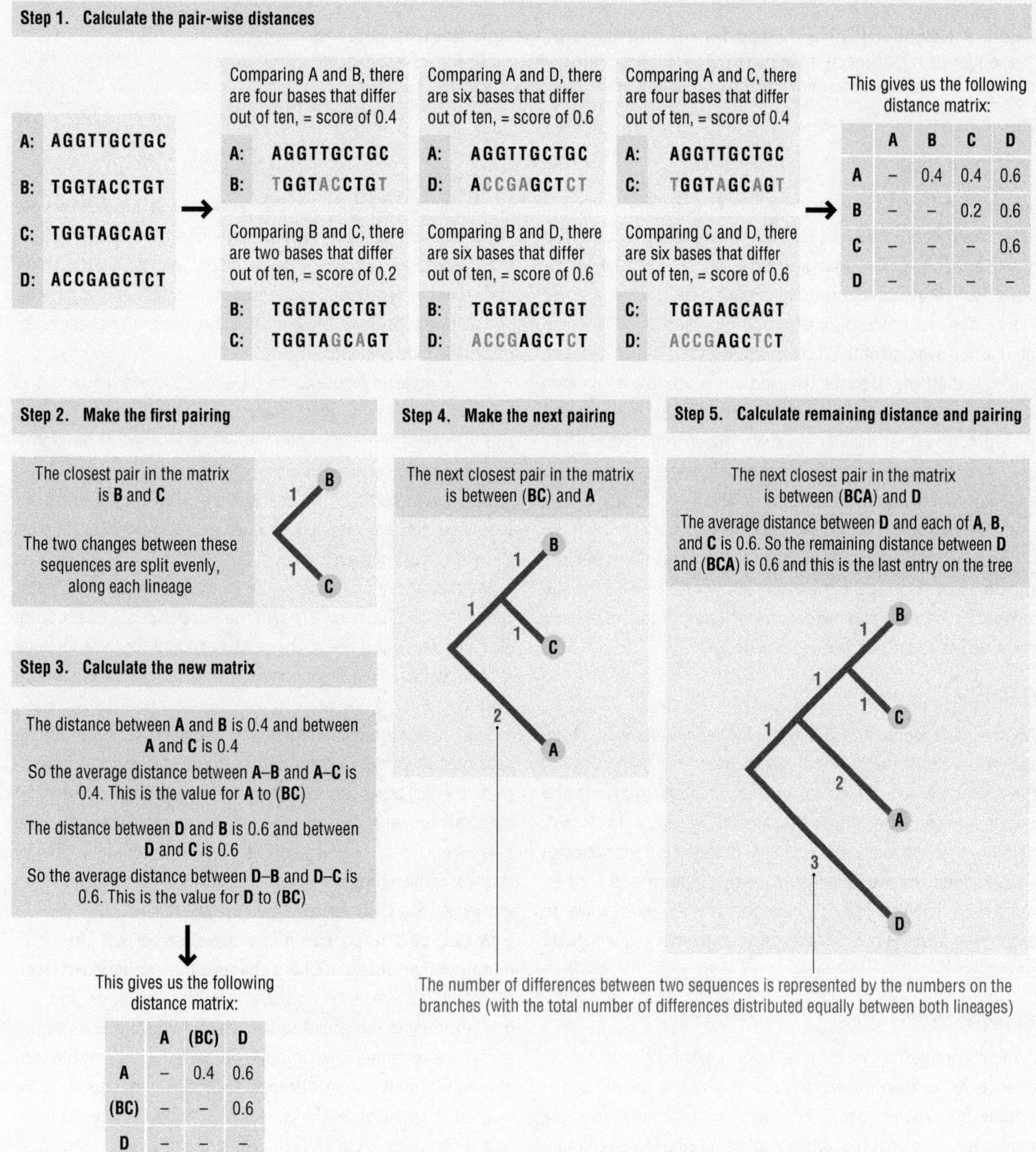

Step 1. Calculate the pair-wise distances
A: AGGTTGCTGC
B: TGGTACCTGT
C: TGGTAGCAGT
D: ACCGAGCTCT
Comparing A and B, there are four bases that differ out of ten, = score of 0.4
A: AGGTTGCTGC
B: TGGTACCTGT
Comparing A and D, there are six bases that differ out of ten, = score of 0.6
A: AGGTTGCTGC
D: ACCGAGCTCT
Comparing A and C, there are four bases that differ out of ten, = score of 0.4
A: AGGTTGCTGC
C: TGGTAGCAGT
Comparing B and C, there are two bases that differ out of ten, = score of 0.2
B: TGGTACCTGT
C: TGGTAGCAGT
Comparing B and D, there are six bases that differ out of ten, = score of 0.6
B: TGGTACCTGT
D: ACCGAGCTCT
Comparing C and D, there are six bases that differ out of ten, = score of 0.6
C: TGGTAGCAGT
D: ACCGAGCTCT
This gives us the following distance matrix:

	A	B	C	D
A	–	0.4	0.4	0.6
B	–	–	0.2	0.6
C	–	–	–	0.6
D	–	–	–	–

Step 2. Make the first pairing
The closest pair in the matrix is B and C
The two changes between these sequences are split evenly, along each lineage
Step 3. Calculate the new matrix
The distance between A and B is 0.4 and between A and C is 0.4
So the average distance between A–B and A–C is 0.4. This is the value for A to (BC)
The distance between D and B is 0.6 and between D and C is 0.6
So the average distance between D–B and D–C is 0.6. This is the value for D to (BC)
This gives us the following distance matrix:

	A	(BC)	D
A	–	0.4	0.6
(BC)	–	–	0.6
D	–	–	–

Step 4. Make the next pairing
The next closest pair in the matrix is between (BC) and A
Step 5. Calculate remaining distance and pairing
The next closest pair in the matrix is between (BCA) and D
The average distance between D and each of A, B, and C is 0.6. So the remaining distance between D and (BCA) is 0.6 and this is the last entry on the tree
The number of differences between two sequences is represented by the numbers on the branches (with the total number of differences distributed equally between both lineages)

Figure A Unweighted pair-group method with arithmetic means for constructing phylogenetic trees. This series of panels provides a brief stepwise guide to constructing a tree using UPGMA, one of the simplest methods of building a tree.

TOOL BOX 4.3 Continued

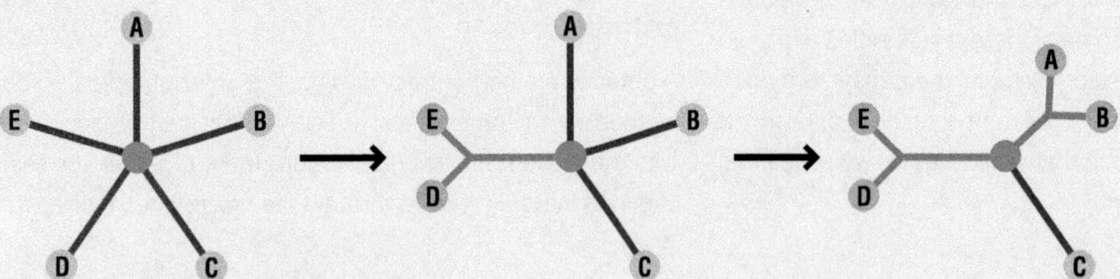

Figure B Neighbor joining method for constructing phylogenetic trees. Neighbor joining displays the sequence in a connected star pattern. The lengths of the branches are adjusted to indicate the distances between the sequences to minimize the total length of the branches in the tree.

first step is to build a pair-wise matrix to tabulate the number of differences between each pair of DNA sequences. The closest pair of sequences B and C is then grouped together. The distance matrix is then recalculated, using the grouped pair (BC) as a single entry in the matrix. The distance to this grouping of (BC) is calculated as the average of the distances to B and C. The next closest pairing is then added to the tree, and this procedure is repeated until all sequences are located on the tree. The final tree is shown in step 4. Note that the tree is usually drawn with proportional branch lengths to reflect the number of differences between each pair of lineages, with the total number of differences split evenly along the two branches that diverge from each node.

UPGMA is based entirely on pairs of sequences taken in order and does not examine the overall topology of the tree or re-evaluate the original pairs as more data are included; distances are simply added as additional pairs are included.

Neighbor joining

Another distance approach, neighbor joining, starts with all the sequences connected in a star arrangement and then pulls out the pairings of similar sequences that ultimately minimize the total length of the tree. This is illustrated in Figure B. As in UPGMA, pairs of sequences are used, but, in this case, the distances are not simply added; the overall length of the tree is minimized. Furthermore, with neighbor joining, sequences that are most similar to each other **and** the most different from the other sequences are paired.

Parsimony

A more complex approach that looks at positions within a sequence, rather than the sequence as a whole, is known as parsimony. "Parsimony" means "economy" or "thriftiness," so these methods are the ones that attempt to identify the fewest changes needed to assemble the tree. Each position in the sequence is examined, one at a time; all the possible trees that can account for the differences between the sequences at this position are deduced, and the resulting tree that involves the fewest possible changes overall is chosen. A nucleotide is ignored if it is uninformative, either because all sequences are identical at that position or if all tree topologies require the same number of changes for that nucleotide position.

For example, in Figure C, we have four aligned sequences: E, F, G, and H. The four sequences are identical at positions 3, 4, 7, 8, and 9, so these sites are uninformative and ignored. At position 1, there are three possible tree topologies for the four sequences. For each tree topology, the changes that would account for the sequence data are mapped onto each of the possible trees, as shown in step 2 of Figure B.

We then move on to the next informative nucleotide—in this case, the one at position 2. Again the three possible tree topologies are drawn, and the changes that account for the sequence data at position 2 are mapped onto these trees, as shown in step 3 of Figure B.

The process is repeated for each of the informative nucleotide positions, as shown in steps 4–6 in Figure B. When this is complete, then all three possible tree topologies are considered, and the total number of changes needed to account for the sequence data for each tree is calculated. The tree with the least number of changes is then chosen as the most parsimonious account of the sequence data, as shown in step 7 of Figure B.

A tree built using parsimony therefore reflects the minimum number of evolutionary changes that would have taken place since divergence from the common ancestor, and not from examining the entire sequence as a unit but considering each base separately. Parsimony approaches are computationally more intense than distance approaches but tend to be favored in terms of reliability. However, they become less reliable when there is considerable variation in branch lengths, in other words, when both closely related and distantly related sequences are used.

TOOL BOX 4.3 Continued

Step 1.
The sequences to be compared (E, F, G, and H) are aligned
Nucleotides that are not informative are ignored

	Nucleotide position									
	1	2	3	4	5	6	7	8	9	10
E	T	C	T	C	C	A	T	G	C	A
F	A	G	T	C	G	A	T	G	C	T
G	A	C	T	C	C	A	T	G	C	A
H	A	G	T	C	G	T	T	G	C	T

Step 2. All possible trees are built for the first informative nucleotide (nucleotide #1)

	1
E	T
F	A
G	A
H	A

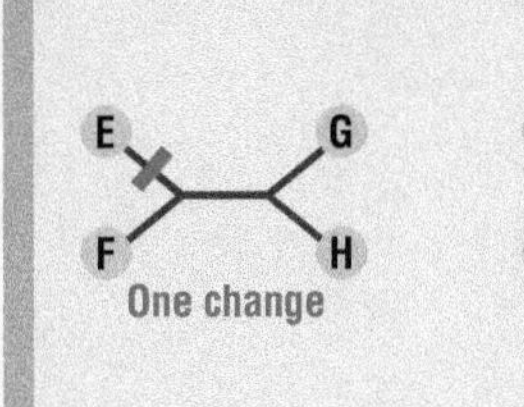

E G H F — One change

E F H G — One change

Step 3. All possible trees are built for the next informative nucleotide (nucleotide #2)

	2
E	C
F	G
G	C
H	G

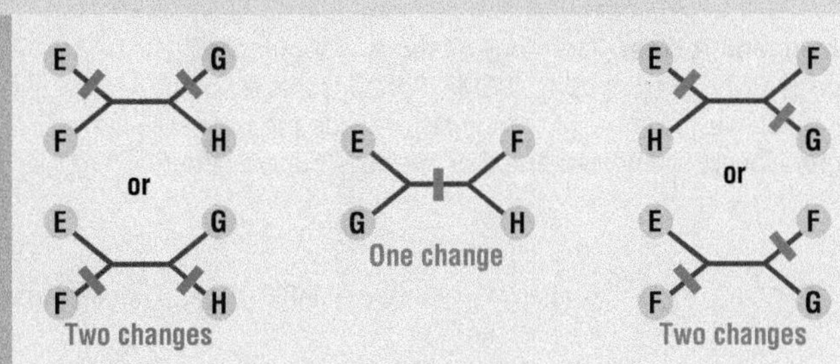

Step 4. All possible trees are built for the next informative nucleotide (nucleotide #5)

	5
E	C
F	G
G	C
H	G

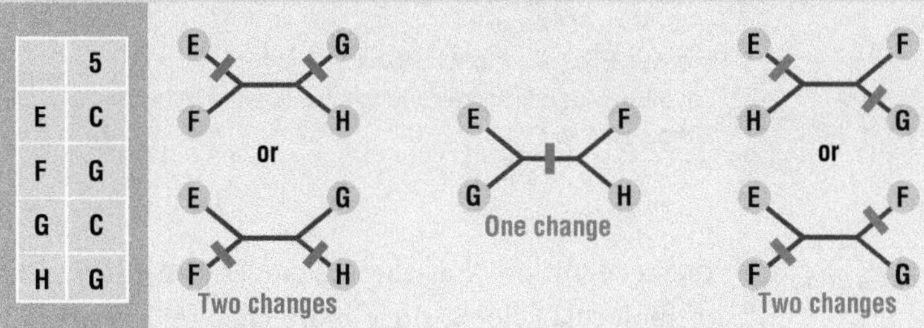

Step 5. All possible trees are built for the next informative nucleotide (nucleotide #6)

	6
E	A
F	A
G	A
H	T

E G F H — One change

E F G H — One change

E F H G — One change

TOOL BOX 4.3 Continued

Step 6. All possible trees are built for the last informative nucleotide (nucleotide #10)

	10
E	A
F	T
G	A
H	T

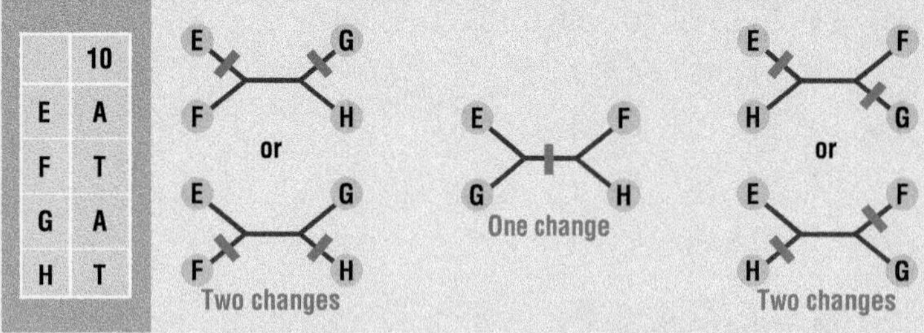

Step 7. Putting all the possible trees together to choose the most parsimonious:

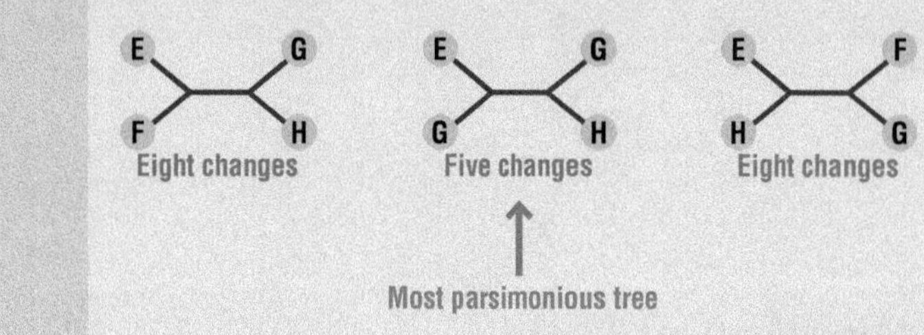

Figure C Parsimony method of constructing phylogenetic trees. Parsimony methods look at individual positions in the sequence, rather than the sequence as a whole string, as in UPGMA. Positions that have not changed are ignored. The selected tree is based on the fewest number of substitutions needed to connect the sequences. Parsimony methods are more computationally intensive than other methods but are generally considered to produce more reliable trees.

FIND OUT MORE

An overview of parsimony methods:

Baldauf, S. L. (2003) Phylogeny for the faint of heart: a tutorial. *Trends Genet* **19**: 345–51

Dowell, K. (2008) *Molecular phylogenetics: an introduction to computational methods and tools for analyzing evolutionary relationships*. Available at: http://www.math.umaine.edu/~khalil/courses/MAT500/papers/MAT500_Paper_Phylogenetics.pdf [accessed 1 April 2016]

National Center for Biotechnology Information. *Parsimony methods.* Available at: http://www.ncbi.nlm.nih.gov/Class/NAWBIS/Modules/Phylogenetics/phylo14.html [accessed 1 April 2016]

A discussion of phylogeny:

Singh, M. (1999) *Phylogeny*. Available at: http://www.cs.princeton.edu/~mona/Lecture/phylogeny.pdf [accessed 1 April 2016]

VIDEO 4.1 demonstrates the creation of a phylogenetic tree using DNA sequences. Find it at **www.oup.com/uk/meneely**

Phylogenetic trees represent descent with modification

The mutations that accumulate in DNA sequences over time can be analyzed and compared to infer evolutionary history and relationships between the sequences. These analyses are based on the process of changes that arise during DNA replication, as well as the recognition that these changes constitute a type of molecular clock, since more changes are predicted to accumulate over time. The construction of realistic phylogenetic trees uses much more data than the simplified examples in Tool Box 4.3. As shown in Table 4.4, the number of trees that can be constructed increases dramatically with the number of sequences that are available. Using only the sequence for a single gene from ten species of mammals, there are more than 34 million possible rooted trees. Thus, even with a relatively small and defined set of data points, it is exceptionally difficult to construct an "optimal tree;" in

Table 4.4 The number of rooted and unrooted trees with different data sets

No. of sequences	Possible unrooted trees	Possible rooted trees
2	1	1
3	1	3
4	3	15
5	15	105
6	105	945
7	945	10,395
8	10,395	135,135
9	135,135	2,027,025
10	2,027,025	34,459,425

Note that the number of unrooted trees for n data points is the same as the number of rooted trees for n − 1 data points because any of the data points can be used as the root.

fact, it has been shown that it may not be possible to construct an optimal tree involving more than 15 sequences in an efficient time interval. As is increasingly true for many questions in biology, acquiring the data is not the limiting step; the challenge is to bring together the data that we have.

Researchers will therefore often make use of several tree-building methods in search of the most "reliable" one, and many of the approaches will produce highly similar trees. In all cases, the trees must be considered as hypotheses for the evolutionary relationships under consideration, because it is impossible to know the actual evolutionary history with absolute certainty. As with many other hypotheses, the reliability of a tree can be tested by statistical methods that are beyond the scope of this book.

How then do we know that the methods used to construct trees are the most reliable? One simple and widely used approach is to use different methods on the same data set. Many parts of the tree are likely to be the same under different methods. This usually provides a consistent and reliable view of the phylogenetic relationships. However, the parts of the tree that vary under different methods are often the ones of greatest interest. One method that has been used to test and validate the methods of constructing trees involves the use of data sets from rapidly evolving organisms, whose evolutionary history is partially or fully known, since it has occurred in a relatively short time span. While most commonly used to infer ancient histories in biology, evolutionary trees can also be used to construct more recent histories such as infectious disease outbreaks or the history of human immunodeficiency virus (HIV) within an infected individual.

The Tree of Life

The ultimate challenge in constructing phylogenetic relationships is to produce a tree that plots interrelationships among all living organisms. This is referred to as the "Tree of Life." In principle, this can be extended back to a predicted last universal common ancestor (LUCA) of all living organisms roughly 3.8 billion years ago. As the LUCA and the vast majority of the ancestors of today's living organisms are extinct, constructing the Tree of Life provides an outstanding example of how evolutionary biologists use data from present-day organisms, in conjunction with fossil records, to infer the most likely tree of relationships among living organisms.

The Tree of Life is perhaps the best illustration of the value of phylogenies based on sequences, rather than on phenotypes. For many decades, organisms were grouped into two categories, based on an easily observed phenotype: prokaryotes, which lack a nuclear membrane, and eukaryotes, which have a nuclear membrane. As more organisms without nuclear membranes were found, often in unusual ecological niches such as geysers and deep sea vents, "prokaryotes" were divided into two subcategories: the Eubacteria ("true" bacteria) and Archaebacteria (now called Archaea, from the Greek *archaea* meaning the ancient ones).

The Archaea and Eubacteria have many features in common, in addition to lacking a nuclear membrane. Both archael and eubacterial cells are small, have cell walls, and lack intracellular membranes. Prior to the 1970s, when classifications were made largely on morphological characters and microscopes had very little power to see inside prokaryotic cells, both Archaea and Eubacteria were considered to belong to the same kingdom and share a common ancestor, separate from the eukaryotes.

In the 1970s, however, it became apparent that the resemblance between the two groups is merely superficial. Archael ribosomes and cell membranes more closely resemble those of eukaryotes; the processes of transcription and replication in Archaea are comparable to those of eukaryotes, rather than those of Eubacteria, and many other archaeal processes and enzymes are more similar to those in eukaryotes than those in Eubacteria. Thus, even before molecular sequence data were introduced, questions were being raised about the relationships among Archaea, Eubacteria, and eukaryotes.

The principal methods that allowed reliable sequencing of any DNA molecule were published in 1977 and were based on the sequences of ribosomal RNA molecules. In order to acquire sequence data for Archaea, George Fox and Carl Woese used an enzyme known as T1 RNase to digest ribosomal RNA molecules into very short pieces, which could then be sequenced with the technology available at that time. The sequences of ribosomal RNAs can be aligned and compared across all extant species in all kingdoms of life. In addition to being present in all species, ribosomal RNA is highly abundant and easy to obtain, which was also important for the techniques available at the time.

From their sequence comparisons of ribosomal RNA, Fox and Woese discovered that the Archaea were more similar to eukaryotes than to Eubacteria. The tree generated from their sequence data is redrawn in Figure 4.33; this tree shows that the ancestors of present-day Eubacteria (commonly referred to as the bacteria) and Archaea diverged from each other before the eukaryotic ancestor branched out of the archaeal lineage. The three oldest major branches on the tree are referred to as the three **domains** of life: Bacteria, Archaea, and Eukarya. Although their techniques were limited and only ribosomal RNA sequences were used, Fox and Woese's seminal data and their Tree of Life have withstood the test of time. A global phylogeny of 191 fully sequenced organisms from 2006, shown in Figure 4.34, confirms the evolutionary history proposed by Fox and Woese; subsequent analysis of many more species and many more genes has re-enforced their hypothesis. As a result of their work, sequence comparisons are now the principal tool for inferring evolutionary lineages.

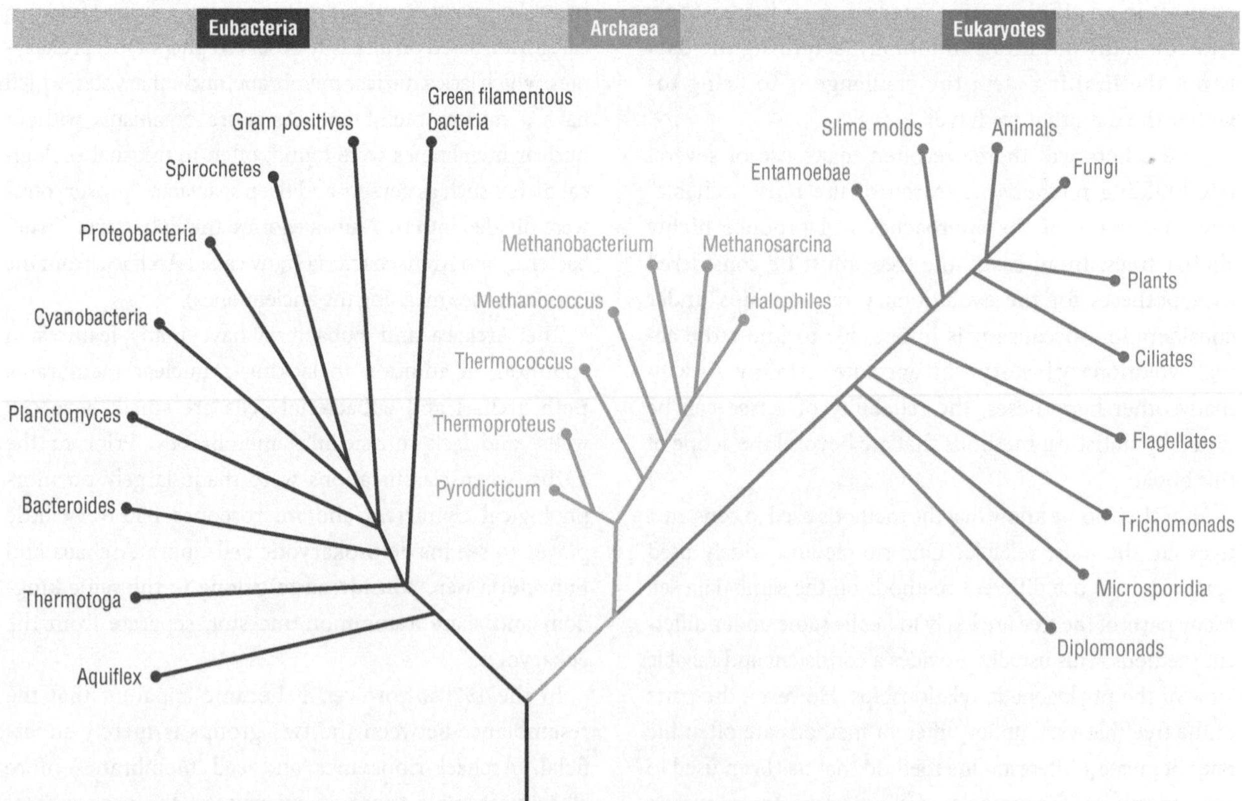

Figure 4.33 Tree of Life. Ribosomal RNA sequences have been integral to building the current view of the phylogenetic tree of all living organisms. Some of the information is summarized in this dendrogram. The Tree of Life can be divided into three domains, with the Archaea (formerly known as Archaebacteria) and eukaryotes more closely related to each other than either is to bacteria. Previous groupings classified organisms as either prokaryotes or eukaryotes only, but the sequence data allow a more nuanced view.

Source: Redrawn from http://www.sciencepartners.info/?page_id=348.

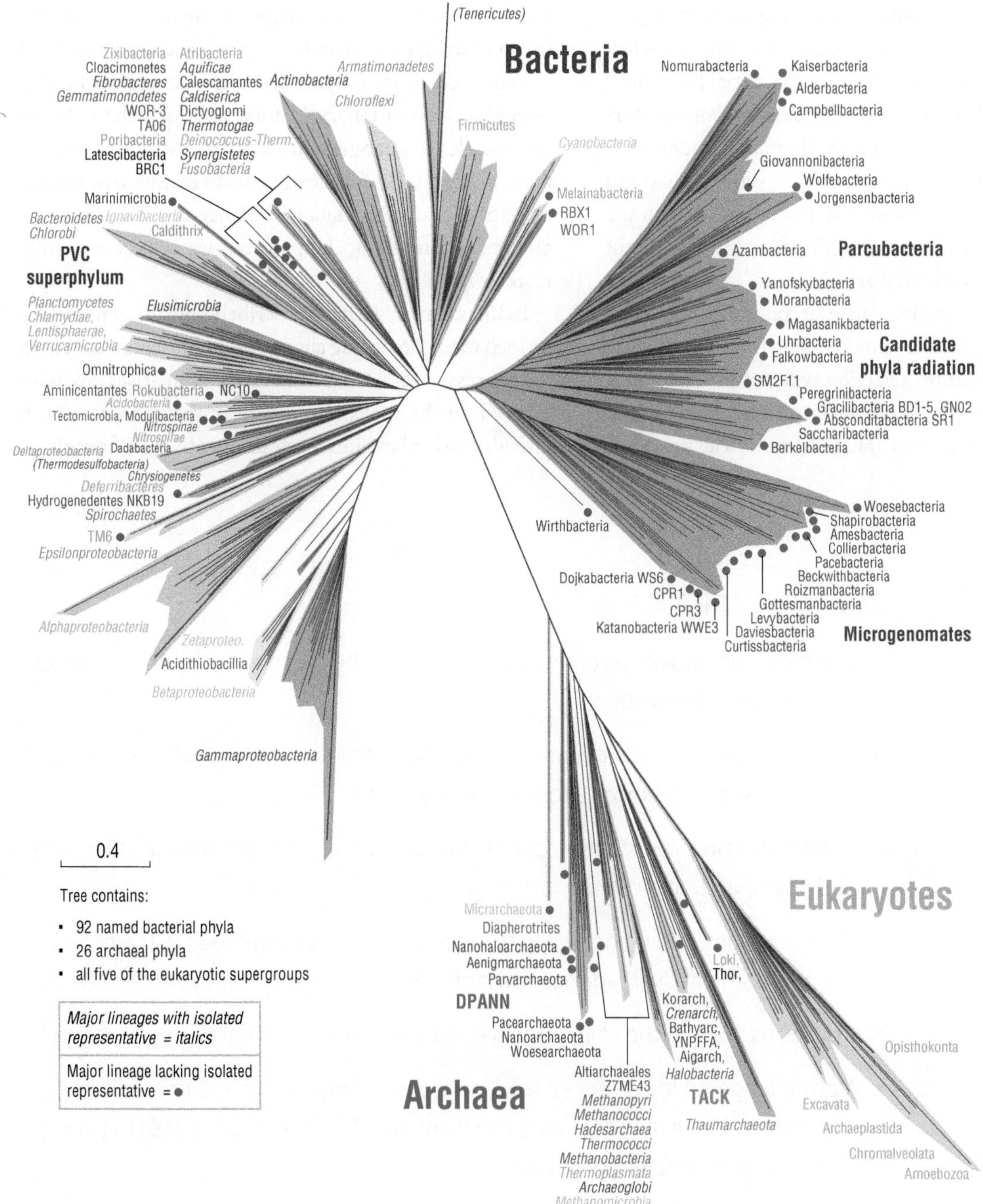

Figure 4.34 Phylogeny of organisms built using data from whole-genome sequencing. Whole-genome sequencing was used to identify and align 16 different ribosomal protein sequences. Comparisons of these sequences were used to build this tree. The topology of the tree agrees with that constructed with the single ribosomal RNA gene sequences depicted in Figure 4.33.

Source: Reproduced from Hug, L. A., Baker, B. J., Anantharaman, K. et al. (2016). A new view of the tree of life. *Nature Microbiology*. Article number 16048 under the terms of a Creative Commons Attribution 4.0 licence. © Macmillan Publishers Limited 2016.

4.8 Summary: descent with modification

Darwin's principle of descent with modification relies on twin processes to preserve continuity and to create diversity. These processes are both inherent within the template-driven replication of DNA. The use of each DNA strand as a template to make a new strand during replication provides the continuity necessary for descent.

The unavoidable errors that arise during replication provide the modification. Most of these errors are corrected, but some remain and provide the raw material for natural selection to sort through. It is fortuitous that not all errors are repaired. To paraphrase Lewis Thomas, if not for mistakes, we would still be living in a swamp and there would be no music. DNA replication teaches not only about the hierarchical nature of biological diversity but also about the mechanisms by which it arises.

The changes themselves arise at random, but selection provides for the non-random survival or extinction of the changes. Natural selection will not produce optimal results or "perfect" biological processes. It cannot. It relies on the basic principle that biological processes cannot be perfected—a principle borne out by the chemistry of biological molecules. We can use the relatedness of sequences from different species and populations to reconstruct evolutionary history, although the data are too complex to guarantee that any reconstruction is completely accurate. In fact, it will always be this way, and no amount of additional data will change this. We can refine our understanding of the past, but we cannot ever fully recreate it.

In this chapter, we have described how the genome described in the preceding chapters is replicated. Inherent in the process of replication is that variation arises. We have focused on the molecular nature of the genome, which provides an explanation for descent with modification.

CHAPTER CAPSULE

- The molecular processes of replication describe the mechanism by which descent with modification occurs.
- Descent implies continuity, which comes from how the DNA sequence of each strand is used as a template to make a new strand.
- Modification implies variation, some of which comes from the changes that arise during and after replication.
- These changes, known generally as mutations, can be deleterious to the fitness of the organism or advantageous; most of them are probably neutral in their effects.
- Mutations occur in the cells of the germline and in somatic cells.
- Many changes in the DNA sequence, whether arising during and immediately after replication or at other times, can be repaired. Different types of DNA damage are repaired by different processes.
- Changes in the nucleotide sequence accumulate over time. The number of changes when nucleotide sequences are compared reflects the time since the sequences were present in a common ancestor.
- The similarities in the sequences can be used to determine evolutionary relationships among sequences and organisms, and to estimate the most recent time when they shared a common ancestor.
- Evolutionary relationships can be depicted in the form of phylogenetic trees.
- Phylogenetic trees built using sequence data can provide a reinterpretation of our understanding of evolutionary relationships that were previously based on phenotypic characteristics.

STUDY QUESTIONS

Concepts and Definitions

4.1 How does the process of DNA replication explain Darwin's principle of "descent with modification"?

4.2 Define how the following molecules or activities help to address one of the "challenges" faced by DNA replication *in vivo*.

a. Telomerase

b. Okazaki fragment

c. Primase

d. 3′ to −5′ exonuclease activity

e. Topoisomerase

f. MutS

4.3 How does DNA replication in bacteria differ from DNA replication in eukaryotes?

4.4 What are some of the ways that organisms use to ensure the fidelity of DNA replication? Why is it important that the fidelity of DNA replication is an evolutionary balance between faithful replication and the existence of some errors?

4.5 How does the recognition that mutations arise at random, with respect to fitness, affect our view of evolutionary change?

4.6 What are some of the concepts and assumptions illustrated in a phylogenetic tree?

Beyond the Concepts

4.7 Origins of replication are typically AT-rich (that is, they contain more As and Ts than Gs and Cs). What do you predict would happen in a hypothetical organism in which replication began at a GC-rich region?

4.8 DNA polymerases and RNA polymerases carry out fundamentally similar reactions, albeit with different nucleotides for their substrates. Yet RNA polymerase can initiate the reaction without a primer, whereas DNA polymerase cannot. Why is this an important distinction between the two types of polymerases?

Looking ahead **4.9** When gene sequences from closely related individuals are compared, transitions are both more common and less deleterious in their effects than transversions. Propose an explanation for why transitions are both more common and generally less harmful than transversions. Figure 13.16, which presents the genetic code based on the chemical similarities of amino acids, could be helpful in thinking about the harmful effects of different types of mutation.

4.10 In 1995, Hamilton Smith, Craig Venter, and co-workers published the first complete genome sequence of a self-sustaining organism, that of the bacterium *Haemophilus influenzae*. Analysis of the genome sequence revealed that the organism does not have a gene that could encode a telomerase enzyme. Explain why this is unlikely to have any consequences for the genome of this organism.

4.11 *Escherichia coli* and other bacteria methylate adenines on the original strand to distinguish the original strand from the newly replicated strand of DNA. Why is this distinction important?

4.12 Rare genetic diseases in humans arise from mutations in components of the mismatch repair system, the base excision repair system, and the nucleotide excision repair system. Individuals with these mutations are highly "cancer-prone," that is, they have a very high rate and recurrence of different types of cancer. Why is this so?

4.13 Some cases of basal cell carcinoma ("skin cancer") in humans arise from somatic mutations in a gene called *patched* (*Ptch1*). One study sequenced the *Ptch1* gene from basal cell carcinoma cells found on the nose or cheek of different patients. About half of these samples affected adjacent thymines. Explain this result.

4.14 We cannot be certain that a particular phylogenetic tree is the "optimal" tree, as the data set becomes larger. Since we cannot know the "right answer" for a tree for sure, why are trees useful? What types of evidence give us confidence that a tree is accurate or would cause us to doubt the accuracy of a particular tree?

Applying the Concepts

4.15 Figure B in Box 4.1 illustrates the results of the Meselson–Stahl experiment after a single cycle of replication in ^{14}N.

a. Explain the results they observed after two rounds of replication in ^{14}N medium.

b. Draw out the expected results if a third round of replication were allowed in ^{14}N medium.

c. Two other models for template-directed replication were considered as alternatives to semi-conservative replication. One of these was conservative replication, in which the parental strands were unpaired, replicated, then reannealed such that the parental strands stayed together and the newly synthesized strands were together. The second model was dispersive replication, in which one strand was used as the template for polymerization, then the polymerase switched to using the other strand as the template, and subsequently switched back and forth between the two strands until both were fully replicated. Each of these models is ruled out by one of your results from (a). Which results from (a) rule out these alternative models? (Hint: Different results from (a) rule out different models.)

4.16 Mutagens are chemical or physical reagents that increase the rate of mutation. Different mutagens work by different mechanisms to produce different types of mutations. Each of the following mutagens is used in some experimental organisms. Postulate what type of mutation each mutagen is likely to induce at the highest frequency.

a. Ethyl methanesulfonate (EMS) attaches an alkyl group to guanine, which increases the frequency at which it pairs with thymine rather than cytosine.

b. Ethidium bromide inserts between the base pairs

c. Nitrous acid converts cytosine to uracil.

d. Ethylnitrosourea (ENU) attaches ethyl groups, primarily to thymines. This results in more frequent pairing with guanine rather than adenine.

e. 5-bromouracil (BrdU) is incorporated into DNA in place of thymine.

4.17 The DNA polymerase *Taq*, commonly used for PCR, as described in Tool Box 4.1, is an error-prone polymerase.

a. Define what "error-prone" means.

b. What is one practical reason to be aware that *Taq* is an error-prone polymerase when working with PCR fragments?

Challenging **c.** It is possible to use "high-fidelity" polymerases, such as *Pfu* or *Tth*, in PCRs, rather than *Taq*. Most high-fidelity polymerases, however, require longer than *Taq* to complete the elongation step in PCR. Why is that?

Looking ahead **4.18** Conditional lethal mutations are useful in analyses of complex essential processes, as will be discussed more fully in Chapter 15. Temperature-sensitive mutations, which are one form of conditional lethal mutations, allow growth at one temperature (for example 30°C) but not at a higher temperature (for example 42°C).

a. Most temperature-sensitive mutations are missense mutations. Postulate an explanation for this observation.

b. A large number of temperature-sensitive DNA replication mutants have been isolated from a bacterium. The mutants are unable to replicate DNA at 42°C but can do so at 30°C. If the bacteria are grown at 30°C for several generations and, then the temperature is raised to 42°C, two categories of mutants can be distinguished, referred to as "quick-stop" and "slow-stop". Quick-stop mutants stop replicating DNA immediately once the temperature is raised. Slow-stop mutants replicate DNA for a few more minutes after the temperature is raised and then stop. Give one example, with a brief justification, of a specific cell component that could be disabled in a quick-stop mutant and another similarly justified example of a component that is defective in a slow-stop mutant.

Challenging **4.19** DNA is very stable, allowing us to determine partial genome sequences from organisms that are extinct. However, because DNA can often become fragmented over time, it is virtually impossible to get complete sequences from DNA derived from ancient archeological specimens. Most methods for capturing DNA into sequencing libraries depend on ligating double-stranded DNA fragments to a double-stranded adapter. When DNA fragments have 3′ and/or 5′ overhangs, as is common in ancient DNA, this ligation is inefficient. To increase the amount of sequence data that can be obtained from ancient DNA, Meyer *et al.* (*Science* 2012, **338**: 222–6) "... devised a single-stranded library preparation method wherein the ancient DNA is dephosphorylated, heat denatured, and ligated to a biotinylated adaptor oligonucleotide [this is a single-stranded adapter with a hook on it], which allows its immobilization on streptavidin-coated beads [beads with an "eye" that the hook can hold on to]. A primer hybridized to the adaptor is then used to copy the original strand with a DNA polymerase. Finally, a second adaptor is joined to the copied strand by blunt-end ligation, and the library molecules are released from the beads. The entire protocol is devoid of DNA purification steps, which inevitably cause loss of material." The process is summarized in Figure Q4.1. This innovative protocol allowed them to obtain DNA sequence from fragments that were single-stranded in the archeological specimen or did not have "polished" blunt ends, ultimately leading to a 6-fold or greater yield in the DNA sequence obtained.

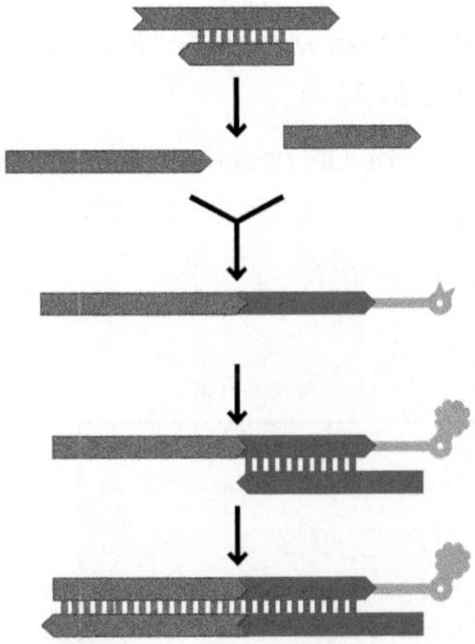

Figure Q4.1 For single-stranded library preparation, ancient DNA molecules are dephosphorylated and heat-denatured. Biotinylated adaptor oligonucleotides are ligated to the 3′ ends of the molecules, which are immobilized on streptavidin-coated beads and copied by extension of a primer hybridized to the adaptor (essentially by a single round of *in vitro* replication akin to the polymerase chain reaction, using DNA polymerase). Finally, the beads are destroyed to release the library molecules (not shown).

a. Why is a single-stranded adapter of known sequence required in this protocol?

Challenging **b.** Guanine is the base most susceptible to depurination. Meyer *et al.* noted that, compared to traditional double-stranded sequencing library preparation, their method was particularly valuable for obtaining sequence of relatively G + C-rich DNA. Provide a plausible explanation for why this might be the case.

4.20 Positions 43 to 51 of the 159-base RNA that is an essential component of the enzyme telomerase in *Tetrahymena*, a eukaryote, have the sequence CAACCCCAA. Scientists in Elizabeth Blackburn's lab mutated this sequence to C**G**ACCCCAA. How will this change affect the nature of *Tetrahymena* telomeres?

4.21 It is now feasible to sequence DNA from normal cells and tumor cells of a patient rapidly, at a relatively low cost (that is, a few thousand dollars). However, it is still difficult to find cancer cells early, and most of the sequencing is done at a later stage on rapidly dividing cells of a tumor.

a. Outline an experimental strategy that could be used to distinguish passenger from driver mutations in later-stage tumor cells.

b. Why might cancer biologists want to be able to distinguish driver from passenger mutations?

4.22 A still ongoing cholera outbreak in Haiti began after an earthquake. This is the first cholera outbreak in Hispaniola (the island comprising Haiti and the Dominican Republic) in over a hundred years. One hypothesis is that the earthquake caused ocean perturbations that allowed cholera strains residing close to other North and South American shores to reach Haiti. An opposing hypothesis is that *Vibrio cholerae* was accidentally brought to Haiti by United Nations (UN) peacekeepers flown in at short notice from cholera-endemic countries in Asia. The second hypothesis caused the victims to blame and attack UN personnel. To determine which hypothesis is most likely to be correct, scientists decided to compare the sequences of Haitian isolates with those from isolates from other parts of the world. Investigators identified and sequenced 1588 genes that were present in a selection of pandemic *V. cholerae* strains, including the Haiti 2010 outbreak strains. They used the resulting nucleotide sequence to produce the phylogenetic tree shown in Figure Q4.2. The year and country of isolation were noted for each strain. Very closely clustered branches are magnified, so that you can read the text. (The highlighted boxes are of no significance; they are strains selected by the investigators for other studies.) The phylogenetic tree is rooted with three pre-1920 strains that are not shown in this figure because they are very distantly related to the strains shown.

a. Explain why the investigators thought it necessary to root the tree with strains isolated prior to 1920, even though the strain of interest was isolated in 2010.

b. Do the data more closely support the ocean perturbation or the UN peacekeeper hypothesis, and why?

c. Do the data effectively rule out either hypothesis? Why or why not?

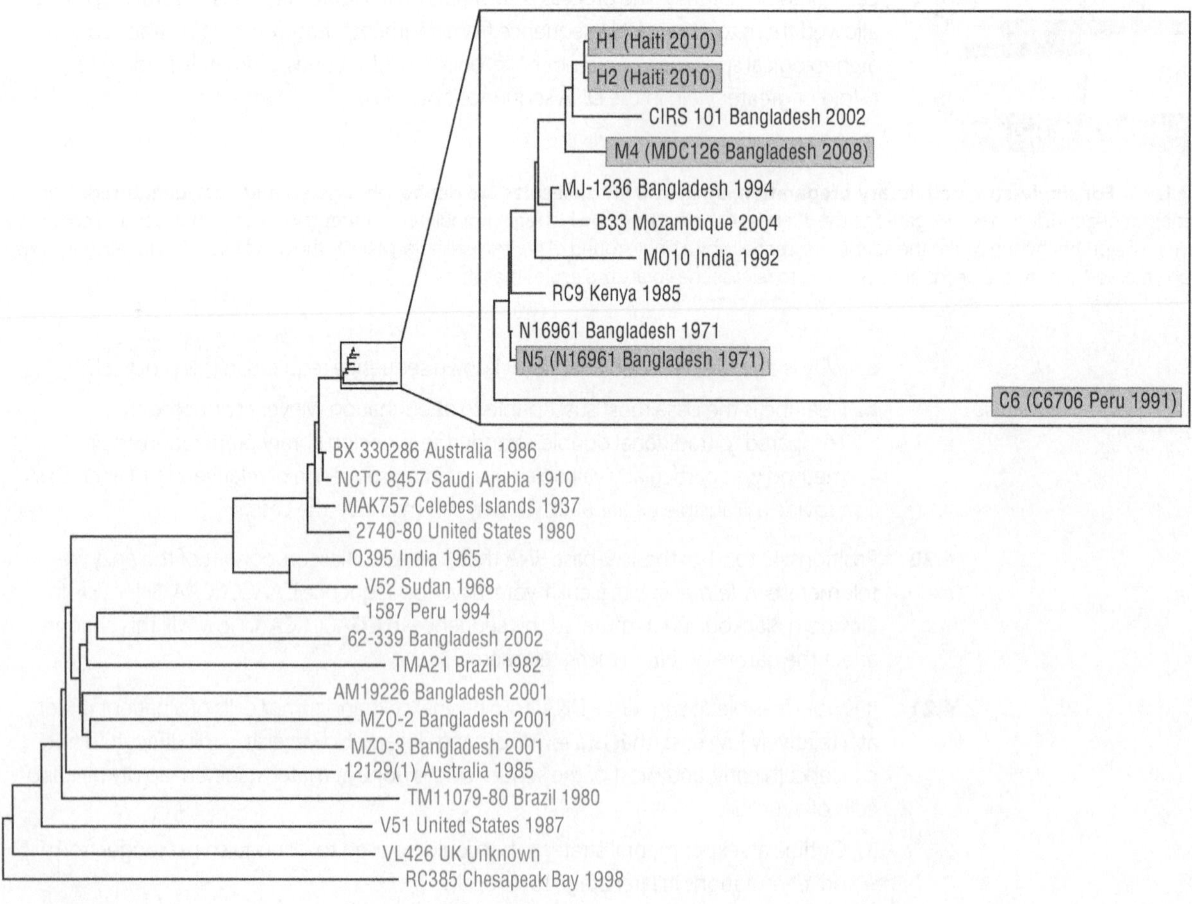

Figure Q4.2

Source: Reproduced from Chin, C.S. et al. (2011). The origin of the Haitian cholera outbreak strain. *N Engl J Med.* Vol. 364: 33-42. doi: 10.1056/NEJMoa1012928.

CHAPTER 5

The Inheritance of Single-gene Traits

5

IN A NUTSHELL

Gregor Mendel's experiments with true-breeding pea varieties revealed the basic principles that underlie the inheritance of biological traits in eukaryotes. In this chapter, we explore how Mendel's work gave rise to the concepts of "discrete units of inheritance" (now known to be genes), dominant and recessive traits, and the Law of Segregation. His insights also made it possible to calculate the probability of outcomes in a cross. Although first determined in pea plants, Mendel's laws of inheritance can be applied to all genes in any eukaryotic genome.

5.1 Overview: genes and inheritance

In our discussion of genes and genomes in Chapter 2, we noted that genes have two necessary functions: inheritance (which is connected to replication and variation, as covered in Chapters 3 and 4) and expression. We also said that these two functions are connected to the molecular biology of genes and genomes, that is, to DNA sequences. In this chapter, we will begin to develop these topics, particularly the concept of inheritance. Indeed, the inheritance of genes and genomes will be our primary topic for the next several chapters.

Before considering inheritance, however, let's return for a moment to gene expression. Genes are expressed to produce phenotypes, as we summarized at the end of Chapter 2. We will explore the **mechanism** of gene expression in more detail in Chapters 12 and 13. For now, we will consider gene expression in its most general sense, that is, a gene produces a product—RNA or protein—that results in a change within a cell. This cellular activity, in turn, influences the physical traits or the morphology of an organism. The expression of genes provides the heritable component of the morphological diversity necessary for evolutionary change as well. Similarly, it is the expression of genes that produces the physical traits or phenotypes that are often used as tools to monitor inheritance. It is therefore important to understand that, when we consider inheritance and selection, we are often dealing with outcomes of gene expression. Fortunately, we can consider gene expression for these purposes without having to fully understand the mechanism by which it occurs. Sometimes, this more general sense of how a gene produces a phenotype is referred to as gene activity.

KEY POINT A gene must be expressed to impart a physical outcome—a phenotype or trait—upon the individual. That phenotype can be monitored to track the inheritance of the gene.

Let's now consider some additional background concepts from the previous chapters that will be particularly relevant here.

- Genes are located at defined positions on chromosomes. Each position is often called a **locus**, so the terms "gene" and "locus" are frequently used interchangeably. This idea will be developed more fully in Chapter 9.
- Changes in the DNA sequence of genes—that is, mutations—can result in changes in morphology and phenotype.
- Diploids, which include most multicellular organisms, have two copies of every chromosome, one inherited from each parent. The most notable special case involves the sex chromosomes, which are discussed in Chapter 7.

In the following sections, we will build upon this foundational understanding to explore the relationship between genes and inheritance in some additional ways. First, the inheritance patterns of individual genes are predictable and can be tracked by well-chosen phenotypes. Because the inheritance pattern of the gene is predictable, the frequency with which certain phenotypes are expected to be observed can be calculated. We will discuss this for one gene in this chapter and for more than one gene in Chapter 8. Second, the predictable inheritance pattern of genes arises from the behavior of chromosomes during meiosis, the topic of Chapter 6.

5.2 Mendel: a man and his garden

Inheritance from one generation to the next has been studied for as long as there have been historical records and clearly began well before we have any documentation by humans. Parents have long recognized that their children resemble them and each other and are different from the children of other parents. Our forebears grew plants for food and raised animals for food and labor. Even more significantly, they intentionally selected desirable characteristics in their crops and animals because they recognized that the offspring of these artificially selected plants and animals were more likely to have the same desirable characteristics as their parents.

Many ideas and guesses about how inheritance worked emerged over time, but the turning point in our knowledge of genetics came in the 1860s. Gregor Mendel, a monk at the Abbey of St. Thomas in the city of Brno, which is now part of the Czech Republic, was conducting a lengthy series of plant breeding experiments. The best known photograph of Mendel is shown in Figure 5.1(a).

Mendel worked with garden peas, shown in Figure 5.1(b), and his first important scientific breakthrough was to use **true-breeding** varieties. An organism is said to be "true-breeding" for a trait if it will pass that trait to all offspring of all subsequent generations when crossed to an organism that is true-breeding for the same trait. The concept of being true-breeding, which is applied to breeds and varieties of individual animals and plants, is the same as a trait being fixed, as introduced in Chapter 1, which is applied to the genetic structure of populations.

Prior to Mendel's work, most plant and animal breeders began with diverse starting material and attempted to make sense of what they observed. Mendel used pea varieties that always produced offspring that looked like their parents for the traits he considered. He analyzed the inheritance of seven different characteristics in these peas, shown in Figure 5.2. Each of these characteristics had two easily distinguishable morphological phenotypes. For example, he used varieties that produced peas that were either wrinkled or round. If he crossed a plant that had wrinkled peas with another plant with wrinkled peas or if he self-fertilized a plant with wrinkled peas, he got only wrinkled peas in the offspring. When he crossed a plant with round peas with another plant with round peas, he only got offspring with round peas, as illustrated in Figure 5.3. To the best of our knowledge, no one before Mendel had begun with such true-breeding varieties. By doing this, Mendel controlled the amount of variation that he was analyzing, which allowed him to focus on a few discrete characteristics.

Simple and complex traits

Of course, many phenotypes do not fall into two discrete categories like these. Even with garden peas,

Figure 5.1 Gregor Mendel, an Augustinian monk at the Abbey of St. Thomas in Brno, now in the Czech Republic, is considered the founder of the science of genetics. The most familiar picture of Mendel is shown in (a). He used true-breeding varieties of peas (b) growing in the abbey garden to study inheritance. His meticulous records and quantitative handling of data were essential to understanding patterns of inheritance.

Source: Image of Mendel from the frontispiece of *Mendel's Principles of Heredity: A Defence*. 1902. Image of peas from Prof. Dr. Otto Wilhelm Thomé, *Flora von Deutschland, Österreich und der Schweiz* 1885, Gera, Germany. Permission granted to use under GFDL by Kurt Stueber.

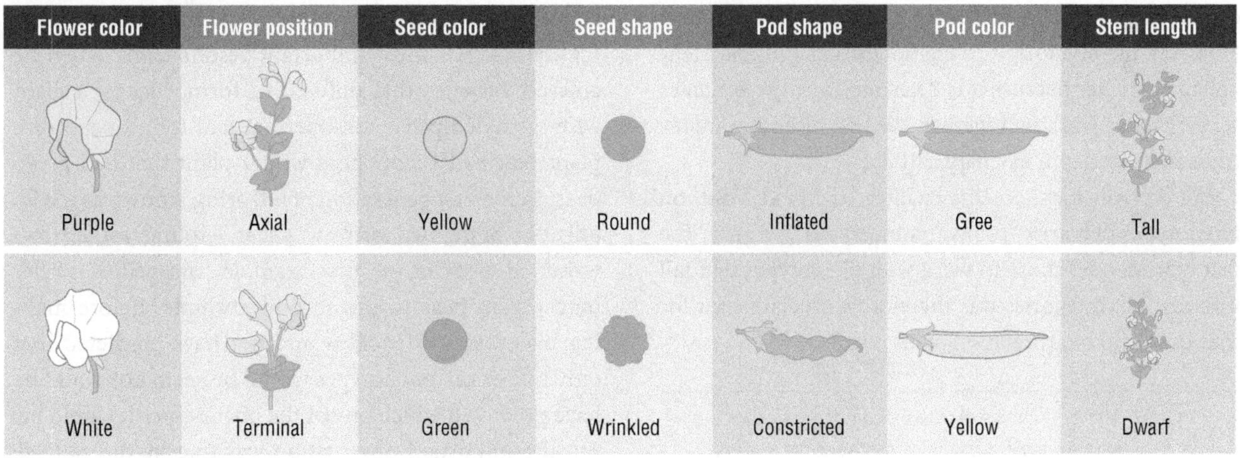

Flower color	Flower position	Seed color	Seed shape	Pod shape	Pod color	Stem length
Purple	Axial	Yellow	Round	Inflated	Gree	Tall
White	Terminal	Green	Wrinkled	Constricted	Yellow	Dwarf

Figure 5.2 The seven traits that Mendel used and their alternative phenotypes.

phenotypes, such as the number of pods and the days from germination to flowering, are inherited but do not fit into two separate categories. Mendel probably recognized this and intentionally chose phenotypic characteristics that could be easily distinguished and classified. For example, one of the traits that he chose was plant height or stem length, and he categorized plants as "tall" and "dwarf." Although "tall" and "dwarf" are discrete and non-overlapping categories, plants would certainly vary within these categories, so that not all tall plants had exactly the same height, but all would be taller than any dwarf plant.

Those phenotypic traits that cannot be readily separated into two distinct categories often show continuous variation (like height or weight in humans) and are frequently examples of multigenic or complex traits—traits

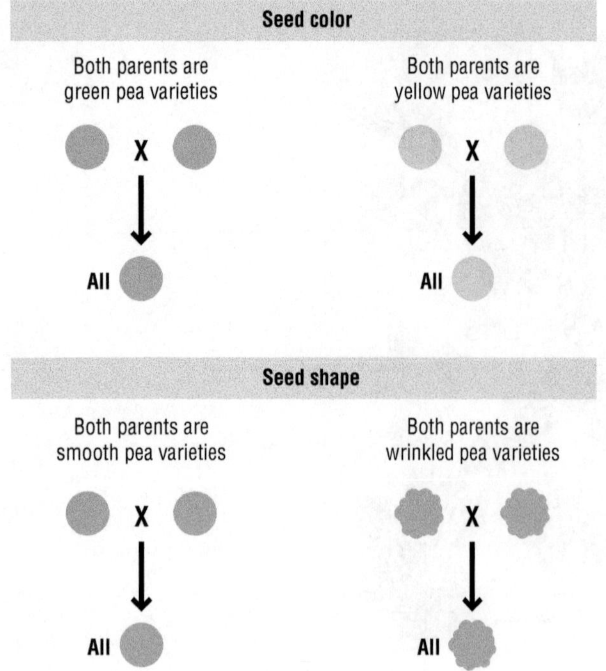

Figure 5.3 True-breeding varieties. In a true-breeding variety, all of the offspring are like the parent. This is illustrated for two of the traits Mendel used: green and yellow seed colors, and round and wrinkled seed shapes.

that are the outcome of the actions of many genes. For complex traits, the **inheritance** of each individual gene behaves like a simple Mendelian unit, following the laws of inheritance that we will cover in the next few chapters. However, the **phenotype** might not exhibit a simple Mendelian character because it is determined by the action of several genes working together. We will discuss complex traits in more depth in Chapter 10.

But we will focus in this chapter, as Mendel did, on phenotypes that arise from variation at a single gene. The fact that Mendel chose to work with phenotypes that fall into distinct categories was therefore a crucial factor for the successful interpretation of his results.

KEY POINT Mendel's choice of true-breeding varieties and simple, discrete phenotypes was critical for the success of his experiments.

Dominant and recessive traits

Looking back at his experiments, we can say that Mendel's pea plant was a model organism—a research organism that has particular advantages for study and for which the conclusions can be generalized to other organisms. Mendel himself may not have realized that his results with peas could be applied generally to inheritance of every gene in every other plant and animal; in fact, he probably believed that his results did not apply to all plants. The recognition that he had found nearly universal rules for inheritance in eukaryotes did not come for another 35 years, well after Mendel had died.

Mendel kept meticulous records on everything—he was extremely quantitative in his approach and counted everything. His original notebooks on his plant breeding experiments have been lost, but the Abbey of St. Thomas does have Mendel's notebooks on meteorology, one of his other interests. These notebooks are incredibly detailed, with more than 20 climate and weather characteristics meticulously recorded daily for years. His published paper on genetics shows a similar level of quantitative detail. It was this level of detail that was instrumental for those who went on to verify (or question) his results.

We are first going to summarize Mendel's results for one characteristic, since this underlies most of what will be discussed in the next six chapters. Communicating Genetics 5.1 describes some of the notation and defines some terms important for understanding this discussion.

Every variety of peas that Mendel used was true-breeding for each of the seven characteristics noted in Figure 5.2. His most important results came when he crossed varieties that differed in form, that is, a plant with wrinkled peas with one that had round peas, or a plant that had yellow peas with a plant that had green peas. In the first generation of offspring, known as the F_1, only one of the two forms was seen—round, rather than wrinkled, peas in the first example, and yellow, rather than green, peas in the second example. Before making his crosses, Mendel would not have predicted that only one of the phenotypes would be seen, nor could he have predicted which one of the phenotypes it would be. But the important observation was that he did not see a blended inheritance, such as slightly wrinkled or yellowish-green, or even a plant that had some pods with wrinkled peas and some pods with round peas. The F_1 was uniform and indistinguishable from one of the two parents.

When Mendel took the F_1 plants and generated an F_2 generation (by self-fertilizing a plant or crossing the F_1 plants among themselves), he found another unexpected

COMMUNICATING GENETICS 5.1 Genetic notation for genes and alleles

The nomenclature and symbols that are used to represent genes and alleles can be among the most confusing topics in genetics. However, following the correct conventions and nomenclature is an important aspect of communicating genetics clearly to others. In this box and in Communicating Genetics 5.2, we will review some of these conventions and definitions.

Indicating the generations

The parental strains, that is, the ones that are true-breeding and are used to initiate the mating, are called the P_0 generation—the parents represent generation zero. The offspring of the P_0 generation are referred to as the F_1 (for first filial generation), and the offspring when members of the F_1 generation are mated to each other are referred to as the F_2, etc. When you write out a cross, each generation is denoted by including the corresponding nomenclature ("P_0" or "F_1," etc.) to the left-hand side of the page, as shown in Figure 5.5. When the mating strategy does not involve mating individuals among themselves to produce the next generation, it is convenient to refer to the G_2 or G_3 generation.

Dominant and recessive alleles

Throughout Chapter 5, we used the common convention of capital and small letters to indicate the different alleles of a gene. We also use ***R*** because the phenotypes are round or wrinkled, but you can use any letter that is clear to you for working the problems and following the inheritance patterns. In addition to being easy to use, the capital and small letters system, such as ***R*** and ***r*** or ***A*** and ***a***, has the advantage that it allows an easy designation of which allele is dominant. When using capital and small letters to write the alleles in a genetic cross, you should always follow the convention of designating the dominant allele as the capital letter and the recessive as the lowercase, whenever possible. This will help everyone to follow your cross more clearly. Note that we will also write the gene names in italics, as previously discussed in Communicating Genetics 1.1.

Designating multiple alleles for a gene

While it is an appropriate convention for learning inheritance with single genes, the ***Aa*** system of nomenclature also has some disadvantages when more realistic cases are considered. The most important of these is that this system only easily allows for two alleles, whereas a gene can have many different alleles, as will be discussed in Chapter 8. That is an enormous shortcoming for experimental genetics. For some organisms in which some of the research predated our understanding of the possible complexities of genes and alleles, a version of the Aa system still shows up occasionally; this is most apparent among some historic gene names in mice, corn, and *Drosophila melanogaster*, as well as some domesticated species and occasionally in humans.

Whenever more than two alleles of a gene need to be represented, it is possible to use superscripts to differentiate them. For example, a gene named *agouti* (a gene whose symbols in mice genetics actually are ***a*** and ***a***) is one of the principal genes affecting coat color in mammals. More than 370 different alleles at the *agouti* locus have been experimentally isolated and characterized in mice, many with different phenotypes. A few of these phenotypes are shown in Figure A. All of these alleles need symbols to identify them, and it would be convenient to know which alleles

Figure A Since genes can have many different alleles with different phenotypes, the nomenclature has to be more complicated than the simple uppercase vs. lowercase system. Shown in this figure are the phenotypes of mice with the following genotypes for *agouti*, for which many alleles have been identified. The mouse on the left is ***a/a***, the one in the middle is A^{y}/a, while the one on the right is $A^{w\text{-}J}/A^{w\text{-}J}$.

Source: Image 1: Jackson Mouse Model KK. Cg-AY/J ID 002468. Image 2: Jackson Mouse Model C57BL/6J-Aw-J/J 00051. © 2016 The Jackson Laboratory - used with permission.

COMMUNICATING GENETICS 5.1 Continued

are recessive and which are dominant. Consequently, these alleles have symbols such as A^y, a^u, A^w, A^{vy}, and so on. The ***A/a*** symbols can't represent all of the different possible alleles and their phenotypes, so we don't know if A^y is dominant to A^w or if either of them is dominant to ***A***. This topic of inferring the dominance relationships among different alleles of a gene will arise again in Chapter 8.

Wild-type

For experimental organisms used in genetics research, a different notation for alleles and genes is frequently used. The naming of genes depends on the concept of a **wild-type**. The wild-type is the reference standard that is used by research laboratories and can be generally thought of as "normal" or what would be found in the wild. A typical situation in the history of genetics was that one prominent researcher or research laboratory began working with a particular research organism and chose one bacterial culture, plant, or worm, or a pair of flies to initiate the culture in the laboratory; some of these are shown in Tool Box P.1, Figure A. These organisms might have been brought in from nature and grown in the laboratory. All of the growth media, handling techniques, and general laboratory procedures were developed for this initial organism and its offspring. The phenotypes that this founder organism displayed are called the wild-type, and all of the alleles are defined to be the wild-type alleles.

In an abstract sense, the wild-type is what we imagine *Escherichia coli*, a fruit fly, a mouse, a nematode, or the *Arabidopsis* plant looks like and how it normally behaves. In reality, the designated laboratory strain often has some unusual genetic properties of its own, so most model organisms have several different "wild-type" strains that are used in the laboratory. Documentation from the laboratory has to indicate which wild-type it is using. For example, for ***Drosophila*** *melanogaster*, the two most commonly used wild-types are Oregon-R and Canton-S. Most of the mutations that have been studied were first identified beginning with the Oregon-R line.

All of the alleles found in that original individual or population are called the "wild-type alleles" and are designated by a + sign. Depending on the nomenclature system for that organism, the + sign might be a superscript or it might not. The wild-type allele is usually one that allows the gene to be functional in nature and in the laboratory.

Nearly all mutations that are found in research organisms are recessive to the wild-type allele. In fact, it is **generally assumed** that mutant alleles are recessive to wild-type. So the wild-type allele ***a+*** is equivalent to writing ***A***, and all of the other recessive alleles are versions of a. For every organism, there is an associated system for writing the recessive alleles, and it is not easy to master or summarize these. For our purposes, we will use ***a+*** or ***A*** to indicate the dominant wild-type allele of a gene, and ***a*** to indicate a recessive mutant allele. When we write **+** without a corresponding gene symbol or letter, we will mean the wild-type allele for any gene that is being considered.

Heterozygotes

Another possible confusion can arise in the way that we write homozygotes and heterozygotes. Again, the conventions are a bit different for various organisms, so we will describe the systems you are likely to encounter. We might write either ***Aa*** or ***A/a***, with the / symbol to separate the two different alleles. When using ***a+/a*** or ***+/a***, we will almost always use the / for clarity in the text. A homozygote is written as ***AA***, ***a+/a+*** or possibly just **+/+**. You will also see the notation of ***a+/_*** or ***A/_***. The _ (underscore) means that the phenotype is the same, no matter which allele is present, because the other allele is dominant. This comes in handy when we don't know the entire genotype for sure and have to figure it out by doing a test cross, as earlier, for example.

Alleles dominant to wild-type

Occasionally, there are mutant alleles that are dominant to wild-type. We use different conventions for denoting such dominant mutant alleles in different organisms, but usually there is some obvious way to make this clear. In *Drosophila*, there is a gene called *Notch* (abbreviated *N*), named because the mutant flies have notches in the margins of their wings. Some alleles of *N* (such as N^8) are dominant to wild-type, whereas other alleles are recessive to wild-type, so the use of the capital letter to indicate the dominant allele can be confusing. For other organisms, geneticists might use the symbol (*d*) or (*dom*) or an asterisk (*) to indicate a mutant allele that is dominant to wild-type. There are different conventions about italics, boldface, superscripts, capital and small letters, and so on. Model organisms have genome databases that list the mutants, their phenotypes, and other information; these genome databases also have the guidelines for nomenclature. In this book, we will use forms that should be clear in the context of the chapter, even if these are not precisely the way that gene symbols are used for that particular organism in the research laboratory.

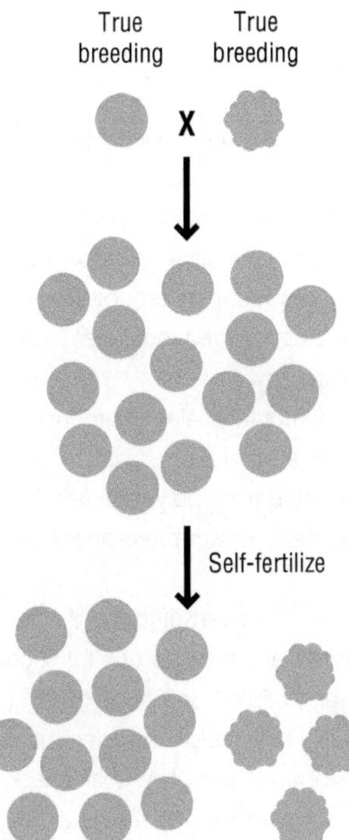

Figure 5.4 Mendel crossed a true-breeding pea plant with round peas with a plant with wrinkled seeds. All of the F_1 offspring had round seeds. When the F_1 plants were mated to each other or self-fertilized, a mixture of plants resulted, three-fourths with round seeds and one-fourth with wrinkled seeds.

result, as shown in Figure 5.4. Self-fertilization of the F_1 produced pods that contained both round and wrinkled seeds. So the phenotype or the morphological variant that disappeared in the F_1 was not lost from the population of pea plants. It was simply masked by the other I"dominant" form.

One of the most biologically important observations that Mendel made is that some traits were recessive and some were dominant. This has become so familiar that it is easy to overlook some of its implications about gene activity, gene dosage, and evolution. Some of the implications about dominant and recessive traits are explored in more depth in Box 5.1.

Instead of trying to continue this as a mystery story, let's describe these experiments with more contemporary language and explanations. Many of the observations and inferences made by Mendel have been incorporated into the modern language of genetics; we have collected some of the most important definitions together in Communicating Genetics 5.2. Once you have reviewed this panel, return to the cross described above and shown in Figure 5.4 to see if you can describe the cross using each of the relevant genetic terms.

BOX 5.1 *Going Deeper* Recessive and dominant alleles

Among the first things that students learn about genetics is the concept of dominant and recessive alleles. Because we learn these ideas so soon in our understanding of genetics, we may not think about them again as we learn more. Yet it is worth reflecting on these concepts and what they tell us about biology.

First, it should be recognized that the "dominant allele" or the "recessive allele" is only a relative statement. Each allele has to be compared to another allele of the gene, in the form of a heterozygote. Let's say that there are two alleles of the *A* gene, ***a1*** and ***a2***. Which is dominant, and which is recessive? In order to determine that, we make a heterozygote that is ***a1/a2***. (In technical jargon, a combination involving two alleles of the same gene is more properly termed "heteroallelic.") What is the phenotype of this ***a1/a2***? If it looks similar to ***a1/a1***, then ***a1*** is the dominant allele; if it looks similar to ***a2/a2***, then ***a2*** is the dominant allele. Until we do the experiment and make the ***a1/a2*** combination, we don't know which one is the dominant allele. Even if ***a1*** is dominant to ***a2*** in this combination, there could be another allele that is dominant to ***a1***, or another allele that is recessive to ***a2***. The dominant allele is defined as the one whose phenotype appears in the heterozygote, no more or less. As we will discuss further in Chapter 8, the concept of recessive and dominant also refers to a specific phenotype.

Let's return to the concept of wild-type alleles, introduced in Communicating Genetics 5.1. As a rule of thumb, the wild-type alleles are dominant to almost all other alleles of the genes. This rule of thumb is true because, in general, the wild-type organism was originally found living successfully in nature, so most of its alleles must have been functioning. Most mutant alleles that we generate in the laboratory will not function as well as the wild-type allele, simply because they have not been subjected to generations of natural selection. Since these other mutant alleles do not function as well as the wild-type allele, they are often called **loss-of-function** (LOF) alleles. With only occasional exceptions, LOF alleles are recessive to wild-type, and recessive alleles are usually thought of as loss-of-function alleles.

BOX 5.1 Continued

Let's think a bit about the fact that most LOF alleles are recessive to wild-type alleles. Loss-of-function alleles contribute nothing, or next to nothing, to the function of the process. The fact that they are recessive to wild-type implies that, in general, a diploid organism makes much more functional product from one wild-type allele that it needs to carry out the process. We can use a specific example. The *white* gene in *D. melanogaster* encodes a protein called an ATP transporter, a broad class of proteins that use ATP to generate the energy necessary to move a cargo across a membrane. In the case of the white protein, its cargo is pigment granules. A fly with white eyes (that is, ***w/w***) cannot transport these pigment granules across the membrane of eye cells (called ommatidia), so the cells are unpigmented—they are white in phenotype. A heterozygote ***w+/w*** has wild-type eyes (the white protein works correctly), so we know that the ***w+*** allele is dominant to the w allele. This also means that the one ***w+*** allele makes enough of the functional gene product to carry out the transport process, as well as a fly with two functional ***w+*** alleles.

Because most wild-type alleles are dominant to most mutant alleles, organisms apparently produce more of the gene product than is actually necessary for nearly every biological function. The exceptions to this rule of thumb are uncommon, and nearly all of them represent some version of a special case. In general, the uncommon dominant mutant alleles fall into one of three categories.

First, and most often, a dominant mutant allele can arise because the mutation results in an overproduction of the gene product or in a gene product that cannot be easily "turned off." Dominant mutant alleles involved in cancer (called oncogenes) usually fall into this category.

Second, a dominant mutant allele can produce a gene product that interferes with the normal function of the gene. This occurs for proteins that assemble from polypeptide subunits. Having a mutant subunit can interfere with the assembly and function of the normal subunit.

Third, a few genes are "haplo-insufficient," that is, the biological process is dose-sensitive and requires two functional alleles. The biological roles of such haplo-insufficient genes are unpredictable, in part because so few haplo-insufficient genes are known. Two of the genes involved in the DNA repair checkpoint that assesses the genome for damage and regulates the cell cycle in humans—*p53* and *PTEN*—appear to be haplo-insufficient, such that individuals who are heterozygous for mutations in either of the genes have an elevated cancer risk because their cells do not arrest at the DNA repair checkpoint and continue the cell cycle with damaged DNA.

COMMUNICATING GENETICS 5.2 The language of genetics

Many of the observations and inferences made by Mendel have since been incorporated into the modern language of genetics. Below is a list of important words and concepts that will be needed, as we explore Mendel's experiments and beyond.

Phenotype—the appearance. Anything that you could possibly measure by any device that you have (a ruler, a color wheel, a written test, the appearance of a band on an electrophoretic gel, a nutritional requirement, or a peak on a spectrophotometer, etc.) is a phenotype.

Genotype—the genetic constitution that underlies the phenotype. The phenotype is the outcome of the genotype, the environment, and the interaction between them. For some phenotypes, particularly the ones that we will discuss in the next few chapters, the environment plays no role (or a minimal role), so the phenotype depends completely on the genotype. Most of the phenotypes that we will consider (such as ABO blood type in humans or white eyes in *Drosophila*) are determined by the action of a single gene, with no contribution from the environment. Other phenotypes (such as height in humans) are the outcome of many different genes, as well as the environment. We will consider the inheritance of some of these complex traits in Chapter 10.

As a historical reality, the phenotype has often been what geneticists have been able to measure directly. The genotype, on the other hand, often has to be inferred from the phenotype and its inheritance pattern in particular crosses. Problems at the end of this chapter will ask you to infer the genotype, based on the results of certain crosses and the appearance of particular classes of offspring. However, with the advent of DNA sequencing and other molecular methods, we now have the ability to look directly at a DNA molecule and its base sequence. Remember, we have just defined "phenotype" as appearance or anything that can be measured or read. So when we are looking directly at the base sequence of a stretch of DNA, this can be considered a phenotype because we have "measured it" by sequencing it. Thus, in experiments where we are following the inheritance or the function of a particular section of DNA sequence itself (rather than following a physical trait such as eye color, etc.), the DNA sequence at that particular location in the genome is both a genotype and a phenotype.

COMMUNICATING GENETICS 5.2 Continued

Gene—for Mendelian genetics, the gene is the unit of inheritance and is usually described in terms of the phenotype that is seen. So we can talk about a gene for pea texture, a gene for pea color, a gene for flower color, etc. As we discussed in Chapter 2, a precise and inclusive definition of a gene is much trickier. As used in Mendelian or classical genetics—that is, most of the material that we will discuss in Chapters 5 through to 9—the gene is a unit of inheritance and is encoded in DNA. More nuances to this definition will be added in other chapters.

Alleles—the different and alternative forms that exist for a gene. The gene for pea texture has alleles that give rise to round or wrinkled peas. Mendel worked with two alleles of each gene that he studied. However, a gene can have a nearly infinite number of possible alleles, which might or might not have different phenotypes. We will encounter the enormous diversity of possible alleles again in Chapter 9. Different alleles arise from different DNA sequences at a given genetic locus. "Alleles," "variants," "polymorphisms," and "mutations" are closely related and overlapping concepts, although with slightly different connotations. For now, we can say that an allele is a form of the gene that confers a recognizable phenotype.

Homozygous—describes the situation where a diploid organism contains two matching alleles for the gene in question.

Heterozygous—describes the situation where a diploid organism contains two different alleles for the gene in question.

Dominant—the phenotype that appears in the F_1 of a cross of two different true-breeding parents. This is a very important concept. The dominant phenotype is not necessarily the "normal" phenotype or the common one. It is simply the one that appears in the F_1 of a cross between two true-breeding strains. This concept is developed more fully in Box 5.1.

Recessive—the phenotype that is masked in the F_1 of a cross of two true-breeding parents and reappears in the F_2.

Carrier—is used most often in human genetics to describe an individual heterozygous for a recessive allele, so the allele is "carried" in the individual's genome without being detected at the phenotypic level. The phenotype of the carrier is that of the dominant allele.

5.3 Predicting the outcome of crosses

Let's consider Mendel's F_2 round and wrinkled peas again. Both morphological phenotypes were seen. Mendel's second significant contribution to the study of inheritance was to tabulate how many of each phenotype he observed. His quantitative skills and meticulous record-keeping were central to the way the results could be interpreted. Figure 5.4 shows that three-fourths of the F_2 generation had the dominant phenotype, and one-fourth had the recessive phenotype. This ratio was observed no matter which dominant and recessive phenotypes Mendel started with, whether it was round or wrinkled peas, yellow or green peas, white or purple flowers, or any of the other traits he studied. Thus, the ratio has nothing to do with the phenotypes themselves and must be explained by how the genes for these phenotypes (or the alleles that affect them) are being inherited.

KEY POINT For a single gene with two alleles, when two different true-breeding lines are crossed together and the F_1 is self-crossed, the dominant phenotype always appears in three-fourths of the F_2, and the recessive phenotype always appears in one-fourth. This gives the characteristic phenotypic ratio of 3:1 in the F_2 generation.

By observing these ratios in dozens of crosses and thousands of peas over several years, Mendel was able to deduce what was happening. He postulated that each parental plant had two alleles for each gene. In a true-breeding plant, these two alleles are the same—so, for example, one parent would have two "round" alleles, and the other would have two "wrinkled" alleles. Wrinkled and round are two alternative alleles of the same gene. Because we have seen that the round allele is dominant to the wrinkled one, we will write it as an upper case ***R***, with the recessive allele being the corresponding lower case ***r***. We will represent the genotype of a plant, showing the wrinkled phenotype by rr and a true breeding plant with round peas as ***RR***. When they are crossed, the F_1 receives one allele from each parent via the **gametes**, as shown in Figure 5.5. (In animals, the gametes are the sperm and the ova. In plants, the gametes are the pollen and the egg cells.)

Now each individual F_1 plant has two alleles of each gene, but these two alleles are different because they were inherited from different parents. The genotype of the F_1 will therefore be represented as ***Rr***. However, while both alleles are present in the **genotype**, only one of the two alleles, the dominant allele, shows up in the **phenotype** of

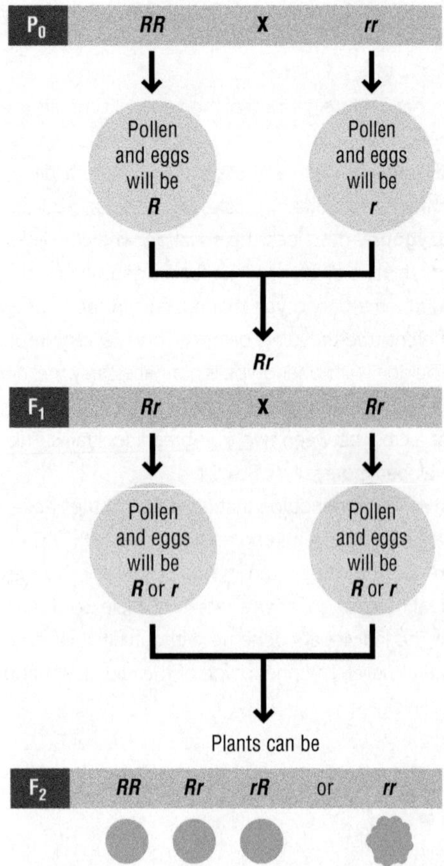

Figure 5.5 When a true-breeding individual carrying two alleles for a gene (*RR*) is crossed to a true-breeding individual containing two different alleles for that gene (*rr*), the F_1 offspring will receive one allele from each parent via the gametes. Because they are all ***Rr*** heterozygotes, they will all display the dominant phenotype. The offspring in the F_2 generation will receive one allele from each parent via the gametes to produce combinations of alleles that show a 3:1 ratio of dominant to recessive phenotypes.

the F_1 plant, while the recessive allele is masked. Each pollen grain or egg cell made by the F_1 plants inherits only one of the two alleles, but it can be either one ***R*** or ***r***. In fact, half will get one allele, and half will get the other allele.

When the F_1 is then self-crossed to produce the F_2, the pollen grain and the egg could have either of the two alleles. These unite with each other to give a genotype in the F_2 that again has two alleles, as shown in Figure 5.5. We now say that, if the two alleles for the gene are the same, the individual is **homozygous**. If the two alleles are different, the individual is **heterozygous**.

Punnett squares

We can represent such a cross in a more diagrammatic fashion by producing a Punnett square—a way of depicting crosses that was first developed in the early twentieth century by the British geneticist Reginald C. Punnett, who introduced it in one of the first widely used textbooks on Mendelian genetics, appropriately entitled *Mendelism* (1905). His method takes the genotype of each parent and divides it into the gametes that the parent would make. The gametes made by one parent are written on one side of a square, while the gametes made by the other parent are written on the other side. The square is then filled in by combining the gametes from each side of the square to give all possible outcomes of progeny. Each entry in the Punnett square represents an equally likely outcome from the cross, although versions will be used in Chapter 9 in which the squares do not represent equally likely outcomes. Writing out a cross in the form of a Punnett square can therefore be a very helpful way of keeping track of the numbers of possible combinations that might be seen in the offspring.

A Punnett square depicting Mendel's cross of wrinkled peas and round peas is shown in Figure 5.6. The Punnett square displays the outcome from the F_1 plants mating among themselves or self-fertilizing to produce an F_2. Notice how the completed square shows the genotype of each F_2 individual and the corresponding phenotype. When producing a Punnett square, the best approach is to work with genotypes first and then determine the

		Male gametes	
		R	***r***
Female gametes	***R***	***RR*** round	***Rr*** round
	r	***rR*** round	***rr*** wrinkled

Figure 5.6 The Punnett square of a cross between two heterozygous round pea plants illustrates the Law of Segregation. The two alleles of each parent ***R*** and ***r*** segregate from each other when the gametes are made and are inherited separately. This is shown diagrammatically in the Punnett square by writing the two possible gametes produced by one parent along the top of the Punnett square (shown in blue) and writing the two possible gametes produced by the other parent down the left-hand side (shown in pink). The gametes are then combined into all the possible combinations of offspring by filling in the Punnett square. The result is one offspring with the ***RR*** genotype, two offspring with ***Rr***, and one with ***rr***, giving a 1:2:1 genotypic ratio. The phenotypic ratio is 3:1—three round-seeded plants to one wrinkled-seeded plant.

phenotype that corresponds to each genotype in the square. As we shall see, the inheritance of the genes, and thus the genotype, is quite predictable, while the inheritance of the phenotype may be much less so. The distinction between genotype and phenotype is very important, and clarifying which is being considered is a critical step in understanding any cross.

Look again at Figure 5.6, and notice that the genotypes ***RR*:*Rr*:*rr*** are found at a ratio of 1:2:1. But because ***R*** is the dominant allele, the phenotypes are found at a ratio of 3:1 round:wrinkled. If we saw a plant with round peas in the F_2, we would not know if its genotype is ***RR*** or ***Rr***. On the other hand, we know that the wrinkled peas must have the genotype ***rr***.

Mendel's familiar 3:1 ratios were carefully calculated from many hundreds of plants. Some have looked at his data and wondered if perhaps his data are a little too carefully calculated—in other words, if Mendel might have selected or altered his data in some way. We discuss this question in a bit more detail in Box 5.2.

KEY POINT Punnett square diagrams show the gametes of each parent on the sides of the square and the possible genotypes of their offspring in the resulting grid. This method of analyzing a simple cross gives the genotypic ratios of the offspring, from which phenotypic ratios can be deduced.

BOX 5.2 *An Historical Perspective* Did Mendel cheat?

In 1936, the statistician and geneticist Ronald A. Fisher wrote that Mendel's results from his experiments with peas were "too good to be true" in how well they fit his hypothesis about the inheritance of traits, that is, any experiment that involves quantitative samples also requires some margin of error in the sampling. If we flip a coin, we expect that it will come up heads half the time and tails half the time, but we are not concerned if, after 20 flips, we got 11 heads and nine tails, rather than ten of each. Data do not always exactly conform to our expectations. Fisher concluded that, in effect, the variations in Mendel's data were too small, so that Mendel must have biased his results—or worse. This created quite a controversy then and is still debated today, although there is no evidence that Mendel deliberately altered his results. Still, his data are statistically closer than expected to the 3:1 ratio Mendel anticipated from a cross of two heterozygotes.

At a distance of 150 years and with no access to the original experiments and methodology other than the data that Mendel presented in his paper, it is impossible to know why his data fit so close to his hypothesis. Perhaps he saw ratios close to 3:1 in some pilot experiments and stopped counting when he got a 3:1 ratio in his follow-up experiments. Perhaps he included only some of his data and ignored the ones that did not fit his hypothesis. These are not acceptable practices nowadays, but Mendel lived in a different time and culture. It is also possible that his data really did show the ratios that he presented; we don't expect to get exactly 50 heads when we flip a coin 100 times, but it does happen. Thousands of geneticists since Mendel's time have done similar experiments with tens of thousands of genes and hundreds of thousands of offspring, so we know that Mendel's 3:1 ratio is not a statistical anomaly. We even know now the biological basis for the ratios that Mendel saw, and we will explain this in Chapter 6. So we can't question the validity of his conclusions, even if we suspect that his published data might be too good to be true.

Thinking about Mendel's data leads us to consider how scientific data are selected and presented in a more general sense. When should data be included in an analysis, and when can they be excluded? Is it acceptable to present only those results that support a given hypothesis? For one example, when selecting an image for a scientific publication, is there bias when the image chosen most clearly depicts and supports the hypothesis of the study? When is it appropriate to say that certain data points are "outliers" due to experimental error and to ignore them? When is this considered reasonable, and when does this constitute scientific fraud? These are not easy questions to answer, for Mendel or for us. In fact, if Mendel had not been so careful about presenting so much of his data, no one could have suspected that his errors were suspiciously small.

FIND OUT MORE

Much has been written about Mendel and his data. A general source on Mendel and his influence is:

Carlson, E. A. (2004) *Mendel's legacy. The origins of classical genetics*. Cold Spring Harbor Press, Cold Spring Harbor, NY

A recent discussion of this controversy with a somewhat different perspective is offered by:

Radick, G. (2015) History of Science. Beyond the "Mendel-Fisher controversy". *Science* **350**: 159–60

For more details on this topic and a more in-depth statistical analysis, refer to:

Pikas, C. (2012) *Mud sticks, especially if you are Gregor Mendel*. Available at: http://scientopia.org/blogs/guestblog/2012/08/03/mud-sticks-especially-if-you-are-gregor-mendel/ [accessed 2 August 2016]

Inferring the genotype using test crosses

Let's consider another hypothetical example. Suppose that Mendel, despite his care and meticulous record-keeping, finds a plant on the edge of his garden that produced round peas. He does not know what the ancestors of this plant looked like, so he does not know its genotype. Based on his previous findings, what could he do to determine the genotype of the plant that has this phenotype?

Although we do not know the parents of this plant, its round peas tell us that it has to have at least one copy of the ***R*** allele. However, the genotype could be either ***RR*** or ***Rr***. We can write this as ***R_***, meaning that it has one dominant ***R*** allele that masks the other allele. How could we determine the other allele? We could let our undetermined plant self-fertilize. If the plant is ***RR***, none of its many offspring plants will produce wrinkled peas. If it is ***Rr***, then about a fourth of the offspring plants will produce wrinkled peas, as shown in Figure 5.7(a).

This method of controlled crosses works well for peas and other organisms that have a large number of offspring, but it does not work as well for those with smaller numbers of offspring—say, a litter size of three or four puppies, or in humans in which usually only one offspring is produced during each pregnancy. If only one-fourth of the offspring are expected to be homozygous recessive, then it is possible that, by chance, none of them in a small litter or family will be homozygous recessive. Thus, relying on the phenotypic ratios to infer the genotype might not work very well if the number of offspring is small. As we will see in Chapter 7, this can be very important for the inheritance and occurrence of recessive genetic diseases in humans. Parents can both be heterozygotes—"carriers" in the common vernacular—but not realize it because we have small families. We return to this in Section 5.5.

In addition—and unlike peas and many other plants—most organisms cannot self-fertilize. The unknown individual or plant will have to be mated with another individual in order to produce offspring, so it makes sense to perform a cross that gives an unambiguous result and requires the fewest offspring. The simplest approach is to cross our ***R_*** plant with a homozygous recessive plant, one that is ***rr***, which, in this case, has wrinkled peas. The plant with wrinkled peas contributes no dominant alleles, so whatever allele is present in the other parent plant with the unknown genotype

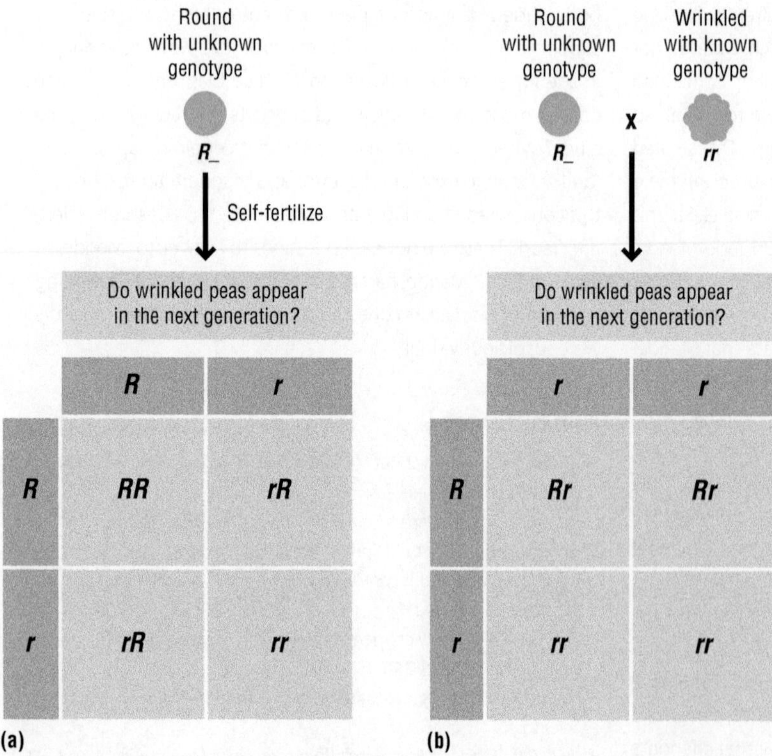

Figure 5.7 Crosses can be performed to distinguish whether an individual showing a dominant phenotype has a homozygous or heterozygous genotype. A plant can be self-fertilized for this purpose, as shown in (a). Alternatively, the plant can be test-crossed to one that is homozygous recessive, as shown in (b). Both examples illustrate the outcome if the unknown individual is heterozygous.

will be seen in the offspring. Thus, if the unknown parent plant is ***Rr***, half of the offspring are expected to have wrinkled peas, as shown in Figure 5.7(b). As soon as we see a plant with wrinkled peas, we know that the unknown parent had to be heterozygous. If we do not see any wrinkled peas among the offspring, even among five or six offspring, it is unlikely to be a heterozygote. We discuss the concept of "do not see any wrinkled peas among the offspring" in more detail in Section 5.4 when we introduce calculating probabilities, which reconsiders the question of whether five or six offspring is enough to draw this conclusion.

This type of mating, in which a parent of unknown genotype is mated with one that is homozygous recessive, is called a **test cross**, as illustrated in Figure 5.7(b). We are testing the genotype of the unknown parent by using another parent that contributes no dominant alleles that could mask the effect of the unknown alleles. Test crosses are one of the most useful diagnostic tools in genetics and will be employed repeatedly in the next few chapters.

Let's return to the crosses in more detail. Notice that the F_1 is giving only one of its two alleles of a gene to each of the F_2. This has come to be known as the **Law of Segregation**. The two alleles of the gene, ***R*** and ***r***, **segregate** from one another. We will return to this behavior of the two alleles when we describe meiosis in Chapter 6.

KEY POINT In a diploid organism, the two alleles of a gene separate from one another during gamete formation, and each gamete receives one or the other but not both alleles. This is known as Mendel's Law of Segregation.

From all of these crosses with seven different traits, Mendel concluded that genes are discrete units with different forms or alleles. They affect different characteristics of the plant, and they are inherited in predictable ways. Let's now develop this concept that inheritance is predictable in a little more detail.

5.4 Probability methods for calculating the expected ratios

Punnett squares, as we drew in Figure 5.6, always work if they are set up correctly and can always be used as a reliable method to solve a difficult problem. However, Punnett squares take time to draw and are prone to human error, particularly when more than one gene has to be considered. In addition, in many cases, we may need to know the frequency of only one class of offspring, so calculating the frequency of all of the classes of offspring is unnecessary. A faster way to solve problems is to use the rules of probability—the area of mathematics dedicated to determining the likelihoods of particular events.

The concept of probability is familiar to us from everyday examples such as weather forecasts. When the forecasters say that there is a 30% chance of rain, we understand that the weather conditions are such that rain is possible but not likely. Probability does not tell us if an event will occur or not but rather how **likely** it is to occur. For example, forecasters usually do not tell us that it will definitely rain tomorrow in our location, but they tell us how often similar conditions have produced rain. Some people, hearing that there is a 30% chance of rain, will bring an umbrella; others will pack sunglasses. Either of these two choices might be the better one once the day is under way, but, before it occurs, probabilities require judgments. In genetics, probability might not tell us what an individual offspring or seedling will look like, but it can tell us how often a particular genotype or phenotype is found among a very large population of offspring or seedlings.

The product rule to calculate genetic ratios

A simple rule of probability is called the **product rule**. The product rule says that:

> The probability that two or more independent events will occur together is the product of their individual probabilities.

Let's suppose that the probability that it will rain tomorrow is 30%. The probability that you will wear a blue shirt tomorrow is 25%. Now, let's assume that you don't pick your shirt based on the likelihood of rain and that the color of your shirt does not influence the probability that it will rain. Thus, raining and the color of your shirt are independent events. So the probability that tomorrow will be a rainy day when you are wearing a blue shirt is $0.30 \times 0.25 = 0.075$, or 7.5%. In contrast, the likelihood of rain and the color of your jacket are probably not independent events, since you may choose which jacket to wear based on the likelihood of rain.

We can illustrate the use of the product rule with our genetic crosses. From the Punnett square involving two heterozygotes, as in Figure 5.6, the probability that the male parent will donate an ***R*** is one-half. The probability that the female parent will donate an ***R*** is one-half. These are independent events, since the allele that one parent donates does not affect the allele that the other contributes. So the probability of the offspring being ***RR*** is one-half × one-half, that is, one-fourth. We can do this calculation for any combination and include more than two events, as long as we know the probability of each event.

This product rule shows up in many other examples. Have you played board games that involve dice? Suppose that one die is red, and the other one is gray, which makes it easier for us to discuss them. What is the probability that you will throw the two dice and get a 12? We can answer using the product rule. In order to get a 12, each die has to come up 6. Since a standard die has six sides, the probability that the red die will be a 6 is 1/6. The probability that the gray die will be a 6 is also 1/6. The number that comes up on the red die is independent of the number that comes up on the gray die, which is important for applying the product rule. Thus, the probability of obtaining a 12 on the two dice is 1/6 × 1/6 = 1/36.

What is the probability that you will throw a 12 on the first toss and a 2 on the second toss of the dice? By applying the product rule, we know that the probability of getting a 12 on the first toss is 1/36. By applying the product rule again, we know that the probability of getting a 2 on the second toss (1 on both dice) is 1/6 × 1/6 = 1/36. Thus, the probability that we will get a 12 on the first toss and a 2 on the second toss is 1/36 × 1/36 = 1/1296.

The product rule can be used to find the likelihood that any number of independent events will occur together. We simply multiply the probability of each separate event. Suppose that you have three coins: a penny, a dime, and a quarter. What is the probability that you will throw three heads when you flip the coins? The probability that the penny is a head is one-half; the probability that the dime is a head is one-half, and the probability that the quarter is a head is one-half. Thus, the probability of getting three heads is 1/2 × 1/2 × 1/2 = 1/8.

As an alternative to drawing Punnett squares, it is sometimes preferable to diagram a problem as a branching diagram and then use the product rule. An example of a branching diagram for our three coins is shown in Figure 5.8. In this type of diagram, the outcomes of the separate independent events are represented as branches or forks in the path. A probability is assigned to each branch (particularly if the two outcomes are not equally likely), and the final phenotype is then computed by tracing the path that led to it. Almost every practicing geneticist switches back and forth among Punnett squares, branching diagrams, and simple probability calculations, depending on the complexity and intent of the question to be answered. If only a single class of progeny is the one of interest, a probability calculation may be the easiest; if all of the possible outcomes are important, branching diagrams and Punnett squares are more widely used.

Let's use the product rule again to explore the statement we made about determining the unknown genotype of a plant with round peas. In that example, we test-crossed a plant with the dominant round pea phenotype to a plant with wrinkled peas and said that, if no wrinkled peas were seen in the next generation, we could be sure that the plant with round peas was ***RR***, rather than ***Rr***. The judgment needed here is how many plants will have to be examined to be confident of our conclusion about the genotype of the round plant. As soon as a plant with wrinkled peas is seen in the next generation, we know that the parental plant with round peas has the genotype ***Rr***—it only takes one offspring with the recessive genotype to prove that. But how many plants with **round** peas do we need to see in the next generation before we are confident that the parental plant with round peas is ***RR***?

If the parental plant is RR, then we could count thousands of peas in the next generation without seeing a wrinkled one, so that is not very informative. We would rather know about the **fewest** number of plants with round peas we need to see in the next generation before we are confident the parent is RR—after all, why do more work than necessary?

There are other ways to solve this problem, but we can reason as follows, as shown in Figure 5.9. In order to calculate how many total plants we need to count, we need to calculate first the probability of the round pea phenotype for an individual plant. By knowing the probability of a round pea phenotype for one plant and then applying the product rule, we can determine how many plants we need to examine. If the plant with round peas of unknown genotype is heterozygous, then half of the offspring of the cross will have wrinkled peas and half will have round peas. Thus, the likelihood that a single plant among the offspring has round peas is one-half. The likelihood that the first two plants to be produced will both have round peas is one-half × one-half, or $(1/2)^2$, or one-fourth, so long as these are independent events.

We can apply this rationale for each subsequent plant. For five plants to all have round peas, the probability is $(1/2)^5$, or 1/32, or about 3%. In other words, if the parental

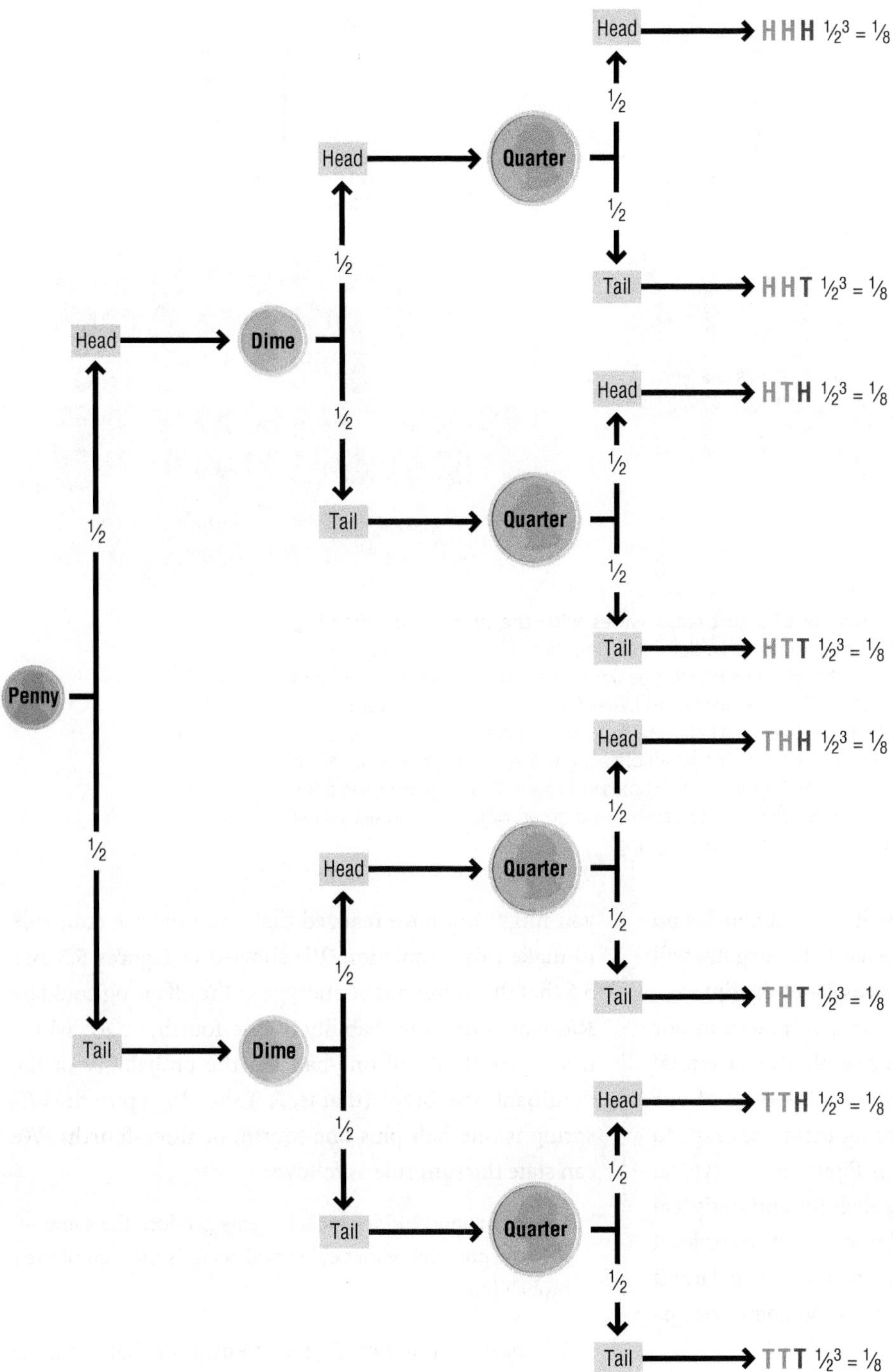

Figure 5.8 Branching diagrams. Another method to solve questions involving probability is to construct a branch diagram with the possible outcomes. Each branch is assigned a probability, which are multiplied to get the probability of the final outcome.

plant is a heterozygote and produces five F_1 pea plants by self-fertilization, there is a 3% chance that all five plants will have round peas.

More generally, because each new plant is the outcome of an independent fertilization event, the probability of seeing only round peas among the offspring is $(1/2)^n$, where **n** is the number of plants we examine.

To figure out how many plants with round peas we must count before we are confident that the parent was ***RR***, we must first make a judgment call about how "confident" we'd like to be. This is no different in principle from deciding if a 30% chance of rain calls for an umbrella or sunglasses. For example, if we count five plants and see no wrinkled peas, we may decide that the parent is likely to be

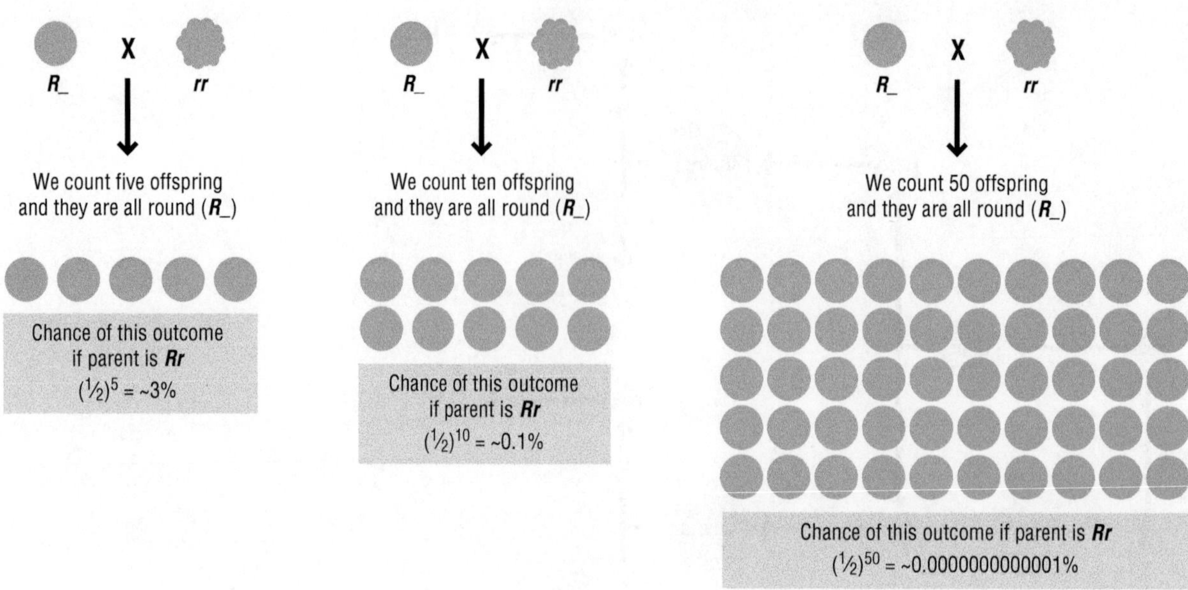

Figure 5.9 The significance of the outcome of a test cross varies with the number of offspring screened. In this figure, we have taken an individual with a dominant phenotype but unknown genotype (***R_***) and test-crossed it to an individual with the recessive phenotype (***rr***). In the first case, we have counted five offspring, all with the dominant phenotype. This may lead us to believe that the parent must have been homozygous for the dominant allele (***RR***). However, even a heterozygous parent (***Rr***) could produce five offspring with the dominant phenotype. There is a half chance that an offspring will inherit the dominant allele from a heterozygous parent, and the chance of this happening five times in a row is $(0.5)^5$. However, the odds of seeing only offspring with the dominant phenotype if the parent were heterozygous decreases, as we count more offspring (as calculated for the cases of ten and 50 offspring).

RR; we now know that the probability of that conclusion **not** being true is about 3%, since even heterozygotes will produce five offspring with round peas 3% of the time.

We may decide we want to be more confident in our answer. If we count an additional five plants (for a total of ten) and see no wrinkled peas, then the odds of our conclusion that the parent is ***RR*** being incorrect drops to $(1/2)^{10}$, or about 0.1%, as shown in Figure 5.9. A typical level of confidence used in many probability and statistical tests is 5%, but sometimes a level of confidence (or level of significance) of 1% is more appropriate, particularly if it only involves examining a few more outcomes or pea plants.

The sum rule

The product rule is one of the two fundamental rules for probability that are widely used in Mendelian genetics. The other rule is called the sum rule and also shows up in the Punnett square.

We can illustrate the sum rule with the following question. What is the probability of seeing the dominant phenotype from a cross of two heterozygotes? We know from the Punnett square that the answer is three-fourths, but you might not have realized that you used the sum rule to make this calculation. We showed in Figures 5.5 and 5.6 that the dominant phenotype in the offspring could be ***RR***, which has a probability of one-fourth, or ***Rr***, which has a probability of one-half. So the probability of the dominant phenotype (that is, ***R_***) showing up in the offspring is one-half plus one-fourth, or three-fourths. We can state the sum rule as follows:

> If two (or more) independent events produce the same result, the probability that either will occur is the sum of their probabilities.

We used the notation ***R_*** here to indicate that the same phenotype will occur, no matter what the other allele is, since ***R*** is dominant. But what is the probability of seeing the recessive phenotype? We can say with absolute certainty (a probability of 1) that we will see either the dominant phenotype **or** the recessive phenotype. So we can write:

Probability of dominant phenotype + probability of recessive phenotype = 1

We just showed that the probability of the dominant phenotype is 3/4. So the probability of the recessive phenotype is (1 – 3/4), or 1/4.

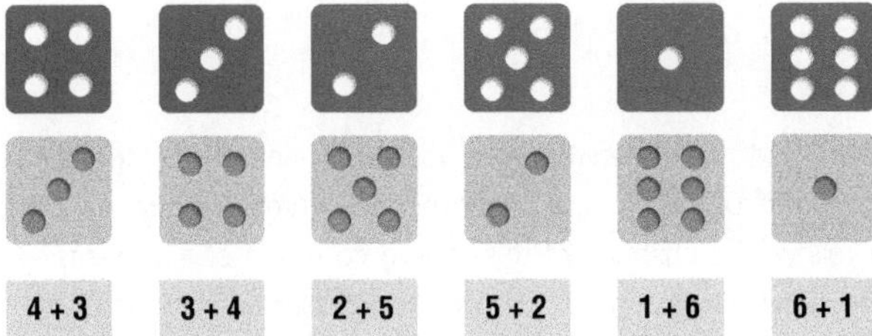

Figure 5.10 Shown are all the possible ways of rolling a 7 with two dice. The probability of each combination is 1/6 ×1/6 = 1/36. Thus, the overall likelihood of obtaining a 7 is equal to 1/36 + 1/36 + 1/36 + 1/36 + 1/36 + 1/36 = 6/36 = 1/6.

We can use the sum rule with our dice game as well. With a standard pair of dice, what is the likelihood of rolling a 7? We show this diagrammatically in Figure 5.10. For example, a 7 can arise if the red die is a 4 and the gray die is a 3. The probability that this particular combination will occur is 1/6 × 1/6 = 1/36. But there are six different combinations that produce a 7, and, if we sum all of these separate probabilities, we get a total of 6/36 = 1/6. As it turns out, although it happens only 16.7% of the time, 7 is the number most likely to occur when you roll a pair of dice.

As we noted in Box 5.2 for Mendel's results, the observed numbers in different phenotypic classes found in an experiment may not match the expected numbers. When the observed numbers are sufficiently different from what is expected, we may need to reconsider our hypothesis about how the trait is inherited or what our expected outcomes might be. One commonly used statistical method to determine if our hypothesis should be reconsidered or rejected is known as the chi-square (χ^2) test. We discuss this test in Box 5.3.

BOX 5.3 *Quantitative Toolkit* The chi-square test

Any genetics experiment counts or samples a finite number of offspring or outcomes from an infinitely large number of possible outcomes. Therefore, the observed number of individuals in each category may be fewer or greater than what you would predict or expect if an infinite number of offspring had been sampled. This is true whether your hypothesis is correct or incorrect. How do you make the judgment about how close your observed number must be to the predicted or expected number in order to support your hypothesis?

For example, you may be testing a hypothesis that a trait is recessive, so that mating of two heterozygous parents would yield 1/4 of the progeny expressing the mutant phenotype. You observed 14 mutants out of a group of 50 offspring, rather than **exactly** the 1/4 or 12.5 mutant offspring you predicted. What's important here is not that there is deviation from the expected numbers, but how large that deviation is and how much importance or significance is placed on that difference. The greater the difference of observed from expected, the less likely it is that your hypothesis is correct. Suppose that you found only three mutant offspring in the group of 50, when 12.5 mutants are expected. Your intuition is such that you would question your hypothesis if only three mutant offspring are observed in such a cross but probably not question your theory if 14 mutant offspring are observed. You are recognizing that it is not the number of offspring but rather the difference between the number of offspring you expected and the number you observed.

When experimental results fall into discrete categories, such as white eyes vs. red eyes or round peas vs. wrinkled peas, geneticists can use a statistical test called χ^2 (chi-square) to test the "goodness of fit" of the observed data to the expected numbers based on the hypothesis. Using the results of the χ^2 analysis, the geneticist can make a decision as to whether to accept the original hypothesis or to reject it in favor of an alternate hypothesis.

To use the test, you must first work out your expected results. Based on the hypothesis about how traits are inherited, you can make a prediction about how often each of the traits should be found among the offspring, or the frequency of a particular outcome. For each category of expected results, that is, for each phenotypic class, you then calculate the expected number of individuals based on this hypothesis. It is very important to note that, while your hypothesis may well be about the fraction of each type of offspring among the total, the data used in the χ^2 test are the raw numbers, rather than the fractions.

The difference between the observed and the expected number is then calculated. This number is squared and divided by the expected value. This calculation is repeated for each phenotypic class. The numbers calculated for each phenotypic class are added together to obtain a χ^2 value. Therefore, the formula for calculating the χ^2 value is:

χ^2 = the total of (**observed − expected**)^2 divided by the expected number for all categories of data expected.

Each χ^2 value has an associated probability, which is the likelihood that the deviation or difference from the expected values would be this amount or greater in repeated experiments. Table A lists χ^2 values and their associated probabilities.

The other information needed to determine the probability associated with a particular χ^2 value is called degrees of freedom. For our purposes, the number of degrees of freedom is the number of different categories minus 1. Therefore, if there are two expected

BOX 5.3 Continued

phenotypes of offspring in a mating—such as males and females—the degrees of freedom is two categories minus 1 = 1. If an experiment had eight categories of data, the number of degrees of freedom would be 7.

By reading across the row with the appropriate degrees of freedom, you can determine the probability for any value of χ^2. The number at the top of each column is the probability of obtaining that χ^2 value with that number of degrees of freedom by chance, that is, if the hypothesis is correct and if the experiment were repeated many times, you would have seen this amount of deviation by chance in that fraction of repeated tests.

By convention, if the probability for a given χ^2 value is greater than 0.05 (5%), the data support the hypothesis; the appropriate way to state this is that the data do not allow you to reject the hypothesis. If you determine the probability for a given χ^2 value to be 0.05 or less, the data are said not to fit the model or hypothesis. There is little likelihood (≤5%) that the deviation from predicted results occurred by chance. Setting the rejection level at 0.05 allows a large amount of deviation before you reject a hypothesis and reduces the chance that you will reject a true hypothesis. If the χ^2 value has a probability of 0.05 or less, your model may not be correct, or else there is another factor that is skewing your data. If this occurs, you will want to test your results using another hypothesis or look for other factors that may have affected your experiment.

Example

The χ^2 test can best be illustrated with an example. Suppose that, in one of Mendel's experiments, he self-fertilized plants that were heterozygous for green and yellow pods; these heterozygotes have green pods, so green is dominant to yellow. The offspring of this mating consisted of 428 plants with green pods and 152 yellow-podded plants, for a total of 580 plants.

Mendel's hypothesis is that three-fourths of the plants should have green pods and one-fourth should have yellow pods. (If this is not clear to you, be sure to work it out.) Out of a total of 580 plants analyzed, 145 (1/4) are expected to have yellow pods and 435 (3/4) are expected to have green pods. We summarize the χ^2 calculations in Table B.

The χ^2 value is therefore 0.451, and there is one degree of freedom in this experiment. From the table of χ^2 values, the probability of this much variation is about 50%. In other words, if this experiment were repeated many times, 50% of the trials would have this much deviation from the expected values. Because this is greater than the 5% cut-off, these data support the hypothesis that the green and yellow traits follow the Law of Segregation and that green is dominant; more formally, we cannot reject this hypothesis.

There are a few important points to note in this example. First, we used the raw numbers of observed and expected plants, rather than the ratio or fraction. This is how the χ^2 test takes sample sizes into account. Second, the χ^2 value increases when the observed and expected values are further apart. This check of your understanding of the test can help to make sure that you are reading the table of values correctly.

Table A χ^2 values and their probabilities

Degrees of freedom	Possibility of chance occurrence as a percentage (5% or less considered significant)								
	90%	80%	70%	50%	30%	20%	10%	5% (sig.)	1%
1	0.016	0.064	0.148	0.455	1.074	1.642	2.706	3.841	6.635
2	0.211	0.466	0.713	1.386	2.408	3.219	4.605	5.991	9.210
3	0.584	1.005	1.424	2.366	3.665	4.642	6.251	7.815	11.341
4	1.064	1.649	2.195	3.357	4.878	5.989	7.779	9.488	13.277
5	1.610	2.343	3.000	4.351	6.064	7.289	9.236	11.070	15.086
6	2.204	3.070	3.828	5.348	7.231	8.558	10.645	12.592	16.812
7	2.833	3.822	4.671	6.346	8.383	9.083	12.017	14.067	18.475
8	3.490	4.594	5.527	7.344	9.524	11.030	13.362	15.507	20.090
9	4.168	5.380	6.393	8.343	10.656	12.242	14.684	16.919	21.666

BOX 5.3 Continued

Table B Data from a hypothetical cross

	Observed	Expected	(Ob – exp.)²/exp.
Green	428	435	0.113
Yellow	152	145	0.338
Total	580	580	0.451

The sum rule and product rule together

The sum rule and the product rule cover most cases of probability that we will encounter in genetics. Here is one more application that uses the product rule and the sum rule together to simplify a problem. We can illustrate with a variation on our coin problem.

If three coins are flipped, what is the probability that at least one of the coins will **not** be a head? This situation is depicted in Figure 5.11. We could work out all of the ways that at least one coin is a tail (that is, not a head) based on the product rule, and then use the sum rule to calculate the probability that at least one will be a tail. But the easiest way to solve this problem is to reason backwards. We know that the probability that all three coins will be heads is 1/8. In every other combination, at least one of the coins is not a head. There are many ways that this could happen, and we could write them all down and work it out. But more simply, if the probability that all three coins show a head is 1/8, then the probability that **at least one** of them is not a head is 1 – 1/8, or 7/8. Either all coins are heads or at least one of them is not a head.

Often, we can use subtraction to find the probability of some combination of events. For example, once we determined from earlier that there was a 3% chance that a heterozygote would produce five plants with round peas, we could have also concluded that there is a 97% probability that it will **not** produce five offspring with round peas—in other words, at least one plant will have wrinkled peas. If there are many different ways that the event could happen, it may be simpler to find the probability of the rare chance that the event **does not happen** and subtract this from 1. A good rule of thumb is that problems with wording such as "at least one" or "not affected" can often be figured out using this approach.

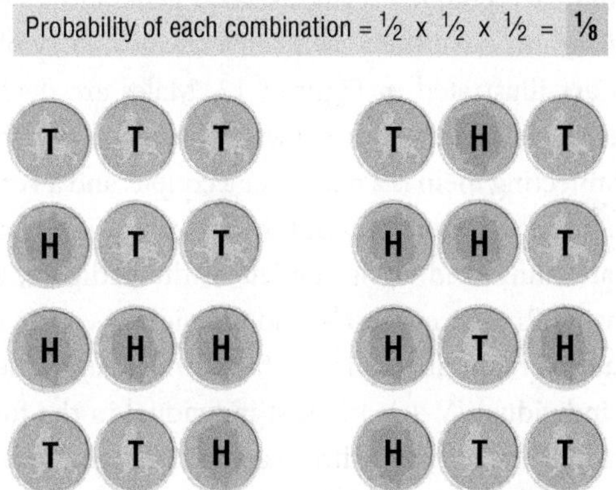

Figure 5.11 If three coins are flipped, what is the probability that at least one of the coins will **not** be a head? Heads are depicted as H and tails as T on the coins in the figure. The probability of flipping three tails is 1/2 × 1/2 × 1/2 = 1/8, so the probability of NOT getting this is equal to 1 – 1/8 or 7/8.

5.5 Using pedigrees

Even very large families in humans do not have enough children to make phenotypic ratios meaningful. Nonetheless, because we know the principles of inheritance from other organisms in which a large number of F_1 offspring is feasible, we can apply these same principles in human genetics, even without using the expected ratios of phenotypes. In studying inheritance patterns in humans and some other animals that have low numbers of offspring, a family tree or a pedigree can be very useful.

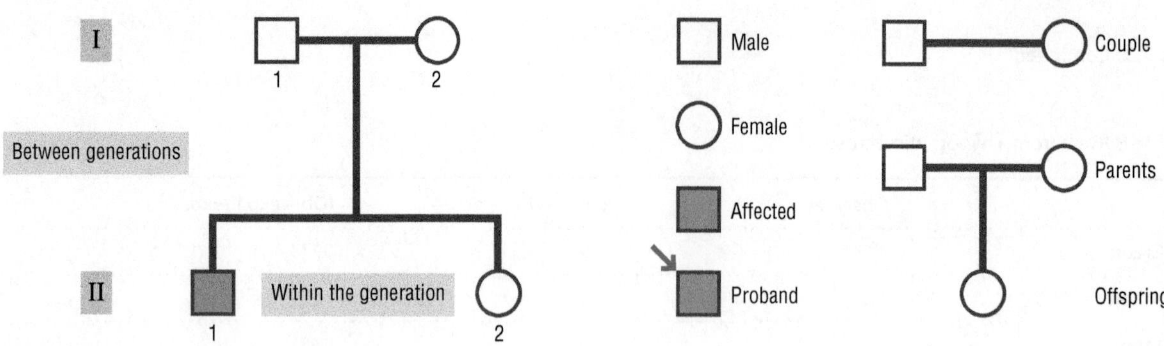

Figure 5.12 A summary of the most commonly used symbols in pedigree analysis.

A pedigree diagrams the matings and offspring arising from each couple through several generations. Pedigree diagrams have their own conventions and symbols, which are illustrated in Figure 5.12. Males are depicted as squares, and females are shown as circles. A horizontal line connecting them is a mating or a couple, and a vertical line indicates a generation and the offspring. The generations are numbered along the left of the pedigree, typically using Roman numerals, and individuals within each generation are numbered in Arabic numbers underneath. Thus, individual IV-1 is the first individual in the fourth generation. The number within each generation is primarily used to designate the person of interest, and may or may not be related to relative age or birth order. Typically (although not always), a filled symbol is an affected individual, and the unfilled symbols are individuals who are not affected. "Affected" in this sense means that this individual shows the phenotype in question, whether or not this phenotype might be considered a genetic disorder.

A frequent situation in human genetics is that an individual affected by a genetic disorder appears in a clinic or hospital. The first individual in a pedigree known to be affected is referred to as the **proband**, and the appearance of the proband is usually the impetus for pedigree analysis of the family history. Many single-gene disorders in humans—that is, those that follow simple inheritance patterns—have severe phenotypic consequences, so the proband is often an affected infant or young child, although this is not always the case. Some single-gene disorders manifest themselves at puberty or even at post-reproductive ages—Huntington's disease is an autosomal dominant neurological disorder in which the average age of onset is among people in their 40s. Other single-gene traits are not diseases at all, and these may not be recognized or diagnosed as often unless there is some other interest; a few of these traits will be used in study problems in this chapter, Chapter 7, and Chapter 8. Nevertheless, most of the identified single-gene traits in humans are disorders that manifest themselves in children, and many of them are rare. The authoritative source for human genetic traits is known as OMIM (Online Mendelian Inheritance in Man), found at **www.ncbi.nlm.nih.gov/omim**.

Determining inheritance patterns in pedigrees

When viewing a pedigree, the basic inheritance pattern is usually not that difficult to determine. Because most genetic disorders are rare, it is usually assumed that people who are not directly descended from individuals I-1 and I-2, that is, people who marry in from outside, are not carriers for the trait. This assumption can be revised if the inheritance pattern warrants it, but this is the initial assumption. It is also important to determine if both males and females are affected by the trait and if they are affected in about equal numbers; in Chapter 7, we will return to pedigrees in which males are affected much more often than females.

If the trait is rare and both males and females are affected, then the inheritance pattern can be inferred from our knowledge of inheritance patterns of other traits in other organisms, as illustrated in Figure 5.13. If an affected child has two unaffected parents, then the trait is very likely to be recessive and the parents are very likely to be heterozygous. In the Punnett square, this shows up as ***Aa*** × ***Aa*** producing ***aa*** offspring. If, on the other hand, affected children always have an affected parent, then the trait is likely to be dominant. In the Punnett square, this shows up as ***Aa*** × ***aa*** producing ***Aa*** offspring.

Determining probabilities in pedigrees

A typical situation in a pedigree, and one that will occur in many genetics questions, will require that you determine

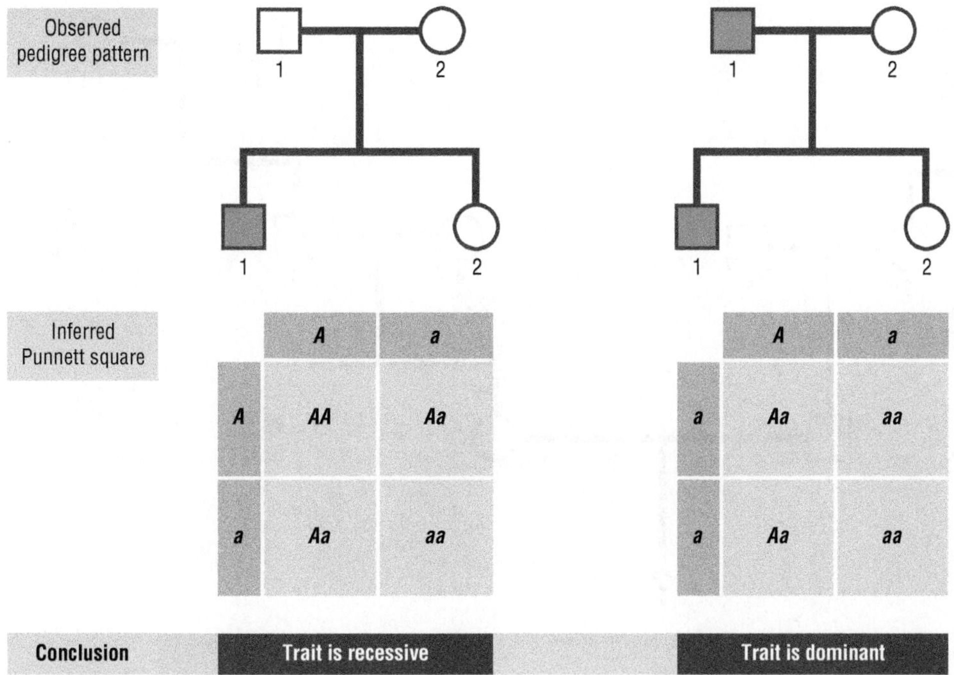

Figure 5.13 The relationships between two commonly observed patterns in pedigrees and the Punnett square that would give rise to them. A recessive trait is shown on the left, with a dominant trait on the right.

the probability that an unknown person will be affected by the trait in question. Perhaps the woman is pregnant, and her brother and her aunt, as well as the father's uncle, have the same genetic disorder; she may want to know the probability that her unborn child will be affected. This requires a careful stepwise analysis of the phenotypes and likely genotypes in the pedigree, and the application of the rules of probability discussed in Section 5.4.

To illustrate this process, let's work out the pedigree in Figure 5.14, which presents the situation we have just discussed. A woman (III-2) is pregnant with her first child, and neither she nor her spouse (III-1) is affected by the genetic trait in question. The trait is rare, but her husband's uncle (Individual II-1 in the pedigree), her aunt (Individual II-6 in the pedigree), and her brother (Individual III-3 in the pedigree) are affected. We have extended the pedigree back by an extra generation to show the couple's grandparents (Individuals I-1 through I-4) in case we need information on them. What is the probability that her child, represented by the "?", will be affected by the trait?

First, we have to determine how this trait is inherited. We notice that both II-1 and III-3 are affected by the trait but do not have affected parents. This indicates that the trait is recessive. Notice that II-6 is also affected, but her mother I-4 was also affected, so this is not helpful and can be ignored. Since the trait is recessive and neither of the couple III-1 and III-2 is affected by the trait, the child could only be affected if both are heterozygous. So our next step is to figure out the probability that III-1 and III-2 are each heterozygous.

Let's begin with III-2. Her parents must both be ***Aa*** because her brother is affected by the trait. So since her parents are both ***Aa***, what is the probability that she is also ***Aa***? This is a bit more subtle that you may first realize. You may have answered one-half, since half of the offspring from ***Aa*** × ***Aa*** are ***Aa***. But that omits an important piece of information. The woman herself is not affected, so she cannot be ***aa*** and must be either ***AA*** or ***Aa***. Among those who are ***A_***, two-thirds are ***Aa***, as shown in the Punnett square involving two heterozygous parents in Figure 5.13. So her probability of being ***Aa***, knowing that she is not affected by the trait, is two-thirds. Notice that, even though we were told that the trait is rare and we assumed that the people from outside the family were not carriers, her mother II-4 must be a carrier for her brother to be affected. When there is clear evidence that an individual from outside the family must be a carrier, we have to drop our assumption.

Now we turn to her spouse, III-1, and we have to go back further in the pedigree. He is not affected, so he is

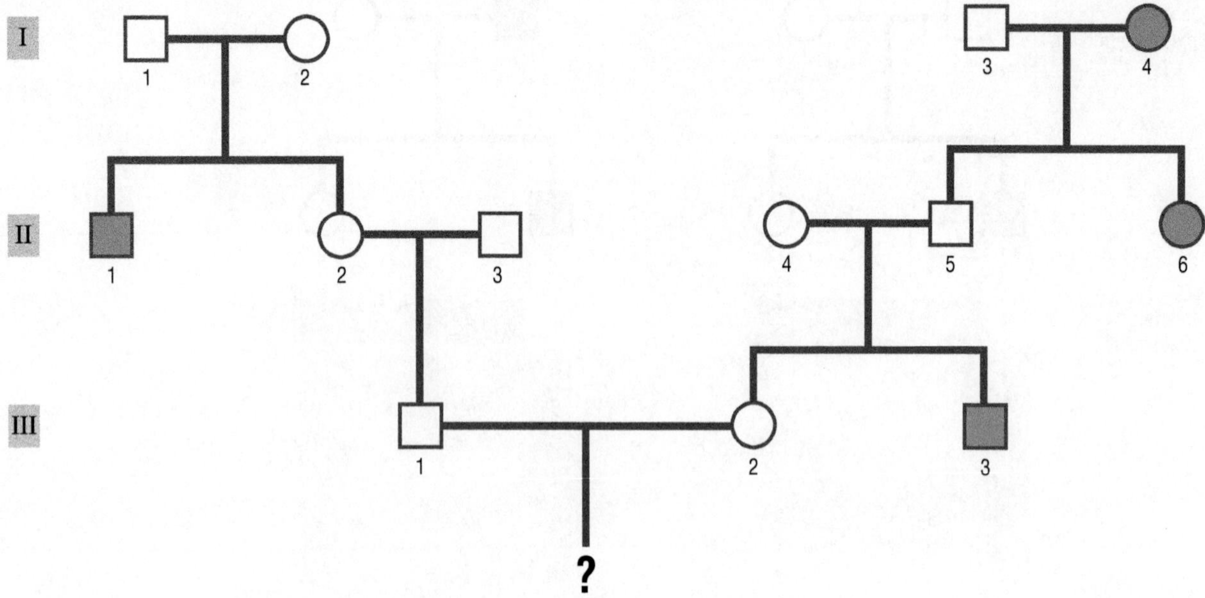

Figure 5.14 A pedigree that is solved in the text.

A*_**. His mother II-2 is not affected either (and his father II-3 is assumed to be ***AA, since there is no contrary evidence). The probability that his mother is a carrier is two-thirds, by the same reasoning used for III-2. His mother must be ***A*_**, but her parents I-1 and I-2 must both have been ***Aa***. If II-2 has a two-thirds probability of being ***Aa***, then III-1 has a one-third probability of being ***Aa***. Half of the children from a mating of ***Aa*** × ***AA*** are ***Aa***. We have just used the product rule—the probability that his mother is ***Aa*** times the probability that he inherited the ***a*** allele from her.

We are not quite done yet. The probability that III-2 is ***Aa*** is two-thirds. The probability that III-1 is ***Aa*** is one-third. So the probability that both of them are ***Aa*** carriers (which is the only way they could have an affected child) is 2/3 × 1/3, or 2/9. We used the product rule again. But, even if both of them are carriers, the probability that their child is affected is one-fourth. So the total probability that her child will be affected is 1/3 × 2/3 × 1/4, or 2/36 or 1/18. It is the probability that her spouse is a carrier times the probability that she is a carrier times the probability that her child will be affected, even if both are carriers.

Let's ask one more question. What is the probability that her child **will not** be affected? This could be worked out by carefully determining all of the possible ways that the child inherits one ***A*** allele (which is all that is needed to be unaffected), but that would be tedious and there are numerous possibilities for mistakes. It is simpler to work out the probability that the child will be affected, as we did above, and find that it is 1/18. Then the probability that the child is not affected must be 17/18.

This is a relatively complicated pedigree question, but by methodically working through the possibilities and recognizing the probabilities from Mendelian ratios, we arrived at a solution. We worked out the genotype of people only as necessary to solve the line of direct descent; in this particular example, that ended up being most of the people in the pedigree, but that is not always necessary. This problem also illustrated two common pitfalls in working pedigree questions. First, the probability that III-2 is ***Aa*** is two-thirds, not one-half, since we know she is not affected. Second, even if both parents are affected, the probability of having an affected child is only one-fourth. The study questions offer some additional examples.

WEBLINK: The analysis of this pedigree question is reviewed in Video 5.1. Find it at **www.oup.com/uk/meneely**

5.6 How do phenotypes arise from genotypes?

With practice, a person can do many genetics problems using Punnett squares, branching diagrams, and probability, without thinking very much about the underlying biology of what is occurring. In the next chapter, we will describe how inheritance patterns, such as the 3:1 ratio, arise from the biological process of meiosis. But now we want to put some of the concepts in this chapter, together with some concepts from previous chapters, and discuss phenotypes and genotypes in terms of molecules and genomes.

How does a particular genotype give us a particular phenotype? This is a big question, with many possible answers, often specific to each genotype and phenotype; most commonly, the answer is not entirely known. The most fundamental answer about the connection between the genotype and the phenotype is that the product of a gene is often a protein, as discussed in Chapter 2. (Exceptions in which the product of a gene is a functional RNA, rather than a protein, will be discussed in Chapter 12, but all of the examples we have been describing involve proteins.) The phenotype arises from the activity of one protein, or perhaps many different proteins, encoded by different genes. It could also arise from the activity or inactivity of a protein that is missing or altered. Let's go back to our most familiar example to discuss this question.

Why then were Mendel's peas wrinkled? How did that phenotype arise from the ***rr*** genotype? This question was answered about 20 years ago, or more than a century after Mendel's experiments, and is illustrated in Figure 5.15. Peas, like many vegetables, get their shape from the synthesis and storage of starch; starch storage also affects how much water is retained, or the osmolarity. The gene that affects pea shape encodes an enzyme involved in the biosynthesis of starch. If this enzyme works normally, large amounts of total starch and amylopectin (the branched form of starch) are made and stored, so the pea is round. However, the wrinkled allele makes an inactive form of that enzyme, so less amylopectin is created, which affects the osmolarity of the pea. Since less amylopectin is made, water diffuses out as the seed matures, and the pea is shrunken or wrinkled. We return to this example in the questions at the end of the chapter.

Of course, wrinkled and round peas are not the only example to illustrate how genotypes and phenotypes are connected to one another, and many others will be developed in other chapters. Here is another example, involving a gene with a familiar effect in humans. You may realize that different coffee drinkers can have very different responses to caffeine; you may not realize that, to a large extent, these responses are affected by a single gene. Caffeine, found in coffee or many energy drinks, is broken down in our liver by an enzyme known as cytochrome P450 1A2, which is encoded by the gene *CYP1A2*. This enzyme is the product of a single gene, and the gene has different alleles. To connect this a little more fully to concepts in Chapters 2 and 3, we know that different people have slightly different nucleotide sequences for this gene, which means that they make cytochrome P450 1A2 enzymes with slightly different amino acid sequences. All of these variations, or alleles, produce a functional enzyme, but the different forms of the protein do not function equally well. We will call two of the most common forms of the gene the ***A*** allele and the ***C*** allele, referring to the nucleotide at a particular position in the gene.

These naturally occurring variations affect how quickly a person breaks down caffeine. Homozygotes for the ***A*** allele break down caffeine much more quickly than ***A/C*** heterozygotes or than ***C/C*** homozygotes. As a result, people with the ***A/A*** genotype are able to drink coffee later in the evening without being kept awake, unlike a heterozygote or a ***C/C*** homozygote who will still be experiencing the effects of caffeine. Genotypes produce the phenotypes; in this case, the nucleotide sequence of the *CYP1A2* gene and the corresponding amino acid sequence of the cytochrome P450 enzyme affect how rapidly caffeine is broken down and whether coffee keeps you awake at night.

For the sake of completeness, the metabolism of caffeine is a bit more complicated than we have just suggested. While *CYP1A2* is the primary gene responsible for breaking down caffeine and many other drugs, its

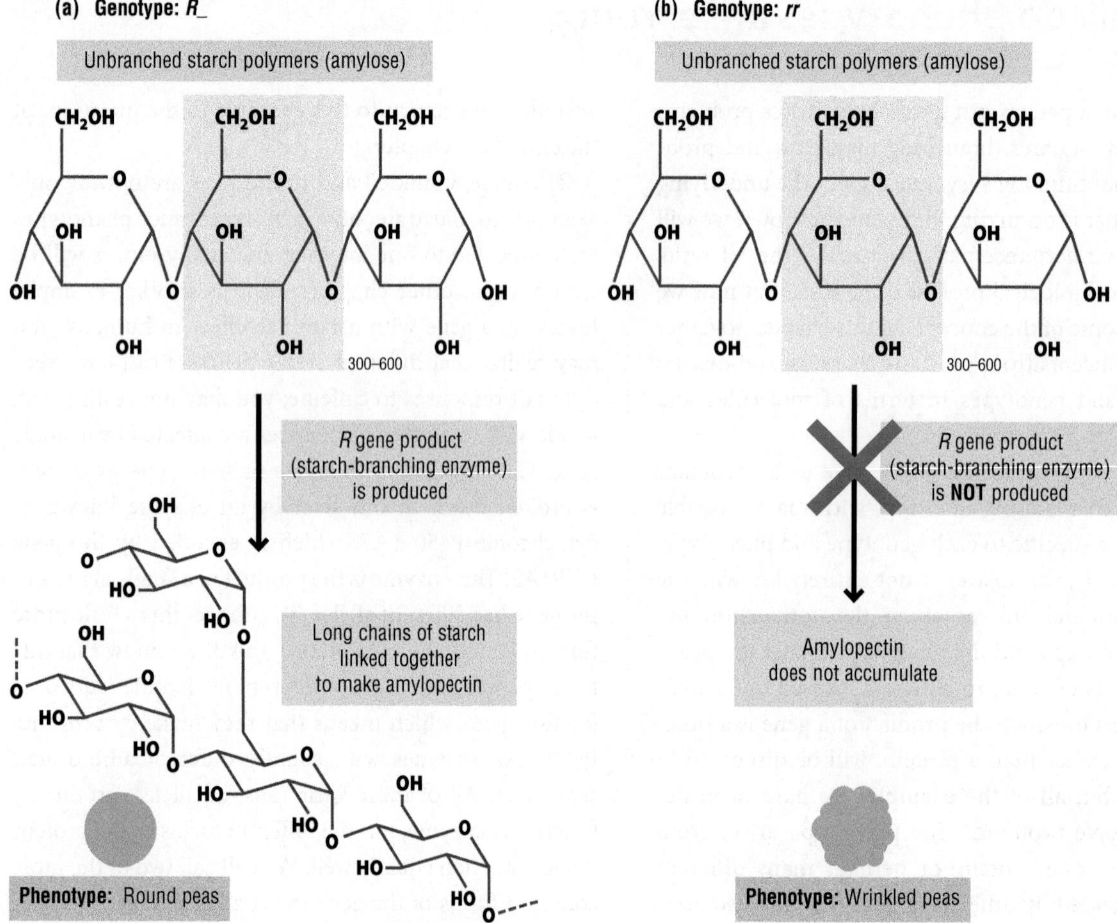

Figure 5.15 The basis for round and wrinkled peas. Starch is a polymer of glucose subunits, some of which are linear and some branched. The starch-branching enzyme (SBEI) is the protein that is altered in Mendel's wrinkled peas, leading to reduced concentrations of amylopectin, the branched form of starch.

expression is also affected by regular caffeine consumption. In a concept that we will develop more fully in Chapter 14, the transcription of *CYP1A2* is induced by the presence of caffeine. Thus, heterozygotes who regularly drink coffee can break down caffeine more rapidly than *A/A* homozygotes who do not drink coffee because chronic caffeine consumption results in a higher level of transcription of the *CYP1A2* gene. To oversimplify, what the heterozygotes may lack in the quality of their cytochrome P450 is being made up for by its quantity.

5.7 Summary: Mendel and his peas

Gregor Mendel developed the universal rules of genetics in eukaryotes by using true-breeding varieties to study inheritance in pea plants. From these studies, he observed that traits were inherited beyond the first generation, even if they were not observed in the F_1. From this observation came the idea of units of inheritance, which we now call genes and alleles, as well as the concepts of dominant and recessive traits. Mendel deduced that each parent carried two alleles for each trait, and these separated before fertilization, what we refer to as the Law of Segregation. By understanding this segregation of alleles during a cross, we can predict patterns of inheritance and apply rules of probability to determine the likelihood of specific outcomes. We can then begin to think about how a particular phenotype results from a given genotype—the underlying changes in protein structure or activity that result from a change in the gene.

CHAPTER CAPSULE

- A gene must be expressed to impart a physical outcome or phenotype upon the individual.
- Changes in the DNA sequence of a gene can result in changes in the phenotype associated with that gene. Different variants of the same gene are referred to as alleles.
- Most inherited traits are the product of combined inputs from the genome and the environment. However, there are some traits where there is a direct and simple relationship between a single gene and a corresponding phenotype.
- Following these simple phenotypes in crosses between individuals allows us to make deductions about the inheritance patterns of the associated gene.
- Mendel's use of simple, discrete phenotypes and true-breeding lines in his crosses of pea plants was critical to the success of his experiments.
- Mendel's observation that traits are inherited beyond the first generation, even if they were not observed in the F_1, led to the idea of genes and alleles and to the concepts of dominant and recessive traits.
- Mendel deduced that each parent carried two alleles for each trait, and these separated before fertilization. This is called the Law of Segregation.
- A test cross is a mating in which a parent of unknown genotype is mated with one that is homozygous recessive.
- By understanding this segregation of alleles during a cross, we can predict patterns of inheritance and apply rules of probability to determine the likelihood of specific outcomes.
- Punnett squares, branching diagrams, and pedigrees are used to determine the likelihood of particular outcomes.
- Applying Mendel's laws of inheritance can help us establish links between genotypes and phenotypes. We can then begin to think about how a particular phenotype results from a given genotype, such as the underlying changes in protein structure or activity that result from a change in the gene.

STUDY QUESTIONS

Concepts and Definitions

5.1 Define the following terms, and discuss how they are used in this chapter.

- **a.** Locus
- **b.** Diploid

c. Recessive and dominant

d. Proband

5.2 Why is a test cross useful to infer the genotype of an individual who might be heterozygous or homozygous? Are there situations when observing the ratios in the next generation might be as informative as, and preferable to, doing a test cross?

5.3 It is easier to establish true-breeding lines for recessive traits than for dominant traits. Why?

5.4 We stated that Mendel was able to deduce the rules of inheritance because he used true-breeding lines, he concentrated on a few traits, and he kept accurate and quantitative records. Why was each of these experimental strategies important?

5.5 What is a "wild-type"? The concept of a wild-type is similar to one of Mendel's experimental strategies. Which one?

5.6 The concept of "wild-type" is not typically applied to most human traits, but there are some situations in which the idea of a wild-type is helpful in human genetics, even if the term is not used. What might be such a situation when the concept of a wild-type is useful for human genetics?

5.7 Without doing the maths, what parameters are important in deciding if experimental results match the expected results?

Beyond the Concepts

5.8 Mendel presented his results to his scientific society in Brno and also communicated them to colleagues in Vienna and other places. Suppose that you heard his results and saw his data. What questions might you have raised about his findings? (Assume that he was honest and careful and he did, in fact, accurately record what he found.)

5.9 Consider a trait that you think might have a genetic basis. What information would be helpful in determining which phenotype is dominant to the other?

5.10 Many traits that people think about as "running in families" or as being heritable in plants or animals are not inherited in the simple patterns that Mendel observed. What are some of the reasons that a trait could be strongly affected by genes but still not show the simple ratios that Mendel observed?

5.11 Why are pedigrees a better method for analyzing genetic results in humans than in fruit flies?

5.12 People can be carriers (heterozygotes) for traits for many generations and not be aware of it. In fact, all of us are heterozygous for some severe or life-threatening genetic diseases. Using the principles you learned in this chapter, why do most of us not know what genetic diseases we carry? What might be some cultural or social conditions that affect how often a child is born with a rare genetic disease?

5.13 Wrinkled peas arise from a mutation that inactivates a gene that encodes an enzyme involved in the production of starch. Why should a mutation that inactivates the gene be recessive? What does this tell you about how much of the enzyme is needed by the plant?

Applying the Concepts

5.14 In tomatoes, the shape of the leaf called potato is recessive to a leaf shape called cut. A true-breeding variety called Mortgage Lifter with cut leaves is crossed to a true-breeding variety called Hillbilly with potato leaves, and seeds are collected. These F_1 plants are grown.

a. What is the genotype of these F_1 plants? What is the phenotype of their leaves?

b. The F_1 plants are self-fertilized, and their seeds are collected. These seeds are planted to create an F_2 generation. What fraction of the F_2 plants are expected to have potato leaves?

Challenging **c.** Approximately how many seeds have to be planted in order to be confident (at a 95% level) that at least one seedling has a potato leaf?

5.15 In many mammals, including rabbits, there is a hair texture known as angora, in which the hair is long and soft. Angora rabbits are highly prized for the quality of their fur. A rabbit breeder has four rabbits: an angora male, an angora female, a short-haired male, and a short-haired female. (The sex of the rabbits is not relevant to the solution.) He made the following crosses with the outcomes shown.

Angora female × angora male = all angora

Angora female × short male = all short

Short female × angora male = 4 short and 5 angora

a. Which phenotype is dominant?

b. What are the genotypes of each of the parents?

c. If the short female and the short male were mated to each other, what fraction of the offspring is expected to have short hair?

d. Even for experts, the sex of rabbits is very difficult to determine, but the breeder wants to keep the males and females separate as much as possible. When the mating in (c) is performed, a litter of five rabbits is born. What is the probability that at least one of the five is a female?

5.16 In certain games that use dice, rolling doubles (both dice show the same number) gives an advantage to the player such as an extra turn. In any given roll, what is the probability of rolling doubles of any number?

5.17 In Chapter 4, we saw that the small probability for error during replication can result in mutations that change the function of a gene. Certain mutations in the *Escherichia coli rpsL* gene result in the phenotype of resistance to the antibiotic streptomycin. Similarly, specific mutations in the *rpoS* gene yield rifampicin-resistant varieties. The antibiotics have distinct mechanisms of action, so that a bacterial cell might be resistant to one, both, or neither. When 10^8 of a certain strain of mismatch repair-deficient *E. coli* bacteria from a rapidly dividing culture are plated onto nutrient agar plates containing streptomycin, 119 resistant colonies (each arising from a single cell) are obtained. When the same number of bacteria are plated out on plates containing rifampicin, 35 colonies are counted. What is the probability of obtaining a colony resistant to both streptomycin and rifampicin?

5.18 The nematode *Caenorhabditis elegans* can reproduce by self-fertilization or cross-fertilization. Hermaphrodites are capable of self-fertilization or cross-fertilization by mating with a male. A wild-type male with normal movement is mated to a paralyzed hermaphrodite. All of the F_1 offspring are capable of movement. One of the F_1 offspring is allowed to self-fertilize to produce an F_2 generation.

a. Which allele is dominant, and which is recessive?

b. Among the F_2 generation worms, what fraction of them do you predict will be paralyzed?

Looking ahead **c.** When an observant student did the cross, she noticed that some of the worms in the F_2 generation moved normally, some were paralyzed, and some moved sluggishly, although they could move. She decided to count the number of worms in each category, with the following results.

Normal movement = 73

Paralyzed = 66

Sluggish but moving = 139

How might you explain these results? What do these results suggest about the F_1 worms that are sluggish but capable of movement?

d. Test your hypothesis about the results in (c) by using a χ^2 test.

5.19 A family has three children, none of whom are twins.

a. What is the probability that they have two daughters and a son?

b. What is the probability that they have at least one son and one daughter?

Challenging **c.** Suppose that we know that their oldest child is a girl. Based on this additional information, what is the probability that they have two daughters and a son?

5.20 The ability to taste bitterness in certain foods, such as Brussels sprouts, depends on the ***T*** allele, which is dominant to non-tasting (***tt***). A couple has a daughter and a son. Both parents can taste bitter, but their daughter cannot taste bitter.

a. What is the probability that their son can taste bitter foods?

b. How does the phenotype of the daughter help you answer this question?

c. When he is tested, their son can taste bitter foods. He marries a woman who cannot taste bitter foods, and they have a son. What is the probability that their son (the grandson of the original couple) will be able to taste bitter foods?

Looking ahead **d.** Speculate about the cellular and molecular biology of the ability to taste bitter chemicals. What function might the *T* gene encode? How might this explain which allele is recessive and which allele is dominant? What type of protein could have this function?

Looking back and ahead **e.** Check your speculation from (d) about the molecular biology of this trait by doing an Internet search for the ability to taste bitter chemicals in humans, and see if you can answer the following questions. Do gorillas and chimpanzees (our closest living relatives) have the ability to taste bitter chemicals? Why might the ability to taste bitter chemicals provide a selective advantage in humans that might not have made so much difference in gorillas and chimpanzees?

5.21 Figure Q5.1 shows a pedigree for a rare trait in humans. Affected individuals are indicated by the filled-in symbols. The woman II-2 is pregnant, with II-3 being the father. Her unborn child is represented by the "?"

a. What is the relationship between the unborn child and individual I-2?

b. What is the relationship between the unborn child and individual II-4?

c. Is this trait inherited as a dominant trait or a recessive trait? What is the best evidence for your answer?

d. What is the probability that the unborn child will be a boy and not affected by the trait?

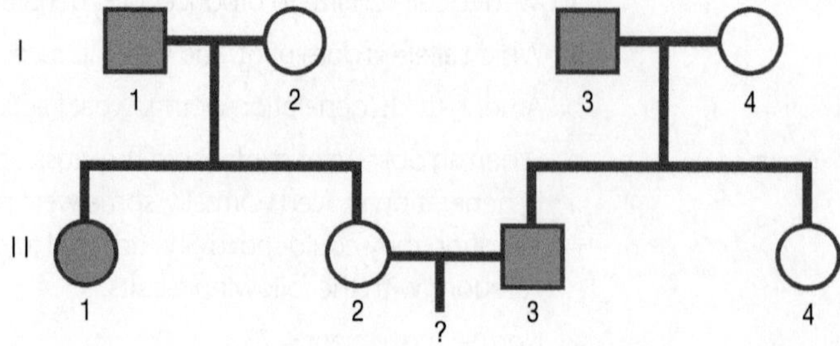

Figure Q5.1

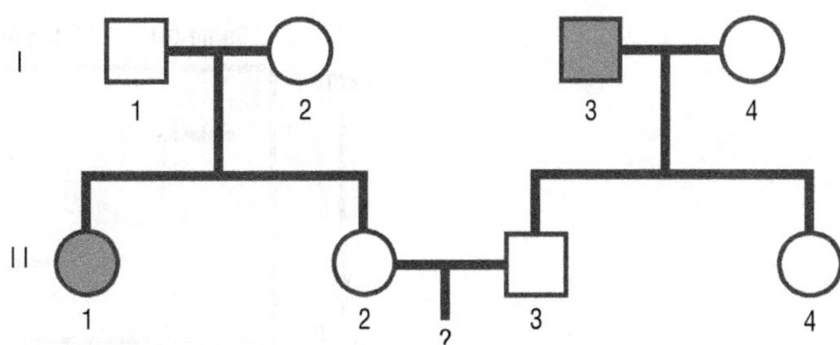

Figure Q5.2

5.22 Figure Q5.2 shows a pedigree for a rare trait in humans. Affected individuals are indicated by filled-in symbols. The woman II-2 is pregnant, with II-3 being the father. Her unborn child is represented by the "?".

a. Is this trait inherited as a dominant trait or a recessive trait? What is the best evidence for your answer?

b. What is the probability that both II-2 and II-3 are heterozygous for the trait?

c. What is the overall probability that the unborn child will be unaffected by the trait and be a girl?

5.23 The Rh blood factor is a single-gene trait, with Rh-negative recessive to Rh-positive. Two Rh-positive parents have an Rh-negative child.

a. What must be their genotypes?

b. What is the probability that their next child will be Rh-positive?

Challenging **5.24** A research laboratory is studying a gene that encodes a protein involved in the flower colors in petunia. They have a DNA fragment that encodes the gene from true-breeding red petunias and white petunias and use this DNA to make the protein *in vitro*—precisely how they do this is not important for this question. One tube has the reaction to make the protein for red flowers, while another tube has the reaction to make the protein for white flowers. The protein made from each reaction is run on a gel that separates the proteins, based on their size. When the flower protein extracts are run on gels, the two homozygous individuals displayed different size bands, as shown in Figure Q5.3.

a. Based on what you know (from Chapters 2 and 3) about genes and proteins, what molecular change in the gene could produce a protein in white flowers that is smaller than the one in red flowers?

b. Based on this, do you predict that red or white will be dominant, and why do you predict this?

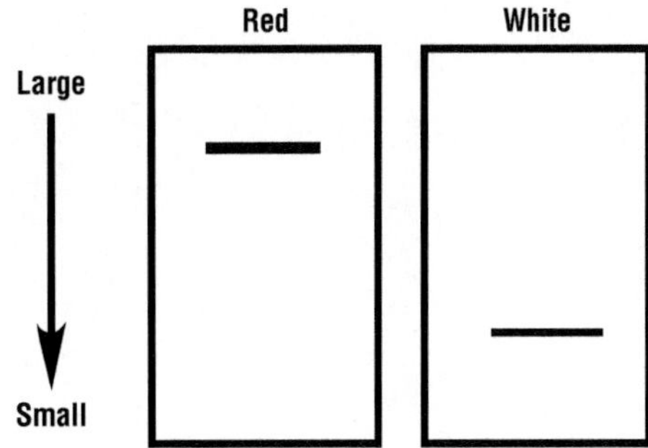

Figure Q5.3

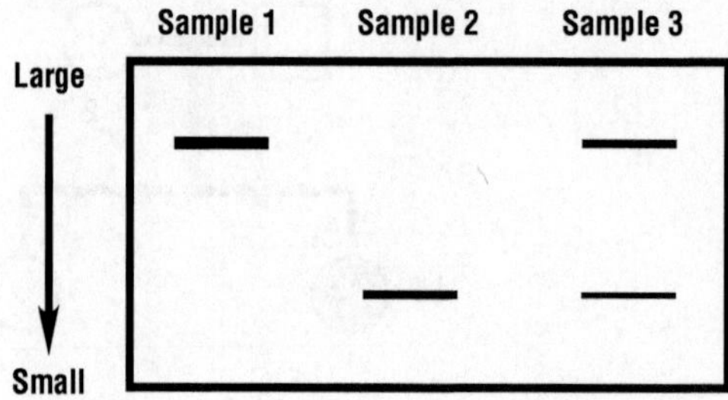

Figure Q5.4

c. A true-breeding red petunia is crossed to a true-breeding white petunia. Seeds are isolated and grown, and the DNA is isolated from F_1 plants. This DNA is used to make the protein as before, and the protein is run on a gel. Draw the pattern of bands that you would expect to see.

d. One of the F_1 petunias in (c) is self-fertilized to produce an F_2 generation. Seeds are planted and grown. You carefully sort the F_2 generation into plants with red flowers and those with white flowers. From each single plant, you isolate the DNA fragment and make the protein *in vitro*. You then run the product of each individual reaction on a protein gel. The results from three different plants are shown in Figure Q5.4. Label each of the three lanes with the color of the flower that produced this result.

CHAPTER 6

The Cellular Basis for Mendelian Genetics

6

IN A NUTSHELL

The process of meiosis helps us understand on a cellular and molecular level the inheritance patterns that Mendel observed. During meiosis, the genome of the cell is halved, with each resulting cell containing one of the two members of a homologous pair of chromosomes. Meiosis is the biological process that results in the phenotypic ratios Mendel observed in his crosses of peas, as described in Chapter 5.

6.1 Overview of Mendel's contributions

Mendel's research was not discovered or widely read for about 35 years after he completed his work with peas, and Mendel himself went on to study things besides genetics, including meteorology and bee-keeping. Remarkably, three other biologists who were studying inheritance around 1900—Carl Correns, Hugo de Vries, and Erich von Tschermak (shown in Figure 6.1)—independently realized that the inheritance patterns found in their data had been previously recognized and explained by Mendel. It is to their credit that these three men acknowledged and celebrated Mendel. Carl Correns was a German botanist working in Tubingen. His mentor was Karl Nägeli, a prominent botanist with whom Mendel had an extended correspondence, but Nägeli failed to see the significance of Mendel's work. Erich von Tschermak, an Austrian agronomist, was trained at the University of Vienna (as was Mendel), and his grandfather was Mendel's professor of botany. Hugo de Vries was a Dutch botanist who was the most scientifically prominent of the three men. de Vries gave us many of the terms that we currently use in genetics—gene, allele, and mutation, among others.

During the time when Mendel's work was largely unknown and Darwin's work was gaining acceptance, cytologists (or what we now call cell biologists) were developing techniques to look at the structures inside cells. Chromosomes were identified and described, and their movements during cell division were recorded and categorized. Some cytologists looked at testes from grasshoppers and locusts, among many other organisms, and used the study of these specimens to describe the process of meiosis.

By the time Mendel's work was rediscovered in 1900, meiosis had been well described for many organisms, and the differences in shape and size of chromosomes within a genome or karyotype were noticed. The German biologist August Weismann had even realized that the products of

(a) Carl Correns (b) Hugo de Vries (c) Erich von Tschermak

Figure 6.1 The rediscovery of Mendel. Mendel's work was largely ignored until it was independently discovered in the early 1900s by these three men: Carl Correns, Hugo de Vries, and Erich von Tschermak.

Source: Image of Hugo de Vries from *Popular Science Monthly* Vol. 67 (1905). Image of Erich von Tschermak from *Acta horti bergiani bd. III*, no.3 (1905).

meiosis had half the chromosome number of the somatic cells and that this was necessary to maintain continuity of the germline from one generation to the next. A number of biologists, including de Vries, the British geneticist William Bateson, and the American cytologist Edmund B. Wilson, recognized that Mendel's patterns of inheritance and the segregation of chromosomes during meiosis were two sides of the same biological coin—Mendel's rules and patterns of inheritance were indicating how the chromosomes moved at meiosis; meiosis explained how Mendel's rules came about.

KEY POINT When cell biologists rediscovered Mendel's research in 1900, they were able to relate the patterns of inheritance that he observed in peas to the changes they observed in chromosomes during meiotic cell division.

6.2 Mitosis preserves chromosome number during somatic cell division

In order to explain meiosis and its connection with Mendel, we will begin with a quick summary of mitosis and cell division. Eukaryotic organisms begin as a single cell called a zygote containing the complete genome. Indeed, many organisms, including yeasts, the green algae *Chlamydomonas*, and others, remain single-celled; the cell divides to reproduce, and the genome size stays constant with each division. By contrast, many other organisms develop to become multicellular; each of the cells in such an organism has the same DNA sequences and the same genome.

In order for cells of a multicellular organism to possess the same genome, even after they have divided, they must first **replicate** or copy their entire genome before division occurs. We noted in the Prologue and in Section 4.4 that the cell cycle is divided into several phases, shown in Figure 6.2. These include G_1 in which cells increase in size and then commit themselves to going through division at a time referred to as START, S in which DNA is synthesized, G_2 the second growth phase, and M during which mitosis occurs. Genome replication occurs during the S phase (synthesis phase) of the cell cycle. Then, during mitosis or M, the duplicated genomes separate, yielding two cells, each with a complete copy of the genome that was present before the S phase.

Not all cells in a metazoan are actively dividing; in fact, relatively few cells in an adult are in G_1 and committed to the cell cycle. Rather, most cells are in a phase called G_0 and can continue to grow and function in G_0 without dividing for a long time, even years in some tissues in humans, for example. Other cells, such as our skin cells, divide more often, so they are rarely in G_0. G_0 and G_1 are often called interphase, when the cells are carrying out their normal functions: expressing their genes, transporting their proteins, communicating with one another—in short, doing everything, except replicating their DNA or dividing.

KEY POINT During mitosis, the chromosomes that were replicated in the S phase of the cell cycle separate, resulting in two cells with the same genetic information.

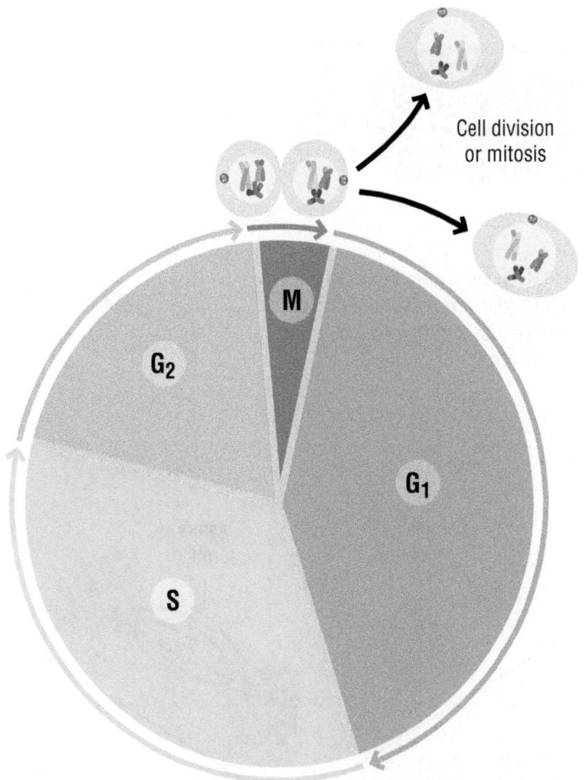

Figure 6.2 The eukaryotic cell cycle. The stages of the eukaryotic cell cycle are shown; cell division or mitosis is represented by the dividing cells during the M phase. The chromosomes in these cells are shown in more detail in Figure 6.3. The lengths of G_1 and G_2 are highly variable. Cells that are not in the process of dividing are in the G_0 phase, which is not shown; G_0 and G_1 are also referred to as interphase.

Mitosis, or M phase, can itself be divided into four stages, as seen by cytologists more than a century ago—prophase, metaphase, anaphase, and telophase, which may be familiar to you; these are shown in Figure 6.3. For genetics, the focus when describing mitosis is on the behavior of the chromosomes. But the subcellular aspects of dividing cells that do not directly involve the chromosome are also important. We will first pay particular attention to two subcellular structures and their behavior: the spindle and the kinetochore, which are also shown in Figure 6.3. In doing so, we will focus on one particular family of proteins: the tubulin family, which is important in the functions of each of these structures.

Tubulin forms the wall of a structural tube in the cell called the **microtubule**, as shown in Figure 6.4. Tubulin is a protein family, with different family members having very similar, but not identical, amino acid sequences. Two family members, α-tubulin and β-tubulin, are the structural components of microtubules; these are shown in Figure 6.4(a). A microtubule consists of alternating subunits, composed of α-tubulin and β-tubulin, and has a polarity or directionality, with an α-tubulin molecule at the negative end and β-tubulin at the positive end, as shown in Figure 6.4(b). Microtubules are found throughout the cell as part of the cytoskeleton where they serve many important other roles. For example, they are involved in moving objects, such as chromosomes and vesicles, around the cell, as well as being involved in the formation of cilia

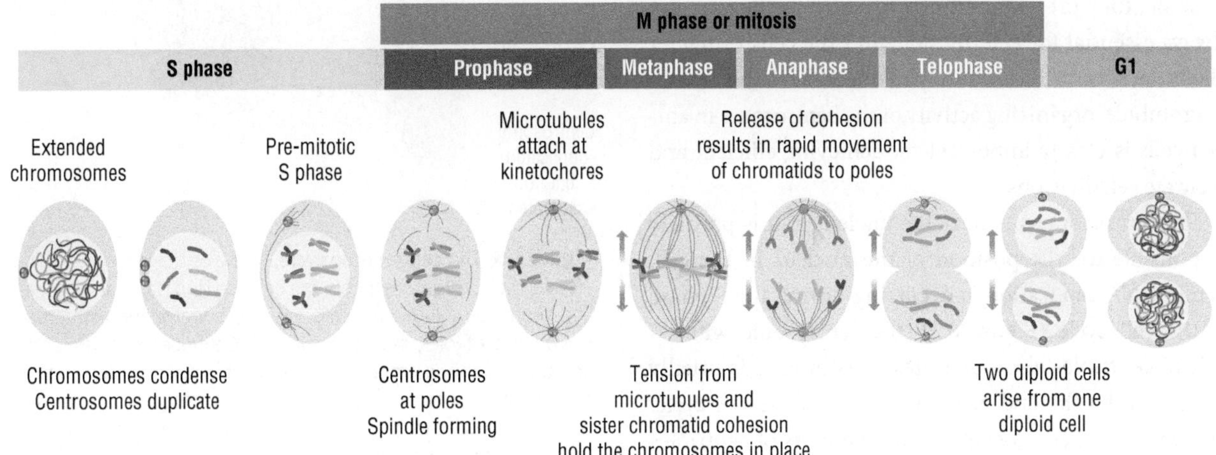

Figure 6.3 Mitosis. Mitosis, the M phase of the cell cycle, is traditionally divided into stages: prophase, metaphase, anaphase, and telophase. The processes are described in sequence, and the distinction between stages is somewhat arbitrary, since the process is continuous. In this drawing, G_2 is not shown, and the duration of each stage is not to scale. For example, anaphase for some cells lasts a few minutes, while prophase and metaphase can be much longer.

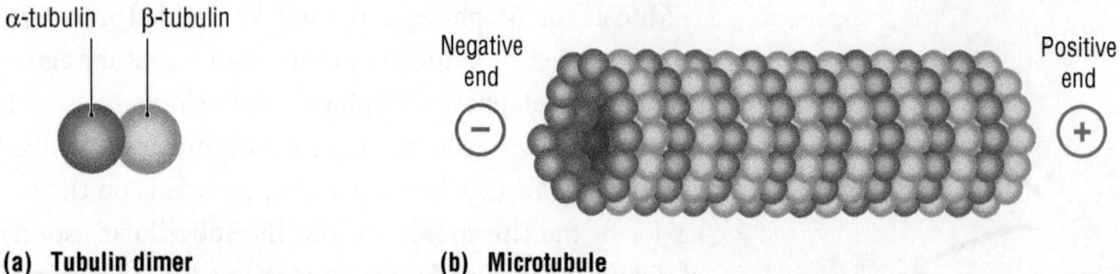

Figure 6.4 Tubulin proteins and microtubules. Tubulin is a family of proteins that are highly similar in their amino acid sequence and functions. α-tubulin is represented in orange, while β-tubulin is represented in gold. (a) α- and β-tubulin form a dimeric protein. (b) These dimers from (a) can assemble into a cylindrical structure called the microtubule. The ends of the microtubule are different, with the end culminating in α-tubulin called the – (negative) end and the end with β-tubulin called the + (positive) end.

and flagella. The microtubules emerge from structures known as microtubule-organizing centers, or MTOCs; for mitotic and meiotic cells, the MTOC is also known as the centrosome. A third member of the tubulin family, known as γ-tubulin, is found at the MTOC.

The centrosome

The centrosome in an animal cell is composed of a pair of structures known as centrioles, oriented at right angles to each other and embedded in a matrix of proteins; these other proteins include rings of γ-tubulin. These structures can be seen in Figure 6.5. Microtubules grow out from the centrosome, using the γ-tubulin rings as their starting points, by adding new α–β subunits to the plus end. As a result, there is an inherent directionality to both microtubules and the structures made from them. Interestingly, although centrioles lie at the heart of the centrosome, they present a bit of a mystery, as they are not always essential for cell division, and the cells of higher plants lack centrioles entirely. Nonetheless, the overall microtubule-organizing activity of the centrosome in animal cells is clearly important for achieving efficient and accurate cell divisions.

The centrosome sits near the nucleus and is partially responsible for the position of the nucleus in the cell. While DNA replication is occurring during the S phase of the cell cycle (Figure 6.2), the centrosome with its pair of centrioles also duplicates. This process is usually completed by the end of the G_2 phase of the cell cycle. The duplicated centrosomes with two pairs of centrioles remain attached to each other, until the beginning of mitosis when the pairs of centrioles are split apart by an enzyme called separase. (We encounter separase in a different role in Box 6.1.) The newly separated centrosomes initiate the growth of microtubules away from them; a

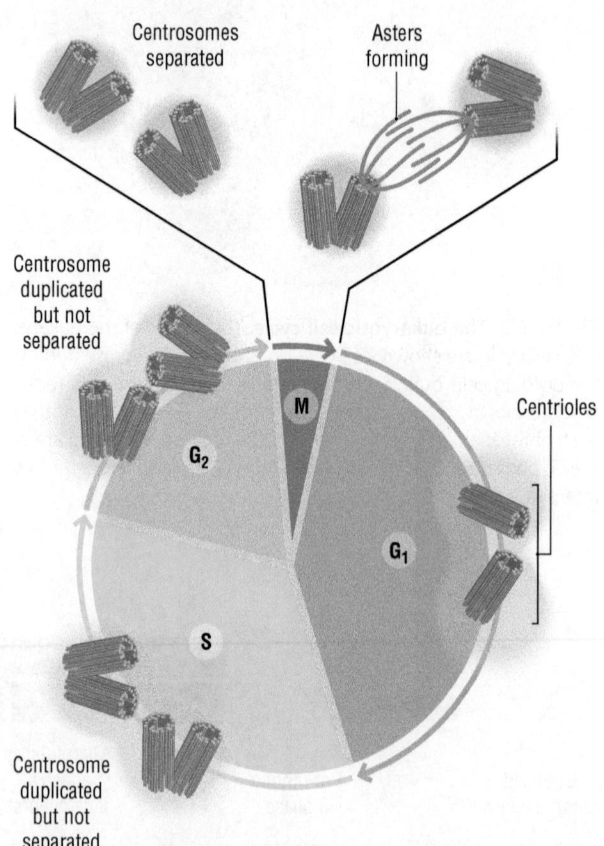

Figure 6.5 The centrosome cycle. As the cell itself is undergoing a cycle, the centrosome within the cell also duplicates and divides. The centrosome in animal cells has two barrel-shaped structures known as centrioles, shown here in pink. Centrioles include a different tubulin family member, known as γ-tubulin, which serves as the nucleation center for microtubule growth. During the S phase, the centrosome doubles; the two centrosomes do not separate from one another, however. This structure remains until the beginning of mitosis when the centrosomes separate from one another, migrate apart, and microtubules begin to grow from their centrioles as a pair of asters. Eventually, the asters form a spindle that spans the dividing cell from one end to the other, as shown in Figures 6.3 and 6.6. After mitosis, each daughter cell has its own centrosome which can resume the cycle.

centrosome with microtubules radiating away from it is called the aster.

The formation of the spindle

The asters migrate to opposite ends of the cell and continue to nucleate the growth of microtubules towards the center of the cell, as shown in Figure 6.6. The growth of microtubules is highly dynamic, with subunits being added and lost and with the microtubules becoming longer and shorter as a result. This variation in microtubule length is referred to as dynamic instability. However, when the microtubules from one aster contact the microtubules from the other aster, the dynamic instability in microtubule length decreases, and the microtubules radiating away from the asters become longer and longer. Eventually they form a spindle that spans the cell from two opposite poles, as shown in Figure 6.6. Each centrosome forms one pole of the spindle, and each half of the spindle has microtubules extending towards the middle of the cell.

The growth of the asters occurs during prophase of mitosis, while the spindle is formed by the end of prophase and the beginning of metaphase, a stage often called prometaphase. The spindle has been assembled outside of the nucleus, but the spindle is not a fixed and stationary structure. Its microtubules are still growing and shortening, a dynamic behavior that is crucial to its next function.

The role of kinetochores in spindle attachment

During prometaphase, the nuclear membrane breaks down into small lipid vesicles. Now the chromosomes are exposed and can come in contact with the microtubules of the spindle. Each chromosome has a structure known as a **kinetochore**, which provides a site of attachment to the spindle. The kinetochore comprises a complex of proteins and has a distinct inward and outward face, that is, there are certain proteins on one side of the structure and different proteins on the other side. The proteins on the inward face contact the DNA and chromatin; the kinetochore and the underlying DNA are referred to as the **centromere**. The microtubules of the spindle attach to the kinetochore of each sister chromatid, as shown in Figure 6.7, and pull the chromosomes to opposite poles of the cell. The terms centromere and kinetochore are nearly synonyms, depending on whether the focus is on the region of the chromosome (centromere) or the proteins associated with the site of microtubule attachment (kinetochore).

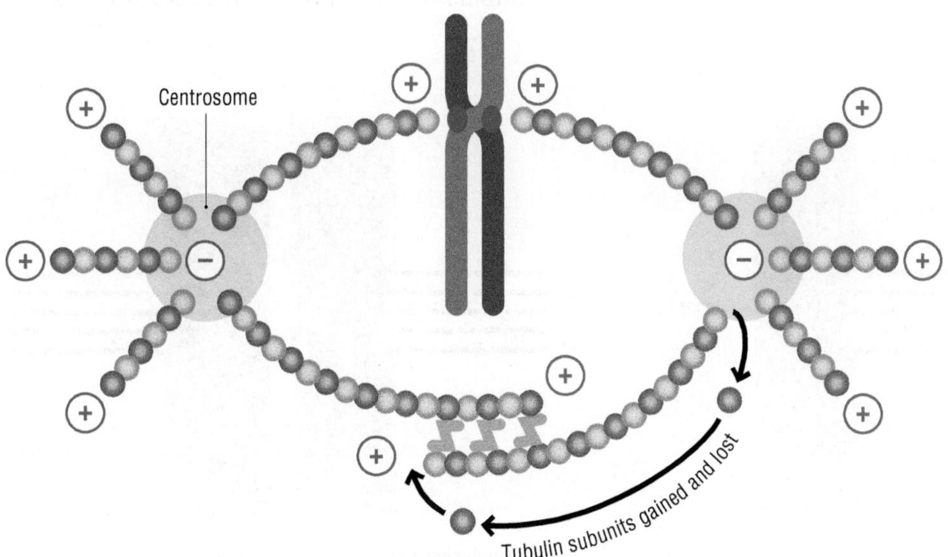

Figure 6.6 Asters and microtubule growth. The centrosomes migrate to the poles of the dividing cell, with microtubules growing out from them, a structure referred to as the aster due to its similarity to the flower. The negative end is at the centrosome, while new tubulin subunits are added at the positive end, resulting in dynamic growth as new subunits are added and lost. Eventually, the microtubules contact the kinetochore on the two sister chromatids, and the spindle forms by this bipolar attachment.

Source: Based on Becker (2015) *World of the Cell*, 9th edition. Pearson.

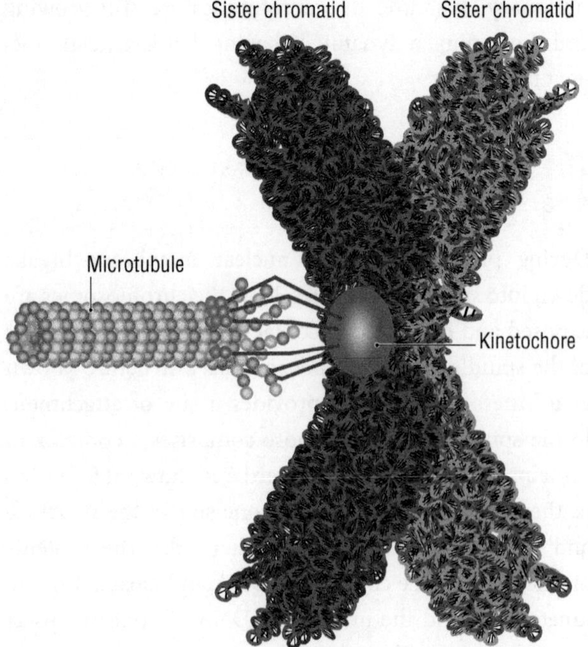

Figure 6.7 Microtubules attach at the kinetochore. One chromosome with its two sister chromatids is depicted here in purple. The chromosome has a kinetochore to which microtubules attach. The kinetochore is found at the centromere and consists of approximately 20 different proteins, most of which are found in all eukaryotes. The term "centromere" connotes the region of the chromosome, while "kinetochore" is more often used to describe the structure, but the terms refer to the same feature of chromosomes.

The kinetochores have an inner and an outer face, as illustrated in Figure 6.8(a), that is, certain proteins in the kinetochore are found on the same side of the chromatid as the spindle attachment, and others are found on the side of the chromatid away from the site of the spindle attachment, associated with the DNA of the centromeric region.

During mitosis, kinetochores on the two sister chromatids face in opposite directions on the sister chromatids, as shown in Figure 6.8(b). This bipolar orientation—with attachment to the spindle on both sides—is important for the kinetochore to function correctly and hence for the processes of mitosis and meiosis. The spindle fibers attached to the kinetochores pull the sister chromatids towards the poles of the cell. At the same time, sister chromatid cohesion acts to hold the chromatids together. These opposing forces hold the chromosomes in the middle of the cell during metaphase.

KEY POINT Kinetochores are complexes of proteins on the chromosome that interact with, and attach to, the spindle fibers that pull the sister chromatids apart during mitosis. The tension between the pull of these spindle fibers and the cohesion of the two sister chromatids holds the chromosomes at the metaphase plate.

Chromosome segregation in mitosis

Now that we have described two of the important subcellular structures, let's go back to the S phase and focus on the chromosomes. Notice how the chromosomes are

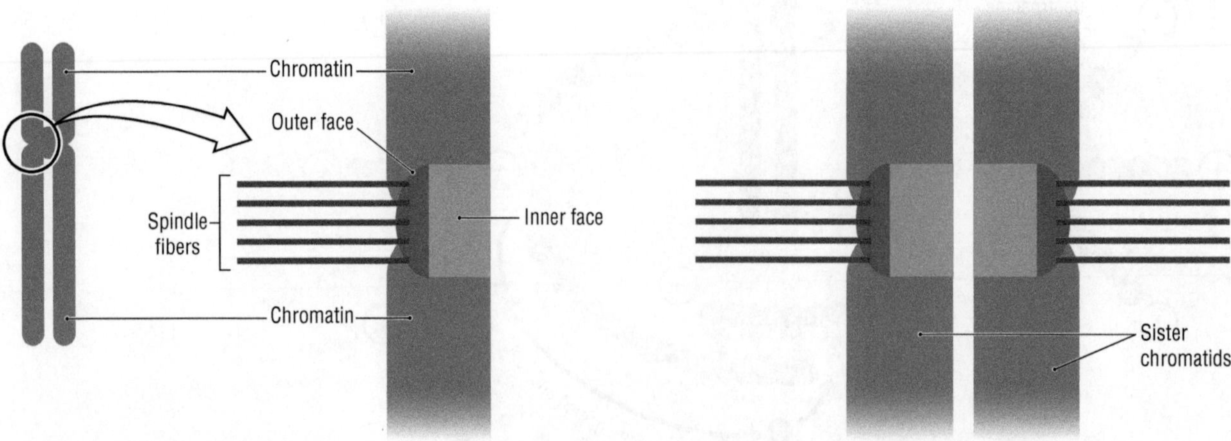

Figure 6.8 Kinetochore orientation. (a) The kinetochore has two sides, and its orientation is important to its functions. The outer side has proteins that attach to the microtubules, while the inner side is associated with the DNA of the centromere. (b) During mitosis, the kinetochores of the two sister chromatids are facing in opposite directions. This bipolar orientation allows the microtubules of the spindle to pull the sister chromatids to opposite ends of the cell during anaphase.

depicted in Figures 6.3 and 6.6. Prior to the S phase, the chromosome consists of one long double helix of DNA with its associated proteins that make up chromatin. During the S phase, the DNA molecule replicates to yield two DNA molecules; these are still tangled around and attached to each other, particularly at the region of centromeric DNA. These two double-stranded DNA molecules make up the **sister chromatids**, which are held together by cohesion along the arms and at the centromere.

The sister chromatids are held together both at the centromere and along the arms by a complex of proteins known as **cohesin**. In the absence of such sister chromatid cohesion, the tension applied by the spindle during metaphase would cause the sister chromatids to separate prematurely or even cause them to break apart. The cohesin protein complex and the process of sister chromatid cohesion during mitosis and meiosis are discussed in Box 6.1.

Thus, during prometaphase, the nuclear envelope has broken down, and the kinetochores have formed, with their outer faces oriented away from each other. Microtubules attach themselves to proteins on the outer face of the kinetochore, which means that the chromosome becomes attached to the spindle. The microtubules on the spindle begin to shorten, which pulls each sister chromatid (via its kinetochore) towards the corresponding pole. However, the sister chromatids are still held together, and the two sister chromatids are being pulled in opposite directions. The so-called bipolar orientation, in which the spindle from each pole is attached to the kinetochore on each sister chromatid, creates tension on the chromosome, as shown in Figures 6.6 and 6.8(b). This tension serves to line up the chromosome at the center of the cell, and the lining up of chromosomes across the center of the cell defines metaphase.

From images of mitotic metaphase, you may have the idea that the chromosomes are sitting in the middle of the cell during metaphase, patiently awaiting the next step, like passengers queuing at a bus stop. In reality, the chromosomes are constantly moving back and forth in small rapid movements, pulled towards the pole by the microtubules attached at the kinetochore and held back by the cohesion to the sister chromatid, which is being pulled in the opposite direction. If these are passengers awaiting a bus, they are very impatient passengers, pacing back and forth, stepping out into the street to see if the bus is approaching, checking their phones, and so on.

As soon as the cohesion between the sister chromatids is released through the action of separase, as described in Box 6.1, the sister chromatids are pulled

BOX 6.1 *Going Deeper* Sister chromatid cohesion and the cohesin complex

The importance of sister chromatid cohesion during mitosis and both meiotic divisions is introduced in Section 6.2. If the sister chromatids are not held together along the arms or at the centromere, the tensions applied by the spindle fibers will cause them to separate prematurely, and orderly segregation to opposite poles of the cell does not occur. Then, after being held together during the S phase, G_2, prophase, and metaphase of mitosis, the sister chromatids must rapidly separate from each other in the transition from metaphase to anaphase. A complex of approximately 12 proteins, known as the anaphase-promoting complex (APC), is responsible for the loss of sister chromatid cohesion, and thus the rapid transition from metaphase to anaphase. One key mechanism by which the APC triggers this transition is by mediating the activity of a protease known as separase. Separase cleaves one of the proteins involved in sister chromatid cohesion, and the chromatids are released, as shown in Figure A. Separase is itself regulated by a protein inhibitor called securin, which sequesters separase to prevent it from cleaving proteins at inappropriate times.

Cohesin during mitosis

The protein complex responsible for sister chromatin cohesion during mitosis and meiosis is called **cohesin**. Cohesin comprises four protein subunits, all of which are found throughout the kingdom of eukaryotes and some of which have orthologs in bacteria (which, of course, do not have sister chromatids). Additional proteins are needed for loading cohesin onto the chromosomes and for its establishment and release, but these four proteins make up its core. Cohesin has been best studied in budding yeast, so we will use those names of the yeast proteins and the corresponding genes in the following description.

Two of the proteins, Smc1 and Smc3, are highly similar to each other in sequence and function. Each of these polypeptides consists of a globular and flexible "hinge" at one end, and part of a nucleotide binding domain (NBD) at the other end. Curiously, the domain structures of the proteins are anti-parallel to each other, so that the NBD of Smc1 is found at its C-terminus while the one for Smc3 is found at its N-terminus. The NBD is responsible for the binding and hydrolysis of adenosine triphosphate (ATP); in order

BOX 6.1 Continued

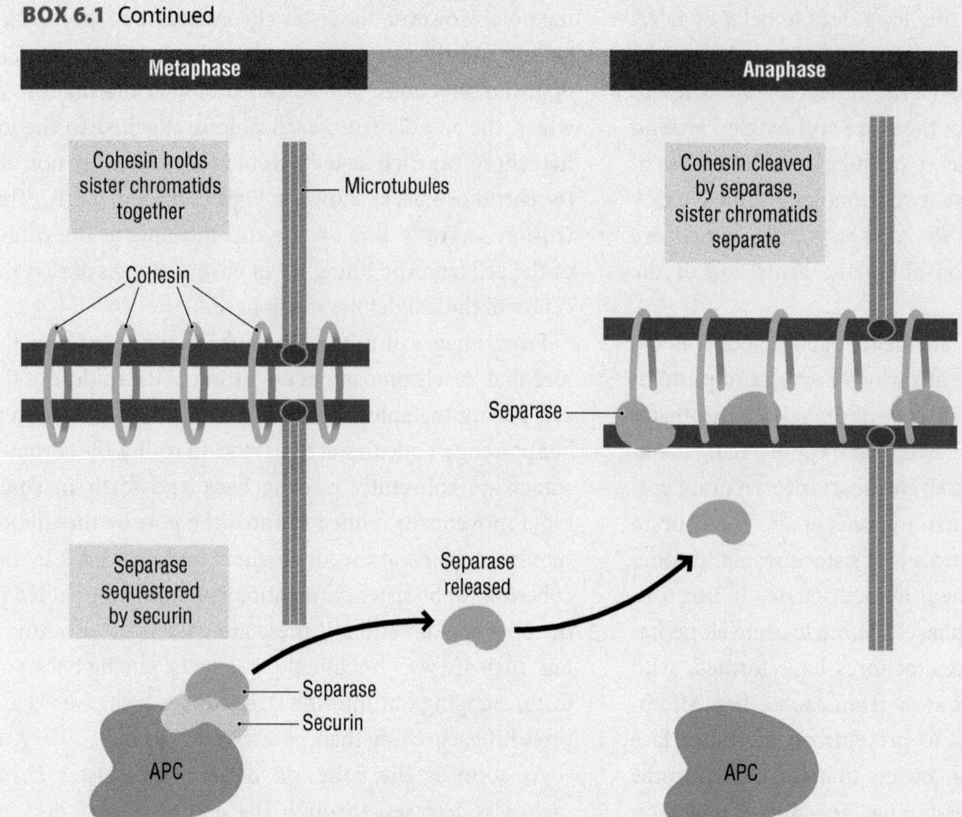

Figure A Sister chromatid cohesion and the metaphase to anaphase transition. At metaphase, cohesin rings (in light blue) hold the sister chromatids (in dark blue) together. Microtubules from the spindle attached at the kinetochore pull the sister chromatids to opposite poles, while cohesin holds them in place. A protein known as securin is part of the anaphase-promoting complex (APC). Securin is a protease inhibitor that blocks the protease separase (in green). When securin is degraded, separase is released from the inhibition, as shown by the arrow at the bottom. This triggers or promotes the transition from metaphase to anaphase. Separase cleaves one subunit of the cohesin ring, which releases the constraint on the sister chromatids and allows them to move to the poles.

to bind ATP, the ends of the two proteins must be in connection with each other, while the hydrolysis of ATP results in the separation of the two ends.

Smc1 and Smc3 form part of a ring structure, as shown in Figure B. The third protein in the ring during mitosis is Scc1, which plays the key role of holding the ring together; Scc3 is more weakly associated with the ring than Scc1, and, while it is necessary for this structure, its exact role is less well understood. Note the structure of the ring, which provides a striking example of how the biochemical structures of molecules determine their functions. The DNA molecules that make up the two sister chromatids pass through the middle of this ring. Picture the two sister chromatids as pencils, and cohesin as a rubber band used to hold them together. The diameter of the ring depends upon the interaction between the NBD and ATP; the structural integrity of the ring depends on Scc1. As the spindle is applying force to pull the kinetochores and the sister chromatids apart, the ring is holding them together.

When the S phase begins, Smc1 and Smc3, and possibly all of the cohesin subunits, are already associated with the replicating DNA molecules. Part of their role appears to be keeping the replicating strands from becoming wound around each other, what we called the "unwinding problem" in Section 4.3. The cohesin ring forms during prophase and stays in place throughout metaphase in mitosis. By holding the sister chromatids together, cohesin facilitates, and may help to ensure, the bipolar attachment to the spindle—that is, it may help to determine the orientation of the faces of the kinetochore.

BOX 6.1 Continued

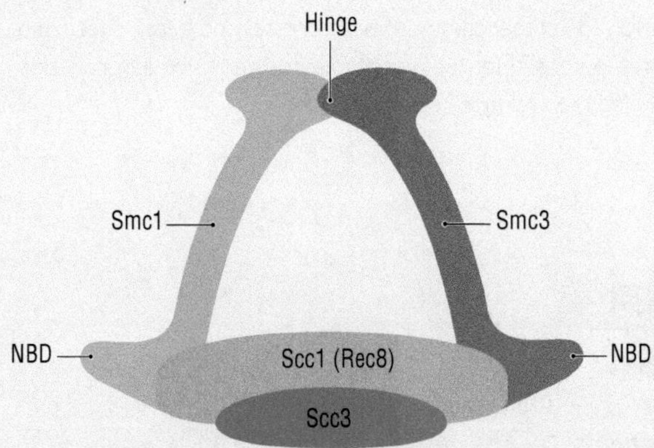

Figure B The cohesin ring. The cohesin complex consists of four conserved proteins that make a ring; this figure shows the names of the proteins in *Saccharomyces cerevisiae*, which are slightly different in other species. The sister chromatids pass through the middle of the ring. Smc1 and Smc3 are highly similar proteins, although with their domain structures oriented in opposite directions. At one end (the top in the figure) is a flexible hinge region. The other end includes half of a nucleotide binding domain (NBD). The binding of ATP brings the two halves together, with Scc1 apparently looping out and reducing the diameter of the ring. Thus, the shape of the ring depends on ATP hydrolysis. Scc1 is associated with all three of the other proteins. The Scc3 protein is necessary for the ring but is more loosely associated. Scc1 is found in mitosis; in meiosis, it is replaced by another member of the same gene family known as Rec8. Scc1 (or Rec8) is cleaved by separase at the metaphase to anaphase transition.

Cohesin does not form a continuous wrap around the sister chromatids but is instead found at discrete sites at the centromere and along the arms, as shown by the drawing in Figure A. It is not clear what determines the locations of the cohesin rings on the sister chromatids; their locations also appear to be dynamic and are possibly capable of sliding along the sister chromatids at different times. Cohesion between the chromatids is released in two phases, with loss of cohesion from the chromosome arms occurring before loss of cohesion at the centromere.

The loss of cohesin occurs when separase cleaves Scc1, as shown in Figure A. This opens the ring and releases the sister chromatids to segregate to the opposite poles. The function of separase is inhibited by a protein known as securin; securin binds to separase and sequesters it from cleaving Scc1. Securin is part of the APC, and its degradation triggers sister chromatin release. Securin is degraded by proteins in the APC, which releases the inhibition of separase. Separase cleaves Scc1, opening the cohesin ring and allowing the sister chromatids to be pulled to the poles by the spindle.

Cohesin in meiosis

Cohesin plays a similar, but possibly more complicated, role during meiosis, again helping to ensure that the kinetochores are oriented appropriately—the kinetochores on sister chromatids are oriented in the same direction during meiosis I (MI), but in opposite directions in meiosis II (MII). In this case, Scc1 is replaced by a meiosis-specific member of the same protein family known as Rec8; the difference between Scc1 and Rec8 lies in their time of expression, since they can functionally replace one another if expressed at the other time. Rec8 is also part of the lateral element of the synaptonemal complex that forms during prophase I; thus it lies both between sister chromatids and between homologs.

While the ring structure of cohesin during mitosis is widely accepted, the configuration of cohesin during meiosis is not quite so well established. One model is that it resembles a set of handcuffs, with one ring surrounding both sister chromatids of a homolog and the two rings held together in the middle. While the handcuffs seem a likely structure, a ring surrounding all four chromatids or a more complicated structure is also possible.

Sister chromatid cohesion is clearly necessary for the orderly segregation of homologs during MI and of sister chromatids during MII, but again the locations of cohesin must be dynamic. As noted in Section 6.3, homologs always cross over during prophase I. As illustrated in Figure C, this requires the loss of sister chromatid cohesion distal to the crossover, while retaining it at the centromere and nearby regions. It must then be lost at the centromere for MII to occur normally. Precisely how this happens is not known.

Other roles of cohesin

As a key structural component of eukaryotic chromosomes, cohesin has been implicated to have numerous other roles, in addition to sister chromatid cohesion. As we have just noted, it plays roles in preventing the sister chromatids from becoming tangled during replication and in ensuring that the kinetochore makes its appropriate attachments at different stages of division. It is involved in DNA repair during mitosis and meiosis and in the repair of double-stranded breaks that initiate recombination. Its location may be tied to the locations of double-stranded breaks, and thus the locations of crossovers.

Degradation of cohesin has been hypothesized to contribute to the maternal age effect in human trisomy. It has also been hypothesized to be involved in the processes of pairing and synapsis

BOX 6.1 Continued

of homologs and has been shown to play important roles in chromosome condensation and in transcriptional regulation. The orthologs of Smc1 and Smc3 found in bacteria clearly were involved in other chromosomal processes and became recruited or re-purposed for the crucial role in sister chromatin cohesion during the evolution of eukaryotes.

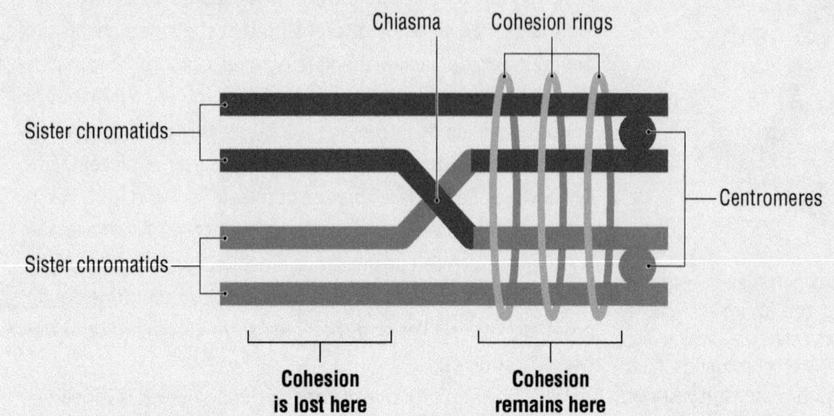

Figure C Sister chromatid cohesion and crossing over. The formation of a chiasma during meiosis I involves the loss of sister chromatid cohesion. As illustrated here, in order for the chiasma to resolve itself into four chromatids at meiosis, sister chromatid cohesion has to be lost at locations distal to the chiasma (that is, on the side away from the centromere), while it remains intact closer to the centromeres. The mechanism by which this occurs is not known.

immediately and quickly to opposite poles from the metaphase plate to the poles during anaphase. In most cells, anaphase occurs rapidly and involves two separate activities. First, the microtubules shorten and pull the chromatids along. Second, the centrosomes themselves move further apart. Microtubule shortening occurs through the loss of tubulin subunits from the ends, while the movement apart of the centrosomes at the poles requires particular motor proteins. Once the chromatids have separated, the nuclear membrane re-forms around them; the spindle disassembles into microtubule subunits, and the centrosome settles next to the nucleus. With the sister chromatids separated into two distinct nuclei, the cell completes division by pinching into two separate cells at telophase, a process called cytokinesis.

A variety of checkpoints and regulatory steps make sure that the process of mitosis occurs normally and without mis-segregation of the chromosomes, as summarized in Figure 6.9. **Checkpoints** are molecular events that monitor a biological process, such as DNA replication or microtubule attachment to the kinetochore, and keep the cell at a particular stage until the process that precedes it has been completed; these were introduced in Chapter 4.

The G_1 checkpoint, also called the restriction checkpoint or START, is the point at which the cell commits to entering the cell division cycle; cells that arrest at this point pass into the G_0 phase and exit the cell cycle, either temporarily or permanently. The G_2/M checkpoint ensures that the S phase is completed and all DNA damage is repaired before mitosis begins. A third checkpoint, the spindle attachment checkpoint, makes sure that all chromosomes are attached to microtubules at metaphase before allowing cohesion of the sister chromatids to be broken. Additional checkpoints occur during meiosis, as discussed in Section 6.5.

Cell division is a highly organized system, and the orderliness of the process comes from the biochemistry of the molecules involved. This description of mitotic cell division is a quick overview to set the stage for a discussion of the more elaborate type of cell division that occurs during meiosis. We will return to the mitotic cell cycle in Chapter 15, as an example of a complicated cellular process that has been well studied through genetic analysis.

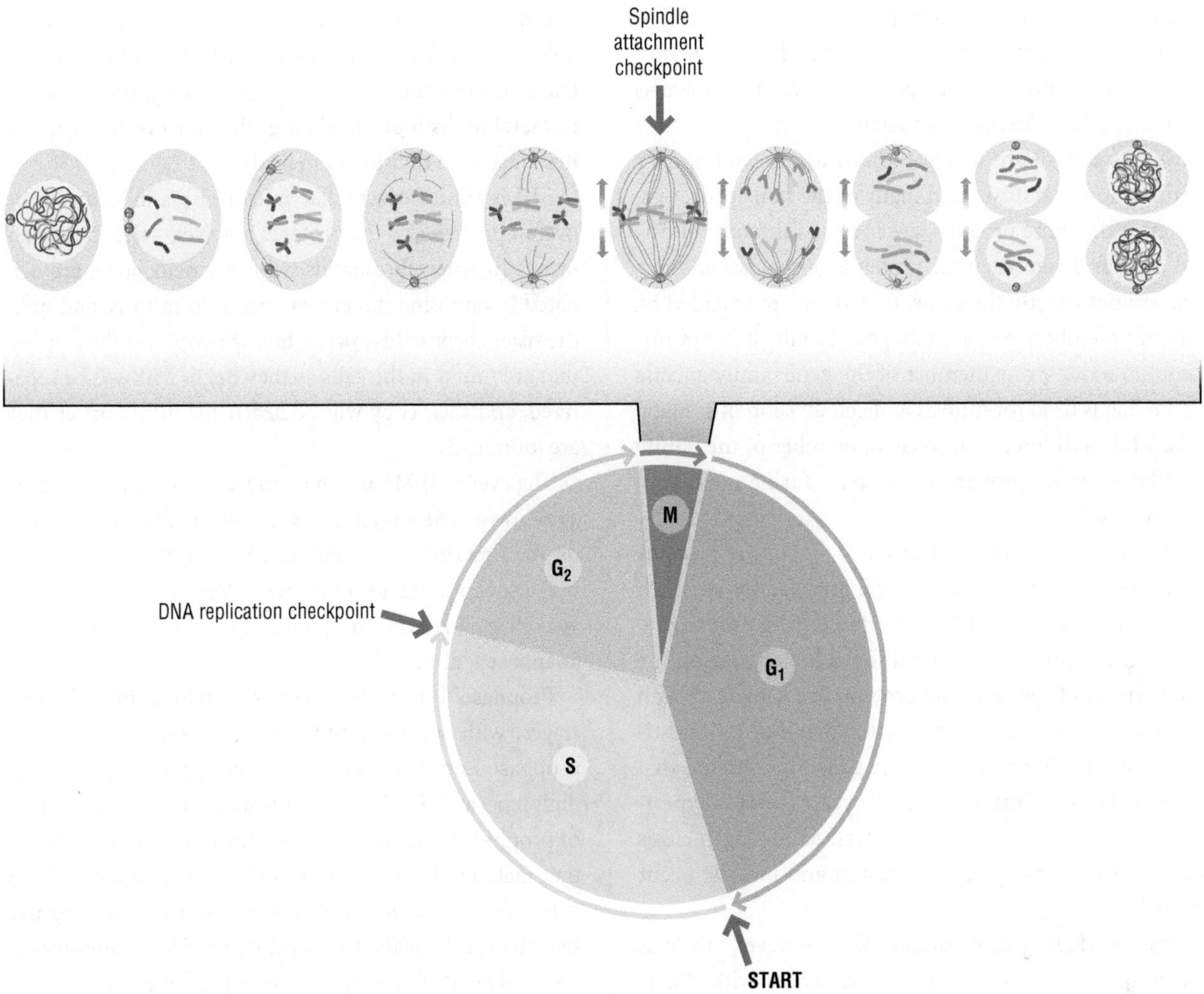

Figure 6.9 Checkpoints during mitosis. A mitotic cell must pass through at least three different checkpoints. The G_1 checkpoint, also referred to as START, allows the cell to progress in the cell division cycle; cells that do not pass this checkpoint return to G_0. After the S phase, a second checkpoint ensures that DNA replication and repair have been completed. Finally, the spindle attachment checkpoint at metaphase ensures that all kinetochores are attached to the spindle before anaphase begins.

6.3 Meiosis reduces chromosome number during germ cell division

Mitosis consists of one round of DNA replication during the S phase, followed by one round of chromosome segregation. Consequently, the size of the genome and the number of chromosomes stay the same. Mitosis occurs in somatic cells in animals, as well as during vegetative growth in fungi and plants. If cells resulting from mitosis were to fuse with one another following cytokinesis, or if the mitotic cells at telophase fail to separate from one another during cytokinesis, the resulting cell would have twice as many chromosomes and a genome twice as large as the parental cell; in fact, as discussed in Box 3.1, this is one mechanism by which polyploid tissues arise.

But mitosis does not generate the gametes or cells involved in reproduction, and the genome does not double in size each time a sperm and an egg fuse at fertilization. Instead, the gametes have only half of a complete genome. When gametes fuse, the genome of the resulting cell or zygote is the same size as the genome in all other cells. Thus, in organisms with a sexual cycle—which include nearly all eukaryotic organisms—a second type of cell division is needed to generate a gamete with only half of the complete genome. This type of cell division, known as **meiosis**, divides the genome in half.

Meiosis is thought to have arisen as an evolutionary modification of mitosis, and many of the same processes, and even the same proteins, are involved in both mitosis and meiosis. Even when the proteins for meiosis and mitosis are not exactly the same, they often are encoded by different members of the same gene family. It is not uncommon to have one member of the gene family encode a gene that is used for mitosis and cell division in somatic cells, while a different, but related, member of the family encodes a similar protein that is used during meiosis in the germ cells.

For example, as noted in Box 6.1, one of the proteins of the cohesin complex is part of a gene family in which one member is required for cohesion during mitosis and a different member is required for cohesion during the first division of meiosis. The proteins are similar enough in amino acid sequence and function that they can substitute for each other in yeast. If the gene for cohesion that is normally used during mitosis in yeast (*scc1*) is experimentally altered, so that it is expressed during meiosis instead of its paralog, *rec8*, cohesion and meiosis occur normally.

Despite their many similarities, however, meiosis and mitosis do have important differences with different consequences. An overview of meiosis is shown in Figure 6.10. Meiosis includes a single round of DNA replication in order to make a copy of the complete genome. However, the S phase is followed by two separate, but sequential, divisions, which reduce the size of the genome that goes into the gametes. The two divisions are called meiosis I (MI) and meiosis II (MII). The phases in MI and MII are named as in mitosis but are given a Roman numeral to distinguish whether they are occurring in the first or second division of meiosis.

MII is almost exactly like mitosis, both in terms of the overall process and the proteins involved—the spindle attaches to sister chromatids, whose kinetochores are oriented in opposing directions, just as in mitosis, and pulls the sister chromatids apart. Thus, the copies of the genome that are found in the cells as they begin MII will be separated, and each copy will be nearly identical, just as they are for mitosis.

The events of MI are much more significant for understanding genetics than the events of MII, so we will focus on the first division of meiosis. In fact, the most important processes for genetics occur during prophase I, so we will spend almost no time on events in metaphase I or anaphase I, or on MII.

Prophase I has also been subdivided into distinct stages, with specific events in each stage. The stages of prophase I are known as leptotene, zygotene, pachytene, diplotene, and diakinesis. While it may be helpful to think of prophase I (or any of the processes in cell division for that matter) as occurring in distinct stages inferred from cytological descriptions, it is also important to recognize that the cytological events and stages do not always correspond neatly to the molecular or genetic events.

Pairing and synapsis

One of the most important events during prophase I from a genetics perspective is chromosome pairing. Recall that a

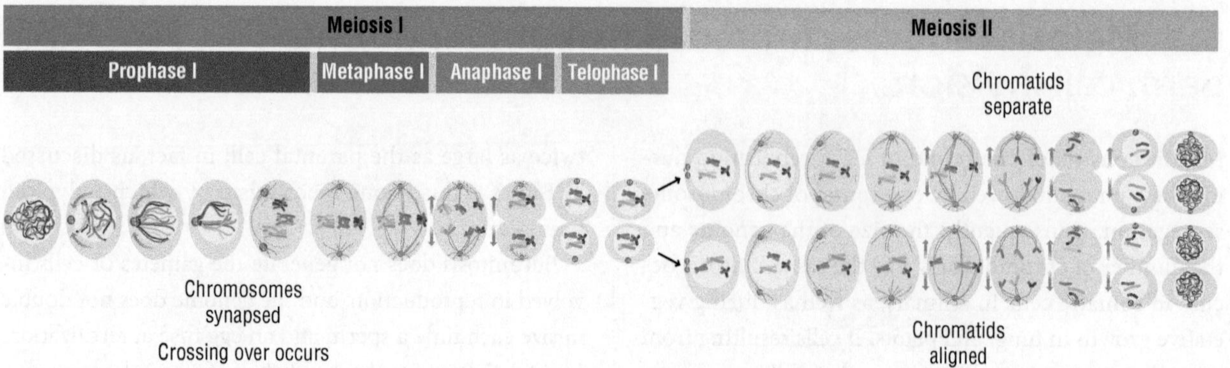

Figure 6.10 Overview of meiosis. Meiosis consists of a single round of DNA replication followed by two cell divisions, referred to as meiosis I and meiosis II. The names of the stages are the same for each meiotic division as for mitosis, but the processes that occur during prophase I and metaphase I are different from mitosis and will be elaborated in subsequent figures. Whereas mitosis generates two daughter cells with the same DNA content as the mother cell, meiosis generates four cells with half the DNA content of the mother cell.

diploid cell has two copies of every chromosome, referred to as homologs. After replication at the S phase, there are two copies of every chromosome, each with two sister chromatids. The homologous chromosomes have the same size and shape and the same genes, although not necessarily the same alleles. One of the homologs came from the mother originally, while the other homolog came from the father.

Homologous chromosomes have no special role in mitosis or in MII. Each pair of sister chromatids lines up on the spindle by itself, with no regard for what any other chromosome is doing. Homologs on the metaphase plate during mitosis or MII are no closer to one another than they are to a non-homologous chromosome. However, in MI, the homologs "find one another" and pair at the beginning of the process. Homolog recognition and pairing are mysterious processes. Somehow, each chromosome is able to identify sequences or structures on its homologous chromosome and pair with it. Not much is known about the cellular and molecular processes involved in such pairing, and it is not yet clear if it occurs in the same way in different organisms. However, it may involve some combination of DNA sequences, RNA molecules, and proteins. Pairing in *Caenorhabditis elegans* is explored in more detail in the online resources.

KEY POINT Chromosome pairing in MI is essential to the separation of homologous chromosomes, although we don't yet know how homologous chromosomes "recognize" each other.

Either before or during this homolog recognition process, one end of the chromosome is attached to the nuclear membrane, as shown in Figure 6.11. As a result, the chromosomes have one end anchored to the nuclear envelope and one end extended into the nucleoplasm. (The nuclear envelope does not break down until the end of prophase I and the beginning of metaphase I, so it is well after this very early step in the process.) This structure was first observed in plant chromosomes during the 1930s and is now thought to occur in all eukaryotes, although we cannot always see it clearly. The structure is known as bouquet formation, since the ends of chromosomes held together at the nuclear membrane look like a collection of stems.

It is not certain if the same end of a chromosome is always attached to the membrane, although most available evidence suggests that it is. Because the ends of the chromosomes are together, homolog recognition and pairing are effectively simplified. If we think of pairing and homolog recognition as being analogous to making a new friend in a social situation, bouquet formation serves like the opening reception that brings a group of people together in one place.

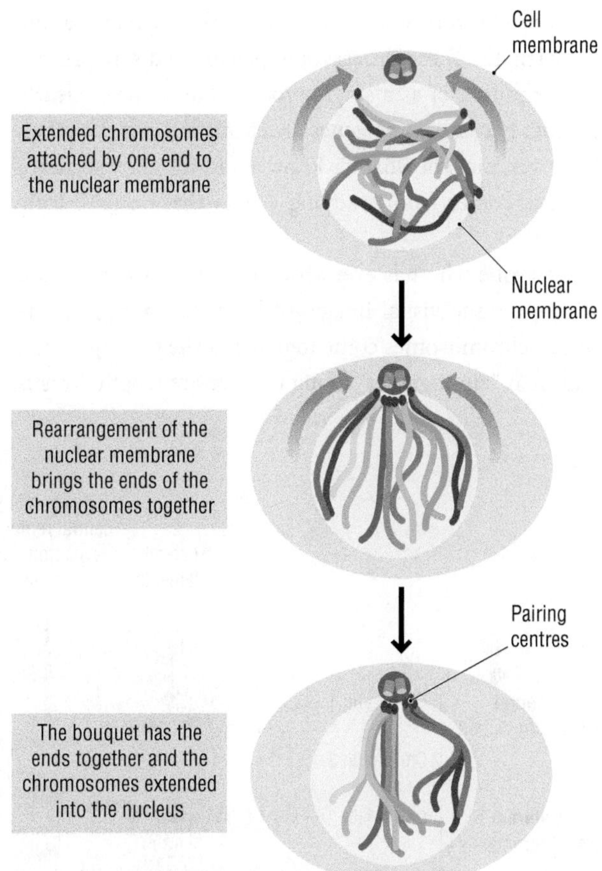

Figure 6.11 Bouquet formation. A crucial early step in pairing is bouquet formation. One end of each chromosome attaches to the nuclear membrane, which rearranges to bring the ends of the chromosomes together. The resulting structure, with the ends held together and the rest of the chromosome extended into the nucleus, was thought by cytologists to resemble a bouquet of flowers. The ends attached to the nuclear membrane are probably the pairing centers, as shown, although this has not been demonstrated for most organisms.

Pairing apparently occurs quite rapidly, and it is not easy to observe among the entangled and extended mass of chromosomes that are present during early meiosis; the nucleus at this stage is sometimes described as looking like a bowl of spaghetti. Identifying and following the specific interaction that is important for pairing is challenging when it is difficult to keep track of individual chromosomes. However, this elongated state is also thought to be important in pairing, since it extends the chromosomes and potentially increases the number of interactions they have with one another. If bouquet formation is analogous to going to an opening reception, the extended chromosomes are analogous to introducing yourself to many people there, whether the interaction proves to be long-lasting or short-lived.

Once chromosomes have paired, they synapse along their length. These processes of pairing and synapsis are often thought of as a single process, but they are clearly distinct because some genes affect one but not the other. However, most of what we know about the early stages of meiosis comes from studying events that occur during synapsis.

Everyone who has ever observed synapsis describes it with the same visual image—it looks like a zipper. The paired chromosomes come together at the pairing region and then "zip together" along their entire length. This is summarized in Figure 6.12(a). Once synapsis begins to occur, as the chromosomes "zip up," the chromosomes become thicker and shorter, and the number of interactions between chromosomes appears to decrease. The distinctive appearance of the fully synapsed homologs identifies this as pachytene.

As the homologs synapse, a structure, known as the synaptonemal complex (SC), forms between them. The precise role of the SC is not clear, but it forms in nearly all eukaryotes (except for a few species of fungi) and probably helps to regulate the positions and

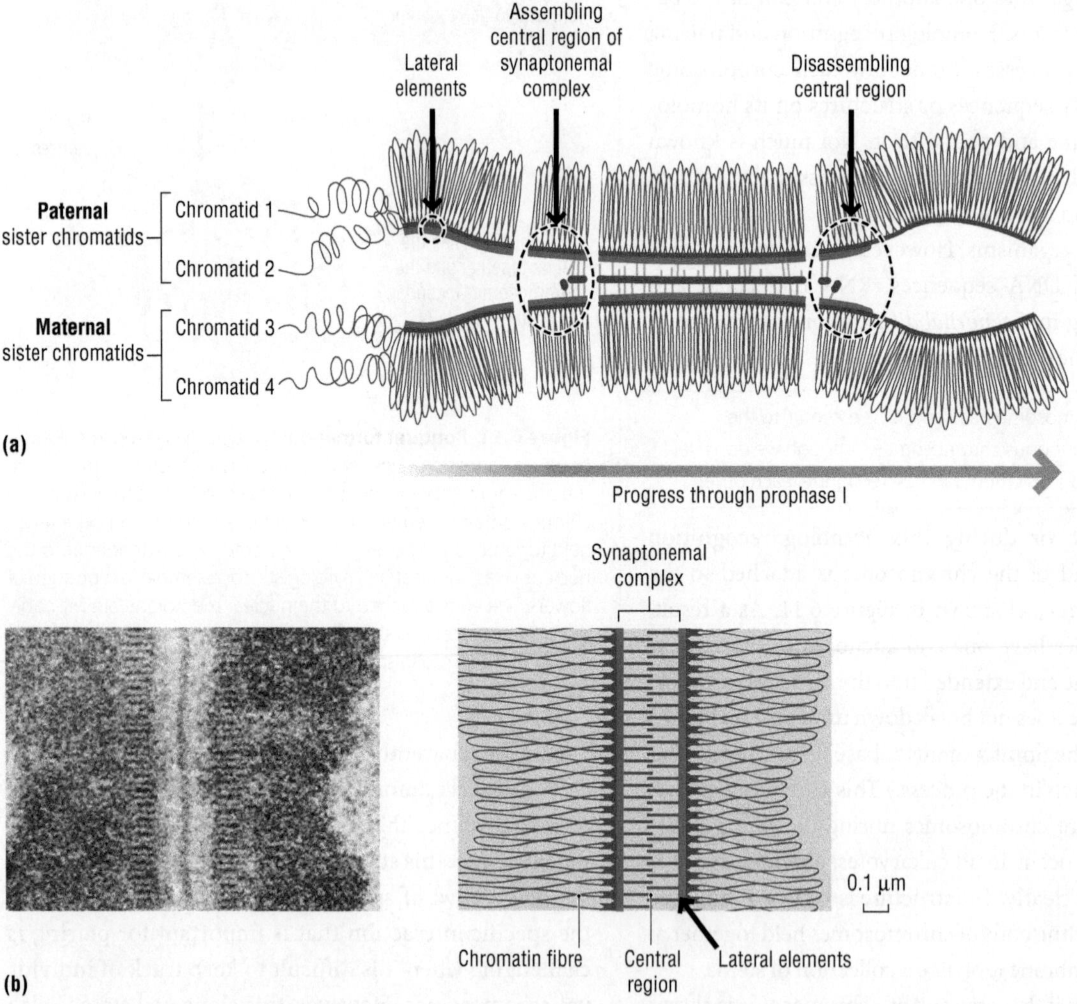

Figure 6.12 Synapsis and the formation of the synaptonemal complex. Once homologs have paired, they synapse along their length during prophase I in a process usually described as resembling the action of a zipper. The synaptonemal complex looks like a ladder, with lateral elements along the chromatin and a central region. (a) A pair of homologs is shown, each with two sister chromatids. The progression of synapsis is shown in the drawing. Some of the proteins that comprise the lateral elements are associated with chromatin before synapsis begins. The central region is assembled, and full synapsis occurs along the length. The central region later disassembles as the homologs begin to separate. (b) An electron micrograph of the synaptonemal complex, with a diagram labeling the various elements. The synaptonemal complex is nearly identical in all eukaryotes.

numbers of crossovers occurring between the paired chromosomes, as discussed in further text. The SC has the structure of a ladder, as shown in Figure 6.12(b). The lateral elements are made up of chromosomal proteins associated with chromatin and DNA, including at least one member of the cohesin complex; the proteins known to lie in the lateral elements have orthologs in many eukaryotes. By contrast, the protein components of the central element are not evolutionarily conserved and are, in fact, different in each group of organisms that have been studied. However, the width of the SC is highly conserved; despite varying protein compositions, the lateral elements are about 200 μm apart, so that the homologs lie the same distance apart in all eukaryotes. In other words, the width of the ladder seems to be functionally important, but the composition of the rungs that determine that width is less so.

Visualizing pairing and synapsis in *C. elegans*

To be clearly understood, the events of meiosis need to be seen, as well as described. Figure 6.13 shows the progression of pairing and synapsis during prophase I for chromosome II in *C. elegans*; we will describe this figure in detail. It seems likely the pairing and synapsis of this chromosome in this organism are similar to those for other chromosomes in other eukaryotes.

The experiment depicted in Figure 6.13 began by using the genome sequence of the worm to identify DNA sequences from different regions of chromosome II, and labeling these DNA sequences *in vitro* with different fluorescent tags. The fluorescent labeling of DNA probes is described in Tool Box 2.2. The labeled DNA probes were then hybridized to the chromosomes *in vivo*, so each band of color corresponds to a region of the chromosome within a nucleus undergoing meiosis.

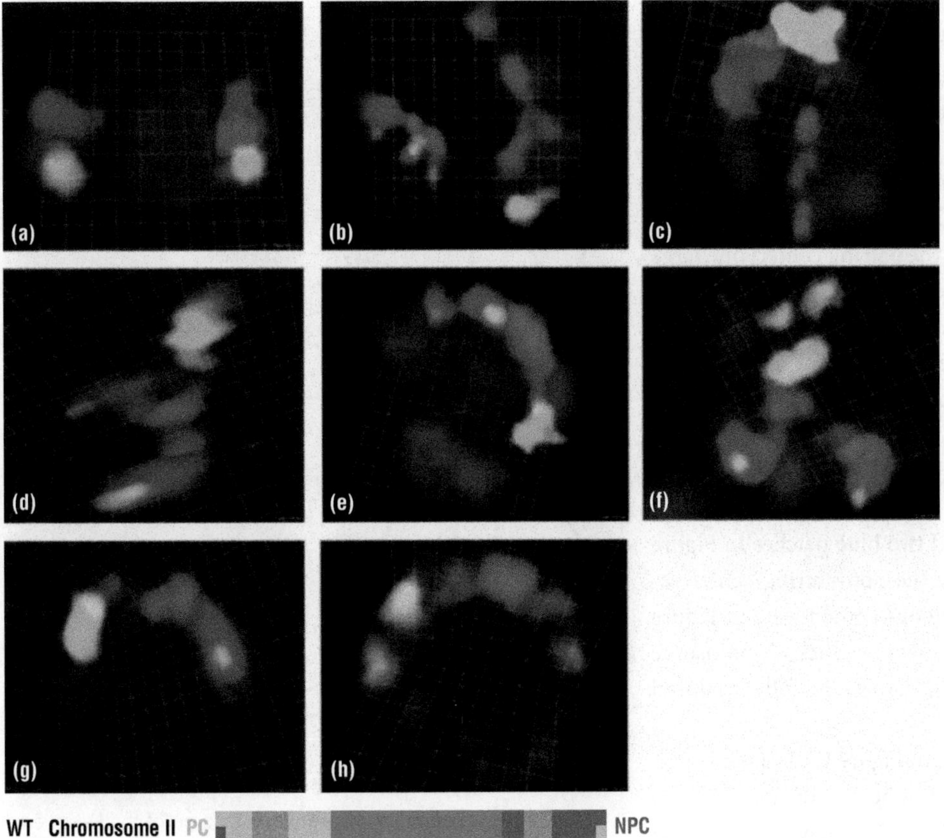

Figure 6.13 Pairing and synapsis during prophase I visualized. The processes of pairing and synapsis of chromosome II in *Caenorhabditis elegans* are shown, using labeled probes to highlight the behavior of different regions of the chromosome. The figure is described in detail in the text. The green, red, and blue colors are false color images from different fluorescent dyes, with the wavelengths analyzed computationally and projected into different colors. The patches in yellow, magenta, and light blue arise from the overlap of different dyes.

Source: © 2011 Nabeshima et al. Chromosome Painting Reveals Asynaptic Full Alignment of Homologs and HIM-8–Dependent Remodeling of X Chromosome Territories during *Caenorhabditis elegans* Meiosis. *PLoS Genet* 7(8): e1002231.

As shown in the diagram of the chromosome below the images in Figure 6.13, sequences from the pairing center (PC) region at one end were labeled with a fluorescent dye that appears green, those from the middle of the chromosome with a dye that appears red, and those from the end opposite the pairing center (non-pairing center or NPC) with a dye that appears blue. Each of the panels in the figure shows the same pair of chromosome as pairing and synapsis progress. The entire process can be followed, since the ovary in *C. elegans* has the stages of prophase I laid out in sequence, making it possible to observe the different stages at the same time by looking at nuclei at different locations in the same ovary.

At the beginning of prophase I, in Figure 6.13(a), two distinct homologous chromosomes are seen, each compacted but not paired; all three fluorescent labels on each homolog are readily seen. Although not shown in this figure, the green end (the pairing center) is likely attached to the nuclear envelope. Note that the chromosomes in Figure 6.13(b) and (c) are elongated, compared to those in Figure 6.13(a); the compacted blue regions (as well as the red and green regions) in (a) have become much longer and thinner in (b) and (c), that is, the colored regions have changed dimensions because the chromosome has changed in shape. In (c), pairing has clearly begun at the pairing center; one large green patch is seen because the pairing regions of the homologs are so closely associated that they cannot be resolved from each other; the red and blue regions are still distinct for each homolog, so these regions are not yet paired or synapsed.

Synapsis progresses in Figure 6.13(d) through (h); the middle region synapses before the other end (as indicated by one red patch and two blue patches in Figure 6.13(d) and (f)), and finally the non-pairing center region synapses as the blue regions come together. Figure 6.13(g) and (h) appear to show one thick chromosome with three bands of color; this shows the fully synapsed homologous pair.

The progression of images in Figure 6.13 suggests that the process of synapsis may not be exactly analogous to the closing of a zipper—the homologs may come apart in some regions before continuing synapsis, or synapsis may not be strictly linear from end to end.

WEBLINK: You can watch a more in-depth analysis of Figure 6.13 in Video 6.1. Find it at **www.oup.com/uk/meneely**

Crossing over as part of recombination

Once the homologous chromosomes have paired and synapsed, they appear as thick threads across the nucleus when viewed under the microscope; we see this appearance in Figure 6.13(g) and at the pachytene stage of prophase I, summarized in Figure 6.14. At this point, a second necessary process in meiosis known as crossing over occurs.

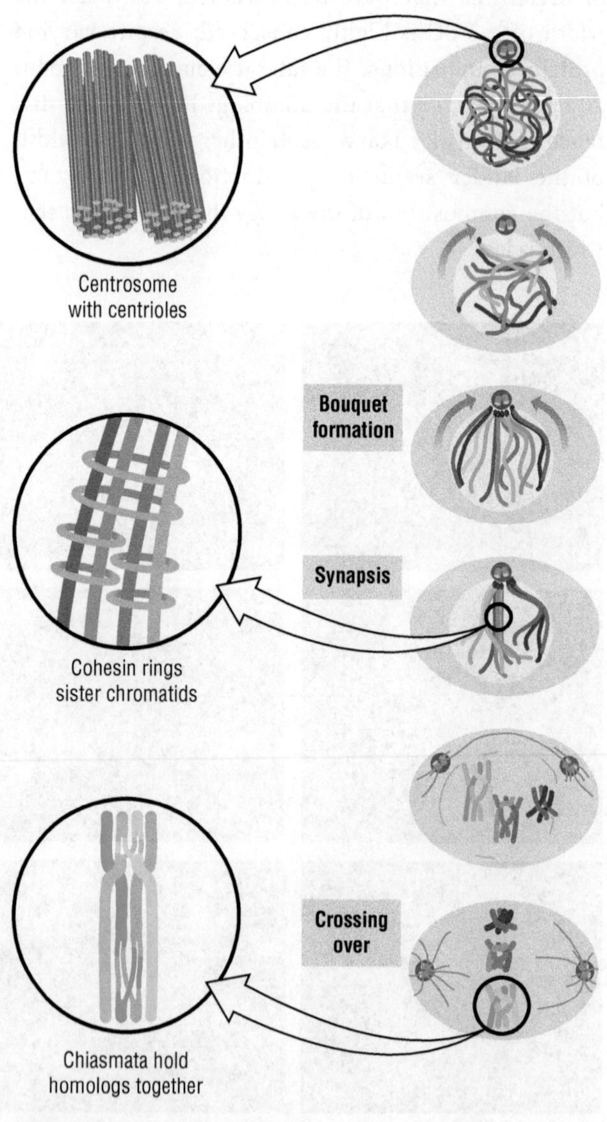

Figure 6.14 Summary of prophase I. Some of the key events during prophase I are shown in summary, with the inset figures identifying particularly significant structures. After bouquet formation and synapsis, the sister chromatids are held together by the cohesin complex. Cohesin forms a ring around the two sister chromatids but the precise structure is not known. Crossing over occurs during synapsis, as evidenced by the formation of a chiasma, which holds the homologs together.

During pachytene, one homolog acquires double-stranded breaks (DSBs) in the DNA sequence of one sister chromatid. (Remember that each homolog comprises two sister chromatids.) These breaks and the ensuing process are illustrated in Figure 6.15. Dozens of DSBs form in the nucleus in mammals, and many more than that are made in flowering plants. The enzyme SPO11, which is evolutionarily conserved among eukaryotes, creates the DSBs at sites throughout the chromosome, more or less at random. (We will see later in this chapter and in Chapter 9 that the locations of breaks and crossovers are not completely random, but this makes a useful approximation for now.)

DSBs in the DNA are generally deleterious and are recognized as DNA damage, as described previously in Section 4.4 and Table 4.3. Thus, most of these breaks are immediately and quickly repaired, so that no damage to the chromosomes occurs. However, a few of these breaks result in one DNA strand of one sister chromatid on one homolog being exchanged with one DNA strand of one sister chromatid of the other homolog, that is, the sister chromatid from one homolog crosses

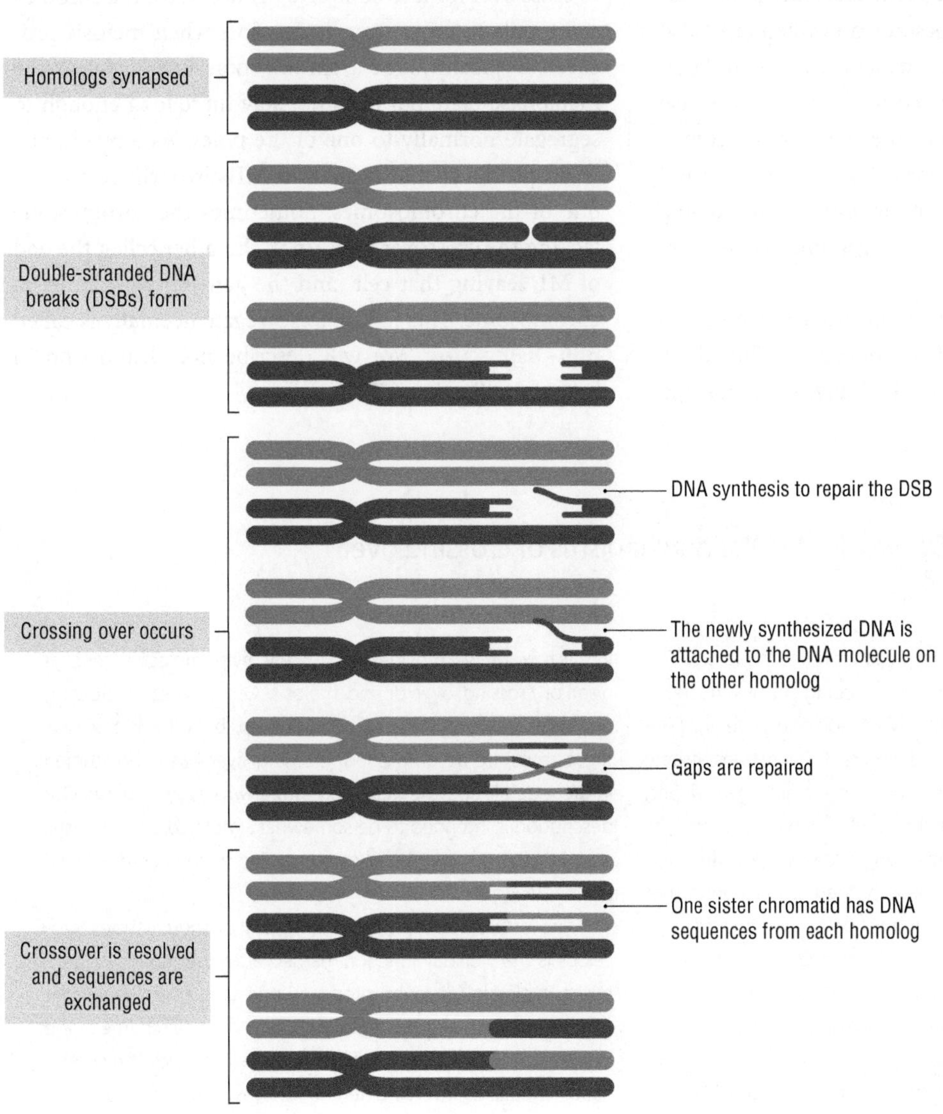
Homologs synapsed
Double-stranded DNA breaks (DSBs) form
Crossing over occurs
DNA synthesis to repair the DSB
The newly synthesized DNA is attached to the DNA molecule on the other homolog
Gaps are repaired
Crossover is resolved and sequences are exchanged
One sister chromatid has DNA sequences from each homolog

Figure 6.15 Crossing over. Once the homologs, one shown in orange and one in blue, are synapsed, double-stranded breaks (DSBs) are made in the DNA sequences. The break is repaired, and, during the repair process, an exchange of DNA strands between the two homologs occurs. As a result, some length of sequence from each homolog is crossed over or recombined to the other homolog; note that two of the chromatids have both blue and orange sequences. The rest of the DSBs are repaired without a crossover, and the two outside chromatids are not directly involved.

over with a sister chromatid from the other homolog. As a result of this crossover, one sister chromatid in each homolog becomes a composite of maternal and paternal sequences, as shown in Figure 6.15. This exchange of DNA sequences is known as **crossing over**. We describe the molecular process of crossing over in Box 6.2.

Crossing over is only one of several processes that produce a new arrangement of the sequence of genome in the next generation. All of these processes together can be thought of as **recombination**. In addition to crossing over, fertilization and independent assortment of non-homologous chromosomes (described in Chapter 8) also recombine the genome; horizontal gene transfer of DNA from other organisms (described in Chapter 11) is also a type of recombination. It is common and often convenient to use "crossing over" to mean "recombination," and vice versa, but it is important to remember that crossing over is only one of the ways that the genome becomes recombined during inheritance.

There are two very important and under-emphasized facts about crossing over during prophase I. While these were known to geneticists and cell biologists as long ago as the 1930s, they have not always made their way into biology textbooks. Now that we have the ability to study many regions of the genome at the same time, we have a much more complete picture of the genome-wide events that occur during the early stages of meiosis, so these should be emphasized.

First, with some unusual and notable exceptions, crossing over **must** occur for the chromosomes to segregate normally during MI. (One exception is male fruit flies, as discussed in Chapter 9.) Crossing over is not an occasional or optional process—every chromosome has to cross over for it to complete MI normally. If a chromosome fails to cross over, it gets lost when meiosis proceeds. Typically, such a chromosome lines up near the metaphase plate but does not stay there long enough to segregate normally to one of the poles. As a result, one of the products of MI (and thus MII also) will be missing one of the chromosomes. Sometimes the chromosome lacking the crossover ends up in the other cell at the end of MI, leaving that cell (and the gamete) with an extra chromosome. This failure to segregate normally is called **non-disjunction**. We will describe non-disjunction in humans in Section 6.5.

BOX 6.2 *Going Deeper* Molecular mechanisms of crossing over

Recombination between two different DNA molecules is one of the most fundamental and widespread activities in biology. The process involves one or both molecules being cut by a nuclease and rejoined with itself or another DNA molecule. Different DNA molecules are joined or rejoined during replication, transposable element movement, viral integration, and crossing over during gene transmission in both bacteria and eukaryotes, as well as during other processes. Recombination is closely associated with the processes of DNA repair—many of the enzymes and other proteins that are involved in recombination are also important in DNA repair. DNA repair is essential; recombination, which creates genetic variability, may have been a very fortuitous by-product of repair that has been widely advantageous during evolution.

Crossing over results in a change in the DNA sequence of the DNA molecules involved; more specifically, it joins DNA sequences that were previously separate. Thus, recombination creates genetic diversity, which is necessary for evolutionary change. The processes of repair and recombination are evolutionarily conserved, and very similar or closely related enzymes and proteins involved in these processes are found in bacteria, archaea, and eukaryotes.

The Holliday junction

In this box, we will provide an overview of the molecular mechanism of crossing over during meiosis. Other types of crossing over usually involve many of the same steps but differ in the ways these steps are accomplished. Crossing over has been studied in great detail, and the details are not always easy to grasp. Our description is intended to be somewhat simplified, although still accurate; many more details can be added by the interested and diligent reader.

A key structure that arises during any of the crossover processes is the Holliday junction, named after the molecular geneticist Robin Holliday who proposed one of the early molecular models by which recombination occurs. A general scheme of the Holliday junction is shown in Figure A; different types of recombination processes involve different versions of the Holliday junction, and meiosis involves two Holliday junctions. We will describe this simplified version as background for meiotic recombination.

The process of recombination begins with a nick in the single strand of one or both DNA molecules, or with a double-stranded break that cuts both strands of one DNA molecule; Figure A(i) shows nicks occurring in one strand of both molecules, which

BOX 6.2 Continued

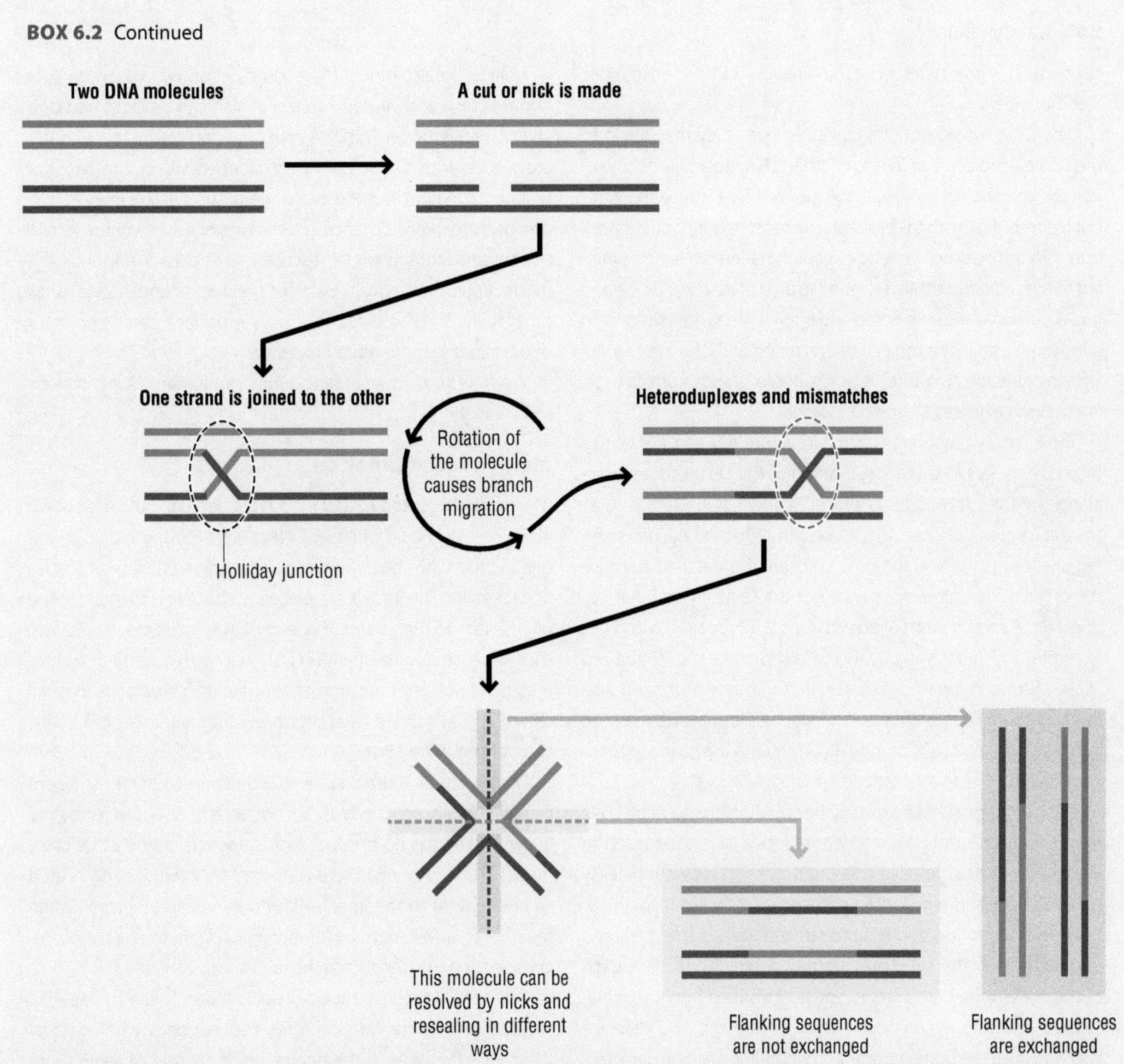

Figure A The Holliday junction. In nearly every form of recombination between two different DNA molecules, a cross-shaped structure known as a Holliday junction forms. The Holliday junction is the region circled with the dotted line. Two double-stranded DNA molecules are shown, one in orange and the other in blue. The orientation of the strands is not shown in this overview; it can be different for different types of recombination. One DNA strand of each molecule is nicked; in order to repair the nick, the strands are rejoined. A Holliday junction forms when the nicked strands are rejoined to the other DNA molecule. When the DNA molecules rotate, the junction migrates to an adjacent region of DNA. In the region "behind" the migration of the Holliday junction, the two DNA strands have mismatches and heteroduplexes (with G:T base pairs, for example), shown with the dotted red circle where blue and orange sequences would be paired with each other. Note that recombination has occurred for sequences here, since blue sequences are connected to orange ones. The Holliday junction can be resolved in different ways, depending on where the nicks and unwinding occur. Resolution in one form results in the reciprocal crossing over of the flanking sequences, while resolution in the other form does not exchange the flanking sequences. See the text in Box 6.2 for a more detailed description.

BOX 6.2 Continued

happens in some types of recombination, but not in meiotic recombination.

When the nicked or cut strands reanneal, they may rejoin to the original molecule or join to another DNA molecule. If they rejoin to their original molecule, the nick or break is repaired without recombination; this is probably the most frequent outcome. If they join to the other molecule, recombination occurs; this joining to the other molecule is referred to as **strand invasion**. Strand invasion usually requires some additional DNA synthesis to fill in gaps that arise as the molecules are joined and gives rise to the Holliday junction structure, in which single strands from two different double helices are joined together.

The Holliday junction is circled in Figure A(i) and (ii). As this molecule rotates in space, the location of the junction can move along the DNA molecules, a process known as **branch migration**, illustrated in Figure A(ii). As branch migration occurs, some regions form, in which the base pairs are mismatched because the original DNA sequences were not the same. In such heteroduplexes, A ends up mispaired with C or G is mispaired with T (for example). In Figure A(ii), this region is shown by the dotted red circle. Mismatch repair is discussed in Chapter 4 and often also involves some DNA synthesis. Potential heteroduplexes are seen in Figure A as regions in orange on one strand are paired with regions in blue on the other strand.

The next step in the process is the most difficult to depict. We have shown the Holliday junction molecule as cross-shaped in Figure A(iii) to help visualize this, but the results may not be easy to see. (In fact, at professional meetings during which recombination is discussed, it is common to see attendees get out colored pencils to draw the structures.) It may be helpful to trace each of the lines in the branch migration figure to see how the cross-shaped structure is composed, to recognize the two different ways in which the same molecular configuration can be depicted.

This junction has to be resolved for the DNA molecules to separate from one another, which involves nicks and unwinding by topoisomerases, as was discussed for replication in Chapter 4. The nicks, unwinding, and resealing can occur in different orientations. The sequences in the middle are exchanged, no matter how the structure is resolved. But let's now focus on the sequences on the outside. If the nicks and unwinding occur along the vertical gray dotted line, the sequences on the outside are exchanged. The two DNA molecules on the left have blue sequences combined with orange sequences in the flanking regions. By contrast, if the nicks and unwinding occur along the horizontal (blue) dotted line, the flanking sequences are not exchanged, and orange still appears with orange and blue with blue. A region is exchanged and recombination occurs, but the outcome does not result in reciprocal products with respect to the outside sequences.

Holliday's basic model for recombination was based on data from bacteria and viruses, some of which can be explained by a version of this model. The DNA structures depicted in these figures are probably less than 1 kb in length (depending on the extent of branch migration), but the exchange of the outside markers can be observed over a distance of kilobases, or even over the length of a chromosome, as will be discussed in Chapter 9. Although this model could not explain data for meiotic recombination or all other types of recombination, it is an important starting point for understanding these other processes.

We now turn to what is thought to happen during meiotic recombination.

Meiotic recombination

A generally accepted model for meiotic recombination is shown in Figure B. We begin with two DNA molecules, one in orange and one in blue, with the different strands shown in different shades of each color. The arrows are placed at the 3′ end to indicate the orientation of the strands on each DNA molecule. In meiosis, each of these double-stranded DNA helices would be one chromatid, but not sister chromatids; they are the chromatids that will recombine, which we have drawn as chromatids 2 and 3 in this chapter and in Tool Box 6.1.

On one chromatid, a double-stranded break is made by the enzyme SPO11. In other words, both strands of DNA are cut by the enzyme, whereas no break or nick is needed on the other chromatid. SPO11 is found in all eukaryotes; it is evolutionarily related to one type of topoisomerase found in archaea, indicating that meiotic recombination is evolutionarily related to replication, recombination, and repair in archaea.

Once the double-stranded break is made, it can be repaired without recombination occurring; this happens with most of the breaks that form at the beginning of meiosis. At some break sites, one of the strands crosses over (or "invades") the helix on the other chromatid; this is the light orange strand in the figure. In doing so, it displaces the corresponding strand of the DNA molecule on the other chromatid (which has the same orientation). This interaction of an invading strand and a displaced strand forms a structure that has many of the features of a Holliday junction.

The process continues in Figure B(ii). The formation of this structure shown Figure B(i) involves new DNA synthesis and probably some branch migration, which is shown in Figure B(ii) as the dashed lines. The colors of the newly synthesized strands indicate which strands are being used as the templates and have the same sequence as one of the original strands. In connecting these newly synthesized strands, a structure known as a double Holliday junction forms. The locations at which an orange line is

BOX 6.2 Continued

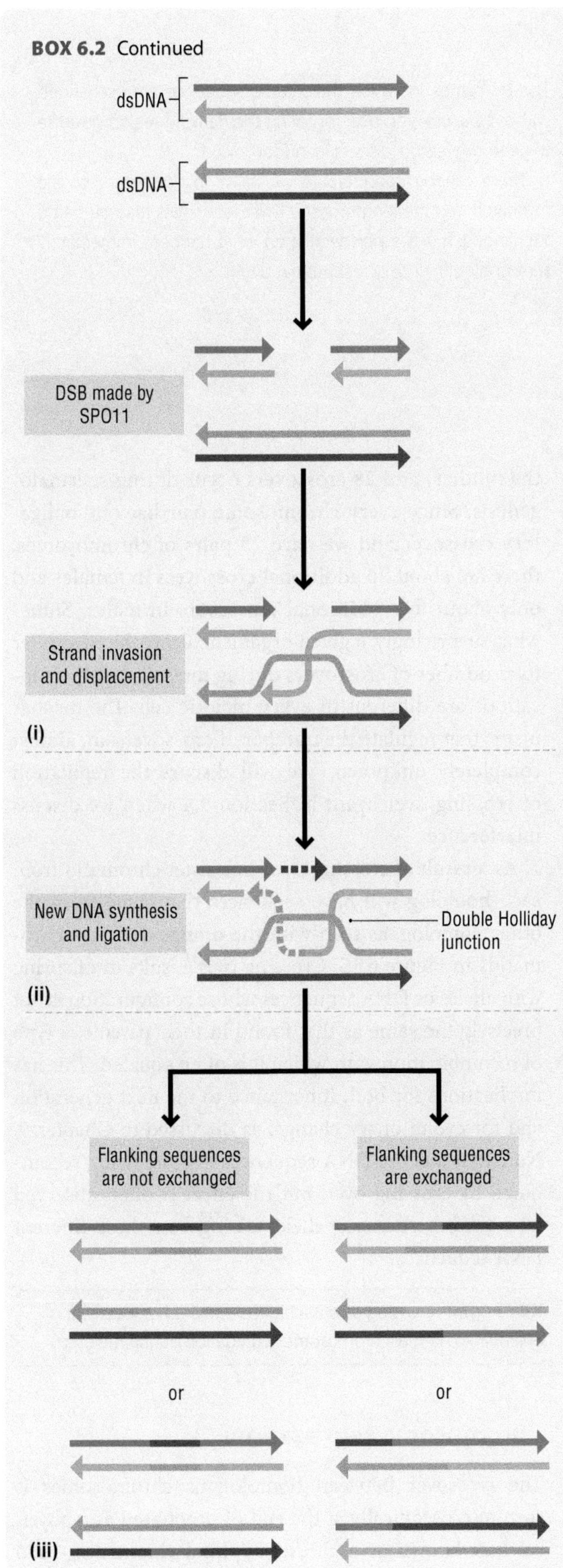

Figure B A model for meiotic recombination. (i) Two double-stranded (ds) DNA molecules are shown, one in orange and one in blue, each with two strands in different shades. These are the two chromatids that will exchange and thus are on homologous chromosomes. Meiotic recombination is initiated by a double-stranded break (DSB) in one of the molecules, catalyzed by the evolutionarily conserved enzyme SPO11. One strand that is cut can invade or base pair with the other DNA molecule, displacing its other strand; this makes a structure similar to a Holliday junction. (ii) Continued DNA synthesis, using one of the strands as a template, followed by joining of the newly synthesized DNA, results in a double Holliday junction structure with sequences recombined in the middle. (iii) This structure can be resolved in four different ways, depending on the locations of the nicks and how it is unwound. Two of the ways it can be resolved result in the reciprocal exchange of the outside flanking sequences, while the other two ways do not result in the exchange of flanking sequences. Recombination without the exchange of flanking sequences accounts for gene conversion, as discussed in Tool Box 6.1.

BOX 6.2 Continued

paired with a blue line indicate possible sites of heteroduplexes or mismatched bases.

This complicated double Holliday junction structure is also resolved by nicks, unwinding, ligation, and resealing, as shown in Figure B(iii). Again, the orientation of the nicks determines what happens with the flanking sequences—but with some sequences in the middle being exchanged in all cases.

The structure can be resolved in four ways—two of these involve exchange of the flanking sequences; two of them do not. The outcomes in which the middle sequence are exchanged without the crossing over involving the flanking region give rise to gene conversion, discussed in Tool Box 6.1.

There can be considerable variability in the extent of the mismatch and the length of the DNA sequences that are being exchanged; this is a perspective on what happens immediately surrounding the site of recombination.

Second, both the locations at which crossovers occur and the number of crossovers in the genome are very highly regulated. We do not know how this crossover control occurs; indeed, the mechanisms may not be the same in all organisms. It is likely to involve some modifications in the structures of chromosomes, and a few specific modifications in the structure of chromatin have been associated with the sites at which crossovers occur. The sites of DSB formation in all eukaryotes for which this has been examined have a very high level of the specific histone modification H3K4me3, that is, the lysine found at position 4 (K4) in histone H3 in nucleosomes around the sites of DSBs has three methyl groups attached to it. (Histone modifications were introduced in Chapter 3 and will be described in Chapter 12 in connection with their role in gene expression.) Other specific proteins also seem to be found at the sites of DSBs in several different eukaryotes, possibly involved in recruiting SPO11 to that location.

The processes that specify the locations of crossovers are an area of active investigation, greatly aided by the availability of genome sequences for all model organisms. For now, it is still easiest to think of crossovers occurring at random sites, although we must recognize that this randomness is partly a reflection of our lack of more detailed knowledge.

While we have some molecular information about how the locations of crossovers are controlled, we have much less knowledge about how the number of crossovers is regulated. For most animals, each chromosome has only two or three crossovers (or fewer). Plants have more than that, and budding yeast has more than 20 crossovers per chromosome. Genome-wide analysis of recombination in humans indicates that an average of 49 total crossovers occur during oogenesis (that is, in the mother) and 28 crossovers occur during spermatogenesis. Since every chromosome pair has one obligatory crossover and we have 23 pairs of chromosomes, there are about 26 additional crossovers in females and only about five additional crossovers in males. Somewhat surprisingly, a given organism always has a similar total number of crossovers during meiosis, but their locations are different in every meiotic cell. The mechanisms that regulate the number of crossovers are almost completely unknown. We will discuss the regulation of crossing over again in Section 9.4 when we discuss interference.

As a result of crossing over, one sister chromatid from each homolog will have sequences that come from the other homolog, as seen with the orange and blue chromatids in Figure 6.15. Crossing over results in offspring with alleles or DNA sequences whose configuration is not precisely the same as that found in their parents, a type of recombination with which it is often equated. This has implications for both inheritance to the next generation and for evolutionary change, as discussed in Chapter 9. Note that it is the DNA sequences that are being recombined by crossing over, but the effect is often observed from the inheritance of alleles arising from these different DNA sequences.

KEY POINT Crossing over is a highly regulated process that is essential to normal chromosome segregation during meiosis I.

Chromosome segregation

The crossover between homologous chromosomes is seen microscopically at the end of prophase I as a physical link known as a **chiasma** (plural chiasmata). Held together by the chiasmata, the chromosomes line up on the metaphase plate, as shown in Figure 6.16. It may help to think of this stage in mechanical terms. The kinetochores from the two sister chromatids are facing the same direction in MI, whereas the ones from the different homologs are facing in opposite directions, as shown in Figure 6.17. The spindle microtubules attach to the outer face of the kinetochore, with tension from the opposing spindle poles; the physical connection between the chromosomes, which is provided by the chiasmata, is holding the chromosomes in an equilibrium position. As soon as the crossover is completed and the connection between the homologs is cleaved, the homologs move rapidly to opposite poles.

When cytologists were looking at chromosomes during meiosis (especially in plants), they named the four stages—prophase, metaphase, anaphase, and telophase. Even though we think of each division of meiosis (and of mitosis) in these four distinct stages, the stages are far from being of the same length. Prophase I is much longer than any other stage, and anaphase I is extremely short. During oogenesis in women, for example, prophase I can last for decades, as discussed in Section 6.4, whereas anaphase I in many organisms often lasts a few minutes.

Following anaphase I, two separate cells are formed at telophase I. Each of these cells has one copy of every chromosome—that is, they are haploid—but has two sister chromatids for each homolog. There is no additional DNA synthesis, and MII proceeds exactly like mitosis. The microtubules of the spindle attach to the kinetochore of each sister chromatid and pull them to opposite poles of the cell in MII, resulting in four haploid cells, each with one of the four sister chromatids. These four products are called a **tetrad**, as described in further text.

We have described the separation of the two homologs in terms of the chromosomes, but remember that every sister chromatid consists of DNA and that the DNA can encode different sequence or different alleles for the gene. This separation of homologs and their corresponding alleles is often referred to as Mendel's First **Law of Segregation**. The two alleles for each gene are on different members of the same homologous pair of chromosomes. These separate or segregate from each other during MI, so that each gamete that results has only one allele for each gene; because there are two sister chromatids, there are two copies of the allele, however. Mendel's First Law of Segregation is really a way of stating what happens during

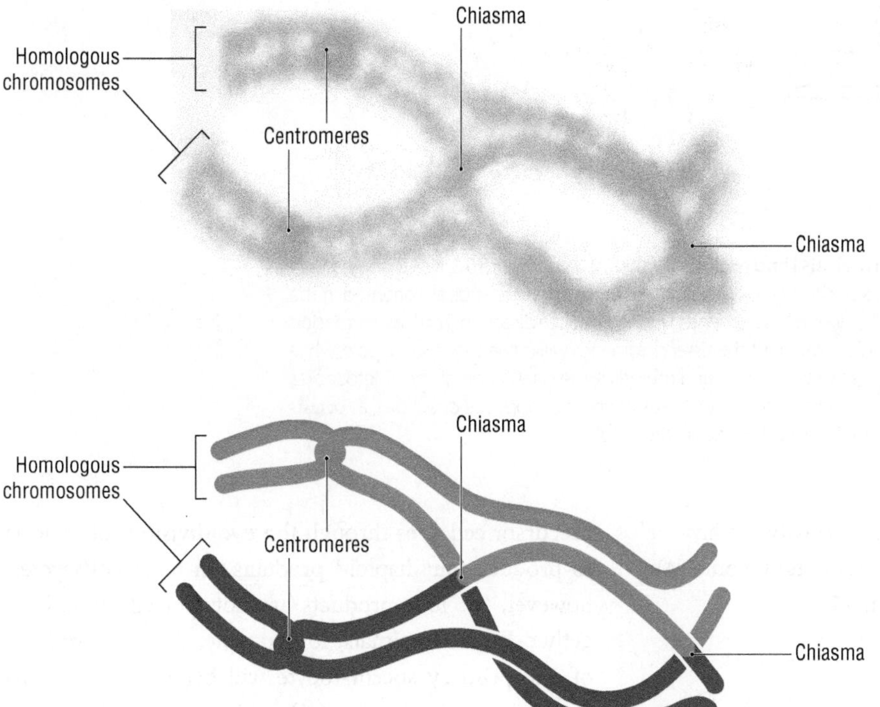

Figure 6.16 Chiasmata showing crossing over. A chiasma is the physical manifestation of the crossover event in Figure 6.15 and holds the two homologs together as they line up on the metaphase plate. The electron micrograph shows a pair of homologs with two chiasmata, as diagrammed below.

Source: Electron micrograph courtesy of James Kezer, University of Oregon. Illustration reproduced from Sadava et al. *Life: The Science of Biology* 8th edition. With permission from Sinauer Associates, Inc.

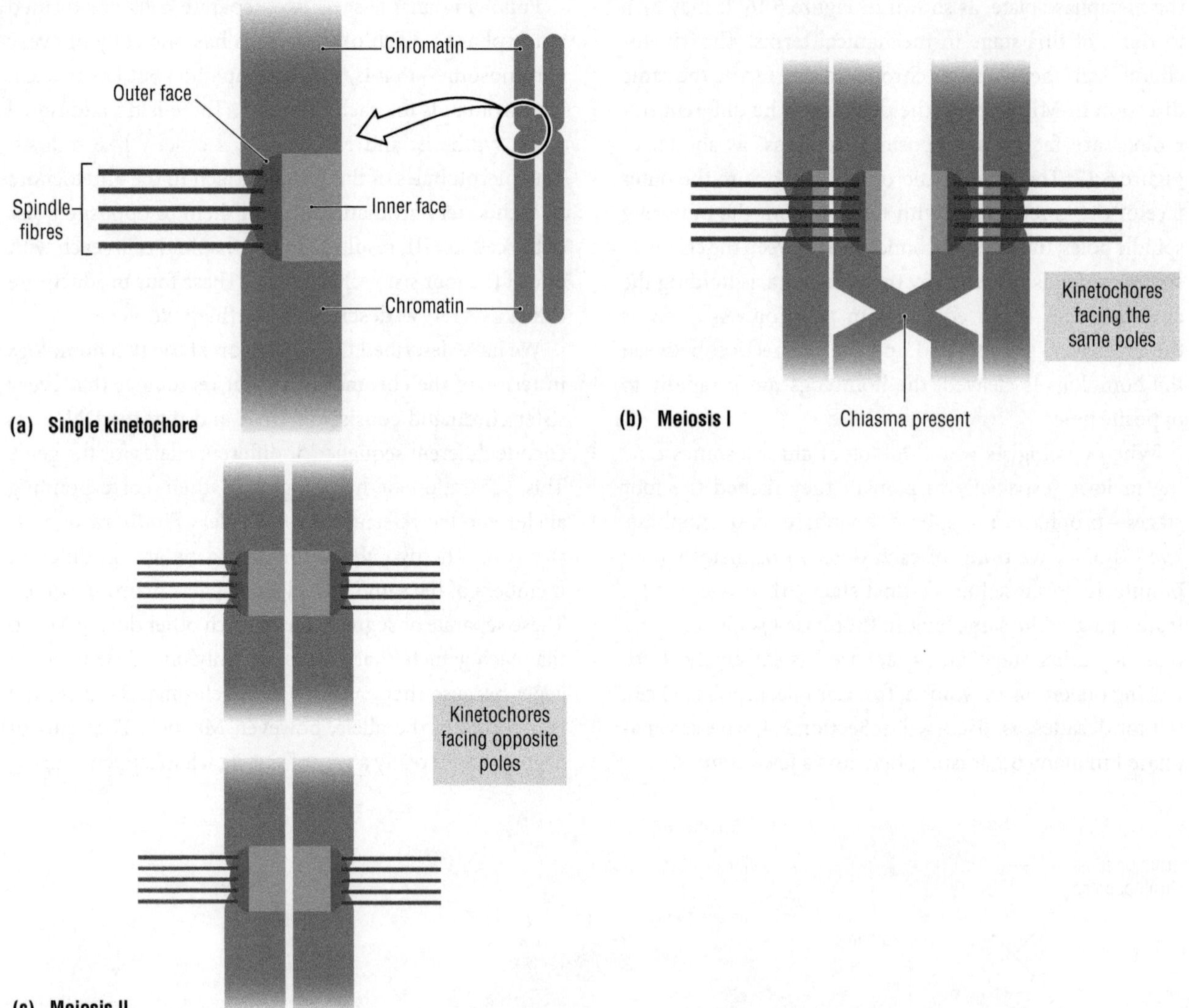

Figure 6.17 Kinetochore orientation in meiosis I and meiosis II. (a) As noted in Figure 6.8, kinetochores have an inner and an outer side. (b) During meiosis I, the kinetochores of the two sister chromatids are oriented in the same direction, while the homologs have the kinetochores oriented in the opposite direction. Thus, when tension is applied from the microtubules of the spindle in meiosis I, the sister chromatids will move to the same pole, while the homologs move to opposite poles. (c) Between meiosis I and meiosis II, the orientation of the kinetochores flips, so that, in meiosis II, the sister chromatids have their kinetochores oriented in opposite directions, as occurs during mitosis. The sister chromatids will move to opposite poles in meiosis II.

meiosis, so it is a law in the sense that gravity is a law—a simpler way to state a natural process and its outcome. We will return to this concept in Section 6.7.

KEY POINT Mendel's Law of Segregation can be explained at the level of the chromosomes by the separation of homologous chromosomes during meiosis I, which generates gametes containing only one allele for each gene.

The sister chromatids at the end of MI then separate from each other in MII, just as sister chromatids separate in mitosis. As a result of MI and MII, one diploid meiotic precursor cell goes through the two divisions of meiosis to produce four haploid products. In most eukaryotes, however, the four products are not typically found together. In male animals, for example, the four products of one primary spermatocyte will become sperm cells (as discussed in Section 6.4), which are found together in the testes, along with the haploid products of many other primary spermatocytes, all of which have also gone through meiosis. Depending on the animal, hundreds or thousands of sperm cells occur together, and it is not possible to determine which sperm cells are derived from the same precursor cell. However, in some fungi known as *ascomycetes*, the four products of a single meiosis are found together in a sac structure called an **ascus**. Since an ascus has all of the products that made up a single tetrad, analysis of tetrads has played a major role in enhancing our understanding of the relationship between meiotic events and genetic ratios. We discuss tetrad analysis further in Tool Box 6.1.

TOOL BOX 6.1 Tetrads and tetrad analysis

In most eukaryotic organisms, the products of different meiotic cells are released together into one organ, so it is not possible to identify which gametes arose from a particular meiotic cell, or meiocyte, at the beginning of prophase I. However, for fungi known as ascomycetes, the four haploid products of meiosis are all found together in a common sac, called an **ascus**. Our ability to recover all of the meiotic products together and know that they occur as a result of the same events in meiosis has greatly furthered our understanding of meiosis. The topics in this Tool Box have a close connection to Chapter 9 when we will discuss how recombination is used to determine the location of genes on chromosomes. We introduce them here to re-enforce your understanding of the behavior of homologous chromosomes and sister chromatids at MI and MII, without focusing on how these events are used to create genetic maps.

The four products of meiosis correspond to the four chromatids—two sister chromatids from a pair of homologous chromosomes—as shown in Figure A. These four chromatids form a **tetrad**. Each chromatid in ascomycetes will become one ascospore; the analysis of these ascospores is known as tetrad analysis. Since each ascospore corresponds to one chromatid, each ascospore is haploid. (The ascospore has one chromatid for each of the chromosomes in the haploid set, but typically only one or a few chromosomes are considered in tetrad analysis, which is how Figure A shows it.) Most eukaryotes can only be grown as diploids, with the haploid phase limited to the gametes. By contrast, ascospores grow into colonies or filaments as haploids, so their genotype can be directly scored from the phenotypes.

While many different types of genetic variants are known, the most widely used are nutritional variants or **auxotrophs**. These are mutants that have lost the ability to produce certain nutrients, and thus they cannot grow unless that nutrient is added to their growth medium. For example, *ade* mutants (called *ad* in the ascomycete *Neurospora*) cannot grow, unless adenine is added to the medium, since the mutation eliminates the ability of the cell to make its own adenine. This concept was introduced in Chapter 2 with the experiments of Beadle and Tatum who used auxotrophic mutants of *Neurospora* that could not synthesize arginine.

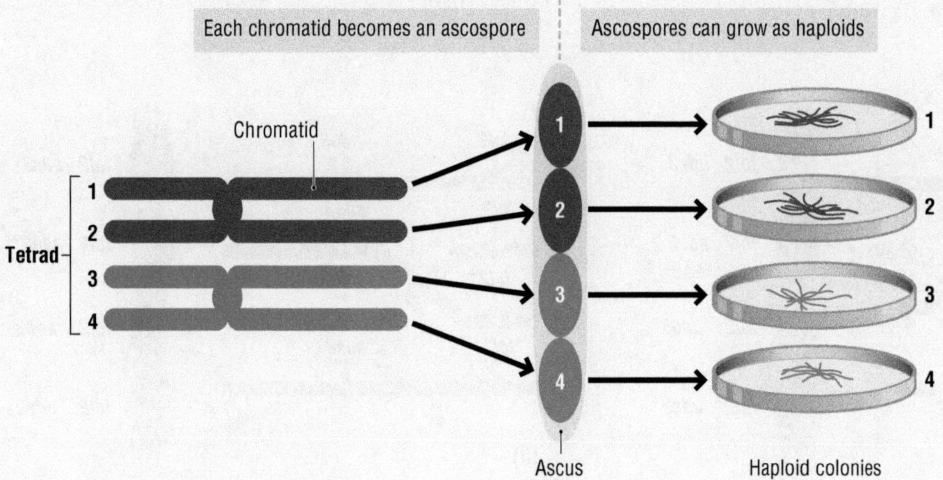

Figure A Tetrads and ascospores. In fungi known as ascomycetes, each of the four chromatids of a tetrad at meiosis becomes one ascospore, which can then grow into haploid colonies or filaments, revealing their genotypes. While this drawing has the chromatids in the tetrad and the ascospores in the ascus in the same order (chromatid 1 gives rise to the ascospore at the top, for example), this only occurs in fungi with ordered tetrads and is presented in this way for clarity.

TOOL BOX 6.1 Continued

Ditypes and tetratypes

Let's consider an example using the budding yeast *Saccharomyces cerevisiae*, one of the best studied eukaryotes and the most familiar ascomycete. A haploid strain that cannot grow without added folic acid (*fol2*) and without additional adenine (*ade3*) is crossed to a haploid strain that grows without any additional nutrients; the two genes are on the same chromosome. The resulting diploid cell is allowed to undergo meiosis, called **sporulation** in yeast and other fungi. The ascus containing the ascospores is collected; the ascus is cut open, and the ascospores are grown on media with different added nutrients to determine their genotypes.

The *fol2* and *ade3* genes are located near each other on the same chromosome, as illustrated in Figure B [B6-7]. Note that one of the parents lacked both the ability to make folic acid and the ability to make adenine, so one of its chromosomes is labeled *fol2 ade3* (on both chromatids), while the other parental strain could make both chemicals and is labeled $fol2^{+}$ $ade3^{+}$. Each homolog is shown in Figure B(i). This diploid cell then goes through meiosis, and a crossover forms at some location between these homologs. If the crossover is not between these two genes but occurs somewhere else on the chromosome, as shown in Figure B(i), two types of spores will be found in the ascus—ones that can make neither folic acid or adenine, and ones that can make them both. Because there are only two types of ascospores, this is known as a ditype ascus. Because the two types have the same genotypes as the two original parental strains, this is known as a **parental ditype**.

Now let's see what happens if the crossover occurs between the *fol2* and *ade3* genes. This is shown in Figure B(ii). Because the crossover happens between the two genes, each of the chromatids (and thus the ascospores) has a different genotype. This type of ascus is known as a **tetratype**. We can begin on the left in Figure B(ii) and trace each chromatid to see the genotypes. Note that two of the chromatids are non-recombinant and two are recombinant. This has important implications that we will discuss again in Chapter 9. For now, we will note that the presence of the tetratype ascus was an important part of the evidence that crossing over occurs when four chromatids are present and involves two of the four. One of the study questions at the end of the chapter returns to the concept of using tetratypes to determine when crossing over occurs during the cell cycle.

Ordered tetrads and the location of the centromere

In *Saccharomyces*, the ascospores from one meiocyte are in the same ascus, but they can be found in any order when the ascus is cut open, that is, the ascospore at the top could have chromatid 4, the second one could have chromatid 2, and so on. This is known as an unordered tetrad. In other ascomycetes, such as the red bread mold *Neurospora crassa*, however, the location of the ascospore in the ascus indicates the order of the chromatids from top to bottom; in other words, sister chromatids always become ascospores that are adjacent to each other. These are known as ordered tetrads, and they allow an additional level of understanding about the behavior of chromosomes during meiosis. For clarity,

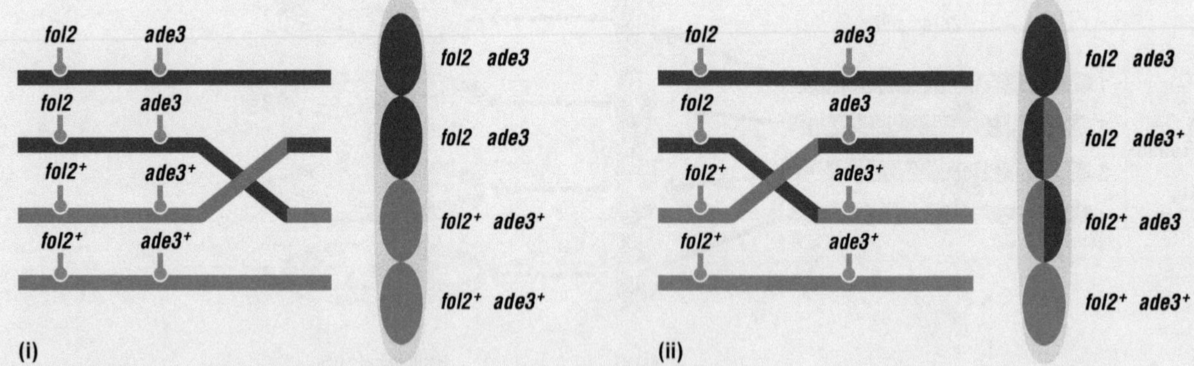

Figure B Tetrad analysis and the location of the crossover. Because all of the four haploid products are found in one ascus, tetrad analysis has been important in understanding the events in meiosis. An example of a diploid that is heterozygous for two genes (*fol2* and *ade3*) on the same chromosome is shown. (i) If the chromatids cross over in any region of the chromosomes other than between the two genes, two types of ascospores are found; the crossover here occurred to the right of the *ade3* gene. These ascospores have the same genotypes as the haploid parents, so this ascus is known as a parental ditype. (ii) If the chromatids cross over between the two genes, four different ascospores are found; this is suggested by the color of the ascospores but is written in the genotypes to the right. This configuration is known as a tetratype ascus.

TOOL BOX 6.1 Continued

we drew Figures A and B as if these were ordered tetrads, even though the ascospores might not occur in this way in yeast.

Before going further, we need to note that both *Neurospora* and *Aspergillus*, the other widely used fungus with ordered tetrads, have an additional feature that will be important in the next section. We introduce it now for accuracy. After the ascospores form, they go through one round of mitosis. Thus, there are eight ascospores in an ascus from *Neurospora* or *Aspergillus*.

Look at Figure C, and recall that the centromeres of the two sister chromatids stay together during MI but separate from each other during MII. Note the round of mitosis represented by the site at which the dotted arrows branch with a red dot. Because the tetrads are ordered, it is possible to determine where the crossover occurred not only with respect to the gene, but also with respect to the centromere. In other words, the centromere can be treated like any other genetic locus. (The location of centromeres can also be determined with unordered tetrads, but the process is a little more complicated.)

In Figure C(i), the crossover occurred between the centromere and the gene *ad-6*. The ascospores have the arrangement 2:2:2:2—two with the ***ad-6*** allele, two with *ad-6*$^+$, two with *ad-6*, and two with *ad-6*$^+$. This is known as a second division segregation because the *ad-6* alleles (on chromatids 1 and 2) separated with their centromere at the second meiotic division; in other words, when the centromeres divided at MII, the *ad-6* alleles divided as well. In Figure C(ii), the crossover did not occur between the centromere and *ad-6*, so the ascospores have the arrangement 4:4. This is called a first division segregation, because the *ad-6* allele and the *ad-6*$^+$ allele segregated from each other at the first meiotic division.

The actual process of tetrad analysis is, in fact, done in the reverse order of how we have just described it, that is, the genotypes of the ascospores are observed and the location of the crossover is inferred from the pattern in the ascus. When the molecular composition of centromeres was not known, it was sometimes difficult to locate them on the chromosome with respect to any of the genes. Crosses in these organisms helped to demonstrate the behavior of the centromere at MI and MII and ultimately led to an understanding of their molecular composition.

Tetrads and gene conversion

Let's return to a single gene and the centromere to illustrate another important point about meiosis that was found from tetrad analysis. With ordered tetrads, a crossover between the gene and the centromere results in the 2:2:2:2 ascus, shown in Figure C(i).

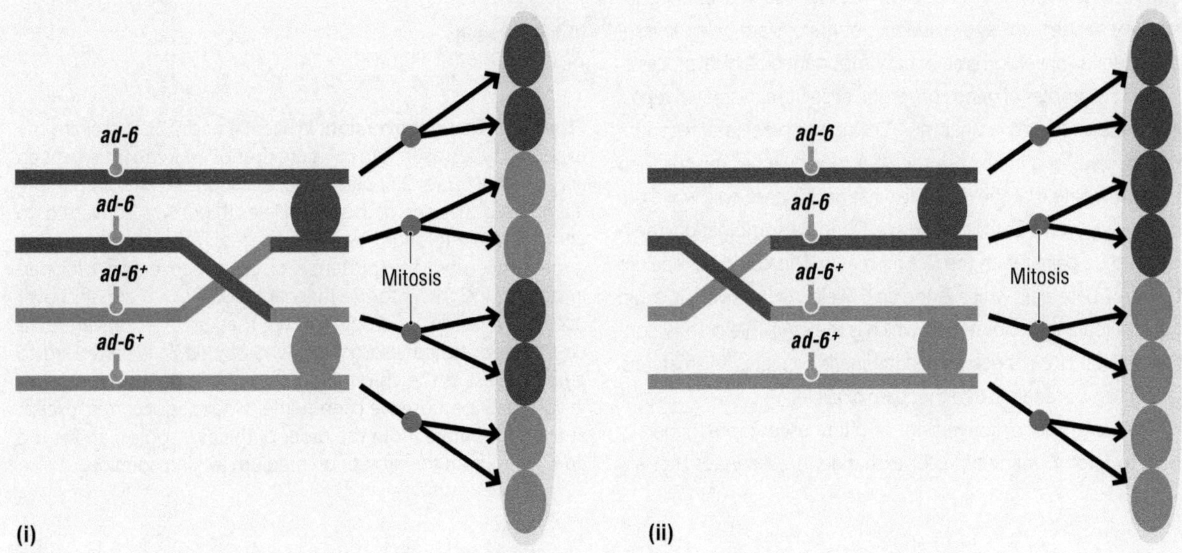

Figure C Ordered tetrads. This figure show ascospores and tetrads as they would occur in *Neurospora*, in which the diploid cell is heterozygous for the *ad-6* gene. The colors represent the genotype at the *ad-6* locus. (i) The crossover has occurred between the *ad-6* gene and the centromere. Two important features are shown. First, the ascospores always form in the same order as the chromatids in the tetrad. To see this, the chromatids can be traced, beginning at the centromere on the right. This results in ordered tetrads. Second, the haploid product of meiosis undergoes a mitotic division, as indicated by the red circle and the branching arrows, so there are eight ascospores. The pattern of the ascospores is 2:2:2:2 when the crossover occurs between the gene and the centromere. (ii) An ascus in which the pattern of the ascospores is 4:4 is shown. This arises when the crossover does not occur in the region between the centromere and the gene, shown here to the left of *ad-6*.

TOOL BOX 6.1 Continued

However, when such experiments were performed in the 1950s, a different type of ascus was also often found. This ascus had a 6:2 arrangement of the ascospores; rather than having four ascospores of each type, there were six of one type and only two of the other type, as shown in Figure D. This was unexpected when it was first encountered because it requires the crossover to be non-reciprocal between the two chromatids, so that more of one type were found than the other type. The terminology is that the *ad-6* allele was "converted" to an *ad-6*⁺ allele, so a 6:2 ascus is called a **gene conversion**. The frequency of gene conversion varies with the organism, but it is found in all eukaryotes and is not rare; it is often as common as reciprocal exchange.

Gene conversion, which can occur from the + allele to the mutant allele, or vice versa, was important in constructing models by which recombination occurs. The Holliday junction model for recombination that is described in Box 6.2 was designed specifically to account for the high frequency of gene conversion. If the Holliday junction diagrammed in Box 6.2 Figure A is resolved by "cutting" the strands horizontally, rather than vertically, gene conversion occurs. Thus, gene conversion, like crossing over itself, is not an unusual feature that occurs only occasionally during meiosis. Rather, it is an inevitable outcome of the process by which reciprocal exchange occurs and was important evidence for understanding how recombination itself happens. While we usually focus on reciprocal exchanges, unless the specifics of meiosis are being investigated, gene conversion is an equally important biological process.

Most examples of gene conversion arise from mismatch repair, a process discussed in Chapter 4. During the process of recombination, discussed in Box 6.2, the DNA molecule has mismatched bases. Imagine that the *ad-6* allele has an A:T base pair, while the *ad-6*⁺ allele has G:C at the same site. During recombination, the A can end up paired with the C, a mismatch. This could be repaired to an A:T base pair, which is the *ad-6* allele. Alternatively, it could be repaired to a G:C base pair, which is the *ad-6*⁺ allele. Thus, conversion will occur because recombination creates mismatches, which can be repaired to either base or allele.

The term "gene conversion" is often used more generally now to refer to many types of exchange in eukaryotes that are directional and not reciprocal, resulting in more of one type of haploid product than the other. However, the original (and still primary) use of the term came from an analysis of tetrads and the unexpected occurrence of asci with a 6:2 arrangement of ascospores.

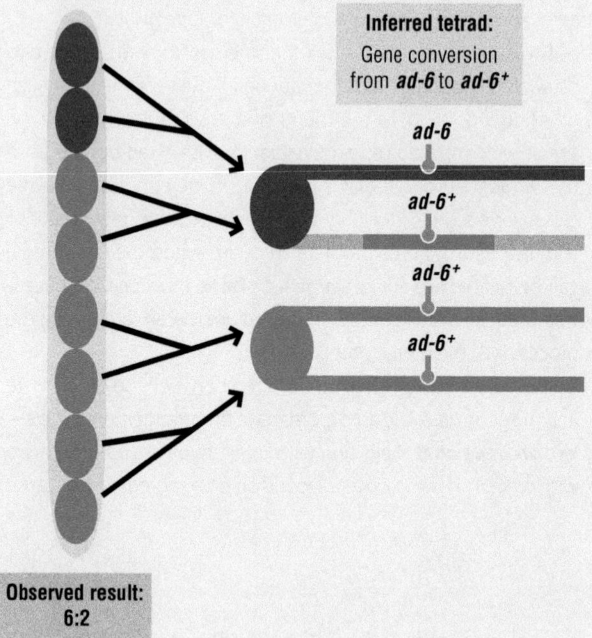

Figure D Gene conversion. While 4:4 and 2:2:2:2 asci are the expected types, as shown in Figure C, another frequent type is an ascus with a 6:2 (or 2:6) pattern of ascospores. From this pattern, it is inferred that one of the *ad-6* alleles (blue) was "converted" to an *ad-6*⁺ allele (orange) by some process. This process is called gene conversion; the presence of gene conversion provided important insights into the mechanism by which crossing over occurs. While this example shows that *ad-6* is converted to *ad-6*⁺, the opposite also occurs, whereby *ad-6*⁺ is converted to *ad-6*. The rest of the chromatid involved in conversion is shown in light gray because we cannot infer what has occurred, except at the centromere (from the order of the ascospores) and at the *ad-6* locus (from the growth on medium lacking adenine).

6.4 Spermatogenesis and oogenesis

In the preceding sections of this chapter, we have set out a general overview of meiosis; this process is essentially what happens during spermatogenesis, pollen grain formation, or meiosis in single-celled organisms such as yeast. In these processes, all four haploid cells at the end of MII survive and are about the same size. The process is more or less continuous from the beginning (S phase) to the end (the completion of MII), with no lengthy pauses or interruptions. (In spermatogenesis, the four cells—called spermatids in animals—go through an additional cellular differentiation process to become functional sperm, but we have described the process as it occurs for the chromosomes themselves.)

Polar bodies

Oogenesis in animals is different from spermatogenesis in two key ways, with several important consequences. First of all, the cells formed at each division in oogenesis are very different in size. MI generates one large cell called the **secondary oocyte**, and one very small cell, the first **polar body**, as shown in Figure 6.18(a). The chromosomal content is the same in each cell, but their size differences are due to differences in the way the cytoplasm is divided between the two. The secondary oocyte gets nearly all of the cytoplasm, and the polar body gets almost none; the differences in size for humans are shown in Figure 6.18(b).

The same unequal division of the cytoplasm occurs during MII. The secondary oocyte undergoes MII, eventually becoming the ovum. Thus, the cytoplasm of the secondary oocyte, with its mitochondria, ribosomes, stored proteins and RNA, and so on, will become the cytoplasm of the fertilized egg and the early embryo. The first polar body may or may not go through MII, depending on the species; in most mammals, it often does not undergo MII and is eventually degraded. At the end of MII, the mature ovum is a large cell, and the second polar body is very small.

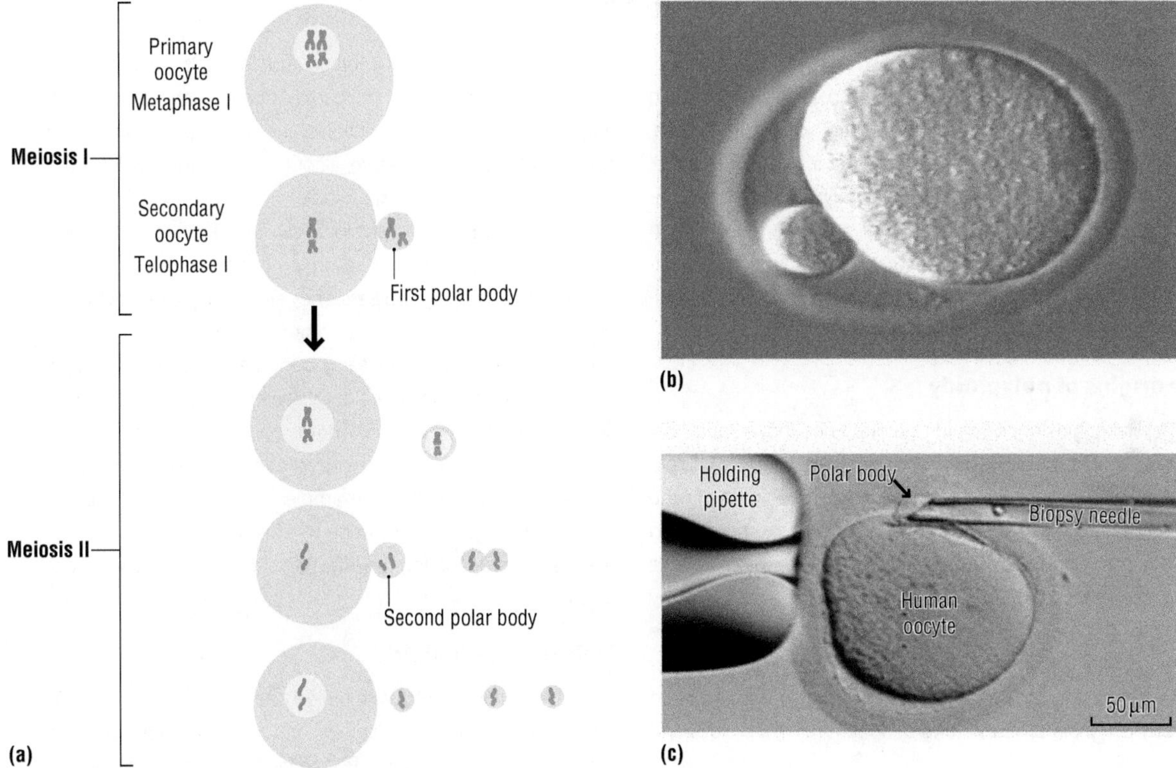

Figure 6.18 Oogenesis and polar bodies in animals. While chromosome segregation is the same in spermatogenesis and oogenesis, the cells arising during oogenesis are not of the same size. (a) The two meiotic divisions during oogenesis, summarized. The cell undergoing meiosis I is known as the primary oocyte, shown here at metaphase I. When the chromosomes segregate at anaphase I to create two cells (at telophase I), the secondary oocyte has nearly all of the cytoplasm, while the other cell, known as the first polar body, has a haploid set of chromosomes but almost no cytoplasm. The same process occurs during meiosis II, producing a second polar body. In this diagram, the first polar body is shown undergoing meiosis II to produce two additional polar bodies. In many species, the first polar body degenerates and does not progress into meiosis II. (b) A micrograph showing the secondary oocyte and the first polar body in humans. (c) During a polar body biopsy, the polar body is removed with a small needle. Polar body biopsy can be used to determine which oocytes from *in vitro* fertilization have the normal allele during pre-implantation genetic diagnosis. If the polar body has the allele responsible for the disease, the secondary oocyte must have the normal allele.

Source: Part (b) courtesy of Douglas Kline, Kent State University. Part (c) courtesy of Gary Wessel, Sandra Carson, Peter Klatsky, and Julian Wong.

If a primary oocyte undergoes MI *in vitro*, the first polar body can be removed by a needle in a process known as **polar body biopsy**, as shown in Figure 6.18(c). This can be done as part of pre-implantation genetic diagnosis during *in vitro* fertilization in humans. If a woman is heterozygous for a life-threatening disease allele, for example, the polar body can be assayed to determine which allele is present. The secondary oocyte and the first polar body have the two different homologs, so if the polar body has the disease-associated allele, the secondary oocyte has the other allele. Pre-implantation genetic diagnosis can be done for dozens of different genetic diseases and for many meiotic errors. An advantage of using the polar body for such analysis is that it does not develop further, so no potential damage is done to the embryo itself in the process of taking the biopsy.

On rare occasions, a polar body may fuse with the oocyte to produce a diploid ovum, or non-disjunction might occur for an entire set of chromosomes, rather than only one pair of chromosomes. These result in diploid gametes, and in **polyploid** zygotes when these gametes are fertilized—"poly" here referring to more than two. Polyploidy has been an important process in genome evolution and is discussed in Chapter 3. We discuss meiosis in polyploids in Box 6.3.

BOX 6.3 *Going Deeper* Meiosis in polyploids

While most animals are diploids with two copies of each chromosome, many plants are polyploids with more than two haploid sets of chromosomes. We introduced polyploids during our discussion in Box 3.1 about mechanisms that increase the size of the genome; both recent and ancestral polyploidy has been important during evolution and in artificial and natural variation in plants. It is important to remember that polyploidy involves complete sets of chromosomes, unlike aneuploidy, which affects individual chromosomes.

The origins of polyploidy

The natural origins of polyploidy are often not known, but most of the mechanisms probably involve an unreduced or diploid gamete, the fusion of two diploid cells, or the fusion of two of the meiotic products such as the polar body with the ovum. Polyploid plants are typically larger and often more vigorous than diploids; thus, they have often been selected for, either naturally or by plant breeders. When a polyploid shoot arises, the plants can be grown vegetatively and propagated, whether or not they produce viable seeds. This has allowed the survival and propagation of many polyploid plants, among both common food crops and ornamentals.

The most common experimental method to produce a diploid gamete (and hence a polyploid offspring) has involved treatment that causes the breakdown of the meiotic spindle. Because the spindle comes apart and the chromosomes (either homologs at MI or sister chromatids at MII) cannot separate at anaphase, the meiotic cell might mature into a gamete with two sets of chromosomes. The most common way to induce polyploidy experimentally has been to treat plants with the chemical colchicine, originally extracted from members of the crocus family. Colchicine binds to tubulin and prevents the assembly of the microtubules that make up the spindle; it has some medicinal uses at low doses to block cell division, in addition to its value to plant breeders as an inducer of polyploidy. Other methods to induce polyploidy experimentally involve temperature shocks (either heat or cold) or high pressure; all of these treatments also interfere with the assembly of the spindle.

As noted in Box 3.1, polyploids can be classified as triploids, tetraploids, hexaploids, and so on, according to the number of chromosome sets they possess. They are also classified by which species contributed the extra chromosome sets. **Autopolyploids** have multiple copies of the same genome; this could arise from any of the mechanisms mentioned earlier, including a spontaneous polyploid shoot. **Allopolyploids** have the genomes of two or more different plants; different varieties of wheat are allotetraploids or allohexaploids, for example. These arise from crossing an unreduced gamete from one species with one from another species and propagating the offspring either vegetatively or from seeds.

Meiosis in polyploids

The behavior of polyploids at meiosis depends on the origins of the chromosomes. Allotetraploids and allohexaploids have two copies of each chromosome from different genomes; these two copies are a pair of homologs, which means that they can usually synapse and segregate correctly, allowing meiosis to occur normally. Thus, most allotetraploids or allohexaploids are fertile and set seeds as well as a diploid plant. Allopolyploids are useful to plant breeders because they can often be selected to combine the most desirable characteristics of each of the individual species in a single new plant.

Autopolyploids, by contrast, have multiple copies of the same chromosomes. This creates a problem when the chromosomes attempt to pair, synapse, and segregate normally; three copies of a chromosome cannot easily make the stable pair needed to

BOX 6.3 Continued

progress through MI. Thus, autopolyploids are usually sterile or nearly sterile, and produce few or no seeds. However, this situation is often desirable for consumers since seedless varieties may be easier to eat, so autopolyploids are common among our foods. Seedless watermelons and bananas are autotriploids, while potatoes are autotetraploid; all of these plants are propagated vegetatively, rather than from seeds. (The few seeds that arise in a seedless watermelon are probably aneuploids and can only rarely be grown into a plant.)

Allopolyploidy in *Raphanobrassica*

In the 1920s, a Soviet agronomist named Georgi Karpechenko attempted to use polyploidy to create a hybrid between cabbages of the genus *Brassica* and radishes of the genus *Raphanus*. Since allopolyploids can have the desirable characteristics of each parental species, Karpechenko apparently envisioned a plant with the leaves of a cabbage and the roots of a radish; such a plant could provide a significant food source in Russia. However, his hybrid plant had the leaves of the radish and the roots of the cabbage instead, as shown in Figure A. In other words, hybridization can sometimes combine the least desirable characteristics of both species as well. The plant, originally referred to as *Raphanobrassica* but now renamed *Brassicoraphanus* and familiarly called radicole, can be used as animal fodder but is not useful for human consumption.

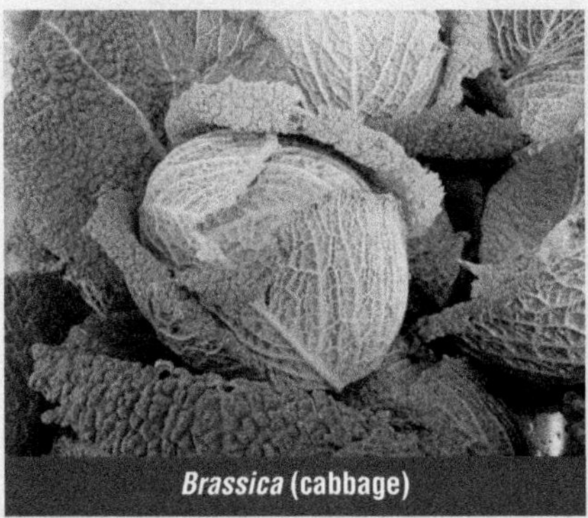

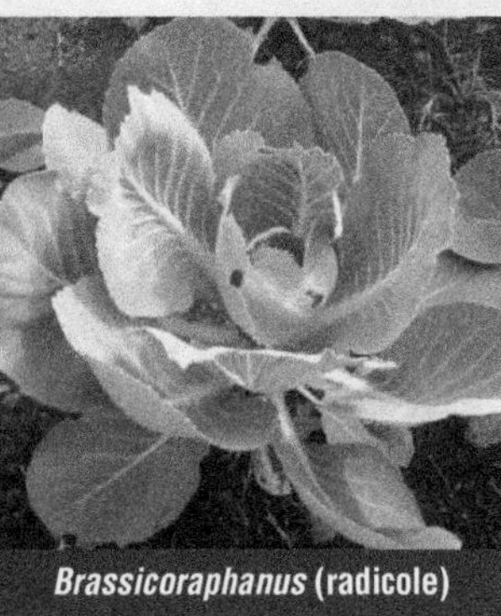

Figure A The allopolyploid *Brassicoraphanus*. Compared to its diploid parents, the cabbage and the radish, the tetraploid plant *Brassicoraphanus*, or radicole, lacks both edible leaves and roots.

Source: Image of cabbage courtesy of Dinkum/ CC0 1.0. Image of radish courtesy of Chixoy/ CC BY-SA 3.0.

Meiotic arrest

The second way that oogenesis in animals differs from the general overview of meiosis is that the process of oogenesis is not continuous. Nearly all vertebrates pause or arrest oogenesis during prophase I, in what is known as the **dictyate arrest**. In humans, all primary oocytes are formed before birth, but arrest after pachytene until ovulation begins at puberty, as shown in Figure 6.19. At ovulation during the reproductive years, the primary oocyte completes MI and arrests again as a secondary oocyte at metaphase of MII. If it is fertilized, the secondary oocyte completes MII; if it is not fertilized, it is degraded. The arrest at prophase I can last for decades in humans, whereas the arrested cell at metaphase II lasts a week or less. Some other animals have the prophase I arrest but don't complete MI at all until after fertilization and thus might not have a pause at metaphase II.

From a broader perspective, the vastly different cellular processes of spermatogenesis (or pollen grain formation) and oogenesis have led to a range of hypotheses and models about different evolutionary pressures on the two sexes. During their respective reproductive lives, males make a very large supply of sperm, whereas females make relatively few oocytes. This has led to models involving differences between maternal and paternal behaviors. Without discussing any of these hypotheses in detail, it is worth noting that males and females are each passing on their genomes to the next generation, but using somewhat different strategies to do so.

KEY POINT Oogenesis differs from spermatogenesis in that the cells resulting from meiosis I and meiosis II often differ in size. The smaller polar bodies receive little cytoplasm and usually degenerate. Oogenesis is not continuous, and the process can pause for weeks or years. Thus, males and females help to ensure the transmission of their genomes to the next generation by somewhat different evolutionary strategies.

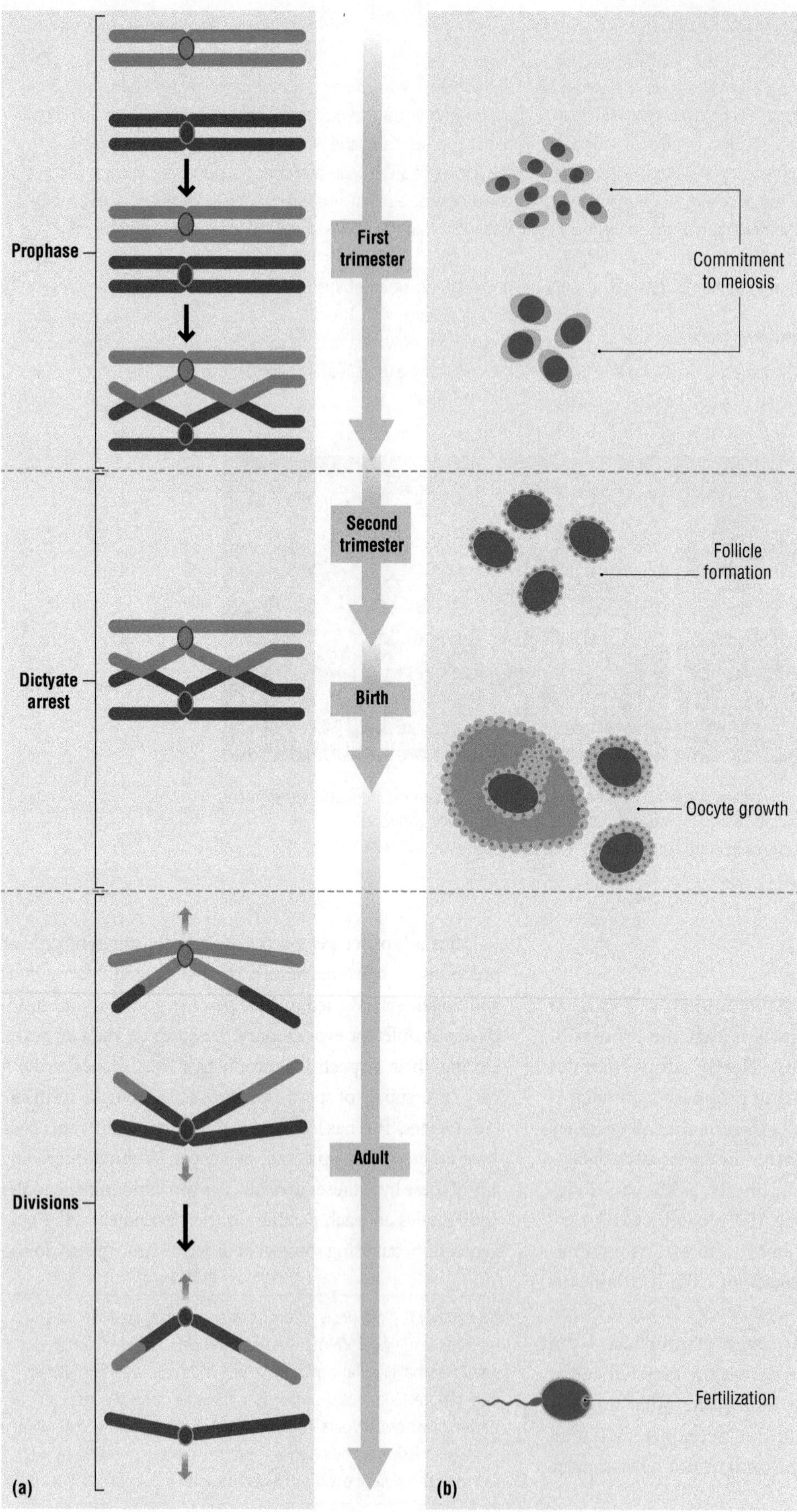

Figure 6.19 Oogenesis arrest cycle. Meiotic prophase begins, and the cell progresses to the formation of crossovers after pachytene in prophase I. In humans, this occurs during the first 12 weeks of gestation. The cells then arrest and continue in this stage—called the dictyate arrest—so girls are born with all of their oocytes committed to meiosis. In humans, this stage can last for years. With the onset of hormonal changes that occur at puberty, the oocyte grows, and ovulation occurs with the release of one (or more) egg. The ovulated egg resumes and completes meiosis I and begins meiosis II, before arresting again temporarily at metaphase II. If fertilized, the secondary oocyte completes meiosis II and becomes a zygote; if it is not fertilized, the secondary oocyte is shed or degraded.

Source: Reprinted by permission from Macmillan Publishers Ltd: *Nature Reviews Genetics*. So, I., et al. Human aneuploidy: mechanisms and new insights into an age old problem. © (2012).

6.5 Meiotic errors

The outcome of meiotic segregation is exceptionally precise, such that nearly all of the viable gametes produced by an individual have one copy of each chromosome, while nearly all of their offspring have two copies of each chromosome, one from the meiosis occurring in each parent. This precision is evolutionarily essential, of course, since meiosis is the process in eukaryotes that ensures continuity of the genome from one generation to the next; an imprecise meiosis would be quickly selected against. This precision arises from a series of regulated steps and checkpoints, each of which eliminates or arrests some of the meiotic errors that do occur.

These meiotic checkpoints are summarized in Figure 6.20. In addition to the G_1 or START checkpoint that initiates the cell division cycle, a further checkpoint ensures that premeiotic DNA replication has been completed normally, as in mitosis. Additional checkpoints also ensure that synapsis has occurred, that

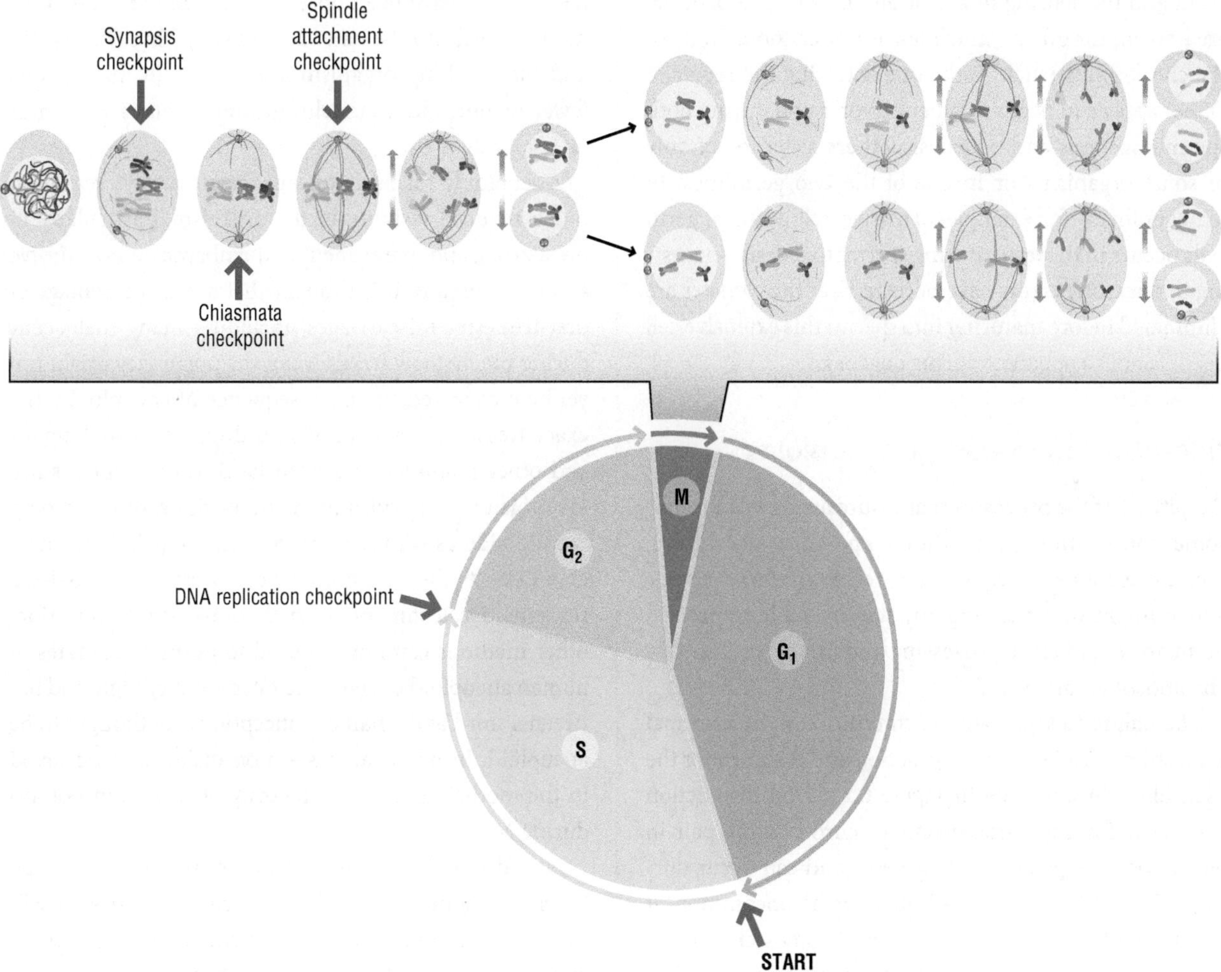

Figure 6.20 Checkpoints during meiosis. A meiotic cell has to pass through at least five different checkpoints, two of which are found only in meiosis and three of which are similar in mitosis and meiosis. The G_1 checkpoint, also referred to as START, allows the cell to progress in the cell division cycle; the genes that regulate this checkpoint appear to be different for mitotic and meiotic cells, but the function of the checkpoint is similar. After the S phase, a checkpoint ensures that DNA replication and repair have been completed; this is common to both mitotic and meiotic cells and appears to be regulated by the same genes. Once a cell enters meiosis, at least three additional checkpoints occur. There is a checkpoint to ensure that synapsis has occurred, as well as one to ensure that crossing over has occurred and all chiasmata have been resolved. Finally, in both mitotic and meiotic cells, there is a spindle attachment checkpoint at metaphase to ensure that all kinetochores are attached to the spindle before anaphase begins.

recombination has been completed, and that all DSBs have been repaired; finally, in both mitotic and meiotic cells, a checkpoint ensures that the spindle has attached to each chromosome. A checkpoint may also confirm that each chromosome has received a crossover; additional checkpoints have also been suggested but not yet proved to occur in all eukaryotes.

If a cell fails at any of these checkpoints, its progress through meiosis is arrested until the particular molecular process can be completed. Alternatively, the cell might exit from meiosis to re-enter a non-dividing state or undergo the programmed cell death process known as apoptosis. For some of these checkpoints, the triggering event and the sensing molecule and its effective pathway are known; for others, much less is understood. Some of the checkpoints (such as the one detecting the repair of DSBs) appear to occur in all eukaryotes and during both spermatogenesis and oogenesis; others seem to act only in some organisms or in one of the two germlines. In many animals, for example, extensive cell death of germ cells occurs in the testes during spermatogenesis, suggesting that meiotic errors have occurred but the gametes are eliminated before maturing into sperm; this extensive cell death is not so apparent during oogenesis.

Non-disjunction leads to aneuploidy

Despite all of the processes that ensure a normal chromosome constitution (a condition known as **euploidy**), meiotic errors do occur. As a result of these errors, which might occur during any step in the long process of meiosis, gametes possessing too many or too few chromosomes arise.

The failure to segregate the chromosomes in a normal 1:1 fashion is called **non-disjunction** and is shown for the light blue chromosome in Figure 6.21. Non-disjunction can occur for any chromosome or chromosome pair in either spermatogenesis or oogenesis, and can occur during MI or MII. Since non-disjunction is uncommon, it is very unlikely that more than one chromosome or one pair of chromosomes will be affected during one complete meiotic process, and even more unlikely that it will occur at both MI and MII. The outcomes of non-disjunction at MI and MII are different; the different outcomes can be most readily appreciated by following the segregation of sex chromosomes, so we will discuss this in more detail in Chapter 7. In addition, as discussed in Chapter 7, sex chromosome non-disjunction has been important in understanding the chromosomal basis for sex determination.

As a result of non-disjunction, a gamete will be missing one chromosome from the haploid set—a condition known as **nullisomy**—or will have two copies of a chromosome—a condition known as **disomy**. If all gametes survive and can be observed, a non-disjunction event is expected to produce equivalent numbers of nullisomic and disomic gametes, since they are the two outcomes of the failure of a chromosome pair to segregate normally. If the gamete from the other parent has a euploid set of chromosomes (as nearly always occurs), the resulting zygote after fertilization will be either **monosomic** with one member of one chromosome or **trisomic** with three members of a chromosome. (Other possibilities, such as tetrasomics, have been observed but are even less common and will not be considered here.) Thus, the zygote and the resulting organism are **aneuploid**, having one fewer or one additional chromosome; this is written as $2n - 1$, or $2n + 1$.

Aneuploidy for an autosome nearly always results in much-reduced viability in the organism, particularly in metazoans, and many aneuploid embryos do not survive at all. In humans, it is thought that a high percentage of first-trimester miscarriages, including many that occur during the first few weeks when the pregnancy might not yet be recognized, are a consequence of aneuploidy. The exact frequency of natural non-disjunction in humans and other mammals has been hard to determine, since investigators are typically in the position of examining the individuals who survive in order to make inferences about events they are not able to observe. With assisted reproductive technologies, such as *in vitro* fertilization, other methods can now be used to estimate the rates of human aneuploidy. Again, the rates are very high, and between a third and a half of conceptions are thought to be aneuploid. In nearly all cases, non-disjunction occurred in the mother, and in the majority of cases (but not all) during MI.

Non-disjunction in humans is most commonly seen for chromosome 21 and for the sex chromosomes; we will discuss sex chromosome non-disjunction in Sections 7.3 and 7.4. For the autosomes, the only aneuploid children that regularly survive to birth are those with trisomy 21, also known as Down syndrome. Children who are trisomy 13 or trisomy 18 are sometimes live-born but die in infancy. Trisomy 22 children have also been observed but again rarely survive infancy. No autosomal monosomic individuals ($2n - 1$) have been observed; while it may be assumed that the occurrence of a disomic gamete should also result in a nullisomic gamete, monosomic embryos

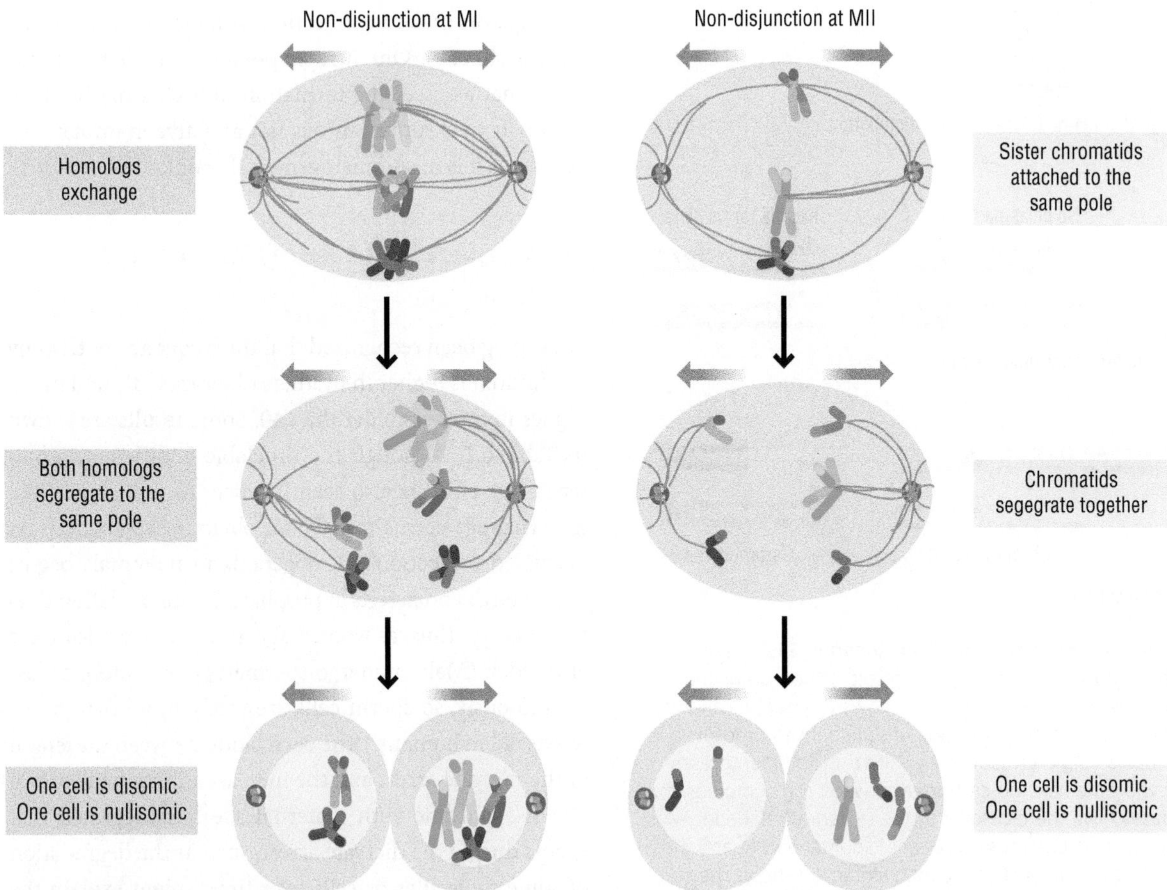
Non-disjunction at MI
Non-disjunction at MII
Homologs exchange
Sister chromatids attached to the same pole
Both homologs segregate to the same pole
Chromatids segegrate together
One cell is disomic
One cell is nullisomic
One cell is disomic
One cell is nullisomic

Figure 6.21 Non-disjunction and aneuploidy. Non-disjunction can occur for any chromosome and during either meiotic division. During meiosis I non-disjunction, shown on the left, both homologs in light blue segregate to the same pole; meiosis II occurs normally. During meiosis II non-disjunction, shown on the right, both sister chromatids in light blue, segregate to one pole. In either case, one meiotic product has two copies of the chromosome and is disomic, while the other meiotic product has no copies of the chromosomes and is nullisomic.

do not survive, so only disomic gametes (and thus trisomic children) are observed. Trisomy 21 has been extensively studied, and while much is known about its origins and contributing factors, some of this information raises more questions.

Trisomy 21 is usually the result of a failure to cross over in the mother

With the availability of DNA sequence variants and analysis of the human genome, it is possible to examine trisomy 21 children and their parents to understand the origin of the non-disjunction event; these are summarized in Figure 6.22. Chromosome 21 usually has a single crossover, which can occur at any position along its length. Figure 6.22(a) shows the pair of homologs with a single chiasma near the middle, which allows normal disjunction. If this crossover does not occur and the chiasma does not form to hold the homologs at the metaphase plate, non-disjunction occurs. More than 80% of the spontaneous cases of trisomy 21 that can be attributed to non-disjunction in the mother are also associated with a failure to cross over on chromosome 21, as shown in Figure 6.22(b).

A minority of cases of maternal non-disjunction are associated with a chiasma that forms right at the end, opposite the centromere, as also shown in Figure 6.22(b); it may be that a chiasma at this location cannot provide enough tension to hold the homologs in place. Most of the remaining cases of trisomy 21 are the result of a translocation in which parts of chromosome 21 have become attached to another chromosome, so that the child

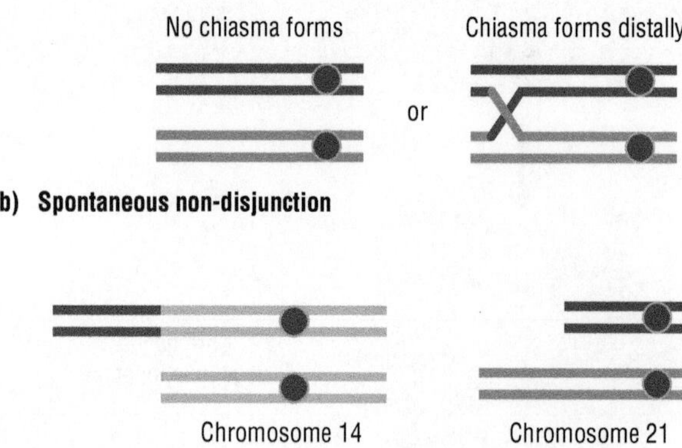

Figure 6.22 Non-disjunction in trisomy 21. Trisomy 21, or Down syndrome, is usually the result of non-disjunction of chromosome 21 in the mother. (a) shows a typical location for the chiasma that results in normal disjunction; chromosome 21 usually has a single crossover. As shown in (b), many cases of Down syndrome have no crossover between the homologs, but some cases have the crossover at the extreme distal end, away from the centromere. These cases arise spontaneously, and there is no increased risk for subsequent children. (c) shows the chromosome translocation that can lead to familial Down syndrome. One parent has a large fragment of chromosome 21 attached to another chromosome, shown here as chromosome 14, the most common location; there may also be a small part of chromosome 14 attached to chromosome 21. Either parent can have this translocation. This parent has a normal chromosome 21 and a much shorter version of chromosome 21 that is missing the portion translocated to the other chromosome. A child who inherits both the normal copy of chromosome 21 and chromosome 14 with the additional part of chromosome 21 from this parent, as well as a normal chromosome 21 from the other parent, is partially trisomic for chromosome 21 and shows the characteristic Down syndrome phenotype. In this case, future children are also at risk of inheriting the translocation.

has two copies of chromosome 21 plus the translocated fragment of chromosome 21 attached to another chromosome, as shown in Figure 6.22(c). Individuals with this translocation are partial trisomics; this condition is sometimes called familial Down syndrome, since subsequent children are also at risk of inheriting the translocation. For most spontaneous or sporadic cases of trisomy 21, the birth of one affected child does not increase the risk that a subsequent child will be affected.

Why do most cases of trisomy 21 involve maternal non-disjunction, rather than paternal non-disjunction? The answer is not entirely known, although many investigators make a connection with maternal age (as we discuss shortly). One further possibility is that a checkpoint that assesses the formation of a chiasma between each pair of chromosomes is not as active in human oogenesis as it is in spermatogenesis, for unknown reasons.

The frequency of trisomy 21 increases with maternal age

It has long been recognized that the frequency of trisomy 21 children is higher in mothers older than 35, and much higher in mothers older than 40. Some results are shown in Table 6.1. Although the timetable is different, a maternal age effect is also seen in mice. At first glance, the maternal age seems to have an obvious explanation. As discussed in Section 6.4, oogenesis in mammals begins before birth and arrests at prophase I until ovulation during puberty. Thus, as women age, the ova they release are also older. (Males undergo spermatogenesis more or less continuously, so sperm cells are rarely more than a few weeks old in humans.) The correlation between the length of the dictyate arrest and the increase in the incidence of Down syndrome with maternal age has suggested that events during the dictyate arrest, such as the degradation of some molecules or cells over time, might explain the maternal age effect.

However, this explanation also leaves some additional unanswered questions. For example, it predicts that non-disjunction events for every chromosome should increase with maternal age, at about the same rate. This is only partially true; while non-disjunction of autosomes increases with maternal age, non-disjunction of the X chromosome has little or no maternal age effect at all. In addition, Figure 6.23 shows how three different autosomes have different rates of increased non-disjunction with maternal age, which suggests more than one explanation is needed.

Furthermore, the time at which dictyate arrest occurs during prophase I does not quite fit this model. As noted previously, oogenesis arrest occurs after crossing over has commenced and probably about the time crossovers have been resolved. Unless there is a mechanism by which oogenesis can assess which oocytes have had a crossover and have those mature first, it is not clear why there should be a difference based on the length of the dictyate arrest. It is possible that some additional surveillance or checkpoint system exists and that this checkpoint begins to fail with maternal age.

Another component of the maternal age effect may be the influence of environmental factors. Ovulation is

Table 6.1 The effect of maternal age on the incidence of trisomy 21

Maternal age	Incidence of Down syndrome	Maternal age	Incidence of Down syndrome	Maternal age	Incidence of Down syndrome
20	1 in 2000	30	1 in 900	40	1 in 100
21	1 in 1700	31	1 in 800	41	1 in 80
22	1 in 1500	32	1 in 720	42	1 in 70
23	1 in 1400	33	1 in 600	43	1 in 50
24	1 in 1300	34	1 in 450	44	1 in 40
25	1 in 1200	35	1 in 350	45	1 in 30
26	1 in 1100	36	1 in 300	46	1 in 25
27	1 in 1050	37	1 in 250	47	1 in 20
28	1 in 1000	38	1 in 200	48	1 in 15
29	1 in 950	39	1 in 150	49	1 in 10

With permission from the National Down Syndrome Society. To learn more, visit www.ndss.org.

dependent on hormonal signals. Some of the chemicals to which we are exposed routinely, such as the plastic additive bisphenyl, are thought to mimic some hormones and may be interfering with checkpoints or the function of cohesin, in addition to possibly affecting meiosis itself. In any event, the maternal effect is a bit more complicated than might first have been imagined.

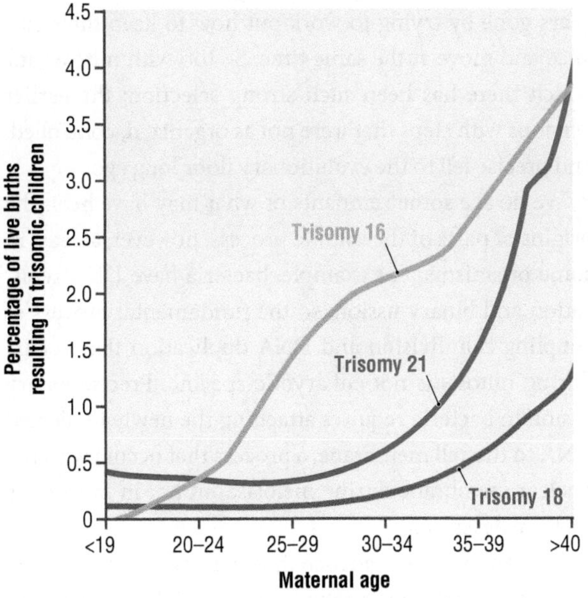

Figure 6.23 Trisomy and maternal age. The incidences of trisomy 16, trisomy 18, and trisomy 21 all increase with maternal age. However, the curves are not the same, suggesting that no single explanation accounts for the maternal age effect for all chromosomes. The data are redrawn from Nagoaka *et al.* 2012 and are based on fertilized eggs from fertility clinics.

Trisomy and aneuploid syndromes

Why should trisomy 21 be the most common autosomal trisomy and the one that survives the best? The simplest explanation is also probably the most likely; chromosome 21 is the smallest human chromosome and apparently has the fewest protein-coding genes on it—about 250 is the latest estimate. Thus, the average number of crossovers on chromosome 21 is low—because it is the smallest chromosome—and the impact of an extra copy of the chromosome is reduced because it has the fewest genes on it.

But even this likely explanation also raises additional questions. Chromosome 22 is only slightly larger than chromosome 21, and trisomy 22 fetuses only rarely survive to birth. In addition, trisomy 13 and trisomy 18 children survive better than trisomy for any other chromosomes, including some that involve chromosomes that are smaller and have fewer genes than chromosome 13 or chromosome 18. Thus, there must be some additional explanations.

The effects of trisomy for different chromosomes also lead to the question, "Why is there a distinct trisomic phenotype?" and again the answers may not be so obvious. Most of us can recognize a person with trisomy 21, and some of their physical features are more similar to other trisomy 21 people than to their disomic siblings, for instance. Experienced human geneticists can recognize trisomy 13 and trisomy 18 infants, again because they have distinctive morphological features. The most likely explanation for these recognizable trisomic phenotypes is that hundreds of genes on each chromosome are being expressed at higher levels than in disomic individuals. We pointed out in Box 5.1 that most genes are relatively insensitive to an increased dose or expression level, but perhaps the cumulative effect of over-expressing many genes leads to the recognizable phenotypes. But this does not help us understand why other trisomics survive less well or not at all.

A second and related explanation for the phenotypes is that, while most genes are relatively insensitive to increased dose, some genes are very sensitive, and an increase in the level of their expression or dose might be responsible for the phenotypes. Some studies have attempted to identify genes on chromosome 21 responsible for particular phenotypes. For example, trisomy 21 individuals have an elevated frequency of Alzheimer disease, which also begins much earlier, on average, than in disomic siblings. One explanation is that one of the genes associated with an elevated risk for Alzheimer disease encodes a protein known as β-amyloid, and this gene is located on chromosome 21. The change in the dose and expression level of this gene may contribute to the elevated risk for Alzheimer disease. The combination of a few genes that are dose-sensitive and contribute to specific phenotypes with the general over-expression of hundreds of genes may provide the explanation for the recognizable trisomic phenotypes.

6.6 Evolutionary origins of mitosis and meiosis

Our understanding of evolutionary biology adds nuance, and even grandeur, to our understanding of mitosis and meiosis. For instance, nearly all eukaryotic cells carry out mitosis in very similar ways, using very highly similar proteins. This indicates that mitosis, a defining characteristic of eukaryotic cells, occurred in the ancestor of all extant eukaryotes. While refinements have been made, many steps in the process are likely to be the ones that were present in that long-extinct ancestor. Similarities between processes of cell division indicate that meiosis is almost certainly an evolutionarily derivative of mitosis that occurred in specialized cells and that MII is essentially mitosis, only with one chromosome of each set.

Furthermore, the amino acid sequences of tubulin proteins, such as those found in the spindle, in different organisms are among the most similar for any protein. This indicates that selection has strongly and consistently eliminated most variations in the tubulin genes and proteins. Changes in tubulin sequence definitely arise, just as changes in every gene and amino acid sequence arise. But those changes must impair the function of tubulin so severely that they are not retained. This level of sequence conservation may not be surprising, given how closely the functions of tubulin are tied to its structure, in other words, to its biochemical properties.

There is another, somewhat more speculative and philosophical, question: how did mitosis arise evolutionarily? When we look at mitosis now, we see a highly organized, tightly controlled, and incredibly precise and complicated process, sometimes referred to as "the dance of the chromosomes." But remember that we are looking at the current version of mitosis, the one that has survived and been modified during billions of years of evolution, and not at any of the many variations that did not survive or that were refined by evolution and natural selection into what we see. A dancer is not born dancing but must first move with stumbling steps, holding onto furniture, and tumbling to the floor, like any other toddler. When we watch a dancer, we might not recognize the toddler from years gone by trying to work out how to keep their balance and move at the same time. So too with mitosis, for which there has been such strong selection; the earlier versions with steps that were not as organized, controlled, and precise fell to the evolutionary floor long ago.

We do see some remnants of what may have been the origins of parts of the mitotic process, however, at least in some organisms. For example, bacteria have DNA replication and binary fission, so the fundamental process of coupling cell division and DNA duplication that occurs during mitosis is not eukaryotic-specific. Precise binary fission in bacteria requires attaching the newly replicated DNA to the cell membrane, a process that occurs with the nuclear membrane during meiosis, but not in mitosis, as we have described.

We can also see evidence in other organisms for how mitosis may have become so complicated. Dinoflagellates, unicellular organisms prominent in aquatic habitats, are considered to be among the most evolutionarily ancient eukaryotes but lack many typical eukaryotic features such as histone proteins; some investigators do not consider dinoflagellates to be eukaryotes at all, although they do have a nucleus. They also have a different type of cell division that is not exactly like binary fission in bacteria and not exactly like mitosis. The division does not have a true spindle, for example, and the chromosomes are attached to the membrane instead, similar to what happens in bacteria (and in bouquet formation in MI). Dinoflagellates have microtubules that play a role in cell structure, but microtubules seem not to be involved in chromosome segregation; furthermore, the chromosomes of dinoflagellates appear to lack kinetochores. So three cellular features that are important in mitosis in other eukaryotes—the centrosome, the spindle, and the kinetochore—are not used in cell division in dinoflagellates.

But these are not the only aspects of mitosis that appear to have been refined during eukaryotic evolution. In many single-celled eukaryotes, such as yeast cells, which have kinetochores and centrosomes, the nuclear envelope does not break down; this is known as having a closed mitosis, rather than the open mitosis found in multicellular organisms. Nuclear membrane breakdown allows the microtubules to attach to the kinetochores in the form of open mitosis we have described, but this may have been a refinement, rather than part of the original process. (We also should point out that dinoflagellates, yeast cells, and other eukaryotes have all continued to evolve from the shared common ancestor, so none of the current processes is likely to be identical to what occurred in that common ancestor.) In other words, there is clear evidence that the process of mitosis did not all appear at one time, fully organized like the process we observe now; like other evolutionary modifications, it occurred in stages, which were subjected to natural selection. Organisms with variants that did not work as well could not survive once alternatives that allowed a more precise division of cells arose.

One more clue about the evolution of mitosis lies in our description above. Dinoflagellates and many single-celled eukaryotes are flagellated cells, and many cells in our body are likewise flagellated or ciliated. It is not far-fetched to speculate that the common ancestor of all eukaryotes was a flagellated cell. The eukaryotic flagellum is made of microtubules, indistinguishable from the microtubules from which the spindle is composed. The basal body that lies at the base of the flagellum, from which the microtubules grow, is highly similar in organization, function, and protein composition to the centrosome. While dinoflagellates do not have a centrosome or a spindle, they do have basal bodies. It is not hard to imagine that the spindle is derived from flagella and that a flagella-derived spindle became attached to the kinetochore to pull the chromosomes apart. Of course, "not hard to imagine" is far from the evolutionary evidence of what happens; that evidence may lie in the evolutionary laboratory notebook of eukaryotic genomes, and not all of the failed experiments in those genomic laboratory notebooks have been saved.

6.7 Summary: Mendel, meiosis, and molecules—putting things together

One of the challenges in biology is being able to interpret one set of findings in terms of another. The preceding chapters have described three related topics that apply to the same phenomena, but at different scales:

- Mendelian inheritance (Chapter 5), which involves observations at the level of the organism
- Meiosis (Chapter 6), which involves observations at the cellular level
- DNA sequences and the genome (Chapters 2 through 4), which include information at the level of the molecule.

These are integrated and summarized in Figure 6.24, which shows diagrammatically how the topics in these chapters fit together. Let's begin our explanation with the topics that we have just been considering, that is, the segregation of chromosomes during meiosis. A pair of homologous chromosomes from a heterozygote ***A*/*a*** individual is shown. Because each chromosome consists of two sister chromatids, this must be after the S phase has completed but before the M phase has begun—in other words, the G_2 phase of the cell cycle. Note also that the two alleles **A** and ***a*** lie on the different homologs and that the sister chromatids have the same allele.

The lowest arrow pointing in Figure 6.24 shows how the chromosomes at meiosis are connected to the rules of Mendelian segregation. During MI, the homologs segregate to different poles, while the two sister chromatids separate during MII. The four chromatids become four rows in a Punnett square, as shown. However, since the two sister chromatids are the same (either *A* or *a*), we can simplify this to the more familiar Punnett square from Chapter 5 with two rows. (In Chapter 9, where we will return to the genetic effects of crossing over, the two sister chromatids may not carry the same alleles.) These rows are the gametes contributed by one parent to the next generation; the other parent in our Punnett square is not shown.

Now, let's return to our chromosomes at meiosis and trace back the arrows to see how this configuration arose. We will begin at the top of Figure 6.24. Within the nucleus of the cell, there are two closely related, but slightly different, DNA molecules that make up this chromosome (along with the associated proteins that comprise chromatin). In our example, one of those DNA molecules has the sequence GGTCAATACCGA (reading only one of the two strands) while the other DNA molecule has the slightly different sequence GGTCAGTACCGA—that is, the base pair at position #6 (reading from left to right) is an A:T pair in one DNA molecule and a G:C pair in the other. This base pair change could be naturally occurring in the population, or it could have arisen during laboratory investigations, but the sequences are slightly different. Now that we have the complete sequence of many genomes, it might be possible to write out the entire DNA sequence for the chromosome; instead, we summarize these as the *A* allele (with the A:T pair) and the *a* allele (with the G:C pair).

Let's pause to think about the two alleles and the phenotypes here. Recall that a phenotype is any feature that can be measured. The DNA sequence itself can serve as a phenotype if we use an assay that looks directly at the base sequence; we will encounter this again in Chapters 9 and 10. When we are looking directly at the DNA sequence, the phenotype and the genotype are precisely the same. More commonly and conveniently, the phenotypes are the outcomes of these sequence changes as observed in an organism; the sequence changes are the genotypes. The phenotype could be round and wrinkled peas, pointed or blunt beaks in finches, nutritional requirements, coat colors, or blood types in humans, for example. In any case, the underlying DNA sequence alteration is the inherited change that results in the differences in the phenotype, even if we are not looking directly at the DNA sequence itself. (We also want to point out that this example uses a single base pair change in the DNA sequence, but Chapter 3 described many other types of molecular variation in the genome that also serve

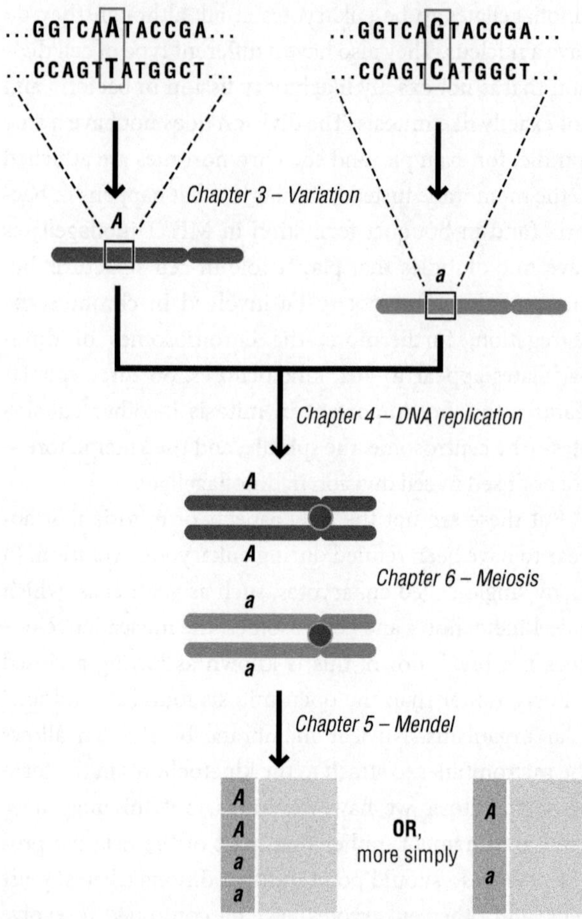

Figure 6.24 The relationships among Mendelian genetics, meiosis, and DNA molecules. The behavior of chromosomes at meiosis can account for the segregation ratios observed by Mendel. This can also be connected to variation in the genome. The figure is explained in detail in the text.

as genotypes and phenotypes.) We have not depicted in this figure how the DNA sequence variation—the genotype—results in different phenotypes, but that topic was covered in Chapter 2.

By continuing to follow the arrow in Figure 6.24 down from the DNA sequences, we encounter DNA replication, as discussed in Chapter 4. (This occurs during the S phase of the cell cycle.) Replication results in two nearly identical copies of each DNA molecule—the sister chromatids. This produces the pair of chromosomes that enter meiosis.

The cellular process of meiosis fully describes Mendelian inheritance. Homologs pair and synapse with each other, cross over and exchange stretches of DNA sequence, and then segregate from one another during MI; at MII, the sister chromatids separate. The haploid products of meiosis, known as spores or gametes, have one of the four chromatids that were paired at prophase I, so the chromosome number and amount of DNA are halved.

The next chapters will describe other aspects of Mendelian inheritance, including genes on the sex chromosomes in Chapter 7, genes on two different chromosomes in Chapter 8, and genes on the same chromosome in Chapters 9 and 10. We will then return to the effect of meiosis and gene transmission in populations and during evolution in Chapter 16. These additional principles of inheritance in the subsequent chapters are not unexpected or novel consequences, or aberrations in meiosis. They simply elaborate some of what has already been seen for meiotic chromosomes in this chapter. Meiosis fully explains Mendel's observations.

CHAPTER CAPSULE

- Prior to cell division—either mitosis or meiosis—the DNA of the cell is replicated during the S phase of the cell cycle.
- Mitosis results in daughter cells that each contains the complete genome of the parent cell.
- During mitosis, the sister chromatids are held together at the centromere, line up on the metaphase plate, and then are separated from each other by the spindle fibers that attach to the kinetochore on each chromatid.
- During meiosis, the genome is duplicated, after which two sequential divisions give rise to cells that contain half of the genetic information found in the original parent cell.
- Crossing over is a highly regulated process that is required for normal segregation of the chromosomes during MI. Crossing over leads to the recombination of maternal and paternal genetic information.
- The behavior of chromosomes during meiosis is the cellular basis of Mendel's Law of Segregation, which states that alleles for each gene separate from each other during the production of gametes.
- Our understanding of Mendelian inheritance can be integrated with the observed movement of chromosomes in meiosis and with our knowledge of the changes in DNA itself during the process of gamete production.

STUDY QUESTIONS

Concepts and Definitions

6.1 Define the following terms as they are used in this chapter.

a. START

b. Centrosome

c. Centromere and kinetochore

d. Cohesion

e. Synapsis

f. Synaptonemal complex

g. SPO11

h. Non-disjunction

6.2 Mitosis and meiosis are both processes of eukaryotic cell division, with the same names used to describe the different stages.

a. What are the significant cellular and chromosomal events that occur during each stage of mitosis?

b. What are the similarities and differences between the events that occur during each stage of mitosis and the corresponding stage of meiosis I?

c. What are the similarities and differences between the events that occur during each stage of mitosis and the corresponding stage of meiosis II?

6.3 What is the most common underlying cause of trisomy 21 in humans?

Beyond the Concepts

6.4 Compare the orientation of the kinetochore during mitosis with the orientations during meiosis I and meiosis II. How does this change in the orientation affect these divisions?

6.5 The dimensions of the synaptonemal complex are similar in different organisms, but the amino acid sequences of proteins that make up the central element (the "rungs of the ladder") are not similar. What does this suggest has been the evolutionarily important feature of the synaptonemal complex?

6.6 What are two important differences between the processes of spermatogenesis and oogenesis in animals?

6.7 Every eukaryote that has been studied has an ortholog of SPO11, but it is very difficult to obtain and maintain a mutation that knocks out the function of SPO11. What do you predict is the phenotype of a *spo11* mutant that will make it so hard to maintain, and why?

6.8 Triploid plants are typically seedless, while tetraploid plants often produce seeds. Explain why this is true.

6.9 Figure 6.24 is one of the most important unifying figures in the book. Explain this figure to a classmate.

Applying the Concepts

6.10 While cleaning out the storage room in the basement of a biology department, you encounter a box of microscope slides. Most of the labels are illegible, but you can read that the slides are taken from ovaries of some animal that has only two pairs of chromosomes ($n = 2$). The slides appear to be stained with some fluorescent dye that allows you to see the chromosomes. Three of the slides are shown in Figure Q6.1. What stage of meiosis is shown in each slide, and how can you tell?

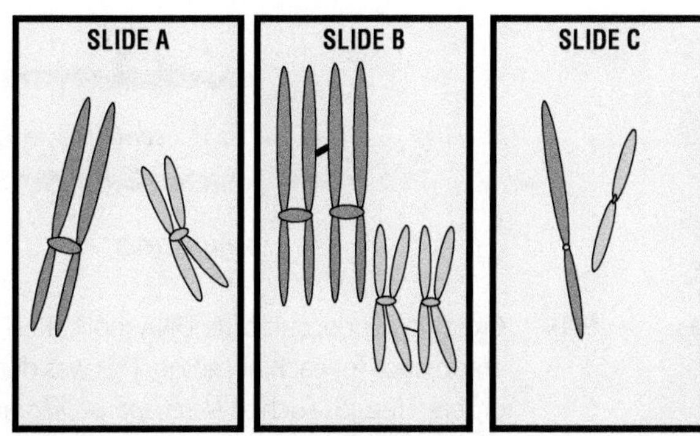

Figure Q6.1

6.11 *Arabidopsis thaliana* is the most widely used flowering plant for genetic and genomic studies. The graph in Figure Q6.2 shows the DNA content per cell at different times points during the life cycle of the plant, beginning with somatic cells of the flower and ending with the single cell seed (after fertilization). Indicate where the stages of premeiosis, meiosis, and fertilization would be found on the X-axis.

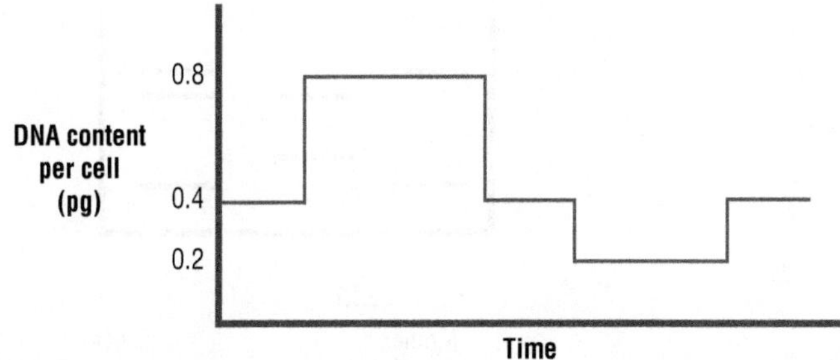

Figure Q6.2

6.12 When chromosomes are examined cytologically, it is often possible to determine if the homologs are paired or unpaired. Paired homologs are called bivalents, while unpaired homologs are called univalents. *Caenorhabditis elegans* has six pairs of chromosomes. Mutants in the gene *him-5* in *C. elegans* affect some process in pairing and synapsis in prophase I, but precisely what process is affected is not clear. Different oocytes from a *him-5* mutant hermaphrodite can have different karyotypes. How can the following karyotypes at the end of prophase I be explained?

a. Six bivalents and no univalents

b. Twelve univalents and no bivalents

c. Five bivalents and two univalents

d. Three bivalents and six univalents

6.13 Tetrad analysis in *Neurospora* is described in Tool Box 6.1. In *Neurospora*, *ser-6* is located to the left of *ad-8* on the left arm of chromosome VI, as shown in Figure Q6.3; the centromere is shown as the circle at the right end, but the right arm of this chromosome is not shown. A heterozygote is set up, as shown, and is allowed to sporulate. Diagram the ascus and the ascospores that will occur under the following circumstances.

a. A crossover occurs between the centromere and *ad-8*

b. A crossover occurs between *ser-6* and *ad-8*

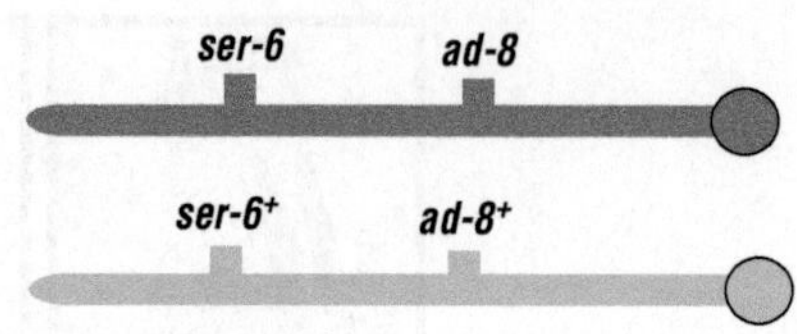

Figure Q6.3

Challenging **6.14** Crossing over occurs after DNA replication has occurred, when there are two sister chromatids for each homolog. This was demonstrated by tetrad analysis in fungi with ordered tetrads such as *Neurospora*. Diagram the expected result if crossing over instead occurred before DNA replication, when only a single chromatid is found for each homolog. From this result, explain the finding that indicated that crossing over occurred after DNA replication occurs.

6.15 A couple has a child with trisomy 21, that is, with three copies of chromosome 21, rather than two. Geneticists analyzed chromosome 21 of both parents and the child in order to understand the origin of this non-disjunction event. Some of the results are shown in Figure Q6.4.

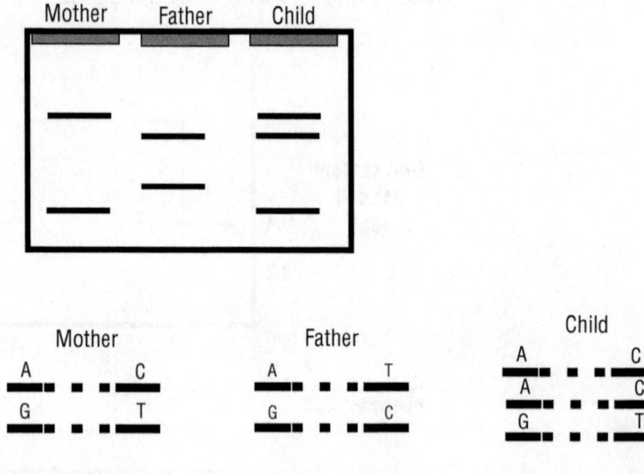

Figure Q6.4

a. The gel at the top of the figure shows a polymorphic locus on chromosome 21 referred to as a copy number variation (CNV). At this locus, different individuals have a different number of copies of a repeated sequence; a higher copy number results in a larger band (closer to the top of the gel) than a low copy number. The locus itself has nothing to do with meiosis; it is simply a marker to track the parents' and child's chromosomes. Based on this gel, which parent experienced the non-disjunction, and how can you tell?

b. The bottom part of the figure shows two different single-nucleotide polymorphisms (SNPs) at different loci on chromosome 21 in each parent and the child. The line is a chromosome from each parent, with the dashed line indicating that the two loci are far apart on chromosome 21. The chromosomes are arranged in the same order as on the gel, so that the top chromosome shown on the gel (in the mother and the child) has the A at the first locus and the C at the second locus. What does this sequence information tell you about the "cause" of the non-disjunction event?

Looking ahead **6.16** *C. elegans* has two sexes—males and hermaphrodites. Hermaphrodites are essentially females that make sperm for a few hours and then shut off spermatogenesis for the remainder of their life. Thus, a hermaphrodite can either reproduce by self-fertilization of its own sperm with its own ova or by cross-fertilization with a male. Hermaphrodites have five pairs of autosomes and a pair of X chromosomes (XX). As described in Chapter 7, males have five pairs of autosomes and a single X chromosome (X0); there is no Y chromosome in nematodes, so males have 11 chromosomes, rather than 12. This mode of reproduction and sex determination has made meiotic mutants particularly easy to identify in *C. elegans*.

a. Most of the offspring of self-fertilization by a hermaphrodite are also hermaphrodites. However, approximately one in 500 offspring of self-fertilization is a male. Explain these results.

b. Mutations that result in non-disjunction in the *C. elegans* hermaphrodite are referred to as Him mutations. "Him" is the phenotypic designation for "high incidence of male progeny." Rather than having one male in 500 offspring, Him mutants have 3% or more male offspring, with the exact percentage depending on the gene and the mutant allele. Explain why mutations that affect meiosis have a Him phenotype.

c. What are some of the biological processes that might be encoded by a *him* gene?

d. Although the Him mutants have a high incidence of male progeny, most of them also lay many eggs that do not hatch. Explain why most Him mutations have so many non-viable offspring.

Looking back **6.17** The woman II-2 shown in the pedigree in Figure Q6.5A is pregnant with her first child. Both her family and her husband have a sister who is affected by an autosomal recessive trait as shown.

a. What is the probability that their child will not be affected by this trait?

b. A geneticist uses RFLP analysis (described in Tool Box 3.1), in combination with a polar body assay, to determine if their child is likely to be affected by this trait. In this analysis, the size of the band on the gel represents a chromosome that has either the "normal" allele or the "disease" allele. The gel is shown in Figure Q6.5B, with each lane of the gel showing the results for one member of her family. The lane labeled "PB" is the analysis of the first polar body. Based on this analysis, what is the probability that their child will not be affected by this trait?

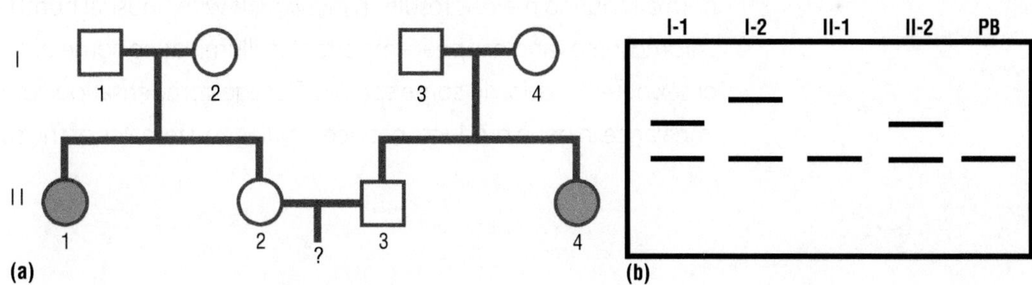

Figure Q6.5

Challenging **6.18** Regions known as pairing centers are thought to occur for all meiotic chromosomes but have only been defined for a few chromosomes in a few species. Pairing center regions in *C. elegans* were identified in part because of the properties of a deletion that removed all or part of the pairing center.

a. What is the expected property when a worm is heterozygous for a normal chromosome and a chromosome that is deleted for the pairing center?

b. *zim-2* is a gene that is thought to be involved in pairing center functions in *C. elegans*. When *zim-2* is deleted, chromosomes I, II, III, IV, and X pair normally, but chromosome V does not pair. List some of the inferences that can be drawn about pairing centers and the functions of *zim-2*.

c. How many univalent and bivalents do you expect to see in a *zim-2* mutant at prophase I?

d. What do you predict will be the phenotype of worms that are homozygous for a *zim-2* mutant?

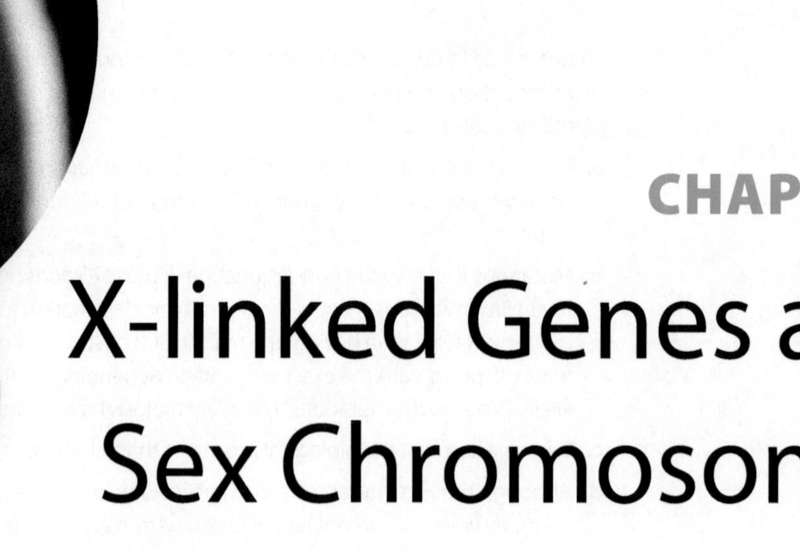

CHAPTER 7

7 X-linked Genes and Sex Chromosomes

IN A NUTSHELL

In many familiar species, the presence or absence of particular chromosomes determines the sex of the individual. These chromosomes are known as the sex chromosomes. One sex is defined by having two different sex chromosomes—we say that it is heterogametic. For example, human males are XY. This has consequences for the inheritance of genes on the X chromosome, such that different patterns of inheritance are observed for X-linked traits in males and females. Non-disjunction of the sex chromosomes or a failure to segregate normally during meiosis results in individuals with unusual numbers of X or Y chromosomes and has been important in illuminating some of the different modes of sex determination in some species. Dosage compensation corrects for potential imbalance in gene products between males and females of the same species.

7.1 Overview: sex linkage and sex determination

Most of the principles of gene inheritance in eukaryotes that we discuss in this chapter and in Chapters 8 and 9 are the logical outcomes of the events that occur during meiosis, described in Chapter 6. Although the processes by which genes give rise to phenotypes and both genes and genomes evolve in populations are not the outcome of meiosis, and bacterial gene inheritance does not involve meiosis, much of the rest of what we will discuss in the next few chapters can be inferred from a careful study of this process.

One of the most apparent phenotypes that we inherit is our biological sex: female or male. Many species exhibit **sexual dimorphism**, that is, physical differences between the sexes in anatomy, physiology, coloration, and behavior. Of course, not all members of one sex look alike, and there is much overlap between the phenotypes of females and males, but, at the level of the chromosomes (and even the genes for the most part), biological sex is often a relatively simple outcome of meiosis and inheritance.

For many animals (and a few plants), sex is determined by the presence or absence of particular chromosomes, known as the **sex chromosomes**. In many animals, including mammals, the sex chromosomes are known as the X and the Y chromosomes. The fact that the two sexes had a different

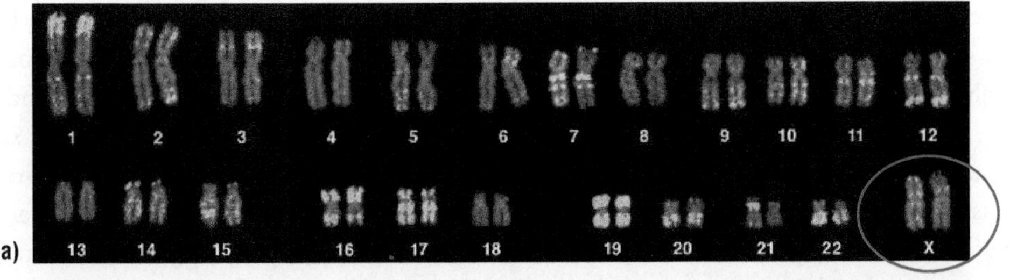

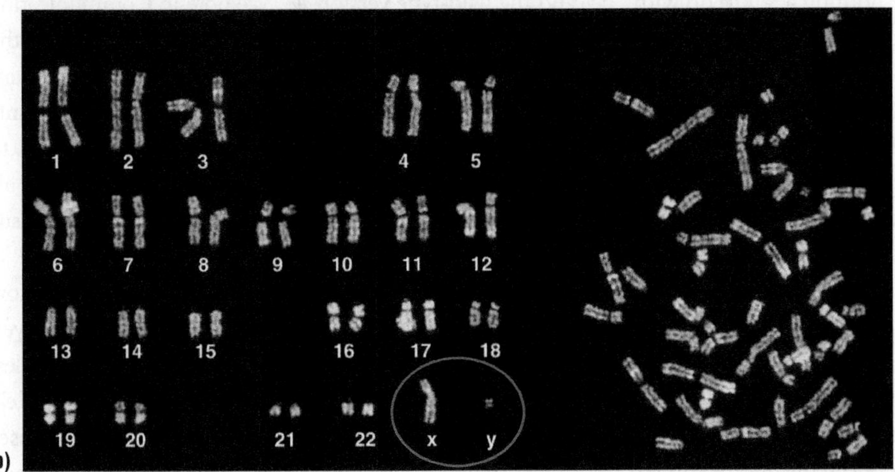

Figure 7.1 The human karyotype, showing the sex chromosomes. These two panels show the chromosomes or the karotype of a human female (a) and male (b). The chromosomes have been fluorescently labeled to make their structures easier to see (different labels are used in (a) and (b)). Note that, for 22 pairs of chromosomes, the autosomes numbered 1 through to 22, males and females have the same karyotype. The 23rd pair is the sex chromosomes, designated either XX or XY. In females, the two sex chromosomes are identical to each other in size, morphology, and staining pattern. In the male, the Y chromosome is much smaller than the X chromosome and has a different morphology and staining pattern.

Source: Part (a) reproduced from Bolzer, A. et al. (2005). Three-Dimensional Maps of All Chromosomes in Human Male Fibroblast Nuclei and Prometaphase Rosettes. *PLOS Biology* 3(5): e157. doi:10.1371/journal.pbio.0030157 © 2005 Bolzer et al. Part (b) CNRI/SCIENCE PHOTO LIBRARY.

chromosome configuration was first recognized more than 125 years ago. The different karyotypes of men and women are shown in Figure 7.1. This difference between the genetic content of the sex chromosomes and the effect of that difference on the inheritance pattern of traits found on the sex chromosomes is the main topic of this chapter.

Erasmus Darwin, Charles' grandfather and a noted naturalist in his own right, referred to sexual reproduction and sexual dimorphism as "the masterpiece of nature," and it is hard to disagree. But it may surprise you to learn that sex determination occurs by many different mechanisms, with very few steps in common. Nature has used many different ways to accomplish this masterpiece. While the XY system of sex determination is the most familiar to us, it is not the only chromosome-based system of sex determination, and several variations are known. In fact, many species have no sex chromosomes at all, and environmental factors often form the basis of sex determination. Nonetheless, most of our attention in this chapter will be on the most familiar system—the X and Y chromosomes.

7.2 Inheritance of the X chromosome and X-linked genes

Let's return to thinking about meiotic chromosomes, and particularly about sex determination in mammals. We said in Chapter 3 that humans have 23 pairs of chromosomes, including one pair of X chromosomes in females. Males have an X and a Y chromosome, which pair and segregate from one another during meiosis, although they are quite different in size and shape, as seen in the karyotypes in Figure 7.1. In addition to the sex chromosome, there are 22 pairs of non-sex chromosomes, referred to as **autosomes**, which are also shown in the karyotypes. Genes on the autosomes are inherited by the patterns developed in Chapter 5.

The Y chromosome in humans is the smallest of the chromosomes, comprising just 59 million base pairs, which makes it about a third of the size of the X chromosome. Unlike the X chromosome, which has thousands of genes on it, most of which have nothing at all to do with sex determination and sex differences, the Y chromosome has very few genes on it—around 200 genes in total. The most important of these (in mammals) is the gene responsible for testis determination.

All mammals and many other animals have XX females and XY males. Since the male has two different sex chromosomes, it is referred to as the **heterogametic** sex; males will make gametes (sperm) with either an X or a Y chromosome. The female with two similar sex chromosomes is called the **homogametic** sex; all of her ova will have an X chromosome. Although the genetic and molecular mechanisms of sex determination are not the same in all animals, the chromosomal mechanisms are similar among species with XX females and XY males. In almost all cases, the Y chromosome has few genes on it. We will talk about the mechanism for mammalian sex determination later, but, for now, let's focus on the inheritance of the X chromosome with its thousands of genes and the Y chromosome with very few genes.

Drosophila males are also XY. Again, the X chromosome has many genes on it—called **X-linked** or **sex-linked** genes—but only a few of them are related to sex determination itself. (The reason that "sex-linked" implies "X-linked" in XX/XY species is that the Y has almost no genes on it.) This description of the inheritance of X-linked genes takes us back to the foundational history of genetics, and a paper published in the journal *Science* in 1910 that studied the inheritance of one of the genes on the X chromosome. While we will describe the inheritance of one gene—the ***white*** gene—the inheritance principles derived for this gene are true for any X-linked gene.

KEY POINT Genes located on the X chromosome are called X-linked genes, while genes located on the other chromosomes,—the autosomes—are referred to as autosomal.

The X chromosome

Thomas Hunt Morgan was an embryologist working at Columbia University in New York. He began working with the fruit fly *Drosophila melanogaster* because it grew so easily in the laboratory and he believed—correctly as it turned out many decades later—that it would be useful for understanding embryology. Fruit flies normally have brick-red eyes, so the wild-type phenotype is "red eyes." One of the very first mutants, discovered in 1908 by Morgan and his wife, Lilian, was a male fruit fly with white eyes. The gene for this trait is known as *white*, with the mutant version abbreviated ***w*** and the wild-type version $\boldsymbol{w}^{+}$. We note in Communicating Genetics 7.1 that geneticists often name genes based on the mutant phenotype, and the *white* gene is one of the primary examples. The gene is named *white* because the mutants have white eyes, but the normal function of the gene is to produce brick-red or wild-type eyes. By definition, the alleles present in the wild-type fly are all denoted with a superscript +, with the mutant version as the gene name.

As diagrammed in the crosses in Figure 7.2(a), the Morgans mated this white-eyed male to a wild-type red-eyed female. All of the F_1 offspring, both males and females, had the wild-type red eyes. This told them that the ***w*** allele is recessive to $\boldsymbol{w}^{+}$. So far, this looks exactly like the crosses done in Chapter 5. But the subsequent crosses, diagrammed in Figures 7.3 through 7.5, gave different results.

Let's write out the initial cross that the Morgans did, using a Punnett square, as shown in Figure 7.2(b). Because the *white* gene is on the X chromosome, which was not known when these experiments were begun, we are going to use a notation for X-linked genes to make this explanation easier to follow. We will write X^{w+} to indicate an X chromosome carrying the wild-type red eye allele and will write the mutant white eye allele as X^{w}.

Notice that the white-eyed male has the genotype X^{w}/Y. The male has only one X chromosome, and the Y chromosome has no corresponding gene, as is true for nearly all X-linked genes. Thus, every allele on the X chromosome appears in the phenotype in males, regardless of whether it is dominant or recessive. The male is said to **hemizygous** for genes on the X chromosome; in a hemizygote, only one copy of the gene is present, so that the allele is seen in the phenotype. The female parent is wild-type with red eyes, and we will note that it is X^{w+}/X^{w+}. The use of a "/" to separate the two chromosomes in a genotype is optional but is sometimes helpful for clarity.

All of the ova produced by the female have an X chromosome with the $\boldsymbol{w}^{+}$ allele. By contrast, half of the males' sperm have an X chromosome with the ***w*** allele and half have the Y chromosome, as indicated in Figure 7.2(b). All of the F_1 **male** offspring get a Y chromosome from the father (or else they would not be males) and an X

COMMUNICATING GENETICS 7.1 Gene names

As we noted in Communicating Genetics Box 1.1, no other topic in genetics has more potential for confusion than gene names and allele designations. The potential for confusion arises from many sources. First, different organisms have different systems of nomenclature. In *Escherichia coli*, gene names are three lowercase letters, followed by a capital letter, such as *lacZ* or *recA*. Gene names in *Drosophila* consist of one or more uppercase or lowercase letters and rarely involve a number (e.g. *w*, *ptc*, *Antp*), whereas gene names in *Caenorhabditis elegans* are three lowercase letters long and always include a number (e.g. *tra-1*, *dpy-10*). In *Arabidopsis*, gene names are usually three lowercase letters but can vary from two to six letters. They may include a number or even two numbers (such as *bri1-1*) in which the second number refers to the allele.

Second, many of the nomenclature systems are products of a long history of research in the organism. A gene name that made sense to an investigator working on a particular mutant allele years ago may be quite obscure to a modern investigator. For example, the *Notch* gene in *Drosophila* encodes a receptor involved in cell–cell signaling; orthologous genes are found in nearly all animals. The gene name arises because the first mutation resulted in notches in the wing margin, a relatively insignificant defect in light of all the other roles of this gene. However, the gene name indicates its history.

Third, a gene name reflects a particular viewpoint on the nature of genes. This last point will provide our brief guide to how genes are named.

Names based on mutant phenotype

In an era of genetics that predated genome projects, or even molecular biology, genes were recognized solely by mutant alleles and their phenotypes. This began with one of the first genes to be named, the *white* gene in *Drosophila*, discussed in the chapter. The gene was called *white* because flies with a mutation in the gene had white eyes. Thus, the gene was named for its mutant phenotype—in effect for what happens when the normal function is missing or altered.

This method of naming genes is contrary to the way everyday objects are named. For example, when I am driving and want my car to stop, I press a particular pedal on the floor and the car slows down. The normal function of this pedal and its associated parts is to brake the car's momentum, and we call the pedal the brake after its normal function. However, with genes, especially in the earlier days of genetics, the normal function was usually unknown; the best clue to the normal function was to see what phenotype resulted when the function was disrupted. Thus, when this particular gene was disrupted, it was observed that the fly had white eyes. The normal function of the *white* eye gene is not to make white eyes—it is to make wild-type red eyes. However, we name a gene for what happens when the gene malfunctions, not what happens when the gene functions normally, so the gene is called the *white* gene. If we were to use a similar system to describe the brake pedal, we would name it for what happens when it is non-functional; thus, it would be called "no stopping," or some other name indicating that the car fails to slow down and stop.

Naming genes by mutant phenotypes also occurs with human genes. Sometimes we speak of "disease genes" when we actually mean that a known disease arises when the gene is **mutated**. The cystic fibrosis gene (called *CFTR*) does not cause cystic fibrosis. It is a **mutation** in the *CFTR* gene that results in cystic fibrosis—again, the gene is named for its mutant phenotype.

This type of shorthand confusion is rampant, particularly when the news media report that a gene for some behavioral trait, such as schizophrenia, has been identified. It must be recalled that the gene is being named (or nicknamed) for the phenotype that arises when the gene is altered. The normal function of the gene is not to cause schizophrenia any more than the normal function of the *white* gene is make white eyes. However, we may not know the normal function of the gene. We only know what happens when the gene is altered, so the gene is being named for its mutant phenotype.

Names based on gene product

The potential for confusion becomes exacerbated because not all genes are recognized first by their mutant phenotype; this is especially true, since biochemical and molecular genetics and genomics are used, in addition to traditional mutant analysis, to identify genes. In humans, the *β-globin* gene is very well studied. Notice how the gene is named—for its polypeptide product, and not for the diseases that arise from mutations. Hundreds of naturally occurring mutations in the *β-globin* locus are known in humans, which result in a group of diseases called the β-thalassemias, including sickle-cell anemia. But the gene is not called the thalassemia gene or *thal-1* or something similar. It is named instead for its normal function and is called the *β-globin* gene. The same is true for many genes that encode metabolic enzymes or familiar proteins such as β-tubulin or collagen.

Names based on homology

Another source of confusion arises from the fact that many genes are evolutionarily conserved and have orthologs in other species. For example, the cyclin B protein regulates the cell cycle universally among eukaryotes, but the orthologous genes in different organisms often have different names. In *Schizosaccharomyces pombe*, the cyclin B protein is encoded by a gene called *cdc2* where *cdc* stands for cell division cycle, the mutant phenotype that is defective when the gene is altered. In *Saccharomyces cerevisiae*, the cyclin B protein is encoded by a gene called *cdc28*.

COMMUNICATING GENETICS 7.1 Continued

Although the proteins are similar enough that one can substitute for the other, the gene names are different because the relationship between the two genes was not known until well after the mutations had been identified and characterized. To make matters even more confusing, *S. cerevisiae* does have a gene named *cdc2*, but it encodes a different protein, and not cyclin B. As cyclin B genes have been found in other organisms, they are often called *cdc2* genes, although not usually because of the mutant phenotype in that organism.

A project called HomoloGene, administered by the National Center for Biotechnology Information (NCBI) and found on their website, attempts to identify homologóus genes in different organisms and organize them together in one place. It also allows the investigator to learn what an orthologous gene is called in other organisms and provides links to the appropriate databases. Although the confusion over gene names probably cannot be completely eliminated, HomoloGene can be a useful tool for what orthologous genes are named in different organisms.

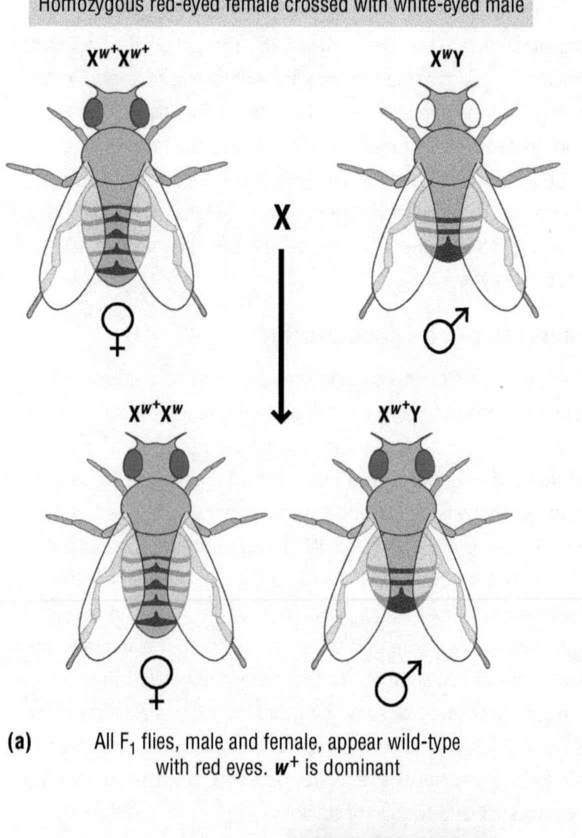
Homozygous red-eyed female crossed with white-eyed male
$X^{w^+}X^{w^+}$
$X^{w}Y$
X
$X^{w^+}X^{w}$
$X^{w^+}Y$
(a)
All F_1 flies, male and female, appear wild-type with red eyes. w^+ is dominant

(b)

		Male gametes	
		X^{w}	Y
Female gametes	X^{w^+}	$X^{w^+}X^{w}$	$X^{w^+}Y$
	X^{w^+}	$X^{w^+}X^{w}$	$X^{w^+}Y$

Figure 7.2 The inheritance of *white* (*w*) eyes in *Drosophila*. (a) A spontaneously arisen white-eyed male fly was mated to a wild-type red-eyed female fly. All of the offspring have red eyes, as expected if the trait is recessive. (b) A Punnett square that diagrams the results. For clarity with what lies ahead, we have written this with *w* and w^+ on the X chromosome, although this cross does not demonstrate this X-linkage yet.

chromosome from the mother, which has the w^+ allele. As a result, all of the male offspring have red eyes. All of the F_1 **female** offspring get the X chromosome with the ***w*** allele from the father and an X chromosome with the w^+ allele from the mother. So the F_1 females are all heterozygotes with X^{w+}/X^{w}. These have red eyes because the wild-type red eye allele is dominant to the ***w*** allele.

Now, let's see what happens when we mate that F_1 female to the F_1 male. This is diagrammed in Figure 7.3. The male parent in this cross has the genotype X^{w+}/Y. Half of its sperm have an X chromosome with the w^+ allele, while the other half of the sperm have the Y chromosome. The female parent has the genotype X^{w+}/X^{w}; it is heterozygous on the X chromosome, with one copy of the wild-type allele and one of the white mutant allele. Consequently, half of its ova will have the w^+ allele, and half will have the ***w*** allele.

In the Punnett square in Figure 7.3(b), there are four genotypic categories of offspring, each occurring as one-fourth of the offspring:

- X^{w+}/Y (red-eyed males)
- X^{w}/Y (white-eyed males)
- X^{w+}/X^{w} (red-eyed females)
- X^{w+}/X^{w+} (red-eyed females)

All female offspring have red eyes because they received the w^+ allele from the father; whichever X chromosome allele is inherited from the mother is masked by the dominant allele from the father's X chromosome. The male offspring reveal the genotype of the mother—half of them are white-eyed, and half are red-eyed.

We will not continue this description, but you should trace the chromosomes in the crosses shown in Figures 7.4 and 7.5. Notice the characteristic inheritance pattern

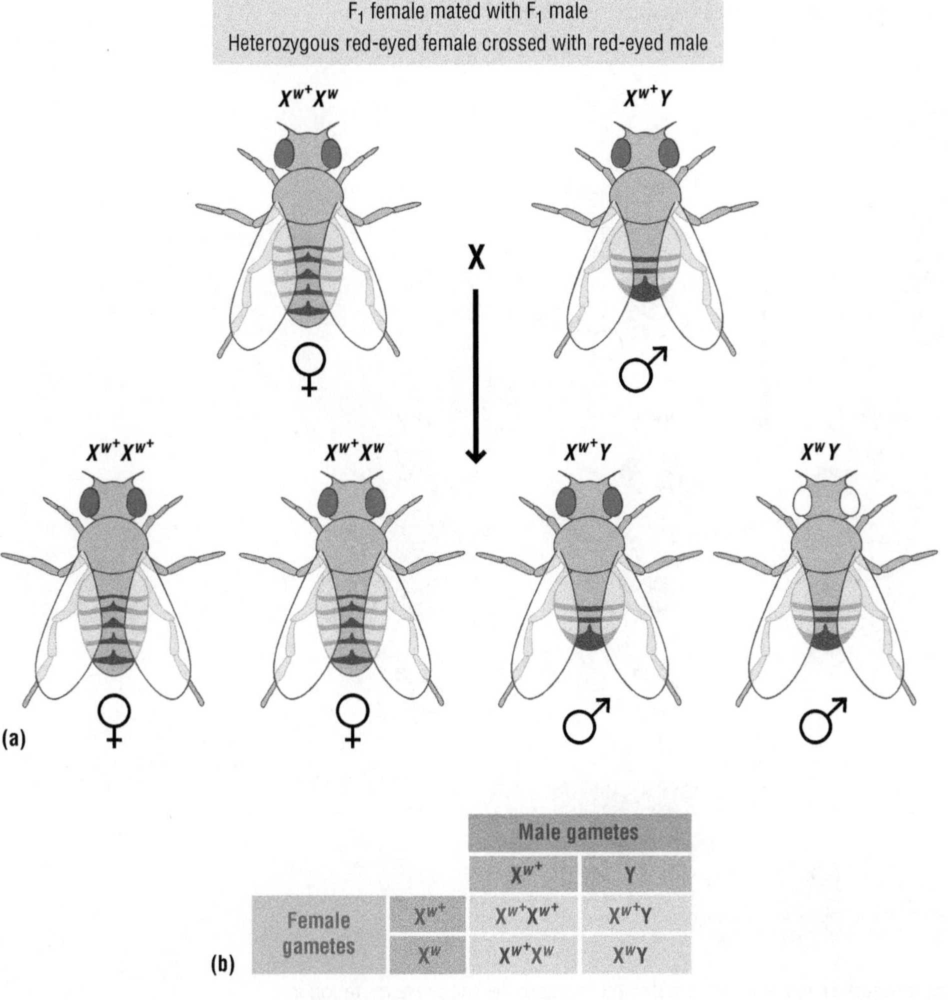

		Male gametes	
		X^{w+}	Y
Female gametes	X^{w+}	$X^{w+}X^{w+}$	$X^{w+}Y$
	X^{w}	$X^{w+}X^{w}$	$X^{w}Y$

Figure 7.3 The *white* gene is X-linked. Heterozygous F$_1$ females from the cross in Figure 7.2 are mated to red-eyed F$_1$ males, as shown in (a). All of F$_2$ females have red eyes, while half of the F$_2$ males have red eyes (X^{w+}/Y) and half have white eyes (X^{w}/Y). These results are different from those expected for a typical autosomal recessive gene. The genotypes of the F$_2$ flies are listed in the Punnett square in (b).

by which the recessive phenotype appears in the male offspring of unaffected, but heterozygous, females.

WEBLINK: Video 7.1 illustrates patterns of inheritance observed with X-linked genes. Find it at **www.oup.com/uk/meneely**

KEY POINT In organisms with XX/XY sex chromosomes, the patterns of inheritance for phenotypes from X-linked genes differ between male and female offspring.

The X-linked inheritance of traits encoded by genes on the X chromosome was one of the key pieces of information that showed that Mendel's rules could be explained simply by meiosis, thus anchoring the gene concept in its cellular context and forging the connection that was made in Chapter 6. In organisms with an X and a Y chromosome or a similar system in which the male is heterogametic (which includes mammals, as well as the model organisms fruit flies and *Caenorhabditis elegans*), any X-linked gene shows this pattern of inheritance, which can be worked out simply by following the inheritance of the sex chromosomes. The key indicator that a gene is X-linked is that males and females show different patterns of inheritance; typically, a recessive phenotype will appear in males when it does not show up in females.

Rather than continue to go through more cases that illustrate these same principles, the Study Questions at the end of the chapter emphasize the inheritance patterns seen with X-linked genes. Exercises like these questions are by far the best ways to understand this topic and to test your understanding.

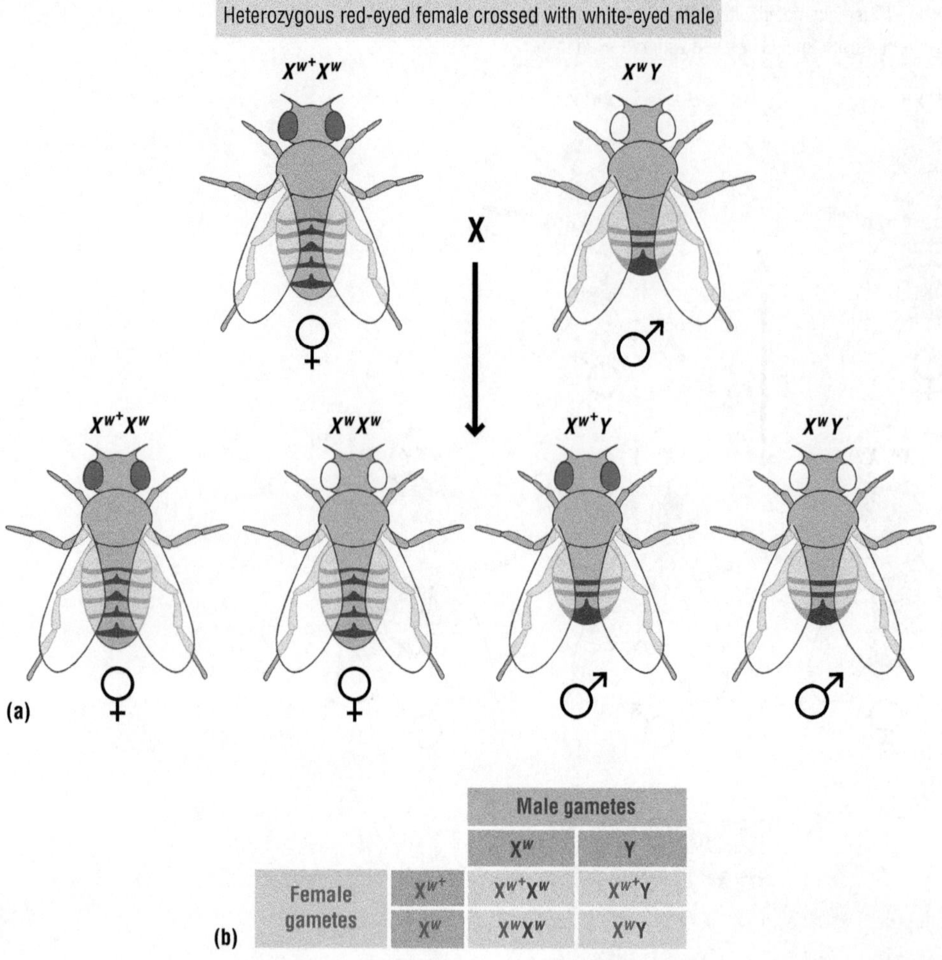

		Male gametes	
		X^w	Y
Female gametes	X^{w^+}	$X^{w^+}X^w$	$X^{w^+}Y$
	X^w	X^wX^w	X^wY

Figure 7.4 White eyes can appear in either sex. When a female heterozygous for the *white* mutation is mated with a *white* male, half of the males are white-eyed and half of the females are also white-eyed, as illustrated in (a). This result illustrates that white eyes can occur in female flies if they are homozygous for the recessive ***w*** allele, so nothing about the *white* gene itself depends on sex. The genotypes of the F_2 flies are listed in the Punnett square in (b).

Some X-linked genes in humans

Hundreds of genes are found on the X chromosome in mammals, almost none of which affects conditions related to sex. A few of these may be familiar to you because of the phenotypes that arise from them in humans, but there are many others with a similar pattern of inheritance. Among the genes on the human X chromosome are the ones that encode light receptor proteins known as opsins; individuals with non-functional versions of one of these genes are red–green color-blind. Because the opsin genes are X-linked, color blindness is more common in males than in females, since males need only one copy of the mutant gene to have this phenotypic trait. The opsin genes are a gene family, as defined in Chapter 3 as genes with closely related functions. In the case of the X-linked opsin genes, the opsin responsible for red light detection and the opsin responsible for green light detection are immediately adjacent to each other on the chromosome, and most cases of red–green color blindness affect the green light receptor. We will discuss the origin of red–green color blindness in Box 9.2. Red–green color blindness is not the only familiar X-linked trait in humans. Two of the most common forms of muscular dystrophy—Duchenne and Becker—are the result of different mutations in the X-linked gene dystrophin, a major protein in muscle; as expected from being recessive conditions on the X chromosome, these diseases appear much more commonly in boys than in girls.

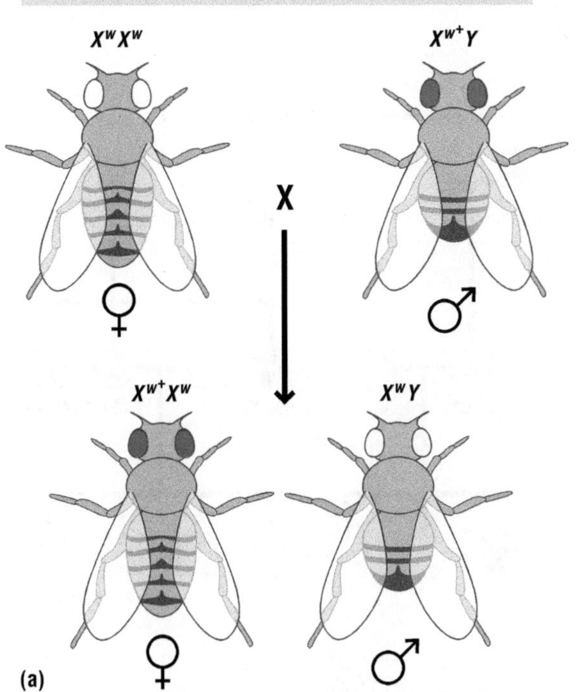

(b)

		Male gametes	
		X^{w+}	Y
Female gametes	X^w	X^wX^{w+}	X^wY
	X^w	X^wX^{w+}	X^wY

Figure 7.5 X-linkage and "criss-cross inheritance." When a white-eyed female is mated with a wild-type male, all of the female offspring have red eyes because they get one X chromosome from the male, while all of the male offspring have white eyes because they get their only X chromosome from the female. These crosses demonstrated that males are the heterogametic sex and that white eyes is an X-linked recessive trait. The term "criss-cross" is sometimes used to describe this pattern in which the trait appears to be passed from mother to son; this simply reflects the inheritance of the X chromosome.

Two more X-linked genes encode the blood clotting proteins factor VIII and factor IX. The most common forms of hemophilia are due to mutations in one of the genes that produce either factor VIII (in hemophilia A) or factor IX (in hemophilia B). These mutations result in slow or reduced blood clotting following an injury. Hemophilia is found in about one out of every 5000 live male births, and hemophilia A accounts for about 80% of cases. As shown in Figure 7.6, Queen Victoria, the queen of Great Britain from 1837 to 1901, was a carrier of the gene leading to hemophilia—at least two of her daughters were also carriers, and one of her sons had the disease. The effect of this genetic disorder on history in the twentieth century, particularly in Russia where the young heir Alexis, Victoria's grandson, was affected, has been the subject of many books and movies. Communicating Genetics Box 7.2 addresses pedigree analysis for an X-linked trait, of which hemophilia is an example. Note the inheritance pattern in which an unaffected female has affected male offspring.

KEY POINT Two well-known X-linked conditions in humans are red–green color blindness and hemophilia, recessive traits that occur more often in males than in females, since males have only one X chromosome.

The Y chromosome

It is generally believed that the X and the Y chromosomes arose from a pair of homologous chromosomes that originally—in evolutionary terms—had the same genes. As with any other pair of homologs, the X chromosome and the Y chromosome pair at meiosis, although this pairing in mammals is limited to a specific region of the two chromosomes and does not extend to synapsis along the entire length like other chromosome pairs. (In females, the two X chromosomes pair, synapse, and undergo recombination just like any pair of autosomes.)

The Y chromosome has many residual and non-functional sequences that are similar to the sequences of functional genes found on the X chromosome, but it has very few intact genes. The Y chromosome is a major repository of apparently non-functional sequences in the genome, such as degenerated viruses and transposons, as discussed in Chapter 3. The very few genes that are on the Y chromosome in mammals—the most important of which is a gene that determines testis formation—show a distinctive inheritance pattern and are transmitted strictly from father to son. This inheritance is known as Y-linked or **holandric**. Other than testis formation itself, none of these Y-linked genes encodes a familiar or easily observed trait. In addition to these few traits, sequence differences on the Y chromosome have been very useful in tracking migration during human history and determining parentage, as will be discussed later in Box 16.3.

Traits that are limited to one sex or the other, or that are more common in one sex or the other, definitely exist. These are known as sex-limited traits if they appear only in one sex, or sex-influenced traits if they appear more commonly in one sex than the other, but the genes that affect them are encoded on the autosomes. Y-linked, sex-limited, and sex-influenced traits are discussed in Box 7.1.

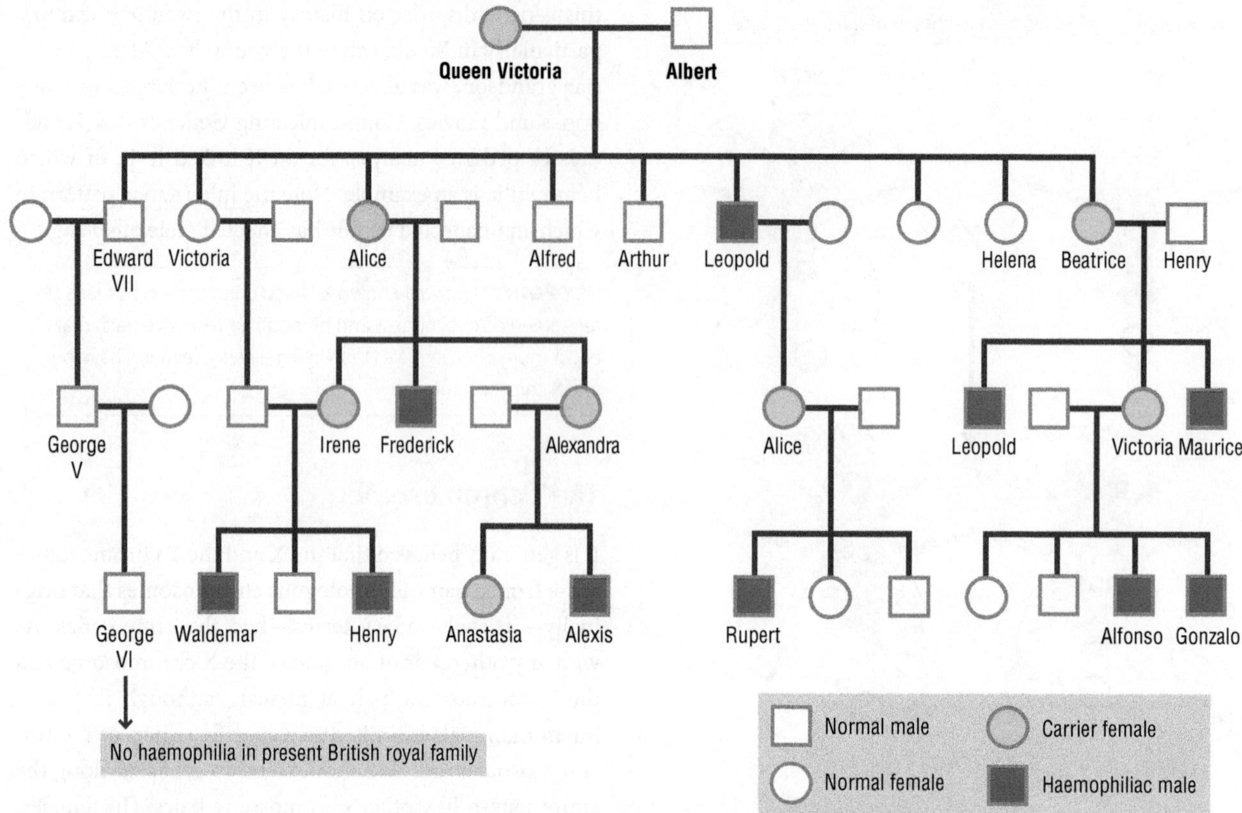

Figure 7.6 The pedigree for Queen Victoria and her offspring. Queen Victoria and Prince Albert of England had nine children, many of whom became members of the other royal families of Europe in the late nineteenth and early twentieth centuries. Queen Victoria's son Leopold, Duke of Albany, had hemophilia; this was the first occurrence of hemophilia in the royal family, indicating that Queen Victoria was a heterozygote and suggesting the mutation arose in her father, Prince Edward. Two of her daughters, Alice and Beatrice, must have been heterozygotes because they had affected sons.

Source: © John W. Kimball/ (CC BY 3.0) and made possible by funding from The Saylor Foundation (www.saylor.org).

COMMUNICATING GENETICS 7.2 Pedigree analysis for X-linked genes

In Communicating Genetics 5.1, we introduced pedigree analysis as an approach for analyzing modes of inheritance. Pedigree analysis is often used when the number of offspring is low but multiple generations can be followed. Thus, it is particularly used in studies of humans and domesticated animals for which extensive records are available. The examples used in Chapter 5 involved autosomal genes, and the key was to determine if the trait was recessive or dominant.

Pedigree analysis can be helpful to quickly illuminate the mode of inheritance for sex-linked genes as well. We will assume that the trait is recessive, as nearly all X-linked traits are. Look at the pedigree in Figure A, and compare the number of affected males and females. You can see that the trait is found much more often in males; in fact, in this case, the trait is found only in males. If a trait shows up in males but rarely in females, particularly if it shows up in a son from an unaffected mother, the trait is most likely to be X-linked and recessive. We can then trace this back through the generations to be sure.

When approaching a pedigree and attempting to understand the pattern of inheritance for a trait, a few simple and quick steps are helpful.

- Are there examples in which unaffected parents have an affected offspring? If so, the trait is recessive. If not, and all affected individuals have an affected parent, then the trait is inherited as a dominant trait.
- Are males affected much more often than females? Although this by itself does not prove that the trait is X-linked, since the number of offspring is small, this is usually an indication that the trait is X-linked. This can be confirmed by working back through the pedigree. If a female is affected, she must have had an affected father.

COMMUNICATING GENETICS 7.2 Continued

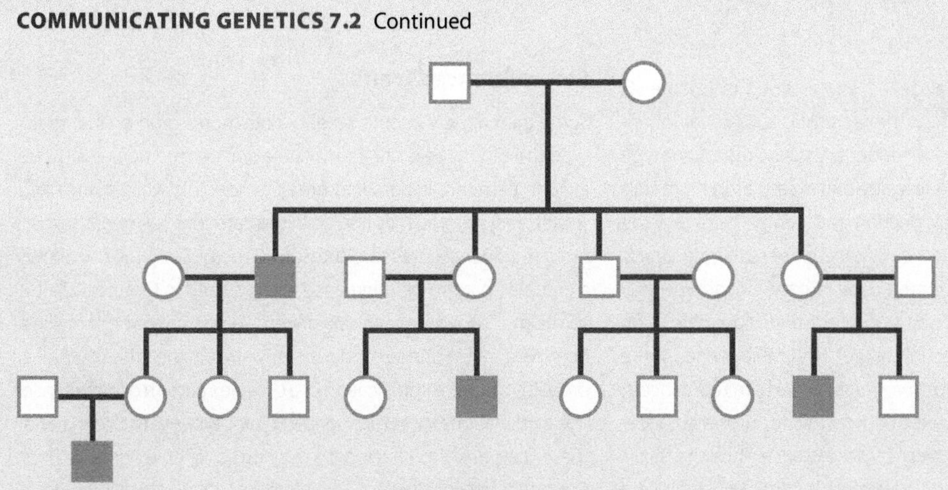

Figure A A pedigree for an X-linked trait. X-linked traits show up much more frequently in males, as shown here; affected individuals are blue. The woman in the first generation must be a heterozygote for this trait, since she has an affected son. Furthermore, this affected son has an affected grandson, although his daughter was not affected. The woman in the first generation also has two daughters who must also be heterozygotes, since they have affected sons as well.

BOX 7.1 *Going Deeper* Sex-limited and sex-influenced traits

Some traits have differences in the phenotypes between males and females, or the frequencies at which the traits are seen differ between males and females. This can occur even if the genes are not on the X chromosome. In this box, we briefly discuss some ways that a trait may show up differently in the two sexes, even if the gene is not X-linked.

Y-linked or holandric traits

While it is convenient, and reasonably safe, to ignore the possibility that a gene might be on the Y chromosome when considering inheritance patterns, there are a few examples of Y-linked traits. In some older literature, these were known as holandric traits, but they are more typically termed Y-linked. The most obvious and important example in mammals is the gene for testis determination itself, *Sry*, which encodes the gene for testis-determining factor, TDF. A few other examples are known in humans, including *AZF1*, *AZF2*, and DAZ1, which are known to be important for spermatogenesis. Thus, the phenotypes (and the genes) are only found in males, and the biological process occurs only in males.

However, not all genes on the Y chromosome are unique to this chromosome. Several others, including *ASMT*, *AMGL*, and *IL3RA*, are genes that have both an X-linked copy and a Y-linked copy. The X-linked copy is usually excluded from X inactivation, consistent with the gene product being needed in two doses in each sex. Usually, the phenotype for these genes is not sex-dependent.

Genes present on both the X and the Y chromosome are known as **pseudoautosomal genes**, since their inheritance pattern looks like an autosomal gene in a pedigree, but the gene actually resides on both the X and Y chromosomes. These genes lie close together in a region that is involved in the meiotic pairing between the X and the Y chromosome. The pseudoautosomal region proved to be important in identifying the gene for the TDF, as described in Box 7.3. These genes support the hypothesis that the X and Y chromosomes were originally a pair of homologs, as developed in Box 7.4. In *D. melanogaster*, the ribosomal RNA genes (encoded by the *bobbed* locus) are located on both the X and the Y chromosomes and also show pseudoautosomal inheritance in a cross.

Sex-limited traits

Some autosomal genes affect biological processes or organs that are found in only one sex. Thus, while these are inherited in the

BOX 7.1 Continued

pattern typical of autosomal genes, as described in Chapter 5, their phenotypes do not show characteristic autosomal inheritance because they are sex-dependent. Usually, the phenotype itself is sexually dimorphic. A few obvious examples in mammals are genes that affect milk production and genes that affect the risk for ovarian, prostate, or testicular cancer. For sex-limited traits, both sexes have two alleles for the gene, but the phenotype may simply not be expressed in one sex. Interestingly, genome-wide association studies, described in Chapter 10, have identified some genes that affect the risk for two different sex-limited cancers; some of the variants identified as risk factors for prostate cancer are also risk factors for ovarian cancer, for instance. Thus, the trait is sex-limited because of the organ affected, but the cellular process involved is not sex-limited, and the genes are clearly expressed in both sexes.

Sex-influenced traits

Some traits appear in both sexes but are more common in one sex than the other. These are known as sex-influenced traits. Examples include pattern baldness and breast cancer risk, while autism and Tourette's syndrome may also be in this category. The genes that affect sex-influenced traits are autosomal, and typically the trait shows a complex pattern of inheritance (as discussed in Chapter 10). It is sometimes thought that a sex-influenced trait might behave like a dominant trait in one sex and a recessive trait in the other, but this is probably an over-simplification of a complex pattern of inheritance. The explanations for sex-influenced phenotypes vary, but many of these traits are also affected by the hormonal environment. Thus, the genes responsible for sex-influenced traits are expressed in both sexes, but, due to some other factor, such as testosterone or estrogen levels, the phenotypes differ in the two sexes.

7.3 X chromosome meiotic errors

The inheritance pattern of the *white* gene in *Drosophila* convinced Morgan that Mendel's rules could be explained by meiosis, but this was not the only line of evidence. Further evidence came from the occurrence of meiotic errors, known as **non-disjunction** events. We discussed non-disjunction in Section 6.5; it results in one meiotic product—the egg or sperm—having an extra chromosome or missing a chromosome, a condition known as aneuploidy. Since one of the gametes has an extra chromosome or is missing a chromosome, an offspring will be either trisomic or monosomic when the aneuploid gamete is fertilized by a euploid gamete from the other parent. This section will take us back to thinking about meiosis and highlight how the process of sex determination was clarified by observing the results of non-disjunction of the sex chromosomes.

Non-disjunction and sex determination

Note that XX females and XY males actually have two differences in their sex chromosome constitutions. One sex has a Y chromosome, while the other sex has no Y chromosome. One sex has one X chromosome, while the other sex has two. (As will be discussed in Section 7.4, males in *C. elegans* have only a single X chromosome and no Y chromosome.) In normal meiosis and sex determination, these two differences go together—the animal with one X chromosome also is the one with the Y chromosome, and the one with two X chromosomes is also the one with no Y chromosome. But which of these two differences, the presence of a Y chromosome or the number of X chromosomes, is responsible for determining the sex?

The answer in *Drosophila* (and other organisms) was deduced from an analysis of X chromosome non-disjunction in flies, in which the normal configuration of sex chromosomes is altered. In particular, non-disjunction in one parent (the female in the crosses in Figures 7.7 through 7.9) resulted in XXY flies and X0 flies (in which "X0" means that there is a single X chromosome and no other chromosome). Let's follow this in more detail for a deeper understanding of both meiosis and sex determination.

We wrote in Chapter 6 that non-disjunction can occur in either sex, for any chromosome, and at either meiotic division. The outcomes are different when non-disjunction occurs at meiosis I or at meiosis II. This can be seen most clearly if we follow an X-linked trait such as white eyes, as can be seen here.

For example, imagine that non-disjunction occurs in a white-eyed X^w/X^w *Drosophila* female during the first

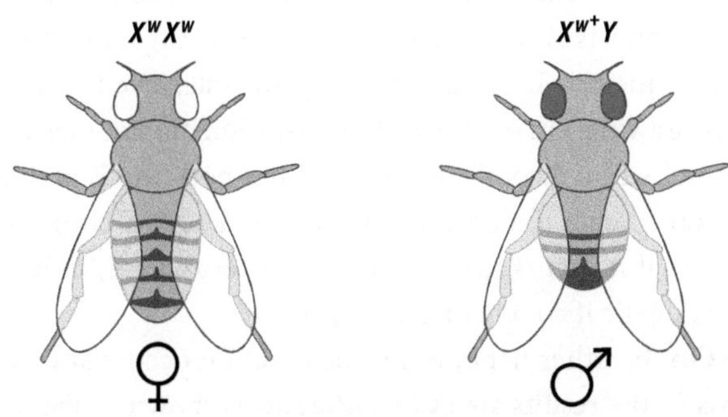

Non-disjunction during meiosis I in homozygous X^wX^w female

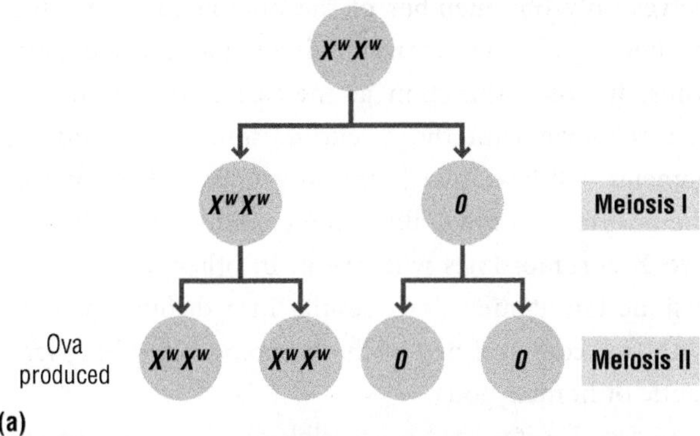

		Male gametes	
		X^{w+}	Y
Female gametes	X^wX^w	$X^wX^wX^{w+}$ Red-eyed female	X^wX^wY White-eyed female
	0	$X^{w+}0$ Red-eyed male, sterile	0Y Dies

(b)

Figure 7.7 X chromosome non-disjunction and sex determination. (a) Non-disjunction occurs during meiosis I in a *Drosophila* female with white eyes. This produces an ovum with two X chromosomes, both with the ***w*** allele, and an ovum with no X chromosome. Meiosis II proceeds normally. (b) A Punnett square showing the results when these ova are fertilized by a sperm from a wild-type red-eyed male. This XX ovum can be fertilized by a sperm with an X chromosome, resulting in an XXX female fly with red eyes, having received the w^+ allele from the father. More significantly, note that, when the ovum with two X chromosomes is fertilized by a sperm with a Y chromosome, it produces an XXY fly, which has white eyes. This fly proved to be a white-eyed female, which can only arise if sex determination depends on the number of X chromosomes, and not on the presence of the Y. If the ovum with no X chromosome is fertilized by a sperm with an X chromosome, the fly will have red eyes. It is also a male whose X chromosome came from the father. (The other offspring, which is Y0, dies because it lacks all of the functions from genes encoded on the X chromosome.) The appearance of white-eyed females (XXY) and red-eyed males (X0) indicates that sex determination in *Drosophila* depends on the number of X chromosomes, and not on the presence of a Y chromosome.

meiotic division. This situation is illustrated in Figure 7.7. As shown in Figure 7.7(a), the female will make a few ova that are XX disomics and a few ova that are 0 nullisomics with the full set of autosomes but no X chromosome. Most of her ova will still have one X chromosome, so these nullisomic and disomic gametes are rare events. When these rare ova are fertilized by a normal sperm from a red-eyed male, trisomic or monosomic offspring result, as illustrated by the Punnett square in Figure 7.7(b).

We can follow the X chromosomes in this example because of the presence or absence of the w^+ allele, which determines the eye color of the offspring. Most of the offspring of this cross will be the normal red-eyed females (X^+/X^w) and white-eyed males (X^w/Y) expected from this cross and are not shown in the figure. The white-eyed female also occasionally (but rarely) produces an X^w/X^w gamete—both X chromosomes have the recessive ***w*** allele. If this ovum is fertilized by a Y-bearing sperm from a red-eyed male, it produces an XXY fly. Because it has only the ***w*** allele and no wild-type w^+ allele, we know it has white eyes.

But what is the sex of this XXY fly? When this experiment was done in *Drosophila*, this XXY fly was found to be a white-eyed female. In fact, this is the only way to explain the occurrence of a white-eyed female in this cross. Thus, female sex determination occurs when the fly has two X chromosomes, even if it also has a Y chromosome.

KEY POINT Analysis of non-disjunction of the X chromosome in flies elucidated mechanisms of sex determination.

We should discuss the other trisomic offspring in this cross. If the XX disomic ovum is fertilized by the X-bearing sperm, the resulting fly is XXX. This fly is female but has red eyes, since the X chromosome from the father has the dominant w^+ allele. Such trisomic females, or triplo-X females, have reduced survival and viability, compared to their XX siblings, for reasons that are not entirely clear.

The event that produces an XX ovum also produces an ovum that has no sex chromosome at all. If the nullo-X ovum is fertilized by a Y-bearing sperm from our red-eyed male, the resulting embryo is 0Y (no X chromosome and only a Y chromosome), which is inviable because it is missing all the genes found on the X chromosome. If the nullo-X ovum is instead fertilized by an X-bearing sperm with the wild-type w^+ allele, the fly is X0. It will have red eyes, since it is hemizygous for the X chromosome with the w^+ allele. More significantly, from the perspective of understanding the mechanism of sex determination, this fly is a male.

In *Drosophila*, the chromosomal signal for sex determination is the number of X chromosomes; 1X embryos develop as males, while 2X embryos develop as females, regardless of the presence or absence of the Y chromosome. In *Drosophila*, X0 males are also sterile, but that is not true in all organisms. In *C. elegans*, no Y chromosome exists, and X0 is the chromosome constitution of normal males. Thus, *C. elegans* also "counts" its X chromosomes to determine the sex. Many other invertebrates also use the number of X chromosomes as the chromosomal signal for sex determination.

Non-disjunction at meiosis I or meiosis II

Non-disjunction is the final outcome of a meiotic error, but the actual error could have occurred at any earlier step in either meiosis I or meiosis II. In Section 6.5, we wrote that the outcome of non-disjunction is different if it happens at meiosis I and meiosis II, which can be illustrated most effectively with sex chromosomes. To show this, we can begin with a male that is XY or with a female that is heterozygous on the X chromosome for ***A*/*a***. This is shown in Figure 7.8.

What happens if non-disjunction of the sex chromosomes occurs during meiosis I? Note that, if non-disjunction occurs at meiosis I, both homologs go to the same pole.

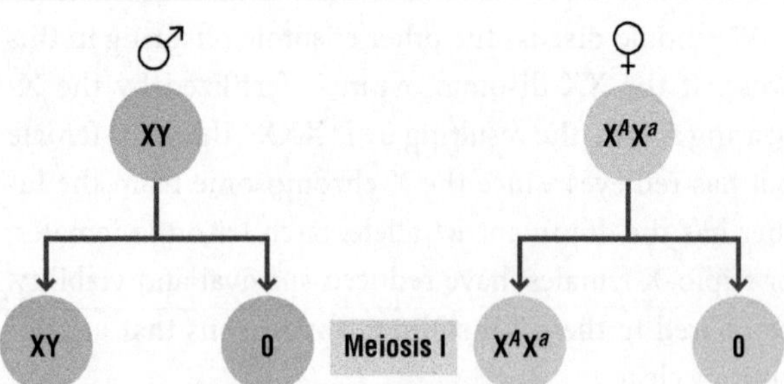

Figure 7.8 Non-disjunction at meiosis I. While non-disjunction of the X and the Y chromosomes is being used in this figure for clarity, non-disjunction can occur for any chromosome with analogous outcomes. We illustrate this with a male (XY) and a heterozygous female (***Aa***). Meiosis II occurs normally in this example and is not drawn. Non-disjunction that occurs during meiosis I results in a gamete that has both homologs, as well as one that has neither. Thus, in a male, the gametes have both the X and Y chromosomes. In the heterozygous female, the gamete will have both the ***A*** and the ***a***. In short, non-disjunction at meiosis I results in gametes that are heterogametic or heterozygous.

Thus, the sperm is XY and has both sex chromosomes. The ovum is ***A*/*a*** and has each of the X chromosomes. (Since meiosis II occurs normally, we will leave it out of our drawing.) In other words, non-disjunction at meiosis I results in a disomic gamete that has both the X and Y chromosomes if it occurs in a male, and both X chromosomes if it occurs in a female. It is heterozygous, or heterogametic if it comes from a male.

On the other hand, if non-disjunction occurs at meiosis II, the results are quite different, as shown in Figure 7.9. Meiosis I occurs normally, and the two homologs go to opposite poles. Non-disjunction at meiosis II involves only one member of the homologous pair, but it affects both sister chromatids of that chromosome. Thus, if non-disjunction at meiosis II occurs in the spermatocyte with the Y chromosome, the resulting gametes will have two Y chromosomes. If it occurs in the oocyte with the ***a*** allele, the resulting ova will have two X chromosomes with the ***a***. In other words, non-disjunction at meiosis II results in a disomic gamete with two copies of one of the homologs. It is homogametic or homozygous.

In practice, it is often not feasible to follow the segregation of the chromosomes during meiosis directly, at least in a sample size large enough to observe the rare non-disjunction events. Aneuploidy is observed in the offspring, and the source of non-disjunction is inferred from the phenotypes of the offspring. As discussed in Chapter 6, non-disjunction at meiosis I (which is more common) is usually associated with a failure of the homologs to cross over and form chiasmata. Non-disjunction at meiosis II is usually associated with a failure in sister chromatid cohesion.

Mendel, sex chromosomes, and meiosis

The final proof that Mendel's rules could be explained by meiosis and that genes were found on chromosomes was the presence of these unusual phenotypes arising from non-disjunction such as those seen in crosses of white-eyed females and red-eyed males. Meiosis explained not only the normal results from genetic crosses but also the unusual results in some crosses arising from meiotic errors. In 1916, Morgan's student Calvin Bridges published the first paper in the first issue of *Genetics*, entitled "Non-disjunction as proof of the chromosome theory of heredity," using the *Drosophila* data and the explanation you have just read. He inferred that the chromosomal signal for sex determination in *Drosophila* was the number of X

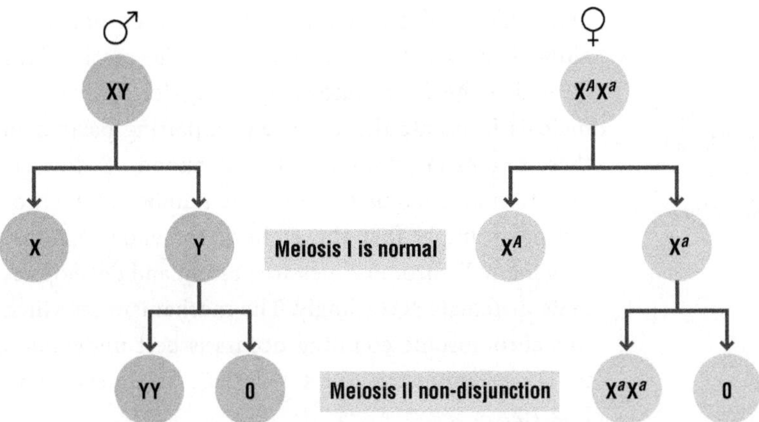

Figure 7.9 X chromosome non-disjunction at meiosis II. The homologs have segregated normally from one another at meiosis I, so each cell (the secondary spermatocyte or secondary oocyte) has only one of the two homologs. The same result arises with either homolog, but only one is shown. Non-disjunction at meiosis II results in gametes that are homozygous. These can have two copies of the Y chromosome or two copies of the same homolog of the X chromosome (***aa*** here). Non-disjunction at meiosis II, like non-disjunction at meioisis I, results in nullisomic (0) gametes, so these are non-informative. However, the disomic gametes can be used to determine which meiotic division has been affected.

chromosomes, and he showed that the unusual results of his crosses could be explained by errors occurring during meiosis.

KEY POINT The study of meiotic errors, such as non-disjunction, in *Drosophila* confirmed the link between meiosis and Mendel's rules, as well as the signal for sex determination.

We pointed out in Section 6.5 that non-disjunction can occur for any chromosome, but that trisomic offspring in animals (arising from non-disjunction) are typically only observed with the smallest autosomes and the sex chromosomes. We also discussed why trisomics for small autosomes might survive better than trisomics for medium-sized or larger autosomes. Likewise, monosomic offspring with only a single chromosome are only observed with the X chromosome in animals, and monosomy for autosomes almost never survives. In nearly all animals, the X chromosome has hundreds of genes on it, so it might seem that genetic imbalance for the X chromosome should have the same dire consequences as genetic imbalance for an autosome. However, the reason that aneuploidy for the X chromosome is more readily tolerated lies in the process of dosage compensation, as discussed in Section 7.5.

7.4 Mechanisms of sex determination

We are going to step away from meiosis and the inheritance of X-linked genes to talk about the developmental biology of sex determination. Because sex determination is directly tied to reproduction and fitness, it might seem that the mechanism for sex determination should be evolutionarily conserved. Thus, you might think that the mechanism for sex determination should be similar in many different animals. In fact, this is not true; sex determination mechanisms in animals are remarkably diverse. In this section, we will briefly summarize some of the different mechanisms involved in sex determination among animals, but this is far from an exhaustive list of the known mechanisms. Among the systems we are not considering are the ones that involve multiple X chromosomes, parthenogenetic organisms, or environmental signals other than temperature.

XY/X0 systems

In *Drosophila*, *C. elegans*, and many other invertebrates, sex is determined by the number of X chromosomes; the presence or absence of the Y chromosome does not matter. So X0 is male, and XXY is female, as indicated in Figure 7.10. In fact, many species, including most

XY = Male
XX = Female
XXY = Female
XO = Male

(a) ***Drosophila***

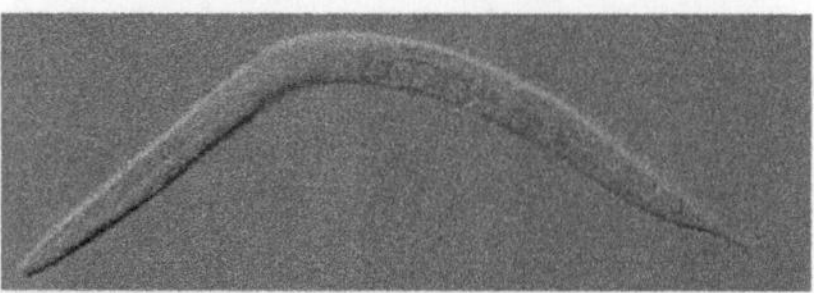

XX = Female
XO = Male

(b) **Nematodes and grasshoppers** (and a number of other species)

Figure 7.10 The number of X chromosomes determines sex in *Drosophila* and many other invertebrates. In *Drosophila* and some other invertebrates, sex is determined by the number of X chromosomes, and the Y chromosome plays no essential role in sex determination. Thus, XXY flies become females, and X0 flies become males. In grasshoppers and nematodes, among other species, sex determination is similar, except there is no Y chromosome. XX is a female and XO is a male. As noted in Box 7.3, the signal for sex determination in *Drosophila* and *Caenorhabditis elegans* is not the number of X chromosomes, strictly speaking, but the X:autosome ratio; thus, this drawing only pertains to normal diploids.

Source: Grasshopper courtesy of Ryan Wood. Nematode *(C. elegans)* courtesy of National Human Genome Research Institute.

nematodes and grasshoppers, lack a Y chromosome altogether, so males are normally X0. (The mechanisms by which the X chromosome segregates normally at meiosis I, despite the absence of a pairing partner in these organisms, are not well understood.) Sex determination systems that rely on the number of X chromosomes imply that the embryo somehow "counts" how many X chromosomes are present and develops as male or female accordingly. The mechanism by which this chromosome counting occurs is best understood in *D. melanogaster* and *C. elegans*, as described in Box 7.2.

The XY/XX sex determination system in mammals works by a different and somewhat conceptually simpler strategy than the ones in flies and worms, as shown in Figure 7.11. (We want to note here that we are using "mammal" in this chapter in the familiar sense to refer to placental mammals. Sex determination in marsupials, such as kangaroos, koalas, and opossums, is a bit different, and, in monotremes such as the platypus, it is quite different.) This is conceptually simpler in that it relies on the presence or absence of one gene, rather than a quantitative difference. In other words, mammalian sex determination is analogous to an ON/OFF switch, while sex determination in *Drosophila* is analogous to a dimmer switch.

In mammals, the Y chromosome has a gene—known as *SRY* in humans and *Sry* in mice—that encodes a protein called testis-determining factor, or TDF. TDF triggers testis differentiation in the immature gonad of the developing embryo; the testes produce testosterone, which then stimulates the complex cascade of differentiation events that produces males. If no Y chromosome is

BOX 7.2 *Going Deeper* How do organisms count X chromosomes?

Drosophila melanogaster and *Caenorhabditis elegans* are among the organisms that determine their sex based on the number of X chromosomes, rather than the presence or absence of a Y chromosome, as happens in mammals. As a conceptual process, it seems simpler to rely on the presence of a Y chromosome as the signal for a developmental event, rather than a quantitative difference between a signal that exists in both sexes. If the Y chromosome trigger is present, as is true in mammals, testes formation occurs; on the other hand, if the Y chromosome trigger is absent, ovarian development ensues. The identification of this testis-determining gene on the Y chromosome in mammals is discussed in Box 7.3.

However, in *Drosophila* and *C. elegans*, the sex determination signal is present in both sexes but at different levels. The trigger is then a quantitative difference in the level of some gene, sequence, or gene product on the X chromosome. Furthermore, the quantitative difference is apparently quite small, only one vs. two in diploids, which is not much of a difference for most biological processes. In addition, as noted in Section 7.5, the difference between 1X and 2X dosage is compensated at the level of gene expression, so that males and females make the same levels of X-linked gene products. How then do these organisms count the number of X chromosomes and use that difference to determine

BOX 7.2 Continued

the profound developmental, genetic, and evolutionary differences between being a male and being a female?

The genetic analysis of the chromosomal signal for sex determination was done first for *Drosophila* in the late 1970s and 1980s, primarily by Thomas Cline, his students, and colleagues, and slightly later for *C. elegans*, primarily by Barbara Meyer, her students, and colleagues. It involves remarkable insights, creative thinking, and an expertise in manipulating genes and chromosomes—as well as hard work—that few other analyses have surpassed. Rather than attempt to recapitulate the details of how this was done, we will summarize the results, based on a few key observations and experiments. Our focus will be on *Drosophila*, since it was studied first, with a comparison to *C. elegans* at the end.

A word of warning. This is a complicated process, so even a simplified version requires some familiarity with transcription initiation, transcription factors, gene dosage, and dosage compensation. The mechanism by which *Drosophila* counts X chromosomes is intricate and beautiful and could probably be taught as a graduate-level seminar in genetics, but it is not particularly simple. The mechanism is summarized in Figure A.

The X chromosome number is "counted" relative to something else, rather than absolutely

Although we have regularly referred to the number of X chromosomes as being the signal for sex determination, that only refers to diploids and is a bit of an over-simplification. It was recognized in the 1930s from experiments with triploids that the actual signal for sex determination is the ratio between the X-linked signals and some signals from the autosomes. This is known as the X:A ratio. The number of X chromosomes—or, more significantly, the **expression** of the key X-linked genes (which might be different in 1X and 2X animals)—is calibrated against the expression of certain autosomal genes that are expected to be the same in both sexes. The autosomal component was referred to as the denominator of the X:A signal, while the X-linked component was referred to as the numerator.

One of the major denominators was recognized, by Cline and others, to be the gene *daughterless* (*da*). The *da* gene is transcribed and translated in the mother, and the protein product is then allocated to her ova. Since the ova develop without regard to what their future sex might be, the amount of Da protein is

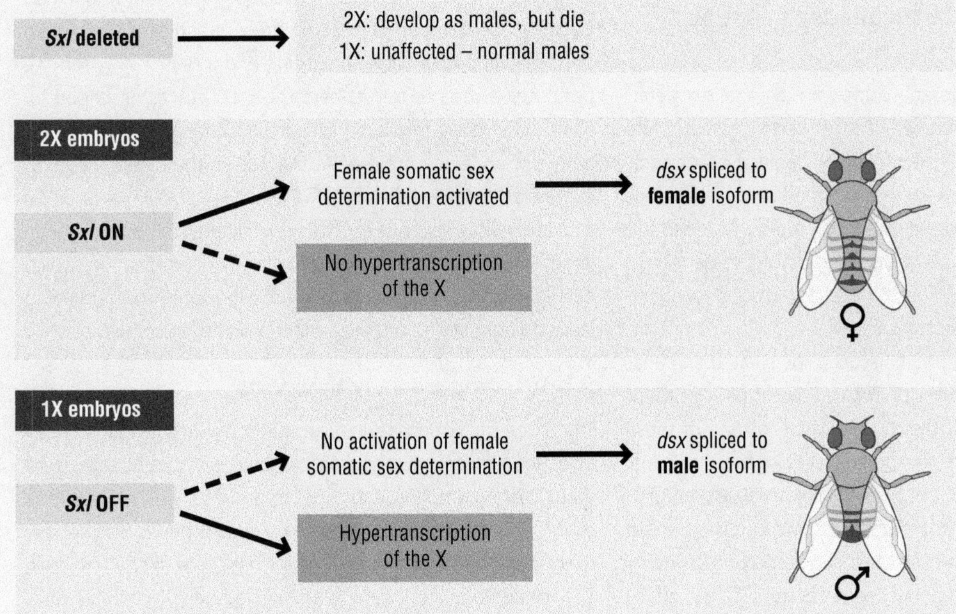

Figure A A summary of the functions of *Sex-lethal* (*Sxl*). *Sxl* is the key regulatory gene for sex determination in *Drosophila melanogaster*. If *Sxl* is deleted, 1X embryos develop normally as males, but 2X embryos usually die due to defects in dosage compensation. If capable of developing, these 2X embryos have male sexual differentiation. Thus, the normal role of *Sxl*$^+$ is to turn on female development in 2X embryos, while also shutting off hypertranscription of the X in 1X embryos. These roles are summarized in diagrams for 2X and 1X embryos. The last gene in the sex determination pathway is *doublesex* (*dsx*), which is spliced to either the female or the male isoform, depending on the activity of *Sxl* and the upstream genes. *doublesex* is referred to again in Box 7.5.

BOX 7.2 Continued

approximately the same in all embryos. This serves as the calibration factor in the denominator. Subsequent analysis showed that the product of the *da* gene is a transcription factor of the basic helix–loop–helix type. The Da protein forms one half of a dimeric protein with other proteins, while the numerator's genes encode the other half. The complete dimeric transcription factor is the active form.

The master regulatory gene Sex-lethal affects both sex determination and dosage compensation

The key gene for understanding the process of sex determination in *Drosophila* is the X-linked gene *Sex-lethal* or *Sxl*. Recessive alleles of *Sxl* are lethal in XX flies but viable in XY flies, suggesting that the gene is needed for some process in females but is non-essential in males, as summarized in Figure A. Cline had the insight that the basis of this lethality is that *Sxl*$^+$ not only is the key gene for sex determination, but it is also the key gene for dosage compensation. The XX *Sxl/Sxl* flies develop as males but die because they also carry out dosage compensation like normal XY males do and hypertranscribe their X-linked genes. When dosage compensation was also compromised or eliminated, these XX *Sxl/Sxl* flies now survived and showed male development.

Regulation of Sex-lethal depends on specific X chromosome genes

Since *Sxl* regulates both dosage compensation and sex determination, the genes that regulate it must be expressed before dosage compensation or sex determination begins. Dosage compensation initiates at a very early stage in embryonic development, so these regulators of *Sxl* expression must be among the first genes expressed in the embryo. Furthermore, these regulators—the numerators—must be X-linked and dose-dependent. This is an unusual set of characteristics.

To identify these X-linked numerators, Cline created and analyzed flies that had extra doses or duplications of parts of the X chromosome. He reasoned that a duplication of one of these numerators would "fool" the fly and activate *Sxl*$^+$ function in 1X embryos. The 1X flies would become females because the duplication would act like a second X chromosome, but they would also die because dosage compensation would occur as if the fly were XX when, in fact, it had only one X. Several different genes with these properties were found, called *sisterless-a* (*sis-a*), *sis-b*, and *sis-c* (although only *sis-a* and *sis-b* were found originally and were sufficient to carry out most of the process). It was usually necessary to change the dose of both *sis-a* and *sis-b* simultaneously, which made the experiments even more challenging. Both *sis-a*$^+$ and *sis-b*$^+$ encode transcription factors, also of the basic helix–loop–helix type, the same type as Da.

The names of these genes help us understand their roles. *Sxl*$^+$ is necessary for 2X embryos, but non-essential in 1X embryos. The products of *da*$^+$, *sis-a*$^+$, and *sis-b*$^+$ are the necessary transcription factors for *Sxl* expression. When *da* is mutated and the protein is not made, the mother cannot allocate the denominator protein into the ova. Thus, *Sxl* cannot be transcribed in the embryo, and the 2X embryos die. The mother has sons—1X embryos that do not need to activate *Sxl*—but not daughters. When *sis-a* or *sis-b* is mutated and the proteins are not made, the embryo also cannot activate *Sxl* expression. Males survive, but they have no sisters.

The simplest version of the model (simplified by omitting some components) is that, when two X chromosomes are present and both *sis-a*$^+$ and *sis-b*$^+$ are active and present at higher levels, they make proteins that form a dimer with Da. This is summarized in Figure B. The Sis/Da transcription factor dimers activate the transcription of *Sxl*, and female development ensues. If not enough *sis-a*$^+$ or *sis-b*$^+$ is made, few dimers with Da can be made, and *Sxl* transcription is not activated; this is what happens when only one X chromosome is present and there is only one dose of *sis-a* and *sis-b*.

Sex-lethal regulates its own activity

Sex-lethal is itself X-linked, so the dose of *Sxl* also can play a role in assessing the level of Sis/Da dimers. More importantly, once *Sxl* begins to be transcribed in 2X embryos, it activates its own expression, via both effects on splicing and a switch in which the promoter is used to transcribe the gene. Thus, as described for some other gene regulatory circuits in Chapter 14, *Sxl* is an example of an "auto-amplifier." At the outset, the difference in *Sxl* expression between 1X and 2X flies is very small, but once *Sxl* activates its own expression, the signal difference becomes quite large. Thus, after the very first steps in sex determination, which culminate in *Sxl* regulating its own expression, the remaining steps do not depend on the number of X chromosomes at all.

A brief comparison with worms

The process in *C. elegans* is analogous, but with a few differences. In effect, similar events occur, but in the opposite sex. Recall that dosage compensation in worms occurs when XX animals down-regulate the transcription of X-linked genes, so the sex-specific lethality is observed in X0 animals, instead of XX animals. The master regulatory gene is an X-linked gene, known as *xol-1*, that is needed for normal male development; mutants in *xol-1* develop as females and down-regulate X-linked transcription. (Strictly speaking, the XX animals are hermaphrodites because they have both a male and a female germline, but, for our purposes, we will refer to them as females.) Since these embryos have only a single X chromosome, this down-regulation results in their death. The X-linked and autosomal regulators of *xol-1* activity are

BOX 7.2 Continued

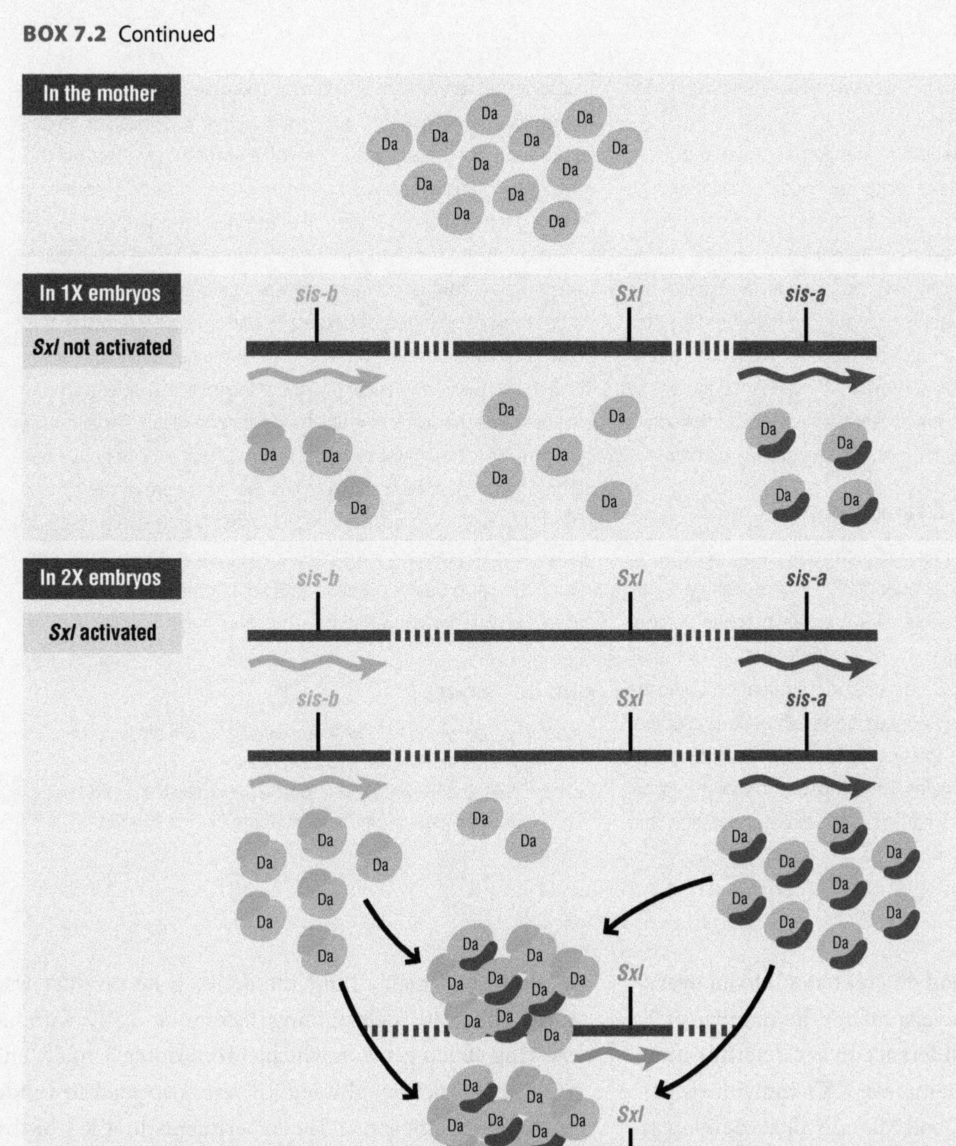

Figure B Sis and Da activate *Sxl* transcription in 2X embryos. This figure is a simplified representation that omits some steps and other genes. The mother makes a protein known as Da, shown here in orange. Da is half of a dimeric transcription factor that activates the expression of *Sxl*. Since the protein is made by the mother, all embryos have about the same amount of Da protein, regardless of their own sex chromosomes. The other half of the transcription factors are encoded by the X-linked genes *sis-a*$^+$ and *sis-b*$^+$. These are expressed very early in the embryo, with *sis-a* depicted in blue and *sis-b* depicted in green. When the *sis* genes are present in a single dose, that is, in 1X embryos, only a small amount of the proteins is made, only a few dimers between Sis and Da are made, and *Sxl* is not activated. When the *sis* genes are present in two doses, that is, in 2X embryos, twice as many of the protein dimers are made, which results in enough Sis–Da dimeric transcription factors to activate the expression of *Sxl*.

not definitively identified, and the activation could be both transcriptional and post-transcriptional. *xol-1* itself down-regulates or shuts off the expression of X-linked genes *sdc-1* and *sdc-2* and the autosomal gene *sdc-3*, which are needed for female development and dosage compensation. Thus, while there are conceptual similarities between *Sxl* and *xol-1* as X-linked genes that are the master regulators of both sex determination and dosage compensation (albeit acting in opposite sexes), the details of how they are

BOX 7.2 Continued

regulated and how they act on the downstream genes are probably not similar.

Both *Sxl* and *xol-1* are related to genes that have important functions in the cell that are not related to sex determination. The *Sxl* gene is thought to have evolved from the kind of gene duplication and diversification event we described in Chapter 3. The duplication event is thought to have occurred at the base of the *Drosophila* lineage when a gene encoding an RNA-binding protein in the fly genome became duplicated. *xol-1* encodes a protein that is related to kinases important in fundamental metabolic processes in the cell; kinases are enzymes that regulate the activities of other proteins by attaching phosphate groups to them.

Sex determination and canalization

One recurring theme in the many varied methods of sex determination is that a relatively small quantitative difference in a signal results in the significant developmental, genetic, and evolutionary difference between female and male differentiation. The small quantitative difference that triggers sex determination could be levels of transcription factors, X-linked gene expression, temperature, or something else, but the final outcome is distinctly different between males and females. This amplification of the signal appears to involve some very early step in sex determination that is auto-regulatory and, in effect, not reversible. There are certainly examples, in all animals, of individuals with ambiguous sexual development, but these typically involve something affecting the downstream steps, such as hormone production or a hormone receptor, rather than an alteration at the initial triggering step.

Although the initial triggering difference may be small, once the embryo commits to one sex determination pathway, it follows that pathway. The biologist Conrad H. Waddington referred to this process as "canalization," so that, despite small and random variations that arise in any naturally occurring process, selection has produced mechanisms to ensure a single end result. Canalization is a form of robustness, as discussed in Chapter 3. Because sex differentiation is so well known to us, we are aware of the canalization that occurs to ensure a reproductively successful male or female. But the general process of canalization is very common in evolution and development and is an essential feature of many processes with natural variation.

FIND OUT MORE

Cline, T. W. and Meyer, B. J. (1996) Vive la différence: males vs females in flies vs worms. *Annu Rev Genet* **30**: 637–702

Parkhurst, S. M. and Meneely, P. M. (1994) Sex determination and dosage compensation: lessons from flies and worms. *Science* **264**: 924–32

present, the immature gonad differentiates into an ovary, which results in female development. The number of X chromosomes makes no difference in sex determination in this process. Thus, in mammals, XXY individuals develop as phenotypic males, and X0 individuals develop as phenotypic females.

The names for the gene and the gene product may be a bit confusing, but they are historical. The gene is named for its mutant function, originally found in mice. It is a Y-linked gene that, when mutated, results in sex reversal. Thus, it was first identified in sex-reversed mutants in mice; the chromosomes in these mice were XX, but the individuals developed as males. Hence, the gene was named *Sry* for "sex reversal on the Y chromosome," but it was not known for sure what the gene product did. It was recognized that these XX mice had a duplication of a small piece of the Y chromosome. In other mutant mice, this piece of the Y chromosome was deleted. These mice were also sex-reversed, but the small-deletion mice were XY females.

It was also known from transplantation experiments in many mammals that, when the embryonic gonad becomes a testis and produces testosterone, the embryo develops as a male. Thus, the active gene product was called the testis-determining factor, or TDF, without knowing which gene on which chromosome actually encodes this function, although it was suspected to reside on the Y chromosome. The experiments that led to the identification of the *SRY* gene as encoding the TDF in mammals are described in Box 7.3.

XXY males and X0 females, and other sex chromosome aneuploids, such as XXX or even XXXX, are not particularly rare in humans, although the exact frequency is hard to estimate. The phenotypes of XXY males (a condition known as Klinefelter syndrome) and X0 females (known as Turner syndrome) in humans vary widely. Some XXY and X0 individuals are probably undetected, while others are easily recognized by geneticists and clinicians. It is estimated that at least half of the cases of XXY and X0 are not detected until puberty, which is often delayed in these people. Since XXY and X0 individuals have reduced fertility or are sterile, they are sometimes identified in fertility clinics.

Many different explanations have been suggested for the range of phenotypes observed from these and other sex chromosome syndromes. One explanation, which

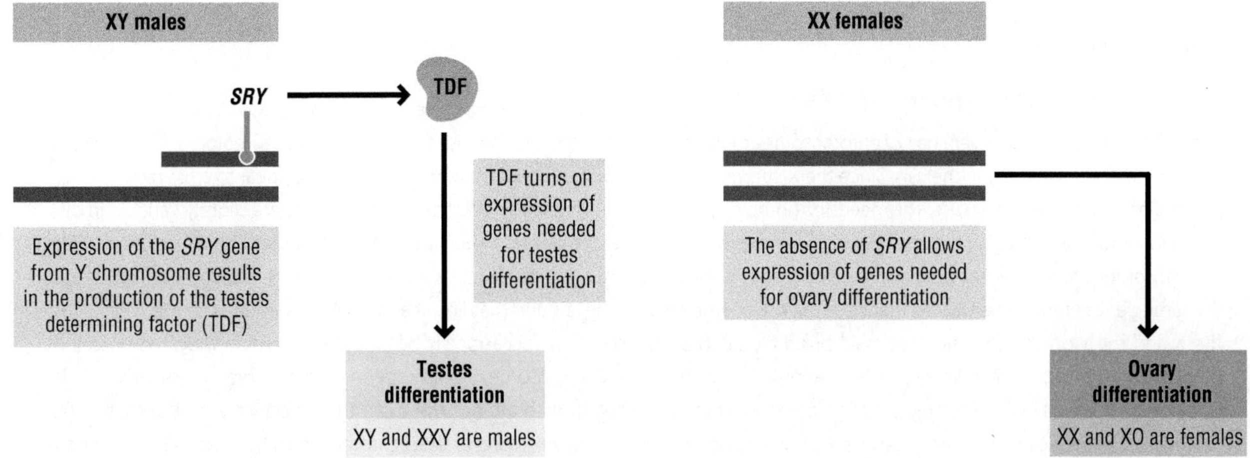

Figure 7.11 SRY triggers testes formation in mammals. In mammals, a gene on the Y chromosome known as *SRY* encodes a protein called testis-determining factor, or TDF. Thus, the presence of the Y chromosome, or even of only the *SRY* gene, results in testis differentiation of the immature gonad and subsequent male sexual development. The absence of SRY allows the immature gonad to differentiate into an ovary. Thus, XXY individuals exhibit male sexual development, while X0 individuals show female sexual development. The phenotypes of XXY and X0 individuals vary widely.

BOX 7.3 *A Human Angle* *Sry*, testis determining factor, and sex determination in mammals

Few phenotypes are as familiar—and as interesting—to us as mammalian sexual differentiation; nearly every identification form that we fill out from childhood asks us to check the box marked "M" or "F." Direct experiments investigating the biological basis for mammalian sex determination have been done for more than two centuries, and speculation about the biological origins of mammalian sex differences are far more ancient than that. We want to point out that we are again using "mammals" here to refer to placental mammals, and are not including marsupials or monotremes in this description.

One of the most important early observations about mammalian sex determination came from cattle. It was observed that a cow sometimes gives birth to a calf that is essentially female (and chromosomally XX once the sex chromosomes were recognized) but has masculinized features and is infertile; these are known as freemartins. Freemartins are always one of a twin, and the other twin is always a male. In the early twentieth century, it was demonstrated that freemartins occur because twins in cattle usually share a placenta and chorions; the testes of the male twin produces a substance, now known to be testosterone, that circulates freely through the shared chorions and placenta to the other twin, resulting in its masculinization, despite being chromosomally XX. Thus, in mammals, testosterone induces male features and inhibits some female features, and testosterone is produced by the testes.

What then induces the sexually indifferent gonad in the embryo to develop into testes? This was called testis-determining factor, or TDF; its biochemical and molecular nature was unknown for decades. Note that, in most mammals, including humans, twins arising from separate fertilizations like this do not have shared chorions, and thus testosterone does not circulate from one twin to the other, as happens in cattle. Human freemartins are found only in fiction and folklore, and not in nature.

It was also recognized more than a century ago that males have a Y chromosome, and it was hypothesized that a gene on the Y chromosome encodes TDF. Many candidate genes were initially proposed and then subsequently rejected as the gene encoding TDF, before the proper gene was found. Among the candidates was a gene in mice known as *Sry*. *Sry* was defined by the mutant phenotype—the mice were XX males and were sex-reversed. This sex reversal was due to the sex-determining region, so *Sry* is also used to refer to both the sex reversal and the region on the Y chromosome responsible for sex reversal. *Sry* is one of the few functional genes that was mapped to the Y chromosome, although the number of Y chromosome genes was not yet fully known when it was mapped. In addition, in mice, *Sry* maps very close to the region that crosses over between the X and the Y chromosome. This proved to be very important and requires a bit more explanation.

BOX 7.3 Continued

The pseudoautosomal region and XX males

Recall from Chapter 6 that every pair of chromosomes must cross over to segregate normally in meiosis. Since the X chromosomes in females can pair and synapse, this need for homology is no different from that for an autosome. However, in males, the Y chromosome is not of the same size as the X and does not have the same genes or the same DNA sequence as the X. Nonetheless, the X and the Y do pair in male mammals, but the pairing and synapsis are limited to a relatively small region at one end. Crossing over occurs in this small region, allowing normal meiotic segregation of the X and the Y chromosome, as shown in Figure A.

Sequences from the X chromosome in this region are crossed over, or translocated, to the Y chromosome, and sequences from the Y chromosome are crossed over to the X chromosome. Since there are few or no genes in this region (for most mammals at least), chiasmata are observed, but recombination is detected only from analysis of the sequences, and there are few or no functional consequences. This region that crosses over between the X and the Y chromosome is called the **pseudoautosomal region**, or more precisely the pseudoautosomal region 1 (PAR1), since two other regions of the X and Y also appear to cross over.

The precise boundaries of the PAR1 are slightly variable in different laboratory strains of mice, and probably in nature as well. Thus, while no genes are affected by the crossovers most of the time, a gene that lies near the PAR1 boundary on the Y chromosome will occasionally be crossed over or translocated onto the X chromosome in a few sperm cells of some mouse strains. This sperm with a bit of the Y can then produce a mouse that is XX, but in which one of the X chromosomes has a small translocation of part of the Y chromosome, as shown in Figure A. The mouse is, in effect, XXY, but the Y is only a small fragment of the total Y chromosome.

Recall that, in mammals, XXY develops as a male. Thus, it was recognized that, in some strains and some mutant lines, there were XX mice that were sex-reversed because a male-determining fragment has been moved onto the X chromosome; this locus

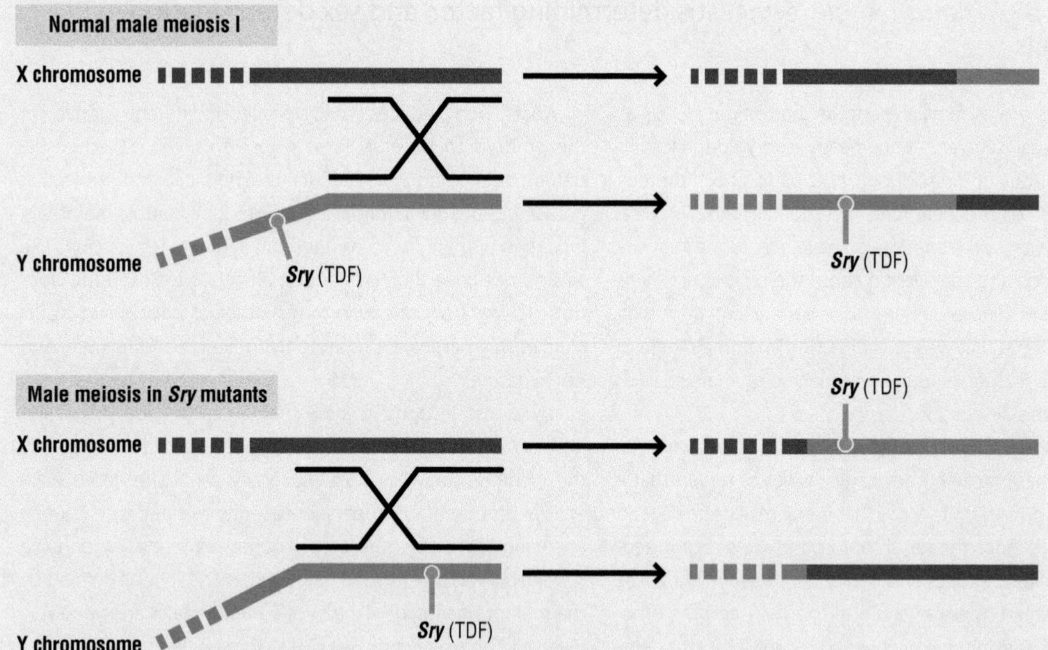

Figure A ***Sry*** **and the pseudoautosomal region.** The X and the Y chromosomes in males normally pair and cross over in a region known as the pseudoautosomal region; the two chromosomes do not synapse over their entire length as homologs do. This is shown at the top. The outcome is that sperm with an X chromosome have a small fragment of the Y chromosome, but typically this region does not have any genes in it. The *Sry* gene is adjacent to this region but is not usually involved in the crossover. In the *Sry* mutant strains of mice, the pseudoautosomal region is slightly extended and includes the *Sry* locus, which gives rise to the mutant phenotype. Thus, some sperm have an X chromosome with a fragment of the Y, and this fragment that is translocated includes *Sry*. When such a sperm fertilizes an ovum, the embryo will be XX but will have the *Sry* gene on its paternal X chromosome; these embryos are sex-reversed and develop as males, as shown in the lower panel.

BOX 7.3 Continued

was named *Sry*. These *Sry* sex-reversed XX mice all had the same fragment of the Y chromosome, so somewhere in this fragment lay the gene encoding TDF.

Meanwhile, studies in humans were identifying the same small region. Translocations between the X and the Y chromosome with functional consequences are fairly uncommon, but such individuals are males with greatly reduced fertility because of the failure of the gonad to develop. Some of these males visited fertility clinics, and the region of the Y chromosome that had been moved to the X could be mapped. Again, this was identified as a small region near PAR1, the same region being identified in mice by the *Sry* translocations. Furthermore, a few of the women who came to fertility clinics were identified to be XY, but with a deletion of part of their Y chromosome; they also had a failure of the gonad to develop. The deletion in these XY women must be removing the gene encoding TDF.

Both the translocations in XX males and the deletions in XY females were of varying sizes, many of them encompassing many millions of base pairs of the Y chromosome, so the key was to find the smallest region that was common to both and was also found in *Sry* XX mice. It also turned out to be important that the alteration affected only this region, and not some other region as well; one false lead had a small change here and a larger and more easily detected deletion elsewhere on the Y chromosome. This region included only two or three protein-coding genes, one of which was *Sry*, but any of these could have been the gene for TDF.

The proof that the *Sry* gene did encode TDF came from several lines of evidence. First, the gene is transcribed in the embryonic gonad during the time that testis determination occurs, and it is transcribed in the appropriate cells of the gonad. The other candidate genes did not have this precise pattern of transcription. Second, the *Sry* gene was conserved among placental mammals and always Y-linked; this also was not true for all other candidate genes. Third, each of the 14 different women who were chromosomally XY with a failure in gonad development had a mutation in this same gene. Fourth, and most compelling, the *Sry* gene from mice was cloned, and the DNA sequence of only this gene, and no other, was inserted into mouse embryos. Embryos that were XX but had only this additional gene developed testes and male sexual differentiation. This then proved that *Sry* encodes the TDF.

Sry and testis determination

What then is the molecular function of *Sry*, and how does the TDF actually trigger testis formation? *Sry* encodes a transcription factor of the SOX family; the family is named for the mechanism by which it binds to the DNA, as discussed in Chapters 2 and 12, and Sry was the member by which the family was named. (SOX is a contraction of "Sry-like box.") The Sry protein forms a heterodimer with another transcription factor called SF1 (for steroidigenic factor 1, since it regulates the expression of genes needed for steroid production, including *Sry*). This Sry/SF1 dimer regulates the expression of several other genes, most notably the gene *SOX9*, which itself encodes a transcription factor of the SOX type. It is the SOX9 protein—as a dimer in combination with SF1—that directly regulates most of the differentiation of the gonad into the testis. (This includes the *DMRT* genes described in Box 7.5.) Thus, *Sry* sets off a cascade of transcription factors, which then activate the expression of genes involved in testis formation, testosterone production, and so on.

Interestingly, the sequence of the Sry protein is not highly conserved among mammals. The main regions of conservation are found in the regions that bind to DNA and regulate the expression of the *SOX9* gene; the *SOX9* gene is more highly conserved. This lack of conservation suggests that the *Sry* gene is needed primarily for one function—to turn on the expression of *SOX9*.

FIND OUT MORE

Kashimada, K. and Koopman, P. (2010) Sry: the master switch in mammalian sex determination. *Development* **137**: 3921–30

works well to explain the range of phenotypes of sex chromosome aneuploids in mice, is **mosaicism**. Mosaicism occurs when an individual has cells with different genotypes—in this case, when not all cells in the organism have the same chromosome number. While the one-celled embryo is XXY (and a similar explanation works for X0 individuals), chromosomes get lost occasionally during mitosis, as summarized in Figure 7.12. Thus, most cells in the adult mouse are XXY, but some are XX, some are XY, and so on, depending on which chromosome is being lost and when during development this loss occurs.

If some cells early in development lose one of their X chromosomes during mitosis, most of the cells in an individual will be XY and follow a male pathway for differentiation. If, on the other hand, it is the Y chromosome that is lost mitotically, most of the cells will be XX and will follow a female differentiation pathway. Some of the phenotypic range observed in XXY and X0 people can probably be explained by mosaicism, but the exact extent is not known. It should also be recognized that the chromosomal signal for sex determination is only the first step in a complex developmental pathway. Just as a wide range

XXY | XXY
Mitotic divisions

XXY XXY XX XXY | XXY XY XXY XXY
XY XXY XX XXY | XXY XY XXY XXY
XY XX XX XXY | XX XY XXY XXY
XY XX XX XY | XX XY XY XX
XY XX XX XY | XX XY XY XX
XY XX XX XY | XX XY XY XX

Figure 7.12 Mosaicism and phenotypic variability. One explanation for the phenotypic variability observed in XXY and X0 individuals is that there is mosaicism due to chromosome loss during mitosis. Examples are diagrammed for XXY. Each fertilized egg is XXY and goes through a series of mitotic divisions. The four cells in the top row continue to divide by mitosis; the vertical column indicates mitotic descendants of that cell. XXY cells that lose an X chromosome become XY, the normal male karyotype, while cells that lose the Y chromosome become XX, the normal female karyotype. This loss can occur in any mitotic division, but once the chromosome is lost, its mitotic descendants retain the same karyotype. XX cells are shown in yellow, while XY cells are shown in green. Note that these two individuals vary in the proportion of XX and XY cells, as well as which particular cell lineages are XX or XY.

of sexual characteristics exists among individuals who are XX or XY, a similarly wide range is likely to exist among individuals who are XXY, XO, or XYY because the developmental pathway involves so many different steps. Most cases of indeterminate sex determination in humans, or probably most cases in which the chromosomal sex does not agree with the development of secondary sex characteristics, are not because of changes in the sex chromosomes. Most such phenotypes appear to be due to the genes that respond to the chromosomal signal and *Sry*.

ZZ/ZW systems

While the XX/XY systems for sex determination are the most familiar to us and the most thoroughly studied experimentally, many other sex-determining mechanisms are found in nature. The most similar to the XX/XY system is the ZZ/ZW system. In birds (as well as some fish, some reptiles, and some invertebrates), the female is heterogametic (ZW) and the male is homogametic (ZZ); this is shown in Figure 7.13(a). In other words, it is a chromosome-based system like an XY/XX system, but it is

(a)

~82.3 Mb and ~770 genes (chickens)
Z
(b) W ~1.3 Mb and nine genes

Figure 7.13 ZW/ZZ sex determination in birds. (a) Many birds are clearly sexually dimorphic; male and female cardinals and chickens are shown here. In birds, the male is the homogametic sex referred to as ZZ, while the female is the heterogametic sex referred to as ZW. Like XX/XY systems, the Z chromosome is fairly large and has many genes on it; in chickens, the Z chromosome is roughly 83 Mb in length and has more than 770 annotated protein-coding genes on it. (b) The W chromosome is usually much smaller and has very few genes on it; in chickens, the W chromosome is 1.3 Mb in length and has nine annotated genes on it. It is not known if sex determination is due to the number of Z chromosomes or the presence of a W chromosome, since ZZW and Z0 birds have not been found.

Source: Image of cardinals courtesy of Ken Thomas. Image of chickens courtesy of Sini Merikallio/ CC BY-SA 2.0.

the female that has two different sex chromosomes and is heterogametic. In order to distinguish these situations when the female is heterogametic from the better known XY systems, the sex chromosomes are referred to as Z and W, with the W chromosome having the sex-determining locus on it. As with an XX/XY system, the Z chromosome is larger and has many genes on it, while the W chromosome has relatively few genes and is smaller in size, as shown in the diagram in Figure 7.13(b).

In birds, any allele that is found on the Z chromosome shows up in the phenotype of females, just as any allele that is found on the X chromosome in humans shows up in the phenotype of males. The XX/XY system in mammals and the ZZ/ZW system in birds apparently arose independently from one another, since genes that are X-linked in mammals are not Z-linked in birds, and vice versa. In fact, the Z chromosome in chickens is most similar in gene identity and arrangement to chromosome 9 in humans. Just as not all systems of XX/XY sex determination rely on the same cellular and molecular mechanisms, it also seems likely that different mechanisms of sex determination may be found among the ZZ/ZW systems.

Mating types with no distinct sex chromosomes

Many organisms have separate sexes and sexual dimorphism but do not have morphologically distinct sex chromosomes. For example, many different vertebrates, such as some reptiles, fish, and amphibians, and a very few plants, such as gingko trees, have separate sexes, and the sexes arise in a 1:1 ratio from a cross. However, a cytological examination of the chromosomes does not reveal any pair of homologs that are clearly differentiated as sex chromosomes. Some of these organisms may have a mating-type locus—that is, a gene with two different alleles, one that determines one sex and one for the other sex. Sex determination in these organisms behaves like a simple Mendelian trait with two alleles, in which one sex is a heterozygote.

For example, in the tropical fish known as a swordtail, shown in Figure 7.14, there is a single gene with two alleles, which we can call ***A*** and ***a***. Fish that are ***AA*** are females, whereas those that are ***Aa*** are males. Thus, in inheritance, this resembles an XX/XY system, but the sex chromosomes are not structurally distinct from one another. It has been postulated that this may be the ancestral condition for chromosomal sex determination and that distinct sex chromosomes arose from a mating type. A model for the evolution of sex chromosomes from a mating-type system is discussed in Box 7.4.

Another genetic system of sex determination, known as haplodiploidy, is found among bees, wasps, and ants, as well as some other insects. A female honeybee has 16 pairs of chromosomes, all of which pair, recombine, and segregate like autosomes during meiosis. Males (or drones) arise from unfertilized ova and thus are haploid, with 16 individual chromosomes. A drone produces

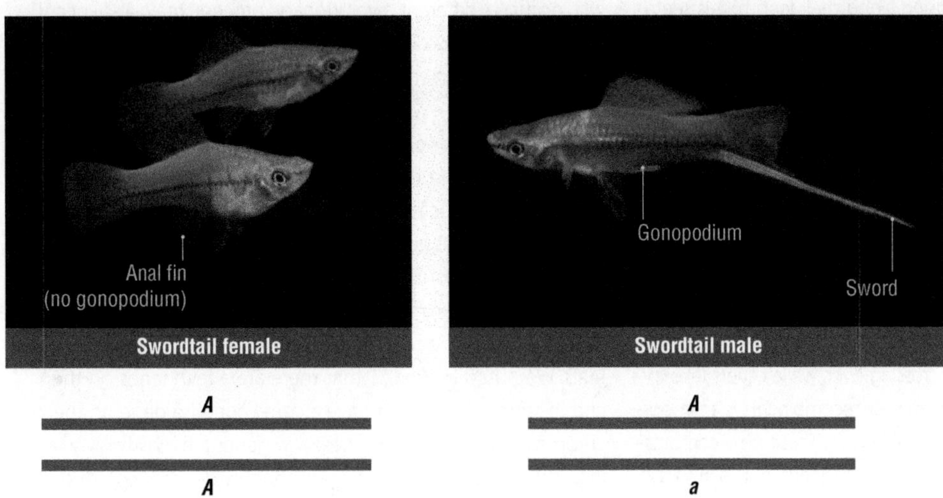

Figure 7.14 Mating types determine sex in swordtails. Swordtails (*Xiphophorus* species) have clearly defined sexual dimorphism between males and females but do not have sex chromosomes. Instead, a mating-type gene with two alleles determines sex, with females being homozygotes and males being heterozygotes

BOX 7.4 *Going Deeper* The evolution of sex chromosomes

Many different mechanisms of sex determination exist among animals, and several of these depend on the existence of sex chromosomes that differ in size, shape, and gene content. The evolution of sexual reproduction is itself a fascinating topic, but we will focus here more narrowly on the evolution of distinct or heteromorphic sex chromosomes. In light of the diversity of sex determination mechanisms in vertebrates (and invertebrates), it seems that heteromorphic sex chromosomes have evolved many types.

Acquisition of a sex-determining locus

The most widely accepted model for the evolution of sex chromosomes begins with a species that has a mating-type locus with two alleles, ***A*** and ***a***, on one of its pair of chromosomes; this type of sex determination is found in many fish species and some snakes, as well as elsewhere. In this model, the male is ***Aa***, while the female is ***AA*** (and dominance is ignored, so the female could also be written ***aa***). In mammals, this is thought to have occurred approximately 300 million years ago when one homolog in the pair acquired the *Sry* gene. This sex-determining locus could be any of a number of upstream regulators of a key gene needed for sexual or gonad development, as we note in Box 7.5. Once one chromosome has acquired a sex-determining locus, several critical steps must then occur for heteromorphic sex chromosomes to evolve from this system.

Suppression of recombination

Next, recombination between the two chromosomes must be suppressed in the ***Aa*** sex, so that the chromosome with the ***A*** allele does not cross over with the chromosome with the ***a*** allele. Perhaps this happens because a series of rearrangements prevent the chromosomes from synapsing along their length; this seems likely to be what has happened in vertebrates. The pair of chromosomes must retain the ability to segregate normally at meiosis, possibly through a mechanism that does not require chiasmata (which happens in *Drosophila*) or through a mechanism in which crossovers are confined to a particular small region (which happens in placental mammals).

In addition to its role in holding the homologs at the metaphase plate, recombination has an important evolutionary consequence—it produces new combinations of alleles on each homolog. (Many students and some biologists believe that the generation of recombinants is the principal role of crossing over, while others take the view, as we have, that the generation of recombinants is an important, and sometimes favorable, by-product.) These new combinations could be advantageous or disadvantageous, but, in any case, recombination will reshuffle them in the next generation, so that disadvantageous combinations are not retained.

Accumulation of deleterious mutations

While recombination will produce new combinations of alleles in the homogametic sex (***AA*** or ***aa***), it has been suppressed in the heterogametic sex (***Aa***). As a consequence, when a deleterious mutation arises on the same homolog as the sex-determining allele—and the majority of new mutations are deleterious or neutral—it cannot be crossed over onto the other homolog. New combinations of alleles cannot be generated for the chromosome with the sex-determining locus, so this chromosome will begin to lose functions. In short, it will begin to degenerate, and, once degeneration begins, it continues. This process is known as Muller's ratchet, named after the geneticist Hermann J. Muller who proposed this for populations that reproduce asexually, as will be discussed in Chapter 16. (In case you have not made a recent trip to a hardware store, a ratchet is a type of wrench that only applies torque when turned in one direction; the term is appropriate here, since the chromosome cannot go backwards and lose the deleterious mutations.)

Degeneration does not happen all at once, and recombination need not be lost all at once for Muller's ratchet to play a role. These deleterious mutations will include viruses and transposable elements (as discussed in Chapter 3), which insert on the chromosome and cannot be removed by recombination. Viruses and transposable elements are repeat sequences, since they can insert more than once, which the cell identifies as a heterochromatin and further inactivates.

Selection for sex-limited functions

While Muller's ratchet postulates how degeneration of the sex-determining (male-determining, in our scenario) chromosome will occur, a different evolutionary process may also occur. Genes on the sex-determining chromosome that contribute to reproduction in that sex—in our model, genes needed for spermatogenesis—will be strongly selected for. The origins of these genes could vary, and they could be any gene with a specialized function in reproduction. Perhaps they are autosomal genes that translocate to the Y; perhaps they are alleles of the X-linked gene that acquires sex-limited expression, or perhaps they arise from horizontal gene transfer (discussed in Chapter 11). Whenever such genes occur, there will be very strong selection in favor of them. This seems to have happened in mammals. We noted in Box 7.1 that there are a few genes on the Y chromosome that do not have a corresponding gene on the X chromosome in mammals; these few genes are mostly involved in spermatogenesis. Evolutionary models do not require that these genes arise but that there will be very strong selection for them once they do occur.

BOX 7.4 Continued

Dosage compensation occurs

The last step of the process could be the most variable in evolutionary and mechanistic terms. The homolog with the sex-determining locus (the Y in mammals) no longer has functional copies of the genes on its homolog. Thus, some mechanism occurs that regulates the expression of genes on the functional homolog (the X in our description), so that 1X and 2X animals make equivalent amounts of X-linked gene products. This could occur by a down-regulation mechanism in 2X animals (as happens in worms and mammals, although by different mechanisms) or by an up-regulation mechanism in 1X animals (as happens in *Drosophila*), and the processes need not be similar. However, if dosage compensation does not occur, one sex will be at a selective disadvantage, as we know from mutants that fail in dosage compensation. Thus, there is strong selective pressure for a compensating mechanism, once it arises.

The genomes of mammals and other vertebrates provide support that this theoretical model, proposed in some form long before sequence analysis was possible, has occurred. It is not the only model for sex chromosome evolution, and the evolution of dosage compensation mechanisms seems the most nebulous. On the other hand, regulating levels of gene expression can occur by many different mechanisms, as we describe in Chapter 12, so there are many possible ways that dosage compensation could have arisen. This too is precisely what we see from the organisms in which this has been analyzed—no two unrelated species use the same mechanism.

sperm cells via a meiotic process that contain one haploid set of chromosomes, although it is not clear how normal segregation occurs in the absence of pairing and recombination. These sperm fertilize ova from the queen to produce diploid females that are the workers; unfertilized ova from the queen give rise to more drone males. Deleterious mutations that arise will result in the death of the haploid drone and thus will not be passed on to the next generation.

Environmental sex determination

Although we have not covered all of the known systems of chromosome-based or genotypic sex determination, we note that many other species have no sex chromosomes and no mating-type locus and use environmental signals, such as the temperature, to determine their sex. Thus, in these organisms, sex is not a genetically inherited trait at all. This is actually quite common, even among vertebrates. In fact, there may well be more species of animals that have some form of environmental sex determination (ESD) than there are species with genetically based sex determination, although this is by no means definite.

Among the most interesting species that exhibit ESD are different species of sea turtles. The system is diagrammed in Figure 7.15. A female turtle crawls up the beach, digs a nest, and lays eggs. The embryos in the eggs are sexually indeterminate and can differentiate as either male or female. The tides wash over the nest, so that eggs in a nest low on the beach develop at a slightly lower temperature than those in a nest high on the beach. In many species, the warm eggs hatch and develop as females, while the cool eggs develop as males, whereas in other species, the situation is reversed. Because nests are made at different locations on the beach, depending in part on the tides, both males and females will occur. A temperature-dependent mechanism like this is also seen in crocodiles and alligators, among other species.

One of many interesting features of this system is that a nest laid at an intermediate temperature has both males and females but rarely has intersexes, that is, individuals that show both male and female development. This suggests that there may be some feedback or amplification system. Once an embryo begins to develop as one sex, which may be initiated by a small stochastic variation in temperature, some key signal is irreversible or amplified in such a way to allow it to continue along that sex differentiation pathway and not to activate the differentiation pathway of the other sex. Feedback systems and genetic amplifiers are discussed briefly in Chapter 14.

Temperature is probably the most common environmental variable used as a signal for sex determination, but it is not the only one. Other environmental signals, such as crowding or external food sources or pH, are also important for sex determination in some species,

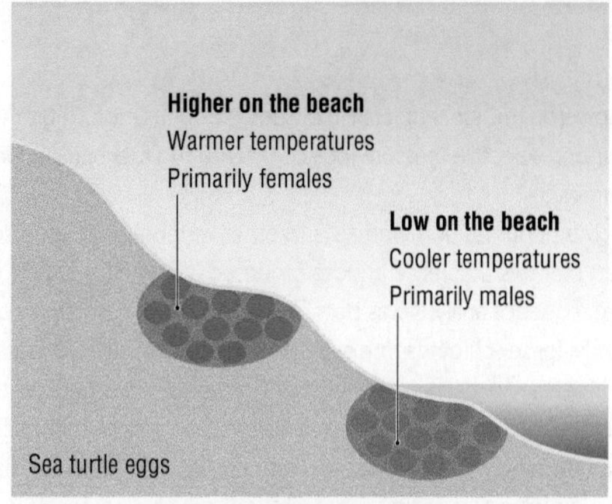

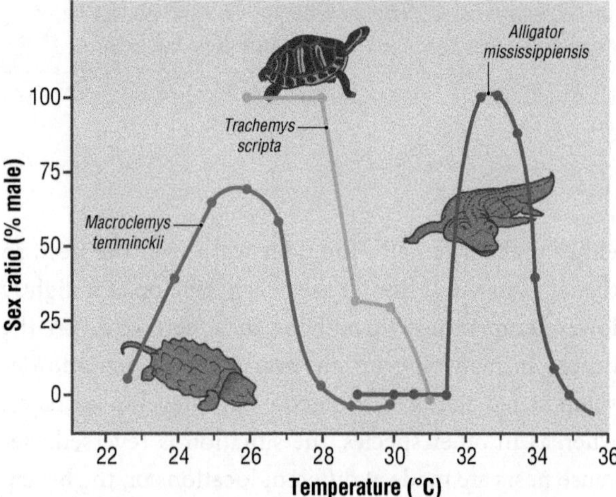

Figure 7.15 Environmental sex determination. Many species with separate sexes use environmental signals to determine sex. Among the best known is temperature-dependent sex determination in different species of reptiles, including turtles. Female sea turtles crawl up the beach, dig a nest, and lay their eggs. Eggs laid in nests low on the beach will be cooled by water at high tide, whereas the ones high on the beach will be warmer from the sunlight. In many species, eggs at low temperatures, 27°C or below, develop as males, while eggs at higher temperatures, above 27°C, develop as females. As shown in the graph (redrawn based on the diagram by Scott Gilbert, *Developmental Biology*, 7th edn, 2003), some species have this pattern reversed, with males developing at high temperatures. In other species, such as alligators, males develop at intermediate temperatures, while females develop from eggs at low and high temperatures.

Source: Top illustration adapted from Crain and Guillette (1998). Reptiles as models of contaminant-induced endocrine disruption. *Animal Rep Sci.* Vol. 53 Iss. 1-4 pp 77-86. Graph reproduced from Gilbert, S.F. *Developmental Biology* 10th edition. With permission from Sinauer Associates, Inc.

particularly fish. A range of the most familiar mechanisms employed in different species is summarized in Figure 7.16.

KEY POINT Several different mechanisms for sex determination are known, including the XY system found in mammals and some invertebrates, the ZW system found in birds, and mechanisms that are environmentally influenced.

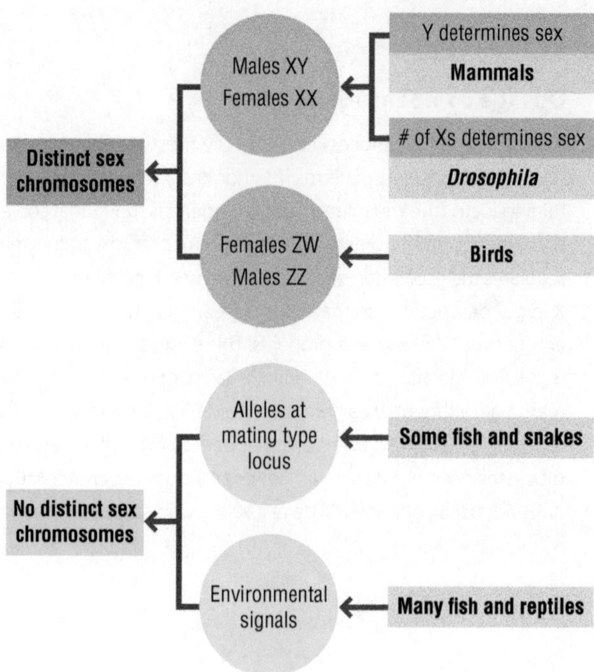

Figure 7.16 Summary of sex determination mechanisms. Sex determination occurs by many different mechanisms in animals. A key difference is the presence or absence of distinct sex chromosomes. Among those with distinct sex chromosomes, some groups have an XX/XY system in which males are heterogametic. In mammals, the Y chromosome is the male-determining signal, while in *Drosophila* and *Caenorhabditis elegans*, the number of X chromosomes is the signal for sex determination. Other groups with distinct sex chromosomes have a ZZ/ZW system in which females are the heterogametic sex; this is seen in birds. Those groups that do not have distinct sex chromosomes may have a sex-determining or a mating-type locus with two alleles; many fish and some amphibians have this system of sex determination. Many other groups use environmental signals for sex determination and have no genetic basis for sex determination. Environmental sex determination (ESD) is widespread among reptiles and fish, with environmental signals, such as temperature, pH, or crowding, being responsible for triggering the sex differentiation pathway.

The examples we have just discussed clearly demonstrate how there is no simple or universal mechanism of sex determination. Indeed, the diversity of sex-determining mechanisms is one of the more fascinating problems in evolutionary and developmental biology. Even among vertebrates, sex chromosomes and sex-determining mechanisms clearly evolved many different times. However, it appears that some of the genes involved in sex determination are highly conserved across animal phyla, as we discuss in more detail in Box 7.5.

BOX 7.5 *Going Deeper* DMRT and evolutionary tinkering with sex determination

As discussed in the chapter, the mechanisms of sex determination among animals are diverse. Among the different signals used to initiate sex determination in different species are the number of X chromosomes, the presence of a gene on the Y chromosome, a single sex-determining locus, and the temperature of the nest. While these may be the best studied examples of sex determination signals, this list does not exhaust the mechanisms that nature has used. In short, the initial trigger for sex determination is not highly conserved. Even among vertebrates, chromosomal sex determination appears to have evolved more than once. This diversity of signals may seem surprising, since sexual dimorphism and sex determination are so important in reproduction and fitness.

The very diversity in the types of signals used to initiate sex determination reveals important points about the evolution of genetic regulatory systems and developmental pathways. In Box 7.2, we introduced the concept of canalization, a form of robustness that helps to ensure a specific and discrete outcome from a variable signal. Here, we introduce another evolutionary concept—in many biological processes, some steps or components are highly conserved during evolution, while other steps and components are much more divergent. Evolutionary constraints are not equally applied to all steps and components in a process, nor should we expect them to be. Sex determination pathways illustrate this very well.

The key to normal male development

Most of the genes involved in sex determination in one organism are also found in other organisms but are not involved in sex determination, for example, *Sxl* and several of its downstream target genes needed for female differentiation in *Drosophila*. Orthologous genes are found in other organisms where they are important in RNA splicing, but splicing is not a critical feature of sex determination in these other organisms. However, nearly all (if not all) animals share one key gene that appears to always be necessary for normal male development. The versions of this gene in different organisms have different names, but the proteins that these genes encode are clearly orthologs of each other. In *D. melanogaster*, the gene is called *dsx*. In *C. elegans*, the gene is called *mab-3*. These were the first two genes that were recognized as encoding orthologous proteins important in sex determination in more than one species, so the other members of the family are named *DMRT* for "*double-sex* and *mab-3*-related transcription factor."

As the name implies, these genes encode transcription factors and thus regulate the transcription of other target genes. Many of these *DMRT* genes regulate the transcription of many targets, a concept described in more detail in Chapter 14. The target genes for the *DMRT* genes are typically those needed for the specific phenotypic features by which the sex is recognized: sex combs, pigmentation patterns, behavioral differences in some species, and, probably most significantly, the development of the gonads.

While *dsx* and *mab-3* are single *DMRT* genes in the fly and the worm genome that are involved in sex determination, a *DMRT* family of genes in the genomes of vertebrates is involved in gonad development, the most important of which seems to be *DMRT1*. The similarity among the proteins in this family, as with all transcription factors, lies in the portion of the amino acid sequence that binds to DNA. The parts of the DMRT proteins outside of the DNA-binding domain are not very similar to one another. The DNA-binding domain is a type of zinc finger protein (introduced in Chapter 2) and is typically found at the C-terminus of the protein.

In nearly all cases, at least one member of the *DMRT* family is expressed in the male and often (but not always) in the developing testis, although they may also be expressed in other tissues. Examples of the male-specific expression of a *DMRT* gene are found in placental mammals, marsupials, chickens, different species of snakes, many species of fish, the frog *Xenopus*, turtles, alligators, mollusks, crabs, and wasps, as well as *Drosophila* and *C. elegans*; a *DMRT* gene is even expressed in coral during its seasonal reproductive cycle. These examples come from every class of vertebrates and diverse phyla of invertebrates. These results indicate that the functions of the *DMRT* genes in sex determination and sexual reproduction are evolutionarily very ancient.

Different signals, same response

Just as significantly from our point of view in this chapter, these examples represent every different form of sex determination signal that we discussed and several that we did not—the number of X chromosomes, the presence of a Y chromosome, ZW chromosomes, multiple X chromosomes, a single sex-determining locus, temperature-dependent sex determination, and even a seasonal cycle regulated by light and temperature.

It is an over-simplification perhaps, but it seems that the role of all of these different signals is to activate expression of the *DMRT* gene in one sex, which becomes the male. In the examples in which results are known, a mutation in the *DMRT* gene is known to affect sex determination, often resulting in intersexual development or sexually ambiguous development. Examples are found in flies, worms, and mammals, so this also suggests an ancient role in sex determination. On the other hand, while the role of the *DMRT* gene is ancient, the means by which the activation of the *DMRT* gene occurs is itself diverse. For example, *mab-3* appears to be transcribed only in males, while *dsx* is expressed in both sexes but spliced differently in the two sexes. In most

BOX 7.5 Continued

other species, the processes that regulate expression of the *DMRT* gene(s) are not known.

What does this tell us about the evolution of genetic pathways? Suppose that, in some ancestral animal, a *DMRT* gene was the key male-determining signal, as the evidence suggests. As long as the role of this gene in sex determination was conserved during evolution, the mechanisms that regulated its expression had less selective pressure and were free to diverge. To use a metaphor from automobiles, so long as the steering mechanism remains connected to the wheels, the shape and composition of the steering wheel and the tires can be more variable.

The genes upstream of the *DMRT* gene, including *Sxl*, *xol-1*, and *Sry*, and most of the somatic sex determination genes in worms and flies, are genes with everyday biological functions that have been "recruited" to be involved in sex determination in that species. These other functions include splicing factors, ligands and receptors, chromosomal proteins, and transcription factors. Any of these common biological processes could be used for sex determination and apparently are used to regulate the expression of the critical *DMRT* gene. To return to Jacob's metaphor introduced in the Prologue, all of these initiating signals are examples of evolutionary tinkering; the *DMRT* gene is the conserved function.

FIND OUT MORE

Kopp, A. (2012) *Dmrt* genes in the development and evolution of sexual dimorphism. *Trends Genet* **28**: 175–84

7.5 Dosage compensation and X chromosomes

The existence of differentiated sex chromosomes introduces a conundrum in genetics and evolution. For many genes, the level of expression roughly correlates with the number of copies in the genome; if there are twice as many copies of a gene, we expect to observe about twice as much RNA transcript. Females have two copies of the genes on the X chromosome, while the male has only a single copy of these genes. There are hundreds of genes on the X chromosome in mammals whose functions are not related to sex differences and which do not have a corresponding allele on the Y chromosome. Thus, if gene dosage and expression were directly correlated for all genes, including X-linked genes, males would make half of much of the X-linked gene products as females make, as summarized in Figure 7.17(a). This implies either that males make half as much of these gene products as they need, which would be evolutionarily deleterious, or that females make twice as much of these gene products as they need, which would seem to be metabolically wasteful.

But when the level of gene expression is actually measured for X-linked genes, males and females in many different species with sex chromosomes make about the same amount, as summarized in Figure 7.17(b), (c), and (d). How does this happen? The imbalance in gene dosage does not result in an imbalance in the level of gene products, because of a process known generally as **dosage compensation**. While dosage compensation probably occurs in nearly all, if not all, species with sex chromosomes, it does not occur in the same way in all organisms. In fact, the three groups in which this has been well studied—*D. melanogaster*, *C. elegans*, and placental mammals—all exhibit dosage compensation, but the mechanisms are different among the three groups.

These differences have been revealed in *Drosophila* and *C. elegans* by the isolation of mutations that alter the process of dosage compensation. In both species, such mutants have sex-specific mutant phenotypes. In *Drosophila*, dosage compensation mutants are lethal in 1X males but have no effect in 2X females. This led to the model summarized in Figure 7.17(b) that, in *Drosophila*, the proteins and RNA molecules responsible for dosage compensation (referred to in the figure as the dosage compensation complex, or DCC) assemble on the X chromosome in males and result in the hypertranscription of X-linked genes. In *C. elegans*, mutations that affect dosage compensation are lethal or subvital in 2X animals (that is, hermaphrodites) but have no effect on 1X animals or males. Thus, in worms, the dosage compensation complex assembles on the X chromosome in 2X animals to reduce the transcription of X-linked genes. Although the mechanisms are different, as revealed by the phenotypes of dosage compensation mutants, the outcome is the same in both species—1X and 2X animals transcribe X-linked genes at similar levels.

Mammals have a different mechanism for dosage compensation from either flies or worms, as summarized in

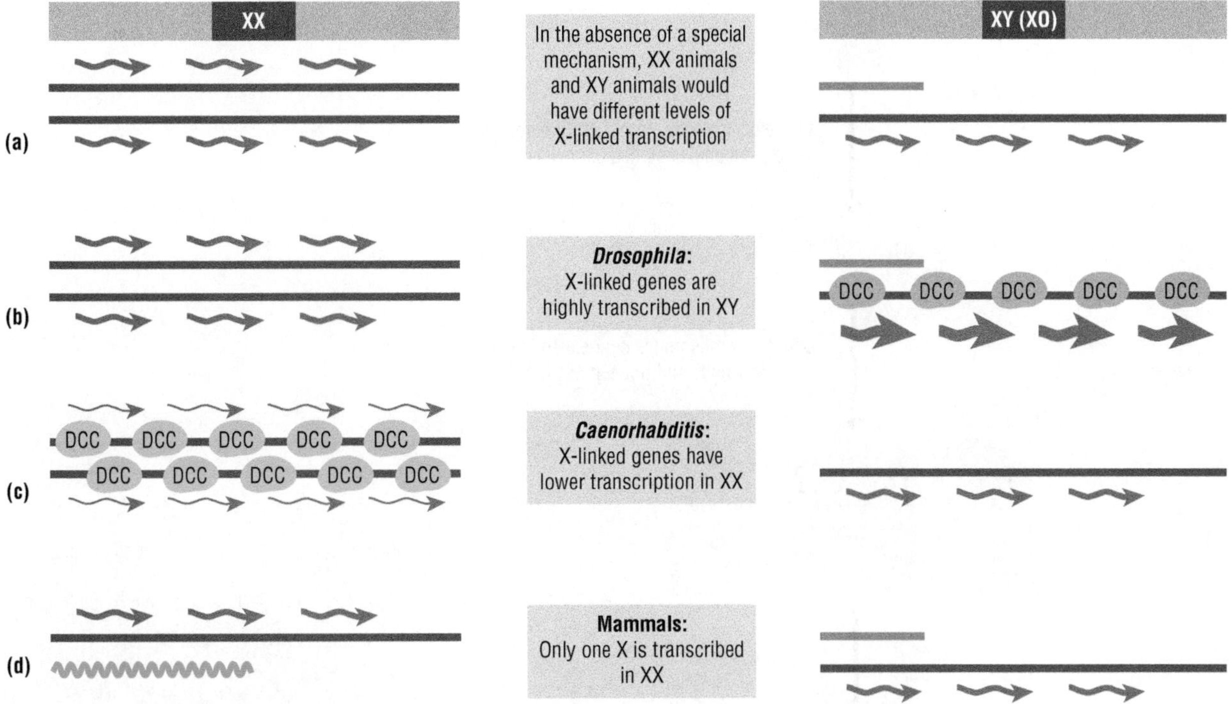

Figure 7.17 An overview of dosage compensation. As shown in (a), in species with genetically distinct sex chromosomes, XX and XY individuals would express their X-linked genes at different levels, unless a special mechanism exists. The mechanism of dosage compensation varies in the three species in which it has been extensively studied. (b) In *Drosophila melanogaster*, XY flies have an elevated level of X-linked gene transcription, compared to XX flies; this is indicated by the heavier arrows for transcription in males. A complex of proteins and RNA molecules, abbreviated here as the dosage compensation complex or DCC, is recruited to the X chromosome in males and is responsible for the hypertranscription of X-linked genes. (c) In *Caenorhabditis elegans*, in which males are X0 rather than XY, the XX worms have a lower level of X-linked transcription, compared to X0 worms, as indicated by the thinner arrows for transcription in XX. The DCC of proteins and RNAs is recruited instead to both X chromosomes in XX worms. The proteins and RNA that make up the DCC in worms are unrelated in sequence to the ones in *Drosophila*. Note that both X chromosomes are transcribed, although at a lower rate. (d) In mammals, only one X chromosome is transcribed in XX individuals, as indicated by transcription arrows occurring from only one chromosome. The other X chromosome is condensed into heterochromatin and is not transcribed. Thus, in contrast to *C. elegans*, only one X chromosome is expressed in any given cell.

Figure 7.17(d) and explained more fully in Figure 7.18. This does not regulate transcription of individual genes so much as it regulates the structure, and thus the transcription, of the entire X chromosome. Since the mechanism of dosage compensation is the same in all placental mammals, we will focus on this process, known as **X chromosome inactivation**. We need to begin with a brief description of mammalian embryogenesis when dosage compensation begins.

Embryology of X inactivation

At an early stage of mammalian embryogenesis, one of the X chromosomes in each of the cells of XX embryos—that is, those embryos destined to become females—is inactivated and made into heterochromatin. One X chromosome is condensed into heterochromatin and moved to the edge of the nucleus, as shown in Figure 7.18. Figure 7.18(a) shows the process of X inactivation diagrammatically, while Figure 7.18(b) is an electron micrograph showing the inactive X chromosome. The inactive X chromosome forms a structure known as a Barr body or a sex chromatin body, and genes on this inactive X chromosome are not transcribed. (At a finer level of analysis, a few X-linked genes do escape inactivation and are transcribed at higher levels in females than in males, but this seems to have very little effect on the embryo.) This X chromosome stays inactive throughout all development, so that, when a cell divides by mitosis, its daughter cells will have the same X chromosome active and the same one inactive.

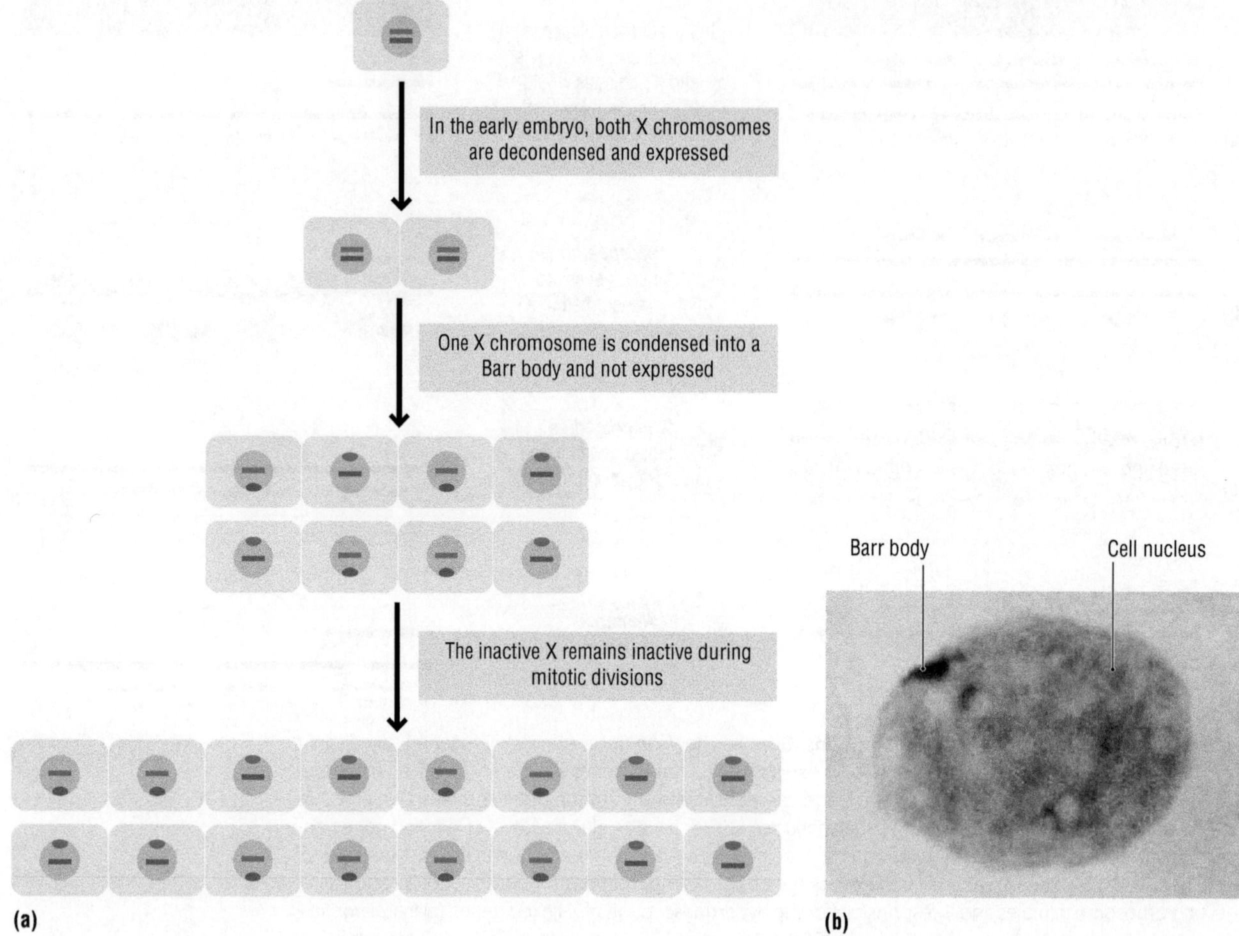
In the early embryo, both X chromosomes are decondensed and expressed
One X chromosome is condensed into a Barr body and not expressed
The inactive X remains inactive during mitotic divisions
(a)
Barr body
Cell nucleus
(b)

Figure 7.18 X chromosome inactivation in mammals. Dosage compensation in mammals occurs by a process known as X chromosome inactivation in XX females. (a) The two X chromosomes are represented as blue or red lines within the nucleus of the cell, and the arrows are cell divisions. In the zygote and early embryo, both X chromosomes can be transcribed. As cells divide, one of the two X chromosomes becomes condensed and localized to the nuclear membrane where it is not transcribed. The inactive X is shown as a ball at the nuclear membrane, while the active X is shown as a line. Inactivation is random, and either X chromosome can be the active or the inactive X. This transcriptionally inactive X chromosome is known as a Barr body or sex chromatin body. Once an X chromosome has been inactivated, it remains inactive through subsequent cell divisions. (b) An electron micrograph that shows the inactive X chromosome as a Barr body at the nuclear membrane.

Source: (b) courtesy of Carolyn Trunca.

In the embryo, the active and inactive X chromosomes are determined at random—about half of the cells in the developing embryo inactivate the X chromosome from the father, whereas the other half of the cells inactivate the X chromosome from the mother. In marsupials, such as kangaroos and opossums, the inactive X chromosome is always the paternal chromosome. Even in placental mammals, the paternal X chromosome is inactivated in extra-embryonic structures, such as the amniotic sac and the placenta, so these cells express only the X chromosome inherited from the mother.

As a result of X chromosome inactivation, male and female mammals make the same amount of gene product from X-linked genes because females only use one of their two X chromosomes in any particular cell. However, since different X chromosomes are inactivated in different cells, female mammals exhibit cell-type differences in expression for all X-linked genes. (Although this is often referred to as another example of mosaicism, the strict use of the term "mosaicism" that we introduced previously requires that the cells have genetic differences, which cells in a female do not have.) We will discuss some of the biological consequences of this shortly.

X inactivation and epigenetics

X inactivation is one of the best examples of epigenetics. Epigenetics refers to a heritable change in gene activity—and thus the phenotype—without a corresponding change in the DNA sequence. In general, epigenetic changes arise because the structure of the chromosome has been altered, and this new chromosome structure is preserved during mitotic divisions. Epigenetic changes underlie a type of "cellular memory," such that, when a cell divides by mitosis, its daughter cells retain the differentiated state of the original cell. In this case, the epigenetic modification is the formation of the Barr body.

The mechanisms responsible for different epigenetic changes are diverse, and many of them are not yet known. While X chromosome inactivation was among the first known examples of epigenetic changes, the mechanism by which it occurs is unique; no other biological process closely resembles the molecular events in X inactivation. The key molecule is an unusual and large RNA that is transcribed from the inactive X chromosome, an RNA of about 17 kb known as Xist. Since other non-coding RNAs are not nearly this large, it is difficult to draw comparisons between Xist and any other molecule or to make inferences about how Xist functions. In general, once the Xist RNA is transcribed, it localizes along the length of the inactive X chromosome and appears to recruit proteins that modify the histones to make an epigenetic code. (The concept of a histone code of epigenetic modifications for other cellular functions will be discussed in Section 12.4.) A brief description of the complicated process of X inactivation and the role of Xist is given in Box 7.6.

X inactivation may have affected the evolution of X-linked genes

X inactivation was discovered in the early 1960s by the British geneticist Mary Lyon; it is still sometimes referred to as "lyonization" in her honor. X inactivation, in which

BOX 7.6 *A Human Angle* X chromosome inactivation and Xist

Dosage compensation in mammals occurs by the inactivation of one of the two X chromosomes in females; the inactive X chromosome is condensed into heterochromatin as a cytologically recognizable Barr body or sex chromatin body. Since trisomic XXX and tetrasomic XXXX cells in culture have two and three Barr bodies, respectively, it might be more accurate to regard the dosage compensation system as keeping one X active, but we will use the more familiar approach and talk about the process of inactivation. Since being proposed by Lyon in 1961, X chromosome inactivation has been repeatedly cited as a prototypical example of an epigenetic effect. While it is an outstanding example of epigenetics and has many features in common with other types of epigenetic effects, as defined in this chapter and in Chapter 3, the process of X inactivation is also quite different from most other epigenetic examples. The process occurs in several distinct steps.

Inactivation initiates at the Xic

In the developing mammalian embryo, one of the two X chromosomes will be inactivated. The earliest differentiation in the mammalian embryo separates the outer layer, known as the trophoblast (and later, the trophoectoderm), from the inner cell mass. The inner cell mass will eventually become the various cell types, tissues, and organs of the organism. The trophoectoderm gives rise to the extra-embryonic tissues such as much of the placenta, the amniotic sac, and the chorion. In these extra-embryonic cells, the paternal X chromosome is always inactivated and the maternal X chromosome is kept active. Thus, there must be some molecular markers on the paternal and maternal X chromosomes that allow them to be distinguished. (In marsupials, the paternal X chromosome is inactivated in all cells, but, in placental mammals, it is only in the extra-embryonic cells.)

The nature of these molecular markers is not clear, but there are many differences between chromosomes that come from the sperm and chromosomes that come from the egg, so there are numerous possibilities. For example, most of the DNA in sperm is condensed with an arginine-rich class of protein, known as protamines, rather than with histones, so sperm chromatin has a different structure from the outset. There is evidence in mice that, at the very earliest stages of inactivation, when the developing embryo consists of only two to four cells, the paternal X is highly condensed and inactive; thus, it may be that the paternal X is kept inactive in those cells that will become the extra-embryonic cells but reactivated in the cells that will become the inner cell mass and the embryo proper. By the time the inner cell mass is distinct from the extra-embryonic cells, both X chromosomes are decondensed and appear to be active, although transcription at this stage is very low.

BOX 7.6 Continued

In the cells of the inner cell mass, the choice between active and inactive X chromosomes appears to be at random; there is no consistent preference for inactivation of the paternal chromosome over the maternal chromosome, or vice versa. The inactivation begins when the inner cell mass has about 8–12 cells, and each cell inactivates its X separately; at the time of inactivation, neighboring cells are no more likely to have the same X inactivated as non-neighboring cells. The active X is designated Xa, while the inactive X is designated Xi. Inactivation does not occur simultaneously in all cells, but the process is complete by the end of gastrulation. The initiating event for inactivation is not known, and the basis for the choice between the two X chromosomes is also not known at this time.

It is clear, however, that inactivation begins at a particular region near the centromere on the long arm of the X chromosome, termed the X chromosome inactivation center, or *Xic*. Some DNA sequence variants at the *Xic* render the chromosome slightly more or less likely to be the inactive X, so there may be a protein that binds at the *Xic* or a structure that forms there to initiate inactivation. (Conversely, something may occur at the *Xic* on the active X chromosome that blocks its inactivation.) It does seem that inactivation continues to completion once it begins. Thus, while the initial choice between the two X chromosomes may be random or involve a very subtle difference, the propagation of that choice into inactivation results in one active X and one inactive X. Once an X chromosome is inactivated, it remains inactivated in its mitotic daughter cells, as described in the chapter; this persistence of inactivation is one reason that calico cats have distinct patches of fur color.

The inactivation spreads from the *Xic* to encompass the entire chromosome. Thus, there is an initiation step, followed by a spreading step. The nature of this spreading step is not known precisely but certainly involves a progressive formation of heterochromatin. Eventually, the entire X chromosome of about 164 million bases is inactivated. As noted in the chapter, some genes, possibly about 5% of X-linked genes, escape inactivation and are expressed from both X chromosomes. The genes that escape inactivation are not precisely the same in all mammals and may not even be precisely the same in different cell types. While it is convenient to imagine that spreading is progressive, like rolling up a rug, a chromosome is a three-dimensional molecule, while a rug is more similar to a two-dimensional object; not all of the genes that escape inactivation are adjacent to each other on the X chromosome, but they may be adjacent to each other on the three-dimensional chromosome.

Structure of the inactive X chromosome

The inactive X chromosome is an example of heterochromatin, as discussed in Chapter 3. Because it is compacted or condensed, Xi stains brightly with DNA dyes. Genes in vertebrates have a collection of C–G dinucleotides in their upstream regulatory region, known as CpG islands. On the inactive X, the CpG islands are heavily methylated, a characteristic of transcriptionally silenced genes. Particular histone modifications that are characteristic of heterochromatin, such as high levels of H3K9 methylation and low levels of H3K4 methylation and H3 acetylation, are also found on the Xi. (Histone modifications are discussed in Chapter 3 and again in Chapter 12.) In addition, a protein known as H2AYF replaces the H2A protein in some of the nucleosomes. Like other regions of heterochromatin, the inactive X chromosome replicates late in the cell cycle. In these ways, the inactive X chromosome is similar to other types of heterochromatin.

None of these modifications or structural changes is likely to be the event that initiates inactivation. Rather, these are the molecular mechanisms that ensure that, once silenced, the chromosome remains heterochromatic and is silenced through subsequent cell divisions. If the initial steps in inactivation are equivalent to closing a door, these steps are analogous to locking it up tightly and keeping it locked.

Xist

X chromosome inactivation is generally similar to the formation of heterochromatin in other regions of the genome and in other organisms. However, the most unusual aspect of X inactivation was the discovery of a non-coding RNA called Xist. Xist (for "Xi-specific transcript" and pronounced "exist") is transcribed from the *Xic* and exclusively transcribed from the inactive X chromosome. Transcription of Xist may be the triggering event or the master regulator for inactivation—both X chromosomes transcribe Xist at low levels before inactivation, and the chromosome that transcribes Xist at high levels first becomes the inactive one—or it may be an early event indicative of X inactivation. Xist is the only transcript that is specific to the Xi. Mice with deletions of the *Xist* gene die as very young embryos and fail to inactivate an X chromosome, indicating that the *Xist* gene has an essential role. Furthermore, if the *Xist* gene is inserted into an autosome in a cell line, that autosome is inactivated like Xi, which suggests that Xist (or something that interacts with the *Xist* gene or its RNA) controls inactivation.

Precisely how Xist is involved in X inactivation is not known, and it may have several roles at different stages of the inactivation process. Xist is a non-coding RNA, so it is not translated into an amino acid sequence and appears not to be capable of being translated. It is spliced and poly-adenylated, like mRNAs and many (but not all) long non-coding RNAs. On the other hand, most non-coding RNAs are relatively short, less than 1 kb in length or shorter than a typical mRNA. By contrast, the Xist RNA is enormous, more than 17 kb long in humans.

BOX 7.6 Continued

The sequence of Xist in humans includes a series of repeated sequences that would be capable of forming complicated intra-strand structures in the RNA, which may play a role in its functions; however, not all of these repeats (and thus, the structures that they can form) are found in the mouse Xist RNA. Certain proteins known to be important in chromatin structure and epigenetic regulation, such as members of the Polycomb family, are recruited to the Xist RNA, so some of its activity may be through its action as a binding site for other factors. It can also form double-stranded RNA hybrids with some other RNA molecules, including an RNA known as Tsix, transcribed using the other strand of the *Xic*; Tsix (Xist spelled backwards) corresponds to the reverse complement of Xist.

Because Xist is a uniquely long non-coding RNA, the mechanisms by which it acts have been somewhat difficult to elucidate; many activities and functions have been found, but the relationship among these and X inactivation are an area of active investigation and speculation.

Once Xist is transcribed from the *Xic*, it spreads along the inactive X, so that eventually the inactive X is coated with Xist. Figure A shows a cell in culture with both an active and an inactive X, and the localization of Xist.

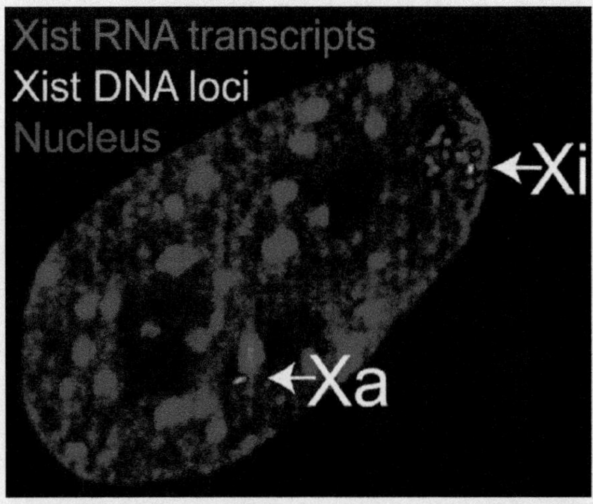

Figure A Xist and the inactive X chromosome. In this image of a fibroblast cell from a female mouse, the DNA is stained blue, while a DNA sequence from the region of the *Xic* is shown in yellow. This can be used to contrast the structure of Xa and Xi. The Xist RNA is labeled in red; note how it is found along the length of the inactive X chromosome.

Source: Reproduced from Reinius, B. et al. (2010). Female-biased expression of long non-coding RNAs in domains that escape X-inactivation in mouse. *BMC Genomics* 11: 614.

X chromosome reactivation

The X inactivation cycle does not end with inactivation, however. While Xi remains inactive in all of the somatic cells of a female, a reactivation process must occur in the germline cells. The mitotically dividing nuclei in the ovary have an inactive X. When these nuclei enter meiosis to form ova, the X must be reactivated or at least decondensed if not actively transcribed. The embryo begins with an active X chromosome from the mother, which implies that her ova had decondensed X chromosomes. This reactivation process has not been studied as intensively as the inactivation process, and not much is known about it.

only one X chromosome is being expressed, has an evolutionary consequence, which was recognized shortly after the process was described. Genes that are X-linked in one placental mammal are almost always X-linked in all other mammals. In other words, their position in the genome—their presence on the X chromosome, as opposed to any other chromosome—has been retained for all of these genes throughout evolution, presumably because of the mechanism of dosage compensation.

X-linked genes are often found in a different order on the X chromosomes in different mammals, but they are still on the X chromosome. This conservation of chromosome location is not true for other chromosomes. As described in Section 9.2, genes that are on chromosome 4 (say) in humans are scattered throughout the chromosomes in mice, although certain blocks of conservation are seen. Likewise, the conservation of X-linkage is not seen for genes in other animals. Genes that are X-linked in *D. melanogaster* are not necessarily X-linked in *Drosophila virilis* or *Drosophila pseudoobscura*, let alone in other insects. This difference between conserved X-linkage in mammals and non-conserved X-linkage in other animals probably arises due to differences in the mechanisms of dosage compensation.

What happens with X inactivation in X chromosome aneuploids? In females or cell lines in culture with multiple X chromosomes, such as XXX, only one X chromosome remains active and all the others are inactive, so an XXX female has two sex chromatin bodies. Similarly, an individual who is XXY has one active X chromosome and one sex chromatin body, whereas an X0 woman has no sex chromatin body. This inactivation process, and the resulting dosage compensation, is the reason that aneuploidy for the X chromosome in humans is much less serious than aneuploidy for an autosome—only one X chromosome is normally expressed in any given cell, and a process exists to inactivate all other X chromosomes. Thus, while we usually focus on one X chromosome becoming inactive, sex chromosome aneuploids suggest that the process might be better thought of as keeping one X chromosome active and all others inactive.

KEY POINT Dosage compensation corrects for a potential imbalance in the amount of gene products made by males vs. females. In mammals, one X chromosome in every cell of the female individual is inactivated, forming a Barr or sex chromatin body.

Biological effects of X inactivation

This mechanism of dosage compensation in mammals also has consequences for phenotypes in heterozygous females. This is most easily illustrated by calico cats, as seen in Figure 7.19. An X-linked gene in cats affects coat color. This gene has two alleles, known as black (***B***) and orange (***b***). Although almost all genes found in one mammal are found in other mammals, and particularly all X-linked genes are found on the X chromosome in most mammals, this particular gene appears to be a curious exception and is only found in cats. There are no calico dogs, horses, mice, or humans, but there have been reports of calico lions.

A male cat will have only one X chromosome, so it will be either black or orange. But think about what happens in a female cat that is heterozygous at this locus. If the X chromosome with the black allele is inactivated in a cell or group of cells, only the orange allele is expressed from the active X chromosome in those cells, and that patch of fur is orange. If the X chromosome with the orange allele is inactivated in those cells, then the fur is black because that is the only X chromosome to be actively transcribed. As a result, the female who is heterozygous has fur with patches of black and orange. Thus, calico cats are nearly always female; the rare calico male cat has to be XXY and is sterile.

While calico cats are the most readily observed example of a phenotype arising from X chromosome inactivation, other examples are known. Some women who are heterozygous for the X-linked muscular dystrophy allele show evidence of muscle weakness, presumably due to inactivation of the functional copy of the gene in some muscle cells. However, X inactivation does not seem to

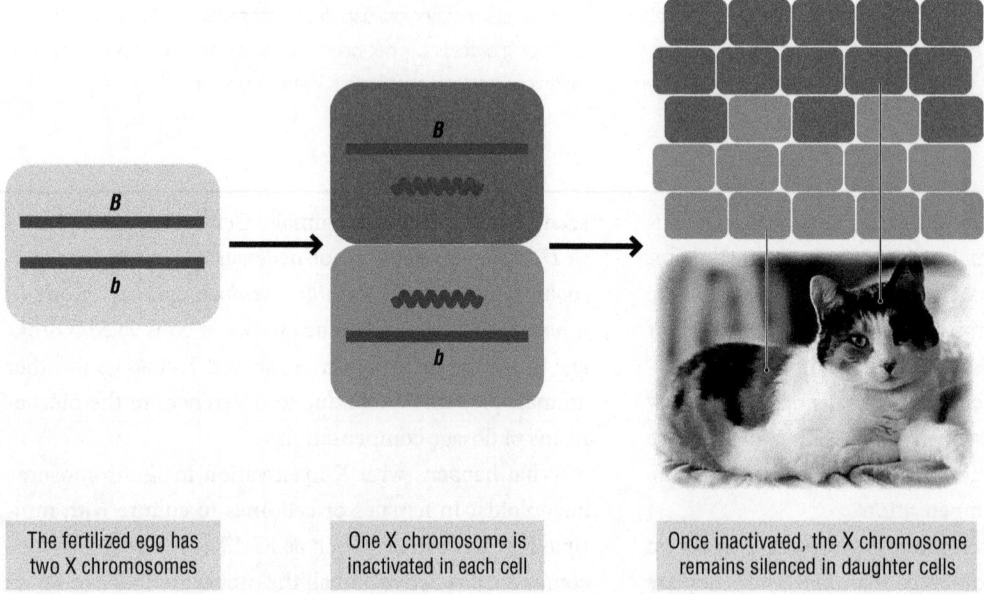
B
b
B
b
The fertilized egg has two X chromosomes
One X chromosome is inactivated in each cell
Once inactivated, the X chromosome remains silenced in daughter cells

Figure 7.19 X inactivation and calico cats. Cats have an X-linked gene with two alleles, ***B*** and ***b***, that affects coat color. In a ***Bb*** heterozygote, some cells have the chromosome with the ***B*** allele active, while other cells have the chromosome with the ***b*** allele active; the other allele is inactivated as a Barr body. Once the chromosome has been inactivated, it remains inactive through subsequent mitotic divisions. As a result, a ***Bb*** heterozygote has patches of black fur and patches of orange or yellow fur. This pattern is known as calico. The white fur on the underbelly is due to the action of a different gene.

Source: Image of cat courtesy of Tatiana Chessa/ CC BY-SA 3.0.

have much of a phenotypic effect on women who are heterozygous for the two most familiar X-linked traits in humans—color blindness and hemophilia. The retina arises from several different cells in the embryo, so it is unlikely that all of these cells would have the same inactive X chromosome, particularly for both eyes. Likewise, factors VIII and IX are diffusible, so which cells produce them is generally not relevant for blood clotting.

7.6 Summary: X-linked genes and sex determination

The inheritance pattern of X-linked genes is one of the most evident consequences of the behavior of chromosomes during meiosis. In species with sex chromosomes, genes occurring on the X chromosomes are inherited in different patterns to genes found on autosomes. Recessive X-linked traits are observed more often in males, as males have only one X chromosome, making dominance irrelevant. Inheritance patterns of X-linked genes provided evidence that meiosis could explain Mendelian genetics. In fact, although it may seem strange to us, genetics textbooks from the 1920s and 1930s often had X-linked inheritance as the first chapter, from which meiosis and autosomal inheritance then logically followed.

While the existence of X-linked genes and their inheritance was a key piece of evidence connecting Mendel's rules with the process of meiosis, additional support came from the study of sex chromosome aneuploids. These arise from meiotic non-disjunction during gametogenesis and either are lacking a sex chromosome or have an additional sex chromosome in their cells. The study of aneuploids also provided information about the various mechanisms of sex determination in many animals.

Despite the differences in gene dosage for X-linked genes in females and males, little or no difference in the levels of X-linked gene products exists between the two sexes. This is important, since so few X-linked genes have functions related to sexual dimorphism. Potential imbalances between males and females in the amount of expression of X-linked genes are corrected by dosage compensation mechanisms. Dosage compensation, like sex determination itself, occurs by different mechanisms in different organisms. In mammals, dosage compensation involves the transcriptional inactivation of one X chromosome in females. It is thought that this mechanism of dosage compensation also imposed an evolutionary constraint, such that an X-linked gene in one mammal is very likely to be X-linked in other mammals.

It may seem like a curious reality that a process such as sex determination, with its fundamental connection to reproduction and fitness, has so many different variations. The evolution of sex-determining systems is a topic that might make us re-think our position that we understand how evolution ought to work. Such evolutionary tinkering with a fundamental process might be unexpected, and the fact that sex chromosomes evolved separately many different times forces us to examine where the actual constraints lie in biological systems. X-linked inheritance is a natural and somewhat obvious consequence of meiosis. Sex determination itself is not such an obvious consequence of evolutionary forces.

CHAPTER CAPSULE

- The chromosomes found in an individual can be either sex chromosomes or autosomes. Humans have 22 pairs of autosomes and one pair or sex chromosomes—either XX in the female or XY in the male.
- In mammals and other animals with X and Y chromosomes, the inheritance of genes on the X chromosome shows different patterns to genes found on the autosomes.

- Males are affected by X-linked traits more often than females, since they are hemizygous for these genes. White eyes in *Drosophila* and red–green color blindness and hemophilia in humans are familiar X-linked traits.
- Errors in the separation of chromosomes or sister chromatids during meiosis, known as non-disjunction, can lead to offspring with too many or too few sex chromosomes. These sex chromosome aneuploids are often viable. Studies of the results of these non-disjunction events have shed light on modes of sex determination.
- Mechanisms of sex determination vary among species. In mammals, the Y chromosome carries genes that initiate the development of the testes and subsequent development of a male individual. In *Drosophila*, the number of X chromosomes determines the sex of the fly.
- In birds and some invertebrates, females are the heterogametic sex (ZW) and males are homogametic (WW).
- Other mechanisms of sex determination include genes for mating type or the effect of environmental cues such as temperature.
- Dosage compensation corrects for the potential imbalance of gene products produced by male and female individuals. In mammals, X inactivation early in development results in a single active X chromosome in each cell, which may vary among cells. As a result, females are mosaics for all X-linked genes.

STUDY QUESTIONS

Concepts and Definitions

7.1 In examining a large pedigree for a human trait, what would be an indication that the trait is X-linked? What would provide the clearest evidence that the trait could not be X-linked?

7.2 Define the difference between heterogametic and homogametic sexes. Give an example of a species in which the male is heterogametic, and one in which the male is homogametic.

7.3 Why is inheritance of X-linked recessive traits sometimes called criss-cross inheritance?

7.4 Why are the terms "X-linked" and "sex-linked" frequently used interchangeably? In at least one widely used domesticated species, the terms are not synonyms. What species is this, and why are these terms not synonyms in that species?

7.5 It was long thought that the appearance of tufts of hair on the outer ear was inherited as a Y-linked trait in humans. What distinctive inheritance pattern would have led to this proposal? Why might there have been evidence that this is not a Y-linked trait?

7.6 Using Online Mendelian Inheritance in Man (OMIM; www.ncbi.nlm.nih.gov/omim) or another source for human genetics, what are examples of X-linked recessive traits in humans?

7.7 Why are male calico cats both rare and usually sterile?

7.8 Individuals who are aneuploid for one of the sex chromosomes usually survive longer and have fewer health consequences than individuals who are aneuploid for one of the autosomes. Explain why this is true.

Beyond the Concepts

7.9 The original white-eyed fly that the Morgans found was a male. Based on what you know about X-linked inheritance and sex determination in flies, why was it much more likely that the first mutant would be a white-eyed male, rather than a white-eyed female?

7.10 Explain why non-disjunction was considered proof of the chromosome theory of heredity.

7.11 The pedigree shown in Figure Q7.1 is an example of an X-linked recessive trait.

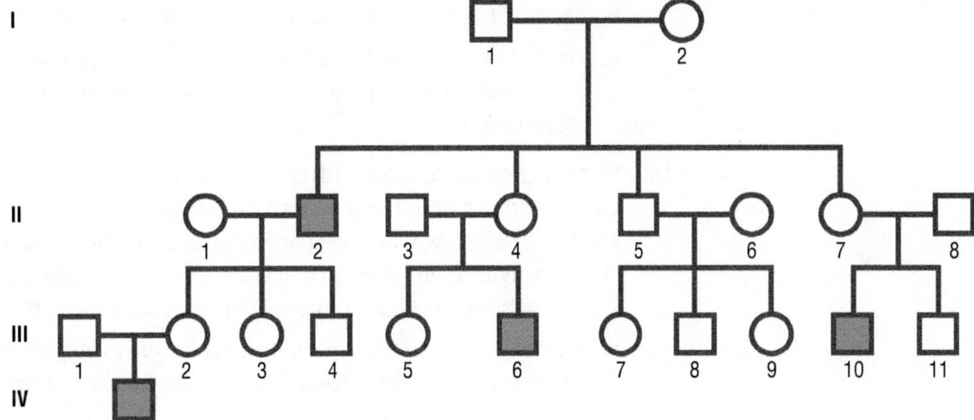

Figure Q7.1

a. Could this inheritance pattern occur if the trait were autosomal? If so, what additional assumptions would need to be made to explain this inheritance pattern?

b. Could this inheritance pattern occur if the trait were X-linked but dominant?

c. For each of the following individuals, calculate the probability that their first son will be affected by the trait.

i. III-3

ii. III-5

iii. III-9

iv. III-10

Challenging **7.12** The genetic analysis of *Caenorhabditis elegans* as a model organism was begun in the early 1960s. Mutations were isolated and numbered in the order in which they could be assigned to a location on a chromosome. Nearly all of the mutants with the lowest numbers are X-linked. Suggest an explanation for this.

7.13 A *Drosophila* female that is missing cross veins on the wing is mated to a male with normal wings. The female F_1 offspring have normal wings, but the male F_1 offspring have crossveinless wings.

a. What two conclusions can you draw so far about the inheritance of crossveinless?

b. One of these F_1 females is mated to a normal male. What are the expected results among the F_2 offspring?

c. Two F_2 females are chosen, and each is put into a vial and mated with a male with crossveinless wings. One of these matings produces only wild-type flies. The other mating yields both crossveinless flies and normal flies. Explain these results. What are the expected proportions of crossveinless and normal flies in this vial?

7.14 A gene known as *lon-2* is X-linked in *C. elegans*; mutations in *lon-2* result in worms that are unusually long. A *lon-2* mutant male was mated to a wild-type hermaphrodite.

a. What will be the cross-progeny of this mating if *lon-2* is recessive?

b. What will be the cross-progeny of this mating if *lon-2* is dominant?

c. In fact, *lon-2* is recessive. One of the F_1 hermaphrodites is mated with wild-type males. What are the expected progeny among the offspring of this mating, and in what proportions will they be found?

7.15 A male mouse with a pale yellow coat called cream is mated to a female mouse with a wild-type gray–brown coat. All of the F_1 mice, both males and females, have the wild-type gray-brown color. One of these F_1 female mice is mated to a cream-colored male.

a. What are the expected results if cream color is X-linked and recessive?

b. What are the expected results if cream color is autosomal and recessive?

c. In fact, cream color is X-linked. When the F_1 female mouse is examined more closely, it is apparent that she has patches of fur that are cream-colored, rather than gray–brown. Explain this result.

7.16 Some parakeets and other birds of the parrot family have a phenotype known as lutino. Lutino is a much prized beautiful coloration pattern, in which the feathers on the body and head are yellow, with silver patches on the cheeks and light yellow tail feathers. (These birds are pretty enough that you might want to look at a picture of them online.) The birds also have red or pink irises in the eyes, rather than the standard black irises.

a. A lutino male parakeet is mated to a normal green female. The female offspring are all lutino, while the male offspring are all green. Explain this result.

b. One of the male offspring in (a) is mated to a lutino female. What are the expected outcomes of this mating?

7.17 As discussed in the chapter, white eyes in *Drosophila* is an X-linked trait. Female flies that are heterozygous for the *white* gene have red eyes, and no heterozygotes have patches of red and white coloration. In fact, of the hundreds of X-linked recessive mutations in *Drosophila*, none of them shows a mosaic or chimeric phenotype in heterozygotes. Why is mosaicism or chimerism seen for X-linked traits in mammals but not in *Drosophila*?

Challenging **7.18** As was shown with calico cats, female mammals that are heterozygous for an X-linked gene have some cells that express one allele and some that express the other allele. However, this difference between the two expression patterns does not make a difference in the phenotypes for most X-linked traits. In other words, calico cats have a very striking phenotype, but this pattern of cell differences is not seen for most other X-linked traits. What are some of the possible explanations for the lack of differences for most X-linked traits?

Applying the Concepts

7.19 Factor VIII and factor IX are encoded by two X-linked genes in humans involved in blood clotting, and mutations in these genes result in hemophilia A or hemophilia B, respectively. Originally, it was not known which form affected the royal families, but the inheritance pattern is the same in either case. Use Figure 7.6 to answer the following questions.

a. Queen Victoria's daughter Alice had two daughters, Irene and Alexandra. Both of these daughters were carriers for hemophilia, as seen from their children. What is the probability that the two daughters born to Alice would both be carriers of the disease?

b. Note that Irene married her first cousin, Henry of Prussia. Normally, the marriage of first cousins results in a higher occurrence of recessive disorders. However, that is irrelevant to the explanation of the occurrence of hemophilia in this family. Why is this information on the family relationship not important?

c. Irene's cousin Alice had a hemophiliac son, Viscount Rupert Trematon, as well as a daughter Mary, and a son who died as a newborn. What is the probability that this son was hemophiliac? (His actual status is not known.)

7.20 A calico cat mates with a black male cat. What phenotypic ratios are expected among her offspring? (Both color and sex of the offspring are relevant phenotypes for this question.)

7.21 Shown below are three pedigrees for rare genetic traits in humans. Affected individuals are shown in blue, while unaffected individuals are shown in white. Although each pedigree has too few members to be absolutely confident about how the trait is inherited in each case, certain modes of inheritance might be likely and some might be impossible. For each of the pedigrees below, which modes of inheritance—that is, X-linked recessive, X-linked dominant, Y-linked, autosomal recessive, and autosomal dominant—can be ruled out? Which is the most likely?

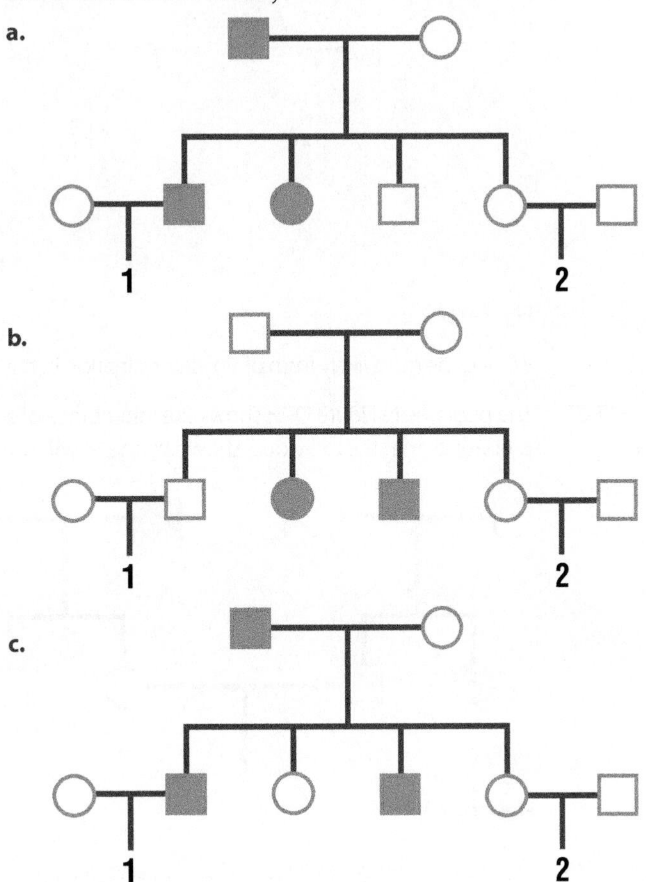

d. For the most likely mode of inheritance in each pedigree, calculate the likelihood that child #1 will be affected by the trait.

e. For the most likely mode of inheritance in each pedigree, calculate the likelihood that child #2 **will not** be affected by the trait.

7.22 You identify a new species of fish, which clearly has two separate sexes—male and female. However, all of the chromosome pairs appear identical in the two sexes, so it is not apparent if sex determination depends on an X/Y system, a Z/W system, or a single mating-type locus. All of these systems of sex determination are known to occur in some species of fish, so any of them could be at work in this new species. You have found a locus that might be able to distinguish these possibilities. Using the polymerase chain reaction (PCR), you amplify this sequence from a male and a female and from two of their offspring—one male and one female. You separate the PCR products on an agarose gel with the results shown in Figure Q7.2.

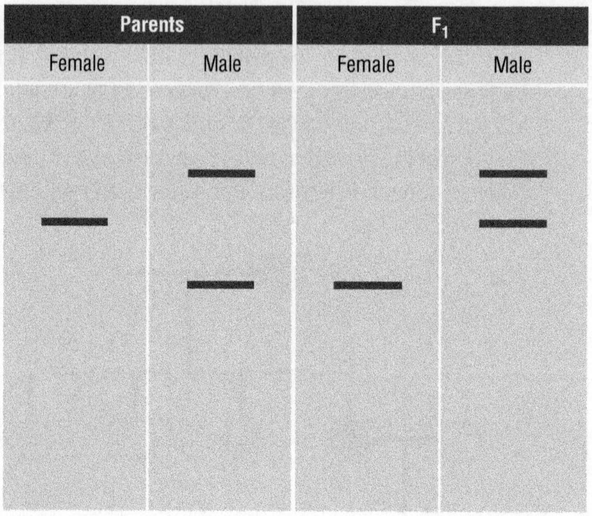

Parents		F_1	
Female	Male	Female	Male

Figure Q7.2

What is the most likely form of sex determination in this fish? Briefly explain your reasoning.

7.23 The pedigree in Figure Q7.3 shows the inheritance of a rare trait in humans. What is the probability that the individual shown by the "?" **will not** be affected by this trait?

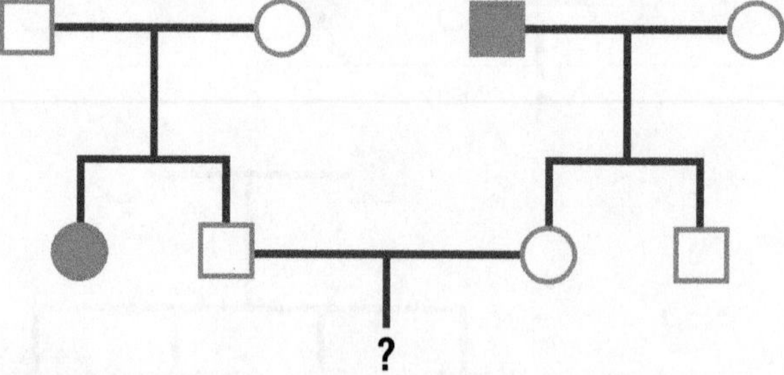

Figure Q7.3

Looking ahead **7.24** In *C. elegans*, the genes *lon-2* and *unc-18* are both X-linked and are very close to each other on the X chromosome. Mutations in the *lon-2* gene result in worms that are unusually long (Lon), while mutations in the *unc-18* gene result in worms that are uncoordinated (Unc).

a. A wild-type male is mated to a hermaphrodite that is both Lon and Unc. The F_1 hermaphrodites are wild-type, while the F_1 males are both Lon and Unc. One of the F_1 hermaphrodites is mated to a wild-type male. What are the expected male progeny from this cross, and in what proportions will they arise?

b. A Lon male is mated to a hermaphrodite that is Unc. The F_1 hermaphrodites are wild-type, while the F_1 males are Unc. One of the F_1 hermaphrodites is mated to a wild-type male. What are the expected male progeny from this cross, and in what proportions will they arise?

Looking back and ahead

c. While nearly all of the F_1 males from the cross in (b) have the expected phenotypes, a few of the males are completely wild-type. Based on what your learned about meiosis in Chapter 6, postulate the origin of these few wild-type males.

7.25 During the Korean War, male soldiers and female nurses were given antimalarial drugs such as primaquine. Some soldiers developed severe and life-threatening hemolytic anemia, in which their red blood cells lysed; other soldiers were completely unaffected. Subsequent genetic analysis showed that the hemolytic response was due to a particular allele of the X-linked gene encoding the enzyme glucose-6-phosphate dehydrogenase (G6PD).

a. Explain why the response of males to this drug was biphasic—some with a severe reaction and others with no reaction, with no intermediate phenotypes.

Challenging

b. Surprisingly, some female nurses also showed this severe hemolytic response, although the allele is rare enough that no nurse was expected to be homozygous. Postulate an explanation for this response.

7.26 The Romanov family, the last czars of Russia, were executed during the Russian Revolution, although there have been long-standing rumors that some family members might have escaped. The last czarina was Alexandra, Queen Victoria's granddaughter in the pedigree, who was married to Czar Nicholas II. They had four daughters and a son, Alexis (or Alexei), who had hemophilia. The relevant part of this pedigree for this question is also shown in Figure Q7.4. The likely grave site for the last of the Romanovs, including Alexis, was identified, and, in order to confirm that the bones were from the Romanov children, the two X-linked genes for clotting factors VIII (responsible for hemophilia A) and IX (responsible for hemophilia B) were analyzed. A small amount of DNA was isolated from the femur of the skeleton of the boy and sequenced. For the factor VIII gene, no sequence changes were found between the DNA from the bones and from sequences in unaffected people. However, when DNA was sequenced from the skeleton of the boy, a single base change was found in the factor IX gene, as shown below the pedigree in Figure Q7.4(A). This sequence change is illustrated in Figure Q7.4 (B) by the arrow on the sequencing scan.

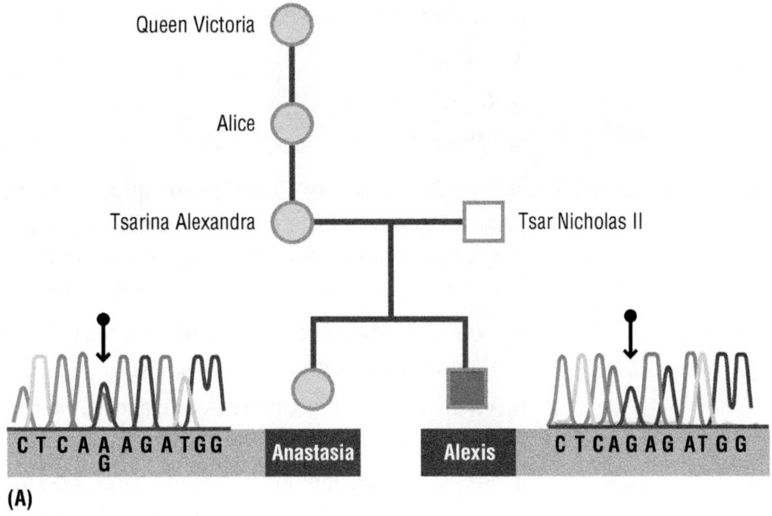

Figure Q7.4

Source: Reproduced from Rogaev, E., et al. Genotype analysis identifies the cause of the "royal disease". *Science*. 2009 Nov 6;326(5954):817.

a. Why were the genes sequenced from the boy, rather than one of his sisters?

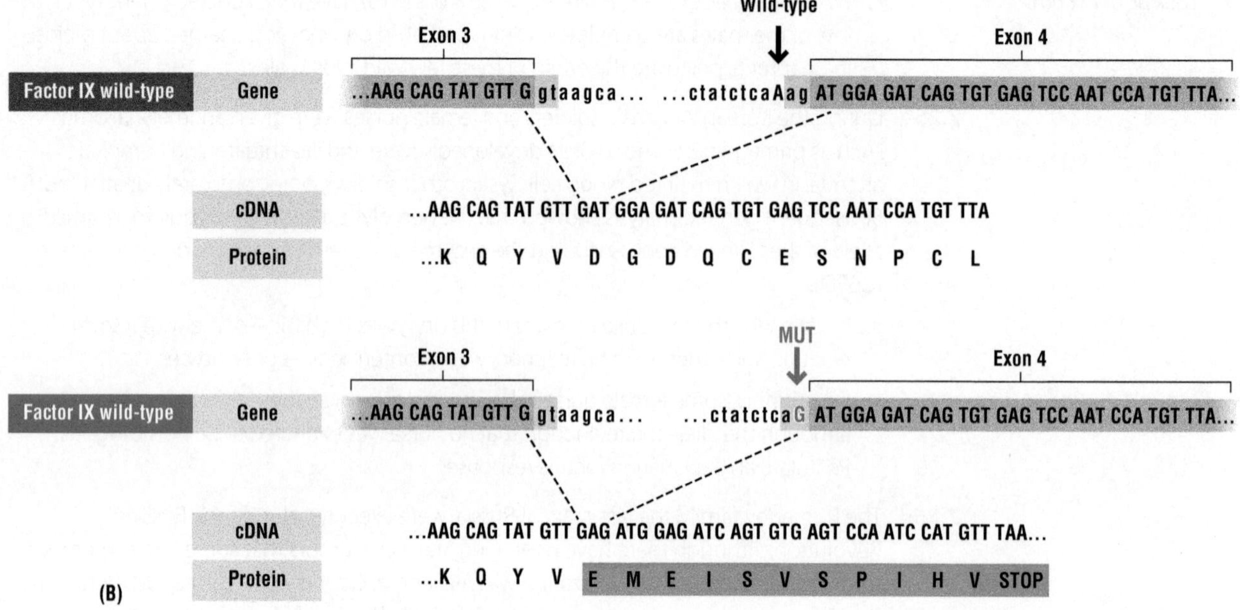

Figure Q7.4 continued

b. The single base change shown above appears to have changed the splicing of the exons in the gene encoding factor IX. How would this mutation alter the function of factor IX and lead to hemophilia?

c. One of the female skeletons was tentatively identified as Anastasia, based on the age of the girl and the presence of some jewelry. Anastasia died before she could have children. Based on the evidence above, what is the probability that her first child would have been a boy with hemophilia?

7.27 Dosage compensation occurs by different methods in different organisms. Suppose that you found a mutation in which dosage compensation did not occur. What is the expected phenotype of a dosage compensation mutant in:

a. *C. elegans*

b. *D. melanogaster*

c. Mice

Looking back and ahead
Progressively challenging

7.28 In *C. elegans*, males are X0, while hermaphrodites are XX. Hermaphrodites have the somatic sex determination of females, but they undergo both spermatogenesis and oogenesis, so they can use their own sperm to fertilize their ovum or they can mate with a male. Hermaphrodites cannot mate with other hermaphrodites.

a. When a hermaphrodite self-fertilizes, what is the expected ratio of males and hermaphrodites among her offspring?

b. When a hermaphrodite mates with a male, what is the expected ratio of males and hermaphrodites among the offspring?

c. Recessive mutations in the X-linked gene *unc-3* result in worms that are uncoordinated. A hermaphrodite that is homozygous for *unc-3* is mated to a wild-type male. What are the expected offspring of this mating? (In *C. elegans* nomenclature, the gene is called *unc-3*, and the mutant phenotype arising from that gene is Unc.)

d. A mating between a male and a hermaphrodite nearly always includes both self-progeny and cross-progeny among the offspring. Postulate at least one method that could be used to distinguish self-progeny from cross-progeny in *C. elegans*. (The most widely used method relies on a topic introduced in Chapter 8 but may well be known to you already.)

Looking ahead

e. A wild-type male is mated to a hermaphrodite that is homozygous for a gene called *dpy-11* on one of the autosomes and heterozygous on the X chromosome for *unc-3* and *lon-2*, that is, it had *unc-3* on one X chromosome and *lon-2* on the other X chromosome. The mutations in *dpy-11*, *unc-3*, and *lon-2* are recessive. What are the expected cross-progeny offspring of this cross? (Hint: This may help you answer (d) above.)

f. Approximately one in 500 of the self-progeny offspring from a hermaphrodite is a male. Explain the process by which such males arise spontaneously.

g. Mutations in genes that affect meiosis in *C. elegans* can be readily identified and have a distinctive mutant phenotype. The mutant phenotype is called Him for "high incidence of male self-progeny," so the genes are named *him-1*, *him-2*, etc. Mutations in many *him* genes result in greatly reduced fertility and many eggs that do not hatch. However, among the eggs that hatch, a relatively high proportion are X0 males. Explain this result in terms of both the Him phenotype and the reduced fertility.

h. A hermaphrodite that is homozygous for one of the *him* mutations and homozygous for *unc-3* was mated to a wild-type male. In addition to the results that you deduced for (c) above, some of the offspring are wild-type males. Explain this result. (Hint: The offspring also include *unc-3* hermaphrodites, but, unless one of the methods from (d) is used, these cannot be distinguished from self-progeny.)

Very challenging

i. In order to investigate which meiotic division was affected by each of the *him* genes, Hodgkin *et al.* (*Genetics*, 1979, **91**: 67–94) did the following experiment. (This has been slightly modified.) A wild-type male was mated to a hermaphrodite that was homozygous for the *him* mutation, homozygous for *dpy-11* on one of the autosomes, and heterozygous on the X chromosome for *unc-3* and *lon-2*, that is, it had *unc-3* on one X chromosome and *lon-2* on the other X chromosome. In other words, it is the same cross as used in (e), but the hermaphrodite is also homozygous for a *him* mutation. The offspring included a few hermaphrodites that were Unc but not Dpy or Lon, and Lon but not Dpy or Unc. Explain this result.

Even more challenging

j. When Hodgkin did the cross in (i) with a meiotic gene known as *him-3*, the progeny also included a few offspring that were Dpy but not Unc or Lon. Explain this result. (Hint: Most of the self-progeny eggs laid by *him-3/him-3* hermaphrodites do not hatch at all, and *him-3/him-3* is nearly sterile, as in (g) above.)

CHAPTER 8

The Inheritance of Multiple Genes

IN A NUTSHELL

The inheritance pattern of eukaryotic genes follows the laws recognized by Mendel. These laws arise from the processes that occur during meiosis in eukaryotes. Even when the inheritance pattern of genes is predictable, the phenotypes that arise from these genes may not be clear-cut. In this chapter, we consider the more complicated phenotypic ratios that can arise when we follow multiple unlinked genes, when alleles interact in unexpected ways, and when additional inputs from the environment influence the phenotype. Yet, even in these more complicated scenarios, the individual contributing genes always follow the fundamental patterns of single-gene inheritance that arise from meiosis. We can often determine the origins of these unexpected phenotypes by working with laboratory organisms, but these same situations prevail in natural environments and so affect evolutionary processes as well.

8.1 Introduction: Mendel expanded and extended

Having understood what happens with the inheritance of a gene on one of the autosomes, as discussed in Chapter 5, and with a gene on the X chromosome, as discussed in Chapter 7, you now have a grasp on the fundamentals of single-gene inheritance. These patterns are true for all genes in all diploid eukaryotic organisms. The rules that Mendel derived from working with peas can be applied to every other diploid eukaryote, underscoring the profound impact of the shared evolutionary history of these basic cellular mechanisms.

In this chapter, we go on to address the question of what happens with two different genes, located on two different chromosomes. In Chapter 9, we describe what happens when the two genes are on the same chromosome. We also consider some examples in which the inheritance of the genes follows Mendelian principles, but the observed phenotypic ratios don't agree with what is expected. Although these are somewhat different topics, we bring them together to emphasize the underlying biological processes connecting gene inheritance with phenotypes. The inheritance of the genes is usually straightforward and predictable. The expression of those genes into a phenotype can be simple and straightforward or can be a bit more complicated. Let's begin with two genes located on different chromosomes and return to talking about Mendel.

8.2 Two genes on different chromosomes

Peas have two copies of each chromosome and don't have X chromosomes, so every gene Mendel tested behaved like a typical diploid autosomal gene. Mendel did experiments involving the inheritance of two genes at the same time, and the results are completely predictable now with an understanding of the process of meiosis and the laws of simple probability.

Suppose that we have a true-breeding round and yellow pea plant and another plant that is true-breeding for wrinkled and green peas. These are two different traits controlled by different genes. We cross them, as shown in Figure 8.1. The F_1 plants produce seeds that are all round and yellow, so those must be the dominant alleles. Now we make an F_2 generation by self-fertilization or by crossing two F_1 plants. What phenotypes do we see and in what frequencies? Looking at Figure 8.1, we can diagram this cross using a Punnett square, writing out all possible gametes from each F_1 plant and combining these in every

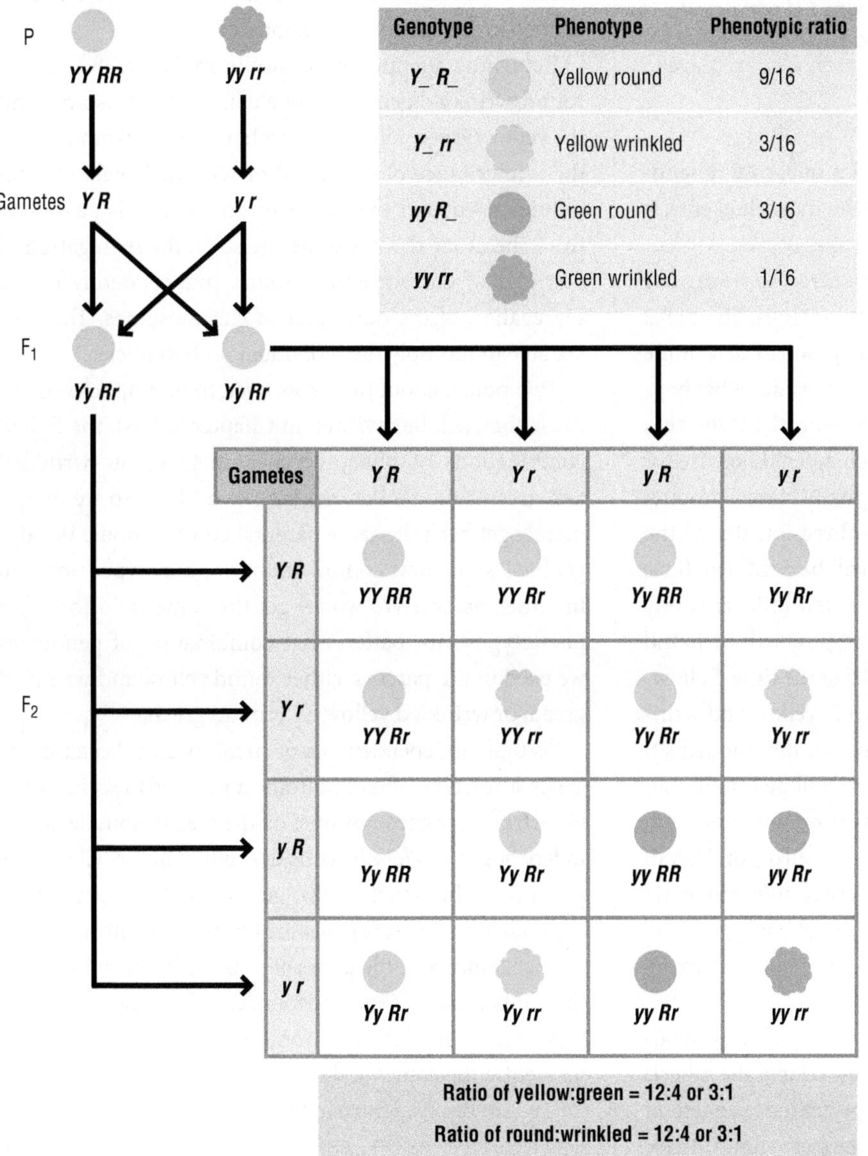

Figure 8.1 The outcome of Mendel's crosses involving two unlinked genes shown using a Punnett square. Gametes from a cross of heterozygotes (***YyRr*** × ***YyRr***) are combined in a Punnett square to reveal the resulting offspring. The phenotypic ratio of the offspring is 9:3:3:1. Looking at each gene alone shows the familiar 3:1 ratio of dominant to recessive phenotypes.

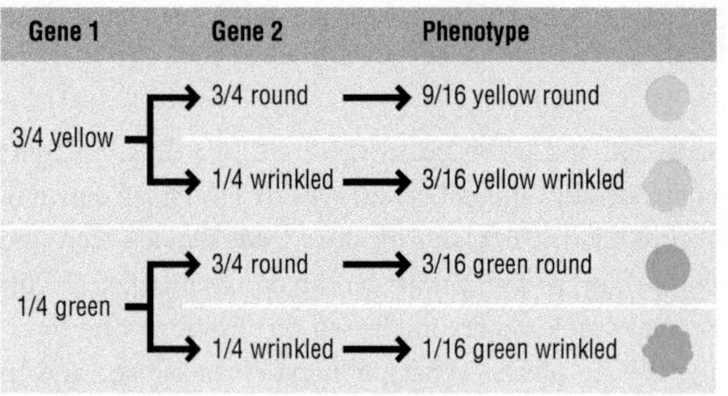

Figure 8.2 Using probability to predict the outcome of a cross involving two unlinked genes. A branching diagram using probabilities is a quick way to obtain phenotypic ratios. Again, heterozygotes for unlinked genes are crossed (***YyRr*** × ***YyRr***). For the colour gene, 3/4 of the peas will be yellow (dominant), while 1/4 are green (recessive) peas. Within each of these groups, 3/4 of the peas will be round, while 1/4 will be wrinkled. Multiplying the probabilities of these independent events yields a 9:3:3:1 phenotypic ratio, as also found using the Punnett square approach.

possible combination. The frequencies of each F_2 phenotype are then found by counting up the individual entries in the Punnett square for each of the phenotypes.

The Punnett square for two genes is large, so we can use the rules of probability to simplify our analysis. Figure 8.2 illustrates the use of a branching diagram to determine F_2 phenotypic outcomes. We will first consider what happens with one gene—seed color. Three-fourths of the peas will be yellow, and one-fourth will be green. Likewise for the second gene—pea shape (or actually starch production, as discussed in Chapter 5)—three-fourths of the peas will be round, and one-fourth will be wrinkled. If we assume that the two genes are inherited independently of each other, then yellow and round peas will be found three-fourths × three-fourths, or 9/16 of the time. Following this logic, as outlined in Figure 8.2, yellow and wrinkled peas will be found at a frequency of three-fourths × one-fourth, or 3/16; green and round will be found one-fourth × three-fourths, or 3/16 of the time; and green and wrinkled will occur one-fourth × one-fourth, or 1/16 of the time. For two genes that are inherited independently of each other, we therefore see a 9:3:3:1 phenotypic ratio. To put it another way, we have two 3:1 ratios occurring simultaneously.

What did we have to assume when we made this calculation to get the 9:3:3:1 ratio? We assumed that the inheritance and appearance of the pea shape are independent of the inheritance and appearance of the pea color. If these two genes are **inherited independently** of each other, we can multiply their probabilities to get the probability of each genotype and then each phenotype.

This illustrates Mendel's Law of Independent Assortment. According to this law, the inheritance of one gene does not affect the inheritance of a second gene. With an integrated understanding of both Mendelian inheritance and meiosis, the biological basis of this law becomes clear. In Figure 8.3, we return to our diagram of meiosis and see two pairs of chromosomes, in this case one pair of long chromosomes and one pair of short chromosomes. Following the chromosomes through meiosis, you can see that one of the pair of long chromosomes is passed on to one daughter cell in the first meiotic division, but this event does not influence which of the two short chromosomes will be passed on with it. Both possibilities are equally likely. Which scenario occurs depends merely on how the pairs of chromosomes happen to line up during metaphase I.

Returning to our pea example in Figure 8.1, if the round/wrinkled gene is on one pair of chromosomes and the yellow/green gene is on another pair of chromosomes, then inheritance of the round or wrinkled allele will not influence whether the yellow or green allele is passed on from the other chromosome. As such, the segregation of one pair of chromosomes occurs independently of the segregation of the other pair of chromosomes. Thus, we are able to multiply the individual probabilities.

Two points about this cross need to be emphasized because they will be revisited in Chapter 9. First, the F_2 had combinations of phenotypes—namely yellow wrinkled and green round—that neither parent had. So we do not merely get back the same parental combinations. We also get back some new combinations of phenotypes not seen in either parent. We would get the same ratio for these phenotypes, no matter what combination of genotypes we used in the parents, either round yellow and wrinkled green, or wrinkled yellow and round green.

Second, independent assortment occurs because the genes are on two different chromosomes and the orientation of each of these chromosomes on the meiotic spindle occurs independently. We will consider situations in Chapter 9 when the inheritance of two genes is not independent of each other because the genes are on the same chromosome. In such situations, the genes are said to be linked because their inheritance is not independent. Once two genes are known to be on different chromosomes and inherited independently, the names of the genes are usually separated by a ";" e.g. ***Aa***; ***Bb***. We return to this in Chapter 9.

KEY POINT The Law of Independent Assortment indicates that the inheritance of one gene or trait does not affect the inheritance of another unlinked trait. The basis of this law lies in the process of meiosis.

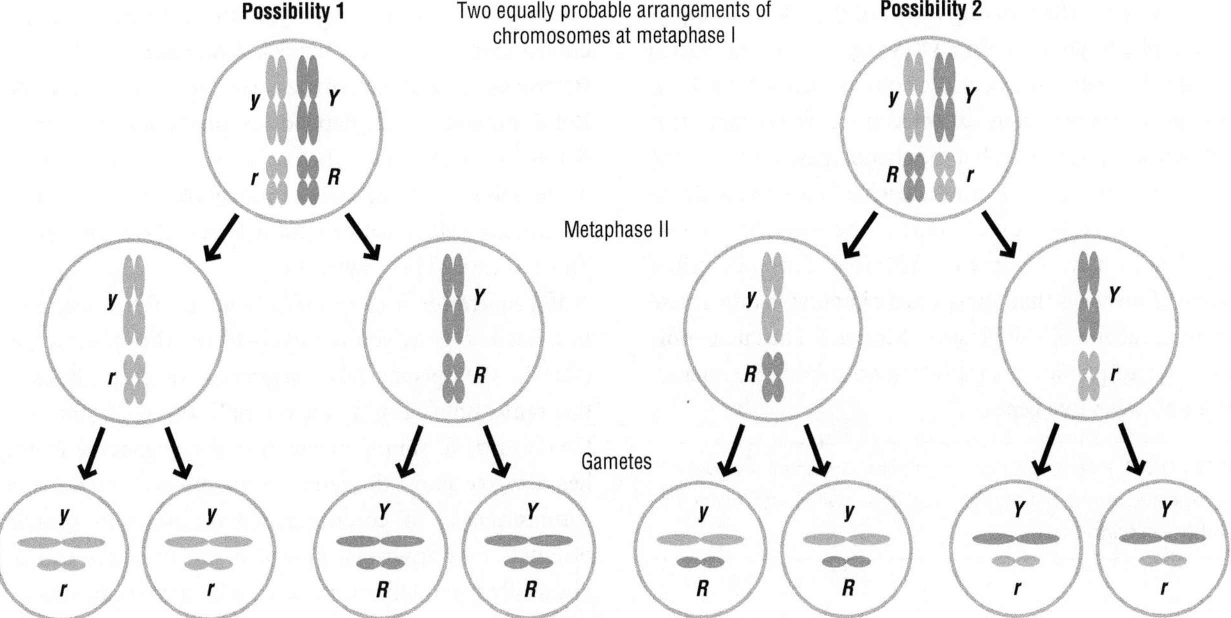

Figure 8.3 Mendel's Law of Independent Assortment is due to the behavior of chromosome pairs during meiosis. Chromosomes separate independently during meiosis. Two chromosomes are shown, one long and one short. The chromosomes from one parent are shown in blue, while the ones from the other parent are in pink. The long pink chromosome will assort with the short pink chromosome, as often as it assorts with the short blue chromosome. Each pair of chromosomes segregates independently, which is the cellular basis for Mendel's Law of Independent Assortment.

The principle of independent assortment can be used to work out any genetics problem that you may encounter, whether in a course, a research or clinical laboratory, or in life outside of academia. The main underlying condition is that the two genes are inherited independently of each other. This includes situations when one gene is X-linked and the other gene is autosomal. It is possible to solve these problems using Punnett squares, but it is much easier to use simple probability or branching diagrams. This is especially true when more than two genes have to be considered and the Punnett square has to be larger to accommodate all the possible combinations of gametes. It is usually much easier, with less possibility of error, if these are treated as separate and independent crosses; the outcome of each of the crosses is multiplied together to get the combined final result.

8.3 Unexpected phenotypic ratios arising from a single gene

Although understanding the processes of meiosis and applying the principles of probability allow us to predict how **genes** and chromosomes are inherited, we cannot always predict how the **phenotypes** will be inherited. There are several reasons that the inheritance of phenotypes may be unpredictable.

First, and particularly relevant for phenotypes in natural populations (including humans), the phenotype might be determined by the actions of multiple genes working together, rather than by a single gene or a pair of genes. The trait could also be affected by the environment. In genetic terms, the phenotype is a **complex trait**, a topic that we will discuss in Chapter 10.

Second, particularly for the controlled crosses of laboratory species and genetics classes, other biological factors are at work. The remainder of this chapter discusses some of these. The unifying theme of these various topics is that phenotypes do not always segregate in quite such predictable patterns as our previous examples have done. The genes segregate, as expected from Mendelian principles based on meiosis, but the phenotypes may not fit the ratios anticipated. These are somewhat loosely related biological topics that indicate some of the ways that phenotypic ratios may be unpredictable. This section describes some of the ways that unexpected phenotypic ratios arise from the alleles of a single gene; Section 8.4 discusses how unexpected phenotypic ratios can arise from the interactions between two genes.

KEY POINT Phenotypic ratios in a cross may sometimes give unexpected results, but the individual genes will always follow Mendelian principles.

Dominance relationships are not always simple

Our first situation with an unexpected phenotypic ratio occurs with the two alleles of a single gene, that is, when attempting to identify which allele is dominant. We remind you that dominance is defined based on the phenotype of a heterozygote. If the heterozygote has the same phenotype as the homozygote for one of the alleles, then that allele is dominant. As we said before, dominance has nothing to do with being common or normal. We discussed this in detail in Box 5.1. Dominance is simply a statement of what phenotype is seen in the heterozygote. But dominance really depends on how the phenotype is defined—for example whether it is viewed at the level of the DNA or at the level of morphology—so different dominance relationships exist between alleles of a gene. This is illustrated in Figure 8.4.

If you are looking specifically at the DNA sequence in a heterozygote, you are likely to see the "phenotype" (that is, the specific DNA sequence) of both alleles at the same time. This is an example of co-dominance. Co-dominance simply means that the phenotype of the heterozygote shows the genotype of both parents. Another common case of co-dominance occurs with protein products or transcripts, in which the products can be physically separated, such as on a gel. But you can also see co-dominance with phenotypes described at other levels, as in the examples below with ABO blood types.

Other types of dominance relationships can exist between the alleles of a gene, however. If you are looking at a color—of a flower perhaps—you may see an intermediate phenotype like pink or lavender. For example, if a true-breeding plant with red flowers is crossed with a true-breeding plant with white flowers, the F_1 plant will be heterozygous and might have pink flowers. Both alleles of the gene are present, but the amount of the product of the

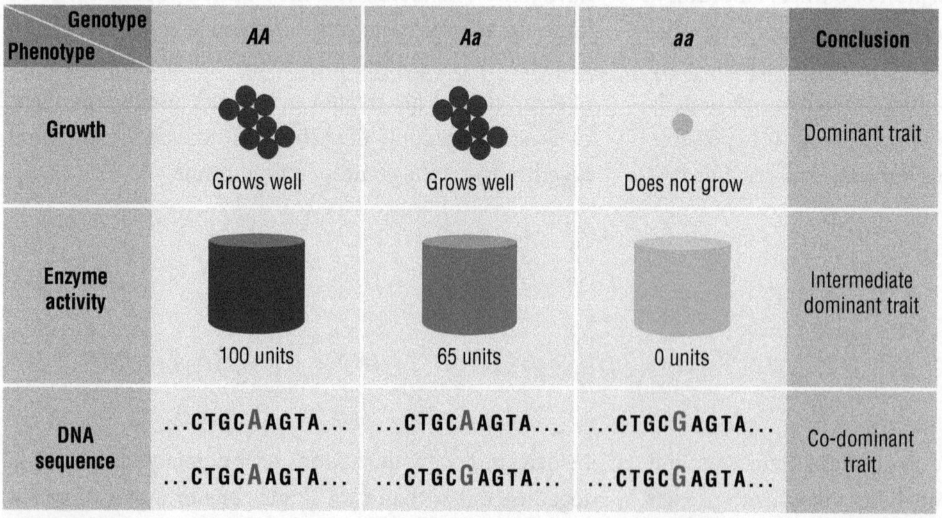

Genotype / Phenotype	*AA*	*Aa*	*aa*	Conclusion
Growth	Grows well	Grows well	Does not grow	Dominant trait
Enzyme activity	100 units	65 units	0 units	Intermediate dominant trait
DNA sequence	...CTGCAAGTA... ...CTGCAAGTA...	...CTGCAAGTA... ...CTGCGAGTA...	...CTGCGAGTA... ...CTGCGAGTA...	Co-dominant trait

Figure 8.4 Levels of dominance depend on the assay being used for the phenotype. A cell or organism that is ***AA*** in genotype grows well, while the ***aa*** homozygote does not grow. The heterozygote ***Aa*** grows as well as the ***AA***, so the ***A*** allele would be considered dominant if the phenotype is growth. However, if the activity of the enzyme is measured, ***Aa*** has a reduced activity, compared to ***AA***, but more than ***aa***. Thus, based on enzyme activity, the ***A*** allele would be considered to show intermediate or incomplete dominance. If the genes are sequenced, both alleles can be observed in the ***Aa*** heterozygote. In an assay based on DNA sequence, the alleles are co-dominant.

red allele plus the amount of the product from the white allele is phenotypically distinct from the amount of the product from the combined effect of two red alleles. This is often called **incomplete dominance**.

Whereas co-dominance results in the phenotypes of both alleles being present simultaneously in the heterozygote, incomplete dominance results in a phenotype in the heterozygote that is intermediate in some way between the two homozygous phenotypes. This typically means that the heterozygote ***Aa*** makes more of the gene product than the homozygote ***aa***, but less than the homozygote ***AA***; the intermediate amount of gene product being made results in an intermediate phenotype for that gene. Nonetheless, the segregation pattern for the two alleles is the same as we have seen before. If that pink-flowered plant is allowed to self-fertilize, what will you see? One-fourth of the plants will be red, one-half will be pink, and one-fourth will be white.

KEY POINT When two alleles don't follow a standard dominant–recessive relationship, the phenotype of the heterozygote sometimes reflects both alleles at the same time (co-dominance), and sometimes a new intermediate phenotype is seen (incomplete dominance).

The difference among dominance, co-dominance, and incomplete dominance may not provide much insight in describing what occurs in producing the phenotype of the heterozygote, so it might not be useful to consider these as fixed and established categories. A more helpful insight is to recognize that phenotypes can be analyzed and described at many different levels—DNA sequences, gene products, cells, particular structures, individuals, and populations.

For example, consider a gene that encodes an enzyme whose activity we can directly measure; let's say that the enzyme is involved in the synthesis (or anabolism) of a macromolecule, and the synthesis produces a pigmented product. While we are not providing data from a specific example, many genes and enzymes could fall into this category. As we have said previously, dominance information depends on the phenotype of the heterozygote ***Aa***. As summarized in Figure 8.4, suppose we look at the overall growth phenotype of the ***AA***, ***Aa***, and ***aa*** individuals or cells. Both ***AA*** and ***Aa*** cells can grow and divide, while the ***aa*** organism or cell cannot grow. Thus, based on this growth assay, ***A*** is a simple dominant allele to ***a***, since ***Aa*** cells can grow and divide like ***AA*** cells.

Suppose instead that we measure the activity of the enzyme from an *in vitro* assay and measure how much of the product is made in a specific period of time; we may be able to detect this using the color of the product or some other assay that measures enzyme activity. In this case, the ***Aa*** individual has an enzyme activity or a phenotype that is intermediate between ***AA*** and ***aa***—not precisely halfway between them, since enzymes don't often work that way, but less than ***AA*** and more than ***aa***. Thus, based on the enzyme assay, this would be considered an example of incomplete dominance.

Finally, instead of looking at growth or enzyme activity, we directly examine the DNA sequence for the gene. By this assay, we can detect the presence of two different alleles—one with a G at position 423 and another with an A at position 423. Thus, using the DNA sequence as our phenotype, the alleles would be considered co-dominant.

The alleles have not changed in this example, and the activities of the alleles have not changed. However, in a formal sense, we could classify these two alleles as having any of three different dominance relationships, depending on the phenotypic assay being used. In other words, dominance refers not only to the phenotype in the heterozygote, but also to the assay that is used to measure that phenotype.

Genes have multiple alleles, not just two alleles

Although we often write alleles as ***A*** and ***a***, a gene theoretically has an extremely large number of possible variants, as we noted in Chapter 3. Alleles have different DNA sequences, which might produce different amino acid sequences in the protein product. It is sometimes useful to group these different alleles into functional categories, based on the phenotypes they produce.

A classic example of sorting different alleles into functional categories is found in the ABO blood antigen in humans, which is critical for ensuring that a patient receives compatible blood during a transfusion. We have dozens of blood antigens encoded by different loci; other than the Rh locus on chromosome 1 and the ABO locus on chromosome 9, most of these other antigens are not important for transfusions, so they are not often discussed. Some of these blood antigens, including the ABO locus, have interesting associations with risks or resistance to various diseases and pathogens, even if they are not important for transfusions, which are discussed in Box 8.1.

Our ABO blood type arises from three different alleles at a single autosomal gene locus. We will call these alleles i, I^A, and I^B, for reasons that will be apparent shortly. As shown in Table 8.1, there are four different phenotypes arising from the ABO blood type locus, based on blood

BOX 8.1 *A Human Angle* ABO blood types

Among genetic traits that we all share, the most familiar single gene trait in humans is probably our ABO blood type. Other familiar shared heritable human traits, such eye color, are not controlled by a single gene. Many traits that are inherited as a single gene are associated with genetic diseases, such as cystic fibrosis and sickle-cell anemia, so they are familiar because their effects have profound, and often tragic, effects in some families, but they do not affect every person. Still other traits that are sometimes taught as being controlled by a single gene, such as tongue rolling and a widow's peak, have a highly questionable genetic basis, since identical twins can differ in these phenotypes. So, when it comes to familiar single-gene traits that every person has, the ABO blood type probably stands supreme; it almost certainly is used more often in murder mysteries and crime dramas than any other genetic trait.

This familiarity does not mean that we know all there is to know about the ABO blood types. While we traditionally (and appropriately) consider three alleles for the gene, based on blood clotting assays or agglutination tests in the clinic, there are many more alleles in natural populations, particularly when the DNA sequences are analyzed. Even based on a simple blood clotting test, subtypes for these three alleles can sometimes be detected, with two common and 18 less common subtypes of the I^A allele, for example. Nonetheless, the general blood clotting phenotype is convenient to assay and well known and is reliable enough for most purposes.

The biological basis of ABO blood types

The underlying biology for ABO blood types is relatively simple, as discussed in the chapter and depicted in Figure A. The protein encoded by the ABO locus is a glycosyl transferase, an enzyme that attaches certain sugars to a polysaccharide stem molecule (the H antigen) found on the surface of red blood cells. The I^A allele attaches the sugar N-acetylgalactosamine, while the I^B allele attaches the sugar galactose. The i (O) allele encodes an inactive form of the enzyme, often because of a deletion in the gene or a premature stop codon, and attaches no additional sugars to the stem. The two sugars that are attached act as antigens in an immune response; we have immunological tolerance for the antigen of our own blood type, and we generate isoantibodies to the antigen with the other sugar during our first year of life. Thus, an individual with the I^A allele produces antibodies against the antigen made by the I^B allele, while a person with the I^B allele generates antibodies to the antigen made by the I^A allele.

Having neither antigen, a person with type O blood produces antibodies against both alleles. As a result, type O people can donate blood to any other blood type, since they are not introducing any antigens that will cause agglutination, so they are sometimes referred to as universal donors. On the other hand, they can't accept blood of any other blood type, since they have antibodies against those antigens. By contrast, a person with type AB blood can receive blood from anyone, since both antigens are

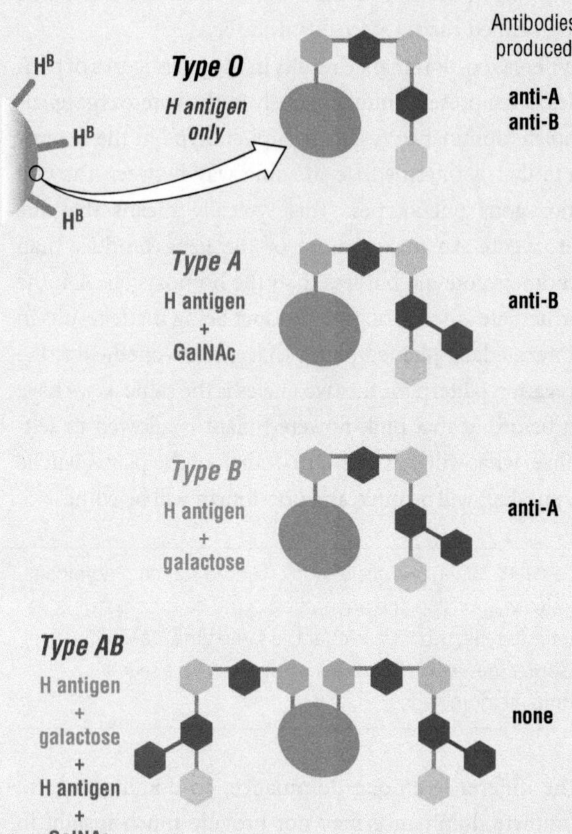

Figure A Antigens and antibodies produced by ABO. The ABO locus encodes a glycosyl transferase that adds sugars to the H antigen. The H antigen is shown in greater detail in Figure 8.5 and summarized here with colored hexagons. In individuals with type O blood, an inactive glycosyl transferase enzyme is made, and no additional sugars are added to the H antigen. Since neither antigen is made, type O individuals make antibodies against both A and B antigens. In individuals with type A blood, the enzyme adds N-acetylgalactosamine (GalNAc) to the H antigen, shown here as the pink hexagon; because they make only the A antigen, people with blood type A make antibodies against the type B antigen. In individuals with type B blood, the enzyme adds galactose to the H antigen, shown as a dark blue hexagon; because they make only the B antigen, people with blood type B make antibodies against the type A antigen. In individuals with type AB blood, the enzyme adds GalNAc to some of the H antigens and galactose to some of the H antigens; because they make both antigens, people with type AB do not make antibodies against either antigen. The presence of the antigen determines if the person can be a donor. Thus, type O blood can be given to anyone, while type AB people can receive any other blood type.

BOX 8.1 Continued

already present and no antibodies are being produced; ABO individuals are sometimes referred to as universal recipients.

Given the existence of this system, we might ask how and why these blood types exist. Blood types were discovered around 1900 when blood transfusions were first being done, which resulted in large amounts of blood from one individual coming into contact with the blood of an unrelated individual. The Austrian physician Karl Landsteiner is credited with identifying type A, type B, and type O blood in 1900 and was awarded the Nobel Prize in Physiology or Medicine in 1930; the less common type AB was discovered a few years after the others. While Landsteiner is credited with this discovery, the Czech doctor Jan Jansky had found all four types at about the same time as Landsteiner, but his work was not widely known until years later. Landsteiner also discovered the Rh blood type and was among the team that discovered the polio virus, so the elucidation of the ABO blood types was not his only contribution to human genetics.

What is the evolutionary basis of the ABO blood types?

The time at which blood types were discovered is significant when thinking about the evolutionary biology of ABO. Different blood types were identified only when blood transfers and blood transfusions were being done regularly, which was millions of years after the blood types themselves arose in the human population. In other words, blood types existed long before we had any routine mechanism to detect these differences.

Based on other primates, the evolutionarily ancestral blood type was probably type A. The I^B allele is thought to have arisen about 3.5 million years ago, while the ***i*** allele is thought to have arisen more than 1.5 million years ago. Since ***i*** is a non-functional version of the gene, it probably arose many times independently, and other higher primates have a type O that is molecularly distinct from ours. Both the I^A and an ***i*** allele are present in the genomes of Neanderthals, indicating that they had at least these two types. In other words, the variation in the phenotype that we know best for the ABO locus—blood clotting during transfusion—was almost certainly not the phenotype under which these different alleles arose. The variation at the ABO locus must be attributed to some evolutionary force other than that arising from blood transfusions.

What then was the evolutionary basis for variation at the ABO locus? Why don't all humans have the same blood type? Have there been specific selective pressures that have led to different ABO alleles? (This is an even more intriguing question for the Rh blood locus, since there has been a selective disadvantage, known as Rh incompatibility, arising when Rh-positive children are born to an Rh-negative mother.)

The answers are not so clear. It has been recognized for nearly a century that human populations differ in the relative frequency of the three alleles. In particular, the I^B allele is absent from some populations such as Native Americans, which has been interpreted by some as evidence for different selective pressures on different blood types. In fact, the evolutionary forces, such as selection, may not have been directed against the blood types at all; approximately 80% of people secrete their ABO antigens into other body fluids, such as saliva and plasma, and the ABO antigens are present on the surfaces of other cells, in addition to red blood cells. Secretion of these antigens is encoded as part of the Lewis blood type on chromosome 19. Thus, it may be that the important biological effects of ABO during human evolution occurred on cells other than the red blood cells.

With such a long history of genetic differences for a familiar trait, many hypotheses have been posed about the effect of our ABO blood type on other phenotypes. Most of the more scientific hypotheses are based on correlations between blood type frequencies and various disease or pathogen frequencies. Some of the well-established correlations are summarized in Table A, but more diseases or pathogen correlations have been proposed; all of these correlations are associated with small changes in risk, typically less than 10% over other blood types.

To stress, these are merely correlations, and very few data exist on a mechanism by which a particular blood type may increase or decrease a cancer risk, for example. In the popular literature, ABO blood type has also been suggested to correlate with beauty, dietary behaviors, personality, and recovery from the effects of alcohol, among other things, but there is no scientific support for any of these claims (even if a subjective phenotype, such as beauty, could be assayed accurately).

One hypothesis based on a possible molecular mechanism arises from proposed structural similarities between the antigens and some infectious agents. It has been proposed that the

Table A

Blood type	Slightly increased risk	Slightly decreased risk
O	Squamous cell carcinoma Basal cell carcinoma More severe cholera symptoms	Pancreatic cancer Gastric cancer Coronary heart disease Deep vein thrombosis
A	Gastric cancer Nasopharyngeal carcinoma	
B	Ovarian cancer	
AB	Stroke Nasopharyngeal carcinoma	

BOX 8.1 Continued

antigen produced by the I^A allele, with N-acetylgalactosamine attached to the stem, is structurally similar to a glycoprotein making up the coat of some subtypes of the influenza virus. Thus, antibodies against the type A antigen (as would be found in people with type O or type B blood) may confer a slight resistance to influenza. Similarly, the antigen produced by the I^B allele with galactose attached to the stem is proposed to be similar to a glycoprotein component of the cell wall found in Gram-negative bacteria, such as *Escherichia coli*, *Salmonella* species, and *Vibrio cholerae*, and may confer some resistance to infections by these bacteria.

Yet another idea, and perhaps the most likely, is that the selective pressures on variation at the ABO locus were small; certainly all of the effects we observe now, other than agglutination, are relatively small in scale, compared to other examples of alleles with selective benefits or risks. With such small effects, mutations in the locus arose and may have persisted or been lost based on evolutionary forces other than selection. In Chapter 16, we will discuss some of these evolutionary forces as they relate to other traits, and it is possible that these also were important for blood types.

clotting assays: type A, type B, type AB, and type O. These represent different dominance relationships among the three alleles, with I^A and I^B showing simple dominance over the ***i*** allele, but co-dominance with each other.

The ABO gene encodes an enzyme known as a glycosyl transferase, whose function is to attach one particular sugar to a stem of other sugars (called the H antigen), which is, in turn, attached to proteins and lipids on the surface of red blood cells, as shown in Figure 8.5. A separate gene on chromosome 19 encodes the enzyme that makes the H antigen to which the particular sugars are attached, and there are many different stem attachment sites on the surface of each cell. (The rare Bombay phenotype arises from mutations in the gene that makes the H antigen, so individuals who are ***h/h*** fail to make the oligosaccharide stem.) The sequences that fall into the O category of alleles, symbolized as ***i***, result in no enzymatic activity from the ABO gene and thus attach no additional sugars to the H antigen stem. Many of these O alleles feature deletions of part of the gene and are recessive to the other types. A person with type O blood has the genotype ***i/i***.

The A category of alleles (denoted as I^A) and the B category of alleles (denoted as I^B) have somewhat different enzymatic activities. The I^A alleles attach the sugar N-acetylgalactosamine to the H antigen stem. The alleles known as I^B attach the sugar galactose. Each of these is dominant to ***i*** (O category), so the phenotype of type A blood, as we can see from Table 8.1, could result from a genotype of either I^A/I^A or I^A/i. Likewise, the phenotype of type B blood could have a genotype of either I^B/I^B or I^B/i.

What happens in an individual whose genotype is I^A/I^B? These two alleles are co-dominant. N-acetylgalactosamine is attached to some H antigens, while galactose is attached to other H antigens on the surface of the same red blood cell. This produces the phenotype of type AB blood.

Table 8.1 Co-dominance in ABO blood types. Four common phenotypes are found for the ABO blood locus, as shown in the table. The phenotypes are based on blood clotting assays and arise from the presence or absence of specific antigens and antibodies. The I^A (A) allele and the I^B (B) allele each produce an antigen found on the surface of red blood cells, while the ***i*** (O) allele produces neither of these antigens. Blood clotting occurs because each type produces antibodies against the antigen made by the other type. The I^A (A) allele and the I^B (B) allele exhibit simple dominance to the ***i*** (O) allele, so type A and type B individuals can be either homozygotes or heterozygotes; type O must be homozygous recessive, as shown. I^A and I^B are co-dominant with respect to each other, so type AB individuals make both antigens and produce neither antibody.

Blood type	Genotypes	Antigens produced	Antibodies made
Type A	I^AI^A I^Ai	A antigen with N-acetylgalactosamine	Anti-B
Type B	I^BI^B I^Bi	B antigen with galactose	Anti-A
Type O	ii	None	Anti-A Anti-B
Type AB	I^AI^B	Both A antigen and B antigen	None

KEY POINT There are many ways the DNA sequence of a gene can vary, and each variant can be considered an allele. But on a functional level, several different DNA variants can alter the function of the gene in the same way and can therefore be grouped together under the same phenotype.

The important focus here is that different genomes may contain different alleles for a locus and that these might

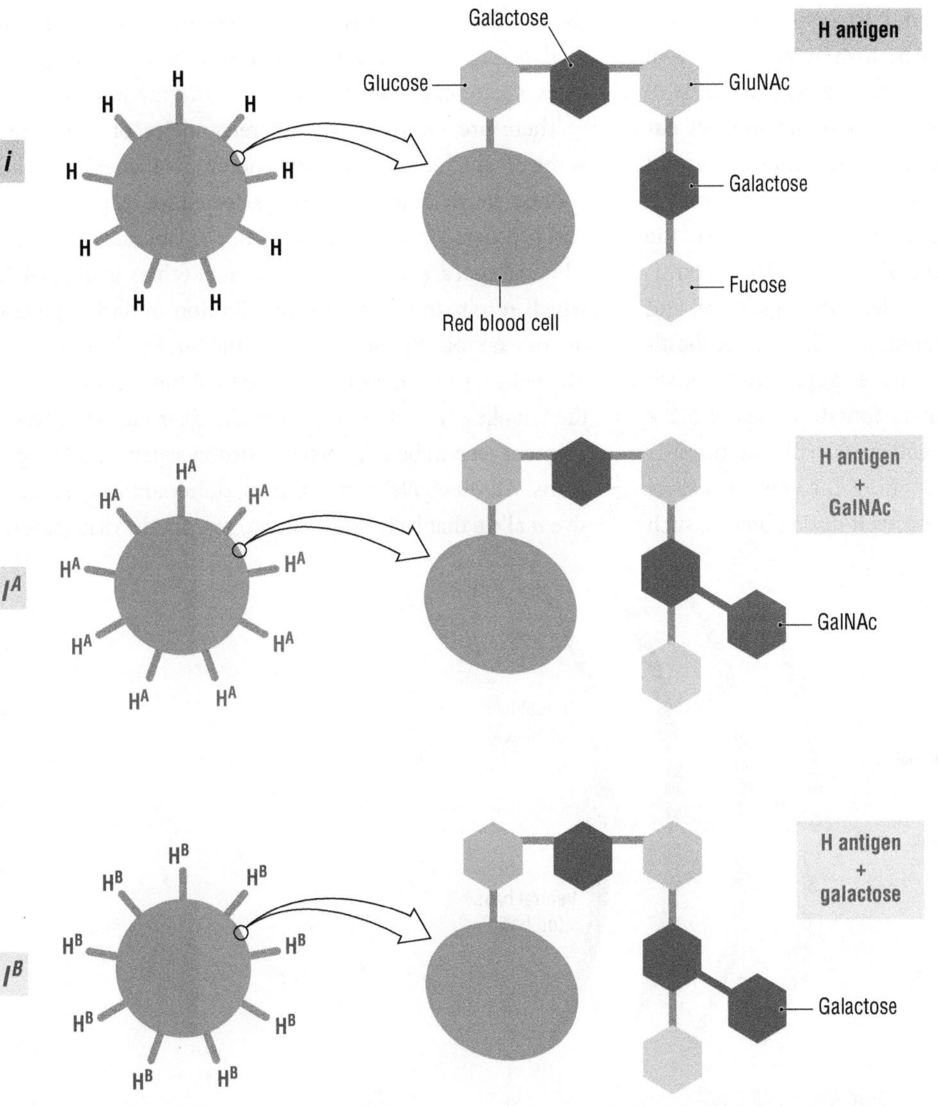

Figure 8.5 The antigens made by the ABO locus. The ABO locus makes a glycosyl transferase that modifies a stem molecule called the H antigen, found on the surface of red blood cells. The H antigen comprises a series of sugar molecules, linked together as shown, and is synthesized by the H locus on chromosome 19; GlucNAc is an abbreviation for the sugar N-acetylglucosamine. People with type O blood make an inactive form of this enzyme and do not modify the H antigen. The enzyme produced by the I^A allele adds the sugar N-acetylgalactosamine (abbreviated GalNAc) to the H antigen, while the I^B allele makes an enzyme that adds galactose to the H antigen.

have functionally different activities. Precisely which alleles are dominant or recessive with respect to each other can often be figured out from working through a scenario, paying close attention to the phenotypes that arise from different allelic combinations, and the particular assay that is being used.

Allelic series can be ranked by dominance

Because there can be many different alleles for a gene, these are sometimes ranked in their order of dominance or by the mutant phenotypes that they produce to give what we call an allelic series. Usually, an allele that completely lacks the activity of a gene, for example, the ***i*** allele, will be recessive to alleles that have some activity for the gene such as the I^A or I^B allele. A completely non-functional allele (such as a deletion of the gene) is generally the one that is recessive to most of the others, although there can be many exceptions to this rule of thumb.

One of the best examples of this type of allelic series comes from the *agouti* gene in mammals. This is one of the major genes that affect coat color (or in humans, hair color and pigmentation), producing patterns that vary, depending on the alleles present at the many other coat color loci. The function of the *agouti* gene product is to limit the extent and timing of the production of black eumelanin pigment; in the regions where eumelanin is not produced, the yellow or orange pigment pheomelanin is deposited instead. The coloration patterns for some of the alleles are shown in Figure 8.6(a); similar, if not identical, patterns are seen for these alleles in dogs, mice, and horses, although different symbols may be used for the alleles. The wild-type allele $\boldsymbol{a^w}$ produces a coloration known as agouti where individual hairs contain bands of black pigment usually at their tip and base, with the number of bands depending on the length of the hair. This allele is responsible for the coloration seen in dog breeds, such as the elkhound and Alsatian or German shepherd, but is also responsible for the coloration seen in the gray–brown color of a field mouse, rabbit, or gray squirrel.

There are hundreds of different alleles of this gene, some of which occur naturally and others that either arose spontaneously in pets and were selected for by pet owners and breeders, or were generated in the laboratory. The allele that appears to be dominant to all others is $\boldsymbol{a^y}$ (sable), which results in the limited distribution of dark pigment in, for example, the noses of pugs and Shetland sheepdogs. The wild-type $\boldsymbol{a^w}$ allele is recessive to $\boldsymbol{a^y}$ but is dominant to the tan allele $\boldsymbol{a^t}$ that directs the wider distribution of black pigment seen in breeds such as Gordon setters and Dobermans. All these alleles are, in turn, dominant to the recessive $\boldsymbol{a}$ allele that lacks the ability to restrict the distribution

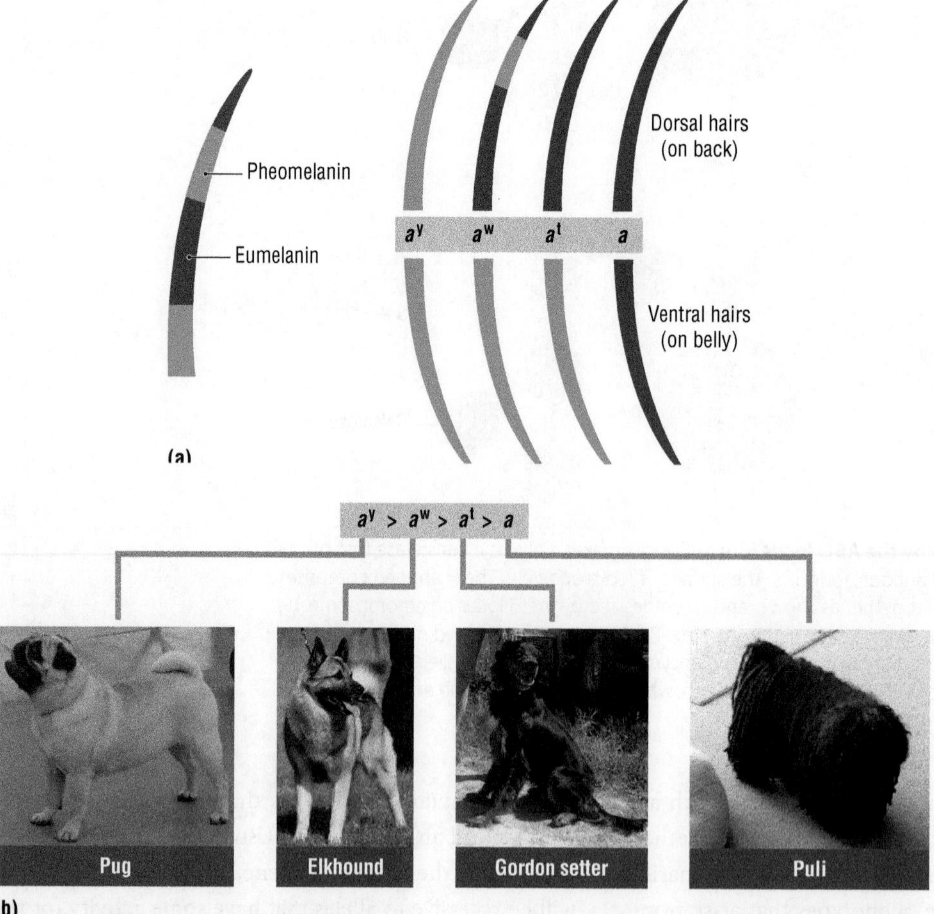

Figure 8.6 Agouti coloration forms an allele series. The *agouti* locus produces a secreted molecule that regulates the synthesis of the black pigment eumelanin. (a) A diagram of a hair shaft with alternating bands of black (eumelanin) and orange (pheomelanin) pigment; the number of bands increases as the hair grows. Different alleles of the *agouti* locus modify this pattern of bands in various ways, with some affecting the dorsal (back) and ventral (belly) hairs differently. (b) The alleles form a series in terms of dominance, as shown. The $\boldsymbol{a^y}$ allele is dominant to all other alleles and completely blocks the synthesis of eumelanin. Allele $\boldsymbol{a^w}$ is recessive to $\boldsymbol{a^y}$ but is dominant to $\boldsymbol{a^t}$, and $\boldsymbol{a^t}$, in turn, is dominant to $\boldsymbol{a}$, which lacks the ability to restrict the distribution of eumelanin pigment, resulting in a completely black coat. Examples of dog breeds with each of these alleles are shown.

Source: Image of pug courtesy of Abuk SABUK/ CC BY-SA 3.0. Image of elkhound courtesy of Dmitry Guskov. Image of Gordon setter courtesy of Pleple2000/ CC BY-SA 3.0. Image of puli courtesy of Puli1989/ CC BY-SA 3.0.

BOX 8.2 *Going Deeper* Pleiotropy and the genome

A gene is pleiotropic when it contributes to more than one phenotype. As we noted in the chapter, most genes are pleiotropic, even if we often only consider one phenotype when we discuss a gene's function. Pleiotropy has been known for many genes since the discovery of mutant phenotypes for a gene, so it is not altogether surprising that genome-enabled analysis of gene functions has enhanced our perspective on pleiotropy.

Pleiotropy arises from either or both of two sources. First, many genes have more than a single function. One example is the gene encoding a protein called β-catenin; the gene has different names in different organisms but encodes a similar protein.

The β-catenin protein plays two distinct roles in animal cells. It was first discovered as one of a group of proteins known as cadherins for their importance in cell adhesion. A few years later, β-catenin was found to have an intracellular and nuclear role in regulating transcription as part of the Wnt signal transduction pathway. These are distinctly different functions encoded by the same gene and performed by the same polypeptide in different subcellular locations. β-catenin is said to be an example of a "moonlighting" protein, but since both of its roles are important in normal cells, it is difficult to determine which function its regular job is and which its moonlighting job is. Its structural similarity to other catenins that are not involved in transcriptional regulation suggests that it is ancestral role was in cell adhesion, however.

Mutations in β-catenin usually affect both cell adhesion and intracellular signaling, since these functions are carried out by the same protein. However, some regions of the protein are needed for one function but not the other, so mutations in such a region will only affect one of the two functions. Mutations that knock out or that cause over-expression of β-catenin in humans are associated with a number of different cancers. These probably involve its role as a transcriptional regulator, since other mutations in the Wnt signal transduction pathway are associated with cancers, while mutations in other cadherins are not considered cancer-causing. In any event, mutations in multifunctional proteins like β-catenin are pleiotropic because the protein is performing several different functions.

Another form and more common source of pleiotropy occurs because the gene is expressed in multiple different cell types. Even though the gene is carrying out the same molecular function in these cells, pleiotropy arises from the expression pattern. This form of pleiotropy is expected to be quite common, since genomic analysis of gene expression patterns in human cell lines has found that only about 6% of protein-coding genes are expressed in exclusively one cell type, and more than 55% were expressed in all cell types surveyed.

No matter the underlying source of pleiotropy, direct analysis of mutations and their effects on phenotypes in various model organisms has consistently shown that most mutations affect more than a single phenotype. One estimate is that every gene has five to ten phenotypic effects; this has to be considered a minimum, since many additional phenotypes could be affected but were not analyzed.

Whether we think of pleiotropy in terms of how many genes are pleiotropic or in terms of how many phenotypes a single gene can affect, we are left with the same conclusion—most genes are pleiotropic. These pleiotropies are often hard to predict. Thus, if we are attempting to use the genome sequence or the genotype to infer the expected phenotype(s), pleiotropy presents a considerable challenge.

FIND OUT MORE

Adams, J. (2008) Obesity, epigenetics, and gene regulation. *Nat Edu* **1**: 128. Available at: http://www.nature.com/scitable/topicpage/obesity-epigenetics-and-gene-regulation-927 [accessed 2 August 2016]

Lobo, I. (2008) Pleiotropy: One gene can affect multiple traits. *Nat Edu* **1**: 10. Available at: http://www.nature.com/scitable/topicpage/pleiotropy-one-gene-can-affect-multiple-traits-569 [accessed 2 August 2016]

Paaby, A. B. and Rockman, M. V. (2013) The many faces of pleiotropy. *Trends Genet* **29**: 66–73

Stearns, F. W. (2010) One hundred years of pleiotropy: a retrospective. *Genetics* **186**: 767–73

of black pigment, as seen in the Puli. It is therefore possible to place these *agouti* alleles in order of the dominance seen when they occur in combination with one another, giving the allelic series shown in Figure 8.6(b).

KEY POINT When different alleles of the same gene produce a variety of phenotypes, it is sometimes possible to rank the alleles relative to one another into an allelic series, from the most dominant to the least dominant allele.

agouti is a well-studied gene in mammals, with effects on phenotypes, in addition to coat color. For example, in mice, different alleles of *agouti* also affect obesity and susceptibility to particular cancers. *agouti* is an example of a gene with pleiotropic effects—a concept we will discuss at the end of this chapter and in Box 8.2.

Alleles, mutations, and polymorphisms are related concepts

We want to return to our knowledge of genomes for a moment here. Alleles arise from the process of mutation, like all genetic variations. So, in some sense, alleles and mutations are simply different ways of referring to heritable changes in the gene. The term "allele" has a slightly different connotation from a "mutation" because, as explained in Chapter 4, many mutations do not result in new alleles that affect the function of the gene. If there is a wild-type form of the gene and variations from that are rare, it is helpful to think of them as being mutations. We might refer to an allele that causes a rare genetic disease as a mutation, for example.

But what terms should we use when the variations are not rare? For example, we know that blue and brown eyes arise from different alleles of eye pigmentation genes in humans. The folklore or "common knowledge" is that blue eyes is recessive to brown eyes, which is generally true if we are discussing alleles for any single gene. However, there are at least seven different genes that affect eye pigmentation in humans, so we cannot say that one eye color phenotype is clearly dominant to the other. Would we say that an allele giving rise to blue eyes is a mutation? Since the African ancestors to all modern humans almost certainly had brown eyes, the change that produced blue eyes could be called a mutation in one sense. (Analysis of Neanderthal genomes has shown that at least some Neanderthals apparently had blue eyes, but this appears to have been a separate molecular and genetic event.)

However, describing blue eyes as a mutation does not really capture the phenotypic diversity that exists among human eye colors, and there is no clear wild-type eye color phenotype that would apply to all human populations. We instead describe these variations as arising from different alleles of the gene. If there is more than one common allele or variant for a gene, the gene is said to be **polymorphic**. We will have a more precise definition of common and polymorphic alleles in Chapter 10, but, for now, we will just say that more than one allele is common. The eye color genes are polymorphic in many human populations because both brown and blue eyes are frequent. The ABO locus, discussed earlier, is polymorphic because there are three common alleles for this gene. On the other hand, most disease-causing alleles are not considered polymorphic in most populations because they are rare. We will return to this principle of common and uncommon genetic syndromes and diseases in Chapter 10.

Some alleles result in lethal phenotypes

Many genes are essential for the viability and fertility of an organism. We know this because a mutation in one of these genes can be lethal. A lethal allele is almost always recessive to other alleles, since a dominant lethal allele with an effect before the age of reproduction could not be passed on. Dominant alleles that result in the premature death or illness of the person can only be passed on if they act after the age of reproduction. Huntington disease is an example of an autosomal dominant disease; part of its tragic effect is that it usually affects people after they have had children, and therefore after they have potentially passed on the disease allele.

Recessive lethal alleles are maintained in the population at low levels in heterozygotes, familiarly known as **carriers**. All of us are carriers for many different lethal genetic conditions, although the exact number is not known. ("Lethal" is being used here in a genetic sense to refer to any mutation that prevents a homozygote from reproducing.) If two parents are carriers for the same lethal condition, some of the children could be affected. More precisely, the probability that a child would be affected is one-fourth. The risk that the parents might both have inherited the same rare recessive mutation is elevated when parents have a common ancestry and is one reason that marriages among first cousins, known as consanguineous marriages, are illegal in many countries. We will talk more about the genetics of human populations, including some genetic diseases, in Chapter 10 and about the effects of consanguinity in Chapter 16.

Lethal mutations can be recognized in a cross because one category of offspring is missing in the surviving offspring. Typically, instead of seeing phenotypic ratios when heterozygotes mate that are based on fourths, the ratios will be based on thirds, as seen in Figure 8.7.

Some unusual mutant alleles are dominant to wild-type, but lethal when they are homozygous. To express this using genetic symbols, if we designate ***A*** as the mutant allele and + as the wild-type allele, ***A*/+** has a mutant

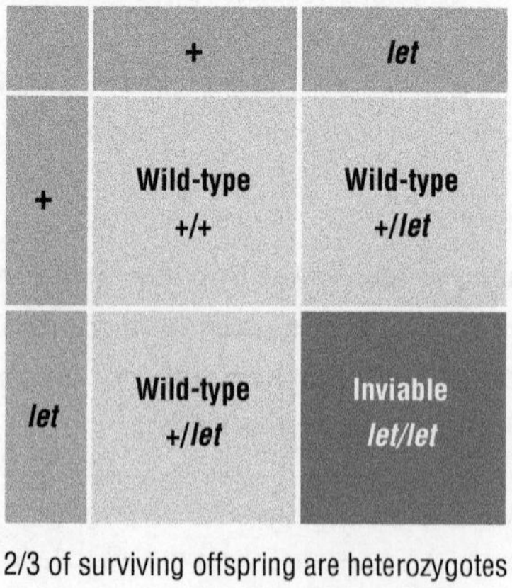

	+	*let*
+	Wild-type +/+	Wild-type +/*let*
let	Wild-type +/*let*	Inviable *let*/*let*

2/3 of surviving offspring are heterozygotes

Figure 8.7 Mendelian ratios with lethal alleles. Many alleles have a lethal phenotype, summarized as *let*. When two heterozygotes for the lethal allele are mated, one-fourth of their offspring will die; since the phenotypes and genotypes ratios are based on surviving offspring, these ratios are based on thirds.

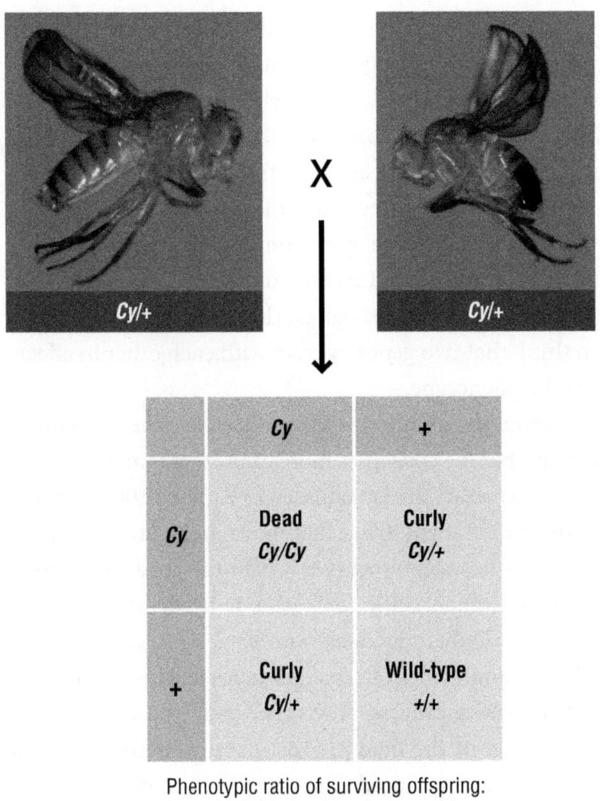

Figure 8.8 *Cy* is an example of an allele that is dominant to wild-type but lethal when homozygous. The Curly wing allele in *Drosophila* is dominant to the allele for straight wings and is also homozygous lethal. When two flies with Curly wings are mated, 1/4 of their offspring die before hatching. As a result, 2/3 of the offspring observed have Curly wings, and 1/3 have wild-type straight wings. Thus, the phenotypic ratio of the offspring is 2:1 Curly:straight.

phenotype, while ***A/A*** is dead. The Curly wings trait in *Drosophila* is one example of this; +/+ is wild-type, ***Cy/+*** has Curly wings, and ***Cy/Cy*** is dead. This can give rise to some unexpected phenotypic ratios. In Figure 8.8, we can see that, when two Curly-winged flies are crossed, the **genotypic** ratios in the resulting Punnett square follow the standard Mendelian pattern. However, the **phenotypic** ratio of the surviving offspring is an unexpected 2 Curly wings:1 wild-type because all *Cy* homozygotes die.

Some dominant genetic conditions in humans, particularly those affecting the skeletal system, are also lethal when homozygous. One example is thought to be achondroplasia, one of the most common forms of dwarfism. The achondroplasia allele is dominant, and most individuals with achondroplasia are heterozygous for an allele that arose from a new mutation in one parent. When two individuals with achondroplasia become parents, some of the children have achondroplasia and some do not, as expected, since both are heterozygous. However, it appears that none of their children is homozygotes for achondroplasia, which is consistent with the hypothesis that the homozygotes are not live-born infants. Further support for this is that a similar dominant mutation in the orthologous mouse gene is lethal when homozygous as well.

KEY POINT Some alleles produce phenotypes that result in lethality. Most lethal alleles in a population are recessive, but dominant lethal alleles also exist. In some cases, an allele can have a viable dominant phenotype that is lethal when homozygous.

8.4 Unexpected phenotypic ratios arising from gene interactions

Almost no gene acts alone in the physiological process that it affects, and only some precisely defined biological processes (such as the ABO blood type) are controlled by a single gene. After all, biology involves organized systems, and the activities of genes are no exception to being part of a system. We have focused on traits affected by a single gene acting alone because these are the easiest to interpret. However, the examples of phenotypes that we use in introductory genetics are more often the specific cases we know about, rather than the general explanation for phenotypic differences between individuals in a population. Again, this emphasis on single genes goes back to how Mendel did his experiments—he was able to interpret his results because he focused on one or two genes acting independently from each other.

But the more common biological situation is that two or more genes interact with each other to produce the phenotype. As we will discuss in Chapter 10, as the number of genes affecting the phenotypes increases, the number of possible phenotypes also increases. We also note in that chapter that the relationship between genotype and phenotype becomes more complex, and the same phenotype can arise from different genotypes. For this chapter, we will limit ourselves to phenotypes arising from crosses involving two genes.

As a result of an interaction between the genes, the phenotypic ratio observed from crossing two heterozygotes is often not the standard 9:3:3:1 shown in Figure 8.9, but rather some modification of that. When heterozygotes for two genes are mated, there are 16 possible genotypes

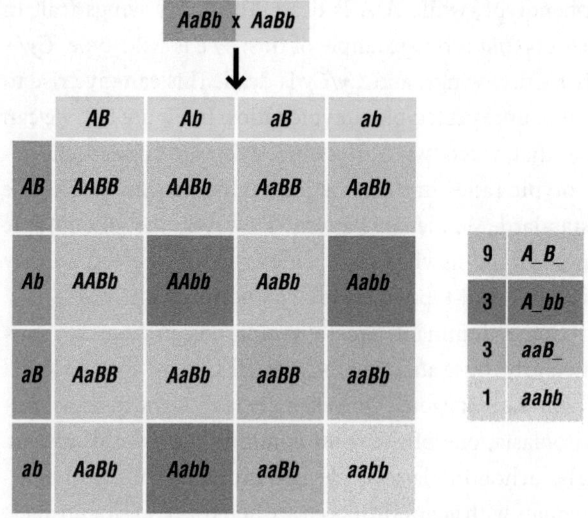

Figure 8.9 Two genes with no interactions. When two heterozygotes ***AaBb*** are mated to each other, the phenotypic ratio in the offspring is expected to be 9:3:3:1. In this diagram, we summarize these phenotype classes by colors. Among the offspring, 9/16 have a dominant allele for both genes, 3/16 have a dominant for one of the two genes, 3/16 have a dominant allele for the other gene, and 1/16 have only the recessive alleles for both genes. This ratio can be modified by gene interactions.

Source: Kristina Yu, © Exploratorium, www.exploratorium.edu.

in the offspring, as seen with the 9:3:3:1 ratio. Interactions between two genes that affect phenotypes can often been observed by noticing the fractions in the phenotypic ratios. If only two genes are involved and the phenotypes are well defined, the phenotypic ratio will still be based on sixteenths, but the ratio might be 9:3:4 or 9:7 or 15:1 or some other combination that adds up to 16 genotypes. When this type of ratio is observed in the mating of two heterozygotes, the immediate reaction is to think that two genes interact with each other in affecting the phenotype.

Various different terms are used to describe these interactions, but the concept is more important than the definitions. Two examples are outlined in Figure 8.10(a) and (b).

Imagine a situation like that shown in Figure 8.10(a), involving a two-step process in which the product encoded by the dominant allele for a gene is needed to carry out each step. Only individuals who have an ***A_B_*** genotype, with the dominant allele for each gene, will be able to carry out the entire process. Any other genotype prevents the production of the final product because both genes and their activities are needed to complete the process. Thus,

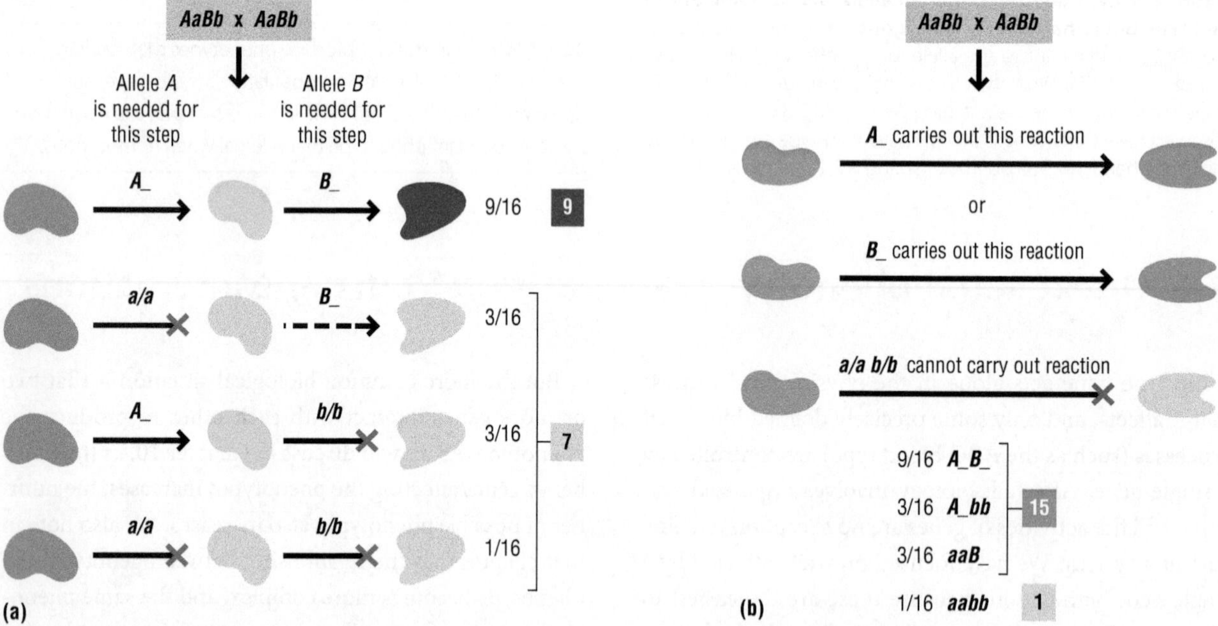

Figure 8.10 Two examples of gene interactions or epistasis. The standard 9:3:3:1 ratio for phenotypes in Figure 8.9 is altered if the genes interact with each other or exhibit a form of epistasis. These phenotypic ratios reflect the biological interactions of the functions of the genes. Several different types of epistasis are possible, of which two are shown here. (a) The activity of each gene is needed for one step that leads to the final phenotype (dark blue). While the process is shown here in two steps, these could be two of many steps in the process and need not occur one after the other. Since the activities of both genes are needed, only ***A_B_*** individuals are dark blue, and all other genotypes fail to carry out one or both steps. This results in a 9:7 phenotype ratio. (b) The genes have duplicate activities, and either one is capable of carrying out the biological process. Thus, only ***aabb*** lacks the activity, and all other genotypes produce the same phenotype. This results in a 15:1 ratio.

when ***AaBb*** heterozygotes mate with each other, there will be a 9:7 phenotypic ratio in the offspring because ***aaB_***, ***A_bb***, and ***aabb*** genotypes are all missing one or both steps in the process and therefore result in the same phenotype.

Now consider the situation shown in Figure 8.10(b), in which either of two pathways can carry out the process, that is, the process can be carried out by the ***A*** allele or by the ***B*** allele, while the normal cell has both—either one is capable of producing the end product because they have duplicate activities. The mating between the two heterozygotes results in a 15:1 ratio because ***A_B_***, ***A_bb***, and ***aaB_*** can all carry out the reaction, and only ***aabb*** individuals are defective in both pathways.

KEY POINT Phenotypes are often produced through the combined action of multiple genes. In dihybrid crosses for such genes, the phenotypic ratios of the offspring may deviate from the expected 9:3:3:1, but the underlying genotypes will still follow this expected ratio.

Epistasis describes some interactions between two genes

The examples above with 9:7 and 15:1 phenotypic ratios are only two of the possible ways that a 9:3:3:1 ratio can be modified by gene interactions; other ratios that add to 16, such as 9:6:1 or 9:3:4, are also possible. These ratios arise between two genes interacting to produce the final phenotype; gene interactions of this type are referred to as **epistasis**. Epistasis involves two (or more) genes whereby the phenotype of one gene masks the phenotype of the other. The gene that masks the phenotype of the other is referred to as the epistatic gene or is said to be "epistatic to" the other gene. Although conceptually analogous to dominance, because one phenotype is seen and one is not, epistasis is not to be confused with the existence of dominant and recessive alleles. The terms "dominant" and "recessive" refer to the relationship between alleles of a single gene. Epistasis occurs between two or more different genes and could involve either dominant or recessive alleles of these genes.

While epistasis is fundamentally important in our thinking about gene interactions, most phenotypes arising from gene interactions in nature are not so simply defined as these examples. However, epistasis has an important role in understanding how genes interact to affect a biological process, as discussed more fully in Chapter 15. Our two examples illustrated two different functional relationships between the genes, one in which both genes were necessary for a function to occur and the other in which either gene is adequate for the function to occur, and was able to distinguish between these two functional relationships, based entirely on the phenotypes that are observed.

In addition, geneticists can use epistasis to determine the order in which genes must act in a complicated biological process. In Tool Box 8.1, we provide an overview with the example of the ***E*** gene acting upstream of the ***B*** gene to affect coat color in Labrador retrievers. We want to point out that "upstream" here does not mean that the

TOOL BOX 8.1 The use of epistasis to construct genetic pathways

The ways that geneticists think about epistasis are a little bit different and more informative than the standard modifications to the 9:3:3:1 ratio that are discussed in the chapter. In controlled laboratory experiments with well-studied organisms, the inheritance patterns of genes are usually known before epistasis is encountered, so the phenotype ratios are not something that needs to be worked out. In fact, the two genes used in the crosses are usually chosen because the investigator is specifically interested in their interaction. Thus, epistasis is an expected outcome of the experiment, and inferences from epistasis provide a useful way of understanding how the genes function together to affect a biological process. While this will be encountered more fully in Chapter 15, a brief introduction here will set the stage.

Recall that the two genes being investigated for a possible interaction are inherited by the established principles of Mendelian genetics. Their interaction is deduced from the interpretation of their phenotypes. Many different kinds of phenotypes can be used to look at gene interactions, including the effects of having mutations in two different genes and the patterns of expression of one or both genes in wild-type and mutant individuals. Let's look at two examples, one that uses one mutant phenotype and an expression pattern phenotype, and a second one that uses two mutant phenotypes.

Using a mutant and an expression pattern

We discussed microarrays in Section 2.5 as a method to examine gene transcription patterns on a genome-wide scale. In wild-type individuals, we extract RNA and hybridize the labeled RNA (or corresponding cDNA) to a microarray to determine which genes are being transcribed and at what levels. The expression profile that

TOOL BOX 8.1 Continued

we observe from a microarray makes a reasonable phenotype, not only for one gene, but also for many genes.

Now suppose that we do the same microarray experiment using a mutation that eliminates the function of a particular gene, called gene *A*, that is, we extract RNA from a population of individuals who are mutant for that gene (***a/a***) and perform the same microarray analysis as we did for wild-type individuals. We show some of these in Figure A. For many of the transcripts that we consider on the microarray, the pattern will be the same in the wild-type and the mutant individuals. We conclude that the transcription of those genes does not depend on the activity of gene *A*.

For another set of transcripts on our microarray, the pattern of transcription is different in wild-type and ***a/a*** mutants. Maybe some of the genes are transcribed at much lower levels, or possibly not transcribed at all, in the ***a/a*** mutant. We infer that these transcripts depend on the normal function of gene *A* because, when the activity of gene *A* is eliminated in the ***a/a*** mutant, the level of expression of these other genes is decreased. Thus, the normal function of gene *A* somehow activates or turns on the transcription of these genes. Notice a very important point. We cannot conclude that

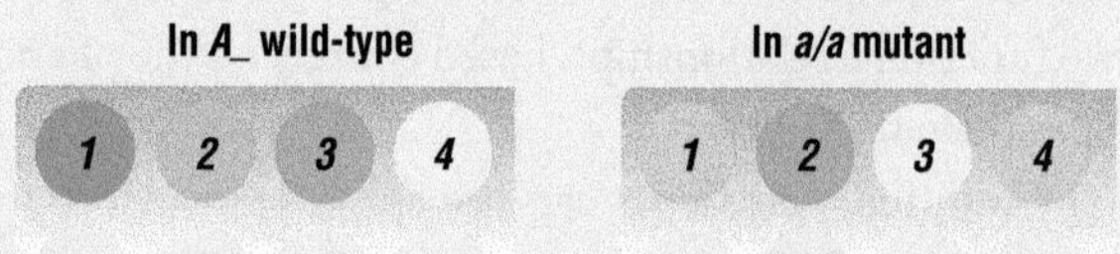

Gene 1: The transcription level goes down when *A* is mutated
Inference: Gene *A* is needed for high levels of gene *1* expression

Gene 2: The transcription level does not change in wild-type and *a/a* mutants
Inference: Gene *A* does not affect gene *2* expression

Gene 3: The transcription is **OFF** when A is mutated
Inference: Gene *A* is needed to turn **ON** gene *3*

Gene 4: The transcription is **ON** when *A* is mutated
Inference: Gene *A* is needed to turn **OFF** gene *4*

Figure A Gene interactions using molecular assays. This assay compares the transcription of genes 1–4 in the presence of wild-type and mutant versions of gene *A*. In this example, the interactions of gene *A* with four other genes is shown, using a microarray as the phenotype for genes *1*, *2*, *3*, and *4*. When gene *A* is wildtype, genes *1*, *2*, and *3* are transcribed while gene *4* is not. A microarray is then done for the same four genes when gene *A* is mutant, shown here as ***a/a***. The transcription of gene *1* has decreased, as indicated by the lighter green signal, indicating that functional gene *A* is needed for high levels of gene *1* expression. The results with the other genes are summarized below the figure. Note that this assay does not mean that gene *A* directly regulates the expression of genes *1*, *3*, and *4*; other genes may be involved, and the effect of gene *A* could be indirect.

gene *A* **directly** controls the transcription of these genes; the effect of gene *A* might be exerted through any number of intermediate processes that we have not defined. But we can say that, in some way, the transcription of these other genes depends on the normal function of gene *A*.

Let's illustrate this with a simple analogy. Suppose that you are reading in your room one evening when suddenly your lamp goes out. You can reasonably conclude that the activity of your lamp has been changed and that something that your lamp needs for its activity has malfunctioned. There might be a malfunction in some component with a very direct effect—the bulb in your lamp has burned out, for example. But there might also be a malfunction in some component far upstream of your lamp—a tree limb has fallen on a power line, for example, so that electricity to an entire neighborhood has been cut off. The transcription of a gene as assayed by the microarray is similar to the reading lamp in this analogy; it has changed, but we don't know if the malfunction (the mutation in gene *A*) occurred directly upstream of the transcript or in some process far upstream. The dependence of the transcript on the function of gene A shows the strategy or the logic of the overall program by which gene A regulates other processes, but it does not provide any of the details.

Before leaving this example, we should consider the transcription pattern of some other genes in our hypothetical microarray. There will undoubtedly be genes that will be transcribed at a higher level in the ***a/a*** mutant than in the wild-type. Thus, our inference is that normal function of gene *A* is somehow to repress or turn off the transcription of these genes. Again, the effect of gene *A* could be direct or indirect. The change in transcription is showing the overall program by which gene *A* affects this process, but not any of the specifics.

We have illustrated this interaction with the changes in transcription, as assayed by a microarray, but this can be generalized to any assay for any process. We could have an assay that looks at changes in splicing pattern, in protein expression, in protein localization, or in any of a number of other processes. The overall strategy for interpreting the results is the same, regardless of the assay or process used.

Using two mutant phenotypes

Sometimes we do not have an easy assay for the transcription or expression of a gene, but we still want to understand how two genes might interact to affect the phenotype. A similar logic of epistasis can still be applied, but we make the inference based on the mutant phenotypes of the two genes, rather than an assay for expression. We can use coat color in Labrador retrievers for this explanation, since it involves a particularly well-studied pair of genes with familiar and recognizable phenotypes.

TOOL BOX 8.1 Continued

There are numerous genes that are responsible for coat color variation in mammals, including the well-studied *agouti* locus discussed in the chapter. In any particular purebred (that is, inbred) breed of dogs or other mammals, some of these genes exhibit no variation, and all purebred dogs of the same breed are homozygous for the same allele. Thus, genes that contribute to the variation in one breed may or may not be responsible for some of the variation in another breed. This is an important concept that will arise again in Chapters 10 and 16; the gene might have a very important function in coat color, but, because there is only one common allele in a particular breed, it does not contribute to coat color **variation** in this breed.

In Labrador retrievers, three coat colors are recognized by kennel clubs and dog breeders: yellow, black, and chocolate brown, as shown in Figure B. Yellow dogs are homozygous recessive for a trait known as the Extension or *E* gene; they are ***ee***. The actual *E* gene is now known to be the melanocortin 1 receptor gene *MC1R*, which also contributes to pigmentation differences in other mammals by regulating the deposition of pigment granules in skin and hair. In humans, one of the most common forms of red hair arises from mutations in the *MC1R* gene, and mutations in *MC1R* were found in the Neanderthal genome as well. Because Labrador retrievers are homozygous for alleles in other genes that regulate or interact with *MC1R* (including the *agouti* locus), mutations that eliminate the function of *MC1R* produce yellow fur.

Dogs with an ***E_*** genotype have a functional *E* gene and deposit pigment granules in the skin and hair. These pigment granules in Labrador retrievers can be either black or brown, depending on the function of the *TYRP1* gene, which encodes an enzyme responsible for the production of eumelanin, the dark pigment in mammalian hair. In standard coat color genetics, this is often symbolized as the *B* gene because of its phenotype; it is the same gene responsible for the black or brown rabbits. Alleles

Figure B Epistasis using color variation as the mutant phenotypes. Epistasis or gene interactions can also be evaluated using visual phenotypes such as coat color. In dogs, a gene known as *E* is needed for the deposition of pigment, while a gene known as *B* is responsible for the synthesis of black or brown pigment. Thus, black Labradors must have the genotype of ***E_; B_***; the ***B*** allele is responsible for the black pigment, while the ***E*** allele allows the pigment to be deposited in the hair. Chocolate Labradors have the genotype ***E_; bb***, since the absence of the ***B*** allele means the pigment is brown. Yellow Labradors do not deposit the pigment. They therefore have the genotype ***ee***, but the pigment genotype cannot be inferred because no pigment is deposited in the coat. The *E* gene is defined as being epistatic to the *B* gene, since the genotype at the *E* gene affects the phenotype controlled by the *B* locus. The *E* gene is considered to act upstream of the *B* gene. Other genes also affect coat color in dogs, but these are not variable in Labrador retrievers, so their effects are not evident.

Source: Image of yellow labrador courtesy of Elf/ CC BY-SA 3.0. Image of black labrador courtesy of Sokova Mary. Image of chocolate labrador courtesy of Ben Swihart/ CC BY 2.0.

TOOL BOX 8.1 Continued

that encode functional enzymes produce black pigmentation, represented as ***B_***. Three different molecular lesions found in dog breeds, including Labrador retrievers, eliminate the function of the gene and produce a brown pigment instead; all three mutations, although different at the molecular level, produce similar hair color phenotypes, at least in Labrador retrievers, which is known as chocolate to dog breeders. Thus, ***bb*** is chocolate brown and ***B_*** is black.

How do these genes interact with each other? Matings between dogs with different coat colors has shown that the *E* gene is epistatic to the *B* gene, so that a dog that is ***ee*** is yellow; no matter which pigment granules are produced by the activity of the *B* gene, those granules are not being deposited in the hair because of the mutant form of the *E* gene. Thus, the *E* gene acts upstream of the *B* gene, although the cellular and molecular effect is indirect.

Let's talk through this example using our desk lamp again. If the power goes out throughout the neighborhood—that is, we have a malfunction in the upstream gene (***ee***)—then it makes no difference if the light bulb is functional (***B_***) or burned out (***bb***). We will only see the effect of the upstream malfunction even if the bulb is working fine, and the lamp will be out. As before, the interaction between the two genes is showing the overall strategy or program by which they act, but not the specifics. With the *E* and the *B* genes, we can also make an analogy to having a functional delivery truck. If the delivery truck is broken down (that is, an ***ee*** genotype), it makes no difference if it is delivering books or television sets; no home delivery will occur.

In either analogy, as in epistasis, the interaction between the upstream and downstream genes could be direct or indirect, and the outcomes would be the same. That is a limitation of using epistasis to build molecular, cellular, or biochemical pathways; there could be many intervening steps between the two being tested. On the other hand, the tremendous advantage of using epistasis to build a pathway is that it provides a logical or functional pathway, even when the biochemical and molecular interactions are not known. Many, if not most, of the cellular and biochemical pathways you may be familiar with were originally constructed using epistasis. The molecular and biochemical interactions were worked out after the logical or functional order was determined.

E locus **directly** controls any of the molecular processes at the *B* locus; in fact, we know from other experiments that it does not directly control the *B* locus. It is rather that the activity of the *E* locus is needed for the activity of the *B* locus to be observed.

KEY POINT When a phenotype of one gene masks the phenotype of a second gene, it is said to be epistatic to the second gene. Such epistatic relationships between genes can be used to determine the order in which gene products act in a biological process.

Epistasis is a powerful, albeit subtle, tool in the genetic analysis of biological processes. The activities and interactions of the genes have been inferred without any knowledge of the molecular function of either gene or the biological process itself. This ability to use epistasis to get a preliminary picture of the strategy by which genes interact to carry out a biological process should be not underestimated; epistasis is often used most effectively when very little is known about the process and has been used as a starting point for the genetic analysis of many cellular processes. We will return to the use of epistasis for genetic pathways in Chapter 15.

8.5 A functional definition of a gene: the complementation test

As we discussed in Chapters 2 and 3, it is not easy to define a gene in a way that encompasses all of its properties. However, by now, we have developed a reasonably accurate working definition of a gene, even if it is difficult to express succinctly. Let's approach the definition of a gene from a different perspective—by describing some experimental crosses and their results that can be used to determine if alleles or mutations fall into the same gene. The procedure that we will describe is called a **complementation test**, and it offers one of the most complete functional definitions of a gene. A complementation test can be applied to any organism and to any gene, with a few exceptions that will not concern us now. The experiment is shown in Figure 8.11.

Consider this conundrum—if multiple genes can contribute to a particular phenotype, then how do we know if

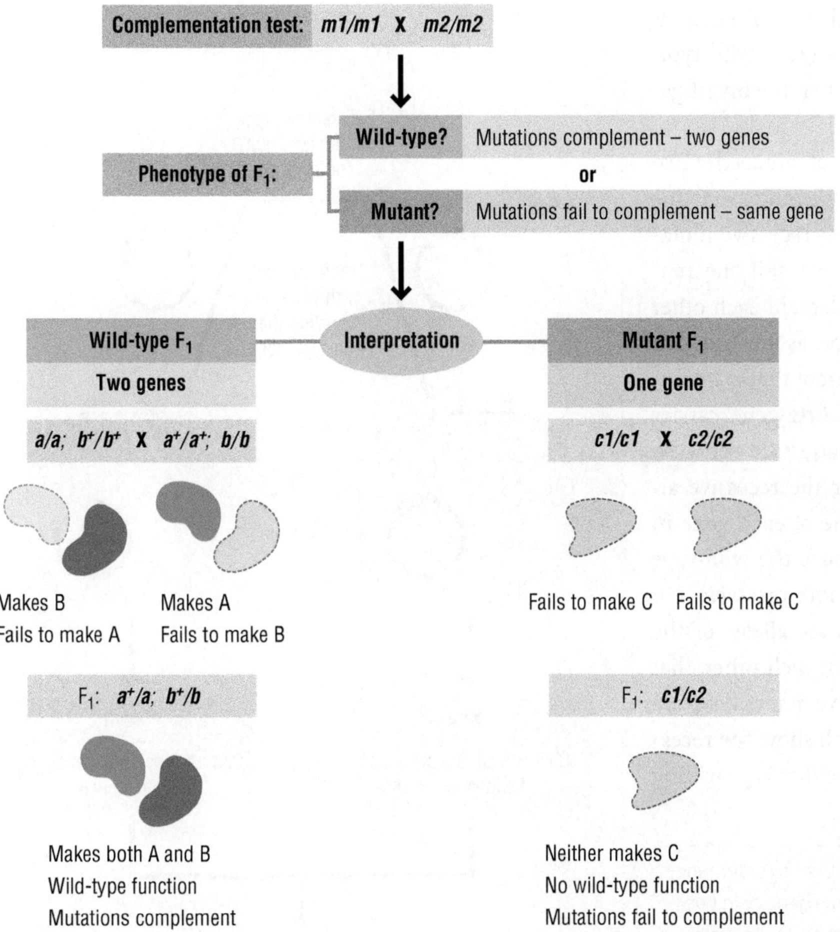

Figure 8.11 A complementation test. The complementation test is a functional test to determine if two mutations are different alleles of the same gene or if they are alleles of two different genes. The alleles must be recessive. A heterozygote with one of each allele is constructed, shown here by the mating of two homozygotes. The phenotype of this heterozygote is scored. If it has a wild-type phenotype, the two mutations are defined as complementing each other and represent two different genes. The interpretation for this result is shown below. The mutant *aa* lacks the A activity but has the B activity, while the mutant *bb* lacks the B activity but has the A activity. Thus, when they are crossed, the heterozygote has both the A activity and the B activity and has a wild-type phenotype. On the other hand, if the heterozygote has a mutant phenotype, the two mutations are defined as failing to complement and represent two different alleles of the same gene. Again, the interpretation is shown below. Both the first mutant (*c1/c1* here) and the second mutant (*c2/c2* here) fail to make the C activity. When they are crossed, neither is able to provide the function that the other lacks, and the heterozygote is mutant. A heterozygote with different alleles for a gene is formally defined as being heteroallelic, which would be the case here.

two individuals showing that phenotype possess specific alleles of the same gene or possess two different genes? This scenario applies to many genetics experiments in which geneticists collect many mutations that have similar phenotypes in relation to the process being studied. The fly embryos have defects in segmentation, the plants have similar flower colors, the worms cannot move normally, and so on. One of the early steps in the genetic analysis of any biological process is to determine how many different genes affect the process. Are all of these different mutant phenotypes the outcome of changes in a single gene, in which case these mutations would be different alleles, or are they the result of the mutation of several different genes with related functions?

We can determine if two mutations are alleles of the same gene by performing a complementation test, as shown in Figure 8.11. A complementation test involves crossing the two mutants and observing the outcome, so it is usually simple to do and to interpret. The one key requirement is that the two mutations to be evaluated must both be recessive to wild-type. If that requirement is met—and most mutant alleles are recessive to wild-type, so this is not a difficult requirement—then the investigator can perform a complementation test.

So how does a complementation test proceed? Two recessive mutations are crossed to each other, and the diploid F_1 offspring are observed. If two recessive mutations are not alleles of the same gene—that is, if they represent different genes—they will complement each other and produce offspring that are wild-type, as illustrated in Figure 8.11. This occurs because the parent that is homozygous for the recessive allele of the first gene carries a wild-type copy of the second gene, and vice versa, so the offspring will be heterozygous for the recessive allele of both genes. One wild-type allele of each gene in the offspring will be sufficient to produce the wild-type phenotype because the mutations are both recessive. On the other hand, if the two mutations are alleles of the same gene, they will fail to complement each other, that is, when individuals with the recessive phenotype are crossed together, all the F_1 offspring will show the recessive phenotype of the parents because neither is providing the function of the gene.

KEY POINT A complementation test can be used to determine whether two different mutations are in the same gene or in two different genes. In this test, two individuals that share the same phenotype, but are each homozygous for a different recessive mutation, are crossed together. If the offspring all show the same phenotype as the parents, we deduce that the mutations were in the same gene.

Complementation tests are fundamental to defining a gene and to thinking about how genes and mutations act. The key concept is that one copy of the gene has the ability to provide the function lacked by another copy of the gene. Thus, a version of a complementation test can also be done in bacteria (which do not normally have two copies of most genes) and in haploid eukaryotes by using a plasmid. Recall from Chapter 3 that a plasmid is a circular piece of DNA that is inherited separately from the main chromosome of the organism and that has its own genes. Most plasmids are dispensable as far as the viability of an organism is concerned—that is, they do not contain essential genes for most environments—but they often contain genes that are helpful, important, or even essential in some environments.

Plasmids provide the necessary tool to perform a version of a complementation test in bacteria and in those haploid eukaryotes containing them such as the yeast *Saccharomyces cerevisiae*. The logic of this experiment is the same as the logic behind a complementation test—the one wild-type copy is providing the function lacking in the mutant, as shown in Figure 8.12. The mutant cell lacks the ability to perform a particular function such as to synthesize the amino acid leucine. Thus, unless the medium is rich in leucine, the cell cannot grow and divide.

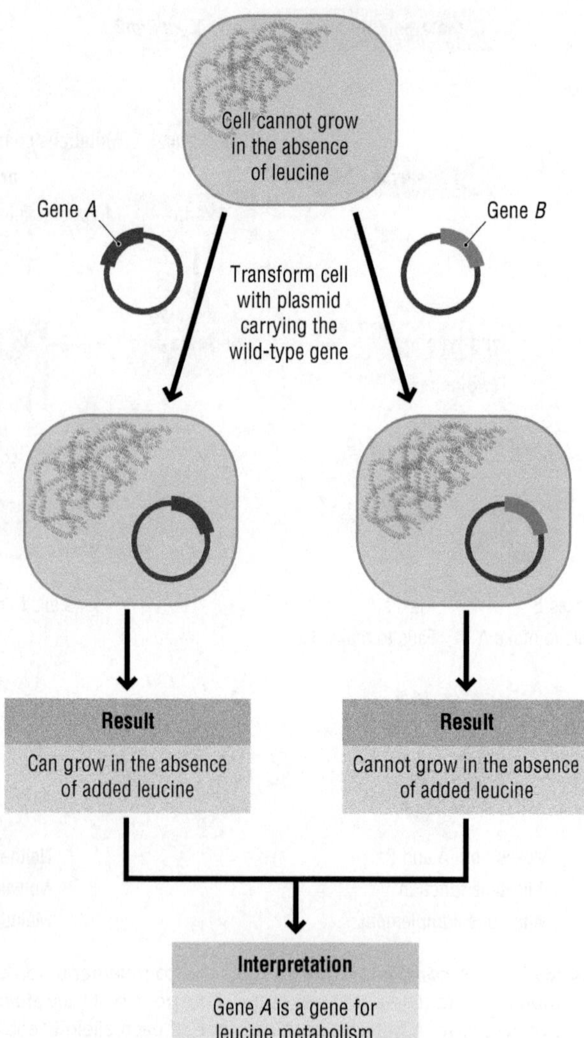

Figure 8.12 Complementation test using plasmids. In this example, the cell has a mutation in a gene affecting leucine biosynthesis and cannot grow, unless leucine is added to the medium. Plasmids are made that have the wild-type versions of two genes affecting leucine biosynthesis, called gene *A* and gene *B* in this example. Each of these plasmids is transformed into the leucine mutant cells. The cells transformed with the gene *A* plasmid are now able to grow in the absence of additional leucine; this demonstrates that gene *A* is able to supply the gene activity that was missing in the original mutant, and thus complements or rescues the mutant defect. On the other hand, cells transformed with gene *B* are still unable to grow, unless leucine is added, so gene *B* is not able to complement the mutant defect and must encode a different activity.

In order to determine if a particular DNA sequence can complement the mutant phenotype, a wild-type copy of a gene is inserted into a plasmid, and the plasmid transformed into a mutant cell. If the plasmid contains the wild-type version of the gene for leucine biosynthesis that is absent in our mutant, then the cell with the plasmid will be able to grow. In other words, the mutant phenotype has been rescued by the addition of the plasmid with the wild-type gene.

On the other hand, if the plasmid has the wild-type copy of a different gene affecting leucine biosynthesis, the mutant will not be rescued and will continue to require leucine to grow. In a complementation test with diploids, the other parent is providing an additional wild-type copy of the gene; in this type of experiment in bacteria and yeast, the plasmid is providing the wild-type copy of the gene.

KEY POINT When a phenotype is believed to be due to a mutation in a particular gene, this can be tested by supplying a wild-type copy of that gene to the mutant to see if this rescues the phenotype.

8.6 More complicated interactions between genes and the environment

Many phenotypes are easy to assess, and we can sometimes make a direct connection from the phenotype to inferring the underlying genotype. For instance, if a person has type O-negative blood by agglutination assays, we can be fairly confident that her genotype is ***ii*** for the ABO locus and ***rr*** for the Rh locus, since these are the recessive traits. But for many phenotypes, it is not so easy to infer the underlying genotype. We have discussed some of the reasons that this connection can be difficult to make. Alleles can have varying degrees of dominance, every gene can have multiple different alleles in the population, some alleles have lethal effects, genes interact with each other in complicated and sometimes unpredictable ways, and so on. To return to the unifying ideas introduced in the Prologue, biological processes arise as a result of organized systems, so phenotypes can be less predictable than expected from the genotypes. In this section, we describe a few more situations that make it more difficult to make direct connections between genotypes and phenotypes.

The first and most familiar of these conditions that affect phenotypes is the environment. Many phenotypes arise from the environment alone or from combined effects of genes interacting with the environment. In this context, the "environment" includes not only factors like temperature, light, and rainfall, but also other non-genetic factors such as nutrition, stress, trauma, infections, and so on. There are also all of the complicated interactions with other organisms living in the same environment, referred to as metagenomics and the main topic in Chapter 17. We now recognize that these interactions can be so intimate and long-lasting that the lines between genotypes and phenotypes, and even between individual genomes, become blurred.

The era of genome sequencing makes clear that knowing the full DNA sequence of a genome has proven to be important knowledge but is not itself sufficient to fully understand or predict all phenotypic outcomes. Some of the challenges are summarized in Figure 8.13.

In fact, successfully navigating this gap between genome sequence or the genotype and phenotype remains one of the current challenges of the "post-genomics" era. A full genome sequence allows us, for the first time, to simultaneously consider the nature of an individual allele for one gene in the context of all the other sequence information present in an individual genome. Nevertheless, there are many subtleties, unknowns, and even puzzling findings that make trying to predict the phenotype from a DNA sequence still a considerable challenge. This is a substantial change in how geneticists work—for many decades, geneticists developed the tools and strategies to find the genes responsible for a phenotype, that is, to work from the phenotype towards the genotype and its underlying DNA sequence. Phenotypes were easy to observe, while DNA sequences were harder to come by. Now with the comparative ease of finding the DNA sequence, and hence the genotype, tools and strategies that work from the genotype towards the phenotype—the reverse of the direction in which we have worked in the past—are needed.

The relative contributions of genes and the environment to a phenotype are often stated as "nature vs. nurture" where nature refers to an organism's genotype and nurture refers to all of the environmental influences

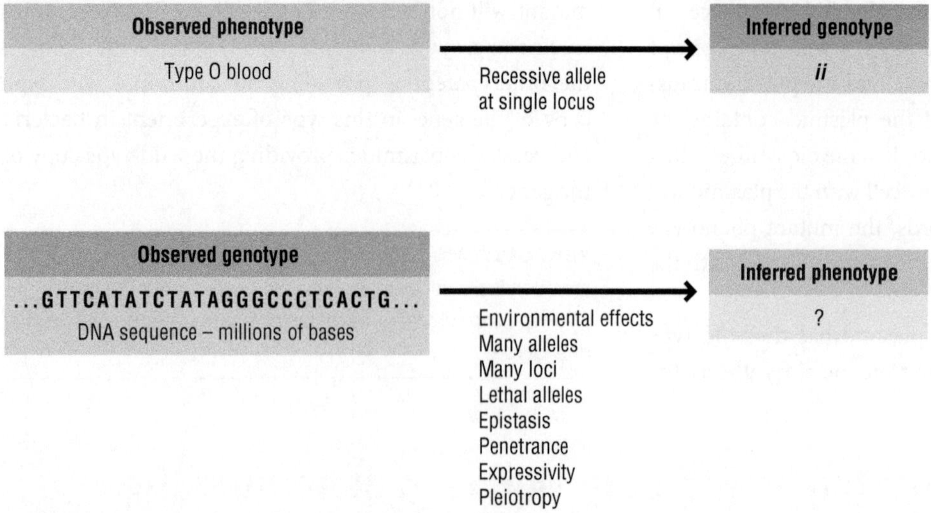

Figure 8.13 Connecting genotypes and phenotypes. Some of the challenges arising in inferring genotypes and phenotypes are summarized. When the observed phenotype, such as type O blood, is due to the effects of a recessive allele at a single locus, it is easy to infer the genotype accurately; it must be *ii*. When genomes have been sequenced, it is often the genotypes that are observed and the phenotypes that must be inferred. This is more challenging because of the effects discussed in this chapter.

acting upon it. But this shorthand greatly understates the complexities of these interactions, as we will discuss more fully in Chapter 10. In fact, thinking of genotypes and environments as separable inputs to a phenotype may be misleading. Environmental inputs can interface with the genome by directly altering the regulation of genes, as occurs for certain toxins, and/or by changing the chemical nature of the DNA and the proteins that associate with it. Furthermore, the same genotype can result in different phenotypes, depending on the environment. Seeds from the same true-breeding variety of a plant do not produce identical plants; purebred dogs from the same litter are not identical, and even identical twins in humans are distinguishable for each other; the reasons for these phenotypic differences among genetically identical individuals are the contributions from the environment, which are not constant, and the interactions between the genotype and the environments. As we will describe in Chapter 15, geneticists have taken advantage of these environmental effects on phenotypes by working with **conditional mutations**. These are mutations whose phenotypic effects are only observed under particular environmental conditions such as elevated growth temperature (temperature-sensitive mutations) or nutritional conditions (auxotrophic mutations). Under other conditions, a conditional mutant has a wild-type phenotype.

Three additional, and possibly related, genetic phenomena also contribute to the challenges of connecting genotypes and phenotypes more directly. These phenomena are reduced penetrance, variable expressivity, and pleiotropy. All are commonly observed, but their underlying causes are not so simple to determine; none of them seems to have a single explanation that accounts for all of the examples, or even for most of the examples. They may be related to one another, and often are treated as being related to each other, but there is not enough evidence to make confident assertions about their relationships. We provide some speculations and ideas in our discussion.

Penetrance

Penetrance is defined by the answer to a simple question—in a population in which all of the individuals carry the same mutant genotype, how many of the individuals are also phenotypically mutant? If all of the individuals who are genotypically mutant are also phenotypically mutant, then the mutation is said to be completely (100%) **penetrant**. All of the situations we have described so far and nearly all of the situations that we encounter in introductory genetics involve phenotypes that are 100% penetrant. Every individual whose genotype at the ABO locus is ***ii*** has type O blood because phenotypes at the ABO locus are 100% penetrant. But many mutations are not completely penetrant, as illustrated in Figure 8.14. For example, it might be that only 18%, or 39%, or 86% of those

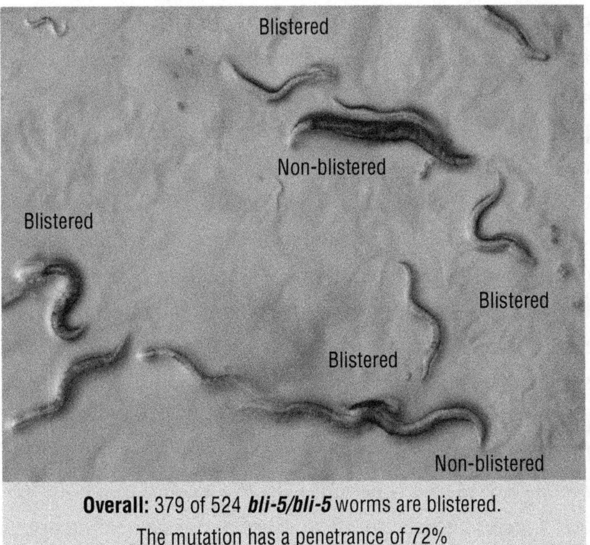
Blistered
Non-blistered
Blistered
Blistered
Blistered
Non-blistered
Overall: 379 of 524 ***bli-5/bli-5*** worms are blistered.
The mutation has a penetrance of 72%

Figure 8.14 Alleles with reduced penetrance. Reduced penetrance means that not every individual with the mutant genotype displays a mutant phenotype. An example for the *bli-5* gene in *C. elegans* is shown here. Recessive mutations in this gene result in the formation of blisters on the head; some worms with blisters are labeled in the micrograph. All of the worms in this picture are homozygous for the *bli-5* mutant allele, but not all of them exhibit the blisters, as shown by the non-blistered worms. In a direct count, 72% of the worms develop a blister, so the mutation is defined as having a penetrance of 72%.

homozygous for a certain allele have a mutant phenotype. We can say that these alleles are incompletely penetrant or have a reduced penetrance, or we can express this as a percentage if that can be reliably determined.

Why do some individuals show a mutant phenotype and some show a wild-type phenotype, even though they have the same genotype? This difference is not completely understood. The most likely explanation is that there is an interaction of the gene product with the environment, including nutrition, light and temperature, pathogens, and some other more subtle factor. However, other explanations are also possible. For example, incomplete penetrance could arise from random fluctuations in the amount of the gene product being made by different individuals. There may be epigenetic differences among individuals arising from differences in their chromatin structure, a topic discussed in Chapter 12, which could, in turn, result in changes in the amount of the gene product being produced. Reduced penetrance could also be due to interactions with other genes in the genotype, which makes the phenotype more or less likely so show up—but no one explanation can fit all examples of incomplete penetrance.

KEY POINT There are some alleles that do not always result in the phenotype associated with that allele. Such alleles are said to have reduced penetrance.

Penetrance becomes especially intriguing when we think about how it might apply to human populations. The concept of "penetrance" might underlie the concept of "risk factors," that is, having a particular mutant genotype places the individual at risk for having a mutant phenotype or a disease, but not all of the people with the mutation will develop the disease or the trait. For example, not every woman who has a particular *BRCA1* mutation develops breast or ovarian cancer, although many do. While we usually say that the *BRCA1* mutation predisposes the woman to developing breast or ovarian cancer or is a risk factor for these cancers, we might also say that the deleterious effects of the *BRCA1* mutation are not completely penetrant, so that some women are unaffected. The biological explanation for why some women develop breast or ovarian cancer and some do not likely involves some combination of interactions with other genes and alleles in the genome and the impact of environmental factors. When we think of risk factors, or reduced penetrance, with human conditions, we usually attribute the phenotypic difference to environmental differences, but genetic differences arising from other alleles in the genome or from random fluctuations cannot be overlooked.

Expressivity

Even when a group of affected individuals have a particular mutant phenotype, they usually do not have exactly the same mutant phenotype; the mutant phenotype may vary. In humans, this could be seen as a difference in the severity of the disease or the trait, the age of onset, other organs that are affected, or other changes. This phenomenon is known as variable **expressivity**; even among the blistered worms in Figure 8.14, the blisters vary in size, for example.

Variable expressivity probably has simple explanations, as well as others that might be more complicated biologically. Most mutations result in a gene that does not work at all or does not work as well as wild-type, so we can draw an analogy with situations in which things do not work. Thus, one way to think of variable expressivity is that, when the function of a process or a gene is reduced, different phenotypic breakdowns can occur. Imagine that the brakes on your car are analogous to the function of a gene. When the brakes on your car are working poorly, they may produce an accident, but they don't always result in the same kind of accident. Sometimes the car hits a tree, and sometimes it bumps into another car. Variable expressivity tends to focus on which object is hit (that is, which tissue is affected), rather than what caused the brakes to fail in the first place (that is, the underlying cellular or molecular defect).

KEY POINT Individuals who share the same genotype for an allele may differ in the severity or strength of the associated phenotype. Such alleles are said to show variable expressivity.

Variable expressivity and reduced penetrance can sometimes be thought of as different positions on a spectrum of effects. We can continue to discuss a risk factor gene such as *BRCA1*; a particular mutation in this gene elevates the risk of breast or ovarian cancer by as much as 8-fold over the risk for women who do not have the mutation. (We want to emphasize that most cases of breast and ovarian cancer are not due to mutations in *BRCA1*, so more authoritative sources should be consulted for precise information on cancer risks.) Among women who have the same molecular change in the gene, the effects or the phenotypes vary widely. Some women develop breast cancer, while their sister develops ovarian cancer, so different tissues can be affected. In addition, the age of onset varies widely, and the biological courses of the cancer (such as rate of growth and metastasis) also vary. Thus, among women with the *BRCA1* mutation, there is variable expressivity. But, in addition, at least 30% of women with this *BRCA1* mutation do not develop cancer at all before age 70; thus, the mutation could also be considered to have reduced penetrance.

In this type of example, which could apply to many other cancers and genetic disorders, such as Tourette syndrome, Marfan syndrome, and many more, variable expressivity and reduced penetrance are closely related concepts and may have related underlying explanations. But this does not explain all examples of variable expressivity and reduced penetrance. Even mutations with 100% penetrance typically have variable expressivity, while some mutations with reduced penetrance do not have much variation in expressivity, at least in laboratory populations.

Pleiotropy

Another factor that makes it more complicated to explain the connections between genotypes and phenotypes is that many genes, and even most genes, affect more than one phenotype if we look closely enough. Genes that affect more than one phenotype or more than one process are said to be **pleiotropic**. Many of the genes we discussed previously for a single-gene trait are actually pleiotropic. For example, white-eyed flies also lack pigmentation in some other organs. The *agouti* gene, discussed in Section 8.3, affects coat color, but some alleles also affect obesity and other phenotypes. As noted in Box 8.1, even the ABO blood type may have pleiotropic effects on the risks for certain types of diseases; it seems plausible that some of the pleiotropic effects were actually the more important phenotypes from the ABO blood type in evolutionary time than agglutination, since blood transfusions are a very recent innovation in the evolutionary life of our species. Although all of these genes are pleiotropic, we could discuss them as examples because we focused on only one of their phenotypes. If we were attempting to infer the phenotype that arises from changes in the gene, these other effects become more important to recognize.

There are at least two ways that a gene can have pleiotropic roles. First, the gene may encode a single product with multiple molecular functions or with separable activities. Thus, changes that affect one region of the gene and protein can have a different phenotype from changes that affect another region. Second, the gene may play the same molecular function but at multiple different times or locations in an organism. For example, the *white* gene in *Drosophila* encodes a transporter protein that transports pigment granules in many different cells, one of which is the eye. Genes and gene products are also parts of organized systems, so pleiotropy from a gene can also arise from its other interactions.

The fact that most genes have pleiotropic roles was recognized even before we knew that DNA was the substance of genes or indeed knew any of the basic processes of molecular biology. However, the ability to determine the sequence of whole genomes has confirmed and deepened our understanding of the causes of pleiotropy and provided some insights into the frequency of pleiotropic effects. When we look at the genome composition of any organism, it is now clear that there are more "things that need to be done" in the life of an organism than there are protein-coding genes to do them. Individual genes must therefore be deployed for multiple tasks.

In many cases, the same molecular function is used to drive multiple biological processes. The gene responsible for that molecular function must therefore be expressed at different times and/or places in the organism. For example, the evolutionarily conserved *sonic hedgehog* gene (*shh*) encodes a molecule that is secreted from cells and acts as a molecular signal to the receiving cells. This *shh*-based form of cell-to-cell signaling is used in vertebrate embryos to establish an important specialized group of cells during the formation of the neural tube—but it is also used in many other instances, including limb formation and axon guidance. The molecular function is to provide a signal from one cell to another, but the outcome of that signal varies, depending on the other genes being expressed at that time and place (along with any environmental inputs). In this way, the limited set of genes present in a genome are used in a variety of different instances, and most genes therefore have pleiotropic roles. Box 8.2 explores the concept of pleiotropy in more detail.

KEY POINT A gene or a mutation in a gene can have multiple effects or phenotypes. When this occurs, the gene or mutation is referred to as being pleiotropic.

Let's now consider some evolutionary consequences of this genomic understanding of pleiotropy. If most genes in the genome have pleiotropic roles, then individual mutations in a gene can differ in how their effects are revealed. For example, a mutation in the *shh* gene that results in a non-functional gene product will affect every process that involves *shh*. On the other hand, a mutation that only changes the expression of *shh* in the developing limbs will have a more specific, less pleiotropic effect—it will affect just limb development, but no other biological event with which *shh* is normally involved.

Interestingly, it turns out that mutations with extensive pleiotropic effects are more likely to have some deleterious effect and therefore be subject to negative selection. This predicts that evolution will often proceed via a series of mutations, each with limited pleiotropic or more specific effects on the gene. As with our example of *shh*, mutations that occur in the genomic regions that control when and where a gene is expressed are expected to have more limited effects than mutations in the actual coding sequence of a gene. We can therefore predict that, when we look at the genomic level, we are more likely to find mutations that contribute to phenotypic evolution in the regulatory regions of genes than in the coding regions of genes.

But if mutations that affect the molecular function of the encoded gene product are likely to be more pleiotropic—and therefore more deleterious—than mutations in the regulatory elements of a gene, then how do new molecular functions ever evolve? Part of the answer here lies in gene duplication. As we noted in Chapters 3 and 4, the duplication of genomic regions containing the coding sequence of entire genes is an important component of genomic change. Once a pleiotropic gene has duplicated, then it is possible for one copy to acquire mutations that alter the molecular function, while the original molecular function is maintained in the other copy. When a pleiotropic gene with multiple regulatory elements duplicates, it is also possible for the duplicated coding sequences to come under the control of a subset of the original regulatory elements, so that both copies are now needed to constitute the full range of expression of the original gene. With both copies now necessary in the genome, they may then differ in their subsequent evolution, and this can play an important role in genomic evolution, as we will discuss in Chapter 11.

So while the term "pleiotropy" was coined to describe a common feature of the first genes to be phenotypically characterized, it also has significant implications for our current understanding of genomes and evolution.

8.7 Summary: the inheritance and phenotypes of two separate genes

The principles for the inheritance of a single gene on the autosomes (Chapter 5) and a single gene on the X chromosome (Chapter 7) can be applied to the inheritance of any two genes that are located on different chromosomes. In fact, these principles of inheritance can be extended to any number of genes, so long as the genes are on different chromosomes. Mendel found this, as captured in a principle referred to as his Law of Independent Assortment.

But like his Law of Segregation, Mendel's Law of Independent Assortment is obeyed because it describes in abstract terms what happens in physical terms to chromosomes when they segregate during meiosis. During prophase and metaphase of meiosis I, chromosomes pair with their homologues. However, the orientation of non-homologous pairs of chromosomes involves independent events. Thus, by understanding the process of meiosis from Chapter 6 and applying simple probability from

Chapter 5, the likely inheritance pattern of any pair of genes can be calculated accurately.

Why, then, do we see so much variation from this predicted pattern when we examine the inheritance of phenotypes in natural populations? Even if we examined a large population of offspring, so large that statistical variation is not a factor, we might not find the ratios of phenotypes predicted from these simple principles. There are several reasons for this variation from the predicted ratios.

1. The genes underlying the phenotypes that we are examining are not on separate chromosomes. Thus, their inheritance is not described by two independent events, but rather by a single event. This common situation, called linkage, is the topic of Chapter 9.
2. The phenotypes that we are examining are the outcome of numerous genes working together in complicated ways. Some of these genes are inherited independently of each other, while others are linked to each other. In addition, environmental factors affect the phenotype. This common situation, referred to as a complex trait, is the topic of Chapter 10.
3. One of the other situations that we describe in this chapter is at work. For example:
    a. There may not be a simple dominance relationship between the alleles, and there may be more than two alleles with different dominance relationships to each other
    b. Some of the alleles might be lethal, in which case some phenotypes are not readily observed
    c. The phenotypes may depend on the functions of two or more genes that interact with each other
    d. Finally, the phenotypes may depend on interactions not only among the genes, but also among the genes and the environment.

The concepts discussed in this chapter bridge observations that often seem disparate from one another. You are likely to learn in the genetics laboratory about convenient and easily calculated inheritance patterns and can often successfully answer the genetics problems at the end of the chapter. But these inheritance patterns don't seem to apply in the real-world situations found outside the laboratory, whether in families, the litters of a pet, the plants in a garden or an orchard, or in nature. But we can come to realize how the same inheritance patterns that occur in these laboratory situations also occur in the real-world situations if we understand a bit more about what other genetic factors are contributing to the phenotypes. In order to make that bridge most effectively, we also need to recall the underlying principles by which genes are expressed and consider how changes in expression also play into these phenotypes. This is one of the powers of genetics—even when the phenotypes are hard to understand, the underlying principles of inheritance are quite predictable.

CHAPTER CAPSULE

- Genes on separate chromosomes are inherited independently of each other, so the inheritance pattern can be predicted from an understanding of meiosis and the application of simple probability.
- Phenotypes that arise from these genes can be affected by:
    - Dominance relationships, such that one allele is not simply dominant to another and multiple alleles exist
    - The occurrence of lethal alleles
    - Epistasis, in which the phenotype of one gene masks the phenotype of another gene or, more generally, by interactions between the genes
    - Interactions between the genes and the environment, often in unpredictable ways.
- Genes can be defined by the inheritance pattern of their phenotypes, but, because these phenotypes can be hard to interpret or the same phenotype

might arise from more than one gene, a gene is functionally defined by the complementation test.

- If two mutations complement each other, each is providing a function that the other lacks, so they define different genes.
- A similar principle of a gene providing the function that a mutation lacks underlies the technique of transformation rescue or plasmid rescue, used in bacteria as well as in eukaryotes.

STUDY QUESTIONS

Concepts and Definitions

8.1 Which of the rules of probability is important in determining the phenotypic and genotypic ratios from crosses involving more than one gene?

8.2 What are the similarities and differences among the terms allele, mutation, and polymorphism? (A more precise technical definition for polymorphism will be given in Chapter 10, so this is asking for a conceptual definition instead.)

8.3 Why are dominant lethal traits not passed on, unless their effect is exerted after the age of reproduction?

8.4 What is a complementation test, and why is a complementation test performed?

8.5 What is epistasis, and how does it differ from dominance?

8.6 What are the molecular or cellular origins of pleiotropy, and why is pleiotropy so common?

Beyond the Concepts

8.7 Diagram how the orientation of two different pairs of homologous chromosomes at metaphase I of meiosis leads to the expected ratios of the possible gametes with two unlinked genes.

8.8 In a cross involving parents who are heterozygous for each of four unlinked genes, how many boxes will be found in a Punnett square?

8.9 Geneticists working with model organisms in the laboratory rarely need to monitor the segregation of three unlinked genes simultaneously.

a. What are some reasons that segregation ratios with more than two unlinked genes are rarely encountered in crosses involving model organisms?

b. What are some experimental situations, possibly not involving model organisms, in which segregation ratios involving three or more unlinked genes might be routinely monitored?

8.10 Explain why nearly every gene has multiple alleles, but how these multiple alleles could result in only two or three different recognizable phenotypes.

8.11 Is it true that every gene, if analyzed fully, has co-dominant alleles? Why is this concept both useful and not always useful?

8.12 Use a pedigree diagram to show why recessive phenotypes appear more frequently among children of first-cousin marriages.

8.13 What might be some reasons that a mutation is not completely penetrant in its phenotype?

8.14 Briefly explain how independent assortment contributes to genomic and phenotypic variation in populations, and why this has evolutionary consequences.

Applying the Concepts

8.15 A female fruit fly that is homozygous for ebony body (on chromosome 3) and brown eyes (on chromosome 2) is mated to a wild-type male fly with gray–brown body and red eyes. All of the F_1 offspring, both males and females, have wild-type bodies and eyes.

a. These wild-type F_1 flies are mated among themselves to produce an F_2. What are the expected phenotypic ratios among the F_2 flies?

b. One of these F_1 male flies is test-crossed to a female with ebony body and brown eyes. What are the expected phenotypic ratios among the offspring of this test-cross?

8.16 For the following crosses, you will need to recall that *Caenorhabditis elegans* can reproduce by cross-fertilization between a male and a hermaphrodite or by self-fertilization of the hermaphrodite. In addition, males are chromosomally X0 (that is, a single X chromosome and no other chromosome), while hermaphrodites are XX.

A wild-type male mates with a hermaphrodite that is both unusually long (*lon*) and has uncoordinated (*unc*) movement. All of the F_1 male offspring are long but with normal movement, while all of the F_1 hermaphrodite offspring are normal in both body length and movement.

a. What can you determine about the linkage or chromosomal locations of the *lon* and the *unc* genes?

b. If the *lon* F_1 males are mated to the F_1 hermaphrodites, what phenotypes are expected among the cross-progeny in the F_2 generation, and in what proportions?

c. If the F_1 hermaphrodites are instead allowed to self-fertilize, what phenotypes are expected in the F_2 generation, and in what proportions?

8.17 In humans, the ABO blood type locus is on chromosome 9, and the Rh locus is on chromosome 1. The dominance relationships among the three ABO alleles are discussed in the chapter; at the Rh locus, Rh-negative is recessive to Rh-positive.

a. A woman whose blood type is O-positive, but whose father was B-negative, marries a man whose blood type is A-negative. His parents' blood types are not known. What are the possible blood types that might be found among their children?

b. Their first child has blood type O-negative. What is the probability that their second child will be type A-positive?

8.18 Summer squash can be round, oval, or long; they can also be rough (covered in wart-like protrusions) or smooth. A true-breeding round rough squash was crossed to a true-breeding long smooth squash. All of the F_1 were oval and smooth. Assuming that each of these traits is due to a single gene, with no interaction between them, what phenotypes and in what proportions would be found in the F_2 if the F_1 were crossed among themselves?

8.19 Bar eye in *Drosophila* is X-linked and exhibits incomplete dominance. Wild-type females have an average of 810 facets in the eye, while homozygous Bar females have an average of 59 facets. Heterozygous females have a kidney-shaped eye with an average of 447 facets. Males have either wild–type-shaped eyes with about 810 facets or Bar eyes with about 60 facets.

a. Why do no males show an intermediate phenotype?

An autosomal recessive allele purple results in purple eyes when homozygous. A female with Bar but red eyes is mated to a male with normal-shaped but purple eyes.

b. What are the expected phenotypic ratios in the F_1 generation?

c. If one of the F_1 males is mated to an F_1 female, what are the expected results in the F_2 generation?

8.20 In humans, the absence of fingernails and knee caps (nail–patella syndrome) is an autosomal dominant trait. Red–green color blindness is an X-linked recessive trait. The family on the left in the pedigree in Figure Q8.1 has a history of red–green color blindness (shown in red symbols) but no history of nail–patella syndrome, while the family on the right has a history of nail–patella syndrome (shown in black symbols) but normal color vision. Individuals from these two families have a child, indicated by the "?".

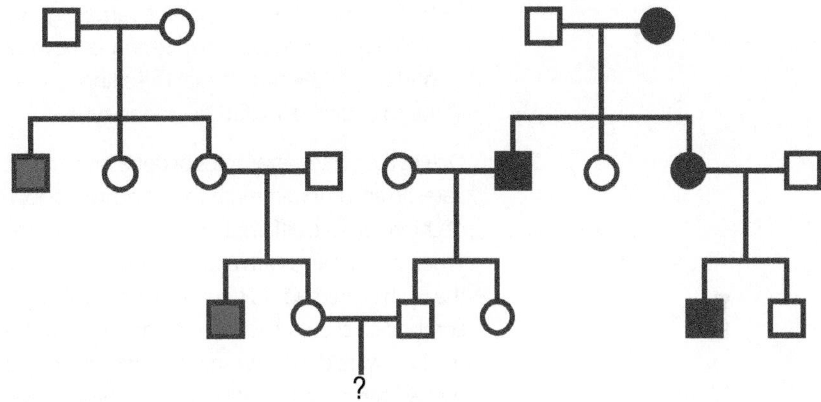

Figure Q8.1

a. What is the probability that this child will be a boy who has normal color vision and with normal fingernails and patellas?

b. In fact, the child is a boy with normal color vision, but he is born with some fingernails missing and one patella misshapen. How do you explain this result?

c. This family has a second son. What is the probability that he will have normal color vision and normal fingernails and patellas?

d. The fingernails of individuals with nail–patella syndrome are reduced in size or missing, but the phenotype is variable. Some individuals lack fingernails completely, while others have fingernails of reduced size. The effects on the patella are also variable among affected individuals. What might account for this range of phenotypes?

e. Many individuals with nail–patella syndrome also have kidney pathologies, but this is not found in all individuals. What might account for the range of different tissues that are affected in these individuals?

Challenging

f. According to Online Mendelian Inheritance in Man (OMIM; http://www.ncbi.nlm.gov.omim) some have speculated that the pathology of the kidneys (mentioned in (e)) is due to a different gene from the one responsible for nail–patella syndrome. What might be one line of evidence that would support the hypothesis that these quite different phenotypes are in fact due to alleles of the same gene? (Hint: The gene for nail–patella syndrome has been cloned, and its DNA sequence and protein product are both known. In addition, nail–patella syndrome has no effect on fertility or survival. Either of these pieces of information could be used to provide evidence in support of this hypothesis.)

8.21 In *Arabidopsis*, the *Dwarf* (*Dw*) gene on chromosome 1 is considered to have incomplete dominance, so that ***Dw***/+ plants are intermediate in height between wild-type (+/+) and *Dwarf* (***Dw/Dw***) plants. One of the genes for resistance to chlorate salts is *chl-2* on chromosome 2, with resistance being the recessive trait. A fully Dwarf plant that is chlorate-sensitive to crossed to a normal height plant that is chlorate-resistant.

a. What are the expected phenotypes of the seedlings in the F_1 generation?

b. One of the F_1 plants is self-pollinated, and its seeds are collected to grow an F_2 generation. What are the expected phenotypes in the F_2 generation?

c. Seedlings can be cultured in the presence of chlorate, so that only the resistant seedlings will grow. A geneticist plants 50 of the F_1 seedlings from (a) above in a chlorate-containing soil. How many of these plants are expected to grow to a full height?

8.22 A student in high school biology class learns that she has type O blood. Her mother has type A blood, but her father's blood type is not known.

a. What are the possible blood types that might be found in her father?

Challenging **b.** Her younger brother takes the same high-school biology class a few years later and also learns his blood type. He comes home and tells his father, "I know my blood type, and I am 100% certain that I also know yours." What must be his blood type for him to be so certain of his father's blood type?

8.23 One of the earliest examples of gene interactions was found by the British geneticists Bateson and Punnett more than a century ago. They noticed that different pure breeds of chickens have different-shaped combs, which they referred to as single, rose, and pea. When pure breeds with a rose comb were crossed to pure breeds with a pea comb, all of the F_1 chickens had a different shape comb, which they referred to as walnut. The F_1 walnut-combed chickens were mated among themselves to produce an F_2 generation; many such matings were done, and several hundred chickens were analyzed, with the results tabulated below. They reported ratios of phenotypes, rather than numbers, but these numbers are consistent with their ratios. Note that neither of their original breeds had either walnut combs or single combs, so both of these phenotypes arose in the F_2 generation.

Single comb 12

Pea comb 39

Walnut comb 113

Rose comb 36

a. What is the evidence that these phenotypes are due to the interactions of two genes, rather than to multiple alleles at a single locus?

b. Explain how these phenotypic ratios arose by writing the genotype responsible for each phenotype.

c. In order to prove their hypothesis about the genotypes giving rise to each phenotype, Punnett and Bateson crossed each of the original breeds (one with pea comb and one with rose comb) to a breed with a single comb. What was their strategy in performing this cross, and what would be the expected outcome?

Challenging **d.** Bateson and Punnett also took **individual** chickens with a walnut comb from the F_2 generation and crossed them to a breed with a single comb. What fraction of these crosses involving a single walnut comb chicken will give all four phenotypes in the offspring?

WEBLINK: An additional example of the solution to such a problem is a classic one involving chicken combs, which we discuss in Video 8.1. Find it at **www.oup.com/uk/meneely**

8.24 Traditionally, blue eye color is considered recessive to brown eyes in humans, although, as noted in the chapter, eye color in humans is not a single-gene trait. It has occasionally been the subject of folklore and "common knowledge" that blue-eyed parents usually have blue-eyed offspring. However, blue-eyed parents do often have brown-eyed children. In addition, when one brown-eyed child is born in this family, it is expected that subsequent children will also have brown eyes. Based on information in this chapter and assuming blue eyes really is recessive to brown eyes for any single gene being tested, how can you explain this result? Be sure to use the correct terminology.

8.25 An individual's ABO blood type can always be determined by an analysis of the blood. About 80% of the people in North America are known as secretors, in which the ABO type is also secreted into other body fluids such as saliva. About 20% of North Americans are non-secretors, so their saliva always appears to be type O, no matter their actual ABO blood type. Non-secreting is a recessive trait. The ABO locus is on chromosome 9, while the secretor locus in on chromosome 19. A non-secretor man with type B blood and a secretor woman with type A blood have a child; the child is determined to be a non-secretor and to have type O blood when the blood is tested. What is the probability that the next child born to this couple will be type A blood but type O saliva?

8.26 When one of the stock vials in the fly room got damp, the piece of laboratory tape with the label for the *Drosophila* stock fell off, so the research student could not tell the genotype of the flies in her vial. She could see that the flies in her unlabeled vial—both males and females—had a black body, among their other phenotypes, so if she can determine which gene is giving rise to this black body color, she can infer the rest of the genotype of her stock. She also knew that she had been working with stocks that had each of three different recessive mutations that gave rise to a black body similar to her unlabeled flies: the X-linked gene *sable*, a gene on chromosome 2 known as *black*, and a gene on chromosome 3 known as *ebony*. She had separate vials with homozygous *sable*, *black*, and *ebony* flies to work with.

a. How should she determine the genotype of the flies in her unlabeled vial?

Challenging **b.** The research student is very clever and figured out that she could get the same information as in (a), but with only a single cross. What single cross does she perform? What will be the result of this cross for each of her three possible outcomes (sable, black, and ebony)?

Looking back and ahead **8.27** A recessive mutation known as ***zb4*** on the left end of chromosome 1 in corn results in zebra crossbands, that is, regularly spaced yellow bands across the leaf. A recessive mutation on chromosome 5 known as ***a1*** results in a colorless aleurone (outermost layer of the kernel), rather than the reddish purple aleurone found in wild corn. That ***a1/a1*** strain of corn is also homozygous for a physical alteration in the structure of the chromosome that produces a knob on the left end of chromosome 1; the knob can be seen when chromosomes are viewed in a microscope but has no effect on growth or any other phenotype. The ***zb4/zb4*** and ***a1/a1*** strains are crossed to produce an F_1 generation, which has no bands on the leaves and a colored aleurone, and which is heterozygous for the knob. The F_1 is bred to produce an F_2 generation.

a. What fraction of the F_2 plants will be zebra-banded with a colorless aleurone?

b. Will these plants have a cytologically visible knob or knobs? Why or why not? (You may want to draw out the chromosomes to answer this question, and you should answer this question before going on to (c).)

Challenging **c.** In fact, a few of the zebra-banded plants with a colorless aleurone do have a knob on the end of chromosome 1. Based on what you learned in Chapter 6, explain this result.

8.28 The Manx breed of cat has no tail or a very short tail. When two Manx cats are bred to each other, the litter size is reduced, some of the kittens have no tails, and some have a full-length tail. Among many such crosses, tailless offspring are about twice as common as kittens with a full tail. No true-breeding Manx cats have been found, despite more than a century of interest by cat breeders.

a. Provide an explanation consistent with these data that explains why no true-breeding tailless Manx cats have been found.

b. In addition to being tailless, the Manx breed is often prone to skeletal problems such as a shortened or poorly formed spinal column. Using the proper terminology, postulate an explanation for this phenotype.

c. Despite the fact that some of the kittens have tails, the Manx tailless phenotype is considered to be highly penetrant. Do you agree with this assessment, or would you think that this tailless phenotype has reduced penetrance? Assuming that you had the space and patience to breed many cats, how could you determine if the tailless phenotype is, in fact, highly penetrant?

8.29 In rabbits and many other mammals, the *C* gene is necessary for colored fur; a ***cc*** rabbit is white. The *B* gene controls the color of the pigment, with ***B_*** being black and ***bb*** being brown. The *C* gene is epistatic to the *B* gene, and all ***cc*** rabbits are white, regardless of their genotype at the *B* locus.

a. Breeder 1 mates a pure-breeding white rabbit to a pure-breeding brown rabbit. All of the F_1 offspring are black. Breeder 2 performs the same mating with rabbits of the same two breeds and colors and obtains the same result—all of the F_1 rabbits are black. Rabbits from Breeder 1 are mated to rabbits from Breeder 2. What is the expected ratio of black, brown, and white rabbits among the offspring of these matings? (Two different breeders and multiple matings are introduced to simplify the question and avoid any complications from inbreeding and the small number of rabbits in a litter.)

b. One of the black F_1 rabbits from Breeder 2 is back-crossed to the white parent from Breeder 1. What are the expected offspring of this mating?

c. Draw the epistatic relationship between the *C* gene and the *B* gene, similar to the diagram in Figure 8.11.

Challenging **d.** In addition to the epistatic relationships described in the chapter (that is, 9:7 ratios and 15:1 ratios) and the one between the *C* gene and the *B* gene in this problem, what other epistatic ratios could occur when two heterozygotes are mated, and how might these ratios arise?

8.30 The location of proteins in an organism can often be determined by immunohistochemistry, that is, by using a tagged antibody that recognizes a particular protein. The fluorescent tag bound to the antibody allows a researcher to see the location of the antibody in the organism, so the presence of the tag indicates the expression and localization of the protein gene product. In wild-type flies, immunostaining shows that the Wg protein is found in the anterior half of the wing blade. In flies homozygous for the mutant gene patched (*ptc*), the Wg protein is found in a much broader region of the wing blade, including in the posterior half. In flies homozygous for the mutant smoothened (*smo*), there is little or no Wg protein found.

a. Diagram a pathway that shows the effects of *ptc* on Wg and of *smo* on Wg in the wing blade.

b. What, if anything, can be concluded about the mechanism by which the *ptc* and *smo* genes regulate the expression of the *wg* gene and its product?

Challenging **8.31** The microarray data in Tool Box 8.1 show that gene *A* is necessary to turn on the expression of gene *3*, that it turns up the expression of gene *1*, and that it turns off the expression of gene *4*. It does not tell us how genes *1*, *3*, and *4* affect one another, however. Suppose that a mutant that eliminated the function of gene 1 was available. This mutant has a functional copy of gene *A*. The results are shown in the microarray in Figure Q8.2. Draw a pathway that shows the sequence of interactions among genes *1*, *2*, *3*, and *4*, assuming that the genes all function in a simple, linear pathway.

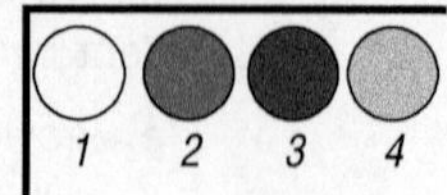

Figure Q8.2

CHAPTER 9

The Locations of Genes on Chromosomes: Linkage and Genetic Maps

9

IN A NUTSHELL

Genes located near each other on the same chromosome are not inherited independently of one another. Rather, alleles of these genes are likely to be inherited together; such genes are said to be linked. Linked genes—or, more accurately, linked combinations of alleles—are not always inherited together, however, because crossing over during meiosis creates new combinations of the alleles. Since genes that are very close together are less likely to have a crossover between them than genes far apart, the frequency of recombination provides one indication of how closely located the genes are on the chromosome, and thus the basis for a genetic map of a chromosome. Maps of the genome based on recombination are among the most important types of maps geneticists use, but not the only type. Other types of maps can be made, particularly since the DNA of genomes can now be sequenced.

9.1 Overview of linked and unlinked genes

The DNA that makes up a eukaryotic genome is found as part of a chromosome in the nucleus. As noted when we discussed meiosis in Chapter 6, the chromosome is the actual vehicle of inheritance; genes can be thought of as the passengers that are followed to monitor the inheritance of the chromosome. In Chapters 5, 7, and 8, a gene was treated as if it were the only occupant of its chromosome, and the connection between Mendelian inheritance and meiosis was clear. In these chapters, the inheritance of one gene could be considered independently of the inheritance of another gene because the corresponding chromosomes for the two genes behaved independently at meiosis. But this was a simplification of inheritance patterns, since a chromosome carries hundreds or thousands of genes. Like passengers in the same vehicle, alleles of genes that are on the same chromosome travel together during meiosis and will usually be inherited together. Such genes are said to be

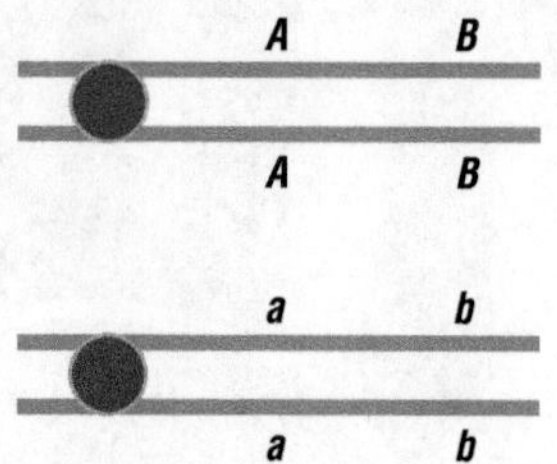

(a) Linked genes are on the same chromosome

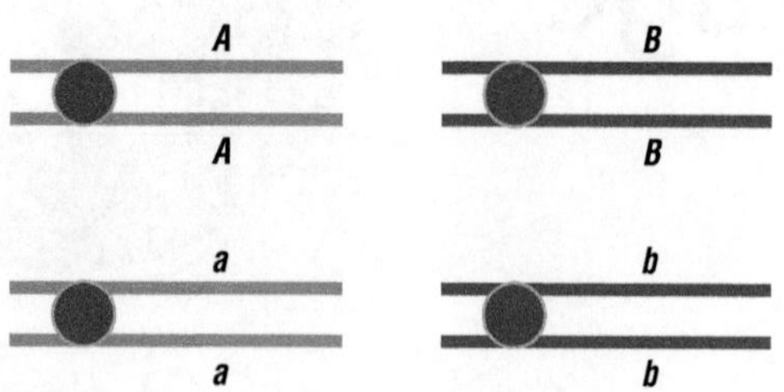

(b) Unlinked genes are on different chromosomes

Figure 9.1 Linked genes are located close to each other on the same chromosome, as in (a). In contrast, unlinked genes are located on separate chromosomes, as shown in (b). The chromosomes are shown at the beginning of meiosis I, when two sister chromatids are present for each homolog.

linked. Linked and unlinked genes are diagrammed in Figure 9.1.

In this chapter, we describe the inheritance of linked genes. Although located on the same chromosome, the combination of alleles for linked genes can be exchanged or recombined during meiosis as a result of crossing over during meiosis I. Recall from Section 6.3 that every pair of chromosomes has at least one crossover between the homologs. Although every pair of chromosomes must have a crossover, the locations of the crossovers are not the same in different meiotic cells, which have observable impacts on the inheritance patterns of alleles.

Like the other properties of Mendelian genetics, linkage between genes is an inevitable by-product of the behavior of chromosomes at meiosis. The crossovers result in combinations of alleles on a chromosome that are different from those of the parent—in other words, there is a recombination of alleles. As we will explain in Section 9.3, the incidence of recombinant chromosomes can provide a map of the locations of genes in the genome, a map based on **recombination**. Maps based on recombination, usually called **genetic maps**, are important tools for understanding the structure and organization of a genome and for predicting how often certain traits will be inherited together. Genetic maps have also formed an essential foundation for finding genes known from a mutant phenotype, that is, for finding the DNA sequence responsible for single-gene traits. Recombination also has important consequences for the inheritance of traits over evolutionary timescales.

9.2 Linked inheritance of genes on the same chromosome

In Chapter 8, we looked at the simultaneous inheritance of two genes, and we treated these as two separate events. Let's quickly summarize our results with the inheritance of two unlinked genes from Chapter 8, as shown in Figure 9.2. We start with two true-breeding strains, one of which is ***A/A; b/b***, and the other being ***a/a; B/B***. The F_1 heterozygote is ***A/a; B/b*** and is **test-crossed** to an individual who is ***a/a; b/b***. The expected results of this cross are shown in Figure 9.2 and summarized below.

In the next generation:

- One-fourth of the offspring will have the genotype ***A/a; b/b*** with a phenotype that is like one of the parents.
- One-fourth of the offspring will have the genotype ***a/a; B/b*** with a phenotype that is like the other parent.
- One-fourth of the offspring will have the genotype ***A/a; B/b*** with a phenotype that is different from either parent.
- One-fourth of the offspring will the genotype ***a/a; b/b*** with a phenotype that is different from either parent.

In other words, the phenotypes of half of the offspring are parental phenotypes, that is, the same phenotypes as those of the two parents (P_0) that were crossed to make the F_1 heterozygote—Ab and aB in this example. (As explained in Communicating Genetics 5.1, we are designating **genotypes** in *italics*, but phenotypes are not italicized.) The phenotypes of the other half of the offspring are recombinant phenotypes—that is, a combination of phenotypes that is different from the original P_0 parents. If we had started with parents who were ***A/A; B/B*** and

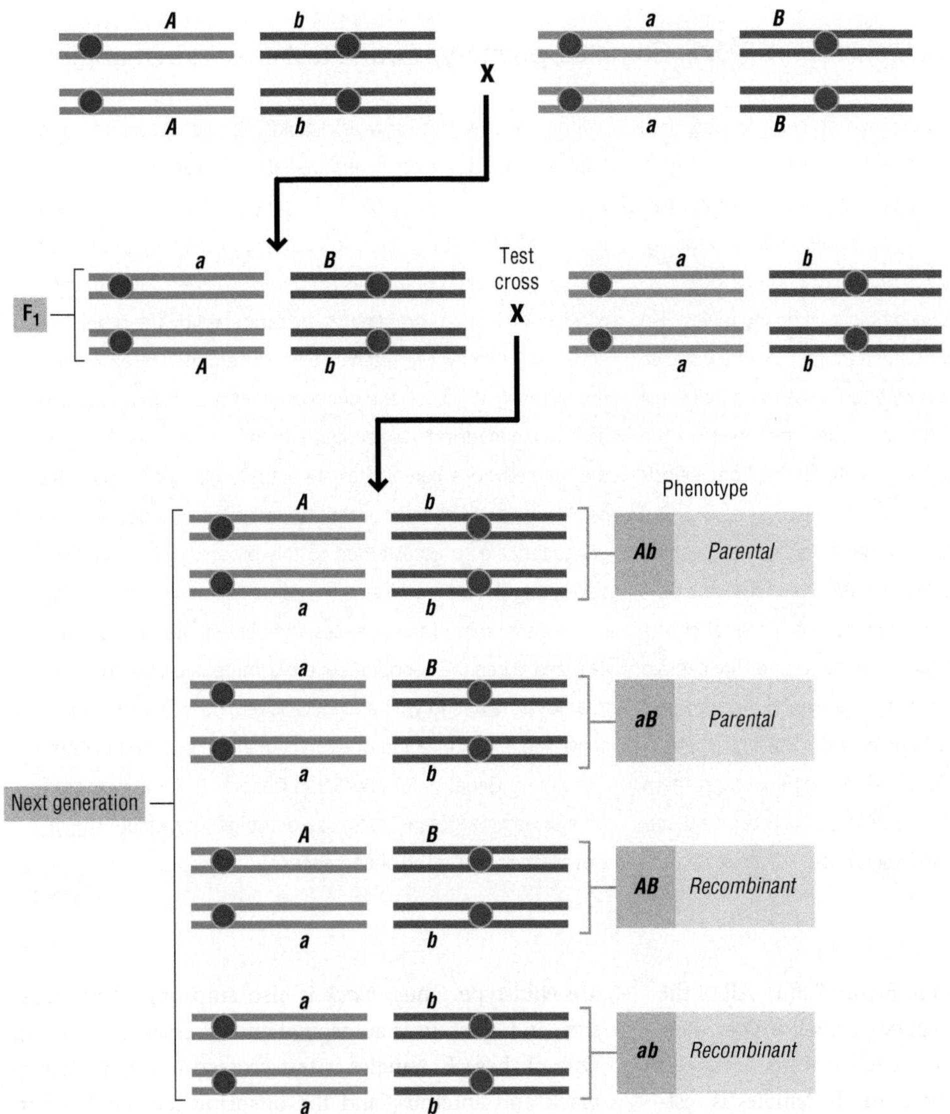

Figure 9.2 A two-factor or dihybrid test cross with unlinked genes. In dihybrid test crosses, an individual heterozygous for both genes is test-crossed to one who is homozygous recessive for both genes. As discussed in Chapter 8, when the two genes are unlinked, the four phenotypes seen in the resulting offspring occur with equal frequency. Note that the "parental" type refers to the phenotypes of the P_0 individuals, the parents of the F_1 heterozygote. Two of the categories of offspring resemble the original parents of the heterozygote, while two of the categories have new combinations of the traits.

a/a*; *b/b (that is, with the same alleles but different parental combinations), we would have obtained the same result with one-fourth of each genotype and phenotype. This ratio arises because the inheritance of the *A* gene is independent of the inheritance of the *B* gene. The ***A*** and ***a*** alleles assort independently of the ***B*** and ***b*** alleles because the two genes are on different chromosomes. Note that we use a ";" to indicate genes on different chromosomes. We are going to work back and forth between the genetic constitutions of the gametes produced by a heterozygote—that is, the genotypes—and the phenotypes of the offspring; this is discussed more fully in Communicating Genetics 9.1.

But what happens if the two genes *A* and *B* are present on the same chromosome? This occurs often, since a chromosome can have hundreds or thousands of different genes on it, so it is very likely that a geneticist will encounter two traits for which the genes happen to be on the same chromosome. Let's show the effects with some examples.

Linked and unlinked inheritance patterns

Suppose that a female fly with an ebony body (with the relevant gene abbreviated *e*) and no wings (a phenotype called apterous, with the gene abbreviated *ap*) is mated

COMMUNICATING GENETICS 9.1 Gametes, phenotypes, and test crosses in mapping

The procedure for determining the map distance between two genes largely follows a standard form; the same procedure is used to predict the frequency of different categories of offspring when the map distance is known. In this procedure, a heterozygote for the two genes is created, usually by mating two homozygous parents if both of these are fertile. Recombination during meiosis I in this F_1 heterozygote is assessed to determine the map distance between the two genes, usually by a test cross to an individual that is homozygous recessive. In other words, the object of the experiment is to survey the haploid gametes arising from meiosis in this F_1 heterozygote.

In some cases, the genotypes of these haploid gametes can be determined directly. For example, it is possible to sequence DNA from sperm or pollen grains; if enough sperm can be sequenced, the map distance can be determined. In many fungi, the haploid products (known as ascospores) can be grown on medium, and their phenotypes determined from growth properties; tetrads and haploid ascospores are discussed in Tool Box 6.1. In these cases where the genotype of the haploid products can be determined directly, the genotypes and phenotypes are the same.

In organisms that grow as diploids, the genotypes of these haploid products are not determined directly. Instead, the heterozygous F_1 is test-crossed. Since the other parent is homozygous recessive for both traits, its contribution to the phenotypes can be largely ignored. It contributes no dominant alleles, so whatever allelic combinations or gametes are produced by the other parent—the heterozygote—can be easily determined by examining the offspring. While it is the phenotypes of the offspring that are being examined, it is the genotypes of the gametes produced by the heterozygous parent that are actually being assessed. This can lead to some confusion, not the least of which arises because geneticists are so accustomed to this process that they themselves might be inconsistent in their terminology. In referring to the genotypes of the gametes, the gene names are routinely italicized. When the genotypes of the gametes are inferred from a test cross, the phenotypes are not italicized. This follows the same conventions for gene and phenotype names described in Communicating Genetics 5.1 and 5.2 in Chapter 5. Remember that, when a heterozygote is test-crossed, the genotypes of its gametes are the same as the phenotypes of its offspring.

to a wild-type male, as shown in Figure 9.3(a). All of the F_1 offspring, both males and females, are wild-type. Thus, both *ebony* and *apterous* are recessive mutations, and both must be autosomal. One of the F_1 females is test-crossed to a male that is both ebony and apterous, and the different types of offspring are tabulated, as shown in Figure 9.3(b).

The test cross results in 154 wild-type offspring, 145 ebony and apterous offspring, 159 ebony flies with wild-type wings, and 149 flies with a wild-type body and no wings (apterous). Since the four categories of phenotypes are about equally frequent, we can conclude that the responsible genes, *ebony* and *apterous,* are inherited independently of each other. Thus, these two genes are both on autosomes, but they are not on the same autosome. This is exactly like some of the crosses in Chapter 8.

Let's work through another example. A female fly with a black body (with the gene abbreviated *b*) and no wings (the *apterous* gene again) is mated to a wild-type male, as shown in Figure 9.4(a). Although black and ebony flies resemble each other in phenotype, the phenotypes are the outcomes of two different genes. All of the F_1 offspring are wild-type. Thus, *black* is also autosomal and recessive, and *apterous* is autosomal and recessive (as before). One of these F_1 females is test-crossed to a male that is black and apterous, and the offspring are tabulated in Figure 9.4(b).

When we count the offspring for this cross, we find something different from what occurred with the genes for *apterous* and *ebony*. Among the 400 offspring, we observe 189 wild-type offspring, 183 black and apterous offspring, 15 offspring that are black with normal wings, and 13 offspring with a wild-type body color and no wings. Clearly, we did not get our expected 1:1:1:1 ratio, and thus the *black* and *apterous* genes are not being inherited independently of each other. A fly that inherited the ***b*** allele from the mother is also likely to have inherited the ***ap*** allele; in fact, of the 183 + 15 = 198 offspring that inherited the ***b*** allele, 183 of them (92.4%) inherited the ***ap*** allele as well, and only 15 (7.6%) inherited the ***b*** allele without also inheriting the *ap* allele. (Since this is a test cross, all of the alleles from the male are ***b ap***, which allows us to focus on the alleles from the mother.)

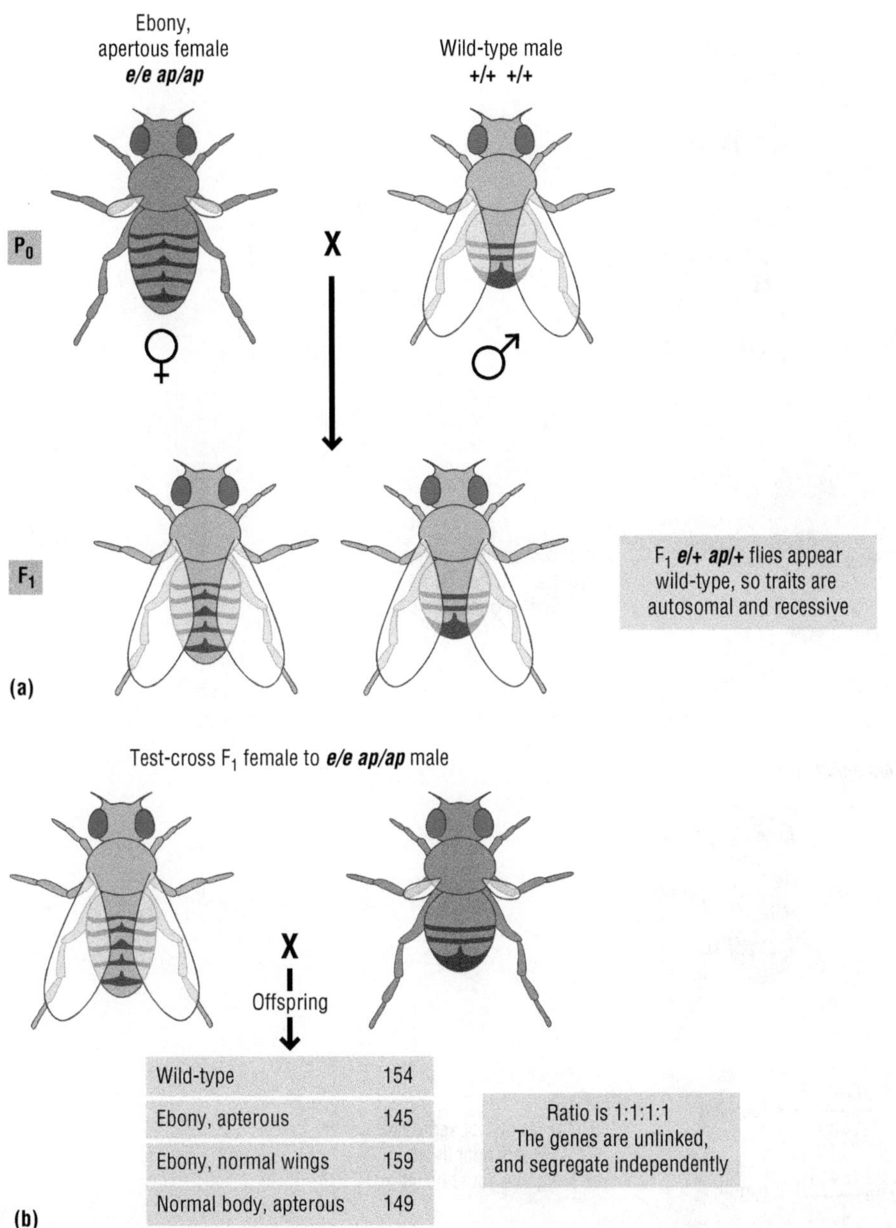

Ebony, apertous female
e/e ap/ap
Wild-type male
+/+ +/+
P0
X
F1
F1 e/+ ap/+ flies appear wild-type, so traits are autosomal and recessive
(a)
Test-cross F1 female to e/e ap/ap male
X
Offspring
Wild-type 154
Ebony, apterous 145
Ebony, normal wings 159
Normal body, apterous 149
Ratio is 1:1:1:1
The genes are unlinked, and segregate independently
(b)

Figure 9.3 Example of two-factor crosses with unlinked genes in *Drosophila melanogaster*. In (a), ebony apterous females are mated to wild-type males; thus, ebony apterous and wild-type are the parental phenotypes. An F_1 female is test-crossed, and the offspring counted and tabulated in (b). The four phenotypic categories are equally frequent, which allows us to conclude that *ebony* and *apterous* are unlinked to each other.

Likewise, we see that 189 + 13 = 202 flies inherited the b^+ allele and have a wild-type body color. Of these flies, 189 (93.6%) also inherited the wild-type ap^+ allele, and only 13 (6.4%) inherited the ***ap*** mutant allele. In other words, the inheritance of the allele for the *black* gene is linked to the inheritance of the allele for the *apterous* gene—alleles for the two genes are not inherited independently of each other. The allele the fly inherited for one of the genes is highly correlated with which allele the fly inherited for the other gene. What is the explanation for difference in the results of these crosses between *ebony* and *apterous* and *black* and *apterous?*

KEY POINT If a test cross involving an F_1 heterozygote carrying recessive alleles for two different genes does not result in equal numbers of four phenotypes in the F_2 generation, these two genes must be linked.

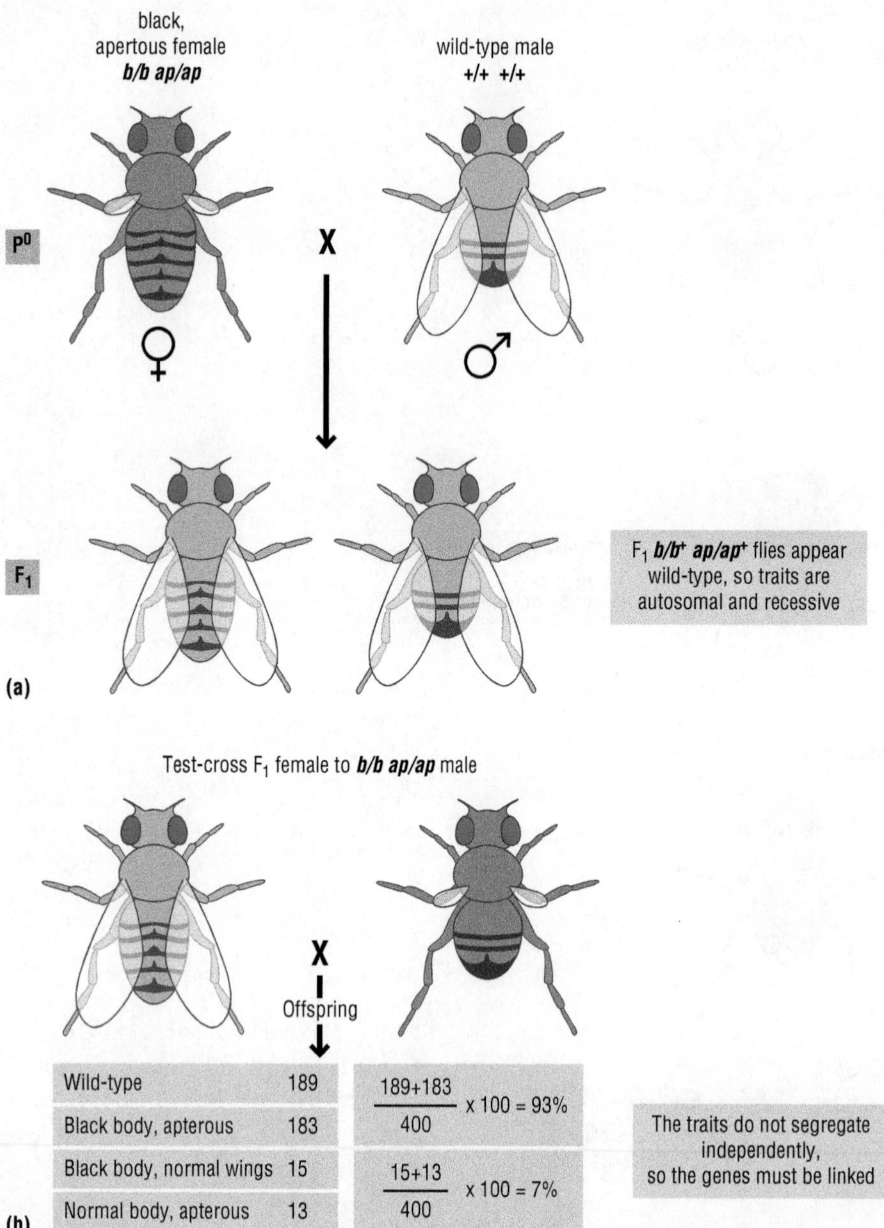

Wild-type	189	$\frac{189+183}{400} \times 100 = 93\%$
Black body, apterous	183	
Black body, normal wings	15	$\frac{15+13}{400} \times 100 = 7\%$
Normal body, apterous	13	

Figure 9.4 Example of two-factor crosses with linked genes in *Drosophila melanogaster*. In (a), black apterous females are mated to wild-type males. An F_1 female is test-crossed, and the offspring are counted, with results shown in (b). In this case, the two parental phenotypes (black apterous and wild-type) are 93% of the total offspring, while the recombinant phenotypes (black with normal wings and wild-type body but apterous) are only 7% of the total. Since the parental types are more frequent than the recombinant types, these two genes are linked to each other.

Linkage and meiosis

In order to explain the results (as with so much of Mendelian genetics), we need to return to our diagram of meiosis, as shown in Figure 9.5. Remember from our discussion of meiosis in Chapter 6 that crossovers between the homologs hold the chromosomes together during prophase I. The chromosome pair must cross over, but the crossover can be at any point along the length of the chromosome pair. If we could examine a large population of gametes that had arisen from meiosis involving these chromosomes and all of their genes, we would find that all of the gametes had at least one crossover, but the location of that crossover would vary.

How does this account for the inheritance of the alleles for *black* and *apterous*? The explanation is diagrammed in Figure 9.6. Since the two genes are not inherited independently of each other, as revealed by the frequencies of

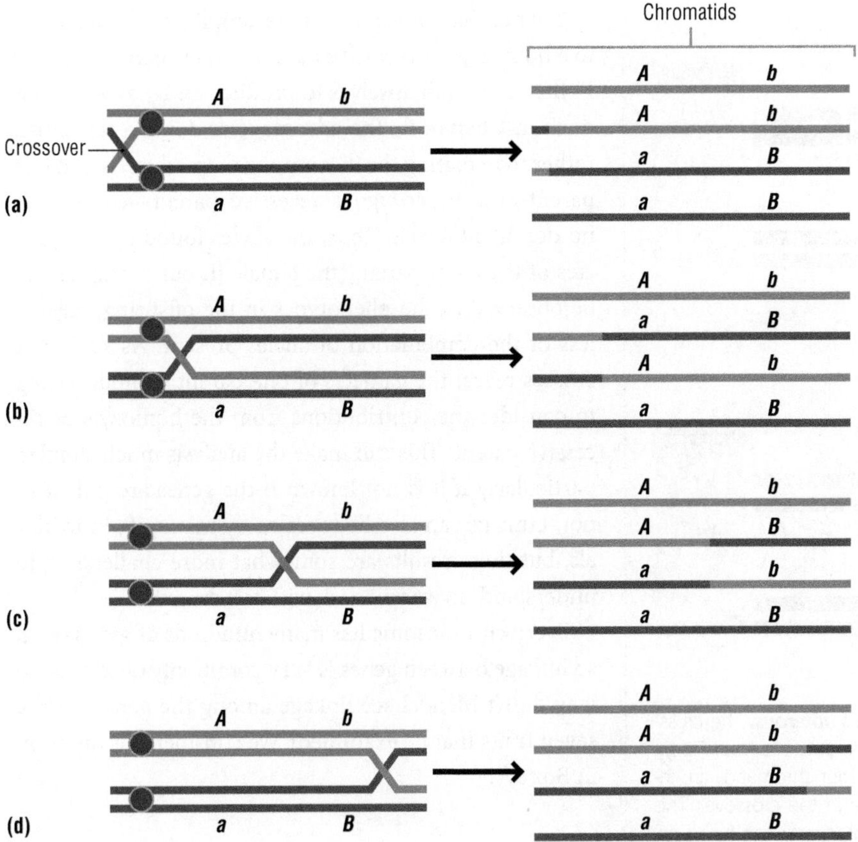

Figure 9.5 Genetic linkage and meiosis. The explanation for genes whose segregation is linked to each other is found in the process of meiosis. Every homologous pair of chromosomes will cross over at least once. If the crossover does not occur between the two genes (as in (a), (b), and (d)), the alleles will segregate together. If the crossover does occur between the two genes (as in (c)), recombinant combinations of alleles are observed. Thus, the frequency of recombinant chromosomes provides one measure of the relative locations of genes on the chromosome.

the different types of gametes, the two genes are inferred to be located on the same chromosome.

The chromosomes in the mother cross over in every meiosis, so she makes a population of gametes with different genotypes. In about 7% of the gametes, the crossover occurred between the *black* locus and the *apterous* locus, such that the gamete had a recombinant combination of alleles at these two loci, either ***b ap***$^+$ or ***b***$^+$ ***ap***, as shown in Figure 9.6(a). Most of the time (accounting for 93% of the gametes, in fact), the crossover did **not** occur between the *black* locus and the *apterous* locus, so the resulting gamete has the same arrangement of alleles at these loci as the parents, either ***b ap*** or ***b***$^+$ ***ap***$^+$, as shown in Figure 9.6(b). Wherever the crossover occurred on this chromosome pair in the mother, it must have been at some other location on this autosome that was not being examined in this cross. Since ***b*** and ***ap*** are on the same chromosome, we write them without a ";" between them.

We want to make a few other points about these data before moving on. Note that black apterous and wild-type flies, the two phenotypes of the original homozygous parents, are equally frequent. Similarly, the two classes of recombinants—black with wild-type wings and wild-type tan with apterous—are also equally frequent. The two classes of parental types (and the two classes of recombinants) are the alternative outcomes of the same crossover event, so the two parental types must be equally frequent. Note also that the frequency of the parental classes of offspring is always greater than the frequency of the recombinant classes of offspring.

KEY POINT The parental combination will **always** be more common than the recombinant combination. If the genotypes of the original parents are unknown for some reason, the phenotypes of the most common classes of offspring must be the parental types, so these must be the phenotypes (and genotypes) of the parents.

Remember from Section 6.3 that, no matter what chromosome pair is being considered, a crossover has to occur somewhere on the chromosome. If this crossover happens

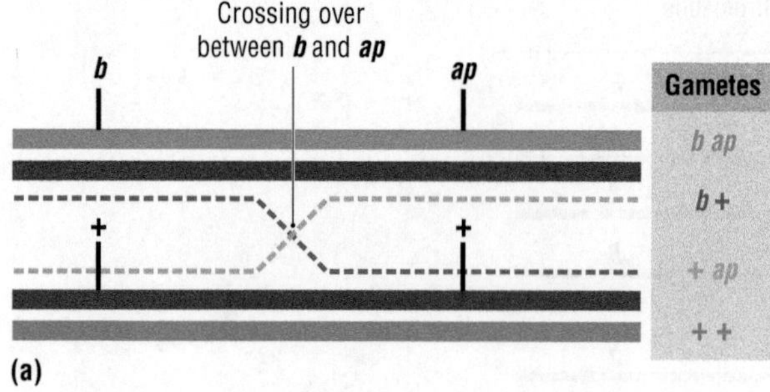

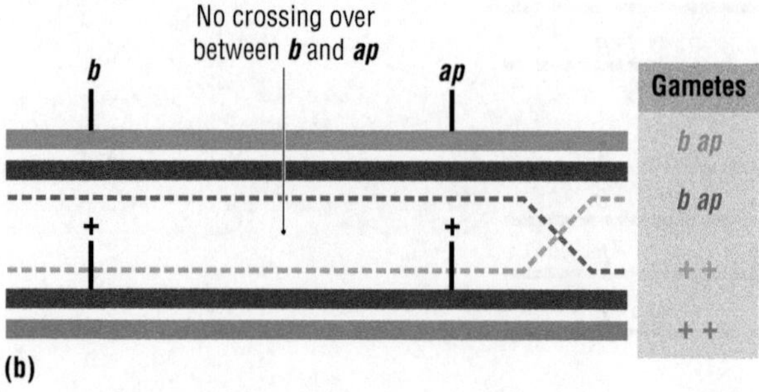

Figure 9.6 Crossing over with *black* and *apterous*. The crossover that explains the results with *black* and *apterous* from Figure 9.4(b) is shown. Note that only two of the four chromatids are involved; two chromatids do not participate in the crossover. This figure is a close-up of the chromosomal region between ***b*** and ***ap***; most of the crossovers on this chromosome will not occur between these two genes.

between the two genes under examination, recombinant combinations of alleles will arise. The frequency of recombinant gametes then provides an indication of how often a crossover occurs at a particular location, and thus a measure of how close the two genes are.

In the case of *black* and *apterous*, 400 gametes (in the form of offspring) were analyzed from the mother, as shown in Figure 9.4(b). The two recombinant classes make up 28 or 7% of the total offspring, so these two genes are said to be 7% apart or 7 map units apart. This distance is sometimes called 7 centiMorgans or 7 cM, in honor of Thomas Hunt Morgan. This is the definition: **one map unit represents 1% recombination between the two genes.**

Note that the map distance is based on the total frequency of recombinant offspring in **both** non-parental phenotypic categories. Sometimes it is only feasible to count one category of recombinants (because one type of offspring dies, for example); however, since the two categories of recombinant gametes are equally frequent, the map distance in those cases can be calculated as twice the number of recombinants observed.

You may have noted that our examples used test crosses to a homozygous recessive parent, rather than mating the F_1 flies among themselves to produce an F_2, as was often done in Chapter 8. The advantage of doing a test cross, rather than mating the flies among themselves, is that one parent—the homozygous recessive parent—contributes no dominant alleles. Thus, the alleles found in the gametes of the other parent (the female in our example) will be observed as the phenotypes in the offspring, regardless of the combination of alleles present. As such, test crosses reveal the gametes of one parent without having to consider the contributions from the homozygous recessive parent. This can make the analysis much simpler, particularly if it is not known if the genes are linked or not. Linkage can also be revealed by mating F_1 individuals, but these results are somewhat more challenging to understand, so we will stick with test crosses.

Every chromosome has many hundreds of genes on it, so linkage between genes is very commonly observed. So why didn't Mendel see linkage among the genes for the seven traits that he examined? We consider this question in Box 9.1.

Map units and recombination

Let's think about the connection between map units and meiosis as it applies to the genome. Genetic loci are spread out along the entire length of the DNA sequence of the chromosome. Now suppose it was possible to test different alleles or sequence variants for each of these loci. In principle, it would always be possible to find the location of the crossover between the two homologous chromosomes, as shown in Figure 9.7. Not only that, it would be possible also to determine how many crossovers occurred on that chromosome during meiosis. There might be chromosomes with one, two, three, or any number of crossovers, but if meiotic disjunction has been normal, it would always be possible to find at least one crossover per chromosome.

Furthermore, alleles or sequence variants that are present together on a homolog will frequently be inherited together as a block in that gamete, and the likelihood that the variants at two different locations are inherited together is an indication of how far apart the locations are on the chromosome. In fact, as we will consider shortly (and again in Chapter 10), it is possible to do this experiment using sequenced genomes. This has led to using genetic maps as a means to identify the DNA sequences for genes, as discussed in Section 9.4.

While crossing over occurs for every chromosome and is necessary for normal meiosis to proceed, there are situations when geneticists want to keep a particular set of alleles together on a chromosome, with no crossing over to generate new combinations. Geneticists have developed special types of chromosomes for which crossing over is suppressed between two homologs, which allows them to maintain a particular combination of alleles together. These crossover suppressors are known as balancer chromosomes; balancer chromosomes are discussed in Tool Box 9.1.

Linkage with X-linked genes

Having considered two genes that are linked on an autosome, we now turn to an example with two X-linked genes. There is no difference in principle from what has been done with autosomal genes, but different crosses are often done. In this case, we will mate a white-eyed female fruit fly with a wild-type body color to a yellow-bodied male that has wild-type red eyes, as shown in Figure 9.8(a). The male offspring have white eyes and wild-type body

BOX 9.1 *An Historical Perspective* Why didn't Mendel encounter linkage?

It is well known that, whenever Mendel crossed two heterozygous F_1 pea plants, he observed a 3:1 ratio among the phenotypes in the F_2 generation. In fact, his ratios are so close to the expected 3:1 ratio that some have suggested that Mendel might have cheated, at least by our current practices of experimentation; this question is discussed in Box 5.2. But another question arises from Mendel's experiments. When he crossed plants heterozygous for two genes, he always found a ratio of 9:3:3:1 in the F_2 generation, as described in Chapter 8, that is, the two traits segregated independently from one another, indicating that the genes are unlinked. In retrospect, working with unlinked genes was a fortunate circumstance. It is likely that the simplicity of Mendel's results was crucial to the rapid spread of genetics upon their rediscovery. More complicated results that included linkage might well have been more difficult to accept. So we can raise another retrospective question about Mendel's experiments—why didn't he encounter linkage? Was it by chance that he worked with genes that were unlinked to each other?

To answer this question, we first need to know how many chromosomes are found in the garden pea, *Pisum sativum*. After all, if the pea had 20 chromosomes in its haploid genome, it is not unlikely that all seven genes would be on different chromosomes. However, we know that peas have only seven chromosomes in their haploid genome, so it is somewhat suspicious that all seven of the genes behaved as if they are unlinked. On the other hand, plant chromosomes are quite large, as noted in the chapter, so genes might be on the same chromosome but far enough apart to segregate as being unlinked. This explanation can be tested if we know the map locations of the genes with which Mendel was working.

We do not know exactly which genes Mendel was using for his experiments, so we can't simply look at the genetic map of peas for the answer. However, for six of the seven traits that Mendel studied, modern geneticists have inferred the gene from his descriptions of the phenotypes and knowledge about what varieties of peas were being grown in Eastern Europe at that time. The most likely candidates for these six genes and their map positions are shown in Table A. Of these six genes, four of them are unlinked to any other, while the ones for pod color (the *GP* gene) and seed shape (the *R* gene) are both located on chromosome V. However, the *GP* and *R* genes are far enough apart on chromosome V that they do not show linkage in most two-factor crosses; thus, they are syntenic but usually not linked. In fact, among his dihybrid crosses, Mendel did not report data for those two genes, so it is possible that he did not test them. Even if he had, he may have found them to be unlinked.

The gene for the seventh trait—inflated and constricted pods—cannot be reliably identified, since several different genes produce this phenotype; the familiar "sugar snap" edible pod peas are examples of constricted varieties, and many different varieties with constricted pods were being grown at that time and place. The two best candidates are the *V* gene and the *P* gene; the *P* gene is on chromosome VI and is unlinked to any other gene with which he worked, so Mendel might have used this one and would not have seen linkage. On the other hand, he may have used a strain with different alleles of the *V* gene, which is on chromosome III, the same chromosome as the *LE* gene he is thought to have used for stem length. These two genes are only 12.6 map units apart, so linkage is definitely observed between these two genes. However, this is another combination of traits that Mendel either did not test or did not report; after all, with seven separate traits, there are 28 pairwise combinations, and Mendel could not be expected to perform and report all of them.

So did Mendel cheat by not reporting linkage? Or was he simply lucky or experienced enough to pick unlinked genes? At a distance of more than 150 years with incomplete records, we cannot know for sure. But we can afford to be charitable and conclude that, while he may have picked one pair of genes on the same chromosome, he did not choose linked pairs of genes. Or if he did pick the one pair that is clearly linked, he did not tell us about it. What we don't know didn't hurt us.

BOX 9.1 Continued

Table A Locations of Mendel's traits in the pea genome: the map positions and inferred functions of the genes with which Mendel is believed to have worked

Mendel's trait	Phenotype	Gene symbol	Chromosome	Gene function
Seed shape	Round vs. wrinkled	*R*	V	Starch branching enzyme 1
Stem length	Tall vs. dwarf	*LE*	III	GA-3 oxidase 1, involved in the production of the hormone gibberellic acid
Cotyledon color	Yellow vs. green	*I*	I	Stay-green gene, due to reduced chlorophyll breakdown in dominant mutant
Seed coat and flower color	Purple vs. white	*A*	II	Basic helix–loop–helix transcription factor
Flower position	Axial vs. terminal	*FA*	IV	Meristem function
Pod color	Green vs. yellow	*GP*	V	Chloroplast structure
Pod form	Inflated vs. constricted	*V* or *P*	III or VI	Sclerenchyma formation

Adapted from Reid and Ross, 2011 Genetics 189: 3–10.

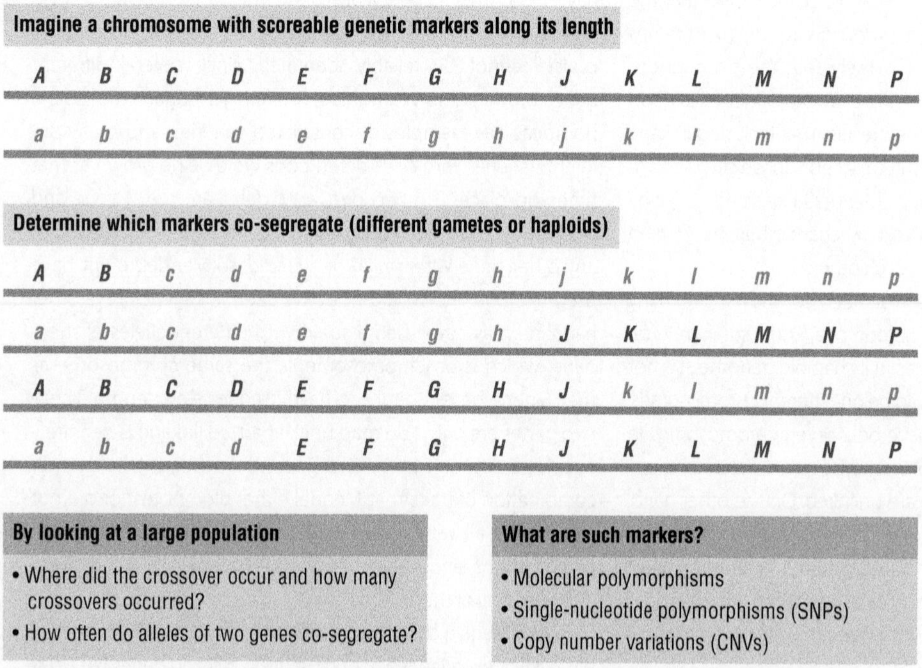

Imagine a chromosome with scoreable genetic markers along its length

A B C D E F G H J K L M N P

a b c d e f g h j k l m n p

Determine which markers co-segregate (different gametes or haploids)

A B c d e f g h j k l m n p

a b c d e f g h J K L M N P

A B C D E F G H J k l m n p

a b c d E F G H J K L M N P

By looking at a large population

- Where did the crossover occur and how many crossovers occurred?
- How often do alleles of two genes co-segregate?

What are such markers?

- Molecular polymorphisms
- Single-nucleotide polymorphisms (SNPs)
- Copy number variations (CNVs)

Figure 9.7 Locating the positions of crossovers. Every homologous pair of chromosomes has at least one crossover. If enough genetic markers on the chromosome can be scored or assayed, the location of each crossover can be determined. Many different types of genetic markers can be used in these experiments, although molecular polymorphisms are typically the most useful.

color. The female offspring have the genotype w^+/w; y^+/y, being heterozygous for both genes and have a wild-type phenotype; we will focus on her male offspring, so the genotype of the male parent isn't considered further.

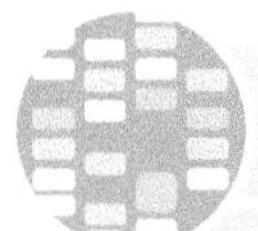

TOOL BOX 9.1 Crossover suppression by balancer chromosomes

Recombination is an inevitable and necessary part of the inheritance of genes. If enough gametes are examined, a crossover will have occurred between any two genes, no matter how close together they lie on the DNA molecule. In fact, recombination also occurs within a gene and has been used to map the locations of different mutant alleles with respect to one another. Like mutation, death, and taxes, recombination always happens.

A geneticist will sometimes want to suppress crossing over in a particular region of the chromosome to ensure that alleles of two different genes will continue to segregate together in subsequent crosses. The most frequent use of crossover suppression arises when one of the genes is essential, and a recessive mutation in it is lethal. Many genes are essential and have lethal alleles, and, naturally enough, many of the most important biological processes are governed by essential genes. If a mutant allele is lethal, its effect is monitored by the **absence** of a certain category of offspring, which can be difficult to detect. Thus, from the early days of genetics, the segregation pattern of lethal mutant alleles was often monitored by the segregation of nearby alleles that were easily visible. An example is shown in Figure A. These visible alleles of other genes could be on the same homolog as the lethal mutation, that is, *in cis* to it, or they could be on the other homolog, that is *in trans* to it; commonly, as shown in Figure A, visible alleles in both configurations were used. In the jargon of genetics, the lethal mutation was kept as a **balanced heterozygote**, the balancing effect being the nearby visible alleles used to monitor its inheritance.

The utility of balanced heterozygotes depends on the recombination between the visible marker alleles and the lethal mutation. In practice, visible markers that were more than 1 or 2 map units away were not very useful because crossing over would separate the marker from the lethal in nearly every generation. To circumvent this limitation, early *Drosophila* geneticists devised a series of specially marked and altered chromosomes that do not cross over with their normal homolog. These were known as **balancer chromosomes**; the lethal mutation of interest was maintained with a visible marker *in trans* to the balancer chromosomes, as shown in Figure B. These balancer chromosomes will be discussed again in Section 15.3 in the genetic analysis of segmentation in *Drosophila*. While balancer chromosomes are most widely used in *Drosophila* genetics, they have also been useful tools for genetic analysis in *C. elegans* and mice.

Properties of balancer chromosomes

A good balancer chromosome has two essential characteristics. First, it must greatly reduce or eliminate crossing over in a region of the genome being investigated. With crossover suppression, the balanced mutation will not be lost, even if the visible marker is not nearby on the normal genetic map. Second, individuals homozygous for the balancer need to be distinguishable from

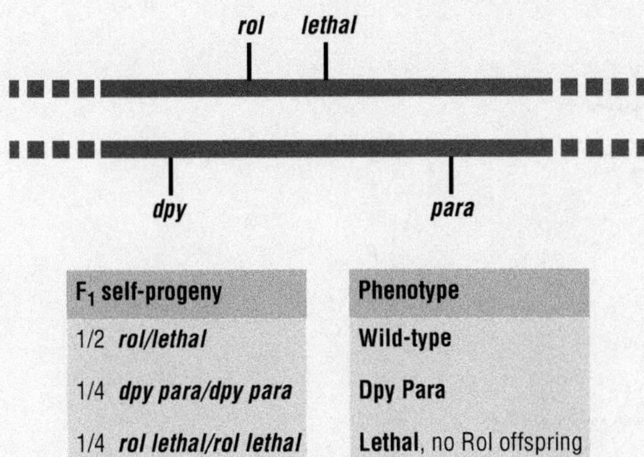

F_1 self-progeny	Phenotype
1/2 ***rol/lethal***	**Wild-type**
1/4 ***dpy para/dpy para***	**Dpy Para**
1/4 ***rol lethal/rol lethal***	**Lethal**, no Rol offspring

Figure A A balanced heterozygote. This example from *Caenorhabditis elegans* shows four linked recessive mutations, all of them heterozygous. As shown here, one homolog has a lethal mutation *in cis* to a *rol* (Roller) visible marker, while the other homolog has a *dpy* (Dumpy) and a *para* (Paralyzed) marker *in cis* to each other. Because all of the alleles are recessive to wild-type, the worm has a wild-type phenotype. Its progeny, upon self-fertilization, are shown in the box. Half of the expected offspring will be balanced heterozygotes for the lethal and wild-type, a quarter of the expected offspring will be Dpy and Para and thus not have the lethal mutation, and a quarter of the expected offspring will be Rol but lethal. Thus, the worm will have no Rol offspring, and the wild-type offspring will be two-thirds of the surviving offspring, while the Dpy Para offspring will be one-third of the surviving offspring.

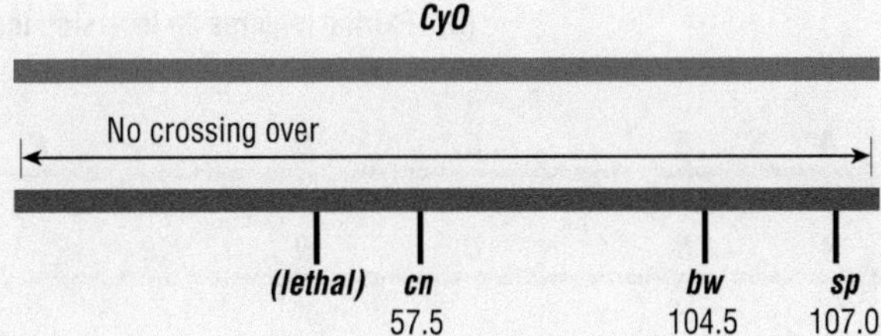

Figure B The *CyO* balancer chromosome in *Drosophila melanogaster*. The rearranged chromosome 2, shown in red, has multiple inversions that prevent crossing over with its normal homolog over most of the length of chromosome 2. It is marked with the dominant *Cy* (Curly wings) mutation, which allows easy scoring of flies heterozygous for the balancer; other recessive mutations on this balancer chromosome are not shown. The other homolog has the normal arrangement of three recessive mutations; these three mutations, *cn* (cinnabar), *bw* (brown), and *sp* (speck), are the ones used in the genetic screen discussed in Section 15.3. The map positions of *cn*, *bw*, and *sp* are given; *CyO* suppresses crossing over for most of chromosome 2 and can be used to balance lethal mutations as heterozygotes for almost any location on chromosome 2.

TOOL BOX 9.1 Continued

heterozygotes, which, in turn, need to be distinguishable from the individuals homozygous for the non-balancer chromosome.

Crossover suppression

The first of these characteristics—crossover suppression—has often been solved with the use of inversion chromosomes, as it was in *Drosophila*. From cytological observations in maize, it is generally believed that an inversion heterozygote can form a loop structure upon meiotic pairing, as shown in Figure C. A crossover that occurs within this loop results in chromosomes that lack a centromere (known as acentric fragments) and ones that have two centromeres (dicentric fragments). Both acentric and dicentric chromosomes fail to complete meiosis normally and thus do not contribute to viable gametes. An inversion

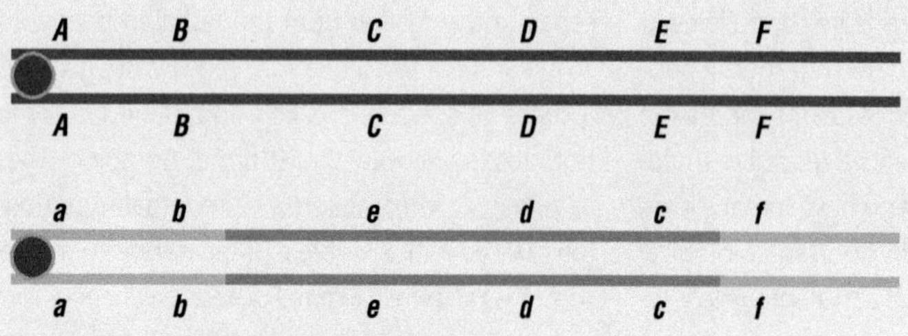

(i) Normal and inversion chromosomes

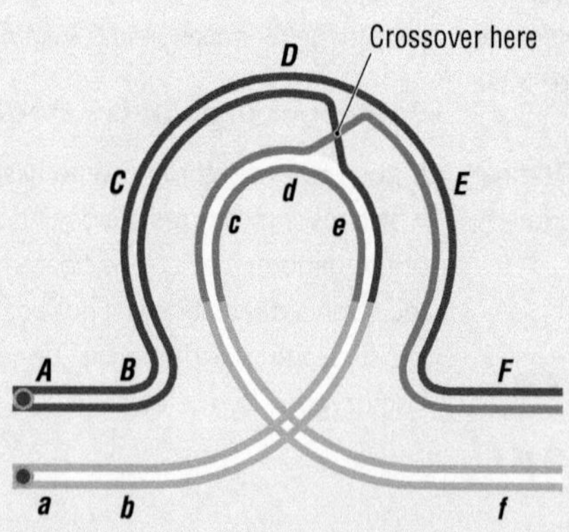

(ii) Pairing requires an inversion loop

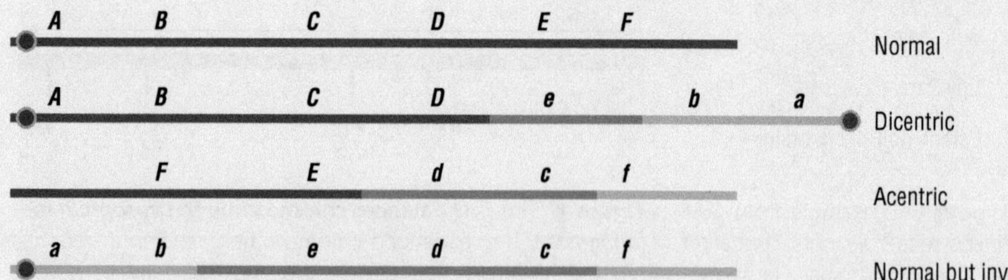

(iii) The recombinant chromatids are acentric or dicentric

Figure C Inversions loops as crossover suppressors. The normal arrangement of the chromosome and the inverted arrangement are shown in (i). The capital and lowercase letters could be considered as different alleles of the same gene or as regions of the chromosome. Note the inversion of the order for C, D, and E. (ii) shows the loop that is required for the inversion heterozygote to pair and synapse its chromosomes, if synapsis occurs over the entire length. The location of a crossover within the inversion loop is shown. (iii) shows the expected gametes that arise from a crossover within this inversion. Note that the two gametes involved in the recombination will not survive; one is dicentric, while the other is acentric. Only the homologs that did not recombine have a single centromere and would complete meiosis.

TOOL BOX 9.1 Continued

effectively reduces or eliminates crossing over because a crossover prevents the completion of meiosis. Plants have particularly large chromosomes, which made it possible to see inversion loops using light microscopy; inversions may also prevent normal pairing, particularly in animal chromosomes, which also suppresses crossing over.

The most widely used balancer chromosomes in *Drosophila* have multiple inversions that eliminate crossing over in a large region of the chromosome. In fact, most of the *D. melanogaster* genome can be balanced by one of three balancer chromosomes, one for the X and one for each of the two autosomes. (The tiny heterochromatic fourth chromosome has very few genes and does not recombine, even in wild-type flies, so no balancer is required, and it is typically ignored in mutagenesis screens.) No other organism has balancer chromosomes for all parts of its genome as *Drosophila* does. The ability to use only three different balancer chromosomes to screen the genome reflects both the years of genetics research using *Drosophila* and the small number of chromosomes.

Distinguishing marker mutations

The second necessary characteristic for a balancer chromosome is the ability to distinguish the balanced heterozygote, which is the genotype of interest for strain maintenance, from either of the two homozygotes. With *Drosophila* balancer chromosomes, this is usually accomplished with a dominant marker on the inversion chromosome. Some of the dominant markers that have been used are themselves homozygous lethal, or the inversion chromosome has other recessive lethal mutations present on it. For example, the balancer chromosome shown in Figure B is marked with the dominant marker *Curly* wings, which is easily identified in heterozygotes; the balancer chromosome with *Curly* also has multiple recessive lethal mutations on it. The *Cy/Cy* homozygotes for the balancer chromosome die because of the recessive lethal mutations, so they can be ignored. The offspring that do not have Curly wings will be the ones in which the lethal mutation of interest is balanced.

A more recent innovation has been to insert the green fluorescent protein (GFP) gene or a modified derivative of GFP as a reporter gene onto the balancer inversion. (The use of GFP as a reporter gene is discussed in Tool Box 12.1.) The fluorescence arising from GFP can be identified in early embryos long before any adult visible phenotype, such as Curly wings, can be spotted. Even more importantly, the embryos can be sorted automatically by fluorescence in a flow cytometer, and the three genotypic classes easily identified and separated from one another at a very early stage. Since balancer chromosomes are used to maintain lethal mutation stocks, it is very helpful to be able to distinguish the inviable embryos arising from the effects of the mutation from those arising from the effects of the balancer chromosome.

Since the males in F_1 have white eyes, we know that *white* eyes is an X-linked gene and that *white* is recessive to wild-type red eyes. We also know that *yellow* body is recessive to wild-type because the heterozygous males have a wild-type phenotype, but we don't know from these results if the *yellow* gene is X-linked or autosomal.

Now let's cross those F_1 flies among themselves. We will consider only the **male** progeny from this cross, since the genes are X-linked. If the ***white*** gene and the *yellow* gene were not linked to each other, we expect a phenotypic ratio of 1:1:1:1, that is, with equal numbers of male offspring who are wild-type for both traits, white-eyed and yellow-bodied, white-eyed and wild-type body, and wild-type red eyes and yellow bodies.

In reality, we are able to count 9026 males, as shown in Figure 9.8(b). (These are the actual data used in the original mapping experiment.) Of these, 4484 males have white eyes with wild-type bodies; 4413 have wild-type red eyes and yellow bodies; 76 are wild-type for both traits, and 53 have both white eyes and yellow bodies. Clearly, this does not equate to the 1:1:1:1 ratio of the four phenotypes that we would expect if the *white* and *yellow* genes are on different chromosomes.

How can we explain this result? Among the 9026 male offspring of the mating, approximately 99% are one of two types: the white-eyed flies with the wild-type body, and the wild-type red eyes with the yellow body. Only about 1% of the male offspring are either wild-type for both traits or mutant for both traits, the recombinant combinations of phenotypes, so the genes are about 1 map unit apart. The parental types are vastly more common than the recombinant types. Remember, by parental types, we mean the two parents (P_0) that were crossed to make the F_1 heterozygote. We are observing the effects of recombination during meiosis I in this F_1 heterozygote.

One other point of terminology that can be helpful in describing the configuration of the alleles in the F_1 heterozygote is illustrated in Figure 9.9. The alleles that are present on the same homolog are said to be *in cis* to

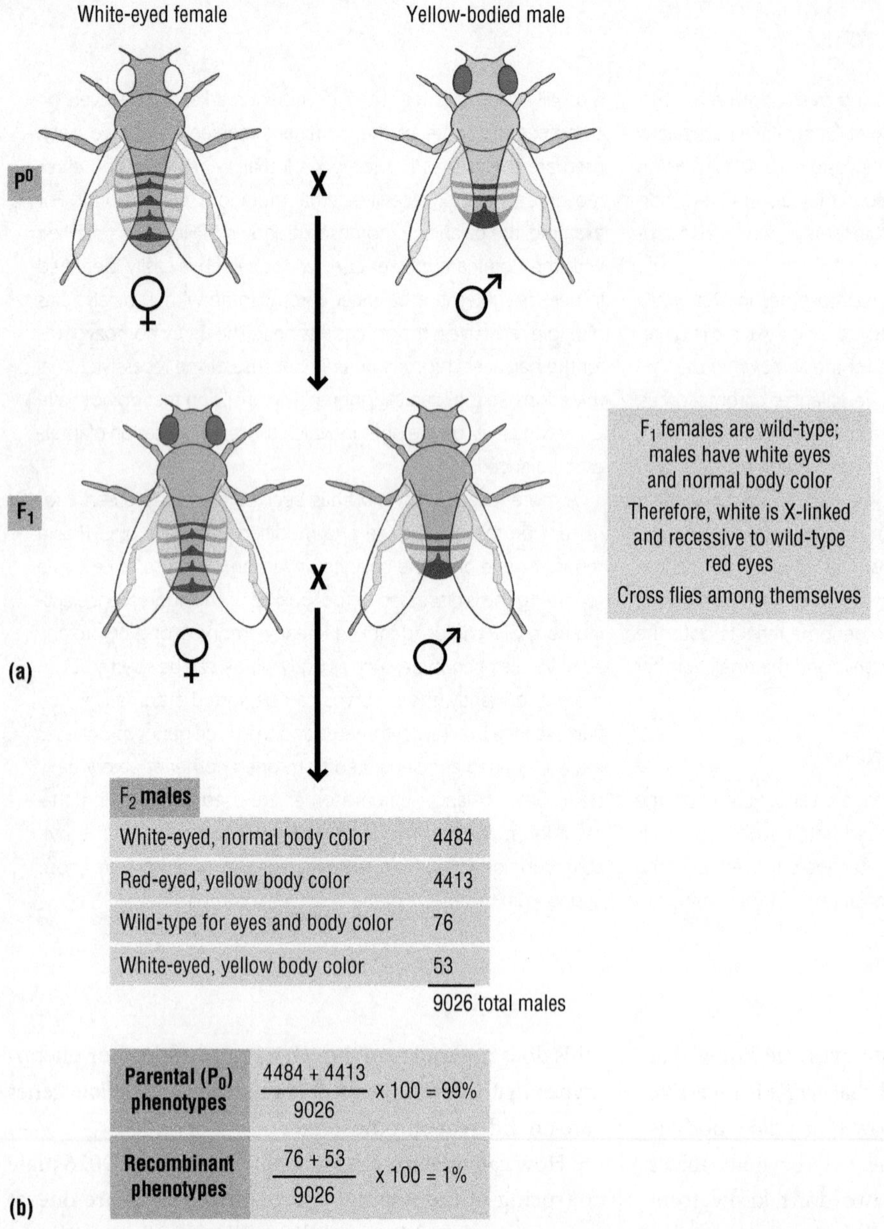
White-eyed female
Yellow-bodied male
P0
X
♀
♂
F1
X
♀
♂
(a)
F_1 females are wild-type; males have white eyes and normal body color
Therefore, white is X-linked and recessive to wild-type red eyes
Cross flies among themselves
F_2 males
White-eyed, normal body color 4484
Red-eyed, yellow body color 4413
Wild-type for eyes and body color 76
White-eyed, yellow body color 53
9026 total males
Parental (P_0) phenotypes $\frac{4484 + 4413}{9026} \times 100 = 99\%$
Recombinant phenotypes $\frac{76 + 53}{9026} \times 100 = 1\%$
(b)

Figure 9.8 Recombination on the X chromosome. The series of matings demonstrating that both *yellow* and *white* are X-linked genes is shown, beginning with a cross between a white-eyed female and a yellow-bodied male in (a). When the resulting F_1 flies are crossed among themselves, four categories of F_2 male offspring are observed, as summarized in (b). The less frequently observed phenotypes are the recombinants and represent 1% of the offspring. Thus, the *white* and *yellow* genes are 1 map unit apart on the X chromosome.

each other. The alleles that are present on opposing homologs are said to be *in trans* to each other.

But remember that the map distance is based on recombination between the two genes, regardless of which alleles were used for the genes. Suppose that we began our cross with a female fly that has a yellow body and white eyes and mated it to a wild-type male, that is, with the mutant alleles in *cis*. This cross is illustrated in Figure 9.10. The *yellow* locus and the *white* locus are 1 map unit apart (based on our previous results), no matter how the cross is done or which alleles are occupying each locus. These are their characteristic locations and the characteristic frequency of recombination between them.

Because this series of crosses began with the mutant alleles *in cis* (that is, with a yellow white parent and a wild-type parent), the recombinant types are those that have yellow bodies and wild-type eyes and wild-type bodies and white eyes. These two recombinant categories are

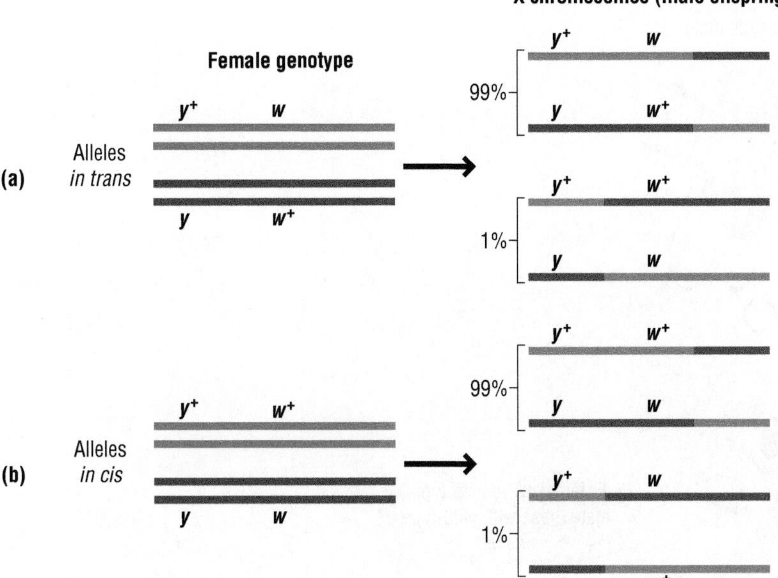

Figure 9.9 Crossing over with *yellow* and *white*. The crossover that gives the results seen in Figure 9.8 is diagrammed. The number of recombinants depends on the positions of the loci, and not on the alleles that occupy those loci. If each homolog has the wild-type allele for one gene (***y*⁺**) and the mutant allele for the other (***w***), as shown in (a), these two alleles are said to be *in trans* to one another. If one homolog has the mutant alleles for both genes and the other homolog has the wild-type alleles for both genes, the alleles are said to be *in cis*, as shown in (b). In either case, 99% of the offspring have the parental arrangement.

expected to be 1% of the total, or ten total male flies, and we expect to see about five male offspring that have yellow bodies with wild-type eyes and about five male offspring that have wild-type bodies and white eyes, since the two classes of recombinants are equally frequent.

KEY POINT Map distances between genes are characteristic for the two loci. The percent of recombinants observed in a dihybrid test cross is defined as the map distance between the two loci.

Map distances cannot exceed 50 map units

Suppose that two genes are unlinked; unlinked genes are as far apart as is genetically possible. What would we calculate the map distance is between them? Look back at the beginning of the chapter, when we test-crossed ***A/a; B/b***, in which the genes *A* and *B* were not linked or between *ebony* and *apterous* which are unlinked. If we had calculated the map distances in these crosses, what would the map distance be?

In a cross involving unlinked genes, the four classes of progeny are equally frequent, so the two parental classes and the two recombinant classes are equally common. Half of the progeny are recombinant, and half are parental. Thus, the maximum map distance that can be measured from a cross involving two genes is 50 map units. This is a very important principle that turns up often and can trip up even experienced investigators at times.

What if the two genes are on the same chromosome, but they are so far apart that there is always a crossover between them? The same 50 map units will be found with a two-factor cross like this. The reason that the maximum recombination frequency is 50%, and not 100%, despite there always being a crossover between the genes, takes us back to meiosis. Look at the diagrams of recombination in Figure 9.11. There are four chromatids, composed of two pairs of two sister chromatids. This four-chromatid configuration is known as a **tetrad**. Two of the chromatids are involved in the crossover, and two are not. Even when a crossover has occurred between the two loci, only 50% of the chromatids are recombinant. For simplicity, we have shown this with only a single crossover involving two chromatids, but the same result holds when more than one crossover occurs and more than two chromatids are involved. Double crossovers are considered in Section 9.3.

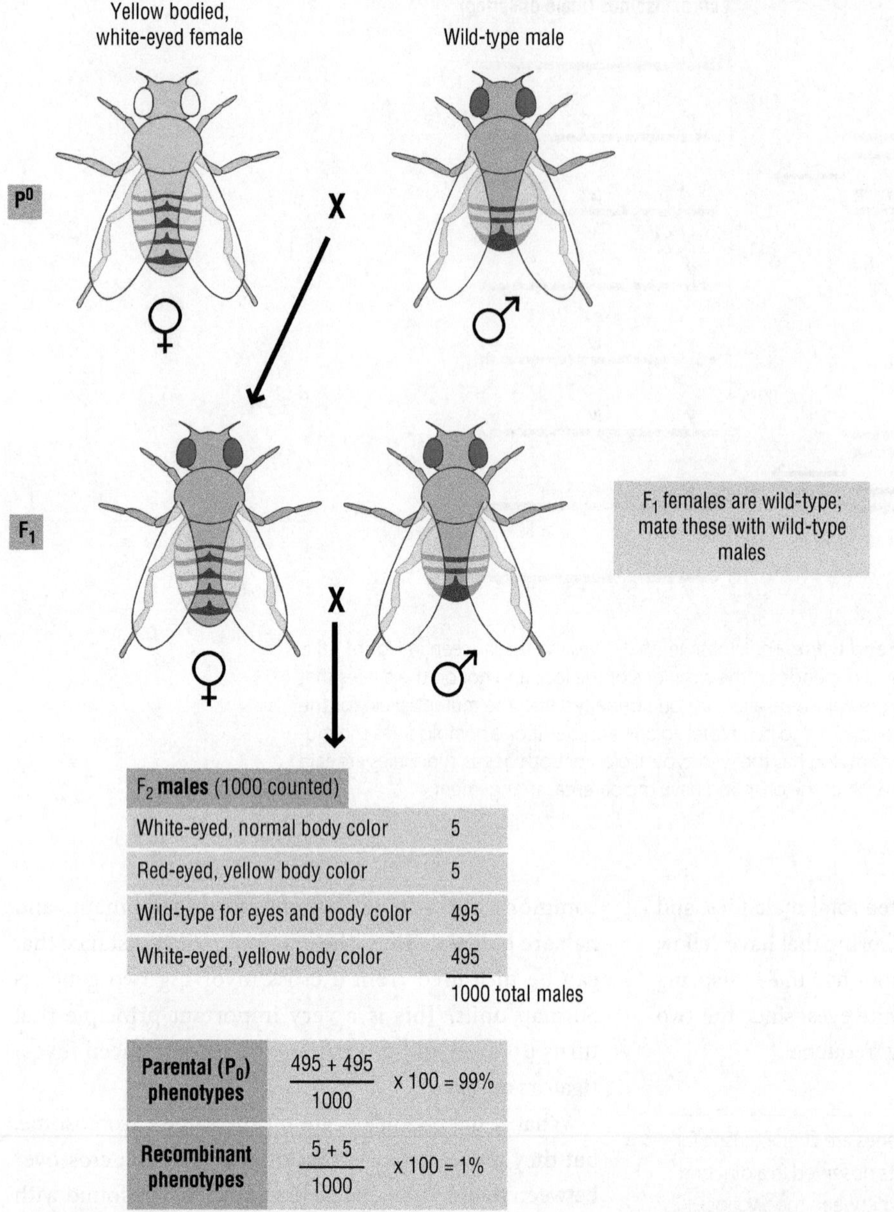

Yellow bodied, white-eyed female
Wild-type male
P^0
X
F_1
X
F_1 females are wild-type; mate these with wild-type males
F_2 males (1000 counted)
White-eyed, normal body color 5
Red-eyed, yellow body color 5
Wild-type for eyes and body color 495
White-eyed, yellow body color 495
1000 total males
Parental (P_0) phenotypes $\frac{495 + 495}{1000}$ x 100 = 99%
Recombinant phenotypes $\frac{5 + 5}{1000}$ x 100 = 1%

Figure 9.10 Recombination when *yellow* and *white* mutant alleles are in the *cis* arrangement. A cross is made between a white-eyed yellow-bodied female and a wild-type male in (a). When the resulting F_1 female flies are crossed with wild-type males, four categories of F_2 male offspring are observed, as summarized in (b). Again, the less frequently observed phenotypes are the recombinants and represent 1% of the offspring, as in Figure 9.8. However, the parental and recombinant phenotypes are different, due to the arrangement of the alleles in the original P_0 generation.

Crossing over during meiosis requires that the two homologous chromosomes align and synapse precisely with one another. Such precise alignments do not always occur, so crossing over can occasionally produce new-variant chromosomes or genes, that is, mutations. The process and consequences of unequal crossing over, and one of its most familiar examples of red–green color blindness in humans, are discussed in Box 9.2.

KEY POINT The maximum map distance that can be observed in a two-factor cross is 50 map units.

Comparing linkage with synteny

Chromosomes always have at least one crossover, so there may be some loci that are physically on the same DNA

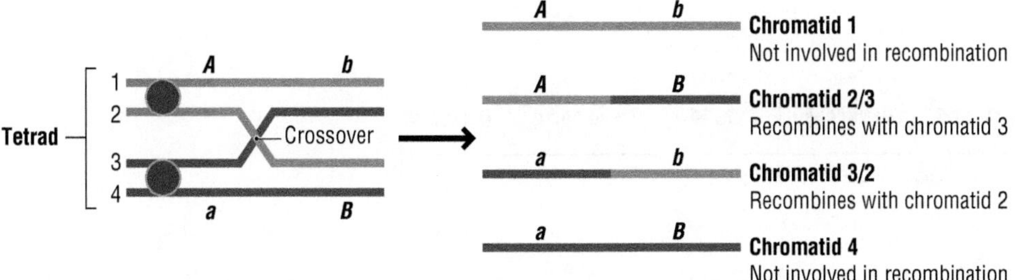

Figure 9.11 Recombinant and non-recombinant chromatids. Recombination occurs during prophase I when the chromosome has doubled to produce two sister chromatids, but before the homologs separate. Since there are four chromatids, this structure is known as a tetrad. Crossing over involves one chromatid from each homologous pair, labeled here as chromatids 2 and 3. The other two chromatids, labeled 1 and 4 here, do not recombine. Thus, even when a crossover has occurred, only half of the gametes are recombinant.

BOX 9.2 *A Human Angle* Red–green color blindness

Mutations that arise spontaneously usually occur as a result of unrepaired errors during replication, as discussed in Chapter 4, or as transposable element movement, as discussed in Chapter 3. However, a few mutant alleles arise during recombination; the most familiar example is red–green color blindness in humans. The same process that produces red–green color blindness has also been an important evolutionary force for shaping genomes and gene families.

The primary photoreceptors in our retina are a family of proteins known as the opsin proteins; the opsins in our cone cells absorb light of different wavelengths and allow us to see colors. Opsins that absorb short wavelengths are the blue opsins; the blue opsin gene is autosomal and is not involved in red–green color blindness. Opsins that absorb long wavelengths are called the red opsins, while those that absorb medium wavelengths are called the green opsins; strictly speaking, the maximum absorbance for the red opsins is now known to be in the yellow spectrum, but they are more sensitive to the long-wave red light than are the green opsins, so the original name has stuck. The genes for the red and green opsins are X-linked and are immediately adjacent to each other on the chromosome. The genes are 96% identical in DNA sequence; in fact, the regions between the genes are also highly similar in DNA sequence. There is usually a single red opsin gene and one to four green opsin genes in tandem duplication, as shown in Figure A(i).

When the X chromosomes synapse, the DNA sequence identity in this region is high enough that the two homologs can misalign, as shown in Figure A(ii). If a crossover occurs with these misaligned chromosomes, the two recombinant chromosomes are not the same as one another; this is known as **unequal crossing over**. Depending on where the crossover occurred in the gene family, the outcome can be a deletion of one red gene or the green gene, or the generation of a hybrid red–green gene. Since these are X-linked genes, males who inherit such an X chromosome from their mothers will be red–green color-blind. The majority of red–green color-blind males have a normal red opsin gene, followed by a hybrid red–green gene, a condition known as deuteranomaly. Smaller percentages have the red gene deleted (protanopia), a hybrid red–green gene with no normal red gene (protanomaly), or a single red gene with the green gene deleted (deuteranopia), depending on where the unequal crossover occurred in these genes.

Because red–green color blindness arises from unequal crossing over, rather than a replication error or a transposition event like most other mutations, the frequency is fairly high; approximately 8% of males are red–green color-blind, a frequency that is similar (although not identical) among most human populations. As many men can attest, the condition is not life-threatening, and most learn to adjust at an early age.

While red–green color blindness is the most familiar example of unequal crossing over for most of us, it is far from the only one, and probably not the most significant one for evolution and genetics. The misalignment of homologs can occur whenever there are duplicate genes or sequences in the genome. This occurs with many gene families like the opsins or the globins, or with many other repeat sequences in eukaryotic genomes, including transposable elements. The genomes of metazoans are littered with repeat sequences that are a source for misalignment and duplication/deletion of chromosomes. Unequal crossing over is the most likely origin of the duplication of chromosomes that result in the expansion of gene families and the duplication–divergence pattern for new genes, discussed in Section 3.6 and Box 3.1.

BOX 9.2 Continued

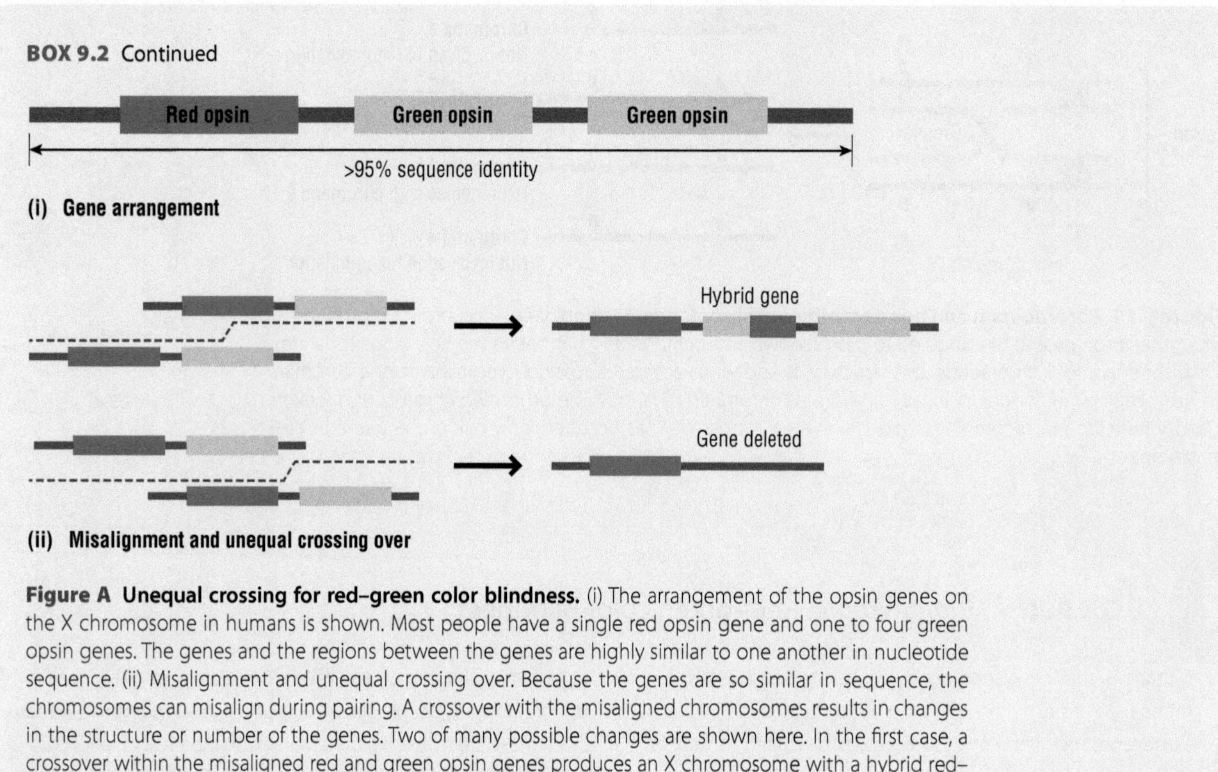

Figure A Unequal crossing for red–green color blindness. (i) The arrangement of the opsin genes on the X chromosome in humans is shown. Most people have a single red opsin gene and one to four green opsin genes. The genes and the regions between the genes are highly similar to one another in nucleotide sequence. (ii) Misalignment and unequal crossing over. Because the genes are so similar in sequence, the chromosomes can misalign during pairing. A crossover with the misaligned chromosomes results in changes in the structure or number of the genes. Two of many possible changes are shown here. In the first case, a crossover within the misaligned red and green opsin genes produces an X chromosome with a hybrid red–green gene. Depending on the location of this crossover within the genes, this male could be red–green color-blind. Although a normal green opsin gene is still present, the hybrid gene next to the red gene affects the color vision. This is among the most common changes. In the second case, a crossover within the region between the genes results in the deletion of the green gene, which also results in red–green color blindness.

molecule but are so far apart that they appear to be unlinked to one another in segregation. For example, genes located at opposite ends of the same chromosome will always have a crossover between them and appear to be on different chromosomes in a genetic cross that considers recombination. Genes that are physically located on the same chromosome (that is, form part of the same DNA molecule) are said to be **syntenic**. For example, all genes on chromosome 7 in humans are syntenic with one another, regardless of the recombination distances between them. Linkage is based on the probability that two alleles will be inherited together. Synteny is based on physical location, regardless of patterns of inheritance.

The concept of synteny is especially important now that it is relatively easy to make evolutionary comparisons between the genomes of different species. It is common to find regions of the chromosome in one species that correspond to a region of a chromosome in other species. These are known as a **syntenic block**, genes that have remained together on the same chromosome during evolution. In fact, the evolution of related species can be considered in terms of the syntenic blocks of chromosomes.

KEY POINT All genes on a single chromosome are said to be syntenic; they are all part of the same DNA molecule. Syntenic blocks of DNA allow for evolutionary comparisons between species.

Figure 9.12 shows the syntenic blocks of human and mouse chromosomes. In this figure, the genes on the mouse chromosomes are colored to match the location of the same genes in humans. Many changes in the linkage arrangement between two genes have occurred, since mice and humans last shared a common ancestor. However, many syntenic blocks can still be found. For example, genes found on human chromosome 4, shown in yellow, appear as syntenic blocks on three mouse chromosomes—numbers 3, 5, and 8. The common ancestor to humans and mice did not have the arrangement of genes that we see for chromosome 4 in humans or the arrangement of genes that we observed for chromosomes 3, 5, and 8 in mice; however, it almost certainly did have these blocks of genes linked to each other on its chromosome.

One of the most interesting examples of synteny is the mammalian X chromosome. Nearly all X-linked genes in

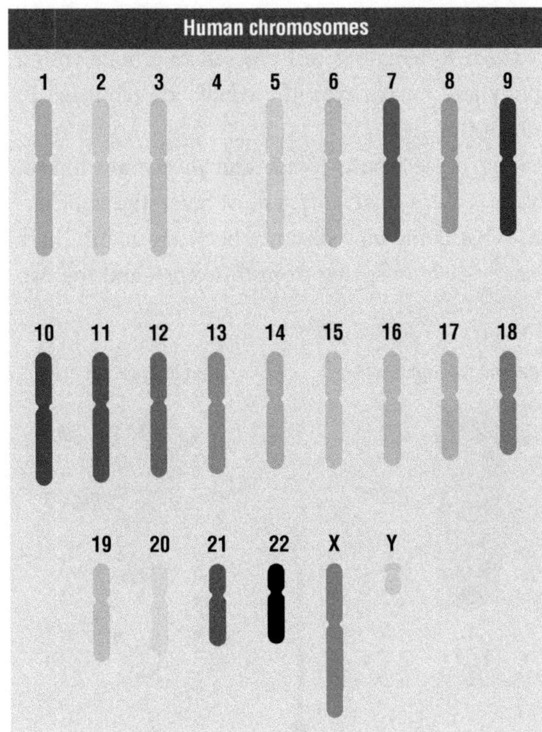

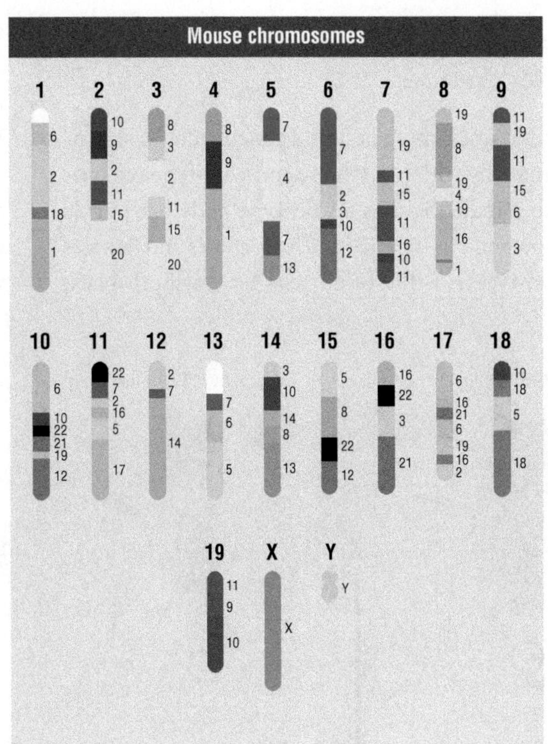

Figure 9.12 Synteny. Genes located on the same physical chromosome are defined as being syntenic. This figure shows syntenic blocks for humans and mice, with the colors based on the human map location. Syntenic blocks show that the common ancestor to humans and mice had many of the same linkage arrangements, but that many chromosome rearrangements have occurred. Note, however, that the X chromosome is largely syntenic for mice and humans, and in fact all placental mammals.

Source: Courtesy of Lisa J. Stubbs, University of Illinois.

placental mammals are syntenic in other mammals—the genes continue to be on the X chromosome, although not necessarily in the same order. This suggests that the common ancestor to all placental mammals had these genes on the X chromosome and that this X-linkage has been retained throughout mammalian evolution; this conservation of synteny is probably because of the mechanism of dosage compensation, described in Section 7.5.

Synteny can be useful in other situations as well. If we find a gene in one species, we can ask if another species has the same gene by using the base sequence of the gene or, more commonly, the amino acid sequences of its encoded protein. Furthermore, we can ask if neighboring genes on the chromosome are also near each other in the second species, that is, we can determine if the genes comprise a syntenic block. We are not measuring recombination in both species, which requires that we can observe and count the offspring of a mating. Instead, we can observe if the genes are physically linked by DNA sequencing and by comparing syntenic blocks. This can be useful for positional cloning, as discussed in Section 9.4.

Syntenic blocks are not confined to eukaryotic chromosomes. The genomes of closely related bacteria also show extensive syntenic blocks or similar gene orders on their chromosomes.

9.3 Genetic maps

The concept of linkage allows geneticists to map the locations of genes on a chromosome. A map distance between two loci is based on how often recombinants arise. These loci can be used to find recombination distances with other loci and thus help us to assemble a map of the relative positions of various genes. A map based on recombination distances is referred to as a **genetic map** or a **recombination map**.

Constructing a genetic map with two-factor data

Let's consider an example of how a genetic map is assembled. A female fly with black body and wild-type red eyes is mated to a male that has a wild-type body but purple eyes, as shown in Figure 9.13(a). The F_1 males and females are all wild-type, so this result tells us that both black and purple are recessive phenotypes and that *black* is not X-linked. An F_1 female is test-crossed to a male with a black body and purple eyes; the results are tabulated in Figure 9.13(a).

Based on these results, *black* and *purple* are linked; since *black* is autosomal, *purple* must be on the same autosome. What is the map distance between them? There were exactly 1000 offspring from this cross, and the two

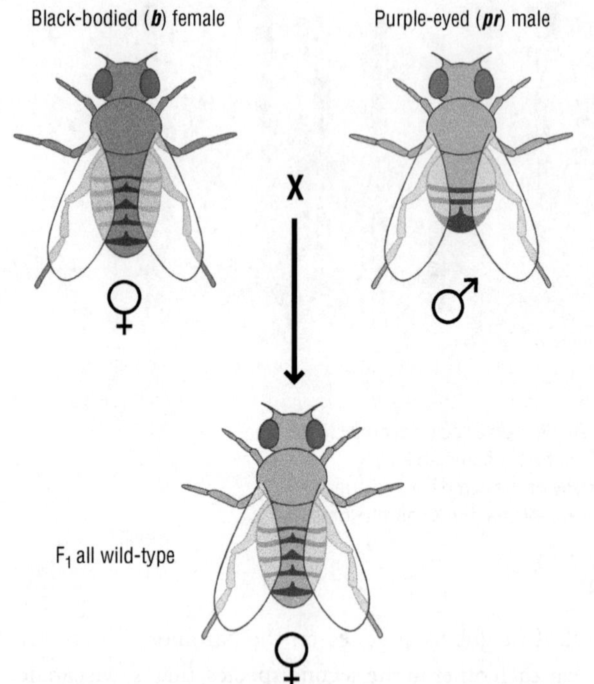

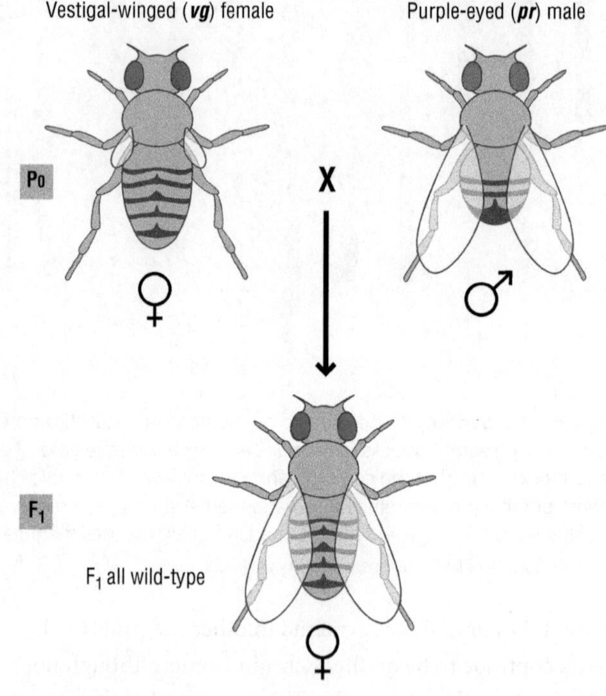

Test cross F_1 female to male with black body and purple eyes

Results	
Black body, red eyes	473
Black body, purple eyes	35
Wild-type body, red eyes	31
Wild-type body, purple eyes	461
	1000 total

$$\frac{35 + 31}{1000} \times 100 = 6.6\%$$

Find that *black* and *purple* are 6.6 map units apart

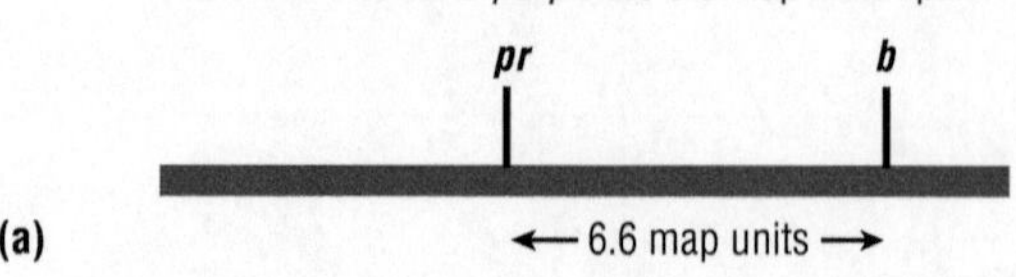

(a)

Test cross F_1 female to male with vestigal wings and purple eyes

Find that *vestigal* and *purple* are 12.1 map units apart

vg   pr

(b) ← 12.1 map units →

Figure 9.13 Building a genetic map. Maps of the chromosomes based on recombination can be assembled from a series of two-factor crosses. In (a), the genes *black* and *purple* are found to be 6.6 map units apart. In (b), *vestigial* and *purple* are found to be 12.1 map units apart. Since both *black* and *vestigial* are linked to *purple*, they must be linked to each other. From a cross involving *black* and *vestigial* in (c), the map distance is found to be 18.7 map units, which allows us to infer the relative positions of all three genes. Based on these experimentally determined map distances, *purple* must be located between *vestigial* and *black*.

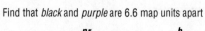

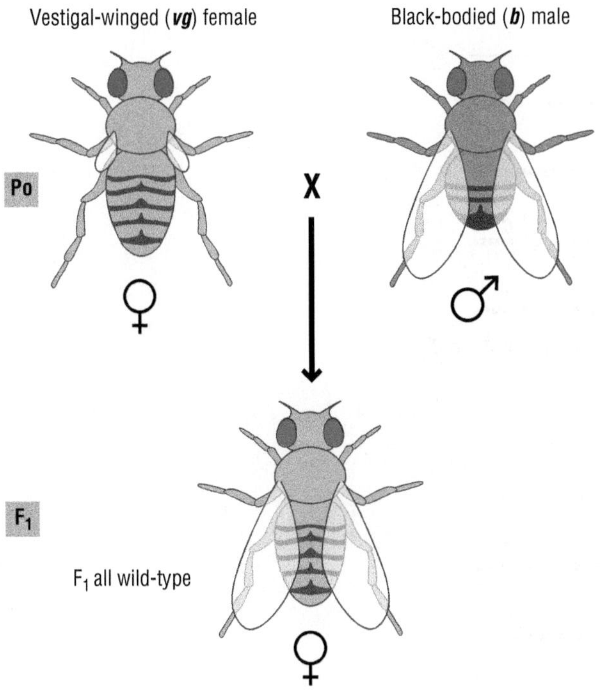

Test cross F_1 female to male with vestigal wings and black body

Find that *vestigal* and *black* are 18.7 map units apart

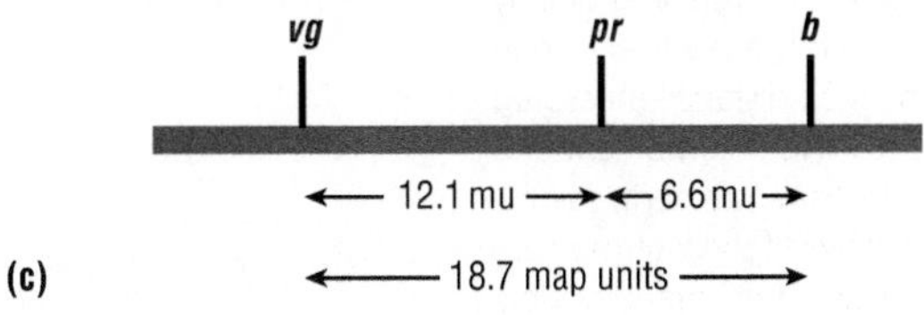

Figure 9.13 Continued

most common phenotypes are black body with wild-type eyes and wild-type body with purple eyes, the parental combinations. The recombinants are the wild-type eyes and body (31 flies) and black body and purple eyes (35 flies), or 66 total recombinants. Thus, the recombinants make up 66/1000 or 6.6% of the offspring, telling us that the two genes are 6.6 map units apart.

The experiment is repeated for the genes *vestigial* wings and *purple* eyes, and a distance of 12.1 map units is found between these two genes, as illustrated in Figure 9.13(b). Thus, *black* is linked to *purple* and *vestigial* is linked to *purple*, so *black* must also be linked to *vestigial*.

There are two possible orders of the three genes *vestigial*, *purple*, and *black* on the chromosome. If the order of the genes on the chromosome is *black–purple–vestigial*, with *purple* being the middle gene of the three, a map distance of about 18.7 map units should be observed between *black* and *vestigial*, the distance from *black* to *purple* plus the distance from *purple* to *vestigial*. On the other hand, if the order of the genes is *vestigial–black–purple*, then the map distance between *vestigial* and *black* is expected to be about 5.5 map units, the distance between *vestigial* and *purple* (12.1 map units) minus the distance between *black* and *purple* (6.6 map units). When the cross is done, as shown in Figure 9.13(c), the distance between them is closer to 18.7 map units. Thus, *purple* is the middle gene between *black* and *vestigial*.

By doing a series of two-factor crosses like this, it is possible (although laborious) to put together a genetic map of the entire genome. In order to get the distance between *black* and *vestigial*, the distances between *black* and *purple* and between *purple* and *vestigial* were simply added. This additive feature of the genetic map usually works best over short distances—a distance of less than 10 map units is typically used for constructing maps, whenever possible, but the effective distance depends on the rate of recombination in an organism.

For most organisms, however, the recombination distances are not completely additive over longer distances because of the occurrence of double crossovers, that is, the chromosomes cross over once and then cross back over in the same interval between the two genes. The effects of a double crossover are shown in Figure 9.14. When looking only at genes *A* and *C*, this double crossover cannot be detected and appears the same as a parental combination; thus, the rate of recombination between *A* and *C* will be underestimated.

For short distances in which one crossover is unlikely, double crossovers are very unlikely. For genetic maps in most animals, double crossovers have proved to be more of a theoretical concern than a practical one, so long as short distances are being compared. In humans, for example, the average number of crossovers is about two per chromosome. Consequently, the probability that both of them would occur between the same two genes is quite small. Nonetheless, there is a long literature on genetic maps and double crossovers.

Part of the fascination with double crossovers has to do with the history of genetic maps. Many early geneticists worked on plants; plants often have multiple crossovers on each chromosome, so the geneticists needed to be concerned about double crossovers. In addition, when genetic maps were first being constructed, there were relatively few mapped genes to use as markers. Thus, geneticists were often forced to work with fairly long genetic distances. Now that many markers are available to examine recombination in most organisms, we rarely need to rely on map distances that are longer than a few map

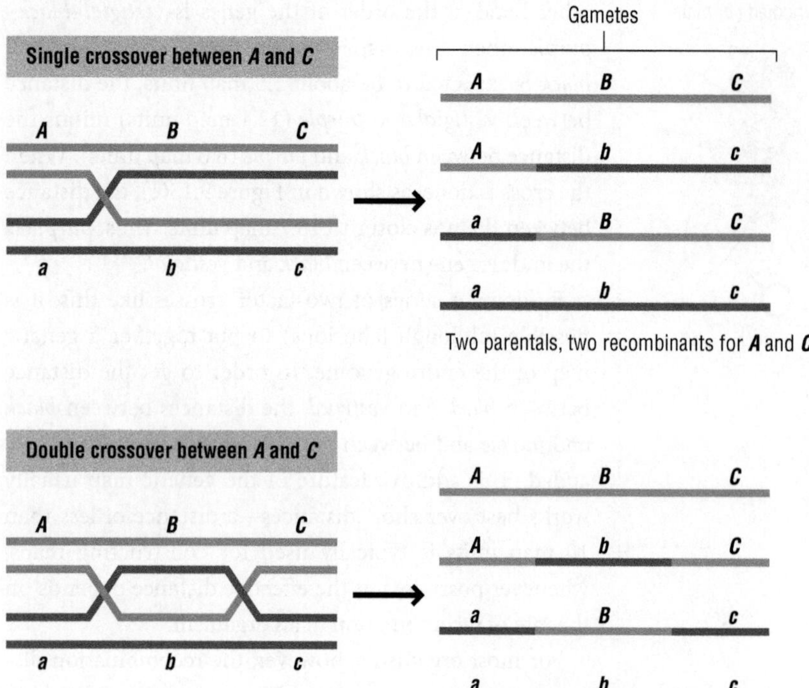

Figure 9.14 Single and double crossovers. Genetic maps are usually constructed from distances of only a few map units because the occurrence of two crossovers between two genes, which is more likely when the genes are farther apart, will result in an underestimate of the map distance. The map distance measured in this cross is between loci *A* and *C*. A single crossover between *A* and *C* produces two recombinant and two parental chromosomes. A double crossover between *A* and *C* restores the original arrangement, so only parental combinations for *A* and *C* are seen; thus, this would be scored as a non-crossover if the intermediate locus *B* were not being examined. In technical terms, this double crossover is known as a "two-strand double crossover," since it involves only two of the four chromatids. Three-strand double crossovers and four-strand double crossovers are also sometimes observed in organisms such as *Neurospora* and *Aspergillus* for which these types can be distinguished.

units, so double crossovers are not a significant concern for constructing most genetic maps or estimating the recombination frequency between two genes.

Genetic maps of *Drosophila melanogaster* and corn were the first to be compiled and have greatly affected the ways that geneticists think about mapping. Both organisms are diploid and out-crossing (as opposed to self-fertilizing), and, in both cases, map data were compiled through extensive test crosses. However, some of the most important information about recombination, and thus genetic maps, came somewhat later from tetrad analysis in ascomycete fungi such as *Neurospora crassa* and *Aspergillus nidulans*, as discussed in Tool Box 6.1. In these fungi, individual haploid ascospores can be grown into colonies, so test crossing was not essential. In addition, all four chromatids arising from a pair of homologous chromosomes undergoing a single meiosis were found together in an ascus. This provided the evidence that crossing over occurs when four chromatids are present (rather than before DNA replication) and that two of the four chromatids are involved in a single crossover. Genetic mapping with tetrads is done similarly to the methods involving recombinant gametes, but the calculations are done a bit differently since all products of meiosis are considered together.

Standard eukaryotic genetic maps

Genetic maps have been compiled for many organisms, with data collated from many hundreds of geneticists and many thousands of experiments. Every model organism used for genetics research has a standard map that is maintained by a group of researchers at its stock center, an agreed-upon reference that all researchers use and to which many people contribute data. These maps can be found at the websites for the different stock centers.

Part of the standard *D. melanogaster* genetic map is shown in Figure 9.15; note the number next to each gene.

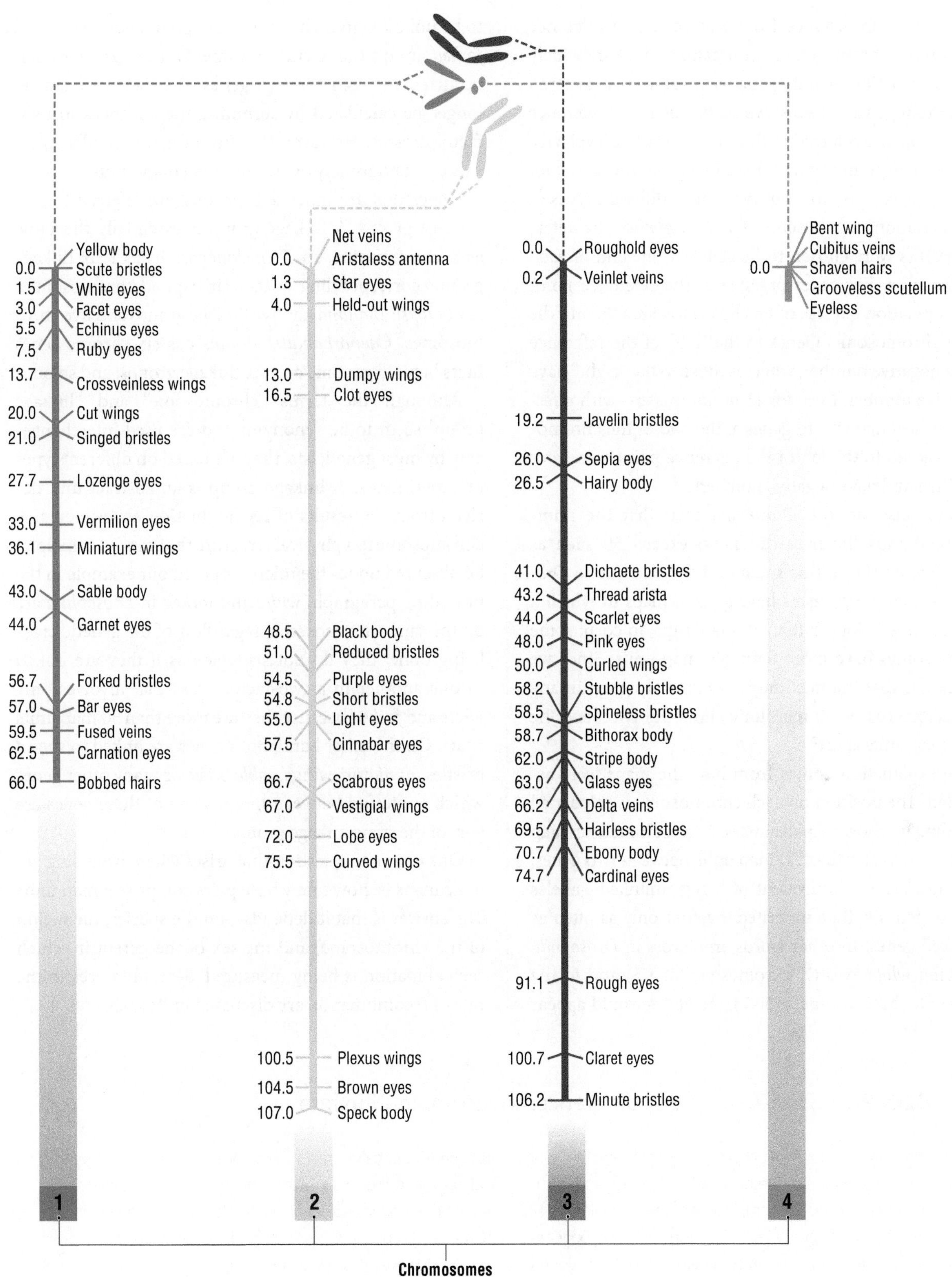

Figure 9.15 The genetic map of *Drosophila melanogaster*. Only some of the thousands of mapped genes are shown. Although the map is often drawn vertically like this to fit on a page, the positions are referred to as left and right, with left at the top and right at the bottom; thus, *brown* eyes gene is on the right end of chromosome 2. The numbers represent the recombination distance from a reference point, as measured by two-factor crosses with closely linked genes.

Source: Griffiths et al. (2002) *Modern Genetic Analysis* 2nd edition, Macmillan.

This is the map distance from that gene to a reference point on the chromosome. For organisms that are widely used for genetic research, geneticists arbitrarily declared that certain genes would serve as the anchor or reference for the map and referred to this position as 0.0. Typically, the reference point was a well-studied gene whose mutant phenotype was easy to identify in many different types of recombination experiments. For *Drosophila*, the reference points were chosen to be genes at one end of each chromosome. For other organisms, the reference point gene at position 0.0 was often chosen towards the middle of the chromosome. Genes to the "left" of the reference have a negative number, whereas those to the "right" have a positive number. Even for *D. melanogaster*—with reference points anchored to genes at the end of the chromosome—genes to the left of the reference point were later identified and have negative numbers.

Notice also on the *Drosophila* map that the numbers (and thus the map distances) exceed 50 such as 57.0, 75.5, or 106.2. This situation is not unique to *Drosophila*—many organisms have genetic maps in which a chromosome is longer than 50 map units; in corn, some chromosomes have more than 200 map units. Since we said earlier that the maximum distance observed in any two-factor cross is 50 map units, how can two genes be 106.2 map units apart?

The explanation comes from how the maps were assembled. The position on each chromosome is calculated by adding up short map distances. If the summed-up map distance is greater than 50 map units between two genes, the genes are the equivalent of being unlinked—alleles of these genes will be inherited together only as often as unlinked genes. In other words, in a cross in *Drosophila* involving *white* eyes (X chromosome at 1.5) and *forked bristles* (X chromosome at 56.7), the genes would appear to be unlinked to each other, even though both *white* and *forked* are on the X chromosome. Once a crossover has occurred between the two genes, the distances can no longer be calculated by summing the distances in two-factor crosses. However, the sum of many smaller intervals can, and usually does, exceed 50 maps units.

Genes that are inherited together are referred to as a **linkage group**. A linkage group is essentially the same as a chromosome, so *D. melanogaster* has four linkage groups corresponding to its four haploid chromosomes, *Arabidopsis thaliana* has five linkage groups and five chromosomes, *Caenorhabditis elegans* has six chromosomes in its haploid genome and six linkage groups, and so on.

Although the terms "chromosome" and "linkage group" seem to be synonyms and are used interchangeably by most geneticists, they are based on different types of experiments. A linkage group is an abstract unit derived from the results of recombination experiments. A chromosome is a physical structure that can (in principle) be observed under the microscope. In our example in the preceding paragraph, *white* and *forked* in *Drosophila* are on the same chromosome, regardless of the genetic cross being done; they do not segregate as if they are linked to one another in a two-factor cross that involves only *white* and *forked* because they are more than 50 map units apart. On the other hand, both genes are linked to *singed* bristles, *miniature* wings, *sable* body, or many other genes which are found between them, so all of these genes are part of the same linkage group.

One common question that arises when discussing genetic maps is: how many base pairs comprise a map unit? The answer is that it depends—on the species, the region of the chromosome, and the sex of the parent in which recombination is being measured. Sex differences in the rate of recombination are discussed in Box 9.3.

BOX 9.3 *Going Deeper* Sex differences in recombination rate

In Section 9.4, we note that the rate of recombination is not constant for all regions of the genome of a species and is quite different when species are compared. Another variable that significantly affects the rates of recombination is whether oogenesis and spermatogenesis are being considered. The explanations for this difference between the sexes include both molecular and evolutionary factors.

As a means of standardization, genetic maps in animals are almost always based on recombination in the female during oogenesis. Spermatogenesis almost always has fewer crossovers on average than oogenesis, and the distribution of the crossovers is different between the two germlines, that is, in general, genes that are 10 map units apart in a female might be only 7 map units apart in a male. However, this overall rate across all chromosomes is not uniform for all genetic intervals; some genetic intervals have more recombination during spermatogenesis than oogenesis, and some are reduced much more than others. (A few domesticated species, such as sheep and cattle, may have slightly more recombination in

BOX 9.3 Continued

males, which has been cited as evidence that the recombination rate is subject to selection.) In humans, spermatogenesis has about two-thirds as many crossovers as does oogenesis, a fraction that is similar in many other animals. Thus, the map of the chromosomes based on crossing over during oogenesis is longer than the map determined through spermatogenesis.

Much speculation and many evolutionary models have considered these sex differences in recombination; the review article referenced below provides an excellent and readable background on these models. While the evolutionary pressures that favor reduced recombination in spermatogenesis are not altogether clear, the molecular factors seem to be a bit more evident. It should be remembered from Section 6.4 that spermatogenesis and oogenesis involve fundamentally different processes at the cellular and molecular level. Most notably, it may be overlooked that chromosomes during spermatogenesis are usually much more tightly compacted than chromosomes during oogenesis. In fact, in many animals (including mammals), many of the histones in the chromatin of spermatozoa are replaced with a small, positively charged protein (or protein family) known as protamine, so that the nucleosome in spermatozoa has a different structure to that described in Chapter 2. This highly compacted structure present during spermatogenesis may make the chromatin less accessible to the machinery of recombination. In other words, the order and the location of the genes on the chromosome or the DNA molecule are precisely the same, but differences in the three-dimensional structure of the chromosome make recombination more likely during oogenesis than during spermatogenesis.

Despite the reduction in recombination overall, sperm chromosomes still have an obligate crossover during meiosis; the reduction in crossover number is related to the occurrence of double and triple crossovers on a chromosome pair.

Drosophila males do not cross over at all

While reduced recombination in males is the general rule for most animals, some species take this reduction to an extreme. The most familiar example of greatly reduced male recombination is also one of geneticists' favorite organisms. *Drosophila melanogaster* (like many other dipteran or two-winged insects) has no crossing over at all in males. Thus, whatever linkage combination of alleles is found in a heterozygous male will be passed on without recombination in the sperm.

The mechanism by which *D. melanogaster* males accomplish normal chromosome segregation in the complete absence of chiasmata is its own active field of research. The loss of recombination in *Drosophila* males is probably not an extreme version of the reduced recombination in spermatogenesis overall, but rather its own independently evolved mechanism. This is one reason that all of our examples in the chapter use a female for the F_1 heterozygote used to examine recombination, and probably contributes to the historical fact that standard genetic maps are based on oogenesis. In *D. melanogaster*, the first organism for which map distances were compiled, the genetic map had to be based on females; genetic mapping in other organisms probably just followed this convention, even though chromosomes in both sexes cross over.

FIND OUT MORE

Coop, G. and Przweorski, M. (2007) An evolutionary view of human recombination. *Nat Rev Genet* **8**: 23–34

Recombination experiments involving more than two genes

For clarity, we have focused on crosses that involve only two genes; this is also appropriate for the way most genetic crosses are done in the laboratory. It is uncommon to do mapping experiments that involve multiple genes, simply because so many different types of offspring would need to be counted to provide statistically reliable results (that is, a geneticist will often have a genetic strain with three or four different mutant alleles in it but will rarely use it to count more than one or two types of offspring). Furthermore, once genomes are sequenced and all of the positions of genes are known, three-factor crosses are not necessary to determine the order and location of genes.

There is, however, extensive literature that used more than two genes to construct genetic maps, most often three-factor crosses, and it is important to understand the principles behind these experiments. The step-by-step directions for solving a three-factor cross are shown in Box 9.4.

Three-factor crosses illustrate an important concept about recombination that cannot be observed from two-factor data. Crossovers do not actually occur at random in the genome. The number and locations of crossovers are regulated by processes that are not well understood. In particular, it was observed from three-factor crosses that the occurrence of a crossover between two genes reduces the probability of a second crossover occurring nearby on the same chromosome; this is known as **interference**.

BOX 9.4 *Quantitative Toolkit* Three-factor crosses

Three-factor crosses illustrate some important concepts, including double crossovers and interference, so it is worth working through at least one problem involving three genes on a chromosome. Fortunately, there is an exact and sure-fire strategy for solving such problems.

Problem

Rollins A. Emerson crossed two different pure-breeding lines of corn and obtained a phenotypically wild-type F_1 that was heterozygous for three alleles that determine recessive phenotypes—***an*** determines anther; ***br***, brachytic; and ***f***, fine. He did not know the order of these three genes or the map distances between them. He test-crossed the F_1 with a strain that was homozygous recessive for all three genes and obtained the following offspring:

Anther	355
Brachytic, fine	339
Wild-type	88
Anther, brachytic, fine	55
Fine	21
Anther brachytic	17
Brachytic	2
Anther, fine	2

What is the order of these three genes, the map distances between them, and the rate of interference over this region of the chromosome?

Solution

The general structure of a three-factor cross is that an F_1 heterozygote whose genotype is not specified is test-crossed and the offspring are counted. (Alternatively, if the genes are X-linked, an F_1 female can be mated to any male and only the male offspring counted.) In any event, one parent contributes no dominant alleles, so we are examining the gametes produced by the F_1 triple heterozygote.

Step 1. Count the number of categories of phenotypes in the offspring, and group them in descending order. There should be eight categories of phenotypes—three genes with two alternative phenotypes each. If there are fewer than eight phenotypes, you need to determine which one is missing; this could be because double crossovers are infrequent or because some classes of offspring die. Notice that the phenotypic categories are paired (for example, one is anther but wild-type for the other genes, and the other is brachytic and fine but wild-type for anther). Notice also that the two phenotypes in a pair are about equally frequent, although Emerson's data were imperfect in this regard, with many fewer triply mutant corn plants (55) than there were wild-type plants (88).

Step 2. Identify the linkage arrangement in the F1 heterozygote by finding the most frequent pair of gametes. Remember that the parental combination will always be more common than recombinants. The most frequent pair has the genotypes of the original parents that were used to make the F_1 heterozygote and thus indicates the linkage arrangement, although not the gene order, in the F_1 heterozygote. In this case, the most frequent pair are anther and brachytic fine. Thus, the original parents were ***an br⁺ f⁺/an br⁺ f⁺*** and ***an⁺ br f/an⁺ br f***. The linkage arrangement in the F_1 heterozygote was ***an br⁺ f⁺/an⁺ br f***, but we don't know the order of the genes.

Step 3. Determine which gene is in the middle by finding the least frequent pair of gametes. Identify the least frequent pair, and compare the phenotypes to the most frequent pair. Note that any pair of gametes is different from any other pair of gametes by a change in one of the three genes. One of the three phenotypes will show a new linkage arrangement. The gene that is different in comparing the least frequent and most frequent pairs must require two crossovers to produce, so this gene is the one in the middle of the three. In our example, we see that fine was originally coupled with brachytic in the parent, but is coupled with anther (and brachytic$^+$) in the least frequent class. This means that ***f*** is in the middle. Rewrite the linkage arrangement of the F_1 to reflect this: ***an f⁺ br⁺/an⁺ f br***.

Step 4. Determine the distances between the middle gene and the two flanking genes. This step is exactly like the two-factor problems in the rest of the chapter. In our case, a crossover between anther and fine produces ***an f*** and ***an⁺ f⁺*** offspring. There are (55 + 2 = 57) ***an f*** offspring and (88 + 2 = 90) ***an⁺ f⁺*** offspring. So the total number of recombinants between ***an*** and ***f*** is 57 + 90 = 147. There were 879 total offspring, so the recombination frequency between anther and fine is 147/879 = 0.167, or 16.7%. The distance from anther to fine is 16.7 map units.

We do the same with the other interval. A crossover between fine and brachytic produces ***f br⁺*** and ***f⁺ br*** offspring. There are (21 + 2) ***f br⁺*** offspring and (17 + 2) ***f⁺ br*** offspring. So there are 23 + 19 = 42 recombinants between fine and brachytic, or 42/879 = 0.048, or 4.8 map units. Notice that the double crossover types (anther fine and brachytic) are counted twice, since they have a crossover in each interval. So we can write the map as:

an——16.7————***f***——4.8—***br***

Step 5. Calculate the amount of interference. In effect, this is asking if a crossover in one genetic interval (*anther* to *fine*) is independent of a crossover in a neighboring interval (fine to *brachytic*). Interference is a measure of the extent to which one crossover prevents the appearance of another crossover nearby. The expected number of double crossovers is the product of their separate map intervals, $0.167 \times 0.048 \times 879 = 7$. The observed number of double crossovers is 4. The coefficient of coincidence (c) is defined as the observed number of double crossovers divided by the expected number. In our case, there were four

BOX 9.4 Continued

observed double crossovers and seven expected double crossovers, so the coefficient of coincidence is 4/7, or 0.57. Interference is defined as 1 – c. In our example, 1 – 0.57 = 0.43, expressed as a percentage, so interference is 43%. This means that a crossover in one interval reduces the probability of a crossover also occurring in the other interval by 43%.

WEBLINK: VIDEO 9.1 illustrates the solution of a three-factor cross. Find it at **www.oup.com/uk/meneely**

The concept of interference is illustrated in Figure 9.16, which illustrates a chromosome on which the distance between genes *A* and *B* is 10 map units and the distance between genes *B* and *C* is 20 map units. Given these map distances, we expect that 0.10 × 0.20 = 0.02, or 2%, of the gametes will have a crossover in the interval between *A* and *B* and a second crossover in the interval between *B* and *C*. In other words, the frequency of *AC* crossovers should be about 2%.

But what happens in reality? The observed occurrence of these double crossover events is almost always **much less** than would be expected; in our hypothetical example, only 1.2% of the gametes had double crossovers. The observed number of double crossovers divided by the expected number is known as the **coefficient of coincidence**. Interference is defined as 1 – the coefficient of coincidence. In the example, the coefficient of coincidence is 0.012/0.02, or 0.6, and the interference is therefore 0.4, or 40%. Interference is usually expressed as a percentage, ranging from 0 (double crossovers occur as often as expected from the independent map distances) to 100% (double crossovers do not occur at all).

Interference is the common manifestation of the more general process by which the locations and numbers of crossovers are controlled, that is, a crossover event at one location on the chromosome reduces the likelihood of crossover event occurring at a different location. All eukaryotes (with the exception of a few fungi) exhibit interference for nearly all chromosomes, although the amount of interference varies with the organism and the genetic interval being tested. For *C. elegans* and some insects, interference is 100%; the presence of one crossover prevents the occurrence of any other crossover on the chromosome, so double crossovers are extremely rare. For most regions of human chromosomes, interference is 60–70%—the rate of double crossovers is reduced, but they do occur occasionally.

The molecular mechanism of interference is still largely unknown. One hypothesis is that the occurrence of a crossover produces a structural change in chromatin that makes other crossovers less likely. We remind you that, while we draw chromosomes as two-dimensional objects with the genes in a straight line, chromosomes in the nucleus are three-dimensional. Genes that lie far apart on

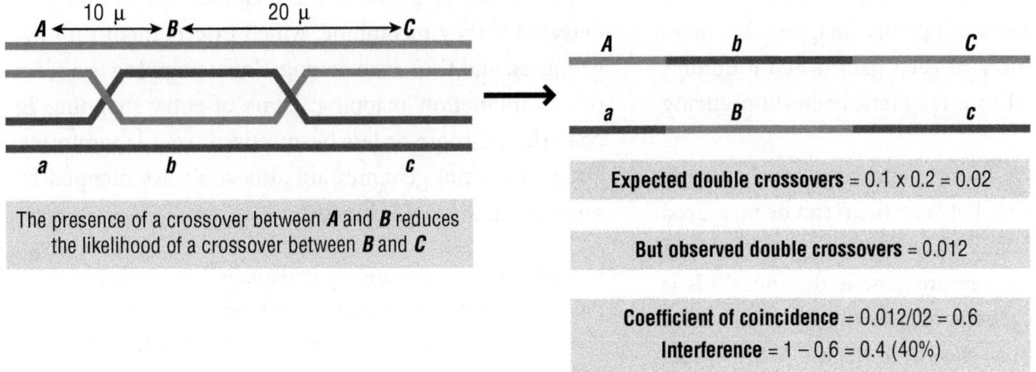
10 μ
20 μ
A
B
C
a
b
c
The presence of a crossover between A and B reduces the likelihood of a crossover between B and C
A b C
a B c
Expected double crossovers = 0.1 x 0.2 = 0.02
But observed double crossovers = 0.012
Coefficient of coincidence = 0.012/02 = 0.6
Interference = 1 – 0.6 = 0.4 (40%)

Figure 9.16 Interference. The presence of one crossover between a pair of homologs reduces the probability of a second crossover between them. The map distance between ***A*** and ***B*** is 10 map units, so the probability of crossover between them is 0.1. The map distance between ***B*** and ***C*** is 20 map units, so the chance of crossover between genes ***B*** and ***C*** is 0.2. The expected frequency of a double crossover between ***A*** and ***C*** is 0.1 × 0.2 = 0.02, the product of the two separate crossover frequencies. The observed frequency of double crossovers between ***A*** and ***C*** is less than this, a manifestation of interference. In this example, only 0.012 double crossovers were found. Interference is usually expressed as a percentage such as 40% interference in this cross. Since every homologous pair must receive a crossover for normal segregation, interference may be part of a mechanism to ensure that crossovers are distributed among the chromosomes.

the DNA sequence and our linear drawings could actually lie near one another in three-dimensional space because of how the chromosome is looped or folded. It is also true that the three-dimensional structure of chromosomes is highly dynamic, so the regulation of crossing over is not a simple spatial problem to solve.

KEY POINT Crossovers on the same chromosome are not independent events; the occurrence of a crossover between two loci decreases the chance of another crossover occurring close by on the chromosome. The amount of this interference varies between organisms and the location within the genome.

Mapping bacterial genomes

Linkage and genetic maps in eukaryotes are analyzed in diploids undergoing meiosis because the relative distance between genes is inferred from the frequency of recombination between them. However, all genomes can be mapped, including bacterial and viral genomes. Maps in viruses and bacteria do not involve meiosis, of course, but the relative positions of genes can be determined by other methods. Many viral genomes recombine within a cell, often by a molecular process similar to recombination during meiosis. Thus, viral genes and genomes can be mapped using analogous techniques based on recombination rates.

The genomes of bacteria, such as *Escherichia coli*, *Salmonella*, and a few other species, have been mapped by different processes. As will be described in Chapter 11, genes in these species can be transferred from a donor cell to a recipient cell via **conjugation**. Thus, these bacterial genomes were originally mapped by measuring the time of entry of each gene when a donor's genome is transferred to a recipient bacterium during conjugation.

Because genes move linearly and sequentially during conjugation, the distance between them can be measured in units of time, and thus early bacterial genomes used minutes as the unit to measure genetic distance. This is seen in the standard genetic map of the *E. coli* chromosome, part of which is shown in Figure 9.17. Genes needed for the synthesis of the amino acid leucine (*leu*) are found at 2 minutes, whereas genes needed for utilization of the sugar lactose (*lac*) are found at 8 minutes. This means that *lac* genes are transferred to the recipient 6 minutes later than *leu* genes in standard strains. (In bacteria, genes with closely related functions are often found closely linked on the chromosome in the form of an operon, so the map shows the location of the cluster

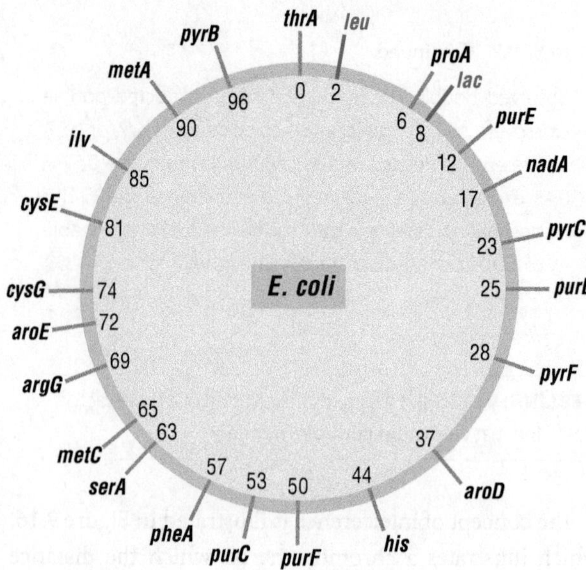

Figure 9.17 This genetic map of *E. coli* K12, a standard laboratory strain, is based on the time necessary to transfer a gene from one strain to another by the process of conjugation, which will be discussed in Chapter 11. The numbers on the circle are minutes; it takes approximately 100 minutes to transfer the entire chromosome if the transfer is not interrupted. If *thrA* is defined as the first gene to be transferred in a standard strain and is set at time 0, then *leu* is transferred 2 minutes later, and *lac* is transferred about 6 minutes after *leu* (shown in red). Time of transfer experiments for mapping are no longer commonly done in bacteria, since it is faster to sequence the DNA.

Source: Courtesy of Stanley Maloy, Center for Microbial Sciences, San Diego State University.

of *leu* genes and *lac* genes. Operons will be described in Chapter 14.)

The arrival of genes in the recipient cell can only be detected if they recombine, which inserts them into the genomes, and thus even conjugational mapping is a form of recombination mapping. Time-of-entry mapping in bacteria has more or less been retired from laboratories. Today, bacterial genomes are almost always mapped by sequencing the DNA directly.

KEY POINT Genetic maps in bacteria were originally based on the time needed for genes to be transferred during conjugation. Most maps in recent years are based on the DNA sequence itself, rather than genetic exchange.

Markers used for genetic maps

The key idea on which the construction of a genetic map is based is the concept of a locus. In order to keep track of the inheritance of the locus, some genetic variation is needed at that site. In the examples so far in this chapter, the variation at each locus was an easily observed phenotype arising from different alleles of genes such as purple vs. wild-type eyes or yellow vs. wild-type body.

But the same concept of a genetic map works with any phenotype and any locus. A cytological feature observed in the microscope, such as a knob found at a particular location on the chromosome, can be used as a phenotype, and recombination distances can be calculated from the knob. Most significantly, the DNA sequence itself can provide the phenotypes for mapping.

There are several different ways to use variation in the DNA sequence directly as the phenotype for mapping. Some methods for easily examining the DNA sequence at a locus include direct sequencing, cutting with restriction endonucleases that digest DNA at specific sequences (described in Tool Box 3.1), and using the different features on a microarray (described in Sections 2.5 and 10.3).

Regardless of the method used to obtain the sequence phenotype, molecular differences at the same locus on homologous chromosomes can be used in mapping just like any other phenotype we have described, as shown in Figure 9.18. In fact, for most mapping experiments in organisms with well-studied genomes, molecular markers

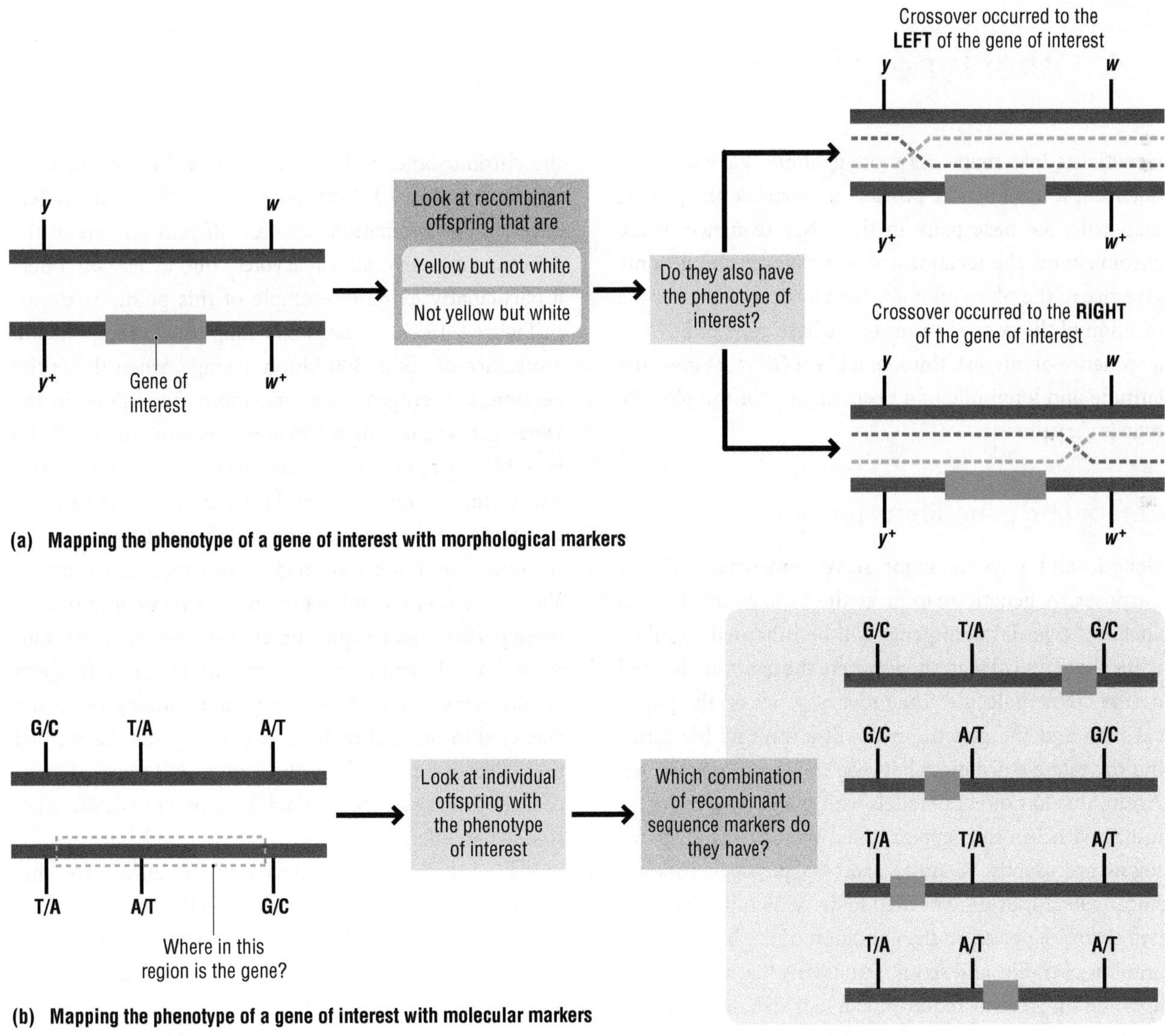

Figure 9.18 Mapping with morphological and molecular markers. One difference between mapping with morphological markers and molecular markers is the order in which the steps are done. With morphological markers, as shown in (a), the recombinant classes are selected and then tested for the presence of the phenotype of the gene of interest. For molecular markers, as shown in (b), the phenotype for the gene of interest is observed first, and then tested for the presence of molecular markers.

are much more widely used for mapping than the eye color or wing shape phenotypes described here.

Molecular marker phenotypes or sequence differences have several advantages over morphological phenotypes arising from different alleles. First, it is often easier to examine many molecular markers at the same time than it is to score many morphological phenotypes simultaneously, and scoring one marker does not prevent the scoring of another one. For instance, it is hard to score more than one allele affecting eye color in *Drosophila* in the same cross, but relatively easy to examine many molecular polymorphisms. Second, there are many more molecular markers available than there are morphological markers; hundreds of thousands of DNA sequence markers exist for a well-researched species. Third, molecular markers are less likely to reduce fertility or viability than are morphological markers. Fourth, as described in Section 9.4, molecular markers directly connect a genetic map to a physical map.

KEY POINT Any type of genetic variation at a locus can be used as a marker phenotype for mapping experiments. Many genetic maps use molecular markers, such as DNA sequence differences, as well as observed phenotypes.

9.4 Other types of maps

Geneticists love maps. As more genomes have been sequenced, it has become possible to correlate the genetic map with the base pairs in the DNA sequence of the chromosome; the locations of base pairs in the genome give rise to the **physical map**. The physical map gives the location of the gene in terms of the base pairs of DNA in a sequence of interest; think of it like a GPS that gives the latitude and longitude of a position, so that the physical map is very precise.

Comparing genetic and physical maps

Genetic and physical maps serve somewhat different purposes. A genetic map gives the probability that the alleles of two different genes will be inherited together, while the physical map shows where the genes are located on the DNA molecule. The order of genes on the physical map and the genetic map will always be the same, but the relative distances between the genes may not be. Again, this has direct parallels with maps that we are familiar with in other applications. We can say that two towns are exactly 28 miles apart—a physical distance. But the length of time needed to drive 28 miles between two towns depends on the condition of the highway, the amount of traffic, and so on. Two towns that are 28 miles apart on the freeway require about half an hour of driving time. But two towns 28 miles apart on rural roads might require more than an hour in driving time. The time to drive between them is analogous to the recombination or the genetic distance.

The exact connection between recombination distance and physical distance depends on the organism, the chromosome, and the location in the genome (and as noted in Box 9.3, probably on the sex). We see differences in recombination rate for different regions of the genome in nearly all eukaryotes, but *C. elegans* offers a particularly extreme example of this point, as shown in Figure 9.19. In *C. elegans*, 1 map unit corresponds to a distance of about 300 kb on average, when the entire genome is averaged. However, there are regions in the worm genome in which 1 map unit is more than 1300 kb (1.3 Mb). Crossovers do not occur frequently in this part of the chromosome, and thus the alleles of genes in this region are very likely to be inherited together. On the other hand, there are regions in which 1 map unit is about 70 kb, so recombination occurs very commonly in these parts of the chromosomes. The data from chromosome I are illustrative; crossovers are far more frequent on the arms of the chromosome than among the genes clustered in the middle in *C. elegans*. We emphasize that these data are drawn from the genetic and physical maps of *C. elegans* and that the data from other organisms are somewhat different.

It is clear from direct assays of the chromatin that regions with high rates of recombination differ in their chromatin structure from regions with low rates of recombination, but the origins of those chromatin differences and the explanations for such a difference in recombination rates are speculative. Nonetheless, differences in the rate of recombination in different parts of the genome are a common feature of many organisms. One common feature is that nearly all organisms have greatly reduced rates of recombination near the centromere.

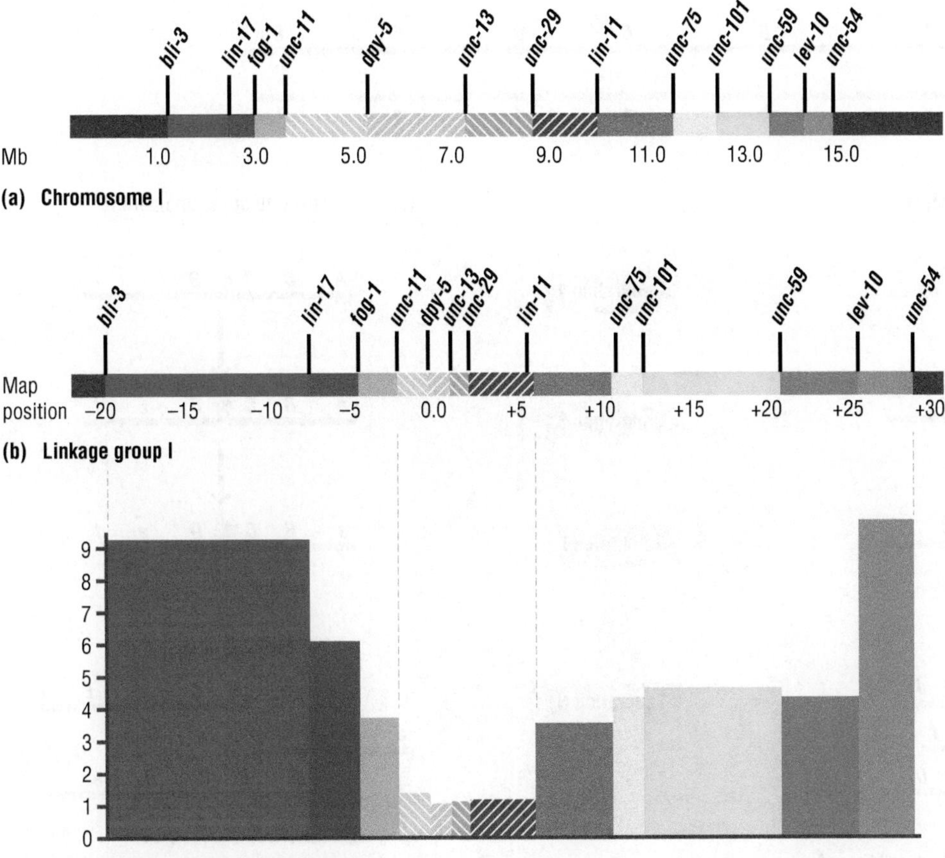

Figure 9.19 The correlation between the recombination distance and the physical distance varies for different regions of chromosomes. Virtually all eukaryotes have regional differences in recombination rate, but an extreme example is found for the autosomes in *Caenorhabditis elegans*, summarized here for some genes on chromosome I. The base pair positions of some known genes along the chromosome are shown in the physical map in (a), in which the numbers refer to the base pairs on the DNA molecule; *dpy-5* includes base pair 5,500,000, for example. The map based on recombination for each of the same intervals is shown in (b). Note that the genes in the cross-hatched intervals in the center of the chromosome are very tightly linked. In (c), the two maps are combined in a graph showing the recombination rate for each interval, that is, the number of base pairs (in millions) corresponding to 1 map unit. Note from (a) and (b) that the 2 Mbp between *bli-3* and *lin-17* have a map distance of 12 map units, whereas the 2 Mbp between *unc-13* and *unc-29* have a map distance of less than 1 because recombination occurs less often in this region. Data for this figure have been taken from Wormbase.org and redrawn.

Recombination rates and genetic diversity

It is informative to think about the recombination rates in different species and the effect this has on their evolution. Although we have presented recombination rates as they relate to inheritance in the next generation, recombination rates also affect the probability that a particular combination of alleles will segregate together over evolutionary time as well. Recombination is a primary source of genetic diversity since it produces new combinations of alleles.

Consider what happens when a new mutation arises on a chromosome, as diagrammed in Figure 9.20. It arises on a chromosome with a particular combination of alleles and linked sequence variants, suggested by the capital letters in Figure 9.20(a). Over time, recombination changes the alleles and sequence variants that are still linked to the newly arisen mutation. If the rate of recombination is high, as in Figure 9.20(b), the region that remains linked to the new mutation is relatively small, and many new combinations of alleles and sequence variants arise. Conversely, when recombination rates are low, as in Figure 9.20(c), the alleles and sequence variants that were linked to the new mutation when it arose will still be linked to it many generations later; fewer combinations of alleles, or less diversity among the alleles, will be observed. The combination of polymorphisms or alleles that

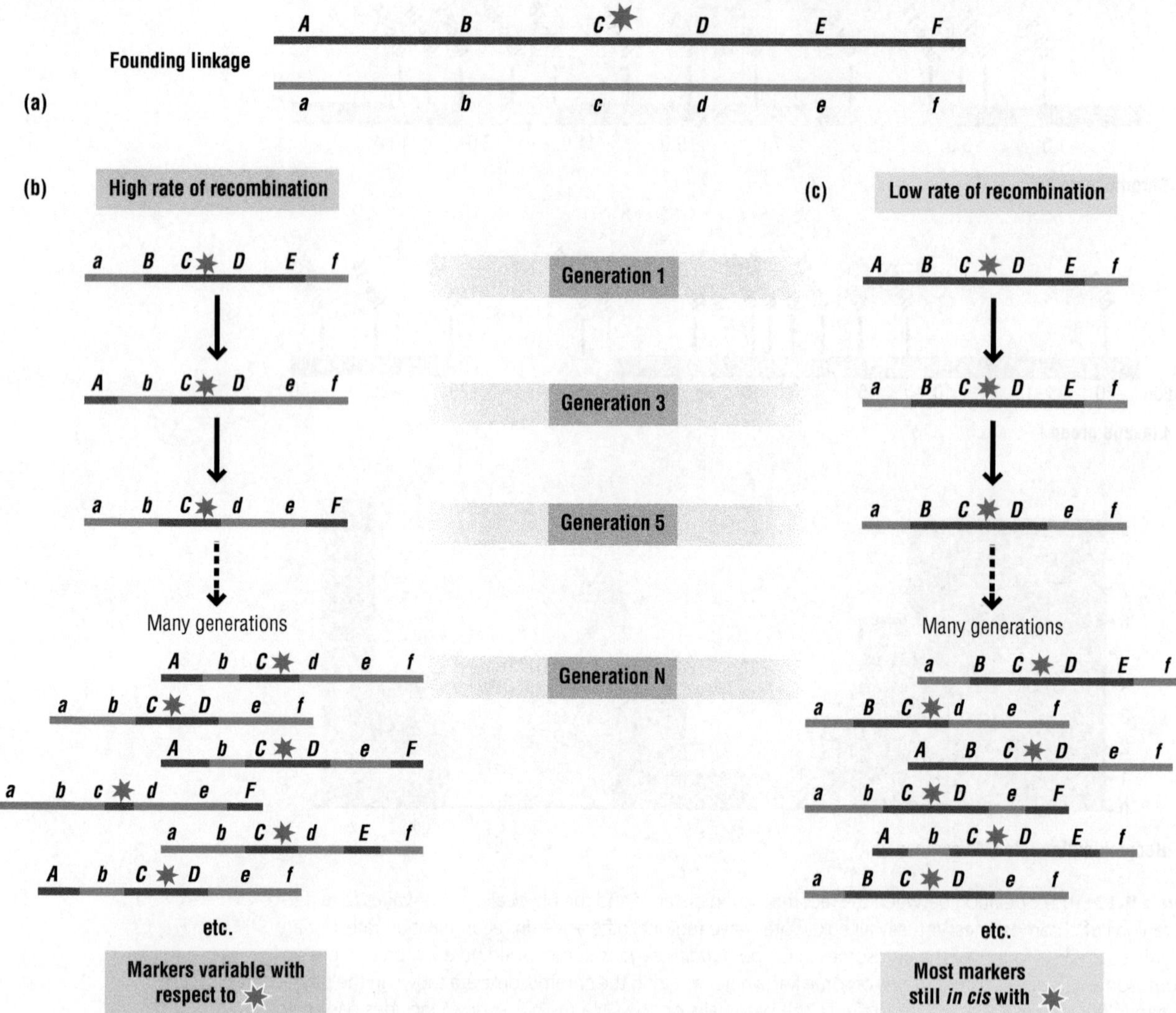

Figure 9.20 Recombination rates and genetic diversity. Variable rates of recombination will also affect the amount of genetic diversity in a population. A founding member of the population has the arrangement of markers shown in (a); the red star indicates the gene whose phenotype will be monitored in subsequent generations, or a newly arisen mutation. In organisms with a high rate of recombination, summarized in (b), the alleles linked to the red star marker are highly variable, and many new linkage combinations are found. In organisms with a low rate of recombination, as in (c), the original linkage arrangement of alleles and polymorphisms persists.

are observed to be inherited together over time is referred to as a haplotype.

The rates of recombination are so different in various organisms that it invites speculations and hypotheses. As a general rule, mammals typically have lower rates of recombination than invertebrates, and plants have higher rates than animals. In the yeast *Saccharomyces cerevisiae*, 1 map unit is only about 20 kb; yeast cells have a very high rate of recombination, so alleles that are linked in one generation are not likely to remain linked over many generations. Thus, the meiotic descendants of a yeast cell in nature that is heterozygous at many loci will show many different genetic combinations, so the population will consist of genetically diverse cells. By contrast, 1 map unit in humans is typically several million base pairs. Thus, the meiotic descendants of a person (more familiarly called a genealogy) will have many fewer different genetic combinations. For genes that are relatively close together, alleles that were linked in your great-great-grandmother are probably still linked in you. This concept is fundamental to Chapter 10 and will arise again in Chapter 16.

Recombination-based maps clearly have evolutionary significance because linked alleles are inherited together in subsequent generations. But the bigger picture connecting physical locations, recombination rates, and evolution is complex. It is clear that the rate of recombination is itself an example of genetic variation, since it is possible to select laboratory lines (of yeast or flies) with higher or lower rates of recombination than the standard wild-type rate. This ability to select for higher or lower rates suggests that the recombination rate in a natural population is the result of a balance of forces. If the rate is too low, unfavorable combinations of alleles will persist in the population; on the other hand, if the rate is too high, favorable combinations of alleles will be lost from the population.

It may be helpful to draw a parallel between the genetic diversity that arises from mutation at a single locus (discussed in Chapter 4) and that arising from recombination between loci. Mutations result in diversity in the DNA sequence and occur because the repair processes in the cell cannot repair all of the damage that may be present. It is possible to select cells with higher or lower rates of mutation than wild-type. Both of these variants have low viability, the high mutation rate lines because of the number of genes being mutated and the low mutation rate lines because the cells grow very slowly. It is not accurate, in a strict sense, to say that the role of mutation is to generate diversity; diversity at the sequence level arises as an outcome of other biological processes, none of which is absolutely precise, and which we recognize as a mutation.

Similarly, recombination results in new combinations of those sequence variations. Diverse combinations of alleles arise from the balance of separate forces that contribute to the rate of recombination. Genetic diversity is not the "goal" of the process of recombination, but rather an inevitable outcome of a complicated biological process that is under evolutionary pressure. Sometimes that diversity proves to be advantageous, and sometimes it is not, but it will always occur. We can speculate that plants gain an evolutionary advantage from a higher rate of recombination than do animals because they are more subject to the effects of their immediate environment; a genetically diverse plant population is likely to have some combinations of allele that will thrive when the environment changes, while animals are more able to move with the changing environment. But this is speculation. Recombination rates vary widely, and the explanation may not be simple.

Cytological maps

Recombination rates and physical positions are not the only types of data that geneticists use to make a map. For some organisms, such as *Drosophila*, corn, mice, and humans, we can also produce **cytological maps**. Cytological maps exploit the size of some chromosomes, as the chromosomes of many organisms are large enough that their physical characteristics, such as knobs or constrictions, can be observed under the light microscope. When stained with particular dyes, the chromosomes in many organisms, including humans, have a reproducible pattern of bands, which are numbered and named;

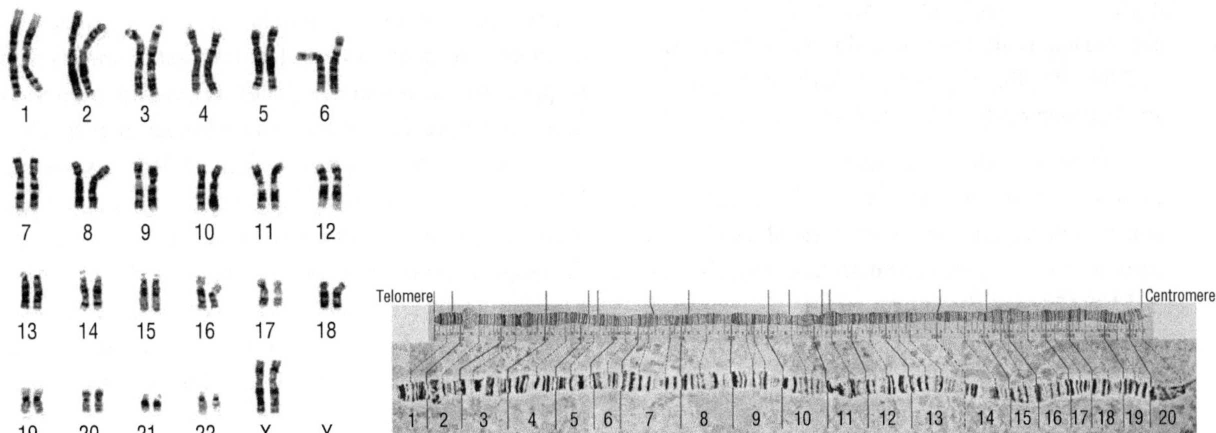

Figure 9.21 Cytological maps. Cytological maps are based on the physical structures of chromosomes and rely on banding patterns (as shown here for the polytene chromosomes of *Drosophila melanogaster*) or knobs and constrictions. Usually these bands are seen when the chromosomes are stained with particular DNA dyes such as DAPI or Giemsa; the G bands in the human karyotype are based on Giemsa staining.

Source: Reproduced from Schaeffer et. al. (2008) Polytene Chromosomal Maps of 11 Drosophila Species: The Order of Genomic Scaffolds Inferred From Genetic and Physical Maps. *Genetics* 179: 1601–1655

these bands create the cytological map. Two examples are shown in Figure 9.21. The G bands in the human karyotype arise from staining metaphase chromosomes with the dye Giemsa. The polytene bands in *Drosophila* salivary gland chromosomes arise from any DNA dye; the salivary glands replicate their DNA many times, without separating the strands, to create polytene chromosomes with as many as a thousand DNA molecules. In each case, the bands can be used as landmarks for viewing and analyzing the chromosomes.

Chromosomes with the centromere near the center are referred to as metacentric, while the ones with the centromere off-center are referred to as sub-metacentric or acrocentric. In humans, the short arm of the chromosomes is called p, and the long arm is called q; the bands are numbered based on the chromosome. Thus, the cytological location in humans is written like 7p12 or 10q35—band 12 on the short arm of chromosome 7 or band 35 on the long arm of chromosome 10. These are neither physical locations (like base pairs) nor recombination distances, but they do provide another landmark to locate genes on chromosomes.

To illustrate these different types of maps and how the information appears, here are the map locations of two genes commonly used as examples in the preceding chapters.

- The *white* gene in *D. melanogaster* is found at 1-1.5 on the genetic map, that is, it is on chromosome 1 (which, in *Drosophila*, is the X chromosome), one and a half map units to the right of the reference marker. Its cytological position is 3B6, indicating its position with respect to a series of bands on the chromosomes. Its physical position is X:2,684,632 ... 2,690,499, that is, it is on the X chromosome and between nucleotides 2,684,632 and 2,690,499.
- In humans, the ABO blood locus is found at cytological position 9q34, that is, at band 34 on the long arm of chromosome 9. Its physical location on chromosome 9 is between nucleotides 136,130,563 and 136,150,630. The human genetic map does not use fixed reference points for recombination, so it is hard to determine the recombination position from data in the Human Genome database. In fact, for many genes in humans, including the ABO locus, the recombination distance to other genes has not been accurately determined because of the small number of children born to any informative family and the effort required to compile recombination data.

The utility of maps

Geneticists put a lot of effort into creating different types of maps, so it is worth asking how these maps are used for experiments. Cytological maps in humans are widely used in amniocentesis and other procedures that involve the number and gross structure of chromosomes. Syntenic blocks can be used to reconstruct evolutionary linkage relationships between genes. Both physical and genetic maps provide important and helpful information, but they are used in different ways. This is no different from the way we use other maps; for example, sometimes we want to know the exact physical location, and sometimes we want to know how long it will take to drive there.

The most common use of a genetic map is to predict how often alleles at two different loci will be inherited together when performing or analyzing a cross or mating. Physical maps are not so useful for this purpose because the experiment is asking about inheritance, rather than about position.

The most common use of a physical map is to locate the position of a gene within the DNA molecule. A gene is often initially defined by its mutant or variable phenotype—white eyes in *Drosophila* or a gene for a genetic disease in humans, for example. Thus, a physical map is used to identify the DNA sequence that corresponds to a particular gene, as defined by its phenotype—that is, "to clone the gene" or to isolate the DNA sequence that encodes the gene. This is shown in overview in Figure 9.22(a) and in more detail for cystic fibrosis in Figure 9.22(b).

Many genes are cloned on the basis of their map positions, particularly for human genetic diseases. For example, the gene responsible for cystic fibrosis was mapped by recombination with respect to numerous molecular markers, and the two molecular markers closest to the gene on each side were identified, as shown in Figure 9.22(b). By knowing that marker A is on one side of the CF gene and marker B is on the other, and knowing the DNA sequence or phenotype of marker A and marker B, it was possible to isolate all of the DNA between A and B; the gene for cystic fibrosis must lie somewhere in that sequence. This type of **positional cloning** was, in fact, how the DNA sequence for the CF gene (formally known as the cystic fibrosis transmembrane conductance regulator gene or *CFTR*) was identified.

From within that region or DNA sequence, the investigators identified transcripts that were expressed in tissues known to be affected by cystic fibrosis to find a candidate gene, as shown in Figure 9.22(b). This

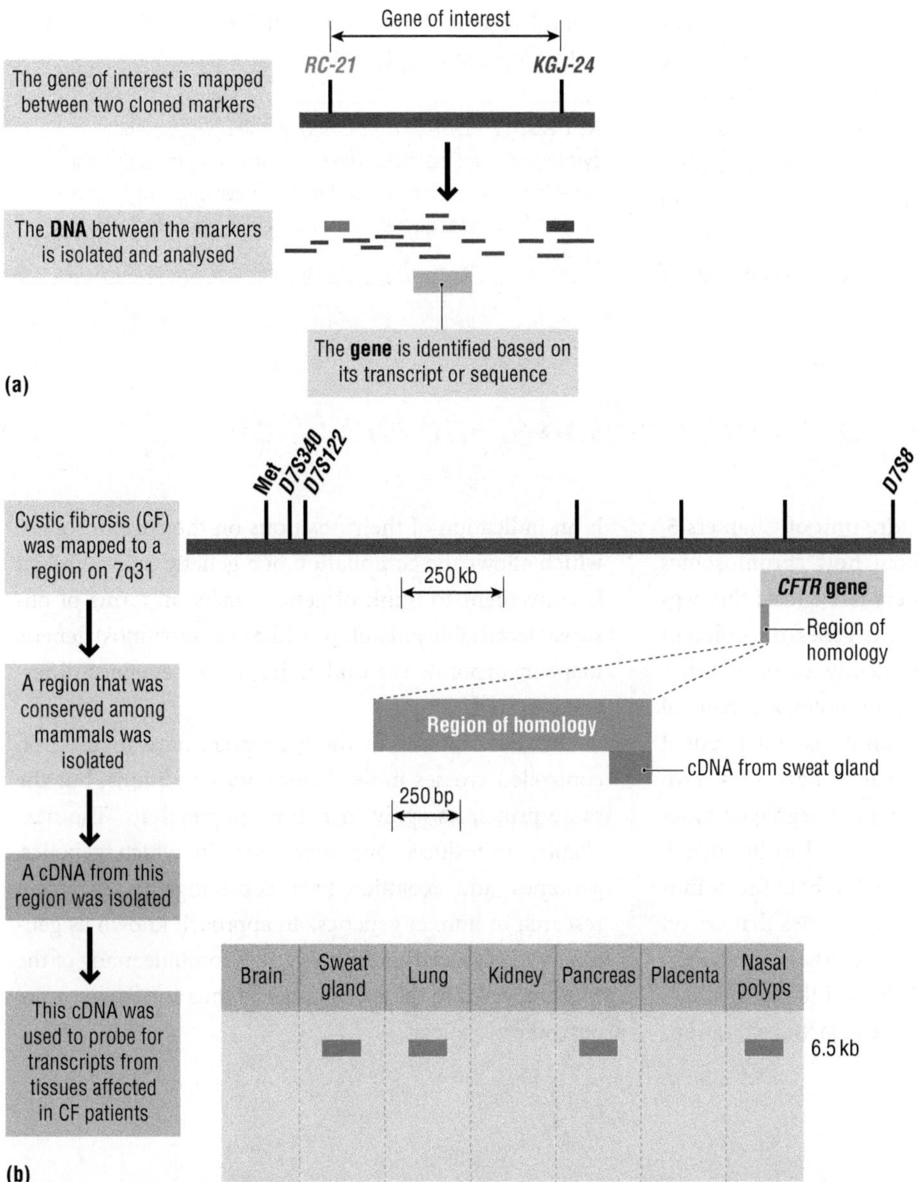

Figure 9.22 Positional cloning of a gene relies on a union of the genetic and physical maps. (a) presents an overview of positional cloning. The gene responsible for the phenotype, such as a disease, has been mapped between two markers whose molecular character is already known; these could be single-nucleotide polymorphisms (SNPs) or other sequence variants. The DNA between these two markers is then isolated and analyzed, a process made easier by genome sequencing. The gene of interest can be identified in this DNA sequence. (b) summarizes how *CFTR*, the gene responsible for cystic fibrosis in humans, was cloned based on its position. By examining many children with cystic fibrosis and their unaffected parents and siblings, *CFTR* was mapped with respect to several molecular polymorphisms on chromosome 7q31. The region between the markers, more than 1.5 Mb in length, was isolated. On the assumption that the gene would be evolutionarily conserved among mammals, a region of high sequence conservation was found and used to identify transcripts from different tissues on a gel. One transcript that was expressed in the tissues known to be affected by cystic fibrosis was found. This transcript corresponded to the gene responsible for cystic fibrosis, as shown by the fact that affected children had a change in its sequence that was not found in unaffected relatives.

candidate gene was then sequenced in children with cystic fibrosis and also in their heterozygous parents; all of them had a sequence change in the candidate gene, and the homozygotes had a sequence change in each copy of the candidate gene, as expected for a recessive trait like cystic fibrosis.

There are many molecular markers that can be used to position a gene and many ways to assay the candidate genes, including transcription patterns, evolutionary conservation, and ultimately the determination of the DNA sequence from affected and unaffected individuals. For cystic fibrosis and many other human diseases, the most time-consuming step in this process was to isolate the DNA between the two flanking markers. The Human Genome Project has performed this step for every region of the genome, so the identification and analysis of candidate genes has become much easier. We will return to this topic in Chapter 10.

KEY POINT Physical maps based on the DNA sequence, cytological maps based on the structure of chromosome, and genetic maps based on recombination have different purposes but can be used together to connect DNA sequences with phenotypes.

9.5 Summary: genetic maps, genomes, and evolution

Nearly all of the topics in the entire unit of Chapters 5, 6, 7, 8, and 9 are primarily about how chromosomes segregate at meiosis and the observable effects this segregation has on the offspring. Two genes on different chromosomes segregate independently of each other because the orientation of one homologous pair of chromosomes on the meiotic spindle is independent of the orientation of a different pair of homologs. Two genes on the same chromosome do not segregate independently of each other, so they are said to be linked. However, since crossovers are needed to hold the homologous pair of chromosomes together, alleles that are on the same homolog may end up on the other homolog in a subsequent generation. The likelihood that two alleles continue to segregate together in subsequent generations is an indication of their positions on the chromosomes, which allows the compilation of a genetic map. While it is convenient to think of genetic maps in terms of observable variable phenotypes, like eye color, most genetic maps incorporate the underlying DNA sequence differences as well.

Our descriptions in these chapters have focused on controlled crosses under laboratory conditions, but the same principles apply in natural populations. The next chapter introduces one more way in which genetics, genomes, and evolution have come together in recent research in human genetics, an approach known as genome-wide association studies. They combine many of the topics covered in these chapters and thus are yet one more outcome of meiosis.

CHAPTER CAPSULE

- A test cross involving an F_1 heterozygote carrying two recessive genes that are unlinked results in a phenotypic ratio of 1:1:1:1. A dihybrid test cross that does not result in equal numbers of four phenotypes in the F_2 generation indicates that the two genes must be linked and therefore are more likely to be inherited together.
- A test cross of an F_1 heterozygote with linked genes yields a large number of parental phenotypes and a small number of recombinant phenotypes.
- The frequency or percentage of recombination observed between two genes is often given in map units, also called a centiMorgan (cM). One map unit is defined as 1% recombination between two genes.
- The expected number of double crossover events is almost always **less** than would be expected, based on the frequency of each crossover individually; this phenomenon is known as interference.

- By using the relative positions of genes or molecular sequence variants with respect to each other, a genetic or recombination map of the genome can be compiled.
- Genetic maps are based on recombination frequencies of genes, while cytological maps rely on knobs or bands seen on chromosomes under the microscope. Physical maps give gene positions based on actual locations within the DNA sequence and thus are quite precise. Genetic and physical maps are often used to locate the DNA sequence corresponding to a gene affecting a phenotype of interest.

STUDY QUESTIONS

Concepts and Definitions

9.1 Define the terms "parental" and "recombinant" as they pertain to linkage.

9.2 For X-linked genes, an F_1 heterozygous female might be mated to a wild-type male, rather than test-crossed to a mutant male.

a. How is this conceptually similar to a test cross, and what has to be changed about how the data are collected?

b. What might be an advantage to using a wild-type male for mapping X-linked genes, rather than a mutant male?

9.3 In a two-factor cross, the longest possible map distance is 50 map units.

a. Explain why this is true.

b. How then can genetic maps show two genes that are 70 map units or more apart?

9.4 What is the difference between "linked" genes and "syntenic" genes, and how are both important in genetic analysis?

9.5 What is interference, and what does that tell us about crossing over?

9.6 Define the term haplotype.

9.7 What roles do genetic and physical maps play in genetic experiments?

Beyond the Concepts

9.8 For the past two decades, genetic mapping has usually involved genetic markers with molecular phenotypes, rather than genes with more "traditional" phenotypes. Discuss some of the advantages of using molecular markers for mapping that have made them the predominant ones used for mapping. Besides the assay used to determine the phenotype, what other part of the standard mapping procedure might be different when molecular markers are used?

9.9 For some organisms, linkage analysis is often done using an F_2 ratio from two heterozygotes, rather than doing a test cross. Consider a recombination experiment that mates F_1 offspring among themselves to produce an F_2, rather than using a test cross.

a. Which classes of offspring would be more and less common in such an experiment?

b. Can you think of one reason that an investigator would use an F_2 ratio, rather than a test cross? (Hint: This is commonly done with *Caenorhabditis elegans* and plants such as *Arabidopsis*, but rarely done with *Drosophila* or mice.)

9.10 In many organisms, it is possible to see the chiasmata that form during meiosis I. Suppose that a chromosome has two easily observed physical landmarks such as a knob at one end and a constriction that forms nearby. When the gametes are examined cytologically, 10% of them have a chiasma between the knob and the constriction. What is the approximate map distance between the knob and the constriction? (Be careful—the answer is not 10 map units.)

9.11 Consider three genes in *Saccharomyces cerevisiae*; the mutant alleles are ***a***, ***b***, and ***c***, while the wild-type alleles are a^+, b^+, and c^+. a^+ and b^+ are on different chromosomes, while a^+ and c^+ are linked. Draw the tetrads that arise under the following situations.

a. Strains that are mutant for ***a*** but wild-type for ***b*** are crossed to the ones that are wild-type for ***a*** but mutant for ***b***. The F_1 is allowed to sporulate, and the tetrads are isolated and analyzed. Diagram the expected result.

b. Strains that are mutant for ***a*** but wild-type for ***c*** are crossed to the ones that are wild-type for ***a*** but mutant for ***c***. The F_1 is allowed to sporulate, and the tetrads are isolated and analyzed. Diagram the expected result if the crossover does not occur between the ***a*** and ***c*** genes.

c. Strains that are mutant for ***a*** but wild-type for ***c*** are crossed to the ones that are wild-type for ***a*** but mutant for ***c***. The F_1 is allowed to sporulate, and the tetrads are isolated and analyzed. Diagram the expected result if a crossover occurs between ***a*** and ***c***.

9.12 How are balancer chromosomes (discussed in Tool Box 9.1) and haplotypes related to each other? What are some ways that they are different?

Looking ahead **9.13** *dpy-18* and *unc-32* are closely linked autosomal genes in *C. elegans*. A student crosses *dpy-18* males to *unc-32* hermaphrodites and then picks and cultures the wild-type F_1 hermaphrodites. She has no time to finish the experiment before the term ends, so her plates sit in the incubator for several weeks. What will be the most common phenotypes when she returns? Assume that all genotypes and phenotypes have equal numbers of offspring.

9.14 Recombination mapping was a common procedure in model organisms, such as flies, worms, mice, and *Arabidopsis*, about 20 years ago, and nearly all genetics graduate students mapped at least one gene in the course of their thesis. Nowadays, most genetics graduate students working with these model organisms do not map genes. Why is mapping no longer a common laboratory practice?

Challenging **9.15** The frequency of red–green color blindness (Box 9.4) is similar in most human populations. It also occurs spontaneously (that is, in families with no previous incidence of it) much more frequently than most other phenotypes. Why is the frequency of red–green color blindness much higher than most other spontaneously arisen traits?

9.16 What role does recombination-based mapping play in each of the following?

a. Predicting the outcome of a cross

b. Making a double mutant strain to work with in the laboratory

c. Positional cloning of a gene

Applying the Concepts

9.17 In *C. elegans*, *lon-2* and *unc-2* are recessive mutations that are 8 map units apart on the X chromosome. A hermaphrodite who is Lon and Unc is mated to a wild-type male. An F1 hermaphrodite is mated to a wild-type male. What are the expected percentages of the different phenotypes among the male progeny?

9.18 In corn, a colored aleurone is due to the presence of an ***R*** allele; ***r/r*** is colorless. Another gene controls the color of the plant, with ***g/g*** being yellow and ***G_***being green. A plant of unknown genotype is test-crossed, and the following progeny plants were obtained.

Colored green	89
Colored yellow	13
Colorless green	9
Colorless yellow	92

a. What was the phenotype and genotype of the plant used for the test cross?

b. What were the phenotypes of the plants used to produce the plant with the unknown genotype? (Assume that the parentals were homozygous.)

c. What is the approximate map distance between the *R* locus and the *G* locus?

9.19 In *C. elegans*, a recessive mutation that causes the worms to twitch is 6 map units from a recessive mutation that results in a dumpy phenotype. A twitcher dumpy hermaphrodite is mated to a wild-type male. A wild-type F_1 male is found and test-crossed to the twitcher dumpy hermaphrodite.

a. What fraction of the offspring is predicted to be twitcher, but not dumpy, males?

Challenging

b. The actual frequency of wild-type males from this cross is lower than expected, and lower than would be found if a wild-type F_1 hermaphrodite had been crossed to a twitcher dumpy male. (This cross can't be done, since twitcher dumpy males don't mate.) Explain why the number of twitcher males is lower if the heterozygote is a male than if the heterozygote is a hermaphrodite. (Hint: Similar results would be observed in most animals for most genetic intervals.)

9.20 Three genes in *Drosophila*—*black*, *dumpy*, and *vestigial*—are linked to each other. From a series of two-factor crosses, the following map distances are obtained.

dumpy–black	34 map units
dumpy–vestigial	50 map units
vestigial–black	19 map units

a. What is the order of these three genes?

b. *dumpy* is found at position 13 on the standard map of the chromosome. What would be the positions associated with *black* and *vestigial*, based on these data?

9.21 (Modified from Srb, Owen, and Edgar, *General Genetics*, 2nd edition, 1965) In tomatoes, round fruit is dominant over elongated fruit, and smooth skin is dominant over fuzzy or peach skin. An F_1 plant with round fruit and smooth skin is test-crossed. The following plants were found in the next generation.

Round and smooth	14
Round and peach	128
Elongated and smooth	124
Elongated and peach	13

a. What were the phenotypes of the true-breeding parental plants used to make the F_1 plant?

b. What is the map distance between these two genes?

9.22 Figure 9.6 reports the actual numbers for recombination between *yellow* and *white*, as done almost a century ago. While nearly equal numbers of the two parental phenotypes were found (4484 and 4413), the numbers for the two recombinant classes seemed a bit different from one another—76 wild-type and 53 both yellow and white.

a. Using χ^2 tests, as discussed in Chapter 5, test the hypothesis that the two classes of parental offspring are equally frequent. Repeat this test for the two classes of recombinant offspring.

b. Based on your χ^2 tests, what is an explanation for the results shown in Figure 9.6?

c. Could you devise another genetic cross, or another form of this cross, that tests your hypothesis in (b)?

9.23 In *Drosophila*, yellow body and crossveinless wings are X-linked recessive mutations, located 14 map units apart. You have a stock of yellow flies and a stock of crossveinless flies, and you want to make a double mutant that is both yellow and crossveinless. Describe the crosses that you would do to make this stock, including an estimate of what percentage of flies would be of the desired phenotype at each step.

Challenging **9.24** A male fruit fly with a black and hairy body (phenotypes due to two different genes) is mated to a female that has scarlet eyes. All of the F_1 flies, both males and females, are wild-type in phenotype. The F_1 females are test-crossed to a male that is black, hairy, and scarlet. The following offspring are found in the next generation at the indicated frequencies.

Phenotype	Frequency
Black, hairy, scarlet	2%
Wild-type	2%
Black and hairy	21%
Scarlet	21%
Black	2%
Hairy and scarlet	2%
Black and scarlet	21%
Hairy	21%

a. Draw a map showing the positions of these three genes with the map distances indicated.

b. Suppose that an F_1 **male** fruit fly had been test-crossed to a female that is black, hairy, and scarlet, rather than using a female F_1 fly. What phenotypes are expected in the next generation, and at what frequencies?

9.25 In *Arabidopsis thaliana*, resistance to chlorate is recessive to sensitivity, and a compact plant is recessive to normal height. The genes are 1.5 map units apart. A chlorate-resistant plant of normal height is crossed to a compact plant that is sensitive to chlorate.

a. The F_1 plants are grown and test-crossed. Among the plants arising from this test cross, what fraction will be compact and resistant to chlorate?

Challenging **b.** *Arabidopsis* can reproduce by self-fertilization. Suppose that the F_1 was allowed to self-fertilize. What fraction of the F_2 plants will be compact and resistant to chlorate?

9.26 In *Drosophila*, the gene for *yellow* body is found at position 0.0, and the gene for *forked* bristles is found at position 56.7 on the X chromosome. A female that is yellow and forked is crossed to a wild-type male; all of the female offspring are wild-type, while all of the male offspring are yellow and forked. An F_1 female is crossed to an F_1 yellow forked male. What percentage of the **males** in the next generation will be yellow but not forked?

9.27 In grapes, the wild-type *DND1* gene confers resistance to certain viruses; *dnd1* strains are susceptible to viruses. The *EDS1* gene confers resistance to powdery mildew, a type of fungal infection. The genes are linked, 8 map units apart, on chromosome 17. Different varieties of grapes have different pathogen resistance, but all original strains used in wineries are inbred and homozygous. A Pinot Noir grape that is resistant to viruses but sensitive to powdery mildew is crossed to a Cabernet variety that is susceptible to viruses but resistant to powdery mildew. The F_1 grapes are test-crossed to a variety that is susceptible to viruses and sensitive to mildew, and the seeds are collected and germinated into seedlings in a greenhouse. When the seedlings are small, the gardener exposes them to a virus and to mildew, so that only those that are resistant to both pathogens will be planted outside and grown into vines for grapes. He wants to have at least ten vines that are resistant to both pathogen. If the gardener tests 200 seedlings, is he likely to have at least ten plants that are both virus-resistant and mildew-resistant?

9.28 The fission yeast *Schizosaccharomyces pombe* is one of very few unusual organisms in which interference is zero. *S. pombe* can be grown as either a diploid that goes through meiosis normally or a haploid since the products of meiosis (known as spores in this yeast) can be plated onto medium and grown directly without the need to do a test cross. A haploid cell that is mutant for the *ade* gene cannot grow, unless adenine is added to the medium; a cell that can grow without added adenine is denoted *ade*+. A haploid cell that is mutant for the *leu* gene cannot grow, unless leucine is added to the medium; a cell that can grow without added leucine is written as *leu*+. On the standard genetic map, *ade* is located to the left of *leu* on chromosome 2.

An *ade* haploid mutant cell is crossed to a *leu* haploid mutant cell, and the diploid products of the cross are isolated. These diploid cells can grow on minimal media without additional adenine or leucine. The diploid cells are allowed to go through meiosis (sporulate), and the haploid spores are plated on various media to test their ability to grow in the absence of adenine and in the absence of leucine. The following results were found.

Phenotype	**Genotype**	**Number of cells**
Grows on minimal media	***leu+ ade+***	101
Require adenine, not leucine	***leu+ ade***	406
Requires both adenine and leucine	***leu ade***	97
Requires leucine, not adenine	***leu ade+***	396

a. What is the map distance between the *leu* gene and *ade* gene?

b. A *his* mutant cannot grow, unless histidine is added to the medium. On the standard genetic map, *his* is located 5 map units to the left of *ade*. A haploid cell that is mutant for *his* and *leu* (but wild-type for *ade*+) is crossed to a haploid cell that is mutant for *ade* (but wild-type for *his*+ and *leu*+). The diploid products of the cross are able to grow on minimal medium with no added leucine, adenine, or histidine. This diploid cell is allowed to go through meiosis. You isolate 1000 haploid spores from this meiosis and plate them on minimal medium with no added leucine, adenine, or histidine. **Since interference is zero**, how many of the 1000 haploid spores will be able to grow?

9.29 In wheat, the dominant alleles of **either or both** of two linked genes are needed for normal plant height. Plants that are homozygous recessive for both genes are dwarf. (In other words, only ***a b/a b*** is dwarf. Any other combination has normal height.) A true-breeding plant with normal height of genotype ***A b/A b*** is crossed to one that is ***a B/a B***. All of the F_1 plants have normal height. One of these F_1 plants is test-crossed to a dwarf plant, with the following results.

Normal height 230 plants
Dwarf 20 plants

What is the map distance between the two genes controlling height?

Challenging **9.30** A particular family has a multi-generational history of an unknown and painful pancreatic disease. So far, as family history can determine, the first person known to suffer from the disease was William, who lived from 1889 to 1972; William's wife Alma (1895–1979) did not suffer from the disease. A few years ago, one member of the family who suffered from the disease contacted a geneticist in an attempt to understand the disease and learn about possible therapies. In reconstructing the family history, the geneticist concluded that the disease was probably an autosomal dominant trait. The affected woman organized a large family reunion of all known descendants of William and Alma, who numbered in the hundreds, and included sons, daughters, grandchildren, and great-grandchildren, with many complicated relationships. The geneticist collected DNA samples from all of the cooperating descendants, classified them according to whether or not they were affected by the disease, and genotyped them for single-nucleotide polymorphisms (SNPs) at four different locations that had previously been associated with pancreatic diseases. Table Q9.1 shows the SNPs and their locations, and the fraction of people who had that SNP and who had the disease.

Looking ahead and back

A modified version of a true story

Table Q9.1

Location	SNP	Frequency of this SNP among the family	Fraction of people with the disease who have this SNP
1q17	G	0.08	0.47
	A	0.92	0.53
2p10	C	0.60	0.49
	T	0.40	0.51
3q5	A	0.27	0.13
	G	0.54	0.15
	C	0.19	0.72
4q12	G	0.43	0.35
	A	0.32	0.31
	T	0.25	0.34

a. Based on data from this family, which of these SNPs is most closely linked to the mutation that causes the disease?

b. Does this SNP cause the disease?

c. What was William's haplotype at that location?

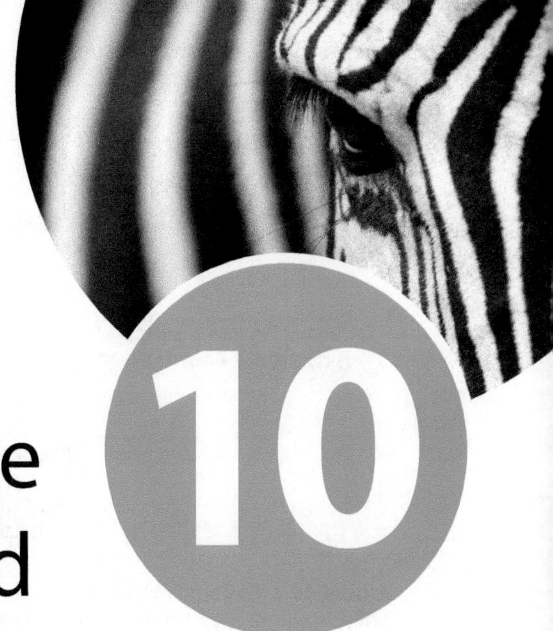

CHAPTER 10

Human Genetic Mapping, Genome-wide Association Studies, and Complex Traits

IN A NUTSHELL

Many human genetic traits and diseases are affected by more than one gene, as well as being influenced by environmental factors. These complex traits can be studied using genome-wide association studies, in which the genomes of thousands of individuals are analyzed for blocks of candidate genes whose presence is associated with the inheritance of the trait.

10.1 Overview of human gene mapping and complex traits

At the end of Chapter 9, we asked why geneticists make maps. The first, and probably most obvious, answer is that genetic maps and linkage are important tools that help us to predict the outcome of crosses and matings. A second answer is that genetic maps can be used in combination with physical maps to identify a gene and corresponding DNA sequences that affect a phenotype. For example, we might ask, "How was the gene that is mutated in Huntington disease identified? What sort of mutation has occurred in individuals who have Huntington disease?" The answer is that the gene and its causative mutation were identified by mapping. Identifying the gene and the causative mutation for this disease—and for dozens or hundreds of other diseases and traits—required both genetic and physical maps.

The use of genetic maps for gene identification has been at the core of molecular genetics for more than 25 years. In this instance, the "identification" of a gene refers to finding the DNA sequence that corresponds to a gene and using that DNA sequence to understand the phenotype. In the jargon of molecular genetics, making the connection between the gene that affects a phenotype (that is, the gene as Mendel identified it) and the DNA sequence (the gene as Watson and Crick identified it) is known as **positional cloning** of the gene, which was described in Section 9.4 and is summarized again in Figure 10.1. To

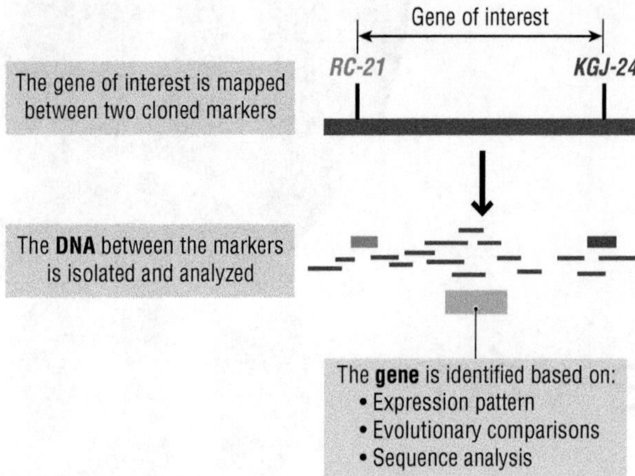
Gene of interest
RC-21
KGJ-24
The gene of interest is mapped between two cloned markers
The DNA between the markers is isolated and analyzed
The gene is identified based on:
• Expression pattern
• Evolutionary comparisons
• Sequence analysis

Figure 10.1 Positional cloning. A gene affecting a trait of interest is first mapped between two markers on the chromosome (here designated as *RC-21* and *KGJ-24*) whose DNA sequence and position are known. These markers could be genes that have been previously cloned or molecular polymorphisms, or a combination of both. All of the DNA between the two markers is then isolated and analyzed. The gene of interest must lie in this region, but there may be many candidate genes. Some other property of the gene of interest, such as its expression pattern, an evolutionary comparison with another species, or the sequence from affected and unaffected individuals, is used to determine which of the candidate genes is most likely to be the one affecting the trait.

return to the themes in the Prologue, identifying the DNA sequence for a phenotype connects the Great Idea on the Gene Concept with the Great Idea on Life as Chemistry; it opens up the possibility that a phenotype can be understood in terms of the underlying molecules that affect it.

Even before the Human Genome Project had started, dozens of genes in humans had been identified using genetic maps, that is, we knew the DNA sequence corresponding to a gene that gave rise to a particular phenotype. Most often, this was a disease phenotype, such as cystic fibrosis, Huntington disease, or sickle-cell anemia. Each of these diseases is the outcome of a mutation in a single gene, analogous in terms of basic biological principles to wrinkled peas or white eyes in *Drosophila*. We do want to always be aware that, while we can discuss traits like wrinkled peas and white eyes in terms of genetic principles, genetic diseases often have heartbreaking effects on people and families. While the genetic principles may be the same for humans and other species, the moral and ethical effects are quite different.

The Human Genome Project has made it possible to identify hundreds (or even thousands) of individual genes that affect disease phenotypes. These are catalogued in a vast and comprehensive database known as Online Mendelian Inheritance in Man (OMIM), the single most important resource for human genetics, found at **http://www.ncbi.nlm.nih.gov/omim/**.

Much of what we know about human genetics comes from genetic diseases rather than common phenotypic traits, but a few other examples, such as the ABO and Rh blood types, have also been important in the study of human genetics. The analysis of non-disease traits, such as eye color, and of traits with medical implications, such as responses to therapeutic drugs, are also now much more completely understood as a result of our analysis of the human genome. In addition, important and interesting information about human origins and human history has emerged from the Human Genome Project, an initiative that sequenced and mapped the human genome by 2003 (**http://www.genome.gov/10001772**). The Human Genome Project informs academic fields like anthropology, archaeology, and linguistics. So while our focus in this chapter will be primarily on human genetic diseases, the analysis of the human genome has allowed a much deeper understanding of many aspects of the overall biology of *Homo sapiens*.

WEBLINK; You can view a timeline of the Human Genome at **http://unlockinglifescode.org/timeline?tid = 4**.

KEY POINT The Human Genome Project resulted in a map of all known human genes, many of which affect disease phenotypes. Information about these genes is collected in the database Online Mendelian Inheritance in Man and is freely available.

OMIM deals principally with diseases inherited as **monogenic traits**, that is, those controlled by a single gene. All of us are carriers for single-gene diseases, but, because most of these diseases are rare, carriers for the same disease are unlikely to have offspring together. (There are many exceptions to this principle, which will be discussed in Chapter 16.) Thus, many of these monogenic traits remain rare.

We may not ever know anyone affected by some of these single-gene diseases, but all of us know about other traits or diseases that "run in families." Such so-called "familial" traits—traits that affect most of us such as cancer risks, cholesterol levels, hypertension (high blood pressure), and adult-onset (or type 2) diabetes—seem to be different somehow from these rare single-gene traits.

For instance, we can't describe easily the risk for hypertension in terms of being dominant or recessive or of being X-linked or autosomal like we can for monogenic traits. We also recognize that hypertension, unlike eye color in *Drosophila* or blood types in humans, is also affected by diet and other environmental factors. Hypertension is an example of a **complex trait**—a trait that clearly is familial and affected by our genotype but whose overall phenotype is affected by several or many genes, as well as by the environment. Since these genetic traits do not seem to be inherited as single genes, how can we use linkage analysis to identify any of the underlying genetic causes? That is one of the primary topics of this chapter.

KEY POINT A complex trait is a trait that is inherited but whose phenotype is affected by multiple genes, as well as by environmental factors.

Most of the material in this chapter has arisen from the Human Genome Project, with results compiled only in the past 10 years or less. Despite this very recent heritage, the use of linkage analysis to find the causative genes for many diseases and traits, referred to as a **genome-wide association study** or **GWAS**, now dominates the literature in human genetics. To describe this new approach, however, we need to go back to some traditional genetic principles that you have already encountered and reapply them with a contemporary genomic perspective.

Let's start with a single-gene trait whose inheritance is already clear to you, so that we can demonstrate this new approach.

10.2 Linkage and genome-wide associations

In linkage analysis, as described in Chapter 9, a heterozygous individual is test-crossed to one that has no dominant alleles, and the offspring are analyzed for the presence of recombinant phenotypes. By determining how often two phenotypic traits are inherited together—that is, how often they are linked or associated with each other—we can determine the map distance between the genes. This works very well in laboratory organisms that have many offspring in short periods of time, which are homozygous wild-type for every trait, except the ones being analyzed, and for which the matings are controlled. But none of these properties can be used for gene mapping in humans.

Mapping in humans

Genetic mapping in humans relies on some of the same concepts that underpin linkage analysis but uses a completely different method. Because each person has a small number of offspring, we cannot obtain the large sample sizes that are needed to detect linkage using a single set of parents and their offspring. In addition, humans are not homozygous for other traits in the genome like laboratory organisms are, and, of course, we do not want the matings to be controlled. These qualities make genetic mapping in humans difficult.

On the other hand, it is easy to get a lot of data from one person; every person's phenotype can be thoroughly described, and we have long historical records of families to allow the analysis of data for many generations.

So, although the methods are different, the ideas behind human genetic mapping are the same as those used for laboratory organisms—how often are two phenotypic traits (or the alleles that cause them) inherited together and associated with each other in the next and subsequent generations?

The most widely used method for genetic mapping in humans is called a **genome-wide association study** (**GWAS**). The principle behind a GWAS is simultaneously incredibly simple and too ambitious to be believed. A GWAS is essentially the following experiment. The investigators find thousands of people from the same population with the same phenotype which the investigators believe to be affected by genes, as well as others who do not show this phenotype. For our purposes, we will call the people who have the trait to be "affected" and those who do not show the trait to be "unaffected." While these terms sound like they refer to diseases (and they often do), they could be used more generally to include a trait such as "affected by red hair" or "unaffected by red hair."

The two groups with and without the trait are compared. The investigators test genetic differences—that is, alleles and polymorphisms—throughout the genomes of these affected and unaffected people. In fact, they test genetic variation in every part of the genome and compare all of this variation between people who do and do not have the trait (that is, non-red hair with red hair). By "every part of the genome," we mean about 500,000 or more different locations; in fact, the original experiments (done in 2007) tested more than 2 million locations in the genome to find the sites that had enough variation to be useful. The data are then used to ask, "What genetic variation or alleles are shared among people who show the trait but are not shared with those who don't have the trait?"

Or to put this in the language of some of the experiments, "How often is variation in this particular region of the genome associated with the occurrence of this trait?"

KEY POINT Genome-wide association studies compare the genomes of affected and unaffected individuals, looking for alleles that are shared among affected individuals but not among those without the trait.

Before going on to discuss how this is done in humans, let's imagine this in terms of mapping white eyes and yellow body in *Drosophila melanogaster*, genes that were mapped in Chapter 9. This situation is depicted in Figure 10.2. Suppose that we began with one fly that had white eyes and a wild-type brown body and with the other parent that had wild-type red eyes and a yellow body, just as we did in Chapter 9. We let them mate, then let their offspring mate, and so on for numerous generations, before we examine our flies. Even though we have not controlled the matings or even counted the number of flies, we would find that most of the flies with white eyes also have a wild-type body, whereas those that have red eyes have a yellow body, that is, the two traits have remained associated with each other, just as they were in the founding population.

The number of white and yellow flies—that is, the association between the alleles for *white* and *yellow* genes—will depend on the recombination rate between the *white* and *yellow* genes and the number of generations since the founding parents. But we will still find more flies with white eyes and a wild-type body (and correspondingly, more flies with red eyes and a yellow body) than white-eyed yellow flies and non-white-eyed non-yellow flies—parental phenotypes will always be more common than recombinant ones.

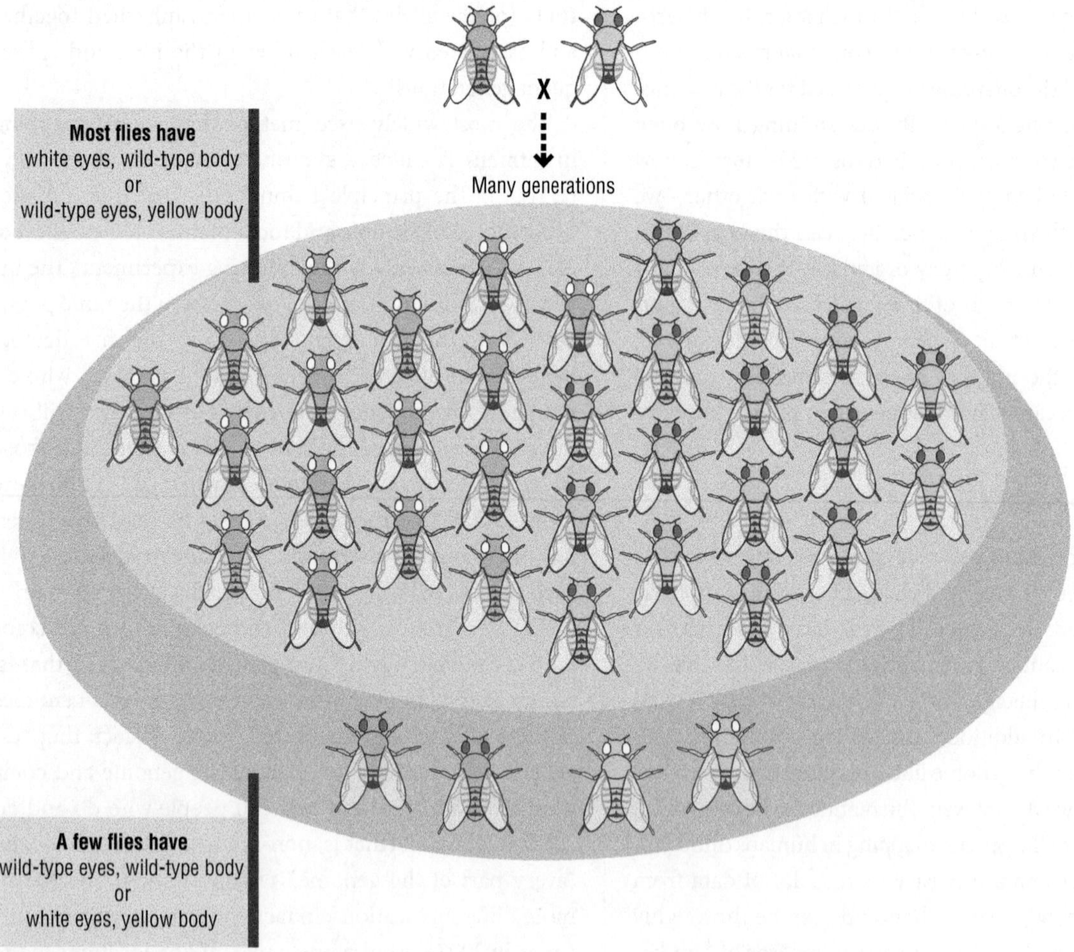
X
Many generations
Most flies have
white eyes, wild-type body
or
wild-type eyes, yellow body
A few flies have
wild-type eyes, wild-type body
or
white eyes, yellow body

Figure 10.2 Association and linkage. As in Figure 9.7 in Chapter 9, white-eyed flies with a wild-type (WT) body are mated to wild-type red-eyed flies with a yellow body. Because these genes are linked, the traits will remain associated with each other many generations later, even if the individual matings are not known. The association between the traits allows us to recognize that the two genes are linked, as well as the genotypes in the founding parental population.

In fact, we can go a bit further with our thought experiment. Suppose that you found a crowded vial of flies in the back of an incubator. You don't know the genotypes of the flies in the original cross or when the cross was done, and you certainly can't account for all of the possible matings that have occurred since the first one was done. Nonetheless, when you count the flies, you find that most of the ones with white eyes have a wild-type body, whereas most of the ones with red eyes have a yellow body. Thus, by reasoning backwards from the generation that you can see, you can infer the genotypes of the original parents in the founding population, without knowing them, and detect an association between the two genes. The technical (and somewhat confusing) term for this is **linkage disequilibrium**, that is, the two genes and their phenotypes have a persisting association—a "disequilibrium"—due to the fact that the genes are linked. This is an important idea for understanding GWAS and is explained in more detail in Box 10.1.

KEY POINT Associations or linkage between genes can still be detected after many generations have passed.

BOX 10.1 *Going Deeper* Linkage disequilibrium

Linkage disequilibrium (LD) is one of the most fundamental concepts in the analysis of the genomes of populations. The concept itself arises directly from the fact that genes are linked to one another, and it is not so difficult to grasp; the name probably makes the underlying idea seem more difficult.

Imagine two genes *A* and *B*, each with two alleles ***A/a*** and ***B/b***. (While we know that a gene has multiple alleles, we can always consider it as having two alleles, one of which is ***A*** and the other is all other alleles, except for ***A***.) We are going to ask about the association in inheritance between the *A* gene and the *B* gene. We sample the genome and determine the number of gametes that have ***A***, the number that have ***a***, the number that have ***B***, and the number that have ***b***. These are known as the allele frequencies and will be discussed in Chapter 16. We are using gametes because they are haploid and have only a single allele for each gene.

Suppose in our test that we find that 70% of the gametes have ***A***, while 30% have ***a***; likewise, our testing shows that 60% of the gametes have ***B*** and 40% have ***b***. We expect that, if *A* and *B* are genes on different chromosomes, that 42% (0.7 × 0.6) of the gametes will be ***A B***, that 28% (0.7 × 0.4) of the gametes will be ***A b***, that 18% (0.3 × 0.6) will be ***a B***, and that 12% (0.3 × 0.4) will be ***a b***. These will be the frequencies of these gametes, and it will not change from generation to generation; the frequencies are at equilibrium. Table A summarizes this information.

Table A Expected frequencies (no LD)

	A 0.7	*a* 0.3
B 0.6	*A B* 0.42	*a B* 0.18
b 0.4	*A b* 0.28	*a b* 0.12

But our observed frequencies of each of the gametes are different, as shown in Table B.

Table B Observed frequencies (LD present)

	A 0.7	*a* 0.3
B 0.6	*A B* 0.56	*a B* 0.04
b 0.4	*A b* 0.14	*a b* 0.26

Note that our frequency of the ***A*** allele is still 0.7 and that of the B allele is still 0.6, but the frequencies of the ***A B***, ***a B***, ***A b***, and ***a b*** gametes are not what we expected. In other words, the frequency of the ***A*** allele is not independent of the frequency of the ***B*** allele—we are more likely to find ***A*** associated (or inherited) with ***B*** and ***a*** associated with ***b*** than would otherwise have been expected. We encountered this exact situation at the beginning of Chapter 9, and we recognized its origin—the ***A*** gene and the ***B*** gene are linked to one another physically, so they are not inherited independently.

Furthermore, we can conclude that the original heterozygote must have had one homologue with the ***A B*** alleles on it, and the other homologue with the ***a*** and the ***b*** alleles on it, because those are the most common types.

Because the *A* and the *B* genes are linked, we cannot know precisely what will happen to the frequencies of the different combinations in subsequent generations. The frequencies will depend on how closely linked the two genes are and how many generations have passed. (If we know one of these two parameters, we can estimate the other. In Chapter 9, we knew how many generations had occurred—only 1. So that allowed us to calculate the

BOX 10.1 Continued

linkage. If, on the other hand, we had known the map distance between them, we may have been able to estimate the length of time since the founding population. This is the type of calculation that allowed the estimate of when the Δ-F508 allele that caused cystic fibrosis occurred because the approximate recombination distance to some nearby markers was known.) The frequencies of the different combinations of alleles will change over time because of linkage and crossing over—their frequencies are not at equilibrium.

The term linkage disequilibrium might be more clearly described as the "unexpected observation of certain allele combinations because the genes themselves are linked." While that might be a clearer description, it would not be nearly as easy to say and write, so LD persists.

While the preceding paragraphs described the logic of the experiment, the design of a GWAS itself might seem more complicated. After all, we don't know the map location of the genes, so all chromosomes have to be tested. Not only that, but **all regions** on all of the chromosomes have to be tested. That requires many markers and, just as with two-factor mapping in Chapter 9, the markers have to be heterozygous. In fact, we don't even know for sure if there is a gene that affects this phenotype, only that people differ in this trait.

Thus, in order to understand how a GWAS works, we need to explain some key facts about the genetic structure of human populations and the human genome. Some of this background was introduced in Chapter 3 and some appeared in Chapter 8, so it may be familiar to you. We also need to reintroduce you to a method to quickly assess genetic variation at many hundreds of thousands of loci throughout the genome—namely the microarray, a method that you have already encountered being used for a different purpose in Chapter 2.

The amount and location of genetic variation in humans

Recall from Chapter 9 that mapping begins with heterozygotes. Thus, the first fact that makes the audacity of GWASs even possible is that our genome has many heterozygous loci that can be used for mapping. As we noted in Chapter 3, most of these loci are not actually genes, however; for example, less than 2% of the human genome encodes genes that produce proteins. Most human DNA sequences are not parts of a gene but the regions between the genes, or what we called "other" in Chapter 3. Molecular differences in these "other" intergenic regions of the genome are the most widely used markers for mapping in humans.

Variations in the DNA sequence are known as **polymorphisms**, whether they occur in genes or in "other" regions of the genome. There are many different kinds of molecular variations or polymorphisms, including copy number variations (CNVs) and insertion/deletions (indels), but the ones used most commonly for mapping currently are known as **single-nucleotide polymorphisms** or **SNPs**, located in regions of the genome that are not genes, as illustrated in Figure 10.3.

An SNP simply means that, at a given location in the genome, some people in the population have one base pair (A:T, for example), while others have a different base pair (C:G, for example). Of course, if the genomes of two unrelated people are compared to each other, SNPs are found at millions of different locations. However, when the genomes of a population of people from the same geographical region are compared to the others, polymorphisms are often found at the same sites, that is, there are certain locations in the genome that are polymorphic within a particular population, a concept introduced in Section 8.3 when we described multiple alleles.

Remember that, for a locus to be classified as polymorphic, the second most common polymorphism must be found at that location on 1% or more of chromosomes. (In population studies, this parameter is often abbreviated as **MAF** for **minor allele frequency**.) Many of the millions of SNPs that would be found in a person's genome are so-called "private polymorphisms"—rare variants found only in that person and immediate relatives. Because their frequency is less than 1%, rare variants would not be useful for a GWAS in the population, although they may be very important for pinpointing the best candidate genes later in the analysis.

But many other SNPs are found at common sites that are shared among unrelated people from the population. In fact, as much as 90% of our polymorphisms are found at sites that are polymorphic among people who originated on the same continent. In other words, if you are Han Chinese, you have a million or more polymorphisms

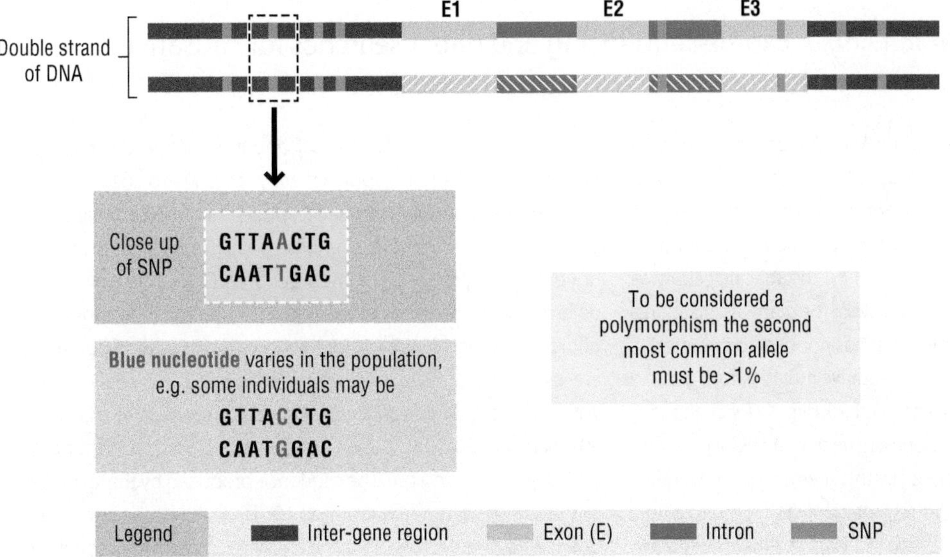

Figure 10.3 Single nucleotide polymorphisms (SNPs) used for mapping are generally not within genes. For a polymorphism to be a useful mapping tool, the second most common variant must be present in more than 1% of the population. The gene is shown with three exons and two introns. Polymorphisms, in blue, are found in many regions, but most are not within the gene itself. In this illustration, most individuals have an A (and complementary T) at the site indicated, but some have a C (and complementary G). Individuals with a G or a T at the variant position would comprise less than 1% of the populations, since these are rare variants.

in your genome that differ from Caucasian genomes. But only about 100,000 of these are found in you and your immediate family members; the other 900,000 are shared with other Han Chinese.

KEY POINT SNPs—single-nucleotide polymorphisms—are single-base changes in the DNA and are the molecular changes most commonly used for genetic mapping in humans.

One advantage of using intergenic regions as our mapping markers, rather than genes, is that these polymorphisms are often selectively neutral; since these regions may have no functions, changes in the sequences make no difference to the health or well-being of the person that might otherwise affect their frequency (as will be discussed in more detail in Chapter 16). A catalog of more than 10 million polymorphisms whose locations are known and that can be used for mapping has been developed, and corresponding catalogs have been used for many different human populations that may not have the same polymorphisms as each other. Thus, the phenotype that we are using for mapping is the DNA sequence itself, rather than white eyes and yellow body, but the inheritance and linkage principles are exactly the same as for conventional alleles.

10.3 Microarrays and polymorphisms

It is one thing to be able to say that millions of polymorphisms exist and that many people are heterozygotes, but now we have to ask how the sequence differences can be scored or assessed. After all, we can't simply look at a person and tell if he has an A or a C at a polymorphic locus in the same way that we can tell if a fly has red or white eyes. In addition, it is not enough to score one polymorphic locus—in order to find an association, the investigators will want to score **all** of the genome, or at least as many different polymorphic loci as possible. Furthermore, they will want to do this as rapidly and as cheaply as possible.

It probably will not be very long until we will be able to sequence the genomes of different people rapidly and cheaply enough to be able to use those sequences directly. Box 10.2 describes one of the first diseases to be analyzed by such direct sequencing methods. Indeed, a stated goal of the Human Genome Project is to make it possible to

BOX 10.2 *A Human Angle* Exome sequencing and direct searches for causative mutations

Genome-wide association studies (GWASs) currently dominate the research literature in human genetics, but they have some limitations, as noted in the chapter. Foremost among these is that, while a GWAS identifies regions of the genome associated with a trait or a genetic disease, it does not find the specific causative gene. That is done by follow-up analysis; for many associations, the actual causative gene or mutation has not been pinpointed.

An alternative approach to find a causative mutation for a genetic disease is to bypass linkage analysis and look directly for changes in the DNA sequences of affected and unaffected individuals. At first glance, a direct sequence analysis like this seems to be a daunting task for the identification of human mutations giving rise to either single-gene or complex traits. Two particular problems immediately come to mind. First, the human diploid genome comprises 6 billion base pairs, so it is much larger than the genome of model organisms such as worms and yeast. Second, unlike laboratory organisms, genomic variation between unrelated people can be expected at hundreds of thousands, or even millions, of sites in the genome, most of which have nothing to do with variation in the trait being analyzed. While we have a reference human genome—in fact, several individual genomes—none of them could be considered a "wild-type" from which other genomes have been derived, as is done in model organisms. Even if the genomes of affected and unaffected individuals could be sequenced, identifying the one causative mutation among all that genetic variation is a formidable task.

Obtaining the sequences

Although these are huge challenges to overcome, the analysis of genetic diseases based on direct genome sequencing is progressing quickly. As the costs of sequencing have decreased dramatically, the feasibility of rapid genome sequencing at relatively low cost has improved. Furthermore, improved filtering methods have been developed to identify the best candidate mutations among the vast array of genetic variation when human genomes are compared.

The initial approaches to directly compare the DNA sequences focused on sequencing the exons of affected people, rather than the entire genome, a procedure known as **exome sequencing**. Exome sequencing offers a direct connection between molecular changes in a gene and the disease syndrome. Exons comprise less than 2% of the total genome sequence, so focusing on them greatly reduces the amount and the complexity of sequence information that is compiled and analyzed. The trade-off for this reduced complexity is that any mutations affecting regulatory regions are missed. For many single-gene genetic diseases with severe phenotypes, this is an acceptable compromise. Mutations in exons are likely to be the mutations with the most profound effects on the function of the gene's protein product, since they can cause amino acid changes or polypeptide chain termination.

The human exome consists of about 30 Mb in the haploid genome, spread among approximately 180,000 exons, so the first stage in exome sequencing is to "capture" the exons from among the entire genome of 3 billion base pairs. Exon capture can be done by hybridizing fragmented genome DNA to two or more different exon microarrays specifically constructed to represent all of protein-coding regions in the genome. Hybridizing fragments are extracted, purified, and amplified by PCR for sequencing. With current high-throughput sequencing methods, as much as 30 gigabases (Gb; 30×10^9) can be generated from one machine in a 10-day sequencing run. The sequence produced by these methods is very short, typically less than 100 bp in length, and each sequence has some errors. Nonetheless, because so many sequence data are generated so rapidly, it is feasible to have each fragment in the exome sequenced 30 to 40 times. While any given exon might be sequenced fewer than 30 times, the multiple passes minimize the impact of errors in any single sequence fragment.

Filtering the sequences

The first use of exome sequencing to identify the causative gene for a genetic disorder was for Miller syndrome. This procedure will be described in some depth. Miller syndrome is a rare genetic disease characterized by craniofacial abnormalities. Only about 30 cases are known worldwide, so the exact mode of inheritance is not clear. The inheritance is consistent with the mutation being an autosomal recessive trait, but an autosomal dominant trait with reduced penetrance could not be absolutely ruled out. Thus, the analysis had to include both possibilities. There are only three known families with two affected siblings, so these families formed the core group for sequence analysis.

The first step was to obtain the exome sequence of two individuals in one family; about 164,000 regions could be sequenced, representing 27.9 Mb and about 96% of the exons. This amounted to 5.1 Gb of DNA sequence per individual, in lengths of about 76 base pairs, or an average 40-fold coverage of their exons.

Many different types of polymorphisms are seen when the two siblings are compared. Because this is a severe disease with profound phenotypic consequences, the authors made the reasonable assumption that the causative mutation in the gene must substantially alter or disrupt the function of the protein product. Thus, mutations that resulted in amino acid substitutions (that is, non-synonymous variants), that altered a splice site, or that inserted or deleted part of the coding region (that is, indels), were compared between the siblings. Each affected child in this study had about 4600 such variants in his genome. The flow chart for identifying the causative mutation from among these variants is shown in Figure A, which has been reconstructed from the original data.

BOX 10.2 Continued

Sequence exomes ~40X coverage

↓

Filtering sequence	IF recessive			IF dominant		
	One child	Two siblings	Plus unrelated	One child	Two siblings	Plus unrelated
Identify non-synonymous mutations, splice variants, indels	2860	2360	1810	4600	3900	3100
Filter out ones found in common polymorphism database	31	9	1	456	228	26
Predict which ones would be damaging	6	1	0	204	83	5

Not this disease gene

Figure A A flow chart summarizing exome sequencing and the causative mutations for Miller syndrome. The exons of affected children were sequenced, with fragments sequenced an average of 40 times to minimize the effects of sequencing errors; while the average coverage was 40 times, the minimum coverage for each fragment was at least seven times. The sequence variants were then filtered in various ways, as shown in the box on the left, with the results from each filter shown by the numbers in the tables on the right. Two different tables are shown, depending on whether the trait is inherited as recessive (as it subsequently proved to be) or autosomal dominant. By reading down the columns in the tables on the right, the effect of each filter on the data can be seen; by reading across the rows, the effects of additional affected individuals can be seen. Note that the program used to predict the most damaging mutations pointed to a mutation in a different gene in the original family, a mutation not shared by other affected children. This mutation was in a different gene for an unrelated phenotype that the siblings shared. More recent analyses have used refined versions of the protein structure prediction program and have been more successful.

Now the question of whether the disorder is inherited as a dominant or recessive trait becomes an important part of the analysis. If the disease gene is an autosomal dominant trait, then the affected siblings only need to share one mutation; in this study, 3900 of the 4600 sequence changes were shared between the two affected siblings. On the other hand, if the disease is an autosomal recessive trait, then an affected individual has to have mutations in both alleles of the gene; furthermore, these variants would be shared between the affected siblings. Note that the mutations need not be the same mutation in the gene; in fact, since the parents were unrelated, the mutations are expected to be different. (In standard genetic terminology, the affected individuals are said to be **heteroallelic** or **compound heterozygotes** for two different recessive mutations, but you can think of them as being homozygous.)

In this study, each child was homozygous for 2860 variants, of which 2360 were shared between the two siblings, as summarized in Figure A. This large number of common variants reflects that siblings have half of their genome in common on average. One of these shared variants is the Miller syndrome disease gene, but there are many other shared variants as well.

The next step was to apply some filtering steps to the data; the effectiveness of these filtering steps is diagrammed in Figure B(i) and (ii). Panel (i) summarizes the effectiveness if the trait is inherited as a dominant, while Panel (ii) summarizes the effectiveness if the trait is recessive. There are databases of known common polymorphisms for particular populations; any polymorphism found in these databases can be ruled out as a candidate mutation, since Miller syndrome is so rare. Of the 3900 single variants shared by the affected siblings (Panel (i)), only 228 variants were not found among the population at large; among the 2360 variants for which the siblings are homozygous (Panel (ii)), only nine were not found in the population at large, that is, almost 95% of the single-gene variants shared by the affected siblings are also shared with people who are not affected; more than 96% of the homozygous variants shared by the siblings are also shared with people who are not affected. Thus, filtering out the common polymorphisms is an extremely effective means to reduce the number of candidate mutations. This supports the concept discussed in the chapter that most of the variation in a person's genome is shared with people of the same ancestry.

In addition to these two affected siblings, the exomes of two unrelated children affected by Miller syndrome were also sequenced and summarized in Figure B(i) and (ii). If only a single mutation is required to cause the disease, that is, if the disease is a dominant trait (Panel (i)), there were 3100 shared variants between the two affected brothers and this unrelated child; if the syndrome is recessive and affected individuals have to be homozygous (Panel (ii)), 1810 possible candidates were identified as being shared among the two affected brothers and the unrelated affected child. By filtering out the polymorphisms that occur commonly in human populations, nearly all of these candidates could be eliminated. Only 26 variants in single genes were shared

BOX 10.2 Continued

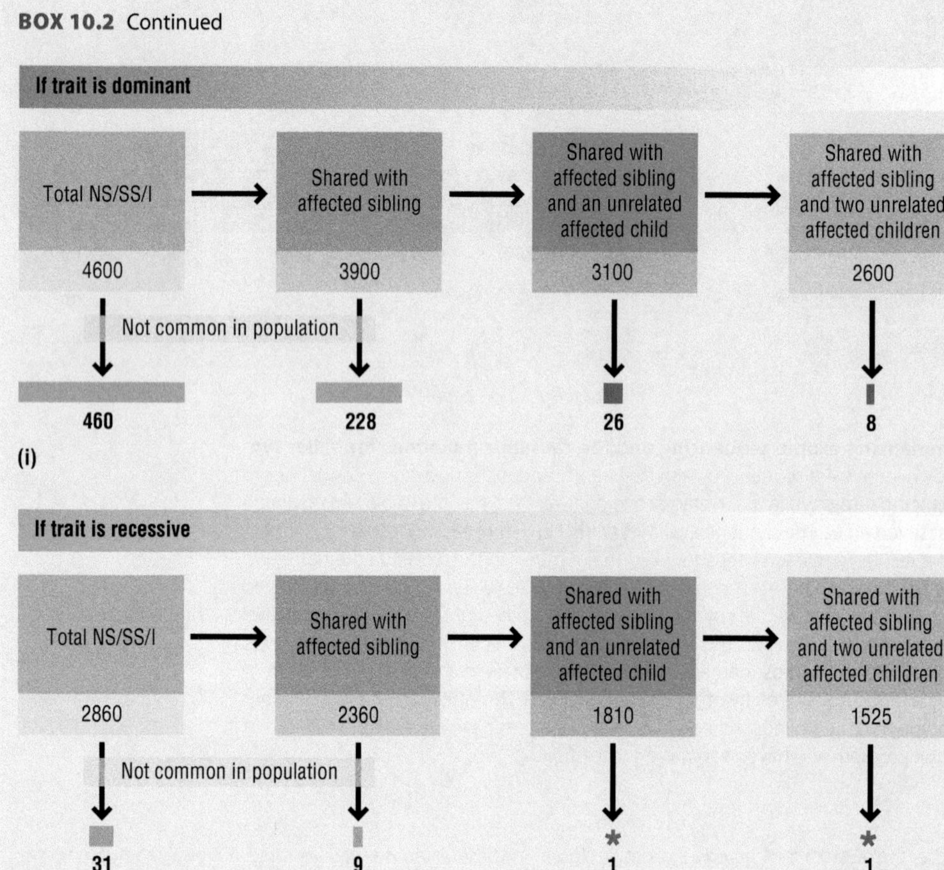

Figure B The effectiveness of filtering steps. Two major approaches were used in this study to filter all of the variants found by exome sequencing. The variants are abbreviated as NS (non-synonymous amino acid substitutions), SS (splice site variant affecting the amino acid sequence), and I (indel or insertion/deletion). The relative sizes of the boxes reflect the effectiveness of each filtering method, and the numbers in each box are the number of remaining candidate variants, rounded off. By following the horizontal arrows across the diagrams, the effectiveness of analyzing additional affected children is seen. By following the vertical arrows down the diagram, the effectiveness of filtering out the common polymorphisms in the population is seen. (i) shows the results of the filtering if the trait is dominant, while (ii) shows the results of the filtering is the trait is recessive. The stars in (ii) indicate that variants in a single gene were found with two affected siblings and one unrelated person, and the same gene was found, regardless of which unrelated person was analyzed.

between the unrelated affected individuals but not in the population at large, and only one homozygous variant was shared by the two affected families but not found in the population at large. This gene is a very strong candidate, and, if the syndrome is inherited as a recessive trait, then this gene is the only candidate.

In fact, when the exome from the other unrelated individual, the fourth person from the third family, was included with the others, only eight dominant variants were in common, and only this same one homozygous variant was shared among the three families. The mutations in this gene, called *DHODH*, made it the best candidate for being the cause of Miller syndrome. The inheritance of unrelated mutations in both alleles of this gene in three unrelated families is very strong evidence that this is the correct gene.

In order to confirm that this gene is the cause of Miller syndrome, this gene (rather than the entire exome) was sequenced from four more affected individuals, one of them an affected sibling of the affected child in family 2 above and three of them the only affected member of their family. All of these individuals were heteroallelic for mutations in the *DHODH* gene. In addition, the parents of the affected individuals were all found to be heterozygous for the mutations found in their children. In total, 11 different mutations in the *DHODH* gene were found in six different families, and none of them was in common, except within the same family.

BOX 10.2 Continued

None of these mutations in *DHODH* was found among 200 unaffected control individuals.

Other means to filter the data

This initial study showed that removing common polymorphisms from consideration was by far the most effective way of filtering out candidate variants. Subsequent exome sequencing studies do this routinely among their first steps. Other filters have also been found to be useful. For example, the study of Miller syndrome also used a program to predict which sequence variants would have the severely damaging effects on the protein, as summarized in Figure A. In the first study, this filter was misleading, since one of the mutations in *DHODH* was not predicted by the program to be damaging and was eliminated from consideration. In fact, this program pointed to another gene for which the two brothers also shared mutations as the best candidate, but this gene was not mutated in affected children from other families. In subsequent studies, more refined versions of protein structure predictions have been used and proved to be helpful.

Even in this initial study, exome sequencing proved to be an exceptionally effective method to find the causative mutation for a genetic syndrome of unknown cause. The exome sequences of only four individuals in three families were necessary to identify the gene, with four more affected individuals tested for verification. In fact, if the investigators had been more confident about the mode of inheritance and simply tested the gene as an autosomal recessive trait, only one unrelated affected individual would have been needed to find the gene. At least 75 other studies using exome sequencing to find the gene responsible for a rare genetic syndrome have been reported in the past 5 years, with similar approaches.

Whole-genome sequencing

The rationale for sequencing exomes, rather than entire genomes, is that most of the severe genetic diseases will be due to mutations in the exons that produce deleterious changes in the amino acid sequence. This is also its primary pitfall—some diseases will be due to changes in the regulatory regions and will be missed by this approach. In order to identify all of the possible causative mutations, it would be better to examine the sequence of the entire genome, both the exons and the regulatory regions.

Some whole-genome sequencing studies have been done, and it is certain that this approach will be increasingly common. While sequencing exomes is much cheaper and faster than sequencing entire genomes, exome sequencing requires an additional step to capture the exons; sequencing technology has improved enough that whole-genome sequencing, which does not require this variable step, is nearly as effective. The primary limitation in using whole-genome variation is that the databases of common polymorphisms focus primarily on exons; thus, it is more difficult to filter out common polymorphisms in other parts of the genome, at least until many more entire individual human genomes are sequenced.

FIND OUT MORE

Ng, S. B., Buckingham, K. J., Lee, C., *et al.* (2010) Exome sequencing identifies the cause of a mendelian disorder. *Nat Genet* **42**: 30–5

sequence a person's genome for less than $1000 in less than 2 weeks. (However, the analysis of the sequence data will take longer and cost more.) The technology to acquire and analyze this much sequence data at this cost is not quite available yet. In the meantime, the method to score many polymorphic loci simultaneously and inexpensively is to use a microarray.

A microarray does not require us to actually sequence the person's DNA. Instead, the investigators can take advantage of the knowledge that there are regions of the genome that are known to be polymorphic and look at those regions specifically, rather than the entire DNA sequence. By focusing on those regions with known variation, the investigators infer or impute the genotype of the individual. In our previous fruit fly example, geneticists can use the observation that the flies have white eyes to infer or impute, with some statistical confidence, that many of them will also have wild-type bodies and that the founding parents had the ***w*** allele and the y^+ allele *in cis* to each other, and the w^+ allele and ***y*** allele *in cis* with each other.

Microarrays were introduced in Chapter 2 as a method to look at expression profiles. They were first used to look at transcription profiles—the application that probably remains the most common to this day. Remember that each spot on a microarray, called a feature, has a short specific DNA oligonucleotide sequence, perhaps 40–60 nucleotides long. The microarray is printed (like a circuit board) with these sequences in a defined and known pattern. With current technology, a microarray chip can hold more than a million features.

In an expression profile, each of the features corresponds to one or more sequence tags from various genes or regions of the genome that are believed to be transcribed. For making an expression profile using a microarray, the investigator isolates mRNA from different tissues, labels it (possibly as cDNA) with a fluorescent tag, and hybridizes it to the microarray chip. Any feature that hybridizes, that is, which finds its complementary sequence on the labeled cDNA, will fluoresce. This technique is described in Section 2.5.

Microarrays used for GWASs employ the same general principles as the ones used for expression profiling but with some very important differences. First, the features on the chip are not sequences from genes (as they are for expression profiles) but are instead sequences that are known to be polymorphic in the genome. Remember that these need not be sequences or alleles from within a gene; in fact, they are probably polymorphic sites that occur in the regions between genes, as we noted earlier. Nonetheless, they are inherited in the same way as standard alleles.

Because a microarray can hold so many features, each polymorphic locus can be represented by more than one feature on a microarray. Suppose, as before, that the common variant at our polymorphic locus has an A, but the less common variant has a C at a particular nucleotide, as shown in Figure 10.4. Individuals with a G or a T at this locus are rare enough that we don't expect to encounter those sequences in the population, so no features are made for them. This polymorphic locus would be represented by two different features on the microarray. The sequences of these two oligonucleotides are identical, except for this one base change. One feature uses an oligonucleotide with an A at the polymorphic site, while the other feature uses an oligonucleotide with a C at the polymorphic site (or if complementary sequences are used, one feature has a T and the other has a G).

The second major difference between using microarrays for expression profiles and using them for genome-wide associations lies in the nucleic acid that is used for hybridization. For expression profiles, mRNA is extracted, labeled, and hybridized. For GWASs and the analysis of genotypes, **genomic DNA** is extracted, labeled, and hybridized to the microarray, as illustrated in Figure 10.4. Each feature on the microarray that has

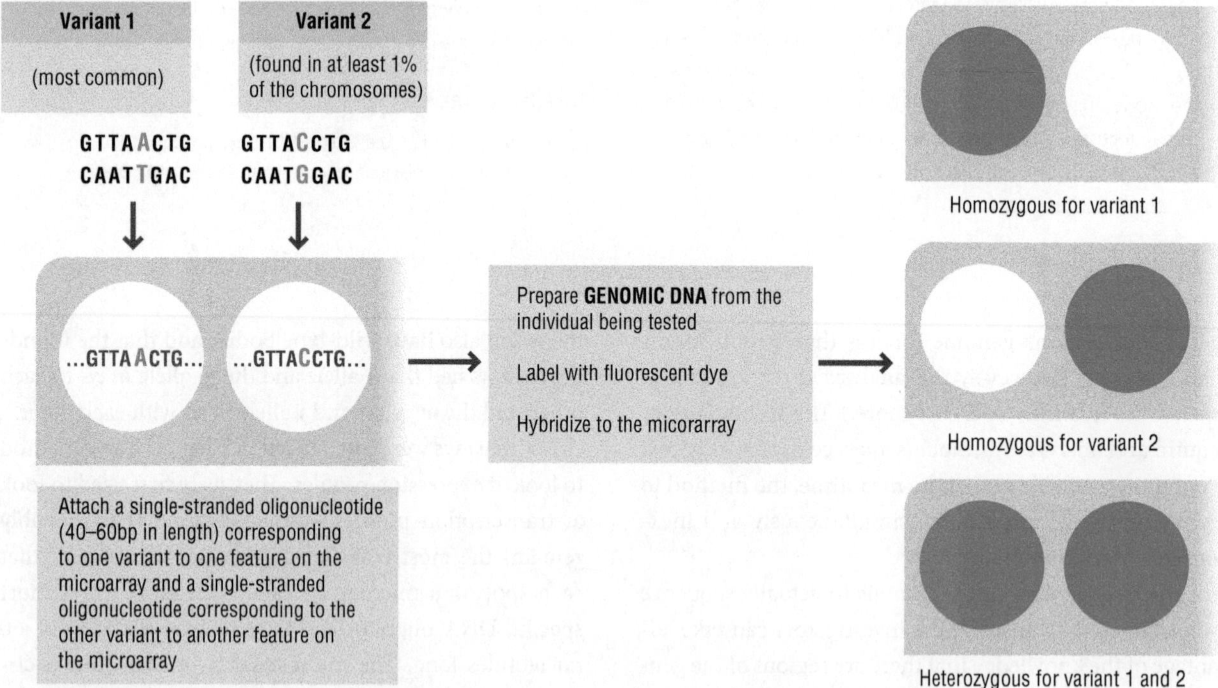

Figure 10.4 Using microarrays to read SNPs. The most common variants, in this case GTTAACTG and GTTACCCG, are represented as features on the chip. Rare variants (in this example, containing a G or a T at the polymorphic site) are usually not represented. For illustration, short sequences are depicted, but each "dot" on a real microarray used for GWAS will contain many copies of a stretch of single-stranded DNA that is 40–60 bp long, and the microarray chip will have as many as a million different features. Genomic DNA from each individual is labeled and hybridized to the array, which can then be used to detect whether the individual is heterozygous or homologous for one of the variants. The usual basis for distinguishing the signals is that the feature with the exact match will hybridize more strongly than the one with the mismatch, so that the intensity is used, rather than the presence or absence of a signal.

a complementary sequence in the sample DNA will fluoresce, and every feature that does not have a complementary sequence will not fluoresce. If the sample DNA that we have labeled has an A (or its complement T) at that site, the feature with an A at the polymorphic site will fluoresce brightly, while the feature with a C (or its complement G) has a mismatch with the sample DNA and will not fluoresce (or will not fluoresce as brightly).

KEY POINT Microarrays used for identifying different polymorphisms in a DNA sequence differ from expression profiling arrays through their use of genomic DNA, rather than mRNA, as the features on the microarray chips. In addition, this genomic DNA comes not from genes but from sequences known to be polymorphic.

Now recall that a microarray can have more than 1 million features on it. Let's do an imaginary experiment to illustrate what this number means for analyzing the genome. Consider a polymorphic site in the genome. We have shown a few nucleotides that surround such a polymorphic site in Figure 10.5. At this site, most people have an A, some have a G, and a few have a T or a C. An oligonucleotide of about 50 bases with one of these four SNPs is placed as a feature on a microarray; each of the four possible features is depicted in Figure 10.5 in the top row of the microarray, with the different SNPs indicated in blue. The second row of features show two completely different polymorphic sites, just to remind you that there can be more than 1 million features on the microarray; the features we are depicting as large circles would only be visible to a laser scanner, and not to your eye.

Now, let's imagine that we isolate DNA from an individual, label it, and hybridize it to the microarray where it is scanned and the signal is recorded. The DNA from the person hybridizes to the site where the sequences (or the complementary sequences) match. Because we are diploids, everyone has two copies of each of the polymorphic sites, so some people are A/G, some are A/A, some are A/C, and so on. The genotypes at the other sites in the genome are scanned at the same time, and all three separate loci are imputed in the genotypes at the bottom. With 1 million features, hundreds of thousands of different polymorphic sites in the genome can be tested, or thousands of sites and haplotypes **per chromosome**.

All of these hundreds of thousands of loci don't represent all of the millions of SNPs that are present in a person's genome, but they represent the genotype at the

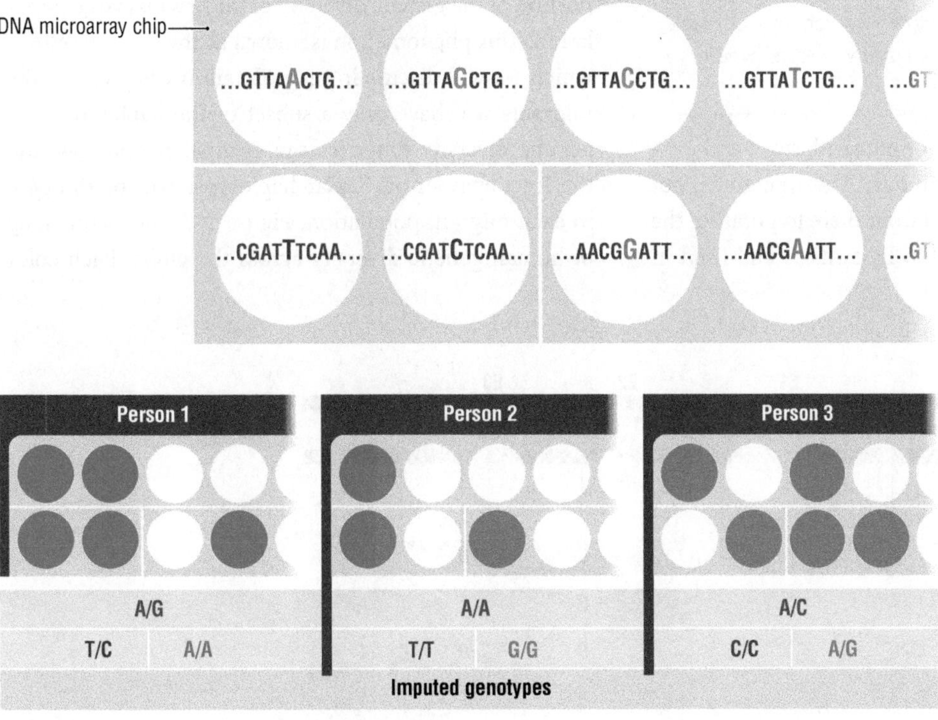

Figure 10.5 Imputing a genotype from microarray results. Three different loci are shown on the microarray chip at the top. The four features in the top row show the four SNPs present at this site, while the second row shows two SNPs present at two other unrelated loci. DNA from an individual is isolated, labeled, and hybridized to the chip, as shown in the middle, and the genotypes at each of these three loci are imputed from the hybridization pattern, as shown in the green boxes at the bottom.

sites that vary most often among humans. Furthermore, since the sequence of our genome in the germline does not change appreciably during our lifetime, once an individual's genotype at each of these sites has been imputed, it can be used repeatedly for different experiments. Thus, all of our heterozygous and homozygous loci are recorded with one experiment, and the results do not change, no matter what future analysis is done with the data.

10.4 Haplotypes and human history

So far, we have described the genome-wide assay for polymorphisms, but we have not yet described how these are used to map a gene that contributes to a phenotype. The key is to remember that we inherit a set of alleles or polymorphisms together in a linkage block known as a **haplotype**, which we introduced in Chapter 9 and show in Figure 10.6. Thus, although we are monitoring an individual site in each spot on our microarray, we can use that polymorphism to indicate which haplotype is present not only for that locus but also for the surrounding loci on both sides. Depending on recombination, this means that one polymorphism is an indicator for a haplotype that ranges in size from a few kb to tens or possibly hundreds of kb. The 200,000 or so different loci being assessed on the microarray represent haplotypes that cover nearly the entire genome.

KEY POINT A block of alleles that are inherited together is referred to as a haplotype. A particular polymorphism in one area of the genome serves as a marker or a tag for the haplotype in which it occurs.

This brings us to the topic of human history and evolution. Humans evolved first in East Africa, migrated out to the Middle East, and spread from there to populate the world. Because humans arose in East Africa and lived in Africa for tens of thousands of years before migrating to the Middle East, Africans are by far the most genetically diverse people. There are vastly more genetic differences among ethnic groups of Africans than exist between any other groups. In fact, all other human populations are a subset of the genetic diversity that is found in Africa, with some new variants that have arisen by mutation along the way. Our ancestors usually lived in small groups, estimated to consist of fewer than 2000 adults at any one location. As these people migrated, they brought along all of their haplotypes.

However, the haplotypes (and the genetic diversity that they represent) are only a subset of the ones that were present in the larger ancestral population from which they came. Thus, the groups of migrants become the founders of a new population in a repeated pattern across the globe, each smaller group containing only a portion of the genetic diversity of the original large population. This phenomenon is known as the **founder effect**, which we will discuss in more detail in Chapter 16. The migrants will have only a subset of the haplotypes and genetic diversity found in the original population, and the frequency of particular haplotypes will be different in each migrant population. Figure 10.7 shows one map of human genetic diversity across the globe. Each color

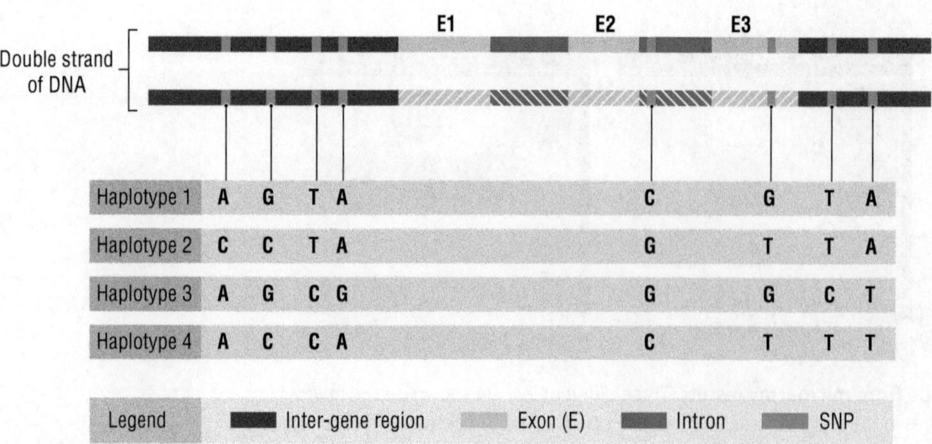

Figure 10.6 Haplotypes. A haplotype is a set of SNPs. A single polymorphism can be used to infer or tag the genotype for the haplotype.

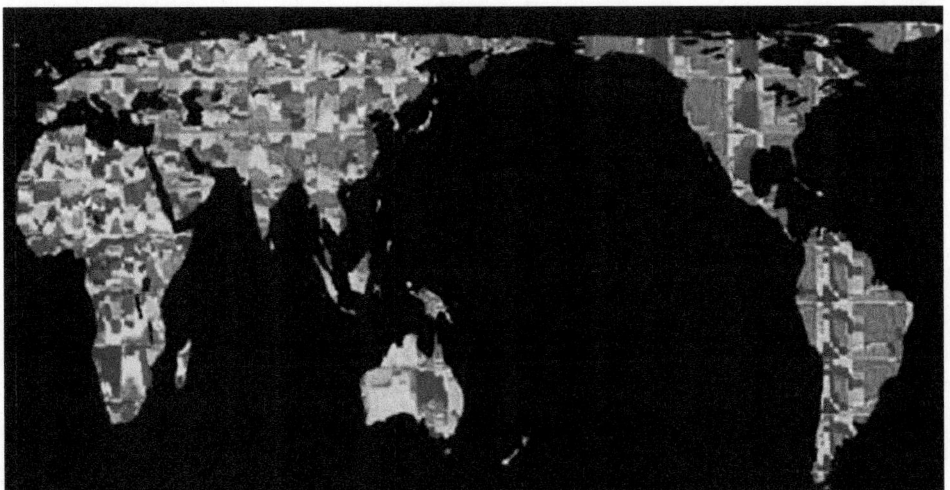

Figure 10.7 Migration out of Africa. A haplotype genomic map of the world shows that genetic diversity decreases outside of Africa. Each color represents a collection of common haplotypes. Africa has more haplotypes, represented by more different and smaller-sized color blocks, than found on other continents, as well as haplotypes found nowhere else. The larger blocks of color on other continents indicate decreased genetic diversity in these locations.

Source: Artistic representation of the patterns of haplotype variation reported by Jakobsson et al. (2008). Image by Martin Soave, University of Michigan Marketing and Design, reproduced with permission from M Jakobsson et al. (2008) *Nature* 451: 998–1003.

represents a common haplotype. Africa has more haplotypes than are found on other continents and specific haplotypes that are found nowhere else. Genetic diversity decreases outside of Africa, as seen by the larger blocks of color on other continents. We discuss how to read haplotype pie charts in Communicating Genetics 10.1.

Here is a thought experiment to illustrate this effect. Imagine that all of the 100 students in a genetics class decide that they will only marry and reproduce with another student from this class. The genetic structure of the population descended from the offspring of the class will be very different from the population of the rest of the world. Some alleles and polymorphisms that are common in the world at large would, by chance, be completely absent from the genetics class' descendants because no one in the class happened to be heterozygous for it. On the other hand, there is certain to be at least one person in the class who is a carrier for some rare genetic disease. Perhaps the worldwide frequency of carriers for this rare genetic disease is one in 20,000 people. But, by chance, among the class population, the frequency of carriers is one in 100. Even though the disease is rare in the worldwide population, it might not be rare among the descendants of the class founder population.

KEY POINT The founder effect refers to the reduction in genetic diversity that occurs when a small group migrates and establishes a new population, separate from the original group. Traits present in the small group will predominate in subsequent generations that arise in the new location.

This repeated pattern of migration, population expansion from a few thousand people, then further migration is why most of the polymorphisms are shared among people from the same geographical group. The main exception comes from African populations, where many more features are necessary to test the entire genome on a single microarray because there is so much more genetic variation arising from the longest human history.

The "common disease, common variant" hypothesis

Now let's consider that a new mutation arose in one of those smaller founder populations, as summarized in Figure 10.8. The mutation occurred in a single individual, and anyone who has the mutation in later generations will probably be a descendant of that long-forgotten founder. (The founder with the mutation may not always be forgotten. Among the isolated religious group known as the Amish in Lancaster County, Pennsylvania, approximately one in eight people is heterozygous for an extremely rare recessive form of dwarfism known as Ellis-van Creveld syndrome. All of these carriers can trace their ancestry back to a man named Rufus King who immigrated to the United States in 1742.) Since all of the people who have the mutation are descendants of the same founder, all of these people will also share the same haplotype, namely the haplotype that was present in the founder with the original mutation, as diagrammed in Figure 10.8. To go

COMMUNICATING GENETICS 10.1 Displaying haplotypes as pie charts

One of the most common ways to depict the frequencies of different haplotypes or genotypes between populations is to use a pie chart. In the pie chart, each haplotype, genotype, or phenotype is represented by a different color, and the area of each colored sector or slice shows the frequency of the haplotype or genotype. This can allow for a quick comparison of the genetic composition of each group.

Figure A shows the frequency of the lactase persistence phenotype and genotypes among different populations in eastern Africa. We discuss lactase persistence (sometimes known as "lactose intolerance") in more detail in Section 16.5, with an emphasis on European populations, but some of the data summarized in Figures 16.21 and 16.22 are also shown here. The location of the pie charts on the map corresponds to where the population sample was taken. Figure A(i) depicts three phenotypes—lactase persistence (LP) in light blue, intermediate levels of lactase persistence (LIP) in medium blue, and lactase non-persistence (LNP) in dark blue. By examining the size of the light blue sector in the pie charts from different populations, it can be concluded that the Beja population from a region in north-eastern Sudan has the highest frequency of lactase persistence, while the Sandawe population in western Tanzania has the highest frequency of lactase non-persistence among the populations examined. (It should be noted that only populations in which lactase persistence is

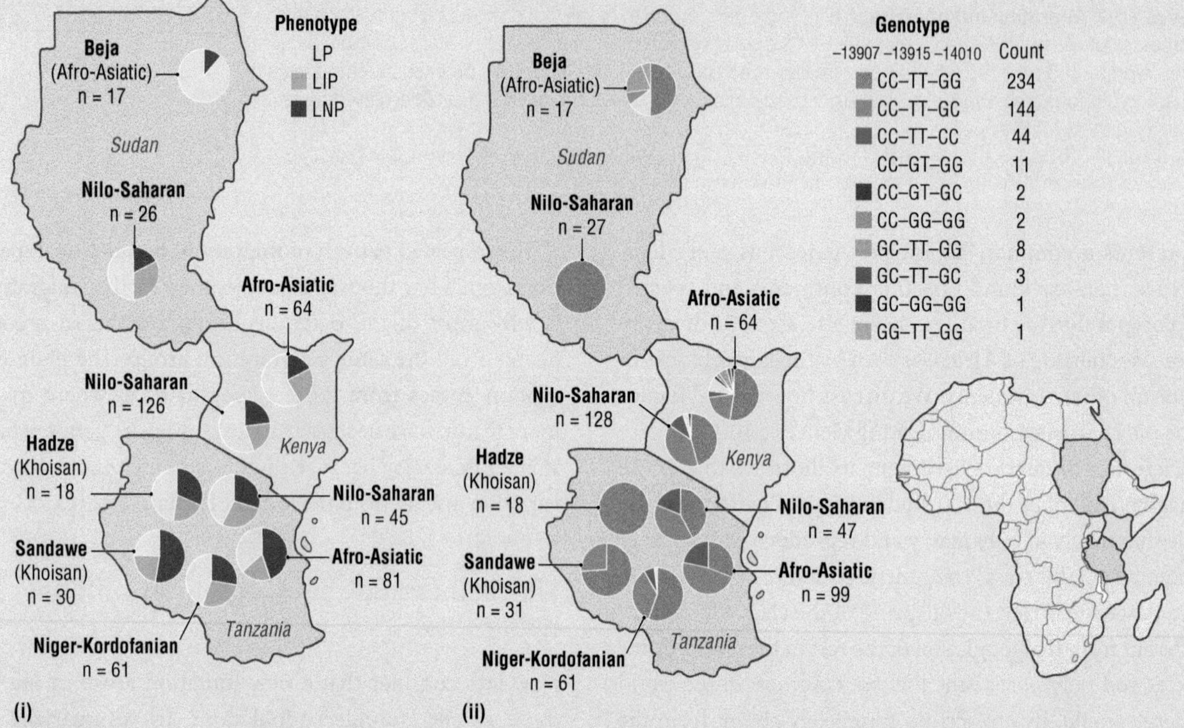

Figure A Lactase persistence among some eastern African populations. The pie charts represent the frequency of each phenotype (in (i)) or genotype (in (ii)). In (i), LP is lactase persistence, LIP is intermediate levels of lactase persistence, and LNP is lactase non-persistence, which is sometimes called lactose intolerance. Most people among these populations exhibit lactase persistence or intermediate levels of lactase persistence, as shown by the size of the light blue and medium blue sectors, but the frequency of this phenotype varies in different populations. The table on the right indicates the SNP at three different positions upstream of the lactase gene (*LCT*), represented by negative numbers. Only one base of each pair is given but for two different (diploid) chromosomal numbers. The map in (ii) shows the frequency of these genotypes—that is, the two haplotypes arising from diploidy—in different populations, color-coded to match the table. Note that the gold genotype, which is common among lactase-persistent people in Tanzania, is absent from the lactase-persistent Beja people in north-eastern Sudan, while the green genotype common among the Beja people is absent from all of the populations in Tanzania and is found at a relatively low frequency in one other population in Kenya. The pie charts illustrate that, although the frequency of a lactase persistence phenotype is high in all of the populations (i), it is due to different genotypes and haplotypes (ii). Note that since the study was done, the country labeled "Sudan" has divided into two countries, Sudan and South Sudan. The lower bar chart in each figure comes from the present country of South Sudan.

Source: Reproduced from Tishkoff et al. (2007). Convergent adaptation of human lactase persistence in Africa and Europe. *Nature* Genetics 39.

COMMUNICATING GENETICS 10.1 Continued

relatively common were tested, and many other populations would have very few individuals who are lactase-persistent.)

The haplotypes surrounding the lactase gene (*LCT*) on chromosome 2 are shown in the table in Figure A. Three positions have SNPs that distinguish the related haplotypes upstream of the start of the *LCT* gene, as represented by the negative numbers. These are shown in more detail in Figure B. Position -13907 has a G/C or C/G polymorphism, position -13915 has a G/C or a T/A polymorphism, and position −14010 has a G/C or a C/G polymorphism; only the first base of each base pair is given in the table on the right of Figure A for simplicity.

Each person has two copies of the haplotype, so ten different haplotype combinations or genotypes were found among these people. For example, as shown in Figures A and B, 234 individuals were homozygous at position −13907 for the C, homozygous at position −13915 for the T, and homozygous at position −14010 for the G. Looking at the next row in the table, 144 people were homozygous for the C at position −13907, homozygous for the T at position −13915, but heterozygous at position −14010 for the G or the C, and so on for each polymorphism at each of the three positions.

Figure A(ii) represents each of the genotypes—that is, the pair of haplotypes—in a different color in a pie chart. This chart shows that, even among those individuals who are lactase-persistent in each population, the genotype that gives rise to this phenotype can be different. For example, among the lactase-persistent people of the Beja population in Sudan, the green genotype is relatively common; only another Afro-Asiatic population in northern Kenya has this genotype at all, and it is less common there. The gold genotype that accounts for most of the lactase persistence in Tanzania is completely absent from the lactase-persistent populations in Sudan; among Tanzanian people, the frequency of the genotype varies, although it is always the most common.

These pie charts show that lactase persistence has probably arisen independently in the Sudanese populations and Tanzanian populations, since polymorphisms that are found in one are absent in the other. The frequency of various genotypes can also provide some clues about the relatedness in the histories of the various populations. Note that the lactase-persistent people from the populations in Kenya have both the Tanzanian haplotypes (in gold) and the Sudanese haplotypes (in green), suggesting that this may be a region where the populations mingled with one another. Obviously, much more evidence than the haplotypes for one gene would be necessary to confirm these inferences, but the pie charts are suggestive.

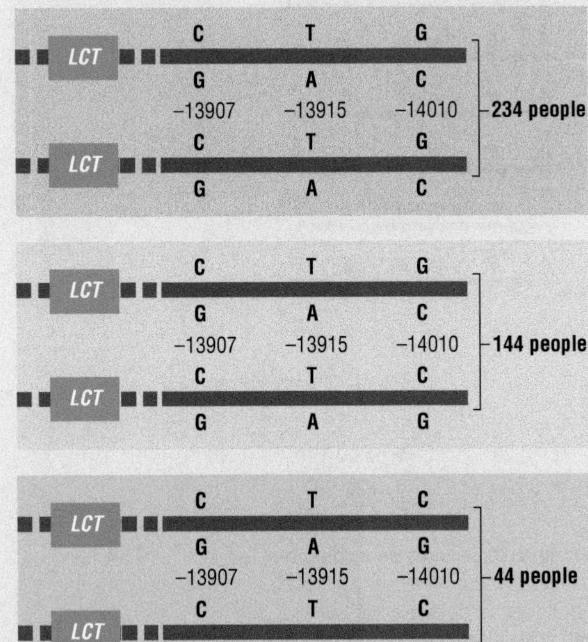

Figure B A summary of the haplotypes for three polymorphisms shown in Figure A(ii). While these polymorphisms are upstream of the *LCT* gene, the numbering indicates that the gene is transcribed from right to left on chromosome 2, so the coding portion of the *LCT* gene is drawn on the left. Both bases of the base pair from the table in Figure A are shown, and each homologous chromosome is shown separately. The color of the three boxes corresponds to the colors of the first three genotypes in the table in Figure A.

back to our earlier example, all of the flies with white eyes in the vial were descended from the same fly with white eyes, and all will have the same molecular change in the ***w*** gene as that original fly had.

This may seem improbable to you, since the modern human population is enormous, diverse, and mobile, but this is precisely the pattern of genetic variation that we see for human populations and haplotypes. In fact, the pattern was recognized when geneticists began to study certain genetic diseases that are common in one population. Among the first and best-studied examples is cystic fibrosis (CF), which we discussed at the end of Chapter 9. CF is inherited as an autosomal recessive trait in standard Mendelian fashion on chromosome 7, and is the most common single-gene disease in people of European descent; approximately one in 25 Europeans (or Americans

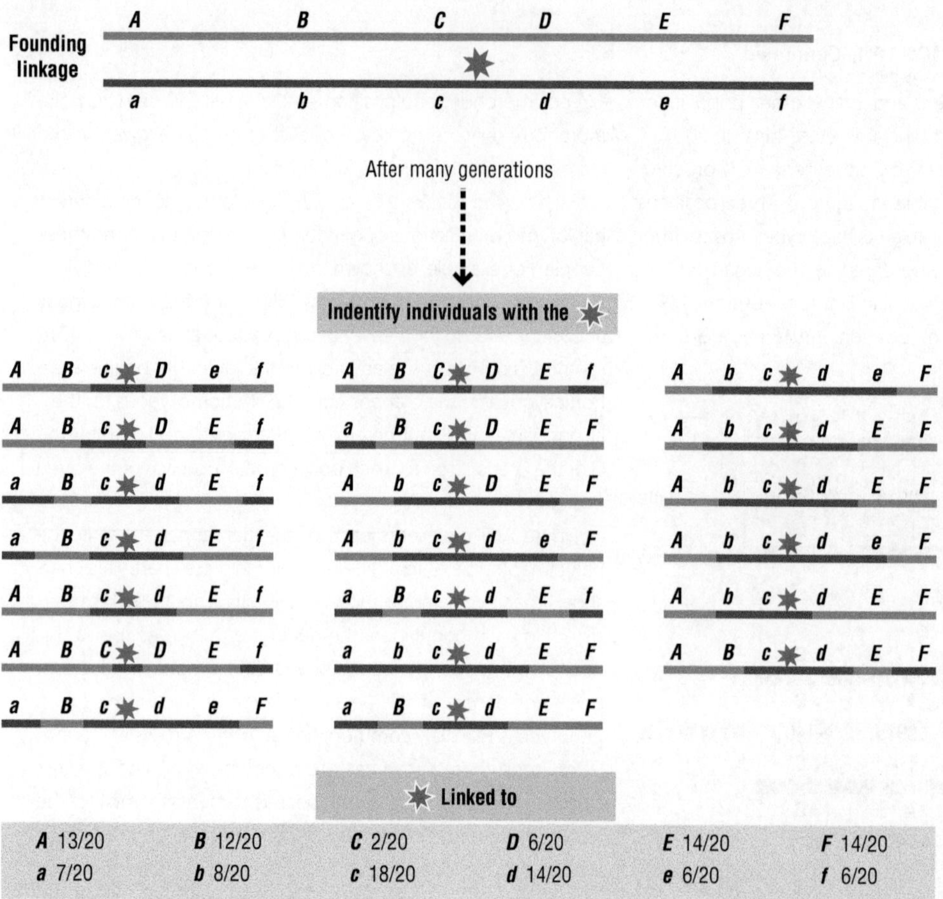

Figure 10.8 Shared linkage from a founder. A mutation, indicated by the red star, arises on a chromosome or a haplotype with a particular set of SNPs, represented by the lowercase letters. After many generations, most of the descendants of that founding member who share the mutation or disease allele will also share most of the surrounding SNPs or haplotype. As represented in the green box at the bottom, the SNPs or haplotypes closest to the location of the mutation or disease allele will show the strongest association.

of European descent) is heterozygous for the disease, although the exact frequency varies across Europe. In the jargon of medical geneticists, CF is a **common disease** among Europeans. (You should realize that every ancestral population has such common diseases, and many others could be used as our example.) In fact, since one in 25 European people is a carrier for the disease-causing allele, the CF gene would be considered a polymorphic locus among Europeans, since the frequency of the second most common allele is about 4%. On the other hand, CF is extremely rare among non-European populations. In Chapter 16, we will discuss a hypothesis for how CF, which has been a fatal disease throughout history, has become so common among Europeans.

As medical geneticists were trying to understand the molecular basis for CF—that is, to identify the DNA sequence corresponding to the CF gene—they realized that more than 70% of Europeans who have CF have exactly the same unusual molecular change that results in the disease. Once the DNA sequence for the gene was known, this mutation proved to be a deletion of three nucleotides that removes a phenylalanine codon at position 508 in the predicted protein, a change called Δ-F508; we will refer to it by that name, although the fact that affected people had the same change was recognized before the nature of the change was identified.

Remember that many molecular changes in the gene are likely to cause the disease phenotype, so it was unusual that so many cases had precisely the same molecular change. In other words, there was a **common variant** for the disease as well. This suggested that the CF mutation found among Europeans arose in a single person. The founder was probably a male, since males have a higher mutation rate than females, who lived in or around Turkey about 10,000 to 12,000 years ago. In fact, this single origin of most cases of CF among Europeans was confirmed when geneticists isolated and sequenced the CF gene (called *CFTR*) from unaffected and affected individuals in 1989. Not only did most people with CF have the same molecular change—the common variant Δ-F508—they also shared many of the polymorphisms flanking the *CFTR* gene as well, that is, they all had a common haplotype, or, more precisely, a series of closely related haplotypes which themselves can be traced back to common origin. (Some of the polymorphisms shown in Figure 9.22 on the positional cloning of *CFTR* are the ones that were found.) The mutation causing CF in most Europeans where the disease is common arose on a single chromosome in a single person.

This result, and similar results for other common genetic diseases in other populations, has led to the **"common disease, common variant" hypothesis**, which states that:

> For a genetic disease that is common in one population but uncommon in other populations, most of the cases are due to a common molecular variant. People with the same disease have the same mutation, and will also have the same haplotype.

There are many examples that support the "common disease, common variant" hypothesis, from many different populations. It certainly cannot explain all monogenic diseases, but it has been shown to be an effective starting point for finding the genetic basis for many of the ones that are common in a particular population. People from the same ancestral population who have the same common disease are likely to share a haplotype as well. This is also seen for many non-disease traits. For example, most Europeans who are lactase-persistent—that is, who can continue to drink and digest milk as adults—have the same mutation near the lactase gene and share one of a set of related haplotypes, suggesting that lactase persistence (the "mutant" or derived condition of what is sometimes called lactose intolerance in the United States) also arose from a mutation in a single individual; we discuss this example in more detail in Chapter 16.

10.5 Complex traits

We have described how the haplotype structure of the human genome, which arises from our history, can be used to map the gene affecting a trait. Let's return to our discussion from the beginning of this chapter and consider complex traits. Recall that a complex trait is one whose phenotype is affected by several or many genes, as well as by the environment. What can GWASs tell us about such traits?

For our first example of a complex trait, let's consider a non-disease trait—the height of human adults. There are some known loci that behave as a single-gene trait and have an extreme effect on height; typically, an individual who has a mutation or change at one of these loci is a dwarf.

Mutations at more than ten different loci result in different types of dwarfism; interestingly enough, several of the dwarfism mutations are inherited as dominant alleles. These types of dwarfism are inherited in the standard fashion of other single-gene traits such as the ABO blood type. These genes are not part of our discussion for this section.

We will focus on the typical height variation among a group of people, none of whom has any known mutations that affect height. They are simply of different heights. Because men tend to be taller than women, we will consider height variation only in one sex. Adult men show a range of heights, as illustrated in the sample shown in Figure 10.9. We can now ask ourselves a number of questions. Is this range of heights due to genetic differences among different men? Or is it strictly due to differences in nutrition or other aspects of the environment? What is the evidence for thinking that there may be genetic differences? What are the relative contributions of genetic differences (that is, genotypes) and nutritional and other environmental differences to the range of heights?

One line of evidence that there may be genetic differences that contribute to variation in adult height is that tall parents tend to have tall children. Siblings tend to be similar in height, and certainly much more similar than non-siblings from the same population. Furthermore, adopted children are closer in height to their birth parents than to the adoptive parents; this suggests that environmental differences alone cannot explain the variation in height. People from different ethnic and geographical origins can be of very different heights, even if they are now living in the same environment—Japanese Americans tend to be shorter, on average, than Swedish Americans, for example, and at least some of that difference is

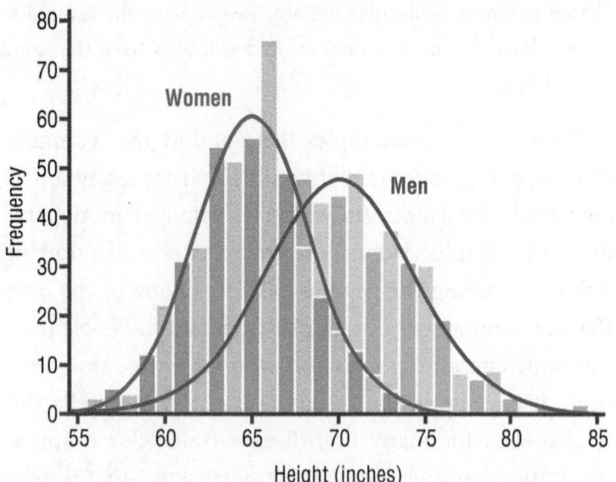

Figure 10.9 Height distributions. The heights in inches of a sample of adult females (in red) and males (in blue) are plotted along the X-axis, with the relative frequency of each height plotted on the Y-axis. The bar graph for each sex can be described by a line that summarizes the distributions.

undoubtedly due to the fact that the ancestral Japanese and Swedish populations have different alleles for some of the genes that affect height.

Height is an example of a **complex trait**. Complex traits are sometimes known as **polygenic traits** or **multifactorial traits** because more than a single gene is involved in the phenotype. They are also known as **quantitative traits**, since the phenotype can often be expressed numerically (such as height or a blood pressure measurement). Genes that affect the phenotype of a complex trait are sometimes referred to as **quantitative trait loci** or **QTLs**. The important thing to note is that all of these terms are essentially referring to the same phenomenon. The phenotype of a complex trait will be affected by the action of genes and the environment, as well as any interaction between genes and the environment.

If only a single gene affects a trait, then the number of possible genotypes and phenotypes is limited. However, as summarized in Figure 10.10, as the number of genes that affect a trait increases, the distinction between phenotypes decreases, and the overall distribution of the phenotype becomes more continuous. Furthermore, the relative effect on the phenotypes that can be attributed to each gene decreases, that is, each allele or gene that affects the trait contributes less to the overall variation. While there is variation among the phenotypes of children affected by a single-gene disorder, such as CF, due to effects like variable expressivity and environmental differences, most of the variation in this trait can be assigned to the

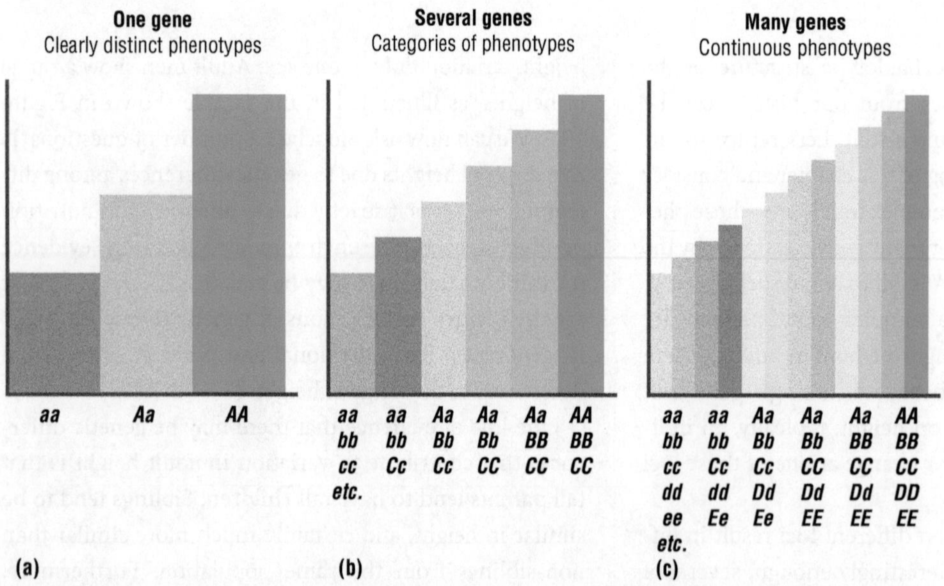

Figure 10.10 Variation in many genes approaches a continuous distribution. In our imaginary example, suppose that each capital letter (***A***, ***B***, or ***C***, etc.) represents an increase in the trait—a centimeter of height, a higher yielding crop, and so on. If only a single gene affects the trait, as in (a), three distinct phenotypes are evident. With a few genes affecting the trait (b), categories of the phenotype are seen, but different genotypes may give similar phenotypes and the categories may blend into one another. Pigmentation in humans might be an example in this category. As the number of genes that affect the trait increases, the distinction between the categories is lost, and the phenotype can be described as a continuous distribution (c). This example treats every allele for every gene that affects the trait as co-dominant or intermediate dominant, so the results are approximately additive. In the terms of quantitative genetics, this is called additive genetic variance. Dominant alleles and interactions between the genes make it even more difficult to assign a genotype to a particular phenotype.

effects of a single gene. As the number of genes that affects a trait increases, the relative impact of each individual gene decreases. This is the effect seen with a trait like height; since alleles at many loci affect height, any single allele might have only a very small effect.

KEY POINT Complex traits are also known as polygenic or multifactorial traits, as multiple genes affect the phenotype. These traits are often quantitative traits, that is, the trait can be measured or expressed as a number. Quantitative trait loci (QTLs) are DNA sequences that contain, or are linked to, genes that affect such a complex trait.

Nature and nurture

In common literature and speech, complex traits are the ones people are usually talking about when they mention the notion of "nature vs. nurture." In more scientific terms, "nature" refers to genetic differences between individuals or between populations; in other words, people have different alleles for some of the genes that affect this trait. "Nurture" refers to environmental differences. Even within a population, we eat different diets, come from different cultures, are infected by different microorganisms, and have different upbringings—all of which can affect the phenotype. Among geneticists, the discussion or debate is not really "nature vs. nurture", as it is so often phrased, but is about the relative contribution of genetic and environmental differences to the different phenotypes that we observe. Of course, this phrasing is not nearly as catchy as "nature vs. nurture," but at least we can understand more specifically what this familiar expression means.

We can go a bit further with this topic and become a bit more technical. Shortly after Mendel's work was rediscovered in the early part of the twentieth century, as described in Chapter 5, genetics research split into two branches. Most of what we described in Chapters 5 through 9 can be traced back to the branch that is referred to as Mendelian genetics. This research focused, as had Mendel, on phenotypes that fall into discrete categories (such as wild-type and mutant) and on inheritance patterns that produce predictable ratios of those phenotypic categories (such as 3:1).

The other main branch of genetics research from the early days of the twentieth century was the quantitative genetics of complex traits; scientists in this branch did not necessarily dispute Mendel's results, but the traits that they were studying, such as crop yield, did not easily fit into Mendelian models. Consequently, those who studied quantitative genetics were often at odds with those who studied traits with simple Mendelian ratios. In fact, the debate between the Mendelists and the quantitative geneticists is one of the most famous academic quarrels in history.

Geneticists studying quantitative traits, who often came from backgrounds in plant and animal breeding, did not see the clear-cut phenotypic categories for their traits that the Mendelists would talk about. To explain their results, some of the quantitative geneticists had to devise new methods to analyze their data and to interpret their meanings. These quantitative geneticists were among the early contributors of the academic field of statistics; one of the most prominent quantitative geneticists, Sir Ronald Fisher, for example, is considered a founder by the professional societies of statistics, of evolutionary biology, and of genetics.

The analysis of complex or quantitative traits is inherently statistical. A thorough description of the experiments and their interpretations for some of the topics in this chapter requires a solid background in statistics, and we understand that not all students reading this book have such training. With this in mind, we will provide a brief review and use the proper terms that statisticians use, without going into the details of the underlying numerical foundation.

KEY POINT The study of complex traits requires the use of statistics to determine the relative contribution of different factors, such as genes and environmental differences, to an observed phenotype.

Distributions, means, and variance

When we take a series of measurements from a population, such as the height of adult men in the United States, we obtain many different values. We can tabulate these, counting how many men are 63″ tall, how many are 68″ tall, or how many are 72″ tall, and all of the values in between. As we tabulate these values, we can represent the data as a graph or a chart. Such a tabulation or its plotted version is called a distribution.

There are many, many different types of distributions, each with its own characteristic shape and described by its own graph. One of the most familiar is a Gaussian distribution, also called a "normal distribution" or a "bell curve," shown in Figure 10.11(a). Many human traits, such as height and diastolic blood pressure, follow a Gaussian distribution, or nearly so. However, there are many other types of distributions that are also important

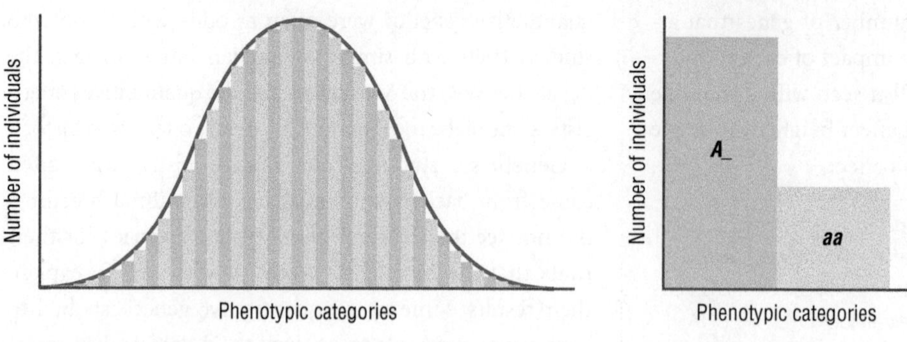

Figure 10.11 Two distributions commonly encountered in genetics. A distribution plots the phenotypic categories on the X-axis and the number of individuals in that category on the Y-axis. (a) shows a Gaussian or normal distribution, which can be represented by the red line or the bell curve. (b) shows a binomial distribution such as arising from a 3:1 ratio with a single-gene trait.

in biology. For example, when we classify flies from a mating into red eyes and white eyes or into males or females, we are using a binomial distribution with two discrete categories, of the type shown in Figure 10.11(b). The rate of mutations, described in Chapter 4, follows yet a different distribution called a Poisson distribution. We will use the Gaussian distribution for our discussion because it is familiar, but you should be aware that there are many other different probability distributions, each with its own properties, and an investigator may not know which one best describes the trait in question.

All distributions are characterized by two key statistical parameters, the mean and the standard deviation (or the variance). These are illustrated for a Gaussian distribution in Figure 10.12 and can be calculated from simple formulas, which we will not discuss here. The **mean** of the distribution is what most of us are referring to when we talk about the average. In a normal distribution, the mean is the height to the top of the bell curve at the middle; the data are symmetrical about the mean, so that there are roughly as many individuals two inches taller than average as there are who are two inches shorter than average. The **variance** refers to how scattered the individual data points are around the mean. If the variance is large, the data are widely spread out. If the variance is small, the data are tightly clustered. The square root of the variance is called the **standard deviation**; this is more readily plotted than the variance, since the standard deviation has the same units as the mean. The standard deviation is plotted for the distributions in Figure 10.12. A defining property of a Gaussian distribution is that it is symmetrical about the mean, such that two-thirds of the data points lie within one standard deviation above or below the mean and 95% of the data points lie within two standard deviations above or below the mean.

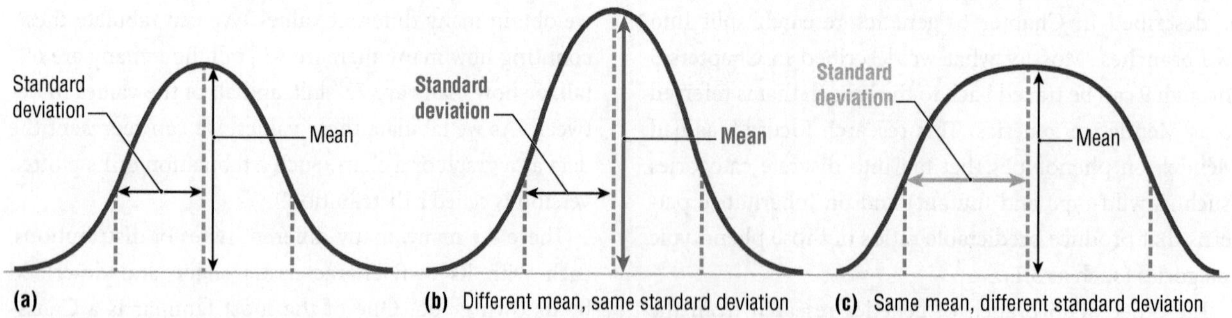

Figure 10.12 Means and standard deviations for a Gaussian distribution. The mean of a Gaussian distribution is the height at the top of the bell curve; (a) and (c) show distributions with the same mean, while the distribution shown in (b) has a different mean, larger than the means in (a) and (c). The standard deviation is a measure of the width of the distribution, or the width of the bell in a Gaussian distribution. The distributions in (a) and (b) have the same standard deviation, while the distribution in (c) has a different standard deviation, greater than in (a) and (b), since the curve is broader.

For quantitative traits, we are often most interested in the variance, rather than the mean. In our hypothetical example of the height of adult men, we would ask, "How much of the variance for this trait can be attributed to genetic differences among the men in our population? By contrast, how much can be attributed to environmental or non-genetic differences?" Note that we are not asking about how genes contribute to the mean for the phenotype, although there are ways to estimate that from the genetic variance. We are not asking, "Why did 5′10″ turn out to be the mean height of the men in Figure 10.9?" nor are we asking about an individual such as, "Why is this particular man the height that he is?" We are asking why some men vary from that average height. In other words, we are attempting to determine how much the genetic differences within a population—the different alleles that people in the population will carry for some genes—contribute to the overall variation in the phenotype.

KEY POINT Because complex or multigenic traits often have a phenotype that is measured quantitatively, the phenotypes are distributed across a spectrum. The distribution of the phenotypes is analyzed by methods of statistics such as the means and standard deviation.

The word **heritability** is often used in the discussion of how much of the variation in phenotype can be attributed to genetic differences. Heritability has a complicated, somewhat confusing, and sometimes disreputable role in the history of quantitative and human genetics. Part of the confusion surrounding heritability is that the same term is used to describe two different genetic situations. We discuss heritability in Box 10.3.

BOX 10.3 *Quantitative Toolkit* Heritability

The concept of heritability, or how much of our variation in phenotypes can be attributed to differences in genotypes, is central in the so-called "nature vs. nurture" debate that engages the popular imagination. One does not need to have a degree in genetics or to read a chapter on complex traits in a genetics textbook to become familiar with the notion that biological factors and environmental factors shape the differences we observe. Anyone who tries to grow a garden, train a pet, or work with children quickly realizes these influences.

The term heritability has acquired different meanings for different types of investigations, which have often been confused with each other. Both meanings define heritability as what fraction of the phenotypic variation can be attributed to genetic variation. The difference between the two meanings lies in the way that genetic variation is defined.

Broad-sense heritability

The more familiar usage in human genetics is technically called the **broad-sense heritability** and is abbreviated H^2. The broad-sense heritability for a trait in a population is the ratio of the total genotypic variance to the total phenotypic variance. Thus, broad-sense heritability asks, "What fraction of the total variation that is seen for this trait can be ascribed to all of the genetic differences among the individuals in the population?" Broad-sense heritability is most applicable when we think of risk factors or familial traits, and paints a rough picture of the contributions of genes and the environment.

Notice that we referred above to all of the genetic differences, so it is worth thinking about the types of genetic differences that are included. Genetic differences arise because there are different alleles that affect the trait, some of which are "favorable" and some of which are "unfavorable;" in this context, favorable and unfavorable refers to whether they increase or decrease the trait, so an allele that produces blonde hair is an unfavorable allele for red hair. Although favorable and unfavorable are terms that are often used, they are only appropriate if there is agreement about the preferred phenotype. This can easily, and sometimes unconsciously, become culturally biased.

But total genetic variation includes much more than the different alleles that affect the trait. It also includes variation that arises because some alleles are dominant; it includes variation due to the interaction of genes and alleles with each other, and it includes the very complicated interactions between particular alleles and the environment. In statistical terms, the total genotypic variance can be partitioned into different components. These components include the variance due to the additive effects of genes and alleles (that is, favorable and unfavorable alleles), dominance, epistasis or gene interactions, and the interactions between the genes and the environment.

Let's express these concepts as equations.

$$H^2 = V_G/V_T$$

and

$$V_G = V_A + V_D + V_I + V_E$$

in which:

- H^2 is the broad-sense heritability,
- V_T is the total phenotypic variance,

BOX 10.3 Continued

- V_G is the total genotypic variance,
- V_A is the additive genetic variance,
- V_D is the dominance variance,
- V_I is the variation from gene interactions, and
- V_E is the variance arising from the interactions between genes and the environment.

(Different sources may use different notations for the variance due to gene interactions and the covariance between genes and the environment.)

In most situations with human populations, the total phenotypic variance can be measured; the total genotypic variation is difficult to measure, although estimates can be made, and the components of genotypic variation are nearly impossible to determine.

Additive genetic variance and narrow sense heritability

Plant and animal breeders have traditionally wanted to select for varieties with particularly desirable traits to improve their stock. Since they can use genetically identical stock or grow their stock in uniform environments, such as greenhouses, to control the amount of genotypic or environmental variance, plant and animal breeders are able to subdivide the genotypic variance into different categories. The ability to subdivide the genotypic variance based on the actions of different genes has allowed breeders to define a different parameter known as the **narrow-sense heritability**, abbreviated h^2. The narrow-sense heritability is defined as the fraction of the total phenotypic variance that can be ascribed to additive genetic variance; in other words, $h^2 = V_A/V_T$. The distinction between broad-sense and narrow-sense heritability lies in which part of the genotypic variance is included; broad-sense heritability includes all components of the genotypic variance, while narrow-sense heritability includes only the genotypic variance that can attributed to favorable or unfavorable alleles.

To illustrate the importance of this, we can describe a standard scenario. When the breeder begins the experiment to improve his stock, he has no idea how many genes are involved and how much of the genotypic variance is due to additive genetic variance, dominance variance, or variance due to gene interactions because these parameters cannot be easily measured in advance. However, the selective breeding program can be used to calculate the additive genetic variance because this is the component that affects how much the stock can be improved, which can then be applied in other breeding programs. In effect, it is asking how much the stock can be improved by replacing unfavorable alleles with favorable alleles.

Again, this might be simplest to illustrate with an equation. It has been shown that:

$$R = h^2S$$

where:

- R is the response to selection or how much the stock can be improved,
- S is the selection differential or the difference between those selected to produce the next generation and the mean of the population, and
- h^2 is the narrow-sense heritability.

While the narrow-sense heritability is not known initially, it can be easily calculated from parameters that can be determined by the breeder (S) or that can be directly measured in the experiment (R). Since the narrow-sense heritability can be calculated, the additive genetic variance can also be calculated. So narrow-sense heritability is important for determining how much the stock could be improved by selective breeding; this is quite different from broad-sense heritability because it is calculated using different parameters.

Considerations about heritability

Our examples allow us to illustrate and clear up some of the potential confusion surrounding heritability. First, whether we are talking about H^2 or h^2, the heritability depends on the population being sampled or (more directly) on which alleles are present in the population being surveyed. An allele that contributes to the trait could be fixed in one population but not in another population. In the population for which the allele is fixed, it will not be recognized as contributing to the phenotype at all because it is not contributing to the variation in the phenotype.

More fundamentally, this means that estimates of heritability from one population cannot be extended to other populations indiscriminately. If we say, for example, that the heritability of colorectal cancer in a Swedish population is 43%, we cannot say that the heritability of colorectal cancer in a Japanese population is also 43%. Some genes affecting colorectal cancer might be fixed for one allele in the Swedish population but polymorphic in the Japanese population, or fixed for a different allele.

Second, high heritability for a complex trait does not mean genetic destiny. There could always be other alleles that affect the trait. There are very complicated and significant interactions between genotypes, environments, and other organisms that we have completely ignored in this chapter, but it is a concept discussed in Chapters 16 and 17. Our simplest experiment to support this point is familiar. Height has a broad-sense heritability estimated at 70–80% in different human populations. Yet we

BOX 10.3 Continued

know that humans in nearly every population have gotten taller in the past four or five generations. This does not imply that humans will always become taller, such that our great-great-great-great-grandchildren will have an average height of 6′ 10″. Alleles will become fixed, which results in no increase in the next generation. There are also selective pressures (such as the size of the uterus) that prevent humans from grower taller in each generation.

Third, broad-sense and narrow-sense heritability cannot be used interchangeably. Every few years, some policy maker or pundit will talk about the heritability of some human trait and use this as data to support an opinion. A common one in the United States has been the heritability of the intelligence quotient (IQ). The IQ, as determined by scores on a battery of tests, is a complex-trait phenotype; after all, anything that can be measured is a complex-trait phenotype, even if the phenotype is not heritable at all. Leaving aside the complicated questions about what IQ tests are actually measuring, we could treat them in a culturally neutral fashion and calculate a heritability of IQ test scores for a particular population. This has been done—many times, in fact. This is broad-sense heritability, of course, although that term is never used in the debates.

One version of the IQ debate posits that, since a particular population has a lower score on IQ tests than another population, the two populations differ in their ability for academic or intellectual achievement. This overlooks the problems of extrapolating heritability estimates from one population to another, of ignoring the interplay of genotypic and environmental variances, and of interpreting the value of an IQ score on the test. Yet it continues to occur.

An even more insidious version of this argument continues that we cannot expect to improve achievement by improving the environment. This argument was used a few years ago in the United States to defend the elimination of early childhood learning programs, for example. Again, leaving aside all of the moral and ethical considerations, this argument fails on genetic grounds. It says, in effect, that response to selection—that is, improvement in the performance for the trait being measured—depends on the broad-sense heritability. It ignores the vast literature on heritability from decades of work by geneticists working with human populations and with plant and animal populations. It ignores our own common-sense experience that, even for traits with high heritability like height and some cancer risks, the phenotypes of the population do continue to change over time.

FIND OUT MORE

Heritability has a long literature. Some recent reviews on broad-sense and narrow-sense heritability include:

Tenesa, A. and Haley, C. S. (2013) The heritability of human disease: estimation, uses and abuses. *Nat Rev Genet* **14**: 139–49

Visscher, P. M., Hill, W. G., and Gray, N. R. (2008) Heritability in the genomics era—concepts and misconceptions. *Nat Rev Genet* **9**: 255–66

Inheritance and complex traits

Hundreds of different genes that contribute at least somewhat to height differences in human populations have been identified. We will describe later in the chapter how genes that affect complex traits can be identified. At least 20 of these genes have been reasonably well characterized as having more significant contributions; as expected for a trait affected by so many different genes, none of these genes affects height variation by more than a few percent. (Since we are discussing height within the usual range for adult humans, genes with major effects on height are probably not polymorphic and would not be detected.) That is to say, different common alleles are known for these genes in humans, and these different alleles affect the distribution of the variation that occurs for human height. These variations or alleles could be in the coding region of the genes, thus producing a protein with slightly different effects, or in the regulatory region of the gene, affecting the level of expression of the gene, or have some other effect on the expression or function of the gene. At this point, all we know is that there are dozens of genes for which variations affect height.

The locations of some of these genes are plotted in Figure 10.13(a). Each of these genes is inherited in a standard Mendelian fashion, just like the genes we have considered in earlier chapters. Some of the genes are X-linked and some are autosomal, as can be seen from the locations in the figure. Most of these independently assort from each other, and there is no particular clustering of genes affecting height. Alleles at each of these loci might be dominant, recessive, or co-dominant. The genes might also exhibit interactions among themselves, although accounting for these interactions requires a more sophisticated type of analysis than we can offer here. These genes are standard genes; they are transcribed at certain times and places, their transcripts are translated to make proteins, and so on.

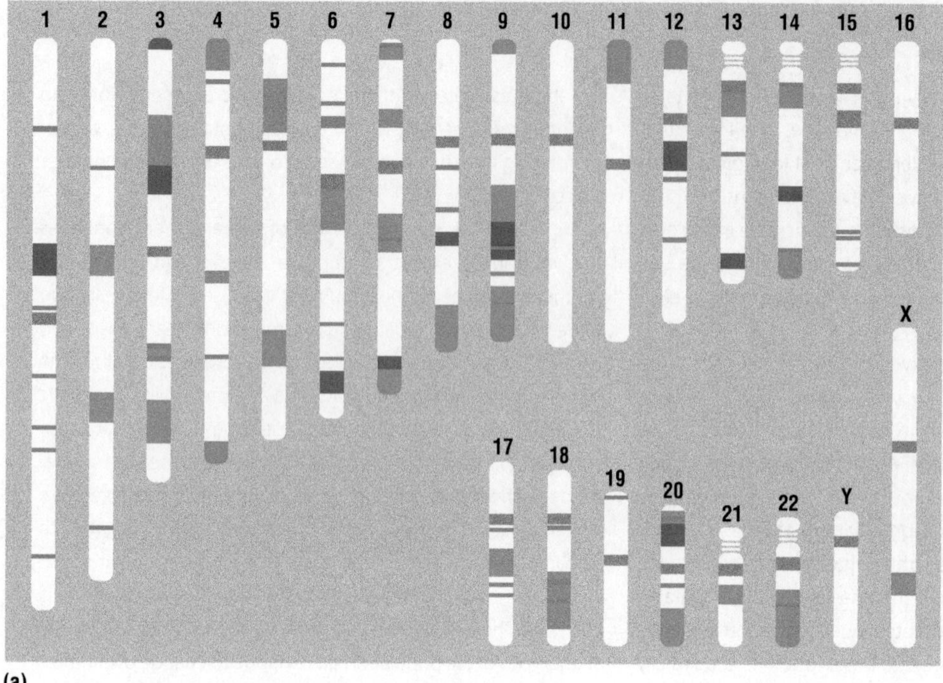

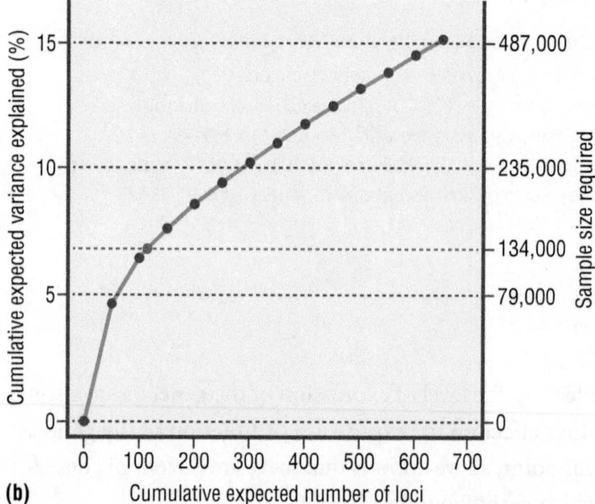

Figure 10.13 Genes affecting height in humans. (a) Regions of the human genome for which variation has been associated with differences in height. The red and pink bands are regions of the chromosome in which a gene has been associated with an effect on height. The blue lines represent loci that were convincingly associated with height from GWASs performed before 2008. The gray bands show the locations of the centromeres. (b) The cumulative number of loci associated with height that are projected to be discovered and their expected contributions to variance. The plot includes already identified loci and as yet undetected loci. The dotted red line corresponds to expected phenotypic variance explained by the 110 loci that reached genome-wide significance in a 2010 study.

Source: Part (a) reproduced from Weedon, M.N. and Frayling, T.M. (2008). Reaching new heights: insights into the genetics of human stature. *Trends Genet.* Dec; 24 (12). Part (b) reproduced from Allen, H.L. et al. (2010). Hundreds of variants clustered in genomic loci and biological pathways affect human height. *Nature.* Oct 14; 467 (7317).

To say they are standard genes is not the final analysis of genes that affect height in humans, however. Polymorphisms in these genes do not account for all of the variation in human height, and each gene alone has a very small effect. For reasons discussed in further text, finding all of the polymorphisms that affect human height requires testing variation in all human populations, which is not feasible. Figure 10.13(b) estimates how many additional genes remain to be found, as more individuals and populations are tested. Note that it is estimated that even if half a million people were tested and 700 genes were found, these would still only account for about 15% of the total variance in human height.

The important message here is that the difference between standard Mendelian traits and complex traits comes not from the genes themselves but from the **phenotypes** they produce. A complex trait phenotype is the outcome of many different genes, each of which makes its own contribution. As we identify genes that affect the trait, we want to ask how much of the variance can be explained by the allelic differences for this one particular gene. Allelic difference at one gene might be responsible for, say, 13% of the total genetic variance for the trait. That gene would be a major contributor. Allelic differences for another gene might be responsible for only 3% of the total genetic variance; each of the genes would affect the phenotype to a different degree.

Let's think a bit more about the genotypes of complex traits. Individuals who have the same phenotype—men who are 70″ tall, for example—might have completely different genotypes for these genes that affect height. One man might be ***AA*** at the *A* locus, ***Bb*** at the *B* locus, and ***cc*** at the *C* locus (and so on for each of the 20 or more genes in Figure 10.13(a)). Another man might be ***Aa***, ***BB***, and ***Cc***, and a third man might be ***aa***, ***Bb***, ***CC***, and so on. Hundreds of different genotypes can give rise to the same phenotype.

Furthermore, different populations will be genetically different. Not every man in the UK who is 5′10″ has the same genotype, even though he has the same phenotype as other British men who are 5′10″. But if we compare British men who are 5′10″ with, say, men in Italy who are 5′10″ or Saudi men who are 5′10″, we would find even more genetic differences. Some genes that are polymorphic in British men will not be polymorphic in Italian men, and vice versa.

There are almost certainly genes affecting height, or any other trait, for which Italian men all have the same allele; the allele is said to be **fixed** in the population. An allele that is fixed in the population could contribute to the mean of the distribution but will not contribute to the variance; every man has the same allele, so this is not a genetic difference among the members of that population. The same gene might be polymorphic in British or Saudi men, or the allele that is fixed among Italian men might be completely absent from Saudi men. Thus, alleles that contribute to the variation in one population may or may not contribute to the variation in another population. Different genotypes can produce the same phenotype both within a population (that is, British men) and between populations (that is, British men compared to Italian or Saudi men). The fact that only genes that are polymorphic are detected when we examine variance for a complex trait illustrates why estimates of the number of genes and the contribution of each one are specific to a particular population.

KEY POINT Genes are said to be fixed in a population if all individuals in that population carry the same allele for a given gene.

If we could look at their individual genotypes—that is, if we could look at their DNA sequences directly and use them to examine the differences in phenotype, rather than height—we could see that these men do, in fact, have different genotypes and phenotypes. The phenotype in this case becomes not the height itself but rather the DNA sequence of genes that affect the height. We might be able to tell that Man 1 who is 70″ tall has one genotype, while Man 2 who is also 70″ tall has a different genotype. So these men who are 70″ tall are not really the same genetically; we just lack an assay to be able to see the phenotype in enough detail.

As a result of the Human Genome Project, our recent analysis of the genetic history of human populations, and the development of microarray technology and DNA sequencing, we now have the tools to see the phenotypes of complex traits in more detail, down to the level of differences in the DNA sequences. As a result of this confluence of advances, we can now use GWASs to identify particular genes and genotypes that give rise to complex traits. In principle, this opens up new avenues for diagnosis and therapies for many of the genetic diseases that plague adults, including heart attacks and strokes, cancer risks, obesity, type 2 diabetes, and many more. The same phenotype (that is, a disease) can have different genetic bases, which could respond differently to therapies or have different prognoses.

Many other complex traits that are not "diseases" but that do affect our health, well-being, and self-image can be analyzed; these include myopia, hair loss, susceptibility to kidney stones, ability to taste or smell certain chemicals, and responses to certain therapeutic drugs. Other complex traits affect our behavior and personality such as our willingness to take risk or our introversion or extroversion. But GWASs all begin with linkage.

10.6 Complex traits and genome-wide association studies

Let's continue this discussion of complex traits and genome-wide associations with an example that was used in one of the earliest GWAS projects. Crohn's disease is an inflammatory bowel disease that is known to be familial. It clearly has a genetic component, but it is not inherited in any simple Mendelian fashion. There is almost certainly an environmental component as well, which could include diet, infections, the bacterial inhabitants of the intestinal tract, as well as other factors, many of which have not been characterized. Crohn's disease is fairly common among white Americans, western Europeans, and Ashkenazi Jews, with about one person in a thousand being affected. People from other geographical origins are less likely to be affected. It is estimated that genetic differences account for about 35% of the total phenotypic variance associated with Crohn's disease.

GWASs were done using information from 3320 affected individuals and 4829 unaffected control individuals. The data were drawn from British, US–Canadian, and French–Belgian databases of populations. A total of 635,547 SNPs were tested. These data are shown in Figure 10.14, referred to as a **Manhattan plot** since it resembles a city skyline. This type of plot is used for nearly all GWASs, with only minor variations in the way the data are presented, so it is worthwhile discussing this one in detail.

The X-axis on the figure shows haplotypes from each region of the genome that was tested, organized by chromosome. (The regions that could not be tested, such as one on chromosome 9, are seen as small gaps.) Although the data from a chromosome look like a block with an irregular surface, this block actually is composed of thousands of dots, with each dot being one haplotype; the apparent vertical lines arise from haplotypes that cover the same genetic location. The haplotypes are plotted in the figure as they occur from left to right on the chromosome, so the first line represents the left-most haplotypes at the left end of chromosome 1 and the last line represents the haplotypes at the right end of X chromosome.

For each haplotype, the relative association with Crohn's disease is plotted as the height of the line (in other words, the Y-axis). The Y-axis represents the relative frequency or association of a haplotype in that region in affected individuals, compared to control individuals. This number on the Y-axis is not always computed in precisely the same way, but the data are interpreted in the same way—the higher the peak, the stronger the association with the trait. The haplotypes that have a statistically significant association with Crohn's disease are shown as green dots in this particular figure. From a careful scrutiny of the plot, haplotypes in 21 different regions of the genome are found to be associated with an increased occurrence or risk for Crohn's disease, as detected for the population in this study: two different regions on chromosome 1, a very strong association with a region on chromosome 5,

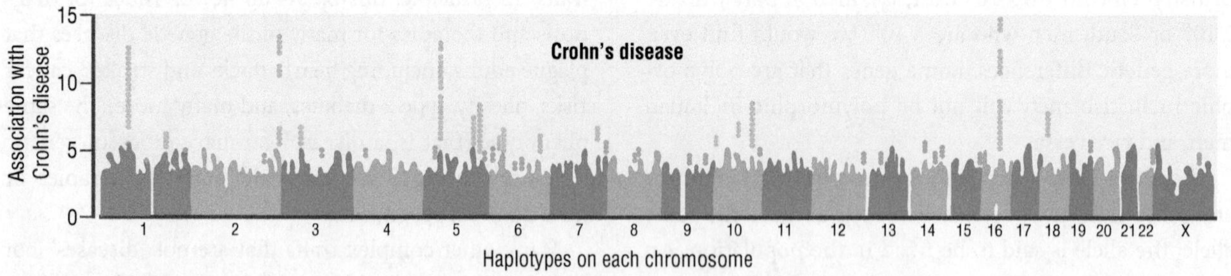

Figure 10.14 A Manhattan plot of the GWAS for Crohn's disease. This was among the first genome-wide associations to be performed. The X-axis shows haplotypes from each region of the genome that was tested, arranged by chromosome. The height of each line on the Y-axis represents the relative frequency or association of a haplotype in that region in affected individuals, compared to control individuals. The higher peaks indicate a stronger association with the disease trait. The haplotypes that have a statistically significant association with Crohn's disease are shown as green dots.

Source: Reproduced from Alexander D.H. and Lange, K. (2011). Stability selection for genome-wide association. *Genetic Epidemiology* 35 (7).

and another strong association with a region on chromosome 16, for example. Additional loci have been found to be associated with Crohn's disease in GWASs done since this original study, particularly as more populations are tested.

In the language of statistics, the experiment is testing the hypothesis that a particular locus is **not** associated with an increased occurrence of the disease or that the haplotype or locus is segregating independently from Crohn's disease. The Y-axis is the P value or the probability that the association was observed by chance. The plots are often done as a negative log scale, so a peak on chromosome 1 with a score of about 12 has a probability of occurring by chance of about e^{-12}. In other words, it is very unlikely **not** to be associated with the increased occurrence of Crohn's disease.

KEY POINT Manhattan plots used in GWASs look at the frequency or association of each haplotype in affected individuals, compared to those without the condition. This type of analysis can be used to find loci on the chromosome that are likely to have genes affecting a particular condition.

What does this tell us about the genetics of Crohn's disease? Most significantly, it tells us that many different genes can contribute to the disease, and these genes are found throughout the genome. Two individuals affected with Crohn's disease could have completely different underlying molecular and genetic causes for the disease—one affected person could have the disease variant in the center of chromosome 1, while the other affected person has the disease variant on chromosome 16. (Recall that the nature of a complex trait is that the same phenotype can arise from different genotypes.) This difference in the underlying molecular basis for the disease could have a dramatic effect on the recommended therapy or course of the disease. Perhaps affected people with the loci on chromosome 1 might respond positively to a drug or dietary change that has no effect at all on the disease in people with other variants.

If we think about this a bit more subtly, with so many different apparent causative genetic factors, perhaps "Crohn's disease" is really not a specific enough description of the phenotype. Until now, it is the best label that we have had because affected people had similar symptoms overall. But Crohn's disease is extremely variable in its manifestations, including differences in the parts of the digestive tract that are affected, the severity of the symptoms, the age of onset, and more. The GWAS suggests that there are at least 21 different loci in the genome that can cause Crohn's disease. Some of the alleles at these loci could be dominant in their effects; others could be recessive. The loci could interact with each other in complicated ways (as we saw for epistasis in Section 8.4) or have complicated interactions with different environmental factors, including the bacteria in the gut. If we want to understand the genetic, cellular, and molecular basis for Crohn's disease, these 21 loci are the places to start.

The "common disease, common variant" hypothesis and GWASs have come to dominate human genetics. Almost all of these used data from the same pre-existing data set of known polymorphisms for that population and did not require additional DNA extraction or hybridization. A complete list of GWASs that meet certain standards for population size and strength of association is maintained at http://www.genome.gov/gwastudies/; additional information, including GWASs done for organisms other than humans, can be found at http://www.gwascentral.org/. An examination of the information at this site will provide the most up-to-date and comprehensive information on traits that have been studied by GWASs and the genes associated with them, as well as links to the original publications. Many hundreds of complex traits and human diseases are cataloged at these databases, and more are added to the site regularly.

WEBLINK: VIDEO 10.1 explains another example of a GWAS for a human disease. Find it at **www.oup.com/uk/meneely**

10.7 Limitations of genome-wide association studies

While GWASs are powerful tools in human genetics, some clear limitations have emerged. The first is that, so far, not all populations can be easily studied. Nearly all of the initial studies listed earlier have been done in American–European–Canadian populations or Han Chinese and Japanese populations. There are many populations for which we simply do not know enough about the existing polymorphisms to make appropriate chips for the microarrays. This limitation is being worked out over time, and examples from different populations appear

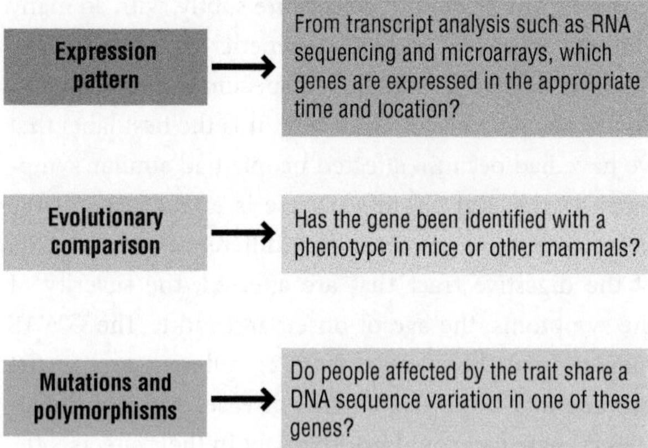

Figure 10.15 Candidate genes from GWAS. Once a region has been associated with a particular trait, it is still necessary to identify the candidate genes for the trait in that region. Three approaches that can be used are summarized here—the expression pattern, known effects in other mammals, and sequence analysis of affected individuals.

regularly in research articles. The populations do matter. As is expected for complex traits, different populations studied for the same trait often do not identify the same regions of the genome. As discussed earlier, a region of the genome that is polymorphic in one population could show an association with the disease, whereas the same region might not be polymorphic in another population and thus would not show an association with the disease.

More fundamentally, GWASs are **association** studies. Although they identify the regions of the chromosomes that are likely to contain the genes for particular traits, they do not identify the specific causative gene within that region. The polymorphisms on the haplotype are not the genes themselves; they are linked to the genes. Other work is usually done to identify the gene within that region, which can take a very long time. Fewer than half of the GWASs listed at the genome.gov website have pinpointed the specific causative gene within that region.

Identifying the causative gene is not a trivial problem. The haplotypes used in these studies typically encompass about 1 map unit on the human genetic map, which can represent a physical region of 500 kb to 10 Mb. There could be many genes within that haplotype, only one of which is responsible for the trait. Furthermore, the variation in the phenotype might be associated with the regulatory region of a gene, which is more difficult to identify computationally and can be quite distant from the protein-coding portion of the gene. How then can the investigators find the best candidate genes within a haplotype? Again, we can use the analysis of CF as an intellectual template.

So far, the best approaches have taken advantage of gene expression patterns and the evolutionary conservation of genes, as shown in Figure 10.5. The gene expression pattern is somewhat easier to determine, although it may not be as informative. Imagine that an investigator has found a haplotype that is strongly associated with a particular dental disorder such as delayed time of tooth eruption. It is reasonable to assume that a gene that affects time of tooth eruption will be transcribed in the jaw or in the bone precursor cells in that region. For most genes in humans, transcription profiles (of the type described in Section 2.5) have been done, although these are not comprehensive for all genes in all tissues at all times. The investigator can look up a database of genes found in that haplotype and determine which ones are expressed in the jaw or bones of the face. These would be good candidates for the gene that affects the disease or the trait, although they would not be the only candidates. For identifying the CF gene within the haplotype, the correlation between the expression pattern and the tissues affected by the disease was a crucial indicator.

A second powerful approach is to take advantage of the evolutionary conservation of genes and investigate similar genes in mice or other mammals. Many different mutations have been identified in mice, and there are ongoing projects to knock out every gene in a mouse one at a time and observe the effects. Perhaps some of the genes in the haplotype have equivalent genes in the mouse that have been analyzed in this way. If one of these genes causes a defect in tooth formation when mutated in mice, then it becomes a very strong candidate as the causative gene in humans. Less was known about genes in other mammals when *CFTR* was analyzed, but the recognition that the gene had similarity to chloride ion channels was reassuring.

A third approach is probably the most powerful of all. In this approach, genes located within the haplotype are directly sequenced from several affected individuals (Figure 10.5). If two individuals are affected by the same disease or trait, and the gene for that gene or trait has been mapped to the same location, then the affected individuals are expected to have mutations in the same gene. This has been done for some single-gene traits in humans and appears to work well. The major challenge is sorting through all of the sequence differences that have been found between the affected individuals to find the one that is causative for the trait.

This third approach has been done with single-gene traits mapped by GWASs but has only been minimally used for complex traits. With cheaper technologies available for sequencing and more sophisticated analysis software, a whole-genome sequencing approach will likely become the preferred method for associating loci with phenotypes in populations.

GWASs and Darwin's finches

You may recall that Chapter 1 begins with the analysis of the genomes of different species of Darwin's finches in order to find genes that affect the shape of the beak. Now that we have described how a GWAS is done in humans, we can revisit that example. Sites that were polymorphic among the genomes of the finches were found, and, because the species are closely related and recently descended from a common ancestor, the genomes of different species could be directly compared. The data were sorted, based on whether the species had a blunt or a pointed peak, and the regions of the genome that were associated with blunt beaks were evaluated. Because they were comparing different species, they did not use polymorphisms within a species like a GWAS does but instead examined regions that were fixed for different haplotypes in different species. In other words, they did a genome-wide association to find 15 regions of the genome that had a gene or genes affecting the shape of the beak.

Having found these regions, they focused on the genes in one particular region. All three of the approaches outlined earlier were used to find the best candidate gene within this region. They used the expression pattern. In other vertebrates, *ALX1*, the gene of interest, was expressed in a pattern consistent with a role in beak morphology. Second, they asked about the role of *ALX1* in mice and humans, finding that mutations of the *ALX1* gene in mice and humans result in profound changes in craniofacial structure. Finally, they directly sequenced the *ALX1* gene and surrounding region from species with broad or pointed beaks and identified eight specific molecular changes between broad- and pointed-beak species in and near the *ALX1* gene.

KEY POINT GWASs are limited by the fact that only some populations have enough data for analysis and that these studies can only highlight associations between regions of the chromosome and a trait. Other techniques must be used to identify the causative gene or genes for each condition.

10.8 Summary: genome-wide association studies combine evolution, genomes, and genetics

GWASs represent a powerful fusion of evolution, genomes, and genetics. They rely on linkage, one of the most basic principles of genetics. Linkage shows up as haplotypes, which arose during human evolution and the dispersal of humans across the Earth. Finally, we can only do these experiments because our genes and our evolutionary history are recorded in our genome, the molecular and biochemical ledger of our past and present. Max Delbrück said, "Any living cell carries with it the experiences of a billion years of experimentation by its ancestors." GWASs are attempts to read the laboratory notebooks of that experimentation, as recorded in our genomes.

CHAPTER CAPSULE

- Gene mapping in humans has been important in positional cloning of many genes that affect human traits and diseases.
- The Human Genome Project has enabled the identification of many millions of sites in the human genome that are polymorphic, that is, where more than one sequence or allele is common.
- A block of the chromosome that is inherited together is a haplotype. The human genome can be subdivided into about 200,000 haplotypes.

- Individuals whose ancestors came from the same geographical region share many polymorphisms and many haplotypes because the current population arose by rapid expansion from small founder populations.
- Alleles and polymorphisms for traits and diseases that were linked result in a persisting association that can be detected many generations later.
- The polymorphisms that are associated on haplotypes from these founding populations can be found by microarray analysis.
- Genome-wide association studies (GWASs) have used the haplotype structure and microarrays to identify associations and linkage for thousands of genes.
- The "common disease, common variant" hypothesis states that individuals with a common genetic disease or trait whose ancestors came from the same geographical area are very likely to have the same molecular variant that produces the disease or trait.
- Complex traits are heritable traits whose phenotypes depend on the actions of many genes and the environment. Many adult-onset diseases, such as cancers and cardiovascular diseases, are examples of complex traits.
- GWASs have been very productively applied to identify many genes responsible for complex traits.

STUDY QUESTIONS

Concepts and Definitions

10.1 What is a complex trait? Give an example of a complex trait other than any of the ones used in the chapter.

10.2 Define linkage disequilibrium, and discuss why it is important in natural populations such as humans.

10.3 What is a haplotype?

10.4 Describe the "common disease, common variant" hypothesis.

10.5 What is a QTL, and how does it compare to the genes discussed in previous chapters?

Beyond the Concepts

10.6 Give at least two ways that genetic maps are used by geneticists working with model organisms. How is this similar or different from how genetic maps can be applied to humans?

10.7 Infants in the United States and the United Kingdom are tested at birth for the genetic disease phenylketonuria (PKU). Use the information about PKU found at OMIM (**http://www.ncbi.nlm.nih.gov/omim/**) to answer the following questions. You will need to access the records on both the PKU disorder (#261600) and the causative gene (#612349).

a. How is the disorder inherited, that is, is it autosomal or X-linked, recessive or dominant? What does this mode of inheritance suggest to you about the normal function of the gene product?

b. In your own words, what is the nature of the biochemical defect in PKU children, and how does this biochemical change lead to phenotypic changes?

c. Is there an available therapy for PKU children, and, if so, what does it entail?

Challenging **d.** What is the most likely basis for the phenotypic variability in PKU patients?

e. Is there any population for which PKU could be considered a common disease?

f. Briefly explain the underlying molecular basis for one or more of the allelic variants that has been described. Is there a common variant for any population?

10.8 Genome-wide association studies (GWASs) are widely used for identifying the genes for human diseases and genetic traits.

a. Outline the steps performed in a GWAS.

b. How does knowing the sequence of the human genome make a GWAS feasible?

c. What is the primary limitation in using a GWAS to find a causative gene for a human trait?

10.9 The data shown in Box 10.2 Figure B(i) and (ii) summarize the variants found in one child's genome by exome sequencing and how these variants were filtered to find the causative gene. How do these data support one assumption of the "common disease, common variant" hypothesis?

Challenging **10.10** Sickle-cell anemia is due a specific mutation in the β-globin gene in humans. As discussed in Chapter 16, sickle-cell anemia is common in parts of the world where malaria is also common because heterozygotes for this specific β-globin mutation are resistant to malaria. Thus, sickle-cell anemia is a common disease among many populations. However, although it is a common disease with a common molecular variant, individuals with sickle-cell anemia from different populations do not have the same haplotype. How can this be explained?

10.11 Explain how GWASs combine evolution, genomes, and genetics.

Applying the Concepts

10.12 As noted in Chapter 9, black body and purple eyes are linked autosomal traits in *Drosophila*, with black and purple being recessive to wild-type body and eyes. One laboratory in an introductory genetics class used flies with the traits to illustrate mapping. When cleaning out the fly incubator at the end of the term, you find several vials that students forgot to throw out. You count the number of flies of each phenotype class in some of the vials; the results are shown in Table Q10.1. (These results are intended to be illustrative and might not represent the actual number of flies from such an event.)

Table Q10.1

Vial	Wild-type	Black body, wild-type eyes	Wild-type body, purple eyes	Black body, purple eyes
1	2454	631	594	2131
2	893	3298	3415	771
3	4632	1181	1098	4254

a. What were the parental phenotypes that were used to set up each vial, assuming that the parents were homozygotes?

b. Note that, in all three vials, the number of wild-type flies is greater than the number of black and purple-eyed flies. Using a χ^2 test (from Chapter 5), is the difference between wild-type and black purple-eyed flies in vial 2 statistically significant?

c. Postulate a reason that the number of wild-type flies is greater than the number of black and purple-eyed flies in every vial.

Challenging
Looking back

d. In one vial, which appears to have come from a different experiment altogether, there were about equal numbers of wild-type flies, black-bodied but not purple-eyed flies, wild-type body but purple-eyed flies, and black-bodied and purple-eyed flies. How could this result be explained? (Hint: Look back at Chapter 8, Study Question 8.26.)

Challenging

e. In light of your answer to (d) above, why is body color in *Drosophila* not considered a complex trait? Or should it be considered a complex trait?

10.13 The length (in Mb) of haplotypes in humans can vary by a factor of 10 or more. What are some of the reasons that haplotypes in different regions of the genome vary in length so much?

10.14 An allele that contributes to a complex trait might be polymorphic in one population but not in another population.

a. What impact does this have on estimates about how many genes affect a complex trait?

b. An allele affecting a complex trait is fixed in one population but not in another. In the population in which it is fixed, what statistical parameter is affected?

10.15 An extensive study published in *Nature Genetics* did a GWAS to identify genes or genetic regions associated with a particular measure of intraocular pressure (IOP) and glaucoma. These are two different, but related, eye conditions; most individuals with high IOP eventually develop glaucoma, but not all cases of glaucoma include high IOP. The study involved more than 35,000 people of either European or Japanese heritage and tested more than 1.2 million SNPs on all 22 autosomes. The results are presented in the Manhattan plot shown in Figure Q10.1.

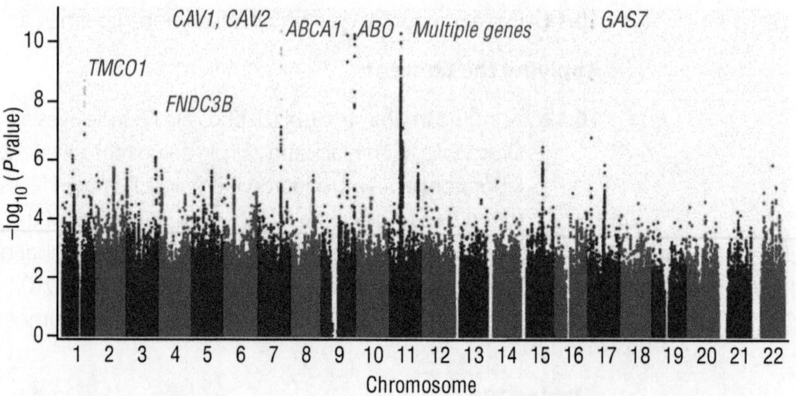

Figure Q10.1

Seven distinct genetic regions were found in this study. Previous studies had identified the loci at *TMCO1*, *CAV1* and *CAV2*, and *GAS7* as being associated with high IOP and glaucoma, so those are shown in light gray. The newly identified regions are *FNDC3B*, *ABCA1*, *ABO*, and "multiple genes." (It may not be easy to tell from the plot, but *ABCA1* and *ABO* are distinct from each other.) Based on this plot, answer the following questions.

a. What does each of the dots represent?

b. Why are there vertical stacks of dots in the regions where candidate genes are found?

c. What does "$-\log_{10}$ (P value)" on the Y-axis mean?

d. Why is there a white line in the middle of chromosome 9 (and also on chromosomes 1, 16, and 19)?

10.16 The named genes in Figure Q10.1 are the ones that the investigators think are the best candidates for being involved in high IOP and glaucoma. The region marked "Multiple genes" includes the genes *AGBL2, SPI1*, and *PTPRJ*, but the investigators could not distinguish which of these three genes was the best candidate (or if more than one is a candidate). They did not sequence the genes from any affected patients, but they did use two other types of evidence to determine the best candidate genes in each region. What **two types of evidence**, other than the linkage association and direct sequencing, could the investigators use to determine which genes are the best candidates for being involved in glaucoma?

10.17 The locus on chromosome 9 is the ABO blood type locus, and the SNP is actually an I^B allele. The physiological reason for this association between type B blood and glaucoma is not known, but it was also previously suspected from clinical information. No similar association with glaucoma is found for type A or type O blood. The information that the I^B allele is also associated with an increased incidence of glaucoma is probably an example of which of the following genetic phenomena? (multiple choice)

a. Epistasis

b. Complementation

c. Pleiotropy

d. Linkage disequilibrium

e. Reduced penetrance

10.18 Imagine that we are 10 years in the future, and you are an ophthalmologist working with glaucoma patients. When a patient with glaucoma arrives in your office, you are able to examine their DNA sequence as easily and cheaply as any other vital sign to look for these "risk polymorphisms." Which of these statements comes closest to what you expect to see, and why? (multiple choice)

a. Most patients will have the risk polymorphism for many of these seven loci, and some will show all seven polymorphisms.

b. Most patients will have one, or possibly two, of the risk polymorphisms, and different patients may have different risk polymorphisms.

c. Since none of the risk polymorphisms is found on the X chromosome, nearly all of your patients will be women.

d. All of your patients will have at least one of these polymorphisms, some will have two or three, a few will have four or more, but everybody will have at least one of them.

e. Even within a family in which multiple siblings have glaucoma, they probably will have different risk polymorphisms.

10.19 The same issue of *Nature Genetics* included two more GWASs for glaucoma. All three studies found the association with *ABCA1*. (Hint: Read both parts of the questions before answering either. The answers for (a) and (b) are not the same, but both relate to different ways in which the studies were done.)

a. A GWAS from an Australian population that used glaucoma, but did not look at high IOP, found associations with two other genes but did not find some of the ones reported here. Very briefly, explain why this study found some different genes for glaucoma.

b. A GWAS from a Han Chinese population that also used a measure of high IOP and glaucoma found one gene that was not identified in the large mixed population study but did not find some of the other genes reported in the large mixed population study. Briefly explain why the GWAS based solely on the Chinese population identified a different gene for IOP but failed to find some that the mixed population study found.

10.20 Ornithologists (bird experts) can score the degree of pointedness of a finch's beak numerically. The gene *ALX1* was recently found to be important for beak pointedness. Investigators measured the pointedness of 62 finch beaks and determined whether each finch was homozygous for either of two *ALX1* alleles, ***B*** and ***P***, or heterozygous. The averages and ranges of pointedness they recorded for each genotype are reproduced Figure Q10.2. While ***PP*** birds showed a higher average pointedness, compared to ***BP*** and ***BB*** finches, beak pointedness did not differ significantly from one category to another, even though other data clearly point to a role for ALX1 in beak shape. What is your best explanation for why there are overlaps in pointedness among the three categories?

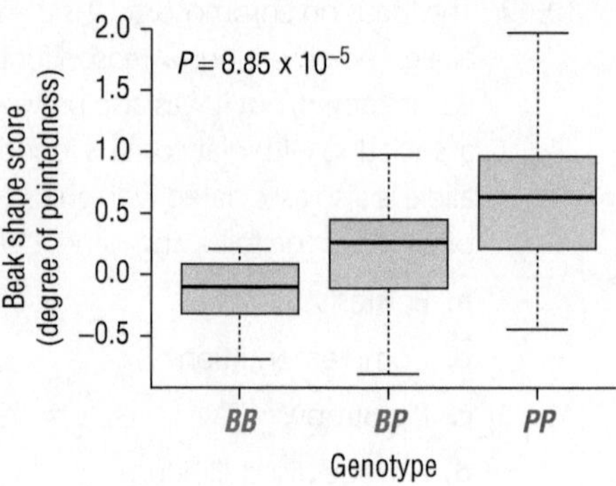

Figure Q10.2

10.21 Figure Q10.3 shows the results of a GWAS of beak shape that was produced by scanning for SNPs in 15-kb windows along alignments of the genomes of two blunt-beaked *Geospiza* finches and two pointed-beaked finches from the same species.

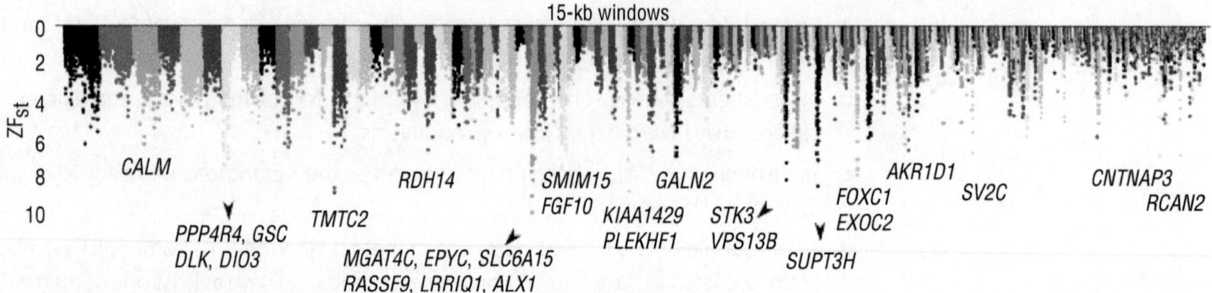

Figure Q10.3

a. There are a wide range of Galapagos finches with blunt and pointed beaks. Why did the investigators limit themselves to closely related finch species?

b. The genes marked on the figure are in regions containing the most variation. (Multiple choice) These genes:

i. Are unlikely to be associated with beak morphology

ii. Are almost certainly associated with beak morphology

iii. May be associated with beak morphology

CHAPTER 11

Exchange and Evolution

11

IN A NUTSHELL

In earlier chapters, we have described the inheritance of DNA from generation to generation. However, organisms can also acquire DNA from other individuals over the course of their lifetime, a process known as horizontal gene transfer. Horizontal gene acquisition is an important force in shaping the content and structure of genomes. Although studied most widely in bacteria, horizontal gene transfer occurs within and across all kingdoms of life. Instances of extensive horizontal gene transfer challenge the traditional view of evolutionary lineages as "branching trees", with even limited horizontal gene transfer having profound evolutionary consequences.

11.1 Overview: vertical and horizontal transmission of DNA

All the genetic processes described in previous chapters have been examples of the **vertical transmission** of genetic material from parents to progeny, that is, through successive generations. We noted in Chapter 2 that one of the essential properties of genes is that they are the units of biological inheritance—vertical transmission of information from one generation to the next. The processes of DNA replication and meiosis, as well as the principles of Mendelian genetics, have all centered on the transmission of DNA to the next generation, with its associated changes. As we have described it so far, variants in the genome, that is, new alleles and polymorphisms, arise primarily from mutations that occur during replication. New combinations of the variants arise as a result of recombination and the independent assortment of chromosomes, which are key aspects of meiosis.

Vertical transmission also occurs in cells and organisms that reproduce asexually, or without meiosis, as shown in Figure 11.1. Asexual reproduction is very common. Eukaryotic cells divide by mitosis, and bacteria reproduce by binary fission; both of these are examples of asexual reproduction. Mutations arise, and these are passed on to the daughter cells, so a small amount of genetic variation occurs. However, since there is no meiosis in organisms that reproduce asexually, the expectation is that, while new alleles will arise, new combinations of variants that arise from recombination or the assortment of chromosomes will not be

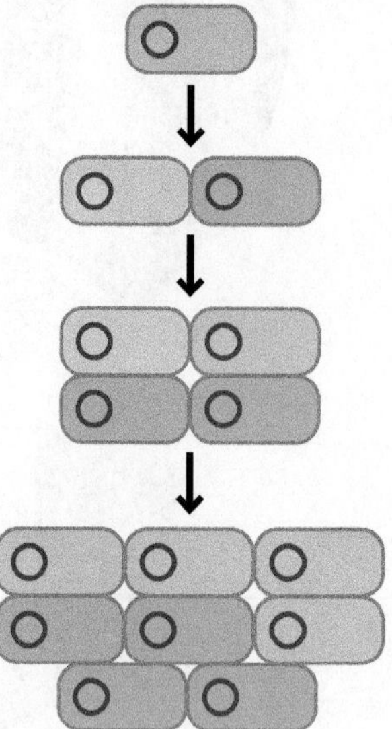

Figure 11.1 Vertical gene transfer. Transmission of DNA from one generation to the next, as shown here with cells dividing by binary fission, is an example of vertical gene transfer. A few mutations arise, so the cells are slightly different shades of blue, but the amount of variation is slight.

seen. Thus, only a few changes are expected during asexual reproduction.

But there are other processes that result in new combinations of alleles and are major sources of genetic change. Genetic information is also transferred between cells within a generation; this is known as **horizontal (or lateral) gene transfer**, as summarized in Figure 11.2. A cell that reproduces asexually can have new combinations of alleles and polymorphisms, and even entirely new genes, because DNA has been acquired from another cell in the population or from the environment. Once this new DNA is stably acquired, it can become a permanent part of the genome of the recipient cell and subsequently be passed vertically to the daughter cells. Thus, one key distinction between vertical and horizontal gene transfer is the **source** of the genetic variation—from the parents in vertical gene transfer or from other cells or the environment in horizontal gene transfer.

Another key difference between vertical and horizontal gene transfer is the **direction** of exchange or transfer that occurs. Vertical gene transfer through sexual reproduction involves the production of gametes, which unite to form a zygote. Both gametes contribute genomes to the zygote, and neither gamete could be considered a "recipient," since a new diploid cell—the zygote—containing both

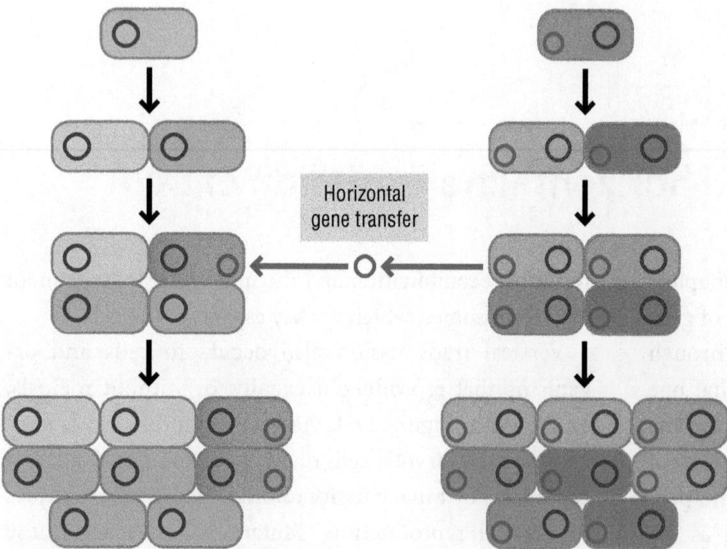

Figure 11.2 Horizontal gene transfer. Horizontal gene transfer occurs between cells in different lineages or populations, as shown by the red arrow and the acquisition of the plasmid (red circle) from the population of orange cells on the right to one cell of the population of blue cells on the left. Acquisition of the new DNA alters the phenotype of the recipient cell, as indicated by the purple color. Once acquired, the new DNA is passed vertically to the next generation, like any other DNA. Although this is illustrated with a plasmid in bacteria for simplicity, many horizontal gene transfer events involve integration into the genome, most are mediated by molecules other than plasmids, and many examples involve eukaryotes.

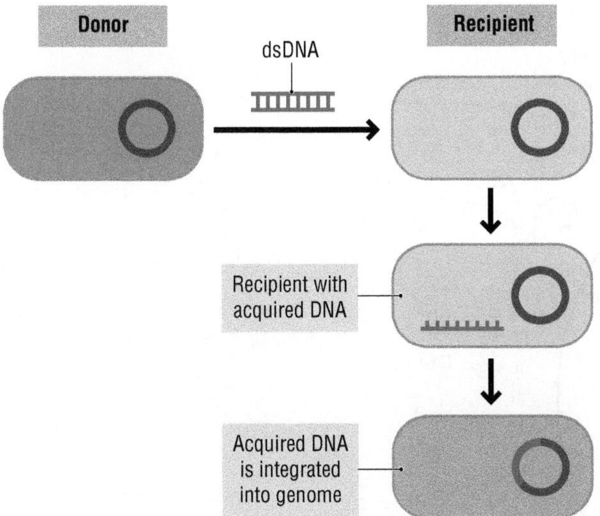

Figure 11.3 Direction of horizontal gene transfer. Horizontal gene transfer involves a donor cell, shown here in red, and a recipient cell, shown in blue. The DNA being transferred is shown as a double-stranded red linear molecule (dsDNA), but different mechanisms of horizontal gene transfer involve circular or linear DNA, and double-stranded or single-stranded DNA, or even RNA. The recipient cell acquires the DNA, here as a single-stranded molecule, and integrates it into its genome. The cell with the newly acquired DNA is shown in purple, indicating that its phenotype has changed.

haploid genomes is produced. In fact, both cells are donors to the genome of the zygote. In horizontal gene transfer, one cell is a recipient that receives DNA from a donor cell directly or indirectly, and donor DNA becomes part of the recipient cell's genome, as shown in Figure 11.3. On the other hand, DNA from the recipient cell is not passed to the donor. Thus, horizontal gene transfer occurs in one direction, from donor to recipient.

Horizontal gene transfer begins with specific mechanisms for importing foreign DNA that nearly all organisms have. For example, viruses and transposable elements, which are found in all organisms, are vectors for horizontal gene transfer. But horizontal gene transfer involves more than simply importing the DNA. Once DNA is acquired horizontally, it must be integrated into the recipient's genome in order to be transmitted vertically to the next generation. If it is not integrated, it is lost by the time the cell divides. Acquisition of new DNA can present a certain hazard to the organism. In evolutionary terms, mechanisms that import and integrate foreign DNA could have enormous selective disadvantages, since it is possible, and even likely, that DNA imported indiscriminately might be detrimental to a recipient cell. Think about picking up random objects found on the sidewalk; occasionally, you will find a coin, but more often you will find something of little value or even something harmful. Thus, most organisms have safeguards that limit the amount and types of horizontal gene transfer. Horizontal gene transfer only rarely involves integration of random DNA into the chromosome because this could be deleterious; the process must be regulated in some way.

While the same evolutionary principles concerning horizontal gene transfer apply to all organisms, they are particularly recognized in bacteria. All bacteria are unicellular and reproduce asexually. Horizontal gene transfer was first seen in bacteria, occurs frequently among them in nature, and has been best studied in these organisms.

KEY POINT Meiosis is not the only means through which different genomes recombine; individuals in the same generation can exchange DNA through other mechanisms. Genetic exchange can also occur in organisms that reproduce asexually if DNA is transferred from one cell to another horizontally.

Most of what we know about horizontal transfer comes from studies with bacteria, but remember that horizontal transfer occurs in, and among, other organisms as well; some of these are described in Section 11.7. Certain species of bacteria exchange DNA horizontally frequently enough to be referred to as "promiscuous." Conversely, other bacterial species do not appear to exchange DNA very much at all. We will focus on one very well-studied species, *Escherichia coli*, with occasional references to other species, but be aware that imagining that *E. coli* is representative of all bacteria would be equivalent to, or even more simplistic than, assuming that fruit flies are representative of all animals. Nonetheless, the use of a model organism to investigate the basic principles and mechanisms of a process like horizontal gene transfer remains a powerful approach.

Chromosomal and extrachromosomal DNA

Before we explore horizontal gene transfer processes, let's review the genomes of typical bacterial cells. As we summarized in Section 3.2, the chromosome of bacteria, such as *E. coli*, is a closed circular molecule that has all of the necessary genes for growth, reproduction, and metabolism. Most known bacteria have a single circular chromosome like *E. coli*, although *Vibrio* species have two circular chromosomes and a few species have linear chromosomes. For a typical *E. coli* strain, the chromosome is 4 to 5 million base pairs and has about 4500 genes; the standard laboratory strain K-12 has a genome of 4.6 million base pairs that encodes approximately 4300

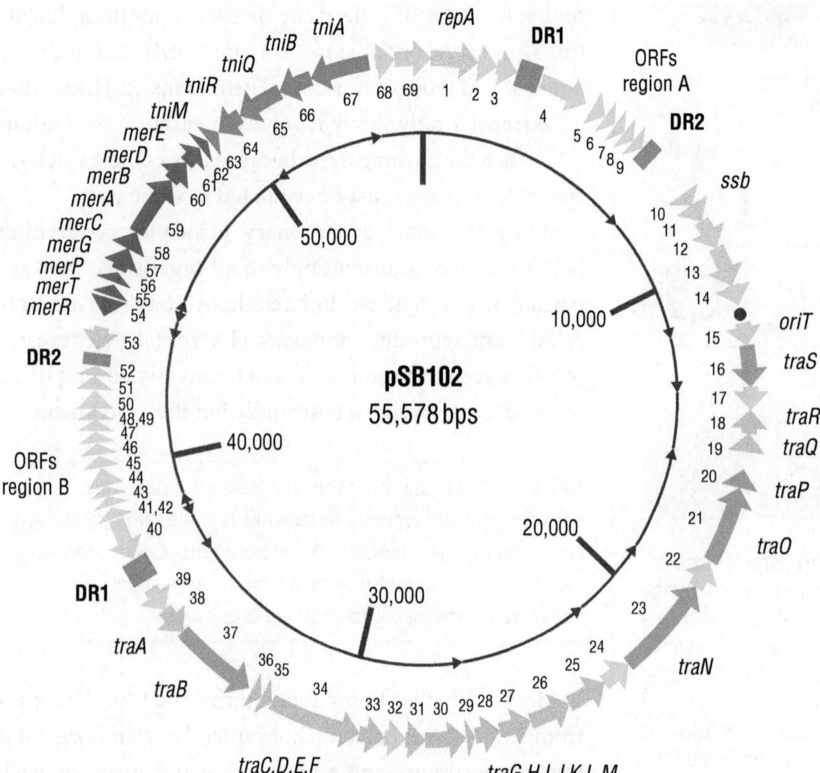

Figure 11.4 A naturally occurring plasmid found in some bacteria. This schematic diagram shows the genes as arrows, with the direction of the arrow indicating the direction of transcription. The genes required for self-transmissibility (discussed later) are shown in green, while the genes required for autonomous replication are shown in yellow. This plasmid encodes genes that confer resistance to mercury (in red), as well as other and unknown functions. This plasmid is 55.6 kb, but plasmids vary widely in size and genetic composition.

Source: Reproduced from Schneiker, S. et al. (2001). The genetic organization and evolution of the broad host range mercury resistance plasmid pSB102 isolated from a microbial population residing in the rhizosphere of alfalfa. *Nucleic Acids Res.* 2001 Dec 15;29(24):5169–81.

protein-coding genes. The chromosome is replicated and segregated into two daughter cells in each generation, a common form of vertical gene transfer.

In addition to the chromosome, bacteria also have smaller molecules of **extrachromosomal DNA**, which can be transient or can be stably transmitted from generation to generation. The most common molecules of extrachromosomal DNA are circular **plasmids**, which replicate independently of the chromosome itself. An example of a naturally occurring plasmid is shown in Figure 11.4. A given bacterial strain may have several plasmids or none at all. Plasmids range in size from about 2 kb to perhaps 2 Mb. They encode genes that can confer a very wide range of functions, including antibiotic resistance, pathogenicity, substrate utilization (such as hydrocarbon breakdown), and DNA mobility. We will consider the role of some plasmids in horizontal genetic exchange in more detail in Section 11.3.

Three different mechanisms for horizontal gene transfer are well known in bacteria: transformation, conjugation, and transduction. These serve as our models for thinking about this process in all organisms, although it is likely that many different mechanisms are involved.

11.2 Transformation

The simplest type of horizontal genetic exchange is **transformation**, illustrated in Figure 11.5. Transformation is the direct uptake of DNA molecules by a cell. This is the process often used in molecular biology laboratories to introduce specific genes—such as those conferring ampicillin resistance carried by a plasmid—into a population of bacteria. Transformation occurs because some bacterial cells, referred to as **competent cells**, have the ability

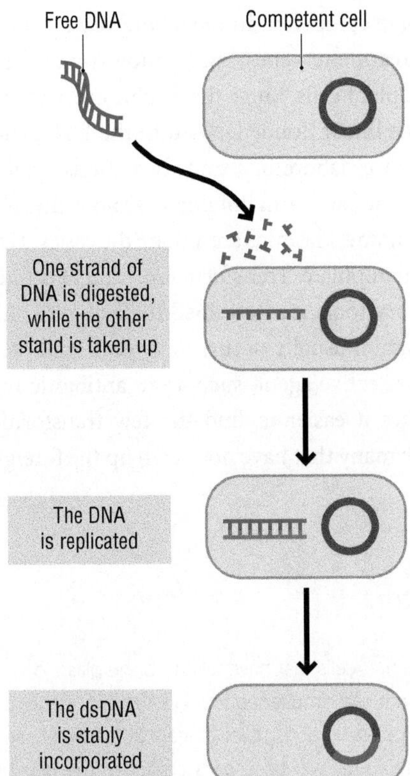

Figure 11.5 The process of transformation. Free DNA in the environment is taken up into competent cells—that is, cells capable of being transformed—with the degradation of one strand. The remaining strand is replicated, and the double-stranded DNA (dsDNA) that results is stably incorporated into the chromosome (shown at the bottom) or kept as a plasmid.

to take up extracellular or naked DNA from the environment. Energy is needed to import DNA, and the DNA often gets degraded upon import. As a result, only very few cells among the billions of bacteria in a culture are transformed, that is, will incorporate extracellular DNA into their genomes.

During transformation, linear DNA is brought into the cells as a single-stranded molecule; the other strand is degraded during uptake. The remaining single-stranded DNA is then replicated, resulting in double-stranded DNA. In order to persist in the cell and be inherited, this newly introduced DNA must become part of the genome of the recipient. This can happen by extrachromosomal maintenance of the DNA as a plasmid or integration into the chromosome via recombination. Although it is used as a standard model for mechanisms of horizontal gene transfer, *E. coli* is not naturally competent and has to be grown and handled using special procedures in the laboratory in order to be transformed. Only about 100 species of bacteria are known to be naturally competent.

KEY POINT Transformation occurs through the uptake of naked DNA.

In nature, transformation is an important mode of DNA transfer in the few bacterial species that are naturally competent. Transformation in bacteria also played an essential role in the recognition that genes are made of DNA. For example, as mentioned in the Prologue, Griffith, and later Avery and colleagues, used naturally competent *Streptococcus pneumoniae* (then referred to as pneumococci) to demonstrate that DNA is the "transforming principle," the molecule of inheritance. Had they chosen one of the many naturally non-competent bacteria for which other DNA exchange processes are more important, such as *E. coli*, it might have taken longer to uncover the central role that DNA plays in biology. Today, this natural competence of *S. pneumoniae* thwarts our ability to control disease-causing strains—this species can acquire multiple antibiotic resistance genes through the process of transformation, allowing the organism to survive conventional drug treatments.

In naturally competent bacterial cells, most transformation involves linear DNA fragments. These linear fragments must recombine with the host chromosome in order to be stably maintained, as shown in Figure 11.6.

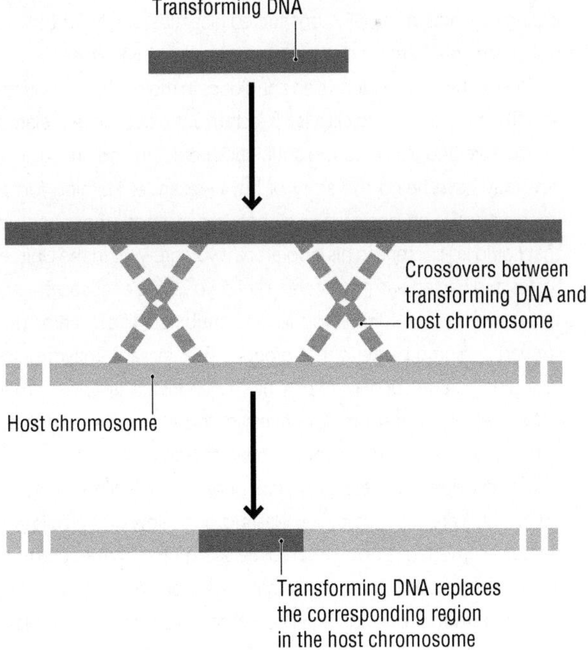

Figure 11.6 Two crossovers are needed to incorporate the transforming DNA. As shown by the dotted lines, the transforming DNA requires a double crossover to be incorporated into the host chromosome. Transforming DNA that is maintained as a plasmid does not require such a double crossover.

If the crossovers with the host chromosome do not occur, the linear DNA fragment will be lost. Note that integration involves two crossovers, one on each side of the DNA fragment being integrated.

Transformation is a common laboratory technique

Transformation is by far the most widely used method for introducing foreign DNA into bacteria in the laboratory. A similar technique is also used to introduce DNA into eukaryotic cells where it is termed **transfection**. Bacterial cells, even from species that are not naturally competent, can be made competent in the laboratory by chemical or osmotic treatment; genes can be moved into these competent recipient cells. Since the double crossovers needed to integrate linear donor DNA into the host genome are relatively rare, laboratory transformations typically use plasmids that can be maintained without integration in the host genome. Plasmids are among the easiest DNA molecules to manipulate. Transformation and plasmid vectors as laboratory tools are described in Tool Box 11.1. Since transformation usually occurs in a very small percentage of cells, a selective agent, such as an antibiotic resistance gene, makes it easier to find the few transformed cells among the many that have not taken up the foreign DNA.

TOOL BOX 11.1 Transformation with plasmid vectors

Transformation is widely used in the laboratory to introduce new or altered DNA molecules into a bacterial cell. This is done most commonly with a plasmid. In order to stably transform a bacterial strain with a plasmid, some method is needed to ensure that the plasmid has been transferred and is being maintained as the cells divide. The most common procedure is to include an antibiotic resistance gene on the plasmid. *Escherichia coli* is naturally sensitive to antibiotics such as ampicillin, tetracycline, and many more—resistance to which is conferred by genes that are often borne on naturally arising plasmids. Artificially constructed plasmids carrying the same resistance genes are used as tools in molecular biology research.

One of the most widely used antibiotics in molecular biology research is ampicillin. Ampicillin is a β-lactam antibiotic, closely related in structure and action to the antibiotics penicillin and amoxicillin. You may have heard the story of how Alexander Fleming fortuitously discovered penicillin as a chemical secreted from a mould that could kill bacteria in his laboratory. If you grew up in the United States and had an ear infection as a child, you were almost certainly treated with amoxicillin. Ampicillin and antibiotics like it can be destroyed by some bacteria that produce β-lactamase enzymes. By including a gene, such as *amp*^{R}, which encodes a β-lactamase that confers ampicillin resistance, on a plasmid, it is possible to identify and select for those cells that have been transformed.

So how does this selection work in practice? Millions of competent cells are exposed to the plasmid, and a few of the cells will be transformed. The population of cells is grown in medium that includes ampicillin. Those cells that have not been transformed die, whereas the few individual cells that have taken up the plasmid with the *amp*^{R} gene will grow into ampicillin-resistant colonies. Thus, cells that include the *amp*^{R} gene have been selected for; the *amp*^{R} gene is referred to as a **selectable marker**. Even if only a tiny fraction of the potential recipient cells are transformed, the powerful selection using ampicillin allows resistant cells to repopulate an entire culture.

The idea is that cells that have taken up the plasmid with the *amp*^{R} gene on it can simultaneously receive other genes on the plasmid. Each plasmid consists of at least 2500 DNA base pairs, of which only about 900 base pairs comprise the *amp*^{R} gene. The rest of the vector must contain genes and sequences that will allow it to replicate in a new cell. However, other genes and functional DNA sequences can then be added to the plasmid and will be transformed into the bacterial cells along with the *amp*^{R} gene. As such, the plasmid is being used as a vector to carry other

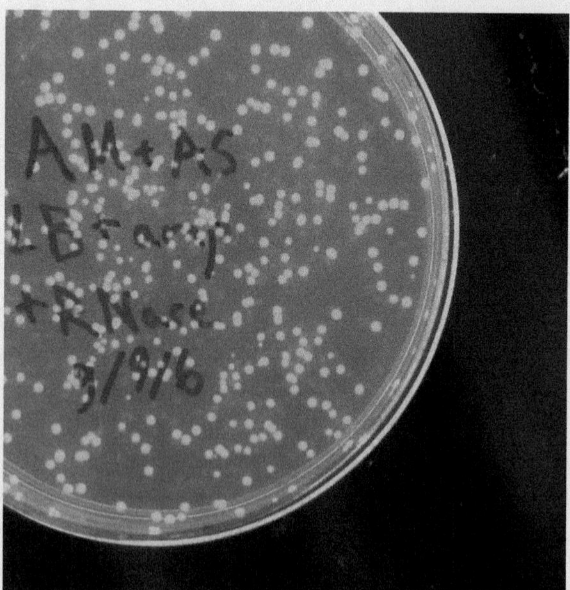

Figure A *E. coli* expressing ampicillin resistance and green fluorescent protein (GFP). A plasmid that included an *amp*^{R} gene and the GFP gene was transformed into *E. coli*, and the bacteria were plated on medium with ampicillin. Colonies that were able to grow because they were ampicillin-resistant also expressed GFP. Source: Courtesy of Lisa Mattei.

TOOL BOX 11.1 Continued

genes, and the selectable marker (in this case, the *amp*^R gene) is being used to select for the transformants.

If the other gene or genes has the normal transcriptional regulatory sequences recognized by *E. coli*, it will be transcribed as if it is a bacterial gene. If it also has the signals needed for translation, the transcript can be translated into protein. This method is often used to produce large quantities of a protein; the gene encoding the protein is inserted on a plasmid that also includes the *amp*^R gene, or another selectable marker, and is expressed in bacterial cells. For example, bacteria can be made to glow green because they have been transformed with a plasmid carrying an *amp*^R selectable marker and a green fluorescent protein (GFP) gene from a jellyfish, as shown in Figure A.

Hundreds of different plasmid vectors are commercially available, with many different selectable markers and signals for expression. Usually, the sequence of the plasmid has been artificially manipulated to make the insertion of other genes or sequences easier. Typically, the plasmid will include one or more sites for insertion of other DNA, known as a **poly-linker** sequence or **multiple cloning site (MCS)**. For example, the pUC18 and pUC19 plasmids have this sequence, as shown in Figure B. Vectors pUC18 and pUC19 replicate very efficiently in *E. coli* from their origin of replication (*ori*), with as many as 300 copies per cell being made. The plasmids carry an ampicillin resistance gene, which serves as a selectable marker, as well as a β-galactosidase gene (*lacZ*) that contains an MCS, which can be cut by several restriction enzymes. When a piece of extraneous DNA is cloned into the MCS site, the *lacZ* gene is inactivated. Therefore, clones containing inserts can easily be distinguished from the vector because they lack β-galactosidase activity.

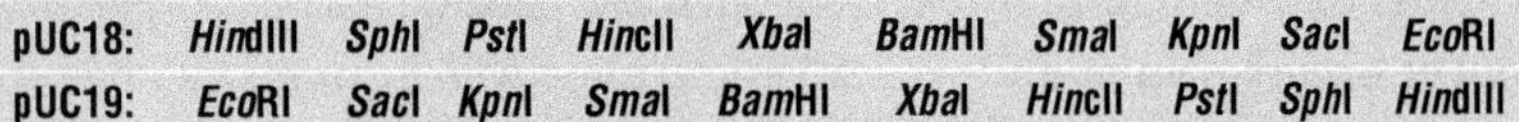

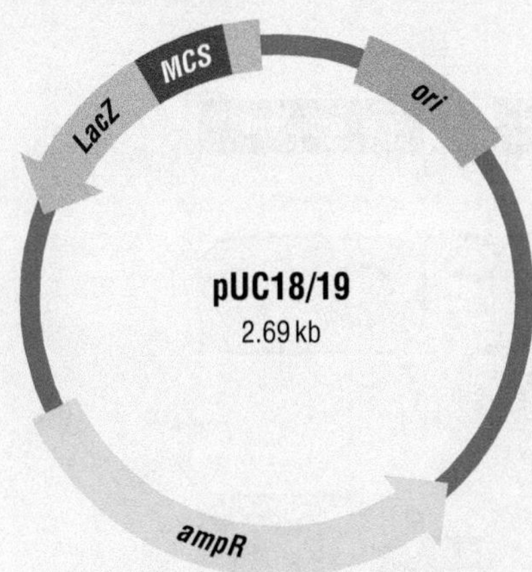

pUC18

399 … 450

GCCAAGCTTGCATGCCTGCAGGTCGACTCTAGAGGATCCCCGGGTACCGAGCTCGAATTC

*Hin*dIII *Sph*I *Pst*I *Sal*I / *Hin*cII / *Acc*I *Xba*I *Bam*HI *Sma*I / *Xma*I *Kpn*I *Sac*I *Eco*RI

pUC19

396 … 447

GAATTCGAGCTCGGTACCCGGGGATCCTCTAGAGTCGACCTGCAGGCATGCAAGCTTGGC

*Eco*RI *Sac*I *Kpn*I *Sma*I / *Xma*I *Bam*HI *Xba*I *Sal*I / *Hin*cII / *Acc*I *Pst*I *Sph*I *Hin*dIII

Figure B Plasmids A plasmid used as a cloning vector. This map shows the commercially available cloning vectors pUC18/19, among the first widely used plasmid vectors. The locations of various restriction sites are shown. In addition to the origin of replication (*ori*), which allows the plasmid to replicate, and the *amp*^R gene, which allows for selection of cells with the plasmid, the plasmid also includes the *lacZ* gene. *lacZ* encodes β-galactosidase, as described in Chapter 14, and is used here as an indicator. As discussed in Tool Box 15.1, if cells are grown on plates with a chemical known as X-gal, the cells expressing lacZ will cleave X-gal to make a blue substrate. Note that the multiple cloning site (MCS) is within the *lacZ* gene. Thus, if a sequence of interest has been inserted at the MCS, the *lacZ* gene is disrupted and not expressed, so the colonies are white. This allows for a procedure known as a blue–white selection. As shown below the plasmid, the MCSs for pUC18 and pUC19 have different sequences and different enzyme restriction sites.

While transformation is widely used in the laboratory, it may not be the most significant form of genetic exchange in natural populations. The natural process by which cells become competent is mechanistically quite distinct from the laboratory process. At least 40 different proteins are needed for natural competence in *Bacillus subtilis*, one of the species in which it is known to occur. From the perspective of bacterial cells, this inefficiency is probably evolutionarily advantageous. Cells that take up foreign DNA too readily would be genetically unstable and could incorporate DNA molecules that are deleterious to their survival. On the other hand, cells that are unable to exchange DNA readily with other cells have reduced genetic variability. The low efficiency of transformation is likely a balance between these two competing evolutionary forces.

11.3 Conjugation

A second method of genetic exchange in bacteria, known as **conjugation**, relies on the direct transfer of DNA from a donor to a recipient cell and is mediated by physical contact between the two cells. Conjugation is best understood in *E. coli*, but it occurs commonly within many other species of bacteria, as well as between different species of bacteria and even between bacteria and some eukaryotes. The ability to serve as a conjugative donor is typically encoded by extrachromosomal DNA such as a plasmid. Genes on that plasmid also typically determine which strains can be recipients for the donated DNA.

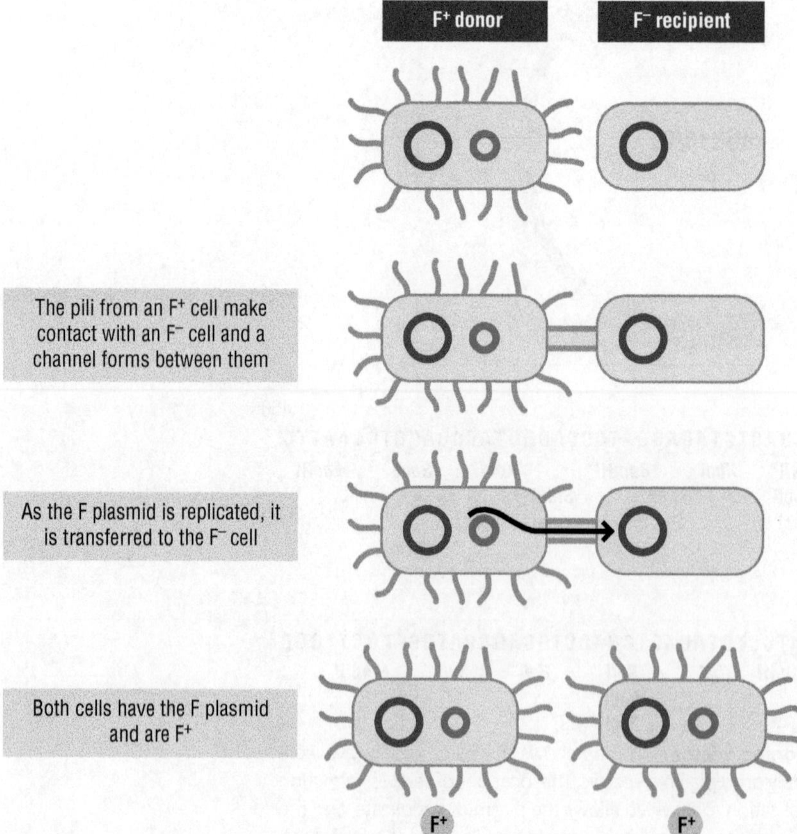

Figure 11.7 An overview of conjugation. In *Escherichia coli*, the F factor or F plasmid (shown in red) encodes the key information for genetic exchange via conjugation. A cell with the plasmid, known as an F^+ cell, makes physical contact with pili to a cell lacking the plasmid, known as F^- cell. The pili form a channel between the two cells. As the F plasmid is replicated, it is transferred from the F^+ donor to the F^- recipient. The plasmid circularizes once the transfer is complete, and both cells are now F^+ and capable of becoming donors.

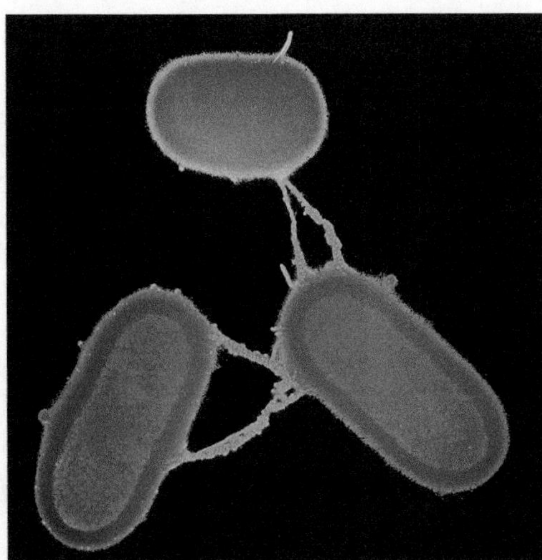

Figure 11.8 An electron micrograph of conjugating cells. False-colour transmission electron micrograph (TEM) of a male *Escherichia coli* bacterium (bottom left) conjugating with two females.
Source: Dr L Caro/Science Photo Library

F plasmids mediate the process of conjugation

In conjugation, as with other forms of horizontal gene transfer, the transfer occurs only in one direction, from the donor cell to the recipient cell, as shown in Figure 11.7. In the best-studied example of conjugation, the donor cell has a plasmid encoding a special transfer function, the best known of which is called the **F factor** or **F plasmid**. "F" originally stood for fertility, since conjugation was first described as bacterial sex in the early 1950s; a cell with the F factor is referred to as F^+, whereas cells that lack the F factor plasmid are called F^-.

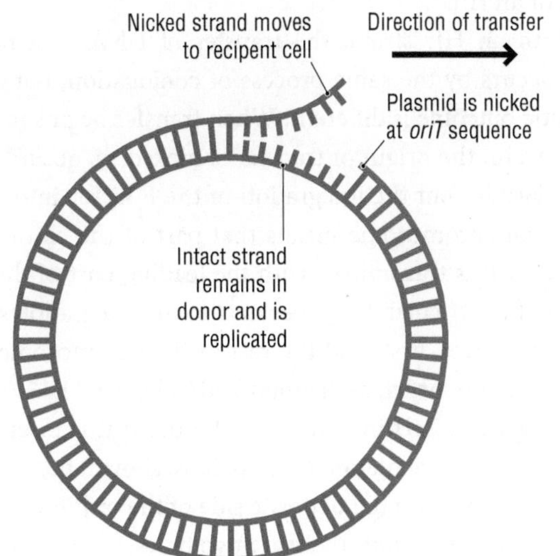

Figure 11.9 The beginning of conjugation. Prior to transfer, the plasmid is nicked at an origin of transfer, *oriT*. The nicked strand is transferred to the recipient cell. The intact strand remains in the donor cell and is replicated to regenerate a double-stranded plasmid.

The F plasmid encodes 28 transfer, or *tra*, genes, most of which are needed to form a long and thick appendage called a conjugative **pilus** (the plural form is pili). The pili form a fragile temporary connection from the donor F^+ cell to the recipient F^- cell and then retract, drawing the donor and the recipient closer together, as shown in the micrograph in Figure 11.8. The F plasmid is transferred to the recipient cell, possibly through the pilus or possibly through a separate mating pore that has not yet been conclusively identified.

Just before transfer, the F plasmid DNA is nicked at a specific site known as the **origin of transfer** (*oriT*), so that one of the DNA strands is broken and the other is left intact. The nicked strand moves to the recipient cell, and the intact strand is left behind. Thus, a single strand is being transferred, as shown in Figure 11.9. Once the recipient cell has received both ends of the plasmid, the split origin of transfer is reconnected or ligated. Both cells synthesize a second DNA strand by replication, so that, at the end of conjugation, both cells become F^+ cells, as shown in Figure 11.7.

Bacteria can carry multiple plasmids of different sequences and types. Some plasmids, such as F, are

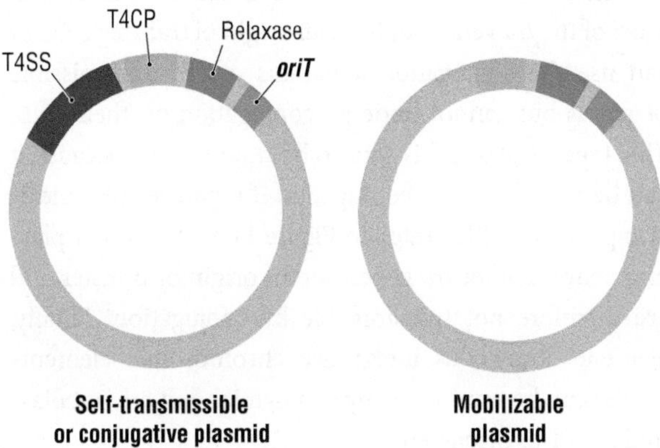

Figure 11.10 Plasmids that can be transferred by conjugation. Four different genetic components are involved in conjugation. The origin of transfer, *oriT* (in blue), is the site where a nick occurs to begin replication and transfer. The type IV secretion system, T4SS (in purple), is a complex of 12–30 proteins that is responsible for the uptake and secretion of many macromolecules by forming the channel between the cells. Both the Ti plasmid of *Agrobacterium tumefaciens* and the F plasmid encode the T4SS proteins. Relaxase (in orange) creates the nick at *oriT* and attaches to the 5′ end of the nicked DNA; it is transferred to the recipient cell, along with the DNA. A type IV coupling protein, T4CP (in green), is involved in the interaction between the relaxase protein and the T4SS. Self-transmissible or conjugative plasmids encode all four of these components. Mobilizable plasmids are missing the T4SS (and T4CP in some cases) but can be moved via conjugation if a conjugative plasmid is present in the same cell.

BOX 11.1 *Going Deeper* Integrative conjugative elements

Integrative conjugative elements (ICEs), also known as conjugative transposons, are DNA elements that can exist extrachromosomally in bacteria and can be transferred from one cell to another but cannot replicate autonomously. They are often thought of as composite mobile elements that encode features of lysogenic phages, conjugative plasmids, and transposons. They also usually encode additional genes, with functions unrelated to integration or transmission, making them similar in this respect to plasmids.

ICEs integrate at specific attachment (*att*) sites in the chromosome. Like lysogenic phages, discussed in Section 11.4, their site-specific integration is catalyzed by an integrase and their excision by a *xis* gene. But unlike phages, ICE genomes are not viruses. Instead, they also encode conjugative systems that are homologous to those on self-transmissible plasmids and are conjugated from one cell to another from an origin of transfer by a process similar to that described in Figure 11.7. Unlike plasmids, however, ICEs do not have an origin of replication and cannot replicate independently. Instead, the genome of an ICE is copied with the chromosome of its host, while it is integrated.

Perhaps the best-studied ICE is the SXT element of *Vibrio cholerae*. This element was first seen in cholera isolates at the end of the 1970s and has since been disseminated so widely that virtually every *V. cholerae* isolate in the past 20 years is SXT-positive. SXT encodes resistance to most of the antibacterial drugs that have been used to treat cholera; its widespread dissemination is linked to human antimicrobial use, since the antimicrobial agents will kill or arrest *Vibrio* strains that lack the resistance gene.

Another example of the effect of ICE on genome evolution is seen with RAGE, an element found in bacteria of the genus *Rickettsiales*, which are intracellular endosymbionts in ticks, mites, and other insects. Roughly a third of the genome of some mites comprises RAGE sequences, which allows for the horizontal gene transfer of other DNA by recombination between RAGE sequences.

FIND OUT MORE

Wozniak, R.A. and Waldor, M.K. (2010) Integrative and conjugative elements: mosaic mobile genetic elements enabling dynamic lateral gene flow. *Nat Rev Microbiol* 8: 552–563.

self-transmissible because they encode all of the information needed for their own mobilization. Others lack most of the *tra* genes but have an origin of transfer, so they can use the conjugation apparatus of self-transmissible plasmids but cannot undergo conjugation on their own. This type of plasmid is said to be **mobilizable** because it can be transferred by conjugation if a partner plasmid is also present, as illustrated in Figure 11.10. Still other plasmids have neither *tra* genes nor an origin of transfer and are therefore not transmissible by conjugation. Finally, elements known as integrative chromosomal elements (ICEs) can be transferred by conjugation but are not plasmids at all, as discussed in Box 11.1.

KEY POINT Conjugation is the transfer of DNA from one cell to another through direct cell contact and is mediated by transfer genes encoded in the donor cell's genome.

HFR and conjugation mapping

As described so far, genetic exchange by conjugation would be limited to plasmid-borne genes. However, plasmids can occasionally become part of the chromosome. For example, the F factor can insert at different sites in the *E. coli* chromosome, as shown in Figure 11.11(a). Certain insertion sites are more likely to be used because a short sequence on the plasmid is similar enough with sequences of the bacterial chromosome to recombine with it, as shown in Figure 11.11(b). Once inserted, sequences on the F factor are replicated, along with the surrounding host chromosomal DNA, and the F factor genes are expressed. A cell with the F factor integrated into the chromosome is known as a **high frequency of recombination** cell, or an Hfr.

With an Hfr strain, the transfer of DNA to a new cell occurs by the same process of conjugation, but the genetic outcome is different. When transfer begins in an Hfr strain, the origin of transfer is nicked, as occurs on the plasmid, but the integration of the F factor into the bacterial chromosome means that part of the chromosome itself is transmitted with the leading part of the F factor. In particular, the genes next to the integration site are transferred first, and the rest of the chromosome is transferred in order, as diagrammed in Figure 11.12. The F factor can insert in either orientation. If it is inserted in the reverse orientation than we have shown, the chromosomal genes on the opposite side of the origin will be transferred first. Thus, there are two types of Hfr strains, which differ according to the genes that are transferred first—the genes transferred first in one type of HFR are the ones that are transferred last if the F factor is inserted in the opposite orientation.

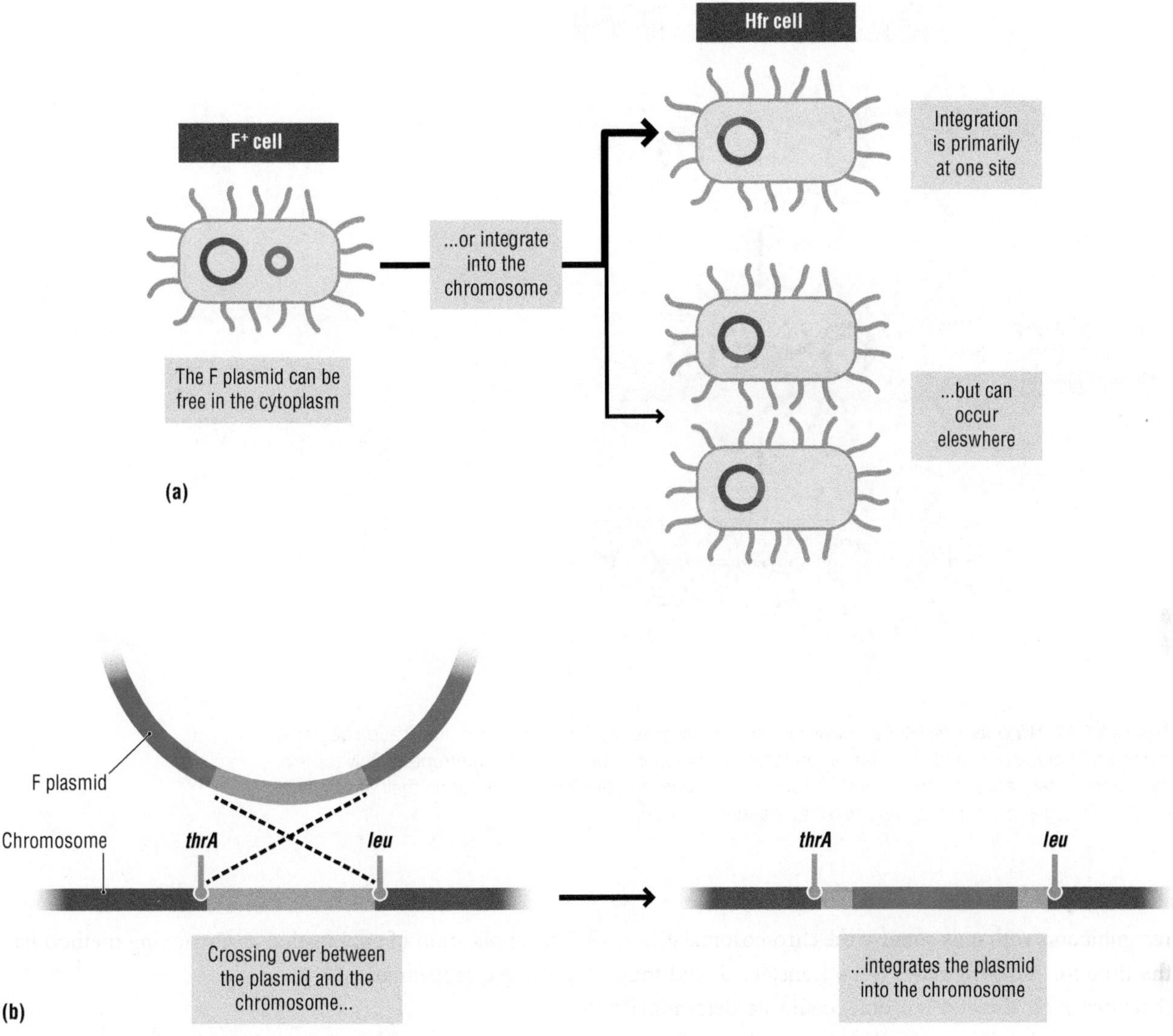

Figure 11.11 F plasmids integrate into the chromosome. The F plasmid (in red) can be free from the chromosome and self-replicating, or it can integrate into the bacterial chromosome (in blue). (a) Integration usually occurs at one site, as shown by the red line, but can occasionally occur at other locations. A cell with the integrated F plasmid is called an HFR cell. (b) Integration occurs by a crossover between homologous sequences (in lighter colors) on the F plasmid and the chromosome. The most common site of integration is between the *thrA* gene and the *leu* gene. Once integrated, the F sequences are part of the bacterial chromosome, so *thrA* and *leu* are further apart. The drawings are not to scale.

Recall that the conjugative pilus is fragile and thus can be broken. It only takes a few seconds to transfer the F factor plasmid, but, when the chromosome is being transferred with the F factor in an Hfr strain, it is common for the pilus to break and for the cells to separate before transfer is complete. In this case, the donor cell is still F⁺, but the recipient cell with a partial F factor has not become an F⁺ cell because the DNA at the other end of the origin of transfer on the F plasmid has failed to enter the recipient cell. As a result, the recipient will have some genes or DNA sequences of the F factor and possibly some chromosomal genes as well. However, the incoming DNA cannot recircularize because one end of the origin of transfer is missing. Most of the time, this DNA is degraded, but sometimes it recombines with the chromosome of the recipient, becoming integrated and replacing some of the original host genes.

Conjugation typically occurs while cells are undisturbed in a liquid medium. By allowing conjugation to begin before disrupting it by mixing in a blender (in a process called interrupted mating) and then looking for

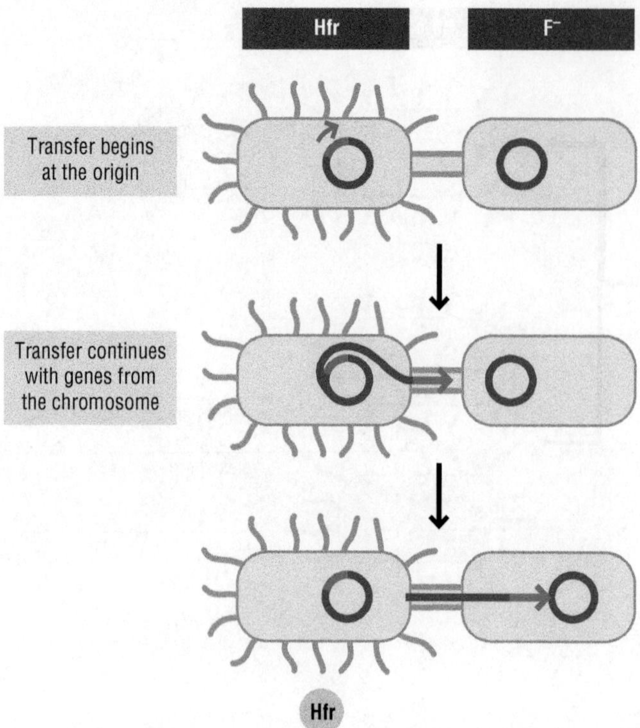

Figure 11.12 Hfr cells. Hfr cells can transfer chromosomal genes during conjugation. Replication begins at the origin (shown in red) and continues around the chromosome (in blue) until it is interrupted. Only some of the F plasmid sequences are transferred, and which host genes are transferred depends on their proximity to the site of integration and the time required for transfer.

recombinants with only some of the chromosomal genes, the time for different genes to be transferred, and thus their order on the chromosome, could be determined. As we noted in Chapter 9, distances on the traditional *E. coli* chromosome are measured in minutes; this is simply the amount of time that it takes to transfer that gene to the F^- cell using the original Hfr strains. Since the *E. coli* chromosome is circular, a map based on time worked particularly well. The location of the F factor in the original Hfr strain that was used has been designated as 0, and the position of every other gene is timed from that. The entire map is about 100 minutes, as depicted in Figure 11.13.

KEY POINT The time needed to transfer genes from a donor to a host via conjugation was the basis for the classical genetic map in *E. coli*.

While conjugation was the basis for the original genetic map in *E. coli*, current researchers rely on direct DNA sequencing to map genes. Conjugation is still studied to understand how bacteria exchange DNA in nature, but the application of conjugation as a mapping method has largely disappeared.

Conjugation across kingdoms

Conjugation donors are bacteria, but the recipients do not have to be. One of most important naturally occurring examples of conjugation occurs between the bacterium *Agrobacterium tumefaciens* and flowering plants. *Agrobacterium* species are generally soil-dwelling, but *A. tumefaciens* and a few other species can also parasitize plants, typically through wounds in the plant cell walls. Once it has infected the plant, *A. tumefaciens* expresses conjugative pili. DNA and proteins are moved from the bacterium into plant cells by means of a special conjugative plasmid known as the Ti plasmid, which is transferred via a mechanism that is similar to conjugation between bacteria.

In addition to conjugation genes, the Ti plasmid also carries genes encoding the ability to make preferred bacterial nutrients, as well as genes that interfere with plant signaling. This interference ultimately leads to the formation of a crown gall tumor on the stem or

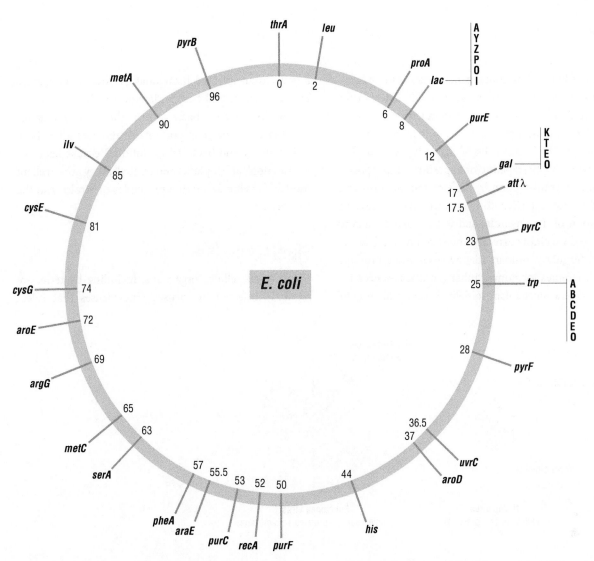

Figure 11.13 The genetic map of the *Escherichia coli* chromosome, in minutes, based on conjugation. The map was based on a particular Hfr strain, with the F plasmid integrated in a certain orientation. In other Hfr strains, other genes may be transferred first, depending on the location and orientation of the F integration, but this map is used as a standard reference strain.

Figure 11.14 *Agrobacterium* causes crown galls. Infection of a flowering or woody plant with *Agrobacterium* causes a localized growth of plant cells known as a crown gall. An example is shown here. Plants can live with crown galls for many years.

roots of the plant, as shown in Figure 11.14. The bacteria live within the tumor, benefitting from a constant supply of nutrients from the host. The ability of *A. tumefaciens* to transfer DNA to plants is the basis for techniques by which plant geneticists introduce foreign DNA into plant cells experimentally. Ti plasmid conjugation is an efficient way to make transgenic plants with specific genes for analysis or desirable characteristics such as disease resistance or improved nutritional qualities.

KEY POINT DNA can be conjugated from bacterial donors to eukaryotic recipients, a process that is typified by conjugation of the *Agrobacterium* Ti plasmid into plant cells.

11.4 Transduction by viruses

A third method of horizontal genetic exchange—**transduction**—occurs via viruses. Virus-mediated transfer is well studied in both bacteria and eukaryotes. It is of continuing interest because of its potential to introduce foreign DNA into cells of many types of organisms for possible therapeutic or experimental purposes. For example, transduction by tobacco mosaic virus can be used to transfer DNA to plants, as an alternative to conjugation of Ti plasmids, and is the basis for many plant-based vaccines currently under commercial development. Similarly, human adenoviruses show promise in gene therapies for human inherited diseases because they provide a mechanism to deliver a normal copy of a gene to an affected individual. As with conjugation, transduction is still best understood in *E. coli*. The relationship between bacteria and the viruses that affect them has been important in understanding both viral biology and bacterial evolution. Because bacterial viruses don't affect plants and animals, they are excellent models for elucidating the principles of virology in the laboratory.

Viruses: a quick overview

Virtually all cellular organisms, including bacteria, are regularly infected by viruses. The viruses that infect

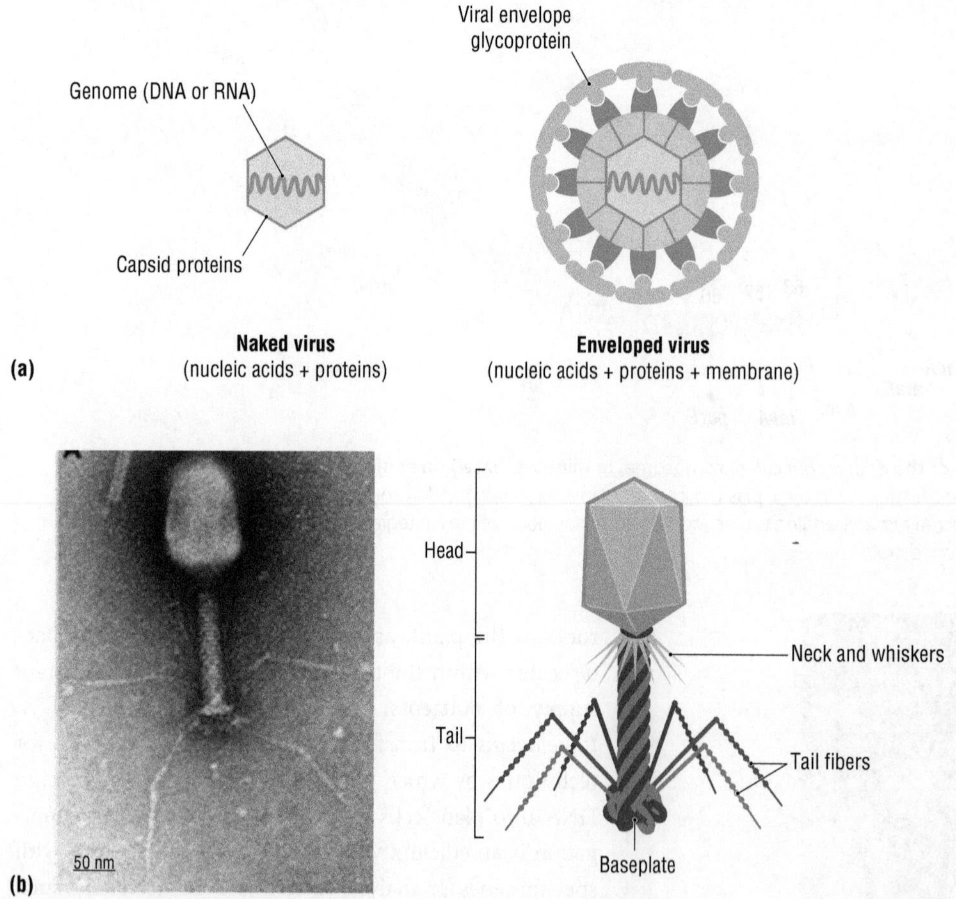

Figure 11.15 The structure of viruses. (a) The general structure of viruses. Viruses have their genome (which can be single-stranded or double-stranded, and either DNA or RNA) encased in a capsid, often referred to as the head. The capsid is made up of proteins encoded in the viral genome. Many viruses also include an envelope, shown in green, in which the lipid bilayer comes from the host cell membrane and the membrane proteins can be from the virus or from the virus and host. (b) A bacteriophage, as seen in an electron micrograph and in diagrammatic form.

Source: Electron micrograph courtesy of M. WURTZ/BIOZENTRUM, UNIVERSITY OF BASEL/ SCIENCE PHOTO LIBRARY.

bacteria are known as **bacteriophages** or, more simply, **phages**. Hundreds, if not thousands, of different phages that can infect *E. coli* have been identified. These phages have a range of life cycles, from the fairly simple to the quite complex, as well as genomes of their own that can encode dozens to hundreds of genes. As with all viruses, the life cycle of a phage depends on interactions between its own genes and proteins and the genes and proteins of the host, in this case the bacterial cell.

Viruses have no cellular machinery of their own and depend on their host cells to replicate their genomes and express their genes. The genomes of viruses can be either RNA or DNA, and the nucleic acid can be either single-stranded or double-stranded. The rest of the viral structure is typically very simple, as illustrated in Figure 11.15(a). It consists of a protective protein coat called a **capsid**, other proteins required for transmission, and, in a few viruses, a small repertoire of enzymes. Some viruses have a membranous envelope as well; the envelope is typically not made by the virus but is instead pinched off from the membrane of the host cell as the virus exits. Some phages have a granular structure, some have a filamentous structure, and many look something like a lunar landing module (used in the US space program), such as the T4 phage similar to that illustrated in Figure 11.15(b).

For T4, the phage structural proteins can self-assemble *in vitro* into three separate components—the head, the tail, and the tail fibers—illustrated in Figure 11.15(b), each of which consists of multiple proteins. (The ability of T4 to self-assemble in the absence of ancillary proteins or other catalysts made important contributions to our understanding of the processes of protein folding.) The head or capsid proteins enclose the nucleic acid, and the size of the head determines its DNA-carrying capacity. The tail fibers attach to the outside of the host cell, while the tail is needed to inject nucleic acid into the host.

In a typical viral infection, as shown in Figure 11.16, the phage docks to a specific bacterial protein on the

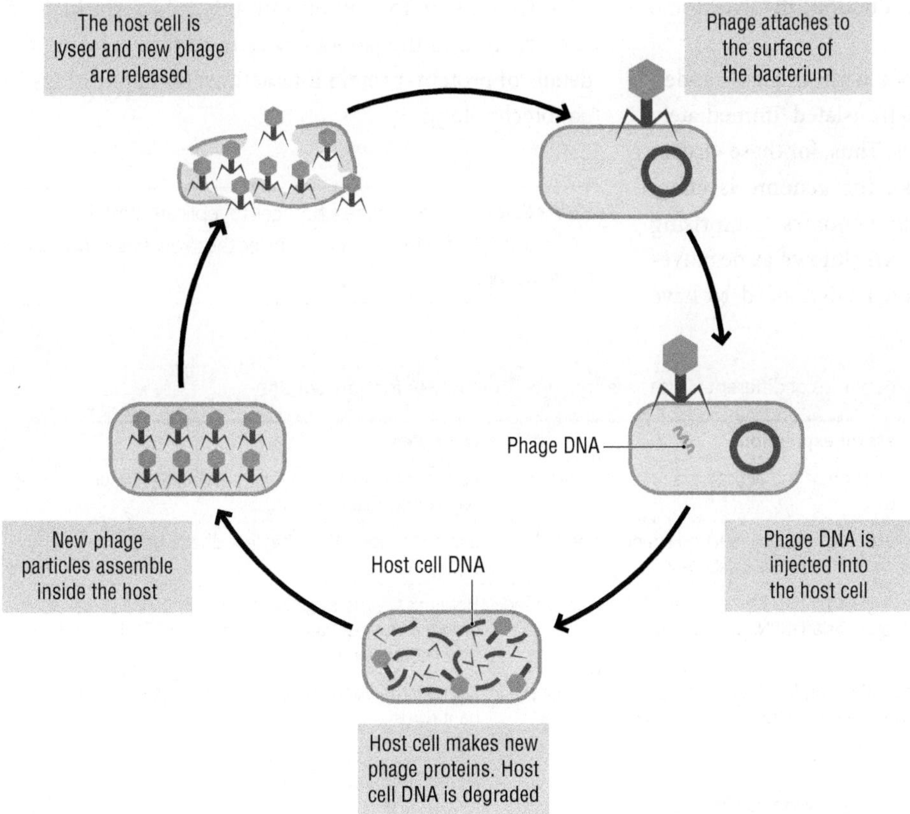

Figure 11.16 Life cycle of a lytic phage. A lytic phage infects the cell, injects its DNA (shown in green), takes over the cellular processes to make more phage proteins from the phage genome, and degrades the host DNA (shown in blue). New phages assemble from the proteins in the host cell and lyse the host cell so they are released to infect a new cell and begin the cycle again.

outside of a susceptible bacterial cell. The bacterially encoded docking protein has other regular functions in the host cell, but the virus can bind to it via one or more attachment proteins. The phage then injects its genome through the cell membrane and into the bacterium, often with a mechanism that resembles a syringe in the tail of the virus. The phage genome then takes over cellular processes to replicate itself.

Some of the phage genes immediately direct the host cell machinery to make phage-specific proteins. These early viral proteins often include those that comprise the capsid or the structural proteins of the phage particle itself. Other phage-specific proteins then direct the remainder of the life cycle, which varies depending on the phage.

The life cycles of different viruses within the host are, in part, dictated by the nature of the nucleic acid comprising the viral genome. Those viruses whose genomes are made up of double-stranded DNA can be transcribed and then translated by the normal host gene expression machinery, like any other DNA. For viruses with single-stranded genomes, the process of making additional viral genomes and particles (that is, virus reproduction) involves some additional steps.

The genomes of some viruses with single-stranded RNA genomes are able to be translated immediately after entry to make viral proteins. Thus, for these viruses (known as plus-strand viruses), the genome is effectively an mRNA. Viruses with genomes comprising RNA complementary to the mRNA (known as negative-strand viruses) or single-stranded DNA need to have their genomes "converted" into a form that the infected cell will use as mRNA. Some of the pathways through which viral nucleic acids are converted into RNA to allow viral reproduction are given in Table 11.1. Since the cell's resources are diverted to produce viral RNA and proteins, viral infection can create an energy burden on the bacterium, slowing its growth or causing cell destruction.

Each of the enormous number of bacterial species is susceptible to several phage types, so there is an immense diversity of bacteriophages, most of which have not been cataloged. A few phages were key experimental organisms in the early days of molecular biology, so these have been very thoroughly studied. Indeed, many of the fundamental molecular processes of cells, such as transcription, translation, mutation, and replication, were first investigated using *E. coli* and its phages.

The enormous unexplored diversity of phages suggests that they may be important in fields of biology yet to be developed. For example, phage and phage lysis proteins are being studied as a non-antibiotic therapy for bacterial infections or for killing unwanted bacteria in food and are used in the laboratory as tools for exploring the details of protein–protein interactions for applications in nanotechnology.

KEY POINT Viruses require a host cell to replicate. When viruses mediate horizontal gene transfer, the process is known as transduction.

Table 11.1 Pathways by which the genomes of different viruses are converted into mRNA for transcription

Genome	Process for expression	Examples
dsDNA	Transcription of one or both strands, like cellular genes	Bacteriophages T4 and λ, herpesviruses, human papillomaviruses, pox viruses
ssDNA	Replication to make the second strand for dsDNA, then transcription like a dsDNA virus	Bacteriophage φX174, bacteriophage M13, many aquatic viruses
ssRNA (+ strand)	The single-stranded RNA genome is the sense strand and serves as an mRNA for translation	Hepatitis C virus, tobacco mosaic virus, Zika virus, severe acute respiratory syndrome (SARS) virus and other coronaviruses
ssRNA (– strand)	The single-stranded RNA genome is the template strand. The sense strand or mRNA is made by RNA polymerase	Rabies virus, Ebola virus, Lassa virus, influenza virus, hantavirus
ssRNA (+ strand, not directly transcribed)	The + strand of the RNA is reverse transcribed to ssDNA by reverse transcriptase. The ssDNA is replicated to produce dsDNA, which is then transcribed into mRNA	Human immunodeficiency virus (HIV) and other retroviruses
dsRNA	One strand is used as the template, and new transcripts are made	Rotaviruses

Lytic phages can carry out generalized transduction

We can identify two major categories of phages, based on their life cycle within the cell; these are known as lytic and lysogenic phages. The life cycle of a lytic phage is depicted in Figure 11.16. Lytic phages multiply in the bacterial cell and then lyse their host cell, releasing hundreds of new phages ready to infect surrounding bacteria. The growth of a lytic phage in bacterial cultures is exponential, and a corresponding exponential death rate is seen in the population of infected bacterial cells. If a single phage is placed on a lawn of bacteria, it soon kills enough bacterial cells to produce a clear hole in the lawn—a plaque—that is large enough to be seen with the naked eye, as shown in Figure 11.17. At least a billion (10^9) bacteria must be lysed to produce a visible plaque.

T4 and similar phages usually incorporate their complete genomes into the new capsids. Phage particles containing the entire genome can then infect other bacterial cells and reinitiate the viral life cycle. However, errors in viral genome packing also occur, which form the basis for transduction. Sometimes the entire phage genome is not incorporated into the capsid, and the resulting phage cannot successfully infect another cell. Two variations in genome packing in T4 allow it to serve as a transduction vehicle for delivering bacterial DNA from a previously infected donor cell to a newly infected recipient. First of all, the process of filling the phage head with DNA is only slightly more orderly than a tourist stuffing clothing and souvenirs into his luggage. Essentially, the capsid is filled with as much DNA as it can hold, which is slightly more than the size of a single phage genome. Occasionally, host DNA can be packaged with phage DNA. In our luggage analogy, this is equivalent to accidently bringing the key or a pen from one hotel room to the next place you stay.

Second, although the entire phage genome is needed to complete another round of infection, it is not needed

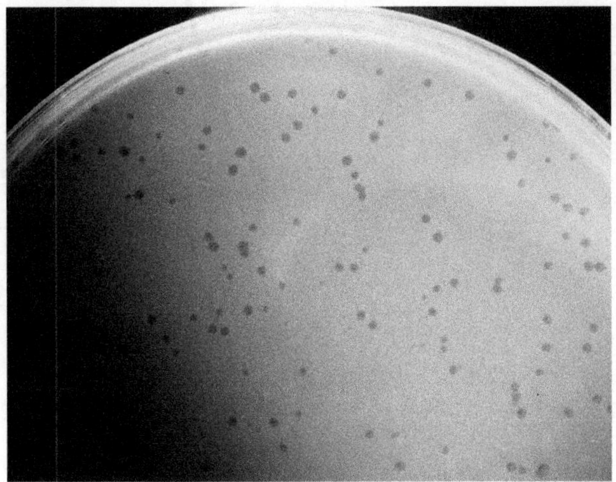

Figure 11.17 Plaques and lytic phage growth. When T4 or a different lytic phage infects a cell, it replicates and lyses (or bursts) the cell. The phages released from the lysed cell infect neighboring cells, and the cycle continues. Bacteria have been spread on this plate as a lawn, which can be seen as the opaque background on the surface of the agar plate. The lysed cells are observed as clear areas, known as plaques, in the bacterial lawn.

Source: Hatfull Lab, University of Pittsburgh, phagesdb.org.

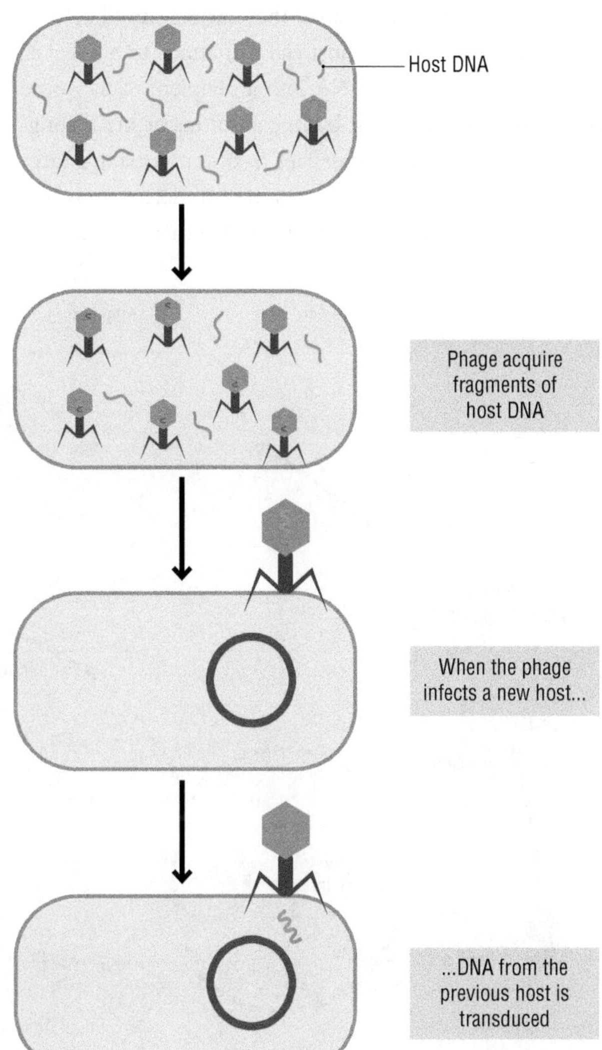
Host DNA
Phage acquire fragments of host DNA
When the phage infects a new host...
...DNA from the previous host is transduced

Figure 11.18 Generalized transduction. The key to generalized transduction is that the amount of DNA that can be packaged into the capsid is greater than the size of the phage genome. Thus, the phage picks up parts of the host cell's genome that was fragmented during lysis, shown here as light blue segments that are part of the genome of the phage. When phages with these bacterial genes infect a new host, host cell genes are transduced to the next host. The genes that are picked up by the phage can come from any part of the bacterial genome.

to begin one. A phage capsid that carries **only** bacterial DNA from the infected host cell can attach to a new cell and inject the foreign DNA into a recipient. This would be analogous to a tourist leaving all of his or her belongings in the hotel room (perhaps in order to bring a large souvenir home). These two features underlie transduction by T4. If a phage particle stuffed with all or part of a previous host bacterium's DNA infects another cell, it mediates DNA transduction from the donor to a recipient, as shown in Figure 11.18. Because T4 picks up host DNA at random, any part of the bacterial genome can be transduced at low frequency when the phage infects another host. If the transduced DNA is self-replicating (as a plasmid) or able to recombine with the host DNA, the newly

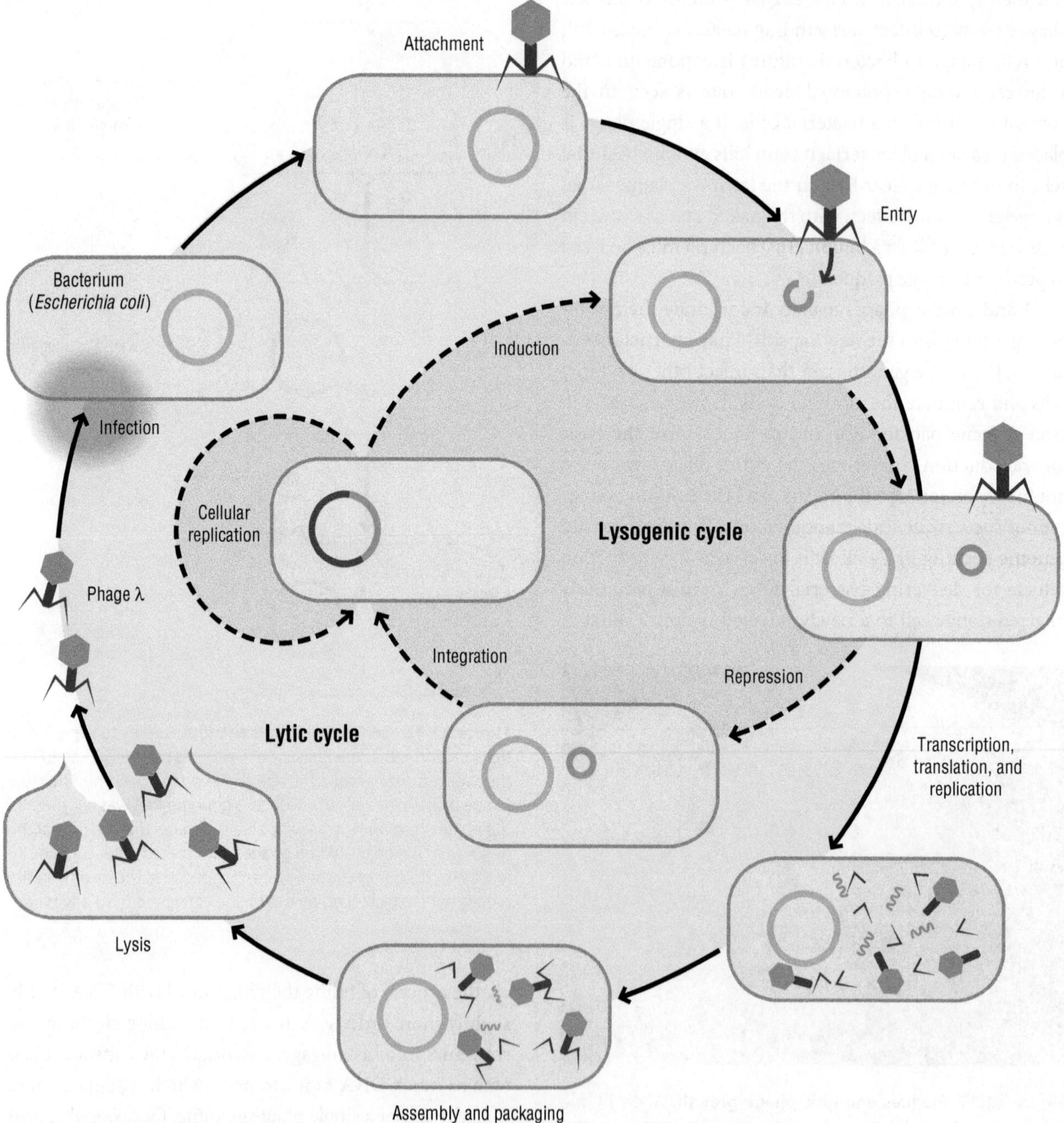

Figure 11.19 The life cycle of a lysogenic phage. As described in the text, a lysogenic phage, such as λ, has two possible life cycles once it infects the host cell: a lytic cycle and a lysogenic cycle. The lytic cycle, which is shown as the outside circle, is similar to that of the lytic phage. During the lysogenic cycle, shown as the inner circle, the phage inserts its genome into the host chromosome as a prophage where it replicates as part of the host chromosome. A prophage can remain integrated indefinitely before excising to initiate the lytic cycle.

Source: Allan Campbell (2003). The future of bacteriophage biology. *Nature Reviews Genetics* 4, 471–477.

acquired DNA is incorporated into the genome and subsequently inherited vertically. Horizontal transfer by this means is termed **generalized transduction** because any part of the genome can be transduced.

Phages can also recombine with each other. Phage recombination occurs when two or more phages infect a single cell. Homologous recombination within the host cell between the DNA of the two phages can occur, or part of one phage's genome can be packed into the head from a different phage, just as your suitcase may end up containing some belongings from your traveling companion. In fact, the DNA in many phages is a genetic composite drawn from different viral genomes. In any event, the genome of one phage has been horizontally transferred to the genome of another.

Lysogenic phages mediate specialized transduction

The plaques made when bacteria are infected with phage λ are cloudy, rather than clear like the plaques made by T4. This difference in plaque morphology led to the discovery of phage **lysogeny**. Lysogenic phages, of which phage λ is the best-known representative, have a much more complex life cycle than T4. When λ or another lysogenic phage infects a cell, two different events might occur. Sometimes λ will immediately enter a lytic cycle similar to the life cycle of a lytic phage discussed previously. Alternatively, lysogenic phages can integrate their genome into the bacterial genome and thus propagate along with the host chromosome, as shown in Figure 11.19. The phage inserted into the bacterial genome, known as a **prophage**, can persist as part of the host genome indefinitely. The host *E. coli* strain is said to be lysogenized by the prophage. The prophage does not lyse the host cell or produce new phage while integrated into the chromosome.

For λ, integration occurs at a specific sequence or site in the bacterial genome, called the *att* site (for attachment). The *att* site has a core of 15 base pairs that is found in both the λ genome and the *E. coli* genome. The sequence on the phage λ genome is called the *attP* site, while the sequence in the bacterial genome is the *attB* site, illustrated in Figure 11.20. *attB* and *attP* share considerable sequence identity, but the length of the identical sequence is too short for crossover to occur by homologous recombination. The integration occurs because phage-mediated **site-specific recombination** occurs between these two sites. The λ genome encodes a protein known as **integrase**, which catalyzes this site-specific recombination. Site-specific recombination based on phage integrases has proved to be a very important tool in the experimental manipulations of DNA in the laboratory, as discussed in Tool Box 11.2.

When λ is an integrated prophage within an *E. coli* genome, it is replicated as if it were a standard part of the bacterial chromosome. A limited amount of transcription and translation of the integrated λ genes produces proteins that specifically repress transcription of lytic genes in the phage. Every λ prophage is also capable of switching from the lysogenic phase into the lytic phase. This is one of most thoroughly analyzed processes in genetics and molecular biology, but the details by which λ represses transcription of most of its genome are beyond our discussion here.

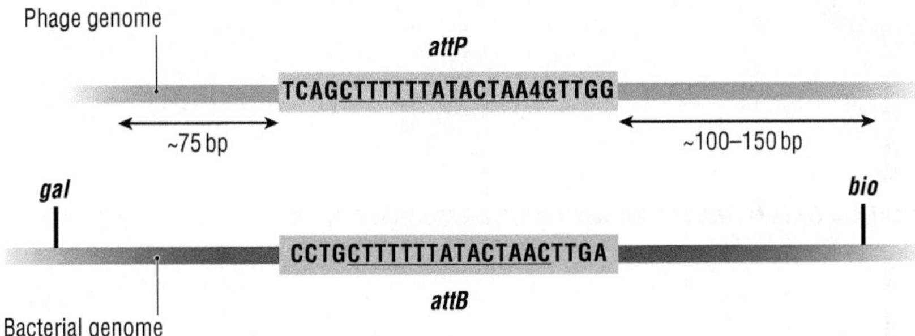

Figure 11.20 The *att* sites and λ phage integration. A 23 bp sequence in the λ genome, known as *attP* and shown in the green box, is highly similar to a sequence in the *Escherichia coli* genome, known as *attB* and shown in the blue box. Recombination between the *att* sites always occurs at the center of the core sequence and is mediated by the site-specific recombination enzyme integrase (*int*), as well as host factors. As shown here, *attB* is found between the galactose (*gal*) utilization operon and the biotin (*bio*) biosynthesis operon, so integration occurs between these two gene regions. After recombination, the flanking regions of the phage genome are recombinants of the two *att* sites and are referred to as *attL* and *attR* (see Figure 11.21).

TOOL BOX 11.2 Applications of site-specific recombination

Homologous recombination occurs at sites throughout the genome but requires the two sites of recombining DNA to share a large region of identical or almost identical sequence. In *Escherichia coli*, this sequence is typically at least 1000 bp; in eukaryotes, the regions of sequence similarity needed for recombination are often many thousands of base pairs. For the meiotic recombination discussed in Chapter 9, this level of sequence similarity is easily accomplished from homologous chromosomes. Certain proteins are required for homologous recombination, but they act on any pieces of DNA in that region of sufficient length, so homologous recombination cannot be easily targeted to specific sites. By contrast, site-specific recombination enzymes target recombination between pieces of DNA harboring very specific sequences. The regions of homology of these sequences can be much shorter than those required for homologous recombination, so recombination between two sequences can be specifically targeted to this location.

Section 11.4 discusses the integration of phage λ into the *E. coli* chromosome with the recombination enzyme called integrase and the regions of homology known as attachment sites or *att*. This is only one of many site-specific recombination systems. Other site-specific recombination systems work by an analogous mechanism with an enzyme that carries out the recombination and a short sequence found both on the phage and the bacterial chromosome, although its name varies. For example, the recombination enzyme from phage P1 is called *Cre*, while the sites of recombination are called *lox*; *Cre–lox* recombination is a widely used tool for working with mammalian genomes.

There are many applications in which an investigator wants to insert or manipulate a sequence at a specific site in a genome; sequence-specific recombination systems provide the tools to do this. For example, any molecule with an *att* site can be recombined with another molecule with an *att* site, so long as the recombination enzyme integrase is provided. Bacterial geneticists use phage integrases to insert sequences that have been manipulated or altered *in vitro* at specific sites in bacterial genomes. Other examples of site-specific recombination are found throughout the biology of bacteria, plants, and animals, including integron systems that allow bacteria to rapidly site-specifically integrate antibiotic resistance genes (as discussed below); rearrangement of the immunoglobulin genes in mammals depends on site-specific recombination catalyzed by enzymes, one of which is evolutionarily related to a bacterial transposase.

While most molecular biologists think of the process of site-specific recombination as a tool for manipulating DNA sequences *in vitro*, it also plays an important role in bacterial evolution. Elements known as integrons are able to integrate sequence cassettes into bacterial genomes at specific locations, by the process shown in Figure A. The integron consists of an integrase enzyme and an attachment site within the integron known as *attI*. Gene cassettes that terminate with a short *attC* attachment site can be recombined into the integron. Bacteria that have integrons adapt very rapidly to changes in their environment because they can relatively easily acquire new genes from other sources. Integron-borne antibiotic resistance genes are almost ubiquitous in Gram-negative bacteria isolated in hospitals where antibiotics are used intensively, and the rapid spread of antibiotic resistance is due, at least in part, to their activity.

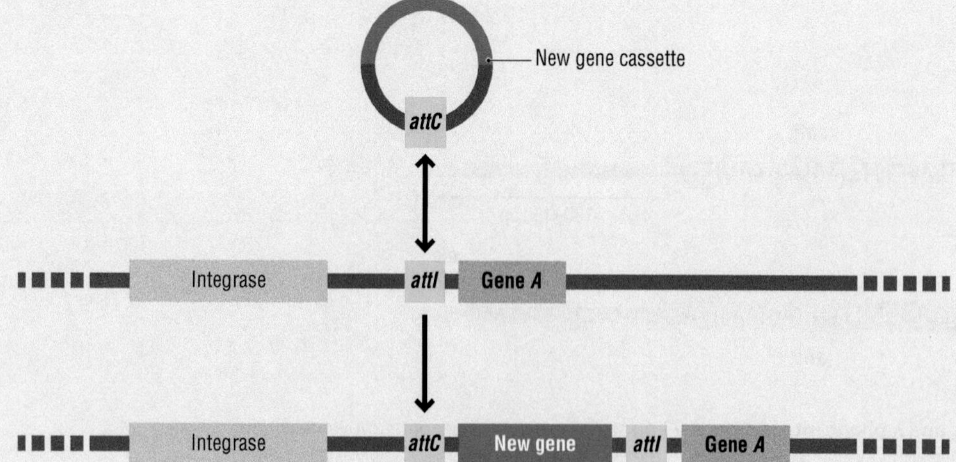

Figure A Integrons integrate into the bacterial genome by site-specific recombination. The integron encodes an integrase enzyme that catalyzes specific recombination between the *attI* site on the integron and *attC* on the gene cassette to insert the new gene.

The genetic switch between lysogeny and lysis requires site-specific recombination between the two ends of the prophage, normally catalyzed by integrase, which reconstitutes *attB* and *attP*. If the excision is not precise, genes from the host bacterium that are adjacent to the prophage insertion site are added and may become part of the phage, as depicted in Figure 11.21. When this happens, this adjacent DNA can be moved to a different cell by **specialized transduction**. Because the portions of the bacterial chromosome that are included are those adjacent to the *att* site, specialized transduction by prophages affects only certain parts of the bacterial genome.

Temperate phages and phage conversion

Some phages are always lytic, while others can toggle between a lytic and a lysogenic life cycle. Still other phages can enter a cell, reproduce, and exit a host bacterial cell without lysing it. These **temperate phages** slow down the growth and metabolism of the host bacterium but do not kill it because they are extruded from the cell without host cell lysis. For example, the phage M13 has a circular genome that can replicate within the host bacterium the way plasmids do and a filamentous morphology that facilitates extrusion. Interestingly, M13 uses the conjugative pilus as its receptor and can therefore only infect cells that have already acquired an F plasmid. Because M13 can exist in an intracellular double-stranded replicative form, but its exported genome is single-stranded DNA, M13 has been used as a vector for producing cloned single-stranded DNA in the laboratory.

KEY POINT Viruses can be lytic, resulting in bacterial lysis, or can integrate into the bacterial genome as lysogenic phages. Other temperate phages can enter cells, replicate, and exit without causing cell death.

Many temperate and lysogenic phages—that is, those that are not lytic—not only have genes that enable them to propagate themselves but also carry genes encoding proteins that could provide a selective advantage to the bacterial cell. The host bacterial cell expressing these genes is said to be **phage-converted** to the phenotype encoded by the bacteriophage. Some well-characterized phage conversions are mediated by lysogenic phages. For example, some phages that are related to λ carry genes encoding a toxin that is deadly to humans,

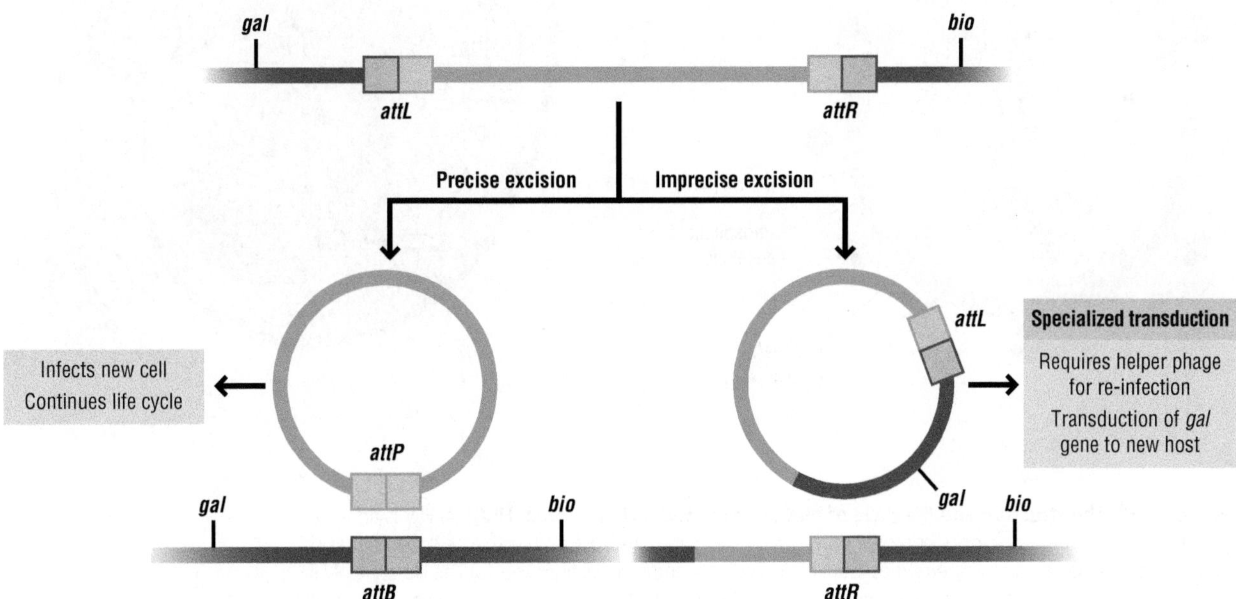

Figure 11.21 Specialized transduction. The λ prophage can excise precisely, as shown on the left, by a crossover involving the flanking sequences. This reconstitutes the *attB* and *attP* loci and produces normal phages, which can reinfect a new host. Alternatively (and rarely), the excision can be imprecise, as shown on the right, and some of the host chromosome (here the *gal* gene) becomes part of the phage genome, with part of the phage genome being left behind in the new cell. Because this phage has an incomplete genome, it requires a helper phage with a normal genome to infect a new host. However, when it does, it will transduce the *gal* gene to the new host. Only genes immediately adjacent to the *att* site can be transduced.

known as the Shiga toxin. When *E. coli* lysogenized for these phages are ingested by a human, the phage-converted bacteria cause diarrhea and life-threatening infections. An evolutionary advantage of producing diarrhea is that it helps to further disseminate the bacterium and the phage it carries.

While the Shiga toxin phage is lysogenic, a similar selective advantage can be seen with other types of phages. The temperate filamentous phage CTXφ carries genes that encode the cholera toxin—an important diarrhea-producing agent. CTXφ mediates conversion of *Vibrio cholerae* cells, such that they produce the cholera toxin. Therefore, infected bacterial strains produce the life-threatening condition known as cholera, while other *V. cholerae* do not. Shiga and cholera toxins act rapidly in humans, so that the bacteria with its phage are dispersed upon infection before the host becomes debilitated.

Eukaryotic viruses and genome evolution

We have focused on viruses that affect bacteria, but viruses that infect eukaryotes can also exhibit lytic, temperate, and lysogenic life cycles. These specific terms are primarily applied to bacteriophages, but the processes are conceptually similar. Just as phages can mediate horizontal gene transfer in bacteria by transduction, analogous virus-mediated genetic exchanges clearly occur in plants and animals, and, based on evidence from genomes, have been occurring for millions of years.

Some eukaryotic viruses complete most of their life cycle in the cytoplasm, but their genome must enter the nucleus in order to be transcribed. This occurs with some of the best-known eukaryotic viruses—the **retroviruses**—whose structure and life cycle is summarized in Figure 11.22. The first retrovirus discovered was a

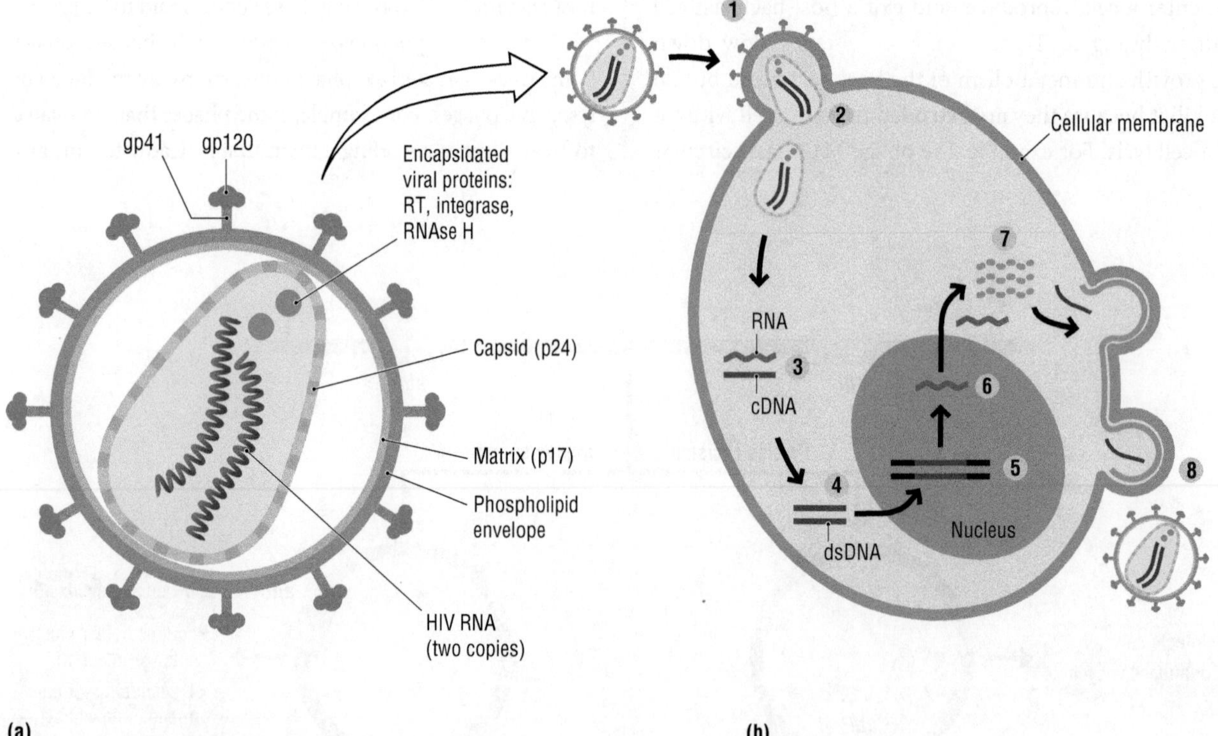

Figure 11.22 The structure and life cycle of human immunodeficiency virus (HIV). (a) HIV is an enveloped retrovirus. The envelope includes proteins and glycoproteins required for fusion with a new cell. The capsid within the envelope contains viral enzymes required to begin the intracellular life cycle. (b) The life cycle of HIV begins with fusion to, and entry into, a susceptible cell (1). The virus uncoats (2), and its RNA serves as template for the viral reverse transcriptase (RT) enzyme, which synthesizes a complementary strand of DNA (3). The RNA in the RNA–DNA hybrid is degraded by RNAse H, and reverse transcriptase then uses the DNA strand as template to produce the second strand of DNA (4). The viral enzyme integrase conveys the double-stranded DNA to the nucleus (5) and catalyzes its insertion into the host genome. Viral RNA can then be transcribed by the host (6). Newly transcribed viral RNA enters the cytoplasm where its translation produces viral proteins (7). Viral RNA and proteins are packaged and bud off the cell surface as new viruses capable of infecting other cells (8).

Source: Reproduced from Simon, V., Ho, D.D., Abdool, Karim Q. HIV/AIDS epidemiology, pathogenesis, prevention, and treatment. *Lancet*. 2006 Aug 5;368(9534):489–504.

cancer-causing virus in birds, but one of the more familiar retroviruses is human immunodeficiency virus (HIV), the virus that causes AIDS.

RNA is the nucleic acid found as the genomes of retroviruses. As illustrated in Figure 11.22(a), the RNA of HIV is packaged into capsids, along with a few enzymes needed to start the life cycle upon infection of a new cell, including reverse transcriptase, a ribonuclease, and integrase. The viral particle buds off the host cell without breaking it open, taking an envelope formed from the cell membrane with it. The new enveloped virus infects a fresh cell by attaching its viral spikes to specific receptors on the cell membrane, as shown in Figure 11.22(b). The viral membrane fuses with that of the new cell, and the capsid and its contents are internalized and the capsid then disintegrates.

In order to integrate into the host genome, the viral nucleic acid must be converted from RNA to DNA before entering the nucleus, as shown in Figure 11.22(b). You will recall from Chapter 2 that the Central Dogma of molecular biology is that DNA is transcribed to RNA, which is then translated into proteins, and a eukaryotic cell does not usually carry out the reverse process. Therefore, it is the retroviral enzyme **reverse transcriptase** that copies the single-stranded viral RNA first to an RNA–DNA hybrid using the incoming RNA as template. (This function of retroviral reverse transcriptase was discussed in Box 2.4 as a method of making cDNAs in the laboratory.) Next, reverse transcriptase destroys the original RNA and uses the remaining single-stranded DNA as a template to synthesize a second complementary strand of DNA. Now that the viral genome is in the form of double-stranded DNA, it is conveyed across the nuclear membrane by integrase, which also inserts it into the host's DNA. Once integrated, the virus is known as a **provirus** and, like a prophage in a bacterial host, can be transcribed and replicated with the host DNA.

Retroviral infections affect the hosts in various ways, depending on which cells are infected and on where the retroviral genome is inserted into the host genome, and what genes are nearby. A retrovirus integrated into a gene, like any sequence inserted into a gene, often inactivates that gene. An insertion upstream of a gene may alter that gene's transcription by separating it from regulatory sequences or creating new or altered regulatory sequences. Retrovirus insertion has effects on the host, like any other insertion or mutagen. Virus-mediated changes in the activity of host genes are one underlying reason why retroviruses can cause cancer in many animals; some of these same mutagenic changes are associated with human cancers, although the causes are typically other mutagenic agents, and not retroviruses.

KEY POINT Retroviruses use the enzyme reverse transcriptase to create DNA copies of their RNA genomes, which can then be incorporated into the eukaryotic chromosome in the cell nucleus via the action of integrase.

When the human genome was sequenced, it became obvious that our evolutionary ancestors were infected with many retroviruses that had lost the ability to replicate. Roughly 8% of the human genome consists of identifiable former retroviruses; other sequences may also have come from retroviruses but have changed so much over time that their origins are not recognized. In some instances, these ancient viral insertions now have a biological effect unrelated to the virus itself. For example, a defunct retroviral prophage is inserted upstream of the copy of the amylase gene that is expressed in the salivary gland human lineage. The insertion results in high levels of transcription of the amylase gene in salivary glands, and much of the starch we eat is broken into sugars by our saliva. Although the amylase gene is present in the genomes of other primates, this viral insertion is absent. The resultant slightly sweet taste may have led humans to favor certain foods, allowing them to utilize foods—like grains—that are high in starch and which other primates rarely eat.

11.5 Transposable elements

Most of the genes and other DNA elements in a genome are found at the same location in all individuals of the species. This has led to our ability to map genes, as discussed in Chapter 9. For example, nearly all humans have the gene encoding the ABO blood type at the same location on chromosome 9, and nearly all humans have the centromere on chromosome 16 at the same location. The exceptions are typically rare and can be recognized because they cause a biological process to go awry. But in every species, some fraction of the genome is not stable, instead being capable of moving from one location to another. The initial observations were made by Barbara

McClintock, who observed that certain genetic elements in maize could move around within the genome. The elements that mediate this type of movement are known as **transposable elements**, "jumping genes," or (particularly in bacteria) transposons.

A functional transposable element is a discrete segment of DNA that is capable of catalyzing its own movement to another location in the genome. The movement may or may not be observable during the life and growth of the individual, but it is evident when different individuals of the same species are compared. Unlike all the mobile elements discussed so far that transfer DNA between genomes, transposable elements mediate DNA transfer **within** genomes. They do not mediate transfer between different genomes, unless they are part of the genome that happens to be moved by one of the other processes. No single transposable element moves very often, but the combined movements of all of the different types of transposons create an enormous amount of sequence diversity and are therefore important in evolution, genomes, and genetics.

Dozens of types of transposable elements have been revealed by sequence comparisons. The general structures of different types of transposable elements are summarized in Figure 11.23. Although there are many different types of transposable elements, some properties are common to all. Transposable elements have defined ends; they encode the functions responsible for their own movement, and they tend to be small. The largest elements are slightly more than 10 kb in size, while the smallest are perhaps 300 bp. RNA elements, often referred to as retrotransposons or Class I elements, have RNA as their mobile intermediate, and DNA transposable elements, or Class II elements, move via a DNA intermediate.

RNA elements

RNA elements, or retrotransposons, are often the remnants of a virus that have integrated into the genome and subsequently (over evolutionary time) lost the capacity for making new virus particles and infecting other cells. The overall structure of retrotransposons is shown in Figure 11.23(a), but there are many variations on this basic structure. Like the RNA viruses from which they were derived, retrotransposons usually encode reverse transcriptase.

The retrotransposon genome is often transcribed at a high rate, and the reverse transcriptase uses those RNA transcripts as templates for making DNA copies of the

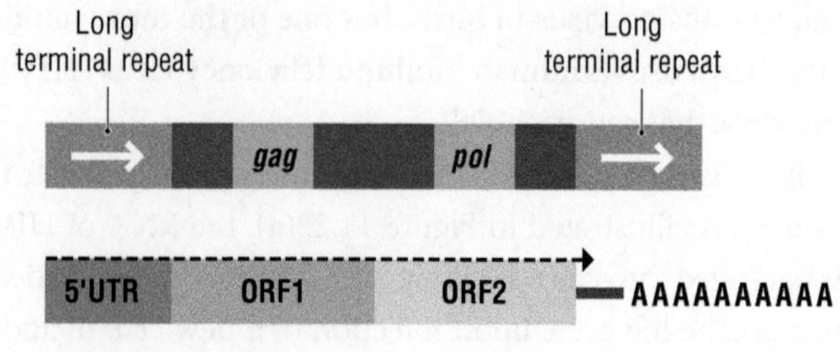

(a) Type I retrotransposons

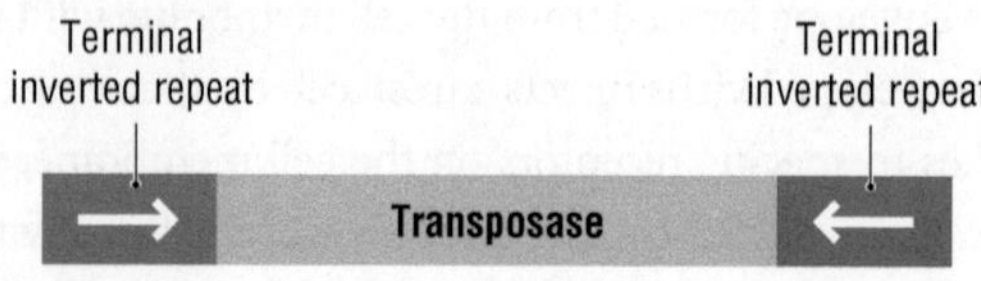

(b) Insertion sequences (ISs) in bacteria

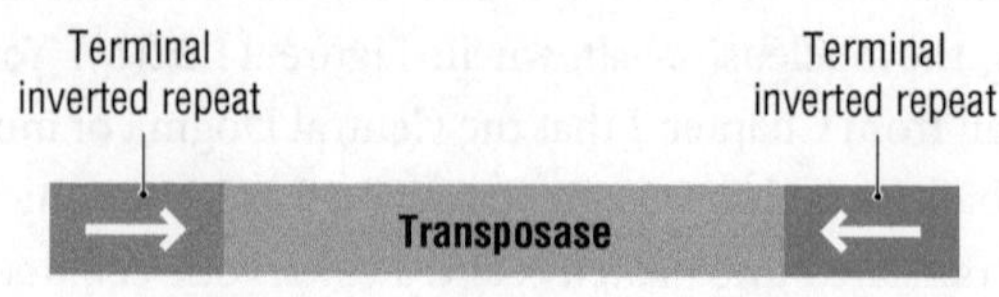

(c) Type II DNA transposable element

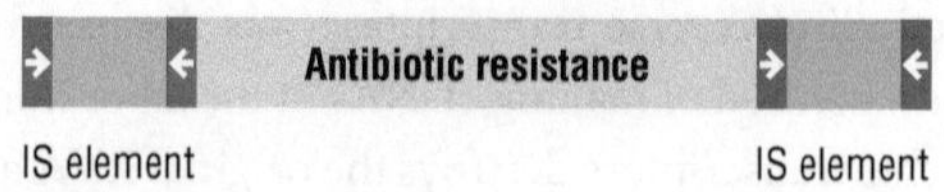

(d) Transposon in bacteria

Figure 11.23 An overview of the structures of transposable elements. There are two general classes of transposable elements based on the mechanism by which they move. (a) Type I elements, or retrotransposons, move via an RNA intermediate. Many type I elements are mutated versions of retroviruses and encode the same genes. The ends are long terminal repeats (LTRs), which direct transcription. One of the genes is usually *pol*, a reverse transcriptase. The *gag* gene encodes part of the capsid from the retrovirus. Type I elements are found only in eukaryotes. (b) Insertion sequences (ISs) in bacteria move via a DNA intermediate and a cut-and-paste mechanism. The ends are inverted repeats; the element encodes a transposable element that makes a cut at the ends to mobilize the element. (c) Type II elements in eukaryotes are analogous to insertion sequences in bacteria and also move by a DNA intermediate and a cut-and-paste mechanism. (d) Transposons in bacteria have a more complicated structure. They move through a DNA element. The ends are IS elements, which may be mutated, one of which encodes the transposase. The middle part of the element encodes one or more genes that confer resistance to an antibiotic. Both the IS elements at the ends and the particular antibiotic resistance are characteristic for each transposon.

element. These DNA copies then integrate into the genome at more or less random locations, often when a break has occurred in the chromosomal DNA. Because the original retrotransposon is still present in its original location, RNA elements typically move by a copy-and-paste mechanism. In other words, the element that moves is a copy of the retrotransposon that is still present at the prior location. Each movement thus generates a new copy of the retrotransposon.

Retrotransposons are especially common in mammalian genomes. More than 40% of the human genome (and nearly all of the Y chromosome) comprises various different types of retrotransposons or the evolutionary remnants of these elements. Not all RNA elements are derived from retroviruses or encode reverse transcriptase, however. The most abundant transposable element in the human genome is the Alu element, which was apparently derived from a small RNA molecule known as 7SL RNA, rather than a retrovirus. There are more than a million copies of Alu elements, comprising about 10% of the sequence of our genome.

DNA elements

The smallest and simplest DNA transposable elements are the insertion sequence (IS) elements found in bacteria, shown in Figure 11.23(b) and Figure 11.24. These elements are grouped into different sequence families and numbered sequentially (IS*1*, IS2, and so on).

The IS element encodes a single protein, an enzyme known as **transposase**. The ends of the IS elements are inverted repeats of 9–40 bp. The sequence can form an intra-strand stem-loop structure, as shown in Figure 11.25. Transposases make an asymmetric cut in the stem of this structure and a corresponding cut in the genomic DNA, which allows the element to move to a new location. Thus, DNA elements often move by a cut-and-paste mechanism; they also often leave part of their terminal sequence behind. The interaction between the transposase and the ends is specific, so the transposase encoded by IS*1* can only act on the sequences found at the ends of IS*1*, and not on the sequences at the end of IS2, for example.

Class II DNA elements in eukaryotes are exactly analogous in structure and transposition to the IS elements in bacteria and are often similar in size, although unrelated to IS elements in sequence. They also have inverted repeats at their ends and encode a transposase that acts specifically on those sequences. Like most other eukaryotic genes (and unlike bacterial genes), a eukaryotic transposase may be encoded in separate exons.

In plants, the most significant transposable element is the Ac/Ds element, which was the first transposable element to be described in any organism. In the 1930s, 1940s, and early 1950s, Barbara McClintock carried out a long series of experiments in corn, remarkable for their detail, insights, and rigor. McClintock received a Nobel Prize in 1983 for her work, when it became clear that transposable elements were common to all genomes.

The most important element in *Drosophila* genetics is the P element, while the most common one in *Caenorhabditis elegans* is known as Tc1. Elements related to Tc1 are found in nearly all animals, so the element must be evolutionarily ancient, whereas P elements appear to have invaded the *Drosophila* genome early in the twentieth century. DNA elements are not as common in the human genome as RNA elements, representing perhaps 2–5% of our genomes.

A more complicated type of DNA element found in bacteria, known as a transposon, is shown in Figure 11.26. Transposons have IS elements at each end and thus have both the inverted repeats of the IS elements and the direct repeat of the entire IS element itself; the inverted repeats at the ends of the IS elements are often mutated, such that the IS ends cannot move on their own. The ends are characteristic for each transposon. For example, Tn*10* has IS*10* elements as its ends, while Tn*5* has IS*50* elements as its ends.

Figure 11.24 The structure of an insertion sequence element. The sequences at the ends of one particular insertion element are shown. Note that the sequences form an inverted repeat, so that the ends read the same (from 5′ to 3′) on each strand. This allows the element to form a hairpin structure by intra-strand base pairing.

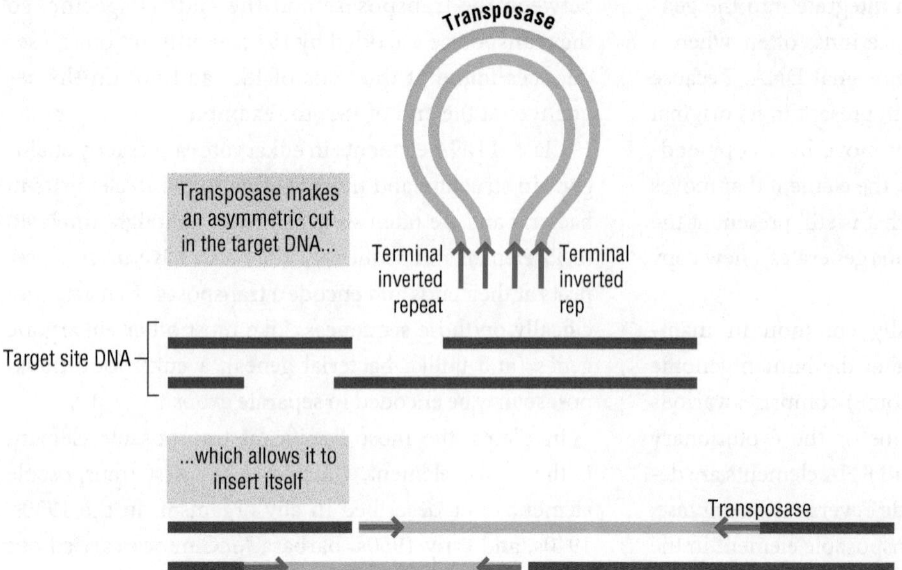

Figure 11.25 The movement of a DNA element. A DNA element has inverted repeat sequences at its ends and encodes transposase. The element can form a stem–loop structure in which the inverted repeats provide a stem. Transposase makes an asymmetric cut to release the element and a corresponding cut in the genome at the site of insertion. The element inserts at the cut target site, and the small gaps created by the asymmetric cut are filled in. Movement of DNA elements often leaves a few bases of the inverted repeat behind; thus, excision is imprecise.

As before, transposase is encoded within one of the IS elements, while the other IS element typically has a mutated and non-functional version of transposase. In addition to transposase, and unlike any other transposable element, bacterial transposons also encode other genes. The earliest transposons to be described contained one or more genes for antibiotic resistance that are also specific to each transposon. For example, Tn*10* encodes the gene for tetracycline resistance, and Tn*5* encodes genes for resistance to ampicillin, bleomycin, and streptomycin. The antibiotic resistance genes are found between the IS elements at each end and are not needed for the normal growth of bacteria. However, transposons frequently integrate into plasmids, as well as into other transposons, creating composite elements that then accumulate genes

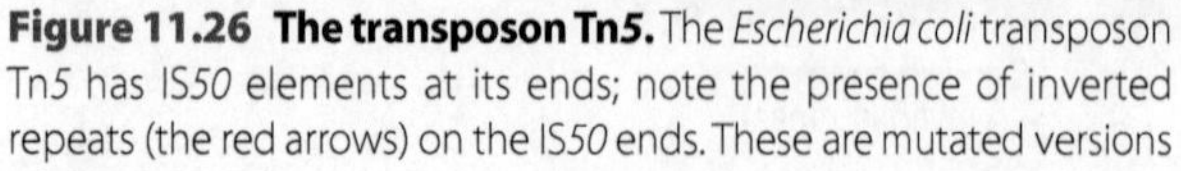

IS*50L* | Antibiotic resistance genes | IS*50R*

Figure 11.26 The transposon Tn*5*. The *Escherichia coli* transposon Tn*5* has IS*50* elements at its ends; note the presence of inverted repeats (the red arrows) on the IS*50* ends. These are mutated versions of IS*50*, which cannot move independently; the transposase for the element is encoded in IS*50R*. The central part of the element encodes genes that confer resistance to the antibiotics streptomycin, bleomycin, and ampicillin. The IS*50* elements at the ends of Tn*5* also encode an inhibitor of transposition, as well as other proteins.

for resistance to multiple, unrelated antibiotics, posing a challenge to the treatment of bacterial infections.

Autonomous and non-autonomous elements

Most transposable elements have become mutated and have lost the capacity to move on their own. For Class II DNA elements, this is often due to mutations or deletions in parts of the transposase gene. However, a functional version of transposase encoded by any element in the genome can move not only that element but also other elements of the same class, whether or not they have a functional transposase. Elements capable of moving on their own are referred to as **autonomous** elements, whereas those that can only be moved by another element are called **non-autonomous** elements. This property has been used as an important experimental tool in *Drosophila* genetics. The transposase encoded by a P element can be replaced by another gene or DNA sequence. While this creates a non-autonomous element, the gene or DNA sequence can be inserted into the genome if an autonomous element is also present.

The Ac/Ds name of the transposable element in corn reflects this difference between autonomous and non-autonomous elements. McClintock, who did nearly all of her work

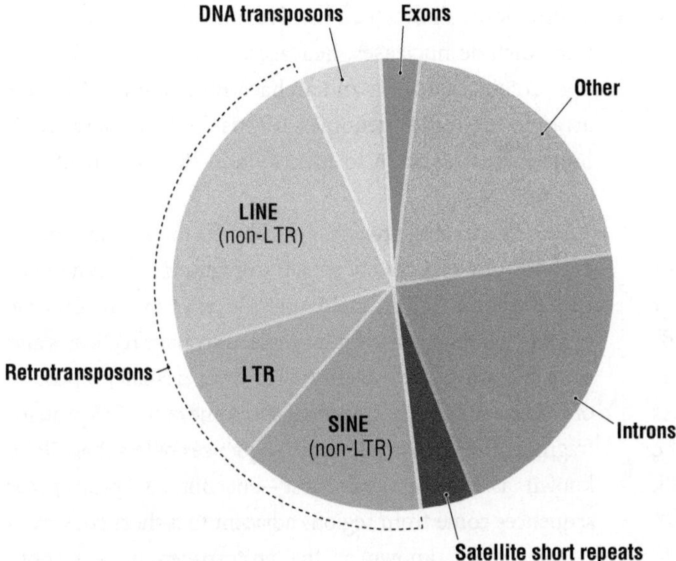

Figure 11.27 Transposable elements in the human genome. Transposable elements or the remnants of transposable elements comprise a very substantial portion of the human genome; different types of retrotransposons are especially common. LTR, long terminal repeat; LINE, long interspersed nuclear element; SINE, short interspersed nuclear element.

before it was recognized that DNA was the genetic material and well before it was possible to determine DNA sequences, realized that Ac elements could move on their own, whereas Ds elements required an active Ac element to move, so she named them as different elements. Ac encodes the functional transposase and is thus an autonomous version of the element, whereas Ds elements have deletions in the transposase and are the non-autonomous version.

Transposable elements and genome evolution

Transposable elements of different classes are common—and, in many cases, the **most** common—genetic elements in a genome. The fractions of transposable elements that comprise parts of the human genome are summarized in Figure 11.27. Transposable elements have a profound effect on the evolution of genome structures, since they are one of the principal sources of mutations and genetic variation. In addition, recombination between transposable elements can rearrange the linear order and organization of chromosome. One of the first and most essential steps in annotating a newly sequenced genome is to document the presence and locations of transposable elements, both active elements and inactive remnants of elements.

In fact, transposable elements are so common and important to genome evolution that some key functions in eukaryotic cells are derived from the components of transposable elements. We noted in Chapter 4 that telomerase, needed to replicate the ends of chromosomes, includes a reverse transcriptase that likely originated from a retrotransposon. A more specialized example is that the vertebrate immune system uses the enzymes RAG-1 and RAG-2 in the rearrangement and recombination of the immunoglobulin genes, which generates antibody diversity. RAG-1 and RAG-2 are apparently derived from the transposase of one or more DNA elements that invaded the ancestral genome of jawed vertebrates.

11.6 Barriers to horizontal gene acquisition

As mentioned previously, the unrestricted acquisition of foreign DNA is more likely to be deleterious than advantageous. Even if transcripts of the foreign DNA are not directly harmful to a recipient cell, there is a metabolic cost to replicating the DNA acquired horizontally, such that, unless the new genes afford a selective advantage, acquisition could reduce growth rates and fitness. For these reasons, most cells have mechanisms for inactivating or degrading foreign DNA. Horizontally acquired DNA can only be incorporated into the genome if it evades these mechanisms.

Clustered interspaced short palindromic repeats (CRISPRs)

One important system in bacteria for recognizing foreign DNA is known as **CRISPR**, the abbreviation for **clustered interspaced short palindromic repeat**. Studies of sequence patterns in genomes of bacteria and archaea identified arrays of direct repeated sequences interspersed with short, variable spacer sequences; the repeats are illustrated by the orange arrows in Figure 11.28. The CRISPR array is specific to different isolates of bacteria, even from the same species of bacteria, so, while different isolates will have a CRISPR array, they will not necessarily have the same sequences within the array. The repeats also have an internal repeat structure that makes a palindrome, a sequence that reads the same in either direction. As shown in the figure, the repeat can fold back on itself into a hairpin structure based on its internal palindromic structure, with the spacer separating adjacent repeats. The variable spacer regions in Figure 11.28 are short sequences sampled from previously encountered foreign DNA.

CRISPRs are usually located close to CRISPR-associated (*Cas*) genes that encode proteins with a wide variety of functions, all of which involve nucleic acids; these proteins include nucleases, helicases, and other DNA-binding proteins. Bacteria and archaea maintain the CRISPR array in their own genomes and use it as a reference library, allowing them to identify and then inactivate invading DNA.

The overarching processes used by the three main types of CRISPR/Cas systems are shown in Figure 11.29. When potentially lethal foreign DNA enters a cell, from the genome of a lytic bacteriophage for example, the phage replicates and usually destroys the bacterium. However, a small proportion of bacteria survive the infection, and some of the Cas proteins fragment the foreign DNA into sequences of less than 30 bp known as protospacers. Most—but not all—protospacer sequences come from regions adjacent to a short conserved 2–3 bp motif, known as the protospacer-adjacent motif (PAM). Due to the proximity to the PAM, the protospacers are not entirely random selections of the foreign DNA. Protospacers are then incorporated without the adjacent PAM nucleotides as a spacer into the 5′ end of the CRISPR array. Thus, the spacer sequences within CRISPRs act as "catalogs" of foreign DNAs that have been encountered previously.

The CRISPR array uses this catalog to detect and degrade foreign DNA from the same source when it is encountered again. When the CRISPR is transcribed, the palindromic sequence within the repeat folds back on itself to form a hairpin structure, with adjacent hairpins separated by the spacers, as shown in Figure 11.30. The

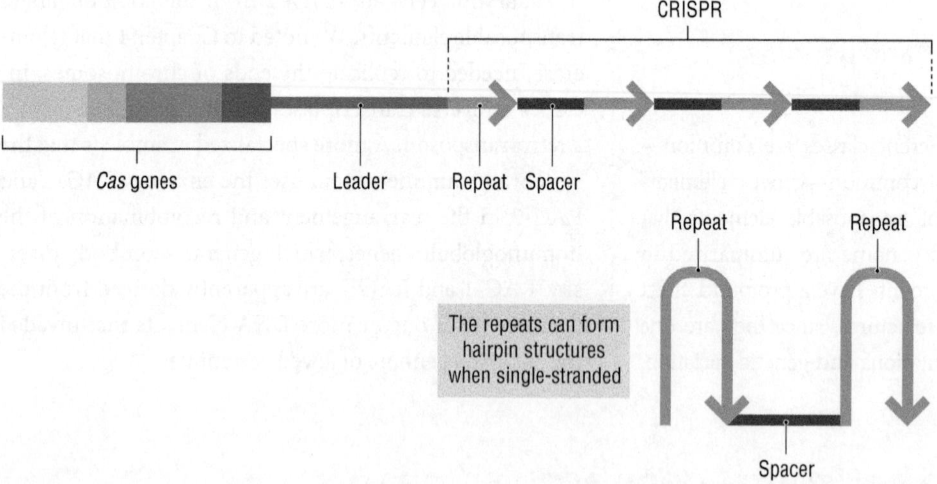

Figure 11.28 The structure of CRISPRs. CRISPRs mediate a mechanism by which bacteria and archaea recognize and destroy invading DNA sequences. CRISPR is an array of short repeats (shown here as orange arrows) that also have an internal repeat or a palindromic sequence. The repeats are separated by spacer regions of variable length and sequence, represented here in orange. The internal palindromes allow the repeats to form a hairpin structure by intra-strand base pairing, with the spacers between the hairpins, as shown below. The CRISPR array lies near a group of genes involved in a variety of different nucleic acid–protein interactions, including nucleases and helicases. These are called the CRISPR-associated or *Cas* genes.

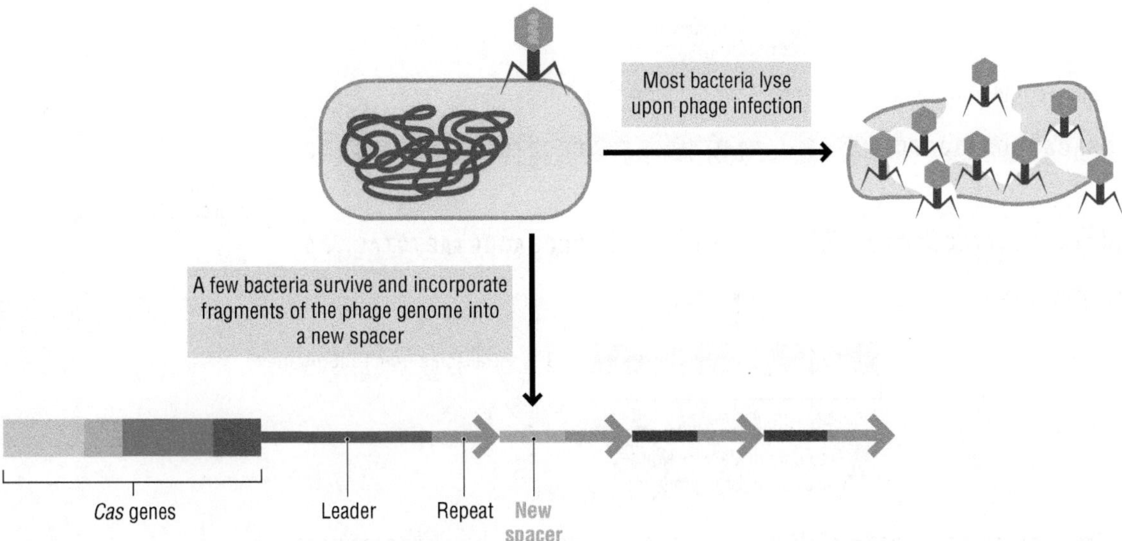

Figure 11.29 Fragments of invading DNA become spacers. When a phage (whose genome is shown in green) or other invader infects a bacterial cell, most of the bacteria lyse. A few survive, however. The *Cas* genes fragment the phage genome, and short segments of the phage genome become new spacers in the CRISPR array, as shown by the incorporation of the green sequence into a spacer.

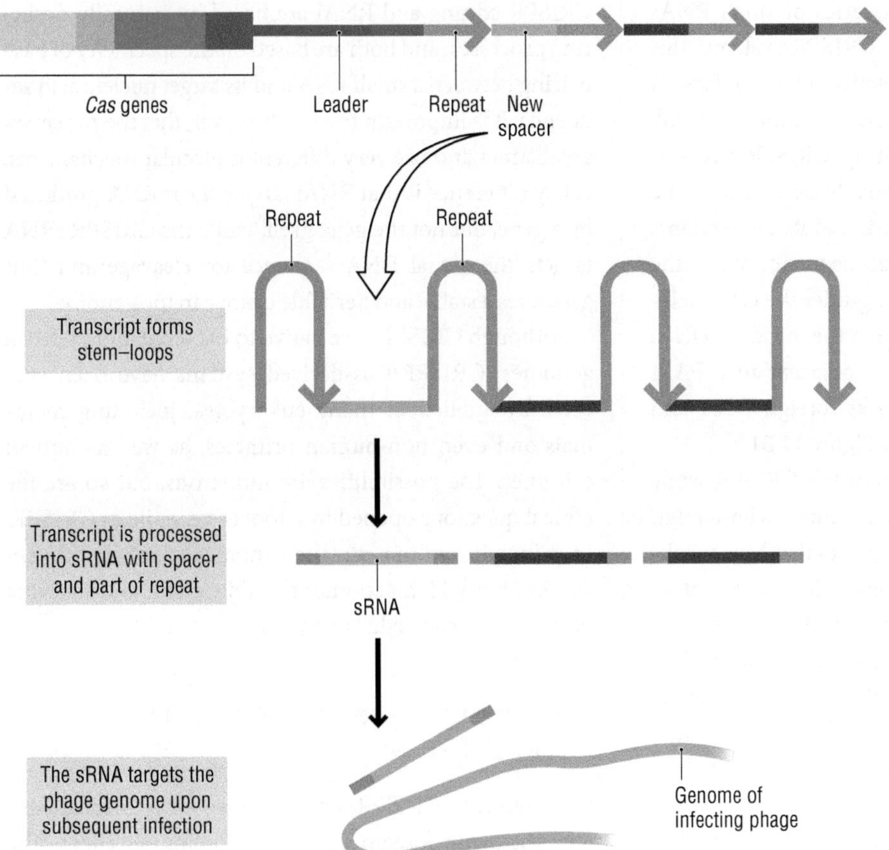

Figure 11.30 Transcription results in stem–loop formation. When the CRISPR array is transcribed, the repeats form a series of stem–loop structures, with the spacers between them. A Cas protein cuts the stem–loops, leaving RNA fragments known as small RNAs (sRNAs), which consist of the spacers with short fragments of the repeat at the ends. Upon infection by a phage or an invader whose genome is similar to a previous spacer, the spacer forms a double-stranded hybrid, which targets the invading phage genome for degradation.

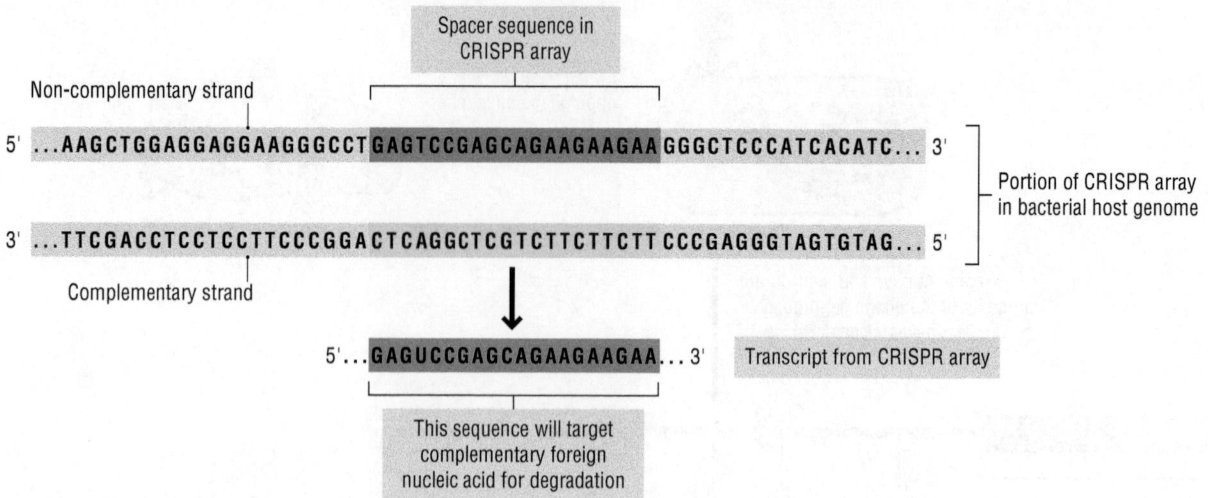

Figure 11.31 Degradation of the invading DNA. If the bacterium is infected by a phage with a sequence complementary to one of the spacers, the invading phage DNA is recognized and degraded.

Source: Adapted from Nishimasu, H. et al. (2014). Crystal structure of Cas9 in complex with guide RNA and target DNA. *Cell.* 2014 Feb 27;156(5):935–49. doi: 10.1016/j.cell.2014.02.001.

resulting RNA is then cut into a series of small RNAs (**sRNAs**); in Type I and Type III CRISPR systems, this cleavage is by one of the Cas proteins, while in Type II systems, a small RNA with partial complementarity to the repeat region (transactivating CRISPR RNA or tracrRNA) guides a cellular ribonuclease to make the cleavage. When the bacterium or one of its descendants encounters invading DNA with this sequence again, the variable spacer region of the sRNA guides the Cas nuclease to its complementary sequence in the invading DNA. If this complementary sequence is adjacent to a PAM sequence, the DNA is recognized as foreign DNA and cleaved, as shown in more detail in Figure 11.31.

The variable regions of an organism's CRISPR array provide a genomic record of past encounters with foreign DNA. Even organisms of the same species that have inhabited different habitats for several generations will contain different CRISPR arrays, each of which will reflect resident populations of bacteriophages, plasmids, and other mobile DNA from that environment. CRISPR/Cas systems provide bacteria with the ability to recognize foreign molecules, analogous to the adaptive immune response in mammals.

The ability of CRISPR to recognize and cut specific DNA sequences has been used as an experimental tool to carry out targeted gene editing or to inhibit transcription, as detailed in Tool Box 11.3. The use of sRNA to target nucleic acid sequences in CRISPR has certain parallels to the use of small interfering RNA used in RNA interference (RNAi) in eukaryotes, as discussed in Chapter 12. Both CRISPR editing and RNAi are based on naturally occurring processes, and both are based on the specificity of base pairing between a small RNA and its target nucleic acid sequence. It is important to note, however, that the processes are distinct and use very different molecular mechanisms. A key difference is that RNAi targets the mRNA produced by a gene, but not the gene itself, while the CRISPR sRNA targets the actual DNA sequence for cleavage and thus generates a stable and heritable change in the genome.

Although CRISPRs are native to bacterial and archaeal genomes, CRISPR/Cas-derived systems have been successfully applied in many eukaryotes, including mammals and even non-human primates, as well as human cell lines. The possibilities are numerous, but so are the ethical questions opened by a tool as versatile as CRISPR. In principle, we can edit the genomes of many different species, but which genomes should we edit and for what purposes are such edits acceptable?

Other systems that protect against foreign DNA

In addition to CRISPR, bacteria have other mechanisms for destroying foreign DNA or reducing its effect. The earliest identified mechanism is also dependent on recognition of specific sequences of bases, although, in this case, the recognition is done by a protein. Intracellular enzymes known as **restriction endonucleases** recognize and cut specific short sequence motifs within the foreign

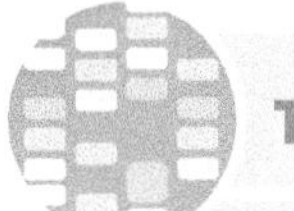

TOOL BOX 11.3 CRISPR applications: *in vivo* DNA editing and more

The CRISPR/Cas system that evolved to protect bacteria and archaea from foreign DNA has rapidly been used by molecular biologists for a number of purposes. As discussed in Box 6.2, recombination in the cell initiates with a double-stranded break in the DNA sequence. Once that break has been made (and not repaired), the sequences at that site can be altered, deleted, or replaced by recombination. As noted in Tool Box 11.2, site-specific recombination is a widely used tool for editing genomes.

Recombination at double-stranded breaks allows a scientist to cleave DNA within a living cell by providing that strain with the appropriate nuclease—most commonly Cas9 from the well-studied CRISPR/Cas system of *Streptococcus pyogenes*—and a complementary RNA designed to look as though it has been transcribed and clipped off a CRISPR array.

The key to the editing is to target double-stranded breaks precisely. Think of how we edit a manuscript in a word processing program. We search for the part of the manuscript we want to edit, possibly by using the "Find" function to locate a particular word or string of characters. We then click on the mouse to define the exact positions of the edits we want to make. We then make the edits—corrections, insertions, deletions, and so on—at that site. This analogy works for understanding genome editing using CRISPR. Cas nucleases are proteins that cleave DNA after being guided to a target site by a complementary RNA, that is, the guide RNA is analogous to the Find function that locates the part of the genome to be edited. The Cas nuclease that makes a double-stranded break is the equivalent of the mouse click.

Cas9 and guide RNAs have been used in this way to edit genes in bacteria, invertebrates, and a wide range of model animals, ranging from nematodes and flies to zebrafish and cynomolgus monkeys. Figure A shows phenotypes affecting body shape, embryo structure, and eye color that were produced in worms, zebrafish, and flies, respectively, using CRISPR.

Cas9 proteins have two catalytic domains, with each one cleaving one strand of the target DNA to produce a blunt-ended cleaved site. The catalytic domains of Cas9 require a lysine (K) at position 10 and a histidine (H) residue at position 840 for activity—a Cas9 derivative with mutations at these sites is unable to cleave DNA to which it is targeted but can still bind it. This "dead Cas9" (dCas9) prevents transcription by sterically hindering RNA polymerase and is employed for this CRISPR interference (CRISPRi) process. This method has been used in bacteria to block transcription of individuals, since RNAi cannot be used.

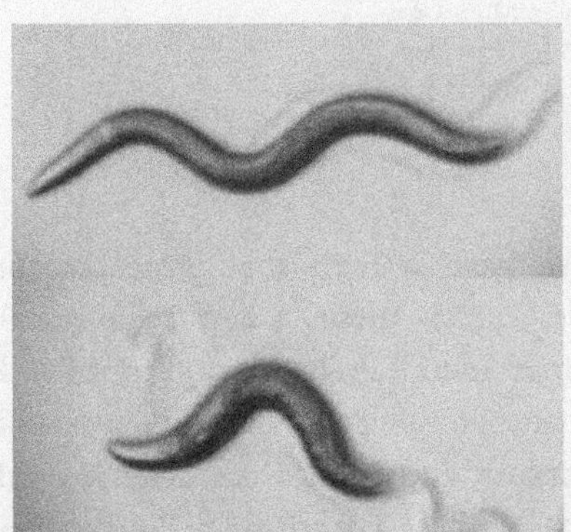

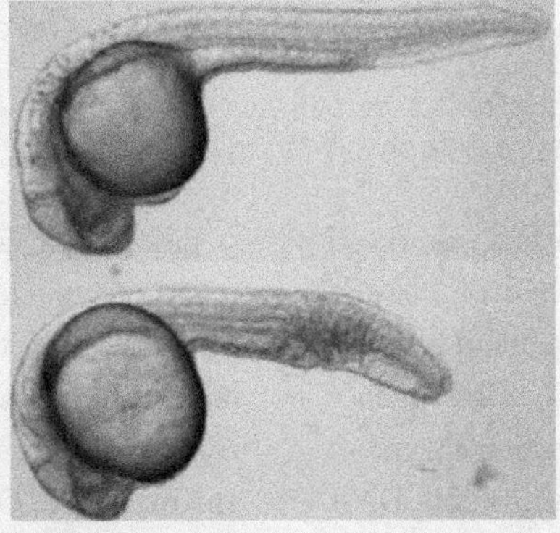

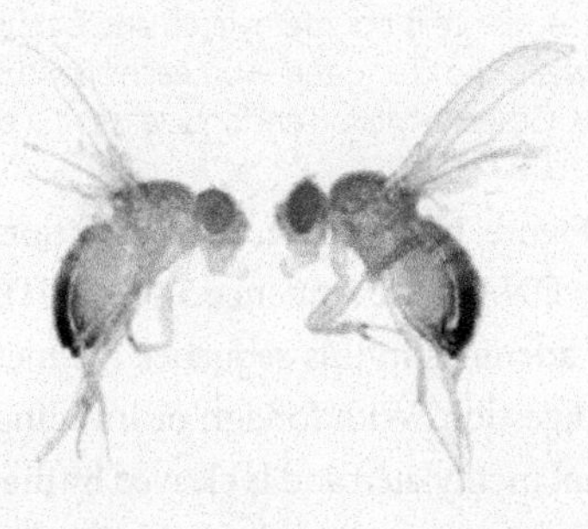

Figure A CRISPR as a gene editing tool. CRISPR is a widely used tool for engineering specific mutations in various eukaryotes. Mutant phenotypes are shown in the images for (from left to right) *Caenorhabditis elegans*, zebrafish embryos, and *Drosophila*.

Source: Image of *C. elegans* courtesy of Gina Broitman-Maduro and Morris F. Maduro. Image of zebrafish courtesy of Joanna Yeh and Andrew Gonzales, Harvard Medical School. Image of *D. melanogaster* courtesy of Dr. Scott Gratz, University of Wisconsin – Madison.

DNA. Restriction endonucleases were introduced in Tool Box 3.1 in our discussion of methods to detect sequence variation in the genome and are also widely used as experimental tools for manipulating DNA molecules. However, their natural function is to protect the bacterial cell against the incorporation of foreign DNA.

As noted in Tool Box 3.1 and shown in Figure 11.32, restriction enzymes recognize a specific DNA sequence and make a double-stranded cut in the DNA. It is possible then that a bacterial cell might encode a restriction enzyme that targets sequences in its own genome for cleavage. However, bacterial species protect themselves from their own restriction enzymes by attaching methyl groups to the sequence recognition site. Thus, the restriction enzyme is part of a larger restriction modification system that includes not only the restriction endonuclease that cuts the DNA but also a methyltransferase that attaches methyl groups to the DNA. For example, as shown

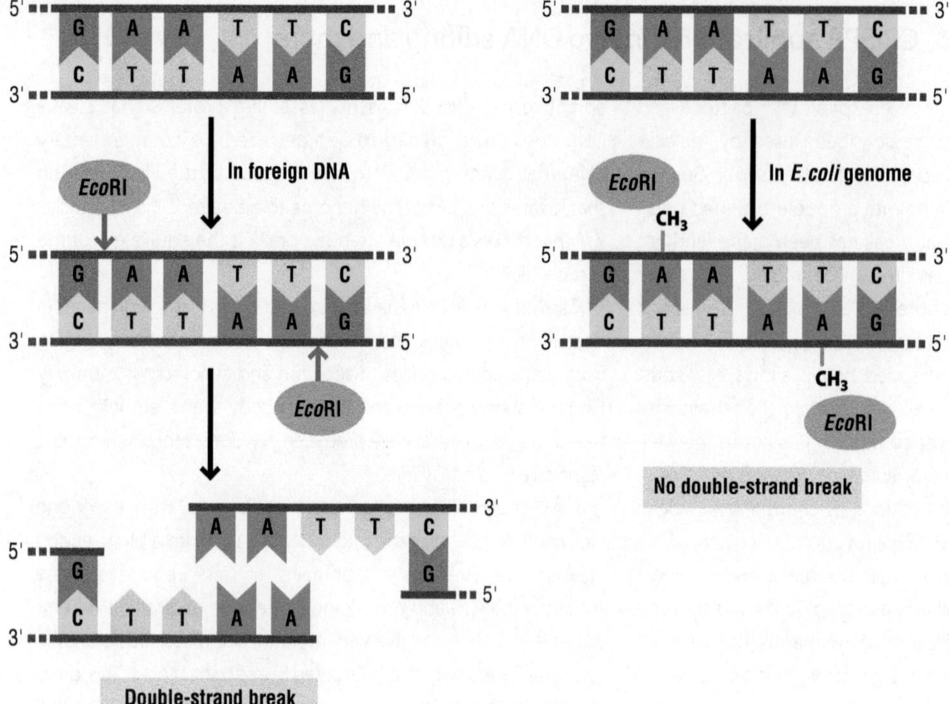

Figure 11.32 Restriction modification systems. Another method that bacteria and archaea use to detect and degrade invading DNA is the restriction modification system. A restriction endonuclease recognizes and cuts at a specific short DNA sequence; in this figure, the restriction endonuclease *Eco*RI from *Escherichia coli* is used as an example. *Eco*RI cleaves the sequence 5′-GAATTC-3′. Within the *E. coli* cell, this sequence is methylated by a methyltransferase on the adenines, and *Eco*RI cannot cleave it. However, on foreign DNA, this sequence is not methylated and is thus cleaved by the restriction enzyme. *Eco*RI makes asymmetric cuts between the G and the A on each strand, as shown; this leaves overhangs on each strand, known as sticky ends. Other restriction endonucleases cut at the same place on each strand and leave blunt ends.

in Figure 11.32, the *E. coli* restriction endonuclease *Eco*RI cuts DNA at the sequence 5′-GAATTC-3′. In the *E. coli* cell, the adenines in this sequence are methylated and resistant to digestion. With foreign or invading DNA, this sequence is not methylated and is cleaved by the restriction enzymes.

Two systems that defend bacteria from invading DNA sequence and do not depend on the recognition of specific sequences are summarized in Figure 11.33. Because most bacteria have circular genomes, the ability to enzymatically degrade linear DNA is one way to target and destroy foreign DNA, as shown in Figure 11.33(a). Bacteria may also be protected against foreign DNA by repressing transcription from invading DNA. The best-studied example of this phenomenon is bacteriophage lysogeny, discussed in Section 11.4, in which phage replication is repressed by a phage-encoded protein. Another example of transcriptional repression is found in *E. coli* and related organisms. Most bacterial genomes are GC-rich, so AT-rich DNA is likely to be foreign. *E. coli* and related organisms produce a protein called H-NS that binds to, and represses transcription from, AT-rich DNA, as shown in Figure 11.33(b).

Eukaryotes also have systems that protect against invading nucleic acids; these are mechanistically distinct from those found in bacteria. Probably the most common types of foreign genomes encountered by eukaryotes are viruses, and many virus defense mechanisms exist. One of these is shown in Figure 11.34. When a human cell is infected by a retrovirus, viral reverse transcriptase converts viral RNA to DNA by producing a cDNA–RNA hybrid. This RNA–DNA hybrid is one of the recognition signals to the mammalian host that a viral infection has occurred. Humans produce a family of enzymes known as APOBEC3 that deaminate cytidine in DNA–RNA hybrids to uracil; the particular family member for this activity is APOBEC3G. Uracil pairs with adenine, rather than guanine, producing G:C to A:U missense mutations when the virus replicates. APOBEC3G therefore causes mutations to accumulate in the viral genome, preventing its replication and integration.

Mechanisms for containing or destroying foreign DNA are abundant in all types of cells, but only a few well-studied

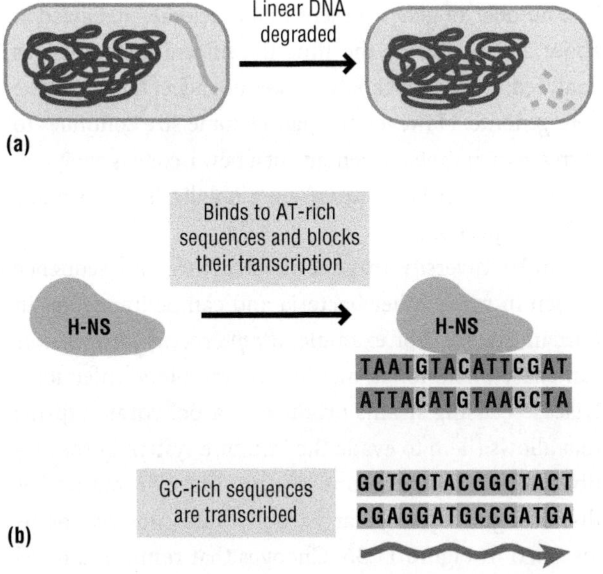

Figure 11.33 Sequence-non-specific defense mechanisms. In addition to sequence-dependent mechanisms, like CRISPR, and restriction modification systems to limit the extent of horizontal gene transfer, bacteria also have mechanisms that do not depend on sequence specificity. Two examples are shown here. (a) Since most bacteria have circular chromosomes, linear DNA is highly susceptible to degradation as a likely invader. (b) Since bacterial genomes are usually GC-rich, AT-rich DNA is a likely invader. Transcription from AT-rich regions is blocked by proteins such as H-NS in *Escherichia coli*.

examples have been discussed. However, they all share some ability to discriminate between foreign DNA and their own DNA, allowing them to target the foreign DNA selectively. The foreign DNA is degraded or its transcription is blocked, or both. But these systems are not flawless. Like their hosts, invading viruses and other sources of foreign DNA also change, and some of these changes allow them to evade the host defense mechanisms, at least temporarily. As we examine the genome of any organism, we are merely capturing a snapshot of the ongoing evolutionary back and forth between the host and invading DNAs.

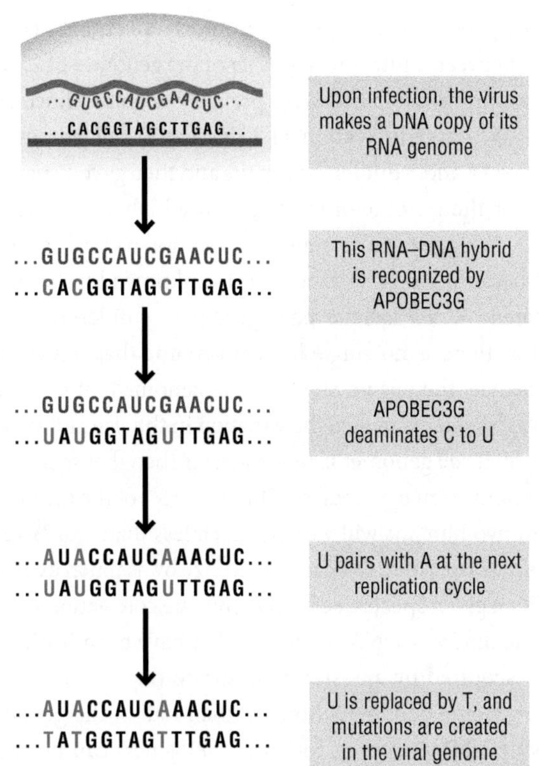

Figure 11.34 An antiviral activity in humans. Enzymes in the APOBEC3 family deaminate cytosines to make uracil; the one responsible for this particular activity is APOBEC3G. When a retrovirus infects a cell, it makes a cDNA copy of its genome. Thus, there is temporarily an RNA–DNA hybrid molecule. This is recognized by APOBEC3G, which deaminates the cytosines in the DNA to uracil. Uracil base-pairs with adenine in the subsequent replication and is replaced by thymine after that. As a result, many G:C base pairs in the retrovirus genome become A:T base pairs, creating mutations that can render the virus inactive.

Source: Reproduced from Simon, V., Ho, D.D., Abdool, Karim Q. HIV/AIDS epidemiology, pathogenesis, prevention, and treatment. *Lancet*. 2006 Aug 5;368(9534):489–504.

KEY POINT Horizontally acquired DNA is sometimes beneficial but can be deleterious to a recipient cell. Cells have defense mechanisms that limit the amount of DNA that can be laterally acquired and incorporated into the genome.

11.7 Horizontal gene transfer and genome evolution

The occurrence of horizontal gene transfer by these, and possibly other, mechanisms has been recognized for many decades. However, the extent of horizontal gene transfer among species and its effect on the evolution of their genomes has been more fully appreciated only as the genomes of individual members of a species or different isolates have been compared. This level of analysis has only been possible with the ability to sequence genomes rapidly. While many examples could be discussed, we will turn again to the analysis of *E. coli* as our model.

Core genomes and pan-genomes

The genomes of several hundred strains of *E. coli* have been sequenced and published. Comparisons among these different genomes reveal that *E. coli* isolates share a core genome that is present in nearly all strains. However, between a fifth and a quarter of the genomes of *E. coli* strains are variable and were probably acquired horizontally. We wrote in Section 11.1 that the *E. coli* genome is between 4 and 5 million base pairs and then gave a specific size for the genome of the widely used laboratory strain K-12. We presented a range of sizes for natural isolates not because the precise number is unknown but because different *E. coli* isolates have genomes of different sizes. Just as there is no single human genome that represents all people, there is no single *E. coli* genome that captures its vast diversity. In fact, the variation in the size and structure of *E. coli* genomes is much greater than that seen with plant and animal genomes. While the sizes of the genomes from two humans will vary by much less than 0.01%, different isolates of *E. coli* can vary in size by as much as 20%.

For a given species, such as *E. coli*, the core genome and all the different DNA sequences that have been horizontally acquired by any strain constitute the **pan-genome**. In a genome of 4.5–5.5 Mb, less than half of the genome (about 2 Mb encoding roughly 2200 genes) can be considered to be part of the core. For *E. coli*, the number of genes known to be horizontally acquired in at least one strain and that are part of the pan-genome is now larger than the core genome of the species and, in some cases, much larger than the core genome.

The effect of sequencing additional genomes on defining the core and the pan-genome is shown in Figure 11.35. The number of genes present in all genomes stabilized at about 2200 genes by the time the fifteenth genome was analyzed; subsequent analysis does not affect the size of the core genome. However, the pan-genome size continues to increase each time the genome of a new isolate is analyzed, since DNA can be acquired horizontally from so many different species.

Similar diversity in genome structure and sequence is seen in many other bacteria and can be important in human disease. For example, *Streptococcus pneumoniae* can cause pneumonia, meningitis, and blood infections. Disease-causing strains produce a carbohydrate capsule that allows them to evade the immune system of susceptible hosts. Host immune systems develop antibodies that recognize particular carbohydrate capsules, as illustrated in Figure 11.36. Changes that result in a novel carbohydrate structure in the capsule arise and confer a selective advantage, since the capsules are not recognized by the host, and the bacteria will reproduce.

When the genomes of closely related *S. pneumoniae* strains were sequenced, it became evident that capsule type switching often occurred by horizontal acquisition of new capsule genes, rather than point mutations in existing ones. Figure 11.37(a) shows the region encoding the capsular gene module (*cps*) from one strain, and Figure 11.37(b) shows the corresponding region from other strains. Although the *cps* genes are located between the *ugl* and *pbpX* genes on one side and the *pbpA* gene on the

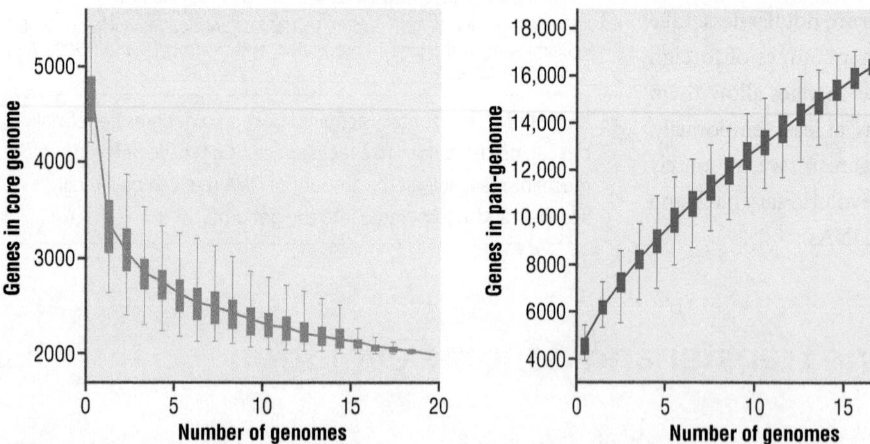

Figure 11.35 Core genomes and pan-genomes in *Escherichia coli*. Hundreds of different *E. coli* strains have had their genomes sequenced, and new isolates continue to define the pan-genome, although the sizes of the genomes are all between 4.5 and 5.5 Mb. (a) The core genome. The graph shows the number of common genes on the Y-axis, as more genomes are sequenced on the X-axis. When 15 genomes have been sequenced, the number of common genes, or the core genome, is observed to have about 2200 genes; this number does not change when additional genomes are sequenced. (b) The pan-genome. On the other hand, the number of genes found only in that strain or in a few strains continues to increase with each additional genome, and almost 18,000 individually acquired genes have been found.

Source: Reproduced from Tenaillon, O. et al. The population genetics of commensal *Escherichia coli*. *Nature Reviews Microbiology* 8, 207–217 (March 2010) doi:10.1038/nrmicro2298.

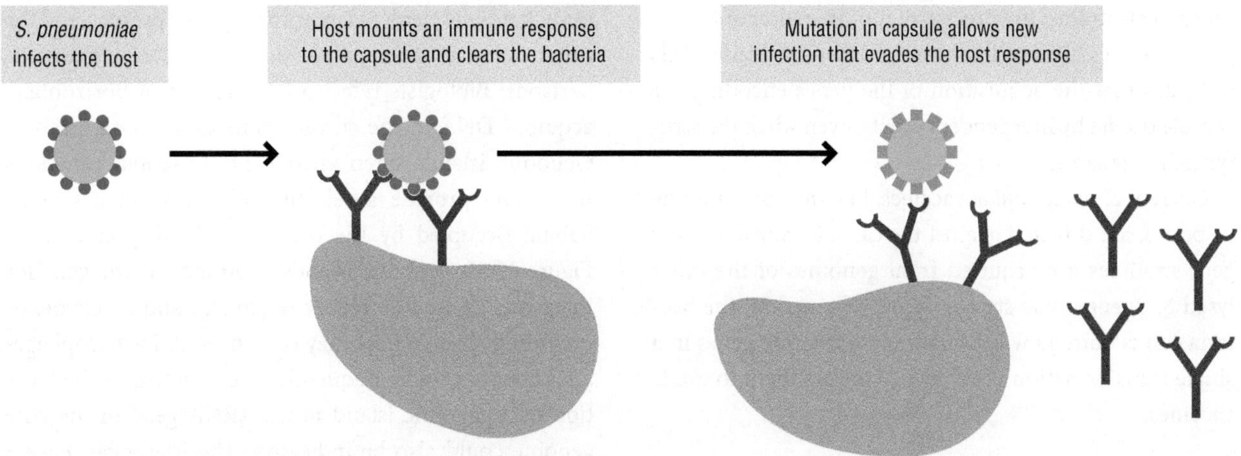

Figure 11.36 ***Streptococcus pneumoniae*** **capsular genes and infection.** *S. pneumoniae* strains have different carbohydrate capsules, as shown here by the colored shapes on the surface. *S. pneumoniae* infects a host and can proliferate until the host mounts an immune response. The immune response is directed against the capsule, as shown here. When the capsule mutates (from a red circle to an orange square), the host is not protected from infection, and the bacteria can proliferate.

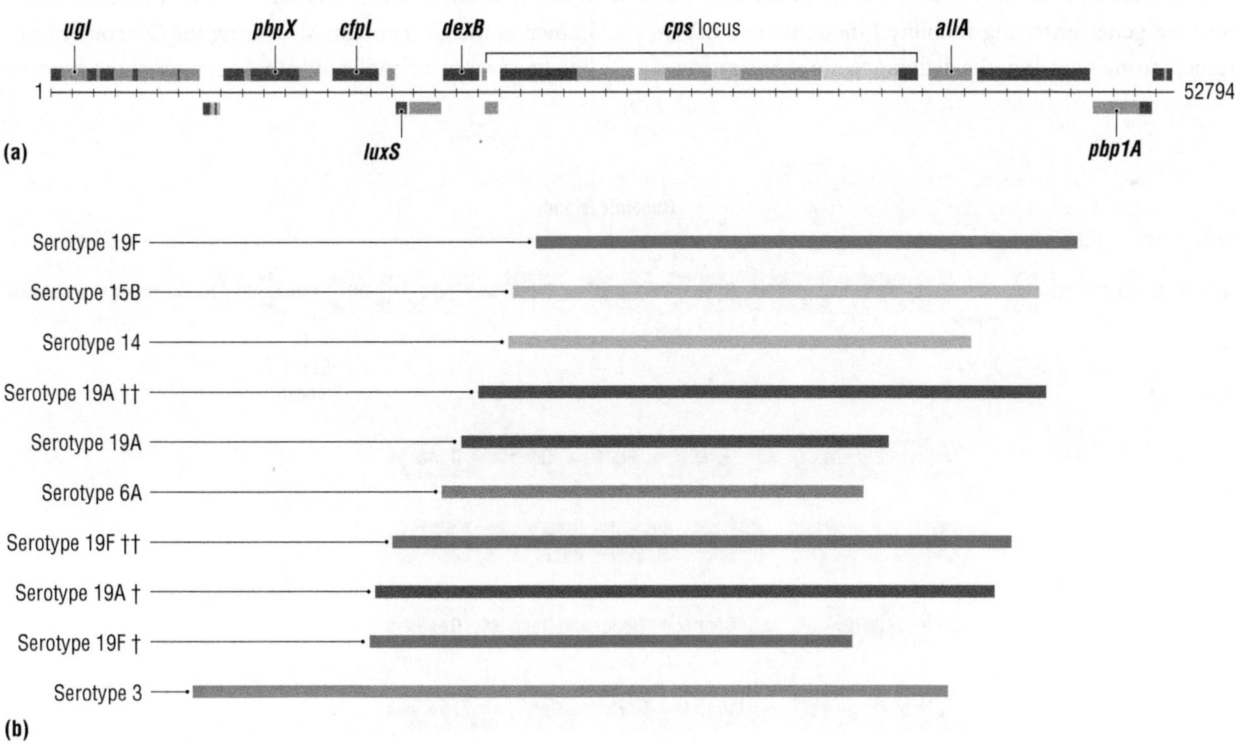

Figure 11.37 Streptococcal pathogenicity and horizontal gene transfer. The *cps* gene is responsible for producing the capsule and thus the pathogenicity of the bacterium. As shown in (a), the *cps* gene is located between the *ugl* and the *pbp1A* genes. (b) shows the location of the *cps* gene in different strains. Different strains have slightly different sites of insertion, as shown by the length and position of the line in different strains. This indicates that these were acquired independently, rather than by base changes from a common ancestor. Note that even the same serotype, which would make the same capsule, such as 19A or 19F (shown by †), differs in their precise position, indicating that even different 19A strains acquired the *cps* gene independently.

Source: Reproduced from Croucher, N.J. et al. Rapid pneumococcal evolution in response to clinical interventions. *Science.* 2011 Jan 28;331(6016): 430–4.

other in all strains, the precise sites of insertion of the *cps* gene modules are not identical in different strains. This indicates that the acquisition of the genes encoding the capsule occurs by independent events, even when the serotypes are the same.

Current *S. pneumoniae* vaccines, like the host immune response, are directed against the capsule. Since new *cps* gene modules are acquired from genomes of the other lysed *S. pneumoniae* strains by transformation, the bacteria can acquire new capsular polysaccharide genes in a single transformation event, which allows them to evade vaccines.

Recognizing horizontal gene transfer

In our example with *S. pneumoniae* above, the differences in the insertion sites of the *cps* genes were used as an indication that horizontal gene transfer had occurred. Horizontally acquired genes are clearly present in many other bacterial genomes but are not always easy to detect. Some acquisitions can be recognized to be prophages or integrated plasmids, based on their sequences, but, because genes encoding mobility functions may not be under strong selection, the defining features or signatures of such mobile elements may no longer be present, and the origins may not be apparent simply from sequence comparisons. Biologists refer to large blocks of horizontally acquired DNA in the chromosome as **genomic islands**. Genomic islands often encode non-essential functions that could provide a selective advantage in a specific habitat occupied by the organism. As diagrammed in Figure 11.38, genomic islands often include the gene for integrase, as well as repeat sequences and IS elements, providing clues of how they once moved. Bacteriophages are known to insert frequently near a tRNA, so the location of a genomic island near a tRNA gene in the core genome could also be indicative. The island can have a variable number of genes among different isolates, which may be involved with pathogenicity and virulence, with drug resistance, or with various metabolic functions, for example.

In addition to sequence features related to mobile elements, another way to recognize the presence of a genomic island is by its distinct base composition. One widely used characteristic of a DNA sequence or a genome is the fraction of bases that are G:C or C:G base pairs, known as the GC content. Measuring the GC content can be a handy tool for confirming that portions of the genome

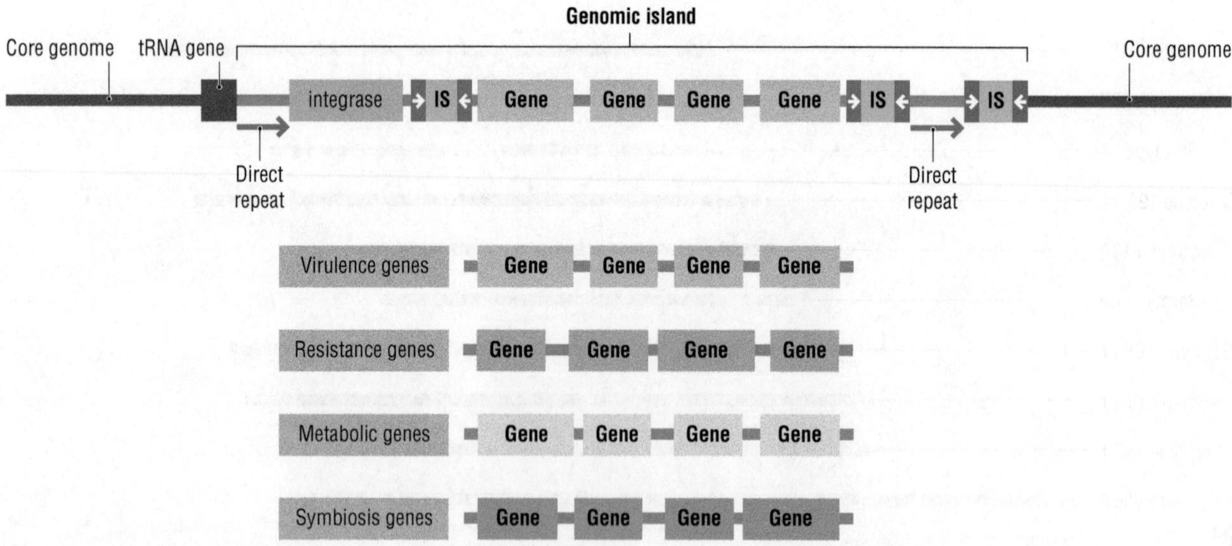

Figure 11.38 The structure of a genomic island. Genomic islands are clusters of genes on bacterial chromosomes that have been acquired horizontally. Genomic islands are often inserted near tRNA genes in the core genome, a site where many prophages also insert, and frequently include genes for integrases, as do lysogenic phages. Genomic islands often include direct repeats (red arrows), IS elements, and other evidence of mobile elements and transposons. They typically encode non-essential genes that are advantageous in particular habitats. Diagrammed here are different islands with genes for virulence, genes for drug resistance, genes for metabolic functions such as the utilization of a particular substrate, or genes that function in symbiosis with other organisms in the environment.

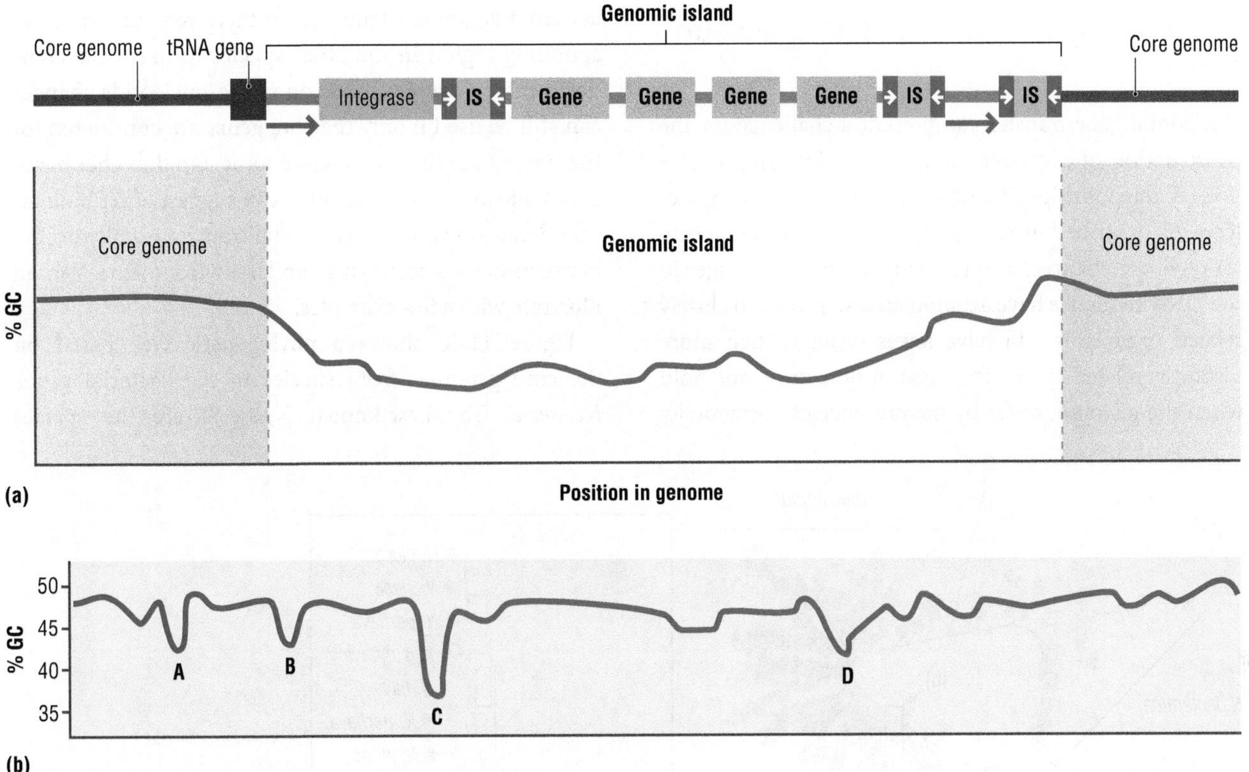

Figure 11.39 Genomic islands and the GC content. In addition to the sequence features such as integrase, IS elements, and repeats, genomic islands often differ from the core genome in the percentage of G:C base pairs. (a) The genomic island from Figure 11.38 is drawn here, with a hypothetical GC content graphed beneath it, showing that the core genome has a higher percentage of G:C base pairs than does this region. (b) Data from chromosome 1 of a strain of *Vibrio cholerae* in which the GC content has been plotted using a sliding window of 20 kb. While most of the genome has a GC content of about 48%, four distinct regions, labeled A, B, C, and DGC, have a lower percentage of G:C base pairs. These are four genomic islands, acquired through horizontal gene transfer.

Source: These data have been redrawn from Karlin (2001) Trends Microbiol 9: 335–343.

of an organism may have come from a different species, since the GC content of an acquired region may be different from the genome as a whole. As shown in Figure 11.39(a), if the GC content is plotted along a sliding window across the genome, peaks and troughs across the genome can often point to sequences that were acquired horizontally from other species. A specific example from chromosome 1 of *Vibrio cholerae* is shown in Figure 11.39(b), with four genomic islands indicated.

WEBLINK: VIDEO 11.1 shows how a plasmid can be analyzed for the presence of horizontally acquired gene clusters. Find it at **www.oup.com/uk/meneely**

KEY POINT The GC content, repeated sequences, and other DNA signatures can be used to identify portions of a genome that were acquired relatively recently in evolutionary time.

While the GC content, sequence features, and other measures can be used to detect the **presence** of DNA acquired through horizontal gene transfer, it is not always easy to determine the **source** of horizontally acquired DNA. Once the new DNA has been acquired and integrated into the genome, it evolves with the genome of its new host. Thus, portions of the acquired DNA may be lost; in addition, horizontally acquired stretches of DNA may themselves be invaded by other pieces of DNA. Furthermore, in unicellular organisms, DNA is probably more commonly exchanged among closely related organisms where the sequence features and the GC content may be similar, so that acquired DNA is not detectable. Some genomic islands can be transferred, but many are no longer mobile. Over time, foreign DNA acquired by transformation, conjugation, or transduction may integrate into the host chromosome, acquire point mutations that, in some cases, alter the function of encoded genes, or even lose parts of the sequence.

Horizontal gene transfer and phylogenetic trees

Horizontal gene transfer can present a challenge for the construction of phylogenetic trees and inferring evolutionary relationships among species, particularly in bacteria. As described in Section 4.7, a common assumption for phylogenetic trees and relationships is that changes in the DNA sequence have accumulated stepwise, so closely related organisms will have fewer changes than more distantly related ones. This assumption does not hold when the genomes differ by the presence of horizontally acquired genomic islands, since there may be no corresponding region in the other species to make the comparison. Phylogenies based on single-nucleotide changes can still be used if only the core genes are considered for the tree, rather than the pan-genome. On the other hand, genes outside the core genome can suggest other biological relationships, since horizontal transfer usually occurs between species occupying similar environments. We can illustrate with a few examples.

Figure 11.40 shows a phylogenetic tree based on the core genomes from species in the bacterial genus *Neisseria*. The best-known *Neisseria* are the species

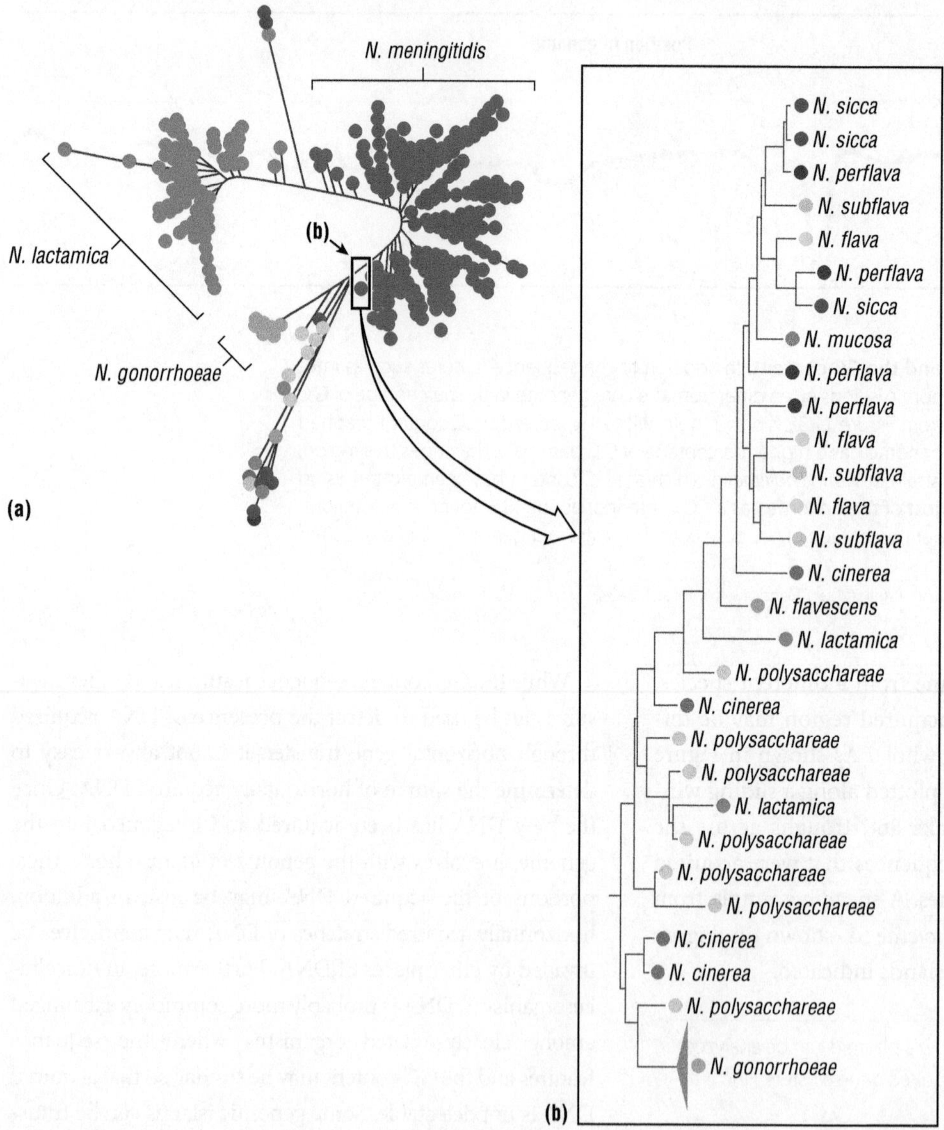

Figure 11.40 Phylogenetic tree constructed from the sequences of seven core genes of members of the genus *Neisseria*. (a) Tree of the genus *Neisseria* depicting a clear demarcation of the species *N. meningitidis*, *N. lactamica*, and *N. gonorrhoeae*. (b) Expansion of a node from (a), demonstrating that the borders among other *Neisseria* are "fuzzy" due to prodigious horizontal gene transfer among them. The *N. gonorrhoeae* clade has been collapsed in this subtree. Notice that strains belonging to different species are within the same clade and that several species are found in multiple clades.

Source: Hanage, W.P. et al. Fuzzy species among recombinogenic bacteria. *BMC Biol.* 2005 Mar 7;3:6.

N. meningitidis and N. gonorrhoeae, which cause meningitis and gonorrhea, respectively, in humans. Although *N. meningitidis*, *N. gonorrhoeae*, and *N. lactamica* form well-defined clusters, the other species do not, as evident from Figure 11.40(a). Figure 11.40(b) expands the tree at node B, which includes these other species. It is evident that the other *Neisseria* species do not form species-specific clusters. This finding is explained by horizontal exchange of DNA in these species. Genetic exchange in *Neisseria* occurs at a high rate, so this is an extreme example, but similar findings have been reported for other bacteria.

Another example is shown in Figure 11.41. These are two trees produced using accumulated single-nucleotide changes from the genomes of 240 closely related *Streptococcus pneumoniae* isolates derived from a single virulent lineage. Each terminal node of the tree represents a single strain, and the color of the lines indicates from which continent the strains were derived. The tree in Figure 11.41(a) uses all the single-nucleotide polymorphisms (SNPs), while the one in Figure 11.41(b) uses data only from the core genomes.

Although both trees represent isolates from the same virulent lineage, these trees have different shapes, which illustrate different evolutionary paths for the isolates. The branches in the tree in Figure 11.41(b) are shorter than those in Figure 11.41(a), indicating that shorter periods of time separate the isolates. The tree in Figure 11.41(b) predicted that this particular virulent lineage emerged around 1970, which agrees with the epidemiological data. The tree in Figure 11.41(a) has longer branches because there are many more changes and suggests that the virulent lineage emerged in 1930; however, many of these changes arose not stepwise from single-nucleotide changes during replication, but from horizontal transfer of sequences from other isolates. The inclusion of horizontally acquired DNA in (a) thus produced an inaccurate picture of the evolutionary history of the *S. pneumoniae* lineage.

KEY POINT Horizontally acquired DNA can make it more complicated to determine phylogenetic relationships.

Horizontal gene transfer involving eukaryotes

We have focused on horizontal gene transfer among bacterial species, but many examples are known for eukaryotes as well, including *Agrobacterium tumefaciens* and retroviruses as

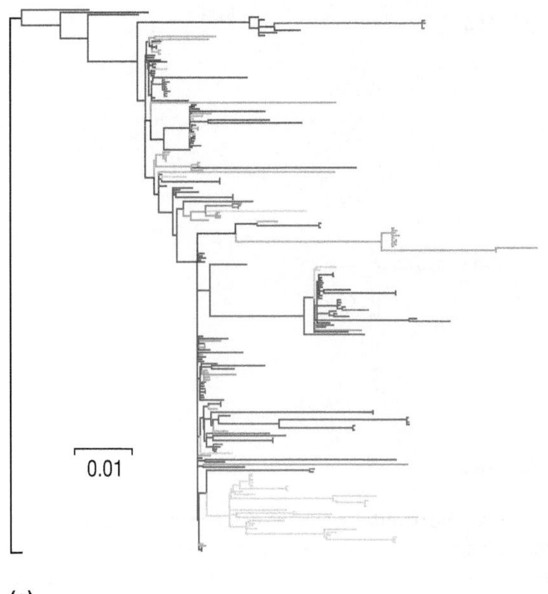

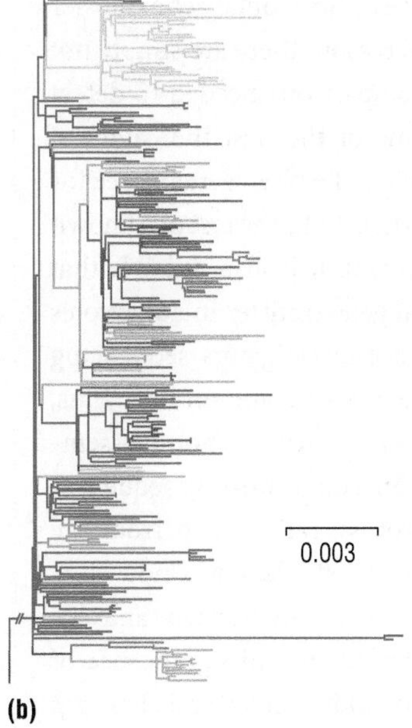

Figure 11.41 Phylogenies *for Streptococcus pneumoniae* PMEN lineage that caused a substantial proportion of infections in humans. The tree in (a) is based on all SNPs, while that in (b) excludes SNPs that are within sequences acquired horizontally in the relatively recent past. The differences in tree topology illustrate how horizontal gene transfer events can obscure phylogeny. Note the difference in the scale bars between (a) and (b); lines of similar length in the drawings are actually much shorter in (b) than in (a). The much more compact tree in (b) suggests a time of divergence that agrees with epidemiological data. The scale bars indicate the respective percentages of the total number of polymorphic sites considered.

Source: Reproduced from Croucher NJ, et al. Rapid pneumococcal evolution in response to clinical interventions. *Science*. 2011 Jan 28;331(6016):430–4.

the vectors by which genes are transferred. As genomes from more species are sequenced and analyzed, even more examples of horizontal gene transfer involving bacteria and eukaryotes are being found. For example, at least a dozen species of plant-parasitic nematodes have genes that dissolve plant cell walls, and more than 50 such genes have been found. Non-parasitic species do not have these genes. These genes apparently came into the nematodes' genomes via horizontal gene transfer from at least 15 different species of soil bacteria that are symbiotic with both the nematode and the plant, or are pathogenic to the plant. Some of these same bacterial genes are also found in the genomes of species of fungi and insects that infect plants, as shown in Figure 11.42, indicating that this transfer has been extremely widespread.

More complex cases of eukaryote-to-eukaryote gene transfer also exist. For example, the genes that give pea aphids their color have their origins in an ancient horizontal gene transfer from fungi of genes that produce carotenoids. Horizontal gene transfer between multicellular organisms is particularly apparent in the sharing of genes between symbionts or parasites and their hosts, which may occur at the time of infection.

As with bacteria, establishing that a gene in a eukaryote's genome has arrived there by horizontal gene transfer can be complicated. For one reason, there are simply not as many genomes for the comparisons needed to detect horizontal gene transfer. One of the best indicators of horizontal gene transfer is when the phylogenetic tree for a particular gene is not congruent with the rest of a known phylogenetic tree for the species. It is also possible that some examples of horizontal gene transfer in eukaryotes have been overlooked because investigators sequencing the genomes were not aware of how common it is. Thus, bacterial sequences encountered during genome assembly were assumed to come from contaminating sequences in the genome assembly process, rather than from the naturally occurring genome. Nonetheless, it is now clear that horizontal gene transfer between bacteria and eukaryotes occurs, and that the extent and significance of horizontal gene transfer in individual eukaryotic lineages vary widely. Horizontal gene transfer is often associated with adaptation to highly specialized niches, and organisms living in close contact increase the opportunity for this exchange.

A particularly interesting example is found in the case of the bacteria in our gut. Some bacteria residing in the gut aid our digestion because they produce

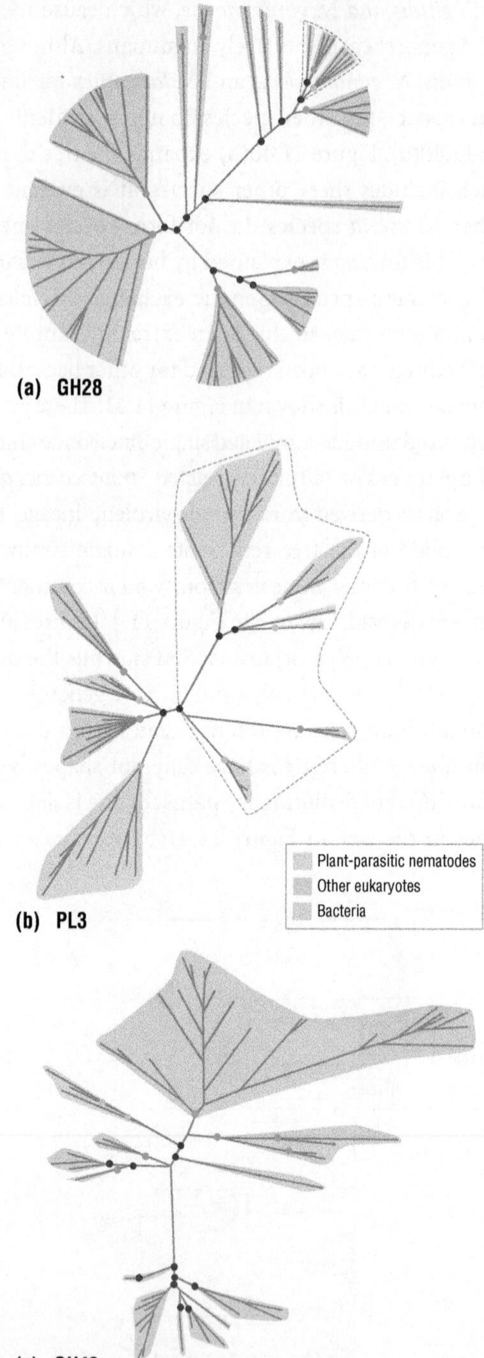

Figure 11.42 Horizontal gene transfer in eukaryotes. Plant-parasitic nematodes often invade plants by degrading or modifying cellulose in the plant cell wall. Phylogenetic trees based on sequence comparisons for three different enzymes involved in cell wall modifications or degradation (a, b, and c) are shown here, with each branch representing the enzyme from a different species of nematode. The colors indicate the source of the enzyme, with light green representing enzymes found in other nematodes, purple representing enzymes found in other eukaryotes (primarily fungi), dark green representing enzymes found in bacteria.

Source: Reproduced from Etienne, G. J. et al. Multiple lateral gene transfers and duplications have promoted plant parasitism ability in nematodes. *PNAS* October 12, 2010 vol. 107 no. 41 17651–17656.

polysaccharide-digestive enzymes that the human genome lacks; one such gut bacterium is *Bacteroides plebeius*. While *B. plebeius* is found in individuals throughout the world, the genes encoding the enzymes for degrading polysaccharides are found only in the *B. plebeius* isolated from individuals in Japan, and not from North Americans.

These genes for digesting polysaccharides appear to have originated in the genome of the marine bacterium *Zobellia galactanivorans*. These marine bacteria live on the red algae found on the edible seaweed *nori* commonly used in making sushi in Japan. One hypothesis is that, when people eat the seaweed with their sushi, the genes for degrading polysaccharides can be transferred from the genomes of *Zobellia* on the seaweed to the genomes of *Bacteroides* in the gut. The presence of the genes from *Zobellia*, in turn, allows humans to digest the seaweed that they have eaten. In populations that do not regularly consume edible seaweed, *Zobellia* is not present, and no horizontal gene transfer to *Bacteroides* has occurred. Thus, horizontal gene transfer, like other types of genetic changes, occurs whenever the opportunity presents itself. We discuss the complex web of connections between horizontal gene transfer, organismal interactions, environmental factors, selective advantages, and evolution in Chapter 17.

The most biologically significant horizontal transfer events were undoubtedly the endosymbiotic origins of mitochondria, chloroplasts, and other DNA-containing organelles. These arose from the transfer of entire genomes into eukaryotic cells (albeit not into nuclear genomes). These transfers, as discussed in Chapter 3, resulted in the evolution of new kingdoms of life.

11.8 Summary: exchange and evolution by horizontal and vertical gene transfer

Being animals ourselves and being familiar with genetic exchange among plants and animals, we may think that meiosis and sexual reproduction are the only means by which genetic exchange occurs and genetic variants are passed from one generation to the next. This is clearly not true. Bacteria do not undergo meiosis but have at least three other mechanisms for genetic exchange: direct uptake of DNA from the environment (transformation), physical contact and DNA transfer between cells (conjugation), and virus-mediated transfer of host and viral DNA sequences (transduction). These processes occur within or between all kingdoms, and they are the basis of many standard laboratory techniques. But while the mechanisms of such transfers may be well known, the evolutionary and biological impacts of these transfers are only beginning to be recognized.

Once transferred horizontally between cells by one of these mechanisms, DNA can be incorporated into the genome where it is transferred vertically to succeeding generations. In fact, genetic exchange is usually only recognized if both horizontal and vertical transmission have occurred. Just as sequence changes arising during replication and repair in the genome have affected the phenotypes and shaped the evolution of all species—bacteria, archaea, and eukaryotes—horizontal gene transfer is now recognized as having a profound impact on the evolution of all species. We recognize the extent of horizontal gene transfer because we have the capacity to sequence genomes rapidly. Bacterial genomes are smaller and easier to analyze, but many more examples are sure to exist among archaea and eukaryotes.

Genome analysis is analogous to exploring items found in an attic where we encounter many unusual or unfamiliar objects. One question we ask is "What does this thing do?" and we can often make inferences about the function based on its structure. When exploring a genome, scientists perform a similar analysis by using the DNA sequences and predicted amino acid sequences to infer functions. Another common question is "Why was this thing saved?" While we can make reasonable guesses, a definite answer is usually elusive for both attic exploration and genomic analysis. Yet another frequently asked question in both attic exploration and genome analysis is "Where did this thing come from?" Perhaps we assumed that most elements in genomes came from ancestral species and that these ancestral relationships could be traced vertically through time, like objects in the attic inherited from parents and grandparents. This is certainly true for many parts of genomes. But we may not have recognized how much of each genome was acquired not only from ancestors but also from neighbors in the community via horizontal gene transfer.

CHAPTER CAPSULE

- Most of an organism's genome is acquired vertically from parents. Organisms can, however, acquire DNA horizontally by mechanisms that require cell-to-cell contact (conjugation), by mechanisms mediated by viruses (transduction), or by uptake of naked DNA (transformation).
- DNA can be transmitted horizontally within and between species and kingdoms.
- Once within the cell, DNA is often degraded but may instead replicate autonomously or become incorporated into the genome by homologous recombination or site-specific recombination.
- In some species, DNA acquired horizontally represents a large proportion of the genome and confers many advantageous phenotypes.
- Horizontally acquired DNA shapes evolutionary history by allowing large-scale sudden changes in the phenotype.
- While horizontal gene transfer is common among bacteria, many examples are being discovered among eukaryotes as more genomes are sequenced and analyzed.

STUDY QUESTIONS

Concepts and Definitions

11.1 How does vertical gene transmission differ from horizontal gene transfer?

11.2 Define transformation, transduction, and conjugation. Describe how these are different from one another and how they are similar to one another.

11.3 Define the following terms:

a. Competent
b. F plasmid
c. Ti plasmid
d. Lytic and lysogenic phages
e. Transposable element

11.4 Briefly explain the difference between the core genome and the pan-genome of a species of bacterium.

11.5 What is meant by the term "genomic island"?

Beyond the Concepts

11.6 Crossing over is needed between the incoming DNA and the bacterial genome for stable maintenance and vertical transmission of DNA by transformation, conjugation, and transduction. Crossing over is not needed when plasmids are transferred by transformation.

a. Why is crossing over not needed for stable maintenance of plasmids?

b. What function must be encoded on a plasmid in order for it to be transferred vertically?

11.7 In early days of *Escherichia coli* genetics, two different strains were widely used for conjugation mapping. In many strains, the genes for leucine biosynthesis (*leu*) and lactose catabolism (*lac*) were transferred early in the process, but the genes for threonine biosynthesis (*thr*) and the genes for isoleucine and valine biosynthesis (abbreviated *ilv* in Figure 11.13) are rarely transferred. In other strains, the genes for threonine biosynthesis and *ilv* were transferred early in the process, while the genes for leucine biosynthesis and lactose catabolism were rarely transferred.

a. Explain this result.

b. On rare occasions, a strain of one type would switch to become the other type. What chromosomal event must have occurred to allow this phenotypic difference?

11.8 Briefly describe the differences between generalized transduction and specialized transduction, including which genes are being transferred in each case.

11.9 Transposable elements are found in the genomes of nearly every species. What sequence features can be used to distinguish RNA-based elements or retrotransposons from DNA-based elements?

11.10 Restriction enzymes are very widely used tools in molecular biology research.

a. What properties of restriction enzymes have allowed them to be so useful in laboratory techniques?

b. What is the natural role of restriction enzymes, and how do bacteria protect themselves from this natural role?

11.11 In what way do CRISPR regions of bacterial genomes provide insights into the history of that bacterial strain?

11.12 In what way do instances of extensive horizontal gene transfer challenge the "branching tree" view of evolutionary lineages?

11.13 Describe a factor or scenario that increases the likelihood of horizontal gene transfer into the genomes of multicellular eukaryotes.

11.14 Why is horizontal gene transfer particularly important in organisms that reproduce asexually?

Applying the Concepts

11.15 The following type of experiment was used to elucidate mechanisms of horizontal gene transfer early in the history of bacteriology. In the experiment, two strains of bacteria in liquid culture were placed in opposite arms of a U-shaped tube. Dividing the two compartments was a glass filter whose pore size was too small to allow bacteria through. However, when pressure was applied at either end, liquid from the medium with any macromolecules it contained could pass from one arm to another. One of the strains was a donor and the other a recipient of DNA that can be transmitted horizontally. Predict if recombinant cells would occur if the process of transfer was each of the following. Explain your answer.

a. Conjugation

b. Transformation

c. Transduction

11.16 Once DNA has been transferred to a recipient cell via conjugation, describe two ways in which the DNA might then be maintained in the recipient cell that would enable it to be subsequently transmitted vertically.

11.17 You are looking at the differences between the genomes of two isolates from the same species of a fungus found in the soil. What types of genomic information could be helpful in determining which differences have arisen by vertical gene transmission and which ones might be due to horizontal gene transfer?

11.18 In the past few years, CRISPR has become a widely used method to "edit" the genomes of any organism. Very briefly, describe how CRIPSR-mediated gene editing is accomplished.

11.19 (The following questions are based on experiments done in *Saccharomyces* by Boeke *et al.*, 1985 (*Cell* **40**: 491–500.) You have identified two different types of transposable element in a poorly studied eukaryotic organism. Although they have the sequence features of typical elements, you want to investigate the mechanism by which each one moves. You call these TE1 and TE2. Their sizes in kb are shown on the agarose gel in Figure Q11.1. You also make a modified version of each of these elements by inserting a known intron from the fission yeast *Saccharomyces bombe* into each one; these are named TE1-A and TE2-A. You insert TE1-A into one strain of the organism and TE2-A into another strain, and allow each of them to move. You then re-isolate each element, after it has moved; these are labeled TE1* and TE2*.

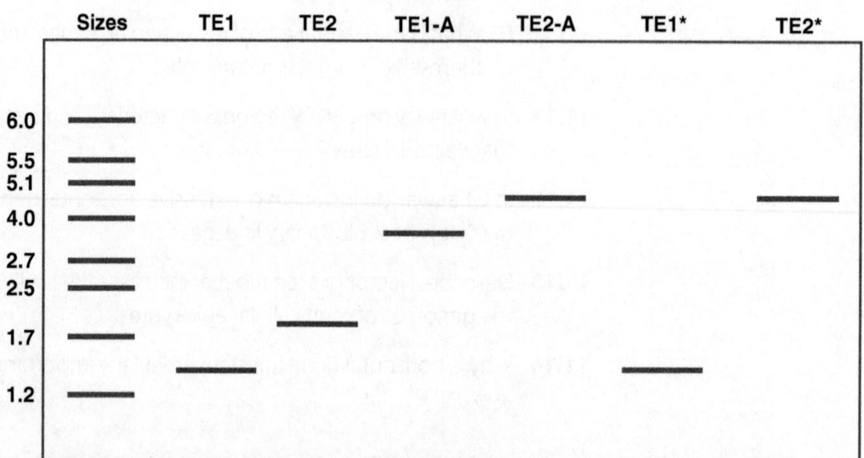

Figure Q11.1

a. What is the approximate size of the intron that you have inserted to make TE1-A and TE2-A?

b. What can you conclude about how each element moves, based on the results with TE1* and TE2*?

c. Although splicing has not been studied in this poorly studied eukaryote, what can you infer about the signals and the proteins needed for RNA splicing?

11.20 Refer to Tool Box 11.1 concerning transformation and plasmid vectors. The first figure (Figure A) is a photograph of colonies of an *E. coli* strain transformed with a plasmid carrying an ampicillin resistance gene and the gene encoding green fluorescent protein. Assuming that the plasmid contains the minimal genetic material needed to produce this phenotype, sketch a map of the plasmid using a circle and annotated arrows, similar to the diagram for the plasmid pUC18/19 in Figure B.

11.21 Undergraduate researchers Daniel and Jenny were interested in identifying the bacteria that live on bamboo leaves and spruce tree needles. They collected samples of the leaves from both plants on their university campus and mixed them vigorously in a wash buffer containing some surfactant. Inevitably, some of the leaves were broken. They then collected the washes (which had a slight green tint) and extracted total DNA from them. When they screened their washes for bacterial ribosomal RNA genes, they did not find any ribosomal RNA gene sequence that was identical to DNA from known terrestrial bacterial species. They did, however, find that each sample contained lots of DNA that was similar to ribosomal RNA from marine cyanobacteria. They considered it unlikely that their samples would harbor marine organisms. What is the most likely explanation for their finding? How might they confirm your hypothesis?

11.22 *Streptococcus pneumoniae* is the best-studied example of a naturally competent organism. Its genome consists of a 2 Mb chromosome and almost never includes independently replicating extrachromosomal elements like plasmids. Over 250 genomes of independent isolates of this species have been sequenced. Considerable variation among *S. pneumoniae* is indicative of extensive horizontal transmission of DNA. The normal habitat for *S. pneumoniae* is in the nostrils of humans and other mammals where it comes in contact with DNA from a broad range of organisms and with diverse potential sources of new DNA. What, in your estimation, accounts for the fact that most of the DNA in *S. pneumoniae* genomes originates from other streptococci?

11.23 Gierer and Schramm (*Nature*, 1956, **177**: 702–703) demonstrated that lesions on tobacco leaves were produced by a virus. They could quantify infectivity by counting the number of lesions and obtained the data shown in Table Q11.1 for tobacco leaves infected with viral particles or with purified RNA (devoid of all other macromolecules) from the virus. From these lesions, they were able to isolate virus particles that could infect new leaves.

Table Q11.1

Sample used for infection	Tobacco mosaic virus	RNA extracted from tobacco mosaic virus
"Normal"	629	488
"Normal" with serum from a rabbit that had never been infected	117	180
"Normal" with antiserum raised in a rabbit that had previously been injected with tobacco mosaic virus	0	145
"Normal" with ribonuclease (RNase) enzyme	473	0
"Normal" stored for 48 hours at room temperature (20°C) before the experiment	130	2

a. In two to four sentences, state what molecule carries the genetic blueprint for tobacco mosaic virus.

b. What role(s) are performed by other macromolecule(s) that are likely present in viral particles?

CHAPTER 12

Transcription: Reading and Expressing Genes

IN A NUTSHELL

The DNA sequence that comprises a gene is expressed by the transcription of the gene into an RNA molecule. Transcription is highly regulated, and cells with the same DNA sequence have different appearances and functions because they transcribe different genes. Transcript initiation is the key step but is only one of several points through which gene expression can be controlled. The points during the transcription process that are regulated by the cell are also those that have been subject to the most evolutionary variation.

12.1 Overview of gene expression and its regulation

To a first approximation, all of the cells in a metazoan have the same genes. In addition, a particular cell has the same genes regardless of what environment it is in. Despite having the same genetic information, however, different cell types do not look the same as each other or perform the same functions. In fact, there can be an incredible diversity in cell structures and functions within an individual, as seen in Figure 12.1. Cells in different regions of a multicellular organism form specialized tissues that look and work differently, even though they contain identical genomes. Additionally, the structure and function of individual cells can also change over time, all while maintaining the same genome. Differentiation of cellular function therefore occurs in both space and time.

Furthermore, scenarios where cells share identical genomes but look and behave differently are not restricted to multicellular organisms. For example, all the cells of a microbial strain of unicellular organisms have the same genes, but different genes are expressed in different environments. This allows the cells to respond to changing environmental conditions and to interact with each other.

How then do cells with the same genome come to differ from one another? The key is that, while all cells from a microbial colony or cells in a multicellular organism contain the same genes, they do not **express** the same genes. Gene expression—that is, the process of transcribing a subset of the genes in a genome into RNA to mediate some kind of biological function—differs across time and space and is highly regulated.

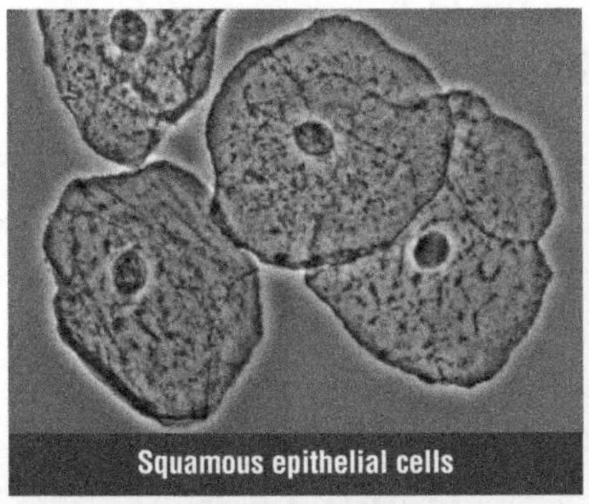

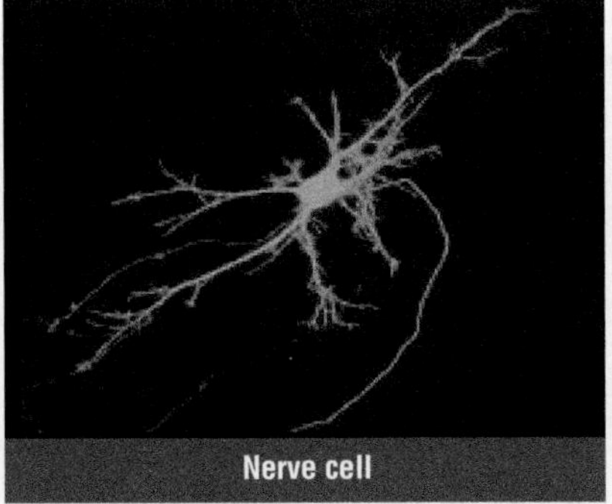

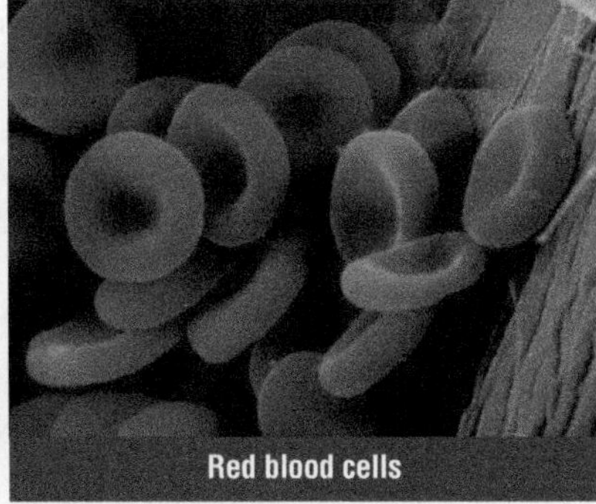

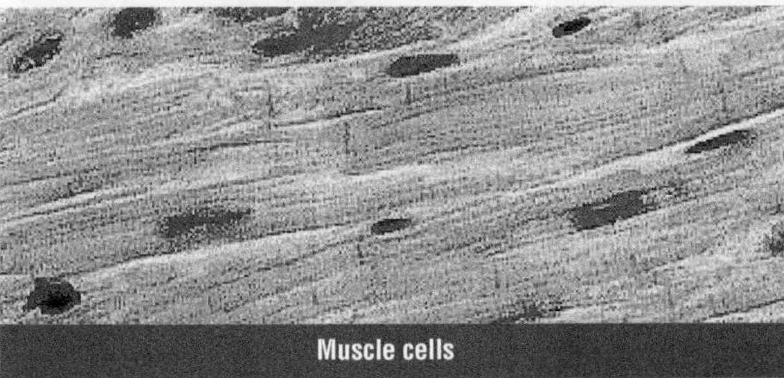

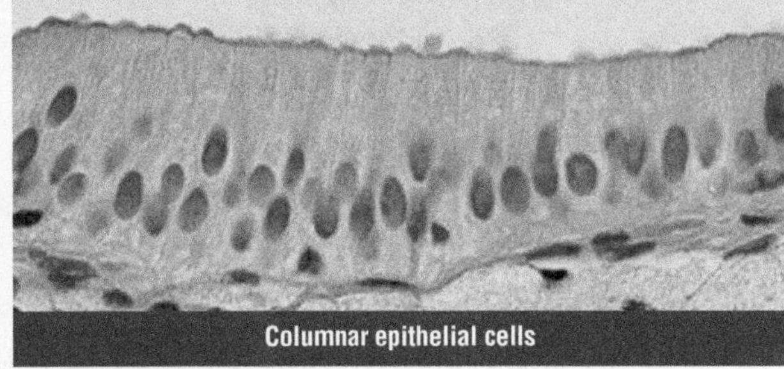

Figure 12.1 Some human cell types. All of these cells started with the same genome but expressed different genes through development and thereafter. As a result of these differences in gene expression, the cells have many structural differences and show different phenotypes.

Source: Squamous epithelial cells courtesy of Alex_brollo/ CC BY-SA 3.0. Nerve cell from Lee, WCA et al. (2006). Dynamic Remodeling of Dendritic Arbors in GABAergic Interneurons of Adult Visual Cortex. *PLoS Biology* Vol. 4, No. 2, e29. Red blood cells courtesy of John Alan Elson/ CC BY-SA 3.0. Muscle cells courtesy of Rollroboter/ CC BY-SA 3.0. Columnar epithelial cells courtesy of Dr. Glenn H. Kageyama, California State Polytechnic University, Pomona.

KEY POINT Groups of cells that contain the same genome do not necessarily express the same sets of genes at the same time. Gene expression is, in fact, tightly controlled in terms of where and when each gene is expressed. As a result, differences in how particular genes are being expressed can lead cells with identical genomes to look and behave differently from one another.

You will recall from our discussion of the Central Dogma in Chapter 2 that there are numerous steps in gene expression, from the transcription of a gene into RNA, to the processing of the RNA molecule, through its translation into an amino acid sequence (for at least some genes), and to its processing and folding into a mature protein. The expression of a gene is regulated at each step by a variety of different mechanisms. For an organism to survive and reproduce, these regulatory mechanisms have to be coordinated with each other to ensure that the gene is expressed in the proper cells at the proper levels and at the proper time. Every step in the regulation of gene expression has been subjected to evolutionary pressures, sometimes to change and sometimes to remain unchanged, since every step provides opportunities for mutations to produce variation. Among the core topics in molecular genetics, the regulation of gene expression has probably generated the most research.

In this chapter, we will review the first steps of this process—transcription from DNA into RNA in bacteria and eukaryotes, focusing on the processes that affect the expression of a single gene. In Chapter 13, we will describe how some RNA transcripts, the messenger RNAs (mRNAs), are translated into a polypeptide sequence. In these chapters, we will view gene expression in terms of information flow from the DNA sequence of the genome to the gene products that will ultimately influence the phenotype and behavior of a cell.

The processes of gene expression—the transcription of the DNA sequence of the gene, the processing and splicing of the transcript, and, in some cases, the translation of that transcript into a polypeptide sequence—provide the framework to highlight the many opportunities for control and evolutionary change. Gene expression is inextricably connected to its regulation, so the processes of gene regulation will also be encountered. Figure 12.2 shows an overview of the various processes of gene regulation that occur in eukaryotes. As we will discuss later, some of these processes do not occur in bacteria or most viruses where transcription and translation are more tightly coupled. However, transcription and translation are universal processes with many points for control in all kingdoms. Changes in gene regulation have played a significant role in shaping the evolutionary differences among

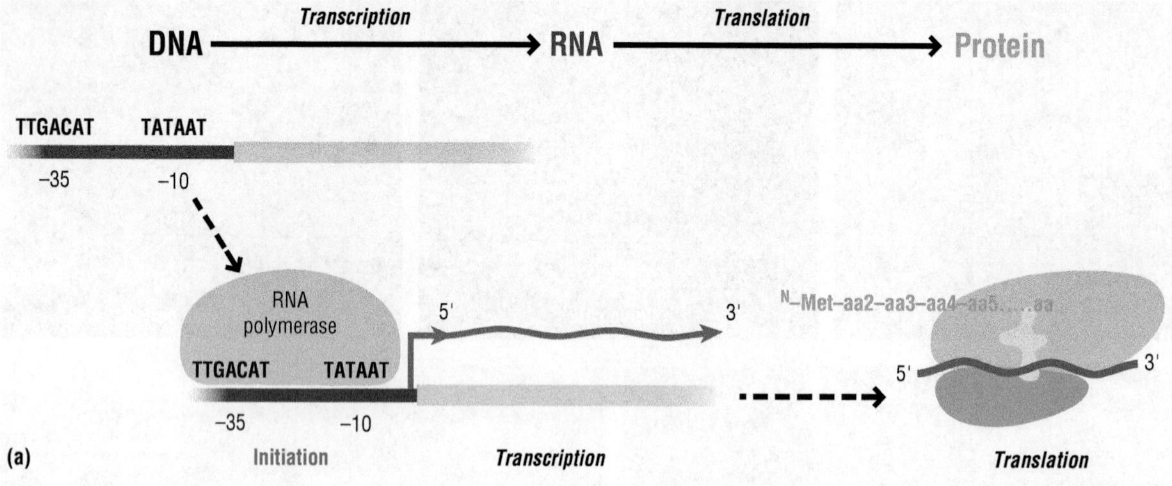

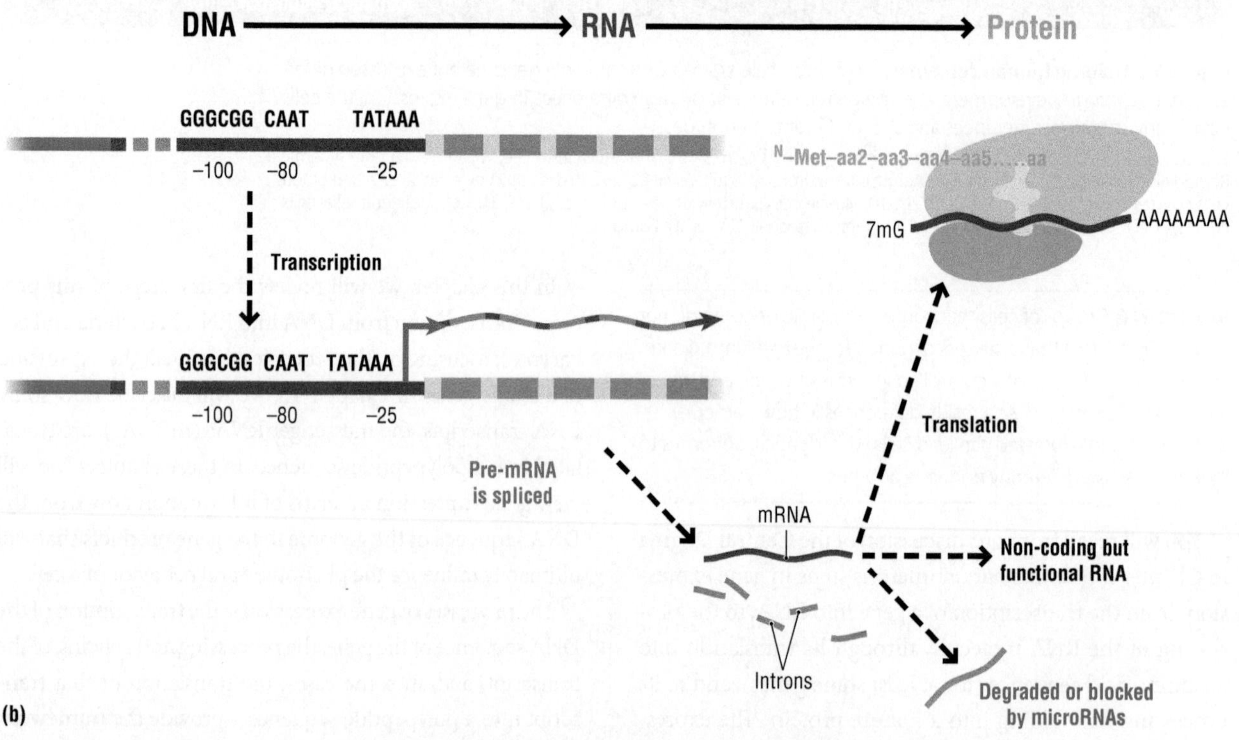

Figure 12.2 Overview of the processes involved in gene expression and its regulation in the Central Dogma. (a) depicts the events in bacteria where nearly all of the regulation occurs at the initiation of transcription. (b) depicts the events in eukaryotes, with regulation occurring not only at the initiation of transcription but also during splicing and with microRNAs. Many transcripts are made but are not translated into proteins, particularly in eukaryotes. These are the non-coding RNAs.

species; in addition, the misregulation of gene expression plays a major role in many diseases, for example in many types of human cancers.

While Chapters 12 and 13 will focus on the expression of individual genes, you should also recognize that the regulated expression of a single gene must be coordinated with the hundreds or thousands of other genes in the genome, each with its own pattern of expression, to produce the complete characteristics of a cell and the organism. The functional and histological differences between a muscle cell and a nerve cell are the outcomes of many dozens or hundreds of genes working together, rather than that of a single gene, although there are "master switch" regulatory genes that set the developmental program of a cell on a particular path. This topic will be discussed in more detail in Chapter 14.

Changes in the coordinated regulation of gene expression among many genes are also important in producing evolutionary variation, as we noted previously in Section 3.6. It is now clear that most of the morphological and behavioral differences between closely related species do not arise because a number of new genes have appeared during speciation (although those are important too) but rather because there have been changes in the expression patterns of genes both species share. Among individuals, changes in gene expression are often associated with changes in the phenotype and, in humans, with many genetic diseases. Thus, throughout this chapter, we will try to note those steps in the process of gene expression that have been subject to the most change during evolution and those that have been more highly conserved.

KEY POINT Changes in the regulation of gene expression play an important role in many diseases and also account for much of the evolutionary diversity among species.

12.2 Initiating transcription

As a very quick reminder of topics introduced in Chapter 2, a gene is transcribed by an enzyme complex known as RNA polymerase into an RNA sequence. Transcription begins at specific sites, known as transcription start sites (TSSs), which are located near the regions where RNA polymerase binds to DNA to begin making the transcripts. In eukaryotes, transcription occurs in the nucleus. The transcript is processed into a mature RNA through a variety of steps, some of which occur in the nucleus and some in the cytoplasm. For genes whose final product is a polypeptide, the RNA transcript (the mRNA) is translated into an amino acid sequence in the cytoplasm by processes at the ribosome. In bacteria, transcription, RNA processing, and translation are tightly coupled in both time and location, since there is no membrane-bound nucleus.

Figure 12.3 shows the structures of a "typical" gene in bacteria and in eukaryotes. Most genes have a structure similar to this, although none matches it exactly. Recall from our description in Chapter 2 that genes have a polarity, with the coding sequence running in a 5′ to 3′ direction. Recall also that transcription always proceeds by addition of the 5′ end of an incoming base to the 3′ end of the base

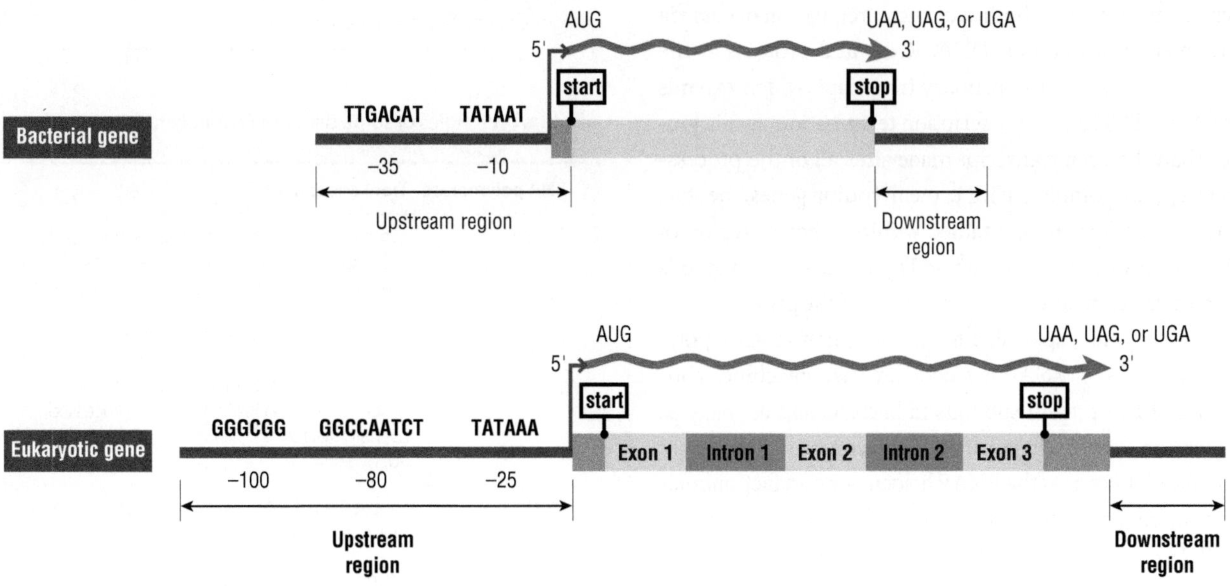

Figure 12.3 A typical gene with its promoter in bacteria and eukaryotes. The upstream region includes most of the signals needed for the initiation of transcription. The downstream region may involve other regulatory regions. Although this diagram of a eukaryotic gene promoter shows a traditional TATA box, the sequence elements of eukaryotic gene promoters are highly variable, and only a subset have the sequences shown here.

preceding it in the transcript. As a result, transcripts also have a polarity with a 5′ and a 3′ end. But how the polarity of the transcript relates to the polarity of the gene can sometimes be confusing, so we will review this quickly here and describe it more fully in Communicating Genetics 12.1.

Although the DNA that comprises a gene exists in the form of a double-stranded helix, most diagrams of genes depict the sequence of just the coding strand, with the 5′ end (the start of the coding sequence) oriented to the left, and the 3′ end (the end of the coding sequence) oriented to the right, as conventionally drawn. During transcription, the other strand of DNA (known as the non-coding, or template, strand) is used as a template to synthesize a complementary RNA transcript, which grows in a 5′ to 3′ direction as it is made. The resulting transcript therefore directly matches the polarity of the coding strand of the gene, with the 5′ end at the start of the transcript and the 3′ end at the end of the transcript.

The DNA sequence located more 5′ than the start of the gene (to the left in our standard drawings) is referred to as the **upstream region**, while the region immediately more 3′ than the end of the gene is referred to as the **downstream region**. The upstream region of a gene usually includes information for the gene-specific regulation of transcription of the gene, while the downstream region may include information for stabilizing the transcript or its translation. The regulatory information may be resident in the DNA sequence itself or in the RNA sequence transcribed from it, or in both sequences; regulation occurs via the interactions among DNA, RNA, and proteins at the **promoter region**. The primary transcript, which extends from the TSS to the transcription termination site, is longer than the final transcript made after all of the processing steps are completed; for protein-coding genes, the start and stop codons are contained within a shorter region of the final transcript, as shown in Figure 12.3. This region is often referred to as the **coding region** of the gene.

The enzyme responsible for transcription is RNA polymerase (RNAP). RNAP is a complex "nanomachine," consisting of five protein subunits in bacteria and as many as 12 protein subunits in eukaryotes; the complete polymerase complex is known as the RNAP holoenzyme. In fact, animals have three structurally related, but distinct, RNAPs (known as RNAP I, II, and III), while plants have these three polymerases plus two additional ones, known as RNAP IV and V. Mitochondria and chloroplasts have their own RNAPs that are somewhat similar to the single RNAP found in some viruses but distinct from the ones in the eukaryotic nucleus.

In addition to the subunits of the RNAP holoenzyme, eukaryotes have additional protein complexes, called general transcription factors, to recruit the entire nanomachine to promoter regions; in bacteria, as we will note later, this recruitment is done by one of the subunits of the RNAP holoenzyme itself. Different general transcription factors play important roles, so it is useful to name them separately. In eukaryotes, the nomenclature is "TF," followed by the Roman numeral of the particular RNAP that it recruits. Thus, the transcription factors associated with RNAP II are all called TFII, whereas those for RNAP III are TFIII. The subunits are then given letters; thus, TFIID is one of the general transcription factors that associate with RNAP II; the important role of TFIID will be described later. Most of these general transcription factors are themselves complexes of several different proteins, so the entire process of recruiting a eukaryotic RNAP to the promoter region involves the functions and interactions of dozens of polypeptides.

The different RNAPs in eukaryotes transcribe different categories of genes, as summarized in Table 12.1; RNAP II is responsible for the transcription of protein-coding genes (and many non-coding RNA genes), so we will focus on it as our example. Many of the protein subunits in different RNAPs are similar or identical to each other in both sequence and function.

KEY POINT RNA polymerase is a multi-protein enzyme that synthesizes the RNA transcript during transcription. The precise protein composition of RNA polymerases can vary, and different RNA polymerases are involved in transcription of different categories of genes in eukaryotes.

Table 12.1 Roles of some different RNA polymerases

RNA polymerase	Transcripts made
RNAP I	Synthesizes the ribosomal RNAs (rRNAs)—5.8S, 18S, and 28S. These are made together as the pre-rRNA of 45S and processed to the final sizes
RNAP II	Synthesizes the precursors to mRNA, as well as microRNAs and long non-coding RNAs, most of the RNAs involved in splicing, and other non-coding RNAs. These are processed to their final forms by capping, splicing, polyadenylation, etc.
RNAP III	Synthesizes transfer RNAs and the 5S rRNA, as well as a few other small non-coding RNAs
RNAP IV	Found only in plants. Synthesizes small interfering RNAs
RNAP V	Found only in plants. Synthesizes small interfering RNAs

Note: Additional and distinct RNA polymerases are also found in mitochondria and chloroplasts.

COMMUNICATING GENETICS 12.1 Drawing genes

It is often helpful to sketch out a picture to illustrate the structure of a gene. Each person making such a sketch will, of course, bring their own style to their drawing. However, there are some basic conventions that are generally followed when illustrating genes. These conventions communicate information about a gene concisely and reduce possibilities for confusion. We will review some of these general conventions here, as they are shown in Figure A.

1. Although the entire gene is composed of a continuous double helix of DNA, the DNA may be drawn in many ways. Often it is depicted simply as a single line (with other features then highlighted on that line), or sometimes the two strands of the DNA may be shown in more detail. Whatever convention is adopted, it is important to remember that the entire gene is a continuous DNA double helix, no matter how complicated—or simple—the annotation of the drawing, with all its boxes and circles and colors, might make it look! The gene is also usually drawn such that the coding sequence runs from left to right.
2. The exons of a eukaryotic gene are often drawn in a way that distinguishes them from the introns. Often the exons will be highlighted and numbered sequentially in boxes, or the exons and introns might be colored differently from one another.
3. Similarly, the introns present in the initial RNA copy of the gene are often drawn in a V-shape to indicate how the process of splicing will proceed.
4. The region of the core promoter where RNA polymerase binds and transcription initiates is often indicated with an arrow pointing in the direction of transcription.
5. Regions of the gene that lie upstream and/or downstream of the coding sequence that are not transcribed but are important for the regulation of transcription are often demarcated in some way—by color, blocks, or circles—on top of the DNA line. Sometimes, the binding sites of key transcription factors or the transcription factors themselves are highlighted in these regions, shown as colored ovals or boxes.
6. Regions of the gene that are transcribed but are not translated (the 5′ untranslated region or UTR that precedes the translation ATG start codon and the 3′ UTR that follows the translation stop codon) are often highlighted differently from the coding sequence. In our example, this is indicated by the gray coloring of the untranslated regions versus the yellow coloring of the coding sequence.

Of course, every gene has its own structure, and every person has their own stylistic preferences. But by following these basic conventions, you will be able to make your sketches more meaningful for others, and you, in turn, will be better prepared to interpret the many different illustrations of genes that you are likely to encounter.

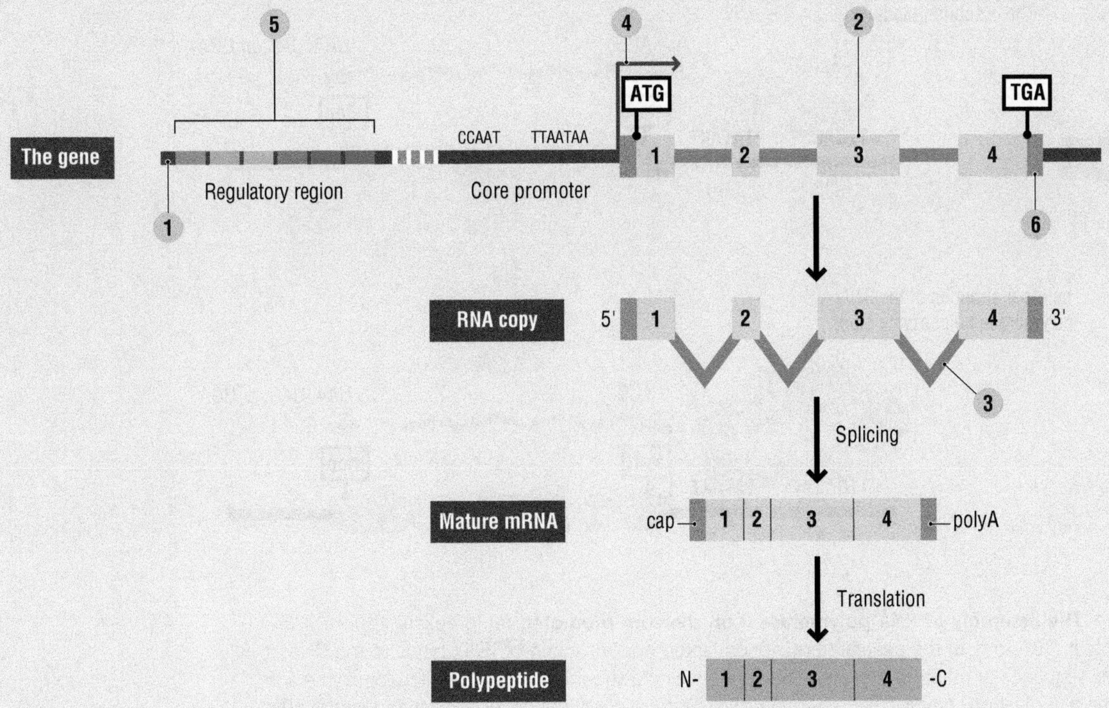

Figure A Conventions for gene expression. Some of the conventions used for drawing the various steps in gene expression.

Initiation and the core promoter

Transcription is regarded as having three stages: initiation, elongation, and termination. However, in reality, this is an over-simplification of the many processes that are occurring. In particular, initiation in eukaryotes includes important pre-initiation steps that are crucial for proper regulation and which are often considered separately from transcription itself.

Initiation occurs when RNAP binds to the region upstream of the gene, known as the **promoter**, to begin transcription of the gene, as diagrammed in Figure 12.4. Since the term "promoter" is sometimes used more generally to refer to the entire regulatory region upstream of

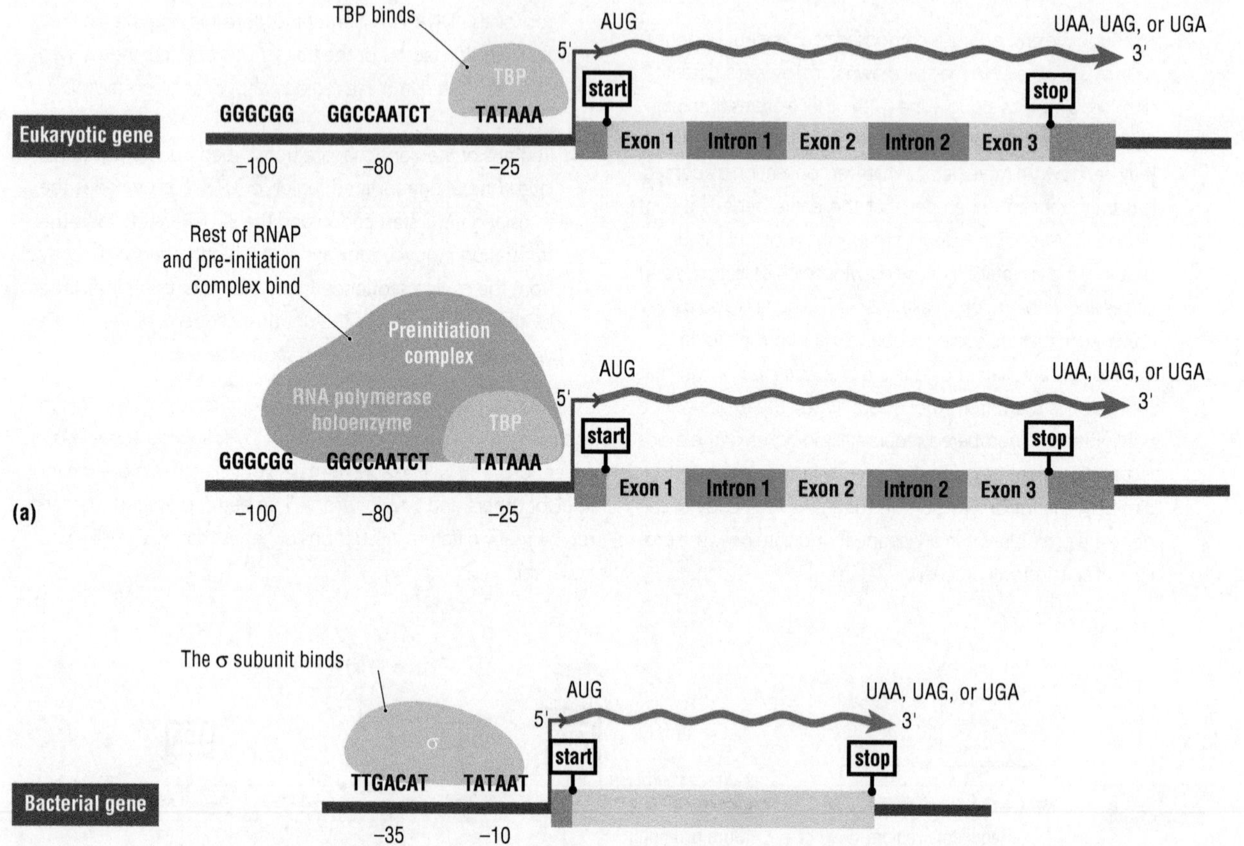

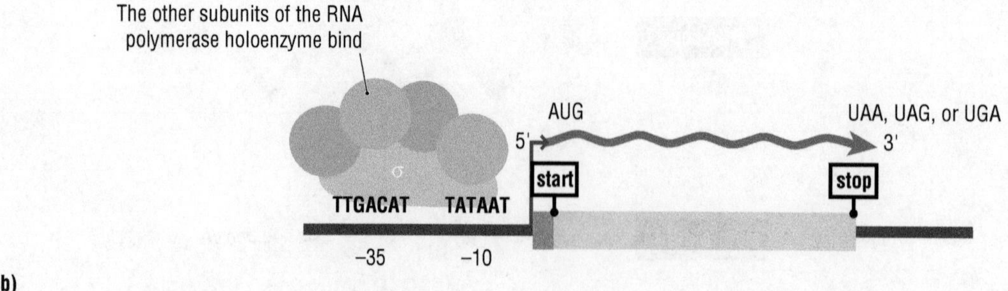

Figure 12.4 The assembly of RNA polymerase II on the core promoter. (a) In eukaryotes, the TATA-binding protein (TBP), one of the general transcription factors and a subunit of TFIID, binds to the DNA at the TATA box sequence or other initiator elements. The remaining subunits of the RNAP II holoenzyme and the rest of the pre-initiation complex then assemble. (b) In bacteria, the σ subunit recognizes and binds to the core promoter and then recruits the remaining subunits of RNAP, permitting transcription to begin.

a gene, the location where RNAP binds is referred to as the "core promoter" or "minimal promoter," to avoid any confusion. Transcription begins about 25–30 base pairs downstream of the core promoter, at the **transcriptional start site** or **TSS**, although this distance varies widely for different genes.

KEY POINT Transcription can be considered in three phases: initiation, elongation, and termination. In the first phase, RNAP binds to a region upstream of the gene known as the core promoter. Transcription then initiates a little further downstream at the transcription start site (TSS).

Initiation in eukaryotes

The core promoter for many genes in eukaryotes includes a 6- to 10-base pair sequence, rich in alternating thymidines and adenines, referred to as the TATA box; the actual sequence is often TATAAA, or some slight variation of this, as shown in Figure 12.4(a). Here is where it becomes useful to know the role of one of the individual proteins in one general transcription factor. Most of the proteins that make up general transcription factors and RNAP cannot bind to DNA on their own. However, a protein known as TBP (TATA-binding protein), a component of the general transcription factor TFIID, binds to the DNA at this sequence. Once TBP and the rest of TFIID are bound to DNA, the other general transcription factors and RNAP assemble into a complex by interactions with TFIID and each other. The complex of these subunits is known as the **pre-initiation complex**.

It was once thought the TATA box was found at the core promoters of all eukaryotic genes, but genomic analysis has shown that only a subset of genes have a traditional TATA box sequence—less than a quarter of the genes in mammals. In the other genes, TBP binds to sequences known as **initiator elements**, which allows the other subunits to assemble and act. The importance, if any, of a gene having a TATA box, as opposed to another initiator element, is not known, and related genes in different species sometimes differ in the kind of sequence found at their core promoters.

Each part of the pre-initiation plays important roles in the assembling of the RNAP holoenzyme and regulating the initiation of transcription. The role of the Mediator complex, which is often considered to be part of the pre-initiation complex, is discussed in Section 12.4.

KEY POINT The TATA box is a sequence motif at the core promoter of many eukaryotic genes that is recognized by the TATA-binding protein (TBP) of the general transcription factor TFIID. Binding of TBP to the TATA box is an important step in triggering the assembly of a full RNAP complex. Other sequence motifs that serve the same function as the TATA box are referred to as "initiator elements."

Many genes have more than one core promoter and more than one TSS. In some cases, these different promoters are active at different times or in different cells, so that the switch from one to the other is important in regulation. In other cases, there seems to be little consequence of having more than one promoter or TSS.

As noted earlier, the promoter and TSS are typically about 25 bases apart, but examples are known in which the distance between them is as much as 90 base pairs. More significant than the distance between the core promoter and the TSS is the distance between the TSS and the ATG codon that marks the start of translation for genes encoding proteins. This distance is rarely less than 30 bases and can be quite long—1000 bases or more. Thus, these parts of the genome are transcribed into RNA but aren't then translated into protein. This region between the start of transcription and the start of translation is known as the 5′ untranslated region or **5′ UTR**.

What is the evolutionary significance of a long stretch of RNA that lies upstream of the protein-coding region? Genes with a long 5′ UTR are frequently subject to regulation at the level of translation, often by proteins that bind to this region on the RNA. We will return to this topic in Chapter 13. Although we are discussing the initiation of transcription in this section, we note that untranslated regions of RNA also lie downstream of the stop codon for a protein; these are referred to as the **3′ UTR**. Like the 5′ UTR, a long 3′ UTR also can be part of a regulatory signal for the expression of the gene, as we note in Section 12.4.

Initiation in bacteria

Transcription initiation in bacteria is conceptually similar to that in eukaryotes, but the particulars are different, as shown in Figure 12.4(b). RNAP in bacteria has multiple subunits: two α subunits, a β and a β' subunit, and accessory subunits, of which the σ subunit is the most important and best understood. The α and β subunits represent the core enzyme that does the actual transcription, but they cannot bind to DNA by themselves. σ subunits bind to the core promoter; when bound, the σ subunit recruits RNAP, analogous to the role that TBP plays as part of the TFIID subunit in eukaryotes. Thus, σ subunits (or factors) act like the general transcription factors in bacteria. However, while TBP and TFIID proteins are used for all genes transcribed by RNAP II in eukaryotes, bacteria have different σ factors that bind to different core promoters and regulate transcription for different genes.

The σ factor recognizes the promoter and unwinds the DNA upstream of the TSS. For many years, it was thought that the binding of the σ factors undergoes a cycle; once transcription has initiated, the σ factor dissociates from the rest of the RNAP and moves to recruit a new enzyme complex to the same, or a different, promoter. However, more recent evidence suggests that a σ factor itself can stay bound to the promoter, and its affinity for binding the other components of RNAP is what cycles as genes are being transcribed. In any event, once bound to the core promoter, RNAP works similarly in bacteria and eukaryotes.

Escherichia coli has seven different σ factors, designated by their apparent molecular weights, with the functional associations shown in Table 12.2; σ^{70} (with a molecular weight of 70 kilodaltons (kDa)) transcribes most of the genes under most conditions in a growing bacterial cell. The σ^{70} factor forms direct contacts with two regions in the core promoter, at −35 and at −10 (where +1 is the first base in the TSS). By contrast, some σ factors are only activated under certain conditions. For example, σ^{38} is predominantly active in the stationary phase, and σ^{32} is activated in response to heat shock.

Table 12.2 Some of the alternative sigma (σ) factors in *Escherichia coli* and their cognate promoter sequences

σ factor	Promoter recognized	Genes transcribed
σ^{70}	TTGACA-17bp-TATAAT	Many and diverse
σ^{32}	CNCTTGAA-14bp-CCCCATNT	Heat-shock response
σ^{54}	CTGGNA-7bp-TTGCA	Many and diverse
σ^{28}	TAA-15bp-GCCGATAA	Chemotaxis, motility, flagellar components

Reproduced from Perry, J.J., Staley, J.T. and Lory, S. (2002) *Microbial life*. Sunderland, Mass.: Sinauer Associates.

As each σ factor recognizes and binds to different sequences at the promoter, they regulate different sets of genes. The sequences listed in Table 12.2 represent consensus sequences of the ideal promoter for each factor. Strong or good promoters with high affinity for the σ factor are close to the consensus. However, in reality, almost all promoters depart somewhat from the consensus so that σ does not bind too tightly.

KEY POINT In bacteria, the σ factor plays an analogous role to eukaryotic TBPs in binding the core promoter and assembling a functional RNAP. However, whereas eukaryotic polymerases operate with dedicated TBPs, bacterial σ factors can vary, and the presence or absence of specific σ factors plays a role in regulating which genes are expressed.

12.3 Regulating the initiation of transcription

In bacteria, the principal role of σ factor binding is to regulate which genes are being transcribed, so these factors differ among genes. In eukaryotes, transcription initiation is controlled by the general transcription factors that interact with RNAP II. The general transcription factors and RNAP II are largely the same, regardless of what gene is being transcribed; in fact, the core promoter for one gene can be substituted *in vitro* for the core promoter of a different gene, with little effect on transcription initiation. So while the general transcription factors direct the transcription of nearly every eukaryotic gene, they do not confer any of the specificity associated with transcriptional regulation. Thus, they are necessary for transcription to occur, but they do not differ among cells.

How then is transcription highly specific for each stage and cell type when the general transcription factors and RNAP itself have almost no specificity? The key step in eukaryotic transcriptional regulation lies in other proteins that control the binding of the pre-initiation complex to the gene. The proteins that confer the **specificity** of transcription in eukaryotes are known as sequence-specific transcription factors, to distinguish them from the general transcription factors, or simply as **transcription factors**. A generalized diagram of transcription factor action is shown in Figure 12.5; you may

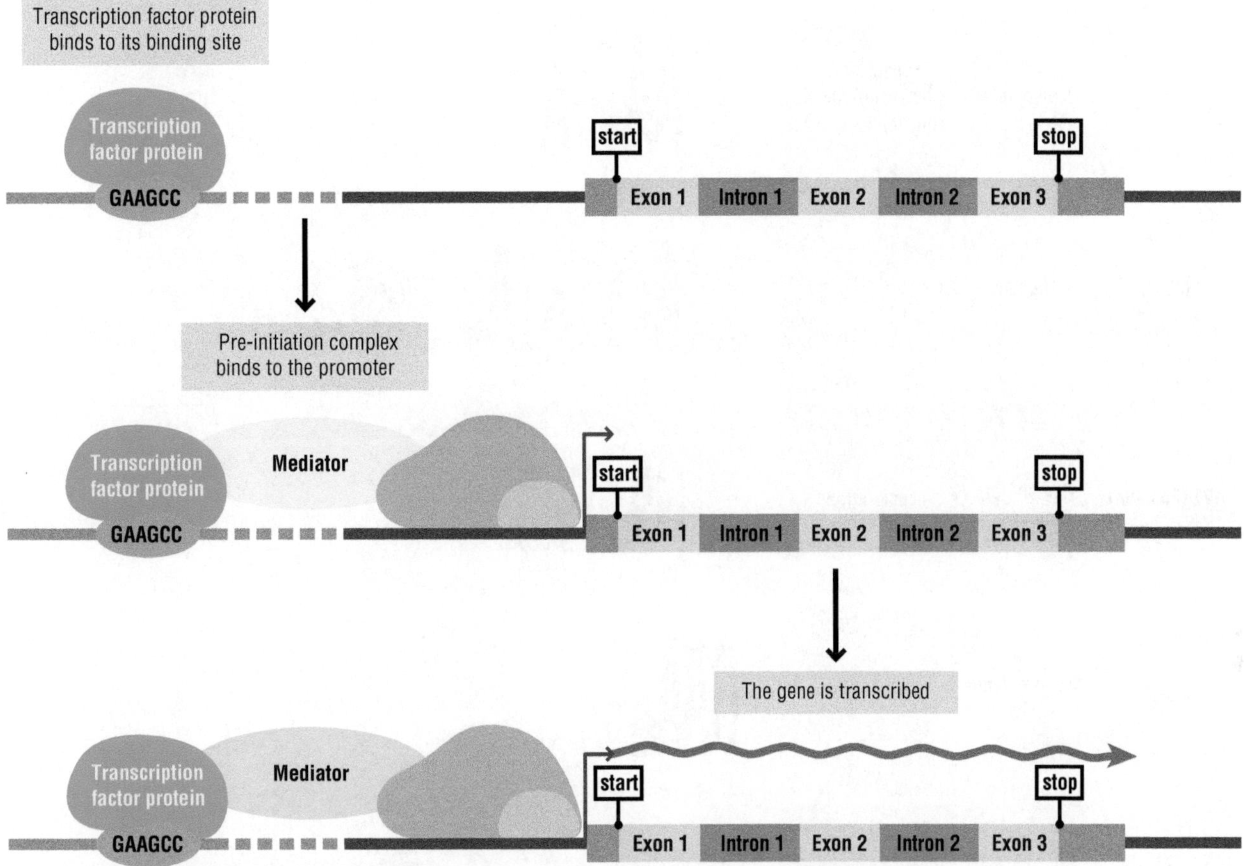

Figure 12.5 Regulation of pre-initiation and the initiation of transcription by RNA polymerase II in eukaryotes. A transcription factor protein binds to the DNA at its binding site. RNAP and the pre-initiation complex bind at the core promoter, and transcription begins. The interaction between transcription factors and the pre-initiation complex, including RNAP, depends on a protein complex known as Mediator. The binding of the transcription factor regulates the transcription of the gene.

recall from Chapter 1 that the ALX1 protein involved in the shape of the beak in Darwin's finches is a transcription factor of this type. The term "sequence-specific" refers to the fact that these transcription factors bind to a specific DNA sequence. In general, when molecular geneticists refer to a "transcription factor," they are referring to a sequence-specific transcription factor in eukaryotes, rather than the general transcription factors. Bacteria also have transcription factors that recruit RNAP to weak promoters, but these function somewhat differently.

KEY POINT In bacteria, the type of σ factor present plays a primary role in controlling which specific genes are being expressed. In eukaryotes, sequence-specific transcription factors, distinct from the general transcription factors that comprise part of the pre-initiation complex with RNAP, play the primary role in regulating the specificity of transcription.

There are hundreds of different transcription factors in a eukaryotic genome, as may be expected if transcription factors are controlling the specificity of transcription. As a rough rule of thumb, about 5–10% of the protein-coding genes in an animal's genome encode transcription factors, a bit less than that in fungi and higher plants. Some examples, with their consensus DNA-binding sequences, are shown in Figure 12.6. Humans have as many as 1000 different genes that encode transcription factors, enough to allow transcription initiation to be very highly regulated for different genes. Each of these transcription factors has its own characteristic pattern of expression, so cells differ from each other in which sets of transcription factors are being expressed.

In fact, one of the primary reasons that eukaryotic cells are different from one another in function and morphology is that they are expressing different sets

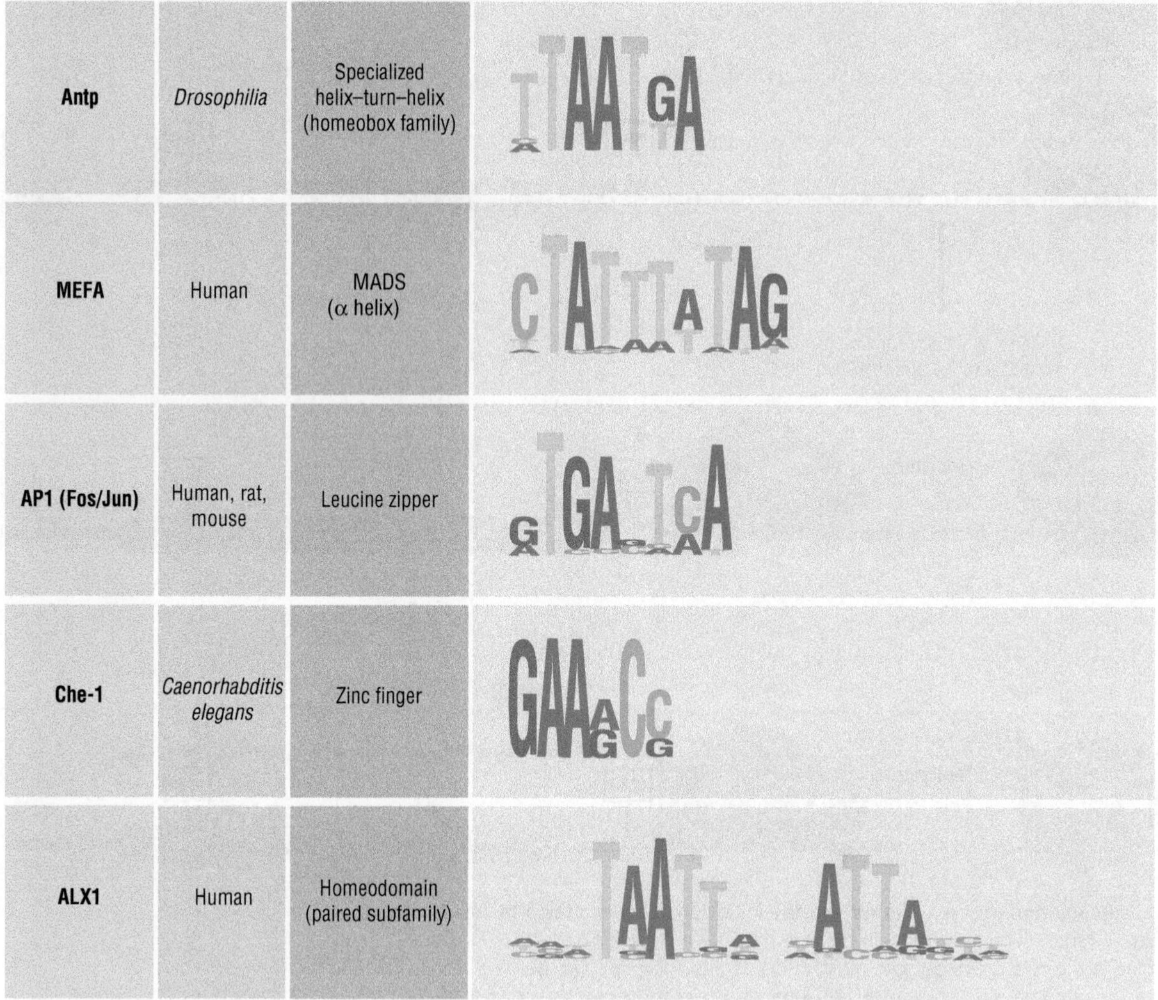

Figure 12.6 Some transcription factors with their consensus binding sequences. The size of the letter in the binding sequence represents the level of conservation of that nucleotide.

Source: Data compiled from Mathelier, A. et al. (2013). JASPAR 2014: an extensively expanded and updated open-access database of transcription factor binding profiles. *Nucleic acids research*, p.gkt997

of transcription factors. Transcription factors are expressed in different cells because their own transcription is regulated by another group of transcription factors, which are, in turn, regulated by other transcription factors, and so on. There are "master regulator" transcription factors that primarily regulate the expression only of other transcription factor genes, and "effector" transcription factors that primarily regulate the expression of the structural proteins and enzymes that are characteristic of the cell type, as well as other patterns of cross-regulation. We will discuss this interaction among transcription factors, the transcriptional regulatory network, in Chapter 14. For now, we will focus on how the transcription of a single eukaryotic gene is regulated.

Transcription factor binding

We noted earlier that eukaryotic transcription factors are sometimes called "sequence-specific" transcription factors because they carry out their functions by binding to specific sequences on the DNA. We say that these are specific sequences, which is true, but they are not always precisely the same sequences, as noted in Figure 12.6. For example, the human transcription factor JUN (related to AP1 in Figure 12.6) binds to the 7-base sequence 5′-TGA(C or G)TCA-3′, the middle base being either C or G, so that TGACTCA and TGAGTCA would both be considered possible JUN-binding sequences. But studies that looked directly at JUN binding *in vivo* and then asked what sequence is bound found that only perhaps half of the sites are actually one of these two sequences, while the rest differ from this sequence by one or two bases. Thus, the binding site is usually considered a consensus sequence, rather than an exact match.

WEBLINK: VIDEO 12.1 provides additional information about consensus DNA-binding sequences and their depiction. Find it at **www.oup.com/uk/meneely**

To make the matter even less predictable, not every TGAGTCA sequence in the genome examined in the study was bound by JUN, even in cells in which JUN was present and active. In other words, while the consensus binding sequence for JUN is 5′-TGA(C or G)TCA-3′, not every JUN-binding site has this sequence, and not every occurrence of this sequence acts as a JUN-binding site. Because this sequence is only seven bases long, it is found in many places in the genome—within exons, in introns, in regions with no genes nearby, and so on. While the consensus binding sequence is important for the function of the transcription factor and for understanding its role in the cell, knowing the binding site is not the entire story.

KEY POINT Sequence-specific transcription factors do not bind DNA randomly. They recognize exact or close matches to specific sequence motifs referred to as binding sites. However, of the many potential binding sites in a genome, not all will be bound by the transcription factor in question.

You may be asking, "How many different transcription factors regulate the transcription of a single gene?" This is a good question, but until recently the experiments needed to answer that question accurately were difficult to do; the answer remains incomplete and depends entirely on the gene whose transcription is being examined. Thus, our discussion will give a general perspective on this question but not much exact information for any individual gene.

A few genes are regulated by only one or two transcription factors; there appear to be more genes with such simple regulation in yeast and single-celled organisms than in multicellular organisms. Conversely, some genes are regulated by dozens, or even 100 or more, of different transcription factors. So it appears that the number of transcription factors per gene can fall anywhere in the range from one to 100. However, the **mode** of this distribution—that is, the number that occurs most often—is about 12 transcription factors per gene in mammals, perhaps a few less than for a gene in *Drosophila* or *Caenorhabditis elegans*. But this "average" number of transcription factors per gene is simply to give a sense of the scale.

While an answer about the average number of transcription factors might be unsatisfying, we should think about what information has to be found in order to make that estimate. The human genome encodes about 1000 transcription factors. Of these, we know the consensus binding sites *in vitro* for perhaps 400 of these, although this is an active area of investigation, so the number changes often. Thus, there are many genes that we know encode transcription factors, but we do not know what sequences these transcription factors bind to *in vitro*. (An experimental approach to identify the preferred *in vitro* and *in vivo* binding sites for transcription factors is known as a ChIP assay and is described in Tool Box 14.1.) Of these 400 transcription factors whose *in vitro* binding sites are known, about 100 or so transcription factor proteins have actually been tested for binding activity *in vivo* with assays that allow us to see which transcription factors co-regulate a gene in the cell. So even with known transcription factors, we might not know their preferred binding site from *in vitro* assays or how often they bind to that site *in vivo*. Clearly, there is much to learn.

Furthermore, one reason that cells are different from one another is because they express different sets of transcription factors. So we really know which transcription factors are regulating a gene only in a given cell type, because a different cell type could have a slightly different set of transcription factors for the same gene. Furthermore, our cells respond differently under different environmental conditions, including age, nutrition, physical stresses, and other environmental factors, in part by regulating transcription.

So how many transcription factors regulate a single gene of interest? In order to answer the question accurately and fully in humans, we will need to be able to test the activity of every one of 1000 transcription factors, in every cell type, under every conceivable condition, including at different times. In sum, our estimate that a typical gene is regulated by 12 transcription factors should be recognized as conveying a sense of the actual level of regulation but not as being the final word. It is almost certainly an underestimate.

KEY POINT The average number of transcription factors that regulate a single mammalian gene is very roughly about 12. However, this number can vary widely from gene to gene and varies for the same gene in different contexts.

A specific example with one gene

One specific example of a gene and the transcription factors that regulate its expression is shown in Figure 12.7. We will discuss the example in some detail to illustrate the general principle of how transcription factors regulate gene expression, recognizing that every gene and every cell type is different; the specifics of this example

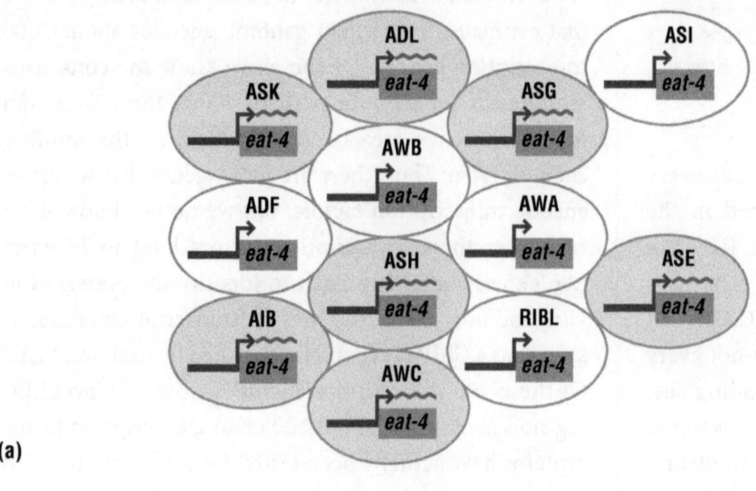

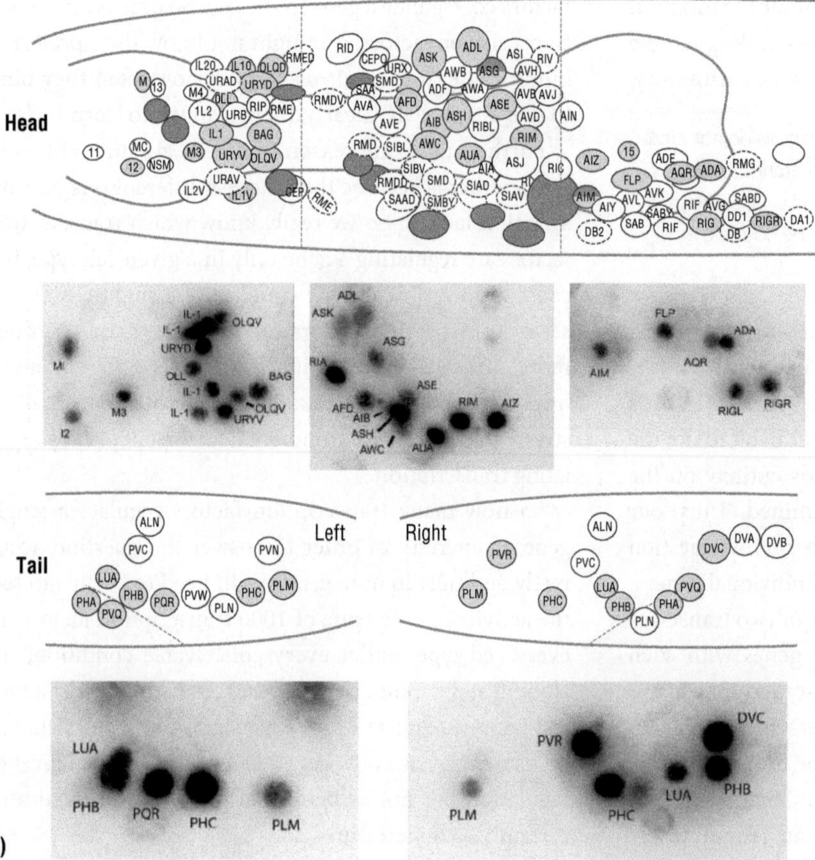

Figure 12.7 Transcription of the *eat-4* gene in *Caenorhabditis elegans*. The *eat-4* gene is expressed in a variety of cells in the head and tail, including 12 neurons from the nerve ring in the head, shown diagrammatically in (a). (b) shows the expression in these 12 cells (in the mid-region of the head) and expression in cells in other regions of the head and tail using a green fluorescent protein (GFP) reporter gene, which was used to dissect the *cis*-regulatory module. Each panel in (b) shows cell staining for the GFP expression and a corresponding diagrammatic representation of the cells for each region of the body shown.

Source: Part (b) reproduced from Serrano-Saiz, E. et al. (2013). Modular Control of Glutamatergic Neuronal Identity in *C. elegans* by Distinct Homeodomain Proteins. *Cell* Vol 155, Iss. 3.

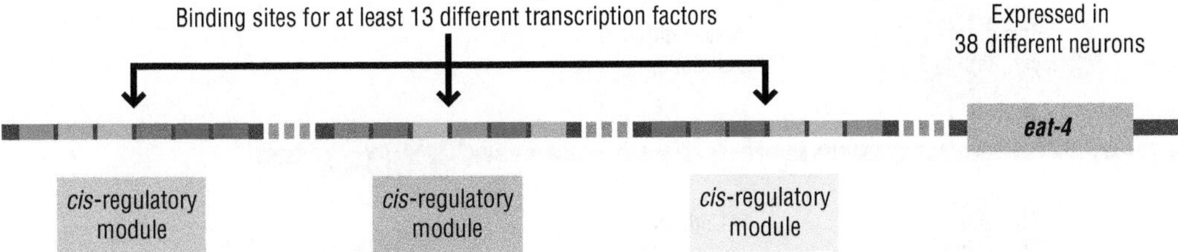

Figure 12.8 ***cis*-regulatory modules for the *eat-4* gene.** The *eat-4* gene has binding sites for at least 13 different transcription factors expressed in 38 different neurons in *Caenorhabditis elegans*, including the 12 neurons in the nerve ring, shown in Figure 12.7(a). Three distinct *cis*-regulatory modules were found in different regions upstream of the gene.

are not as important as the general principles that it demonstrates. In the nematode worm *C. elegans*, a gene called *eat-4* encodes a protein known as a vesicular glutamate transporter, abbreviated VGLUT in other animals. This protein is responsible for the uptake of glutamate, an amino acid used as a neurotransmitter, into the synaptic vesicles in neurons. In other words, the *eat-4* gene is expected to be expressed primarily, if not exclusively, in neurons. This provides a cell type to begin our discussion of *eat-4* regulation.

Worms have 302 neurons, of which 180 form a structure known as the nerve ring. The *eat-4* gene is expressed in 38 different types of neurons, a few of which are diagrammed in Figure 12.7(a); the original published data are shown in Figure 12.7(b) for comparison. Note that, while all neurons, and all cells in fact, have the *eat-4* gene in their genomes, even adjacent neurons in the nerve ring differ in whether or not this gene is transcribed. This difference lies in the transcription factors that regulate *eat-4* expression.

As we noted earlier, the sequence to which a transcription factor binds is usually found upstream of the core promoter, although there are many examples of transcription binding sites within an intron of a gene or downstream of a gene. We will discuss the sites for regulation as being upstream of the core promoter since that is the location of most of them. For *eat-4*, the regulatory region is upstream of the gene, with 13 different transcription factors being known to regulate *eat-4* expression, as summarized in Figure 12.8. Often, the binding sites for the different transcription factors are clustered together in a regulatory region, and there could be several different binding locations for the same transcription factor within a single regulatory region. This region is called the **cis-regulatory module** or **CRM**. We discuss later why this is called a module.

Most transcription factors stimulate or activate the expression of a gene. Thus, the binding region for such a transcription factor is referred to as an **enhancer**. The term enhancer is often used broadly. It is acceptable to refer to the entire region that confers very specific regulation as an enhancer, for example as "the enhancer for pancreatic cell expression," without knowing the specific site within that region that has enhancer activity. It is also possible to refer to one specific sequence within the regulatory region as an enhancer; the binding sequences for the transcription factors in Figure 12.6 would be part of an enhancer. The collection of enhancers and any other functional elements to which transcription factors bind that control the specific expression of a gene is called the *cis*-regulatory module (CRM). The term "*cis*" indicates that it is on the same DNA molecule as the gene being regulated. The transcription factors can be encoded by genes located anywhere in the genome and are usually not near the gene, so they comprise the *trans*-regulatory components.

KEY POINT Transcription factor binding sites are often clustered into *cis*-regulatory modules (CRMs), which integrate inputs from multiple transcription factors and co-factors to regulate transcription of the associated gene. CRMs are usually located upstream of genes but can also be found downstream and in introns, and they comprise part of the modular nature of genes.

For *eat-4*, the entire upstream regulatory region has been subdivided into three distinct regions, as shown in Figure 12.8. Each of these three regions has binding sites for a specific subset of transcription factors, although some transcription factors have binding sites in more than one region, as shown in Figure 12.9. Furthermore, each of these regions directs transcription in a specific and limited subset of neurons. For example, the region in blue in Figure 12.9 has binding sites for seven transcription factors and is necessary for transcription in 12 different neurons, while the region in orange has binding sites for four different transcription factors and directs transcription in ten different neurons. (The specific regulatory

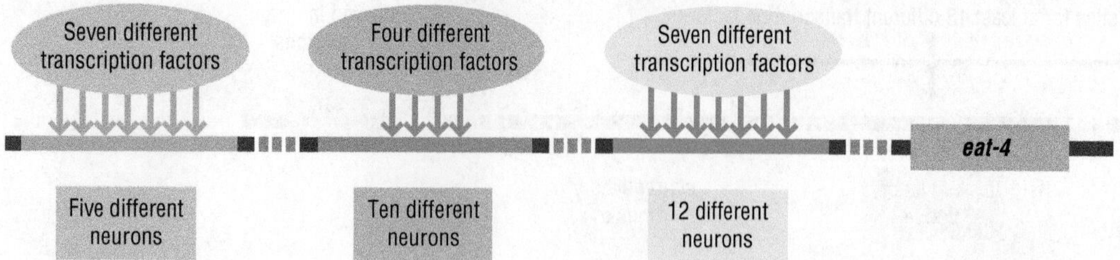

Figure 12.9 The correlations between different transcription factors and different CRMs for *eat-4*. Each of these regions directs transcription in a specific and limited subset of neurons. For example, the region in blue has binding sites for seven transcription factors and is necessary for transcription in 12 different neurons, while the region in orange has binding sites for four different transcription factors and directs transcription in ten different neurons.

modules for some neurons have not been fully defined, so the number of neurons in Figure 12.9 does not add up to the 38 total neurons.)

Genes have a modular structure

An individual CRM and the transcription factors that bind there confer a specific aspect of the transcriptional pattern for that gene, that is, when and where transcription occurs. The full transcriptional expression pattern usually results from the composite contributions of multiple CRMs, as we saw earlier for *eat-4* in *C. elegans*. Let's now focus on the upstream module, shown in green in Figure 12.9, which has the binding sites for seven different transcription factors and directs transcription in five different neurons.

Some of the results for an upstream region regulating transcription of the *eat-4* gene are shown in Figure 12.10. Note that no single transcription factor and no single binding site within this region is responsible for the regulation of this gene. Rather, successful expression in the proper cells is the outcome of the action of a combination of transcription factor proteins binding to different enhancer sequences in this CRM. For example, among the genes that regulate *eat-4* transcription, the neuron type called ADL expresses only the transcription factor gene *lin-11*,

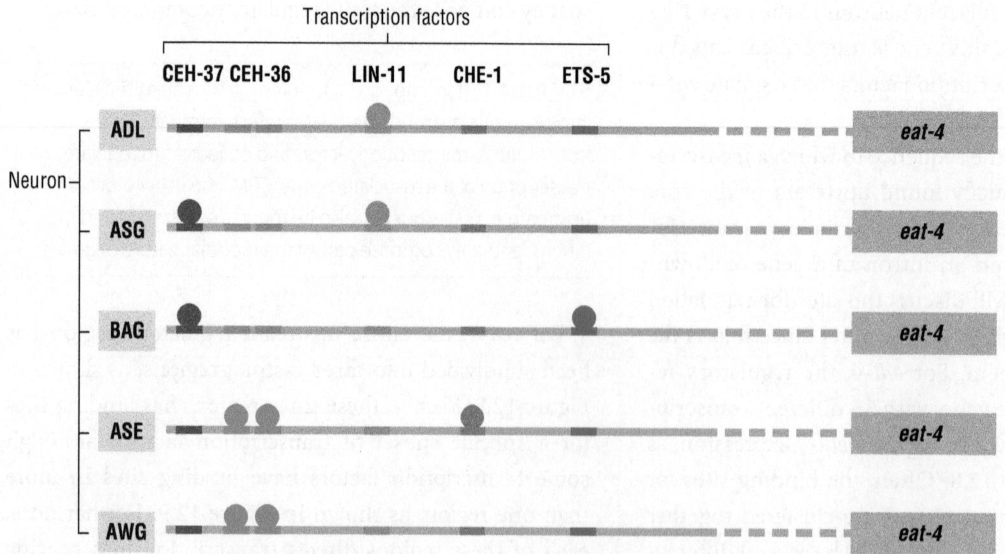

Figure 12.10 The transcription factors that bind to the green *cis*-regulatory module of the *eat-4* gene from Figure 12.9 in five different neurons. The names of the neurons are shown on the left in a three-letter designation. The transcription of the *eat-4* gene is controlled by different combinations of transcription factors in different neurons. For example, the neuron called ADL contains transcription factor LIN-11 but not the other four transcription factors, and this is sufficient to activate transcription of *eat-4* in the ADL neuron, whereas the ASG neuron contains both LIN-11 and CEH-37, and these activate transcription of *eat-4* in this neuron.

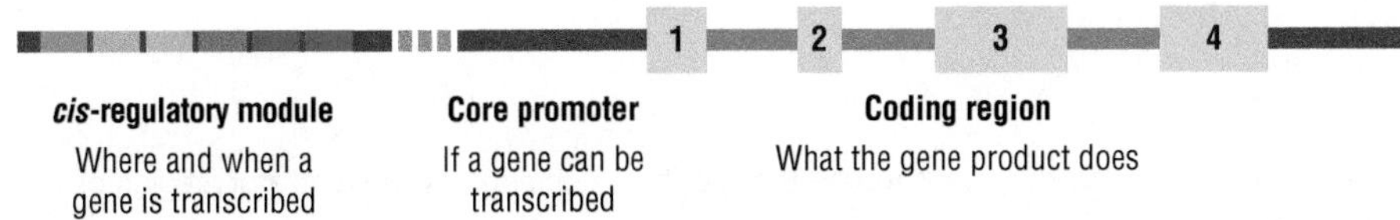

Figure 12.11 The modular structure of a typical eukaryotic gene. The core promoter determines whether the gene can be transcribed, while the *cis*-regulatory region controls where and when the gene is transcribed. The coding region controls what the gene product (protein) does. These different parts of the gene can function separately from other another, which allows these component parts to be modified independently.

whereas the ASG neuron expresses both *lin-11* and *ceh-37* genes. In fact, each of the five neurons depicted in Figure 12.10 expresses a specific combination of these transcription factors.

In general, the three components of the gene that we have discussed—the regulatory region, the core promoter, and the gene body containing the protein-coding region—are modular and separable and can be readily swapped between genes, as summarized in Figure 12.11. The core promoter is very general, so any core promoter can be used to regulate the transcription of any gene body and can be regulated itself by the CRM from almost any other gene. This has important experimental and evolutionary implications—the expression pattern of a eukaryotic gene can be "re-programmed" or regulated either *in vivo* or *in vitro* without changes to the core promoter itself.

The part of the gene that actually encodes biological function lies in the "body" of the gene or the coding region. The sequence in the body of the gene determines if the gene product is an enzyme, a structural protein, or a transport protein, or if it has another function. The body of the gene probably does not have any information that affects the specificity of transcription, unless some of the enhancers happen to lie within it. That said, the function of the gene may be to encode a transcription factor, and it is entirely possible that one function of the transcription factor is to auto-regulate its own expression. As we will discuss in more detail in Chapter 14, auto-regulation of transcription is very common, both in bacteria and in eukaryotes, so in that indirect sense, the body of the gene has information that affects the specificity of transcription. But except for this activity, the body of the gene, like the core promoter, can likely be replaced with another gene body without affecting the transcription of the gene.

You may begin to see why the CRM is referred to as a module. This is the region that **does** control the specific transcription of a gene. As a result, the regulatory region of one gene can be moved (experimentally or evolutionarily) upstream of a different core promoter and a different protein-coding region, and this new gene will now be transcribed under the regulation of the relocated CRM, rather than its original one.

KEY POINT The CRMs, core promoter, and coding sequence of a gene give it a modular nature, in which the individual components can be swapped between genes. This has important implications both evolutionarily and experimentally.

The modular nature of the regulatory region has important implications, not only for the function of genes *in vivo* but also for the techniques that can be used *in vitro* and for the evolutionary changes. Changes in the CRM change the specificity of transcription. In Tool Box 12.1, we describe how the modular structure

TOOL BOX 12.1 Reporter genes for the analysis of gene expression

The expression pattern of a gene is central to its function. "Gene function" and "gene expression" are not synonyms; "gene function" is a broader term that could include an interpretation of mutant phenotypes, while "gene expression" typically refers to transcription and/or translation of the gene. Even "gene expression" includes several distinct, but related, concepts, since the time and location in which a gene is transcribed are not always the same as when and where its product is active. One of the most widely used methods to study the expression pattern of a gene—both its transcription and the location of its protein product—is to use a reporter gene.

The use of **reporter genes** as experimental tools is based directly on the modular structure of genes, as depicted in

TOOL BOX 12.1 Continued

Figure 12.12. Note in this figure that the products encoded by the gene—its RNA and protein—function independently of the mechanism by which the expression pattern is controlled by the promoter and upstream regulatory region. To understand how a gene product functions in a cell or an organism, the coding portion of the gene is critical; to understand how the transcription is regulated, the upstream regulatory region is crucial. Reporter genes, or more formally reporter gene constructs made *in vitro*, separate the regulation of transcription from the function and are therefore used primarily to analyze the regulation of genes.

So how are reporter genes made in the laboratory? In overview, the DNA sequence of a gene of interest and the surrounding region of the genome, which is presumed or known to include the regulatory region, is isolated *in vitro*. This surrounding region is then retained, but the coding region of the gene is replaced by, or supplemented with, the coding region of an unrelated gene whose protein product can be readily visualized, as illustrated in Figure A; this unrelated coding region is the reporter. This reporter gene construct is reintroduced into the organism to make a transgenic organism, in which the expression of the easily visualized reporter is controlled by the regulatory region of the gene of interest. The expression pattern—that is, the time and location—of the easily visualized reporter protein in the transgenic organism thus reveals the specificity conferred by the regulatory region of the gene of interest.

In this box, we briefly consider two main points: the easily visualized proteins (or rather, protein-coding regions) routinely used as reporters; and the difference between using the reporter gene to **replace** the coding region of the gene of interest and using it to **supplement** the coding region of the gene of interest.

A good reporter protein is easy to detect and quantify

In order to be seen *in vivo*, a reporter protein must be easy to visualize and measure accurately. The expression of most common reporter proteins is determined using colorimetric assays, which can also be used quantitatively to compare **changes** in the level of expression. The reporter protein should not be quickly degraded in the organism or during extraction. The size of the reporter protein is also important—it should be large enough that it cannot diffuse out of the cell but not so big that it affects the activity of the protein being studied. Furthermore, the model species cannot have another protein whose properties are similar to the reporter protein.

Although other reporter proteins have occasionally been employed, four reporters are commonly used. Examples depicting

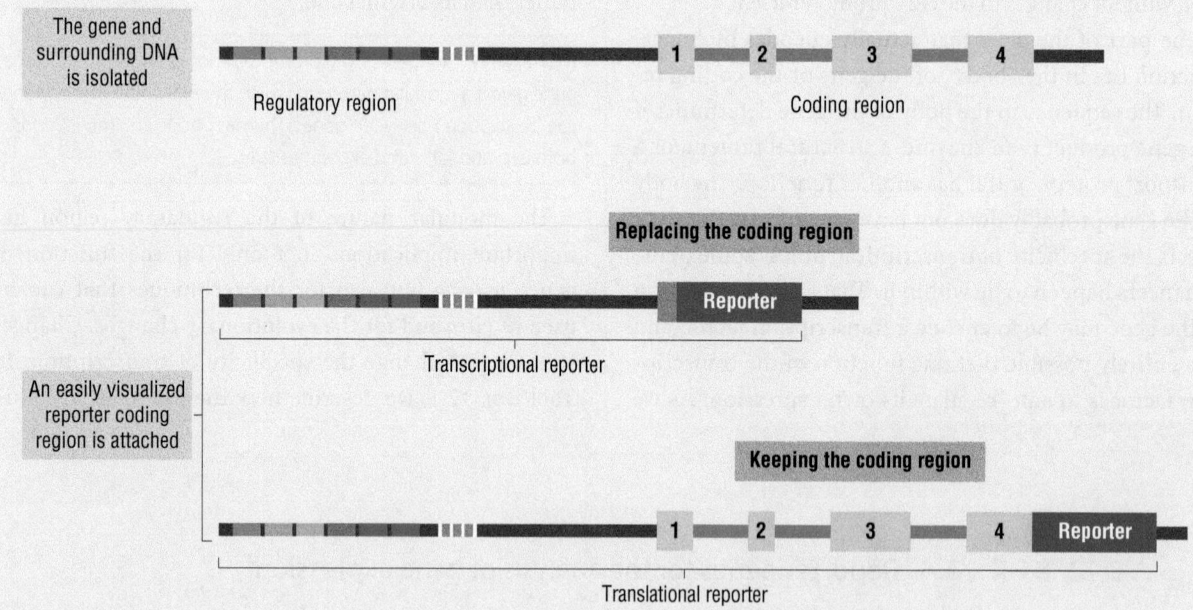

Figure A Making a reporter gene. A gene with its surrounding region, which includes the regulatory region, is isolated. The protein-coding region of a reporter gene, shown here in dark blue, is then attached to the gene *in vitro*, either replacing most of the coding region of the gene, above, or as a fusion to the end of the coding region of the gene, below. These are transcriptional and translational reporter constructs, respectively, which are then reintroduced into the cell or organism.

TOOL BOX 12.1 Continued

the use of each of these reporter genes are shown in Figure B. Green fluorescent protein (GFP) from the jellyfish *Aequorea victoria* and luciferase (LUC) from the firefly *Photinus pyralis* are used in both animals and plants; β-galactosidase, encoded by the *lacZ* gene of *Escherichia coli*, is widely used in animals, while β-glucuronidase (GUS), encoded by the *Escherichia coli uidA* gene, is widely used in flowering plants. All these reporters have been used in bacteria.

GFP has become the most widely used and versatile reporter, and the scientists who developed it for experimental use were recognized by a Nobel Prize in 2008. GFP has natural green fluorescence under blue light, and no additional substrate needs to be added for detection. This provides the important advantage that GFP can be assayed in living cells, unlike the other reporter proteins. Thus, GFP can be used for monitoring dynamic patterns of gene expression and subcellular localization, as well as for separating live cells based on expression patterns.

In general, detection of GFP is somewhat less sensitive than the other reporters, although enhanced GFP with increased fluorescence is commercially available. The GFP protein is small, and its spectral properties are easily modified *in vitro*. Thus, in addition to its natural green fluorescence, the GFP gene has been modified to emit yellow, cyan, or red fluorescence; these modified versions have been given the names YFP, CFP, and RFP. (The original RFP has been largely replaced by an analogous red reporter called mCherry, isolated and modified from the red coral *Discosoma*.) These permutations of GFP allow for the simultaneous labeling and tracing of multiple genes in the same cell or organism.

Reporters other than GFP usually cannot be used in living organisms, but even single fixed cells can be assayed quantitatively by colorimetric assays. LacZ and GUS reporters are especially sensitive methods for studying expression patterns in fixed specimens. The *lacZ* gene will be discussed in detail in Chapter 14 and in Tool Box 15.2, but, for now, the important points are that it encodes an enzyme known as β-galactosidase or β-gal. For β-gal detection, the specimen is fixed, and the substrate 5-bromo-4-chloro-3-indolyl-β-D-galactopyranoside (more commonly called X-gal) is added. β-galactosidase hydrolyzes X-gal to produce galactose and the deep blue precipitate 4-chloro-3-bromo-indigo, so the presence of the blue color indicates the activity of β-gal. This is seen in Figure A (iv). Colorimetric substrates of β-gal other than X-gal are also commercially available, as discussed in Tool Box 15.2.

In addition to its use as a reporter gene in many animals and tissue culture cells, *lacZ* is also used as a reporter gene in some applications of gene cloning, as part of a process known as blue–white selection. During gene cloning, a gene or sequence of interest is inserted into a plasmid vector; the plasmid is ligated,

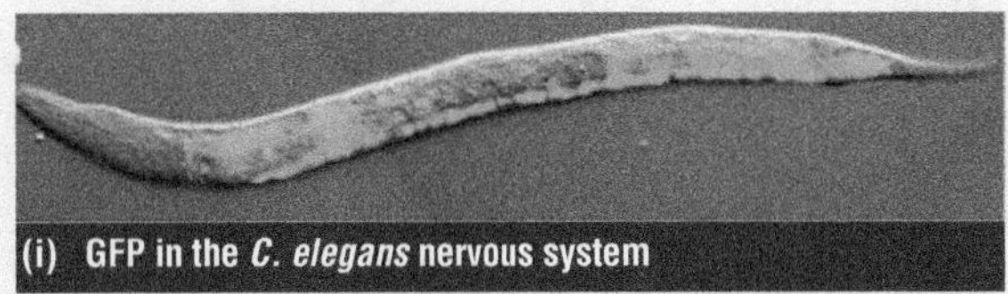
(i) GFP in the *C. elegans* nervous system

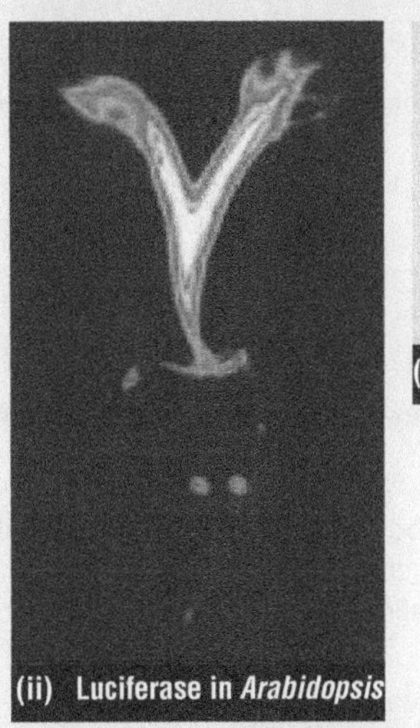
(ii) Luciferase in *Arabidopsis*

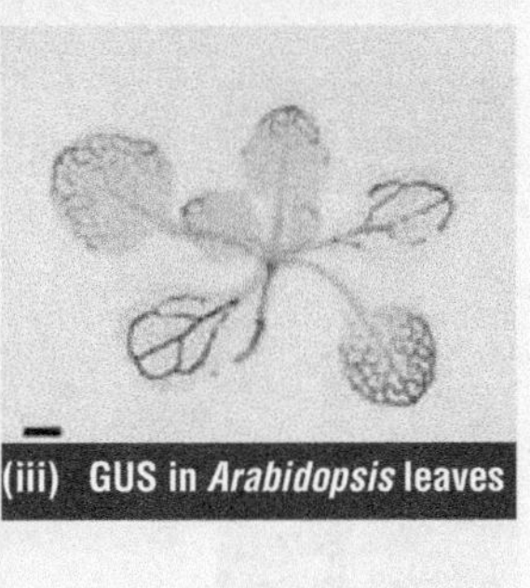
(iii) GUS in *Arabidopsis* leaves

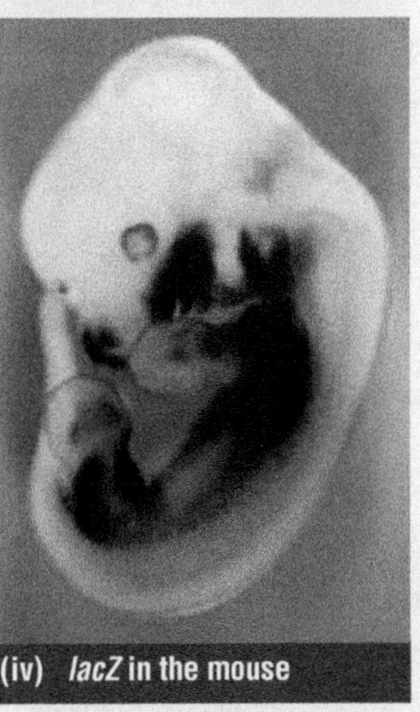
(iv) *lacZ* in the mouse

Figure B Examples of reporter genes. Four widely used reporter genes are shown. (i) GFP. (ii) Luciferase. (iii) β-glucuronidase or GUS. (iv) β-galactosidase (lacZ).

and the new plasmid is transformed into bacterial cells. However, the process is quite inefficient, so most of the time the plasmid is ligated without an insert, and the gene or sequence of the insert is not actually cloned into the vector. In order to improve the process for detecting plasmids with an insert, a plasmid with the *lacZ* gene can be used as the vector. The cloning site for the insert is within the *lacZ* gene, so a successful insertion of foreign DNA (that is, the gene or sequence of interest) disrupts the *lacZ* gene. Thus, bacteria expressing the intact *lacZ* gene with no insert will

TOOL BOX 12.1 Continued

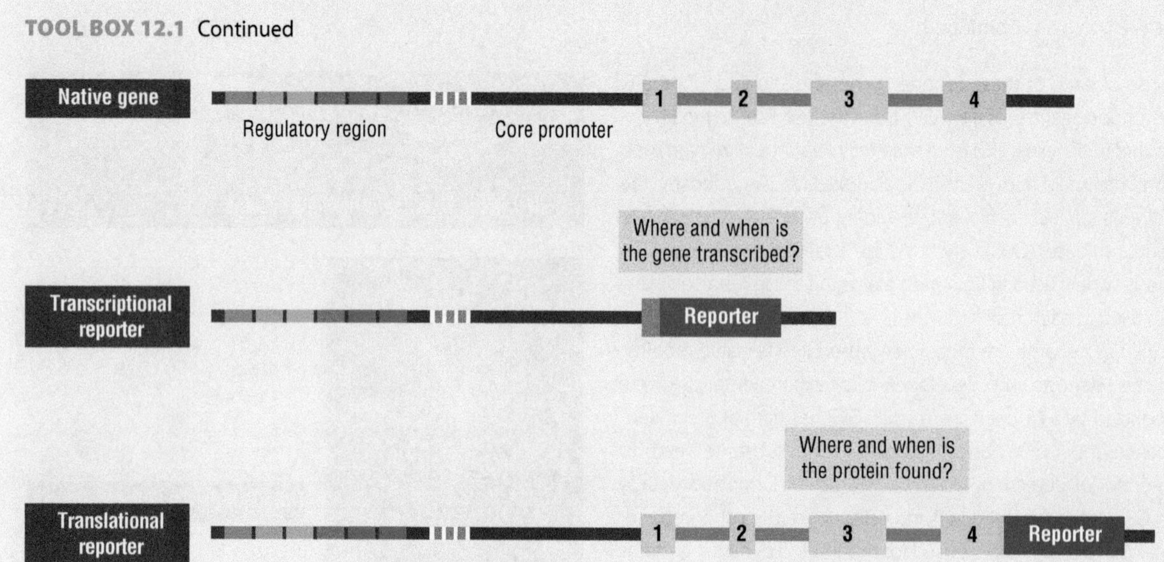

Figure C Information provided by transcriptional and translational reporter genes. The properties of transcriptional and translational reporter genes are summarized. Transcriptional reporters, which do not require a functional gene product, are much easier to construct, while translational reporters offer more information.

be blue, while bacteria with the gene of interest inserted into the *lacZ* gene will be white. By screening for the white colonies in a background when most colonies are blue, the investigator can more readily find the plasmid containing the gene of interest.

GUS assays were developed for use in plants to avoid an endogenous enzyme activity that prevented the use of *lacZ*. They are done in the same manner as for β-gal, with the substrates being X-gluc (5-bromo-4-chloro-3-indolyl glucuronide) or MUG (4-methylumbelliferyl-β-D-glucuronide), with similar spectroscopic properties.

Luciferase was isolated from fireflies as the enzyme that makes fireflies luminesce. As observers who have seen fireflies (or lightning bugs) on a summer evening will realize, a small amount of light is released when luciferase acts on its substrate, luciferin. In luciferase assays, the cells are lysed and combined with luciferin immediately before quantification in a luminometer, which measures the amount of light emitted. Because the light released from the metabolism of one luciferin molecule is small and transient, luciferase is primarily used when rapid detection is important, often in cell cultures.

Transcriptional and translational reporter genes

As noted above, "gene expression" can refer to the pattern of transcription of a gene or to the location of a gene product, or both. While these two patterns are the same for many genes, there are enough exceptions (such as proteins secreted by a cell or tissue) that both transcription products and translation products need to be considered.

Reporter genes can be used to analyze either the transcription pattern or the translation pattern and subsequent protein localization associated with the expression of a given gene. In either case, the regulatory region of the gene of interest is retained. The difference between transcriptional and translational reporters lies in where the protein-coding region of the reporter gene is inserted and how much of the coding region of the gene of interest is retained; these are shown in Figure C.

In transcriptional reporters, the reporter coding region replaces most (or all) of the coding region of the gene of interest. Thus, no functional protein product from the reporter transgene is made, and post-transcriptional or post-translational modifications (which depend on the sequence of the coding region) do not occur. The expression of the reporter protein displays the pattern of **transcription** for the gene.

In translational reporters, the reporter coding region is fused in frame to the coding region of the gene of interest to make a fusion protein that encodes both the reporter and the protein of interest. In an ideal translational reporter, the protein of interest retains its function, as well as all of its post-transcriptional modifications (its protein-trafficking pattern, for example). In this case, the expression of the reporter protein displays the location of the final protein product in the cell. This in-frame fusion between the coding region of the normal gene and the reporter gene makes translational reporters more informative but also more difficult to create than transcriptional reporters.

of genes is used to study the pattern of gene expression. This is a technique referred to as a reporter gene, in which the specificity of the CRM is "reported" by placing it upstream of a coding region that encodes an indicator whose expression can be monitored, such as green fluorescent protein.

A gene's modular structure is important evolutionarily

The separate modules from which a gene is composed have also had important evolutionary implications. Remember from Chapters 3 and 4 that mutations occur randomly throughout the DNA sequence. A mutation in the body of a gene might change the function of the gene or eliminate its function altogether. For a protein-coding gene, a mutation might substitute one amino acid for another (a missense mutation) or produce a stop codon that terminates the protein (a nonsense mutation). Mutations in the protein-coding body of the gene will be subject to selective pressures that limit which of the many possible mutations is capable of surviving in a population. Any base can be changed, but not every change is compatible with survival; many substitutions in the body of a gene will be lost from a population because these substitutions affect the function of the gene product. Other substitutions will produce a gene product that has an altered function but whose function is still compatible with survival.

But what about changes in the DNA sequences that make up the CRM? Mutations arise there as well, but they do not affect the molecular function of a gene. Instead, these substitutions might alter the **expression** of the gene product, so that the equivalent gene product may have different patterns of transcriptional expression in different species, as summarized in Figure 12.12. Because mutations in a CRM may eliminate or create a new binding sequence for a transcription factor—and recall that the binding sequences are short and the binding is often tolerant of substitutions, so regulation can be quite flexible—the gene might be transcribed in cells that have not previously expressed it, or its transcription might be lost from certain cells, and so on. Such changes in the time and location of transcription are likely to have had a significant impact on the evolution of body forms and functions. As genomic sequences of related species are compared, examples of such changes abound.

As we discussed in Section 3.6, the *Hox* genes of animals provide a particularly striking example of the importance of such changes. *Hox* genes are found in the genomes of all animals; their expression in regional domains along the anterior–posterior axis of the animal controls the identity of the structures that form along this axis. Figure 12.13 compares the expression patterns of equivalent *Hox* genes in *Drosophila melanogaster* and humans.

The *Hox* genes are transcription factors that regulate the transcription of numerous genes, so they are examples of the type of "master regulator" genes that directly and indirectly control the expression of many other genes, a concept discussed further in Chapter 14.

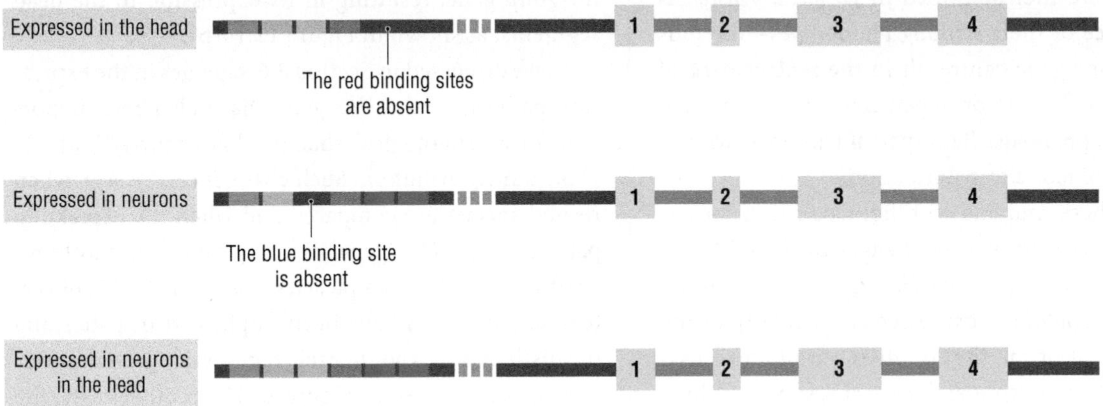

Figure 12.12 Changes in the regulatory region affect evolutionary change. Mutations in the *cis*-regulatory module (CRM) do not affect the molecular function of the gene but alter where and when the gene is expressed. The diagram indicates equivalent genes in three different species, in which slight differences in the sequence of the CRM produce differences in the expression pattern of a gene.

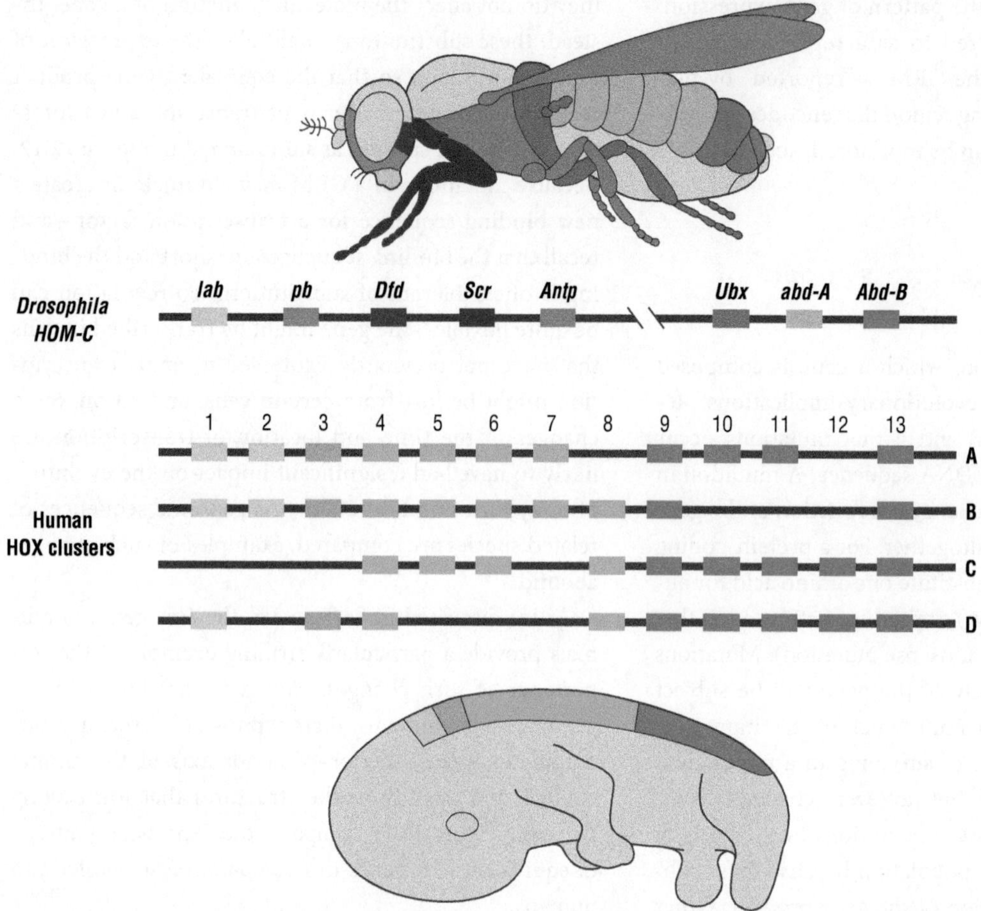

Figure 12.13 The *Hox* genes and evolutionary change. *Hox* genes are expressed in regional domains along the anterior–posterior axis of animals and control the identity of the structures that form in these regions. Note that both *Drosophila melanogaster* and humans have similar *Hox* genes arranged on the chromosome and expressed along the body axis in similar patterns, as shown by the corresponding colors.

Hox genes were first identified in *D. melanogaster* as a consequence of their striking phenotypes—the mutation of a *Hox* gene can result in the replacement of one normal body part or appendage with a different body part or appendage. This type of transformation is known as a homeotic transformation.

Many of these mutants with homeotic transformations have proved to involve changes in the CRM for the genes. For example, the *Hox* gene *Antennapedia* (*Antp*) in *Drosophila* is expressed most highly in the second segment of the thorax where it initiates a cascade of target gene expression patterns, some of which direct the outgrowth from this segment to form a leg. The original mutation that defined the gene results in legs, rather than antennae, being formed in the head, depicted in Figure 12.14(a). Subsequent analysis has shown that this mutation was a change in the CRM for the *Antp* gene, resulting in its expression in the head segments, as shown in Figure 12.14(b).

As we discussed in Section 3.6, changes in the expression patterns of the *Hox* genes have also been important for morphological changes that occurred in body plans during evolution. Such changes can be seen when related species are compared and when the expression pattern of the *Hox* genes is compared in distantly related species. For example, changes to the CRMs of vertebrate *Hox* genes have been implicated in patterning the distinct neck and thoracic regions of different vertebrates, as illustrated in Figure 12.15. In other cases, the expression of the *Hox* genes themselves is broadly conserved, but there are changes in the CRMs of the genes they regulate; changes in the CRMs of the target genes also result in morphological differences between species. Such changes have been implicated in the differing

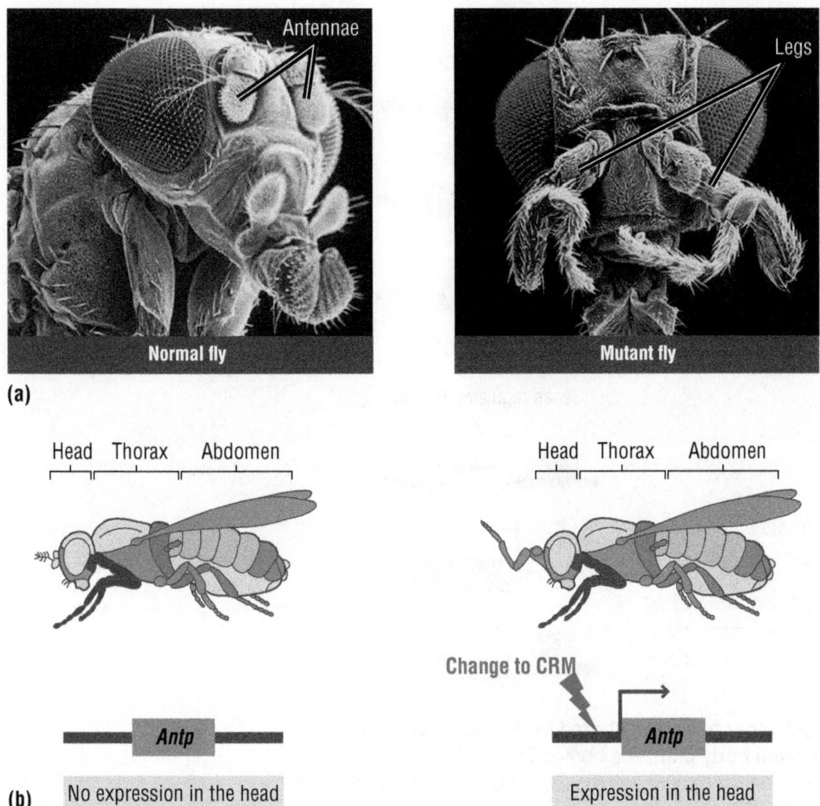

Figure 12.14 Homeotic mutations result in the replacement of one body part with another. (a) The heads of a wild-type (on the left) and an Antennapedia mutant fly (on the right) showing the homeotic transformation from antenna to legs in the mutant fly. (b) The mutant phenotype of Antennapedia arises from changes in the CRM of the *Antennapedia (Antp)* gene, which changes its pattern of expression but not its function.

Source: Wild-type *Drosophila* courtesy of EYE OF SCIENCE/SCIENCE PHOTO LIBRARY. Antennapedia mutant courtesy of Science VU/Dr. F. Rudolph Turner, VISUALS UNLIMITED /SCIENCE PHOTO LIBRARY.

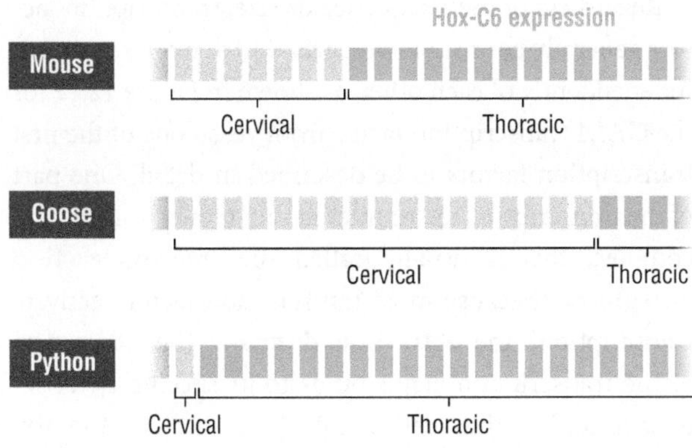

Figure 12.15 ***Hox*** **gene expression and evolutionary change.** Major transitions in body plan organization correspond with major shifts in the spatial boundaries of *Hox* expression, shown here for three different vertebrates and the gene *Hox-C6*. *Hox-C6* is involved in thoracic development. Compared to mice, geese have an expanded region of cervical development, while pythons have an expanded domain of *Hox* genes involved in thoracic development and a reduced region of cervical development.

Source: Reprinted by permission from Macmillan Publishers Ltd. from Cohn and Tickle (1999). Developmental basis of limblessness and axial patterning in snakes. *Nature* 399, 474-479 (3 June 1999). doi:10.1038/20944.

hind-wing morphologies of insects, diagrammed in Figure 12.16.

There have also been changes to the coding sequence of *Hox* genes over the course of evolution, which have resulted in slight, but significant, changes in their functions. Although we cite the *Hox* genes here because of their profound evolutionary effects on animal body plans, thousands of other genes whose expression patterns have changed during evolution could also be used as examples. In fact, we began Chapter 1 by describing *ALX1* in finches, another example of a gene whose expression pattern has changed during evolution. It is clear that the modular structure of genes, in which the CRM and the coding region of a gene have separable activities, has allowed for significant flexibility during evolution.

KEY POINT Mutations in CRMs may result in changes to transcription factor binding sites that alter the expression pattern of a gene without changing the encoded product of the gene. Such changes have been an important aspect of gene evolution.

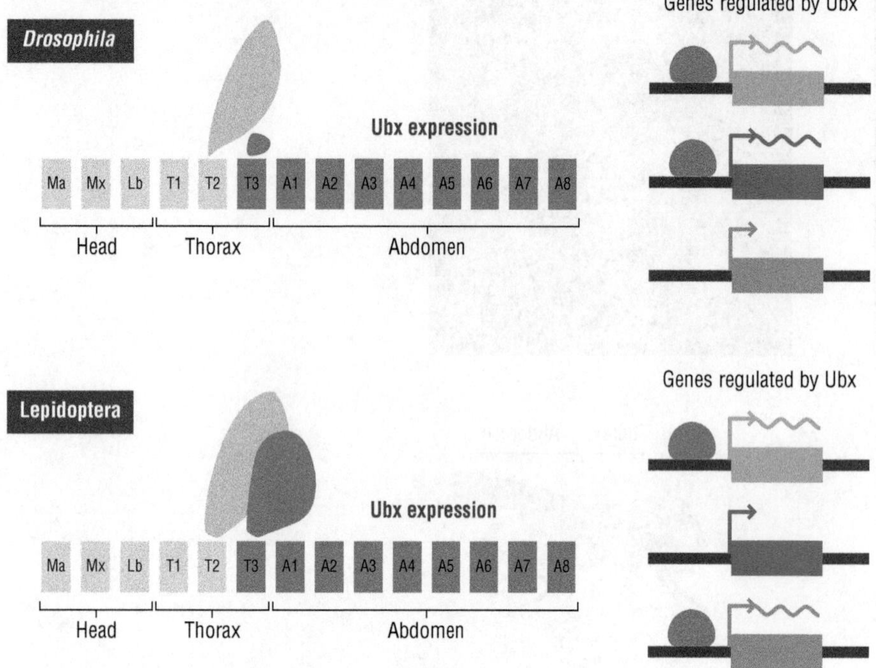

Figure 12.16 Changes in the CRMs of *Hox* target genes have contributed to the morphological diversification between species with a conserved overall body plan. The homeotic gene *Ubx* is expressed in roughly the same regions of the larvae of *Drosophila*, which has two wings, and *Lepidoptera* (butterflies and moths), which have four wings. The differences in wing development are due primarily to changes in the target genes regulated by *Ubx*, rather than changes in *Ubx* expression.

The functions of transcription factors are reflected in their structures

Thousands of different transcription factor proteins exist, each with its own preferred binding site, its own amino acid sequence, and its own set of target genes. Transcription factors from different organisms are listed in multiple online databases. How then can we quickly summarize what we know about their structures and functions? It turns out to be relatively easy, at least as an overview.

Transcription factors have two fundamental biochemical activities—they bind to a specific DNA sequence, and they interact with the pre-initiation complex (or, in the case of bacteria, with RNAP) to regulate transcription. We will discuss these two activities in a bit more depth in the succeeding sections, but, for now, we want to look at how these activities are embedded in the structures of the eukaryotic transcription factor proteins. Recall one of our unifying themes of biology from the Prologue, namely that life depends on chemistry. This is often manifested in the way that molecular structure and function are closely connected—that is, the chemical properties that determine the structure of the molecule directly influence its biological properties. Structure and function are often inseparable concepts. This is certainly true for eukaryotic transcription factors.

Regardless of their sequence or size, transcription factors generally consist of two parts, which work somewhat independently of each other, as shown in Figure 12.17 for the GAL4 transcription factor from yeast, one of the first transcription factors to be described in detail. One part of the transcription factor interacts with the pre-initiation complex; this is usually called the ***trans*-interaction domain** or (because most transcription factors activate transcription) the **activation domain**. The other part of the transcription factor binds to its specific DNA sequence and is thus called the **DNA-binding domain**. The amino acid sequence of the DNA-binding domain determines the nucleotide sequence of the target binding sequence; a change in one of the critical amino acids in the DNA-binding domain might change the preferred binding sequence or (more likely) eliminate the binding to DNA altogether.

We need to make a quick aside about the biochemical vocabulary here. By the strictest biochemical definition, a "domain" is a part of a polypeptide that is capable

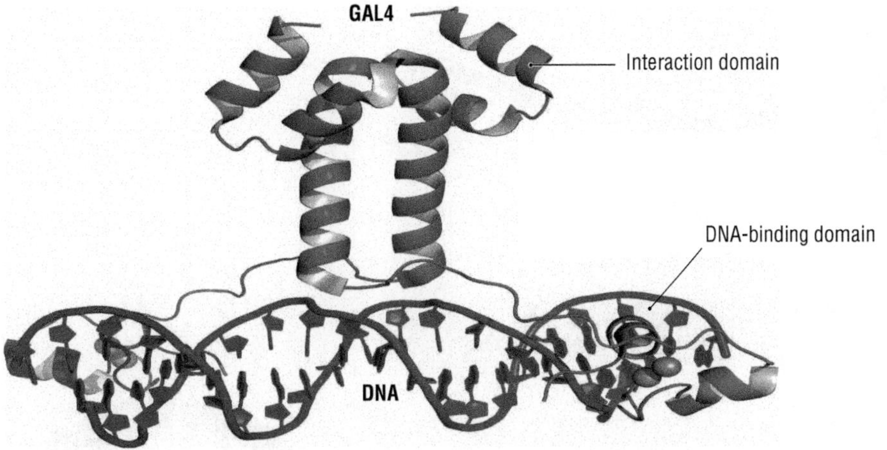

Figure 12.17 Transcription factors have separable DNA-binding domains and interaction domains. GAL4 is a transcription factor in yeast, and is among the best studied. The functional protein is a dimer, with two identical subunits. The DNA-binding domain is shown with its interaction with the DNA sequence (in blue); the interaction or activation domain interacts with RNA polymerase II and the pre-initiation complex.
Source: Generated from PDB ID 3COQ.

of functioning, evolving, and even folding into its three-dimensional structure, independently of the rest of the polypeptide chain. This definition requires specific tests and experiments, so we often do not know all of the information that would be needed to call a particular amino acid sequence a domain. We do, however, have extensive amino acid sequence information for proteins, and we can infer evolutionary relationships between proteins based on these sequences. Recurring patterns of amino acid sequences, as observed from sequence comparisons with related proteins, are called **motifs**. Motifs and domains can range from 12 or so amino acids to more than 300 amino acids.

Motifs and domains are not quite synonyms—several copies of a motif may comprise a domain, for example—but the terms are often used as synonyms when describing evolutionarily conserved recurring patterns in an amino acid sequence. Indeed, a motif is often also shown to be a domain, but the experimental process used to define domains and motifs is different. (It is much easier to spot a recurring pattern of amino acid sequences than it is to do the experiments to make sure that this recurring pattern makes up a domain.)

Transcription factors are classified according to the sequence and structure of their DNA-binding domains, some of which are shown in Figure 12.18(a). There are more than 40 different classifications of eukaryotic DNA-binding domains or motifs, although the exact number depends on who is doing the classification, since some of them are related to one another. This is a bit like asking who the members of your family are; sometimes we include cousins, aunts, and uncles, while other times we refer only to parents and siblings.

For example, one common motif that binds to DNA is the C_2H_2 zinc finger, so called because the amino acid sequence has two conserved cysteine residues (C), 2–4 amino acids apart, and two conserved histidine residues (H), 3–5 amino acids apart; there are 12 other amino acids between the two cysteines and the two histidines, shown in Figure 12.18(a) and in more detail in Figure 12.18(b). This amino acid sequence interacts with a zinc ion to form the structure that binds to a DNA sequence at the enhancer. A typical C_2H_2 zinc finger protein will have two, three, or four or more such zinc finger motifs together. By noting these recurring amino acids in the sequence of an unknown protein, an investigator can assume that this protein binds to a nucleic acid sequence and can plan experiments based on this inference, often without having to directly test its ability to bind to DNA.

KEY POINT Transcription factors are modular, with a DNA-binding domain that recognizes the nucleotide sequence at the binding site and an interaction (or activation) domain that interacts with RNAP and the pre-initiation complex. The amino acid sequence and structure of the DNA-binding domain determine the DNA-binding site specificity, and transcription factors are classified by the type of DNA-binding domain they possess.

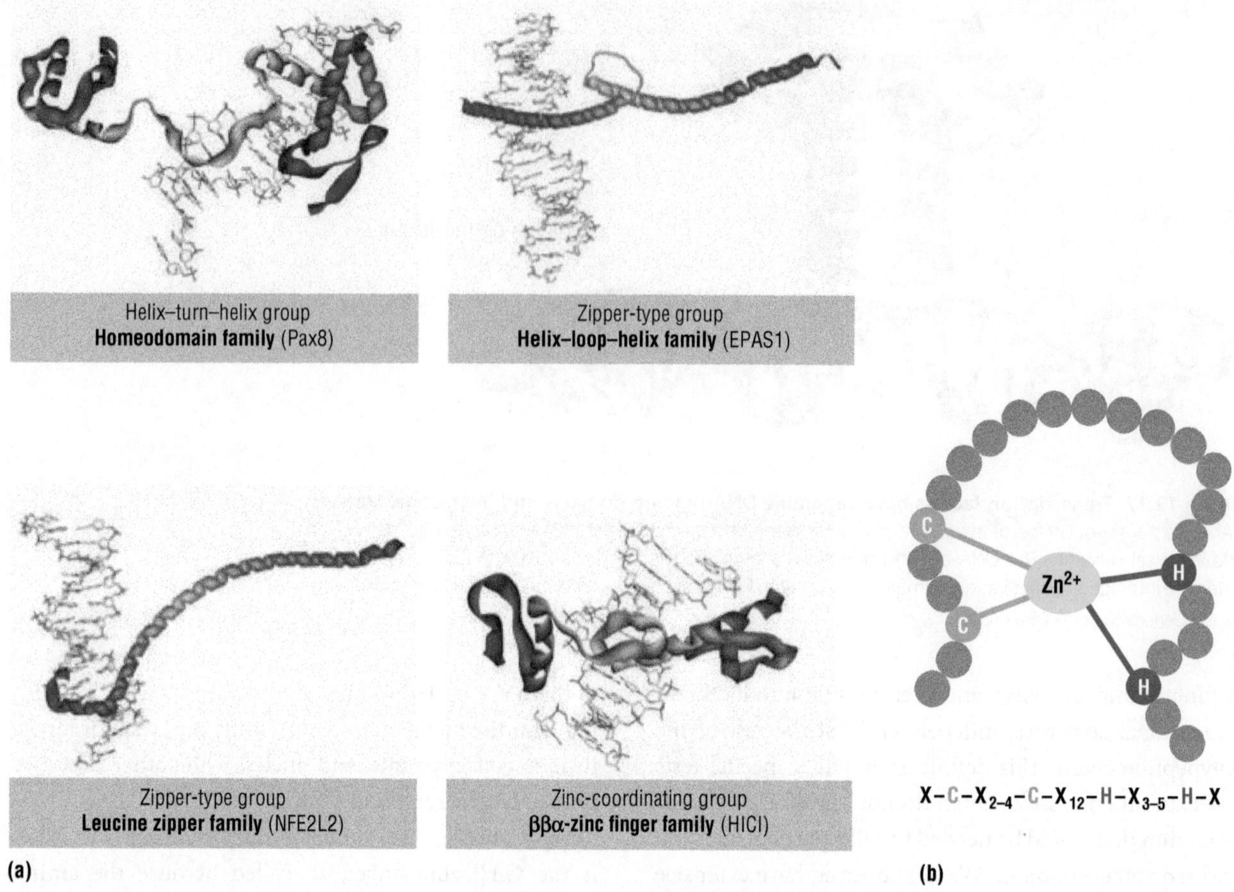

Figure 12.18 The structures of some DNA-binding domains. (a) The proteins are shown as ribbons interacting with the DNA double helix in the center of each diagram. Four different structures of DNA-binding domains are shown. (b) A more detailed diagram of the amino acids that make up the C_2H_2 zinc finger-binding domain. Note the two cysteines (C) with 2–4 amino acids in between, and the two histidines (H) with 3–5 amino acids in between. These interact with a zinc ion to form the finger structure.

Source: Part (a) reproduced from Yusuf, D. et al. (2012). The transcription factor encyclopedia. *Genome Biol.* 2012; 13(3):R24.

Although the DNA-binding domain is used to classify transcription factors, its presence alone cannot be used to determine that a protein acts as a transcription factor. For example, some zinc finger proteins, as well as proteins with other types of binding domains, bind to RNA or to single-stranded DNA rather than to double-stranded DNA, and many zinc finger proteins bind to double-stranded DNA but do not activate transcription. Proteins bind to DNA as part of a variety of biological functions other than regulating transcription—recombination, repair, and replication, for example—and at least some of the proteins involved in these other processes bind to DNA via a zinc finger domain.

The DNA-binding domain can be likened to the plug associated with an electrical appliance. If one sees a plug, it is safe to assume that the device uses electricity for its function. However, just as the same plug structure can be used for a toaster, a vacuum cleaner, a saw, or a hairdryer, one does not know the function of the protein based solely on its DNA-binding domain. The domain allows the protein to bind to DNA, but the function that it carries out once it is bound still has to be determined.

Our analogy about DNA-binding domains and electrical plugs can be extended. As international travelers quickly discover, the same electric appliance requires a different plug in different countries—not because the function of the appliance has changed but because the interaction between the appliance and the outlet is determined by a different structure. Likewise, proteins in different species might have similar biological functions, while using different DNA-binding domains to bind to DNA.

Transcription factors bind to specific nucleotide sequences

We noted earlier that a transcription factor binds to a specific nucleotide sequence, or a consensus sequence, to regulate transcription; a few examples are shown in Figure 12.6. The consensus sequences are typically short, in the order of 9–14 nucleotides, of which some bases are absolutely necessary for binding and some are less necessary and can be variable. As noted, these binding sequences are often clustered in a region upstream of the gene called the *cis*-regulatory module or CRM.

Since there may be multiple binding sites within the CRM, it is reasonable to ask about the relationships and interactions among different transcription factors. In general, different types of interactions are seen. In some cases, transcription factors bind independently of each other, so regulation is combinatorial—the particular constellation of transcription factors, each of which binds independently of the others, determines when and where the gene is expressed. These are often the easiest enhancers to analyze experimentally because each part of it can be studied separately from the others.

In other examples, the binding is cooperative, such that the binding of one transcription factor makes it more likely that another transcription factor will bind. Cooperative binding can arise from protein–protein interactions between transcription factors. For example, there might be two transcription factors, called A and B, that regulate the expression of a gene. Transcription factor A can bind to DNA on its own, but transcription factor B cannot bind DNA unless it is part of a dimer or a complex with protein A. These protein complexes could include several additional proteins as well, all of which play some role in regulating transcription. For any CRM and its set of transcription factors, both combinatorial and cooperative interactions among the transcription factors are likely to be important.

KEY POINT The suite of transcription factors that bind a CRM may bind independently of one another or cooperatively, or a combination of both. Ultimately, the outcome in terms of the effect on transcription is often the result of interactions between the different factors simultaneously bound at the CRM.

Darwin's finches and transcription factors, revisited

We began Chapter 1 with a discussion of the changes in beak morphology that helped to inspire Darwin to propose the theory of evolution by natural selection. Before beginning our description of how scientists identified the *ALX1* gene as being important in this change in beak morphology, we indicated that some of the concepts we introduced would be elaborated in later chapters. This may be a good time to revisit part of our discussion of the role of ALX1 as a transcription factor.

The *ALX1* gene encodes a transcription factor related to the homeodomain class, that is, it binds to DNA using an amino acid motif similar to that used by other homeodomain proteins such as Pax8, shown in Figure 12.18(a); among transcription factors with homeodomains, its binding domain is one of the subcategory called the paired family. *ALX1* is expressed in craniofacial tissues during vertebrate embryogenesis; thus, it is likely to be regulating the transcription of several genes involved in craniofacial functions, including the formation of a beak in birds, although its exact targets are not known. The consensus binding site for ALX1 in mice is shown in Figure 12.6, so we can predict that the genes whose transcription is regulated by ALX1 will have a sequence similar to this in their CRM, but there may also be some slight differences between the binding sequence in mice and that in finches.

As discussed in Section 1.2, when finches with blunt beaks were compared to finches with pointed beaks, eight different sequence changes were found associated with the *ALX1* gene itself. Three changes are predicted to alter amino acids in the DNA-binding domain of the ALX1 protein. It is likely that these substitutions changed the preferred target sequence for ALX1 binding; thus, there may be genes regulated by ALX1 in blunt-beaked species that are not regulated in pointed-beaked finches, and vice versa, because the binding specificity of the ALX1 protein is slightly different. Four changes are found in the likely CRM of the *ALX1* gene and might change its pattern of expression. (The eighth change associated with *ALX1* between birds with blunt and pointed beaks is downstream and near to the *ALX1* gene and could also affect its expression, but this is not so clear.) Thus, the *ALX1* gene may be expressed in slightly different tissues or in slightly different times in blunt-beaked and pointed-beaked species.

In summary, the *ALX1* gene shows the importance of modular gene structure for evolutionary change. In comparing species with different beak morphologies, changes in both the likely CRM and the likely binding domain of the predicted protein have occurred. These may have resulted in small changes in the targets of *ALX1* regulation, both because the protein is expressed in a different pattern and because its binding specificity may have changed. Note that none of these changes is predicted to eliminate, or even significantly reduce, the function of the *ALX1* gene; such changes, which have been found by mutations in mice, are probably lethal and would not have survived during finch evolution. But small changes in the expression pattern and the target genes have apparently had significant effects on beak morphology, such that Darwin could use them as evidence for evolution by natural selection.

12.4 Chromatin and pre-initiation

While we frequently discuss transcription factors in terms of their binding domains and their target sequences and genes, we should keep in mind that DNA sequences *in vivo* are part of chromosomes, rather than occurring in the form of the naked DNA that we typically work with in the laboratory. The organization of the DNA affects how readily transcription factors can access and bind to it, and consequently affects the initiation of transcription. In *E. coli*, for example, the proteins Fis and Integration Host Factor (IHF) regulate numerous genes by altering the number and nature of DNA supercoils in the nucleoid.

Binding sequences occur as part of a structure in chromatin

In eukaryotes, the focus has been on the state of the chromatin, rather than on the conformation of the DNA molecule itself. Recall from Chapter 2 that DNA is in a complex with histones and other chromosomal proteins in a eukaryotic chromosome; the chromatin structure often involves an interaction between proteins and the phosphate backbone of the DNA. A total of 146 base pairs of DNA are wrapped around each nucleosome, as shown in Figure 12.19. In other words, a gene that is 7 kb long (which is not unusual for a gene in invertebrates but is short for a gene in vertebrates) will be wrapped around approximately 35 nucleosomes.

These chromatin complexes of DNA and nuclear proteins vary widely in their accessibility to transcription factors. In some cells, the chromatin is open, and the DNA sequence is more exposed, making it accessible to transcription factors (which can then bind the DNA). In other cells, however, the chromatin is closed, and transcription factors cannot bind to their regulatory sequences easily,

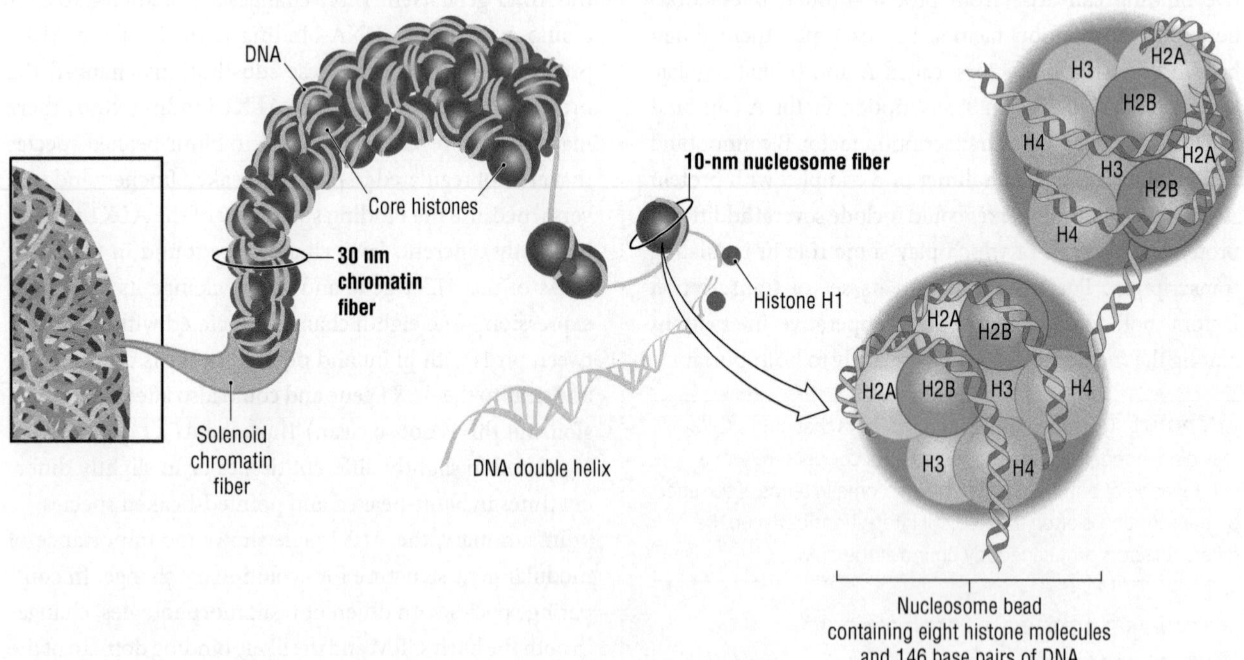

Figure 12.19 The structure of chromatin. DNA is wrapped around the core histones to form the nucleosome or a 10-nm fiber, and the nucleosomes are linked to make the 30-nm fiber. The expanded view shows the structure of two nucleosomes and the DNA in between them. There are 146 base pairs of DNA wrapped around each nucleosome, which is composed of eight histone molecules, two copies each of H2A, H2B, H3, and H4.

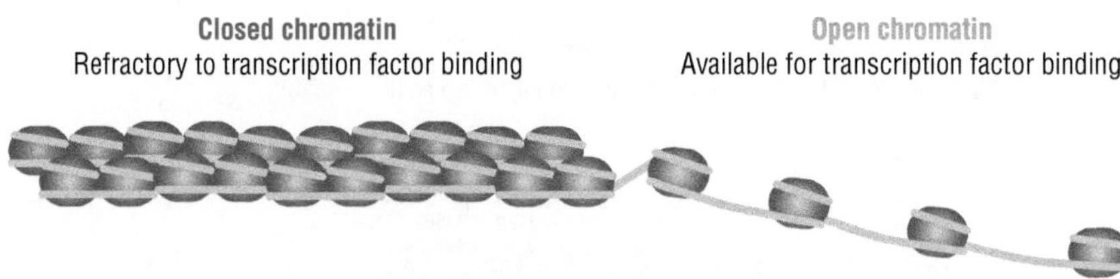

Figure 12.20 Chromatin can be open or closed, depending on how tightly packed the nucleosomes are. In the open state, the DNA sequence is more exposed, so transcription factors can bind more readily. In the closed conformation, transcription factors cannot easily access the DNA sequence.

as shown in Figure 12.20. Accessibility of chromatin is a big topic with its own rich literature; despite how much we know already, there is still much to be learned about the dynamic structure of chromatin in the nucleus and its effects on transcription.

A few transcription factors, sometimes called pioneer transcription factors, can bind even when the chromatin is closed; in some cases, binding of these pioneer transcription factors helps to open the chromatin to allow other transcription factors to bind. These are examples of cooperative interactions among transcription factors, discussed in Section 12.3.

One molecular change that is most strongly correlated with open and closed chromatin occurs on the histone proteins that comprise a major component of chromatin. You will recall from Chapter 3 that there are four core histones—H2A, H2B, H3, and H4—all of which have a tail of amino acids that sticks out away from the DNA molecule and is capable of interactions with other proteins, shown in Figure 12.21. (While these are usually called histone tails, they are actually at the N-terminus of the protein.) These tails can be covalently modified, and dozens of modified histones are known, although these modifications may not represent the full extent of those possible. There are several different components to be considered when categorizing the histone modifications, summarized in Figure 12.22.

- Which histone is being covalently modified? Histone H3 is the one with the most modifications and whose modifications are the best studied. But important modifications have also been found for H4, H2B, and H2A.

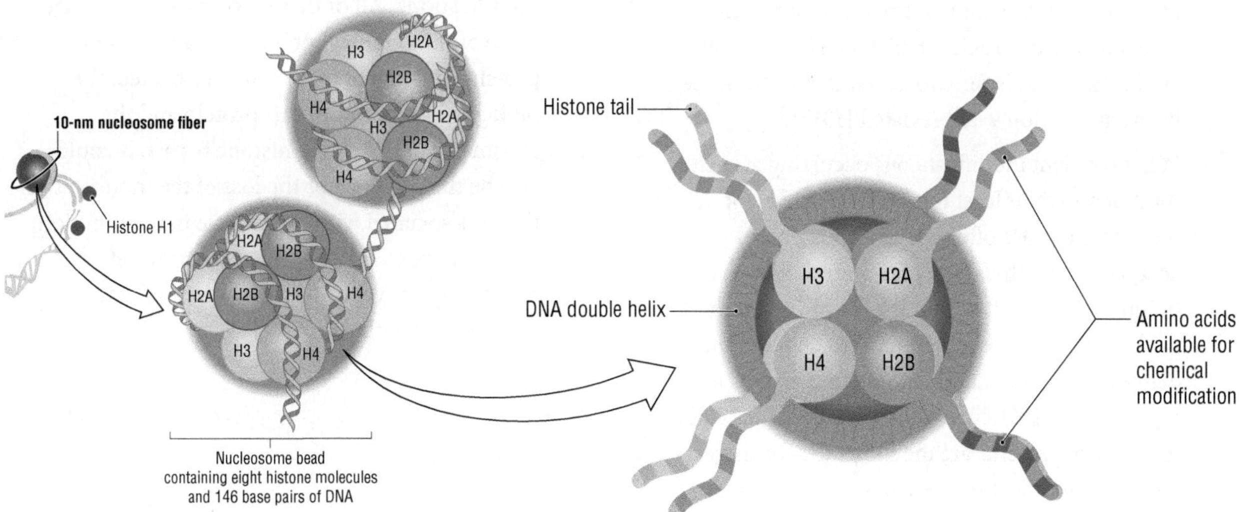

Figure 12.21 The structure of the nucleosome. All four core histone proteins—H2A, H2B, H3, and H4—have a "tail" of amino acids that sticks out away from the DNA molecule. These tails interact with other proteins and can be covalently modified.

Source: Reproduced from Gargalionis, A.N. et al. Histone modifications as a pathogenic mechanism of colorectal tumorigenesis. *The International Journal of Biochemistry & Cell Biology* Volume 44, Issue 8, August 2012, Pages 1276–1289. With permission from Elsevier.

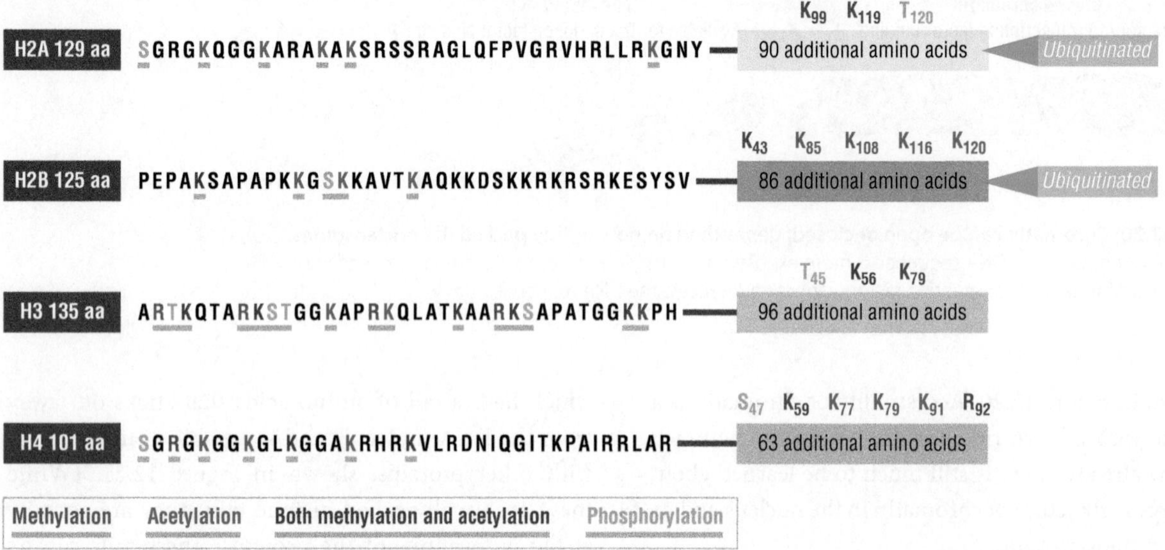

Figure 12.22 Summary of histone modifications. The locations of amino acids that are modified in specific histone tails are indicated.

- Which amino acid within that histone is being modified? Many modifications occur at lysine residues (abbreviated by the letter K), although modifications can also be found on serines (S), threonines (T), and arginines (R). But even this is not specific enough because the same modifications can occur at different locations but with different effects. H3 has nine different lysine residues, for example, for which some modifications are known. The lysine at position 4 (the fourth amino acid) in histone H3 is written as H3K4, which is functionally different from the lysine at position 9, abbreviated H3K9.
- What covalent modification is occurring at each modified site? At least four different covalent modifications are often found on histones: acetylation, methylation, phosphorylation, and ubiquitination. (Ubiquitin is a tag of 76 amino acids that is attached to proteins to target them for degradation.) We diagram these four groups in Figure 12.23. Particular amino acid residues within the histone proteins are the substrates for these covalent modifications. Serines and threonines can be phosphorylated; lysines can be acetylated, methylated, or ubiquitinated; and arginines can be methylated. Thus, we can describe H3S10P (an H3 protein with a phosphate group attached to the serine residue at position 10) as a modification associated with mitotic cells, or H4K16Ac (an H4 protein with an acetyl group attached to its lysine residue at position 16) as one associated with cancer cells.
- But even this is not a sufficient description, because some lysines can have one, two, or three methyl groups attached to them, and the functional consequences can be quite different. Thus, we often need to write H3K4me2 or H3K4me3 to distinguish between different modifications. All of these modifications change some of the chemical properties of the histone protein, and any of them might also affect the biological functions of this protein and the chromatin of which the histone is part. It could also be the presence or the loss of the modification that is associated with the change in the function; the presence of H3S10P is a sign that a cell is mitotically active, while the absence of H4K16Ac is associated with cancer cells.

A histone code underlies many epigenetic changes

With scores of possible modifications, no one can be expected to remember all of them. In addition, new functions are being associated with certain modifications routinely, so any specific modification that we try to remember now is likely to be incorrect or incomplete in a few months. Furthermore, the modifications are

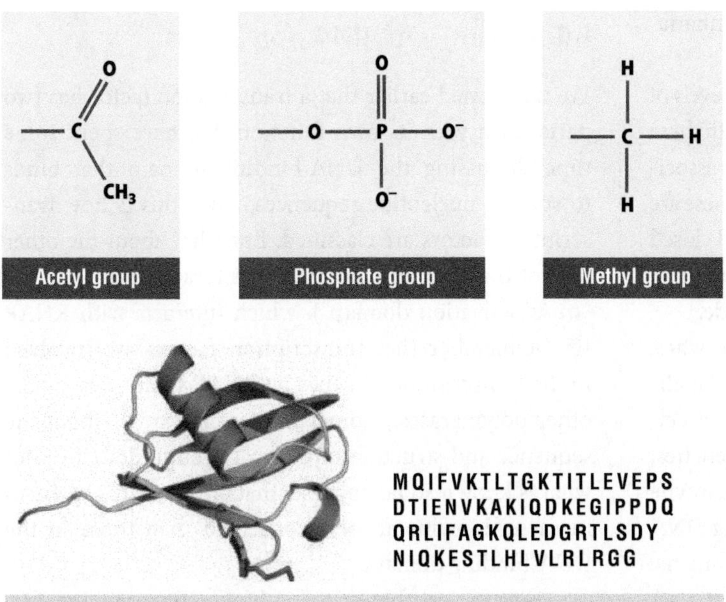

Figure 12.23 The chemical structure of the four common chemical groups that modify histones, resulting in acetylation, methylation, phosphorylation, and ubiquitination of particular amino acids of the histone proteins.

dynamic, so remembering a specific modification might not provide helpful insights.

One insight that **is** helpful, however, is to recognize that specific combinations of histone modifications comprise a snapshot of the state of the chromatin, referred to as a **histone code**. An example of a histone code surrounding a typical gene from *Drosophila* is shown in Figure 12.24; most eukaryotic genes have similar chromatin structures and a similar histone code. Different combinations of histone modifications, that is, different histone codes, are associated with actively transcribed and silenced genes. For example, the lysine at position 4 in H3 has frequently been found to be methylated at sites of active enhancers and promoters but not so often at inactive enhancers and promoters. Active enhancers frequently have a single methyl group (H3K4me1) or two methyl groups (H3K4me2), whereas active promoters and TSSs are characterized by H3K4me2 and H3Kme3. Thus, H3K4me2 is characteristic of active chromatin and is found at both enhancers and

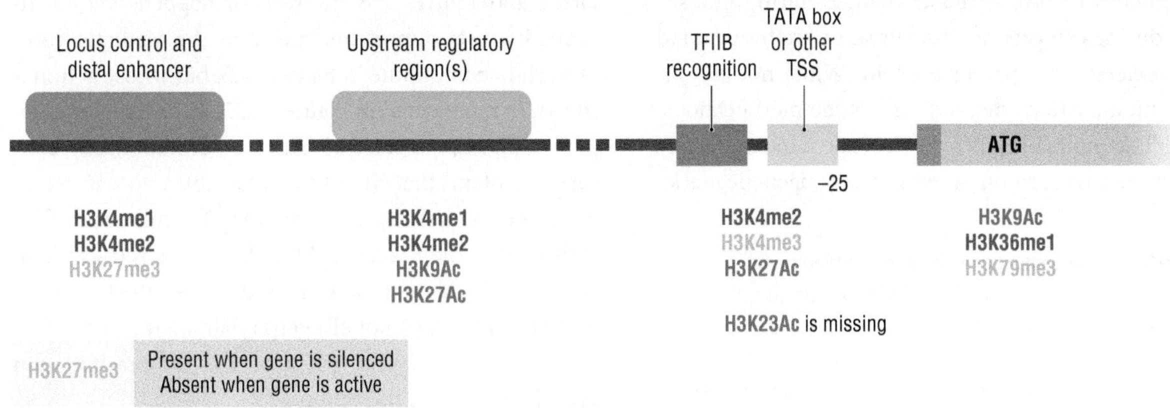

Figure 12.24 An example of a histone code for a typical gene from *Drosophila*. Different combinations of histone modifications are associated with actively transcribed genes. The modifications that are commonly found at each of these locations are indicated, although any specific gene may not have all of these.

promoters, while H3K4me1 is found at active enhancers and H3K4me3 is found at active promoters.

Promoters and regulatory regions also have high levels of H3K9Ac and H3K27Ac. In contrast, H3K27me3 (with three methyl groups attached to the lysine at position 27) is associated with chromatin that is closed to transcription. These are not the only histone marks associated with open and closed chromatin, but these are some of the reliable indicators of the chromatin state or components of the histone code.

Many histone modifications are stably inherited when a eukaryotic cell divides by mitosis, so that the daughter cells often retain the histone code. This type of cellular memory is an important example of epigenetics. **Epigenetics** refers to a heritable change in the phenotype without a corresponding change in the underlying DNA sequence. Consider a differentiated skin cell in humans. It is a skin cell because it is making a characteristic set of proteins, and it is making these proteins because it is transcribing a characteristic set of genes. It is transcribing these particular genes in large part because of the accessibility of sites on its chromatin, which depends on the particular histone modifications of the chromatin.

When a skin cell divides by mitosis, its daughter cells are also skin cells that transcribe this same set of genes and make this same set of proteins. The daughter cells do not undergo the entire differentiation process that gave rise to the skin cell phenotype because they retain much of the histone code found in skin cells from the preceding cell generation. In other words, there is a cellular memory of which genes are being transcribed, although the DNA sequence in skin cells and other cell types is unchanged.

There are many other examples of epigenetics; we introduced this topic in Section 7.5 when discussing X chromosome inactivation in mammals. Most examples of epigenetics involve heritable changes during mitosis, that is, during cell generations, but some are transmitted across meiosis from parent to child. While not all epigenetic modifications **depend** on histone modifications, nearly all examples of epigenetics include some characteristic histone modifications as part of the epigenetic mark.

KEY POINT In the context of a genome, the binding of transcription factors occurs within the DNA–protein complex of chromatin. The core histone proteins of chromatin can be dynamically modified in a variety of ways that influence transcription factor binding and transcription. The particular set of histone modifications in the chromatin around a gene is therefore sometimes referred to as the histone code. The histone code is an important component of epigenetic changes.

Interacting with RNA polymerase

We mentioned earlier that a transcription factor has two parts, each with its own function. We have spent some time discussing the DNA-binding domain that binds to specific nucleotide sequences, since this is how transcription factors are classified. But what about the other part of transcription factors, the interaction domain (or "*trans*-activation domain"), which interacts with RNAP II? (Remember that transcription factors are involved in the transcription of genes by RNAP II, and not by the other polymerases.) Somewhat less is known about the sequence and structure of these activation domains, but what is known so far suggests that amino acid sequences in these domains are less conserved than those in the DNA-binding domain.

One question that arises is how a transcription factor, whose binding site may be thousands of base pairs away from the core promoter, can physically interact with RNAP II, which is found at the core promoter. One key point is that, while a DNA sequence can be thought of as being two-dimensional, a chromosome needs to be considered in three dimensions. As shown in Figure 12.25, the chromatin may be folded or looped in such a way that sites that are distant from each other in the DNA sequence can be quite close together on the chromosome. This appears to be true for the interaction between transcription factors and RNAP II.

KEY POINT Looping of DNA on the chromosome can bring transcription factors bound at otherwise distant CRMs close to RNAP II at the core promoter.

In addition, the interaction between the transcription factor and RNAP II is indirect—it happens via an intermediary set of proteins known as the Mediator complex, shown in outline in Figure 12.5, but whose action is shown in more detail in Figure 12.25. The Mediator complex comprises at least 30 different evolutionarily conserved proteins that collectively assemble into a structure whose molecular weight is more than 1.2 million Da. (For comparison, the *E. coli* RNAP holoenzyme is 0.4 million Da.) The Mediator complex is essential for the transcription of nearly all (if not all) genes that are transcribed by RNAP II in eukaryotes and is part of the pre-initiation complex described in Section 12.2.

This huge macromolecular structure provides a host of opportunities for protein–protein interactions, both within the Mediator complex itself and between the

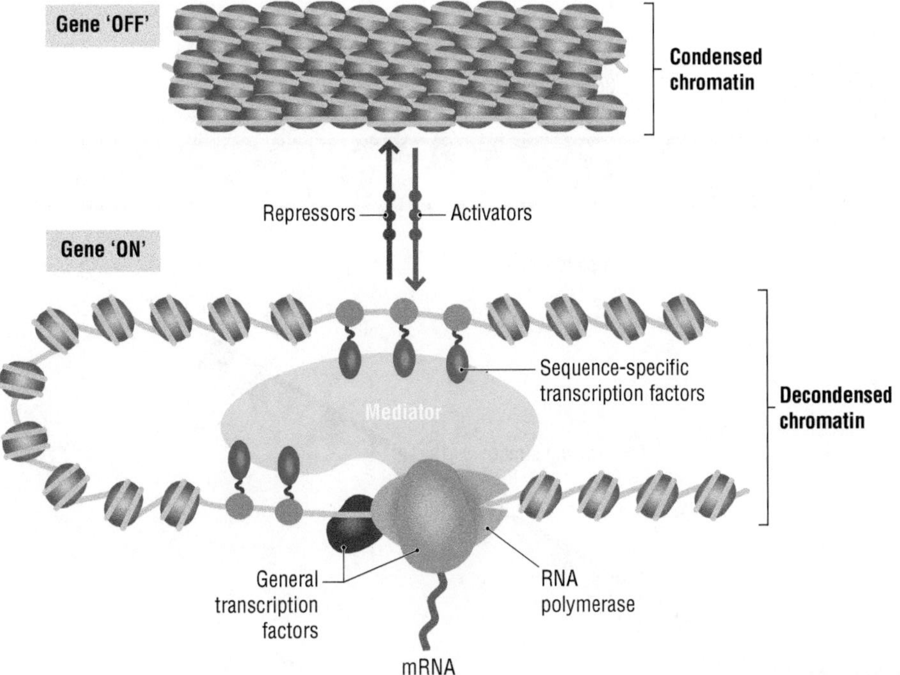

Figure 12.25 The role of the Mediator complex. Chromatin is looped to bring the transcription factors into interactions with the pre-initiation complex, with the Mediator complex forming the bridge.

Source: Reprinted by permission from Macmillan Publishers Ltd: Malik, S. and Roeder, R.G. The metazoan Mediator co-activator complex as an integrative hub for transcriptional regulation. *Nature Reviews Genetics* 11, 761–772 (November 2010) | doi:10.1038/nrg2901.

Mediator complex and the rest of the transcriptional machinery. Part of the Mediator complex interacts with RNAP II and the general transcription factors required for transcriptional initiation at the promoter. Another part of the Mediator complex interacts with transcription factors bound to their sequences at distant CRMs. Thus, the Mediator forms an essential bridge between other components of the pre-initiation complex for genes transcribed by RNAP II and the specific transcription factors that regulate transcription in time and space.

KEY POINT In eukaryotes, a huge multi-protein complex (the Mediator complex) mediates the interaction between transcription factors bound at CRMs and RNAP II at the core promoter.

12.5 Completing the transcript

We have spent a long time describing the initiation of transcription, although many additional details could still be added. One reason that we focus on initiation is that much of the regulation of gene expression occurs at this stage. After all, it is always simplest to stop a process before it starts, and, since transcription requires energy and cellular resources, initiation might be the most efficient control point. But it is far from the only control point, as we shall see.

Elongation includes pauses

Once transcription initiates, as shown in Figure 12.26(a), and RNAP has cleared the TSS, the polymerase often pauses. The duration and importance of this pause, which occurs after about 25–40 nucleotides have been transcribed, is still a matter for investigation. However, it definitely occurs in both bacteria and eukaryotes, and there is evidence that some transcripts are paused for some period of time before resuming transcription.

Following the pause, the transcript is elongated by adding one base at a time to the 3′ end, using the non-coding, or template, strand of DNA as the guide, as shown in Figure 12.26(b). The DNA unwinds ahead of the progressing polymerase, and the transcript grows. Multiple polymerases can work in series from the same unwound DNA template, so multiple copies of the transcript can be

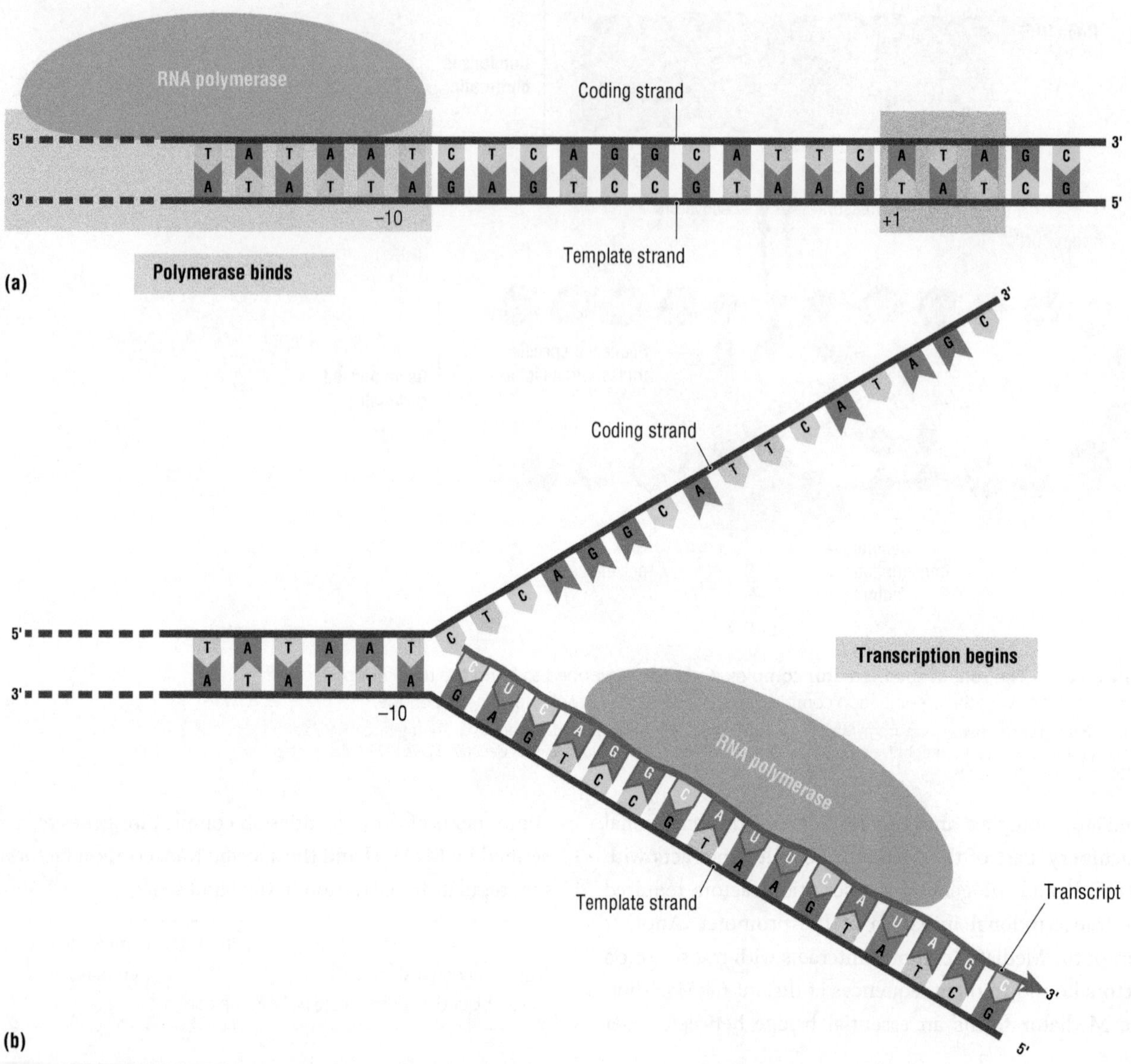

Figure 12.26 An overview of the initiation of transcription. (a) RNA polymerase binds at the core promoter; for simplicity, a bacterial gene is shown, but the process is similar for eukaryotes. (b) The transcript is elongated by adding one base at a time to the 3′ end, using the template strand of DNA as the guide. The DNA unwinds ahead of the progressing polymerase, and the transcript grows.

made at once, as can be seen in the electron micrograph in Figure 12.27.

KEY POINT RNAP pauses briefly after transcription has initiated, and then elongation of the transcript follows as RNAP reads from the unwound template strand of DNA and adds bases to the transcript in a 5′ to 3′ direction. Multiple polymerases can work in series from the same unwound template strand.

Because multiple copies of the same transcript are being made at the same time, and elongation is not completely continuous once it begins, the rate at which a gene is transcribed can only be estimated. One estimate is that transcription elongation in bacteria proceeds at a rate of about 50–60 nucleotides per second, while the rate in eukaryotes is about 15–20 nucleotides per second. But these are estimates, and more precise measurements are hard to obtain. Nonetheless, transcription takes time, particularly in eukaryotes. A bacterial gene that is 1200 base pairs long takes about 20 seconds to be transcribed; a human gene, which might be 40 kb in length, takes about half an hour to be transcribed. We don't usually consider the time that it takes for a biological process, such as transcription, to occur, but this can be critically important if, for example, the cell or the organism needs

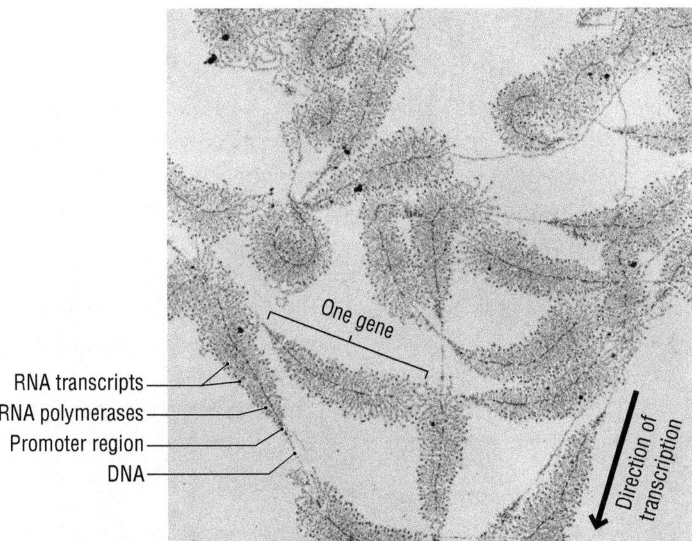

Figure 12.27 An electron micrograph showing chromosomes being transcribed, with multiple copies of the same gene being made simultaneously. Note how the transcripts are shorter near the promoter and grow as transcription proceeds, so this pattern identifies the direction of transcription.

Source: Reproduced from Griffiths et al. (2012). *Introduction to Genetic Analysis*, tenth edition, Macmillan. Image from O. L. Miller Jr. and Barbara A. Hamkalo.

to make rapid responses to environmental changes or divide quickly.

Transcriptional elongation—unlike DNA replication discussed in Chapter 4—has few, if any, proofreading functions, and transcriptional errors in base incorporation undoubtedly arise. However, because there are multiple copies of every transcript on average and transcripts are short-lived, such random base incorporation errors probably do not play a large evolutionary role. A few proofreading or surveillance functions do occur during transcription, but most are poorly understood or are found in specialized cases. Transcription elongation is the target for the toxins made by some of the most common poisonous mushrooms, however; this is described in Box 12.1.

BOX 12.1 *A Human Angle* How toadstools affect transcription

One of the most common and widespread mushrooms in Europe is also among the most toxic toadstools. *Amanita phalloides* (familiarly known as the death cap mushroom or toadstool) grows throughout European forests, typically at the base of hardwood trees such as oaks, and is frequently encountered by picnickers and hikers in late summer or early autumn. The export of hardwoods to other countries has also led to the spread of *A. phalloides* to North and South America, as well as Australia. Accidental ingestion of death cap mushrooms—which resemble many edible mushrooms such as the paddy straw mushroom common in Asia—is responsible for the majority of mushroom-related illnesses and deaths in the world. It is estimated that, without treatment, a person can die from ingesting a single death cap mushroom.

Most of the lethal toxicity of *A. phalloides* is caused by a peptide known as α-amanitin, whose structure consists of eight amino acids that form a ring with a molecular weight of about 900 Da. α-amanitin moves readily across cell membranes and binds at very low doses (1 μg/ml) to the largest subunit of eukaryotic RNAP II. Once bound, it inhibits transcription by preventing RNAP II from translocating along the DNA molecule; in other words, transcriptional elongation is blocked. Since RNAP II is responsible for the transcription of all protein-coding genes, the effects of blocking its activity are expected to be catastrophic to the organism.

In humans, the primary effect of ingesting *A. phalloides* is seen in the liver and kidneys, particularly with the death of hepatocytes or liver cells. These are the organs most responsible for filtering substances we ingest; liver cells are among the most actively dividing cells in adults, which also would contribute to the toxicity from blocking transcriptional elongation. Symptoms usually do not appear until more than a day has passed, when the toxin has been absorbed from the stomach into the bloodstream; at this point, treatment options are limited. By the fourth

BOX 12.1 Continued

or fifth day after ingestion, symptoms of liver and kidney failure become evident, and about 15% of those diagnosed with *Amanita* poisoning die with 10 days. Most of those who survive the poisoning have liver damage, and many require a liver transplant. The delayed appearance of the symptoms makes *Amanita* poisoning difficult to diagnose and treat; this delay probably arises, at least in part, because α-amanitin blocks the **elongation** of new transcripts, so that existing transcripts and polypeptides are not affected.

The specificity of α-amanitin for RNAP II provided an important research tool in the early days of studying transcription. RNAP I, responsible for the transcription of the ribosomal RNA genes, is insensitive to α-amanitin, and RNAP III is less sensitive than RNAP II to its effects. This differential sensitivity to the toxin was early evidence that eukaryotes have different RNA polymerases that transcribe different classes of genes.

Humans have been aware of the effects of toadstools or poisonous mushrooms for a very long time; it is thought that Roman emperor Claudius (in 54 AD) and the Hapsburg emperor Charles VI (in 1740) died from eating *A. phalloides*. With such a long history of human experience comes a similarly long history of folklore about toadstools, much of which is unreliable. However, one possible treatment of *Amanita* poisoning also comes from folk remedies; silymarin, an extract from the seeds of the milk thistle bush (*Silybum marianum*), which has been used for centuries to treat liver ailments, has shown promise as a therapeutic approach to *Amanita* poisoning. Its possible modes of action, as well as any side effects, are not yet thoroughly understood, but it may act by competing with the toxin for uptake into hepatocytes.

FIND OUT MORE

Santi, L., Maggioli, C., Mastroroberto, M., *et al.* (2012) Acute liver failure caused by *Amanita phalloides* poisoning. *Int J Hepatol* 2012: 487480. Available at: http://dx.doi.org/10.1155/2012/487480 [accessed 3 August 2016]

Stickel, F., Egerer, G., and Seitz, H. K. (1999) Hepatotoxicity of botanicals. *Public Health Nutr* 3:113–24

This box was written with assistance from Alison Meneely Wilson, RN, who has been involved with the treatment of families poisoned by *A. phalloides*.

Termination includes several different mechanisms

In most eukaryotes, each gene is usually transcribed separately. In most bacteria and a few eukaryotes, however, multiple adjacent genes arranged in an **operon** can be co-transcribed. We diagram such a structure in Figure 12.28. (Operons will be discussed more fully in Chapter 14 because of their importance in the regulation of bacterial gene expression.) The genes in an operon use the same promoter, located upstream of the gene at the 5′ end. Irrespective of the nature of the transcript, like all good things, transcription eventually terminates. In bacteria, two different processes are used for transcriptional termination, depending on the transcript—Rho-dependent

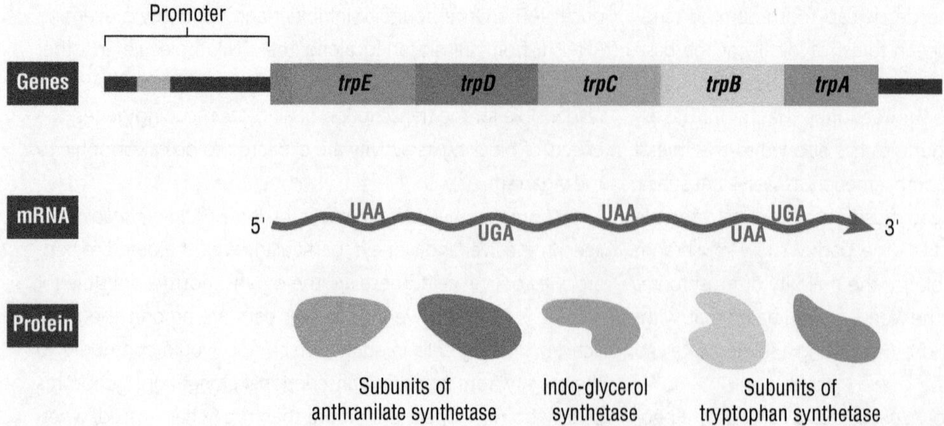

Figure 12.28 An operon structure in *Escherichia coli*. The tryptophan (*trp*) operon is shown as an example. There is a single promoter, so all five of the genes are transcribed on the same transcript. When this is translated, the stop codons (in red) result in five different polypeptides. Operons are described in more detail in Chapter 14.

and Rho-independent termination. These two processes are summarized in Figure 12.29.

For Rho-dependent termination, shown in Figure 12.29(a), an RNA-binding protein called Rho binds to a 70–100 base region of the transcript (called the *rut* sequence) that is rich in cytosines but lacks intra-strand secondary structure. Rho has helicase activity, so it actively unwinds the RNA at the termination site from the template strand of DNA and releases the transcript. Rho-dependent termination of this type occurs for about half the genes in *E. coli*.

In Rho-independent termination, also known as intrinsic termination, a sequence downstream of the transcript determines the site of termination, as shown in Figure 12.29(b). The termination signal is a sequence 7–20 bases in length that is rich in G and C; the complementary sequence is encoded in inverse orientation immediately downstream on the same strand. When this region of the gene is transcribed, the inverted repeat sequences in the RNA can form intra-strand G:C base pairs, resulting in a structure known as a stem–loop or a hairpin. The hairpin

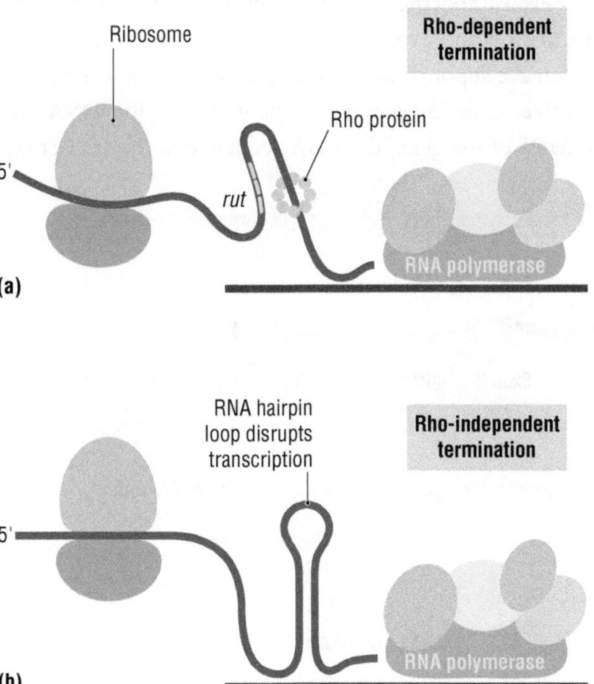

Figure 12.29 Transcription termination in *Escherichia coli* occurs by two different mechanisms. (a) In Rho-dependent termination, a helicase protein known as Rho binds to the transcript and interacts with the *rut* sequence downstream of the stop codon. This dissociates the transcript from the DNA. (b) In Rho-independent termination, the transcript has an inverted repeat structure downstream of the stop codon. When the repeats form a hairpin structure through intra-strand base pairing, transcription terminates.

structure blocks RNAP from carrying out further transcription; instead, RNAP dissociates from the template, and transcription terminates.

Termination in eukaryotes is somewhat analogous but is not so clearly understood or precise. For transcripts made by RNAP I (that is, the ribosomal RNA genes), an RNA helicase protein with some functional similarity to the *E. coli* Rho protein appears to be involved. Transcripts made by RNAP II—which constitutes most transcripts—often extend hundreds or thousands of bases beyond the end of the coding region of the gene. Two factors associated with the C-terminal domain of RNAP II recognize the poly-adenylation signal in the transcript (discussed later) and recruit other proteins to continue and complete RNA processing. In both of these cases, known as factor-dependent termination, the elongation complex remains associated with the DNA, so it is not entirely clear how termination finally occurs.

In contrast to these cases, transcripts made by RNAP III (transfer RNAs and some other short structural RNAs) terminate by a factor-independent process. A stretch of thymines in the DNA downstream of the gene on the coding strand (or uracils in the RNA) is recognized by RNAP III, and termination ensues, but precisely how this occurs is not clear; it does not appear to involve intra-strand secondary structures, however.

Elongation and termination are necessary and essential steps in transcription, of course, but they do not seem to be the principal sites of regulation. Examples can be found in eukaryotes in which pausing seems to be an important regulatory step, and there are examples in bacteria in which termination is an important regulatory step, but most of the regulation of transcription and most of the evolutionary changes associated with transcriptional regulation are found with the processes that lead to initiation.

KEY POINT In bacteria, transcript elongation can be terminated by Rho protein, which recognizes a specific sequence motif in the transcript, or termination may follow the formation of a hairpin in the secondary structure of the transcript. Somewhat analogous, but distinct, mechanisms can be found in eukaryotes. While important, transcript elongation and termination are not common steps regulating a gene's expression.

The initial transcript is processed into its final version

The transcript made by RNAP is still not in its final form, and some important additional processes occur during "post-production"—that is, as part of the transition from a precursor mRNA (pre-mRNA) to a mature mRNA, shown in Figure 12.30. For example, a cap consisting of a modified guanine nucleotide with a methyl group attached at the 7′ position, called 7′-methylguanine (m^7G), is attached to the 5′ of an mRNA in eukaryotes. This 5′ cap is important for transporting the transcript out of the nucleus and for the interaction with the ribosome during the process of translation, as well as other functions. No 5′ cap is added to transcripts in bacteria, which, of course, have no requirement for nuclear transport; likewise, transcription and translation are often coupled in bacteria, rather than occurring as discrete processes separated in time and space.

At the other end of the eukaryotic transcript, a stretch of 100 or more adenines is added to the 3′ end to form a structure known as the polyA tail. The sequence AAUAAA (or a very similar variant) encoded on the RNA is the signal for factors associated with RNAP II to cleave the transcript and add a string of consecutive As. The polyA tail confers stability to the transcript; transcripts that lack a polyA tail or that have a very short polyA tail are rapidly degraded in the cell. Curiously, some transcripts in bacteria are also poly-adenylated, but, in these cases, the polyA appears to target the transcript for degradation, rather than conferring stability. Thus, the enzymatic machinery for adding a polyA tail to transcripts is evolutionarily ancient, but the impact of poly-adenylation is not the same in bacteria and eukaryotes.

KEY POINT The transition from a precursor mRNA to a mature mRNA involves additional steps such as the addition of a modified guanine "cap" to the 5′ end of the transcript in eukaryotes and poly-adenylation at the 3′ end.

Splicing the exons together

While capping and poly-adenylation are important processes, the most significant post-transcriptional step in the transition from pre-mRNA to mature mRNA in eukaryotes is undoubtedly splicing—the removal and rejoining of internal sequences in the transcript, as shown in Figure 12.30. The sequences that are retained in the final transcript (and so will be expressed in the amino acid sequence) are referred to as the **exons**. The sequences that are removed—that is, those that are intervening—are called **introns**. (Introns were originally called intervening sequences, so the connection between "intervening" and "intron" may be a useful memory aid.)

Because splicing is such an important step in both the regulation and evolution of gene expression, there is an extensive supporting literature, which we will only summarize. One important principle is that the DNA sequence of the gene, the RNA sequence of the transcript,

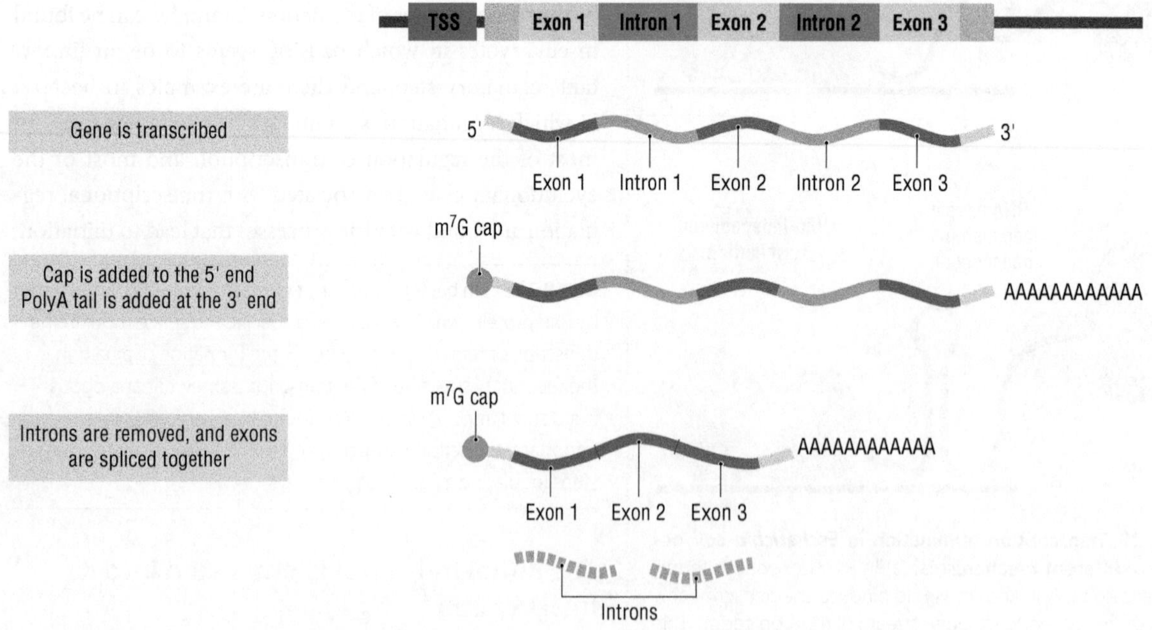

Figure 12.30 The modifications that occur to mRNA after transcription in eukaryotes. The pre-mRNA transcript made by RNAP II is modified by the addition of 7-methylguanine as a cap at the 5′ end and the polyA tail at the 3′ end. The introns, shown as the dashed lines, are removed from the transcript, and exons spliced together as the precursor mRNA is processed into a mature mRNA.

and the amino acid sequence of the polypeptide are in the same order; in the jargon of molecular biology, the gene and the polypeptide are co-linear. In other words, the linear order of the exons in the coding sequence of a gene is the same as their order in the final mRNA, and thus the sequence of amino acids if the mRNA is translated—exons aren't "scrambled" or taken out of order during RNA processing. So, despite the presence of introns, the co-linearity of an mRNA and its corresponding polypeptide is retained.

In some cases, particularly in viral genomes where genome space is at a premium, the same stretch of DNA can be translated in different reading frames, which allows for multiple polypeptides to be produced from the same mRNA. Examples of eukaryotic exons that are translated in more than one reading frame are known and are not rare, but, as we will see shortly, the existence of alternative reading frames is not the principal way in which multiple polypeptides are generated from the same transcript. However, the different exons comprising a single gene may be in different reading frames—but once they are spliced together, the reading frame of the mature transcript will be continuous.

But how are the sequences to be removed during splicing recognized? In the cell, the macromolecular complex of proteins and RNA molecules that carries out the splicing, referred to as the **spliceosome**, uses protein–RNA interactions and base complementarity between its RNA components and sequence signatures in the intron (not all of which are known) for accurate splicing. Some of these signatures can be used to detect the introns in a sequence computationally, so that splicing patterns can be predicted from the gene sequence. However, these predictions are not 100% accurate, so the most reliable way for an investigator to identify which sequences in the DNA comprise introns is to compare the sequences of the mature gene transcript with genomic DNA. Interestingly, and very importantly, however, exons can be included or omitted from different versions of the transcript to generate what are called **alternative splice variants**. Most genes in metazoans (and more than 85% of genes in mammals) have alternative splice variants.

Many types of alternative splicing are known, as summarized in Figure 12.31. On average, a human gene has more than six different splicing variants, four of which affect the amino acid sequence and two of which affect the 5′ UTR or the 3′ UTR. These are probably underestimates, since they only represent the splice variants that have been found in the tissues and conditions tested so far. Of course, there is no "average" number of splice variants, and some genes have only one or two transcripts while others have dozens.

As a consequence of alternative splicing, the human genome, with roughly 22,000 protein-coding genes, is estimated to encode more than 75,000 different proteins. There is a rough correlation between the number of splice variants and what we think of as the biological complexity of organisms. For example, although the estimated number of genes encoding proteins in mammals (approximately 22,000) and nematodes (approximately 21,000) is very similar, the number of proteins is thought to be 2- or 3-fold greater in mammals as a result of more extensive alternative splicing.

KEY POINT The order of exons in a gene is maintained in the corresponding transcript. However, during splicing, not only are the introns removed, but some exons may or may not also be removed, producing a variety of alternative transcripts from the same gene that differ in exactly which combination of exons have been included.

Alternative splicing is so widespread that nearly every biological process in a metazoan offers examples in which it is important. A particularly interesting example of alternative splicing that affects an easily observed biological process involves the *doublesex* (*dsx*) gene in *D. melanogaster*, summarized in Figure 12.32. The *dsx* gene encodes a DNA-binding protein that directly regulates the expression of many genes involved in sexual dimorphism, including genitalia, mating behavior, pigmentation, and morphology; its evolution is the topic of Box 7.5. The differences between males and females in flies depend, to a large extent, on which splice variant of the *dsx* gene is expressed. Some of the exons are common to the transcript found in both sexes, whereas some are sex-specific. In XX flies (normally females), exons 1–3 are spliced to exon 4, and no additional exons are included. In XY flies (normally males), exon 4 is skipped altogether, and the first three exons, which are common to transcripts in both sexes, are spliced to exon 5. Exon 5 is, in turn, spliced to exon 6. Thus, the Dsx proteins in males and females have the amino acids at the N-terminus (from exons 1, 2, and 3) in common but have completely different amino acid sequences at the C-terminus.

At first thought, based on what we have said about DNA-binding proteins and realizing that *dsx* encodes a transcription factor, it is tempting to imagine that the exons that are alternatively spliced encode the part of the Dsx protein that binds to DNA; this hypothesis predicts

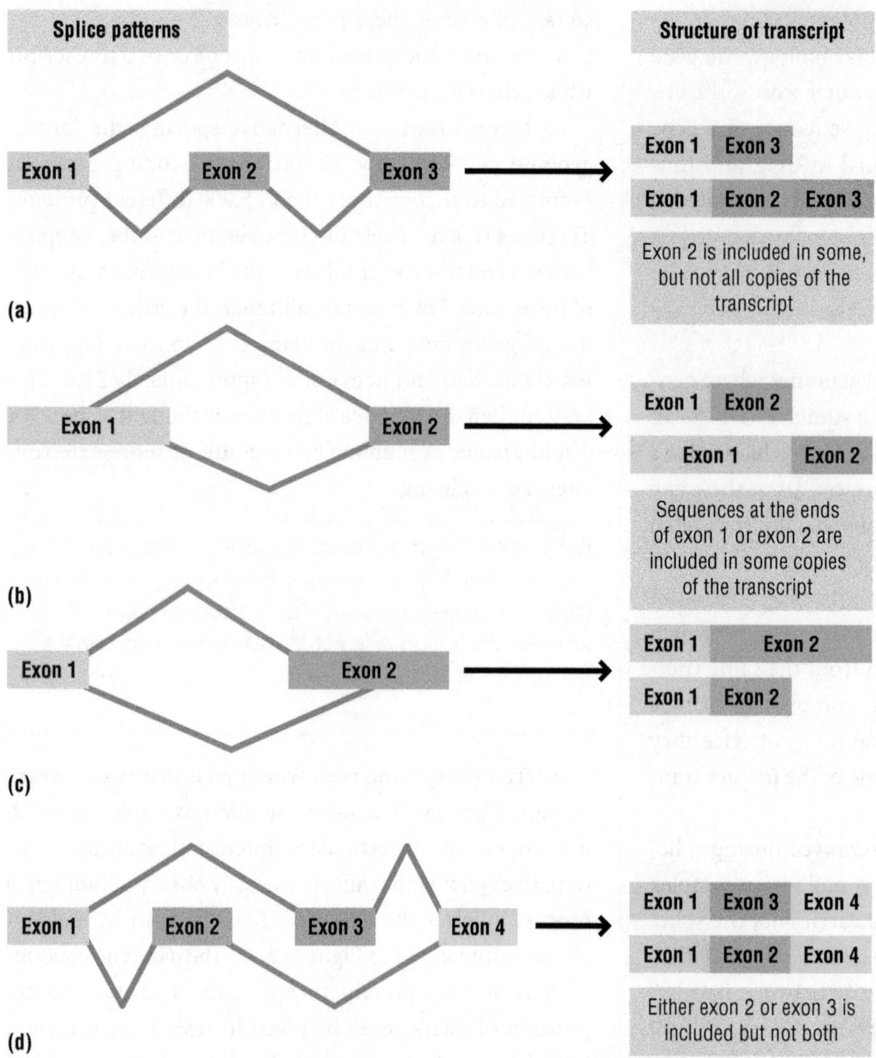

Figure 12.31 Alternative splicing. Many transcripts in metazoans have more than one pattern of splicing. The alternative splices can occur within a cell or, more commonly, in cells of different tissues or at different times. Exons can be skipped, as in (a), or the splicing can occur at different locations in the preceding or following exons, as in (b) and (c). In addition, exons can be mutually exclusive, so that transcripts that include the exon 3 never include exon 2, and vice versa, as shown in (d). Each of these scenarios results in changes in the subsequent amino acid sequence as well, so that multiple amino acid sequences can arise from the same gene.

that males and females would encode proteins with different binding sequence specificities because of how exons 4, 5, and 6 are spliced. However, that seemingly good guess proves not to be true. The exons of the *dsx* transcripts that are the same in males and females actually encode the DNA-binding domain. The exons that are alternatively spliced encode parts of the Dsx protein that is the interaction domain. This serves as a reminder of how biological processes are shaped by evolution, so they might not work in the way we envision by design principles.

When genomes of multicellular organisms are compared, it is apparent that alternative splicing is an important evolutionary mechanism for creating a wider array of proteins and for diversifying the functions encoded in the genome. Evolutionary change depends on molecular differences and variation. In Chapter 3, we considered molecular variation in terms of the DNA sequence itself; this is certainly an important and well-studied source of molecular variation. Much of the variation in the DNA sequence arises from the errors that occur during replication and recombination. Alternative splicing can be

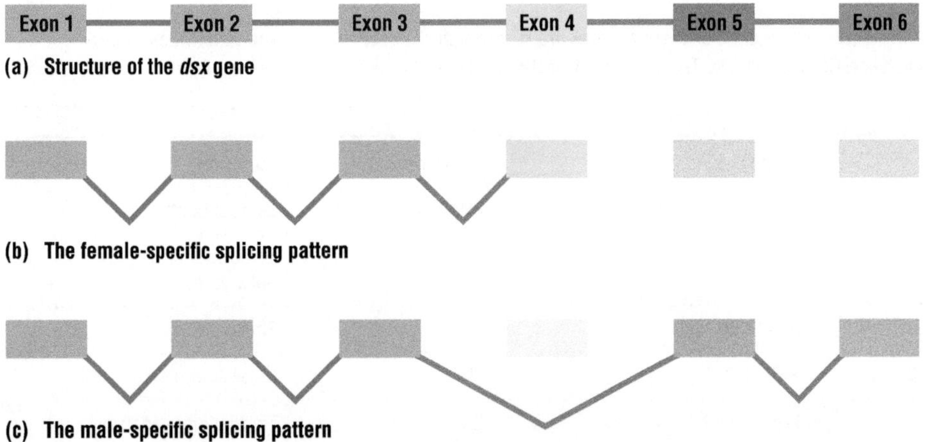

Figure 12.32 Alternative splicing of *doublesex (dsx)* in *Drosophila* regulates sex determination. (a) The *dsx* gene in *Drosophila* has six exons. (b) In XX embryos (normally females), exon 4 is spliced to exons 1, 2, and 3, and exons 5 and 6 are not used. (c) In XY embryos (normally males), exon 4 is skipped, and exons 5 and 6 are spliced to exons 1, 2, and 3.

thought of in much the same way—the process of splicing is precise but not perfect, and molecular variation arises as an inevitable consequence of the imprecision. A few of these splice variants result in a protein that has a selective advantage or a slightly different function, whereas others have little or no functional effects. Molecular variation arising from splicing is subject to the same evolutionary forces that characterize other parts of biology.

12.6 From transcript to function

Transcription occurs by similar processes for all genes in all organisms. RNAP binds to DNA at a specific site, uses one DNA strand as a template, and makes an RNA copy by adding to the 3′ OH of the preceding nucleotide. Different types of RNAPs vary in their subunit composition, which affects the promoter sequence with which they interact, but the process of transcription is the same once the polymerase is bound. However, biologically important differences occur at the level of pre-initiation—that is, in getting the genes transcribed under the appropriate conditions—and (in eukaryotes, at least) at the level of RNA processing, particularly splicing. All of these processes yield a large collection of different RNA molecules—hundreds or thousands in a bacterial cell and up to tens of thousands in a eukaryotic cell. The functions of these RNAs vary widely, and much remains to be learned about their functions.

It was not so long ago that RNA was neatly divided into only three functional categories, as summarized in Table 12.3. Messenger RNAs (mRNAs) are the transcripts that are translated into polypeptide sequences. (Translation is the topic of Chapter 13.) The site at which this happens is the ribosome, a complex comprising both proteins and ribosomal RNA (rRNA). A key molecule in the process of translation is transfer RNA (tRNA).

These categories of RNA molecules are still helpful and appropriate, but they vastly understate the functional complexity of the RNAs in a cell. With genome-based methods of transcript analysis, we now know that there are several more types of RNA in the cell, even if we do not yet know all of their functions, how to classify them, or whether all types of transcripts even **have** a function. The structures of some of the best-characterized RNA types are illustrated in Figure 12.33. Many of the different structures and functions associated with different types of RNAs correlate to different steps in the overall process of gene expression, as summarized in Figure 12.34.

Coding and non-coding RNA

The function that we understand the best for RNA, and the one that most of us think of, is that mRNA is translated into a polypeptide; this is part of the Central Dogma that we discussed in Chapter 2 and is one of the primary functions depicted in Figure 12.34. In reality, less than 5%

Table 12.3 Different types of RNA molecules in bacteria and eukaryotes are summarized. For the rRNAs, the size (measured in sedimentation units, S) of the RNA in bacteria is listed first. The eukaryotic polymerase that makes each type is shown. Bacteria have a single RNAP

Type of RNA		Size (nt)	Transcribed by	Function
rRNA	5S	120	RNAP IIII	Large subunit
	5.8S	156		Large subunit in eukaryotes
	16S/18S	1542/1869	RNAP I	Small subunit
	23S/28S	2906/5070	RNAP I	Large subunit
tRNA		75–95	RNAP III	Translation—"adaptor"
mRNA		Varies from approximately 400 to more than 1,000,000	RNAP II	Information for amino acid sequence
microRNA (miRNA)		70, cleaved to 22 nt	RNAP II	Down-regulates gene expression
Long non-coding RNA (long ncRNA)		>200	RNAP II	Unknown and very diverse

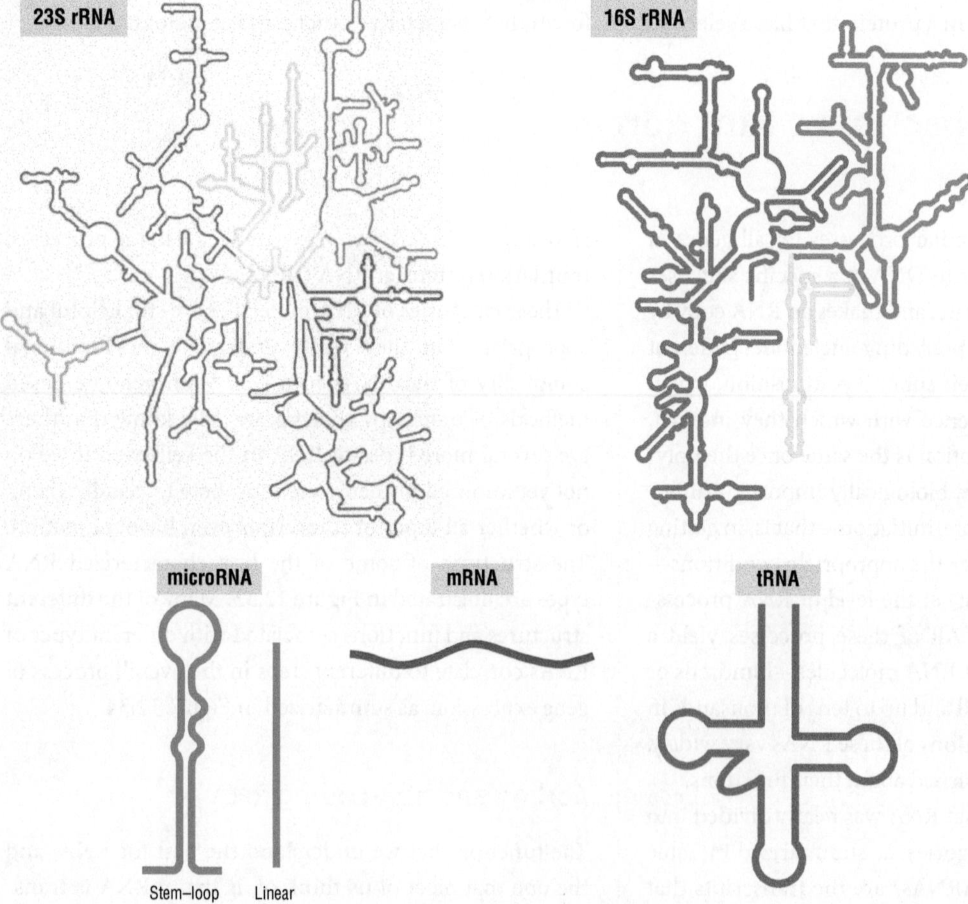

Figure 12.33 The structure of several different types of RNA. In most, if not all, of the cases of non-coding RNA, the folded structure arising from intra-strand base pairing is important for the proper function of the RNA molecule.

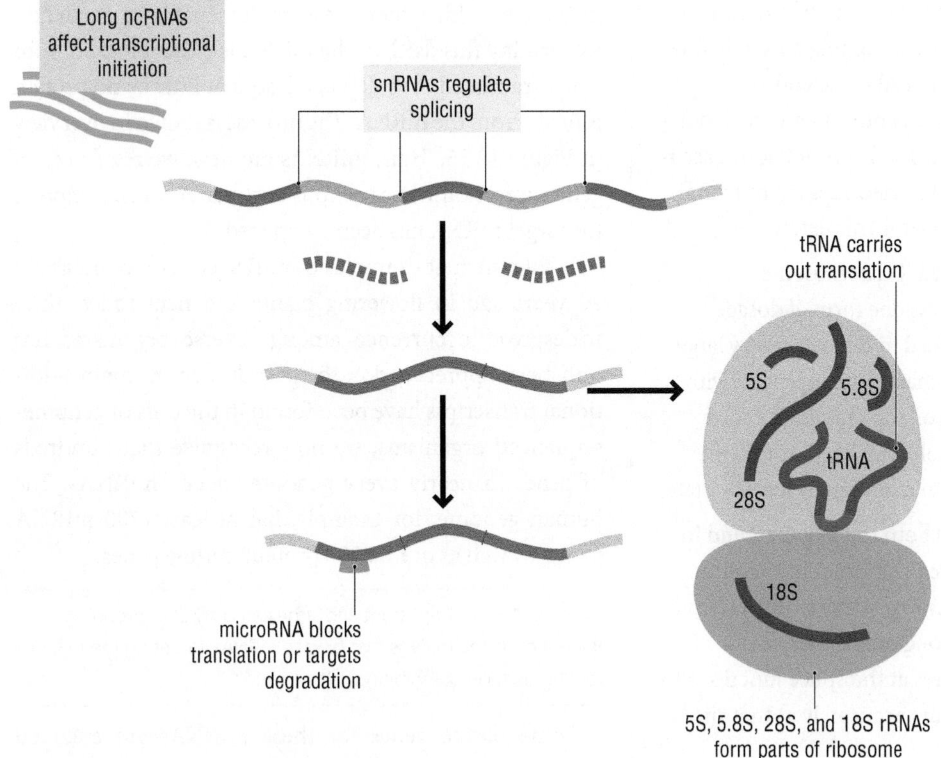
Long ncRNAs
affect transcriptional
initiation
snRNAs regulate
splicing
tRNA carries
out translation
5S
5.8S
tRNA
28S
18S
microRNA blocks
translation or targets
degradation
5S, 5.8S, 28S, and 18S rRNAs
form parts of ribosome

Figure 12.34 The roles of different non-coding RNAs in the regulation of an mRNA. Most of the functional focus is on an mRNA, which will be translated into a polypeptide. The small nuclear RNAs (snRNAs) are involved in its splicing. The rRNAs and tRNAs are important for this translation. The microRNA blocks translation of mRNA or targets it for degradation before it is translated. The long ncRNAs are a diverse group, but at least some have been implicated in the structure of chromatin, which affects transcriptional initiation.

Source: E. coli RNA from H. Noller, Center for Molecular Biology of RNA, University of Santa Cruz http://rna.ucsc.edu/rnacenter/ribosome_images.html

of the total RNA that is produced by a cell (and probably even less than 1% in a cell) is translated into a polypeptide sequence; most of the RNA extracted in bulk from cells is rRNA and, to a lesser extent, tRNA.

Both rRNA and tRNA are critical components of translation. But there are other classes of RNA that are not translated and not involved in translation. These are usually called **non-coding RNA** or functional RNA, since the final functional product of the gene is the RNA molecule itself. We will call them non-coding RNA, meaning that they do not encode amino acid sequences. However, whenever objects are grouped together based on what they are not—that is, as "non-coding" RNA—the distinctions about what they actually are may be lost, so it should be immediately acknowledged that many different functional types of RNA could be grouped together when we refer to them as non-coding RNA.

You will recall that, in eukaryotes, rRNA genes are transcribed by RNAP I, and tRNA genes are transcribed by RNAP III, so these genes encoding these RNAs are transcribed by their own polymerases. Nearly all of the other non-coding RNAs are the products of transcription by RNAP II, as is mRNA.

KEY POINT Genome-based methods of transcript analysis reveal that, in addition to the three well-established categories of RNA (rRNAs, tRNAs, and the protein-coding mRNAs transcribed by RNAP I, RNAP III, and RNAP II, respectively), there are many other non-coding RNAs also transcribed by RNAP II.

There is a conceptual analogy to be drawn between proteins and non-coding RNAs. Just as it is impossible to list all of the functions that proteins play in the cell, it is impossible to list all of the functions that non-coding RNAs play in the eukaryotic cell. Here are a few biological processes that you have already learned about, for which a non-coding RNA plays a critical role; we did not always point out at the time that the process involved a non-coding RNA molecule.

- In Section 4.3, during our description of DNA replication, the problems associated with replication at the ends of linear chromosomes were discussed. This problem has been solved in eukaryotes by an enzyme known as telomerase, which extends the end of the lagging strand before replication. Telomerase is a ribonucleoprotein, consisting of both a protein component and a non-coding RNA component. The RNA in telomerase is a transcript of about 450 nucleotides (in humans) that serves as the template for this extension.

- In Chapter 7, we discussed X chromosome inactivation in mammals as one form of dosage compensation. As described in Box 7.6, a very large non-coding RNA (more than 25 kb in length, with the exact size depending on the mammal) called Xist appears to form a molecular cage around the inactive X chromosome to maintain its inactive state.
- RNA splicing, discussed both in Chapter 2 and in Section 12.5, involves a collection of non-coding small RNA molecules known as the snRNAs (small nuclear RNAs) that, among other roles, form complementary structures at the splice junctions to ensure that splicing occurs accurately at that site.

These examples are well known, but there are dozens, if not hundreds, of other functions involved in a host of other fundamental biological processes. From the perspective of regulating gene expression, we will focus first on one category, called **microRNAs (miRNAs)**.

microRNA regulation has given rise to RNA interference

miRNAs are 22-nucleotide RNAs that form a double-stranded RNA (dsRNA) hybrid with a specific target mRNA, to which they are complementary in sequence. By forming this dsRNA, the mRNA is either blocked from being translated or is degraded, so that little or no protein is made from the mRNA. The process is shown in overview in Figure 12.35. Thus, miRNAs are negative regulators of gene expression whose actions occur after transcription of the target mRNA has been completed.

While the first examples of miRNAs were found about 20 years ago in flowering plants and nematodes, their widespread occurrence among diverse organisms has only been appreciated in the past decade. As many additional transcripts have been found in the cells of genome-sequenced organisms, we now recognize that hundreds of genes in nearly every genome encode miRNAs. The human genome, for example, has at least 1700 miRNA genes, as well as our 22,000 protein-coding genes.

KEY POINT An important and large category of non-coding RNAs are the microRNAs that bind to mRNAs, targeting the mRNA for degradation or inhibiting its translation.

Some of the genes for these miRNAs are encoded within other genes, on the opposite strand of an intron, for example. Others map to their own location and can be mapped just like the protein-coding genes we discussed in Chapter 9; in fact, some of the first ones discovered in nematodes were mapped based on mutant phenotypes several years before their identity as miRNA genes was realized. As will be discussed in more detail in Chapter 14, some miRNAs regulate the expression of one or two other genes, while others regulate the expression of dozens, or even hundreds, of other genes. Indeed, most protein-coding genes, particularly in mammals,

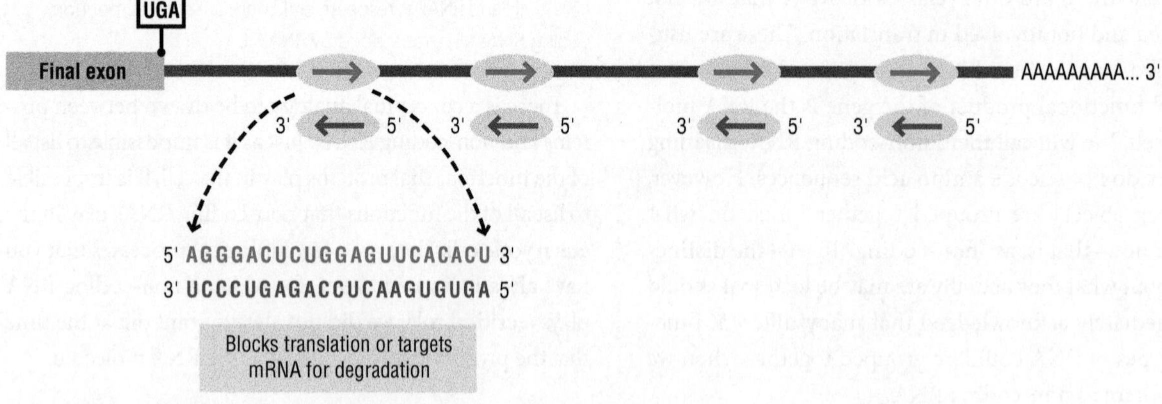

Figure 12.35 miRNAs and gene regulation. miRNAs form a double-stranded RNA hybrid with their complementary sequences in the target mRNA. This double-stranded hybrid blocks the translation of the mRNA or targets it for degradation. In animals, the target sequences for miRNAs are usually located in the 3′ UTR of the mRNA, as shown here. This location is more variable in plants. miRNA regulation is considered in more detail in Chapter 14.

are regulated by one or more miRNAs. In animals, the naturally occurring target sequence for the miRNA is frequently located in the 3′ UTR of the mRNA, whereas the target sequence in plants is often internal, within an exon, for example.

Whenever we are tempted to think that molecular biologists, geneticists, and biochemists have identified all of the important molecular features of the cell, we need to recall miRNAs. Their existence was probably undetected for so many years because an miRNA is so small. Their role was probably not appreciated because the mechanisms of action are different from how proteins work or how proteins and DNA interact. miRNAs and mRNAs form double-stranded RNA molecules, and, for many decades, we did not think of dsRNA as being a normal component of gene regulation. Nonetheless, their function depends on one of the most fundamental unifying principles of molecular biology, the base pair—only this time, the base pair involves two RNA molecules.

In a sense, miRNAs serve as a counterpart to transcription factors; we summarize this in Figure 12.36 with a gene from *Drosophila* known as *snail*. Most transcription factors activate transcription and so turn on or turn up gene expression (though there are well-documented examples of some transcription factors acting as repressors of transcription). For *snail*, the relevant transcription factor proteins include Twist, Tinman, Krüppel, and others. miRNAs work after transcription has been completed but before or at the point of translation, and turn down or turn off gene expression. For *snail*, at least three different miRNAs down-regulate its expression: *mir-1*, *mir-983*, and *mir-1900*. Most genes are regulated by one or more transcription factors, and some are regulated by dozens of them. Likewise, most genes in metazoans are also regulated by one or more miRNA molecules. We will consider this in more depth in Chapter 14.

In addition to their biological importance, the cellular machinery by which miRNAs act has given rise to one of the most powerful experimental tools for genetic analysis, known as **RNA interference** or **RNAi**. RNAi is used to decrease or eliminate the expression of a gene and is broadly applicable to most eukaryotic organisms. This method is described in more detail in Tool Box 12.2. In order to perform an RNAi experiment, the investigator introduces a dsRNA corresponding to any part of the transcript of the gene into the organism or its cells. This introduced RNA, known as a small interfering RNA (siRNA), forms a dsRNA hybrid with the mRNA of the target gene. The dsRNA arising from the interaction between the siRNA and its target mRNA is processed by the cell, resulting in the degradation of the mRNA transcribed from the target gene. RNAi has allowed genetic analysis to be carried out for organisms for which traditional genetics involving mutant alleles cannot be done easily, as will be described in more detail in Chapter 15.

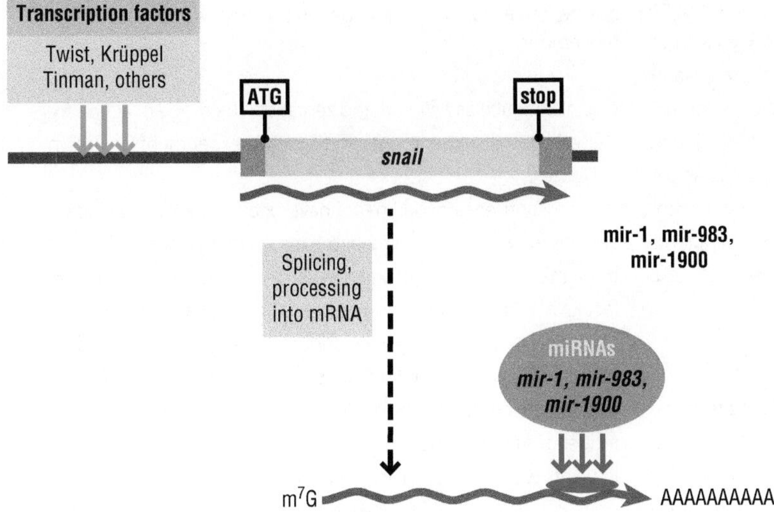

Figure 12.36 The regulation of the expression of the *snail* gene in *Drosophila*. The roles of transcription factors and miRNAs in regulating the expression of this gene are used as an example. A more global perspective will be presented in Chapter 14. Transcription factors, such Twist, Tinman, and Krüppel, activate the transcription of *snail*, while several miRNAs are known to reduce or turn off the expression of the mature *snail* mRNA.

TOOL BOX 12.2 microRNAs as an experimental tool: RNAi

In Section 12.6, we describe how miRNAs regulate gene expression by knocking down the expression of specific genes. This knockdown can occur either by targeting an mRNA for degradation or by blocking its translation, or both. Somewhat independently of the identification of miRNAs as a natural cellular process, a laboratory technique called **RNA interference (RNAi)** was developed that relies on the same biological mechanisms as miRNAs. RNAi and miRNAs are two faces of the same phenomenon, one being a laboratory technique and the other occurring in nature. The connection between miRNAs and RNAi is both mechanistic and personal. Not only do miRNAs and RNAi work by some of the same mechanisms, but the experiments were often done by the same laboratories. The significance of this work was recognized by the awarding of the 2006 Nobel Prize to Andrew Fire and Craig Mello, who were among those who developed the connection between miRNAs and RNAi.

Geneticists are always interested in knocking out a gene or knocking down its expression in order to understand its function. As sequences of individual genes and complete genomes have become available, this knockdown can be specifically targeted to individual genes using RNAi. In a typical RNAi experiment, dsRNA corresponding to a portion of the transcript from a gene is introduced into the cell or organism. The dsRNA specifically blocks or reduces the expression of that gene, thereby producing a phenotype, as illustrated in Figure A.

For example, when dsRNA that corresponds in sequence to part of the mRNA from the myosin heavy chain gene is introduced into *Caenorhabditis elegans*, it causes the worms to be paralyzed. This phenotype resembles that seen for actual mutations in the myosin heavy chain gene, although it is usually not quite as severe. But (with some exceptions) the effects of RNAi are not heritable because it is the mRNA that has been disrupted, rather than the DNA. The DNA sequence remains intact to be passed to the next generation. Strictly speaking, RNAi-treated individuals should not be called mutants (as that implies heritable change), although they are commonly referred to in that way. A non-heritable mutant phenotype induced by environmental agents like RNAi is more accurately, but less commonly, referred to as a **phenocopy**.

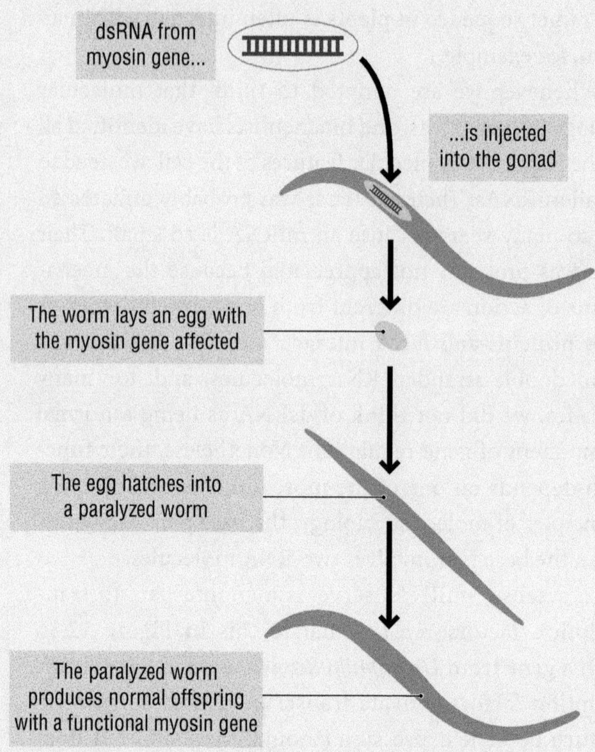

Figure A An overview of RNA interference. RNAi is done by introducing dsRNA corresponding to part of the gene. In this drawing, part of the coding region of the myosin heavy chain gene is microinjected into worms. The dsRNA blocks the expression of the normal gene and thus mimics the mutant phenotype of the gene. In this example, the dsRNA blocks expression of the myosin gene, resulting in a paralyzed worm. The effect is usually not heritable, so, in the next generation, the worm has normal movement.

The antecedents of RNAi lie in other experiments

RNAi is rooted in an older approach known as **antisense** technology. Antisense techniques involve producing a single-stranded RNA, or another modified nucleic acid, that is complementary to the mRNA or sense strand. The single-stranded RNA is predicted to form a double-stranded hybrid with the mRNA and thereby block its translation or splicing. Indeed, antisense techniques using modified nucleic acids known as morpholinos are still used in this way to block expression of mRNAs in a variety of organisms such as *Xenopus* and zebrafish. The mode of action for these antisense experiments and for RNAi experiments is shown in Figure B.

Many antisense experiments have been done over the years, and it is now clear that some of the results that were first thought to be mediated by single-stranded antisense nucleotides were in fact RNAi experiments arising from small amounts of dsRNA in the preparations of antisense RNA. Indeed, careful follow-up experiments in *C. elegans* showed that dsRNA gave stronger effects than the single-stranded antisense RNA. Interestingly, dsRNA-mediated reductions in gene expression had also been previously reported in plants where the process is referred to as **post-transcriptional gene silencing** (PTGS). Despite the different terminology, the general mechanism for gene silencing by dsRNA appears to be the same in plants and animals. Thus, although the experimental history of RNAi is recent, the evolutionary history of the response to foreign dsRNA pre-dates the divergence of plants and animals.

TOOL BOX 12.2 Continued

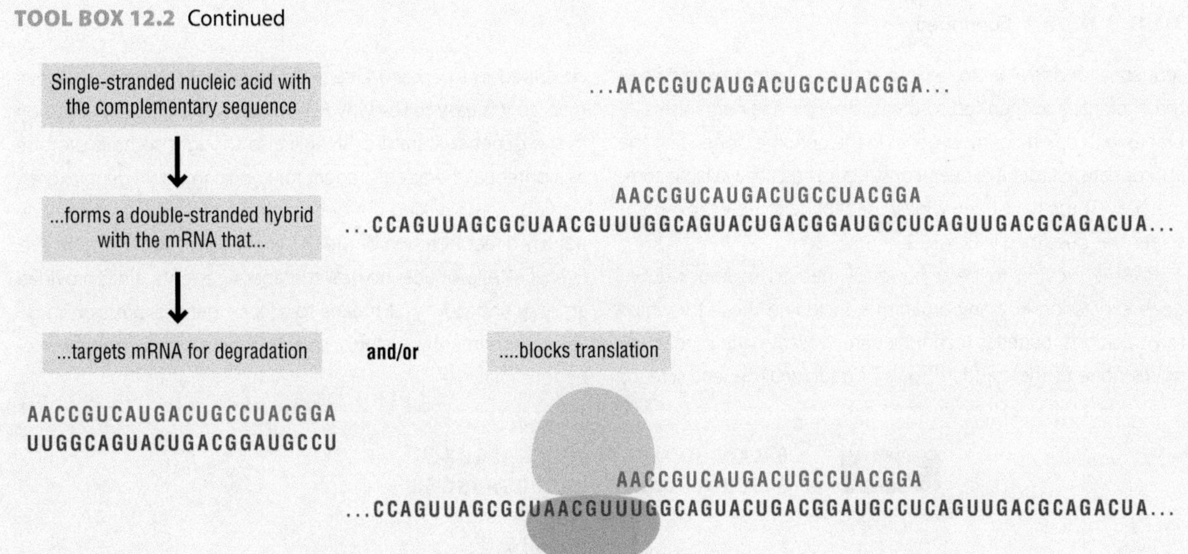

Figure B Modes of actions of dsRNA, antisense RNA, and miRNAs. A single-stranded RNA is generated which makes a dsRNA hybrid with the target mRNA. This dsRNA may target the mRNA for degradation, block its translation, or both.

It has been widely postulated that the RNAi response is an ancient type of immune response, protecting eukaryotic organisms from RNA viruses and the effects of transposable elements, both of which involve dsRNA molecules.

As discussed in Section 12.6, it is now understood that small non-coding RNAs are important naturally occurring regulators of gene expression for nearly all eukaryotes, and certainly all multicellular organisms. The normal cellular response to these **miRNAs** involves many of the same biochemical steps exploited in RNAi experiments. Since the cellular machinery used in miRNA regulation is evolutionarily conserved and found in nearly all eukaryotes, RNAi can also be done in nearly all eukaryotes. In a few organisms, such as *C. elegans*, miRNAs and also the RNAi effect can diffuse between cells, so that the gene knockdown occurs throughout the organism. In most organisms, the effect of RNAi occurs only in the cells into which it was introduced, so the experiments are usually done with tissue culture cells.

The mechanism of RNAi relies on normal cellular functions

How then does the dsRNA produce its silencing effect? Just like miRNAs, the dsRNA molecule for RNAi knocks down the expression of the mRNA, and thus the function of the gene, by two different mechanisms—blocked translation or targeted mRNA degradation. In either case, the introduced dsRNA has one strand identical in sequence to the mRNA (the sense strand), whereas the other is the complementary sequence (the antisense strand). The introduced dsRNA is cleaved into fragments by the enzyme Dicer. As its name implies, Dicer cuts the dsRNA into smaller fragments of 19–24 nucleotides in length.

The RNA fragments produced by Dicer have a characteristic structure—a central dsRNA region of about 19 nucleotides with a 2- or 3-base overhang at each end, as depicted in Figure C. These fragments are called short interfering RNAs (siRNAs), and their appearance in the process is common to all organisms. The sense strand of the siRNA is then degraded, leaving the antisense strand, which can form the dsRNA hybrid with the target mRNA. This dsRNA hybrid may block translation of the mRNA, as antisense molecules, morpholinos, and miRNAs also do.

Alternatively, the antisense strand can become incorporated into a protein complex referred to as RISC (for RNA-induced silencing complex); the presence of the siRNA in this complex targets it to the corresponding complementary mRNA. RISC degrades the corresponding mRNA without degrading the antisense strand of the siRNA, so the same complex can target and degrade many copies of an mRNA; this targeted degradation is also a mechanism used by miRNAs. Because the same RISC is used repeatedly, the pool of mRNA from a gene can be degraded and no protein is produced. Both translational blocks and degradation likely occur for nearly all RNAi experiments.

Introducing the dsRNA into the cells

All RNAi experiments require that dsRNA is introduced into cells or the organism, and the methods to introduce dsRNA depend on the organism. In many cases, the dsRNA is directly injected into the cells or the organism, which is the method indicated in Figure A. In others, such as mammalian cells, the sequence for the dsRNA is cloned into a viral vector that is introduced into the cells where it is expressed; the vector can also be based on a miRNA gene, rather than a virus. In *C. elegans*, which eat *Escherichia coli*, the

TOOL BOX 12.2 Continued

sequence for dsRNA is cloned into an *E. coli* plasmid, expressed in the bacteria, and then fed to the worm. For many organisms, libraries with dsRNA for most genes in the genome cloned into the appropriate vector have been constructed and are available commercially. These have been widely used for genome-wide mutant screens, as described in Box 15.2.

RNAi is proving to be a powerful method to knock down gene expression in many organisms, including those for which few traditional genetics techniques are available. The process requires none of the standard tools of traditional genetic screens described in Chapter 14, such as mutations or even a genetic map, so it is easy to see why RNAi approaches are now common in the genetics literature. RNAi is also thought to hold promise as a potential therapeutic agent for knocking down gene expression, such as blocking the genes involved in macular degeneration by direct injection of the corresponding siRNA into the eye. Even if RNAi produces no new therapeutic agents, it has provided an easy and powerful means to silence gene expression in diverse experimental systems, opening them up to the power of genetic analysis.

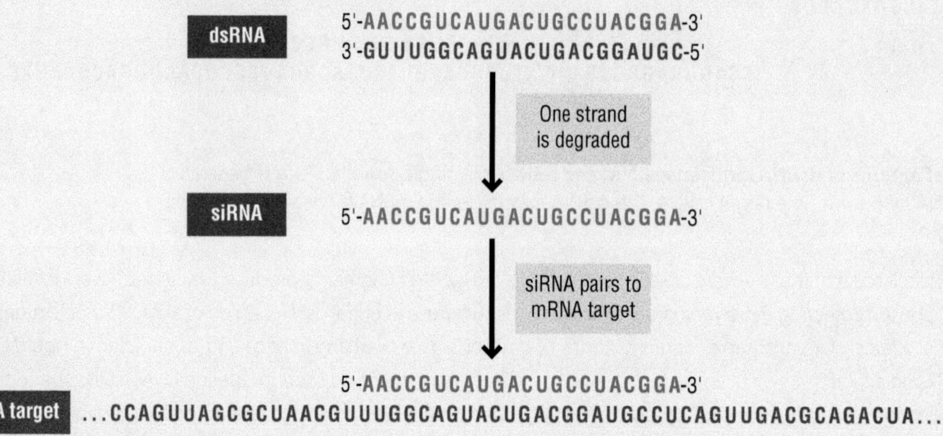

Figure C Short interfering RNAs. When the dsRNA is introduced, one strand is degraded, leaving only a single strand. This single strand is known as a short interfering RNA or siRNA. The siRNA, about 22 nucleotides in length, forms the double-stranded hybrid with the target mRNA.

Long non-coding RNA

While the roles of miRNAs in gene regulation are well established, the roles of another class of non-coding RNAs, known as the **long non-coding RNAs (long ncRNAs)**, are much less understood. In fact, the class is defined by what its members are **not**. They are transcribed by RNAP II, so they are not rRNAs or tRNAs. (Their nucleotide sequences also have no similarity to rRNA or tRNA.) They are at least 200 nucleotides long, so they are not miRNAs. They do not encode the amino acid sequence of a protein, so they are not mRNAs. Since long ncRNAs are currently defined so broadly, there could be many different types that are being considered together. The distinct functions exhibited by the many members of the long ncRNA class promise to be one of the fascinating topics in molecular genetics and genomics in the upcoming years.

Long ncRNAs were recognized from the analysis of transcription patterns in metazoan genomes. Hundreds or thousands more transcripts are being made in the cell than expected or than could be accounted for by conventional ideas. Current evidence suggests that the human genome encodes more than 10,000 different genes that encode a long ncRNA. Like mRNAs, they are transcribed by RNAP II, and they are often capped at the 5′ end, spliced, and poly-adenylated at the 3′ end. Unlike mRNAs, however, they are not translated or found associated with ribosomes. In fact, some are found entirely in the nucleus, some in the cytoplasm, and some in both locations. They are generally shorter than a typical mRNA, usually hundreds or a few thousand bases in length, rather than thousands or tens of thousands of bases. It is widely believed, and shown in some cases, that long ncRNAs are involved in regulating processes in the cell, including chromatin structure, DNA replication, transcription initiation, and others. Compared to mRNAs and miRNAs, long ncRNAs are more highly variable in their expression patterns and are much less evolutionarily conserved between related species. More than 20% of the currently known long ncRNAs in the human genome are found in only one cell type, suggesting that many more could be found as the transcription profiles of more cell types are analyzed.

Let's try to put this transcription pattern into context. For example, while about 2% of the bases in the human genome are transcribed and then translated into protein, more than 65% of the bases are transcribed and then **not** translated into protein. The portion that we have studied the longest is the translated portion—the mRNA. We simply did not know about the other 65% of the transcripts until current methods for transcriptional analysis were developed, and we have much to learn. Some of these long ncRNAs may be present in only a few copies, may be made in only a few cells, and may be very short-lived. Others may be expressed much more widely and have stable transcripts. For most of these transcripts, we have no idea about their functions, and it is possible that some (or even many) of the ones that have been detected do not have a function in the cell. But because these transcripts are made and are part of the molecular variation in the cell, they could also be subjected to the same set of selective pressures as any other molecular variant.

In Chapter 13, we will turn our attention to translation, the process by which the small, but critical, percentage of cellular transcripts—the mRNA—is translated into proteins. Many transcripts are made, but few are chosen to make proteins.

CHAPTER CAPSULE

- Genes are modular in nature with:
  - Distinct coding regions
  - Promoters that bind RNA polymerase and associated factors
  - *Cis*-regulatory modules (CRMs) that bind additional transcription factors involved in directing temporal and spatial regulation of expression.
- This modularity has important implications for evolution and for the development of many genetic tools.
- Gene expression begins with transcription into an RNA molecule, which involves the following steps in both bacteria and eukaryotes:
  - Recruitment of RNA polymerase to the core promoter upstream of the gene
  - Initiation and elongation of the transcript by adding nucleotides to the 3′ end
  - Termination and processing of the transcript.
- Additional steps occur during transcription of eukaryotic genes, particularly for a gene that encodes a protein. These steps include:
  - Establishing an appropriate chromatin context for transcription to occur, including particular histone modifications
  - Binding of the transcription factors to sites at the CRM, which helps to recruit RNA polymerase to the core promoter
  - Splicing of the transcript, most likely with multiple isoforms
  - microRNA interactions to reduce expression.

- Genes need to be expressed in the correct cells at the correct levels at the appropriate times.
- Consequently, transcription is highly regulated, and any of these steps could be a key point for regulation when the transcription patterns of different cells are compared.
- Most of the regulation of bacterial genes occurs in the recruitment of RNA polymerase to the promoter.
- The regulation of eukaryotic genes occurs primarily at the pre-initiation stage, in which the chromatin context is set up to allow transcription factors to bind and specific transcription factors bind to CRMs, but also during splicing and through interactions with microRNAs.
- These regulatory steps that are found by examining transcriptional differences between cells are also the ones found by comparing evolutionary changes in gene expression.

STUDY QUESTIONS

Concepts and Definitions

12.1 In Chapter 4, we discussed DNA replication and mentioned that this could be accomplished *in vitro* with a surprisingly small list of ingredients. DNA can also be transcribed *in vitro*. Make a list of ingredients that would be necessary and sufficient to transcribe a gene *in vitro*.

12.2 Diagram a gene, labeling the upstream and downstream regions, the promoter, and the coding region. Include RNA polymerase at the correct location on the DNA.

12.3 Describe the function of a transcription factor such as ALX1. What is meant by the "consensus binding sequence" of a transcription factor?

12.4 A 90-residue peptide is found to be encoded by a 600-nucleotide gene that contains no introns.

a. What are possible explanations for the small size of the peptide, relative to the gene?

b. The mRNA transcribed from the gene is 680 nucleotides long. The RNA contains 30% adenines, a much higher proportion than predicted from the gene. What explains the large size of the mRNA relative to the gene?

12.5 List the basic steps involved in transcription in bacteria. What are some of the other steps found in eukaryotes that are not present in bacteria?

12.6 What does the term "homeotic transformation" mean? How does the fact that the *Hox* genes encode proteins that bind DNA help to explain homeotic transformations?

12.7 Figure Q12.1 shows DNA wrapped around a set of proteins in a structure known as a nucleosome.

a. What are these proteins called, and what are the four types of covalent modifications that occur on them that can alter their function?

b. Indicate on the figure which area of the proteins is subject to these modifications.

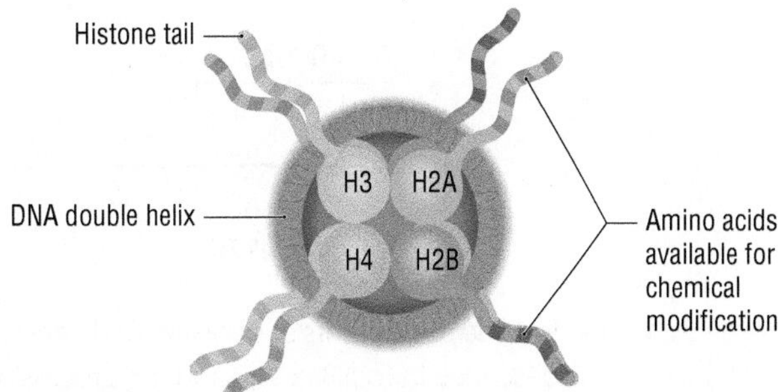

Figure Q12.1

c. In your own words, describe what is meant by H3K27Ac.

d. Very briefly (two or three sentences), explain how modifications to these proteins can alter gene expression levels.

12.8 What is meant by the term "epigenetic," and give an example of an epigenetic modification to DNA.

Beyond the Concepts

12.9 Here is a cartoon schematic of a eukaryotic gene (Figure Q12.2):

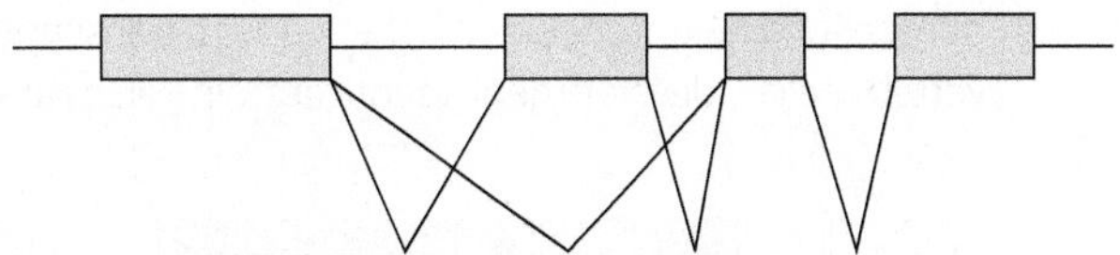

Figure Q12.2

a. How many different mRNA products are indicated as being produced by the process of alternative splicing?

b. Draw one additional potential splicing pattern for this gene.

c. Why is alternative splicing not seen in bacterial genes?

12.10 Protein-coding genes comprise only about 2% of the human genome, but a much higher percentage of the human genome is transcribed. Explain this briefly.

12.11 A group of investigators has cloned a gene encoding a mouse enzyme and attempted to express it in yeast. The gene is functional in yeast but expressed at very low levels. How might the investigators alter their clone to achieve higher expression levels in yeast without altering the sequence of the protein produced?

12.12 A protein that is used to make the bristles on the surface of a caterpillar is found at high levels in one cell, but there are no detectable levels of the protein in a neighboring cell in the epidermis of the caterpillar. This is found to be true in all caterpillars of this same species. Provide a possible explanation for this difference.

12.13 In Chapter 1, we introduced the *AXL1* gene, which may be involved in the evolution of different beak shapes in Darwin's finches. When the DNA sequence of this gene was compared between finches with blunt beaks and pointed beaks, eight sequence differences were found within or near the gene. Figure Q12.3 shows how these can be divided into three groupings: four differences upstream of the coding sequence (Group 1), three differences in the coding sequence (Group 2) and one difference downstream of the gene (Group 3). For each group, briefly speculate how sequence differences: (i) might or might not affect the expression of *ALX1* and (ii) could be affecting the phenotype of the finch's beak.

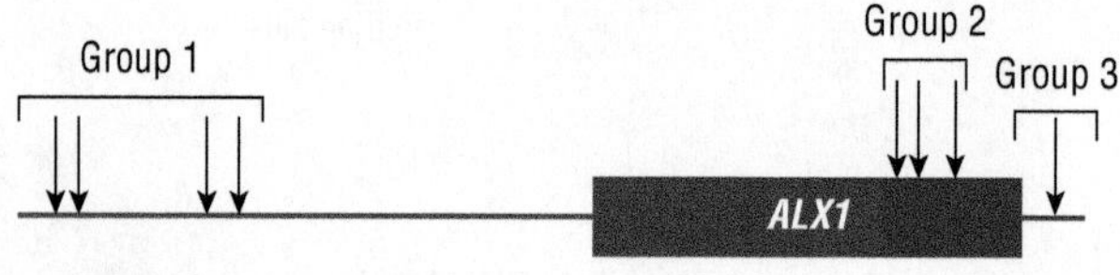

Figure Q12.3

Looking back **12.14** Provide three points of similarity or difference between the CRISPR/Cas system as described in Tool Box 11.3 and the process of RNAi.

Applying the Concepts

12.15 A protein is known to be present in a specific subset of neurons in the brain of the mouse. However, when the gene that codes for this protein is mutated in such a way that it is rendered non-functional and the protein is no longer produced in these neurons, the scientists studying this gene are unable to detect any change in the appearance or function of these neurons. Provide two possible explanations for this result.

12.16 When the influenza virus (the virus that causes "flu") genome enters the nucleus of a human host cell, it cleaves off the 5′ end of host RNAs as they are being made. These "snatched caps" are then attached to the 5′ ends of viral RNAs. Why might "cap snatching" be necessary for this RNA virus to complete its life cycle in a human cell?

12.17 Figure Q12.4 shows the sequence of the C-terminal three-quarters of the *ALX1* gene for a number of different species. *ALX1* encodes a transcriptional regulator that is important in vertebrate jaw development. Identical amino acids are shaded purple.

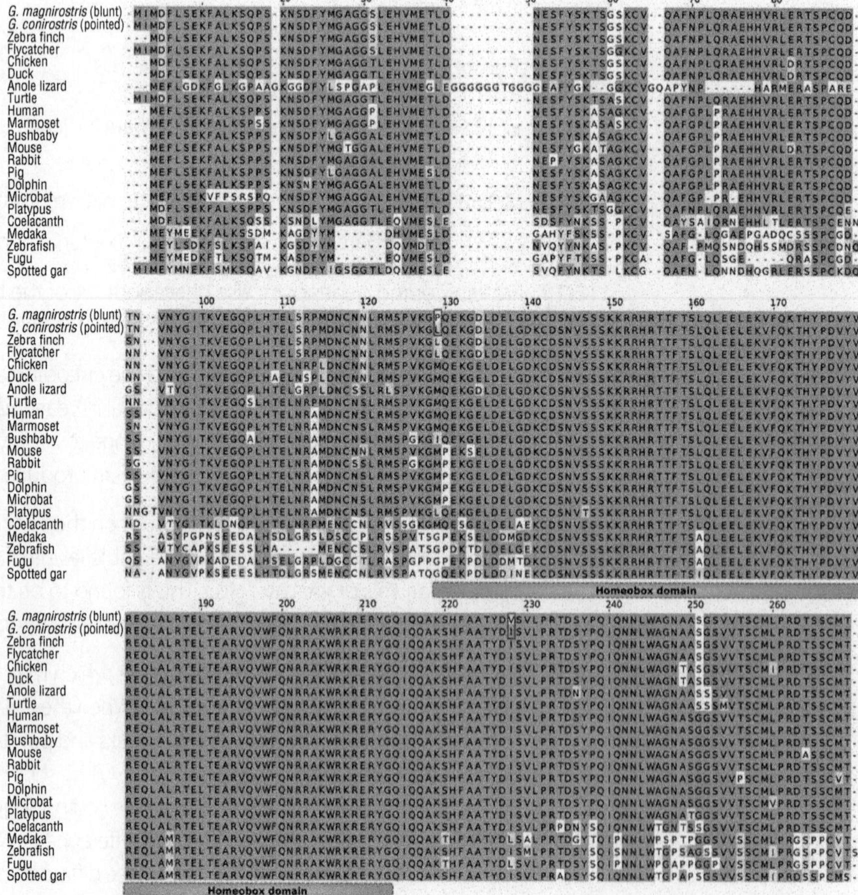

Figure Q12.4

a. Why is the variation across different species not distributed uniformly throughout the molecule?

b. What is your best explanation for the very limited variation within the marked homeobox domain?

c. Specific residues at positions 129 and 228 (boxed) are associated with beak shape in *Geospiza* finches on the Galapagos Islands. Choose one of these sites, and propose a hypothesis for why a change results in a significant alteration in beak shape.

12.18 The expression patterns of two genes—gene *1* and gene *2*—are being monitored using reporter genes in an animal. The structures of the genes and the reporter gene constructs are shown in Figure Q12.5.

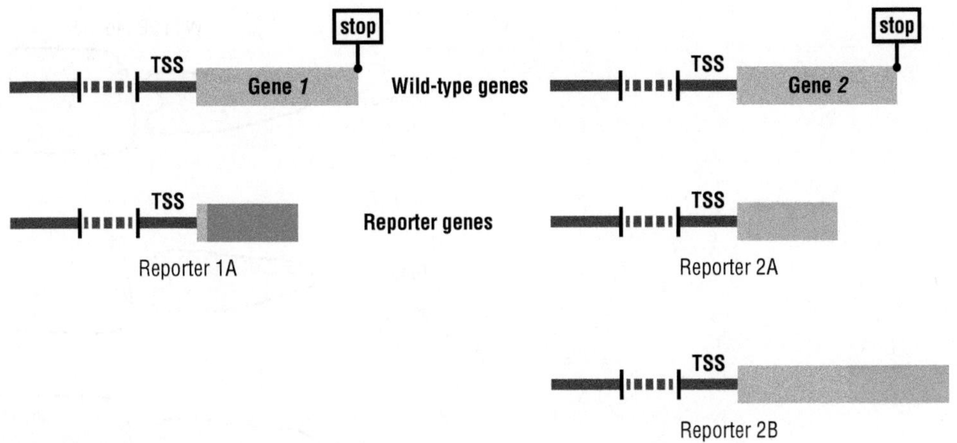

Figure Q12.5

For gene *1*, one reporter construct (reporter 1A) is made, in which the coding region of the gene is replaced by the reporter gene mCherry, which encodes a protein that fluoresces red. For gene *2*, two reporter constructs (reporters 2A and 2B) are made. In reporter 2A, the coding region of the region is replaced with the green fluorescent protein (*GFP*) gene, while, in reporter 2B, *GFP* is placed downstream of the coding region of the gene, but the coding region of the gene is left in place. In each case, the core promoter regions and the upstream *cis*-regulatory module for each gene are used to drive the expression of the reporter.

a. What aspect of gene expression is being monitored by the A reporter constructs reporter 1A and reporter 2A? Explain your answer.

b. What aspect of gene expression is being monitored by reporter construct 2B? Explain your answer.

Looking ahead

c. For which of the reporter constructs will it be important to keep the reading frame intact, and why? Why is the reading frame not important for the other construct or constructs?

d. What are some of the advantages of using GFP and mCherry for these experiments, rather than β-galactosidase or luciferase?

12.19 The reporter gene constructs 1A and 2A from Figure Q12.5 are introduced into an organism, where they are expressed. Two cells are shown in Figure Q12.6, with the expression of the two reporter gene constructs shown beneath the cells with no reporter.

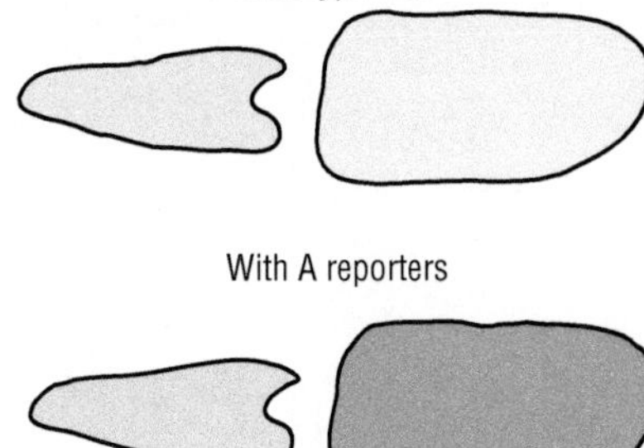

Figure Q12.6

a. Explain this pattern of expression.

b. Suppose that the investigator made a reporter construct in which the upstream *cis*-regulatory module from gene *2* in Figure Q12.5 is replaced with the upstream *cis*-regulatory module from gene *1*; the core promoter and the rest of reporter 2A are left intact. Which cell or cells will express this new hybrid reporter? Explain your answer.

Looking ahead **12.20** The expression pattern over time for reporter construct 2B from Figure Q12.5 over time is shown in Figure Q12.7. Three time points are indicated: time = 0, time = 5, and time = 10; the units of time are not important for the question. Explain this pattern of expression.

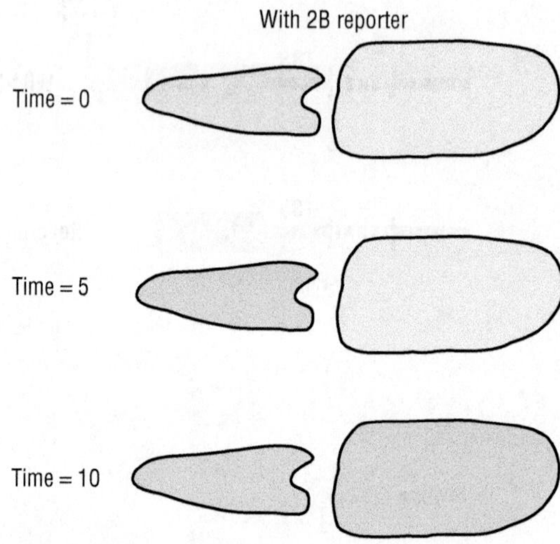

Figure Q12.7

12.21 The reporter constructs from Study Question 12.19 are used for a series of other experiments.

a. It is likely that the expression of genes *1* and *2* is also controlled by microRNAs, but this is not known. Which, if any, of the reporter constructs could be used to monitor regulation by a microRNA? Explain your answer.

b. Reporter 1A is introduced into a mutant that is homozygous for a deletion of gene *1*. Do you expect that this would complement the mutant or restore wild-type function? Explain your answer. Which, if either, of reporters 2A and 2B might restore wild-type function to a mutant in gene *2*?

Looking ahead
Challenging

c. The expression pattern of reporters 2A and 2B was monitored in a mutant that is homozygous for a deletion of gene *1*. Neither reporter 2A nor reporter 2B shows expression. What does this suggest about the function of gene *1*?

CHAPTER 13

Translation: From Nucleic Acids to Amino Acids

IN A NUTSHELL

In keeping with the Central Dogma, some of the RNA transcripts made from the DNA sequence of a gene are translated into the amino acid sequence of polypeptides. While it is now recognized that only a subset of transcripts are translated, most cellular functions in all organisms are carried out by proteins, the products of translation. The process of translation is highly similar in all living organisms and probably arose only once, with relatively small modifications during evolution.

13.1 Translation from nucleic acids to polypeptides

The ability to perform large-scale analysis of genomes, such as finding all of the RNA transcripts made in a cell, has fundamentally changed our perspective on gene expression. Genome-based analysis of transcription has led to the realization that the genomes of nearly all organisms (but particularly metazoans, including humans) are "pervasively transcribed," that is, there are vastly more transcripts than we expected. For many genes—and for **most** genes in multicellular organisms—transcription into RNA yields the final product. The RNA product is termed "non-coding" because it does not code for a polypeptide. It is not yet clear what pervasive transcription means on a functional or an evolutionary level, but surely some of that understanding will arise soon.

The concept of a functional, but non-coding, RNA is not entirely novel. Some examples of functional RNAs, such as the transfer RNAs (tRNAs) and ribosomal RNAs (rRNAs) necessary for translation, have been known for more than 50 years; others, such as the functional RNAs associated with eukaryotic telomeres, splicing, and X chromosome inactivation, were found more recently but have still been known for decades. In these examples and others, the RNA molecule itself performs a function, often arising from its ability to base pair with DNA or other RNA molecules, including itself, or to serve as a binding site for a protein. While the existence of non-coding RNAs was not a new finding, the realization that there are a vast number of non-coding RNAs was surprising.

The Central Dogma, as discussed in Chapter 2, originally did not include such non-coding RNAs. Rather, it posits that the functional molecules in the cell are polypeptides and that a gene encodes the capacity to make a polypeptide—one gene, one polypeptide, as discussed in Chapter 3 and illustrated in Figure 13.1. This remains the dominant, and largely appropriate, framework for thinking about the functions of genes, the processes of genetics, and the genetic basis for phenotypes. Polypeptides carry out most of the functions that we understand best, and nearly all of the functions that we have discussed in other chapters are performed by polypeptides.

Polypeptides are polymers of amino acids. Genes and transcripts of the gene are made of nucleic acids, polymers of the nucleotide bases. Thus, as proposed by the Central Dogma, a process must exist to translate the information from the string of nucleotides in nucleic acids into the string of amino acids in polypeptides. The mechanism of this translation is the topic for this chapter.

More attention has been paid to transcription than translation in the biological literature, but this does not mean that translation is less interesting or less important. One reason that translation has received less attention in recent years is that the process is very similar in all organisms. Unlike transcription, there appears to be very little variation in translation from cell to cell, from individual to individual, and even from species to species. When biologists think about the molecular processes that generate different phenotypes, the immediate thought is that these will involve transcriptional differences; in eukaryotes, the differences may possibly involve splicing and microRNA regulation as well. Translational changes are further down the list when the origins of biological diversity or gene regulation are being considered.

This is not to suggest that translational differences are unimportant or unknown, or that everything about translation has been figured out already. In fact, we will discuss an example of translational regulation in Section 13.3. It is simply that, when we hear the music of living organisms, transcription is the lead guitar or the soloist, while translation is the rhythm section. It works so reliably that we don't always appreciate its essential contributions to the survival of cells and organisms.

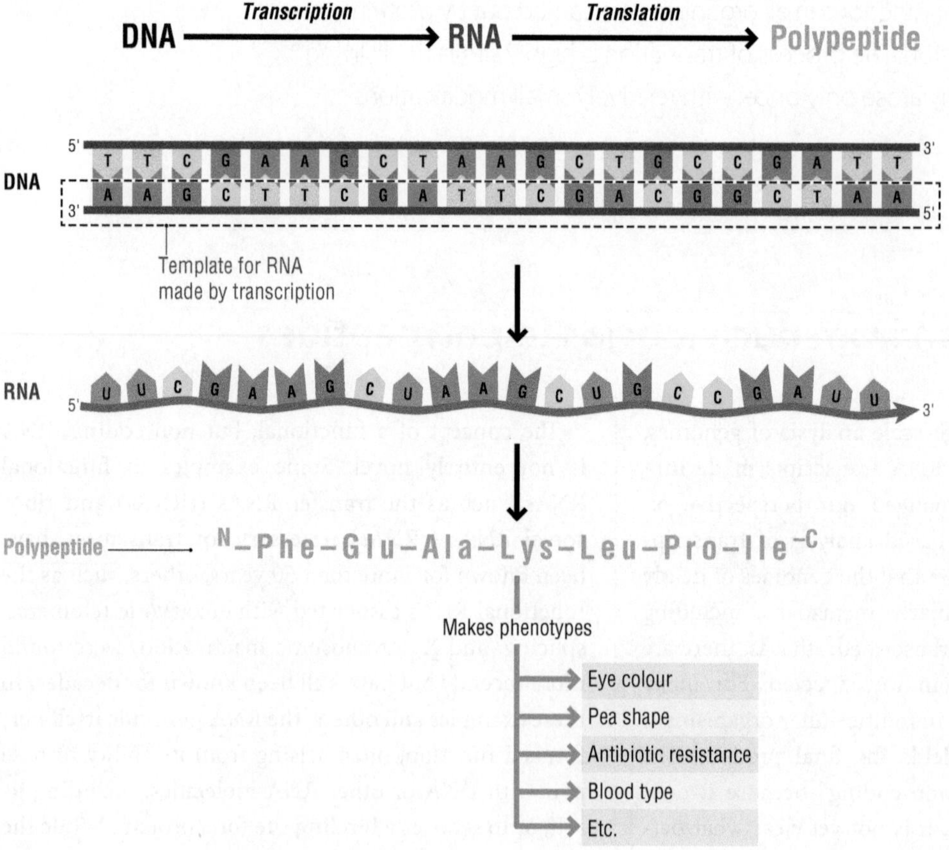

Figure 13.1 The Central Dogma. As originally posed, the Central Dogma of molecular biology is that DNA forms the template for making mRNA by the process of transcription. mRNA is then translated into polypeptides, which contribute to phenotypes.

13.2 The process of translation

As noted above, a small fraction of the total RNA transcribed in the cell is messenger RNA (mRNA), the RNA that will become translated into a polypeptide. In bacteria, mRNA begins to be translated into an amino acid sequence as it is being transcribed. Thus, transcription and translation in bacteria can be coupled in both location and time, which has implications for the coordination of gene expression, as described in Chapter 14 and summarized in Figure 13.2(a). The coupling of bacterial transcription and translation is also an important determinant of gene–product localization. Bacteria have localization machinery for some RNAs and proteins but lack many of the intracellular transport and localization systems present in eukaryotes. For reasons that remain unclear, even when transcription is completed, many mRNAs stay close to their site of transcription. For this reason, placement of some genes on the large circular bacterial chromosome determines where their mRNA and protein products will be at highest density. This could be the evolutionary basis for the functional clustering of genes in bacterial genomes.

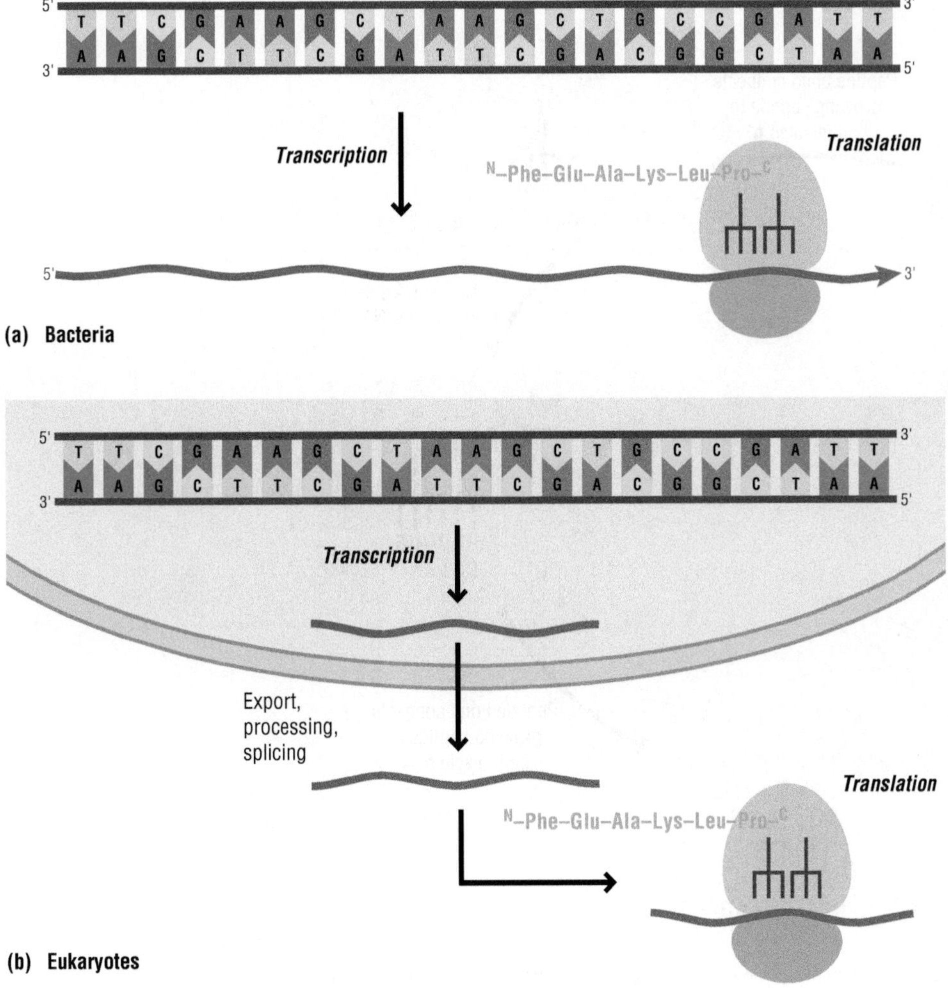

Figure 13.2 Translation and transcription in bacteria and eukaryotes. (a) summarizes the processes of transcription and translation. These processes are coupled in bacteria, such that mRNAs are being translated on the ribosomes as they are being transcribed. (b) summarizes the processes in eukaryotes. Transcription occurs in the nucleus; the transcript is exported out of the nucleus into the cytoplasm, during which it is processed and spliced, and then is translated on the ribosomes in the cytoplasm. As a result, in eukaryotes, transcription and translation occur in different cellular locations and potentially at different times.

In eukaryotes, mRNA is exported out of the nucleus as part of the processing that precedes translation. This means that transcription and translation in eukaryotes occur in distinct cellular compartments and can occur at very different times, as summarized in Figure 13.2(b), that is, an mRNA may be sequestered in the cytoplasm (usually as part of a specific RNA–protein complex) for some period of time before it is translated. This presents important opportunities for the translational control of gene expression. While some eukaryotic genes with important roles are regulated by such translational controls, most genes appear not to be subject to translational regulation. So far as we know, most mRNAs are translated as soon as they become available in the cytoplasm.

KEY POINT In bacteria, the translation of mRNA into protein begins while the mRNA is still being transcribed. In eukaryotes, the possibility of regulating the time and location of translation offers some additional levels of gene regulation. Translational regulation is known for some genes but does not seem to be the principal mechanism by which gene expression is regulated for most genes.

An overview of the process of translation is shown in Figure 13.3. The key molecules for translation are the mRNA and a family of tRNAs, at least one for each amino acid (as we discuss further below). The key molecular interaction, as with so many processes in molecular genetics, involves base pairs, in this case, between three-base codons on the mRNA and three-base anticodons on the

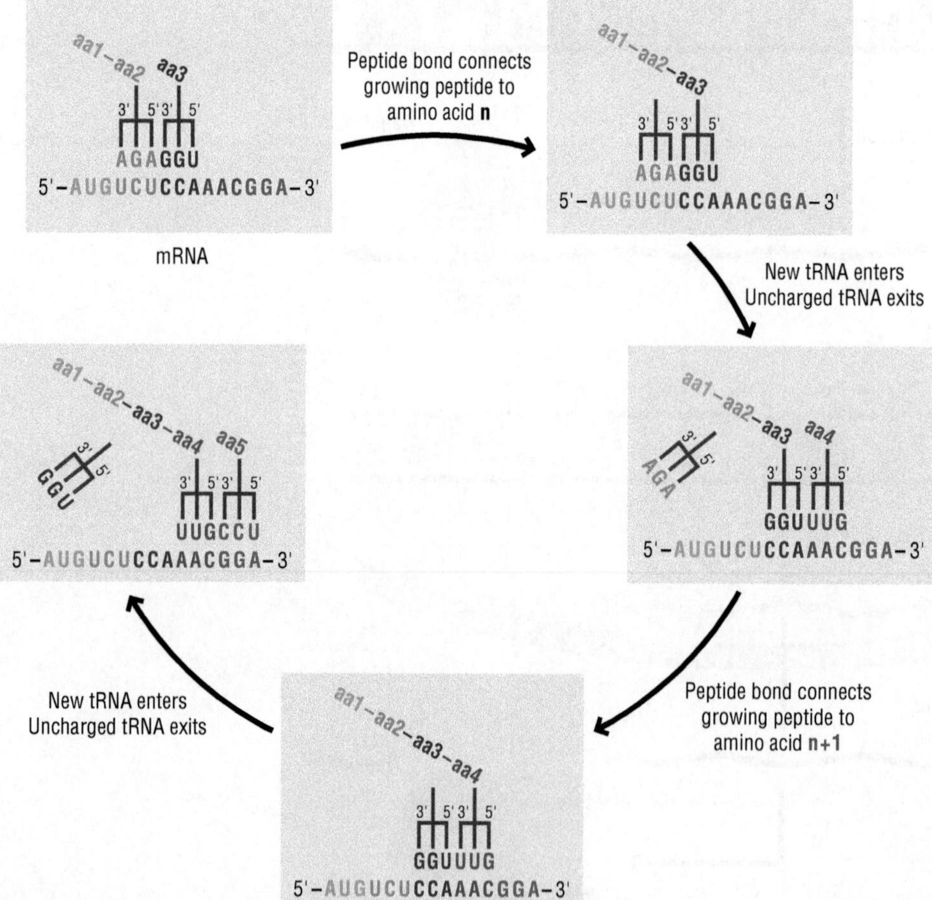

Figure 13.3 The cycle of translation. The key molecule in translation is tRNA shown here as a trident, and the process occurs at the ribosome; for clarity, the ribosome is not shown in this summary figure. An mRNA is brought to the ribosome where the base sequence is exposed in groups of three-base codons. A tRNA molecule with a complementary anticodon sequence and a specific amino acid attached at its 3′ terminus can pair with the codon; this specificity is shown by matching colors. A peptide bond forms between the incoming amino acid and the peptide on the ribosome, and the peptide becomes attached to the new amino acid. The movement of the ribosome and the mRNA exposes a new codon, the new tRNA with the appropriate anticodon and amino acid is brought to the ribosome, and the old tRNA from which the peptide has been removed is ejected. This process continues until the entire mRNA is translated. Note that the mRNA and the tRNA are anti-parallel.

tRNAs. The tRNA molecule has an amino acid attached to its 3′ end, the identity of which depends on the anticodon sequence. This specificity is shown in Figure 13.3 by the colors of the codons, anticodons, and amino acids.

An incoming tRNA base pairs with the codon, and the peptide from the preceding tRNA forms a bond to the amino acid it carries. In so doing, the peptide transfers to the new tRNA and has increased in length by one amino acid. As the next codon is exposed, and a new tRNA is brought in, the old tRNA with no amino acid on it (referred to as "uncharged") is ejected. This cycle continues until translation terminates, amino acids being added one at a time according to the next codon exposed on the mRNA. Translation occurs at the cellular organelle known as the ribosome. Thus, we need to describe the ribosome and the tRNA to describe translation more fully. These are shown in diagrammatic form in Figure 13.4. We will begin with the ribosome.

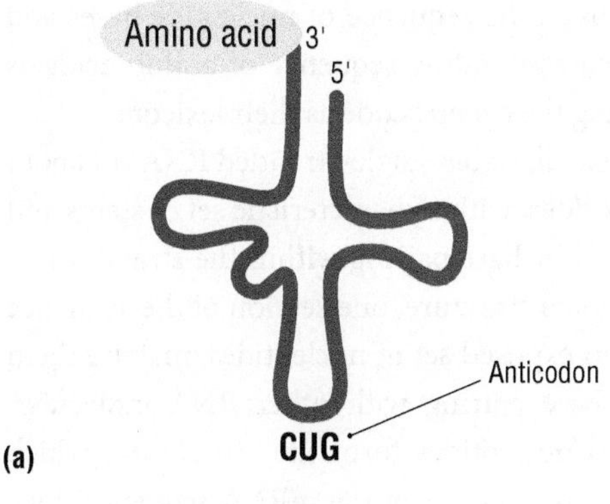

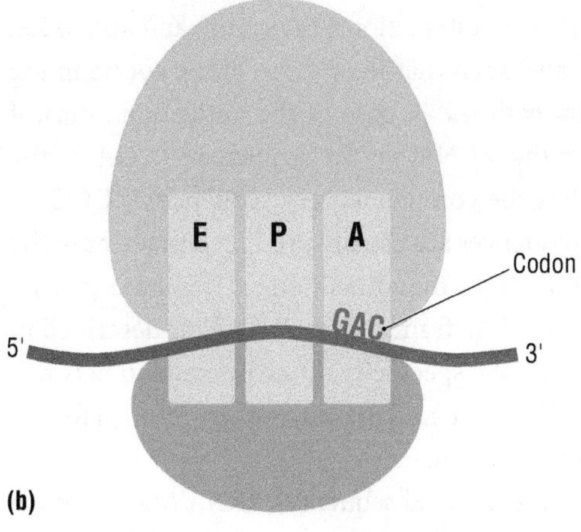

Figure 13.4 The components of translation are the tRNA and the ribosome. (a) diagrams the structure of tRNA, with the anticodon exposed at one side and the corresponding amino acid attached to the 3′ end. (b) diagrams the ribosome, with a large and a small subunit. The mRNA is threaded between the subunits. The ribosome has three functional and biophysical pockets, known as the A site, the P site, and the E site. The codon is exposed at the A site, with the corresponding anticodon entering there.

Ribosomes

In both bacteria and eukaryotes, translation occurs at a large complex of proteins and RNA called the **ribosome**, depicted in Figure 13.5(a) and (b). The bacterial ribosome has a diameter of 20 nm, while the eukaryotic ribosome is slightly larger in size, about 25–30 nm in diameter. The ribosome can be thought of as a "translating machine" with a small and a large subunit, called the 30S and 50S subunits in bacteria and the 40S and 60S subunits in eukaryotes. (The "S" refers to the

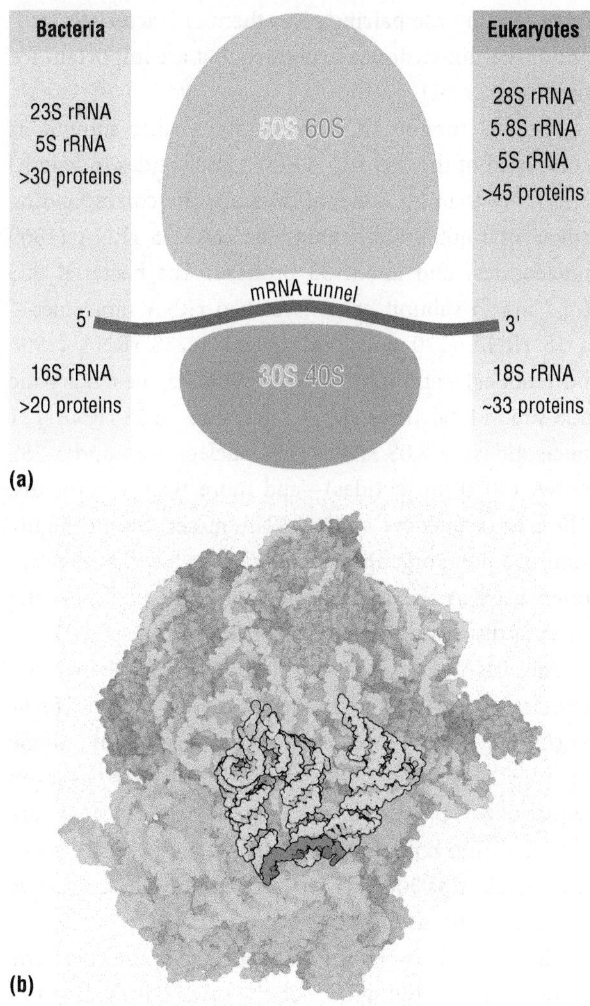

Figure 13.5 The structure of the ribosome in bacteria and eukaryotes. (a) The ribosome is composed of specific RNA molecules and proteins, and has a small and a large subunit. The tunnel for mRNA is formed between the subunits. The structures of bacterial and eukaryotic ribosomes are largely similar, although bacterial ribosomes have three ribosomal RNA (rRNA) molecules and eukaryotic ribosomes have four ribosomal RNA molecules, and some of the ribosomal proteins are different. (b) A space-filling model of the ribosome showing the A, P, and E sites.

Source: Part (b) from the Protein Data Bank. January 2010, David Goodsell, doi:10.2210/rcsb_pdb/mom_2010_1

sedimentation rate during centrifugation, the technique by which the structures of ribosomes were originally analyzed. The S value correlates with size, but it is not a precise linear correspondence; the 30S and 50S subunits combine to make a 70S ribosome.) Each subunit comprises its own characteristic set of proteins and its own single-stranded ribosomal RNA (rRNA) molecule or molecules, which collectively align the mRNA so that translation can occur. The proteins of the ribosome appear to provide a scaffold for the rRNAs, and most do not appear to have a direct role in translation. Although the rRNA molecules are single-stranded, extensive intra-strand base pairing gives them characteristic two- and three-dimensional structures that are important for their functions.

The 30S subunit in bacteria, the smaller subunit, is composed of the 16S rRNA (1542 nucleotides in length) and more than 20 different proteins; the corresponding eukaryotic 40S small subunit has an 18S rRNA (1869 nucleotides) and about 33 proteins. The bacterial 50S (or "large") subunit consists of two rRNA molecules—a 5S rRNA (120 nucleotides) and a 23S rRNA (2906 nucleotides)—and more than 30 proteins; the eukaryotic 60S subunit has three rRNA molecules—a 5S rRNA(121 nucleotides), a 5.8S rRNA (156 nucleotides), and a 28S rRNA (5070 nucleotides)—and more than 45 proteins. The base sequences of the rRNA molecules are highly similar among organisms, so similar in fact that they are often used as the molecule to construct phylo-genetic trees, as discussed in both Section 4.7 and Chapter 17.

The rRNA genes are transcribed by RNA polymerase I and are usually clustered together on the genome. *Escherichia coli* has the three rRNA genes together in a single operon; the genome of the standard strain K-12 has seven copies of this operon. In eukaryotes, the rRNA genes are clustered into repeat units. In humans, the rRNA genes are organized as 300–400 repeats found in clusters on five different chromosomes.

The region between the two subunits of the ribosome forms a tunnel through which the mRNA is threaded as it is being translated. In bacterial ribosomes, the tunnel is about 8 nm long and about 2 nm wide; it has similar dimensions in eukaryotes. As the nascent polypeptide exits the ribosome, it begins to become folded into its characteristic shape, and some ribosomal proteins are thought to play a role in polypeptide folding. While the mRNA is in the tunnel, its sequence is exposed as a codon of three bases at a time. The exposed bases are translated into an amino acid, one at a time, by a tRNA molecule.

KEY POINT Translation occurs at a large protein–RNA complex called the ribosome where each three-base codon in the mRNA is processed and the appropriate amino acid delivered by a corresponding tRNA molecule.

tRNA

tRNA is among the best illustrations of the relationship between the structure of a molecule and its function, as shown in Figure 13.6. In fact, it is such a remarkable fusion of structure and function that Francis Crick postulated the existence of a tRNA molecule as an "adaptor" between RNA and protein several years before it was found. Recall that there are two languages of molecular biology—the language of nucleic acids for which the alphabet consists of nucleotide bases, and the language of proteins for which the alphabet consists of amino acids. tRNAs are the literal "translators" between the two languages—they read the sequence of nucleotide bases and produce the corresponding sequence of amino acids as an output, using the genetic code as their lexicon.

tRNAs are small, stable single-stranded RNAs of about 80–110 nucleotides, with a characteristic set of stems and loops generated by base pairing within the strand. In its three-dimensional structure, one section of the sequence sticks out as an exposed set of nucleotides, making them available for base pairing with other RNA molecules. These exposed nucleotides form the **anticodon**, which base pairs with the codons of the mRNA sequence. Like the codon, the anticodon is three bases long, and like other base pairing interactions, the codon and anticodon are anti-parallel such that the 5′ base of the codon in the mRNA pairs with the 3′ base of the anticodon. Thus, if the codon in the mRNA has the sequence 5′-CGA-3′, the anticodon has the complementary sequence 5′-UCG-3′. (Note that sequences are usually presented in terms of the codons, rather than the anticodons.) This base pairing provides the reading frame of the mRNA, as described in Section 2.3, and the specificity of the interaction between the bases in the codon and the anticodon is critical for the accuracy of translation.

Like other nucleic acid sequences, tRNA has a 5′ and a 3′ end; the bases at the 3′ end are exposed and do not pair with other bases in the tRNA or any other RNA. A family of enzymes called the **aminoacyl synthetases** is responsible for loading (or "charging") the collection of tRNAs with their associated amino acids. Each aminoacyl synthetase recognizes the RNA sequence at the anticodon, together with other features of the tRNA sequence and structure,

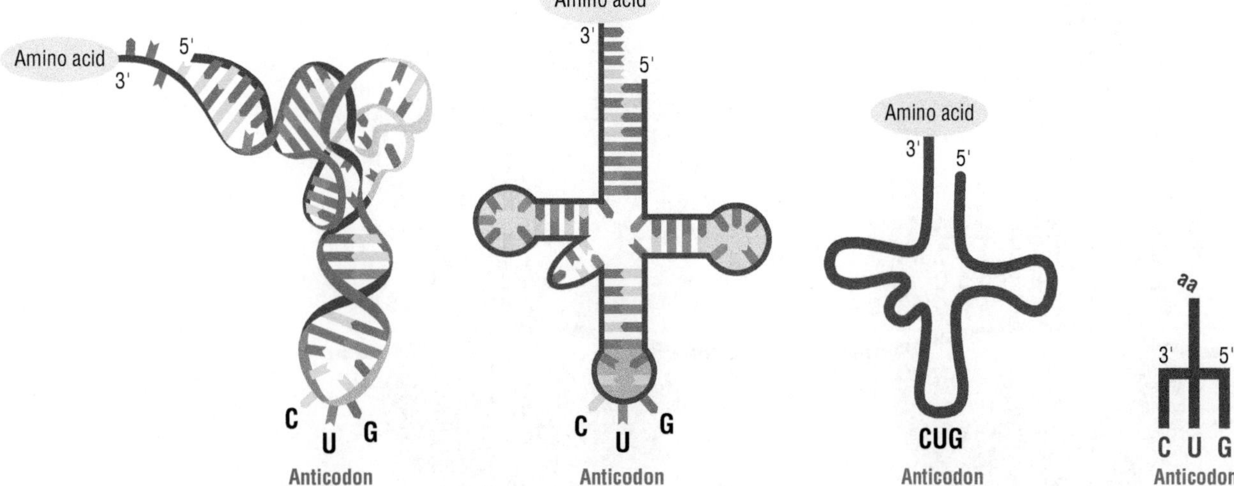

Figure 13.6 The structures of tRNA. Different methods to diagram the structure of tRNA are depicted, each showing the anticodon, which pairs with the codon of mRNA, and the amino acid attachment site. The anticodon sequence is exposed at one side of the molecule, and the corresponding amino acid is attached to the 3′ end. The drawings in this chapter primarily use the trident structure on the far right.

Source: Two left-hand images courtesy of Denise Woodward, The Pennsylvania State University.

and attaches the appropriate amino acid to the 3′ end of the tRNA. There are usually about 20 aminoacyl synthetase genes in a genome, one gene and enzyme for each amino acid. Thus, if the anticodon has the sequence 5′-CCU-3′, an aminoacyl synthetase attaches the amino acid arginine to the 3′ end of the molecule. If the anticodon has the sequence 5′-AGC-3′, a different aminoacyl synthetase attaches the amino acid alanine to the 3′ end of the tRNA.

While there are about 20 aminoacyl synthetase genes in a genome, the number of tRNA genes is larger and much more variable. *E. coli* and most other bacteria usually have about 80 tRNA genes in their genomes, while mammals typically have 400–500 tRNA genes. The number of genes may not be particularly significant; some invertebrate genomes have more than 1000 tRNA genes, and related species can vary by a factor of 5. The tRNA genes are frequently clustered together in the genome, and their transcription can be co-regulated. Transcription of tRNA genes is carried out by RNA polymerase III.

The elongation cycle

With a full collection of charged tRNAs in hand, let's look at the process of translation, as shown in Figure 13.7(a) and (b). We will begin by describing the steps in elongation, that is, the addition of the next amino acid to a growing polypeptide chain already in place. Because the first step in translation—the initiation step—is more complicated to describe (and variable between bacteria and eukaryotes), we will start in the middle of the process—elongation—to provide an overview before filling in some details.

In addition to the tunnel through which the mRNA passes, the ribosome has three pockets at which different steps of translation occur—the A, P, and E sites. These are biophysical pockets arising from the three-dimensional structure of the ribosome and the interactions between the rRNAs and the ribosomal proteins. As shown in Figure 13.7(a), the peptide chain is attached to a tRNA molecule at the P (peptidyl) site on the ribosome during elongation. At the A (aminoacyl) site, the codon on the mRNA is presented and the anticodon on the tRNA base pairs with it. The peptide chain, attached to the tRNA present at the P site, is then detached from its tRNA molecule and forms a peptide bond to the amino acid at the A site. Thus, the polypeptide chain has grown by one amino acid, which is now attached to its C-terminus.

The ribosome and mRNA then move with respect to each other, so that the peptide chain again occupies the P site, and the next codon is presented at the A site. This is shown in Figure 13.7(b). The uncharged tRNA that was formerly at the P site is now at the third pocket—the E (for exit) site—which allows it to leave the ribosome and become recharged.

This cycle then continues; each new codon at the A site interacts with a charged tRNA molecule; the peptide chain at the P site is attached to the new amino acid, and the ribosome and mRNA shift with respect to each other to expose the next codon. In bacteria, there are about 15 cycles per

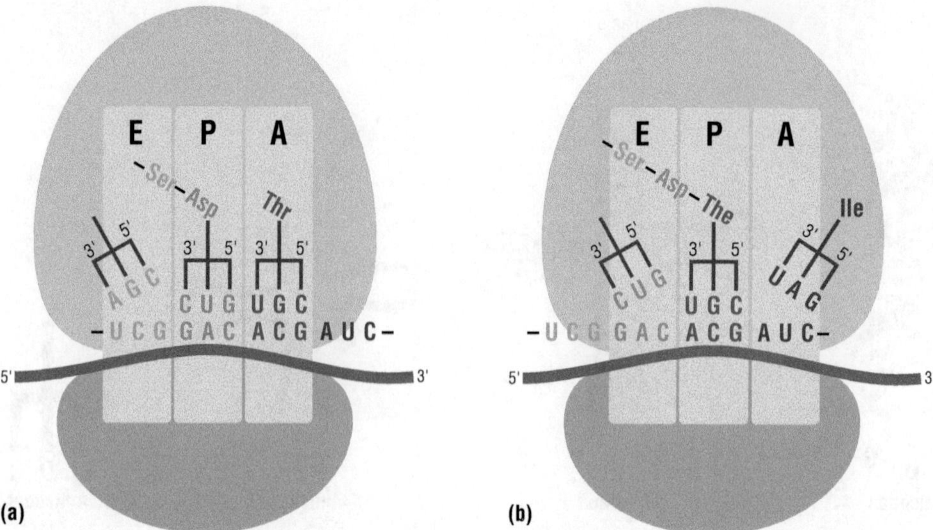

Figure 13.7 A more detailed view of the cycle of translation shown in Figure 13.3. (a) The codon of the mRNA is exposed at the A site, with the peptide on the tRNA at the P site. The E site is where the uncharged tRNA without an amino acid exits. (b) The next codon is exposed; the peptide grows by an amino acid, and the uncharged tRNA exits.

second, so an mRNA of 1800 bases, coding for a polypeptide of 600 amino acids, can be translated in about 40 seconds. Translation in eukaryotes probably occurs at about this rate once it begins, so translation itself is rapid. While we have focused entirely on the interactions between the codon and anticodon for simplicity, proteins known as elongation factors play roles in bringing the newly charged tRNA to the A site, in providing the energy to the ribosome with respect to the mRNA, and in adding accuracy to the process.

KEY POINT mRNAs proceed through three sites on the ribosome where each codon is revealed and bound by the corresponding tRNA (A site), the amino acid on the tRNA is added to the growing polypeptide chain (P site), and the discharged tRNA then exits the ribosome (E site).

Starting and stopping translation

We began our description with elongation in order to show how the peptide chain grows and the mRNA is translated. But how does the process of translation begin? Initiation is fairly complex, and somewhat dissimilar in bacteria and eukaryotes. An overview is shown in Figure 13.8.

The first codon to be translated is AUG (methionine). The AUG codon that initiates translation (called the initiator AUG or initiator methionine) is recognized by a specific tRNA molecule; AUG codons that are internal to the mRNA are not recognized by this same initiator tRNA. In bacteria, this initiator tRNA has been chemically modified and is called the **f-Met-tRNA**. In eukaryotes, the initiator AUG codon is modified in other ways; the cap attached to the 5′ end of mRNA also plays a role in initiation.

The AUG initiator codon is brought to the P site on the ribosome by proteins known as initiator factors; note that all other codons come first to the A site, so the location at the P site is a unique feature of the initiator codon and its interactions with the initiation factors. Once the initiator tRNA with methionine is present at the P site, the next

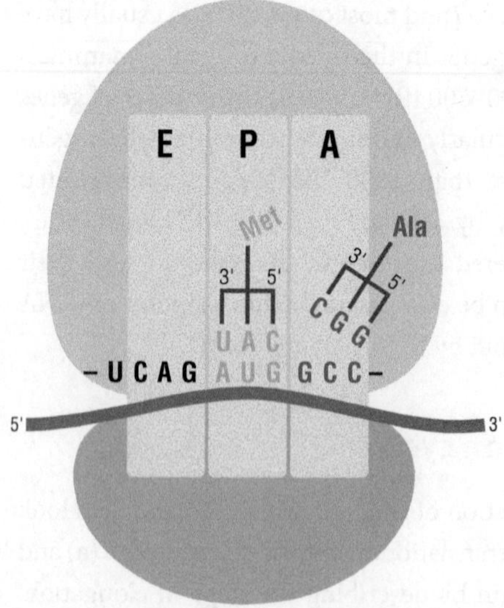

Figure 13.8 Initiation of translation. Translation begins when the initiator AUG is brought to the P site. The initiator AUG and its corresponding tRNA are the only ones that appear at the P site, rather than at the A site. The tRNA for the second codon is then brought to the A site to elongate the chain.

tRNA is brought in at the A site, and elongation commences. A small fraction of mRNAs do not begin with AUG; initiation for these mRNAs is not well understood.

How then does the chain stop? Or, to ask this question another way, why do UAA, UAG, and UGA codons stop translation? The answer is revealed in Figure 13.9 and is surprisingly simple—no tRNA has an anticodon that can base pair with these three codons. Thus, when one of them appears at the A site, no tRNA can pair and the cycle terminates. There are specific proteins known as release factors that recognize the unpaired stop codon at the A site and release the polypeptide chain from the P site, but it is the absence of the corresponding tRNA that stops translation.

Like any other gene, tRNA genes are subject to mutations, which change their base sequences. Among the most significant mutations in tRNA genes for our understanding of translation have been ones that allow the mutated tRNA to recognize a stop codon and insert an amino acid. These are known as nonsense suppressors and are described in Box 13.1.

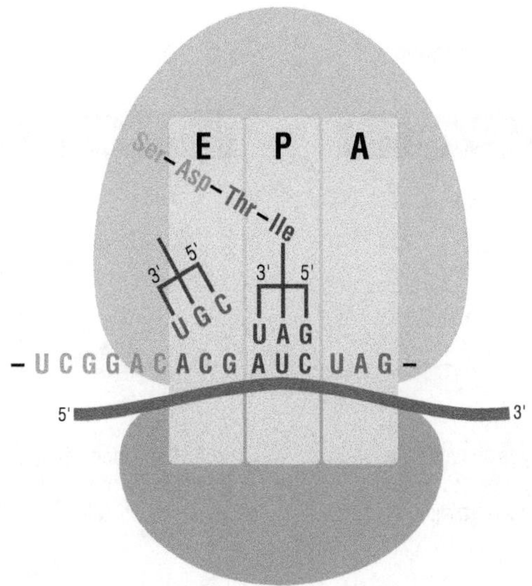

Figure 13.9 Termination of translation. When a stop codon appears at the A site, no tRNA molecule with the corresponding anticodon exists, so translation terminates. In the diagram, this is shown for the stop codon UAG. The tRNA with the peptide on it is released by the action of termination factor proteins.

BOX 13.1 *Going Deeper* Nonsense suppressors

Every gene is subject to mutation, including the genes encoding tRNAs. As will be discussed in more detail in Chapter 15, mutations can be used to understand the biological basis of a process even when the molecules themselves and the overall organization are not known. Mutations in tRNA genes in bacteria were instrumental in understanding the logic by which translation occurs and in recognizing a key aspect of the information flow in all cells.

Let's review some terminology first. A **nonsense mutation** is one in which a codon for an amino acid is mutated so that the codon is now one of the stop codons, UAA, UAG, and UGA. A nonsense mutation is also referred to as a premature stop codon because translation terminates at this point, rather than at the normal stop codon. As we point out in Section 13.2, translational termination occurs because there are no tRNAs that have an anticodon that can effectively base pair with these three codons. (Strictly speaking, the anticodon 5′-CCA-3′, which base pairs with 5′-UGG-3′ to insert tryptophan, can sometimes recognize the UGA stop codon via wobble at the third base.)

A mutation in a tRNA gene can create an anticodon that is able to base pair with one of the stop codons. The mutation could be in the sequence that gives rise to the anticodon itself; more often, the mutation is a few bases away from the anticodon sequence but causes a change in shape or a shift of the base pairing of the tRNA molecule, which exposes different bases at the anticodon site. The tRNA is charged with the same amino acid as before, and only the anticodon sequence is changed. An example from a tRNA gene is diagrammed in Figure A. Such a mutated tRNA now pairs with a stop codon and inserts an amino acid. The premature stop codon no longer results in termination, and translation continues.

Such a mutated tRNA is called a suppressor tRNA. A **suppressor mutation** is a mutation that overcomes the effects of a previous mutation. (If the mutation reverses the original mutation, it is called a revertant. Suppressors, or more accurately extragenic suppressors, are mutations in a different gene to the original mutation.) The function of the original gene was altered because a nonsense mutation arose. The mutation in the tRNA results in the insertion of an amino acid at that nonsense mutation to allow translation to continue, so it overcomes or suppresses the effect of the original mutation.

Nonsense mutations are quite common. The effect of the suppressors was recognized with nonsense mutations in genes from bacteriophage, particularly bacteriophages T4 and λ. Bacteriophage were discussed in Chapter 11; these are viruses that infect *Escherichia coli* or other bacteria and rely on the host translational machinery, including its tRNA genes, to produce viral proteins. As summarized in Figure B, the mutated phage with a nonsense mutation is not able to infect the host bacteria and grow. A suppressor mutation in *E. coli* allows the phage to grow; growth, and thus **suppression**, could be readily seen, even when such mutations

BOX 13.1 Continued

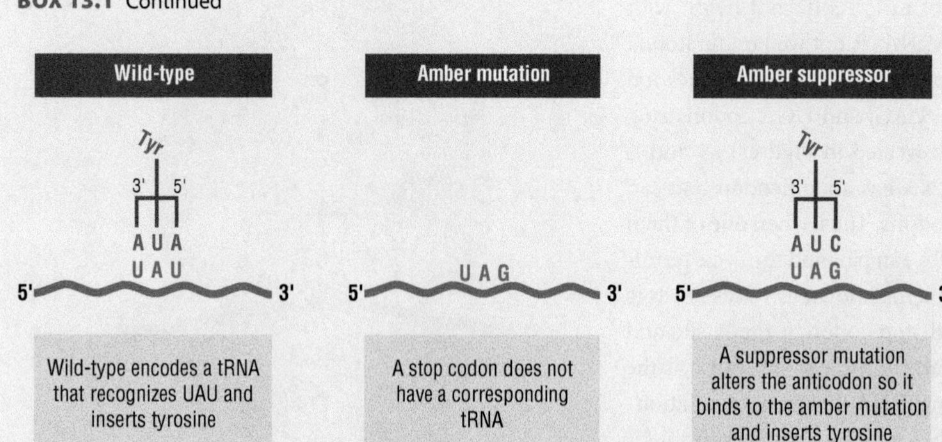

Figure A The mechanism of nonsense suppressors. A nonsense mutation arises when a change in the coding region of the gene creates a premature stop codon. In this example, a UAU codon in a gene is mutated to UAG, known as the amber stop codon. This produces a mutant phenotype for the gene. This nonsense mutation can be suppressed or overcome by a mutation in a tRNA gene, such that the corresponding tRNA can recognize UAG codons and insert an amino acid. The amino acid inserted by the mutated tRNA may or may not be the same as the one mutated in the gene originally.

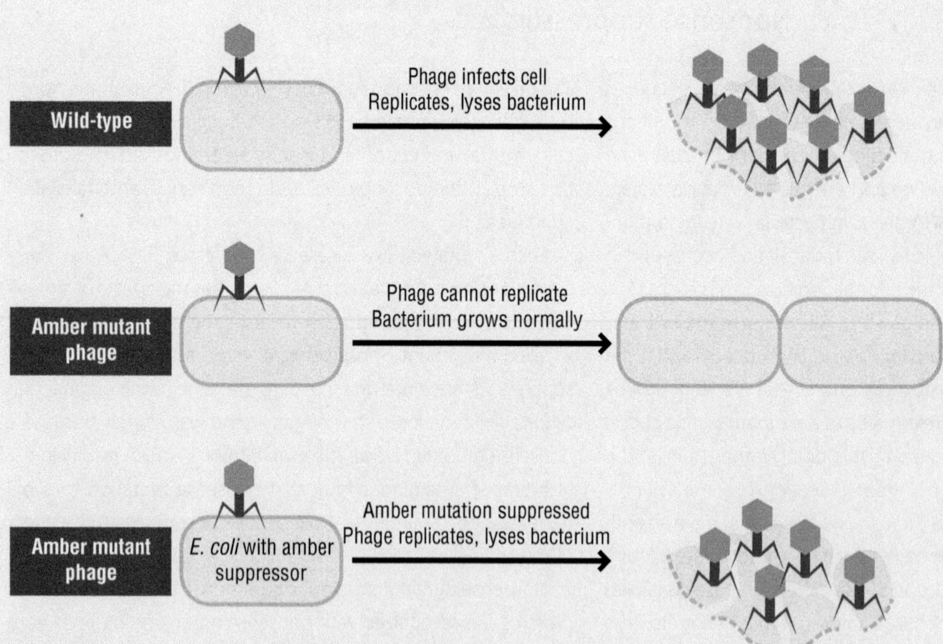

Figure B Nonsense suppressors with bacteriophages. Many nonsense suppressors were found in *E. coli*, recognized by their for their ability to suppress mutations in bacteriophage genes. As discussed in Chapter 11, some bacteriophages infect a bacterium and lyse the cell, as they proliferate. An amber nonsense mutation in a phage gene prevents the phage from completing its life cycle, by blocking phage genome replication, for example. A mutation in a tRNA gene in *E. coli* enables it to recognize the UAG stop codon in the phage gene and insert an amino acid, which restores at least partial function to the phage gene. Thus, the phage can replicate and lyse the bacterium. Nonsense suppressors of this type were important genetic tools in understanding the mechanisms of translation, as well as the biology of many different phages.

BOX 13.1 Continued

are rare, since tens of thousands of bacterial cells are present in a culture. Because these mutations in *E. coli* tRNA genes overcame the effects of nonsense mutations (in phage genes), the bacterial mutations are called **nonsense suppressors**.

Nonsense suppressors were identified before the sequences of the stop codons were known. For historical reasons, the UAG stop codon (the first to be identified) is called the amber stop codon, so nonsense mutations to UAG are also called amber mutations. The suppressors are then called amber suppressors. Once the first stop codon was named for a brown gemstone, the others were as well; UAA is designated the ochre codon, and UGA (the last to be identified) as the opal codon. Amber and ochre suppressors were very widely used in *E. coli* and phage genetics to understand the information flow from transcription to translation to polypeptide in the cell. Thus, these are also known as **informational suppressors**. The various terms encompass one another—informational suppressors include nonsense suppressors, as well as some other types of suppressor mutations that affect ribosomal proteins or rRNA genes, and nonsense suppressors include amber suppressors, as well as ochre and opal suppressors.

Nonsense suppressor mutations and their mode of action are explored more fully in the study questions for this chapter.

13.3 Translational control

The genes whose expression is regulated by translational control don't fall into any convenient categories, either by the functions of the genes or by the mechanisms of regulation. While we can give a general picture of the regulation of transcriptional initiation by transcription factors or σ subunits, or one of regulation by alternative splicing or microRNAs, there seems to be no similar uniformity for regulation by translational control. One broad principle is that, if translational regulation occurs for a gene, it probably occurs during the initiation of translation; once the process begins, it is likely to continue to the end (although at least a few examples do involve pausing or stalling of the ribosome).

A second broad principle of translational regulation is that it likely involves a specific interaction between some sequences in the mRNA and some protein or proteins; the sequences in the mRNA usually are located at the 5′ or 3′ end, possibly more commonly in the 5′ untranslated region (UTR). But these are very broad principles. We will describe two specific examples that are reasonably well understood, but other genes regulated by translational controls may differ from these.

Iron storage in mammals

Iron is a necessary component of many proteins—such as hemoglobin—but also causes cellular damage when circulating at high levels. The regulation of circulating iron levels in many animals, including humans, involves a number of proteins, particularly one known as ferritin. Ferritin forms a complex with iron in the bloodstream to prevent the toxic effects of high iron levels, so ferritin levels are high when excess iron is present; on the other hand, when little iron is present, ferritin levels are low, so that the iron is available to carry out its essential functions.

The level of ferritin expression is regulated at the level of translation and involves interactions among iron, iron-response proteins (IRPs), and a sequence in the 5′ UTR of the ferritin mRNA called the iron-response element or IRE. These interactions are summarized in Figure 13.10. The IRPs are a family of proteins with slightly different functions, but we will summarize their actions by focusing on one member of the family—the IRP. When the IRP binds to the IRE sequence on the ferritin mRNA, translation initiation is blocked, and little or no ferritin protein is produced, as shown in Figure 13.10(a).

The IRP also interacts with iron (or, more accurately, an iron–sulfur complex). When bound with iron, the IRP structure changes such that it cannot bind to the IRE, as shown in Figure 13.10(b). As a result, translation of the ferritin mRNA occurs, and ferritin protein is made. Ferritin then sequesters the excess iron and prevents it from exerting its toxic effects on the cells. When iron levels then drop, the IRP adopts the conformation that allows it to bind to the IRE and block the translation of ferritin mRNA, so that some iron remains available for its essential functions.

The interaction between the IRP and the ferritin mRNA is not the only one between IRPs and mRNAs to regulate iron metabolism. Another example is shown in

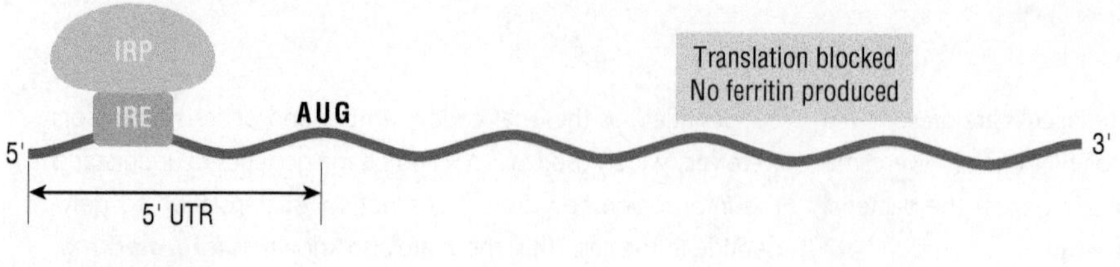

(a) Low iron levels

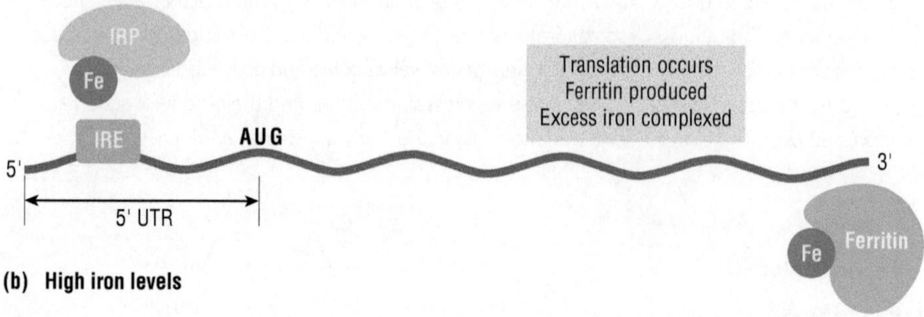

(b) High iron levels

Figure 13.10 An example of translational regulation: ferritin. Ferritin is a protein that sequesters free iron and removes it from the circulation, protecting the cell from the harmful effects of high iron concentrations. The translation of the mRNA encoding the ferritin protein is regulated by the interactions involving a sequence in the 5′ UTR known as the iron-response element (IRE), a binding protein known as the iron-response protein (IRP), and free iron. The interaction between IRP and free iron affects the structure of the IRP and its ability to bind to the IRE. (a) When iron is low, IRP can bind to the IRE, and translation of the mRNA is blocked. No ferritin protein is made, so free iron is not sequestered. (b) When the free iron level is high, it binds to the IRP and prevents it from binding to the IRE. As a result, the ferritin mRNA can be translated; ferritin protein is made, and the free iron is then sequestered by ferritin. When the concentration of free iron drops, IRP again binds to the IRE, and the ferritin mRNA is not translated.

Figure 13.11. A receptor protein known as transferrin is involved in iron transport into the cell. The mRNA of the transferrin gene has three IREs in its 3′ UTR. These are bound by the IRPs when iron levels are low—that is, when most of the excess iron is bound to ferritin and not circulating freely, as shown in Figure 13.11(a). The binding of the IRPs in the 3′ UTR does not affect translation of transferrin directly but increases the stability of its mRNA, so that it is not degraded as rapidly. Consequently, more transferrin protein can be made. As a result, even the low levels of circulating iron are transported into the cell.

When circulating iron levels are high, however, as shown in Figure 13.11(b), the iron interacts with the IRPs. This interaction between iron and the IRPs again prevents the IRPs from binding to the mRNA, so transferrin mRNA is degraded, very little transferrin protein is made, and no iron is transported into the cell.

The intricate balance that exists to ensure that sufficient iron is present, yet prevents too much from entering the cells, arises from these translational controls. These are not the only examples. At least ten other genes have IREs in the sequences of their mRNA; most, if not all, of these genes have a role in iron storage and metabolism. As with ferritin and transferrin, the regulation appears to involve translational control through the interaction of the IRP with IREs in the absence of iron, but the details are not known for most of them.

Anterior–posterior axis specification in *Drosophila*

As we saw with iron storage, translational control mechanisms have often evolved in response to selective pressures in scenarios where tight control of the production of the protein is critical. This tight control can be with respect to the overall levels of the protein being made, as in the example of ferritin described above. But sometimes there can also be a spatial component whereby the production of the protein is needed in one area, such as one side of a cell, but must not be made in another area. One of the best-studied examples of this type of regulation is in the developing embryo of the fruit fly *Drosophila melanogaster*.

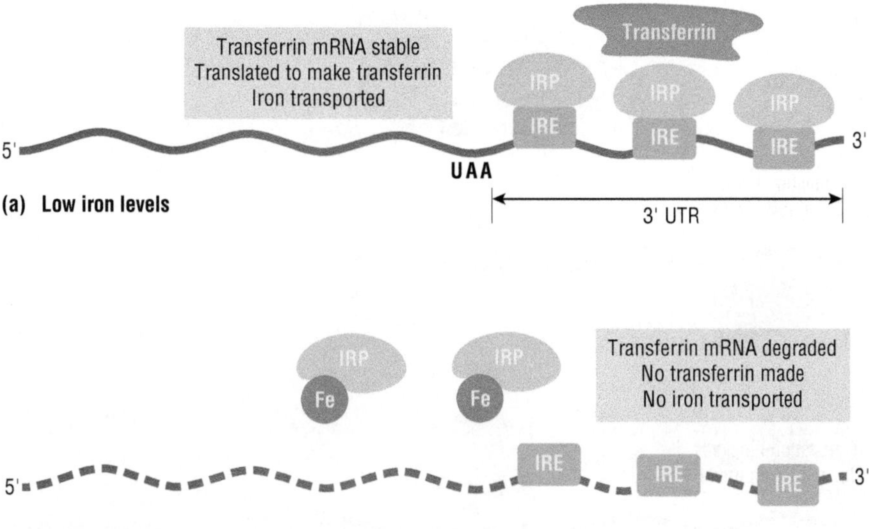

Figure 13.11 An example of translational regulation: transferrin. Transferrin is a protein that transports iron into the cell where it can serve as a co-factor. In this case, three iron-response elements (IREs) are found in its 3′ UTR. (a) When iron levels are low, the iron-response proteins (IRPs) bind to the IREs and stabilize the transferrin mRNA. This allows the translation of transferrin mRNA and the production of transferrin protein, which transports some iron into the cell. (b) When iron levels are high, iron binds to the IRPs, and they do not bind to the IREs. The transferrin mRNA is unstable and is degraded, so no iron is transported into the cell. The balance between the regulation of ferritin in Figure 13.10 and transferrin in Figure 13.11 results in low levels of iron being transported into the cells for its essential function but not so much that the harmful effects occur.

Recall from Section 6.4 that, at the completion of oogenesis, the ovum has a large quantity of cytoplasm. When the ovum is fertilized to become a one-cell zygote, nearly all of the cytoplasm in the zygote (and thus the early embryo) has come from this ovum. In other words, all of the cytoplasmic components in the early embryo—the mitochondria, the ribosomes, the proteins, and the mRNAs, among others—were made by the mother. Insect embryos, such as *Drosophila*, use gene products deposited by the mother into one end or the other of the egg to determine the anterior and posterior ends of the developing embryo. The location and concentration of the gene products are therefore critical in this scenario and are achieved through the combined effects of localizing key mRNAs and proteins to the correct ends and restricting the translation of specific mRNAs to the desired regions of the embryo.

Let us take a look at just a few of the components of this system. Figure 13.12 illustrates how maternally supplied mRNA of the *hunchback* (*hb*) gene is uniformly distributed in the egg; the Hb protein plays a critical role in directing the development of anterior structures. Conversely, maternally supplied mRNA of the *nanos* (*nos*) gene is tightly localized to the posterior pole; the Nos protein is required for the posterior of the embryo to develop correctly. One of the mechanisms in place to ensure that the Hb protein is only made in the anterior end of the embryo is the repression of the translation of *hb* mRNA in the posterior end. It is the presence of the Nos protein in the posterior of the embryo that provides this repression.

The repression of translation of the *hb* mRNA in the posterior of the embryo is dependent on specific sequences present in the 3′ UTR of the *hb* mRNA, but these sequences are not bound directly by the Nos protein. A protein known as Pumilio (Pum) recognizes and binds directly to these sequences (termed "Nanos response elements" or NREs) in the 3′ UTR, as illustrated in Figure 13.12. Pum is uniformly distributed throughout the embryo. The binding of Pum alone, however, does not prevent translation of the *hb* mRNA—the Nos protein must also be present. Nos forms a complex with Pum; when this complex is present on the 3′ UTR of the *hb* mRNA, additional components are recruited and its translation is repressed.

VIDEO 13.1 provides a more in-depth explanation of the anterior–posterior axis specification in developing *Drosophila* embryos. Find it at **www.oup.com/uk/meneely**

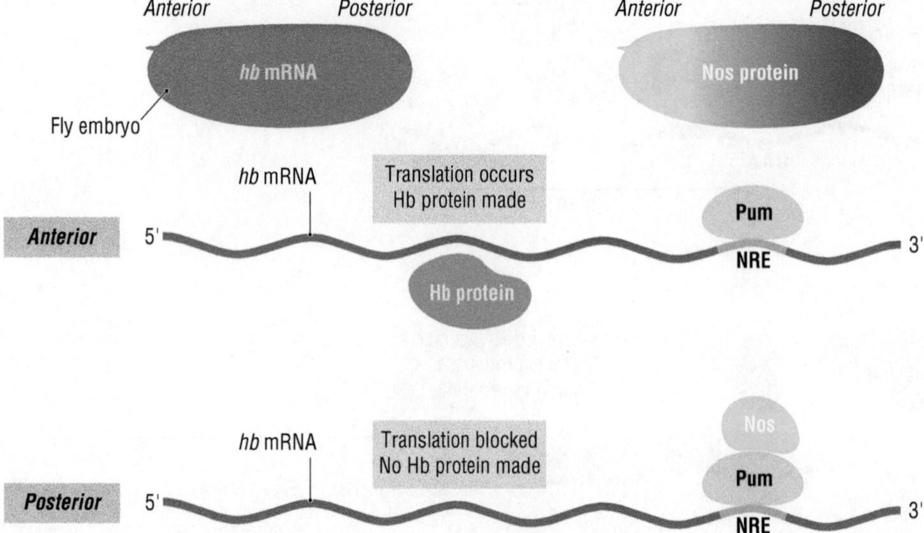

Figure 13.12 Translational regulation in *Drosophila* embryogenesis. The anterior–posterior axis of *Drosophila* is determined by the translational regulation of several key genes; the roles of two of these genes *nanos* (*nos*) and *hunchback* (*hb*) are shown here. The mRNA for each gene is made by the mother and is sequestered without being translated into the embryo. The *hb* mRNA is not tethered and thus is found uniformly in the ovum and early embryo. The *nos* mRNA is tethered at one end, which becomes the posterior. After fertilization, the *nos* mRNA is translated to make the Nos protein, which diffuses to set up a gradient from the posterior towards the anterior. The Nos protein blocks the translation of the *hb* mRNA, so the Hb protein also forms a gradient from the anterior to the posterior, in the opposite direction. These are two of the concentration gradients that determine the body axis. The Nos protein cannot bind to mRNA on its own but interacts with a conserved RNA-binding protein called Pumilio (Pum), which binds to *hb* RNA at the NRE but does not block its translation.

The repression of *hb* translation by the spatial distribution of Nos is only one example of several translational repression mechanisms involved in specifying the anterior–posterior axis of the fruit fly embryo. The mRNAs of several other important genes (including *nos* itself) are also subject to spatially restricted translational repression, as explored more fully in Box 13.2. In all these cases, it should be noted that specific sequences in the UTRs of the mRNAs and/or the secondary structures that form as a result of these sequences play an essential role in providing recognition sites for regulatory proteins to bind to directly, recruiting the machinery necessary for localization and/or translational repression of the mRNA.

BOX 13.2 *Going Deeper* The role of translational regulation in patterning the anterior–posterior axis of *Drosophila melanogaster*

One of the most compelling examples of translational regulation involves the processes to determine the anterior–posterior body axis in *Drosophila melanogaster*. We will describe this example in more detail in Chapter 15 when we discuss the genetic screen that found genes involved in segmentation itself, but the earliest steps in determination of the body axis involve genes subject to translational regulation. Two of the key genes involved are *bicoid (bcd)*, which affects the specification of events at the anterior region of the embryo, and *nanos (nos)*, which affects the specification of events at the posterior region of the embryo. Both of these genes regulate the expression of other genes; two of their important targets for our discussion are the genes *hunchback* (*hb*) and *caudal* (*cad*). Both *hb* and *nos* were introduced in the current chapter.

As we noted in the chapter, all of the cytoplasmic components in the early embryo—the mitochondria, the ribosomes, the proteins, and the mRNAs, among others—were made by the mother. This maternal contribution is essential to understanding the role that *bcd* and *nos* play in the body axis of *Drosophila*.

The second key to early development in *Drosophila* is that, as we will discuss in more detail in Chapter 15, the *Drosophila* embryo undergoes many rounds of nuclear division before it forms

BOX 13.2 Continued

any cellular membranes between the dividing nuclei. The early embryo features many nuclei, all sharing a common cytoplasm, a type of cell known as a **syncytium**. The common cytoplasm came from the mother, so any contribution from the mother that diffuses freely throughout this cytoplasm affects all of the nuclei.

With this brief background, we can now describe the roles played by these four genes in establishing the anterior–posterior axis.

All four of the genes *bcd, nos, hb*, and *cad* are transcribed by the mother in the ovary. The mRNAs for each of these genes are packaged into the cytoplasm of the ovum but are not translated. Instead, the mRNAs for the genes are sequestered in the cytoplasm with translation blocked. For *hb* and *cad*, the mRNAs are found at uniform levels through the cytoplasm of the ovum. However, the mRNAs for *bcd* and *nos* are sequestered at opposite ends of the ovum, with *bcd* mRNA located at the future anterior end and *nos* mRNA located at the future posterior end.

The processes by which these mRNAs come to lie in these locations are more detailed than we can address here, but the localization is crucial to normal anterior–posterior development. Many of the gene products involved in the localization of *bcd* and *nos* mRNAs also play important roles in localizing mRNAs asymmetrically, such as in neural cells and stem cells in many animals. Once the mRNAs of *bcd*, *hb*, *nos*, and *cad* have been deposited and/or localized in the ovum, they are held without translation as maternally contributed mRNAs until after fertilization.

Upon fertilization, these and other maternal mRNAs start to be translated. The anterior and posterior localized mRNAs *bcd* and *nos* are translated to make Bicoid (Bcd) and Nanos (Nos) proteins. While the mRNAs are held in place in the cytoplasm of the ovum and the embryo, the proteins are able to diffuse. Diffusion of the proteins through the cytoplasm forms a gradient—the highest level of the protein is found at the end where the mRNA was held, and progressively lower levels are found further away. Thus, the Bcd concentration is highest at the future anterior, while the Nos concentration is highest at the future posterior, that is, the nuclei in the syncytium experience a different level of BicBcd or of Nos, depending on where they happen to lie in these gradients. The gradients of Bcd and Nan Nos proteins then play an important role in setting up the anterior–posterior axis.

The anterior determinants Bcd and Hb

The mRNA of *bcd* is localized to the anterior of the embryo, while the maternally supplied *hb* mRNA is distributed uniformly. The Bcd protein contains a Hox domain (described in Sections 3.6 and 12.3), which allows it to bind to nucleic acids. While we usually think of Hox domain proteins as being transcription factors that bind to DNA, the Hox domain of Bcd plays two different roles.

The first of these roles is its function as a transcription factor that activates the transcription of *hb* in the embryo. This is the activity of transcription factors discussed in Chapter 12. Embryonic transcription of *hb* results in *hb* mRNA levels increasing beyond the low levels deposited uniformly in the ovum by the mother. The activation of *hb* transcription in the embryo occurs most strongly in the nuclei closest to the anterior where the concentration of Bcd is the highest, and less strongly in the nuclei further down the gradient. Because Bcd forms a gradient from anterior to posterior, the Hb protein also forms a concentration gradient but of a slightly different shape than the Bcd gradient.

Hb itself is a transcription factor, so Bcd and Hb together regulate the transcription of other genes in the embryo, based on their concentrations, with higher concentrations of Bcd and Hb activating the transcription of a different set of target genes than lower concentrations of Bcd and Hb. This triggers a cascade of gene activities that subdivides the anterior of the embryo into specific domains, as will be discussed further in Chapter 15. Substances like Bcd that vary in concentration and drive different cell fates at different threshold levels are referred to as **morphogens**.

Interestingly, the central role of *bcd* in establishing the anterior axis of the embryo is likely to be a more recent evolutionary innovation. More distantly related flies and other insects are not patterned in this way. Ancestrally, it is possible that *hb* may be the central player of anterior patterning, comprising a system into which *bcd* then became co-opted in fruit flies.

The posterior determinants Nos and Cad

The mRNA of *nos* is localized to the posterior of the ovum, while the maternally supplied *cad* mRNA is distributed uniformly. The Cad protein plays an important role in defining the posterior of the embryo. How then is the activity of the Cad protein restricted to the posterior of the embryo when its mRNA is distributed uniformly? This brings us to the second role of the Hox domain of Bcd. In addition to its role as a transcription factor for *hb*, the Bcd protein can bind to specific sequences in mRNA and block translation. An important binding site of Bcd is the 3′ UTR of the mRNA of *cad*. Because the level of Bcd is highest in the anterior end of the embryo, it effectively blocks the translation of *cad* mRNA in the anterior. As a result, the levels of the Cad protein are lowest at the anterior end and highest at the posterior end. Thus, the Bcd and Cad proteins form opposing gradients in the early fly embryo.

As we discuss in the chapter, Nos, the other posterior determinant, exerts its influence on the patterning system by

BOX 13.2 Continued

participating in the translational repression of *hb* mRNA. This function of Nos ensures that the Hb protein is not translated in the posterior region of the embryo, thereby restricting the activity of Hb to the anterior of the embryo. In a further twist to this story, it turns out that *nos* mRNA itself is also subject to translational control, such that only the *nos* mRNA that is localized to the very posterior of the embryo is translated successfully into the Nos protein.

Overall, as a result of both transcriptional activation and translational repression, the genes in nuclei at different positions in the syncytium experience varying levels of Bcd, Hb, Cad, and Nos proteins. Nos blocks the translation of *hb* mRNA in the posterior, while Bcd blocks the translation of *cad* mRNA in the anterior. Because Bcd, Hb, and Cad are also transcription factors, different sets of genes are activated in the embryo, based on the levels of these three proteins. This, in turn, sets up the expression patterns shown by the segmentation genes, discussed in Chapter 15.

Translational control is not the only mechanism used for the anterior–posterior patterning of the *Drosophila* embryo, but it is a critical one. When put together, the anterior–posterior patterning system seems rather convoluted. But this is an outstanding example of the concept of evolutionary tinkering described in the Prologue. There has apparently been strong selection for preventing key mRNAs from being translated in regions of the embryo where the production of those proteins have a deleterious effect; mutants and experimental manipulations that have such ectopic protein expression in the embryo are invariably lethal. There has also been selection for mechanisms that produce gradients of protein production in specific regions of the embryo. Because the specific localizations and levels of proteins have been under such strong selection, multiple steps that regulate the expression pattern have arisen. As a result, mechanisms of localizing, anchoring, inhibiting, or activating translation of key mRNAs are all involved in how a fly embryo sets up heads and tails.

Spatially restricted control of the translation of specific mRNAs is not limited to developing insect embryos. There are many examples in the developing nervous system where the repression of translation of specific mRNAs on one side of a progenitor cell has been implicated in the asymmetric cell divisions that give rise to two different daughter cells, which go on to follow importantly different cell fates. The precise mechanisms of such translational repression are complex and varied, but it is now understood that some overlapping components are used in both the insect embryos and the asymmetric cell divisions within the nervous system. Furthermore, Pum is a conserved protein whose orthologs are also involved in translational controls in the regulation of germline sex determination in *Caenorhabditis elegans* and mammals.

13.4 The intricate beauty of the genetic code

Having described the process of translation and introduced the topic of translational regulation of gene expression, we now turn to the genetic code. We could just point to Figure 13.13(a) which shows the genetic code—the lexicon that provides correspondence between the codon and its amino acid—and move on to our next topic. After all, if biology was "just the facts," the facts of the genetic code are easy to look up—UCA is a codon for serine, GGA encodes glycine, and so on, for each of the 64 codons. It is unlikely that very many biologists "know" this codon table in the sense that we can get it right without looking it up. Most of us have used it enough to know some bits of it—the stop codons are worth remembering—but the rest of it, we look up using an app on our phone.

But the intricate beauty of the genetic code would be lost by merely looking it up. Like a work of art, the genetic code merits our contemplation. Features embedded into the genetic code give us insights into the evolution of molecular processes that lie at the foundation of living systems. Don't memorize the codon table, but contemplate it with us for a moment. Here are some things that you may observe.

Wobble and degeneracy

There are 64 codons, of which 61 code for amino acids. But there are only 20 amino acids. As a result, most amino acids can be encoded by more than one codon—the only exceptions being AUG (methionine) and UGG

		Second position				
		U	C	A	G	
First position	U	UUU Phe UUC Phe UUA Leu UUG Leu	UCU Ser UCC Ser UCA Ser UCG Ser	UAU Tyr UAC Tyr UAA STOP UAG STOP	UGU Cys UGC Cys UGA STOP UGG Trp	U C A G
	C	CUU Leu CUC Leu CUA Leu CUG Leu	CCU Pro CCC Pro CCA Pro CCG Pro	CAU His CAC His CAA Gln CAG Gln	CGU Arg CGC Arg CGA Arg CGG Arg	U C A G
	A	AUU Ile AUC Ile AUA Ile AUG Met	ACU Thr ACC Thr ACA Thr ACG Thr	AAU Asn AAC Asn AAA Lys AAG Lys	AGU Ser AGC Ser AGA Arg AGG Arg	U C A G
	G	GUU Val GUC Val GUA Val GUG Val	GCU Ala GCC Ala GCA Ala GCG Ala	GAU Asp GAC Asp GAA Glu GAG Glu	GGU Gly GGC Gly GGA Gly GGG Gly	U C A G
						Third position

(a)

Amino acid	Codon	Number of codons
Alanine (A)	GC-x	4
Arginine (R)	CG-x	4
	AG-R	2
Asparagine (N)	AA-Y	2
Aspartic acid (D)	GA-Y	2
Cysteine (C)	UG-R	2
Glutamic acid (E)	GA-R	2
Glutamine (Q)	CA-R	2
Glycine (G)	GG-x	4
Histidine (H)	CA-Y	2
Isoleucine (I)	AU-Y	2
	AUA	1
Leucine (L)	CU-x	4
	UU-R	2
Lysine (K)	AA-R	2
Methionine (M)	AUG	1
Phenylalanine (F)	UU-Y	2
Proline (P)	CC-x	4
Serine (S)	UC-x	4
	AG-Y	2
Threonine (T)	AC-x	4
Tryptophan (W)	UGG	1
Tyrosine (Y)	UA-Y	2
Valine (V)	GU-x	4
STOP	UA-R	2
	UGA	1

(b)

Figure 13.13 The genetic code. (a) This table lists the three-letter nucleotide mRNA codons and the amino acid that is encoded by each. In this standard diagram, the first bases of each codon are shown on the left, the second bases across the top, and the third bases on the right. (b) With 20 naturally occurring amino acids and 61 codons that code for amino acids, most amino acids are encoded by more than one codon. The codons that code for the same amino acid are highly related to one another. For example, GCA, GCG, GCC, and GCU all encode alanine, so only the first two bases of the codon matter. This is shown as GC-x, where x can be any base. In the case of asparagine, the codons are AAU and AAC, that is, the third base is specified only as being a pyrimidine. This is shown as AA-Y in which Y stands for either pyrimidine, C or U similarly, R stands for either purine, A or G.

(tryptophan), which we will discuss momentarily. The number of codons that represent each amino acid varies—there are two codons for tyrosine and eight other amino acids; three codons for isoleucine; four codons for valine and four other amino acids; and six codons for leucine, arginine, and serine. Note further that the codons encoding the same amino acid are usually very similar to each other, as summarized in Figure 13.13(b). The tyrosine codons are UA-pyrimidine, the phenylalanine codons are UU-pyrimidine, the lysine codons are AA-purine, and so on. (The symbol Y is used for either pyrimidine, and R is used for either purine.) The valine codons are GU-x, and any nucleotide can be present in the third position. With only two exceptions, the third base in the codon is either not specified at all or is specified only as being either purine or either pyrimidine. This suggests that the interaction between the third base of the codon and the first base of the anticodon is the weakest of the base pairs—that is, that position can "wobble."

Because more than one codon can specify the same amino acid, the code is said to be degenerate. (Many think that this terminology is unfortunate. "Degenerate" implies that specificity has been lost during evolution. It seems more likely that specificity was added, that is, a primordial genetic code might have had two-base codons, which could specify 16 amino acids, or perhaps two-base codons with the third base specified to be either a purine or a pyrimidine, which could accommodate 24 amino acids. The need for specificity in the third position is likely to be a derived feature of the codon.) The important implication of a wobbling degenerate codon, as we noted with the examples above, is that transition mutations that replace a purine with a purine or a pyrimidine with a pyrimidine in the third base make almost no difference, and that any mutation in the third base makes very little difference.

KEY POINT There is degeneracy in the genetic code in that changes to the third base of the codon have little impact on which amino acid is encoded.

Wobbling at the start and stop

We noted that the third position is specifically important only for two pairs of codons AUA/AUG and UGA/UGG. These codons are almost the exceptions that prove the rule for wobble and degeneracy, because each of these pairs of codons has a special property. AUG is the start codon, and UGA is one of the stop codons, so perhaps we can infer why these two are uniquely specified. But what can we say about their partners AUA (isoleucine) and UGG (tryptophan)? Why should these be "wobble-related" to the special start and stop codons? (The other two stop codons UAA and UAG are wobble-related to each other.) We don't have an exact answer, but we can see some interesting properties that might provide insights.

Notice from Figure 13.14 that there are three isoleucine codons—AUU and AUC, as well as AUA. In other words, organisms do not have to use the AUA codon to specify isoleucine, since they can use one of the other two. Or to put this into a more correct language, there may be an evolutionary disadvantage to using AUA to specify isoleucine, rather than using AUU or AUC. In fact, this is probably true. We can use information from codon usage tables to look this up. These tables, compiled for millions of genes from thousands of organisms, list how often a particular codon is used for that organism. Isoleucine is a commonly used amino acid, comprising about 6% of all of the amino acids in proteins. However, in nearly every organism, AUA is among the least used codons; isoleucine is specified by AUA only about 10% of the time, and by AUU or AUC 90% of the time. Thus, there appears to have been selection against using AUA or in favor of using AUU and AUC. This may be a reflection of its wobble relationship with AUG.

For UGG/UGA, the situation is different. UGG is the only codon that can be used to specify tryptophan, so codon usage preference cannot apply here like it can for isoleucine. But tryptophan is the least commonly

First position	Second position: U	C	A	G	Third position
U	UUU Phe	UCU Ser	UAU Tyr	UGU Cys	U
	UUC Phe	UCC Ser	UAC Tyr	UGC Cys	C
	UUA Leu	UCA Ser	UAA STOP	UGA STOP	A
	UUG Leu	UCG Ser	UAG STOP	UGG Trp	G
C	CUU Leu	CCU Pro	CAU His	CGU Arg	U
	CUC Leu	CCC Pro	CAC His	CGC Arg	C
	CUA Leu	CCA Pro	CAA Gln	CGA Arg	A
	CUG Leu	CCG Pro	CAG Gln	CGG Arg	G
A	AUU Ile	ACU Thr	AAU Asn	AGU Ser	U
	AUC Ile	ACC Thr	AAC Asn	AGC Ser	C
	AUA Ile	ACA Thr	AAA Lys	AGA Arg	A
	AUG Met	ACG Thr	AAG Lys	AGG Arg	G
G	GUU Val	GCU Ala	GAU Asp	GGU Gly	U
	GUC Val	GCC Ala	GAC Asp	GGC Gly	C
	GUA Val	GCA Ala	GAA Glu	GGA Gly	A
	GUG Val	GCG Ala	GAG Glu	GGG Gly	G

Figure 13.14 The genetic code: starting and stopping. The third position in the codon matters for only two pairs of codons, highlighted here. AUG (in green) is the initiator codon; its wobble partner AUA is the least used of the three isoleucine codons (in blue). The stop codon UGA (in pink) is wobble-related to UGG, which encodes tryptophan (in purple), the least used amino acid. The other stop codons UAA and UAG are wobble-related to each other.

occurring amino acid in proteins, with a frequency of only about 1–1.5%. Thus, while there is not a selective bias against using UGG because there are no alternatives, UGG simply does not occur that often. Because of codon usage or amino acid frequency, AUA and UGG are two of the least used codons in every organism. Having them wobble-related to the start and a stop codon thus presents less of a problem for the cell. (We should note that the opposite may also have been true, since we do not know how the genetic code evolved. It is possible that AUG and UGA were selected for as start and stop codons because their cognate wobble-related codons are not commonly used.)

KEY POINT The codons that are "wobble"-related to the start AUG and the stop UGA codons are two of the least used codons.

Related amino acids have related codons

Every amino acid has its own unique biochemical properties, and some of them are chemically related to each other. We introduced this notion in Chapter 2 and summarize one important type of biochemical relatedness—hydrophobicity or water-repelling character—using colors in Figure 13.15. Hydrophobic (also called nonpolar) amino acids are shown in light green, whereas hydrophilic (or polar) amino acids are in other colors. Note that the codons for hydrophobic amino acids tend

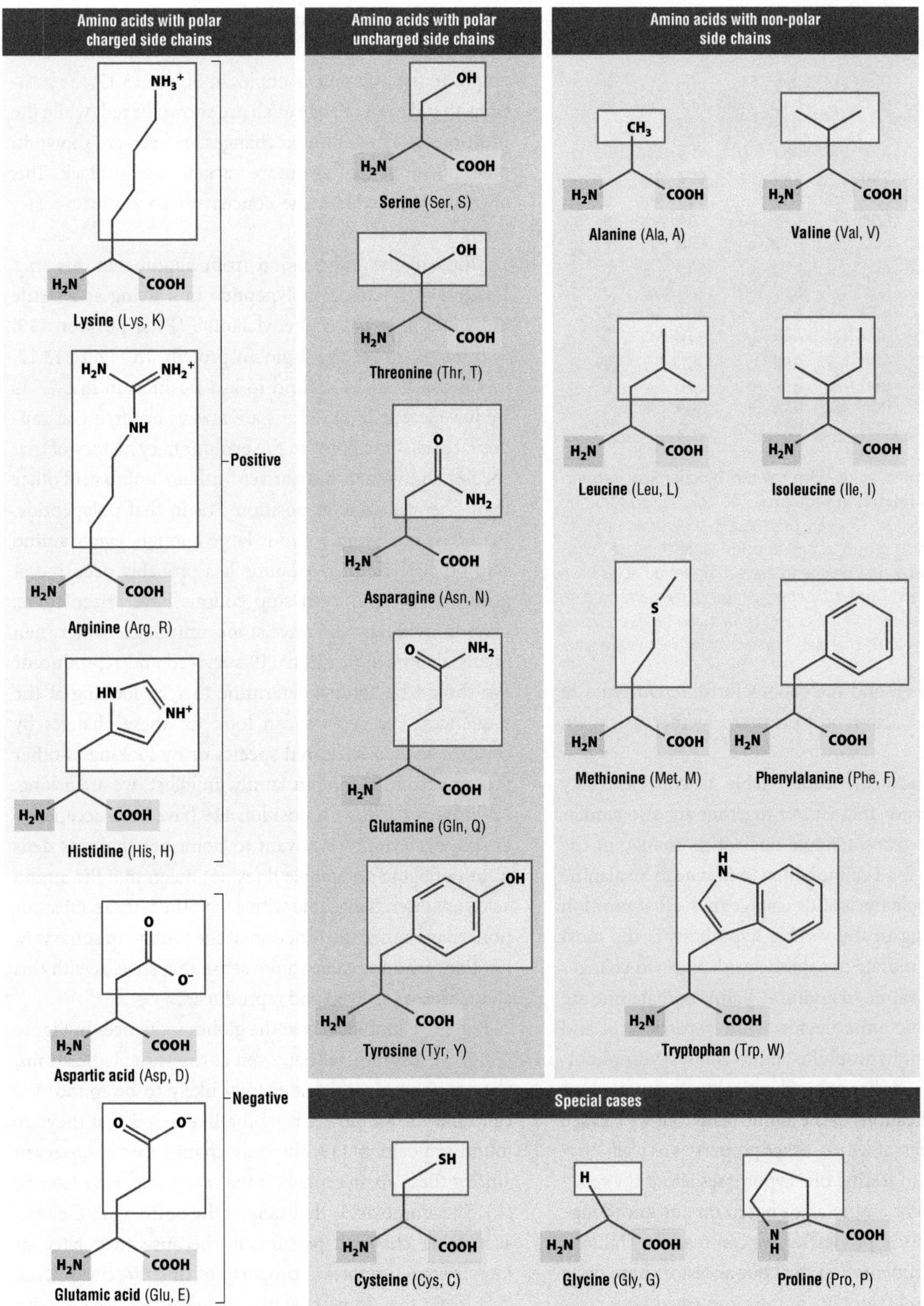

Figure 13.15 The structures of the 20 naturally occurring amino acids are grouped by the properties of their side chains. Other properties for grouping amino acids by their side chains are sometimes used, but this is one of the most common. The amino acids shaded in light green have non-polar or hydrophobic side chains. Among the amino acids with polar side chains, the ones shaded in light blue are charged, at neutral pH while the ones shaded in light purple are polar but not charged. The three amino acids highlighted in light orange have side chains with special properties that often dictate their behavior. The side chain on glycine is exceptionally small; the side chain on proline is particularly rigid, and the side chain on cysteine can form cross-links with another cysteine.

First position	Second position: U	C	A	G	Third position
U	UUU Phe	UCU Ser	UAU Tyr	UGU Cys	U
U	UUC Phe	UCC Ser	UAC Tyr	UGC Cys	C
U	UUA Leu	UCA Ser	UAA STOP	UGA STOP	A
U	UUG Leu	UCG Ser	UAG STOP	UGG trp	G
C	CUU Leu	CCU Pro	CAU His	CGU Arg	U
C	CUC Leu	CCC Pro	CAC His	CGC Arg	C
C	CUA Leu	CCA Pro	CAA Gln	CGA Arg	A
C	CUG Leu	CCG Pro	CAG Gln	CGG Arg	G
A	AUU Ile	ACU Thr	AAU Asn	AGU Ser	U
A	AUC Ile	ACC Thr	AAC Asn	AGC Ser	C
A	AUA Ile	ACA Thr	AAA Lys	AGA Arg	A
A	AUG Met	ACG Thr	AAG Lys	AGG Arg	G
G	GUU Val	GCU Ala	GAU Asp	GGU Gly	U
G	GUC Val	GCC Ala	GAC Asp	GGC Gly	C
G	GUA Val	GCA Ala	GAA Glu	GGA Gly	A
G	GUG Val	GCG Ala	GAG Glu	GGG Gly	G

Figure 13.16 Amino acids with related biochemical properties have codons related in sequence. The colors used to group the amino acids in Figure 13.15 are used here to show which codons correspond to which group of amino acids. Note that the colors form blocks, showing that biochemically similar amino acids have similar codons. Even if other biochemical properties are used to classify the amino acids, or if the order in which the bases are listed on the sides of the table is changed, blocks of color like this are seen.

to cluster together, and the codons for different types of hydrophilic ones also tend to cluster together, as shown in Figure 13.16.

Because of how the codon table is conventionally drawn, amino acids that cluster together are also similar in sequence—it takes a single nucleotide change in the second position for a valine codon to mutate to an alanine codon, but a single nucleotide change in the first position (which according to the wobble hypothesis is the most stable bond) to mutate to aspartic acid, and two changes to mutate from valine to arginine. Valine and alanine are both hydrophobic amino acids, while aspartic acid and arginine are both hydrophilic amino acids. Hydrophobicity or polarity is only one biochemical property used for chemical classifications of the amino acids, but we can see similar relatedness if we use other properties as well.

But rather than testing our hypotheses about the relatedness of different amino acids based on our knowledge of their biochemical properties, we can read the "billions of years of experimentation" that our ancestors have done during evolution by looking at the amino acid sequences of known proteins, as summarized for one protein in Figure 13.17. The amino acid sequences of the β-globin protein from different species of placental mammals are aligned with each other, so that each position in the polypeptide can be directly compared. As we examine this alignment, it is clear some positions in the sequence always have the same amino acid, whereas other positions can have different amino acids in different species. For example, in the β-globin examples in Figure 13.17, the positions that do not change at all are shown in red, while the positions at which limited changes are seen are shown in blue; positions that are more variable are in black. This alignment introduces the concept of an **evolutionarily accepted mutation.**

Think of the information from alignments this way. Imagine a functional polypeptide in a living and fertile individual that has a phenylalanine (F) at position 119, as humans do for the β-globin protein. In Figure 13.17, this is the F in the second to last position in line 2. As we have learned, mutations are always occurring at random, so, at some point in the evolutionary history of that species, individuals have arisen with an amino acid other than phenylalanine at position 119 in that polypeptide. In fact, if our sample size is large enough, **every** amino acid other than phenylalanine has probably occurred at position 119, and even stop codons have arisen there. Have individuals who have some amino acid other than phenylalanine at position 119 survived and reproduced? We should be able to determine that by looking at the sequences. Perhaps we can look for these changes by comparing closely related species or by looking at other proteins from the globin family. In effect, we are asking, "What substitutions at position 119 have been acceptable during evolution?" We want to point out that this definition of being acceptable does not mean that the amino acids are interchangeable with each other. These substitutions may change the function of the protein in some way, but they are acceptable in the sense that species with this change have survived and reproduced.

But as we look closely at the globin sequences in Figure 13.17, we also notice that, even at the variable positions, not every amino acid is equally likely to be found. We can rank the amino acids by the likelihood that they are found at position 119; the only change that is observed among these six mammals is that aardvarks have leucine (L). The amino acids that rank at the bottom are the least acceptable changes, presumably because they alter or impair some important property of this protein. In fact, it is likely that some changes are never (or very rarely) observed. If our sample is large, any mutation that could have occurred has occurred. So if we don't observe that change, that mutation must have so severely impaired the function of the protein that the individual could not survive and reproduce in nature; such a mutation is completely unaccepted evolutionarily. On the other hand,

```
Bat       -VHLTNEEKTAVIGLWGKVNVEEVGGEALGRLLVVYPWTQRFFESFGDLSSPSAIMGNPK
Aardvark  -MVLTADEKALVSSLWCKMNVDEAGAEALGRMLVVYPWTQRFFDHFGDLSSASAVMGNAK
Human     MVHLTPEEKSAVTALWGKVNVDEVGGEALGRLLVVYPWTQRFFESFGDLSTPDAVMGNPK
Rabbit    MVHLSSEEKSAVTALWGKVNVEEVGGEALGRLLVVYPWTQRFFESFGDLSSANAVMNNPK
Whale     -VHLTAEEKSAVTALWAKVNVEEVGGEALGRLLVVYPWTQRFFEAFGDLSTADAVMKNPK
Mouse     MVHFTAEEKAAITSIWDKVDLEKVGGETLGRLLIVYPWTQRFFDKFGNLSSALAIMGNPR
           : :: :**: : .:* *:::::.*.*:***:*:*********: **:**:. *:* *.:

Bat       VKAHGKKVLNSFSEGLKNLDNLKGTFAKLSELHCDKLHVDPENFRLLGYILLCVLARHFG
Aardvark  VKAHGKKVLNSFSDGLKHLDDLKGTFAQLSELHCDKLHVDPENFRLLGKXXVCVMARHLG
Human     VKAHGKKVLGAFSDGLAHLDNLKGTFATLSELHCDKLHVDPENFRLLGNVLVCVLAHHFG
Rabbit    VKAHGKKVLAAFSEGLSHLDNLKGTFAKLSELHCDKLHVDPENFRLLGNVLVIVLSHHFG
Whale     VKAHGKKVLASFSDGLKHLDDLKGTFATLSELHCDKLHVDPENFRLLGNVLVIVLARHFG
Mouse     IRAHGKKVLTSLGLGVKNMDNLKETFAHLSELHCDKLHVDPENFKLLGNMLVIVLSTHFA
          ::******* ::. *: ::*:** *** ****************:***   : *:: *:.

Bat       KEFTPQVQAAYQKVVAGVATALAHKYH
Aardvark  PEFTPQAQAAYQKVVAGVANALAHKYH
Human     KEFTPPVQAAYQKVVAGVANALAHKYH
Rabbit    KEFTPQVQAAYQKVVAGVANALAHKYH
Whale     KEFTPELQAAYQKVVAGVANALAHKYH
Mouse     KEFTPEVQAAWQKLVIGVANALSHKYH
           ****  ***:**:* ***.**:****
```

Figure 13.17 Allowed substitutions among globin sequences. Another method to infer the relatedness of amino acid properties is to determine which amino acids have been able to substitute for one another during evolution. This example shows the sequence of the β-globin protein from six different mammals; the amino acid sequences are continuous but shown in these three blocks due to space constraints. The sequences were aligned using CLUSTALW 2.1, a commonly used multiple sequence alignment program; other multiple alignment programs produce very similar results for these six proteins. Positions in the aligned sequences that are identical in all six species are indicated by a * in CLUSTALW below the sequences and in red for emphasis. Positions in which the amino acids are highly similar in all six species are indicated by a colon (:), whereas positions that have some similarity are indicated by a period (.). Positions that vary are shown without a symbol below them. Other multiple sequence alignments use different properties to determine which amino acids are considered similar, but CLUSTALW is standard.

the amino acids that rank at the top, the ones that are found the most often, must be the ones that are the most accepted evolutionarily. Evolutionary changes are therefore telling us which amino acids are the most and the least related to one another in function.

KEY POINT Comparisons among functionally related proteins from different species can show which substitutions have been tolerated during evolution and which have not survived.

If we then plot the frequency at which one amino acid has replaced another one, it is apparent that evolutionarily accepted substitutions involve amino acids that are close together in the codon table. The program we used to align the sequences, called CLUSTALW, depicts identical amino acids with a "*," those amino acids that are similar in charge and size by a ":," those amino acids that are similar in charge or size by a "."," and those that are not similar with a blank space.

We can use this estimate of biochemical relatedness made by CLUSTALW to illustrate another point about the genetic code. Of the 36 changes in these proteins involving the most similar amino acids, all but one can be accomplished by a single nucleotide change in the corresponding codon. Likewise, all nine of the changes involving similar amino acids could result from a single base change. By contrast, those changes that CLUSTALW considers non-conservative more often involve more than a single base change; in other words, it is likely that these amino acids have changed more than once. Thus, whether we use our knowledge and hypotheses about which amino acids are biochemically related to one another, or if we allow evolutionary history to tell us which amino acids have actually been able to replace one another, we come to similar conclusions—closely related amino acids are encoded by closely related codons. Again, we can infer why this would be an evolutionary advantage to minimize the deleterious impact of mutations.

Asking about acceptable mutations that occurred for one protein is only the beginning of this analysis. We can broaden our lines of enquiry by aligning functionally equivalent proteins from many different species—the amino acid sequences of aminoacyl synthetases that attach phenylalanine to the tRNA from hundreds or thousands of different species, for example. We can then compile a frequency table for each amino acid at each position, called a **position-specific weight matrix**, and determine which positions have allowed regular substitutions during evolution and which positions have not. In addition, we can ask which substitutions have been accepted at each position. We can combine these data with similar substitution matrices for thousands of families of related proteins and produce one 20 × 20 table that shows the frequency with which each amino acid has stayed the same and which ones have replaced which other ones.

The original table of such data, compiled by Dayhoff in 1978, is shown in Figure 13.18. This figure shows, among many families of proteins, how often each amino acid has been replaced by another and how often it has not changed. For example, among 10,000 alanine residues in the original protein used for a comparison, these have remained an alanine 9867 times. Each of the 133 changes to another amino acid is tabulated; an arginine was observed once, an asparagine was found four times, and so on. We discuss this process more fully in Tool Box 13.1.

There are many assumptions about protein evolution that go into this analysis, which go beyond our description here. Nonetheless, the database of substitutions is mammoth—billions of accepted substitutions involving tens of millions of proteins have now been compiled—so the numbers can be considered quite reliable.

KEY POINT Sequence comparisons that analyze which amino acids tend to be found in equivalent positions of a protein in different species can give information about which amino acids are most functionally related to one other. Closely related amino acids are encoded by closely related codons, reducing the deleterious impact of single base changes in the DNA.

		Ala	Arg	Asn	Asp	Cys	Gln	Glu	Gly	His	Ile	Leu	Lys	Met	Phe	Pro	Ser	Thr	Trp	Tyr	Val
		A	R	N	D	C	Q	E	G	H	I	L	K	M	F	P	S	T	W	Y	V
Ala	A	9867	2	9	10	3	8	17	21	2	6	4	2	6	2	22	35	32	0	2	18
Arg	R	1	9913	1	0	1	10	0	0	10	3	1	19	4	1	4	6	1	8	0	1
Asn	N	4	1	9822	36	0	4	6	6	21	3	1	13	0	1	2	20	9	1	4	1
Asp	D	6	0	42	9859	0	6	53	6	4	1	0	3	0	0	1	5	3	0	0	1
Cys	C	1	1	0	0	9973	0	0	0	1	1	0	0	0	0	1	5	1	0	3	2
Gln	Q	3	9	4	5	0	9876	27	1	23	1	3	6	4	0	6	2	2	0	0	1
Glu	E	10	0	7	56	0	35	9865	4	2	3	1	4	1	0	3	4	2	0	1	2
Gly	G	21	1	12	11	1	3	7	9935	1	0	1	2	1	1	3	21	3	0	0	5
His	H	1	8	18	3	1	20	1	0	9912	0	1	1	0	2	3	1	1	1	4	1
Ile	I	2	2	3	1	2	1	2	0	0	9872	9	2	12	7	0	1	7	0	1	33
Leu	L	3	1	3	0	0	6	1	1	4	22	9947	2	45	13	3	1	3	4	2	15
Lys	K	2	37	25	6	0	12	7	2	2	4	1	9926	20	0	3	8	11	0	1	1
Met	M	1	1	0	0	0	2	0	0	0	5	8	4	9874	1	0	1	2	0	0	4
Phe	F	1	1	1	0	0	0	0	1	2	8	6	0	4	9946	0	2	1	3	28	0
Pro	P	13	5	2	1	1	8	3	2	5	1	2	2	1	1	9926	12	4	0	0	2
Ser	S	28	11	34	7	11	4	6	16	2	2	1	7	4	3	17	9840	38	5	2	2
Thr	T	22	2	13	4	1	3	2	2	1	11	2	8	6	1	5	32	9871	0	2	9
Trp	W	0	2	0	0	0	0	0	0	0	0	0	0	0	1	0	1	0	9976	1	0
Tyr	Y	1	0	3	0	3	0	1	0	4	1	1	0	0	21	0	1	1	2	9945	1
Val	V	13	2	1	1	3	2	2	3	3	57	11	1	17	1	3	2	10	0	2	9901

Figure 13.18 The Dayhoff substitution matrix. The biochemist Margaret Dayhoff aligned orthologous proteins of many different types and computed how frequently each amino acid stayed the same and/or was replaced by another. This is the matrix of her original data. The top line is the presence of that amino acid in her reference sequence, while the column on the left shows how often it is found in orthologous sequences. The numbers are based on 10,000 occurrences of the amino acid. The frequencies of replacement are found by reading down the columns. From this, Dayhoff compiled a matrix of the evolutionarily allowed substitutions; this matrix formed the original scoring system for alignment programs such as BLASTP.

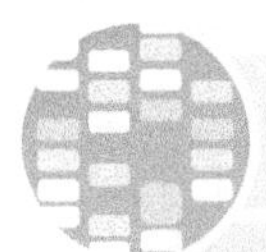

TOOL BOX 13.1 Evolution, genes, and genomes: sequence comparisons

Biological information depends on sequences—sequences of nucleotide bases in DNA and RNA, and sequences of amino acids in polypeptides. Changes in these sequences occur constantly and may result in changes in the functions of the nucleic acids or the proteins. These changes in function might be small or large and may be seen between individuals within a species within generations or between individuals of different species over evolutionary timescales. Genomic methods allow us to obtain vast amounts of sequence information to infer such relationships, as we noted with our discussion of phylogenetic trees in Section 4.7. However, once we obtain the sequence information, methods are needed to compare sequences for similarity and differences and to interpret the similarities and differences that are seen. Methods to compare sequences are widespread in molecular biology; the best known of these is the suite of programs known as BLAST. Interpreting the results of sequence similarity requires a bit more subtlety than simply pressing the button marked BLAST at the bottom of the page; every BLAST search and comparison should be understood as a test of hypotheses about evolution.

Recall from our discussion in Section 4.7 that phylogenetic trees are an attempt to show the evolutionary relationships among a group of individuals (or species or sequences). Since changes in sequences occur constantly, no current sequence (or node on the tree) represents the ancestral state that gave rise to the others. Rather, the similarities between sequences are used to infer their descent from a common, often unknown, ancestor. Descent from a common ancestor is defined as **homology**. We cannot observe homology or descent through evolutionary time. We can, however, observe similarities in the sequences and infer a shared origin and function.

Evolutionarily accepted mutations and the PAM1 matrix

An important principle in using sequence similarity to infer homology is the concept of an evolutionarily accepted mutation, introduced in Section 13.4. Remember Delbrück's statement that every living cell carries in its genome the results of a million years of experimentation by its ancestors. Nowhere is this experimentation more evident than when comparing sequences. Such comparisons can be done using DNA sequences or amino acids, although the theory is better understood for amino acid sequences, which we will discuss here.

Mutation is a constant generator of new sequences and combinations; natural selection provides the experimental test of the viability and fitness of those sequences. Thus, when protein sequences are aligned with each other, the identities, similarities, and differences can be compared, position by position, to learn which amino acids have been "accepted" at that site by evolutionary testing. Some positions in the sequences being compared will have the same amino acid. This suggests either that there has not been enough time for a change to arise or, more often, that any other amino acid at that site resulted in a non-functional protein and so was not accepted by selection. Likewise, some positions will have functionally similar, but not identical, amino acids, and yet others will have functionally distinct amino acids.

Of course, when the analysis is begun, we do not know which sites have tolerated changes or what changes can be tolerated. While these changes are called "accepted mutations"—accepted since the protein is still functional in both species—the changes may have altered some important properties of the protein, that is, they might not be neutral mutations. The accepted mutations are a tally of which changes have occurred and have not been eliminated by natural selection. In addition, when the analysis begins, we do not know which sites tolerate changes and which do not, so the safest and most informative approach is to tally all of the changes.

During the 1960s and 1970s, the protein biochemist Margaret Dayhoff did exactly this analysis for protein sequences by aligning related amino acid sequences from known proteins and determining how often each substitution would occur. (In order to simplify her work for the computers available at that time, she also devised the single-letter amino acid code, which we still use.) This was an incredible body of work, both in terms of the amount of effort and the amount of information; her work is considered to be one of the foundations of the field of bioinformatics.

In order to explain this work, it is informative to look again at one of her data tables, shown as Figure 13.18. This table shows the frequency at which different substitutions occurred when the overall rate of change was one amino acid in 100; Dayhoff referred to this as a PAM1 matrix. PAM is an acronym for "percent accepted mutation," a rough indicator of the evolutionary distance between the two sequences—a time long enough for 1% of the amino acids to be replaced. The numbers in the data table are given per 10,000 amino acids to make them easier to read, with the original amino acid in the row across the top and the replacement amino acid in the column on the left. The terms "original" and "replacement" simply refer to which sequence Dayhoff designated for her reference point and do not imply any information about the ancestral sequence that both of them shared.

The table in Figure 13.18 has a wealth of information, so let's look at the upper left corner to illustrate the data, expanded in Figure A. Among 10,000 positions at which alanine (A) was found in the original sequence, alanine was retained at 9867 of these positions in the corresponding sequence from a related protein, while one had an arginine (R), four had an asparagine (N), six had aspartic acid (D), and so on. Thus, by scanning down the column, we see that, at the PAM1 distance, most alanines do not change to another amino acid; however, when alanine has changed, it is almost ten times more likely to be serine (28 occurrences) than it

TOOL BOX 13.1 Continued

	A	R	N	D	C
A	9867	2	9	10	3
R	1	9913	1	0	1
N	4	1	9822	36	0
D	6	0	42	9859	0
C	1	1	0	0	9973

Figure A The Dayhoff PAM1 matrix. As described in the Tool Box, a substitution matrix was made to show how often one amino acid remained the same or changed to another amino acid when 1% of the amino acids have changed. The complete table is shown in Figure 13.18; a portion is shown here for discussion.

Source: Adapted from Atlas of Protein Structure and Function, Suppl 3, 1978 M.O. Dayhoff, ed.National Biomedical Research Foundation. 1979.

is to be leucine (three occurrences). Is it obvious why this is so? One can speculate about the various biochemical properties of alanine, serine, and leucine, but such speculation is not necessary; evolutionarily change has told us that this is what has occurred.

Figure 13.18 and Figure A are rich in information. For example, not all amino acids are equally likely to have changed; more accurately, changes have occurred but have been lost during evolution. By reading the diagonal, we see that, among 10,000 sites where it is found, 133 alanines have changed while only 87 arginines have changed. Thus, for whatever reason, changes in alanine are more likely to be tolerated than changes that involve an arginine. Overall, we can see that changes that involve cysteine, tryptophan, and tyrosine are much less common than changes involving asparagine and serine. We can also see changes that did not occur, at least at this PAM distance, for these proteins.

As noted in the chapter, a comparison of the changes in Figure 13.18 and the codon table in Figure 13.16 finds that the most evolutionarily accepted changes occur between amino acids whose codons differ by a single nucleotide. This relatedness carries over to the amino acids. The colors used in the codon table in Figure 13.16 indicate some basic biochemical properties of the amino acids. For example, alanine is in green and arginine is in light blue; when these, and other amino acids, are observed to have changed in the PAM diagram, they most often change to one with the same colors, representing the same biochemical properties. Thus, three different pieces of information about evolutionary change in protein sequences—the biochemistry of the amino acids, the sequences of the codons, and the observed substitutions—yield similar conclusions despite their different assumptions.

Substitution matrices, scoring matrices, and protein evolution

The PAM1 matrix sets a foundation for thinking about protein evolution and provides a method to use that understanding for genomic comparisons. Recall that it was based on very closely related sequences in which 1% of the amino acids had changed. To look over longer evolutionary time, Dayhoff multiplied the matrix by itself to extrapolate to PAM10 (10% of the amino acids that could change have changed), PAM100, and PAM250 tables. At PAM250 distances, the amino acids that could have changed have done so multiple times. The probability of a substitution from amino acid 1 to amino acid 2 is shown as the substitution matrix in Figure B.

The raw numbers in Figure B become awkward to work with, so it is convenient to convert them into a scoring matrix. A scoring matrix uses the probability that an amino acid might change—alanine is more likely to change than arginine—and the probability that it will change to a given second amino acid—alanine is more likely to change to serine than to leucine. It then uses these numbers to compute an overall probability of change from amino acid 1 to amino acid 2. These probabilities for each possible change among the amino acids are converted to log-odds scores, as discussed below, to produce a scoring matrix. The PAM250 scoring matrix is shown in Figure C(i).

For a variety of reasons, not least of which has been the explosion of amino acid sequence data in the decades since Dayhoff compiled her original substitution tables, the PAM matrices have largely been supplanted by another family of scoring matrices known as the BLOSUM matrices. The overall principles for the BLOSUM matrices are the same, but the assumptions used to compile the data are a bit different. The BLOSUM matrices begin by identifying "blocks" of amino acids that are similar in different proteins, that is, the proteins themselves may not be similar in overall sequence, but they share a particular sequence of amino acids that is highly related. Hundreds of thousands of such blocks have now been compiled, so the BLOSUM matrices rely entirely on observed data, rather than extrapolation, as had to be done for more distant PAM matrices.

Within that block of highly similar or identical sequences, the amino acid substitutions are noted. For example, in the widely used BLOSUM80 matrix, the blocks of amino acid sequence are

TOOL BOX 13.1 Continued

		Original amino acid																			
		Ala	Arg	Asn	Asp	Cys	Gln	Gln	Gly	His	Ile	Leu	Lys	Met	Phe	Pro	Ser	Thr	Trp	Tyr	Val
		A	**R**	**N**	**D**	**C**	**Q**	**E**	**G**	**H**	**I**	**L**	**K**	**M**	**F**	**P**	**S**	**T**	**W**	**Y**	**V**
Ala	**A**	13	6	9	9	5	8	9	12	6	8	6	7	7	4	11	11	11	2	4	9
Arg	**R**	3	17	4	3	2	5	3	2	6	3	2	9	4	1	4	4	3	7	2	2
Asn	**N**	4	4	6	7	2	5	6	4	6	3	2	5	3	2	4	5	4	2	3	3
Asp	**D**	5	4	8	11	1	7	10	5	6	3	2	5	3	1	4	5	5	1	2	3
Cys	**C**	2	1	1	1	52	1	1	2	2	2	1	1	1	1	2	3	2	1	4	2
Gln	**Q**	3	5	5	6	1	10	7	3	7	2	3	5	3	1	4	3	3	1	2	3
Glu	**E**	5	4	7	11	1	9	12	5	6	3	2	5	3	1	4	5	5	1	2	3
Gly	**G**	12	5	10	10	4	7	9	27	5	5	4	6	5	3	8	11	9	2	3	7
His	**H**	2	5	5	4	2	7	4	2	15	2	2	3	2	2	3	3	2	2	3	2
Ile	**I**	3	2	2	2	2	2	2	2	2	10	6	2	6	5	2	3	4	1	3	9
Leu	**L**	6	4	4	3	2	6	4	3	5	15	34	4	20	13	5	4	6	6	7	13
Lys	**K**	6	18	10	8	2	10	8	5	8	5	4	24	9	2	6	8	8	4	3	5
Met	**M**	1	1	1	1	0	1	1	1	1	2	3	2	6	2	1	1	1	1	1	2
Phe	**F**	2	1	2	1	1	1	1	1	3	5	6	1	4	32	1	2	2	4	20	3
Pro	**P**	7	5	5	4	3	5	4	5	5	3	3	4	3	2	20	6	5	1	2	4
Ser	**S**	9	6	8	7	7	6	7	9	6	5	4	7	5	3	9	10	9	4	4	6
Thr	**T**	8	5	6	6	4	5	5	6	4	6	4	6	5	3	6	8	11	2	3	6
Trp	**W**	0	2	0	0	0	0	0	0	1	0	1	0	0	1	0	1	0	55	1	0
Tyr	**Y**	1	1	2	1	3	1	1	1	3	2	2	1	3	15	1	2	2	3	31	2
Val	**V**	7	4	4	4	4	4	4	4	5	4	15	10	4	10	5	5	5	7	4	17

Figure B The mutation probability at PAM250. By multiplying the PAM1 information by itself to identify greater evolutionary distances, more PAM matrices were constructed. The PAM250 mutation probability matrix, shown here, represents the likelihood that a particular substitution is observed when every position has had the opportunity to change at least two and a half times. Note that, even at this distance, the most likely outcome is that the same amino acid is observed at each position. However, the probability of change varies for each amino acid.

Source: Adapted from *Atlas of Protein Structure and Function,* Suppl 3, 1978 M.O. Dayhoff, ed.National Biomedical Research Foundation. 1979.

80% or more identical; the substitution frequencies are based on the remaining 20% of the amino acids that have changed. In a BLOSUM45 matrix, the blocks are identical in 45% or more of their amino acids, and the substitution frequencies are based on the remaining 55%. (Thus, all of the blocks used in a BLOSUM80 matrix will also be used in a BLOSUM45 matrix.) Again, the BLOSUM matrix provides a measure of evolutionary distance and time; a BLOSUM80 matrix includes only highly related sequences, so it is more appropriate for comparing substitutions that have occurred among more closely related species, such as among mammals. A BLOSUM45 matrix is appropriate for less closely related sequences, and thus for organisms over a longer evolutionary distance, such as bacteria and mammals.

The extensively used BLOSUM62 matrix is shown in Figure C(ii); this scoring matrix is appropriate for comparisons among two different eukaryotic proteins. (You may note in Figure C(ii) that BLOSUM includes some additional abbreviations, such as B, Z, and X, to represent particular pairs of related amino acids. For example, B is used to mean either asparagine (N) or glutamine (D), since these are closely related amino acids.) Note that the scoring matrix is a series of numbers, ranging from –4 to 12, with many of the numbers being 0. These numbers are the log-odds scores for each substitution—that is, the logarithm of the probability of a particular change. These numbers are calculated essentially as they are for the PAM250 matrix, using the probability that an amino acid changes and the overall frequencies of the different amino acids; the key difference with a BLOSUM matrix is that this is based only on observed changes in amino acid blocks.

How do we read such a scoring matrix? If the rate of the predicted change is the same as the observed rate of that change (that is, if a change has occurred as often as predicted by chance), then the BLOSUM score is 0. For example, changes from alanine (A) to threonine (T) occur about as often as expected, given how often changes at alanines occur and the overall frequency of threonine.

Positive numbers in the matrix are changes or relationships that occur far more often than predicted by chance. Recall that, in all cases, the most likely situation is that the amino acid has not changed at all, so a score indicating that an amino acid will stay the same in both proteins is always a positive number. Thus, tryptophan (W) and cysteine (C), the two amino acids that were the most likely to stay the same in the PAM1 substitution table, receive the highest BLOSUM scores, 12 for tryptophan remaining as a tryptophan and 9 for cysteine remaining as a cysteine. Conversely, scores of less than 1 represent changes that occur less often than expected by chance; a phenylalanine (F) in the first sequence is very unlikely to have been replaced by a proline (P) in the second sequence, as shown by the BLOSUM score of –4.

TOOL BOX 13.1 Continued

	A	R	N	D	C	Q	E	G	H	I	L	K	M	F	P	S	T	W	Y	V
A	2	−1	0	0	0	0	0	0	−1	−1	−1	0	−1	−2	0	1	1	−4	−2	0
R	−1	5	0	0	−2	2	0	−1	1	−2	−2	3	−2	−3	−1	0	0	−2	−2	−2
N	0	0	4	2	−2	1	1	0	1	−3	−3	1	−2	−3	−1	1	0	−4	−1	−2
D	0	0	2	5	−3	1	3	0	0	−4	−4	0	−3	−4	−1	0	0	−5	−3	−3
C	0	−2	−2	−3	12	−2	−3	−2	−1	−1	−2	−3	−1	−1	−3	0	0	−1	0	0
Q	0	2	1	1	−2	3	2	−1	1	−2	−2	2	−1	−3	0	0	0	−3	−2	−2
E	0	0	1	3	−3	2	4	−1	0	−3	−3	1	−2	−4	0	0	0	−4	−3	−2
G	0	−1	0	0	−2	−1	−1	7	−1	−4	−4	−1	−4	−5	−2	0	−1	−4	−4	−3
H	−1	1	1	0	−1	1	0	−1	6	−2	−2	1	−1	0	−1	0	0	−1	2	−2
I	−1	−2	−3	−4	−1	−2	−3	−4	−2	4	3	−2	2	1	−3	−2	−1	−2	−1	3
L	−1	−2	−3	−4	−2	−2	−3	−4	−2	3	4	−2	3	2	−2	−2	−1	−1	0	2
K	0	3	1	0	−3	2	1	−1	1	−2	−2	3	−1	−3	−1	0	0	−4	−2	−2
M	−1	−2	−2	−3	−1	−1	−2	−4	−1	2	3	−1	4	2	−2	−1	−1	−1	0	2
F	−2	−3	−3	−4	−1	−3	−4	−5	0	1	2	−3	2	7	−4	−3	−2	4	5	0
P	0	−1	−1	−1	−3	0	0	−2	−1	−3	−2	−1	−2	−4	8	0	0	−5	−3	−2
S	1	0	1	0	0	0	0	0	0	−2	−2	0	−1	−3	0	2	2	−3	−2	−1
T	1	0	0	0	0	0	0	−1	0	−1	−1	0	−1	−2	0	2	2	−4	−2	0
W	−4	−2	−4	−5	−1	−3	−4	−4	−1	−2	−1	−4	−1	4	−5	−3	−4	14	4	−3
Y	−2	−2	−1	−3	0	−2	−3	−4	2	−1	0	−2	0	5	−3	−2	−2	4	8	−1
V	0	−2	−2	−3	0	−2	−2	−3	−2	3	2	−2	2	0	−2	−1	0	−3	−1	3

(i) PAM 250 matrix

	A	R	N	D	C	Q	E	G	H	I	L	K	M	F	P	S	T	W	Y	V	B	Z	X	*
A	4	−1	−2	−2	0	−1	−1	0	−2	−1	−1	−1	−1	−2	−1	1	0	−3	−2	0	−2	−1	0	−4
R	−1	5	0	−2	−3	1	0	−2	0	−3	−2	2	−1	−3	−2	−1	−1	−3	−2	−3	−1	0	−1	−4
N	−2	0	6	1	−3	0	0	0	1	−3	−3	0	−2	−3	−2	1	0	−4	−2	−3	3	0	−1	−4
D	−2	−2	1	6	−3	0	2	−1	−1	−3	−4	−1	−3	−3	−1	0	−1	−4	−3	−3	4	1	−1	−4
C	0	−3	−3	−3	9	−3	−4	−3	−3	−1	−1	−3	−1	−2	−3	−1	−1	−2	−2	−1	−3	−3	−2	−4
Q	−1	1	0	0	−3	5	2	−2	0	−3	−2	1	0	−3	−1	0	−1	−2	−1	−2	0	3	−1	−4
E	−1	0	0	2	−4	2	5	−2	0	−3	−3	1	−2	−3	−1	0	−1	−3	−2	−2	1	4	−1	−4
G	0	−2	0	−1	−3	−2	−2	6	−2	−4	−4	−2	−3	−3	−2	0	−2	−2	−3	−3	−1	−2	−1	−4
H	−2	0	1	−1	−3	0	0	−2	8	−3	−3	−1	−2	−1	−2	−1	−2	−2	2	−3	0	0	−1	−4
I	−1	−3	−3	−3	−1	−3	−3	−4	−3	4	2	−3	1	0	−3	−2	−1	−3	−1	3	−3	−3	−1	−4
L	−1	−2	−3	−4	−1	−2	−3	−4	−3	2	4	−2	2	0	−3	−2	−1	−2	−1	1	−4	−3	−1	−4
K	−1	2	0	−1	−3	1	1	−2	−1	−3	−2	5	−1	−3	−1	0	−1	−3	−2	−2	0	1	−1	−4
M	−1	−1	−2	−3	−1	0	−2	−3	−2	1	2	−1	5	0	−2	−1	−1	−1	−1	1	−3	−1	−1	−4
F	−2	−3	−3	−3	−2	−3	−3	−3	−1	0	0	−3	0	6	−4	−2	−2	1	3	−1	−3	−3	−1	−4
P	−1	−2	−2	−1	−3	−1	−1	−2	−2	−3	−3	−1	−2	−4	7	−1	−1	−4	−3	−2	−2	−1	−2	−4
S	1	−1	1	0	−1	0	0	0	−1	−2	−2	0	−1	−2	−1	4	1	−3	−2	−2	0	0	0	−4
T	0	−1	0	−1	−1	−1	−1	−2	−2	−1	−1	−1	−1	−2	−1	1	5	−2	−2	0	−1	−1	0	−4
W	−3	−3	−4	−4	−2	−2	−3	−2	−2	−3	−2	−3	−1	1	−4	−3	−2	11	2	−3	−4	−3	−2	−4
Y	−2	−2	−2	−3	−2	−1	−2	−3	2	−1	−1	−2	−1	3	−3	−2	−2	2	7	−1	−3	−2	−1	−4
V	0	−3	−3	−3	−1	−2	−2	−3	−3	3	1	−2	1	−1	−2	−2	0	−3	−1	4	−3	−2	−1	−4
B	−2	−1	3	4	−3	0	1	−1	0	−3	−4	0	−3	−3	−2	0	−1	−4	−3	−3	4	1	−1	−4
Z	−1	0	0	1	−3	3	4	−2	0	−3	−3	1	−1	−3	−1	0	−1	−3	−2	−2	1	4	−1	−4
X	0	−1	−1	−1	−2	−1	−1	−1	−1	−1	−1	−1	−1	−1	−2	0	0	−2	−1	−1	−1	−1	−1	−4
*****	−4	−4	−4	−4	−4	−4	−4	−4	−4	−4	−4	−4	−4	−4	−4	−4	−4	−4	−4	−4	−4	−4	−4	−1

(ii) Blosum 62

Figure C Scoring matrices. The mutation probability matrix is converted to a log-odds matrix by identifying the probability that amino acid 1 changes at all and the probability that it changes to amino acid 2. By converting these probability matrices using logarithms, a substitution or scoring matrix is obtained. (i) shows the PAM250 scoring matrix, which used the data in Figure B. (ii) shows the more commonly used BLOSUM62 scoring matrix. In each case, a score of 0 indicates the change occurred as often as expected by chance. A score greater than zero indicates a change that is more likely to be observed than expected by chance, whereas a negative number indicates a change is less likely than chance. The * is the penalty for a gap. These scoring matrices are an important component of the BLASTP program used for sequence comparisons.

Source: BLOSUM matrix from Henikoff, S; Henikoff, JG (1992). Amino acid substitution matrices from protein blocks. *Proceedings of the National Academy of Sciences of the United States of America* 89 (22): 10915–9. doi:10.1073/pnas.89.22.10915. PMC 50453. PMID 1438297.

TOOL BOX 13.1 Continued

We see this same pattern with common substitutions. For example, the PAM1 matrix showed that alanine was most likely to change to a serine, and that this change was much more likely than the alanine becoming a leucine in the second protein; in the BLOSUM62 matrix, alanine to serine has a score of 1 (meaning that it occurs more often than expected by chance), while alanine to leucine has a score of −1 (meaning that it occurs less often than expected by chance).

BLASTing protein evolution

It is interesting and informative to look at these scoring matrices to make inferences about evolutionary changes. However, the most important role of these scoring matrices arises from their use in the alignment program BLASTP. BLAST is a series of programs that aligns a given sequence to all of the other known sequences of the same type and computes a score for each alignment; BLASTP is the program that does this for amino acid sequences. The numbers in the BLOSUM scoring matrix are used to calculate the score for a particular alignment between two sequences; for example, having a tryptophan (W) at the same site in both sequences is worth 12 "points," whereas having a tryptophan at the site in one protein and an alanine at the site in the second sequence deducts 3 points, since those score a −3. The total score for the alignment of two sequences is calculated by simply adding the individual scores. As we noted with the rules of probability in Chapter 5, adding the scores is equivalent to multiplying their probabilities, since these scores are based on logarithms. The best alignment between the two sequences is the one that receives the highest score.

It is important to recognize that this similarity score is based on relationships among known amino acid sequences from functional and existing proteins, rather than some arbitrary standard. The highest score will come from the alignment that has a combination of the most identities, the most accepted mutations, and the fewest unlikely mutations. BLAST is among the most widely used tools in molecular biology. At the core of its functionality are the evolutionary models contained in the BLOSUM scoring matrix—a fusion of evolution, genes and their protein products, and genomics.

The puzzle of serine

Let's look some more at the genetic code in Figure 13.19. Three amino acids are encoded by six codons—serine, leucine, and arginine. All of these are commonly used amino acids—leucine is the most frequently occurring amino acid in proteins, in fact—but amino acid frequency alone does not explain the number of codons for each amino acid. For leucine and arginine, the six codons are related to one another; for leucine, the six are UU-R (that is, either purine in the third position) and CU-x, for instance. But serine is different. Four serine codons are UC-x, while the other two are AG-Y (with either pyrimidine in the third position). In other words, the two sets of serine codons cannot be connected to each other by less than two nucleotide substitutions.

This is a puzzle. All of the serine codons are used in most organisms, albeit with some difference in codon usage, so we cannot use the same argument that we used to account for the AUA codon for isoleucine. In fact, the least often used serine codon in an organism is typically UCG, and the AG-Y pair seems to be used about as often as the UC-R pair. There has been speculation about the lack of relationship between the two sets of serine codons, but it may help to remind us that not all secrets of evolution are given up readily.

First position	Second position: U	C	A	G	Third position
U	UUU Phe	UCU Ser	UAU Tyr	UGU Cys	U
U	UUC Phe	UCC Ser	UAC Tyr	UGC Cys	C
U	UUA Leu	UCA Ser	UAA STOP	UGA STOP	A
U	UUG Leu	UCG Ser	UAG STOP	UGG Trp	G
C	CUU Leu	CCU Pro	CAU His	CGU Arg	U
C	CUC Leu	CCC Pro	CAC His	CGC Arg	C
C	CUA Leu	CCA Pro	CAA Gln	CGA Arg	A
C	CUG Leu	CCG Pro	CAG Gln	CGG Arg	G
A	AUU Ile	ACU Thr	AAU Asn	AGU Ser	U
A	AUC Ile	ACC Thr	AAC Asn	AGC Ser	C
A	AUA Ile	ACA Thr	AAA Lys	AGA Arg	A
A	AUG Met	ACG Thr	AAG Lys	AGG Arg	G
G	GUU Val	GCU Ala	GAU Asp	GGU Gly	U
G	GUC Val	GCC Ala	GAC Asp	GGC Gly	C
G	GUA Val	GCA Ala	GAA Glu	GGA Gly	A
G	GUG Val	GCG Ala	GAG Glu	GGG Gly	G

Figure 13.19 The genetic code: the serine codons. Serine is specified by six codons, UC-x and AG-Y (AGU and AGC). The two sets of codons are not related to one another, which invites speculation about the evolution of the genetic code and serine.

The universal code

The genetic code is universal, and all organisms across all kingdoms of life use the same code to convert a nucleic acid sequence to an amino acid sequence. (Some protozoa and mitochondria have slightly modified versions of the code, typically involving AUA or UGA, or both, but these seem to be secondarily derived from the original universal code.) This fact has both practical and theoretical significance. Let's list a few implications.

- First, all existing life forms can be traced back to a common origin. The universality of the genetic code is strong evidence for common ancestors for all living organisms.
- Second, all species are a finite number of genetic changes from every other species. In Chapters 4 and 11, we demonstrated the construction of tree diagrams that show the relatedness of genes and species to each other. Naturalists and philosophers since antiquity have noted the relatedness of species and constructed diagrams to illustrate this relatedness. The universal genetic code provides one mechanism to explain how this relatedness occurred.
- Third, if the DNA sequence from one species can be transcribed in another species, it will be accurately translated into the same amino acid sequence as found in the first species. In evolutionary terms, this is important in horizontal gene transfer, as discussed in Chapter 11. In broader biological terms, it is also important in infections and parasitism—a virus, for example, transfers its genome into its host, and the host makes viral proteins as a result. Have you had a cold or the flu lately? Blame or credit the universal genetic code because your ribosomes and tRNAs are translating viral mRNAs to make viral proteins in human cells. The universality of the code is also essential for many practical laboratory applications of genetic analysis and genetic engineering. We can easily move DNA from one species into another species—the cDNA sequence of a mammalian gene for insulin can be used to transform *E. coli*. We might take for granted that *E. coli* will accurately produce mammalian insulin. But this would not happen if mammals and bacteria used different genetic codes.

KEY POINT All kingdoms of life are evolutionarily related through their use of the same basic genetic code. This has important and far-reaching implications for the ways organisms evolve and how they interact.

It is nearly impossible to explain esthetics. If you have tried to convince a skeptic why your favorite artist, movie, or musician is under-appreciated, you know that beauty really does lie in the eye of the beholder. But we will assert that the genetic code is one of the most beautiful parts in all of biology. It lies at one of the most scenic intersections of evolution, genomics, and genetics.

13.5 Post-translational regulation

The version of the polypeptide that is released from the ribosome is only rarely the operative protein in the organism. Nearly every polypeptide is modified after it is synthesized, sometimes as part of the process of transporting it to the site of action within the cell, tissue, or organism. All of these post-translational modifications provide further opportunities for the regulation of function. Because our focus is on genetics and genomics, rather than biochemistry and the broader aspects of molecular biology, we will not discuss these in any detail, but we summarize them in Figure 13.20.

Covalent modifications to amino acids

Many covalent modifications are made to amino acids that affect the function of proteins. We noted in Section 12.4 and Figure 12.23 that histones can be methylated,

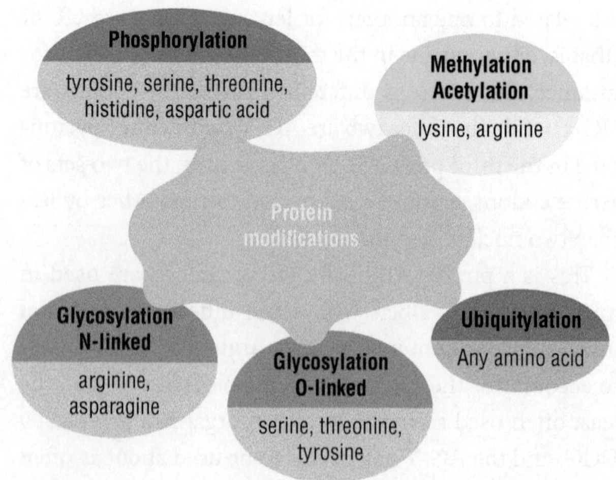

Figure 13.20 A summary of protein modifications. Once a polypeptide is made by translation, numerous post-translational modifications may occur, as summarized here. The amino acids at which the modifications occur are indicated.

acetylated, and phosphorylated, for example. Methyl groups and acetyl groups are usually attached to the positively charged amino acids lysine and arginine; consequently, histones that are rich in lysine and arginine are prime targets for these modifications. Phosphorylation is much more widespread; there are many thousands of phosphorylated proteins, of which histone H3 is only one.

Phosphorylation is carried out by enzymes known as kinases. Phosphorylation can occur on serine and threonine, or on tyrosine; bacteria also have phosphorylation on histidine and aspartic acid, and some plants have histidine phosphorylation as well. These different kinases are classified according to the amino acid they phosphorylate: serine/threonine kinases, tyrosine kinases, or histidine kinases. A genome encodes dozens or hundreds of different kinases. In Chapter 15, we will discuss a particular example with cell signaling involving tyrosine kinases. Changes in the phosphorylation of certain proteins are among the hallmarks of changes that occur in cancer cells, for example.

In addition to these changes, many proteins are glycosylated, with different carbohydrate molecules being attached to them either as they are being translated or as they are being processed and transported in the cell. The two most common forms of glycosylation are N-linked glycosylation (where the modification occurs at the nitrogen of arginine or asparagine) and O-linked glycosylation (where the modification occurs at the oxygen of serine, threonine, or tyrosine—or to modified versions of proline and lysine called hydroxyproline and hydroxylysine). Glycosylation was introduced in Chapter 8 when we discussed the ABO blood types; the ABO locus in humans encodes an enzyme called a glycosyl transferase, which attaches different sugars to a polypeptide stem. Nearly all proteins can be ubiquitylated, which tags them for degradation.

Proteolytic cleavage

For some proteins, the polypeptide arising from translation is processed by specific cleavage reactions. One familiar example is the peptide hormone insulin, which is made as a larger protein called preproinsulin and then cleaved twice, first to produce proinsulin and then to produce insulin, as shown in Figure 13.21.

One common proteolytic cleavage occurs at the N-terminus. Nearly all mRNAs begin the open reading frame with AUG, so nearly all newly translated polypeptides have methionine as the amino acid at the N-terminus. However, this methionine is frequently cleaved off the protein during post-translational processing, so many functional proteins do not have methionine at the N-terminus.

In some cases, multiple products of a proteolytic cleavage are functional peptides. For example, the principal proteins that assemble into a human immunodeficiency virus (HIV) particle are encoded by the *gag* and *pol* genes, which produce

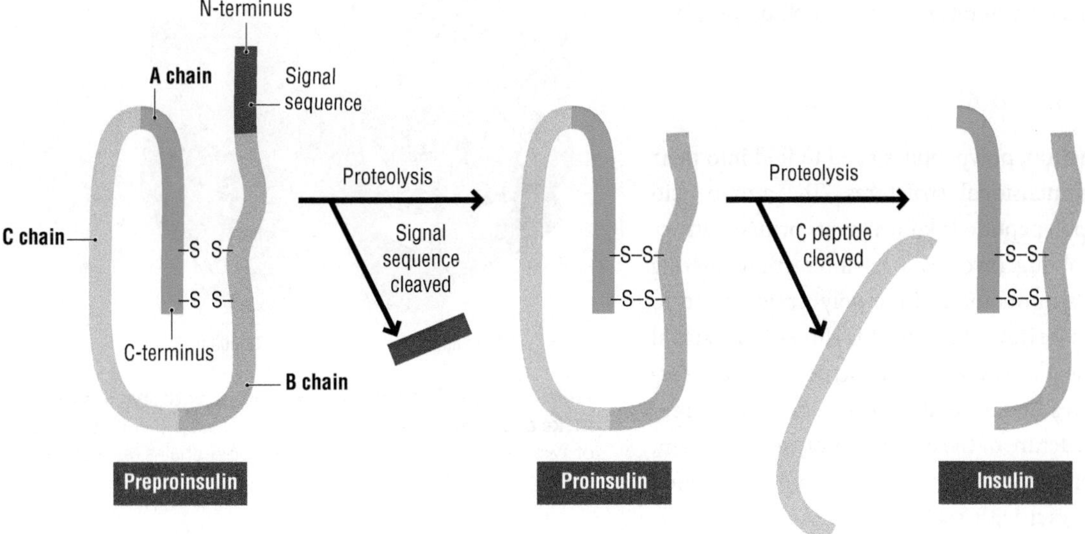

Figure 13.21 Post-translational cleavage of polypeptides: insulin. Insulin is made by two proteolytic cleavage steps. The initial preproinsulin polypeptide of 110 amino acids is metabolically inactive. The first cleavage removes the signal sequence from the N-terminus during synthesis, and proinsulin is trafficked from the endoplasmic reticulum to the Golgi apparatus. In addition, disulfide bonds form between parts of the sequence that will become the A and B chains, as well as some within the A chain that are not shown. The second cleavage releases the C chain of 31 amino acids, and the A and B chains form the active insulin protein of 61 amino acids.

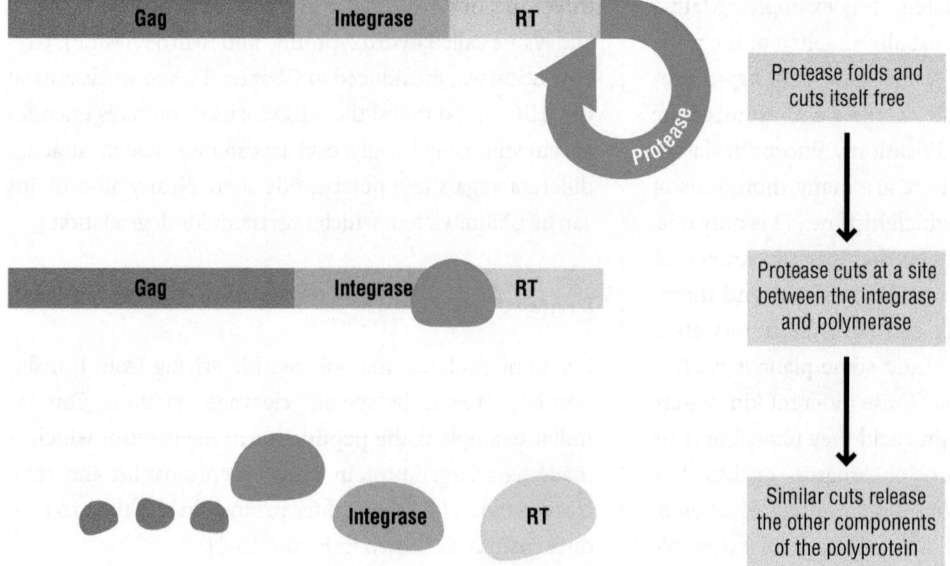

Figure 13.22 Post-translational cleavage of polypeptides: HIV polyprotein. The HIV genome is transcribed and translated to make a single large polyprotein. This polyprotein is cleaved in two steps to make the other components of the virus. The first cleavage frees the protease from the rest of the polyprotein. The protease then cleaves to release integrase, reverse transcriptase (RT), and the Gag polypeptides; while many retroviruses have a single large Gag protein making up the capsid, in HIV, the Gag protein is cleaved into smaller polypeptides that make up the capsid.

a transcript that is translated into a single polypeptide chain, as shown in Figure 13.22. The GagPol polyprotein is then cleaved by protease, one of the proteins found in this polyprotein, to produce final versions of protease, reverse transcriptase, and integrase, which are packaged into the virus, and Gag subunits, which are structural proteins for the virus capsule.

Assembly into a multimeric protein

In order to function, polypeptides need to fold into their proper three-dimensional structures. The amino acid sequence of a polypeptide is known as its primary structure. Protein folding involves local interactions among the amino acids in one part of the polypeptide chain to form regions possessing characteristic three-dimensional structures, known as the secondary structure; these secondary structures then assemble to give the larger three-dimensional structure of the complete polypeptide chain, known as its tertiary structure. Two or more polypeptide chains may interact with each other or with other molecules to form a larger complex with multiple subunits, known as the quaternary structure; the quaternary structure for hemoglobin is diagrammed in Figure 13.23. All of these interactions occur after translation is completed (or during translation in some cases), and all are critical for the proper functions of the polypeptide.

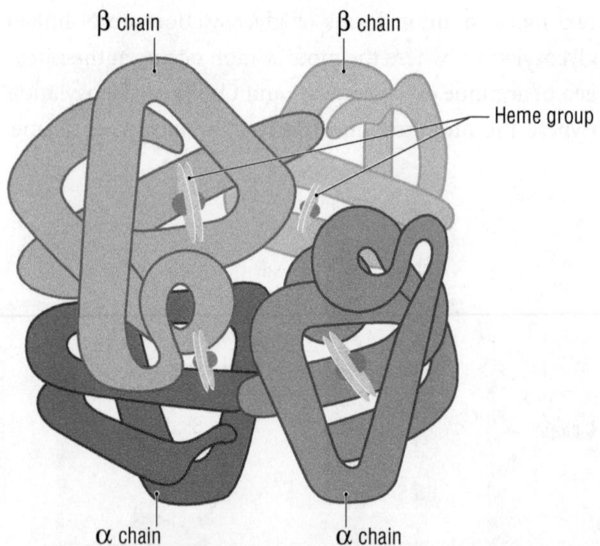

Figure 13.23 Polypeptides assemble with each other to make a functional protein. The hemoglobin protein is composed of two α-globin chains and two β-globin chains, as well as the heme group.

KEY POINT Many proteins undergo additional modifications after translation, including covalent modifications (such as methylation, phosphorylation, and glycosylation), proteolytic cleavage into smaller subunits, and/or assembly into higher-order structures and complexes.

13.6 Summary: gene expression and its regulation

Although Chapters 12 and 13 have covered a lot of information, there is much more that would take us beyond the scope of this book. In this summary section, we return to an overview, not only of the processes described in Chapters 12 and 13 but also as they relate to other chapters in the book. We summarize this in Figure 13.24.

Phenotypes arise from the complicated interplay among the functions of many proteins and RNA, and the impact of environmental factors such as nutrition, climate, and many others. In Chapters 5 through 10, we described the patterns by which single gene and complex phenotypes are inherited without much consideration given to the biological processes that generate such phenotypes. Whether the connection between a gene and a phenotype is complex or relatively simple, the gene must first be transcribed. The regulation of transcription therefore lies at the heart of how genetics contributes to phenotypes, both at the level of the individual and across the course of evolution.

In order to produce a phenotype, or some component of a complex phenotype, genes have to be expressed at the proper time and location. Figure 13.24 summarizes many (but not all) of the ways that organisms regulate the expression of genes. The first and critical step in that expression is transcription into RNA. Species vary in how much of their genomes are transcribed; for viruses and bacteria, almost all of the genome is transcribed or is involved in the regulation of transcription, as summarized in Figure 13.24(a). For eukaryotes, much of the regulation of gene expression occurs at the pre-initiation of transcription, but additional common regulatory steps occur by RNA splicing and via interactions with microRNAs, as shown in Figure 13.24(b). We also have come to recognize that many more transcripts are produced than are transcribed; the functions of these long non-coding RNAs are largely to be determined, but their regulation also appears to occur primarily at the pre-initiation of transcription, as shown in Figure 13.24(c). While most of the attention in research laboratories has focused on the pre-initiation and initiation of transcription, splicing, and microRNAs as the most important aspects of gene regulation, none of the processes to regulate gene expression is unimportant.

In addition to the importance of regulating gene expression for cellular functions and differentiation,

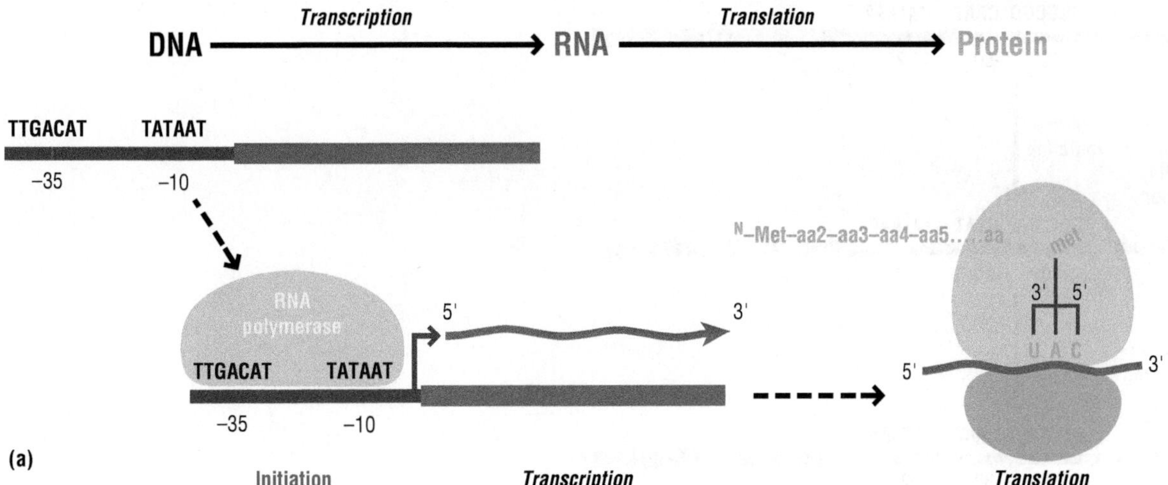

Figure 13.24 Regulating gene expression. Variation and regulation of gene expression can occur at almost any step in the processes shown in the Central Dogma, but steps indicated in red are those that are most likely to vary between cells during development and differentiation, between organisms in population diversity, and between species during evolution. (a) Bacteria. In bacteria, most of the regulation of gene expression occurs at the initiation of transcription. (b) Eukaryotes. For protein-coding genes in eukaryotes, most of the regulation occurs during pre-initiation of transcription with the binding of transcription factors. Significant regulation also occurs during splicing and with microRNAs. (c) Non-coding RNA genes in eukaryotes. A significant number of the transcripts made in a genome are never translated into proteins. The regulation of these transcripts occurs primarily at the pre-initiation of transcription. RNAP, RNA polymerase; TBP, TATA-binding protein.

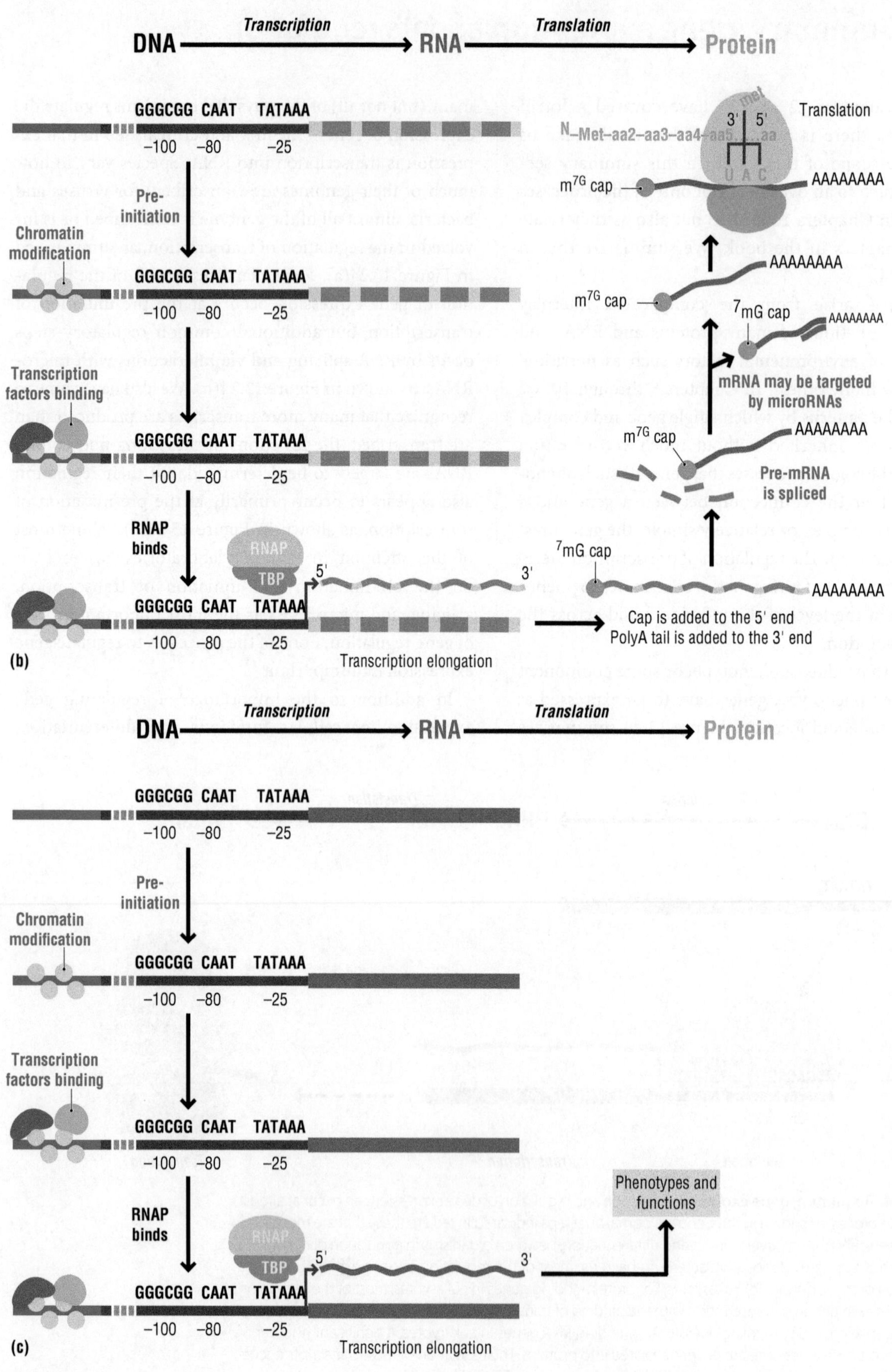

Figure 13.24 *(Continued)*

regulation of gene expression is also important for evolutionary change. We have known for a long time that changes in the protein-coding regions of genes that result in changes in their amino acid sequences are crucial for evolutionary change. Other than cases of recent horizontal gene transfer discussed in Chapter 11, it is very unlikely that a gene has the exact same nucleotide sequence in two different species, no matter how closely related the species are. Evolution requires variation. These changes in the amino acid sequences are the modifications that comprise an essential component of evolution.

But evolutionary change also involves modifications in all of the other processes in gene regulation and function. Typically, the steps in the regulation of gene expression that have changed during evolution are the same as those that vary between different cells or at different times. Even when two species have highly similar genes with similar functions, differences in the time, location, and amount of transcription are usually seen; these arise from differences in the transcriptional regulatory region of the genes. For metazoans, there will likely be differences in the splicing patterns of the genes in related species, which may be small or large. Evolutionary differences may also be seen in the processes of transcriptional elongation, termination, or translation, but these are less likely to change. For metazoans, evolutionary changes in the regulation by microRNAs are also common, possibly as common as differences in transcriptional initiation and splicing. Differences in any of the processes of post-translational modifications are also widespread.

For at least three decades, genetic analysis has investigated the molecular changes associated with changes in gene expression, as these occur between different tissues or at different times or conditions. Much of what we know about the regulation of gene expression has come from these studies. Genome analysis is now giving us the ability to examine the experiments on gene regulation that evolution has already tried.

CHAPTER CAPSULE

- While most of the genome is transcribed in both bacteria and eukaryotes, only some of these transcripts become the mRNAs that are translated into proteins.
- In bacteria, transcription and translation are coupled in both location and time, so that an mRNA is translated as it is being transcribed. In eukaryotes, transcription occurs in the nucleus while translation occurs at the ribosomes in the cytoplasm, so the processes are not coupled.
- Translation occurs at the ribosome and involves tRNA molecules to "adapt" the nucleotide sequence of the mRNA to the amino acid sequence of the protein.
- Translation is highly similar in bacteria and eukaryotes, suggesting that the process evolved only once in a common ancestor to all living species.
- The universality of translation and the genetic code has important implications for evolution and for experimental biology but also indicates that most genes are not subject to translational regulation.
- Some examples of translational regulation are well known, although translational regulation is not as common as regulation by transcriptional initiation, by splicing, or by microRNAs.
- Once a polypeptide is made, it may be modified post-translationally before it is in its functional form.

STUDY QUESTIONS

Concepts and Definitions

13.1 Define each of the following terms as used in this chapter.

a. Aminoacyl synthetase

b. mRNA tunnel

c. Suppressor mutation

d. Anticodon

e. Wobble

f. Nonsense mutation

g. Evolutionarily accepted mutation

13.2 The sequence of an mRNA is 5′-AUGGCUACUUGCAG-3′.

a. Using the codon table in Figure 13.13, translate this into an amino acid sequence in all three reading frames.

b. Is one of these three reading frames more or less likely to be the correct one for this gene? Explain your reasoning.

13.3 Before tRNA was identified, Crick postulated the existence of an "adaptor" molecule. Why is this an appropriate name for the role of tRNAs?

13.4 Could eukaryotic ribosomes accurately translate an mRNA from bacteria? Explain your reasoning.

Beyond the Concepts

13.5 Describe what happens at the A, the P, and the E sites of the ribosomes.

13.6 How do stop codons stop the polypeptide chain?

13.7 What are the similarities and differences between the processes of translation as it occurs in bacteria and eukaryotes?

Looking back and ahead **13.8** As noted in Chapter 4 and as will be discussed in Chapter 17, the sequence of 16S rRNA is one of the most widely used methods for looking at the phylogenetic relationship among bacteria. Give some of the reasons that 16S rRNA is an appropriate choice for identifying sequence relationships.

13.9 While translational control of gene expression is relatively uncommon for eukaryotic genes, it does occur for some genes. However, translational control of gene expression is very rare for bacterial genes. Explain why this is so.

13.10 The GC composition of bacterial genomes varies widely, from less than 35% to nearly 70% of the bases being G:C base pairs. However, the frequencies at which particular amino acids occur are similar in most organisms. Explain how highly similar amino acid frequencies can arise from such different base frequencies.

13.11 With some exceptions, the genetic code is universal. Explain some of the practical and evolutionary implications of this.

13.12 What might be an evolutionary advantage to a genetic code in which closely related amino acids have closely related codons?

13.13 Certain dyes, such as acridine orange, insert between the bases of DNA during transcription and result in the addition or loss of a base in the mRNA sequence. The resulting change is known as a frameshift mutation. Why are these called frameshift mutations?

13.14 In sickle-cell anemia, a glutamic acid in β-globin is replaced by a valine. What is the minimal number of base changes that is required to produce this change, and what are they?

Applying the Concepts

13.15 Design a single-stranded oligonucleotide that could hybridize with all mRNAs that encode the peptide sequence Arg-Lys-Ser-Thr-Gly. (Be sure to pay attention to the directionality of the peptide, the mRNA, and the oligonucleotide.)

13.16 The earliest experiments to decipher the genetic code used an *in vitro* translation assay. Different mixtures of ribonucleotide triphosphates were added along with RNA polymerase to make a RNA molecule, which was then translated. The resulting peptide sequence was then determined. What would be the peptide or peptides that would arise from the following? Assume that initiation could occur normally in all cases.

a. Only G was added to the mixture

b. Only A was added to the mixture

c. An RNA of alternating A and G was added to the mixture

13.17 Many different antibiotics affect steps in translation. Based on the mode of action of each antibiotic, suggest a possible mechanism by which it works.

a. Chloramphenicol blocks the peptidyl transferase activity

b. Tetracycline blocks the A site

c. Streptomycin binds to the 30S subunit and results in misreading the mRNA. (The precise mechanism by which this occurs is not known, but a reasonable hypothesis can be formulated.)

13.18 Cfr is a protein encoded on plasmids in some bacteria. Bacteria harboring the *cfr* gene are rendered simultaneously resistant to a large number of structurally different antibiotics, all of which are known to interfere with protein synthesis and to act at the P site of the ribosome. Strains carrying *cfr* remain sensitive to antibiotics that block other ribosomal sites or have mechanisms of action not involving the ribosome. *In vitro*, Cfr can catalyze the methylation of select cysteine residues in RNA. What is a potential explanation for the antibiotic resistance that Cfr confers?

13.19 Nonsense suppressors are discussed in Box 13.1.

a. Describe a nonsense suppressor, particularly focusing on an amber suppressor.

b. Which tRNAs could mutate to produce an amber suppressor tRNA by a single base change in the anticodon?

c. Strains of *E. coli* with an amber suppressor grow more slowly than wild-type strains, even when the amber suppressor is the only mutation in the genome. This is also true for amber suppressor strains in other organisms. Explain why amber suppressor strains usually grow more slowly and are less vigorous than wild-type strains.

d. Amber suppressors are sometimes referred to as "allele-specific, gene non-specific" suppressors, since they suppress only certain mutant alleles of a gene. On the other hand, they suppress alleles of genes with many different functions. Explain why amber suppressors are allele-specific but gene non-specific in their action.

13.20 An investigator has a sequence of 700 bases that includes the beginning of a gene. However, the precise start of the gene is not known.

a. Outline the logical steps that the investigator might perform computationally to identify the most likely beginning of a gene. (Don't forget that she would not know if this is the template strand or the coding strand.)

b. How would these steps differ for genes from bacteria and genes from eukaryotes?

13.21 Although the genetic code is universal, a few organisms, such as *Paramecium*, have a slightly modified version, in which UGA, a stop codon for most organisms, codes for tryptophan. A researcher identified a gene from *Paramecium* and ran both its mRNA and its protein on denaturing gels (which separate based on size, rather than structure). The results are shown in Figure Q13.1 in the lanes labeled "*Paramecium*". She also inserted the DNA sequence from this gene into a plasmid and transformed the plasmid with the insert into *E. coli*. *E coli* both transcribed and translated the gene; fortunately, the gene has no introns.

a. Draw the approximate locations of the mRNA and the protein as they would be found when expressed in *E. coli*. You cannot know the exact size, so draw this relative to its size when expressed in *Paramecium*.

b. Suppose that the researcher wanted to make an *in vitro* translation system using all of the components from *Paramecium*. Which of the components, if any, would she need to replace in order to have an *in vitro* system that was universal?

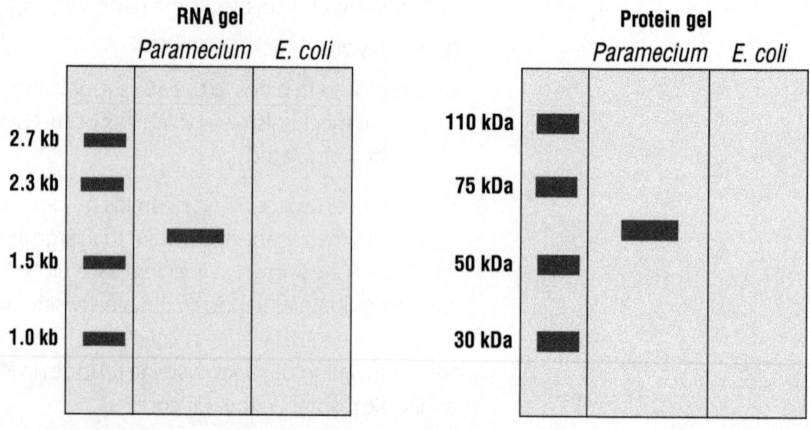

Figure Q13.1

Challenging **13.22** The BLOSUM Scoring Matrix, shown in Tool Box 13.1 Figure B, is a critically important component of amino acid sequence comparisons and alignments in programs such as BLASTP. This matrix shows the likelihood (on a log scale) that an amino acid in one sequence will be replaced by another in another sequence. Based on this scoring matrix, answer the following questions.

a. What does a score of 0 represent?

b. What amino acids are the most likely to be conserved when two related sequences are compared?

c. The score for a lysine (K) to remain a lysine is 5. The score for a lysine to be replaced by an arginine (R) is 2. Explain what this means.

13.23 Use Online Mendelian Inheritance in Man (OMIM; www.ncbi.nlm.nih.gov/omim), and look up the *CFTR* gene, entry #602421. Mutations in the *CFTR* gene are responsible for cystic fibrosis, as discussed in Chapter 10. Follow the link for "Allelic Variants" to answer these questions. More than 135 allele variants are listed from patients who have been diagnosed with cystic fibrosis (CF). Of these, the most common is variant 0001, the in-frame three-base deletion described in Chapter 10 as Δ-F508; here, this is abbreviated as PHE508DEL to indicate the amino acid (Phe), the position (508), and the change in the gene (DEL). Use the codon table in Figure 13.13 to answer the following questions.

a. What single nucleotide change could produce variant 0011, SER54ILE?

b. What change in the DNA sequence occurred in variant 0050, and how did this affect the amino acid sequence?

c. What change in the DNA sequence occurred in variant 0053, and how did this affect the amino acid sequence?

d. What change in the DNA sequenced occurred in variant 0008, and how did this affect the function of the gene?

Challenging *Looking ahead* **13.24** The clinical symptoms in patients with CF are somewhat variable but generally very similar.

a. In particular, the symptoms in a patient with the most common variant PHE508DEL are very similar to the symptoms in variant 0047, which is CYS524TER, and to many other variants that result in stop codons in the gene. What does this suggest about the functional activity of the PHE508DEL variant?

b. Variant 0024 is ILE506VAL. This variant appears to cause no major change in the function of the gene and was detected only because the PHE508DEL variant was also present. Speculate, based on the genetic code, why variant 0024 may be benign.

c. SER1455TER results in chloride ion changes but not in clinical CF. Speculate about why this variant has a much milder effect than PHE508DEL and most other variants that produce stop codons. (Hint: The full-length CFTR protein is 1480 amino acids.)

13.25 Researchers in the department of genetics at Harvard University have been engineering genomically recoded organisms. In one experiment, they replaced all 321 UAG (stop) codons in *E. coli* with UAA (stop) codons and deleted release factor 1, the UAG translational terminator, to produce strain "C321."

a. Do you expect the proteome (full complement of proteins produced by the organism) of strain C321 to differ from that of wild-type *E. coli*? Why, or why not?

Challenging **b.** How might the investigators use strain C321 to produce proteins that contain 2-naphthylalanine, an amino acid that is not one of the 20 required for life?

c. The genome *of E. coli* bacteriophage T7 carries 60 stop codons, six of which are UAG. What are your expectations with respect to susceptibility of C321 to T7?

13.26 Mutations in the β-globin gene, one of the two polypeptides that make up hemoglobin in humans, result in β-thalassemia, a form of anemia that can be life-threatening. (Sickle-cell anemia is one form of β-thalassemia.) Many thousands of individuals with β-thalassemia have been identified in clinics worldwide, and physicians and geneticists have thoroughly studied the molecular basis for the disease in many of these individuals. The β-globin gene and the location of some mutations are shown in Figure Q13.2. The diagram at the top shows the normal gene with its promoter and three exons. The locations of five mutations in the gene are shown by red arrows; for two of these, the actual base changes are shown as well. The DNA, RNA, and protein from an unaffected person (Unaff) and from the five patients (numbered 1 to 5) with β-thalassemia are shown in the DNA gel (a Southern blot), an RNA gel (a northern blot), and a protein gel (a western blot). The lanes in each gel correspond to the results from that particular mutation, such that variant 1 is shown in lane 1, and so on. Based on the locations of each mutation in the gene and the results on the gels, what is the most likely molecular explanation for thalassemia in each of the five affected individuals?

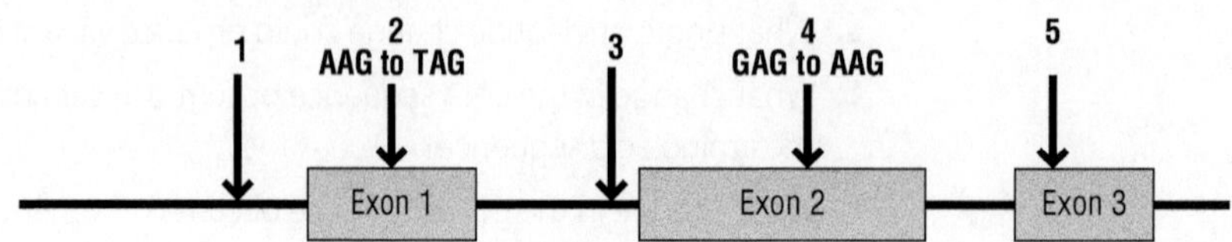

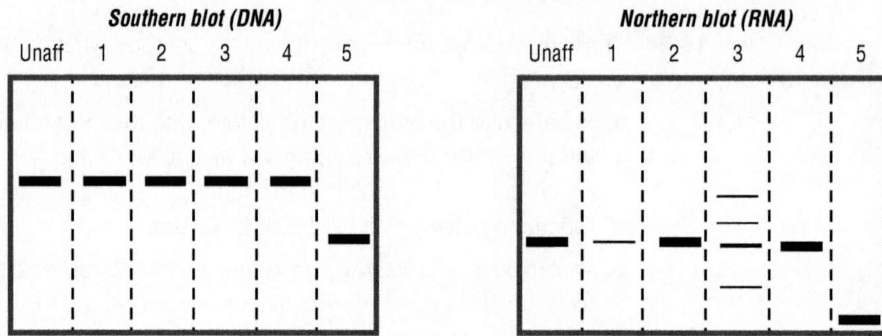

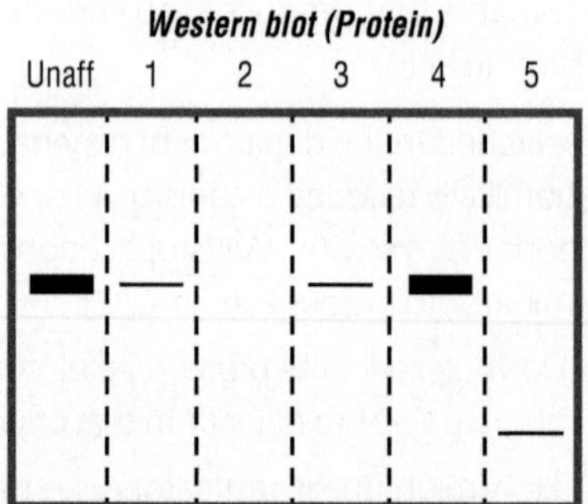

Figure Q13.2

CHAPTER 14

Networks of Gene Regulation

IN A NUTSHELL

While the regulation of gene expression occurs at many different steps in the information flow from DNA to the protein or RNA product, it is the regulation of transcription that lies at the heart of gene regulation. Every gene has its own transcription pattern, and changes in that pattern can have significant effects on phenotypes and morphological variation. However, all of those individual patterns of gene transcription must be coordinated, so that the organism is transcribing all of the correct genes at the correct times and places. This chapter moves beyond the regulation of the expression of an individual gene to the orchestrated regulation of many genes. The processes by which gene expression is coordinated are different in bacteria and eukaryotes, although some similar principles can be found. It is now possible to look at transcription throughout the genome and across time.

14.1 Overview: networks of gene regulation

In Chapters 12 and 13, we described how the expression of an individual gene occurs and some of the ways that expression is controlled. This control can occur at many different levels in the information flow from DNA to RNA to protein. Regulation can occur at any step from the initiation of transcription to covalent modifications of the polypeptide product. Important examples of gene regulation can be found at any one of these steps, but the critical regulatory point for most genes occurs at the initiation of transcription. After all, it is usually easier to stop a process before it starts than it is to control it once it has begun. Thus, when we talk about the regulation of gene expression, we are usually thinking about the regulation of transcription, albeit with mental reminders that the other steps are important as well.

As you will remember from Chapter 12, transcriptional regulation of an individual gene occurs by the recruitment of the RNA polymerase holoenzyme (that is, the complete enzyme with all of its subunits) to the core promoter. In bacteria, this recruitment is regulated by the binding of the σ subunit of RNA polymerase itself, as illustrated in Figure 14.1(a). In eukaryotes, recruitment of

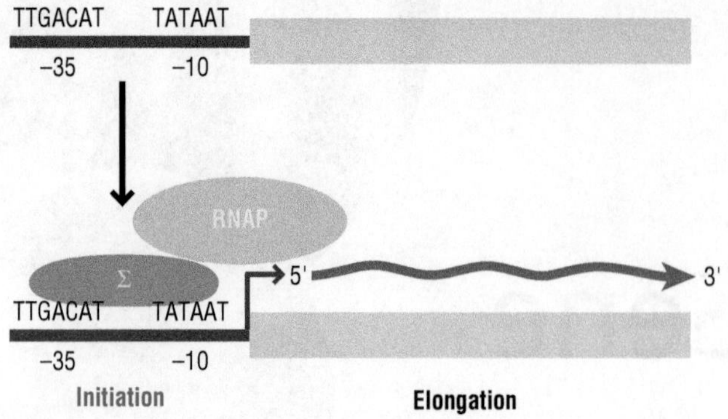

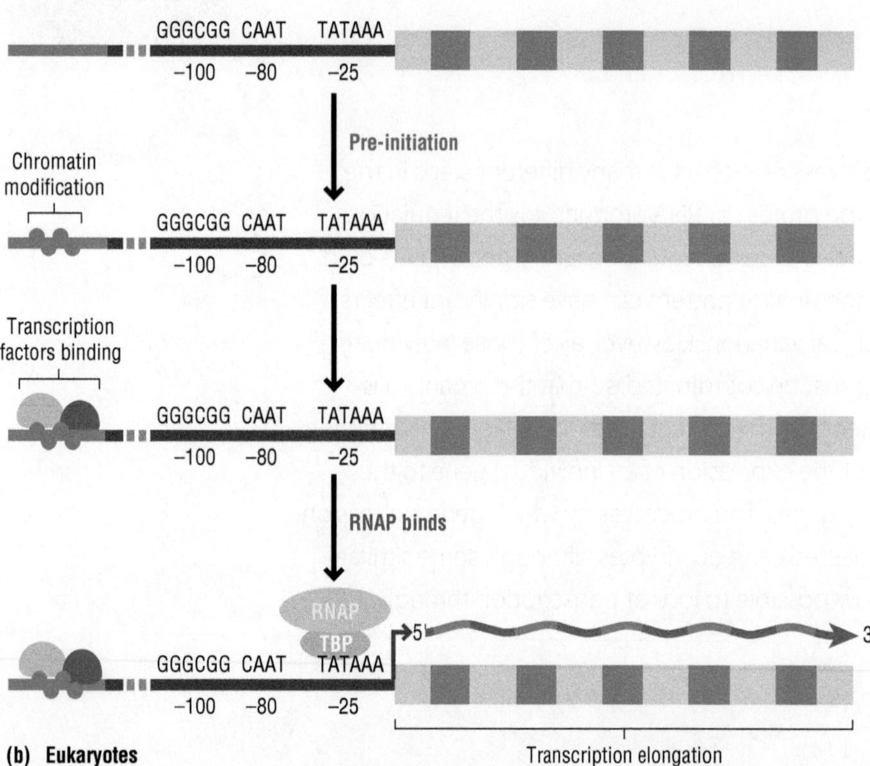

Figure 14.1 An overview of transcriptional regulation in bacteria and eukaryotes. As described in Chapter 12, the initiation of transcription in bacteria is regulated by σ factor proteins, which bind to the promoter to recruit the rest of the RNA polymerase (RNAP) holoenzyme. In eukaryotes, recruitment of RNA polymerase to the core promoter involves interactions with a suite of transcription factor proteins. These transcription factor proteins bind to sequences in the *cis*-regulatory modules, initiating transcription. The binding of the transcription factors then recruits RNA polymerase to initiate transcription. TBP, TATA-binding protein.

RNA polymerase to the core promoter involves interactions with a suite of transcription factor proteins. Recall that an individual gene is regulated by multiple transcription factors that bind to the sequences in the *cis*-regulatory module, which brings RNA polymerase to the core promoter, initiating transcription. This process is also shown in Figure 14.1(b). The regulation of the initiation of transcription of an individual gene in a eukaryote is a complicated, multistep process.

Yet any particular cell, whether in bacteria or in eukaryotes, is transcribing only a subset of its genes at any given time or under any set of conditions. It is essential

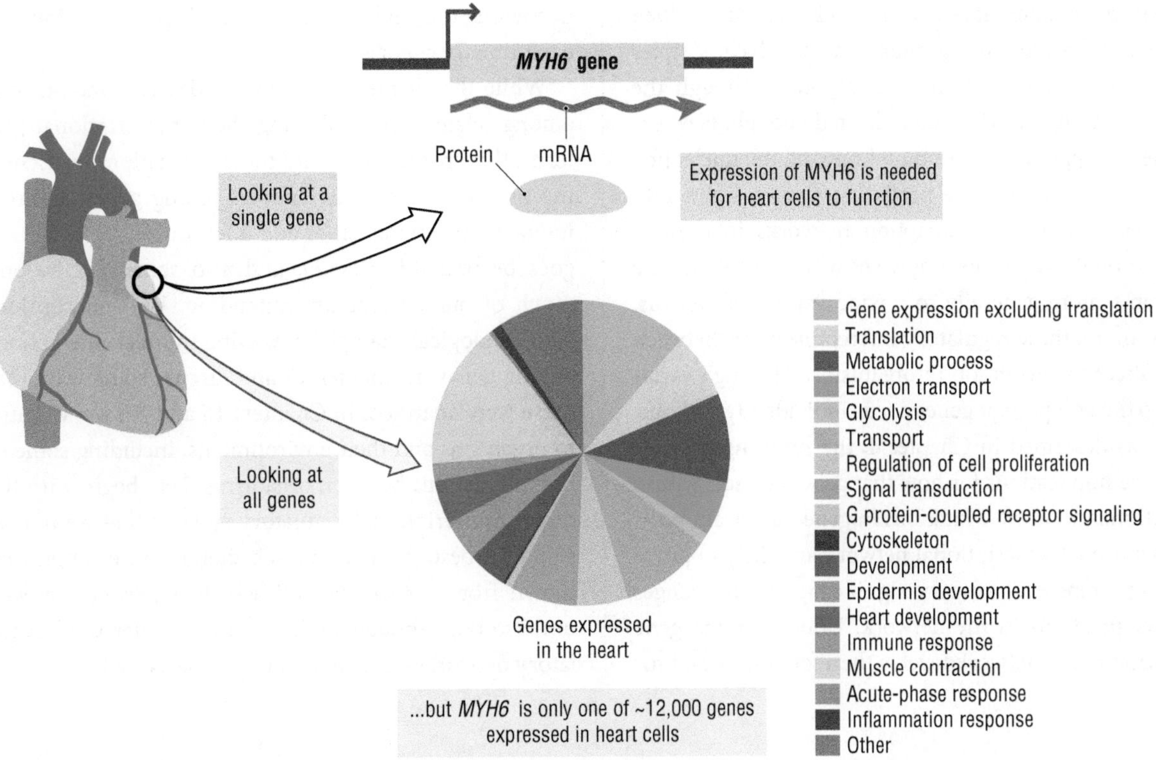

Figure 14.2 Transcription is coordinated among many genes. An example is shown for the mammalian heart. The *MYH6* gene encodes cardiac myosin, which is essential for the normal function of the heart. However, the expression of *MYH6* is coordinated with more than 12,000 other genes that are also expressed in the heart.

Source: Pie chart from Ramsköld, D., et al. (2009). An Abundance of Ubiquitously Expressed Genes Revealed by Tissue Transcriptome Sequence Data. *PLoS Comput Biol* 5(12) © 2009 Ramsköld et al.

to the survival of the organism, whether a single-celled organism or a multicellular organism, that it transcribes the proper set of genes at the right time or place; failure to organize gene expression properly will affect viability and function. In addition, almost every critical biological process is affected by multiple genes rather than an individual gene. Thus, the expression of each individual gene, whether in bacteria or in eukaryotes, has to be coordinated with the expression of dozens or even hundreds of other genes, as illustrated with the mammalian heart in Figure 14.2. In order for the heart to function, the cardiac myosin gene *MYH6* needs to be expressed properly; however, the expression of *MYH6* is only one of more than 12,000 genes transcribed in the heart, all of which need to be regulated coordinately and expressed in the proper time and location.

One of the Great Ideas that we introduced in the Prologue is that biology consists of organized systems. There are many such systems occurring at every level of biology, from the subcellular to populations and ecosystems. Organized systems are inherently more complicated than the individual components, and are likely to be harder to study. In order to understand organized systems, we often have to begin by understanding the individual components and combining them as we piece together the complex system; this can be thought of as a "bottom-up" approach. (Another approach, working from the process itself to the genes, or from the "top down," will be described in Chapter 15.) We may not know the rules by which the system is organized, let alone the mechanisms by which these rules are executed, so we look for the interactions among the components for clues. For gene expression, some of these organizational principles and mechanisms are reasonably well understood. As genomes have been sequenced, it has become increasingly routine to know about the functions and expression of the individual components, namely the genes. With that knowledge of the component parts, it has become more feasible, if formidable, to describe how the expression of those parts is organized.

We have much to learn in this area. It is clear that the basic organizational principles for networks of gene expression are different in bacteria and eukaryotes, albeit with a few features in common. Bacteria have simpler systems of organization than eukaryotes, although the organization in bacteria is complicated enough. New experimental approaches based on knowing the nucleotide sequences of genomes are providing some insights into the organization of transcription networks in eukaryotes, although we do not know yet which principles are applicable more generally to a wide variety of systems. Furthermore, these regulatory interconnections between genes affect the evolution of organisms. Although variations to the component genes occur individually and randomly, as described in Chapter 4, the resulting changes may have implications for how that gene interacts within a functional network. Mutations in genes at certain positions within a transcriptional network are likely to prove to be of more evolutionary significance than changes at other positions in the network. Understanding gene regulation at the network level is therefore important for deepening our understanding of both gene function and evolutionary change.

While the chapter will begin to describe interactions among related genes affecting their transcriptional patterns, this chapter simplifies the organization by discussing only the **direct** interactions among genes. Indirect interactions among genes add a level of complexity that goes beyond this chapter, and also somewhat beyond much of our current understanding. We will consider some biological examples involving indirect interactions among genes in Chapter 15 and discuss some ways that these were analyzed. In Chapters 16 and 17, we will turn to organisms and their environments, including some of the interactions between organisms. Let's begin with the direct transcriptional regulatory systems that we understand the best, those found in bacteria. Our examples are drawn from *Escherichia coli*, which is particularly well understood, although similar, but less understood, regulatory networks occur in other bacteria as well.

14.2 Operons in bacteria

The key biological feature that coordinates transcription among functionally related genes in bacteria is their physical location on the chromosome; genes that are regulated together, or **co-regulated**, and involved in the same biological process are often found very near to each other on the chromosome. This is a distinctive and important feature of transcriptional regulation in bacteria. In eukaryotes, co-regulated genes are usually located throughout the genome, rather than next to each other, and genes that are adjacent to each other on the chromosome often have no function or regulation in common. We will discuss two particularly well-studied examples of co-regulated genes in *E. coli*. There are many details of these examples that are being omitted in the main text, some of which are included in boxes. In Chapter 15, we include some information about the experimental methods and results that led to the picture we present here.

Genes required for the biosynthesis of tryptophan are only transcribed when tryptophan is absent

Tryptophan is one of the 20 naturally occurring amino acids and so is essential for all life processes. Normal *E. coli* cells use tryptophan from the environment when it is present, but when tryptophan is lacking, the bacteria can make their own because their genome encodes the enzymes necessary for tryptophan biosynthesis. Thus, when grown in rich medium with an abundance of tryptophan, the cells do not transcribe the genes for tryptophan biosynthesis but turn these genes on only when the cells are starved of tryptophan.

It is possible to find mutants that cannot grow without added tryptophan; these are cells that cannot produce their own tryptophan. Mutants like this that lack the ability to synthesize an essential macromolecule are known as **auxotrophs**, and the wild-type cells are called **prototrophs**. When these tryptophan auxotrophic mutants were mapped, nearly all of the mutations mapped extremely close together on the chromosome. In fact, these mutants mapped so close together that, if map position alone had been the sole criterion, they might have been classified as alleles of the same gene. However, complementation tests among the mutants, of the type described in Chapter 8, ruled out this possibility and clearly showed that these mutants define five distinct genes, named *trpA* through *trpE*. The arrangement of these five genes on the chromosome is shown in Figure 14.3.

As we previously noted in the Prologue, the biochemical process of synthesizing a more complex macromolecule,

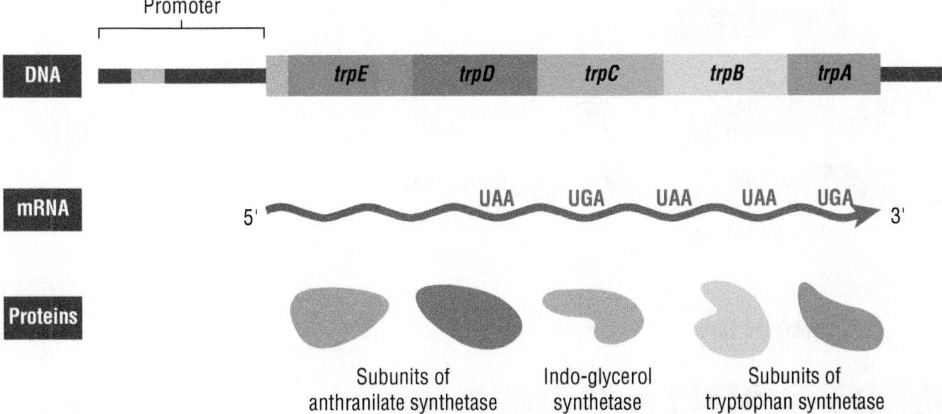

Figure 14.3 The tryptophan (*trp*) operon in *E. coli*. The five genes *trpA* through to *trpE* are located adjacent to one another in the *E. coli* genome. Each gene encodes one of the enzymes necessary for tryptophan biosynthesis. A single promoter is located upstream of the *trpE* gene, and a single mRNA that encodes all five polypeptides is produced. The promoter is bracketed; note that the blue box within the promoter is the operator sequence, to be discussed. Note the presence of the stop codons UAA and UGA, which mean that the transcript is translated to make five different polypeptides. This structure, in which multiple genes with related functions are transcribed as one transcript from a single promoter, is known as an operon. The pale gray box to the left of *trpE* represents the leader sequence, discussed in Box 14.1.

such as tryptophan, from simpler molecules is called **anabolism**. Each of the genes *trpA* through *trpE* encodes the information to make one of the enzymes necessary for tryptophan biosynthesis. Thus, it is easy to see why mutations in these genes are tryptophan auxotrophs—each mutant lacks an enzymatic step necessary to make tryptophan. But what is not so clear from this picture is the importance of these five genes mapping so close together. The answer was provided by many genetics experiments using these mutants in different combinations and conditions, but rather than work through the analysis, we will simply provide the results of many experiments.

Note from Figure 14.3 that the five genes are transcribed from the same promoter, which is located upstream, that is, to the 5′ side, of the *trpE* gene. The termination signal for transcription for these genes is downstream of the *trpA* gene. Thus, RNA polymerase binds to the promoter upstream of *trpE* and transcribes all five genes as a single long transcript, terminating downstream of *trpA*. This is called a **polycistronic transcript**; "cistron" is an old term for a gene as defined by complementation tests, which is now used only in this one context to describe a transcript that encompasses more than one gene. When this polycistronic transcript is translated, five different proteins are produced, since the coding region of each gene begins with a ribosome-binding site and ends with a translation stop codon. Therefore, a single transcript produces five distinct proteins with closely related functions in tryptophan biosynthesis. This arrangement of the genes is referred to as an **operon**, in this case the *trp* operon. Because these genes encode the enzymes that synthesize tryptophan, this is an example of an anabolic operon.

KEY POINTS An operon consists of multiple genes that are transcribed as a single transcript and code for proteins with related functions. Operons are a distinctive feature of transcriptional regulation in bacteria.

The operon arrangement, with multiple genes encoded on a single transcript, guarantees the coordinated regulation of all five genes. When one gene is transcribed, all of them are transcribed, and every step in the anabolic process occurs. Furthermore, because transcription and translation are often coupled in bacteria, all the proteins required for the process are produced in the same location in the cell. This localization of the gene products may be important for bacterial cells, which lack the elaborate transport systems used in eukaryotic cells. Many bacterial genes that encode products that contribute to a function or pathway are organized in operons, many of which—like the *trp* operon—offer the added advantage of coordinated regulation.

E. coli does not synthesize the *trp* enzymes all the time, that is, the genes are not expressed **constitutively**. As we

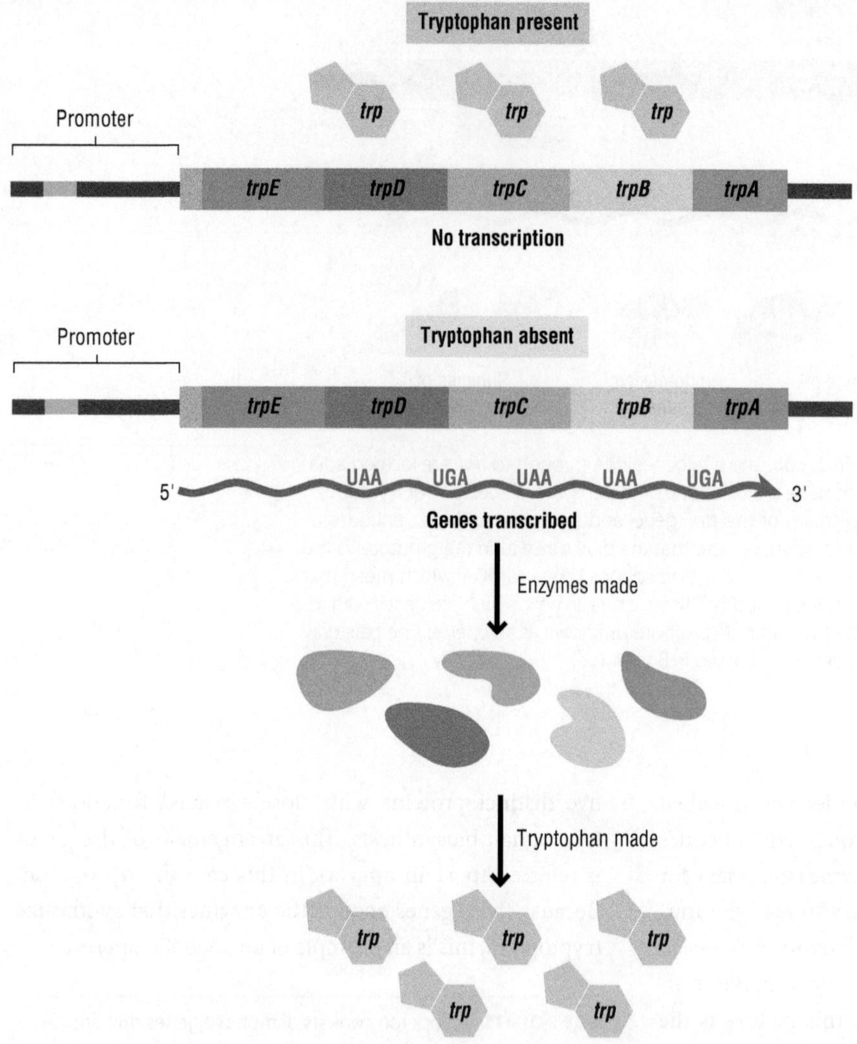

Figure 14.4 Tryptophan regulates the transcription of the *trp* operon. When tryptophan is present, as shown at the top, the operon is not transcribed, none of the enzymes is produced, and no tryptophan is made. When tryptophan is absent, as shown on the bottom, the operon is transcribed, all five enzymes are produced, and tryptophan is synthesized. Note the presence of the operator sequence, represented as the blue box, within the promoter region, which is important in the regulation of the transcription of the operon.

noted previously, these enzymes are only needed when tryptophan is absent, so they are not produced when the medium or the environment has an adequate supply of tryptophan. It would be metabolically wasteful for bacteria to transcribe genes and synthesize proteins that are not necessary, so *E. coli* has adapted to transcribe and translate the genes only when needed, as illustrated in Figure 14.4.

The *trp* operon also provides the means for coordinated expression of these genes or, more accurately, for the coordinated **repression** of these genes when their products are not necessary. Thus, the operon is said to be **repressible**. Repression occurs as transcription is being initiated, mediated by molecular activities occurring at the promoter region. Within the promoter region lies a sequence called the **operator**, shown in Figure 14.4, which is the key for the regulation of the *trp* operon. The operator does not make a gene product; instead, it serves as the binding site for a protein known as the tryptophan **repressor**. The tryptophan repressor is encoded by a gene known as *trpR*, which maps to a different location on the chromosome, separate from the *trpA* through *trpE* genes. With the tryptophan repressor bound at the operator, RNA polymerase cannot

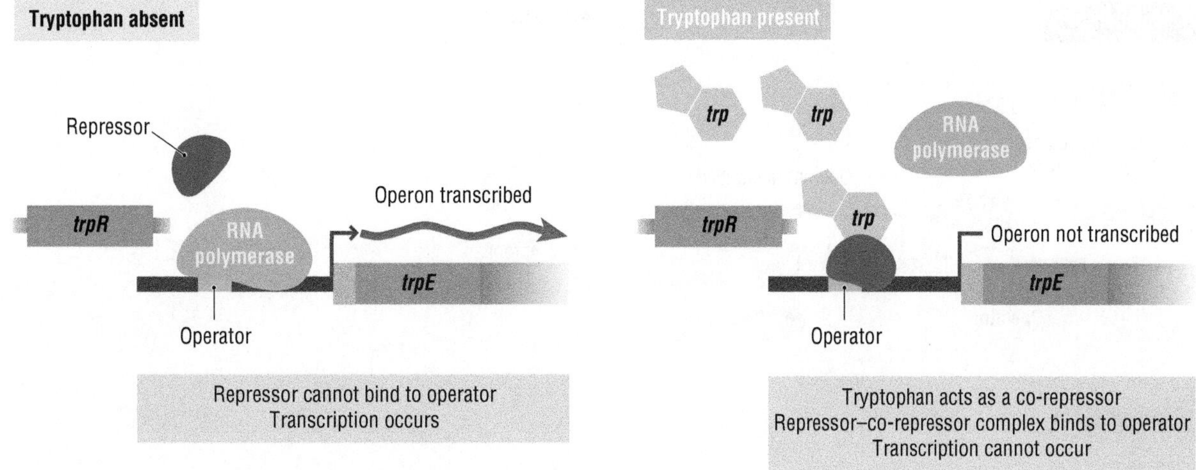

Figure 14.5 The Trp repressor blocks transcription. A protein known as the Trp repressor and encoded by the *trpR* gene is capable of binding to the operator sequence within the promoter. When tryptophan is absent, the repressor cannot bind to the operator. RNA polymerase can bind to the promoter, and the *trp* operon is transcribed. When tryptophan is present, the repressor interacts with free tryptophan, leading to a change in its structure. The tryptophan–repressor complex can then bind to the operator, preventing the binding of RNA polymerase to the promoter, so no transcription occurs and no tryptophan is made.

bind to the promoter, so no transcription occurs and no tryptophan is made. This is what happens when the environment is rich in tryptophan or the tryptophan concentration is high.

KEY POINTS Operator sequences are binding sites for repressor proteins, which block the binding of RNA polymerase to the promoter, thus inhibiting the transcription of the operon.

How then does the presence (and absence) of tryptophan regulate the expression of the *trp* operon? The solution is shown in Figure 14.5. The key feature is that the repressor cannot actually bind to the operator on its own. However, the repressor also interacts with free tryptophan, and the interaction with tryptophan changes the three-dimensional structure of the repressor protein. Tryptophan itself acts as a **co-repressor**. The repressor in a complex with its co-repressor tryptophan now can bind to the operator sequence. The fact that the binding of repressor requires the product of the activities encoded in the operon is the key for the regulation of tryptophan biosynthesis.

As summarized in Figure 14.5, when present at high levels, tryptophan interacts with the repressor, changing its three-dimensional shape, so that the resulting complex binds to the operator. With this complex on the operator, RNA polymerase is blocked from binding to the promoter, so no transcription of the genes occurs. When tryptophan is absent or present at very low levels, the repressor cannot bind to the operator. Thus, RNA polymerase can bind to the promoter, and the *trp* operon is transcribed. From the transcript, the enzymes necessary for tryptophan biosynthesis are produced, and tryptophan is made until it builds up to a level high enough to shut off its further synthesis.

This system of regulation was solved by identifying mutants that either could not make tryptophan, even when tryptophan was lacking, or that always transcribed these genes, even when tryptophan was present, that is, a constitutive mutant. A few examples of such mutants are shown in Figure 14.6, with some further examples as part of the questions at the end of the chapter. While the regulation of the *trp* operon is conceptually simple, some of the details are much more complicated. We describe some further regulation of the *trp* operon, a process known as **attenuation**, in Box 14.1.

KEY POINTS The tryptophan operon is turned off in the presence of tryptophan, which itself serves as a co-repressor of the *trp* operator. When tryptophan is absent, the repressor alone cannot bind to the operator, permitting transcription of the *trp* operon and the biosynthesis of tryptophan.

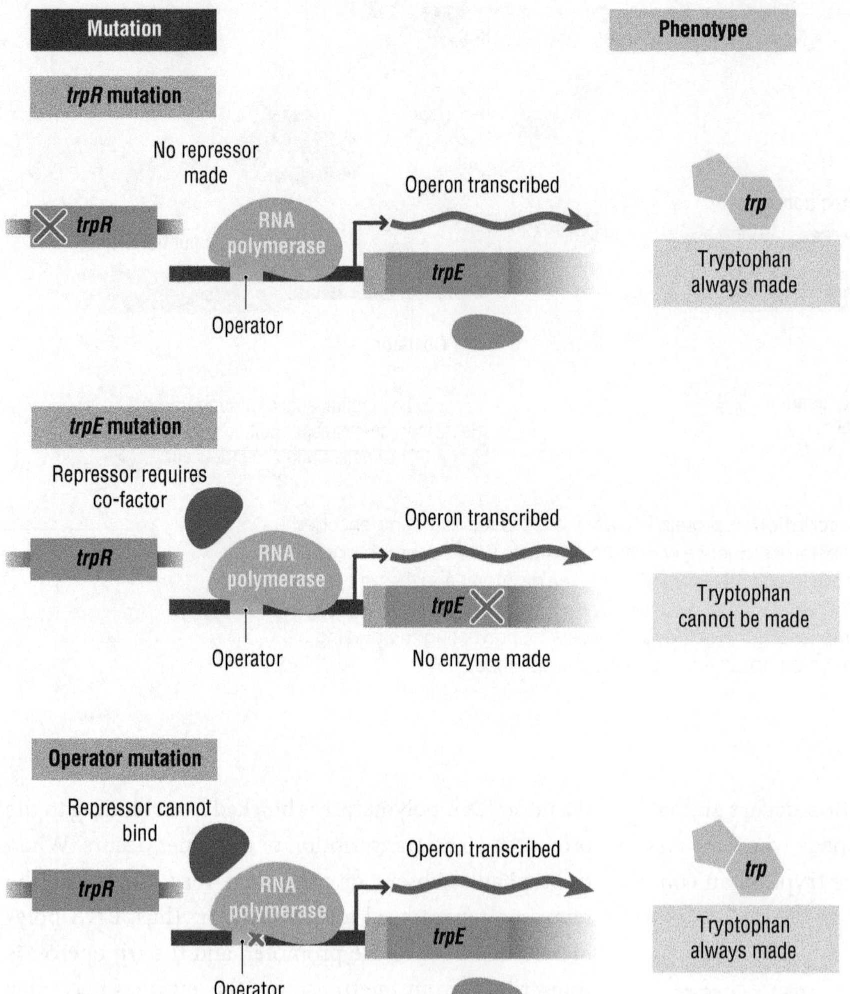

Figure 14.6 ***trp*** **operon mutants were used to infer the regulation.** The isolation of mutants either that made tryptophan constitutively or that could never synthesize tryptophan allowed the regulatory circuit to be analyzed. Three types of mutants, among the many that were isolated, are shown. If no repressor is made (top panel), or if the repressor cannot bind to the operator (bottom panel), tryptophan is made constitutively. Mutations in the *trpE* gene, or any of the other structural genes for the enzymes, are tryptophan auxotrophs and cannot grow in the absence of tryptophan.

BOX 14.1 *Going Deeper* Attenuation and the *trp* operon

Expression of the *trp* genes is largely controlled by the repressor protein, TrpR, binding at the operon's promoter upstream of *trpE*, as described in the text. However, there are additional points for control in this operon that we should note.

First, there is a second and much weaker promoter located between *trpD* and *trpC*, which allows the genes *trpC*, *trpB*, and *trpA* to be expressed at different levels from *trpE* and *trpD*. Box 14.3 discusses internal promoters in operons and the related phenomenon of genetic polarity.

Second, a common type of post-translational control known as feedback inhibition occurs with the first enzyme in the biosynthetic pathway, anthranilate synthetase, the product of the *trpE* gene. Tryptophan binds to the subunit encoded by *trpE* and inhibits its enzymatic activity, thus preventing the synthesis of more tryptophan when the amino acid is present.

A third type of regulation known as transcriptional attenuation or, more often, simply **attenuation** is provided by the leader region of the operon upstream of *trpE*, referred to as *trpL*. This leader is shown as a small gray box to the left of *trpE* in Figure 14.3, while the nucleotide and amino acid sequence features are depicted in more detail in Figure A(i) and (ii).

BOX 14.1 Continued

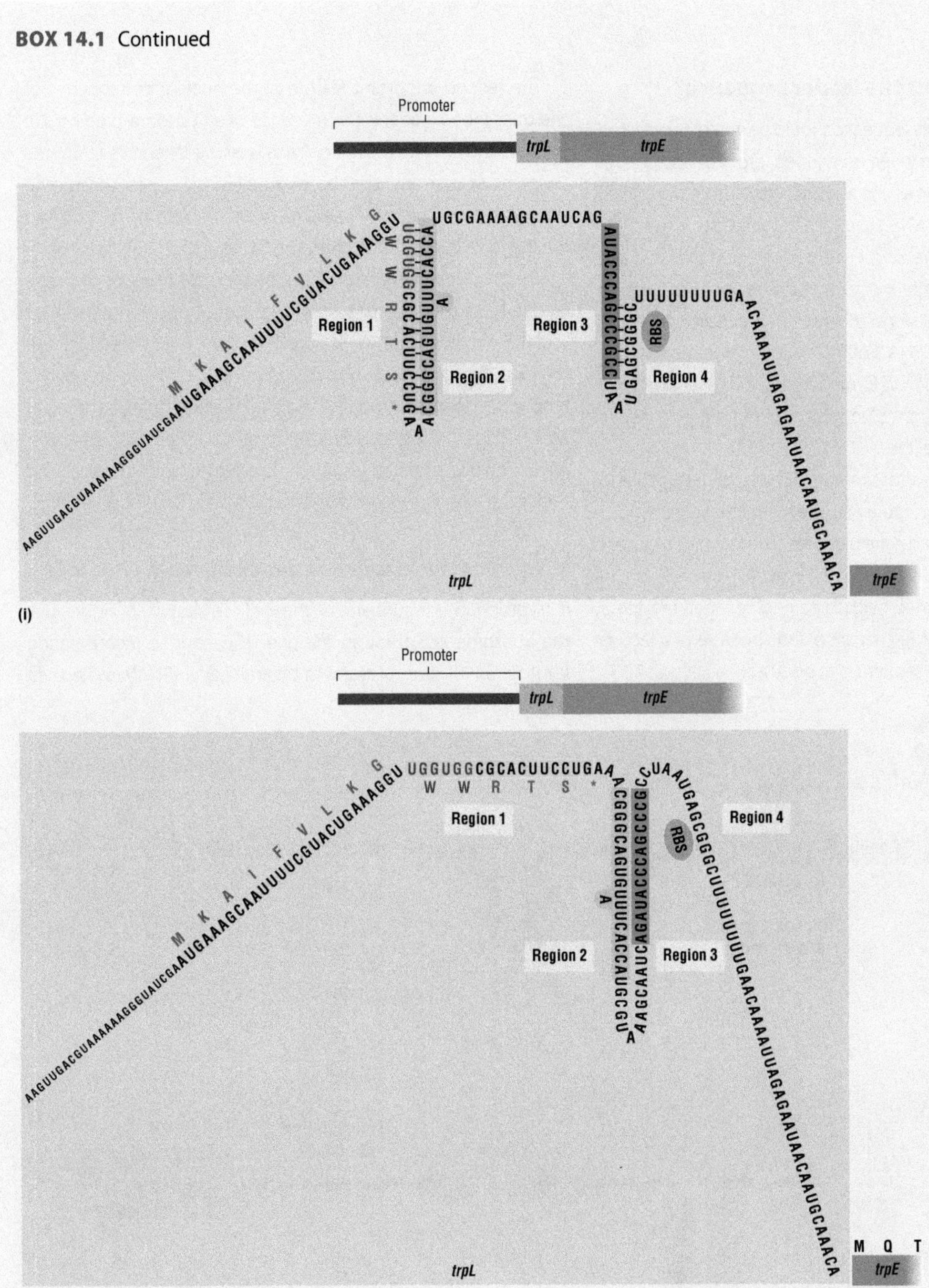

Figure A The leader sequence (*trpL*) can attenuate *trp* operon transcription. The *trpL* gene, shown as a gray box immediately upstream of *trpE*, encodes an RNA and a very short peptide that is involved in additional regulation of the *trp* operon. The nucleotide sequence of the RNA encoded from the *trpL* locus is shown. This leader has three features that contribute to attenuation. First, the RNA has four regions with base complementarity to each other that can form different combinations of hairpins. Second, a ribosome-binding site (RBS) is found at the end of region 4. Third, the leader can be translated into a short peptide, with the amino acids indicated next to the sequence for region 1. Note the presence of the two consecutive UGG codons that specify tryptophan (Trp, W), and the UGA stop codon (*); the tryptophan codons are shown in purple. (i) Regions 1 and 2 form a hairpin, as do regions 3 and 4. Among other effects, this blocks the RBS at the end of region 4. (ii) Regions 2 and 3 form a hairpin, while regions 1 and 4 remain unpaired, leaving the RBS accessible.

BOX 14.1 Continued

What is special about the leader sequence?

Three features of this leader sequence should be noted, two related to its nucleotide sequence when transcribed into RNA and one related to the sequence of 14 amino acids translated from the mRNA.

The first feature of the RNA sequence of the leader is the presence of regions that have sequence complementarity to each other; this complementarity means that the regions are capable of forming hairpins by base pairing with one another once the leader is transcribed into RNA. For illustration in Figures A and B, the portions of RNA that pair to form these hairpins are numbered 1–4; region 1 is shown in light blue, region 2 in light green, region 3 in red, and region 4 in yellow.

In Figure A(i), regions 1 and 2 form a hairpin, and regions 3 and 4 form a second hairpin. In Figure A(ii), the hairpins involve regions 2 and 3, while regions 1 and 4 are unpaired. The formation of hairpins by the *trpL* transcript is an essential feature of attenuation. In particular, the hairpin formed by base pairing between regions 3 and 4 is highly similar to the sequence and structure of a factor-independent termination sequence, as discussed in Section 12.5.

The second important feature of the nucleotide sequence of the transcribed leader is the presence of a ribosome-binding site overlapping the end of region 4; this is also referred to as a Shine–Delgarno sequence and is abbreviated RBS in Figures A and B. The ribosome-binding site helps to recruit the ribosome to mRNA to initiate translation, as discussed in Chapter 13.

The third feature is found in the amino acid sequence of the leader. Note the presence of two consecutive UGG tryptophan codons in region 1, shown in purple in Figure A. As noted in Chapter 13 during the discussion of the genetic code, tryptophan is among the least commonly used amino acids; only about 1% of amino acids in *E. coli* proteins are tryptophan residues. The presence of two tryptophan residues in tandem is highly unusual, and their presence is key to how attenuation is accomplished by *trpL*.

How does the leader sequence have its effect?

If tryptophan is abundant in the cell, as shown in Figure B(i), the ribosome zips through the two UGG codons, incorporating tryptophan residues brought in by aminoacyl tRNA^{Trp} molecules,

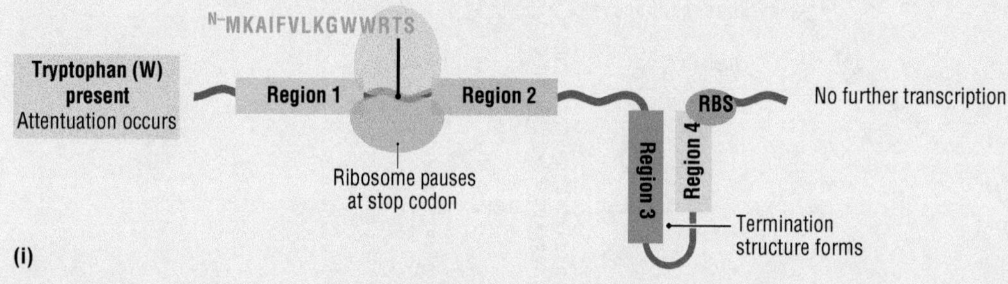

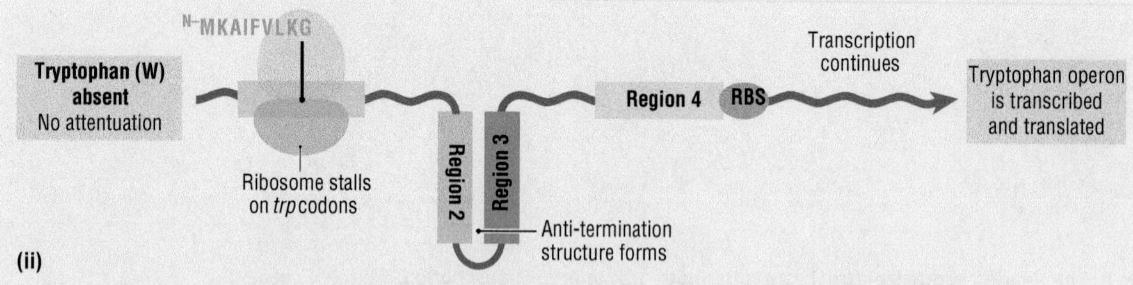

Figure B Attenuation, termination, and anti-termination of the *trp* operon. The leader RNA encoded from *trpL* can form either of two structures, based on intra-strand base pairing. (i) When tryptophan is present, the ribosome is able to translate beyond the tandem tryptophan (W) codons in *trpL*. The ribosome completes the transcription of the tryptophan codons in region 1; by the time the ribosome has reached the *trpL* stop codon in the RNA, a terminator has formed from the hairpin involving regions 3 and 4. In the terminator conformation, the ribosome-binding site (in purple) is inaccessible. The terminator hairpin ends or attenuates transcription and translation. (ii) If no tryptophan is present, the ribosome stalls at the tryptophan (W) codons. When the ribosome is halted at that position, the RNA folds into the anti-terminator structure involving regions 2 and 3, leaving the ribosome-binding site (RBS) in region 4 accessible. The anti-terminator prevents the terminator structure from forming and allows transcription and translation of the *trpEDCBA* genes.

BOX 14.1 Continued

completing the synthesis of the peptide and coming to rest at the UGA stop codon. This leaves region 3 in the RNA available to pair with region 4 to form a factor-independent transcription termination signal, which attenuates further transcription of the operon; the hairpin also blocks the ability of the ribosome to bind at the ribosome-binding site, which prevents translation.

As the leader RNA is transcribed, it is also being translated. If tryptophan is absent or scarce, as shown in Figure B(ii), the cell cannot recruit aminoacyl $tRNA^{Trp}$ to synthesize the peptide, and the ribosome pauses at these tryptophan codons. The stalled ribosome blocks region 1, which then cannot base pair with region 2. This means that region 2 will invariably pair with region 3. As a further consequence, since region 4 is not paired, the transcriptional termination signal formed by the hairpin between regions 3 and 4 does not form. The ribosome-binding site in the unpaired region 4 is accessible, and the operon's mRNA will be transcribed and translated. (The hairpin formed by base pairing between regions 2 and 3 is referred to as an anti-terminator, since it prevents the termination from occurring.)

Thus, the presence of tryptophan shuts off the operon in a variety of ways. The primary action of tryptophan is to act as a co-repressor, as part of the complex with the repressor protein that binds at the operator to prevent RNA polymerase from binding; this is the action described in the chapter. But beyond that, the presence of tryptophan also results in the attenuation of the transcription of the operon and serves as a source of feedback inhibition to the product of the *trpE* gene.

Attenuation occurs in a number of other bacterial operons, as well as in bacteriophage λ, but has been best studied for the *trp* operons of *E. coli* and *Bacillus subtilis* and the *Salmonella* histidine synthesis (*his*) operon. Attenuation is an effective, but somewhat costly, mechanism for control because at least some RNA is synthesized before the operon is shut off. However, it offers a second level of control, ensuring that the biosynthetic pathway is only used when necessary. These two points could suggest that the selective advantage of attenuation is perhaps most pronounced with more fine-tuned control, as in the *trp* operon. However, this does not necessarily allow us to predict which control mechanism evolved first. Charles Yanofsky, whose laboratory outlined the function of the *trp* operon, has suggested that attenuation is an ancient form of regulation that evolved when life was catalyzed and perpetuated mainly or solely by RNAs. If that is true, it will have preceded activator and repressor controls at operons.

Growth curves provide important information about bacterial gene regulation

Our next example of coordinated gene expression in bacteria requires a brief introduction to bacterial growth in culture. The time it takes a single cell to complete its cycle and divide into two daughter cells is referred to as the **doubling time**. Given constant conditions (such as nutrient supply and temperature), a typical cell lineage will show the same doubling time from one generation to the next, producing an exponential relationship between cell number and time. Cell number can be measured by actually counting the cells or, because the cloudiness of cultures is proportional to the number of cells, by using optical density as a surrogate measure. When data from such an experiment are graphed, the result is a plot known as a **growth curve**. A diagram of a growth curve is shown in Figure 14.7.

There is a lag time at the start of most experimentally derived growth curves, during which the bacteria are turning on (that is, transcribing and translating) the enzymes

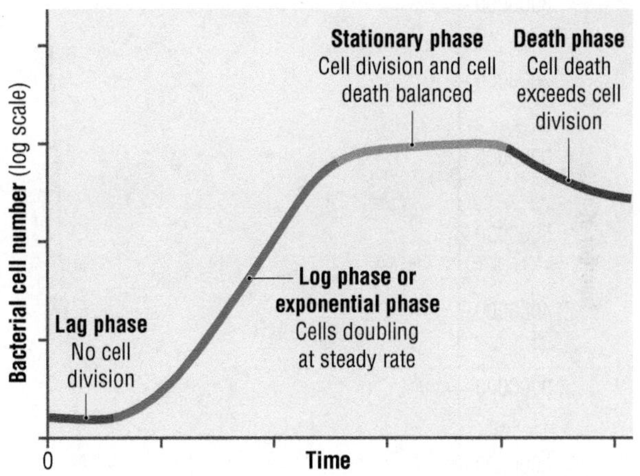

Figure 14.7 Bacterial growth curves are an important method for elucidating gene function and regulation. Growth curves are plotted by the change in the number of bacteria, often measured by optical density and plotted on a logarithmic scale on the Y-axis, over time on the X-axis. A typical bacterial growth curve begins with a lag phase, during which the cells begin to produce the enzymes required to utilize nutrients in the medium. This lag phase is usually followed by a period of exponential growth. Once nutrients are exhausted, growth slows, yielding a stationary phase. Eventually, the rate of cell death exceeds the rate of cell division, and the cell population declines.

required to utilize nutrients present in the medium. As shown in Figure 14.7, the lag phase is typically followed by a period of exponential growth. When nutrients are exhausted, growth rates typically slow until they do not exceed death rates, and a stationary phase of the curve is seen. If the experiment is pursued long enough, a death phase showing a decline in cell numbers follows. Some of the mathematics of growth curves are worked out in Box 14.2.

The doubling time of *E. coli* in rich medium is typically about 20 minutes, whereas, in medium containing only a simple carbon source and an inorganic nitrogen source, the doubling time lengthens to twice this time or more, and the slope of the growth curve is diminished. Similarly, growth curves for different bacterial species show different slopes in rich medium because their doubling times are different. Addition of a chemical that inhibits growth can also alter the structure of a growth curve. The slope and

BOX 14.2 *Quantitative Toolkit* Bacterial growth curves: a quantitative perspective

Bacterial growth involves the analysis of populations. Bacterial cells do grow in size, but many studies of "bacterial growth" are focused on the number of cells in a population. The number of cells is determined, in turn, by how fast the cells are dividing. Growth of a cell population is easily documented, typically by measuring the optical density of the culture using a spectrophotometer, and plotted as a growth curve. Growth curves contain a lot of information, much of it quantitative.

During what we term the exponential phase of growth, each bacterial cell in a culture divides into two; after a set generation time, each of the resultant daughter cells will again divide. The number of cells in the culture therefore increases geometrically, so that, by the third generation, each single cell would have become 2^3 or eight cells, and by the 24th generation 2^{24} cells. This growth is depicted in Figure A. Plotting out the number of descendants of one hypothetical starter cell over 24 generations on a graph yields an exponential curve, as shown in Figure A(i), or, if the axis with the number of cells is transformed logarithmically, a straight line as shown in Figure A(ii).

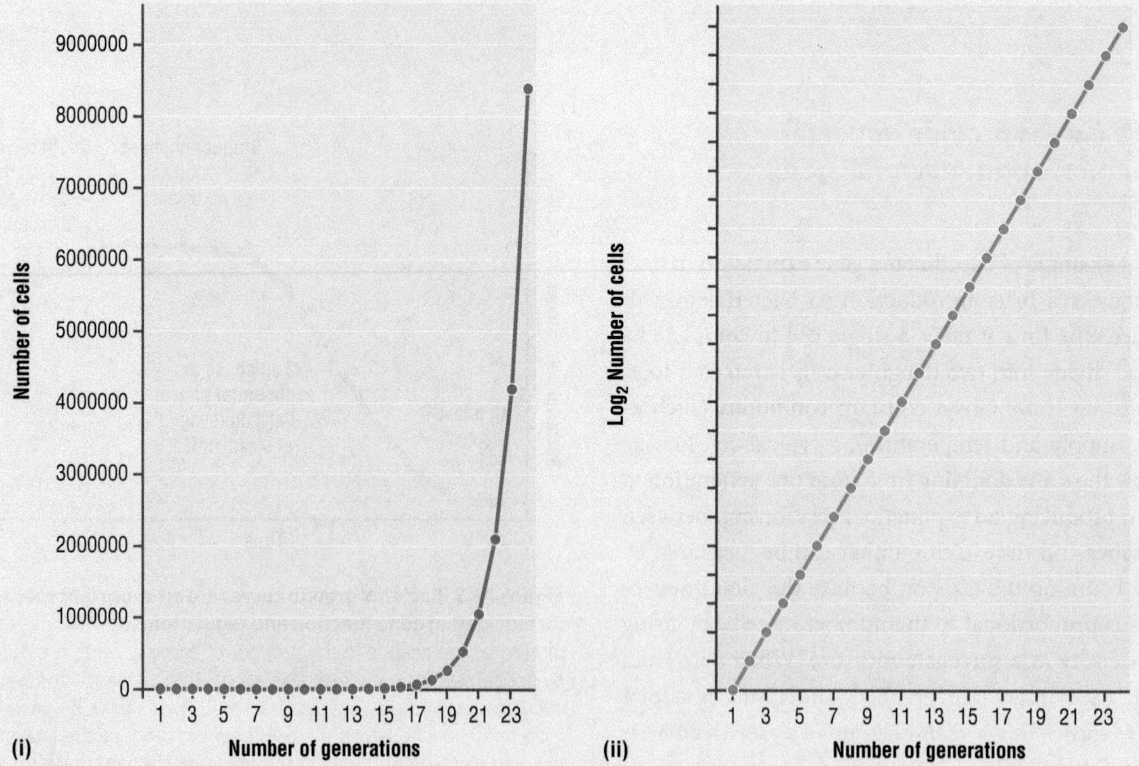

Figure A Plotting bacterial growth. A single cell was allowed to double for 24 generations, and the number of cells at each generation was plotted. The plot shown in (i) is a simple linear plot, which yields an exponential curve, while the plot in (ii) uses a logarithmic scale for the Y-axis. Note that the semi-log plot in (ii) produces a straight line.

BOX 14.2 Continued

We can express this mathematically. If medium containing n_1 cells, is cultured for x generations, the number of cells in the culture would have increased to n, where:

$$n_2 = n_1 \times 2^x$$

$$\text{If } n_2 = n_1 \times 2^x$$

Now suppose we don't know how many generations have gone by. Perhaps we don't know how fast the bacteria divide under the test conditions; indeed, these are the very parameters that scientists who generate growth curves often want to know. However, we can state that:

$$x = t/t_d$$

where:

- t_d is the doubling time
- $1/t_d$ is the reciprocal of the doubling time or the doubling rate constant, *k*.

Therefore:

$$n_2 = n_1 \times 2^{t/t_d}$$

and:

$$n_2 = n_1 \times 2^{tk}$$

Solving for *k* in this equation would give you the doubling time. This is much easier to do by taking logarithms:

$$\ln n_2 = \ln n_1 + tk$$

$$k = (\ln n_2 - \log n_1)/t = \ln(n_2/n_1)/t$$

Or if you are calculating in $\log_{10}$, instead of in base 2:

$$k = \log_{10}(n_2/n_1)/t \times 0.301$$

k is a constant as long as the nutrients in the medium are not limited—that is, all along the portion of the growth curve where dN/dt produces a straight line in the semi-logarithmic plot.

In the language of differential equations, you would describe the exponential graph shown in Figure A(i) by an equation describing the rate of change of cells (the dependent variable) over the independent variable of time:

$$kN = dN/dt$$

where:

- N is the number of cells
- t is time
- *k* is a first-order rate constant (in time −1) that is proportional to the cell division rate (or doubling time) of the organism under the test conditions and can then be computed from the equation or the graph.

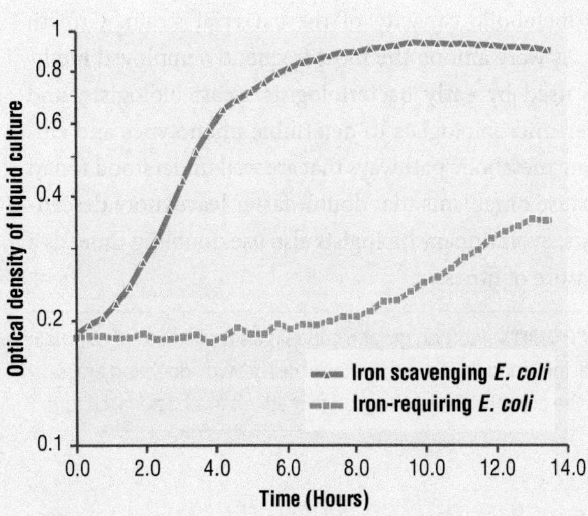

Figure B Experimental use of growth curves. Growth of two *E. coli* strains in minimal media depleted of iron.

Here are just a few of the very many things microbial geneticists can do with growth curves:

- Determine the optimal medium and conditions for growth.
- Determine the effect of medium constituents, environmental conditions, stress, or even specific genes on the rate of cell division. For example in iron-limited media, the iron-scavenging *E. coli* strain tested in Figure B divides more rapidly than more common iron-requiring *E.coli*.
- Compute how quickly growth will be visible in a culture.
- Determine which lineages are evolutionarily fitter. As will be described in Chapter 16, this can be best done by growing the strains together to identify the better competitor.
- Dissect regulator programs. So-called diauxic curves are produced when bacteria exert regulatory control over use of one or more substrates contained in a mixture.
- Study cell populations more broadly. The dynamics of bacterial growth curves are true for all cells dividing by binary fission and are used for similar purposes in studies of unicellular eukaryotes, such as yeasts, or cultured eukaryotic cells.

FIND OUT MORE

Monod, J. (1949) The growth of bacterial cultures. *Annu Rev Microbiol* **3**: 371–94

Neidhardt, F. C. (1999) Bacterial growth: constant obsession with dN/dt. *J Bacteriol* **181**: 7405–8

topology of a growth curve can therefore provide many clues about the composition of the growth medium and the metabolic capacity of the bacterial strain. Growth curves were among the most frequently employed methods used by early bacteriologists, yeast biologists, and other microbiologists to determine phenotypes and elucidate metabolic pathways that are well understood today. Because organisms that double faster leave more descendants, evolutionary biologists also use doubling times as a measure of fitness.

KEY POINTS Plotting growth curves is a simple method to obtain initial information about the rates of cell growth, doubling times, and the composition of the growth medium in a bacterial culture.

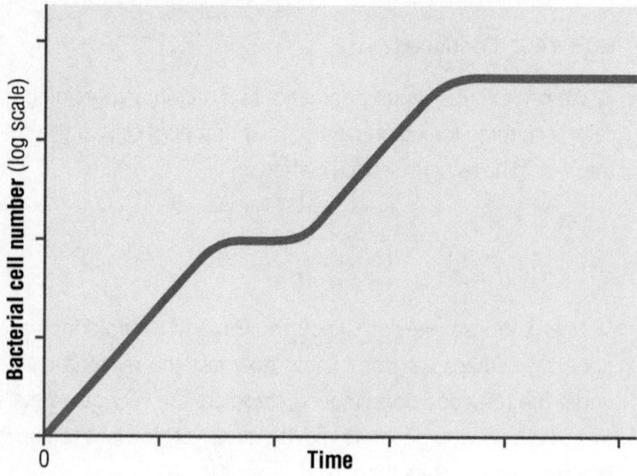

Figure 14.8 Diauxic growth curves were used to understand the genetics of lactose utilization. When grown in a mixture of sugars, such as glucose and lactose, bacteria such as *E. coli* exhibit as two-phase growth curve.

The lactose operon is both repressible and inducible

Growth curves provided important insights into how bacteria utilize various nutrient sources and, in turn, into the regulation that breaks down these nutrients. Prototrophic or wild-type bacteria break down sugars as a source of carbon to produce other macromolecules. The biochemical process of breaking down a complex molecule into its simpler components is referred to as **catabolism**, as noted in the Prologue previously. *E. coli* exhibits a metabolic preference for some sugars, rather than others, as illustrated by a classic growth curve experiment done by Jacques Monod in 1941. Experiments of this nature formed the basis for Monod's lifetime of work with *E. coli*, which illustrated many general principles of gene regulation, demonstrated the existence of mRNA, and greatly increased our understanding of the biochemistry and molecular biology of the interactions between proteins and nucleic acids. Monod and his long-time colleague François Jacob (whose essay on evolution as a tinkerer was cited in the Prologue) shared a Nobel Prize in 1965 with André Lwoff, who studied the response of bacteria to viruses known as bacteriophages and introduced Monod to microbiology.

Monod cultured *E. coli* in medium that had a mixture of the sugars glucose and lactose and analyzed the growth rate over time. Glucose is a monosaccharide, or simple sugar, which bacteria can readily use as a carbon source; lactose is a disaccharide or complex sugar that is broken down into two simple sugars, glucose and galactose. The growth curve he obtained, known as a **diauxic curve**, is shown in Figure 14.8. The diauxic curve shows that the bacteria divide rapidly for a time before the division rate slows or even stops. Then, after a lag time, the bacteria begin to divide rapidly again, before finally exhausting all of the sugar in the medium.

It took Monod and others more than 15 years to understand the process that was occurring during diauxic growth in glucose and lactose. Among the keys to the experiments were mutant cells that could not use lactose as a sugar, mutant cells that always used lactose as a sugar, even when glucose was present, and mutants that could not grow well, even when only glucose was present. Some of these mutants and their effects are described in Chapter 15. For now, we will focus on the overall picture, which turned out to be relatively simple—at least in concept. When provided with both sugars, *E. coli* uses glucose and divides rapidly. Eventually, the glucose is exhausted, so the cells slow or stop dividing. After the lag time, the bacteria begin to use lactose as their energy source, which continues until lactose is also depleted. These stages are illustrated in Figure 14.9.

KEY POINTS *E. coli* preferentially use glucose as an energy source but can turn on genes to produce the enzymes necessary to metabolize other sugars in the absence of glucose.

The regulatory circuit for utilizing these two sugars must then have multiple components and controls. One component includes the genes that are needed to break down lactose into glucose and galactose; these genes must be active during the lactose utilization phase of the diauxic growth but inactive or repressed during the glucose phase so that lactose is not used then. During the intermediate lag phase, these genes are being turned on.

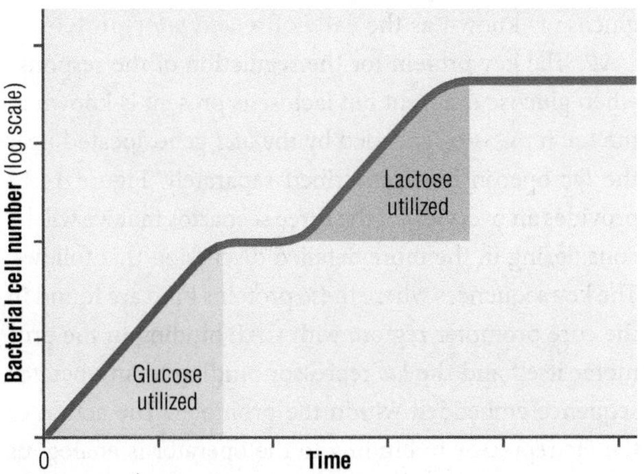

Figure 14.9 Sugar utilization during diauxic growth. *E. coli* breaks down the simple sugar glucose more readily than it does the disaccharide lactose. When provided with both glucose and lactose, *E. coli* uses glucose and divides rapidly, showing an increase in the population size. This is the first log phase of the diauxic curve. When the glucose is gone, the cells slow or stop dividing before beginning to use lactose as the carbon source. This is the second log phase of diauxic curve.

Cells with mutations in these genes cannot use lactose as a sugar at all; in other words, they fail to catabolize lactose. Two such genes were identified, named *lacZ* and *lacY*, with a third gene called *lacA* that is also involved in lactose catabolism but not essential for it. (The third gene was not named *lacX* because X is often associated with an unknown, rather than with an identified gene.) These are three distinct genes, which map close together on the chromosome. Thus, these genes form the structural genes of the *lac* **operon**, as illustrated in Figure 14.10.

The *lacZ* gene encodes the enzyme β-galactosidase, the enzyme that actually carries out the breakdown of lactose into glucose and galactose. We encountered *lacZ* and β-galactosidase in Tool Box 12.1 in our discussion of reporter genes. *lacZ* is one of the best-studied genes in all of biology, and β-galactosidase is a very well-studied enzyme. The *lacY* gene encodes β-galactoside permease, which is necessary for the transport of lactose into the cell where it can be broken down. The *lacA* gene encodes β-galactoside transacetylase, which transfers an acetyl group to β-galactosides; the physiological importance of this activity is still not entirely clear, and it is not essential for the breakdown of lactose. The promoter for the three genes is upstream of *lacZ*, and the genes are part of a polycistronic transcript, as with other operons.

KEY POINTS The *lac* operon in *E. coli* controls the expression of three genes—*lacZ*, *lacY*, and *lacA*—involved in the metabolism of lactose. The *lacZ* gene encodes β-galactosidase and is also often used as a laboratory tool.

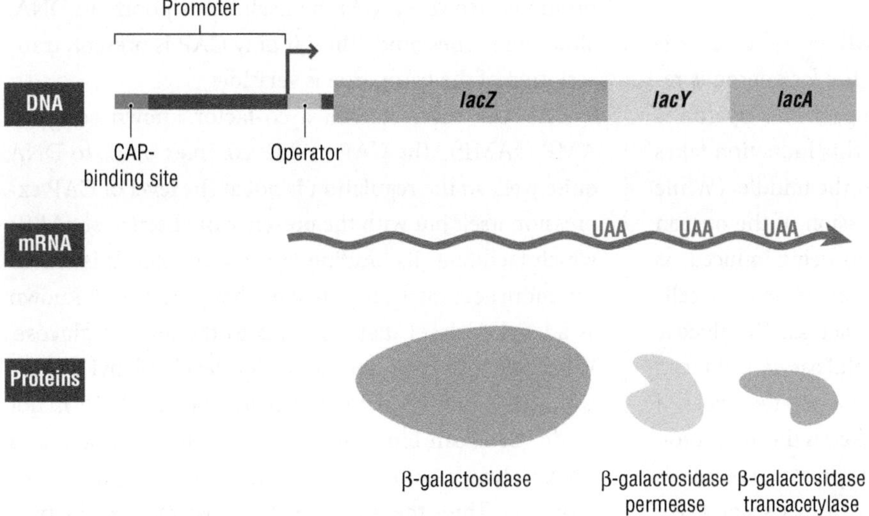

Figure 14.10 The lactose (*lac*) operon is needed for lactose utilization. The structure of the *lac* operon consists of a promoter region and three structural genes, *lacZ*, *lacY*, and *lacZ*, as shown. *lacZ* encodes the enzyme β-galactosidase, the enzyme that actually carries out the breakdown of lactose into glucose and galactose. The *lacY* gene encodes β-galactoside permease, needed for the transport of lactose into the cell. The *lacA* gene encodes β-galactoside transacetylase, which transfers an acetyl group to β-galactosides. The promoter for the three genes is upstream of *lacZ*, and the genes comprise a polycistronic transcript. Regulation of transcription occurs via two sequences in the promoter region—the CAP-binding site, shown in orange, and the operator, shown in blue.

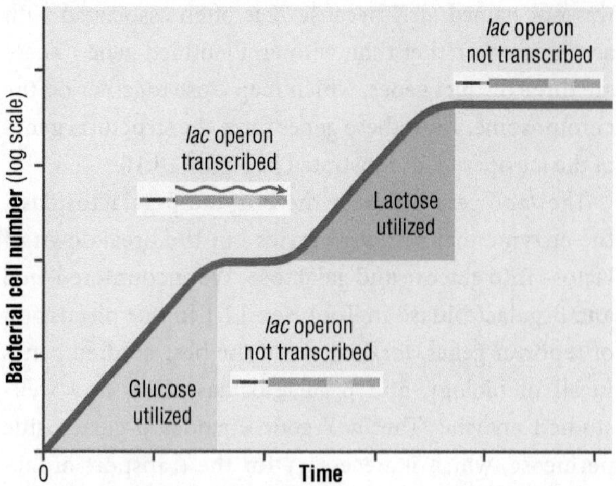

Figure 14.11 ***lac*** **operon transcription and the diauxic growth curve.** During the first phase of the growth curve, glucose is being used as the carbon source, and the *lac* operon is repressed. When the glucose is used up, the *lac* operon is de-repressed; this induction takes time, resulting in the intermediate lag phase. When the *lac* operon is fully expressed, the enzymes necessary for the breakdown of lactose are transcribed and translated. The cells can begin dividing again, using lactose as their primary sugar.

But the *lac* operon is not constitutively transcribed; the genes are not expressed at all when glucose is present because lactose is not being catabolized then. Glucose is the preferred sugar when both are present. The dynamic relationship between growth conditions and gene expression status is illustrated in Figure 14.11.

During the first phase of the growth curve, glucose is being used as the carbon source and the *lac* operon is repressed. When the glucose is used up, the *lac* operon is activated, induced, or de-repressed; this induction takes time, which explains the lag phase in the middle. (While the most accurate term is that expression of the operon is de-repressed, it is often described as being induced, as we shall see.) When the *lac* operon is expressed, the cells begin dividing again, this time using lactose. The glucose utilization genes remain on, since glucose is produced from lactose catabolism. For the *lac* operon, two types of control circuits must exist—one that keeps the *lac* operon from being expressed when glucose is present, regardless of the presence or absence of lactose, and a second one that allows the *lac* operon to be expressed when glucose is absent but lactose is present. We will describe the circuitry of these three scenarios in more detail in the following sections.

The circuits are remarkably sophisticated, but the coordination between them occurs at the promoter region. The key protein for the regulation of the response in glucose is known as the **catabolite activator protein or CAP**. The key protein for the regulation of the response when glucose is absent but lactose is present is known as the ***lac*** **repressor**, encoded by the *lacI* gene, located near the *lac* operon but transcribed separately. Figure 14.12 provides an overview of the three scenarios that we will be considering in the more detailed discussion that follows. The key sequences where these proteins bind are found in the core promoter region, with CAP binding in the promoter itself and the *lac* repressor binding to an operator sequence embedded within the promoter. The action of the *lac* repressor in binding to the operator is analogous to the role of the *trp* repressor in binding to the operator of the *trp* operon, since, in either case, the repressor binds to the operator and blocks RNA polymerase from binding to the promoter. However, it is important to recall that the *trp* operon is off when tryptophan is present, while the *lac* operon has to be on when lactose is present. The role of CAP is a bit different.

CAP binding is indirectly regulated by glucose

Let's begin with CAP and the regulation of the *lac* operon by glucose. The process is diagrammed in Figure 14.13. The main point is that CAP binds to the promoter region, which facilitates the binding of RNA polymerase to the promoter. However, CAP by itself binds poorly to DNA, although it does bind. Thus, if only CAP is present, transcription of the *lac* operon is very low.

CAP can interact with a co-factor known as cyclic AMP (cAMP). The CAP–cAMP complex binds to DNA quite well, so the regulation is not at the level of CAP expression itself but with the presence or absence of cAMP, which facilitates its binding to the promoter. It is cAMP (or, more accurately, the enzyme that produces it, known as adenyl cyclase) that responds to the level of glucose. When glucose is present at a high level, adenyl cyclase is inactive, and cAMP is not made. Since cAMP is not made, CAP can bind only poorly to the promoter, and RNA polymerase binds very weakly, if at all, at the *lac* promoter. Thus, the presence of glucose reduces the production of cAMP, which, in turn, results in only a low-level binding of CAP and little transcription of the *lac* operon. This is the situation in Figure 14.13(a). The *lac* operon is not actively repressed in glucose-rich medium; it is just not induced. Glucose, which is metabolized by a different set of genes altogether, is used as the carbon source.

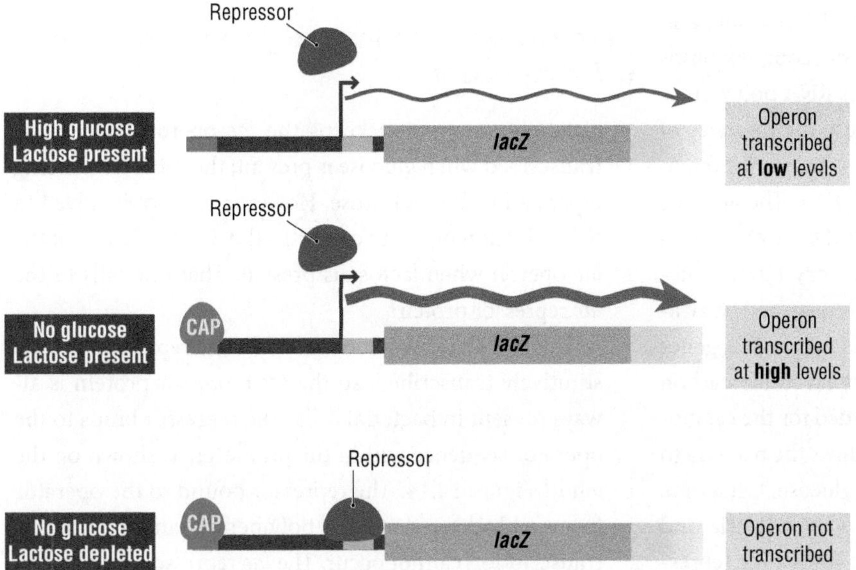

Figure 14.12 The two exponential phases are regulated by two different proteins. Catabolite activator protein (CAP) and the *lac* repressor bind to distinct regions in the promoter region and play distinct roles. When glucose levels are high and lactose is present, neither the CAP protein nor the repressor is bound, and the transcription of the *lac* operon is very low. As glucose is exhausted and lactose is present, the CAP protein binds to the CAP-binding site in the promoter, RNA polymerase is recruited, and the operon is transcribed. As lactose is exhausted, the repressor binds to the operator sequence, and the operon is no longer transcribed.

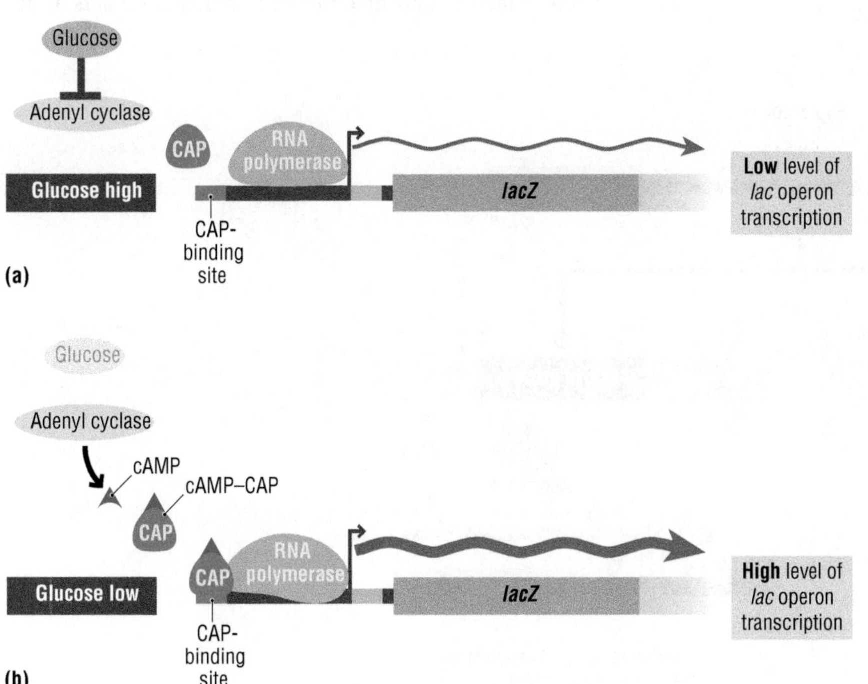

Figure 14.13 Positive regulation of the *lac* operon is regulated by the CAP protein. The CAP protein facilitates the binding of RNA polymerase to the promoter, resulting in high levels of transcription. CAP requires the co-factor cyclic AMP (cAMP) for maximal binding. CAP binding is regulated indirectly by glucose through its effect on the enzyme adenyl cyclase, which makes cAMP. (a) Glucose inhibits adenyl cyclase, so no cAMP is made when glucose is present, and CAP binds very poorly to the promoter region; as a result, transcription of the *lac* operon is low. (b) When glucose is absent, adenyl cyclase produces cAMP, which interacts with the CAP protein. The CAP–cAMP complex binds maximally to the CAP-binding site, which recruits RNA polymerase, and transcription is high.

When glucose is low or absent, cAMP is made, and so the CAP–cAMP complex forms. This complex binds strongly to the promoter, which recruits RNA polymerase to bind to the promoter as well. Because the presence of CAP–cAMP at the promoter induces transcription to occur, this is known as positive regulation. The absence of glucose induces the transcription of the *lac* operon, as shown in Figure 14.13(b). This regulatory circuit, mediated by glucose, is called **catabolite repression**. Catabolite repression occurs in bacteria when the genes necessary for the breakdown of a less favorable carbon source (in this case, the *lac* operon needed for the catabolism of lactose) are repressed, which allows the bacteria to use a preferred carbon source such as glucose. Catabolite repression is a rapid response system when bacteria find themselves in a favorable environment. Catabolite repression also occurs in some single-celled eukaryotes, such as yeast, albeit by a different mechanism.

KEY POINTS The *lac* operon is not induced in the presence of glucose, regardless of whether lactose is present or not, due to catabolite repression. When glucose is low, however, CAP–cAMP binding recruits RNA polymerase to the promoter.

The response to lactose is mediated by the *lac* repressor

Catabolite repression keeps the *lac* operon from being transcribed when glucose is present; the catabolism being repressed is that of lactose. However, it is not involved in the activation or, more correctly, the **de-repression** of the *lac* operon when lactose is present. That role falls to the *lac* repressor protein.

The *lacI* gene, which encodes the *lac* repressor, is constitutively transcribed, so the *lac* repressor protein is always present in bacterial cells. The repressor binds to the operator sequence within the promoter, as shown on the left of Figure 14.14. The repressor bound to the operator forms a block, so that RNA polymerase cannot bind, and transcription cannot occur. The *lac* repressor binds to the operator in the absence of other co-factors or molecules and keeps the *lac* operon transcriptionally off. Thus, the binding of the repressor represents negative regulation, since its binding shuts off the transcription.

However, the *lac* repressor also interacts with another molecule—the sugar allolactose, which is structurally related to lactose and made from it. Allolactose acts as an

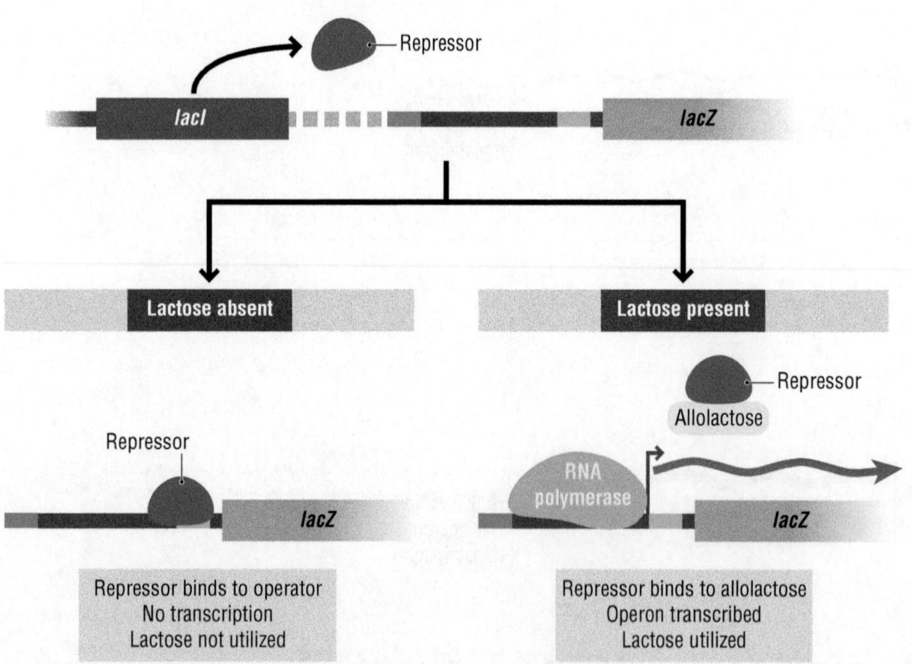

Figure 14.14 The role of the *lac* repressor in regulating transcription of the *lac* operon. The *lacI* gene encoding the *lac* repressor is constitutively expressed at low levels, so the *lac* repressor is always present in bacterial cells. In the absence of lactose, as shown on the left, the repressor blocks the binding of RNA polymerase, and transcription cannot occur. When lactose is present, as shown on the right, allolactose, a sugar derived from lactose, is also present and interacts with the repressor protein. The allolactose–repressor complex cannot bind to the operator. Since the promoter is no longer blocked, RNA polymerase can bind and transcribe the *lac* operon. Thus, allolactose acts as an inducer of the *lac* operon.

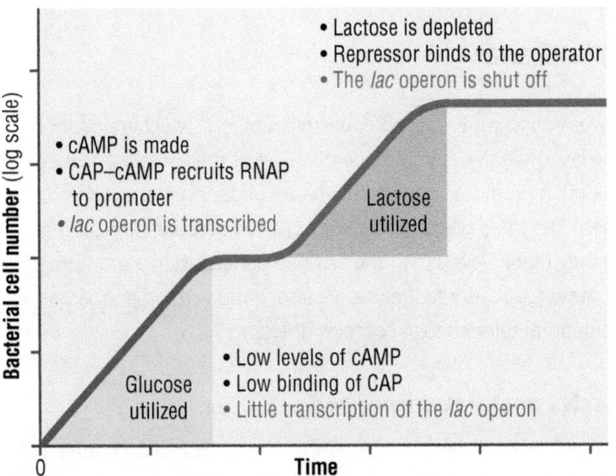

Figure 14.15 A summary of the roles of CAP and the *lac* repressor on diauxic growth and sugar utilization.

inducer of the operon; it binds to the repressor protein, which changes the three-dimensional shape of the protein. The allolactose–repressor complex cannot bind to the operator. Thus, when allolactose is present—that is, whenever lactose is present—it interacts as an inducer with the repressor protein and prevents it from binding to the operator, as shown on the right of Figure 14.14. Since the promoter is now freed of its block, RNA polymerase can bind and transcribe the *lac* operon. As the genes are transcribed and the enzymes are made, lactose is broken down. Eventually, the concentration of lactose (and allolactose, the actual inducer) becomes low, and the repressor binds again to the operator and the *lac* operon is again shut off. This entire process is summarized in combination with the diauxic curve shown in Figure 14.15.

KEY POINTS When the glucose concentration is low and lactose is present, allolactose binds to the *lac* repressor, changing its shape so that it can no longer bind to the operator. Without the *lac* repressor, transcription of the *lac* operon can begin.

The *lac* operon is almost certainly the best-studied gene regulatory circuit in all of biology. It has influenced our thinking not only about the logic of gene regulation but also about how individual subunits of a protein come together to make a functional multimeric protein, since the lactose repressor is as a tetramer of four identical subunits. Our understanding of the interaction of proteins with co-factors and with DNA sequences was also heavily influenced by work on the *lac* repressor and the *lac* operon, since this was one of the first examples known. Likewise, transcriptional start and termination sites and other concepts that are fundamental to molecular biology arose from work on the *lac* operon. For example, one type of mutation, known as a polar mutation, demonstrates the key coupling between transcription and translation in bacteria. Polar mutations are described in Box 14.3.

As with other well-studied biological examples, the *lac* operon has also been the source for several important tools in molecular biology. Some of these were described in Tool Box 12.1, while a few more will be discussed in Tool Box 15.2.

While the *lac* and *trp* operons are the two best-studied examples of coordinated gene expression in bacteria, it should be realized that the *E. coli* pan-genome is estimated to have about 650 different operons, each with its own process of gene regulation. Most of these operons are broadly similar to the two described here, although none is precisely like these and both of these have additional features that we have not discussed. Some operons are inducible; others are repressible; still others (like the *lac* operon) have both sets of controls, and some have no control beyond σ-dependent transcription, as described in Chapter 12. Some operons are catabolic, while others are anabolic. Clearly, the operon structure for gene regulation works very well in bacteria.

While our examples with the *trp* and *lac* operons have the genes mapping very close together, some bacterial activators and repressors regulate multiple genes and operons, which are not necessarily located close to one another on the chromosome. The complete set of genes controlled by a specific activator or repressor is referred to as a regulon.

For example, many *E. coli* carbohydrate metabolism genes are subject to catabolite repression, and most of them are activated by CAP–cAMP in a manner similar to what we have described for the *lac* operon. However, the various genes and operons regulated by CAP do not map together.

Similarly, a number of the rapid, but error-prone, polymerases of *E. coli* discussed in Chapter 4 are typically repressed during normal bacterial growth by a protein known as LexA. When DNA is damaged, these polymerases are activated in a process known as the SOS response. The protein RecA binds to the single-stranded DNA, and this complex interacts with LexA. As a result, LexA autocleaves (essentially destroys itself) and no longer blocks the transcription of the genes encoding these polymerases, as shown in Figure 14.16. Thus, DNA damage causes LexA de-repression of the genes involved in DNA repair. Genes that are repressed by LexA and de-repressed by

BOX 14.3 *Going Deeper* Genetic polarity in bacteria

The coordinated expression of genes in an operon depends on the shared upstream promoter, as discussed in the chapter. However, genetic analysis of particular types of mutations found that expression also depends on changes in the structural genes themselves. Specifically, mutations in the upstream gene in the operon (such as *trpE* and *lacZ*) can reduce the expression of the gene downstream (*trpD* and *lacY* in our examples). Mutations in one gene that diminish the expression of downstream genes are said to be **polar**. Polar mutations occur when the mutated gene and the affected downstream gene(s) are in the same polycistronic operon. Polar mutations arise because the genes are spaced closed together in the operon and because transcription and translation are tightly coupled in bacteria.

While our drawings of operons do not have a sense of the scale, the physical distances and sizes become important for understanding polar mutations and their impact inside the cells. Genes in a simple operon are typically less than 20 bp apart; in many cases, adjacent genes overlap by as much as 4 bp. Polar mutations are usually alterations in the DNA that increase the non-coding distance between two genes in an operon. They can be nonsense or frameshift mutations that introduce a stop codon early in a preceding open reading frame, or large insertions. The premature stop codon arising from such a mutation can cause ribosomes to prematurely disengage from the transcript; as a consequence, downstream open reading frames cannot be translated, so the premature stop codon is polar.

In keeping with the idea that the key to polarity is the distance between the genes, stop codons introduced towards the end of an open reading frame are often not polar or are less polar on downstream genes; they do not increase the distance between genes in the operon very much.

Because transcription and translation are coupled in bacteria, premature interruption of translation of an early gene in an operon can also prompt Rho-dependent termination (described in Section 12.5), so that end genes in the affected operons are not transcribed. The net effect is that expression of genes downstream of the polar mutation is greatly reduced.

Polarity may seem like an unusual feature of certain mutations in operons, but the discovery of genetic polarity provided early evidence that transcription occurs linearly and directionally along a strand of DNA. Studies of polarity were used to physically map genes within operons and to understand the nuances of their coordinated regulation. For example, polar mutations in the *trp* operon revealed that this operon is not quite as simple as we described in the chapter—in fact, it has a second internal promoter. This came to light when it was found that polar mutations in *trpE* prevented TrpD from being produced but did not affect the production of TrpCBA. In other words, not all transcription was beginning at the promoter upstream of *trpE*. Further research identified a weak constitutive promoter upstream of *trpC*, which transcribes the last three genes in the operon at 2% of the levels seen when the main promoter is used. Thus, even under normal conditions, these last three genes (and their polypeptides) are expressed at slightly higher levels than the first two genes. Internal promoters or other regulatory sequences are also found with other operons, sometimes referred to as complex operons.

Koch's postulates revisited

The existence of polarity means that the loss of a function cannot always be simply equated to a mutation in the gene. Recognizing this, Stanley Falkow in 1988 outlined what he referred to as "molecular Koch's postulates." These postulates were modeled on Robert Koch's postulates, which have long been the accepted way to attribute a specific disease to a pathogenic organism. As it became possible to attribute pathogenicity not simply to an organism but also to specific genes within that organism, it became necessary to revisit and amend the original postulates.

Falkow's molecular Koch's postulates are as follows:

1. The gene under investigation should be present in organisms that show the phenotype. The original postulate was that the organism needed to be present in diseased individuals but absent in healthy individuals. The amended version recognizes that pathogenicity might arise from particular genes within the organism, rather than the organism itself. Koch himself came to recognize that non-pathogenic isolates could be present in healthy individuals and stopped using this principle.
2. Inactivation of the gene should abolish or diminish the phenotype.
3. Reversion or allelic replacement of the gene (often effected by plasmid rescue, as described in Chapter 8) should restore the phenotype. The equivalent among Koch's postulates was that the organism should be isolated from a diseased individual, introduced into a healthy individual, and cause the disease. It should then be isolated from the newly infected individual. The molecular version of the postulate bases the same principle on the idea of inactivating or supplying particular genes. Polar mutations are relevant to this postulate, since a gene may be inactivated by a polar mutation in the upstream gene. This will not be rescued when the wild-type copy of the gene is supplied, since the wild-type copy is present but inactivated.

While polarity is not the only reason why a mutation may not be rescued by a plasmid clone, it is the most common one in bacteria. Fulfilling the molecular Koch's postulates increases the

BOX 14.3 Continued

accuracy of gene annotations. However, it requires first the construction, and then the *trans*-complementation, of each mutation. Sequentially deleting and complementing genes was the method of molecular bacteriology for three decades following Falkow's postulates. In practice, these standards are tedious and difficult to apply to genome-wide screens.

Just as Koch's postulates have their limitations, so the molecular Koch's postulates have limitations that extend beyond technical difficulties in implementation. *Trans*-complementation tests are typically performed using plasmids, but the copy number of the cloned gene on the plasmid may be much higher than the copy number in wild-type. This gene dosage change can have a discernible effect on the phenotype under study. This has led some investigators to perform *cis*-complementation tests by placing the reintroduced gene on the bacterial chromosome, so that the gene copy number and dosage are the same as in wild-type. While that addresses the dosage problem with *trans*-complementation, it makes the process of fulfilling the molecular Koch's postulates even more technically challenging. Addressing polarity in bacteria is an important challenge for genome-wide research.

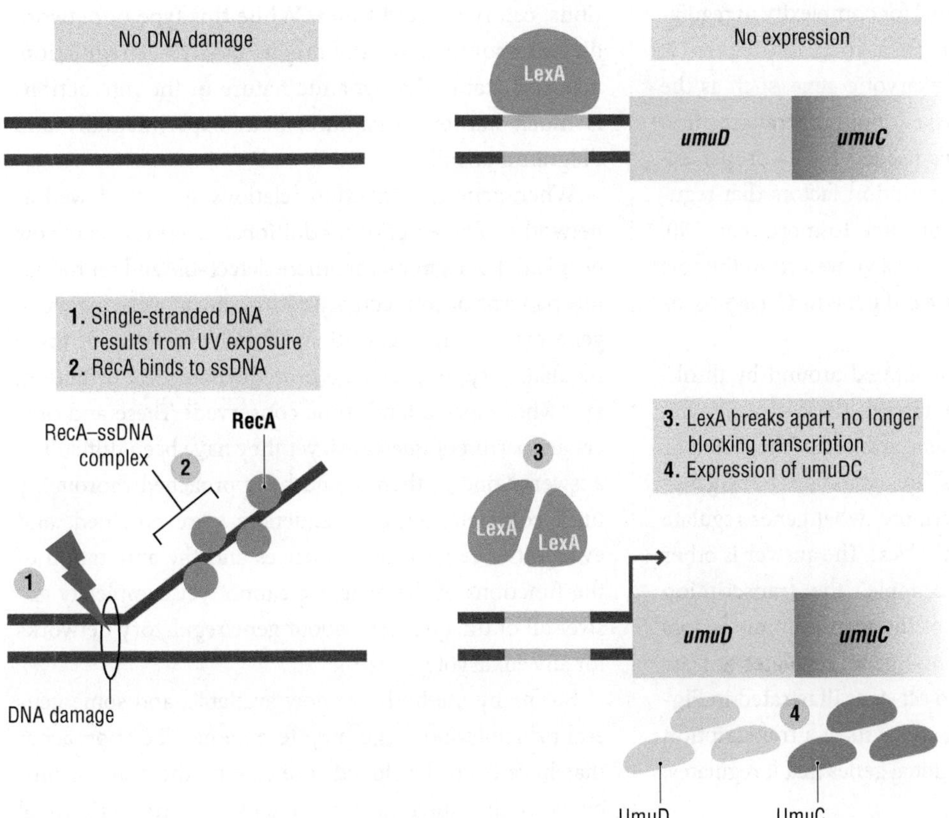

Figure 14.16 The SOS response regulon in *E. coli*. During the SOS response in *E. coli*, the RecA protein binds to single-stranded DNA (ssDNA) produced following DNA damage. The ssDNA–RecA complex causes LexA to autocleave and fall off promoters that it normally represses. This allows expression of error-prone DNA polymerases (such as those encoded by the *umuDC* operon), UV repair proteins including UvrA, RecA, and numerous other proteins not illustrated in the figure. Collectively, the LexA-repressed genes that are de-repressed in this manner comprise the SOS regulon.

single-stranded DNA and RecA also include those that inhibit cell division (including some regulatory genes), as well as *recA* itself. Thus, *E. coli* has evolved a program for responding to DNA damage and ensuring that repair is complete before the cell divides. The genes controlled by LexA and RecA in this manner are said to belong to the SOS regulon.

However, although the operon structure is widespread in bacteria and archaea, it is largely absent in eukaryotes. Some organisms, notably *Caenorhabditis elegans*, have polycistronic transcripts, but the genes encoded on them are often not related in function. The role of polycistronic transcripts in eukaryotes is unclear, but it is evident that gene expression is not coordinated in the same way as the classical operons in bacteria. In eukaryotes, co-regulation of gene expression is done by suites of transcription factor proteins, which appear to provide the additional transcriptional flexibility needed in these organisms.

KEY POINTS While operons are a very widespread strategy for co-regulation of gene expression in bacteria, they are not the only way that gene expression is coordinated.

14.3 Principles of gene regulatory networks

These examples from bacteria demonstrate that the organization of gene regulatory networks can be viewed from different perspectives, which yield different insights. As you might imagine, the potential for complexity in regulatory networks is even greater in eukaryotes. In Chapter 12, we looked at an individual eukaryotic gene, such as the *eat-4* gene in *C. elegans*, and asked about the transcription factors and regulatory regions needed for its proper expression. The number of transcription factors that regulate a gene varies, from two or three to more than 100. This is transcriptional regulation as viewed from the role of one gene, for example, the *eat-4* gene in *C. elegans*, as summarized in Figure 14.17.

But that perspective can be flipped around by thinking about the transcription factors themselves. What genes are regulated by the same transcription factor or factors? These "shared targets" are candidates to be genes that are co-expressed. Furthermore, what genes regulate the transcription factors themselves? The answer is other transcription factors. For example, the transcription factor LIN-11, which is one of the transcription factors that regulate the *C. elegans eat-4* gene, regulates at least ten other genes, in addition to *eat-4*, as illustrated in Figure 14.18. One of these ten genes is itself a transcription factor with its own set of additional genes that it regulates, and is shown in purple.

It is important to note that these types of regulatory diagrams do not necessarily indicate that all the targets are being regulated by LIN-11 in the same way in any individual cell or at a single point in time. Rather, they illustrate a collection of genes known to be regulated by LIN-11 under a variety of different conditions, cell types, and times. While this type of network diagram summarizes the targets of LIN-11 regulation, it is also static; the dynamic nature of the interactions is much harder to capture, both experimentally and diagrammatically.

When gene transcription relationships are viewed as networks of interactions, additional questions can now be asked. For example, are there detectable and reproducible patterns of connectivity within the network by which gene expression is coordinated? What aspects of these regulatory systems change over the course of evolution, and what aspects tend to be conserved? These and others are pertinent questions, yet they have been difficult to answer. None of them could be approached thoroughly until complete genome sequences were obtained, and even with the genome sequences and the annotation of the functions of the gene, we cannot yet completely answer all of the questions about gene regulatory networks for any eukaryote.

But many methods are now available, and some general principles are beginning to emerge. The approaches that have been developed to elucidate the relationships between the different components of a transcriptional network tend to focus on analyzing the sets of genes being actively transcribed or repressed in a particular situation. Usually, this description includes analyzing which genes

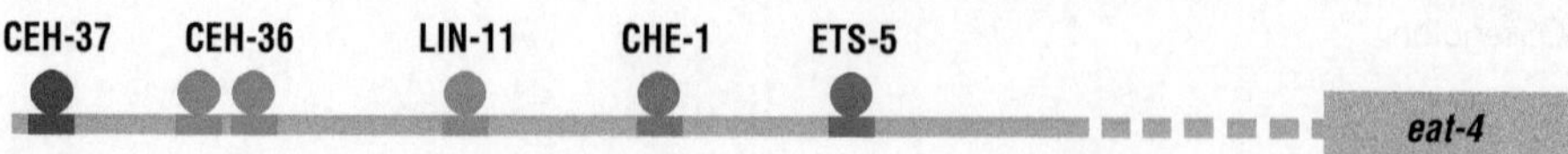

Figure 14.17 An example of eukaryotic transcriptional regulation. Some of the transcription factors that regulate the expression of the *eat-4* gene in *Caenorhabditis elegans* are shown; only one part of the *cis*-regulatory module and five of the transcription factor proteins (colored balls) are shown. Figure 12.10 showed that different neurons express different transcription factors, whereas this version summarizes the overall binding without regard to the specific neuron.

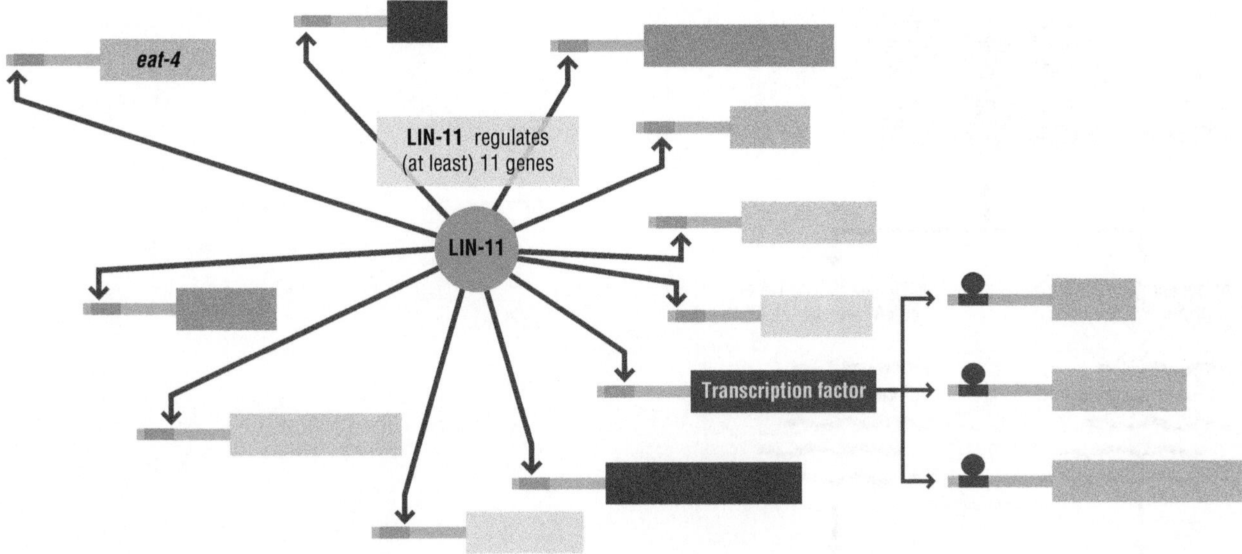

Figure 14.18 The genes regulated by LIN-11. The LIN-11 transcription factor regulates at least ten other genes in addition to *eat-4*. Each of these target genes has a binding region for LIN-11 upstream of the gene. One of these genes itself encodes a transcription factor, which regulates the transcription of another set of genes, shown here in purple. In these and other diagrams, direct regulation is shown by arrows without regard to whether the genes are turning on or turning off the expression of another gene. Since the majority of transcription factors stimulate the expression of other genes, these are assumed to be positive interactions, but that may not always be the case.

and associated regulatory elements a specific transcription factor may be binding to. The goals and the different types of experimental approaches in these studies are considered in more detail in the following section.

Different genome-based methods are used to analyze the networks of gene regulation

The first important stage in understanding how gene expression is coordinated is to determine which genes are being transcribed and under what conditions and in which cells they are being transcribed. This is often done using microarrays, a technique described in Section 2.5, to compile a **transcription or expression profile**. More recently, as discussed in Section 2.5, expression profiles are also done using the more sensitive method of RNA-seq.

Our discussion of microarrays in Section 2.5 described the process by which one-channel arrays are done. A one-channel array compiles a transcriptional profile under one set of conditions or in one cell type. A related method called a two-channel **microarray** is used to determine how the transcriptional profiles change under different conditions; the different conditions could be different cell types, normal cells compared to cancer cells of the same type, the same cells and gene at different times in the life cycle, and so on. For example, an early use of microarrays was to compare the expression profiles of two sexes in worms, while another very early use was to compare the expression profiles of normal mammary cells with breast cancer cells. One-channel arrays (and RNA-seq) are used as sensitive measures of which genes are being transcribed, while two-channel arrays are used as a measure of how that transcription changes.

The process for compiling a two-channel array is very similar to what is done for one-channel arrays, as shown in Figure 14.19(a). A microarray chip is prepared, as described in Section 2.5, with features from potentially every gene, and often with multiple features per gene. Such chips representing the genome of many organisms can be purchased from commercial vendors. The investigator isolates mRNA from each of the two samples and either directly labels the mRNA or produces labeled cDNA from each sample. For a two-channel array, each sample of mRNA or cDNA has a distinct fluorescent label. Typically, one of these is the dye Cy3, which fluoresces yellow–green (which the computer scanner depicts as green), while the other one is the dye Cy5, which fluoresces red.

As shown in Figure 14.19(a), the microarray chip is hybridized with equal amounts of labeled cDNA or RNA

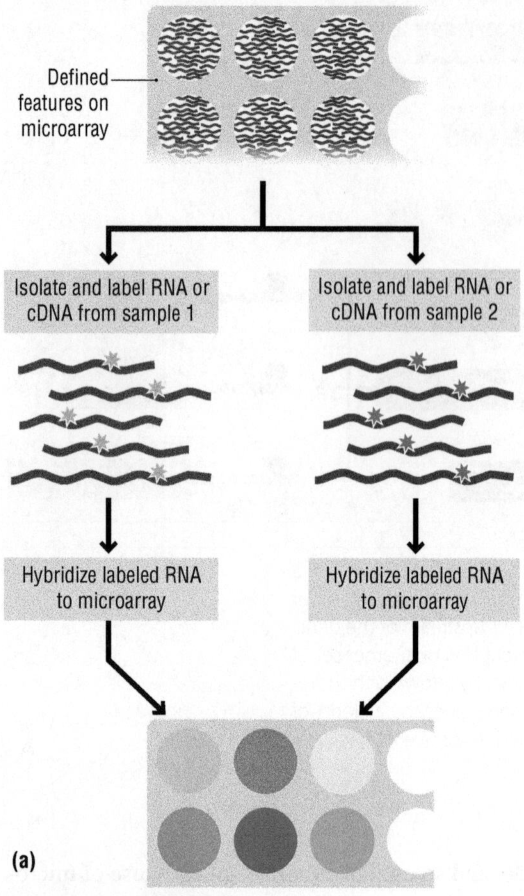
Defined features on microarray
Isolate and label RNA or cDNA from sample 1
Isolate and label RNA or cDNA from sample 2
Hybridize labeled RNA to microarray
Hybridize labeled RNA to microarray
(a)

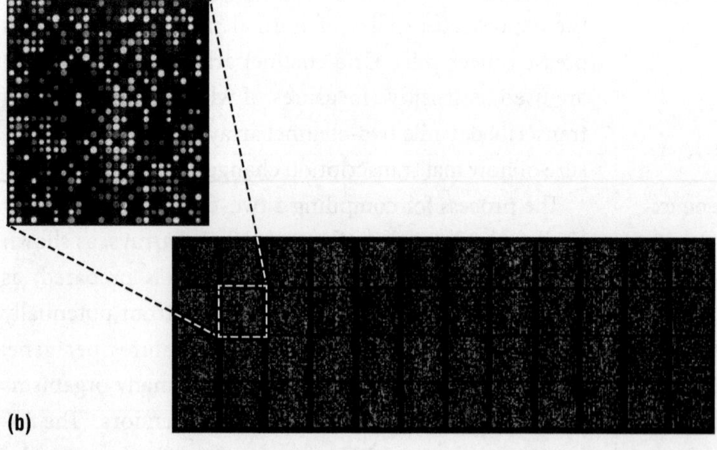
(b)

Figure 14.19 Two-channel microarrays. Two-channel microarrays present a comparison of the expression profiles in two different samples. The features on the microarray chip are prepared as discussed in Section 2.5. (a) diagrams the process. RNA is extracted from each of the two samples and either directly labeled or made into cDNA, which is labeled. In this example, RNA isolated from sample 1 is labeled with Cy3, depicted as green, while RNA isolated from sample 2 is labeled with Cy5, depicted as red. These are hybridized together to the microarray chip, and the colors of the images scanned for the relative contributions of each sample. A feature representing a gene expressed primarily in sample 1 will be green; an example is shown in the lower right corner. A feature representing a gene expressed primarily in sample 2 will be red; two examples, with different levels of expression, are shown in the bottom row. Features representing genes whose expression is the same in the two samples will be yellow. (b) shows data from mRNA isolated from normal mammary cells and breast cancer cells; note that most features are black and do not show either label indicating genes that are not expressed in either sample.

from the two samples. Each of the labeled cDNAs or RNA will make a double-stranded hybrid with complementary sequence features on the microarray chip. Since the number of copies of the complementary sequence on the feature is in vast excess of the number of copies of the labeled cDNA or RNA from the samples, all of the cDNA or RNA samples will hybridize to features, and the quantitative fluorescence signal will show the relative abundance of each sequence in the samples. There may be thousands of features that fluoresce predominantly red, since those genes are expressed only or preferentially in that sample; likewise, features that are predominantly green represent genes that are only or primarily expressed in that sample. Still other features will be yellow, indicating genes that are expressed at similar levels in the two samples. The colors provide a comparison of the gene expression profiles of the two samples.

For example, suppose that the cDNA made from mRNA from a normal tissue is labeled with Cy3, which will be scanned to show up as green, while the cDNA made from the mRNA in a tumor cell is labeled with Cy5, which shows up as red, in Figure 14.19(a). A feature from a gene whose expression decreases when the cells become cancerous will be green or primarily green. A feature from a gene whose expression is activated when the cells become cancerous will be red or primarily red; genes that show up yellow will be the ones whose expression does not change. (Features that are black did not hybridize with either sample and may represent genes that are not expressed in that particular cell type.) An example of a two-channel array is shown in Figure 14.19(b).

The two-channel array is showing a comparison of the transcriptional profiles in the two samples, but it may not be showing only the direct effects. Consider an experiment that asks about the target genes being regulated by a particular transcription factor. One cDNA sample could be prepared from cells in which the transcription factor is active, while the other cDNA sample is prepared from the same cells in which the activity of the transcription factor is knocked out. While the array will show changes arising from the knocking out of the transcription factor, these will include both direct and indirect changes. If one of the direct targets of the transcription factor is another gene that encodes a transcription factor, then the microarray will show changes in the expression pattern not only of the gene encoding the target transcription factor but also of the genes regulated by that target transcription factor. As a result of such effects, many of the changes in transcription that are observed under different conditions or with different transcription factors are due to indirect regulation, although the changes are often important physiologically. A transcriptional profile is an important piece of information but is not the best one for determining the direct targets of a transcriptional factor.

Direct detection of the target sequences

Since transcription factors have preferred binding sites, one could imagine identifying the direct targets of a transcription factor by computationally scanning the genome sequence for candidate binding sites. This method does work, and many different algorithms have been developed to do this. In practice, this is a more difficult problem than it might seem. Recall from Chapter 12 that transcription factor binding sites are often short sequences, with some mismatches allowed. In addition, although most transcription factor binding sites are found in *cis*-regulatory modules (CRMs) upstream of the transcriptional start site (TSS), the distance between the TSS and CRM varies widely for different genes. Consequently, computational searches for transcription factor binding sites produce a very large number of possible sites, only some of which are actually used. Many false positives are found.

The most widely used method to find the direct targets of transcriptional factors is **chromatin immunoprecipitation or ChIP**. The method is outlined in Figure 14.20 and described in more detail in Tool Box 14.1. A ChIP assay directly determines the sites throughout the genome where a transcription factor is bound to DNA in a particular cell or tissue at a specific time or under specific conditions. These sites can be analyzed independently of predictions about what or where the preferred sites might be. ChIP assays detect direct targets of a transcription factor and are also independent of other techniques, although they are usually combined with a transcriptional profile during the analysis.

Thousands of ChIP assays have been done, involving hundreds of transcription factors in many different species under different conditions, and have identified tens of thousands of target sequences. Large-scale projects from many different laboratories have been organized in attempts to annotate all of the transcription factor binding sites in a genome. The projects for humans, mice, *Drosophila melanogaster*, and *C. elegans* are collectively referred to as the ENCODE projects, where ENCODE is an acronym for Encyclopedia of DNA Elements; the ones for *C. elegans* and *Drosophila* are collected as modENCODE for model organism ENCODE. Many similar projects are under

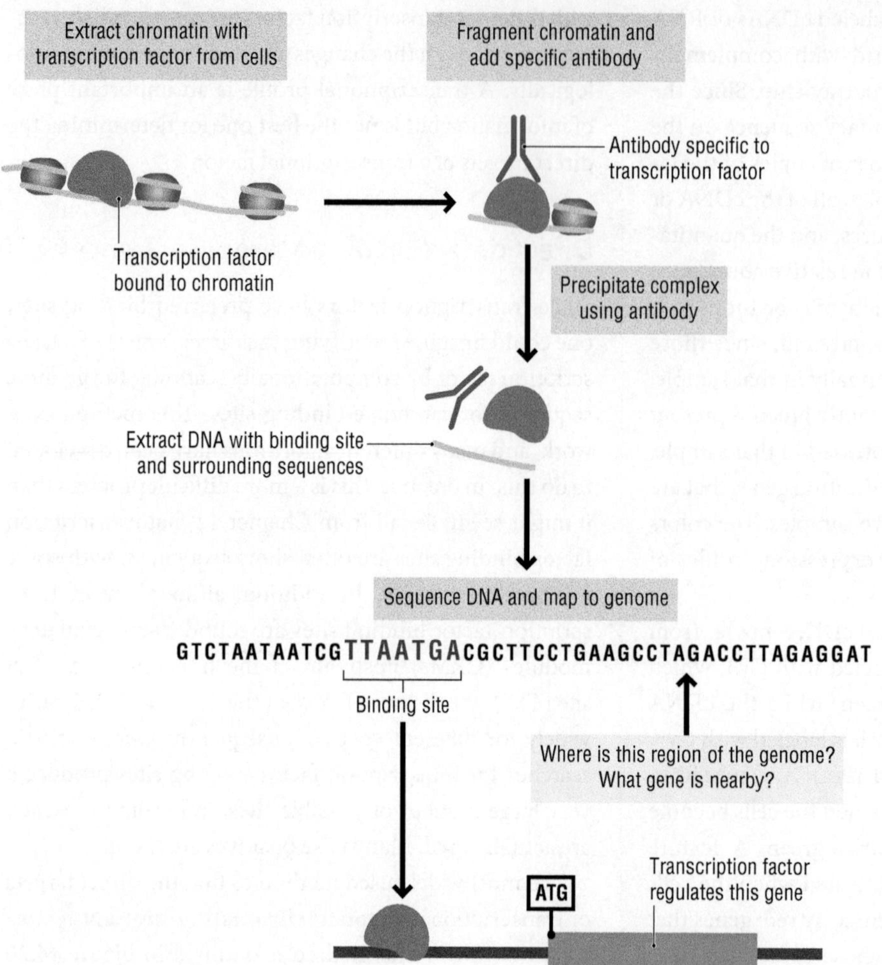

Figure 14.20 ChIP involves isolating chromatin with the transcription factor bound. Chromatin is isolated from a cell in which a transcription factor is likely to be active, and fractionated in solution. The chromatin complex including the transcription factor is precipitated from the fractionated solution by using an antibody specific to that transcription factor. Because the transcription factor is associated with chromatin, approximately 50–200 base pairs of DNA are co-precipitated with the transcription factor. This DNA can be purified and sequenced, and the sequence is compared to the genome to determine the location where the transcription factor was bound.

way for other organisms. All of these genome annotation projects include the identification of transcription factor binding sites as one of the principal goals.

A few observations can be made based on the data from these projects so far. Some transcription factors regulate only one or two genes, whereas others regulate more than 200, as summarized in Figure 14.21; from data from yeast, worms, flies, and *Arabidopsis*, the mode seems to be about 40 target genes per transcription factor, but the range of targets is probably more informative than the mode, since every transcription factor is different. The number of targets per transcription factor is probably a bit higher in mammals than in these model organisms.

The data about transcription factor connections can be presented in tables, but it is often easier to view the information graphically, in which the transcription factor protein is represented as a circle or a triangle. Lines connecting the circles represent a direct regulatory interaction between one transcription factor and another; these lines have arrows to indicate the direction of regulation. We can use the information from the ENCODE project to illustrate this. Figure 14.22 shows the information for the Krüppel (Kr) transcription factor from *Drosophila.* Note that the arrows coming towards the central Kr transcription factor indicate the genes that regulate its expression. The arrows going away from the Kr transcription factor indicate the transcription factor

TOOL BOX 14.1 Chromatin immunoprecipitation (ChIP)

As discussed in the chapter, the most direct method to identify the sites at which transcription factors bind *in vivo* is to use chromatin immunoprecipitation, known as ChIP. The name accurately summarizes the technique, depicted in Figure 14.20, but some elaboration could be helpful. Transcription factors bind to specific DNA sequences, but these DNA sequences are found embedded in the chromosome, as part of the DNA–protein complex known as chromatin. The goal of ChIP is to identify the DNA sequence and its location in the genome; by using this location in the genome, it is possible to infer which gene is regulated by the transcription factor. Thus, ChIP has to be able to isolate one specific protein and its short binding site, amid a complex of many proteins and long stretches of DNA.

The key to ChIP is an antibody, the "immuno" part of the name. Antibodies (or immunoglobulins) are proteins made by the B cells of the mammalian immune system and are used to neutralize proteins from invaders such as bacteria and viruses. The B cells recognize a segment of the foreign protein, known as an antigen, and stimulate the production of antibodies that are highly specific to that antigen. The specificity of the antigen–antibody interaction has provided a versatile and powerful tool for detecting, localizing, and isolating proteins from a cell.

As an experimental tool, antibodies are made by injecting the protein of interest into a mammal, such as a rabbit, which makes antibodies against it; these antibodies can then be purified from the serum of the rabbit. The antibodies themselves have a constant and variable region; the variable region is the part of the antibody that interacts with the antigen, while the constant region, which is common to many different antibodies with different variable regions, is the part used to isolate or detect the antibody. For ChIP, the transcription factor of interest is used as the antigen.

Since production of antibodies specific to each of the hundreds of transcription factors in an organism is often a laborious step, another procedure is to clone the gene for the transcription factor into a plasmid with a common protein sequence, so that every transcription factor is produced as a fusion protein with the same common sequence. This common sequence is known as an epitope. Antibodies to the epitope can be used to isolate any transcription factor of interest; this is known as epitope tagging.

In whichever way an antibody to a transcription factor is produced, these form the basic components of a ChIP experiment. The process itself is summarized in Figure 14.20. Nuclei are isolated, and the protein interactions and protein–DNA interactions in chromatin are stabilized by chemical cross-links. The chromatin is fractionated, such as by sonication, to produce randomly sized fragments in solution. The antibody to the transcription factor is then added to the solution where it interacts with its antigen. The antibody–antigen complex is precipitated out of solution and purified. The cross-links that hold the complexes together are then reversed to release the individual components. Thus, immunoprecipitation is also widely used to study protein–protein interactions, although this is not the goal of ChIP. For ChIP, the associated DNA is isolated, and its sequence is determined.

Originally, the sequence of the DNA was determined using microarrays; since microarrays are nicknamed "chips," this procedure is known as ChIP-chip. It is now more common to sequence the DNA directly, so the procedure is known as ChIP-seq.

The DNA purified from immunoprecipitation is 50–200 bases long. Part of this sequence is the binding site for the transcription factor, which is typically about 8–12 bases. The other part of the sequence is the region flanking the binding site. This can be compared to the genome sequence of the organism to determine the location at which the transcription factor is bound.

Thousands of ChIP assays have been performed, so a large volume of data is available. Nonetheless, they have a few limitations. Since eukaryotes have hundreds of transcription factors, each of which needs to be analyzed individually, many transcription factors have not yet been analyzed. More notably, the binding of transcription factors to their sites is highly dynamic, changing with different cell types and different conditions. ChIP provides a snapshot of where a transcription factor was bound in a cell at a particular time but does not yet capture the changes that occur in all of the many cells under all of the conditions. Even with these recognized limitations, ChIP experiments dominate the literature on transcription factor binding sites and chromatin structure.

genes that Kr directly regulates. Note that any genes that do not encode transcription factors are not included in this figure, and Kr may have many more target genes that these.

Despite the huge amount of work that has been done with transcription factor binding sites, much more remains. For example, in *C. elegans*, more than 16,000 binding sites have been annotated, but only 3–4% of the transcription factors have been analyzed so far. Even for those, only a few of the myriad growth conditions and genotypes have been tested. Since changes in transcription are the key for adapting to different environments, developmental changes associated with the life cycle, and nearly every aspect of morphological diversification, all of these conditions could be relevant for understanding regulatory networks. Nonetheless, recognizing that our

Some transcription factors regulate many genes

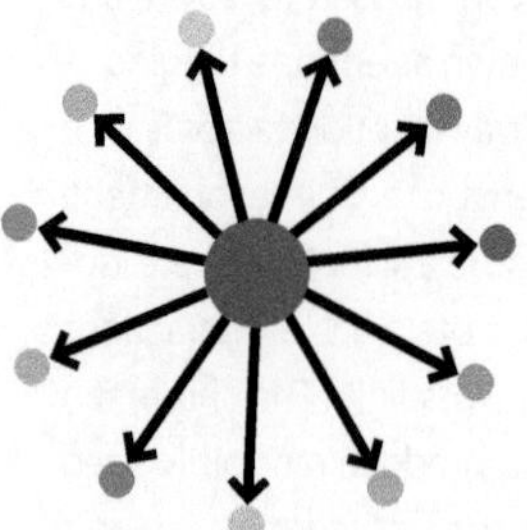

Some transcription factors regulate only a few genes

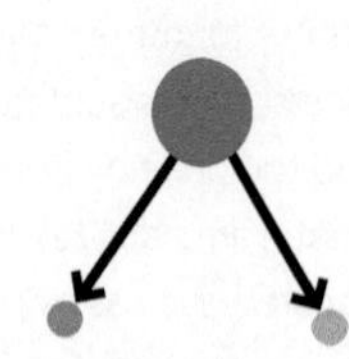

Figure 14.21 The number of genes regulated by a transcription factor varies widely. Some transcription factors regulate only a few genes, while others regulate dozens or hundreds of genes. For many eukaryotes, the mode is about 40 target genes per transcription factor, but the range is probably more important than the mode.

knowledge is limited and based on only a few species, some key principles of transcriptional regulation networks appear to be emerging. It is possible that these may change as more is learned; it seems very likely that more general principles will be found.

Auto-regulation is a common feature in both bacteria and eukaryotes

Many transcription factors have their own gene as a target for regulation. This principle is known as **auto-regulation**. Auto-regulation is so common that it has to be considered a key principle in gene regulatory networks.

Here is an example of auto-regulation involving the *caudal* gene in *Drosophila*, diagrammed in Figure 14.23. As discussed in Section 13.3, the anterior–posterior body axes of the embryo and the fly are largely determined by

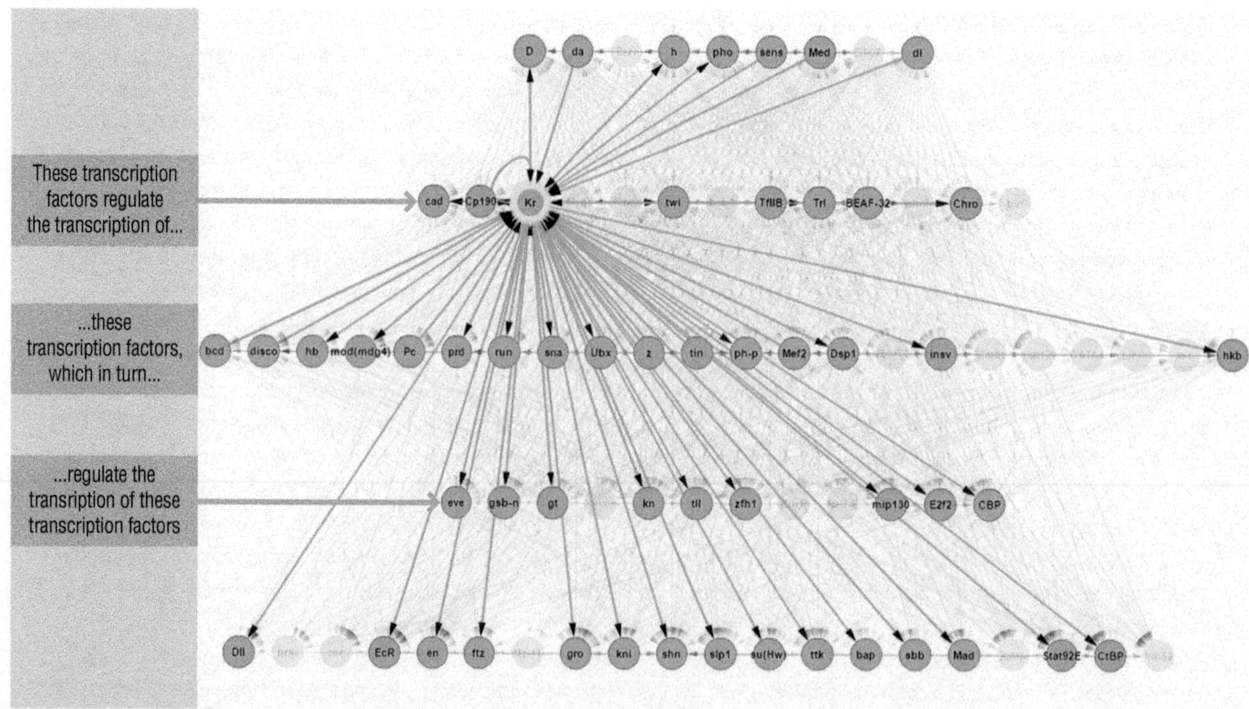

Figure 14.22 Regulation among transcription factors can be depicted and analyzed graphically. Many transcription factors regulate the expression of other genes encoding transcription factors. These interactions can be by a series of arrows indicating which genes are regulating which other genes. This example, based on data from the modENCODE website (http://www.modencode.org/), uses the *Drosophila* transcription factor Krüppel (Kr), shown in the middle. The purple arrows pointing towards Kr are transcription factors that regulate its expression; more than 15 different genes have been identified as regulating Kr expression. Note that many, but not all, of these are genes that occupy higher or equivalent positions in the hierarchy, as will be discussed. The lines with arrows emanating from Kr are the transcription factor genes whose expression it regulates; more than 40 such targets are indicated, with most, but not all, of them occupying equivalent or lower positions in the hierarchy. Only the transcription factor genes interacting with Kr are shown in this figure, and this may not include all of those. Many other transcription factors are present in the background that do not interact with Kr. Likewise, genes that do not encode transcription factors but whose expression is regulated by Kr are not shown.

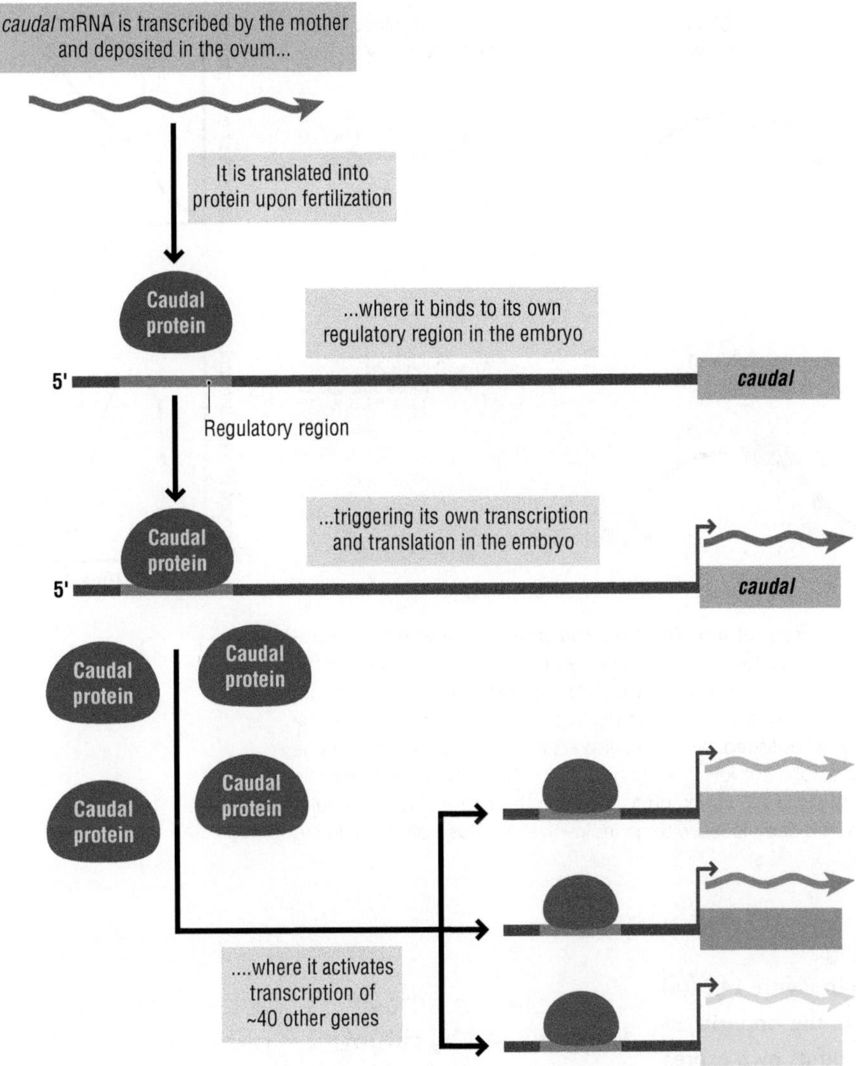
caudal mRNA is transcribed by the mother and deposited in the ovum...
It is translated into protein upon fertilization
Caudal protein
...where it binds to its own regulatory region in the embryo
5'
caudal
Regulatory region
Caudal protein
...triggering its own transcription and translation in the embryo
5'
caudal
Caudal protein
Caudal protein
Caudal protein
Caudal protein
....where it activates transcription of ~40 other genes

Figure 14.23 Auto-regulation is a common feature of bacterial and eukaryotic transcription networks. Auto-regulation, in which a gene regulates its own expression, takes many forms. An example is shown by the *caudal* gene in *D. melanogaster*. The *caudal* gene encodes a transcription factor. The gene is transcribed in the mother and deposited as an untranslated mRNA in the ovum. Upon fertilization, the mRNA is translated into the Caudal protein, which binds to the regulatory region of its own gene in the embryo. The binding of Caudal to its own regulatory region stimulates transcription of the gene in the embryo. Thus, more Caudal protein is produced in the embryo, which stimulates the transcription of at least 40 additional genes.

mRNAs and proteins that are synthesized by the mother and deposited in the ovum. Following fertilization, the mRNAs are translated into protein, which establishes the body axis.

One such maternally synthesized mRNA is from the *caudal* gene. When the *caudal* maternal mRNA is translated after fertilization, some molecules of the Caudal transcription factor protein are made. The Caudal transcription factor binds to the regulatory region of the *caudal* gene in the genome of the embryo, resulting in a higher level of expression of the *caudal* gene and its protein product in the embryo. The Caudal protein then serves as a transcription factor that turns on the expression of more than 40 additional genes in the embryo, many of which are important for the specification of the anterior–posterior body axis. Thus, auto-regulation by *caudal* amplifies the relatively modest signal deposited by the mother in the ovum into a signal with a sweeping scope in the embryo.

The most common occurrence of auto-regulation involves a similar amplification, in which an initially small or stochastically variable signal is intensified to produce

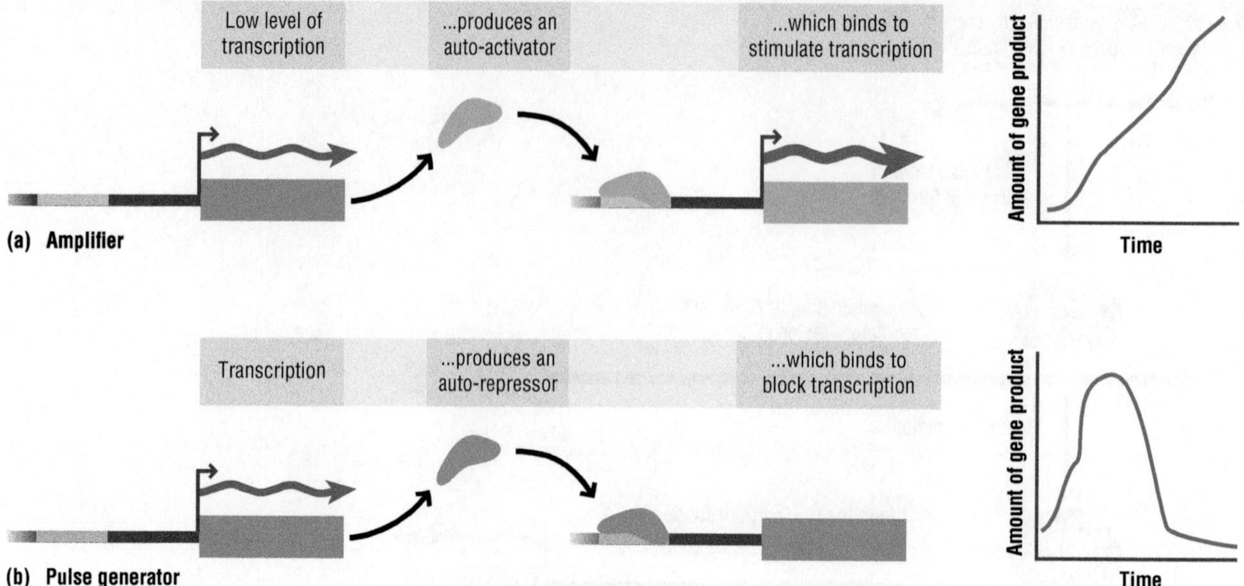

Figure 14.24 Auto-regulation can take the form of amplification and pulse generation. (a) Auto-regulation in which a gene stimulates its own transcription works like an amplifier to greatly increase the expression of the gene. A gene encoding a transcription factor is initially transcribed at a low level. The resulting transcription factor protein stimulates its own transcription. As summarized in the graph on the right, the expression of the gene continues to increase. (b) Auto-regulation can also act as a pulse generator. A gene encoding a transcription factor is initially transcribed at a low level. The resulting transcription factor protein represses its own further transcription. As summarized in the graph to the right, this pattern creates a pulse of gene expression. Pulse generation of this sort is often found with genes involved in developmental timing such as circadian rhythms and molting.

a high-volume signal, as summarized in Figure 14.24(a). A second type of auto-regulation has the opposite effect—a transcription factor may shut off its own expression, as shown in Figure 14.24(b). Thus, after an initial pulse of transcription and protein expression, the protein shuts off the continued expression of the gene, so that the signal does not persist. Examples of pulse generation are found among genes involved in time-related developmental events such as molting and circadian rhythms. Pulse generation is a less common use of auto-regulation in regulatory circuits than amplification but is still quite frequent.

We have focused on auto-regulation of transcription for this chapter, which is common, but other steps in the gene expression pathway may also be subject to auto-regulation. For example, sex determination in *Drosophila* depends on auto-regulation by mRNA splicing, and, as illustrated earlier in Figure 14.16, the SOS response in *E. coli* includes a process in which RecA indirectly autoactivates its own transcription by mediating the proteolytic cleavage of LexA, which represses the *recA* gene.

Different biological processes appear to be governed by different types of regulatory networks

Auto-regulation is only one of many of the different patterns or recurring modules that are found in transcriptional networks. Many other modules involving the connections between more than one gene—with names such as feed-forward loops, single input motifs, and bi-fans that suggest a connection to electrical engineering and signaling circuitry—have also been found, and their frequencies analyzed. Some analyses have indicated that different biological processes are characterized by different types of transcriptional regulatory circuits.

An example is found with two processes in yeast, as diagrammed in Figure 14.25. Yeast responds to nutritional or environmental stress by activating some genes and inactivating others, as do other organisms. The key to a stress response is that it needs to be rapid, but it may not need to be highly coordinated about the various components. The transcriptional network of the stress response shows such a pattern. There are only a few transcription

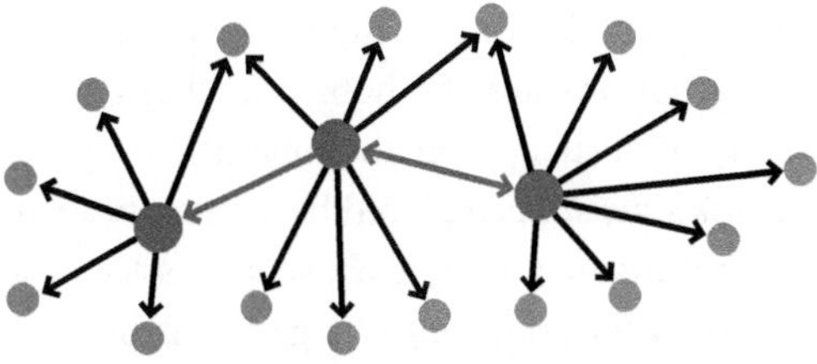

Few interactions among transcription factors
Many targets per transcription factor
Examples: DNA damage repair, stress responses

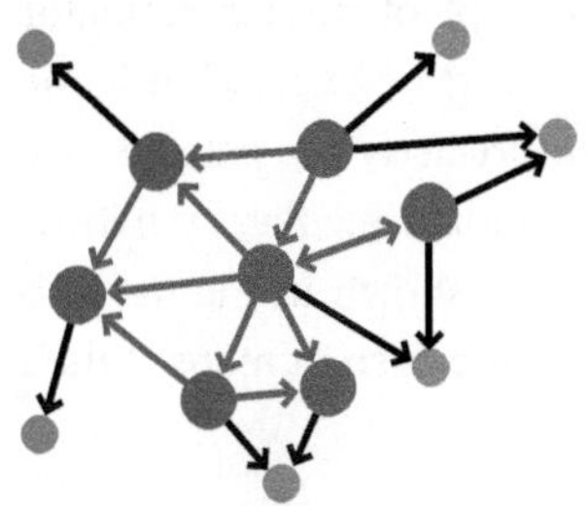

Many interactions among transcription factors
Few targets per transcription factor
Examples: Cell cycle control, sporulation

Figure 14.25 Different biological processes are characterized by different network topologies. In yeast, and probably in other eukaryotes, different types of biological processes are characterized by different structures of the regulatory network. Two different types are shown here. Genes encoding transcription factors are shown in blue, while effector genes encoding other proteins are in orange. For the genes involved in the stress response and repairing DNA damage, shown at the top, the regulatory network is characterized by short paths between the transcription factors and the effector genes and by few interactions among the transcription factors themselves. Such a network is presumed to allow a rapid, if uncoordinated, response. For the genes involved in cell cycle and sporulation, shown at the bottom, the regulatory network is characterized by many interactions among the transcription factors, which results in long paths between transcription factors and the effector genes. Such a network is presumed to allow a tightly coordinated, if slower, response.

factors, but each of them regulates a large number of other genes. Most of these response genes are the ones with an immediate effect on the stress response, with very few steps involving other transcription factors among them. These genes encoding the proteins that carry out a specific and immediate function and produce the phenotype of a cell, including enzymes, structural proteins, and transport proteins, are often called **effector genes**.

There is typically only a single link between the transcription factor and the effector gene; in the terminology of graphs, the path length is very short. Image the stress response as analogous to making an emergency call (to 911 in the United States or 999 in the United Kingdom). The operator immediately contacts the appropriate agency without having to direct the call through other operators or supervisors. The responders try to stabilize the situation as quickly as possible, without worrying very much about the long-term consequences of breaking down a wall or delaying traffic. The stress response pathway has a similar pattern—the response is rapid but not necessarily well-coordinated with other processes, many of which are shut down.

A different pattern is observed with the transcriptional circuit governing cell division control. Cell division occurs in a highly ordered manner when the nutritional and environmental conditions are good. In this case, many transcription factors are involved, with more regulation in which transcription factors control the expression of other transcription factors. Thus, this response is highly organized, with all of the different components "communicating" through transcriptional links to one another, and many alternative paths connecting two points. The effector genes have multiple and coordinated inputs. To continue with an analogy, cell division cycle control is similar to project management. The communication between the project supervisors, analogous to the links between transcription factors, helps to ensure that the project is done on schedule, with very few mistakes or uncompleted tasks. These are characteristics of cell division—on time, highly organized, and highly accurate.

Because it is intellectually appealing to interpret these biological processes in terms of familiar human behaviors, it is easy to fall into thinking of these circuits as "design principles," rather than as evolutionary outcomes. As we noted in the Prologue, evolution "tinkers" with the connections and reuses available components, rather than designing efficient systems from scratch, and the variants arise randomly by chance. Thus, while we can impose our perspective and imagine why some organization "makes sense," we do not usually know the evolutionary challenges that were faced. Our perspective could well be correct, but it comes from looking backward, rather than forward. We have little information so far that allows us to compare transcriptional regulatory circuits in closely related organisms in order to infer the selective pressures. From our point of view, a regulatory network with short paths to the effector genes and few connections among transcription factors seems likely to be able to change the cell or organism and adapt rapidly. A regulatory network with long paths to the effector genes and many connections among transcription factors seems likely to be buffered against small changes in the environment, and possibly more highly conserved when different species are compared. But we don't yet have enough data among related species to know if these perceptions are accurate.

Transcription networks are hierarchical

Much of animal biology is very highly regulated to ensure that body plans are properly organized, tissues and organs are functional and connected, and developmental timing is accurate. We expect to see this high level of organization with the regulation of transcription, although it also occurs at other levels of gene expression such as splicing and microRNA regulation. Many decades ago, genetic analysis identified mutations in which the broad patterns of the organization of the body are disrupted. One example of such an analysis will be discussed in Chapter 15. Among the most striking organizational disruptions are the homeotic changes, in which one body part, such as an antenna in *Drosophila* or a flower in *Arabidopsis,* is replaced by a different relatively normal, but distinct, body part, such as a leg or a sepal. Mutations in the *Antennapedia* gene, introduced previously in Section 3.7, are one such example. Such mutations led to the concept of master regulatory genes that control the large-scale organization of the body. Such master regulator genes ultimately exert their regulation through effector genes, transcribed in particular tissues or at particular times.

Implicit in the concept of master regulator genes is that the organization of gene expression is hierarchical. In fact, this has been precisely the pattern that has been observed for transcriptional regulatory networks in eukaryotes. An example is shown in Figure 14.26, but we encourage you to view the current maps at the ENCODE

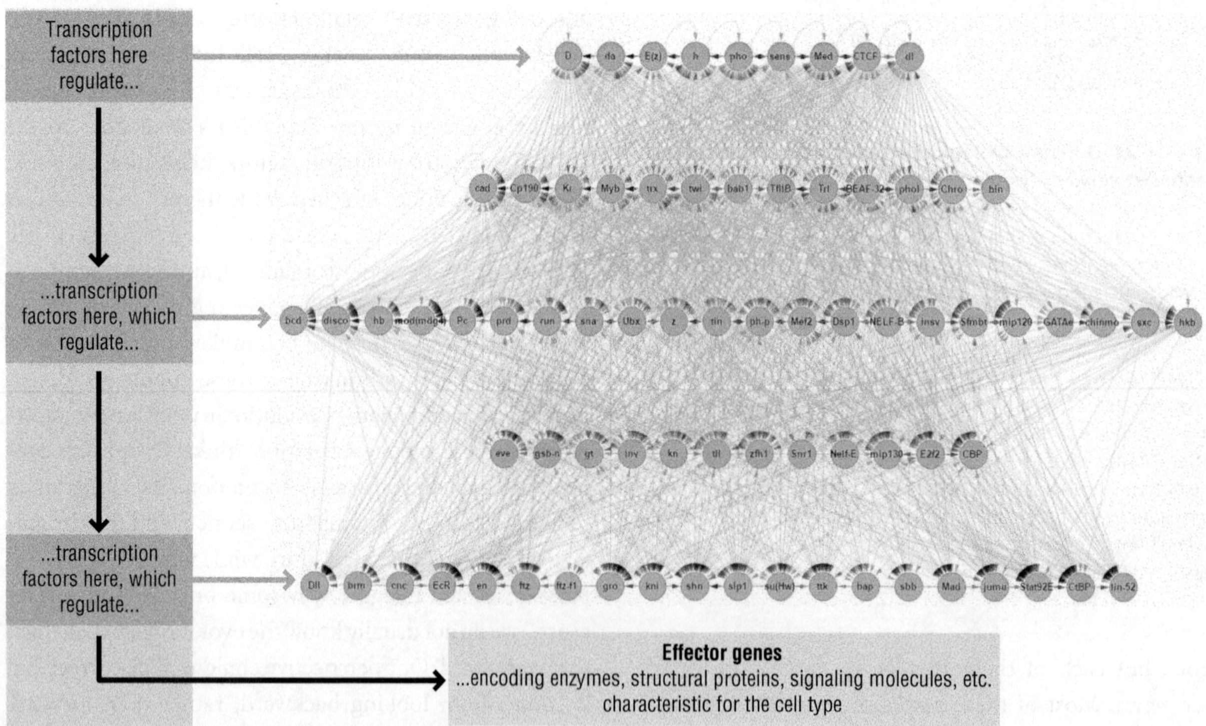

Figure 14.26 Transcriptional regulation in *Drosophila* has a hierarchical structure. The diagrams in this figure and Figure 14.27 show the structure of the known transcriptional regulatory network in *Drosophila* and *C. elegans*, as found on the modENCODE website at **http://www.modencode.org/**. Green arrows indicate the transcription factors that regulate the transcription of other transcription factors; the effector genes are not shown. The genes are organized by which other genes they regulate. Note that the transcription factors in *Drosophila* fall into a clear hierarchy of regulation, with most of the regulation occurring among genes at the same level of the hierarchy or with genes lower in the hierarchy. Genes at the top of the hierarchy include the ones that had been identified by genetic methods as master regulatory genes. Some of the assignments to particular tiers are somewhat arbitrary, so the arrows representing direct regulatory interactions are key.

projects, which provide an interactive perspective on the regulatory hierarchy.

A clear hierarchy is observed among the transcription factors. At the top level are genes that primarily regulate the expression of other transcription factors, with extensive interactions among themselves and down the hierarchy, but rarely with the effector genes at the bottom. Many of these include the master regulator genes previously found by genetic analysis, including the homeotic genes. If we make an analogy to a business, these are the chief executives. At the bottom are transcription factors whose principal targets are the effector genes such as the enzymes, the transport proteins, and the structural proteins that characterize a cell or tissue type. In our business analogy, the transcription factors are the shop stewards and floor managers who oversee floor assistants, loading dock personnel, and cash register operators, effectors that attend the customer or carry out other specific tasks. In between are the "mid-level manager" genes, which are regulated by a constellation of the master regulator genes at the top and relay this organization to the genes at the bottom.

Complex diagrams, such as the one in Figure 14.26, are packed with information. Figure 14.27(a) isolates one

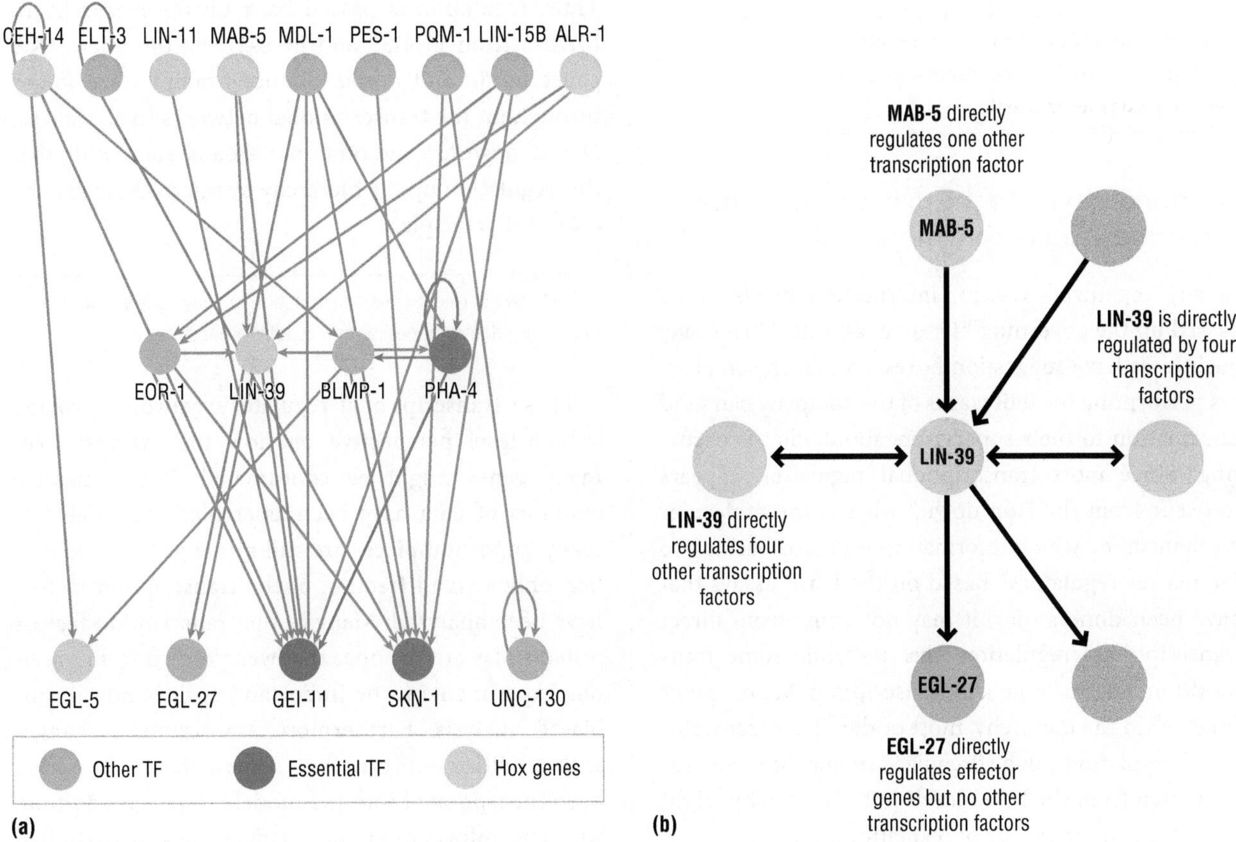

Figure 14.27 The transcriptional regulatory network in *C. elegans* also shows a hierarchical structure. (a) shows the data from the modENCODE website. Different types of transcription factors are shown in different colors, but these are not important for our discussion. Only the regulatory interactions among transcription factors are shown. Note that the transcription factor show a hierarchy of regulation, as has been seen for other organisms. Note also the prevalence of auto-regulation, as represented by the arrows circling back. (b) highlights one particular transcription factor hierarchy, involving MAB-5, LIN-39, and EGL-27. MAB-5, one of the master regulatory genes, regulates the transcription factor LIN-39 but no others among the ones shown. Four transcription factors, including MAB-5, regulate the expression of LIN-39, while LIN-39 regulates four other transcription factors. Note that two of these make up small circuits of mutual regulation, in which LIN-39 and another transcription factor regulate the expression of each other. Among the targets of LIN-39 is EGL-27. While EGL-27 does not directly regulate any other transcription factors among this group, it regulates at least 24 effector genes.

example to illustrate some interactions. The *mab-5* gene in *C. elegans* is one of the master regulator genes and encodes the MAB-5 transcription factor found in the top level of the hierarchy. (In fact, *mab-5* is one of the homeotic genes in *C. elegans.*) Among the investigated transcription factors shown at the modENCODE site, MAB-5 directly regulates the transcription of the gene for one transcription factor, *lin-39*. The LIN-39 protein directly regulates four genes encoding transcription factors, one of which is *egl-27*. The EGL-27 protein directly regulates no transcription factors—but it directly regulates the transcription of at least 24 other genes, most of which are effector genes. This relationship between MAB-5, LIN-39, and EGL-27 is highlighted in Figure 14.27(b) to show the hierarchical nature of their interactions.

WEBLINK: VIDEO 14.1 provides more information about the interpretation of regulatory networks. Find it at **www.oup.com/uk/meneely**

Coordination up to the master regulatory genes may come from microRNAs

In any organized system, information needs to be passed up the governing hierarchy as well. This is why businesses have suggestion boxes, so that the employees performing the daily tasks of the company can send information to their supervisors about what is occurring. Since most transcriptional regulation appears to occur from the "top down," what is the molecular mechanism by which information is passed back up to the master regulators? Based on the ChIP assays that have been done so far, it may not come from direct transcriptional regulation, that is, while some transcription factors regulate transcription factor genes "higher" in the hierarchy, most of the direct transcriptional regulation can be organized in one direction; information from the lower levels in the hierarchy might come from another type of regulation.

One tantalizing hint is that regulation from the bottom up may involve the microRNA genes. Many transcription factors regulate not only other transcription factor genes but also genes encoding microRNAs. microRNAs, discussed in Chapter 12, turn off or turn down the expression of other genes by targeting their mRNAs for degradation or by blocking their translation; they act in the opposite direction of transcription factors, most of which stimulate transcription of their targets. We can include the interactions between transcription factors and microRNA genes in our diagrams as well, using red arrows. An example is shown in Figure 14.28(a) and highlighted in Figure 14.28(b).

Note how the expression of *mab-5* is regulated by the interaction of its mRNA with the microRNAs encoded by *mir80* and *mir90*. The transcription of *mir80* and *mir90* is controlled by the transcription factor LIN-39, a direct target of MAB-5 regulation, as well as by EGL-27, which is one of the direct targets of LIN-39. Thus, regulation is passed back up the hierarchy to MAB-5 from LIN-39 and EGL-27 via the microRNA genes *mir80* and *mir90*. Similar examples are found throughout the transcriptional networks in *C. elegans*, *D. melanogaster*, and other metazoans, suggesting that this regulation up the hierarchy by microRNAs may be a general principle.

KEY POINTS microRNAs may function in relaying feedback to master regulatory genes in multicellular organisms.

These transcriptional regulatory networks provide a high-level perspective on how the expression of many genes might be coordinated. While massive amounts of data have been assembled and analyzed, many important pieces are still missing. As noted earlier, only a small fraction of the transcription factors have been analyzed. Many of the transcription factors tested so far are the ones that were known from previous experiments to be important, so it is not an unbiased analysis. Furthermore, transcription changes with conditions and time; only some of these factors have been incorporated into the models so far. In addition, we have information for relatively few well-studied species. We believe that what occurs in these species is typical of others, but this has not been demonstrated. Nonetheless, these principles do seem to be general, and all of them were also identified through other types of experiments.

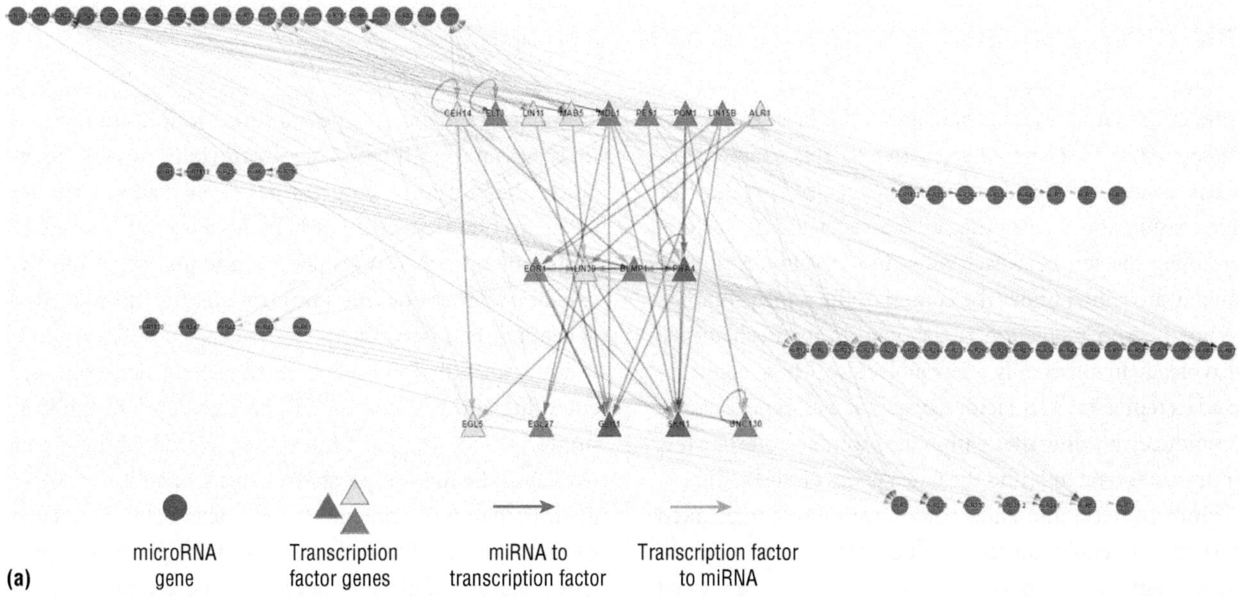

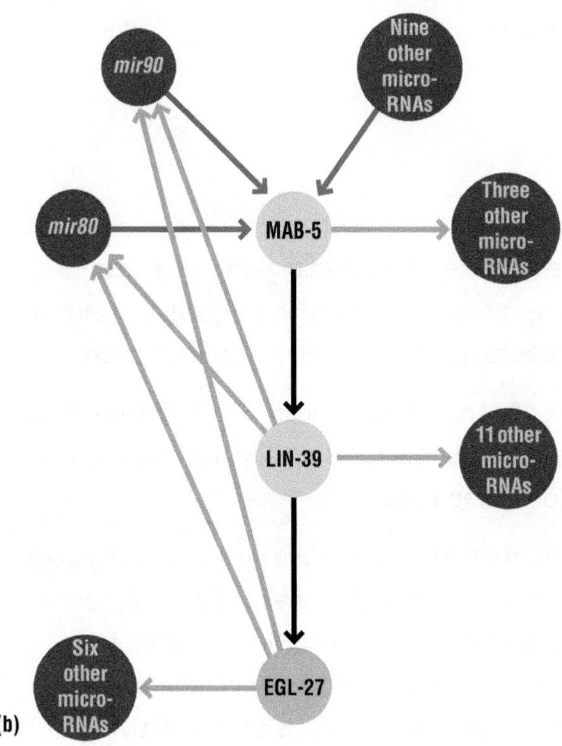

Figure 14.28 Feedback up the hierarchy may be provided by microRNAs. Transcription factors also regulate the transcription of microRNA genes, as shown by the red arrows connecting the transcription factors to the *mir* genes. The microRNA can regulate the expression of its targets through interactions with its mRNA, as described in Chapter 12. While the direct interaction of the microRNA with its target is also depicted by a red arrow, these are negative interactions in which the microRNA is reducing the expression of its target genes. As shown in the data from the modENCODE site in (a), the interactions between microRNAs and transcription factors also have a hierarchy of regulation, but this is somewhat less clear-cut than the ones for transcription factors. In particular, microRNAs often regulate the expression of genes higher in the hierarchy. The particular example involving MAB-5, LIN-39, and EGL-27 is highlighted in (b). Note that the microRNAs *mir80* and *mir90* are regulated by LIN-39 and EGL-27 and then, in turn, regulate MAB-5. This may provide a mechanism for regulation up the hierarchy.

14.4 Summary: transcription networks

One of the Great Ideas of Biology in the Prologue is that biology consists of organized systems. This chapter provides an overview of a few of the organized systems for gene regulation. Genes that are co-regulated in bacteria are often located near each other in the genome and are often transcribed under the control of the same promoter. While operon regulation is elaborate in its mechanisms, it typically involves only a few molecules. These could include a repressor, a co-factor that acts as a co-repressor or a co-inducer, binding sites within the promoter, one or a few transcription factors, and the target genes of the operon.

Both bacteria and eukaryotes have more complicated systems of regulation that include many more components than a simple operon. These systems can be viewed as hierarchical interconnected networks, and different biological processes appear to be characterized by different types of regulatory networks. Genome projects with annotated sequences, genes, and transcripts are increasingly elaborating these general principles of gene regulation. Much recent progress has been made in studies of eukaryotic networks, perhaps because the absence of more obvious operon regulatory circuits mandated the use of system-wide methods to identify co-regulated genes and their networks.

General principles of transcriptional networks are emerging and will continue to be elucidated in the near future. Among the most interesting aspects of these networks will be the comparative analysis when it is possible to draw parallels and make inferences between related species. We now know that many of the genes and proteins, the essential components of these networks, are conserved among species. We do not yet know very much about how their interactions have been rewired by evolution.

CHAPTER CAPSULE

- Operons consist of multiple genes that are transcribed together and code for proteins with related functions. The promoter region of these genes contains an operator sequence to which a repressor can bind, preventing transcription.
- The *trp* operon is turned off in the presence of tryptophan, which itself serves as a co-repressor. Thus, tryptophan is not made when it is already available to the cell. The *trp* operon is therefore considered repressible.
- The *lac* operon contains two types of control circuits. One prevents transcription when glucose, a preferred energy source, is present in the medium. This circuit is controlled by interactions of the CAP–cAMP complex with the promoter sequence. A second circuit permits the expression of the *lac* operon when the glucose concentration is low but lactose is present, and is controlled by interactions of the *lac* repressor protein with allolactose.
- The complete set of genes controlled by a specific activator or repressor in bacteria is referred to as a regulon; the SOS regulon by which *E. coli* repairs DNA damage is an example.
- While the term "regulon" is not used in eukaryotes, it is feasible to use genome-based techniques to identify all or nearly all of the genes controlled by a specific transcription factor in eukaryotes, which yields analogous information.

- These techniques include the use of microarrays and RNA-seq to determine transcriptional profiles of cells under specific conditions and the use of chromatin immunoprecipitation (ChIP) to identify the direct targets of transcription factors.
- The data from these experiments can be represented by graphs of interactions. Some of the principles emerging from these interaction networks include:
  - The prevalence of auto-regulation
  - The different types of networks found for different biological processes
  - The hierarchy among transcription factors.

STUDY QUESTIONS

Concepts and Definitions

14.1 Define auxotroph and prototroph.

14.2 Distinguish between anabolism and catabolism, and give an example of each.

14.3 What are some advantages and disadvantages of operons as a means to coordinate the expression of different genes involved in the same process?

14.4 What is catabolite repression?

14.5 What is a transcriptional profile, and how are transcriptional profiles compiled?

14.6 Why are ChIP assays widely used for determining transcriptional co-regulation? What are some of their limitations?

14.7 On the modENCODE website (**http://www.modencode.org**), go to "Networks" and then "Fly Regulatory Network" and find the transcription factor known as *engrailed* (abbreviated *en*).

a. Approximately how many transcription factors regulate the expression of *engrailed*?

b. Approximately how many other transcription factor genes does *en* regulate?

Beyond the Concepts

14.8 How does the presence of tryptophan repress additional synthesis of tryptophan? In general terms, what types of operons might be subject to similar regulation?

14.9 How does the presence of lactose induce the *lac* operon? In general terms, what types of operons might be subject to similar regulation?

14.10 One feature of the operon structure is that the gene products are produced at the same levels—for example, we expect that, for each molecule of β-galactosidase, there will also be one molecule of galactoside permease and one molecule of transacetylase. In fact, this stoichiometry is not always found, and one enzyme or protein product might be found at much higher levels than the other proteins encoded by the same operon. What is one mechanism (of those discussed in Chapter 12) by which molecules might be present in different amount?

14.11 What are some of the ways in which coordinated expression of genes in eukaryotes differs from that in bacteria?

14.12 What are some of the mechanisms by which a gene can regulate its own expression?

14.13 What is meant by a master regulatory gene?

14.14 Closely related eukaryotic organisms usually have many of the same transcription factors in their genomes. Recent studies have shown that the regulatory networks for these transcription factors are not always very similar. What are two ways that small changes in the genome can "rewire" a transcription factor network?

Applying the Concepts

14.15 The functions of the *lac* operon's components were defined by mutations. Each of the following phenotypes was found for a mutation in the *lac* operon. Describe what kind of change might give rise to this phenotype.

a. A mutation that maps to the repressor in which the *lac* operon is transcribed constitutively.

b. A mutation that maps to the repressor in which the *lac* operon is never transcribed.

c. A mutation that maps to the operator in which the *lac* operon is transcribed constitutively.

d. A mutation that maps to the *lacZ* gene that fails to produce galactoside permease.

14.16 As described in Chapter 11, it is possible to make *E. coli* cells that have two copies of a gene to test for dominance of a mutation. Which of the mutations in Study Question 14.15 is/are expected to be dominant over the wild-type?

14.17 An unusual strain of *E. coli* expresses the *lacY* and *lacZ* genes both in the presence and in the absence of lactose. State two hypotheses that, if correct, could explain this finding. How might you test your hypotheses?

14.18 Operons can be either inducible or repressible (or both). They can also be anabolic or catabolic. Imagine that there are only four combinations that might be found in nature—inducible and anabolic, inducible and catabolic, repressible and anabolic, and repressible and catabolic. For each type of operon, work through the overall strategy of regulation, that is, under what circumstances the repressors are expected to bind to the operator, when the genes will be transcribed, what molecule might serve as a co-repressor or a co-activator, and so on. Is there any of these four combinations you think will not be found, and why?

14.19 In *E. coli*, the arabinose (*ara*) operon encodes the enzymes necessary for the breakdown of the sugar arabinose into a simpler molecule called xylose 5-phosphate. The *ara* genes, as shown in Figure Q14.1, map adjacent to each other on the bacterial chromosome. The genes are *araA*, *araB*, *araC*, and *araD*.

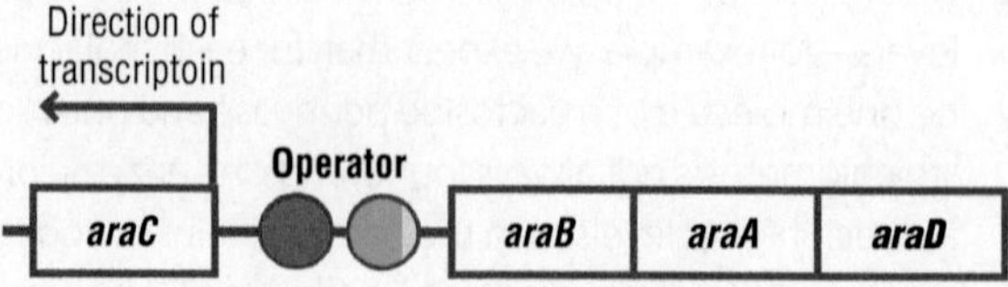

Figure Q14.1

An operator located between *araB* and *araC* has two symmetrical halves, with about 30 bases in between, as shown by the two ovals close together in the figure. The repressor protein has two identical subunits (a structure known as a homodimer), and each subunit binds to one half of the operator. When the repressor binds to the operator, its two subunits come together, which means that the DNA sequence between the two halves of the operator loops out. When the repressor is not bound, the DNA sequence is not looped. In other respects, the *ara* operon is similar to other operons you have learned about.

The direction of transcription for the *araC* gene is shown by the arrow with the dotted line.

a. Is this an anabolic operon or a catabolic operon?

b. Which of these genes encodes the repressor? Very briefly explain your reasoning.

c. Of the four possible choices shown below, which is most likely to be the transcription pattern for the entire operon? Very briefly explain your answer.

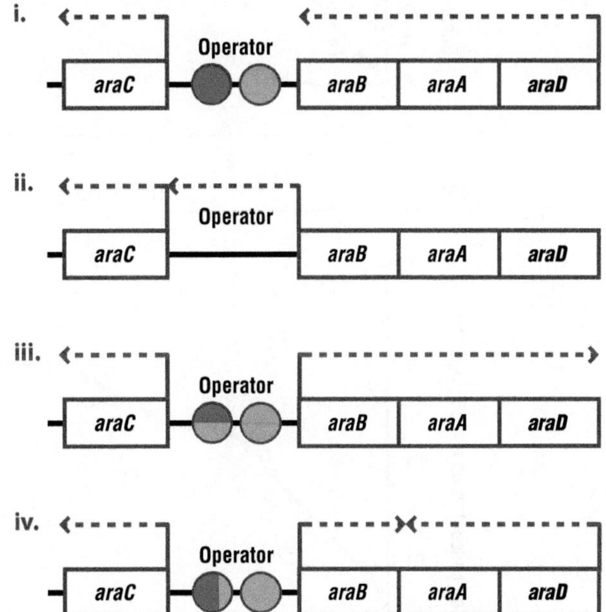

d. What is the expected phenotype of a mutation that deletes the gene for the repressor?

e. Based on your knowledge of other operons and your inferences drawn above, briefly outline how arabinose is likely to regulate the *ara* operon.

f. Like the *lac* operon, the *ara* operon is catabolite-repressed and regulated by CAP and cAMP receptor protein. If *E. coli* are grown in medium containing one part glucose and three parts arabinose as the carbon source and the optical density is taken hourly until both sugars are used up, make a sketch of what you think the resulting growth curve will look like.

14.20 The *E. coli pyr* operon encodes enzymes required for the synthesis of pyrimidines. It is only expressed when the cell is starved of pyrimidines. Early in the sequence of the operon, just downstream of the promoter, is a sequence containing inverted repeats. *In silico* predictions demonstrate that this sequence can form a hairpin by base pairing. The repeat sequence is preceded by a short stretch of AT-rich DNA. Postulate a model for the regulation of the *pyr* operon. What additional information do you require to support the model?

14.21 An *E. coli* culture growing in medium containing 200 µg/ml of glucose is subcultured into fresh tubes of medium. The culture was sampled over time, and bacterial density (BD) was measured in each tube and plotted; the results are shown in Figure Q14.2(a). Tube 1 of the second batch culture contained 200 µg/ml of glucose, while tube 2 contained 200 µg/ml of xylose.

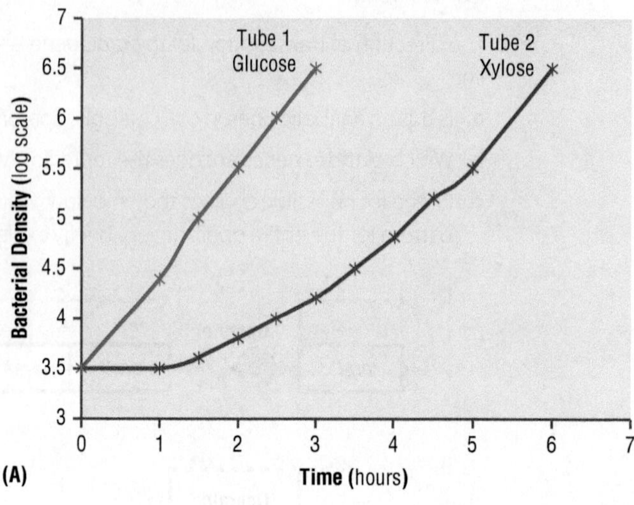

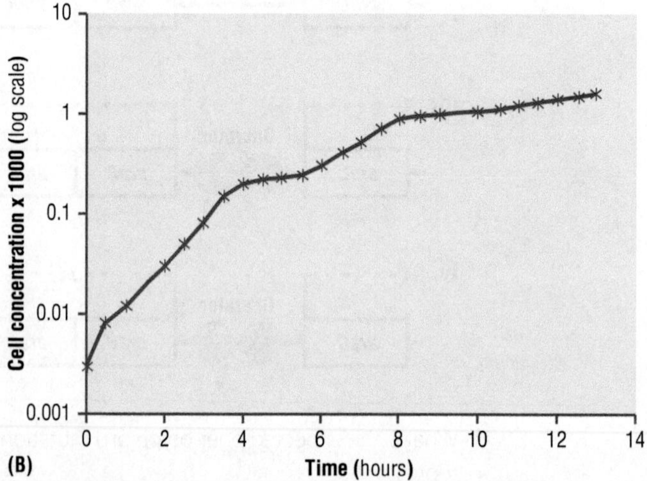

Figure Q14.2

a. If a third tube from the first batch culture is subcultured into a medium containing 100 µg/ml of glucose and 200 µg/ml of xylose, what type of growth curve would you expect to see, and why?

b. The curve shown in Figure Q14.2(b) is obtained if *E. coli* is grown in medium containing 500 µg/ml of glucose, 150 µg/ml of xylose, and 5 mg/ml of lactose, and the cell density is measured over time. How would you describe the curve, and what regulatory mechanisms might account for it?

14.22 In *C. elegans*, *pha-4* encodes a transcription factor that is expressed primarily in the developing pharynx and intestinal cells. *unc-4* encodes a transcription factor that is expressed in certain motor neurons. A gene construct is made *in vitro* that has the upstream regulatory region of *pha-4* and the coding region of *unc-4*. The construct also has the green fluorescent protein (GFP) gene inserted in frame downstream of *unc-4*, so that the activity can be monitored in the reporter gene, as described in Tool Box 12.1. This is inserted back into the worm and is expressed. Which cells are likely to express the GFP reporter? Explain your answer.

14.23 ALX1 is the transcription factor that was implicated as having a role in beak shape in Darwin's finches. However, the downstream target genes that are regulated by ALX1 are not known.

a. Describe how you would identify genes that are likely to be the direct targets of ALX1 regulation.

b. The assay in (a) does not actually demonstrate that the expression of any of these genes is regulated by ALX1. Describe how you might identify genes whose expression is regulated by ALX1, whether directly or indirectly.

14.24 Investigators were interested in understanding the regulatory network for virulence genes in the bacterium *Vibrio cholerae*. They constructed deletion mutants in each of three known transcriptional regulators, *toxRS*, *tcpPH*, and *toxT*. They used microarrays to deduce transcriptional profiles of each mutant, compared to wild-type, and found that transcription of 3678 genes was not affected by any of the three regulators and that the *toxT* mutant was defective in transcribing 13 genes relative to wild-type. For *tcpPH* and *toxRS*, the mutants were respectively deficient in transcription of 27 and 60 genes. They then compared the lists of affected genes for the three mutants and constructed the Venn diagram shown in Figure Q14.3 to illustrate the proportion of genes that were uniquely activated by one of the three regulators or activated by more than one of them. All 13 of the genes activated by all three regulators were effectors, that is, none was a transcriptional regulator. Using this information and the information you can glean from the Venn diagram, sketch out a regulatory network for *V. cholerae* virulence that includes *toxRS*, *tcpPH*, and *toxT* and the 13 listed effector genes.

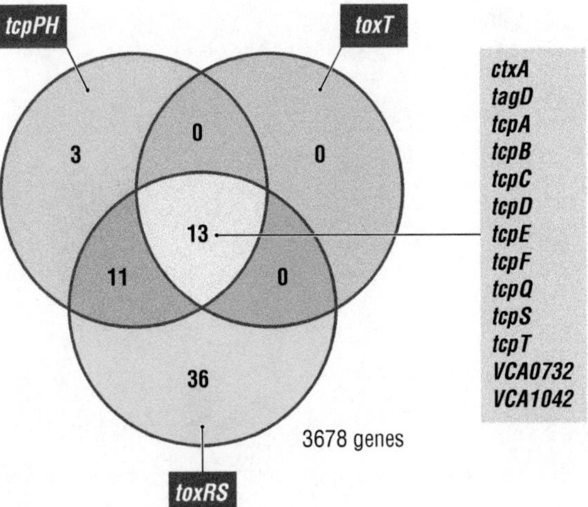

Figure Q14.3

Challenging **14.25** *V. cholerae* is a marine organism that can cause diarrhoeal diseases in mammals. A portion of the genome of *V. cholerae*, known as the TCP island, has been associated with disease. Investigators used RNA-seq to compare the transcriptional profile of genes in this island when the bacteria are grown in rabbits to that obtained when they are grown in laboratory medium *in vitro*. To collect the data, the investigators extract total bacterial RNA under each condition, reverse transcribe it to DNA, and then quantify it by sequencing. Portions of the island that are more heavily represented in the RNA provide deeper reads. Figure Q14.4 depicts the genes in the TCP island, with the gene arrows and the location illustrating whether they are on the forward or reverse strand. Above the gene map are plots of the strand-specific RNA-seq coverage per nucleotide. Data from the rabbit transcripts extend above the midline and are plotted in blue and black for the forward and reverse strands, respectively. Transcripts from growth in laboratory medium are plotted in red (forward) and green (reverse) below the horizontal midline.

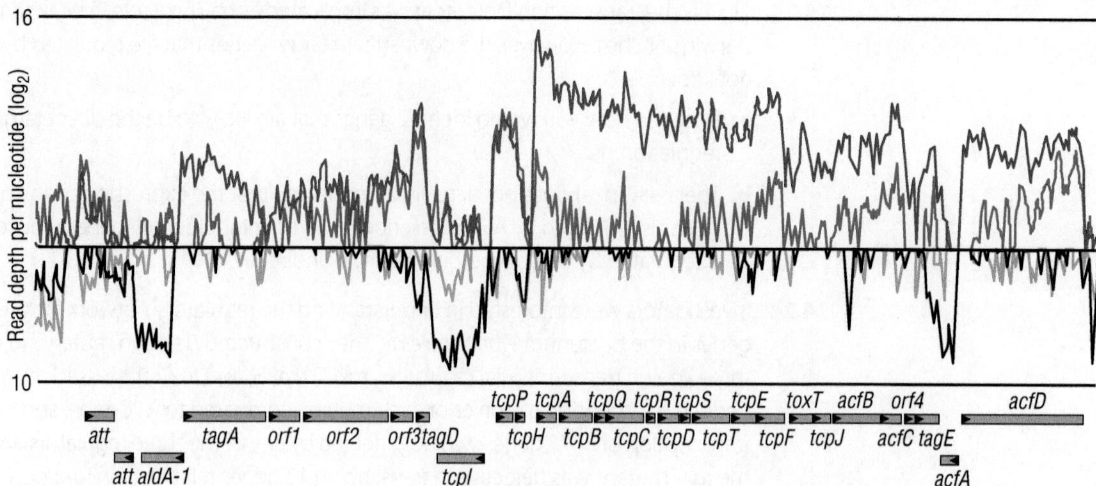

Figure Q14.4

a. Study Figure Q14.4 carefully. Overall, are the island genes more likely to be transcribed *in vitro* or *in vivo*?

b. Which genes are likely in operons, and which ones are almost certainly not? How did you arrive at your conclusion? Which is likely be the largest operon, and how many genes does it contain?

c. What might explain the coordinate transcription of genes in different operons on the island?

CHAPTER 15

Genetic Analysis of Cellular Processes

15

IN A NUTSHELL

The principles of genetics can be applied to identify the underlying genetic basis of complex cellular processes. Such approaches often start with a genetic screen for mutations in the genome that disrupt the process of interest. Characterizing the gene products associated with such mutations can then help identify the molecular pathways involved. The power of this type of genetic analysis is enhanced when conducted in model organisms for which annotated genomic sequence information is available. Shared evolutionary histories—and thus similar cellular and biochemical features—enable many findings from the study of one species to be applied more broadly to other species, providing insights into the underlying workings of many cellular processes.

15.1 Overview: the power of genetic screens

Take a complicated biological process; disrupt it; identify the defects created when a gene was disrupted, and use this information to understand how it all works. Such an approach sounds a little crazy—even audacious—in its scope and aspirations. Yet not only does this strategy work, but it has provided one of the most useful methods of understanding how many biological processes function. Many Nobel Prize-winning discoveries, and much of our knowledge of the fundamental molecular and cellular processes in both bacteria and eukaryotes, have resulted from these kinds of experiments, known as **genetic screens**.

A successful genetic screen can be truly elegant. The precise methodology behind genetic screens and subsequent genetic analyses has been developed and refined over the years to encompass many different techniques. A full description of these methodologies extends well beyond the scope of this book, so we focus here on the underlying logic of the approach, utilizing a few informative examples to illustrate key features of a well-designed genetic analysis. Genetic screens are also called **mutant screens**, and we will use these terms interchangeably.

Genetic screens predate the analysis of genomes by many decades, but the availability of a well-annotated genomic sequence certainly makes genetic screens even more powerful. As we scan through the genome of an organism, we can discern the catalog of components used by that organism over the course of a lifetime; this might be similar to a parts list that comes with a piece of equipment that is delivered unassembled. For any one particular biological process, we need to ask: which of these components are used, and how do they operate together to control the process in question? All of the parts can be listed, but we don't know much about when and how they are used. This is where the art of genetic analysis comes into play to identify which genes are involved in a specific biological process.

In Chapters 5 through 9, we followed the inheritance and functions of simple traits and phenotypes controlled by one or two genes. In Chapter 10, we explored complex traits in which multiple genes contribute to a single specific phenotype. But what if we look beyond a specific trait to the wider genetic network of interactions or the cascade of events behind an entire process? By applying fundamental genetic principles, it is possible to begin to answer these types of questions, even when nothing is known a priori about how the process works and what kinds of genes and their products might be involved. In fact, since little information is needed to begin with, genetic screens have frequently been a pioneering approach to understanding the complicated circuitry of a biological process.

Let's start by describing one of the earliest examples of this approach—the classic experiments of Beadle and Tatum introduced in Chapter 2. By the 1940s, the geneticist George Beadle had realized that the distinction between single-gene traits and biochemical and genetic pathways is not always clear-cut. Working on the fruit fly *Drosophila melanogaster* (first with Thomas Hunt Morgan and then with Boris Ephrussi), he had begun to understand that there was a connection between genes and biochemical processes and that the biochemical process of making the red eye color of the fruit fly was controlled by more than a single gene, that is, while a mutation in a single gene could result in a change in eye color, numerous genes affected the process of producing that eye color, and a mutation in any one of those can alter the eye coloration of the fly. These genes must have biochemical or molecular functions related to one another. But biochemical pathways leading to eye colors are a bit complicated and were unknown at the time. In order to explore the role of the genes involved in pathways controlling biochemical processes, Beadle and his colleague Edward Tatum turned to a simpler system—the orange bread mold *Neurospora crassa*, depicted in Figure 15.1.

The analysis of metabolism using auxotrophs

Beadle and Tatum set up an experimental system using *Neurospora* to identify genes involved in metabolism. Normally, *Neurospora* can grow on minimal medium lacking complex vitamins and amino acids. Beadle and Tatum hypothesized that *Neurospora* must therefore

Figure 15.1 George Beadle and Edward Tatum used *Neurospora crassa*, the orange bread mold, in their experiments to elucidate the genes involved in biochemical pathways. (a) The genus *Neurospora* is named for the features on the surface of the ascospores that were thought to resemble neurons. (b) *N. crassa* growing on sugarcane.

Source: Reproduced from Raju, N.B. (2009). Neurospora as a model fungus for studies in cytogenetics and sexual biology at Stanford. *J Biosci.* 34 (1). Part (a) courtesy of Namboori B. Raju, Stanford University. Part (b) courtesy of Dr. David Jacobson.

contain genes encoding the enzymes that can convert simple nutrients into more complex ones. They further hypothesized that, if one of these genes becomes inactivated through a mutation, *Neurospora* would no longer be able to synthesize a necessary vitamin or amino acid; it could then only grow if the missing component was added to the medium. The concept that the gene encodes the information for an enzyme did not originate with Beadle, as he freely acknowledged. The British medical geneticist Archibald Garrod had introduced this idea decades earlier, shortly after the rediscovery of Mendel's work, based on patients he found with "inborn errors of metabolism." Box 15.1 discusses Garrod's work on this topic.

Beadle and Tatum could not rely on spontaneous mutations that arise during normal fungal growth, since these are rare. Fortunately, however, they had other options available to them. Prior to this work, others had shown that X-rays were an effective means to generate new mutations at a higher rate. Such mutations occur at random—that is, they are not targeted to a particular gene—so irradiated spores could have mutations affecting any cellular process. The "screening" aspect of a genetic screen refers to sorting through the many mutations that were created to find the ones that affect the specific processes being studied.

To identify the genes affecting fundamental metabolic processes, Beadle and Tatum irradiated *Neurospora* conidia (the spores), grew them first in nutrient-rich complete medium, and then grew replicates on minimal medium without all of the added nutrients. Beadle's own schematic diagram for this experimental design is shown in Figure 15.2. Hundreds of irradiated cultures were able to grow on complete, nutrient-rich medium but were no longer able to grow on minimal medium. These cells must have had a mutation affecting the synthesis or utilization of some essential nutrient. Because *Neurospora* can grow as haploids, as discussed in Tool Box 6.1, Beadle and Tatum could identify recessive mutations readily, without having to construct strains with both alleles mutated, as is needed for a diploid.

BOX 15.1 *An Historical Perspective* Archibald Garrod and inborn errors of metabolism

While Beadle and Tatum provided definitive evidence that a gene encodes the information to make a polypeptide, the concept did not begin with them. The British physician Archibald Garrod postulated this relationship in 1902, very shortly after the rediscovery of Mendel's paper and several decades before Beadle's research on flies or *Neurospora*. Garrod recognized that a very rare human disorder known as alkaptonuria usually affected more than one sibling in a family and that the parents are usually unaffected. In addition, eight of the 17 original families in which he observed cases involved first-cousin marriages. Working together with his friend and geneticist William Bateson, Garrod recognized that the pattern of inheritance for alkaptonuria could be explained by a single autosomal gene with a recessive allele that resulted in the disorder.

Alkaptonuria is a non-fatal metabolic disorder in which the urine from affected individuals is initially brown or red before turning black upon standing, such as in a wet diaper. Garrod postulated that a substance he called alkapton accumulates due to the absence of an enzyme (which he referred to as a "ferment") that is responsible for breaking it down. We now know that alkapton is actually homogentisic acid, and the enzyme that is missing is homogentisate 1,2-dioxygenase, also known as homogentisic acid oxidase. This enzyme and homogentisic acid are part of the metabolic pathway that breaks down the amino acids phenylalanine and tyrosine prior to excretion.

Garrod's hypothesis that the condition was inherited as an autosomal recessive condition, that it was recessive because of the absence of the activity of the gene, and that the gene encoded the information to make an enzyme are clearly laid out in his publications. This research was the foundation of biochemical genetics in humans, many years before the nature of genes or enzymes was known. Garrod identified three other human conditions—cystinuria, pentosuria, and albinism—that also behave as single-gene autosomal recessive disorders and that could also be explained by the absence of a metabolic enzyme. (Albinism can arise from the lack of any of at least seven different enzymes in humans. Garrod probably studied tyrosinase deficiency or oculocutaneous albinism type IA, but one review of his work concluded that his inclusion of albinism among his metabolic disorders must have been due to his intuition since, unlike the other three disorders, the evidence available to him was not very convincing.)

Garrod referred to these four conditions as "inborn errors of metabolism," recognizing both their genetic and biochemical basis. They stand as the first four biochemical genetic disorders in humans and were cited by Beadle and Tatum as prior evidence that a gene encodes the information to make an enzyme or, more generally, a polypeptide.

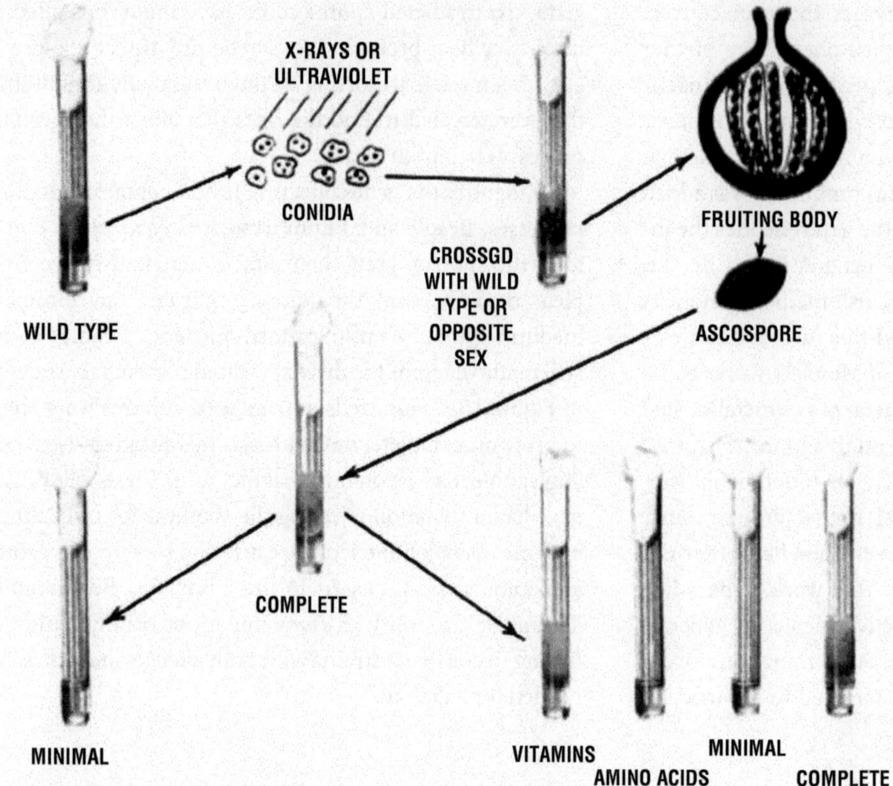

Figure 15.2 A slide drawn and used by Beadle to overview the screens for metabolic mutants in *Neurospora*. Conidia (spores) of wild-type *Neurospora crassa* were irradiated with X-rays or UV light and then crossed with unirradiated wild-type fungi. Ascospores from the resulting cross were cultured and then inoculated into minimal medium and complete medium. Mutants unable to grow in minimal medium were evaluated further.

Source: Reproduced from Singer, M. and Berg, P. George Beadle: from genes to proteins. *Nature Reviews Genetics* 5, 949–954 (December 2004).

They were able to identify which nutrient was lacking by adding individual nutrients to the minimal medium and testing for the ability of *Neurospora* to grow. In other words, they found specific auxotrophic mutants, a concept discussed in Section 14.2. For example, when arginine was added to minimal medium, a few of the mutants were able to grow. These must have mutations affecting genes involved in arginine biosynthesis or utilization, since they could be "rescued" by the addition of arginine but not by any other nutrient. In this way, Beadle and Tatum were further able to identify which specific nutrient was required by each mutant and therefore which metabolic pathway had been disrupted.

KEY POINT A genetic screen is an experimental strategy to find and select for an organism with a mutant phenotype of interest. Mutations are induced in a population that is then surveyed to identify mutants with the altered phenotype.

Building a pathway from individual mutations

With this basic system established, Beadle and Tatum were further able to elucidate multistep metabolic pathways. For example, Figure 15.3 shows the enzymatic pathway through which the amino acid arginine is synthesized. Three enzymes are needed to synthesize arginine in a step-by-step process, but which gene affected which step was not known at that time.

The first step in elucidating this pathway was to screen for mutants that require arginine to be added to the minimal medium in order to grow, as discussed earlier. Note that this results in a collection of mutants that could have alterations in genes encoding any one of the enzymes needed in any of the intermediary steps involved in arginine synthesis.

The next step was to assemble the mutants, and their corresponding genes, into a pathway. The researchers

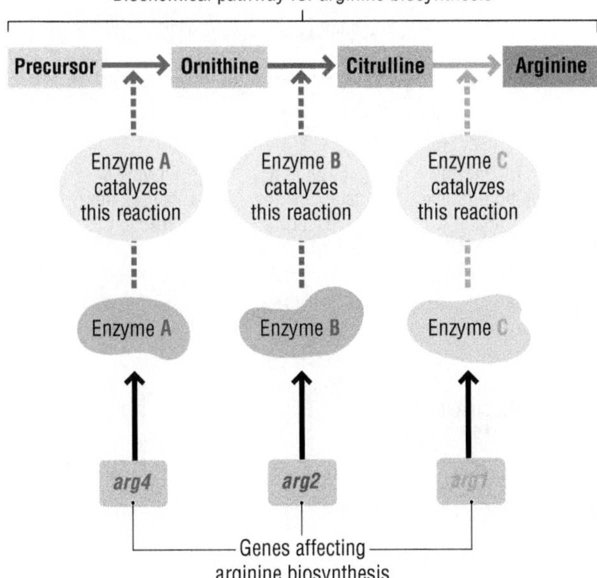

Figure 15.3 A summary of Beadle and Tatum's experiment in the fungus *Neurospora*. *Neurospora* synthesizes the amino acid arginine from a precursor molecule in three steps, with the enzymes responsible for catalyzing each step shown in different colors. This is shown as the biochemical pathway at the top of the figure. Beadle and Tatum identified mutant strains of *Neurospora* that could not synthesize arginine and used these to identify three genes, *arg4*, *arg2*, and *arg1*, shown at the bottom. Each encodes one of the enzymes that catalyzes a step in this biosynthetic pathway, as was demonstrated by adding back the substrate for the reaction, summarized in Figure 15.4.

understood aspects of the metabolic pathway for arginine biosynthesis from biochemical analysis and realized that, when different enzymes are inactivated, different substrates or biochemical products would be missing. Thus, *Neurospora* mutants show different requirements for the added intermediates, such as ornithine and citrulline, depending on which step in the pathway is inactive.

For example, *Neurospora* cultures with a mutation in the gene encoding the first enzyme (*arg4*) will be able to grow in minimal medium when only ornithine is added, as shown in Figure 15.4. The first step in the synthesis of arginine fails to happen, as depicted in Figure 15.4(a), but is bypassed (or "rescued") by the addition of the next intermediate, ornithine, as depicted in Figure 15.4(b). However, addition of ornithine to the medium would not help rescue a mutation in *arg2*, the gene encoding the **second** enzyme, as shown in Figure 15.4(c), since *arg2* mutants can produce ornithine but cannot convert ornithine to citrulline. Using this logic, Beadle, Tatum, and colleagues were able to identify the genes and correlate them with the enzymes that control an essential metabolic pathway.

Key ingredients in genetic screens

The genetic analysis of *Neurospora* conducted by Beadle and Tatum was important in many ways. First, as we discussed in Chapter 2 and depicted schematically in Figure 15.3, their work was central to developing the hypothesis that "one gene encodes one protein." Although we now understand the situation to be more nuanced than this, Beadle and Tatum's hypothesis was nonetheless foundational to our current understanding of the functions of genes and genomes.

In addition to the important biological results from their screen, their work also highlighted the potential for using a genetic approach in a model system to understand basic cellular processes. Although their studies were conducted in the bread mold *Neurospora* and used tetrad analysis to study the outcomes of their crosses, as described in Tool Box 6.1, the implications of their findings extended well beyond this one species. Furthermore, the approach they took in these experiments also illustrates important features for a successful genetic screen. These include:

- **Matching the model organism to the biological question being studied.** We noted that Beadle began this approach with eye color mutants in *Drosophila*, a more complicated and less understood biochemical pathway at the time. By switching to the study of a known metabolic pathway in *Neurospora*, he took advantage of the biological properties of the organism. All of the mutant screens that we discuss in this chapter (and thousands more that we don't) share this feature of asking the right question in an appropriate organism. Since organisms share an evolutionary history, however recent or remote, the roles of the genes can often be extrapolated to other organisms.
- **Setting up a clear and well-defined assay.** Beadle and Tatum established specific conditions for classifying mutant phenotypes—the *Neurospora* cultures had to be able to grow on complete medium but not on minimal medium. Furthermore, each mutant required the addition of one particular nutrient to the minimal medium, focusing the assay specifically on the genes involved in controlling a specific element of the biochemical pathway.
- **Using an unbiased approach.** By employing an unbiased approach—namely, by using randomly introduced genetic mutations—Beadle and Tatum

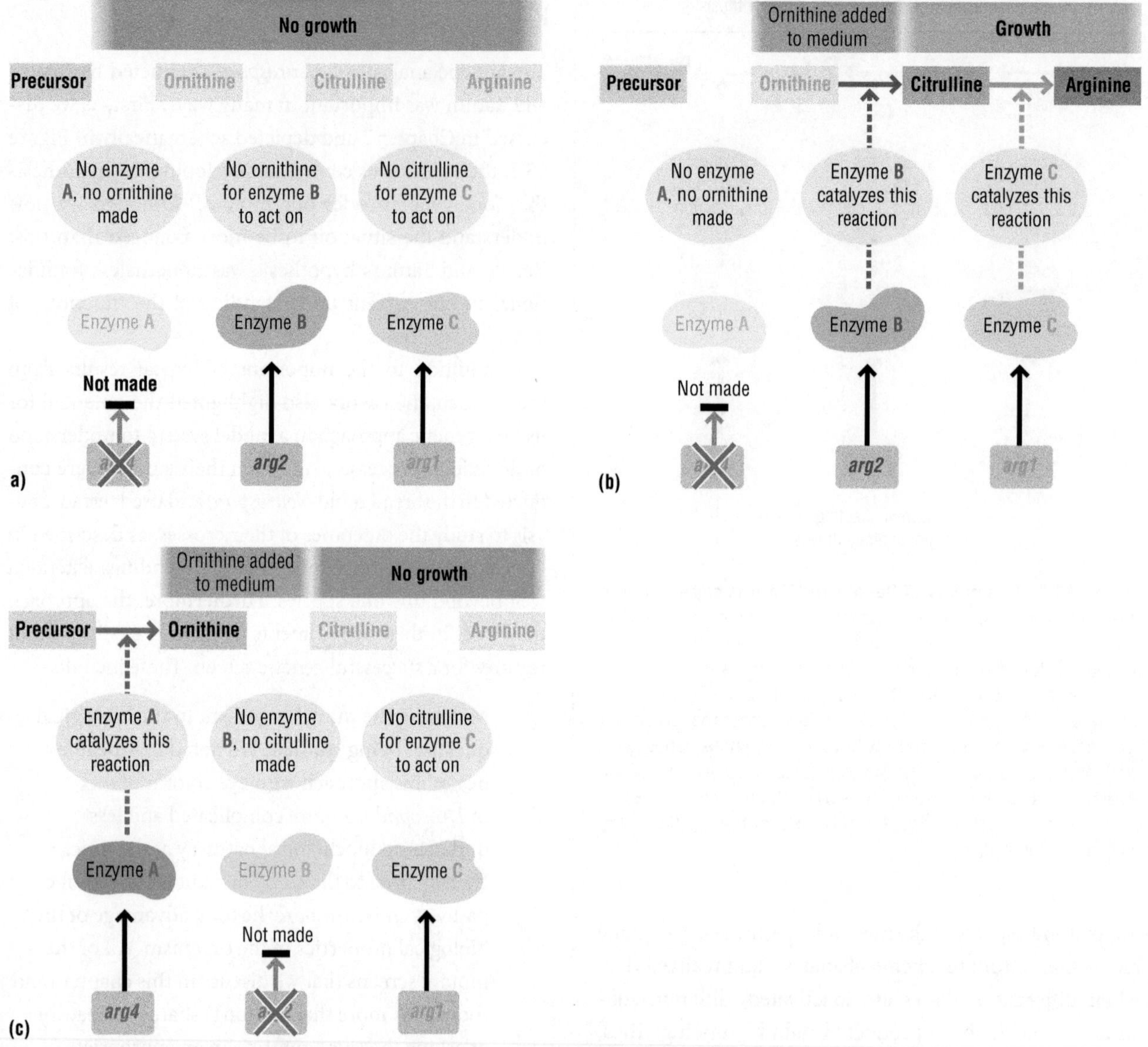

Figure 15.4 Adding back substrates demonstrates the correspondence between a gene and an enzyme. (a) When *arg4*, the gene encoding the first enzyme in the synthetic pathway for arginine, is mutated, ornithine cannot be made, and the cells do not grow on minimal medium. (b) If ornithine is added to the minimal medium, the need for the enzyme to produce ornithine is bypassed, the subsequent steps occur normally, and *arg4* mutants can grow. (c) When *arg2*, the gene encoding the second enzyme in the pathway, is mutated, the addition of ornithine will not help the mutant grow because ornithine is the substrate of the missing enzyme B.

did not make any assumptions about what the components of the process were or what kinds of genes would be found in their screen. Enabling the organism itself to reveal what genes are important through the screening process is one of the most powerful aspects of this type of analysis.

Genetic screens that harness this unbiased approach can lead to important insights. Such methodology is sometimes referred to as "forward genetics," in contrast to a second strategy "reverse genetics," in which a specific gene of interest is mutated directly to observe its potential role in the process. A third strategy, which we discuss in Tool Box 15.1, is to systematically mutate every individual gene in the genome first and then test each mutant for its phenotype. These experimental strategies are illustrated in Figure 15.5. Each of these methods has advantages and limitations. One of the key features of an

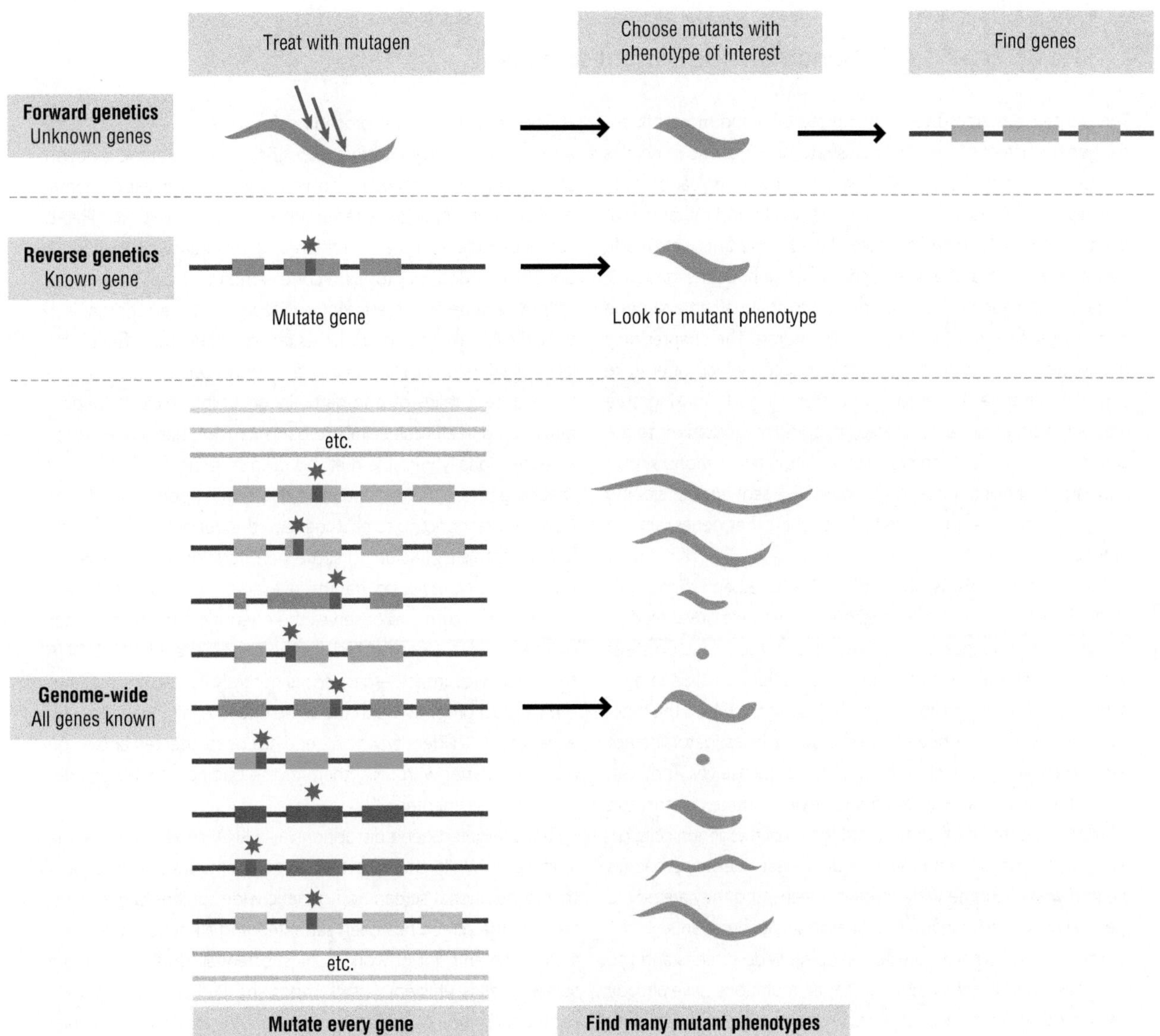

Figure 15.5 Comparison of forward genetic screens with reverse genetic screens and genome-wide screens. In forward genetic screens like those described in this chapter, mutations are induced in a population of wild-type organisms. The mutant organisms with phenotypes of interest are isolated, and the mutation is mapped to a gene on a chromosome. The mutant phenotype and the gene are not known in advance of the study. In reverse genetic experiments, a gene of interest is known and is directly altered or targeted by mutation. The phenotype of mutant organisms is then discerned. This approach is particularly useful to test the effects of a gene found in one organism, such as *Drosophila*, for its effects in another organism such as mice. Genome-wide screens use the genome sequence to individually target mutations to every gene in the genome, regardless of their inferred function. Each mutant is then scored for a phenotype. This method combines many of the advantages of forward and reverse genetic screens but requires a sequenced and well-studied genome, a method to make mutations in the genes, and an efficient screen method to examine each mutant.

unbiased approach is a system that can create **random** mutations in genes across the genome, so there is no bias for a particular class of genes. Not only are the genes mutated at random, but the screening approach should also be as unbiased as possible.

Of course, any screening approach will still have some inherent bias with respect to the types of genes that can be identified. For example, the cultures in Beadle and Tatum's work had to be able to grow on complete medium. Any genes that might have more than one function (that

TOOL BOX 15.1 Genome-wide mutant screens

The chapter describes five of the best-known and most influential genetic screens in order to illustrate how genetic principles can be applied to the analysis of almost any biological process. The key starting point for such screens was to find mutants that failed to carry out the process normally. Such mutants were made by introducing mutations at random throughout the genome, "at random" meaning that the mutations could affect any gene and any process, not just the ones of interest. The chapter only briefly mentions many of the follow-up steps, which often take considerable time. These follow-up steps include making sure that only a single gene is mutated, mapping the mutations to the chromosome, performing complementation tests among similar mutants, identifying the corresponding DNA sequence or cloning the gene, and looking at interactions with other genes, among others.

For model organisms whose genomes have been thoroughly studied, methods are now widely used that take advantage of the genome sequence and bypass some of the follow-up steps required when mutations are introduced at random. In some organisms, resources and libraries of strains exist, in which the function of every gene has been disrupted, so the investigators do not need to begin by inducing the mutations themselves. They can obtain the libraries from a central source or a commercial firm and focus their attention on sorting through all of the genes, cells, organisms, or clones to find the ones of interest. These approaches are known as genome-wide mutant screens, and they are now a method of choice for many investigators and experiments.

The main difference between genome-wide screens and the ones described in the chapter is that all mutations have already been made and assigned to genes in the genome, so the follow-up experiments are usually easier. However, the choice of the organism and the phenotype, the need for a clear-cut assay, and the importance of an unbiased approach are still central to the process.

How are libraries of disrupted genes produced?

Two general procedures have been used to produce libraries in which the function of every gene has been disrupted, and are widely used in different model organisms. In the yeast *Saccharomyces cerevisiae*, collections of strains have been made in which the protein-coding region of each gene has been disrupted by the targeted insertion of a drug resistance marker. These gene disruption strains have been used for hundreds of genetic screens since the collection was made in 2004. Few other organisms have a collection of targeted gene disruptions like yeast does, but less extensive collections of deletion or insertion strains are available in *Drosophila*, *Caenorhabditis elegans*, *Arabidopsis*, and mice, among other eukaryotes.

The second general procedure for a genome-wide mutant screen is to use RNA interference, or RNAi, to disrupt the function of the gene. As described in Tool Box 12.2, RNAi is an experimental method that employs the same cellular machinery as microRNAs, which provide a natural inhibition of gene expression. In RNAi, a double-stranded RNA (dsRNA) corresponding to part of the transcript of a gene of interest is introduced into the cell or organism; the dsRNA makes a hybrid with its target mRNA and either blocks its translation or targets it for degradation, or both.

Because it relies on the naturally occurring microRNA apparatus found in all multicellular organisms, RNAi has proved to be an exceptionally versatile method for disrupting the function of specific genes. Libraries have been made with appropriate dsRNA clones corresponding to each gene in the genome for many different organisms (including some non-traditional model organisms such as *Planaria*), so that the investigator performs RNAi with every gene in turn and observes the phenotype to find the ones of interest. RNAi does not disrupt the gene—the wild-type DNA sequence is left intact—so these are not truly heritable mutations, but they accomplish much the same effect. Again, the follow-up experiments to identify which gene has been affected and so on are much easier with RNAi than for the traditional methods described in the chapter.

Both targeted gene disruptions and RNAi screens have some limitations. While the follow-up experiments involve less work than a traditional screen and genetic-wide screens can guarantee that the screen has been saturated and all genes have been tested, the amount of work is still substantial. Thus, designing a screen that is efficient is still important. Perhaps more significantly, each type of genome-wide screen finds only a particular type of mutation. Traditional genetic screens can identify recessive mutations that knock out the function of the gene or that reduce, but do not knock out, the function of a gene; they can also find dominant mutations that over-express or mis-express the gene. As a general rule, genome-wide mutant screens produce a single type of mutation. Targeted gene disruptions are usually considered gene knockouts, while RNAi effects are usually considered as reductions in the function of the gene, but not knockouts. These are not insurmountable obstacles, and it is common that the genes identified in genome-wide screens are then further analyzed by finding additional and other types of mutations by more traditional methods.

Looking to the future

It seems likely that another method for genome-wide mutant screens will be coming shortly. Within the past few years, CRISPR has been developed as a widely used method to make targeted gene edits throughout the genome; CRISPR is discussed in Tool

TOOL BOX 15.1 Continued

Box 11.3. While no published studies have yet applied CRISPR to make a library of targeted gene edits for every gene, such approaches are certainly being developed and will be used soon. The main requirement to make CRISPR available for a genome-wide screen is to produce a library with guide RNAs that specifically target each gene and to apply these on a wide scale. CRISPR has the potential advantage over other methods in being able to make different types of mutations in the gene—knockouts, reductions of function, mis-expression, and so on.

is, pleiotropic genes) would be missed if one of those functions was needed for general viability. There can be many different biases present in any single screen, but nonetheless the forward genetic approach can remove some of them:

- **Designing for scale and efficiency.** Genetic screens usually involve an assay that can examine individual mutants. In order to be sure that all genes that can be detected in the screen are represented in the resulting mutant collection, especially when a mutation could be anywhere in the genome, the screen must be conducted on a large scale. Beadle and Tatum screened thousands of individual cultures, so efficiency was essential. A well-designed screen might be constructed, such that only the mutants of interest survive, negating the need to examine other mutants at all. This is a powerful variation of a genetic screen known as a **selection**. Since only those cells lacking that specific metabolite would grow when supplemented, and the other cells would not, the screen could be enriched for the mutants of interest. Alternatively, the basic strategy of the experimental set-up might be streamlined, so large numbers of mutants can be observed or tested rapidly when selection is not feasible.
- **Assigning the defects to genes in the genome.** Once a screen generates mutants of interest, the mutated gene needs to be identified. This is usually achieved through a combination of genetic mapping (discussed in Sections 9.3 through 9.5) and complementation testing (discussed in Section 8.5). These approaches are typically easier in model organisms in which previous mapping and complementation studies have been done.

  More recently, the ease of obtaining a genome sequence has eliminated the need for some of the mapping and complementation steps, since it is often feasible to sequence the genome of the mutant to find the newly induced lesion. This is particularly useful in bacteria, whose genomes are relatively small. Even in these cases, however, many mutations and sequence variants can be found, so additional filtering steps are needed. This approach in humans is discussed in Box 10.2.
- **Following up with the analysis of individual mutants and genes.** Once mutants have been found, follow-up experiments are designed in order to understand how many genes have been inactivated in the mutant collection, in what order these genetic components might operate, and how they might interact. We will explore some of these concepts further in Section 15.2. We also note that, based on personal experience, most geneticists who routinely do mutant screens have collections of mutations awaiting analysis to be done "someday."

When the scope and scale of the genetic screen are small and focused, the genetic hierarchy and the biochemical interaction pathway often match up well. This turned out to be the case for the arginine synthesis pathway that Beadle, Tatum, and colleagues elucidated. However, the same genetic approach can still be applied to map out the basic structure underlying a process of significantly greater scale and complexity, even when the underlying biochemical steps are not known or not easily approached. The same principles apply, even if the exact assays are different and even if the genetic pathways constructed from such a screen provide an initial broad outline of the process, rather than a complete biochemical pathway. This will be illustrated in the examples that follow.

KEY POINT A well-designed genetic screen uses an unbiased approach in an assay that is clearly defined and easy to perform on a large scale.

15.2 True in the elephant? A model for gene regulation from genetic screens of lactose utilization in bacteria

Beadle and Tatum's work illustrates how genetic screens can be used to interrogate biochemical pathways. Biology is an organized system of interconnected pathways. Some of these pathways are biochemical, but others represent regulatory networks that can also be pieced together using genetic screens. Perhaps the best-studied regulatory system is the one that controls the use of the disaccharide sugar lactose in *Escherichia coli.* We introduced the workings of the lactose operon in Section 14.2, but let's review its main components, as laid out in Figure 15.6.

The lactose utilization genes in *E. coli* are expressed when needed to derive energy from this disaccharide sugar, but they are repressed in its absence. The two genes directly responsible for lactose import and catabolism are *lacZ*, encoding the β-galactosidase enzyme that breaks down lactose to the monosaccharides glucose and galactose, and *lacY*, encoding a membrane permease that imports lactose into the cell. The *lac* genes also include *lacA*, which encodes a galactose transacetylase, but this gene is not required for lactose utilization, so it was not identified in the original screens. All three genes are expressed from the same promoter, that is, they are within the same operon and regulated by the same mechanisms.

As detailed in Section 14.2 and shown in Figure 15.7, the *lac* operon is repressed in the absence of lactose by the protein LacI, the product of a nearby gene transcribed from its own promoter. LacI (commonly known as the Lac repressor) binds as a multimer to the operator sequence *lacO* or O_1, effectively occluding transcription from the *lac* operon; it is only removed from the operator when its conformation changes in the presence of lactose. This ensures that the lactose catabolism genes can only be expressed when lactose itself is present.

When lactose is absent and the operator sequence is not occupied by LacI—in other words, when the operon is de-repressed—RNA polymerase can transcribe the *lac* operon but requires the catabolite activator protein (CAP) in association with cyclic AMP (cAMP) to do this. Whenever glucose is available for import, CAP–cAMP is not available. CAP–cAMP will only activate transcription of the *lac* operon when glucose is not being translocated into the cell. This ensures that the lactose catabolism genes are not expressed when glucose, a preferable energy source, is present, as summarized in Figure 15.8. The elegance of the regulatory system is only outshone by the screens used to unravel it.

The French bacterial geneticists and Nobel Laureates François Jacob and Jacques Monod performed the original genetic screens that revealed the regulatory structure of the *E. coli lac* operon. One of the important aspects of the genetic screens that were used to understand how the *lac* operon works is that the screens not only asked whether certain strains fermented lactose but also determined the nutritional conditions when lactose utilization took place. As such, lactose fermentation was studied as a step-by-step process that could be turned on or off in response to external conditions, rather than just a simple outcome. This allowed investigators to dissect the regulation of the *lac* genes, as well as the functions of their gene products. In other words, they had a clear and well-defined mutant phenotype.

KEY POINT Genetic screens can be designed to uncover effectors of biological processes, as well as regulators of these effectors.

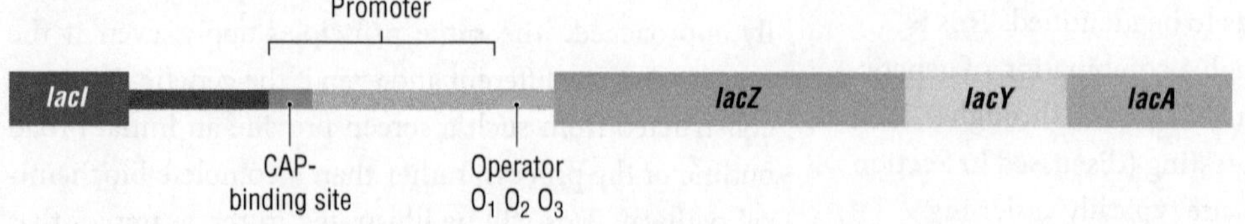

Figure 15.6 An overview of the main components of the *lac* operon in *Escherichia coli*. This is described in more detail in Chapter 14 and is summarized here. The operon has three structural genes involved in lactose utilization, called *lacZ*, *lacY*, and *lacA*. These are transcribed on one RNA molecule from the same promoter. The *lacI* gene encodes the repressor, which binds at the operator in the absence of lactose to prevent transcription of the operon. The CAP-binding site within the promoter is involved in catabolite repression that further ensures that the *lac* genes are only maximally transcribed when glucose (the preferable energy source) is absent.

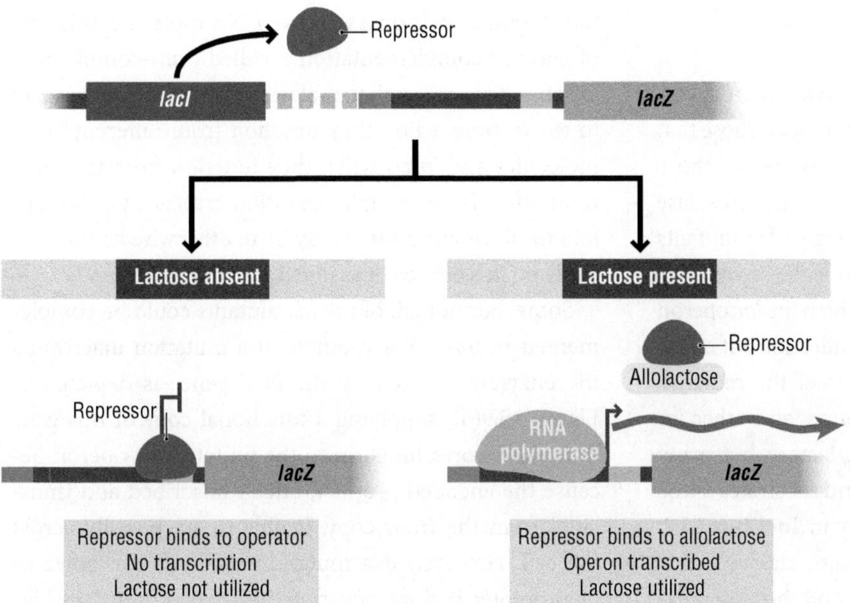

Figure 15.7 Repression of the *lac* operon by the repressor, the product of the *lacI* gene. When lactose is absent, the repressor binds to the operator and prevents transcription of the three enzymes of the *lac* operon. When lactose is present, the repressor does not bind to the operator and the operon is transcribed so that lactose can be utilized. The interaction of the repressor with allolactose, a molecule related in structure to lactose and derived from it, alters the conformation of the repressor to prevent it from binding.

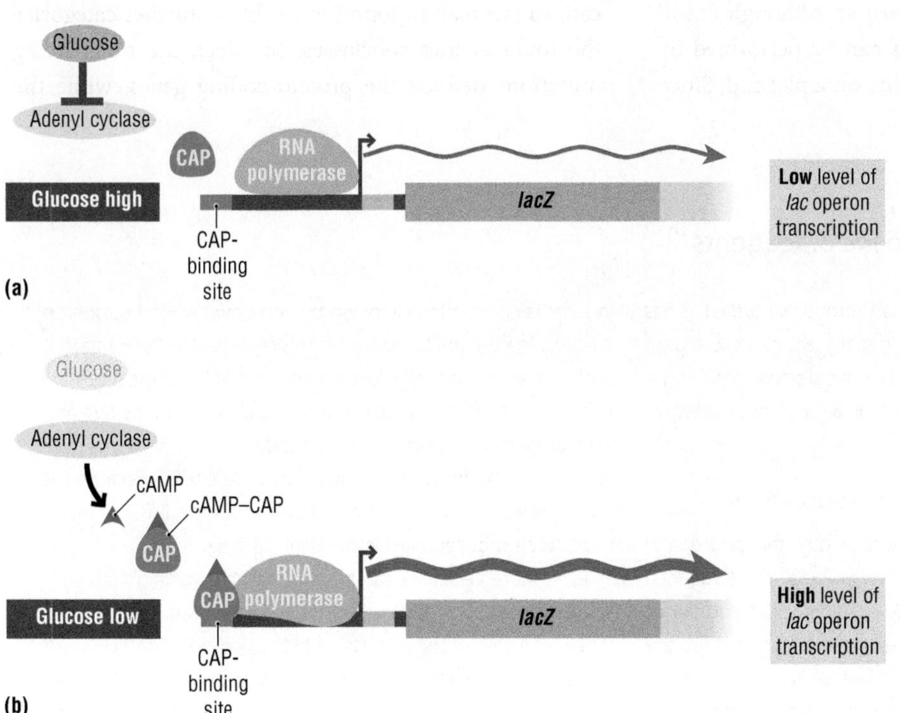

Figure 15.8 The role of CAP in glucose utilization. When glucose is present in the medium, *Escherichia coli* cells do not express the *lac* operon at high levels, which allows utilization of glucose, a better carbon source, instead. When glucose levels are low, a protein known as CAP associates with cAMP and binds to the CAP-binding site within the promoter to recruit RNA polymerase and express the *lac* operon at high levels. More details are provided in Chapter 14.

Different categories of lac mutants

Jacob and Monod initially screened for two different categories of mutants with contrasting phenotypes: those that were unable to utilize lactose as a carbon source and those that did not require the sugar to regulate β-galactosidase production. The first category can be thought of as mutants in which the *lac* operon is always off, while the second category can be thought of as mutants in which the *lac* operon is always on. This is a bit of an over-simplification but might be helpful in thinking about the effects of the mutants. Some screens conducted by Jacob, Monod, and other *lac* operon pioneers were performed using lactose, but many more screens used synthetic inducers and substrates of the *lac* operon that are described more fully in Tool Box 15.2. The screen recovered two broad mutant classes: those unable to ferment lactose at all (OFF), and those that expressed *lacZ* and produced β-galactosidase constitutively irrespective of the presence of lactose (ON). Thus, they had mutations with disruptions in different parts of the process, which was particularly useful for this screen and for many others as well, as we will discuss shortly.

Jacob and Monod then attempted to complement the mutations with a functional version of the *lac* locus in order to categorize the mutants further. Although *E. coli* is haploid, complementation tests can be performed by introducing a second copy of a gene on a plasmid. Since the second copy is on a separate DNA molecule, this type of plasmid complementation is called *trans*-complementation. (Remember that molecular components are said to act *in trans* when they function from different DNA molecules and *in cis* when they function from the same molecule.) *Trans*-complementation creates a partial diploid for the locus under study in an otherwise haploid organism. This approach as illustrated in Figure 15.9.

Some, but not all, of the *lac* mutants could be complemented *in trans*. For example, if a mutation inactivated the enzyme encoded by the *lacZ* gene, as depicted in Figure 15.9(a), supplying a functional copy of this gene *in trans* restores function to the mutated *lac* operon because the encoded product, once transcribed and translated from the *trans* copy, is able to act broadly across the cell. However, if a mutation rendered the operator or promoter regions non-functional, as depicted in Figure 15.9(b), supplying a functional copy *in trans* does not restore function to the mutated *lac* operon because these gene sequences act locally to control expression of adjacent genes on the same DNA molecule.

In general, mutations in the operator or promoter regulatory regions of the *lac* operon cannot be complemented *in trans*, while mutations in the protein-encoding genes can, so Jacob and Monod were able to further categorize the mutants into subclasses. In effect, the *trans*-acting mutations defined the protein-coding genes, while the

TOOL BOX 15.2 No *lac* of reagents

Because the *lac* operon has been so thoroughly studied, it has also been used as an important tool in many experiments in genetics and molecular biology. In this box, we discuss some of the reagents that have allowed the *lac* operon to be used so widely as an experimental tool.

Inducing and monitoring gene expression

The earliest lactose screens used lactose as both the inducer and the substrate. In nature, the lactose isomer allolactose (produced from lactose inside the *Escherichia coli* cell) binds to LacI and causes de-repression or expression of the operon. Since lactose is also the substrate for the *lac* genes, once all the lactose in the medium has been consumed, the inducer activity is also lost. As a result, LacI rebinds to the *lac* operator and prevents further transcription.

While it makes biological sense to have lactose function as both the inducer and the substrate, it makes the experimental analysis simpler if the inducer is not also the substrate—that is, if the inducer is not continually being broken down when the operon is active. To keep the *lac* operon in this repressed (or "OFF") state, it is convenient to use an inducer molecule that binds the repressor to change its conformation but which itself is not a substrate and will not be broken down. The chemical most commonly used as an inducer is isopropyl-β-D-1-thiogalactoside, or IPTG. Because it is not broken down by the activity of the *lac* operon, IPTG is a much more stable inducer than lactose or allolactose.

Even more versatility can be obtained by working with substrates other than lactose. Substrates other than lactose allow for easier assays of the expression of the operon, in particular, for detecting and measuring β-galactosidase activity. These other substrates have the additional advantage of not requiring the LacY permease for entry, so a smaller "cut-down" version of the *lac* operon with only *lacZ* can be used. Thus, both the interaction between the repressor and the inducer and the expression of *lacZ*, both of which require lactose uptake in natural cells, can be done

TOOL BOX 15.2 Continued

independently of LacY and LacI when substrates other than lactose are used.

In addition, the commonly used alternative substrates for β-galactosidase (β-gal) are broken down into products that can be assayed colorimetrically. In combination with IPTG to induce *lacZ* expression, these alternative substrates can provide sensitive assessments of *lacZ* activity without relying on lactose fermentation to see the phenotype. The alternative substrate 5-bromo-4-chloro-3-indolyl-β-D-galactopyranoside, more commonly known as X-gal, turns dark blue when it is broken down by β-gal, while hydrolysis of ortho-nitrophenyl-β-galactosidase (ONPG), a different substrate, produces a yellow nitrophenol. The breakdown of X-gal is easy to see with the unaided eye, so colonies in which *lacZ* is being transcribed can be spotted simply as blue "Lac$^+$" and white "Lac$^-$" colonies on plates. The breakdown of ONPG or X-gal can also be measured colorimetrically and quantitatively in enzyme reactions, so the induction of *lacZ* is easy to detect. The colored products resulting from the action of β-gal on X-gal and ONPG are shown in Figure A

Yet another alternative substrate, phenyl-β-galactoside, is imported by LacY and degraded by β-galactoside but does not de-repress the *lac* operon. In other words, it does not act as an inducer. As a result, organisms that produce functional LacI cannot use phenyl-β-galactoside as a carbon source. Thus, this alternative substrate to lactose was used to select for mutants that express *lac* genes constitutively, which are normally rare. As discussed in the text, screens that include an element of selection are extremely powerful because only the mutants of interest will grow.

IPTG, X-gal, and some of the other *lac* reagents are versatile enough to be employed in organisms other than those that normally display β-gal activity. β-gal can be fused to regulatory sequences and genes in virtually any model organism and then assayed using the β-gal substrates, as described here. As a result, components of the *lac* operon are very often used as tools for monitoring gene expression. We discuss such usage in more detail in Tool Box 12.1.

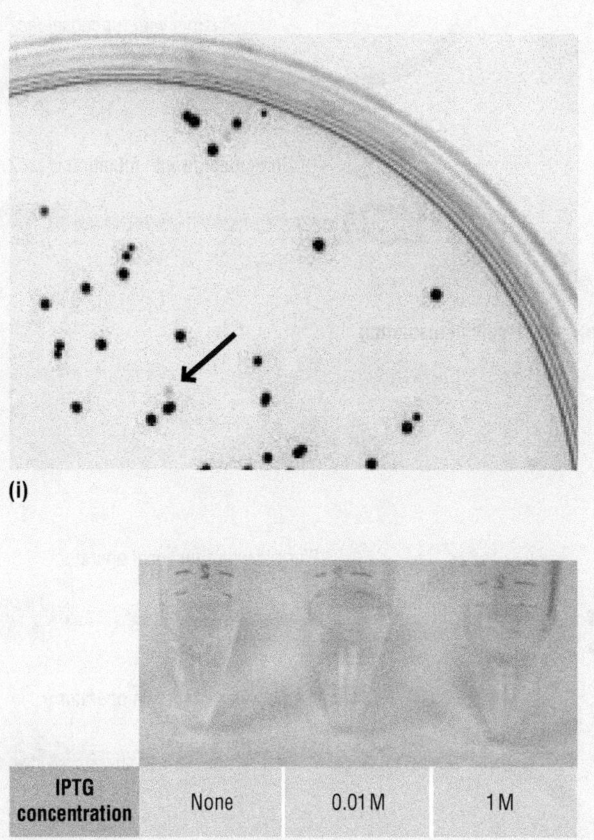

Figure A Alternative substrates to lactose allow for visual and quantitative assays. (i) Tools for lactose regulation screens. X-gal turns blue when hydrolyzed by β-galactosidase, making it possible to identify rare *lac*$^-$ mutants in genetic screens. (ii) Use of ONPG as a substrate allows β-galactosidase activity to be read out colorimetrically. Cell cultures of the *E. coli* strain used for this assay were induced with varying concentrations of IPTG, a synthetic inducer. The cultures were then lysed and then incubated with ONPG. The intensity of the yellow color is proportional to β-galactosidase activity in the cell lysates. While this shows the effects visually, the intensity of the color can also be measured quantitatively.

cis-acting mutations identified the regulatory sequences. The classification scheme Jacob and Monod used to categorize their mutants is outlined in Figure 15.10.

At the time Jacob and Monod were working, there were very few tools for constructing recombinant plasmids *in vitro*, techniques that are now commonplace. To conduct the complementation analysis described above, they therefore took advantage of the fact that the F plasmid occasionally integrates into the bacterial chromosome next to the *lac* locus, as described in Section 11.3; if the F plasmid excises and recircularizes imprecisely, portions of the *lac* operon can sometimes be moved onto the F plasmid. The resulting F′ plasmids containing the *lac* genes could be used to introduce a second copy of *lac* into different *E. coli* strains. It was this early availability of a complementation method that favored the study of lactose fermentation in *E. coli* over parallel studies that were being conducted in other bacteria, another example of matching the model organism to the research question. In addition to classifying mutants, complementation tests

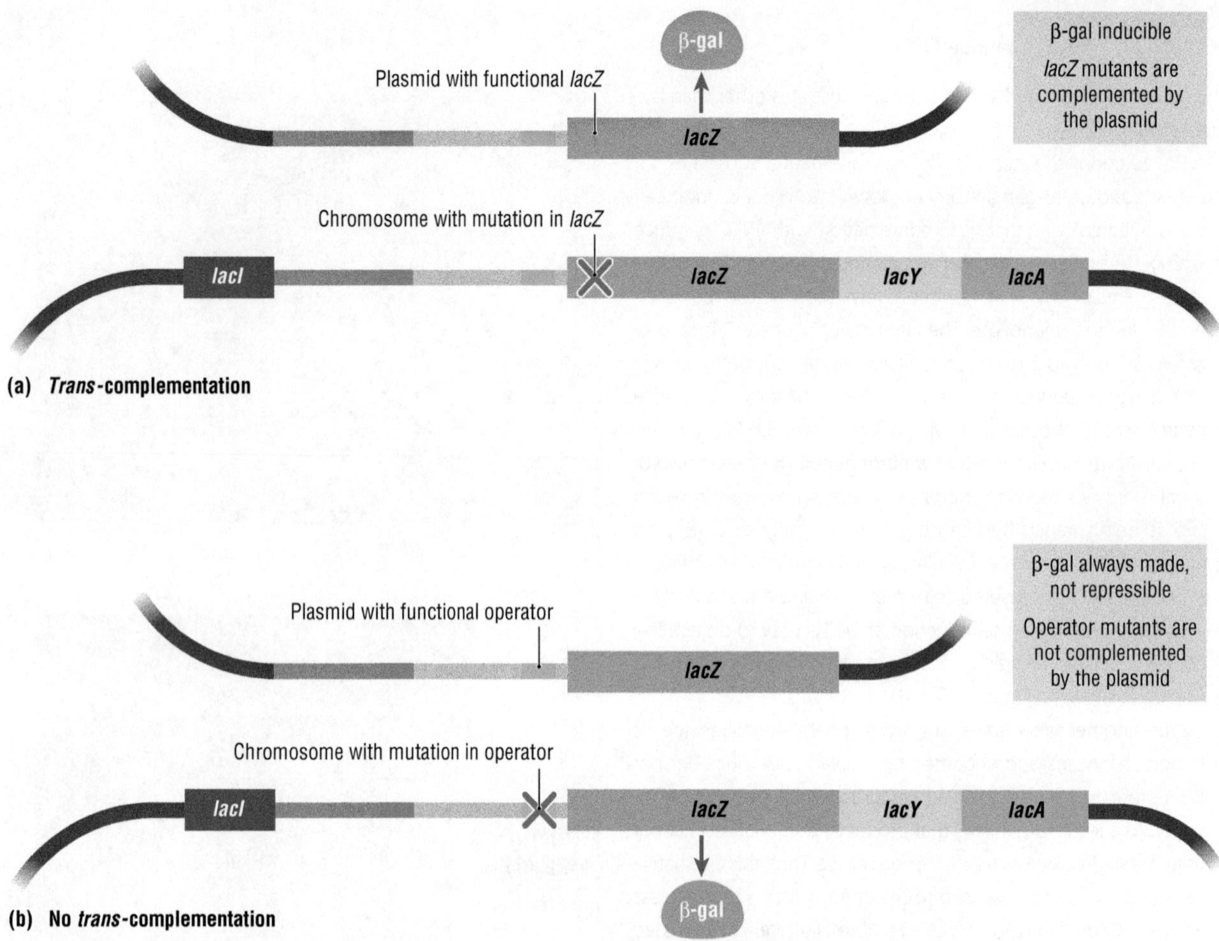

Figure 15.9 ***Trans*-complementation to distinguish different types of *lac* operon mutations.** Part or all of the functional *lac* operon was incorporated into a plasmid, which was then transformed into a mutant cell. If the mutation on the chromosome could be complemented *in trans* by the plasmid, as shown in (a) for *lacZ* mutations, the gene is inferred to make a protein product β-galactosidase (β-gal), which can be supplied by the functional gene on the plasmid. If the mutation on the chromosome cannot be complemented by the plasmid, as shown in (b) for a mutation in the operator, the mutation must affect a binding site, rather than a gene that makes a diffusible protein product.

proved essential for determining the likely functions disrupted in each mutant.

Such complementation tests, together with fine mapping techniques that have been largely supplanted by sequencing nowadays, enabled Jacob and Monod to identify and map *lacZ* and *lacY*, the two essential lactose utilization genes. The existence of these mutants was not unexpected and would be analogous to the arginine biosynthetic mutants in *Neurospora*—they affect genes encoding polypeptides necessary for the process of lactose utilization. However, some categories of *lac*-negative mutants could not use lactose, even when a functional operon was supplied *in trans*. In essence, these are dominant loss-of-function mutants, since a mutant copy eliminates the function of the operon. These mutations were used to define some the regulatory elements of the operon. Insight into the regulatory components of the *lac* operon was particularly useful in highlighting the importance of transcriptional repression. Let's therefore take a look at the genetic analysis of the *lac* repressor gene *lacI* in a bit more detail.

For LacI to function correctly, it must be able to form a tetramer, change conformation in the presence of lactose, and bind to operator DNA. Binding sites in the DNA require specific amino acid residues to be located at certain positions within the binding protein. When lactose (or actually allolactose) is present, there is an interaction with these positions in the *lac* repressor; it changes conformation and cannot bind to the operator; transcription is thus de-repressed.

In the LacI protein, these various functions map to different domains, as shown in the ribbon diagram in Figure 15.11(a) and in the drawing in (b). As such, there

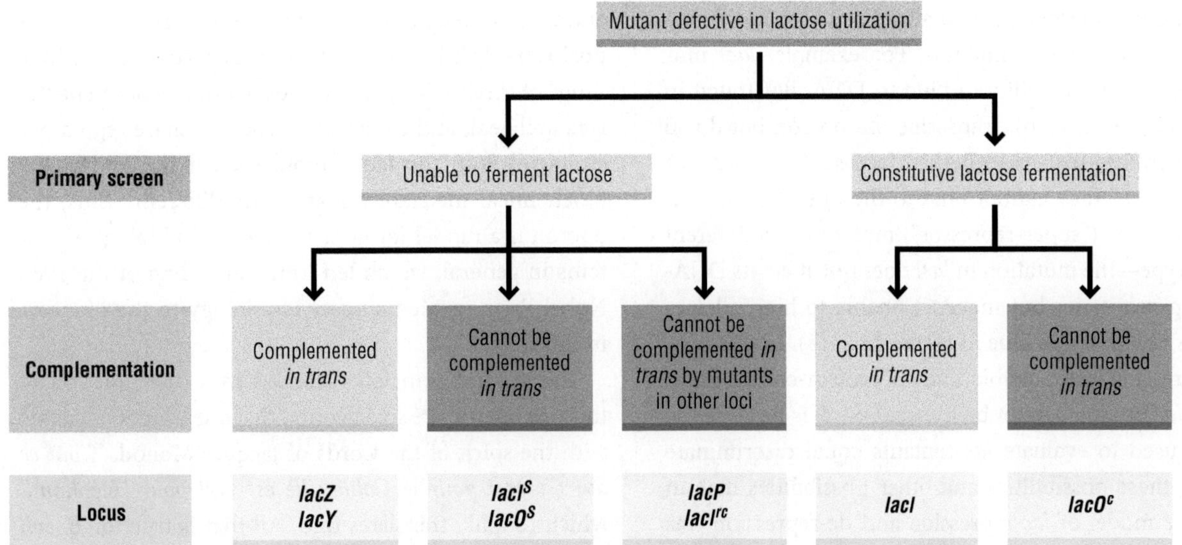

Figure 15.10 A screen for *Escherichia coli* mutants with aberrant lactose utilization phenotypes. Two different types of mutants were found: those that could not ferment lactose and those that constitutively produced β-galactosidase, even when glucose was present. These different types of mutations were then distinguished further by a *trans*-complementation test, done by transforming a plasmid with part or all of the wild-type *lac* operon into the mutant cell. Mutants that could be complemented *in trans*, such as those in *lacZ*, *lacY*, and *lacI*, shown in green, are the ones in which the wild-type genes make protein products. Mutants that could not be complemented *in trans*, such as those in *lacO* and *lacP*, shown in red, are the ones in which the mutation occurred in a binding site on the DNA, rather than in a stretch of the gene encoding a protein product.

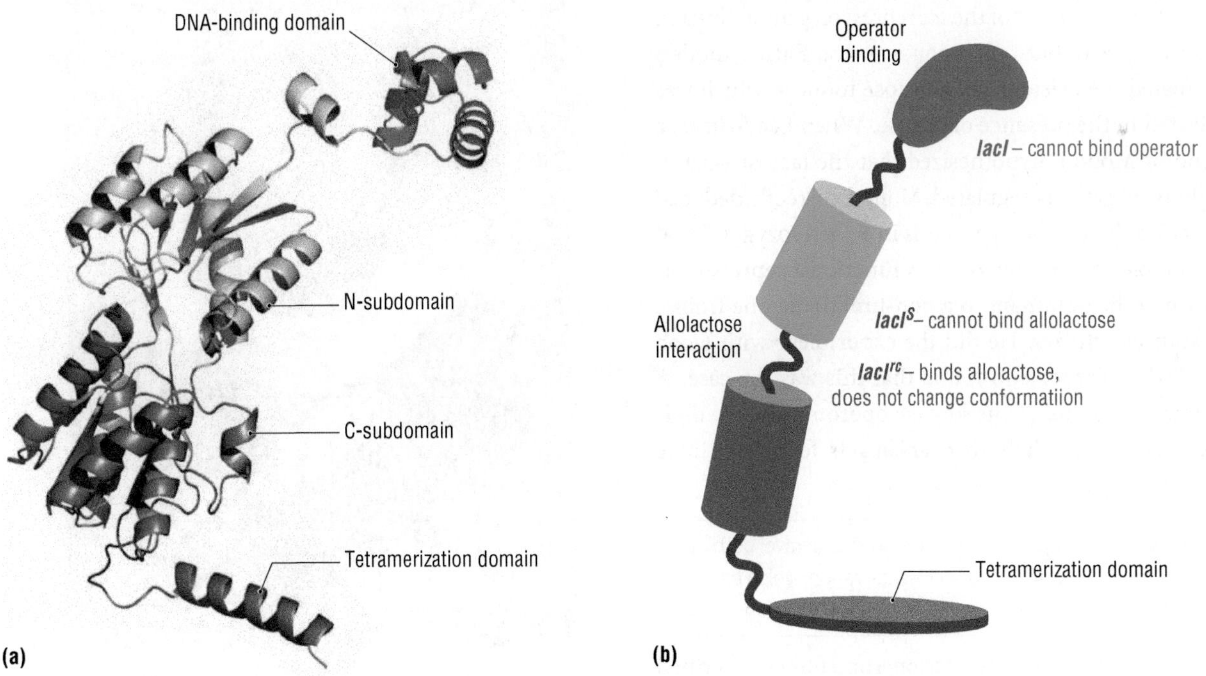

Figure 15.11 Functional domains of the LacI protein. (a) A ribbon diagram of a LacI monomer showing the DNA-binding N- and C-terminal domains that constitute a core to interact with allolactose, and the tetramerization domain. The repressor protein is a tetramer comprising four of these monomer subunits. (b) The locations of some *lacI* mutations and their effects on the protein. *lacI* mutations in the DNA-binding domain cannot bind to the operator. As a result, the operon is not repressed, and the genes are transcribed, regardless of the presence or absence of lactose. *lacI* mutations in the central domain, which interacts with allolactose, have different phenotypes. In *lacI^s* mutants, a "super-repressor" is made that cannot interact with allolactose, and thus the operon is always repressed. In *lac^rc* mutants, the repressor can bind to allolactose but does not change its conformation. Thus, the presence of lactose does not de-repress the operon.

Source: Part (a) reproduced from Wilson, C.J., et al. (2005). The experimental folding landscape of monomeric lactose repressor, a large two-domain protein, involves two kinetic intermediates. *PNAS*, 102(41).

are multiple different ways to alter LacI, each yielding a different class of *lacI* mutants. For example, *lacI* mutants affecting the ability to bind to DNA, illustrated in red in Figure 15.11(b), transcribe the operon but do so constitutively—that is, whether lactose is present or not—because they cannot bind to the operator. In contrast, the *lacI*^{s} ("super-repressor") mutant has a different phenotype—the mutation in *lacI* does not affect its DNA-binding properties but makes it unable to bind allolactose, as illustrated in blue in Figure 15.11(b). In this case, de-repression is impossible, and the protein-coding genes of the *lac* operon cannot be transcribed. It is because the assays used to evaluate *lac* mutants could discriminate among these possibilities and other possibilities that an accurate model of *lac* repression and de-repression was created.

Jacob and Monod's analyses of their early classical genetic screens resulted in a model that correctly predicted the functions of *lacZ*, *lacY*, *lacI*, the *lac* operator, and the *lac* promoter. Figuring out the nature and function of LacI by studying mutants identified in screens for constitutive mutants was the most rewarding challenge associated with studies of lactose utilization.

Until the discovery of the *lac* repressor, gene activation was recognized, but repression was not. Earlier models postulated the existence of a lactose inducer, which was activated in the presence of lactose. When Leo Szilard, a friend of Monod, hypothesized that the lactose operon might be negatively regulated, Monod correctly deduced that he could test the hypothesis in a "heterozygote" that was diploid for the *lacI* locus. A functional repressor allele would be dominant to a non-functional one from a constitutive utilizer. He did the experiment with Jacob and Arthur Pardee and found that this was the case. A summary of some of the key *lac* operon mutants, their phenotypes, and their mechanisms is found in Table 15.1.

KEY POINT Mutant phenotypes were used to unravel both the logic and the molecular basis of lactose utilization, which has influenced thinking about other types of gene regulation.

A written history of the *lac* operon studies described this system as "a paradigm of beauty and efficiency." The study of the *lac* operon affected genetic analysis in ways that extend far beyond the intricacies of sugar utilization by an intestinal bacterium. Through the elegant lactose utilization genetic screens, we have come to know that gene regulation is structured very much like circuitry and that processes can be regulated negatively as well as positively. While such regulation is accomplished by a range of mechanisms, thousands of other genes in bacterial, archaeal, and eukaryotic genomes share regulatory strategies with the *lac* operon, even if the underlying biochemical mechanisms are quite different. Thus, the operon is a model for gene regulation in biological systems in general, which led to the awarding of the 1965 Nobel Prize in Medicine or Physiology to the pioneers in this field.

The general principles learned from the function of this one operon are so profound that one cannot but agree with the spirit of the words of Jacques Monod, "*Tout ce qui est vrai pour le Colibacille est vrai pour l'éléphant*," which roughly translates into "All that is true for *E. coli* is true for the elephant." Monod's pictorial illustration of this idea is shown in Figure 15.12.

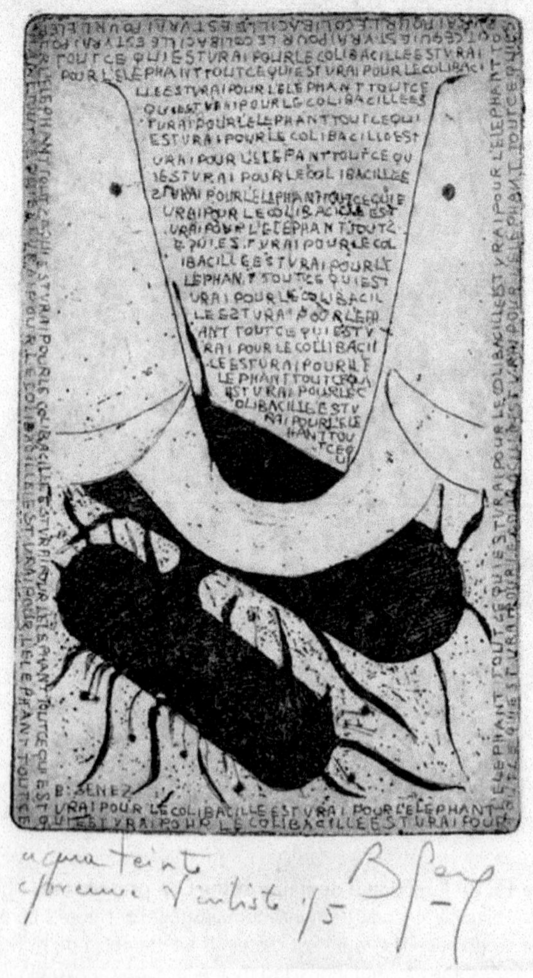

Figure 15.12 Monod's illustration of his dictum about *Escherichia coli* and elephants.

Source: Reproduced with permission from the Pasteur Institute.

Table 15.1 Key *lac* mutants that provided evidence for the now-validated model of lactose utilization in *E. coli*

The mutant phenotype	The gene/locus	The mechanism
Unable to ferment lactose, unable to utilize other β-galactosidase substrates. Can be complemented *in trans*	*lacZ*, β-galactosidase	These mutants do not produce functional LacZ, the enzyme that breaks the β-galactosidase bond. They cannot break down lactose into monosaccharides that can be used as energy sources
Unable to ferment lactose but can use most other β-galactosidase substrates. Can be complemented *in trans*	*lacY*, permease	These are mutants in LacY, the lactose permease necessary for importing lactose into the cell
No phenotype; not discovered in genetic screens	*lacA*, transacetylase	LacA acetylates lactose and IPTG. It is not essential for lactose utilization, and the utility of this function remains unknown
Ferment lactose constitutively, ferment β-galactosidase substrates without lactose or IPTG induction. In essence, these mutants have lost inducibility. Most mutants were selected on phenyl-β-galactoside (they are rare in other screens), which they can use as the sole carbon source, and are complemented *in trans*	*lacI*, the *lac* repressor gene	These mutants have lost the ability to make LacI, the protein that normally represses the lactose operon in the absence of lactose. There are different subclasses of *lacI* mutants. Some fail to make a repressor protein or make a protein that does not bind to the operator, while others are unable to form multimers. Mutants with defects in multimerization can be only partially complemented because they result in a mix of wild-type and defective monomers in the cell
Unable to ferment lactose or other β-galactose substrates. Introduction of a functional *lac* operon on a plasmid does not restore lactose utilization	$lacI^s$ "super-repressor"	These are *lacI* mutants that have lost the ability to bind lactose or allolactose. As such, LacI cannot disengage from the operator sequence. $lacI^s$ mutants are dominant over *lacI* mutants and wild-type *lacI*
Ferment lactose constitutively. Ferment β-galactosidase substrates without lactose or IPTG induction. A loss of inducibility that cannot be complemented *in trans*	$lacO^c$ or O_1, the main *lac* operator	The mutants have lost the LacI-binding site on the operon. This site is called the operator. Unlike *lacI* mutants, they cannot be complemented *in trans* because the defect is in a sequence that must work in concert with the proximal gene, that is, *in cis*. In the absence of selection, operator mutants are recovered relatively rarely because the operator sequence is short compared to the *lac* gene sequences
Unable to utilize lactose. Cannot be used to complement other null mutants, even those from other subclasses	*lacP*	RNA polymerase can no longer bind to the *lac* promoter, and therefore the *lac* genes cannot be transcribed. This is a *cis*-acting mutation arising from a mutation in the promoter sequence
Repressed in the presence of lactose	$lacI^{rc}$	Inducer binding does not result in a conformational change in *lacI*, therefore the operon cannot be de-repressed
A reduction but not an outright loss of inducibility. Not complemented *in trans*	O_2 and O_3	The *lac* repressor LacI works as a tetramer. Two molecules bind the main lac operator (O_1). The other two bind to one or two lesser known weaker operator sites. O_2 and O_3 are redundant, therefore mutants in either of these two operators have no phenotype. A double O_2 and O_3 mutant does show a slight reduction in inducibility. Because the effect is less noticeable than that of the O_1 (*lacO*) mutations, and because of O_2 and O_3 redundancy, the functions of these auxiliary operators were not elucidated until almost half a century after the discovery of O_1

15.3 A foot in the door: *Drosophila* embryonic development

The third of our landmark genetic screens examined an impressively large number of mutants, identified the largest number of genes, and probably asked the broadest biological question: How do animals achieve the body forms that they have? Animals start out as single fertilized cells but develop into "endless forms most beautiful and most wonderful," to quote Darwin. Incredibly, that initial single cell contains all the genetic information necessary to produce these physical forms. What genes are involved, and how do they work to control this intricate process of development? This seems like an impossibly complicated process to try to understand. Nonetheless, these questions motivated Christiane Nüsslein-Volhard and Eric Wieschaus to undertake an ambitious genetic screen of developing embryos. The success of their work has provided important insight into how body plans are patterned and developed from a single fertilized egg in many animal species.

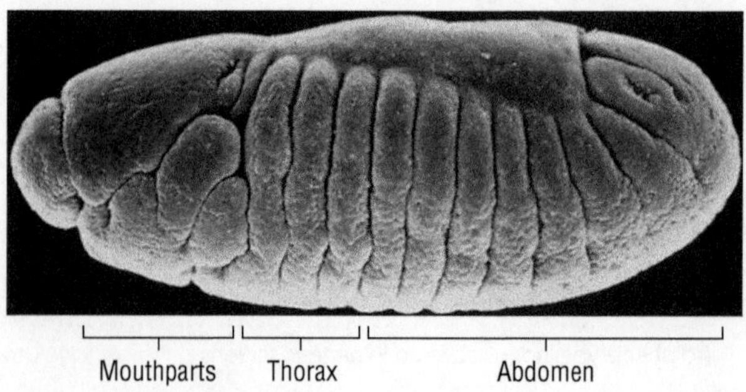

Figure 15.13 Scanning electron micrograph of a *Drosophila melanogaster* embryo. The segments are clearly evident throughout the mid-region, which will become the thorax, and the posterior, which will develop into the abdomen. They are present, but less apparent, in the head region.

Source: Reprinted from Turner, F.R., Mahowald, A.P. Scanning electron microscopy of *Drosophila melanogaster* embryogenesis. II. Gastrulation and segmentation. *Dev. Biol.* 1977, 57: 403–416 with permission from Elsevier.

A crucial contributing factor to the success of Nüsslein-Volhard and Wieschaus's screen lies in their choice of the fruit fly *D. melanogaster* as their model system, since the genetics was well established and the embryology was accessible. As we described previously, the use of *Drosophila* to study genes brings distinct advantages. Of particular importance in this study is the small number of chromosomes they possess and their rapid generation time, along with the fact that many helpful genetic tools and methodologies were already established. However, most of the previous genetic studies had been focused on traits visible in the adult fly, such as eye color. Nüsslein-Volhard and Wieschaus chose to apply and adapt genetic tools to the study of embryonic development.

Drosophila eggs are laid externally and in large numbers, making it possible to easily view embryonic development. Some of the basic embryology of *Drosophila* was being investigated around the time that this screen was being developed, and Wieschaus himself had been involved in these studies. An electron micrograph of a *Drosophila* embryo is shown in Figure 15.13. Note its primary feature—the pattern of segments that make it somewhat resemble a flattened, peeled orange.

An overview of Drosophila embryogenesis

It is helpful to consider some of the features of *Drosophila* embryogenesis before we look at the experimental design of their screen in more detail. The early development of the fruit fly is diagrammed in Figure 15.14(a). The earliest stages of embryonic development in *Drosophila* are characterized by extremely rapid rounds of nuclear division—the single fertilized nucleus undergoes multiple rounds of mitosis to produce thousands of daughter nuclei that migrate to the periphery of the egg. The egg cell and its surrounding plasma membrane have not yet divided, so the result of these nuclear divisions is a single large cell containing thousands of nuclei arranged around the periphery, the stage known as the syncytial blastoderm. In this situation, molecules in one part of the cytoplasm of this cell can affect the gene expression pattern of any of the nuclei lying within range of their diffusion.

The plasma membranes between individual nuclei then form, producing a cellularized blastoderm. By this stage, the germ line has been set apart as a small group of cells (the pole cells) at the posterior end of the embryo.

Subsequent events are shown in Figure 15.14(b). Gastrulation brings some cells to the inside of the embryo and establishes the mesodermal and endodermal cell layers that will give rise to internal cell types, such as cells that form the muscles and gut. The ectoderm that will form the skin and neural tissue derives from the remaining cells around the outside of the embryo. The embryo then divides itself into 14 segments, each of which has its own characteristics and identity.

The genetic control of segment identity is exerted by the homeotic genes that were first characterized in adult flies by Edward Lewis and were described in Section 3.6.

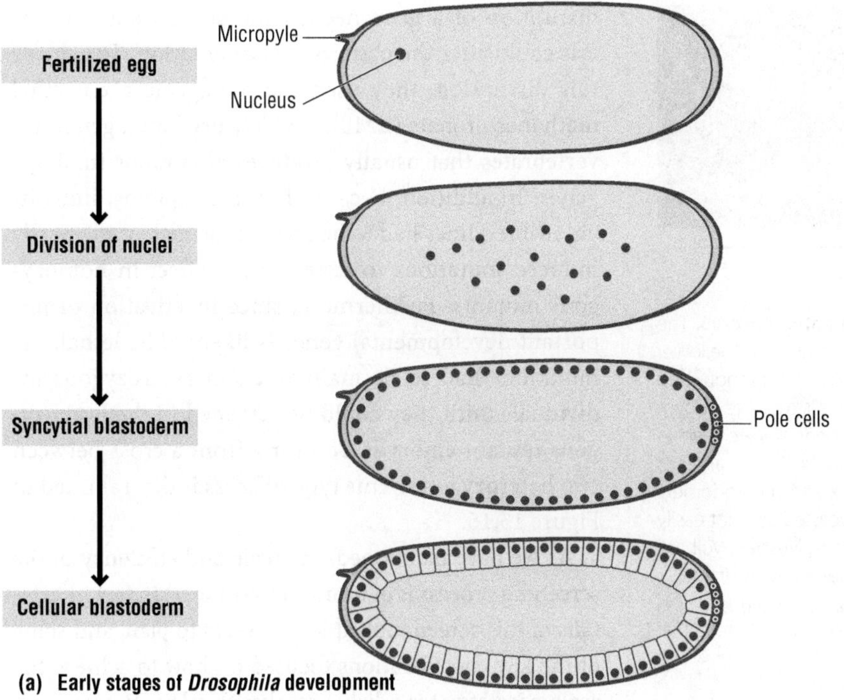

(a) Early stages of *Drosophila* development

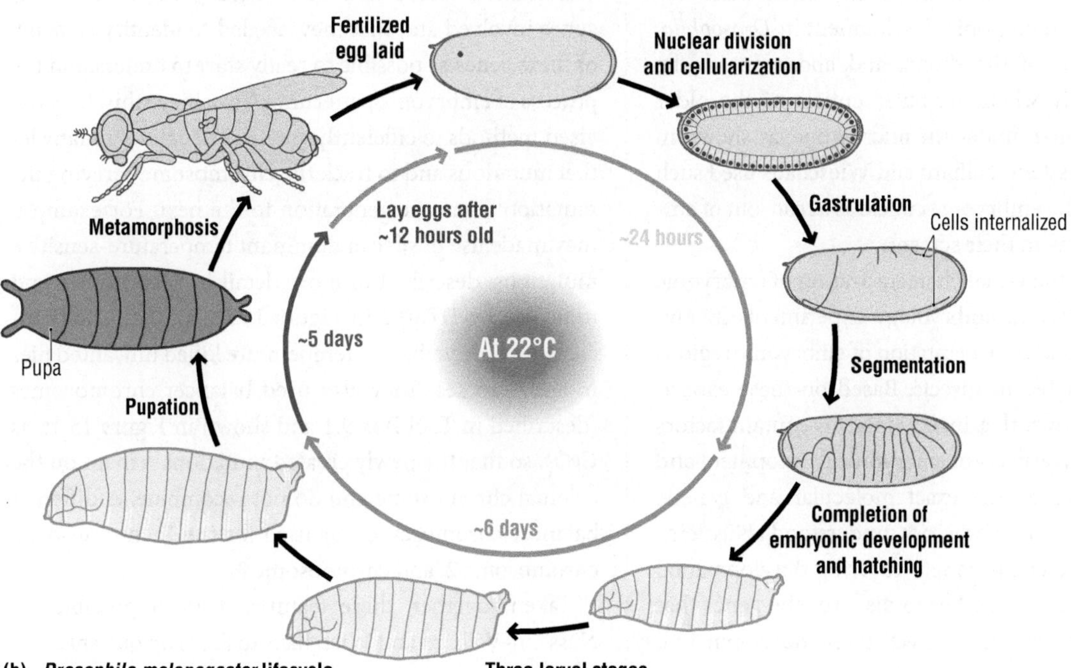

(b) *Drosophila melanogaster* lifecycle

Figure 15.14 Stages in the early development of a fruit fly. The various stages are drawn, with the future anterior region to the left. (a) The early stages of embryogenesis. The micropyle, an opening on the shell of the egg through which the sperm can enter, is positioned at the anterior end. The fertilized nucleus undergoes mitosis multiple times to produce thousands of nuclei, but the egg cell and its surrounding plasma membrane do not divide. This results in a single large cell containing thousands of nuclei, which move to the periphery, forming a syncytial blastoderm. Individual nuclei become surrounded by the plasma membrane, giving rise to a cellular blastoderm. (b) Later stages of embryonic and larval development and completion of the life cycle of the fly. After the cellular blastoderm forms, gastrulation occurs, during which some cells move to the inside of the embryo through specific folds and invaginations. Gastrulation establishes the layers of the ectoderm, mesoderm, and endoderm. The embryo then divides into 14 segments, each of which adopts its own characteristics and identity. This stage corresponds to the one shown in the micrograph in Figure 15.13. These segments can be easily seen in the larval and pupal stages, which follow hatching from the egg.

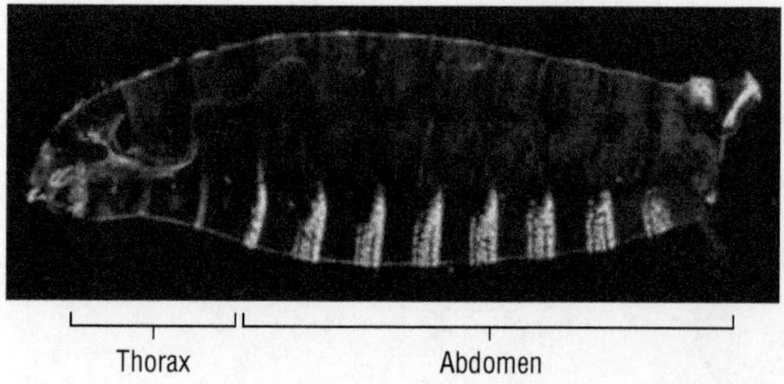

Figure 15.15 Cuticle patterns in the segmented embryo. The anterior is to the left, and the ventral half of the embryo is the lower half of the image. The segments in the embryo can be distinguished from one another by the pattern of bristle-like structures known as denticles. In this dark-field micrograph, the denticles are seen as blocks of bright dots in characteristic patterns along the anterior ventral portion of each segment of the thorax and the abdomen. At this stage, the segments in the head are involuted and not easily seen. Cuticle preparations of this type were used by Nüsslein-Volhard and Wieschaus in their screens for embryonic patterning mutants.

Source: Reproduced from Luschnig, S. et al. (2006). Serpentine and vermiform Encode Matrix Proteins with Chitin Binding and Deacetylation Domains that Limit Tracheal Tube Length in Drosophila. *Current Biology* 16, 186–194.

Segments of the head become mostly internalized towards the end of embryonic development in *Drosophila*, but the segments of the thorax and abdomen can be seen quite clearly when the outer cuticle of the older embryo is examined under the microscope, as shown in Figure 15.15. Nüsslein-Volhard and Wieschaus used such preparations of the embryonic cuticle as a read-out of embryonic patterning in their screens.

In addition to the visual characterization of embryonic development, classical embryology experiments involving the removal and transplantation of embryonic regions had been conducted in insects. Based on these experiments, it was known that insect embryos contain factors that direct certain aspects of embryonic development and patterning. However, the exact molecular and genetic nature of these factors was not understood. Nüsslein-Volhard and Wieschaus sought to bring developmental biology and genetics together to discover the genes that encode these factors and direct these developmental events.

KEY POINT The segmentation pattern of *Drosophila* embryos provided the phenotypes for one of the most influential genetic screens.

An unbiased screen to find segmentation mutants

Using much of the same logic as Beadle and Tatum, Nüsslein-Volhard and Wieschaus reasoned that the disruption of a gene needed for normal embryogenesis could alter the pattern of segmentation. To achieve this disruption, they used the mutagenic agent ethyl methanesulfonate (EMS), a widely used mutagen in invertebrates that usually produces alterations in single genes. In addition, because flies are diploids, individual in-bred lines had to be created for each of the newly induced mutations to examine the effect in homozygous mutants. Furthermore, since inactivation of important developmental genes is likely to be lethal, the mutations had to be maintained in heterozygous individuals until they could be screened in the homozygous mutant embryos, resulting from a cross between two heterozygotes. This type of cross is diagrammed in Figure 15.16.

As we have mentioned, the scale and efficiency of the screening process is of utmost importance. In fact, the details of this screen took at least 2 years to plan, and some of the key considerations focused on how to achieve the scale necessary for reliable results. Nüsslein-Volhard and Wieschaus realized that there were going to be many genes involved and that they needed to identify as many of these genes as possible to really start to understand the process of embryonic patterning. To achieve this, they devised methods to efficiently create and maintain many lethal mutations and to track the chromosome carrying the mutation from one generation to the next. For example, they made use of special dominant temperature-sensitive mutations, described in more detail in Section 15.4 and indicated by *DTS-91* in Figure 15.16, so that briefly increasing the incubation temperature killed unwanted flies in their crosses. They also used balancer chromosomes (described in Tool Box 9.1 and shown in Figure 15.16 as *CyO*), so that the newly created mutations remain on the original chromosome and do not recombine. A different balancer chromosome was used for the X chromosome, chromosome 2, and chromosome 3.

Taken together, these features made it possible for Nüsslein-Volhard and Wieschaus to focus on one chromosome at a time and systematically and efficiently screen through the thousands of mutations created on that chromosome, that is, the mutagenized flies had mutations in many genes throughout the genome, but the mating used in the screen allowed them to focus on only those mutations found on one particular chromosome at a time.

KEY POINT Strategies to identify and maintain mutations in different types of genes are essential for a good genetic screen.

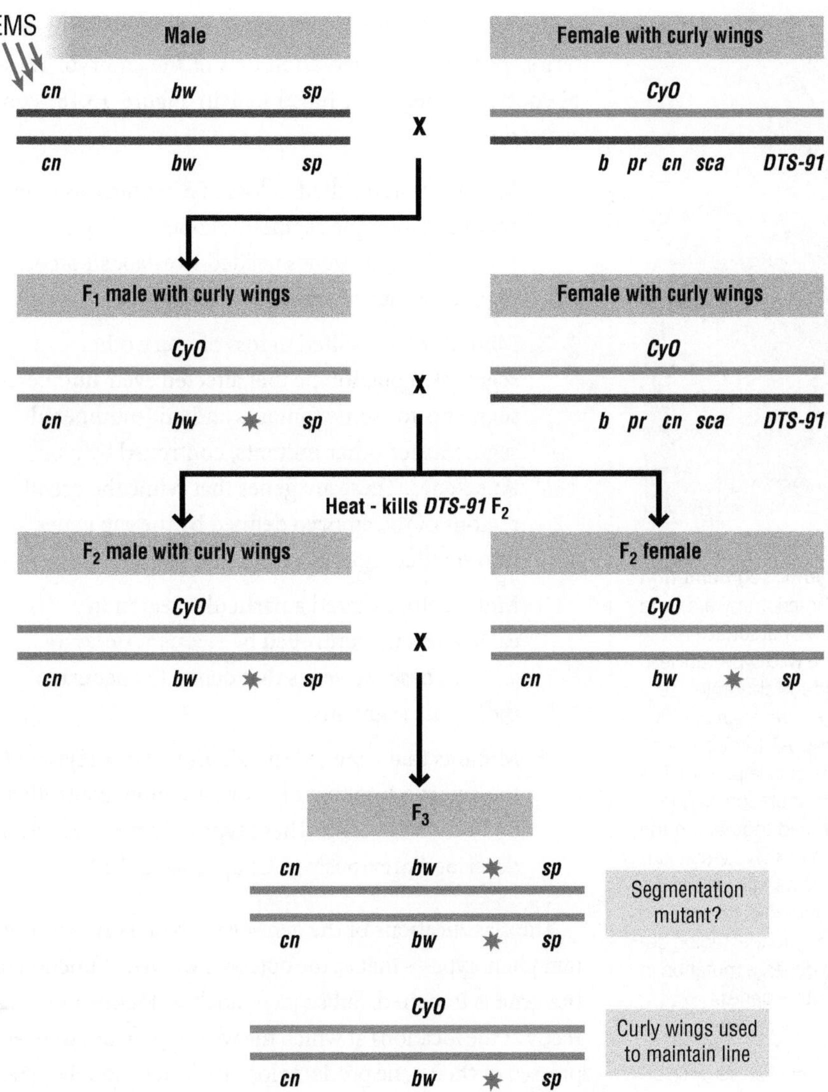

Figure 15.16 Crosses to find mutations affecting segmentation. Since mutations that affect embryonic segmentation are lethal, they must be maintained as heterozygotes. In *Drosophila*, heterozygotes are maintained through the use of a balancer chromosome, which does not cross over with its homologue. These are described in Tool Box 9.1. A balancer chromosome for the second chromosome, called *CyO*, was used in the screen for segmentation mutations, as shown here. *CyO* has the dominant marker Curly wings (*Cy*) and is homozygous lethal, so that only *CyO* heterozygotes survive. A male fly with three easily scored recessive markers (*cn*, *bw*, and *sp*) was mutagenized with ethyl methanesulfonate (EMS) to induce new mutations and then mated to a female heterozygous for *CyO*. The mutagenized chromosome is drawn in purple. In the next generation (F_1), the researchers took individual males with Curly wings. These males were possibly heterozygous for a mutation of interest, represented by the red star, and they were crossed individually to females. These crosses established inbred lines from each male. The temperature was increased to kill any larval offspring carrying the *DTS-91* allele. Since *CyO* homozygotes are also lethal, the only surviving flies were those carrying the mutagenized chromosome inherited from the father and the balancer *CyO* from the mother. These F_2 siblings were left to mate with each other, and the next generation was examined to see whether the mutagenized chromosome was homozygous lethal. All lines that showed homozygous lethality were examined for phenotypes during embryonic patterning, and sibling flies heterozygous for the mutation were used to maintain the stock.

Clear, well-defined phenotypes

Since the outcome of embryonic development is a segmented larva, and not an adult fly, Nüsslein-Volhard and Wieschaus refined methods to isolate and visualize the outer cuticle of the larvae that arose from these mutant embryos. The characteristic patterns of tiny hairs, called denticles, on the cuticle could be used as an indicator of successful developmental patterning, as shown in

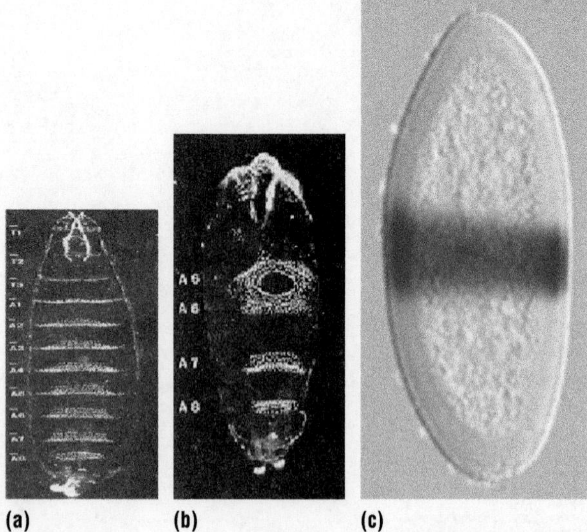

Figure 15.17 Denticle patterns and embryonic segmentation phenotypes. Embryos are oriented with the anterior upwards and looking at the ventral surface (a and b) or with ventral surface to the left (c). (a) shows the patterns of denticles on the wild-type cuticle in a dark-field micrograph. Note that each thoracic and abdominal segment has a distinct pattern. (b) In the mutation known as *Krüppel* (*Kr*), the thoracic segments and abdominal segments A1 through to A5 are missing, creating a characteristic "gap" in the cuticle pattern. The mutant *Kr* phenotype is characteristic of a gap mutation, suggesting that the wild-type gene is needed over a broad segment in the center of the embryo. (c) The expression pattern of *Kr*. Note that *Kr* is expressed in a broad band corresponding to the middle part of the embryo. The expression pattern is similar to the region of the embryo affected by the mutant. Mutants have defects in a broader region because *Kr* regulates the expression of other gap genes; a mutation in *Kr* also affects the segments specified by these other genes.

Source: Parts (a) and (b) reproduced from Nüsslein-Volhard, C. and Wieschaus, E. Mutations affecting segment number and polarity in *Drosophila*. *Nature* 287, 795–801 (30 October 1980). Part (c) reproduced from Haecker, A. (2007) *Drosophila* Brakeless Interacts with Atrophin and Is Required for Tailless-Mediated Transcriptional Repression in Early Embryos. *PLoS Biol* 5(6).

Figures 15.15 and 15.17(a). By contrast, much of the pioneering genetics had been previously focused on the adult fly. The decision to use the larval cuticle was therefore a significant insight that exemplifies the importance of a clearly defined phenotype and assay method.

The next important realization concerned the need to accurately identify discrete and reproducible phenotypes specifically affecting the patterning of the embryonic cuticle. Nüsslein-Volhard and Wieschaus had keen powers of observation and the expertise necessary to interpret the mutant phenotypes they were seeing. They used a dual-headed microscope that made it possible for them to look at a specimen together at the same time, talking through the phenotypes and the significance of their observations.

During the screening process, Nüsslein-Volhard and Wieschaus recognized four distinct categories of mutant phenotype that lacked the correct patterning of the anterior–posterior segmentation axis of the embryo. These phenotypic categories, illustrated in Figure 15.18, consisted of:

1. Mutants that resulted in loss of a contiguous region spanning multiple segments, controlled by **gap genes**. These are genes needed to establish broad regions of the pattern of the embryo.
2. Mutants that resulted in loss of every other segment, a phenotype that affected even-numbered segments for some mutants and odd-numbered segments for other mutants, controlled by **pair-rule genes**. These are genes that refine the broad regions of the embryo defined by the gap genes into smaller regions.
3. Mutants that altered a particular region in each segment, controlled by **segment polarity genes**. These are genes that define the pattern of individual segments.
4. Mutants that changed the identity of one region of the embryo into that of another region, controlled by **homeotic genes**. These types of genes have been described previously in Chapters 3 and 12.

The classifications of the genes were based on their mutant phenotypes—that is, the outcome when the function of the gene is impaired. Subsequent analysis looked more directly at the locations at which the wild-type genes were expressed or their gene products localized. Remarkably, when the expression patterns of various genes in each category were analyzed later, the same patterns were found as seen for these mutant phenotypes, strongly supporting their interpretation of the roles of the different classes of genes.

KEY POINT The key to interpreting a mutant phenotype is often a thorough understanding of the normal organism or, as it has sometimes been called, a "feeling for the organism."

Assigning mutations to genes

The screen itself was an impressive tour de force, but the follow-up analysis required even more work, as can be illustrated by some of the statistics. For chromosome 2, Nüsslein-Volhard and Wieschaus screened about 4600 lines and identified 321 embryonic lethal mutations. Similarly, they identified 122 mutants on the X chromosome (known as chromosome 1 in *D. melanogaster*) and 198 on chromosome 3, totaling 641 different mutations affecting embryo patterning.

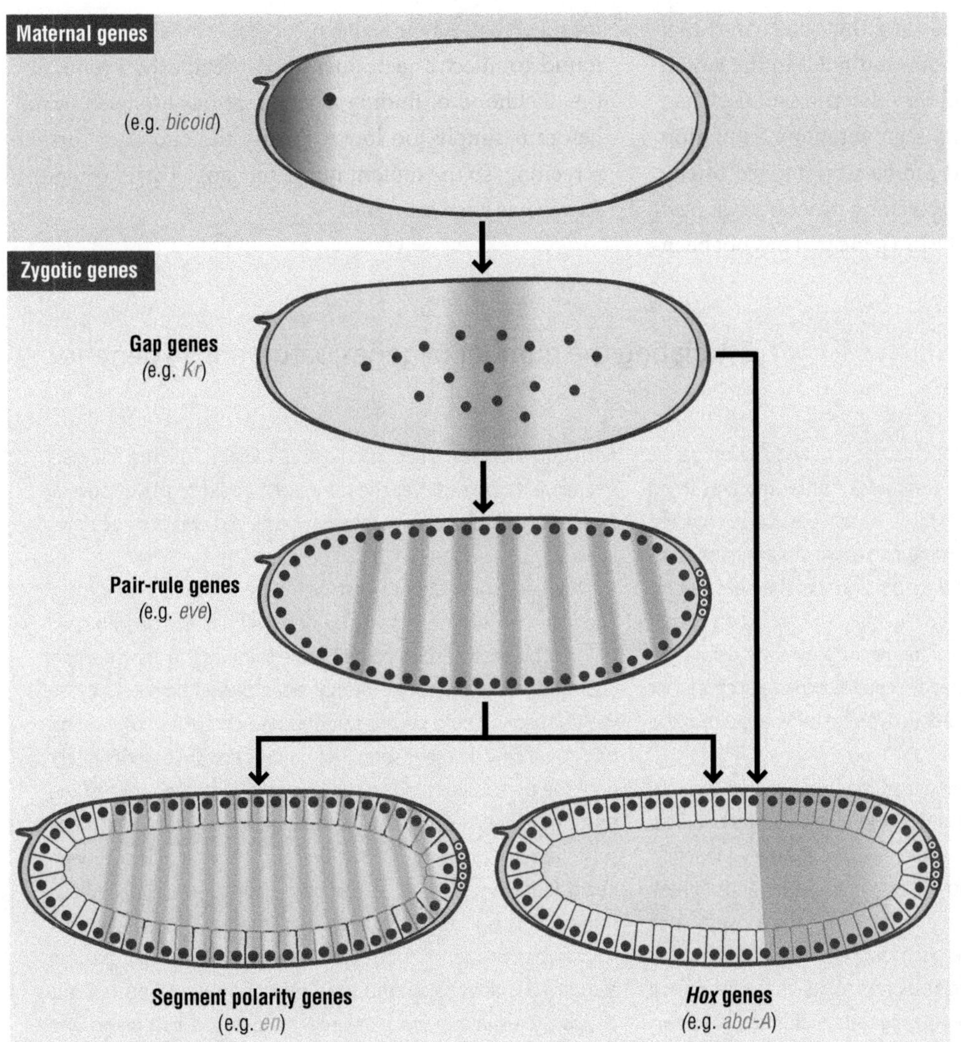

Figure 15.18 Gap genes, pair-rule genes, and segment polarity genes. Different categories of mutant phenotypes were found, as shown here. The mother contributes some gene products in the ovum before fertilization, such as *bicoid*, which is localized to the anterior end of the prospective embryo. Upon fertilization, the rest of the pattern of segmentation is set up by the progressive actions of groups of genes expressed in the embryo (zygotic genes). The gap genes, of which *Krüppel* (*Kr*) is an example, regulate broad and contiguous regions of the embryo. These regions are subdivided by the pair-rule genes, represented here by *even-skipped* (*eve*). Note that *eve* is expressed in seven bands, which correspond to alternating segments later in embryogenesis. The regions defined by *eve* and the other pair-rule genes are then further specified by the segment polarity genes, of which *engrailed* (*en*) is an example. These divide the pairs of segments set up by the action of *eve* and the pair-rule genes into the 14 segments found later in embryogenesis, each with its own orientation and boundaries. The gap genes and the pair-rule genes also regulate the expression of the *Hox* genes (of which *abd-A* is an example) that specify the identities of each segment.

But how many different genes were represented by this collection of mutants? The answer could be determined through complementation tests, and only those located on the same chromosome needed to be tested for complementation. Nüsslein-Volhard and Wieschaus also focused on mutants with similar phenotypes when performing the test crosses, reasoning that mutants with distinctly different phenotypes are less likely to be carrying mutations in the same gene. When the complementation tests were completed, 121 different genes had been identified.

To have identified and characterized 121 different genes needed to correctly pattern the developing embryo was a significant achievement. But do these genes represent most of the genes involved in segmentation or only some of the genes? How did Nüsslein-Volhard and Wieschaus know when they had identified nearly all the genes they were likely to find using this screen and that they could stop searching for new mutants? In the jargon of genetic screens, how could they determine if they had **saturated** the genes affecting segmentation? Saturation is described in Box 15.2 and can be summarized briefly as follows. At the start of the screening process, each new mutant identified is likely to be in a gene that has not been found before. As the screen proceeds, some mutations are found to affect a gene identified previously. Eventually, the likelihood of finding a gene that has not been found before is simply too low to justify the efforts of further screening, so the mutant hunts can stop. This is when the screen has been saturated.

BOX 15.2 *Quantitative Toolkit* Estimating the number of genes: saturation in genetic screens

As they build up a collection of mutations, how do investigators know when they should stop testing for further mutants in an attempt to identify new genes? It is always possible to run the mutagenesis screen one more time, find more mutants, and characterize them. There is always the suspicion that the next mutant could be the most interesting one in the entire collection or that it will identify a crucial gene that has never been seen before. On the other hand, each new mutation requires additional effort that could have been spent more productively on other aspects of the ongoing genetic analysis.

At some point, most of the mutations found are new alleles of previously identified genes and add little new information. This point is known as saturation—when all of the newly identified mutations are alleles of previously identified genes, and no new genes can be found by this method. But how can an investigator recognize when saturation has been reached?

As mutations are isolated and assigned to genes, the frequency with which mutations become assigned to the same known genes can be used to estimate how many additional genes could be found. We cannot know the total number of genes, and the genes that have been identified will inevitably represent only a sample of those that could be found. Therefore, predicting the point of saturation—that is, estimating the number of genes that can be found by a particular mutagenesis procedure—is an inherently statistical concept.

We can illustrate this approach with an analogy. Suppose that I find an old MP3 player, but I do not know how many songs are stored on it. I want to know how many songs are in the library by measuring the frequency at which a given song is played. Like many MP3 players, this one has a function that will select and play a song at random from the library and will count how many times that song has been played. However, the random selection on my hypothetical MP3 player is different from most commercial players in that songs are returned to the library once they are played. (Most actual MP3 players generate a playlist when the random shuffle feature is selected. Thus, once a song is played, it is effectively removed from the library until a new playlist is generated. In statistical terms, this is referred to as sampling **without** replacement. On my hypothetical MP3 player, we are sampling **with** replacement since each song remains on the playlist.) Therefore, with my player, the same song could be played several times while I listen, and might even be played two or three times in a row.

Now let's think about what will happen when I click "play." The first song will necessarily be one that has not been played before. The second song could be the same as the first one, or it could be a new song that has not been played before. Likewise the third song could be the same as one of the first two songs, or it could be a different song. The process continues indefinitely, until the person listening to the song list is satisfied that all the songs have been heard or at least that he can estimate how many songs are stored on the MP3 player. Some songs will have been heard many times, other songs only a few times, and perhaps some songs will not have been played at all. Nonetheless, by knowing how often a song has been played and making some assumptions about how the song selection process occurs, it is possible to estimate how many songs are on the MP3 player. The procedure is saturated when additional song selections will not provide additional information about the total number of songs in the library.

We can express this situation mathematically. Assume that there are N songs in the library. The probability that the first song is the one that has not been played previously is 1. The probability that the second song is the same as the first one is 1/N, and the probability that it will be a different song that has not been played is 1 – 1/N. We can express this more generally if we define h as the number of different songs that have been played before each selection is made. With each selection, the probability of playing a song that has not been heard previously is (N – h)/N.

Let's illustrate this with some numbers. On the first selection, no songs have been heard previously, so h = 0, and the probability of playing a song that has not been played before is N/N = 1, regardless of the size of N. Suppose that N = 3, that is, there are three songs in my music library. After the first selection, h = 1 because one song will have been played. The probability that the second selection will play a different song is (3 – 1)/3 = 2/3. If the second song is different from the first song, then h = 2. The probability that the third selection will play a different song is (3 – 2)/3 = 1/3.

BOX 15.2 Continued

In other words, as the number of selections increases, h gets closer to the value of N, and so N – h approaches zero—most selections will pick a song that has already been heard. Although we do not know N, we can estimate it by using h and the change in h as the number of songs played increases. These are the statistics behind the concept of saturation. This is also the statistical basis of refraction curves used to sample genomes in a population, as discussed in Chapter 17.

Assumptions of a mutagenesis screen

There are several important assumptions in our model for estimating the number of genes, including:

- Every mutation can be identified as an allele of some gene
- Mutations are being induced at random with respect to the genes, without regard to those that have occurred previously; this indicates that mutations can be found in any gene without an intentional bias introduced by the investigator
- All genes are equally mutable.

Let's look at one of these assumptions in a bit more detail—the assumption that genes are equally mutable.

While it seems likely that every mutation can be identified as an allele of some gene and that mutations are being induced more or less at random, most data on large mutant screens indicate that not all genes are equally mutable. Most typically, making this assumption will underestimate the number of genes. For example, data from mutations on chromosome 2 affecting *Drosophila* segmentation, carried out by Nüsslein-Volhard and Wieschaus and tabulated in Figure A, show that one gene is hit 18 times, another one is hit 17 times, and a third is hit 15 times, and yet 13 genes are hit only once. Even without knowing much about frequency distributions, it seems very unlikely that the genes have the same probability of being mutated. Intuitively, it seems clear that the genes hit so much more frequently must somehow present a better target for mutagenesis or that mutations in these genes are more easily recognized and recovered.

Saturation and the *Drosophila* mutagenesis screen

Having examined the underlying assumptions that went into our model, what did the investigators observe and how did they determine if they were near saturation? Among the other attributes of their mutagenesis screen, Nüsslein-Volhard and Wieschaus came closer to saturation than most other geneticists had done

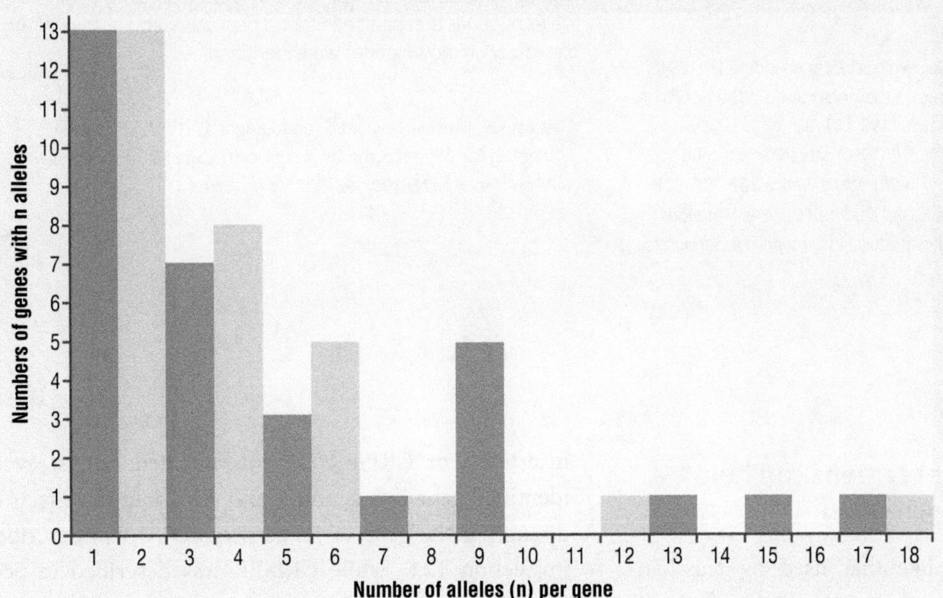

Figure A The distribution of alleles found per gene on chromosome 2 affecting *Drosophila* segmentation. The X-axis shows the number of alleles or hits found per gene, from a single allele to 18 alleles. The Y-axis plots the number of genes with that number of alleles. For example, only one allele was found for 13 genes, and two alleles were found for another 13 genes. Note that 18 alleles were found for one gene, and 17 alleles for another gene. Thus, the assumption that every gene will have the same number of hits is clearly an over-simplification.

BOX 15.2 Continued

for genetic screens of this type. One of the greatest features of this work is the extent of the collection of mutants and the number of genes, which have provided stimulating and informative research projects for many geneticists for almost 30 years.

In order to determine if they were near saturation, Nüsslein-Volhard and Wieschaus plotted the number of new mutations found and the number of new genes found as a function of the number of mutagenized lines. This is reproduced in Figure B and corresponds to a plot of N – h, the frequency with which a new gene is found with each additional mutation.

The data show that more than 5000 chromosomes were scored and the number of mutants increased linearly with the number of chromosome tested. In other words, they were continuing to find new **mutations** each time they did the screen. In contrast, the number of **genes** increased steadily until about 2000 chromosomes had been tested, about 120 mutations had been recovered, and about 50 different genes had been found. Then the number of genes increased much more slowly than the number of mutations, such that, among the final 1500 chromosomes, almost no new genes were found, and N – h was close to zero. This indicated that nearly all the genes that could be found by this method had probably been found. We have the benefit of hindsight to examine this conclusion. More than two decades have passed since these studies were published, and few, if any, additional segmentation genes have been found.

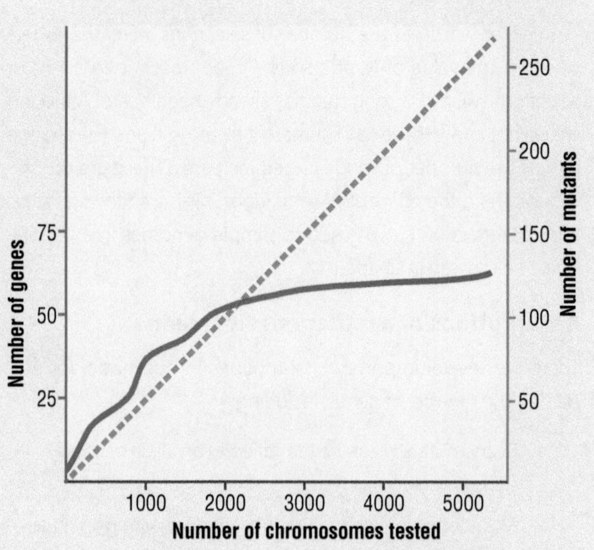

Figure B Estimating the number of genes affecting segmentation. These two curves plot the number of mutants found and the number of genes that these mutations defined as more chromosomes are screened. The number of chromosomes tested is shown on the X-axis. The number of mutants shown as the dashed blue line increases steadily with the number of chromosomes screened; in other words, the more chromosomes they examined, the more mutations they found. The red line shows the number of genes that these mutations fell into. Note that the curve plateaus after about 2000 chromosomes had been screened. While more mutations continued to be found, these mutations were primarily additional alleles of genes that had been found previously, and relatively few previously unknown genes were identified.

FIND OUT MORE

Jurgens, G., Wieschaus, E., Nusslein-Volhard, C., and Kluding, H. (1984) Mutations affecting the pattern of the larval cuticle in *Drosophila melanogaster. Wilhelm Roux Archiv* **193**: 283–95

Nusslein-Volhard, C. and Wieschaus, E. (1980). Mutations affecting segment number and polarity in *Drosophila. Nature* **287**: 795–801

Nusslein-Volhard, C., Wieschaus, E., and Kluding, H. (1984). Mutations affecting the pattern of the larval cuticle in *Drosophila melanogaster. Wilhelm Roux Archiv* **193**: 267–82

Wieschaus, E., Nusslein-Volhard, C., and Jurgens, G. (1984) Mutations affecting the pattern of the larval cuticle in *Drosophila melanogaster. Wilhelm Roux Archiv* **193**: 296–307

Genome-wide mutant screens can help ensure all genes are screened

In mutagenesis screens, like that used by Nüsslein-Volhard and Wieschaus, the mutations are being introduced at random and so run the risk of not finding all the genes involved in a process. The availability of the complete DNA sequence, which encompasses all of the genes in the genome, paves the way for other approaches that help to ensure that no genes are overlooked. For example, researchers can make libraries of RNAi, deletions, insertions, or CRISPR reagents targeted against every identified gene in a genome, and ask about the effects of disrupting the gene or its activity. RNAi was described in Section 12.6, while CRISPR was described in Section 11.6. Tool Box 15.1 discusses **genome-wide mutant screens** in more detail.

While genome-wide mutant screens are unbiased in the sense that every gene is targeted, they do have biases of their own. For example, RNAi induces loss-of-function phenotypes but affects different genes differently; deletions in *Drosophila* usually remove many genes all at once, and some phenotypes are simply easier to see than others. However, genome-wide mutant screens can be done for many organisms and have often become the method of choice for finding the mutant phenotypes to understand a process.

KEY POINT A screen is considered to be saturated when all potential genes that are likely to be found with that approach have been found. Genome-wide mutant screens use the genome sequence to systematically target mutations to each gene, ensuring that no genes go unscreened.

Nüsslein-Volhard, Wieschaus, and Lewis (who studied homeotic genes) were awarded the Nobel Prize in 1995 for their discoveries about the genetic control of embryonic development in *Drosophila*. The value of model organisms is well illustrated by these studies, since many of the genes in this screen are found in most animals. The *Drosophila* studies elucidated many basic genetic pathways and conserved regulatory systems found to be important in the development of other animals, including humans. The genes have also been crucial in comparing the embryonic development of different organisms to understand the evolution of body forms, and many have also been implicated in human genetic diseases, including cancers.

WEBLINK: We explore this groundbreaking genetic screen in more detail in Video 15.1. Find it at **www.oup.com/uk/meneely**

15.4 Ordering genes into pathways

A genetic screen for mutants or a genome-wide mutant screen identifies genes that are involved in a cellular process but does not by itself indicate how these genes work together to carry out this process. An important feature of genetic analysis is its ability to reveal the underlying cellular and molecular logic of a biological process, particularly when the details are not known. In other words, mutants and their phenotypes can be used to place genes into a pathway.

Elucidating genetic pathways can help us understand metabolic pathways

It may be helpful to compare a genetic pathway with a metabolic pathway in which the product of one step becomes the substrate for the next step. In some cases, the two types of pathway are very similar. In the work of Beadle and Tatum in Section 15.1, for example, the one-to-one correspondence between the mutants and the metabolic pathway for arginine biosynthesis allowed them to postulate that each of the genes encoded one of the biosynthetic enzymes. They proved this correlation by using the substrate of one reaction to bypass the mutant defect for the genes, and so placed the genes in the order in which their products acted.

While this is a powerful illustration of how genes can be ordered into a pathway, it is also not the most common situation that is encountered with mutants. We usually do not begin a mutant screen knowing the appropriate pathway or interactions. For example, the organization of the *lac* operon, with the repressor, the operator, and the genes for the necessary enzymes, could not have been solved by this method. Nonetheless, by applying logic derived from the interactions of the mutants, it was possible to work out the cellular logic of lactose utilization.

The screen for segmentation mutants in *Drosophila* also could not have been solved by adding back substrates. The biochemical and molecular pathways were completely unknown; no appropriate biochemical substrates had been identified, and too many genes were found to make for a simple analysis. In this case, the logical pathway of *Drosophila* segmentation was inferred by a careful analysis of mutant phenotypes. On the basis of the mutant phenotypes, it could be hypothesized that the gap genes acted first to organize the embryo into broad regions; the pair-rule genes acted next to subdivide the embryo into groups of segments, and the segment polarity genes acted last to organize each segment. In other words, the genes could be placed into a logical and functional pathway that suggests the strategy by which genetic and molecular processes shape the segmentation pattern. The biochemical or molecular mechanisms by which a gap gene regulates a pair-rule gene were not known; in fact, precisely which gap gene regulated which pair-rule gene was not clear. This knowledge was not necessary for the first insights into the strategy of segmentation. But the value of ordering genes became apparent once the molecular nature of various genes was uncovered, in a satisfying confirmation of the logical pathway inferred from the mutant phenotype.

KEY POINT Genes can be ordered into functional pathways based on their mutant phenotypes, even when the underlying biochemical and molecular mechanisms are not known.

In this section, we will briefly discuss two fundamentally important cellular pathways in eukaryotes—the cell cycle and the Ras signal transduction pathway. Rather than describe the pathways in detail first, we will approach these as the investigators began them—with mutants that did not carry out the processes normally. Each analysis provides not only another example of the power of genetic analysis for understanding complicated cellular processes but also another genetic tool that was important in the analysis.

Elucidating the genetic regulation of the cell cycle using temperature-sensitive mutants

Eukaryotic cells exhibit a well-regulated cell cycle, shown in Figure 15.19, with distinct G_1, DNA synthesis (S), G_2, and mitosis stages. The genetic regulation of the cell cycle was worked out initially by Lee Hartwell and his students. When Hartwell and his co-workers began their research on the yeast cell cycle, it was not clear if each stage began at a set period after time zero, and were thus independent of one another, or if each new stage could begin only when the preceding phase was completed. To put this in terms of pathways—was each stage of the cell cycle a separate genetic pathway, or could the entire cell cycle be described by a single or a few pathways with steps dependent on each other? Or, to use a more familiar analogy, suppose that we have to assemble a complex item of furniture from the various pieces in the packing crate. Which parts can be assembled independently of other parts, and which steps have to be completed before others can be done?

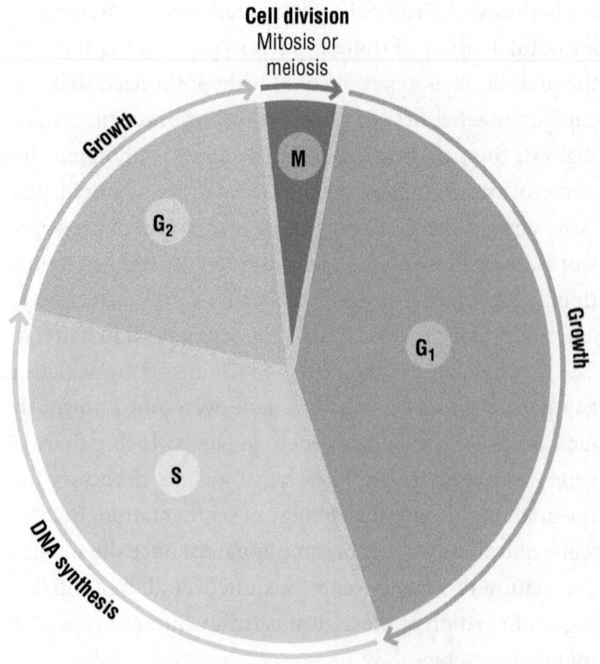

Figure 15.19 The cell cycle in eukaryotic cells.

As with many mutant screens, one of the keys to the success of Hartwell's genetic analysis of the cell cycle was the choice of the organism to be used in the study, in this case the yeast *Saccharomyces cerevisiae*. Yeast can grow as single haploid cells, thereby simplifying the experimental design and the variety of possible phenotypes, while still revealing the workings of a basic eukaryotic cellular process. Because yeast cells can grow and divide as haploids, it means that a single mutant allele is expressed as a mutant phenotype. However, since the cell cycle is essential—it is the process by which cells divide—and mutants affecting it are expected to be lethal, Hartwell needed a genetic tool to work with lethal mutations affecting essential genes.

Yeast can be grown also as diploids. In principle then, Hartwell could have done his experiments in diploids and maintained his cells as heterozygotes. However, this would have made the screen and the analysis more complicated, so Hartwell used a different method to maintain his lethal mutations in haploid strains. This method also provided some other advantages that he used in the analysis of his mutants. Hartwell knew from previous studies in bacteriophage and bacteria, as well as in some eukaryotes, that some mutations have temperature-dependent or **conditional** phenotypes. When grown at low temperature, the cells have a wild-type phenotype and grow and divide normally. A mutant phenotype is observed when the cells are grown at high temperature. The temperature at which the mutant phenotype occurs is termed the **restrictive temperature**, while the temperature at which the process proceeds normally is referred to as the **permissive temperature**. This is summarized in Figure 15.20. Growth temperature is the condition that allows the investigator to observe the mutant phenotypes.

KEY POINT Temperature-sensitive mutations show a mutant phenotype only under restrictive conditions. They can be maintained with a wild-type phenotype under permissive conditions.

Temperature-sensitive mutants thus have the advantage that an otherwise lethal mutation can be maintained simply by growing the organism at the permissive temperature. **Temperature-sensitive mutations** have been found for many genes in many organisms, and, while they

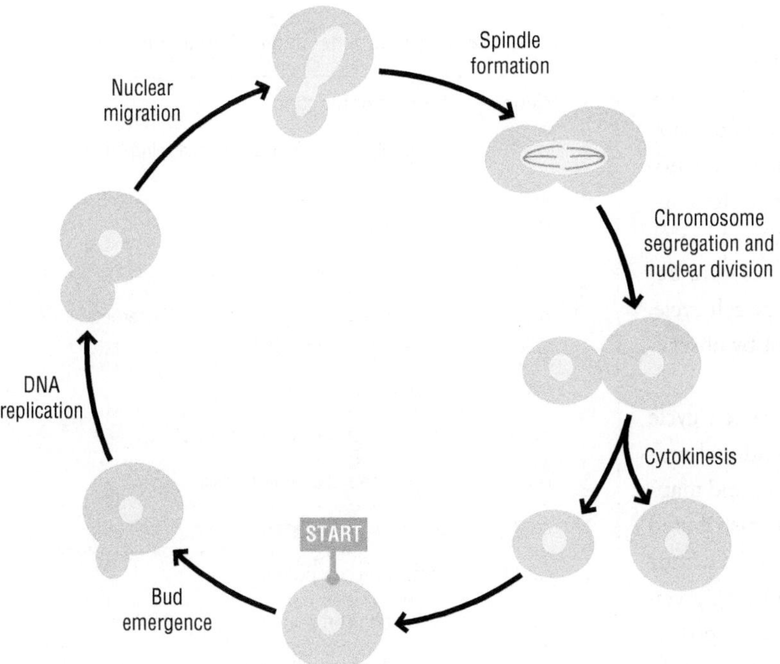

(a) **Permissive temperature: cells grow and divide**

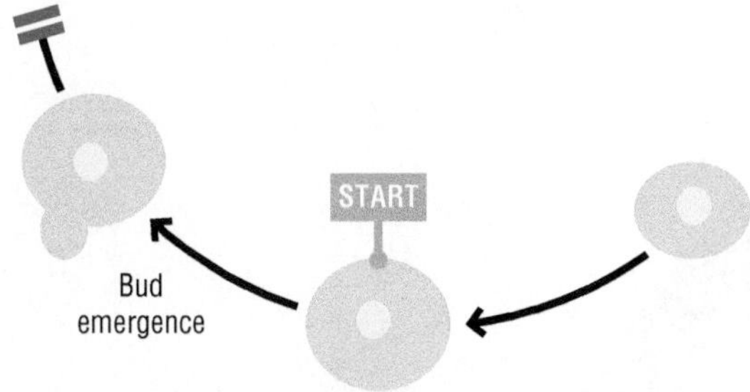

(b) **Restrictive temperature: cells arrest**

Figure 15.20 Temperature-sensitive mutations. Mutations that affect essential genes are often lethal. Lethal mutations can be maintained as heterozygotes, as was done for the segmentation mutations, or as conditional mutations in which the mutant phenotype is seen only under certain conditions. Common examples of conditional mutations are temperature-sensitive mutations, as used in analyzing the cell cycle in yeast. (a) At the low (or "permissive") temperature, the cells grow and divide normally. (b) When the cells are grown at the high (or "restrictive") temperature, the cell cycle arrests and the mutant phenotype is observed; in this example, the mutant causes arrest shortly after bud formation. In addition to allowing lethal mutations to be maintained, temperature-sensitive mutations can be used to analyze the time at which the function of a gene is needed.

are the most widely used example of conditional mutations, they are not the only type. Other types of conditional mutants that have been created display the mutant phenotype only when a particular drug is present or when a co-factor is expressed. A few genes have cold-sensitive mutations, in which the mutant phenotype is exhibited when the cells are grown at low temperature but not when the cells are grown at higher temperatures in which case the restrictive temperature is lower than the permissive temperature. But generally speaking, temperature-sensitive mutations are the most commonly used class of conditional mutants, with the restrictive condition being the high temperature.

Mutant phenotypes used to characterize the yeast cell division cycle

S. cerevisiae divides by budding, with the daughter cell forming on the side of the parental cell. The process of budding has characteristic steps that can be identified either visually, by observing the formation of a mitotic spindle or cytokinesis, as shown in Figure 15.21, or biochemically, by monitoring the replication of the DNA, for example. Hartwell referred to the recognizable steps in cell division in budding yeast as **landmark events**. He and his students screened for mutants that grew normally at a permissive temperature of 22°C but showed a phenotype affecting one of the landmark events when the yeast cells were shifted to the restrictive temperature of 36°C. For example, the cells at the restrictive temperature became stalled or arrested at specific points in the cell cycle, and researchers could infer the point of arrest by observing the landmark event.

Hartwell and his colleagues identified 150 cell cycle mutations in their original studies with conditional mutants in yeast. Following complementation tests and mapping, this set of mutations was found to be distributed across 32 different genes. These genes were named *cdc* for cell division cycle defective, with each gene being given a unique identifying number (*cdc1*, *cdc2*, and so on). A selection of mutants and genes identified in this manner and ordered according to the point of arrest with which they correlate is shown in Table 15.2.

Table 15.2 Some *cdc* genes and their functions

Gene	Product or function
cdc2	DNA polymerase δ involved in lagging-strand synthesis
cdc9	DNA ligase
cdc6	Part of pre-replication complex
cdc4	Part of ubiquitin ligase complex that targets proteins for degradation
cdc13	Telomerase regulation
cdc14	Phosphatase needed for mitotic exit
cdc15	Kinase needed for mitotic exit
cdc3	Septin ring component
cdc10	Septin ring component
cdc11	Septin ring component
cdc28	Cyclin-dependent kinase responsible for START

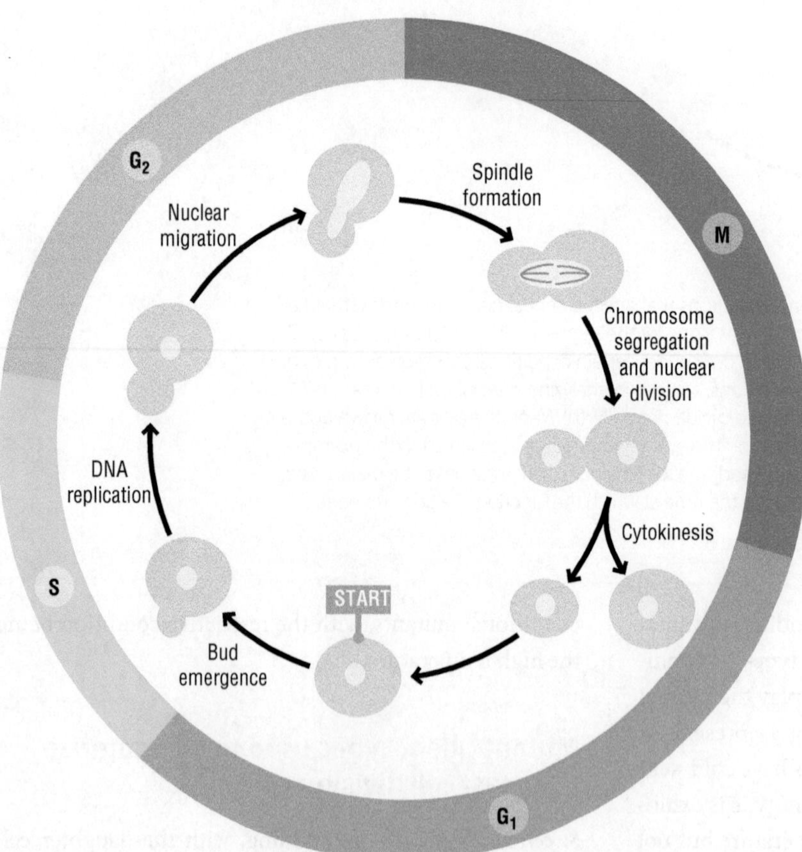

Figure 15.21 The correspondence between the general eukaryotic cell cycle and the budding cycle of the yeast *Saccharomyces cerevisiae*. Characteristic cellular events, known as landmark events, could be used to define and recognize different stages of the cell cycle.

Viewing the process of cell division through the lens of these landmark events also provided a framework for the follow-up analysis of the genes that were identified. A population of yeast cells carrying the same mutation being grown at the permissive temperature will initially have cells at different stages of the cell cycle. However, following a shift to the restrictive temperature, the cells will all arrest at the same landmark event characteristic for that mutant, as shown in Figure 15.22. This suggested that that gene product plays a critical role at a specific step of the cell cycle but might not be needed for the cell to progress up to the landmark event. Shifting from permissive to restrictive temperatures at different points of the cell cycle further defined when the activity of each gene product was needed during cell division, which defined what Hartwell termed the **execution point** for each gene.

KEY POINT Landmark cellular events, such as DNA replication and mitotic spindle formation, provided well-defined phenotypes for the analysis of cell cycle mutants.

We can see from this discussion how the temperature-sensitive phenotypes played two important roles. First, they allowed Hartwell to maintain cultures of cells as haploids at the permissive temperature without a mutant

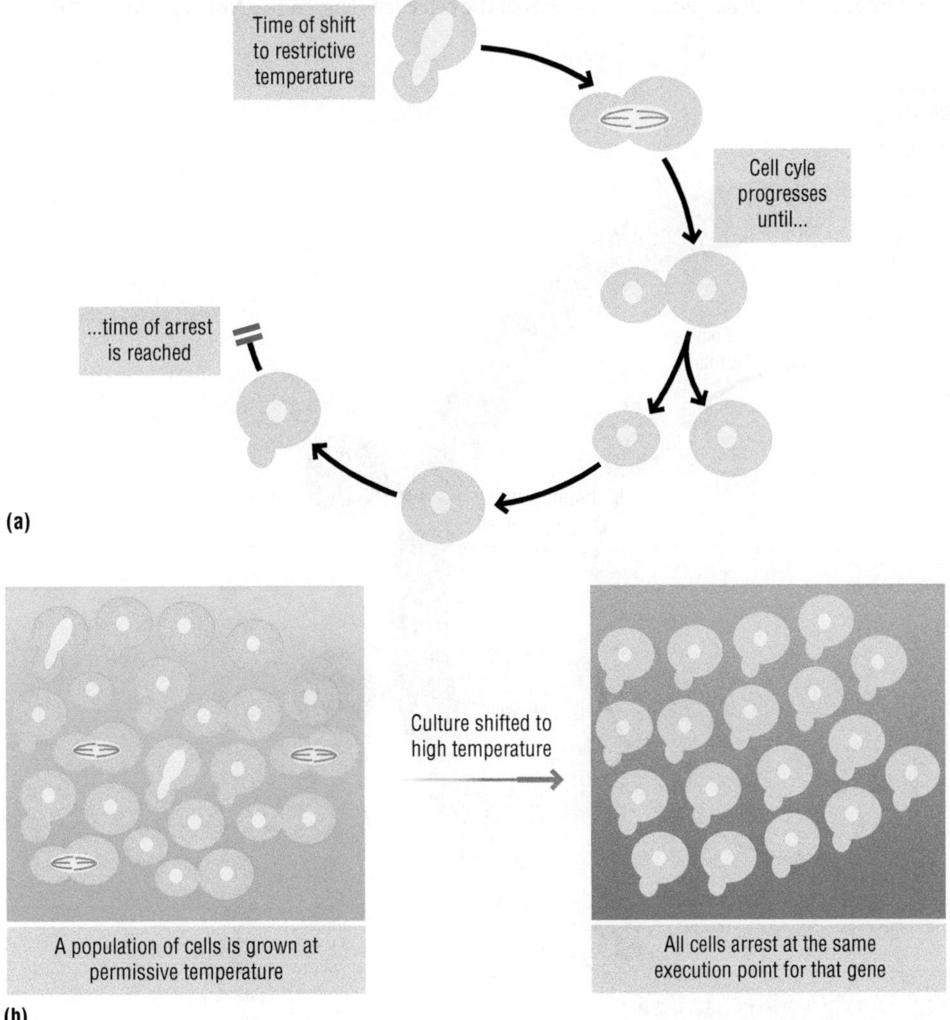

Figure 15.22 Temperature shift experiments and the times of mutant arrest. Temperature-sensitive mutations in genes affecting the cell division cycle, known as *cdc* mutations, were found to have characteristic times of arrest, which could be determined by a temperature shift experiment. (a) When a cell was shifted to the restrictive temperature, it could complete some stages of the cell cycle but would arrest at the time in the next cycle when the gene product was needed. The conclusion for this example is that this gene product is needed after bud emergence but is not needed for the other events of the cell cycle. (b) When a population of cells, originally at different stages of the cell cycle, is shifted to the restrictive temperature, all of the cells arrest at the same cellular event. This implies that this cellular event requires the wild-type function of this gene.

phenotype compromising their growth. Second, they allowed him to determine which genes affected the same landmark events. Thus, by determining when each gene was needed during the cell cycle and correlating the times of arrest with the visible landmark events, he could place them in a functional order or a pathway, as summarized for a few of the genes in Figure 15.23. He could also show which activities were dependent on the successful completion of previous steps and which ones were regulated independently of each other.

The cell cycle mutants and their follow-up analyses have laid the foundation for additional experiments that have led to our current understanding of how cell division in all eukaryotes is controlled. For example, the *cdc28* gene, known as START, is found in all eukaryotes and controls the commitment to entering the cell cycle. It soon became apparent that most of the *cdc* genes have orthologs in all eukaryotes, so the work with temperature-sensitive mutants in yeast became directly applicable to other eukaryotic cells. This work also identified the existence of checkpoints, discussed previously in Chapters 4 and 6, which arrest the cell cycle until the previous steps have been completed. In fact, many cancer cells have mutations in one or more of the *cdc* genes affecting a checkpoint or another aspect of the cell cycle. Hartwell, Paul Nurse, who studied the cell cycle in the fission yeast *Schizosaccharomyces pombe*, and Tim Hunt, who worked out many of the biochemical aspects of the cell cycle, shared the Nobel Prize in 2001.

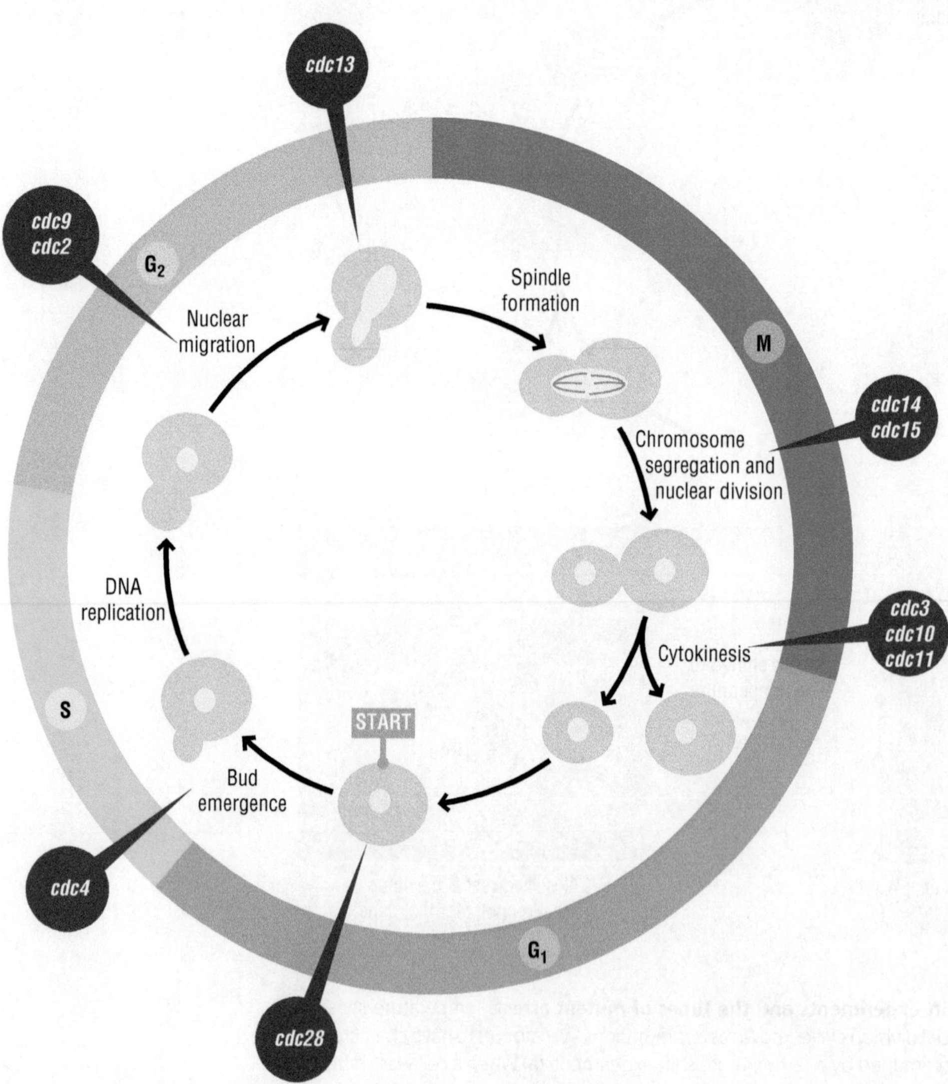

Figure 15.23 Using *cdc* mutations to characterize the cell cycle. Some of the *cdc* mutations used to define events in the eukaryotic cell cycle are shown, with the times when the function of the wild-type gene is needed. For example, the gene *cdc28* defines START, the commitment of a cell to enter the cell division cycle. Most of the *cdc* genes have orthologous genes in all other eukaryotes.

KEY POINT Conditional mutations can also be used to infer an order or a time in which the functions of the genes are needed.

Elucidating the Ras pathway using enhancers and suppressors

The *ras* gene defines one of the most fundamentally important signal transduction pathways in animal and fungal cells. More than a quarter of all human cancers have a mutation in the Ras pathway, and some types of cancer, such as pancreatic cancer, nearly always involve a mutation in the *ras* gene. The pathway is of such fundamental importance that it has become the prototype for the analysis of other signal transduction pathways in eukaryotes, of the sort summarized in Figure 15.24.

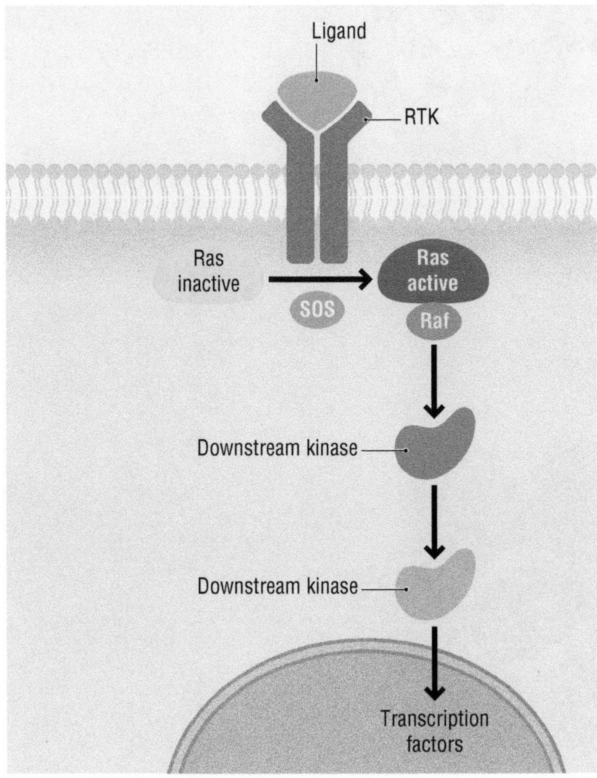

Figure 15.24 An RTK pathway with Ras. This is a general pathway for receptor tyrosine kinases (RTKs), typical of many signal transduction cascades. A ligand or a signal binds the RTK on the surface of the cell. The binding of the RTK converts the inactive form of the Ras protein to its active form; activation of Ras depends on a gene known as *Sos*. The active form of Ras activates another RTK called Raf, which in turn activates a series of kinases downstream. Eventually, these kinases affect the activity of one or more transcription factors to activate the specific transcription of certain genes. The downstream target genes can be numerous and varied, depending on the specific RTK pathway being considered and the cell type.

Receptor proteins are found on the surface of eukaryotic cells and bind to specific signals called ligands. When the ligand binds to the receptor on the cell surface, the receptor relays an activation signal to downstream target proteins within the cell. Ras plays an important role in relaying signals from a particular type of cell surface protein called a receptor tyrosine kinase (RTK). When activated, this type of receptor protein, an enzyme known as a kinase, adds a phosphate group to specific tyrosine residues in its target proteins. When the genetic screens to be described were done, the identities of these downstream target proteins were unknown.

With a pathway of such fundamental importance, it may surprise you to learn that its analysis began with a relatively simple genetic screen based on a very familiar phenotype. You probably know that insects are attracted to light sources; in other words, they exhibit phototaxis. This well-known behavior provided the basis for a genetic screen, done originally by Seymour Benzer and his students, with later analysis by other collaborators.

In essence, *D. melanogaster* were released in a dark box, and a UV light was turned on at one end. (A more complex apparatus was actually used, but this was its underlying design.) As predicted by anyone who has seen a flying insect and a light bulb, the wild-type flies gathered at the end of the box near the UV light. The flies were then mutagenized, mated to produce mutant homozygotes, and released again in the box with the UV light. Those that flew to the light normally were removed from the experiment, whereas those that failed to exhibit phototaxis to the UV light were saved, retested, and examined. The presumption was that these flies must have had a defect in some gene affecting their attraction to UV light.

Many different categories of mutants were found, and many could be quickly disregarded as having non-specific defects in UV light attraction. For instance, some mutants could not fly at all, while others were not attracted to any light source; these could be easily recognized and used for other experiments or discarded. The mutants that exhibited normal attraction to other light sources but lacked the attraction to UV light were analyzed further to identify the genes specifically involved in this behavior.

KEY POINT A simple behavior characteristic of insects—the attraction to light—provided the basis for identifying genes affecting phototaxis.

Drosophila eyes contain special photoreceptor cells

In order to describe the phenotypes of the mutants and the processes they affect, we need to provide some background to the structure of the *Drosophila* eye and light detection. The compound eye of the fly comprises a hexagonal array of hundreds of repeating units called ommatidia, each of which consists of a central grouping of eight photoreceptor cells (named R1–R8) and several associated lens-secreting cone cells, shown in Figure 15.25. The photoreceptor cells R1 through R6 are responsible for the response to blue or visible light, while the photoreceptor cells R7 and R8 are involved in the response to UV light; R8 is also needed for the response to green light, so its role can be distinguished from the role of R7. Since R7 is the cell primarily responsible for detection of UV light, mutants that specifically lack the response to UV are postulated to affect some aspect of the development or function of the R7 photoreceptor cell.

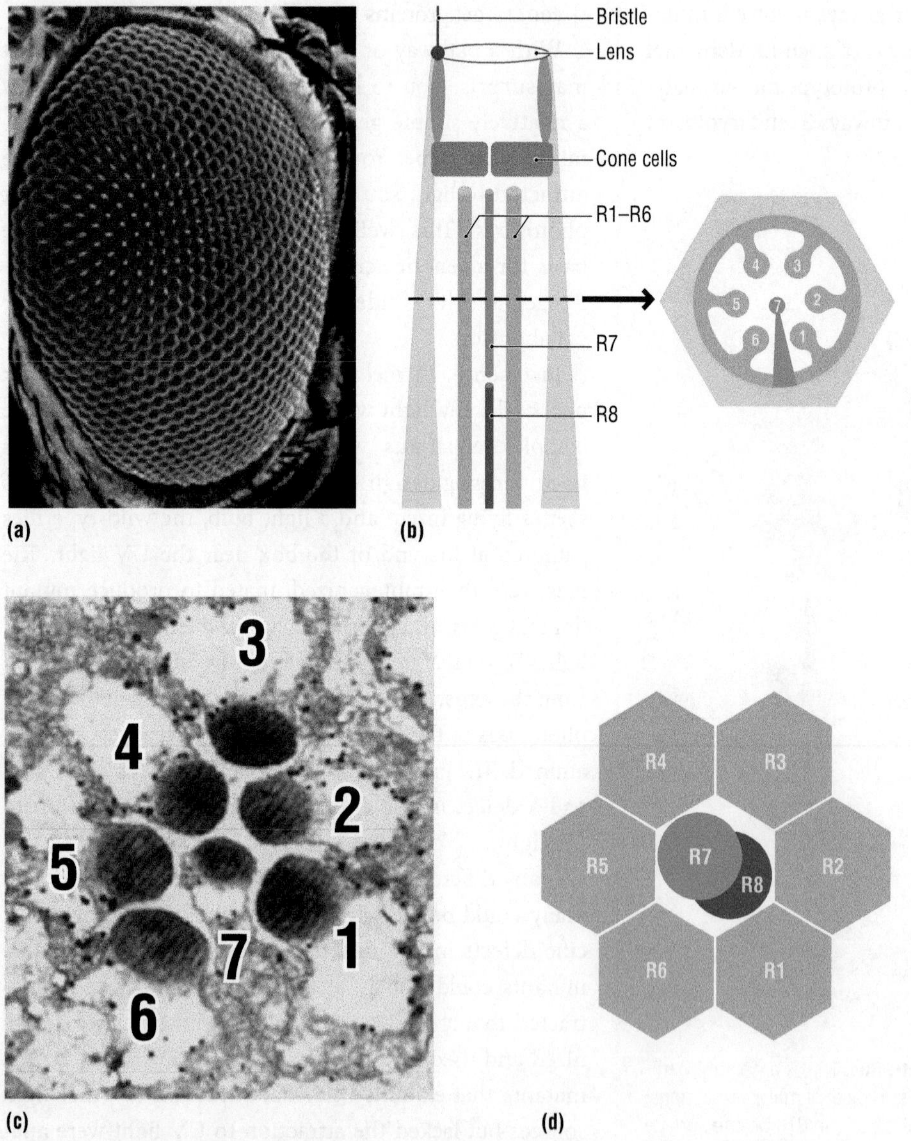

Figure 15.25 The cellular response to light in *Drosophila*. (a) The *Drosophila* compound eye has hundreds of separate light receptors known as ommatidia. (b) The structure of an ommatidium, shown in longitudinal section. Light enters from the top, through the lens. There is a layer of cone cells, with eight photoreceptor cells named R1–R8. The R1–R6 cells respond to visible and blue light, while R7 and R8 are needed for the response to UV light. R8 lies underneath R7; this orientation will be important. (c) An electron micrograph of the photoreceptor cells R1–R7 (with R8 behind R7 and not seen in this section). (d) A diagram of the micrograph from (c).

Source: Tomlinson A., Mavromatakis Y.E., Struhl G. 2011. Three distinct roles for notch in *Drosophila* R7 photoreceptor specification. PLoS Biology 9 (8): e1001132. doi:10.1371/journal.pbio.1001132.

sevenless and related genes comprise a Drosophila version of the RTK Ras pathway

One of the genes identified from this screen was named *sevenless* because mutations that eliminated the function of this gene lacked the R7 cell. When the *sevenless* (*sev*) gene is completely inactive, R7 no longer develops into a photoreceptor but instead adopts the fate of a cone cell, as shown in Figure 15.26. Thus, the wild-type function of the *sev* gene is to direct a precursor cell to become R7, rather than a cone cell. *sev* is expressed in other cells, including the precursors to the R1–R6 cells, but these do not become R7 cells. The molecular and genetic mechanisms by which *sev* specifies this cell fate decision could be clarified by knowing some of the other genes involved in this pathway.

Upon subsequent analysis, the product of the *sev* gene was found to be an RTK. The importance of RTKs in many biological processes was beginning to be appreciated, and there was great interest in understanding the downstream components of these pathways. The *sev* mutants in *Drosophila* provided the perfect opportunity for a genetic dissection of an RTK pathway. By using *sev* mutants with differing amounts of functional gene product, it was possible to identify some of the genes encoding other components of this signaling pathway. This is an experimental application of the principles of epistasis that were discussed in a more formal way in Section 8.4. The logic, which we summarize in Figure 15.27, is as follows.

Many RTKs are essential for normal cellular growth, so a mutation that eliminated one of these would be lethal. Even if *sev*, the RTK in this pathway, is not itself an essential gene, many of the downstream targets could be shared with other RTK pathways and might be essential. The early lethality of mutations in such genes would prevent them from being identifiable with a standard forward genetic visual or behavioral screen of adult flies. An alternative approach is to use *sev* mutations that reduce, but do not eliminate, function. Mutations in other genes in the same pathway or process could then either alleviate the mutant defect of the first mutation or make its effect even more severe. Because the pathway is already so "sensitized" by the first mutation, a second mutation can often exert an effect, even when it is heterozygous.

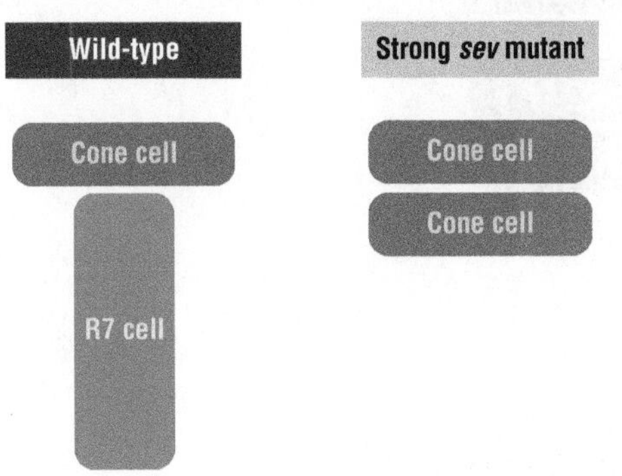

Figure 15.26 ***sevenless*** **mutants lack the R7 cell.** Mutations in the gene *sevenless* (*sev*) were among the mutants that did not respond to UV light. In strong *sev* mutants, in which the activity of the gene is knocked out, no R7 cells form; the cell forms a cone cell instead. The product of *sev* was found to be an RTK.

This approach defines two categories of interacting mutations:

- Mutations that alleviate the defect and make the phenotype of the first mutant more similar to wild-type are called **suppressor mutations**.
- Mutations that aggravate the defect and make the phenotype of the first mutant more severe are called enhancer mutations, or, in order to avoid confusion with the regulatory regions, **synthetic enhancer** mutations. We will refer to these as synthetic enhancer mutations.

KEY POINT Mutations in one gene that alter the mutant phenotype of a different gene can identify other genes in the same pathway or process. Suppressor mutations are mutations in another gene that alleviate the defect of mutations in the first gene. Synthetic enhancer mutations are mutations in another gene that aggravate the defect of mutations in the first gene.

For example, some alleles of the *sev* gene lower the amount of *sev* activity to the minimal level needed for the R7 cell to become a photoreceptor; these mutations reduce the activity of *sev* but do not completely eliminate it. Although weakly mutant in *sev*, these flies still detected UV light; there was no phenotypic effect of the weak *sev* mutation by itself. These mutants provided the genetic tool necessary to find other genes that affect this RTK pathway.

Against this "sensitized" background arising from the mutation that reduces the activity of the *sev* gene, additional perturbations in other genes of the pathway could render it non-functional; as a result, the R7 cell fails to form in these double mutants, as diagrammed in Figure 15.28. Neither mutant alone has this phenotype, so it arises from the combined mutant effects of two different genes. This is why these are called synthetic enhancers—the two mutations need to be put together to show the effect.

Furthermore, even mutations in essential genes that would be lethal as homozygotes can be identified this way

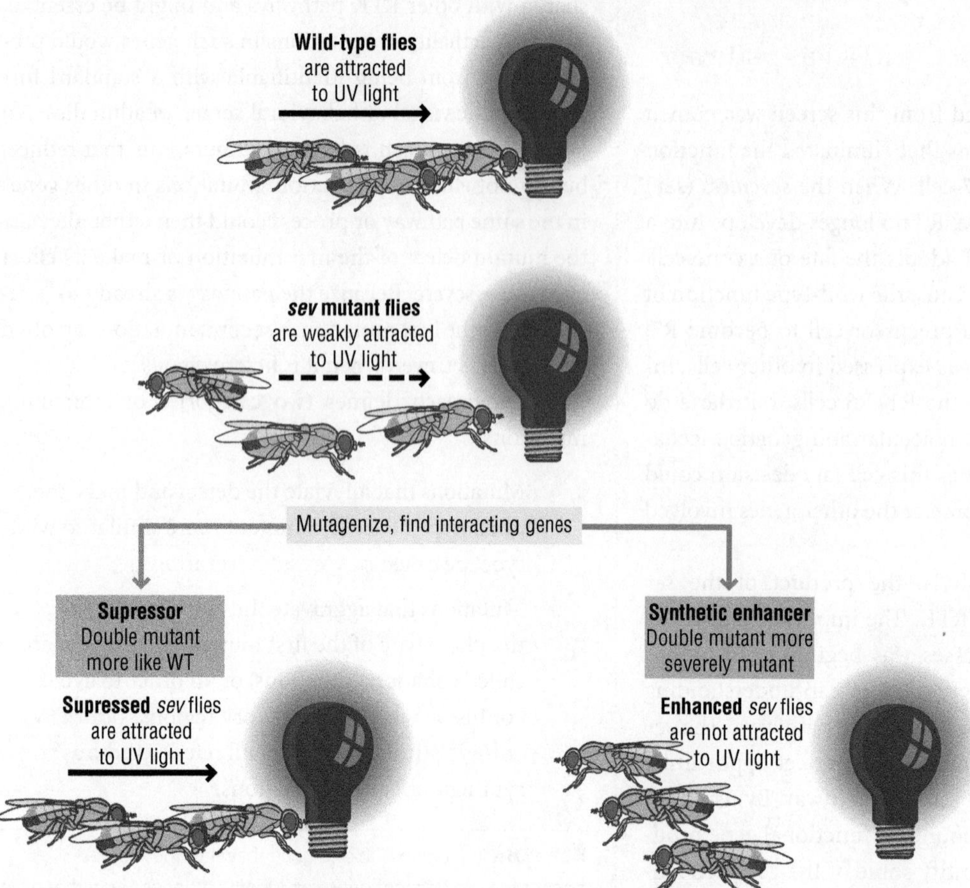

Figure 15.27 Suppressors and synthetic enhancers of *sev*. In order to find other genes in the *sev* RTK pathway, screens were done for mutations that either suppressed or enhanced the *sev* mutant phenotype. A suppressor mutation makes the *sev* mutant phenotype more similar to wild-type, so these flies would respond to UV light. A synthetic enhancer mutation makes the *sev* mutant phenotype more severely mutant. These are double mutant flies, with mutations in one of the interacting genes (a suppressor or synthetic enhancer) as well as a mutation in *sev*. The screens were done in *sev* mutant flies in which the function of the gene was not completely knocked out. While this diagram uses the light attraction phenotype of *sev* and these mutants for simplicity, most of the screens were actually done with other *sev* mutant phenotypes rather than phototaxis.

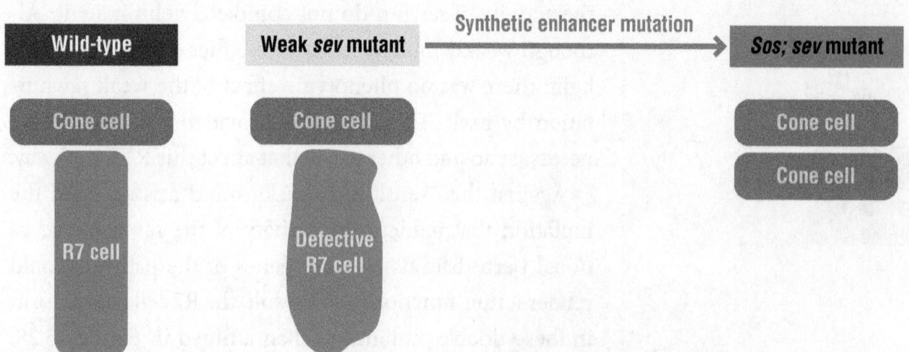

Figure 15.28 Synthetic enhancers of *sev* identified *Sos* and downstream kinase genes. The wild-type pattern of R7 differentiation is shown on the left. A weak *sev* mutant has a deformed R7 cell but has partial *sev* function and exhibits some phototaxis. The weak *sev* mutant fly was mutagenized, and more severely mutant flies that showed no phototaxis were found. These subsequently mutated genes are defined to be synthetic enhancers, since they make the *sev* mutant phenotype more severe. This screen identified the genes *Sos* and *drk*. Subsequent analysis showed that these genes function downstream of the *sev* RTK, as shown in Figure 15.24. However, this pathway was not known at the time and was defined based on these genes. Knocking down *sev* and knocking out the Sos or *drk* gene eliminates the function of the RTK pathway.

if they exert their synthetic enhancement effect as heterozygotes. Compromising the pathway and then screening for synthetic enhancement of the target phenotype allowed the identification of other components of the *sevenless* RTK pathway and avoided the need to rely upon potentially lethal homozygous mutations. Synthetic enhancer screens were used to establish that the wild-type activity of *sev* activates the RTK signaling pathway via multiple intermediates, including Ras and genes known as *Sos* (***So****n* ***o****f* ***s****evenless*) and *drk* (***d****ownstream of* ***r****eceptor* ***k****inase*). Mutations in any of these genes act as synthetic enhancers of the weakened *sev* mutant.

Related, but slightly different, methods using other types of *sev* mutant alleles were used to find suppressors of *sev*, which identified even more genes involved in this pathway. These approaches are outlined in Figure 15.29. For example, in one screen, an allele of *sev* that produced only a few R7 cells was mutagenized, and the flies were then screened for the ones that produced a normal number of R7 cells. In another screen, flies with an over-active *sev* mutant, which resulted in too many R7 cells, were mutagenized and then screened for mutants that reversed that phenotype to yield a normal number of R7 cells. The increased number of R7 cells arises because some of the cells that normally become R1–R6 become R7 instead; this results in the outer surface of the eye having a rough appearance. This rough-eye phenotype could easily be used to screen for suppressors, thereby identifying other components of this pathway.

A gene that was already known to encode an important component of this pathway is the *bride of sevenless* (*boss*) gene. Different mutations in *boss* illustrate the different types of suppression that can occur. The normal *boss* gene is expressed in the R8, rather than the R7, cell and produces the ligand for the Sev receptor; it directs

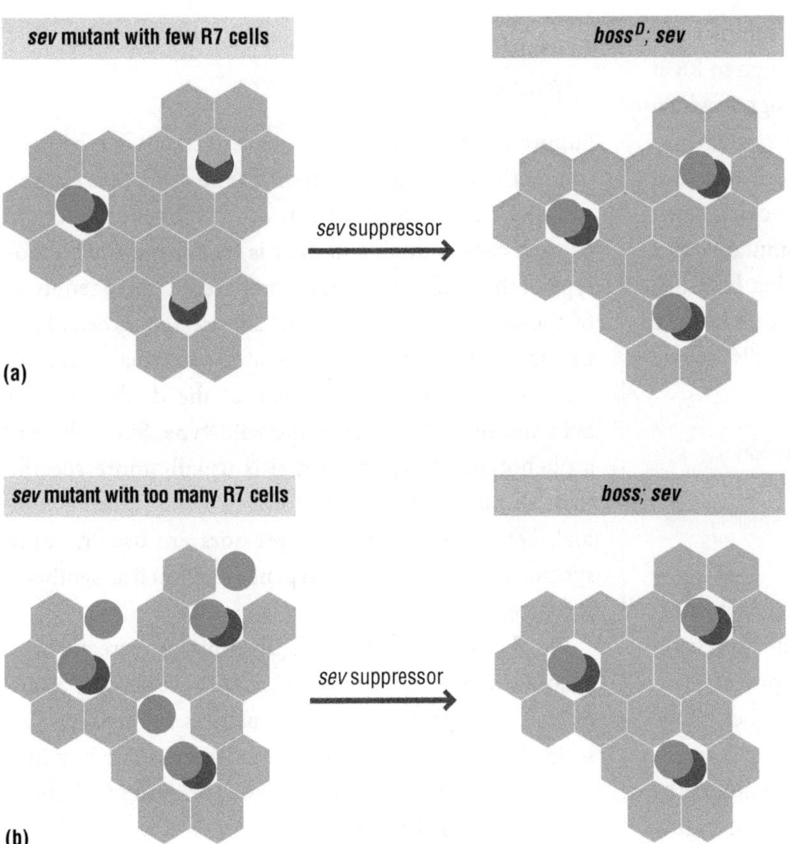

Figure 15.29 Suppressors of *sev* identified *boss*. Two different types of suppressor screens were done, based on the phenotypes of two different types of *sev* mutants. In (a), the *sev* mutant has very few R7 cells. When this fly was mutagenized, double mutants with a more normal number of R7 cells were found. These proved to be mutations that over-expressed a gene called *boss*. The mutations are dominant, as represented by the superscript 'D'. In (b), the *sev* mutant has too many R7 cells. When this fly was mutagenized, double mutants with a more normal number of R7 cells were found. These proved also to be mutations in *boss*. In this case, however, *boss* was under-expressed and the mutations are recessive. Thus, the balance between the level of *boss* expression and the level of *sev* expression determines the differentiation of the cell to become an R7 cell.

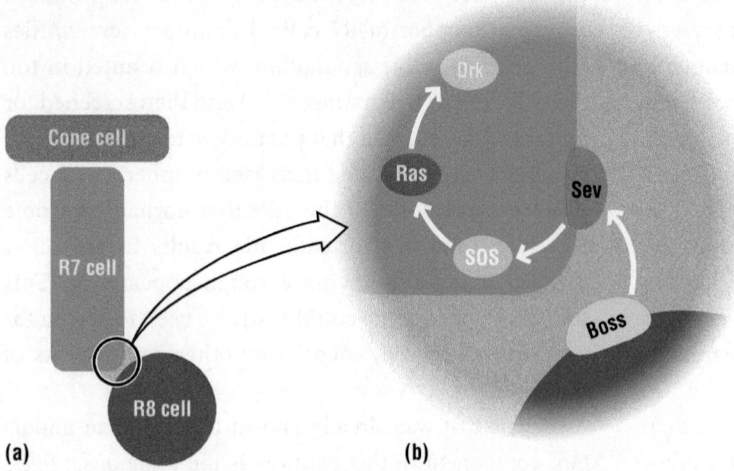

Figure 15.30 Boss is the ligand for the Sev receptor. (a) *boss* is expressed in the R8 cell, which is adjacent to the R7 cell in wild-type flies. (b) Boss is the ligand that interacts with the Sev RTK. The interaction between Boss and Sev activates Sos, which in turn activates Ras, which in turn activates the downstream kinases, such as drk. If *boss* is mutant and Sev is not activated, the cell becomes a cone cell, instead of an R7 cell. Similarly, if *sev* is mutant and cannot respond to Boss, the cell becomes a cone cell.

the precursor cell to differentiate as R7, as summarized in Figure 15.30. Only the R7 cell is close enough to R8 at the right time and place to receive the Boss signal, which allows the other cells to become R1–R6. Thus, when *boss* is a mutant, the corresponding precursor cell does not differentiate as R7, and no R7 cells form. Knocking down *boss* function suppresses the over-active *sev* mutants with too many R7 cells. Conversely, over-expressing *boss* suppresses the *sev* mutants that have only a few R7 cells because some of the cells that normally become R1–R6 cells become R7 instead.

KEY POINT Mutations with different levels of *sev* activity were used to find synthetic enhancers and suppressors. These studies defined other genes that work in this signaling pathway.

While these genes were found on the basis of their effect on UV light detection in *Drosophila*, their expression has proved to impact many fundamental processes in eukaryotic cells. *sev* itself is related to many other RTKs, and many of its downstream targets, including Ras, are the components of important signal transduction pathways in other eukaryotes. Although *boss* is not highly conserved among other eukaryotes, the principles of how it regulates the activity of *sev* have provided insights into how ligands activate RTKs. A seemingly simple behavior—the attraction of insects to a light source—could be used to study one of the most fundamentally important pathways, one that is mutated in human cancer cells.

Considerations in using suppressor and synthetic enhancer screens

Suppressor and synthetic enhancer screens can be used to identify other genes that affect the same process and to help infer a logical order in which the genes function. In each case, a double mutant is made, and the phenotype of the double mutant is compared to the phenotype of the original single mutant. Suppressor screens have the advantage that the effects of the original mutation are silenced, so the phenotype of the double mutant becomes more like that of the wild-type. Since shifting a phenotype closer to normal is usually more specific than shifting a phenotype towards being more abnormal, genes identified as suppressors are usually more specific in their actions than genes identified as synthetic enhancers.

On the other hand, synthetic enhancer screens have become increasingly important as we have become aware of the redundancy of, and robustness in, cells, as described in Section 3.4. Because of the redundancy in a genome provided by gene families, for example, knocking out the function of one gene often does not yield a mutant phenotype; a mutant phenotype arises only when a second gene with a related and overlapping function is also knocked out, as shown in Figure 15.31. Thus, synthetic enhancers can be used to identify not only pathways by which the genes act but also other pathways with related functions.

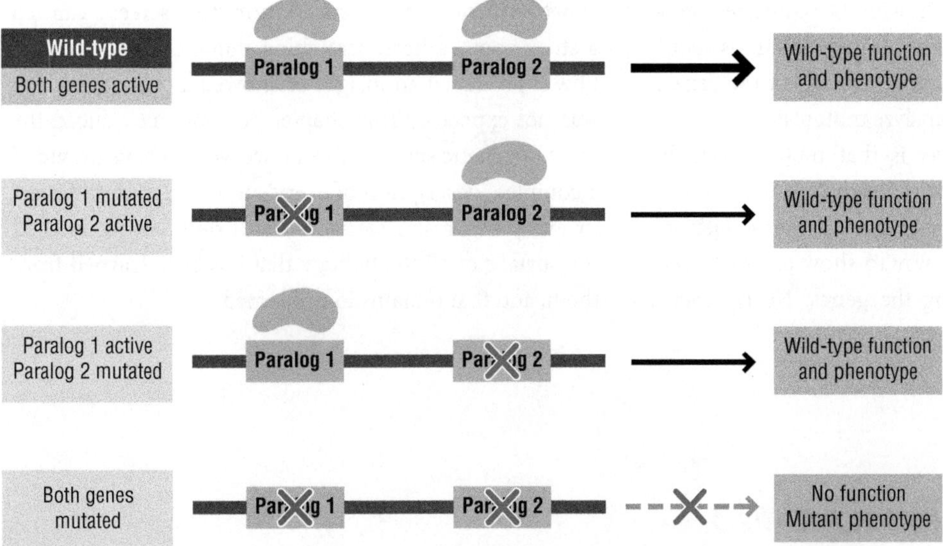

Figure 15.31 Synthetic enhancers can be used to identify redundant functions. Synthetic enhancers that depend on a specific interaction between two genes can reveal functional redundancy. Consider a family with two paralogs with closely related functions; either gene makes enough product to carry out the function. Thus, a mutation in either paralog 1 or paralog 2 will not produce a mutant phenotype, since the other paralog is still carrying out the function. Only a synthetic enhancer interaction in which both paralog 1 and paralog 2 are mutated produces a mutant phenotype. Many examples like this are known, including the ones in which the genes are not paralogs but define functionally overlapping ("robust") pathways, as discussed in Chapter 3.

15.5 Summary: genetic analysis

The ability to identify mutations and subsequently use them to analyze the organization of a complex biological process is the bedrock success of genetics. Sometimes these methods are called "genetic tricks," but there is no trickery involved in using the principles of gene inheritance, expression, and function as an experimental tool. We have focused on five particularly well-known examples of genetic analysis and the biological processes that were studied, but hundreds of other examples could have been used. Likewise, the examples could have been taken from bacteriophages or other viruses, flowering plants, worms, mice, and many other bacteria or eukaryotes. We have also focused exclusively on genetic screens using mutations induced in the laboratory. We could have focused instead on genetic screens based on naturally occurring variation, the type of analysis done with Darwin's finches, which we described in Chapter 1. Most genetic analysis begins with finding a mutant affecting a process, whether in the laboratory, in nature, or in the clinic.

Three of these examples provided insights about genes that have proved to be important in human disease. That was not the motivation for any of the screens, however. None of them was undertaken with the expectation that we would learn about cancer from them, for example. We learned about cancer because these screens were aimed at understanding fundamental processes in eukaryotic cells—the cell cycle, the organization of embryos, and the signaling of one cell to another—those processes whose malfunction can lead to the onset of cancer. The overarching value of mutant screens is that very little of the mechanism or of its components needs to be known in advance, other than having a thorough sense of the process in normal organisms and cells. Nearly every biological process described in this book, or any other biology textbook for that matter, was uncovered in part by a genetic screen.

With the widespread availability of genome sequences, it may seem that genetic screens may be less important than in the past. After all, if we simply gaze at a DNA sequence with appropriately experienced eyes or subject it to the appropriate algorithms, we should be able to find most or all of the genes it contains. By comparing the sequences of these genes and their products with the sequences of related genes, we can even make accurate predictions about the molecular or cellular function of many of the genes. Why then do we still need to analyze mutants?

A short and partial answer is that mutants usually reveal to us how much more we have to learn about a biological process. As valuable as genome sequences are, mutants remain the best way to show us how a cell or organism is actually using the genes. Nearly every geneticist who is experienced in mutant screens can tell a story about a head-scratching mutant phenotype and how it produced an insight or showed a connection that was not expected. This chapter has only introduced the topic of genetic screens. Even if we were to have provided a complete description of every successful mutant screen in every organism. we would still only have scratched the surface of all the biology that has been learned from them, and that remains to be learned.

CHAPTER CAPSULE

- A genetic screen is an experimental strategy to find and select for an organism with a mutant phenotype of interest. Mutations are induced in a population, and individuals with an altered phenotype are then identified for further study.
- Among the key components of a genetic screen are:
  - The choice of the experimental organism for the biological question
  - A well-defined phenotype and assay
  - An approach that is unbiased in terms of the genes and mutants that might be found
  - The scale and efficiency needed to screen many mutants
  - The assignment of mutations to genes
  - The follow-up experiments to understand the genes.
- Genetic screens can be undertaken with little advance knowledge of the molecular and biochemical processes, but a thorough understanding of the process in wild-type organisms and cells is crucial for interpreting the mutant phenotypes.
- Genetic analysis not only finds the genes involved in a process but can also be used to infer a logical order or pathway by which the genes act. Among the methods that can be used to infer such an order are:
  - A prior knowledge of the underlying biochemical pathway
  - Mutations in a gene with different types of phenotypes
  - The mutant phenotypes of different genes themselves
  - Assays that connect the phenotypes to landmark events
  - Temperature shift experiments
  - The use of suppressors and synthetic enhancers.

STUDY QUESTIONS

Concepts and Definitions

15.1 Define the following terms:

a. Selection

b. Restrictive and permissive temperatures

c. Suppressor and synthetic enhancers

d. Saturation

15.2 What are some of the strengths and limitations of using a genetic screen based on mutant phenotypes to analyze a biological process?

15.3 What is a complementation test, and what is its function in a genetic screen?

15.4 What are the advantage(s) afforded by including a selection step in a mutant screen?

15.5 What is meant by referring to a genetic screen as "unbiased?"

Beyond the Concepts

15.6 Make a list of reagents and complementation tools, and sketch out the experimental set-up that could have been used to identify the following components of the *lac* operon.

a. *lacZ* encoding β-galactosidase

b. The *lac* repressor (*lacI*)

c. The *lac* operator

15.7 Bacterial cells do not have an endoplasmic reticulum or known transport systems of the sort that are found in eukaryotes. However, some proteins are known to localize at the cell poles of rod-shaped bacteria. What points would you have to consider in designing a genetic screen that could uncover genes involved in intracellular localization in bacteria?

15.8 Honey bees live in communities in which the queen is fertile and workers are sterile. A saturated genetic screen identified 12 honey bee mutants that were unable to propagate truly social communities because they had lost the ability to restrict reproductive functions entirely to the queen. (These mutants produce an abundance of fertile workers that have been referred to as "anarchists.")

a. If five of the responsible mutations map to chromosome 3 and seven to chromosome 8, what is the minimum number of individual genes that, when disrupted, produce anarchists?

b. How would investigators go about determining precisely how many genes are involved without sequencing the genomes of all the mutants?

15.9 The *Escherichia coli* tryptophan (*trp*) operon was described in Chapter 14. Refer to Section 14.2, and devise a genetic screen that could be used to uncover loci responsible for tryptophan synthesis and activation of the operon by tryptophan.

15.10 Can a mutant screen be used to identify a gene in a specific process if:

a. The product of the gene works with another protein to produce the phenotype?

b. The genome in question encodes two redundant proteins, either of which can produce the phenotype?

c. In addition to the phenotype in question, the gene performs a second function that is essential for viability of the organism?

15.11 What is a temperature-sensitive mutation, and how are temperature-sensitive mutations used in genetic screens?

Applying the Concepts

15.12 An alternative to using a mutagen to produce random point mutations across a genome for a screen is to use a transposable element to mutagenize the genome. Transposable elements that are used for this purpose integrate randomly and disrupt genes and other functional sequences by inserting into them. In bacterial genetics, it is typical to use a transposon that contains an antimicrobial resistance gene or other marker, so that mutants containing insertions can be selected.

a. What other advantages might insertional mutagenesis using transposons offer to the entire process of a genetic screen that chemical mutagenesis or irradiation do not?

b. There are pitfalls associated with using transposons for genome-wide mutagenesis. Can you think of at least one in a specific organism?

15.13 Section 11.4 described the two alternate life cycles of phage λ in *E. coli*: the lytic cycle and the lysogenic cycle. Many of the features of these life cycles were uncovered using simple genetic screens. Design a simple genetic screen to identify genes that are involved in lysogeny that will allow you to screen a very large library of mutants.

15.14 The genetic screens described in this chapter used three different experimental approaches to place the genes in the order in which they act to carry out a biological process. Briefly describe how the investigators used each of these approaches. What are some of the strengths or possible limitations of each approach?

a. Adding back substrates

b. A time of action for the genes based on their mutant phenotypes

c. The phenotypes of double mutants

15.15 Most *E. coli* strains can produce nicotinamide adenine dinucleotide (NAD), which is essential for life. A mutant screen was designed to identify genes required to synthesize NAD from the amino acid aspartic acid. Several mutants that could not grow in the absence of added NAD were isolated. The mutants could be classified into three subclasses, depending on their ability to grow in the presence of other additives:

	Aspartic acid	NAD	Nicotinic acid	Quinolic acid
Class 1	–	+	–	–
Class 2	–	+	–	+
Class 3	–	+	+	+
Wild-type *E. coli*	+	+	+	+

Draw a diagram with arrows that orders the production of each of the additives in the NAD biosynthesis pathway.

15.16 Unlike *E. coli*, the most commonly used bacterial model organism, the crescent-shaped *Caulobacter crescentus* bacterial species has external appendages that change in synchrony with the cell cycle, as illustrated in Figure Q15.1.

a. Relatively few scientists work with *C. crescentus*, but they were able to unravel the details and processes of the bacterial cell cycle in this organism far in advance of *E. coli* researchers. Why might *C. crescentus* be better suited for dissecting the cell cycle genetically?

b. The bacterial cell cycle is very different from that of eukaryotes and does not use homologs of the genes originally identified in yeast. Outline a genetic screen that could be used to identify cell cycle regulators in bacteria.

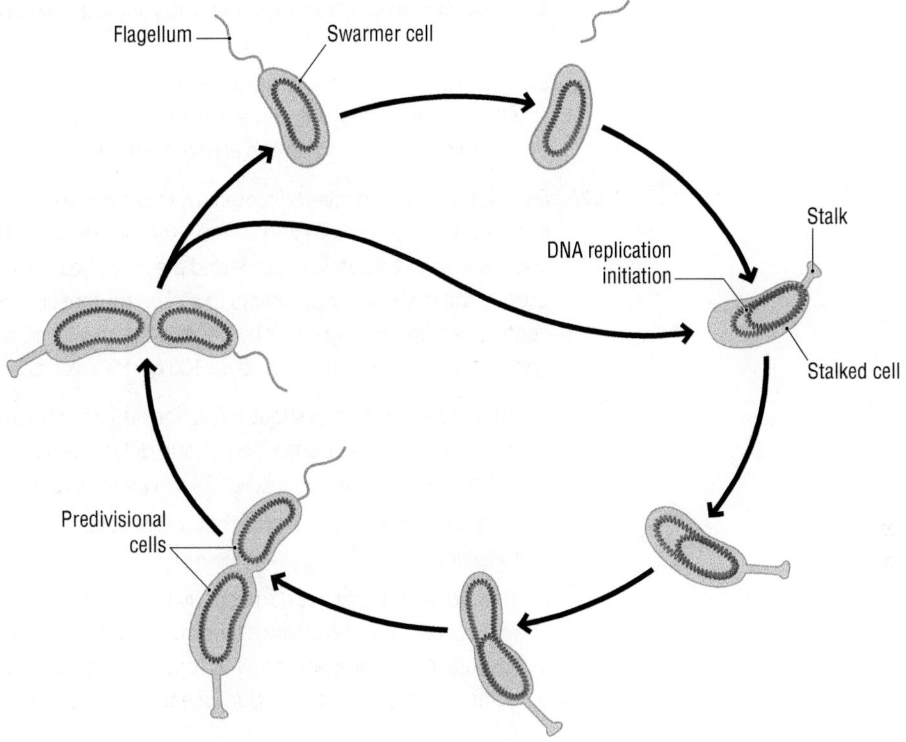

Figure Q15.1

15.17 Imagine that you have domesticated a finch species that can be maintained in a laboratory aviary and that you can obtain ethical approval to perform a genetic screen in this organism.

a. Ideally, what feature(s) would your model finch have to make it suitable for such a screen?

b. Design a screen to identify genes involved in beak morphology.

c. Would you expect your screen to identify the same or different markers from those obtained by genome-wide association studies of wild birds in the Galapagos Islands? Why or why not?

15.18 Retroviruses like human immunodeficiency virus (HIV) contain an error-prone nucleotide polymerase known as reverse transcriptase, which, upon new infection, converts viral RNA to DNA that is incorporated into the infected host's genome (Section 11.4). Because reverse transcriptase is essential for the virus to complete its life cycle, HIV naturally incorporates large numbers of mutations, even within a single host. The HIV genome is small (only about 9 kb), and currently thousands of HIV genomes have been sequenced and are available for researchers to study. Devise an *in silico* experiment to identify essential portions of the HIV genome, without which the virus cannot complete its life cycle.

15.19 Researchers in China have found a strain of bacterium that is resistant to the antibiotic colistin, a last-resort drug for treating antibiotic-resistant infections. The colistin resistance in this strain was mapped to a 64-kb plasmid. Transforming this plasmid into a laboratory strain of *E. coli* renders the laboratory strain resistant to colistin. Since transmissible colistin resistance was previously unknown, the researchers cannot find the colistin resistance gene by performing homology searches against known genes in the database—the gene must be one of the many genes of unknown function present on the plasmid.

a. Design a genetic screen to identify colistin-resistant gene(s) on the 64-kb plasmid.

b. If your screen yields six hits, how will you determine how many genes confer colistin resistance?

Challenging **c.** The plasmid confers a lower level of colistin resistance in the laboratory strain than was seen in the original wild-type resistant strain isolated from nature. Hypothesize why this might be so, and design an experiment to test your hypothesis.

15.20 *Bacillus subtilis* is a species of bacterium that can form a morphologically distinct and heat-resistant cell form known as a spore. Colonies of spores are more opaque than non-sporulating or vegetative bacteria and are more heat-resistant. These properties have been used as the basis for simple assays to screen mutant libraries. In the last half of the twentieth century, random mutagenesis by different methods, followed by a series of *B. subtilis* sporulation screens, uncovered over 100 genes involved in this developmental process.

Much more recently, investigators performed a fresh screen using an ordered mutant library, that is, a library composed of one deletion mutant per open reading frame across the whole genome, and identified 24 new sporulation genes.

a. What are some of the factors that might account for the failure to identify the 24 genes in earlier screens?

b. The ordered library has additionally been used to study the sporulation process microscopically, identifying genes that contribute to the process and timing of sporulation but are not required for this developmental process. Why is an ordered library more useful for this purpose than a random mutagenesis library?

15.21 Genome-wide mutant screens are widely used in eukaryotes, as described in Tool Box 15.1.

a. Describe the basic process by which a genome-wide mutant screen is done using RNAi.

b. What are the advantages of performing a genome-wide screen using RNAi, compared to a more traditional genetic mutant screen using one of the known mutagens?

c. What are some of the disadvantages or limitations of performing a genome-wide screen using RNAi, compared to more traditional genetic mutant screens?

Challenging **d.** It is very likely that genome-wide screens will soon be done using CRISPR, rather than RNAi. (CRISPR is described in Section 11.6.) What will be some of the main differences between mutations made by CRISPR and their phenotypes and mutant phenotypes arising from RNAi?

Challenging **15.22** The various processes of DNA repair, as described in Section 4.4, were worked out largely by genetic screens in bacteria and eukaryotes using mutants that conferred increased sensitivity and resistance to various mutagenic treatments. A chemical known as MMS (methyl methanesulfonate) is not a potent mutagen in yeast because the DNA damage induced by MMS is typically repaired, allowing the yeast cell to continue to grow after application of MMS. One study identified a series of mutants that were sensitive to MMS and also tested the sensitivity of these mutants to UV and to X-rays. These results are slightly modified from this study; other such studies have not always found identical results for all experiments.

a. Outline a process that could find MMS-sensitive mutants.

b. The study identified 29 mutants that were sensitive to MMS and performed complementation tests among these mutants.

i. Describe the basic method by which the complementation test could be done, and the possible results that might be found.

ii. These 29 mutants identified 22 different genes by complementation. One gene had four alleles, one gene had three alleles, two genes had two alleles, and 18 genes were represented by a single mutant allele. Is this screen saturated for MMS-sensitive mutations? Explain your answer.

c. Mutations in six different genes are sensitive to both MMS and UV damage but show the same response to X-ray damage as wild-type strains. What does this result suggest about the normal functions of these six genes?

d. Previous studies had shown that many of the mutants sensitive to UV damage are not also sensitive to X-rays, and vice versa, that is, many mutants sensitive to X-ray damage show a normal response to UV damage. What does this result suggest about the pathways used by yeast to repair UV damage and X-ray damage?

e. This particular study also found mutations in five genes that were sensitive to both X-rays and MMS but not to UV. There were also five genes that were sensitive only to MMS, and not UV or X-rays, and at least two genes that conferred sensitivity to MMS, X-rays, and UV. Based on the results of these mutants and their sensitivities, how many different pathways must exist to repair DNA damage in yeast, and how might they be related to one another?

f. This study found that mutations in *rad6* and *rad18* are sensitive to MMS, UV, and X-rays. However, a yeast strain that is mutant for both *rad6* and *rad18* is no more sensitive than either single mutant. How can this result be explained?

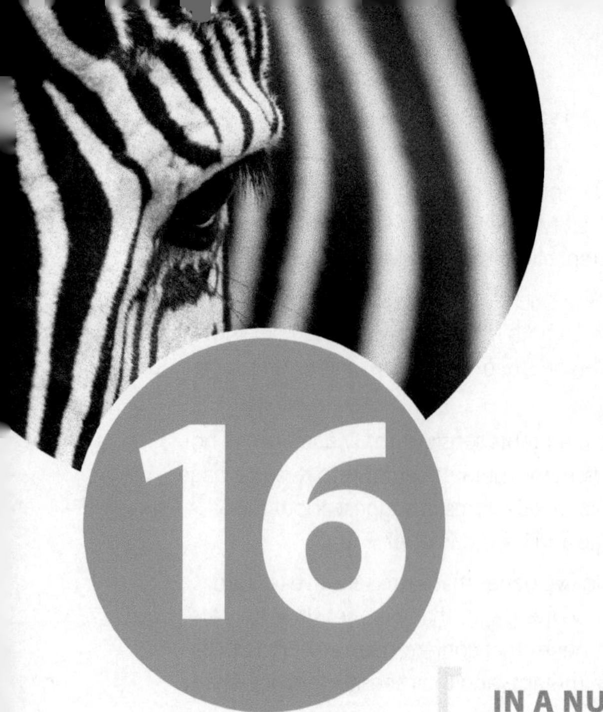

CHAPTER 16

The Genetics of Populations

IN A NUTSHELL

The transmission of alleles and genotypes within a population can be assessed, even when the individual matings cannot be monitored. At equilibrium, the frequencies of the alleles and genotypes will not change from one generation to another. However, no population is at equilibrium for every gene because there are many factors that can change genotype and allele frequencies. The main process that affects genotype frequencies without affecting allele frequencies directly is non-random mating, which can result in population stratification. The factors that change both allele and genotype frequencies—mutations, migration, finite population size, and differential reproductive fitness—make important contributions to the evolution of species. These effects can be seen in all populations, whether they are sexually reproducing species like humans or asexual populations like bacteria. It is the combined effect of all these factors over time that has ultimately shaped how our genomes appear today.

16.1 Overview of population genetics

In earlier chapters, we have explored the composition of genomes and the function and transmission of individual genes. We have also taken into consideration the fact that individual genes and genomes have been shaped by evolution in the context of populations. In this chapter, we will now focus more specifically on what happens to genes and alleles in populations and how the population might change over time. Studying the genetic composition of populations can help us understand the fate of new genetic variants that arise in a population and can reveal the impact and mechanisms of the multiple factors that drive evolutionary change. As we begin this exploration of genes at a population level, we first return to the concept of the gene itself.

While it is easy to talk about a gene and for others to know what we are talking about, it is not so easy to provide a comprehensive definition of a gene. In fact, previous chapters have discussed genes in several different ways, even though all of them refer to the same entity. We summarize these properties in Figure 16.1. For example, in earlier chapters, we have focused on the gene as the unit of inheritance in bacteria and eukaryotes, as shown in

		Male gametes	
		R	r
Female gametes	R	RR round	Rr round
	r	rR round	rr wrinkled

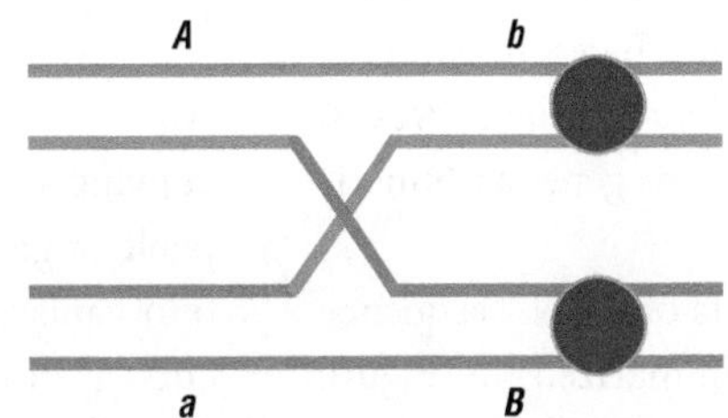

(a) The gene as the unit of inheritance

```
aaaaccgacgagaaaacaacATGAAATCTATTATCATTTTATCAGTTTTCTTAGTTTGTTCCGCGTT
GGCTGCCAGTTCTGATAAGGATACTGAATCAAAAGTGGCCAGATATTTGAAATCCTGTGGAGTTATC
ACTTCATTGAAGACTACTAAATGTGTTAAGgtgggttttataaaattgttcaagcaggcttcaattg
tcatgattgctcatgcagttcttattttggtggaatttcaaacatggagatttatcacagtagtcct
tttttgttgtacttgctgcatttcgaactatttccgtcattttaataaaaaaaacacttttagAGAA
GCGATGAAGTAGATAAGGAGCTCAAAGCTCTCGTAAAGAGCAAAGACAAGGATCTCTCAAAACTCAA
GGCTTCGTGCAAAGACACCATGgtgggtaacatcagtgatattaaaaacttctcaaatgtgtttcag
GAATGCATTAAAGAGCTGAAATGTAAACCACTCGAAGAAGATCACAAGGAGACTCTCGACTTCTGTG
GTCGTGCTCTTTTCCAGGACGACTTCAAGGATTGCAGCAAGAAATTGCAGGCATTGAAGAAAAGTGA
TAAGGATGCTGCAAAGTGTTTGGAAGATTTTAAATCAGAGACAAAGAAGTCGACAAAGGACACTTGC
ACATTTTTGAAGGGCTCGAAAGATTGTATCAAGGCTCATATTAAGAAAGAGTGCGGGGATGATAAGT
TGCAGGGATGGCAGAAGgtaagcggatgatgtcttagaaatcggtttaggcttcggcgcaggcctta
aacttagatgcttaggcttaggcttagagagcctcaaactttcattttcatttccaccaacttttct
aaatttctatttttcagTTTGCTCAAGACTCGTTCAAAGAGAACGAATGCGAGAAGGCAATTGGAGC
AAAATGGAACTAActtcattaaatgttcttcgatttcttgttccttgttacagtaaccctgcaaaac
tgcaataattaatttaaaagtctttaaaattccggttgagacccattcctccctcctacattcttc
atttcttcaaagggtttctctccaggtaccaataaatttttaacgcggtca
```

(b) The gene as a macromolecule

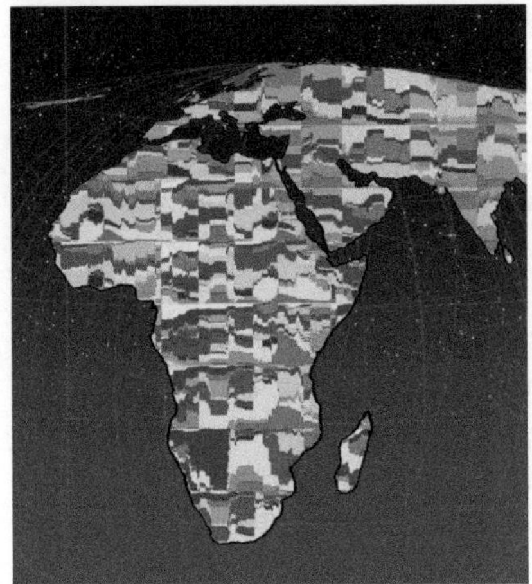

$p + q = 1 \qquad p^2 + 2pq + q^2 = 1$

(c) The gene in the population

Figure 16.1 Different images of genes. Genes are thought of, and depicted, in different ways, but ultimately all of these images describe the same concepts. (a) A gene is the unit of inheritance. This is the aspect when they are depicted in a Punnett square and as a locus on a chromosome. (b) A gene is a sequence of DNA or a macromolecule. (c) A gene is a unit of evolutionary change. This is shown here as haplotype frequencies on a map of Africa and then as allele and genotype frequencies as algebraic equations.

Source: Part (c) shows an artistic representation of the patterns of haplotype variation reported by Jakobsson et al. (2008). Image by Martin Soave, University of Michigan Marketing and Design, reproduced with permission from M Jakobsson et al. (2008) *Nature* 451: 998-1003.

Figure 16.1(a). In this context, a gene is described as a locus on a chromosome, at which different alleles may be found, and which affects one or more phenotypes. In Chapter 8, we also defined a gene as a unit of function by a complementation test. These are properties of genes as they were understood by Mendel, Bateson, Morgan, Beadle, McClintock, Lederberg, and many others. We can think of these as Mendelian genes, or as genes as "inheritance particles."

We also know that a gene consists of a DNA sequence (except in some viruses), as summarized in Figure 16.1(b). As described in Chapters 2 and 12 particularly, this DNA sequence is expressed as an RNA transcript, which is often then translated into a polypeptide or a protein. This is a gene as Watson and Crick, and Avery (among others) defined it—the gene as a macromolecule. All of the components depicted in Figure 16.1(a) and (b) are part of a definition of a gene or its properties, but none of them by itself fully describes a gene.

Another property of a gene has been part of our discussion since the earliest chapters, namely that similar genes exist in different species. This relates to the concept of orthologs—genes with highly similar, but distinct, DNA sequences (and therefore similar or related functions) present in the genomes of different organisms. The inference is that the common ancestor of these species had some version of this gene that has been modified by descent in each lineage. This is a view of a gene through the lens of evolution, as depicted in Figure 16.1(c), in which genes are present in a population and able to change over time. New alleles are created when genes change. Because the genes change, and genes contribute to phenotypes of individuals in the population, then the population also changes. This is a population perspective of genes, as they are often thought of in evolutionary biology as well as by plant and animal breeders.

We have introduced a population concept of genes (or, more accurately, the population concept of **alleles**) in previous chapters. Most notably, the behavior of genes and alleles in a population was a key idea in Chapter 10, but it has been touched on in other chapters as well. As we consider a gene in the context of a population, we still need to think about inheritance patterns. With classical Mendelian genetics, we looked at the mating of two individuals and their offspring. We knew the phenotypes of some of the individuals involved (the parents, the offspring, or both), and we were able to infer the genotypes and phenotypes of others. From these, we could predict the likelihood of a particular genotype or phenotype from a different mating.

But what happens when we consider a gene or genes in an entire large population? We need to begin by defining a population.

Effective populations

An **effective population** is any group of organisms occupying a particular place at a particular time that are capable of genetic exchange. With the important exception of organisms between which horizontal gene transfer occurs, as discussed in Chapter 11, these organisms will be members of the same species. The definition of an effective population includes a time and place, as well as boundaries that define an individual as being part of the effective population (or not). These boundaries may be fixed over time or quite flexible (as we will note), but they exist.

The effective population size, which is important in genetic and evolutionary terms, is a bit different from our usual concept of population size, which is a census of all individuals of a species in a particular time and place. What we think of as a population is larger—sometimes much larger—than the effective population. For example, the human population in the United States includes the effective population of all of the people of reproductive age, as well as many other people who are too young or too old to reproduce, or who cannot reproduce for another reason. These individuals are part of the population in terms of resource utilization and production but are not part of the effective population that is capable of transmitting alleles directly to the next generation.

Among that effective population, many different crosses or matings occur in each generation. These matings involve some combination of heterozygotes and homozygotes for individual genes, but information about the genotype of each parent is not usually readily available or easily tallied. A population is also more complex in its overall genetic structure. When we analyze a single mating involving two parents, we have to consider only four alleles of a gene of interest, some of which could be the same. But in a population, many different alleles might be present, all of which could affect the phenotype. It would be impractical to analyze every separate mating and look at its offspring individually, yet we still want to know about the genetic structure of the population. How can we do that?

A key thing to remember is that the rules of gene transmission are the same as before—for example, a mating involving two heterozygotes is still expected to yield 50% heterozygotes—25% of one homozygote and 25% of the other—but the process of analysis is different because individual matings (or genetic transmission events) cannot be studied separately. Thus, we need to look at the flow of genes (or alleles) without regard to the actual parents. This is the cornerstone of population genetics.

Furthermore, as the changes in the gene flow are measured over time, the genetic structure of populations becomes a key to understanding evolution. The factors that cause those changes are important parameters to understand. Evolution cannot typically be seen in individuals and cannot be reliably seen in the space of just one or two generations; it is best seen in populations over longer periods of time.

The principles described in this chapter apply to any population, and we could use any species or groups of species for this discussion. However, we have made the intentional (and possibly uncomfortable) decision to use humans and human populations for many of our examples. There are a number of reasons to focus on human populations. First, they are probably of the greatest interest to most of us, and we have a more intuitive concept of a human population than of a natural population of, say, fruit flies. Second, the availability and the analysis of human genome data, at a population-level scale, have provided some of the best examples of the concepts we will be discussing.

On the other hand, even a casual study of the history of genetics makes us well aware of how human genetics has been used to reinforce prejudices against certain people or population groups. This should make us all think carefully before discussing genetic differences in human populations. In describing a population as having a particular property—for example, a characteristic height—we need to be constantly aware that not every person in that population has that property. For example, we are factually correct to write that Swedes are taller on average than Italians, but that does not mean that any individual of Swedish heritage is taller than any individual of Italian heritage. There are apparently allelic differences for the genes affecting height between the population of Swedes and the population of Italians, but any person may or may not have those characteristic alleles.

The simple fact remains that, for many phenotypes, human populations are different from one another. We know these differences exist because we can see examples of them constantly; we would be intellectually irresponsible to deny that genetic differences exist. But you should also recognize throughout every example we discuss that we could pick another example from another population that illustrates the same point equally well. In fact, the very biological principles that make human populations differ from one another are the ones that define some of our unique characteristics as a species.

KEY POINT Population genetics considers the transmission of genes or alleles and changes in their frequency over multiple generations.

16.2 Allele frequencies and populations at equilibrium

Each of the approaches to defining a gene that we mentioned in Section 16.1 has helped us to draw out one or more key concepts about genes. In population genetics, the fundamental concept is the **allele frequency**. The allele frequency is simply how often a particular DNA sequence—an allele or a haplotype—is found in a population. We can expand our definition of the allele frequency as follows.

Allele and genotype frequencies

Suppose that it were possible to test every gamete in the population and determine what allele or DNA sequence is present for a particular gene. It is certainly not far-fetched to sequence the DNA in a sperm, a pollen grain, or an ovum to determine what allele is present. But it is hard to imagine extending this to every gamete produced by every individual in the population. However, let's assume that it is possible. We diagram this scenario in Figure 16.2.

In familiar language, the collection of every allele that is present in the population is often called the **gene pool**. We can count up the total number of alleles in the population and ask how many of them are A or a, of course also assuming that only two possible alleles exist for the gene in question. We can then ask about their relative frequencies. Among all of the alleles in a population, what fraction of them are A? We will call the frequency

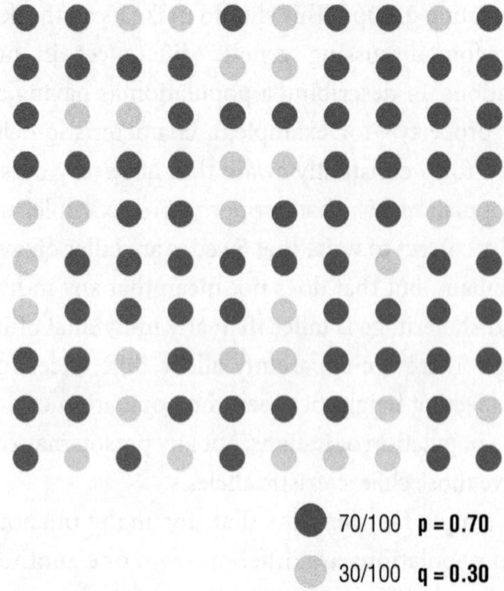

Figure 16.2 Allele frequency. The foundational concept of the allele frequency is shown here, in a format used in other figures in the chapter. Each allele, gamete, or haplotype is represented as a circle, with different colors representing the different alleles. Seventy of the 100 circles shown are blue, so the frequency of the blue allele (p) is 0.70. Thirty of the 100 circles shown are yellow, so the frequency of the yellow allele (q) is 0.30.

of the ***A*** allele p. Among all of the alleles in a population, what fraction of them are ***a***? We will call that frequency q. If we stipulate that there are only two alleles in a population, then $p + q = 1$. Every allele in the population is either one or the other. We can still apply the principle if there are many alleles by classifying them as either ***A*** or **not *A***. The frequency of ***A*** is p, the frequency of not ***A*** is q. It therefore still holds that $p + q = 1$.

While population genetics often simplifies natural events by assuming that there are only two alleles, we can always define the two alleles as being either ***A*** or **not *A***. It is not difficult to expand the discussion to populations with multiple alleles, but the arithmetic gets a little messier, so we will stick with two alleles. Since we are discussing allele frequencies, rather than phenotypes, we don't have to consider if one allele is dominant; ***A*** and ***a*** are being used as indicators for alleles of a gene, and those few cases in which dominance plays an important role will be noted.

Allele frequencies may vary from one population to another, so it is always important to note which population is being considered. In many cases, one allele is more common than the other. If the alleles are known variations in the DNA sequence (rather than being recognized solely by their phenotypes), the alleles are often referred to as polymorphisms. This leads us to a more formal definition of the concept of a **polymorphic locus**, which we introduced in Chapters 3 and 10. A polymorphic locus is one for which the second most common allele has a frequency of 0.01 or greater.

How then can we use allele frequencies to determine the frequency of different genotypes? The answer is diagrammed in Figure 16.3. If we assume that gametes in the population unite at random, an assumption that will be discussed more fully in Section 16.3, we can determine the theoretical **genotype frequencies** by applying the probability rules that we introduced in Chapter 5 as follows. For an autosomal gene, every individual has two alleles. (For X-linked genes, the genotype frequency of males is the same as the allele frequency, as we discuss below.) If alleles in the population combine **at random**—that is, if the choice of the first allele has no effect on the choice of the second allele—the frequencies of the different genotypes would be $(p + q)^2$, or $p^2 + 2pq + q^2$. And together these will total 1 ($p^2 + 2\,pq + q^2 = 1$). That is, ***AA*** would be found at the frequency p^2, ***aa*** would be found at the frequency q^2, and ***Aa*** would be found at the frequency 2pq. The "2" arises because the heterozygote could be either ***Aa*** or ***aA***—but we represent both outcomes in the same way (pq).

We can illustrate this with a simple example, which is shown in Figure 16.3. Suppose that, in a given population, $p = 0.7$, so $q = 0.3$. We expect that the frequency of the ***AA*** genotype will be $0.7 \times 0.7 = 0.49$. Likewise, the frequency of the ***Aa*** genotype is expected to be 2pq or $2 \times 0.7 \times 0.3 = 0.42$. Finally, the frequency of ***aa*** is expected to be $0.3 \times 0.3 = 0.09$. So those are the frequencies of the alleles and the genotypes when we first look at the population.

We can also begin with the genotype frequency and calculate the allele frequency in a population. For example, at the Rh blood locus in humans, there are two functional alleles, with Rh-positive (***R***) being dominant to Rh-negative (***r***). Approximately 16% of the global population is Rh-negative with genotype ***rr***, although individual populations vary widely. In the absence of any other information, we can assume that the human population is at equilibrium with respect to this trait. Thus, $q^2 = 0.16$, and if the population is at equilibrium, $q = 0.4$. This tells us that the frequency of the ***r*** allele is 0.4, and therefore the frequency of the ***R*** allele must be 0.6. We can extend this to other genotypes as well. Since the frequency of the

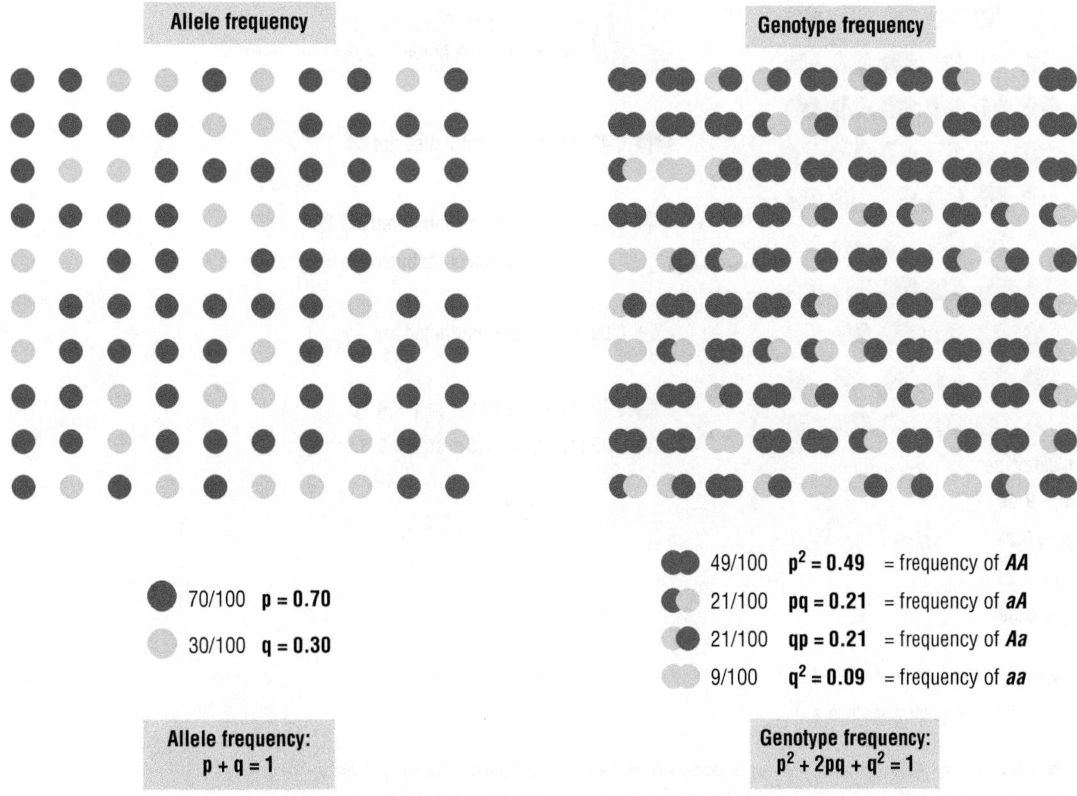

Figure 16.3 Genotype frequency. The alleles depicted in Figure 16.2 are shown in combination in diploid genotypes. The alleles in Figure 16.2 are combined at random, representing genotypes found when mating is at random. Since the frequency of the ***A*** allele is 0.70 and the frequency of the ***a*** allele is 0.30, the frequency of the genotypes can be calculated. The frequency of the ***AA*** homozygote is $p^2 = 0.49$, the frequency of the ***aa*** homozygote is $q^2 = 0.09$, and the frequency of the ***Aa*** heterozygote is $2pq = 0.42$.

R allele is 0.6, the expected frequency of ***RR*** homozygotes is 0.36, and the expected frequency of ***Rr*** heterozygotes is 0.42 at equilibrium.

KEY POINT If the frequency of allele ***A*** is designated as p, the frequency of all other alleles (not ***A***) is then defined as $1 - p$ or q. The genotype frequencies can then be calculated using the equation $p^2 + 2pq + q^2 = 1$, where p^2 and q^2 represent the frequency of the homozygotes and 2pq represents the frequency of heterozygotes.

The allele and genotype frequencies over time

While differences in allele and genotype frequencies may allow us to distinguish one population from another, the most important evolutionary questions arise when changes in the allele and genotype frequencies are compared in the same population over time.

Let's allow the hypothetical population in Figure 16.3 to reproduce into the next generation and see what happens to the allele and genotype frequencies, shown in Figure 16.4. (We are discussing this throughout as if the reproductive ages of different generations do not overlap, which is clearly an oversimplification, but this makes it easier to track generation-to-generation changes.) First, we can look at the allele frequencies.

AA individuals contribute only ***A*** alleles to the next generation. The initial frequency of ***AA*** individuals is 0.49, and the fraction of ***A*** alleles from them is 1, so the frequency of ***A*** alleles in the next generation is 0.49×1 from homozygotes.

Aa individuals contribute ***A*** alleles in half of their gametes. So the frequency of ***A*** alleles coming from ***Aa*** individuals is $0.42 \times 0.5 = 0.21$.

This means that the overall frequency of the ***A*** allele in the next generation is $p = 0.49 + 0.21 = 0.70$.

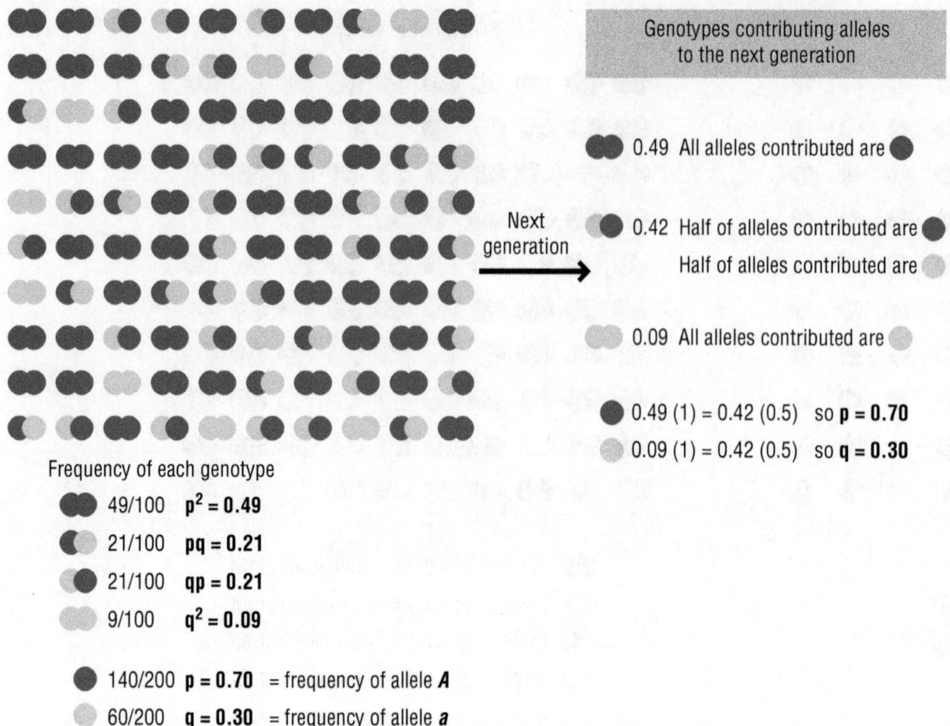

Figure 16.4 Allele frequencies at equilibrium. When a population meets the conditions for a Hardy–Weinberg equilibrium, the frequencies of the alleles do not change from one generation to the next.

Since we said that every allele is either ***A*** or ***a***, and the ***A*** allele has a frequency in this generation of 0.7, then the ***a*** allele must have a frequency of 0.3.

In other words, the allele frequencies did not change from one generation to the next. Furthermore, if we continue to assume that reproduction is random with respect to this gene, the genotype frequencies will also remain the same:

- ***AA*** has a frequency of 0.49.
- ***Aa*** has a frequency of 0.42.
- ***aa*** has a frequency of 0.09.

This simple exercise reveals some important points:

- Allele frequencies did not change in the next generation, and the genotype frequencies also do not change.
- If we know an allele frequency, we can infer a genotype frequency so long as we assume that any allele can unite with any other allele at random.
- Likewise, if we know a genotype frequency, we can infer an allele frequency, and neither the allele frequency nor the genotype frequencies change.

When neither the allele frequencies nor the genotype frequencies change, we can say that the population is at genetic equilibrium. When this equilibrium is reached in the absence of evolutionary influences at play, then it is known as a **Hardy–Weinberg equilibrium**.

Although their names are forever linked in population genetics, Hardy and Weinberg did not know each other. Godfrey H. Hardy was a British mathematician, while Wilhelm Weinberg was a German physician. In fact, for 35 years, no one recognized that they had separately published the same principles in 1908, in part because Weinberg's paper was in German. The geneticist Curt Stern, who was educated in Germany, pointed out in 1943 that the two papers had independently reached the same conclusions, and thus the principle became known as the Hardy–Weinberg (H–W) equilibrium many years later, after Weinberg had died and shortly before Hardy's death. The American geneticist William Castle had also independently reached these conclusions at about the same time, so it is occasionally referred to as the Hardy–Weinberg–Castle equilibrium.

All three men would probably have regarded this principle as a small part of their long and distinguished careers; in fact, Hardy referred to it as a "very simple little point" in his letter to *Nature* that described the principle. Hardy published extensively on number theory and mathematical analysis; Weinberg was among the first to study twinning in mammals and introduced the concept of ascertainment bias in statistics; and Castle, among his many other achievements, inspired Thomas Hunt Morgan to work on *Drosophila melanogaster* and was a founding editor of the journal *Genetics*.

KEY POINT If allele frequencies remain stable through generations over time, and there are no evolutionary influences acting upon them, the population is said to be in Hardy–Weinberg (H–W) equilibrium.

X-linked genes

So far, our discussion has dealt with situations in which an individual has two alleles at a locus, that is, with diploids. For a haploid organism, the allele frequency and the genotype frequency are the same. Similarly, for X-linked genes in a diploid, the allele and genotype frequency of the trait in males are also equal since males have only a single copy of X-linked genes. We also have not considered situations in which allele frequencies are different in the two sexes, but these can be readily solved by elementary algebra.

Alleles and haplotypes

It is convenient and appropriate to discuss allele frequencies, but with our knowledge of genomics, it is also important to point out that these same principles apply to haplotypes. After all, a haplotype is simply the region of the chromosome that is inherited along with an allele. Haplotype frequencies in human populations can be assessed using DNA microarrays, as described in Chapter 10, which makes our original thought experiment about determining allele frequencies in a population a bit more realistic. Haplotype frequencies in various populations are often presented as a pie chart, with each color on the chart (or slice of the pie) representing a different polymorphic haplotype. This is explored more fully in Communicating Genetics 16.1.

COMMUNICATING GENETICS 16.1 Depicting haplotype frequencies

Haplotype information can be collected from the individuals in a population or from individuals in multiple populations. But how can the haplotype relationships and distributions be depicted in a meaningful way? Several approaches are commonly taken, depending on what aspect of the data is being highlighted. Let's consider three distinct approaches.

Depicting the relationships between different haplotypes

The most common way to show how different haplotypes are related to one another is to produce a haplotype network like that shown in Figure A. Several algorithms and software packages exist to do this. In these networks, each distinct haplotype is represented as a circle. The size of the circle reflects the relative abundance of that haplotype in the collection of individuals that are being analyzed. For example, the study in Figure A examined a small stretch of mitochondrial DNA from Barbary macaques, a species of Old World monkeys and the only macaques found in Northern Africa. The most abundant haplotypes are M16 and M02, which have the largest circles. The network of relationships among the haplotypes was constructed using methods described in Tool Box 4.3 for phylogenetic trees, with the connecting lines being single mutational events. When more than one mutational event is involved, each additional mutation is indicated by a dot on the line. For example, there is one change in the DNA sequence that distinguishes haplotype M20 from M19, but there are three differences that distinguish M18 from M19.

Some of the circles are shown as pie charts because that haplotype was found in macaques from different samples. For example, haplotype M09 is found primarily in macaques from the Rif site in Morocco, shown in orange, but also in some macaques from the Gibraltar site, shown in green. This could indicate a past or current movement of the macaques among the different locations. Since the M09 haplotype is found primarily in Morocco and is most closely related to other Moroccan haplotypes, its presence in Gibraltar suggests a movement north from Morocco. This type of figure is useful for showing the relative frequencies and genetic similarities of each haplotype, and can suggest relationships within the populations as well.

COMMUNICATING GENETICS 16.1 Continued

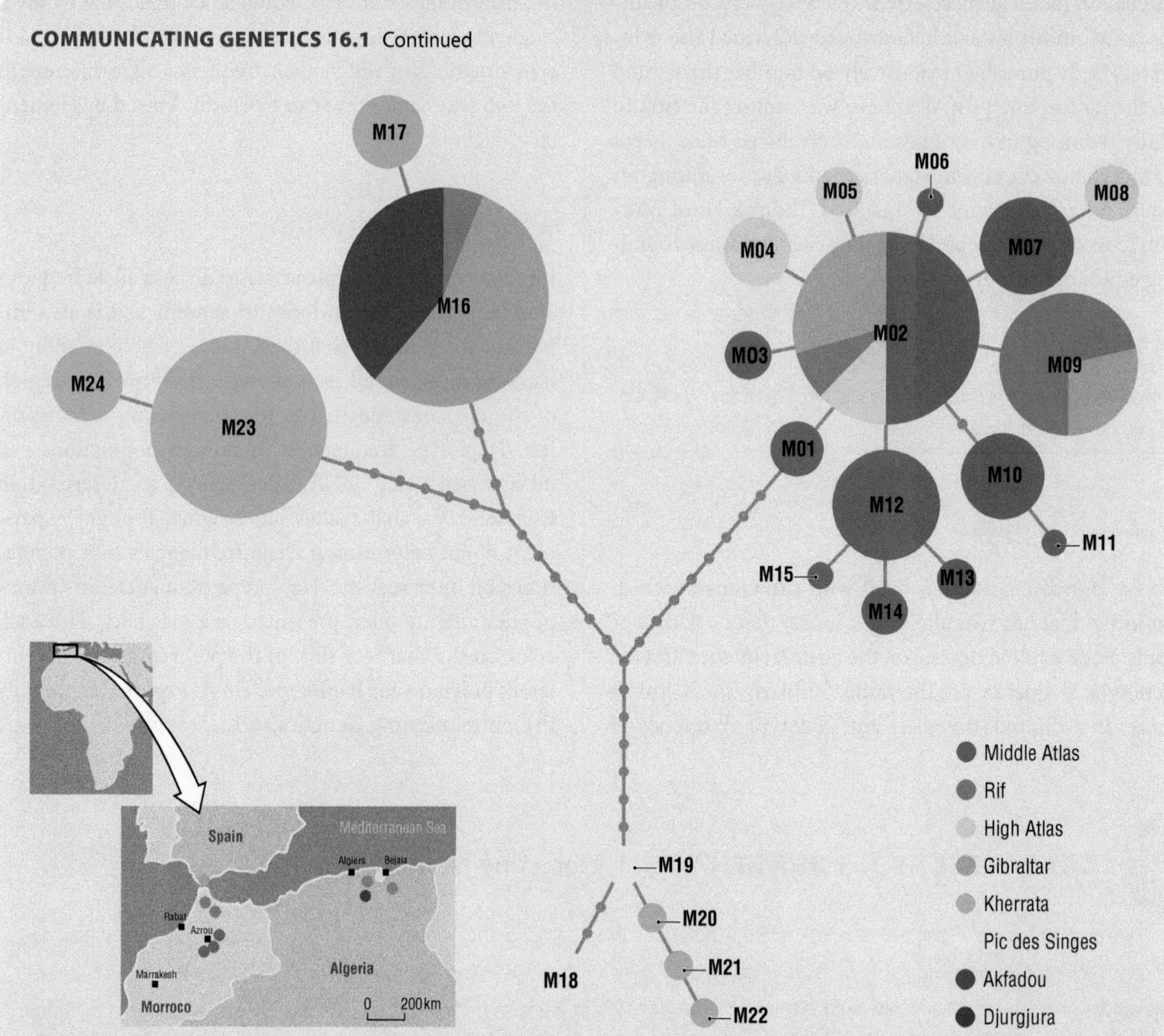

Figure A Haplotype networks. A small stretch of mitochondrial DNA from Barbary macaques was analyzed, and the relationships plotted. Each haplotype is represented as a circle, the size of which indicates the relative abundance of that haplotype. Within each circle, a pie chart shows in which samples that particular haplotype was found and in what proportion. For example, haplotype M09 is found in individuals from two different sample locations: the Rif site in Morocco, shown in orange, and the Gibraltar site, shown in green. The lines between two circles indicate the number of mutational steps that separate those two haplotypes.

Depicting the geographical distribution of haplotypes

In some cases, the geographical distribution of the various haplotypes is of more interest than the relationships of the haplotypes to one another. In these cases, it is common to display the haplotypes as colored circles on a geographical map. For example, Figure B shows the geographical distribution of the five distinct haplotypes associated with the sickle-cell allele ***HbS***, each haplotype named for the country where it was first identified. The colors here represent different haplotypes.

Note that the relative frequency of the haplotype within the population at large is not shown, nor are the sequence similarities among the haplotypes. In this particular figure, the circles are of the same size, so this is showing only those people who have an ***HbS*** haplotype, and not how common the ***HbS*** haplotype is in the population; ***HbS*** is not equally frequent in all of these populations, and other figures may use circles of different sizes to show the frequency of the ***HbS*** haplotype in the population at large. Again, some of the circles are pie charts, but in this case, the colors within the pie chart represent the

COMMUNICATING GENETICS 16.1 Continued

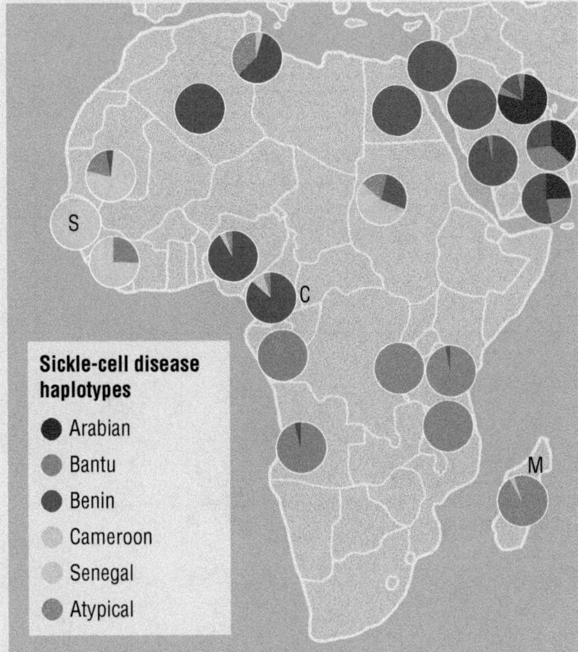

Figure B Geographical distribution of haplotypes for the sickle-cell allele *HbS*. The ***HbS*** allele provides some protection against developing falciparum malaria when heterozygous. Five different haplotypes are shown, named after the country from which each was first identified. Haplotype occurrence is plotted onto a geographical map, with a circle representing the location and a pie chart within the circle indicating the presence and relative abundance of the different haplotypes present in that sample. Since the data were published, the country shown as Sudan has separated into Sudan and South Sudan.

Source: Reproduced from Bitoungui, Valentina J. Ngo et al. (2015). Beta-Globin Gene Haplotypes Among Cameroonians and Review of the Global Distribution: Is There a Case for a Single Sickle Mutation Origin in Africa? *OMICS : a Journal of Integrative Biology* 19:3.

frequencies of different ***HbS*** haplotypes among affected individuals.

For example, we see in Figure B that the Senegal haplotype, represented in yellow, is indeed found in Senegal (S on the map) but is also found in many other areas, including Madagascar (M on the map). We can also see from the circle located in Madagascar that more than one haplotype is found there. Most individuals with the ***HbS*** allele in Madagascar carry the ***HbS*** allele on the Bantu haplotype shown in orange, but some individuals have an ***HbS*** allele associated with the Senegal haplotype (yellow). Still others are atypical. Likewise in Cameroon (C on the map), we can see that some individuals with an ***HbS*** allele carry the Cameroon haplotype (green), but, in fact, most individuals with ***HbS*** alleles in this area carry it on a Benin haplotype (red) and some carry the atypical haplotypes (gray).

When haplotypes are displayed on a geographical map in this way, careful attention must be paid to the geographical location aspect of the figure. For example, if a map was shown with one pie chart pointing to Africa and another to India, does this mean the data in the first pie chart represent the combination of haplotypes found across all of Africa, or is it just the haplotypes found in the particular country in Africa, or even perhaps a particular city? Different conclusions might be drawn in these different scenarios. The only way to be sure is to read the figure legends and study descriptions carefully to find out exactly where the samples were taken and how they might have been combined.

Depicting the chromosomal distribution of haplotype data

The two examples described so far are both scenarios that looked at haplotypes for a single region of the genome. However, some studies involve collecting data for many haplotypes, or even all known haplotypes, for many individuals of a particular species. In this case, the data are collected and displayed in such a way that the entire genome is represented graphically, by featuring pictures of individual chromosomes, for example. All known haplotypes are indicated for all regions of the genome, so that it is possible to look at a particular region of a chromosome and focus on any gene of interest to see:

i. Which haplotypes have been reported for that stretch of DNA, and
ii. Where the individuals with these haplotypes came from.

This interpretation is often done through a web interface, commonly known as a genome browser.

Such studies are large-scale projects that involve the direct sequencing of thousands of single-nucleotide polymorphisms (SNPs) or entire genome sequences. The most extensive project of this type is the human HapMap project. The goal of the International HapMap Consortium is to create a database of the haplotypes present across the entire genome from individuals around the world and to make these data publicly available.

Figure C shows a screenshot captured from the human HapMap genome browser displaying a particular region of the *CFTR* gene. Alterations in the function of this gene underlie the disease cystic fibrosis. On the top of the screen, an overview of human chromosome 7 indicates the location of the *CFTR* gene. Below, we focus on the *CFTR* gene itself. In this region, each small red triangle indicates the position of a documented SNP variant. The cursor is held over one triangle, and the box that is then opened supplies more information about this particular variant in this case, the variation is between a G or an A nucleotide at this position. For example, in the Yoruba population sample from Ibadan, Nigeria, 12% of individuals carry an A and 88% a G, whereas in the Han Chinese population sample from Beijing, China, all individuals carry the G.

COMMUNICATING GENETICS 16.1 Continued

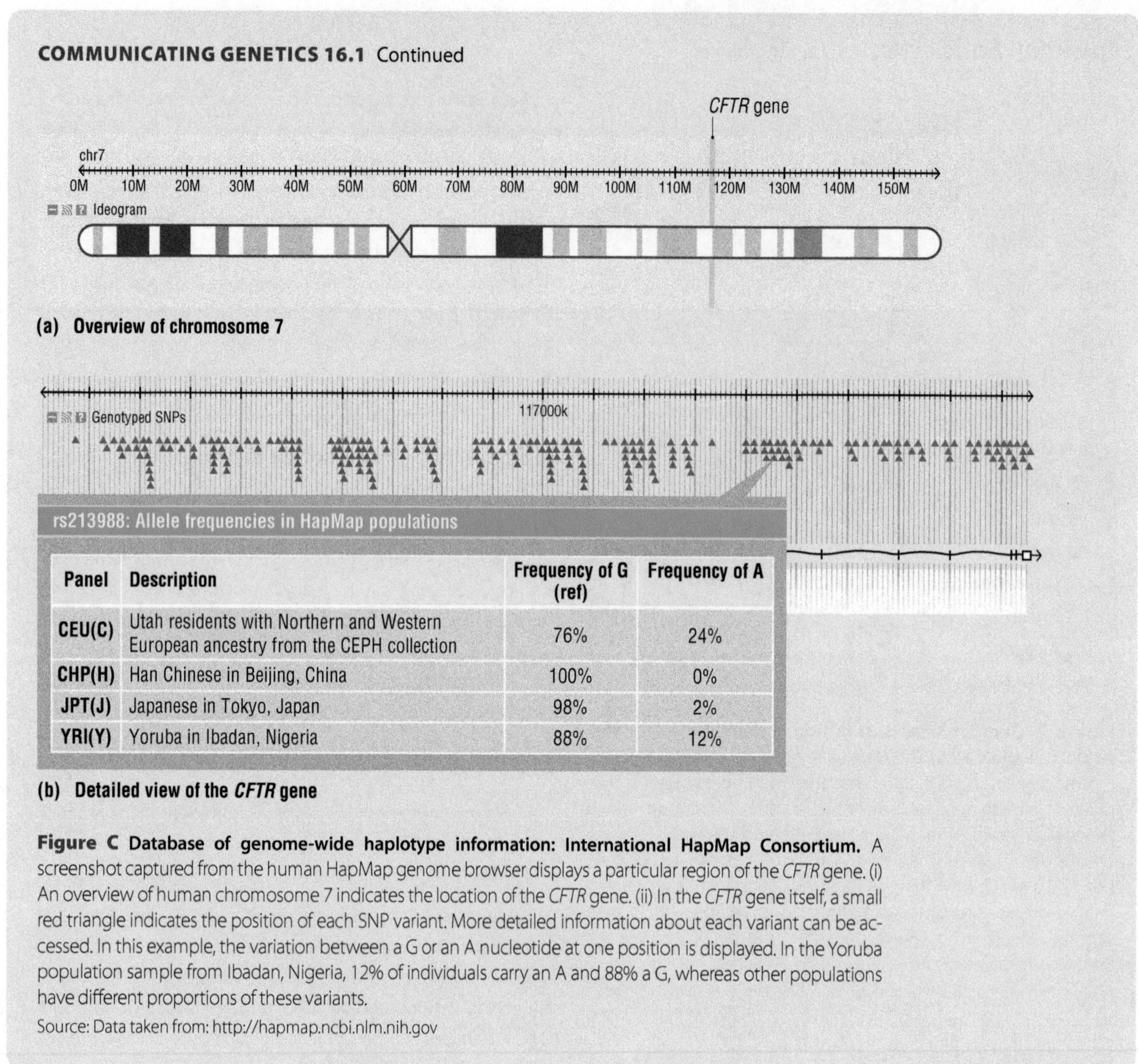

Panel	Description	Frequency of G (ref)	Frequency of A
CEU(C)	Utah residents with Northern and Western European ancestry from the CEPH collection	76%	24%
CHP(H)	Han Chinese in Beijing, China	100%	0%
JPT(J)	Japanese in Tokyo, Japan	98%	2%
YRI(Y)	Yoruba in Ibadan, Nigeria	88%	12%

Figure C Database of genome-wide haplotype information: International HapMap Consortium. A screenshot captured from the human HapMap genome browser displays a particular region of the *CFTR* gene. (i) An overview of human chromosome 7 indicates the location of the *CFTR* gene. (ii) In the *CFTR* gene itself, a small red triangle indicates the position of each SNP variant. More detailed information about each variant can be accessed. In this example, the variation between a G or an A nucleotide at one position is displayed. In the Yoruba population sample from Ibadan, Nigeria, 12% of individuals carry an A and 88% a G, whereas other populations have different proportions of these variants.

Source: Data taken from: http://hapmap.ncbi.nlm.nih.gov

16.3 Changes in genotype frequencies: non-random mating

As shown in Section 16.2 and in Figure 16.4, the frequencies of alleles and genotypes in a population that is at H–W equilibrium will not change from one generation to another. For many genes in humans and other species, an H–W equilibrium applies. But most of the interesting examples in biology come from situations in which the H–W equilibrium does not apply for a particular gene in a population.

WEBLINK: We show how the Hardy–Weinberg equations can be applied to help us understand the frequency of a haplotype in a human population in Video 16.1. Find it at **www.oup.com/uk/meneely**

In order to describe a population as being at H–W equilibrium, five key assumptions have to be made:

- Mating within the population occurs at random
- No mutation occurs
- There is no migration between populations
- The population is extremely large
- There is no selection acting on the population.

You can probably imagine situations in which these assumptions do not apply, and it can be helpful to think through the biological factors that violate each of the

Table 16.1 The assumptions for a Hardy–Weinberg equilibrium

Assumption	Violation (source of change)	Impact on population
Mating occurs at random	Non-random or assortative mating	Genotypes only: stratification or heterozygosity
No mutation occurs	Mutation	New alleles arise
No migration	Migration occurs	If migration is prevented, populations diversify If migration occurs, populations become more similar
Extremely large population	Genetic drift	Allele frequencies change at random
No differential reproduction	Selection	Varies depending on type of selection

assumptions of the H–W equilibrium, thus changing the genetic structure of a population from one generation to the next (at least with respect to one particular allele or genotype). These factors are summarized in Table 16.1. Four of these five factors—mutation, migration, genetic drift, and selection—affect both the allele frequencies and the genotype frequencies; they are shown diagrammatically in Figure 16.5. One of them—non-random

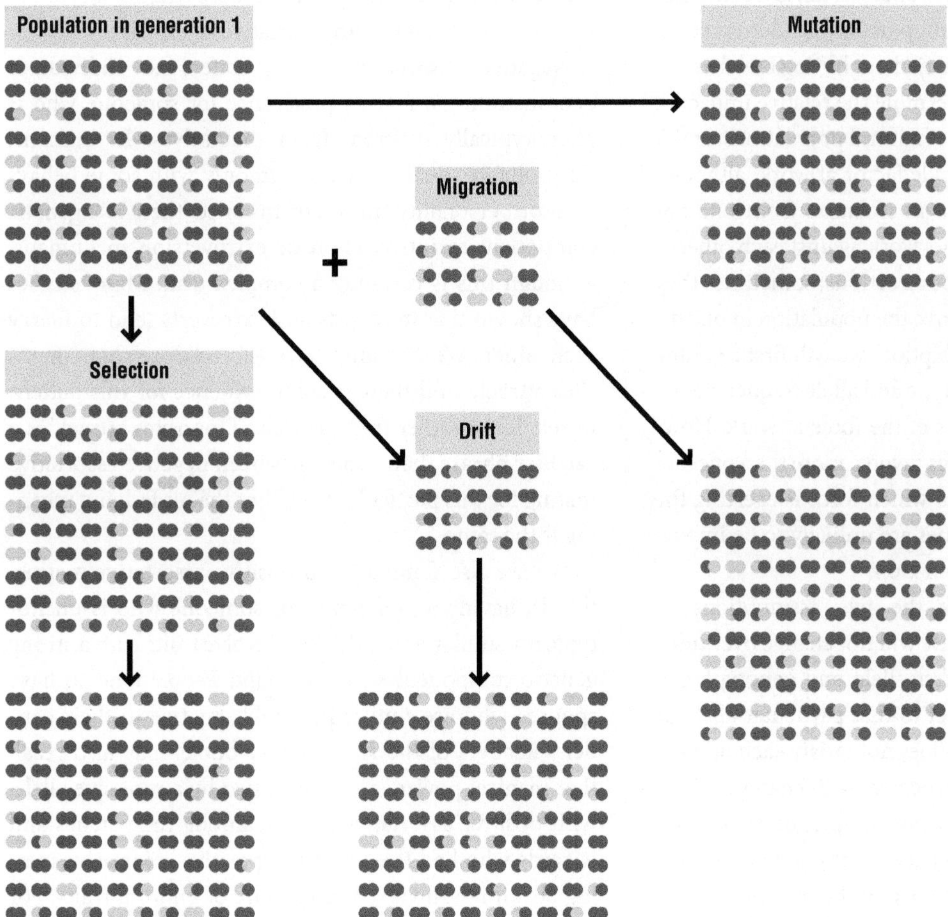

Figure 16.5 Summary of forces affecting allele frequencies. When in equilibrium, allele frequencies do not change from one generation to the next. Four forces that affect this equilibrium are summarized here and developed more fully in subsequent figures. Mutation creates new alleles, shown in red; migration from another population introduces alleles at different frequencies; genetic drift results in random fluctuations in allele frequencies, and selection favors one allele or genotype over the others as shown by the increasing number of blue circles.

mating—affects only the genotype frequencies, and not the allele frequencies. We will discuss each of these factors in this section and in Section 16.4.

Whenever a population changes over time—that is, when it has evolved—some combination of these five factors will have been in effect. If two populations differ from one another in their genetic structure, at least one, and quite possibly more, of these five factors has been at work. In fact, since most populations do differ from one another, the most important thing to determine is which combinations of these five factors have played the greatest roles in shaping these differences.

For example, what is the population genetics explanation that underlies our ability to tell one breed of dog, say a St Bernard, from another, say a Pomeranian? These are two different populations of the same species, but they clearly differ in many phenotypes. Thus, the two breeds must differ in the frequencies of some alleles of the same genes. The explanation for these differences in allele frequencies likely involves some combination of non-random mating, genetic drift, and selection carried out by dog breeders.

It is not always easy to determine the relative impact of each factor on the evolution of natural populations, and it is unusual to have only a single factor at work, although it is helpful to discuss them in isolation from one another. Sometimes these factors work against each other—essentially pulling in opposite directions; sometimes they reinforce one another and drive the population in one direction. In our ensuing description, we will first examine what the factor does to genotype and allele frequencies by itself and then give examples of the force at work. However, nearly all of the examples involve more than one factor, so the focus has to be on which force is exerting the greatest effects and which ones are unlikely to be important for that gene in that population.

If a population satisfies the five assumptions in Table 16.1, its genetic structure will not change over time, and we can accurately predict allele and genotype frequencies, at least with respect to that particular allele or haplotype. If the population does not satisfy each of these assumptions, its genetic structure will change. There will be genetic (and probably phenotypic) diversity both within the population and between that population and a neighboring population, and the population will evolve.

KEY POINT There are five biological factors that affect the frequency of a particular allele or genotype over time. These are mutation, migration, genetic drift, selection, and non-random mating.

We will begin our discussion with the effects of random and non-random mating.

Assortative or non-random mating

In calculating the genotype frequencies from the allele frequencies in Section 16.2, we made the important assumption that mating occurs at random with respect to this trait. This certainly seems like a reasonable assumption for many human traits and for populations in other species as well. For example, we don't choose our mates based on their Rh blood type or their sensitivity to the blood-thinning drug warfarin, so mating is at random with respect to these genetic traits. For these and many traits, the assumption of a random mating is perfectly valid.

But what happens if there is non-random mating for a particular trait? The effect is illustrated in Figure 16.6. The technical term for non-random mating is **assortative mating**, and it can be either negative or positive.

Negative assortative mating refers to individuals demonstrating a mating preference for someone who is phenotypically different from themselves. In humans, we probably see this most commonly with some behavioral or personality traits. For instance, there is a genetic component to introversion or extroversion in humans, although this is certainly a complex trait. Many studies have shown that introverts and extroverts tend to marry each other. We certainly have the folklore that "opposites attract," and there is some evidence for this pattern in relation to other traits as well. "Opposites attract" is a catchier phrase than "There has been negative assortative mating for this particular trait," but the underlying meaning is the same.

We are also familiar with positive assortative mating, that is, having a preference for someone who is phenotypically similar to us. This can involve traits with a strong genetic component such as height. People tend to have spouses who are similar in height, relative to their gender. (Readers of the *Harry Potter* books may also recall that magical powers are determined by a recessive allele, with positive assortative mating among the witches and wizards who had the trait.) But positive assortative mating also might involve non-genetic or cultural traits. For example, German speakers tend to marry other German speakers, people tend to marry within religious groups, and so on. These are not genetic traits—there is no allele for speaking German, for example—but cultural preferences can reinforce traits common to that population that

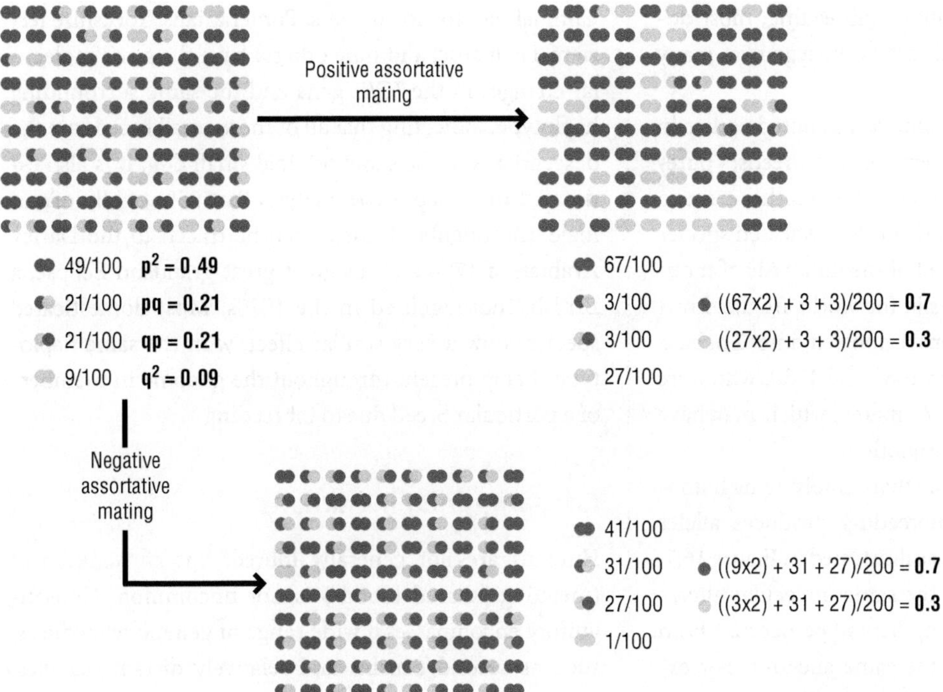

Figure 16.6 Assortative or non-random mating. One assumption of a Hardy–Weinberg equilibrium is that mating occurs at random among the individuals in a population. Non-random or assortative mating occurs when individuals have a mating preference. If the preference is for genetically or phenotypically similar individuals (that is, positive assortative mating), the number of homozygotes (with two blue or two yellow circles) will increase, while the number of heterozygotes (one blue and one yellow circle) will decrease. Under negative assortative mating, individuals prefer genetically or phenotypically dissimilar mates. In this case, the number of homozygotes will decrease, while the number of heterozygotes will increase. In either case, while the frequencies of particular genotypes changes, the frequencies of the alleles do not change.

do have a genetic basis such as height or pigmentation patterns in the hair, skin, and eyes.

KEY POINT Mating is frequently non-random, or assortative. Positive assortative mating indicates a preference for a mate of a similar phenotype, while negative assortative mating means there is a preference for a mate that is of a different phenotype.

Assortative mating changes population structures

Since non-random mating does not by itself change the allele frequencies in a population, how does it affect the genetic structure of the population? Its effects are seen on genotype frequencies, and arise from the principles of single-gene inheritance, outlined in Chapter 5 and shown in Figure 16.6. If ***AA*** individuals tend to mate with ***aa*** individuals (that is, negative assortative mating is occurring for the ***A*** locus), many individuals in the population will be ***Aa***. Thus, negative assortative mating tends to maintain heterozygosity in the population for that trait. On the other hand, positive assortative mating tends to produce more homozygotes and fewer heterozygotes for that trait. If ***AA*** individuals prefer to mate with other ***AA*** individuals, and ***aa*** prefers to mate with other ***aa*** individuals, then the two homozygotes will increase in frequency and the number of heterozygotes for this gene will decrease.

On a simple level, then, negative assortative mating will tend to keep the population more heterogeneous, while positive assortative mating will tend to split the population into subpopulations that mate among themselves but not with the other subpopulations.

The most extreme form of positive assortative mating is **inbreeding**. Inbreeding is defined as the mating of two individuals who are genetically related. Many plants and some animals can self-fertilize, so this is the ultimate form of inbreeding. Many domesticated species of animals are inbred—they are produced by the intentional mating of closely related individuals such as siblings or even parents and their offspring. As a result of inbreeding, most domesticated plants and animals are homozygous at many loci in the genome.

One of the most familiar examples, as noted earlier in this section, is the establishment of dog breeds; "purebred" in dogs and other species is often another term to describe inbreeding. Inbreeding for domesticated species can be quite extreme. Nearly all of the hundreds of thousands of Thoroughbred horses in the world are the direct descendants of three stallions imported into England from the Middle East between 1680 and 1730, who were mated to one of about 50 to 75 mares, which may have themselves been related to one another.

But inbreeding includes more than simply being homozygous at a single locus. Inbreeding produces alleles that are **identical by descent**, as illustrated in Figure 16.7, that is, individuals will have the same molecular alteration and the same surrounding haplotype because both parents are descended from the same ancestor. For example, at least 14 different breeds of small dogs (that is, Chihuahuas, toy fox terriers, Pomeranians, Yorkshire terriers, Pekingese, and nine others) have the same molecular changes in the *IGF1* gene and the same surrounding haplotype, indicating that all of these small breeds are the descendants of the same original small dog. It is also estimated from haplotype analysis that 95% of all current male Thoroughbred horses can be traced to the Darley Arabian of 1704 and his great-great-grandson Eclipse, a British Thoroughbred in the 1770s. Most domesticated species show a very similar effect, with the same haplotypes being present throughout the genome in members of a particular breed due to inbreeding.

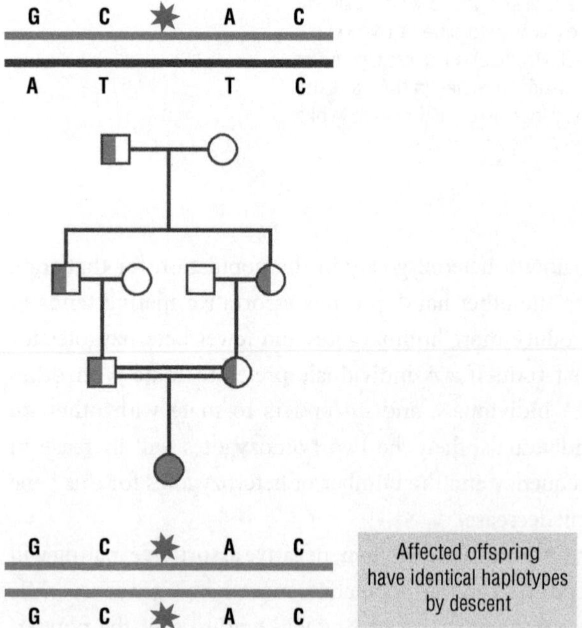

Figure 16.7 Common ancestry results in haplotypes and alleles that are identical by descent. The allele producing the trait is shown as a red star, arising on a haplotype (in orange) with four polymorphic sites shown; the allele arose as a heterozygote, with the other haplotype having other alleles at the polymorphic sites. Heterozygotes for this haplotype or allele are shown as symbols with orange sectors. A pedigree of four generations is shown for simplicity, but the same principle occurs in other populations. In this pedigree, the individuals in the third generation mate and produce an affected offspring. This affected individual is homozygous for the original allele or haplotype because of the shared ancestor.

Consanguinity in humans

Humans are not generally inbred, but **consanguinity** (meaning "shared blood") is not uncommon. Consanguinity encompasses a wide range of genetic relatedness, such as second cousins and relatively distant relatives, and is used commonly in genealogical or legal cases. The degree of relatedness can be calculated by the coefficient of relationship, illustrated in Figure 16.8. The coefficient of relatedness is the probability that two individuals are identical by descent for a particular allele; alternatively, it is the fraction of the haplotypes in the genomes that are likely to be shared because of a common ancestor.

Even if the actual genetic relationship is fairly distant, consanguinity can still have an effect; people in many

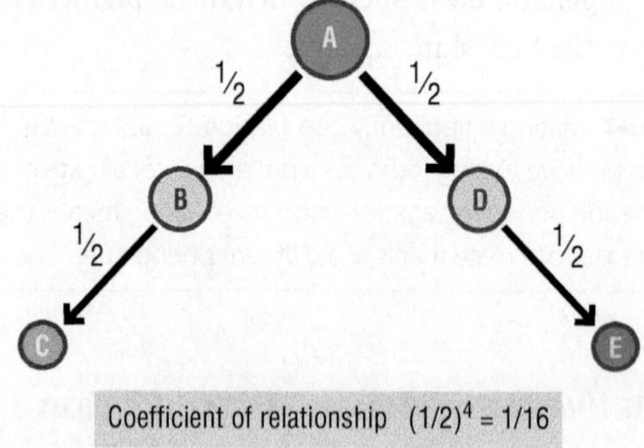

Figure 16.8 The coefficient of relationship. The coefficient of relationship is a measure of the degree of relatedness or consanguinity between two individuals, shown here as the green (C) and purple (E) circles. The common ancestor is shown in blue, and the generations between C and E are counted; the total number of generations between them is four. The coefficient is one-half raised to the power of the number of generations (one-half because each offspring inherits half of their alleles from each parent). Thus, these two individuals are expected to share roughly 1/16 or 6% of their alleles or haplotypes because of their descent from a common ancestor. If multiple common ancestors are found, as frequently occurs, each lineage is computed separately and the results are added.

small populations who are not closely related can still share some genetic heritage. This genetic similarity may arise not from a single shared ancestor, such as a grandparent or great-grandparent, but from multiple shared ancestors, great uncles, second cousins, and so on, who are more distantly related.

In human populations, the closest form of consanguinity that occurs commonly is first-cousin marriages. Many societies allow and encourage first-cousin marriages, but others do not, making the fraction of first-cousin marriages both population-dependent and difficult to assess. First-cousin marriages carry social stigma in much of the Western world and are illegal in 30 states in the United States. But in some countries and cultures, notably among some people of the Middle East, South Asia, and sub-Saharan Africa, first-cousin marriages are both common and preferred, so that as many as 60% of the people in some populations are related by cousin marriages.

On a worldwide scale, it has been estimated by the World Health Organization that approximately 20% of human populations have a preference for consanguineous marriage, and at least 8.5% of children in the world are the offspring of first-cousin marriages. The most frequent form is that a young man will marry his father's brother's daughter, that is, his cousin.

Because there is such a strong social stigma attached to first-cousin marriages in the West, data on the actual genetic consequences are not very extensive. One illustrative study used British Pakistani children in Birmingham, England, in which 25% of all children born in the study were of Pakistani, Bangladeshi, or Middle Eastern origin. A comparison was made between 956 babies of British Pakistani birth and 2432 babies from other British couples. Among the non-Pakistani British couples, only 0.4% of them were related. Among the British Pakistani couples, 69% were related and 57% were first cousins. In fact, because the couples might share multiple ancestors, the actual degree of consanguinity is somewhat higher than that.

Among the non-Pakistani British couples, the prevalence of all congenital and genetic disorders was 4.3% of all births and that of known recessive disorders was 0.28% of births (or 6.5% of those with a congenital disorder). Among the British Pakistani couples, the rate of congenital disorders was 7.9%, not quite twice as high, but the rate of recessive disorders was about 3.3% of births or 41.7% of congenital disorders, more than ten times higher than the rate in the non-consanguineous families. So while the overall frequency of congenital disorders due to all factors increased somewhat, the frequency of congenital and genetic disorders that can be attributed to recessive alleles went up substantially more. This occurs because the individuals become homozygous for recessive traits due to shared ancestry.

The cultural reasons for consanguineous marriages include strengthened family ties for children and the elderly, as well as family support for women (since a bride's in-laws are also her aunt and uncle). The social impact of congenital disorders arising from consanguineous marriages in some cultures may be lower than in the West because strong extended family systems can provide important care at a lower cost. However, recessive genetic diseases do exert a tremendous health impact on any family and tend to have higher rates of mortality than other congenital disorders, so the effects are notable irrespective of where they occur. The World Health Organization has taken the approach that, if the first child from a cousin marriage has a serious genetic illness or dies of such a genetic disease, the couple should be encouraged to have no more children. This approach attempts to recognize the cultural values behind consanguineous marriages while continuing to identify the genetic risk.

KEY POINT Consanguinity refers to genetic relatedness stemming from one or more shared ancestors. Increases in consanguinity in a population can lead to increased frequencies of recessive conditions.

16.4 Factors affecting allele frequencies

As we have just discussed, non-random mating affects genotype frequencies but does not directly affect allele frequencies, at least not if that is the only factor affecting the population. Some genotypes (and phenotypes) become more common as a result of non-random mating, but the frequencies of the alleles do not change. In fact, a single generation of random mating will restore the genotype frequencies to the H–W equilibrium, if non-random mating is the only factor at work in the population.

Four factors change the allele frequency in populations, and thus also affect the genotype frequencies. These factors often work in combination with each other and with non-random mating to shape the genetic structure of a population. Without over-simplifying, all of the genetic variation that we observe in any species, including humans, stems from the biological impact of these factors. Our ability to distinguish one person from another is largely due to the actions of these factors in the past (plus, of course, environmental factors like clothing choices, hairstyles, and nutrition).

More than just recognizing individuals, we can see the effects of these factors in the structure and sequence of the human genome. Our genome, like any genome, is a scrapbook of our history. We can see the remnants of where the species has been and what we have done. Like any scrapbook, we can look at some remnants and wonder why something was saved at all. Some of it seems to make no sense; we may not even know where it came from or what it once represented. Other pieces evoke the great memories that defined us and not only tell us where we have been but also suggest where we might be going. The four evolutionary factors that shape our genome—the scrapbook of our species—are mutation, migration, genetic drift, and selection. Let's now consider each of these in turn.

In the discussion that follows, we have chosen to emphasize the concepts with each of these four factors, rather than provide any information about the quantitative theoretical work that underlies these concepts. Box 16.1 provides a little more information about some of the parameters, assumptions, and quantitative concepts behind our discussion of each of these four factors that affect allele frequencies in a population and thus could result in evolutionary change. Population genetics is a rich field for mathematics and statistics, for which we provide a more qualitative introduction.

BOX 16.1 *Quantitative Toolkit* Quantitative concepts in population genetics

Population genetics is among the most quantitative and statistical fields of genetics; in fact, many population geneticists are also quite accomplished in mathematics and statistics. Our approach in the chapter has been to focus on the quantitative reasoning behind these population concepts without attempting to express these as equations. In this Box, we develop a few of these quantitative concepts a little more fully but will continue to emphasize the underlying concepts, rather than the more rigorous mathematical elements.

The mathematical and statistical analysis of populations is a rich and mature field that deserves to be explored more thoroughly by those with such an interest, and all of the quantitative reasoning concepts that we have included in this chapter have well-developed theoretical foundations.

In addition to the concept of the effective population size (N_e), as defined in the chapter, two other parameters are generally applicable. The first of these is the change in the allele frequency, symbolized as Δq. Mutation, migration, genetic drift, and selection all result in changes in the allele frequency, so Δq is an important parameter to measure or calculate. The second of these is the mean time to fixation and the related concept of the rate of fixation. These are used to calculate the expected number of generations it will take for an allele to become fixed—that is, to reach a frequency of 1.0—under different conditions, or how quickly fixation will occur.

Without going into the mathematical theory, here are some other important quantitative concepts for the specific factors that affect allele frequencies and thus drive evolutionary change.

Mutation rate

A new allele arises by mutation at a rate symbolized by m. If the frequency of an allele at a particular time is q_0, the frequency of the allele in the next generation will be (q_0 + m), that is, it will be the frequency of the allele in the preceding generation plus the rate at which new alleles arose.

This concept makes several simplifying assumptions. First, this is the rate at which a wild-type allele ***A*** will mutate to a non-wild-type allele ***a***. It ignores the possibility of "back mutation" in which a non-wild-type allele mutates to a wild-type allele (***a*** mutates to ***A***) since that is rare. Second, it includes the other assumptions required for a population at Hardy–Weinberg equilibrium, namely that the population is infinitely large, with no migration and no selection for a particular allele.

This last assumption is particularly important. The frequency of non-wild-type alleles (that is, allele ***a***) is expected to increase linearly in each generation at the rate m; under these assumptions, the non-wild-type allele would eventually become fixed, and the wild-type allele would be lost. This does not happen in natural populations because there is selection against most non-wild-type alleles, so they are lost or purified from the population.

There is also an issue of semantics here. If a newly arisen allele increases in frequency enough to become fixed in the population, it is defined to be the wild-type allele.

Migration

The effect that migration has on allele frequency is fairly intuitive. The change in the allele frequency Δq depends on the frequency

BOX 16.1 Continued

of the allele in the migrant population and the number of migrants relative to the effective population size. For example, if the allele is absent from the incoming migrant population, the allele will become less frequent after migration has occurred; the actual value for Δq will then depend on how many migrants entered the population.

Genetic drift

The precise effect that genetic drift has on allele frequencies is among the most difficult to quantify, simply because drift corresponds to a random sample. On the other hand, the factors that affect Δq under genetic drift are relatively easy to imagine and not very difficult to model based on random changes. Drift will have its greatest impact when N_e is small, that is, "sampling error" is largest when the fewest samples are taken.

With genetic drift, the probability of fixation of an allele is equal to the initial frequency of the allele; if the original allele frequency was 0.9 and the population size is reduced at random with respect to the alleles, the probability that the allele would become fixed by chance is also 0.9. The mean time to fixation depends on the initial allele frequency and on the population size; it will take longer if the population size is large than if the population size is small, and longer for rare alleles to become fixed than for common alleles. For a rare allele, that is, when q is very small, the mean time to fixation for q is approximately $4N_e$ or four times the effective population size.

Again, the other assumptions that enter into calculating the effect of drift are the same as for populations at Hardy–Weinberg equilibrium, namely that other mutations do not occur and that there is no selection. But even rare alleles can become fixed by genetic drift if the population size is small.

Selection

There are many types of selection, so we just provide a general perspective here. In order to define the effects of selection, it is first necessary to calculate the mean fitness of a population. Fitness is symbolized by w, and the mean fitness is the sum of the fitnesses of each different genotype. We can say that the allele ***A***, which has a frequency of p, also has a relative fitness of w_1; allele ***a***, which has a frequency of q, has a relative fitness of w_2. Thus, the average mean fitness is ($w_{11} \times p^2 + w_{12} \times 2pq + w_{22} \times q^2$), or the sum of the fitness of the homozygote ***AA***, the heterozygote ***Aa***, and the homozygote ***aa***. The relative mean fitness is defined to be 1.

If w_{11} and w_{22} are not equal to each other—that is, if ***AA*** and ***aa*** do not have the same relative fitness—directional selection occurs; if w_{11} is greater than w_{22}, then the ***A*** allele will increase in frequency at the expense of the ***a*** allele.

The equation for the relative mean fitness also allows us to see the effects in heterozygotes. If w_{12} is less than w_{11}, then the ***A*** allele will increase in frequency by directional selection. If w_{12} is greater than either w_{11} or w_{22}, balancing selection will occur, as discussed in the chapter, because the heterozygote is the fittest.

The one case not discussed in the chapter is when w_{12} is less than either w_{11} or w_{22}, and the heterozygote is the least fit genotype. This is known as disruptive selection or heterozygote disadvantage. Because the heterozygote is the least fit genotype, the population will ultimately split in two to become only ***AA*** individuals in one population and ***aa*** individuals in the other; disruptive selection is one of the processes that could occur during speciation, as discussed in Box 16.2.

Consideration of the relative fitness of various genotypes leads to the concept of the selection differential, symbolized by s. The selection differential is the difference between the mean relative fitness and w_{22}, that is, the relative fitness disadvantage of ***aa***. The value of s ranges from 0 to 1. When $s = 1$, the individual does not survive and reproduce, and severely deleterious alleles have a fairly large value for s, so their fitness is low. With balancing selection in which ***Aa*** is the fittest, both ***AA*** and ***aa*** have an associated selection differential, which need not be the same; ***Aa*** has a fitness of 1, while ***AA*** has a fitness of $1 - s_1$ and ***aa*** has a fitness of $1 - s_2$. With directional selection in favor of the ***A*** allele, the extent of selection against the heterozygote ***Aa*** is represented by hs, where h is the level of dominance with respect to fitness and s is the selection differential. If the heterozygote ***Aa*** is slightly less fit than ***AA*** (that is, if w_{11} is greater than w_{12}), then the fitness of ***AA*** is 1, the fitness of ***Aa*** is $1 - hs$ and the fitness of ***aa*** is $1 - s$.

As discussed in Chapter 3 and this chapter, many alleles are neutral with respect to fitness, so s is often very small and can be difficult to measure in natural populations. Newly arisen alleles with increased fitness are not common and usually have a very small fitness advantage; in human populations, the largest known fitness advantage for any newly arisen allele is for the lactase persistence allele among Europeans, which apparently had a fitness advantage of about 10%.

Selection works in natural populations of finite size, so it is often helpful to try to determine the relative contributions of selection and drift to the genetic structure of the population. One theoretical calculation that combines these two factors finds that, if s is less than $1/N_E$, drift has a greater effect on the population structure than selection. In other words, if s is less than 0.01 and the population size is 100 or more, drift is the more important factor affecting allele frequencies. This may have been the case for many small human populations in which slightly deleterious or nearly neutral alleles arose.

Mutation

Most new alleles began as mutations of some type, so mutations are the primary source of new alleles and new allelic variation in a population, as summarized in Figure 16.9. Some alleles come from horizontal gene transfer or recombination, but most are due to mutations. As we noted in Chapters 3 and 4, while the rate of mutation for a single gene is very low, it is not zero, so new alleles do in fact arise every generation. Recent work on sequencing the entire genomes of families, for which data are not yet extensive, indicates that a child has about 70 new mutations that are not present in either of the parents. Given that coding regions comprise only about 2% of the human genome, most of these new mutations are in non-coding regions, although there are occasionally important exceptions.

Because the rate of mutation is so low, mutation is typically ignored as a factor in changing allele frequencies from one generation to the next. Some genes have a higher mutation rate than average (for reasons we don't always understand), and mutations are always occurring. But mutation by itself still has minimal impact on allele frequencies. On the other hand, mutations occur in individuals that form part of a population, so their effects are usually seen in combination with one or more other factors at play in that population; after the mutation has arisen, one of the other factors may act to allow it to become more common. For example, the combination of a mutation and non-random mating or inbreeding can produce a high frequency of homozygous mutant individuals.

The mutation in the factor VIII gene that gave rise to hemophilia among the descendants of Queen Victoria was shown in Figure 7.6 when we discussed X-linked inheritance. While this is one of the most familiar examples of the combined effects of a mutation and inbreeding, it is far from the only one. Another example emerged from an extensive analysis of families during the nineteenth and early twentieth centuries in an isolated region in the mountains of eastern Kentucky, USA, and involved a condition known as hereditary methemoglobinemia.

Methemoglobin is a form of hemoglobin in which the iron molecule in the heme group is oxidized and cannot release bound oxygen effectively. In normal individuals, methemoglobin comprises about 1% of our total hemoglobin; the blue color that we observe from blood vessels close to the skin surface is due to the presence of methemoglobin. Methemoglobin arises naturally from multiple causes, and levels of methemoglobin up to 10% of the total hemoglobin present are generally asymptomatic. Methemoglobin is converted back to normal hemoglobin by the enzyme methemoglobin reductase, also known as diaphorase.

Elevated levels of methemoglobin, a condition known as methemoglobinemia, arise from the actions of many environmental agents and various drugs, as well as from mutations in the diaphorase gene. While very high levels of methemoglobin can lead to a coma or even death, levels of up to 20–30% result in only skin discoloration, particularly in the mucous membranes and extremities. The phenotype of individuals with a mutation in the diaphorase gene is quite distinctive, however—affected individuals sometimes have a striking blue tinge to their skin.

Because the phenotype is so distinctive and the biochemistry of hemoglobin is well studied, hereditary methemoglobinemia was among the first hereditary conditions in which a specific enzyme defect was identified.

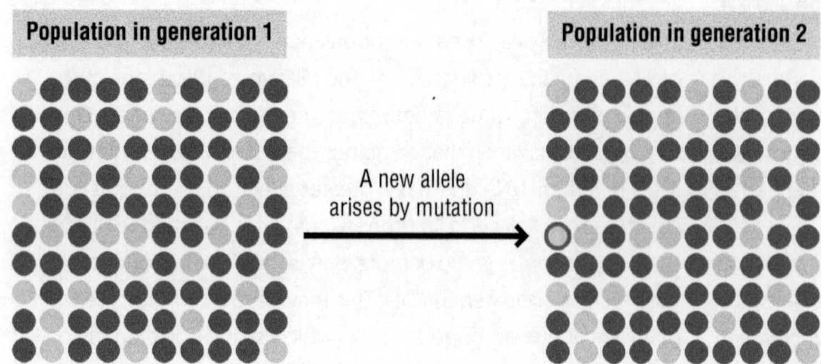

Figure 16.9 Mutation violates one of the assumptions of a Hardy–Weinberg equilibrium and changes allele frequencies. A mutation resulted in one of the blue alleles becoming a yellow allele, represented with a red outline. Thus, the frequency of the blue allele has decreased, while the frequency of the yellow allele has increased.

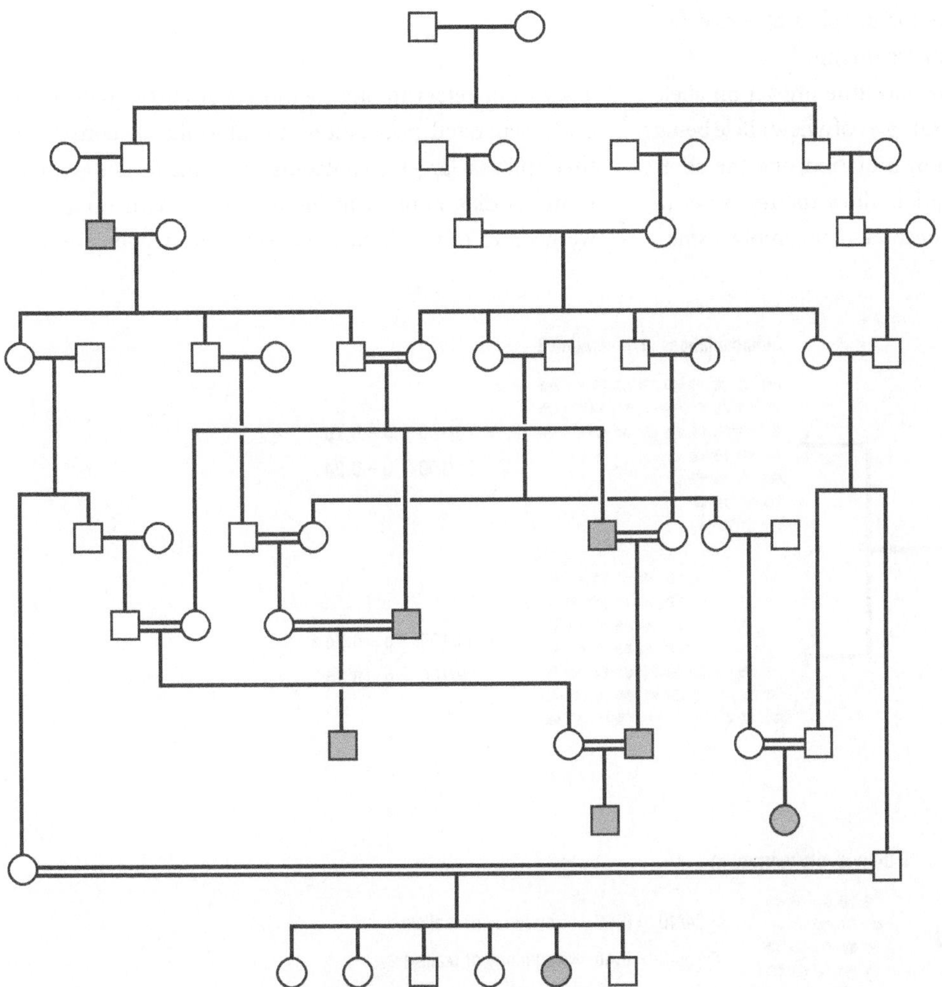

Figure 16.10 In this partial pedigree of these families affected by methemoglobinemia, consanguineous marriages are shown as double lines connecting the parents. Methemoglobinemia appeared in one male but, after two or three generations, occurred frequently within the family, illustrating that a rare trait or recessive mutation can quickly increase in frequency in a local population due to factors such as non-random mating.

Source: Adapted from http://www.indiana.edu/~oso/lessons/Blues/TheBlues.htm. Pedigree information graciously provided by Mary D. Fugate, publisher of The Fugate Family Newsletter.

Individuals with this condition have been reported throughout the world, and it has a particularly high frequency among indigenous populations of Athabaskan Indians in Alaska's interior and Navaho Indians in southwestern USA. In the 1950s and 1960s, hematologists and geneticists at the University of Kentucky Medical Center analyzed the condition extensively after a young boy was brought to a clinic with dark blue skin. Through interviews and family records, the trait was traced back through six generations and more than 150 years to a man named Martin Fugate and his wife Elizabeth Smith; the family became known locally as the Blue Fugates. Part of their pedigree is shown in Figure 16.10.

A high incidence of cousin marriages occurred in these isolated settlements in Kentucky where populations of less than 100 individuals in these settlements were largely derived from four families. Even individuals who were not first cousins often had multiple lines of relationships, so much so that some people joked with the geneticists that they were related to themselves. The condition is considered an autosomal recessive trait, although changes in methemoglobin levels occur naturally with exertion or cold, so some affected individuals may actually have been heterozygotes.

The condition is not generally life-threatening, and many of the "blue people of Kentucky" lived long lives. The condition is also treatable by the administration of an alternative electron donor or reducing agent. Among the most widely used is the dye methylene blue, which is quite effective at converting methemoglobin to hemoglobin, and thus restoring pink pigmentation in the skin. There is a certain ironic symmetry in that the cure for the blue pigmentation was to give the people a blue dye, but this is an effective and inexpensive therapy.

So although mutation alone has little impact on allele frequencies, there are many examples of a new allele being introduced into a population by mutation, and the other factors at play in that population allow the mutation to increase in frequency until it becomes a polymorphism.

Migration: populating the planet

Migration refers to any method that allows two previously separated populations to intermix or, conversely, that splits a larger population into smaller subpopulations, as diagrammed in Figure 16.11. With migration, we need to think about the effects that happen within a

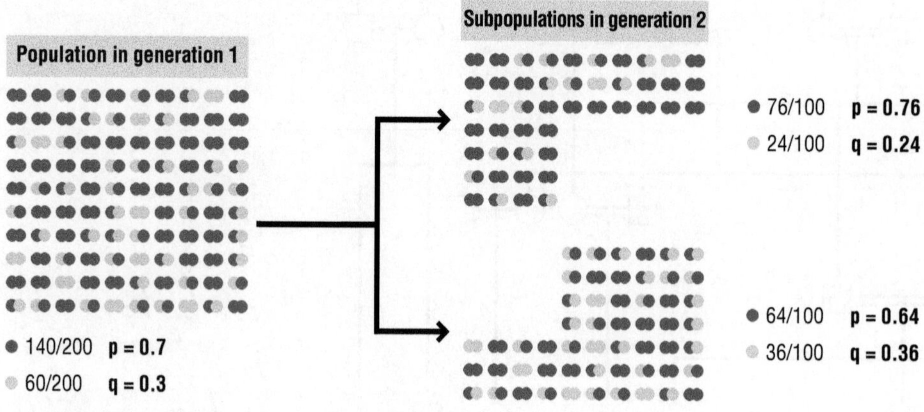

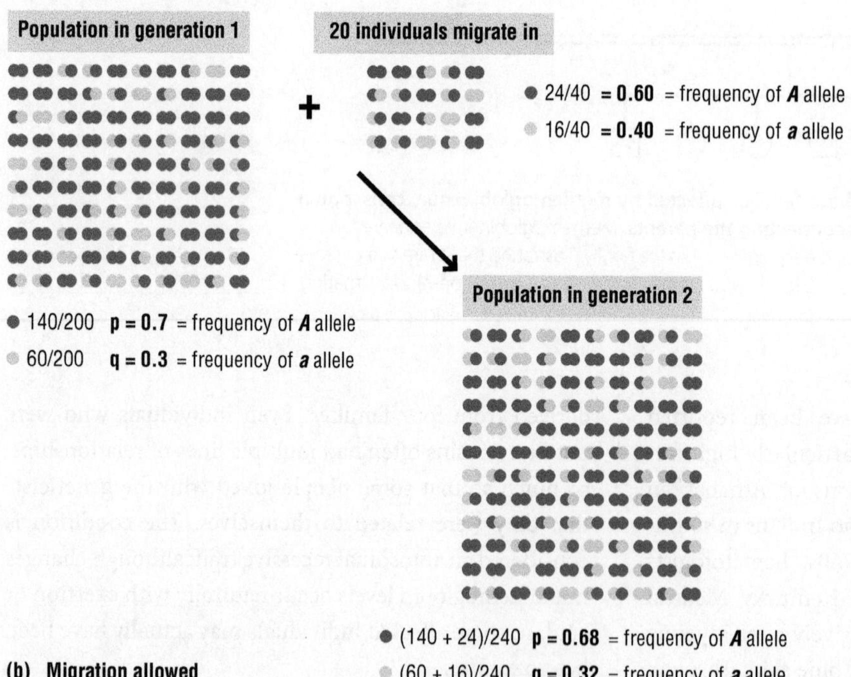

Figure 16.11 Migration changes allele frequencies. Migration refers to the flow of alleles from one population to another, which may or may not involve physical movements. In (a), migration has been blocked between two subpopulations of what was previously one population. When migration is prevented, the two subpopulations will eventually diverge from each other, as shown by the change in allele frequencies for each subpopulation. If migration continues to be blocked and different mutations arise in the two populations, the populations will diverge. In (b), migration of individuals from a different population with a different frequency of alleles ***A*** and ***a*** has occurred. Thus, the allele frequencies of the new combined population become more similar between the populations.

population when migration occurs, and the effects that happen between two populations when migration has been prevented.

We often think of migration in terms of travel or movement of some sort. However, some of the most significant impacts occur when migration becomes limited, such as when geographical or physical barriers separate one population into two. For example, there can be a population of plants in a field that become physically separated when a river becomes diverted through the middle of the field, as suggested in Figure 16.11(a). Over time, subpopulations arise on opposite sides of the river that are distinct from one another, because mutations will occur in each one that don't spread to the other. Many land animals, such as ground squirrels and snakes, on opposite sides of the Grand Canyon show genetic differences in coloration and other traits because the effective populations are separated. The effect is less common in birds, however, which can continue to fly over the Grand Canyon to maintain a single effective population. Such divergence of a population by a geographical or physical barrier is one of the standard models by which two species arise from a single ancestor, as summarized in Box 16.2.

BOX 16.2 *Going Deeper* Speciation

Most of us have at least an intuitive idea of what is meant by the term "species." For example, we understand that a fruit fly and a moth belong to different species. Nonetheless, it can be quite difficult to define the term in a way that is both detailed and all-encompassing. This difficulty is captured in the following quote from Charles Darwin.

> No one definition has as yet satisfied all naturalists; yet every naturalist knows vaguely what he means when he speaks of a species.
>
> Charles Darwin, *Origin of Species*, Chapter 2

The concept of a species has been examined, debated, and challenged extensively over the years, since Darwin first puzzled over how species arose and how they could be identified as such. The concept of a species is one that humans have invented for the sake of classification, so it is not surprising that biology in the real world is more complex. Key contributions to this discussion were made by Ernst Mayr, who proposed that, when a population of organisms becomes separated from the main group by time or geography, the individuals in this population eventually evolve different traits and can no longer inter-breed with the organisms in the main group. Mayr's definition of a species, called the biological species concept, is one of the most widely recognized today and defines a species as follows:

> A species is "a group of individuals that actually or potentially interbreed in nature, and is reproductively isolated from other such groups."
>
> Mayr, 1942

This definition might seem straightforward, but, in practice, it can be difficult to apply. Note that similarity of appearance is not a part of this definition, leading us to ask: what role does morphology play in defining a species?

Other definitions for species concepts have been proposed, including the phenetic species concept that defines species as a set of organisms that are phenotypically similar and distinct from other species. However, at the heart of the biological species concept is the idea of **reproductive isolation**. Understanding how reproductive isolation arises therefore guides our understanding of how different species arise. According to Mayr, reproductive separation between natural populations leads to the creation of new species, so let's briefly consider what processes may give rise to these conditions.

How do new species arise?

The process that gives rise to different species is known as **speciation**, an event represented on a phylogenetic tree by a branching point, as illustrated in Figure A.

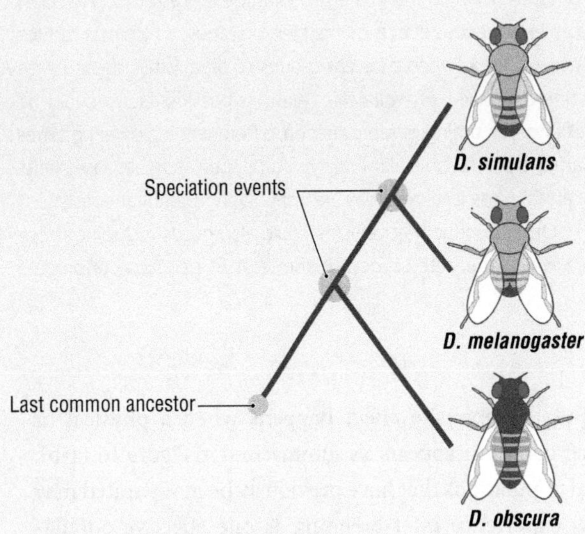

Figure A Speciation. In a phylogenetic tree, speciation events are depicted as branching points, shown here with three species of *Drosophila*.

BOX 16.2 Continued

Mechanisms that give rise to new species generally involve different forms of reproductive isolation, which reduces gene flow among individuals. This then allows for larger genetic differences to accumulate over time between groups within a lineage (probably through mutation followed by selection or drift), eventually causing mating between groups to either not occur or not be successful. Gene flow between different groups of organisms can be reduced in a variety of different ways. For example, in **allopatric speciation**, gene flow is reduced as a result of a geographical barrier that physically separates the organisms, while **sympatric speciation** sees species evolving from a single ancestral species while inhabiting the same geographical region.

Allopatric speciation is widely accepted as being a common way that new species form and is possibly the most common way. The examples in Section 16.4 of a field of plants being separated by a river or of ground-dwelling species on opposite sides of the Grand Canyon provide a framework for thinking about how a geographical barrier could isolate populations and trigger speciation. Sympatric speciation is more frequently contested among scientists, however, as there is much less evidence to support its occurrence.

One way that gene flow could become reduced between parts of a population without a physical barrier is through specialization. For example, if some individuals of a species begin to eat a different type of food, then, over time, there may be preferential mating among individuals using the same food source, eventually leading to reproductive isolation. Another potential driver of reproductive isolation among co-occurring individuals may be through infection by particular microorganisms. The endosymbiont *Wolbachia* (found in many nematode and arthropod species) can exert various effects on host reproduction, which may reduce gene flow between different groups of organisms infected with different strains of *Wolbachia*. Another possible factor could be changes in timing of reproduction; differences in flowering times among plants could allow for sympatric speciation because plants cannot cross-fertilize if they are not flowering simultaneously.

Other genetic mechanisms could also occur, including those in which the heterozygote is the least fit genotype (a process known as disruptive selection, which we discuss in Box 16.5). While we can imagine situations that could result in sympatric speciation, we rarely can watch speciation in progress, so we must make inferences about the sources of reproductive isolation after we observe that it has occurred.

Once groups are separated, mutations will occur randomly in each isolated population, and genetic differences between the populations will accumulate over time, either through natural selection for different variants or through genetic drift in the two different groups, or a combination of the two. At some point, the genetic differences that build up may prevent the two populations from successfully inter-breeding. Such differences can be **prezygotic** (preventing individuals from mating at all) or **postzygotic** (reducing the reproductive success after mating has occurred). For example, hybrid offspring may be produced, but they may be sterile.

If hybrid offspring have any reduction in fitness, compared to either parental form, then further selection may act to increase the reproductive isolation. This occurs because there is increased fitness when individuals from the same group mate with each other, rather than across groups, which results in less fit hybrids. Speciation might then occur more rapidly if natural selection occurs to favor genes that cause individuals to avoid mating with individuals from the other group. Genes involved in the control of mating behaviors in animals are often subject to selection in this kind of scenario. This secondary process is sometimes referred to as reinforcement.

Chapters, or even entire books, can be devoted to the underlying processes of speciation, so this is simply an introduction. Speciation occurs as a natural outcome of the other processes that affect the genetic structures of populations. But precisely what forces were at work are often difficult to know. Typically, by the time biologists identify two populations as being distinct species, the differences between them include geographical separation, behavior, mating preferences, morphology, chromosome and genome structure, and more. In principle, any of these could provide a source for reproductive isolation, and all of them could reinforce reproductive isolation once it has begun.

But the opposite effect happens when a physical or other barrier disappears, as summarized in Figure 16.11(b). Two populations that have previously been separated may now experience inter-breeding as one effective population, perhaps because a land bridge forms (or, in humans, a road is built). As a result, the two populations that were once distinct become more similar to one another; in effect, they become one larger population.

Thus, when migration is prevented, the population can undergo diversification into subpopulations, and separate species could arise. A large population maintains genetic similarity, even when a small amount of migration occurs.

Although it is appropriate to think of migration in terms of physical barriers and movement, cultural and linguistic barriers have also been effective in creating human subpopulations throughout history. These cultural barriers prevent intermixing of populations, thus affecting allele frequencies. The cultural barriers can be religion, class or status, language, or anything else that prevents the free movement of alleles throughout a population. Genetic markers have long been used to try to recreate the migrations by which humans have come to populate the planet. Originally, these markers were blood types, mitochondrial DNA, and Y chromosome DNA. In Box 16.3, we discuss the use of mitochondrial DNA and Y chromosome DNA to study human migration.

Now that we are able to compare haplotypes among different human populations, we can see the impact of migration being restricted by language and culture, as well as by geography. Examples of the effects of cultural barriers on populations who lived in the same location include genetic differences between Jews and non-Jews in Europe (and, to a lesser extent, between Sephardic Jews and Ashkenazi Jews who lived for centuries in different regions and developed linguistic and cultural distinctions), and among people of different social castes in India. The

BOX 16.3 *A Human Angle* Tracking migration with mitochondrial and Y chromosome DNA

Wherever we go, we bring along traces of where we have been. In genetic and genomic terms, this is the process of migration, the leaving of one population and the joining of another, either because of physical movements or because of changes in the reproductive structure of the population. All parts of the genome can now be used to track migration, but some of the earliest evidence about human history and human migration came from two specific components—mitochondrial DNA and the Y chromosome. There is an extensive literature, both in scientific journals and in popular books, about reconstructing human history using mitochondrial DNA and the Y chromosome, which we introduce here only as a brief overview.

Mitochondrial DNA—are you my mother?

Recall from Chapter 3 that mitochondria have their own genomes. The genomes of mitochondria do not mix or recombine with the nuclear genome, and the rate of mutation in the mitochondrial genome is about ten times higher than in the nuclear genome. In addition, since a cell has hundreds, or even thousands, of copies of mitochondria, all with the same genome, it proved easy to isolate and analyze an individual's mitochondrial DNA at a time when the analysis of chromosomal DNA was far less straightforward. Furthermore, because the mitochondrial genome is relatively small—only about 16,600 bp in humans and other mammals—it was easier to work with than chromosomal DNA. Thus, studies about human migration using the mitochondrial genome began in the early 1980s, while migration studies using the nuclear genome are more recent.

The analysis of mitochondrial DNA occurred either by direct sequencing or, more commonly and more rapidly (30 years ago, at least), by restriction fragment length polymorphisms or RFLPs, which is described in Tool Box 3.1. RFLPs are easy to detect, even without sequencing, and certain variable sites in the mitochondrial genome were soon identified. Since there is no recombination with chromosomal genes, the entire mitochondrial genome comprised one haplotype, more properly referred to as a haplogroup.

Mitochondria are inherited exclusively from the mother in the cytoplasm of the ovum, so the haplogroup of your mitochondria indicates your maternal lineage for millennia. In general, changes in the mitochondrial genome are either highly deleterious (in which case, they are quickly lost) or selectively neutral; these neutral changes were the basis for recognizing different haplogroups. The polymorphisms are found primarily in two specific regions of the genome (which have no genes) called hypervariable regions 1 and 2. Based on these regions, there are more than 20 different mitochondrial haplogroups in humans, which have various relationships among themselves that arose by subsequent mutations. The L haplogroup, for example, has at least five subtypes referred to as L1, L2, L3, and so on. Most of these haplogroups are found only in some African populations; as with other types of genomic analysis, we see far more genetic diversity among African populations than in all of the rest of the world.

The differences among these haplogroups have been used to construct a phylogenetic tree, of the type discussed in Chapter 4. The hypothesized phylogenetic tree could be extended back to a single common ancestor (an ancestor of at least the 147 individuals used in the original analysis) who was estimated to have lived about 200,000 years ago, nicknamed Mitochondrial Eve. It is postulated that the early hominids who descended from this common ancestor moved out of Africa and into the Middle East, and subsequently into Europe. All Europeans tested had the L3 haplogroup, so most non-African haplogroups are derivatives of L3.

Since that early study found seven haplogroups among modern Europeans (all derived from L3), these haplogroups were called the Seven Daughters of Eve in a popular book. The "daughters" were given nicknames based on the first initial of the haplogroup; the J1 haplogroup, which is common in the British Isles and Scandinavia, is called Jasmine, for instance. Subsequent analysis with additional data has suggested there may, in fact, be ten

BOX 16.3 Continued

to 18 European haplogroups, rather than the seven originally proposed, which arose at various times and thus have different ages. The total number of surviving haplogroups worldwide could be as many as 29 or 30, depending on how the tree is constructed. There may be nine distinct mitochondrial haplogroups in Japan alone, for example, depending on how much similarity is required within a group.

Mitochondrial haplogroups are most useful for broad characteristics of heritage and have been widely used with old skeletons because mitochondrial DNA is fairly easy to obtain. For example, mitochondrial DNA analysis was used in 2012 to show that skeletal remains found in Leicester, England, probably belonged to King Richard III who reigned only briefly from 1483 to 1485 but was immortalized by Shakespeare. Mitochondrial DNA represents only the maternal line, of course, so mitochondrial DNA analysis is only one a part of a person's genetic heritage.

Y chromosomal DNA: What's your name? Who's your daddy?

The use of the Y chromosome as an analogous method to track paternal heritage began not long after mitochondrial analysis, as the tools for DNA amplification and isolation—particularly the use of the polymerase chain reaction (PCR)—improved. Most of the Y chromosome does not recombine with the X chromosome during meiosis, as discussed in Chapter 7. So, similar to mitochondrial DNA, the Y chromosome is passed to subsequent generations largely intact, with little exchange or mixing with other chromosomes. The differences that do arise are the result of intra-chromosomal rearrangements that can occur within the Y chromosome itself and can be quite extensive, as well as any newly arisen mutations.

Because the Y chromosome is larger (59 Mb in humans) and has more sequence diversity than the mitochondrial genome, Y chromosome analysis provides more specificity than does mitochondrial genome analysis. There are more Y chromosome haplogroups than mitochondrial ones, and more individual variation as well. That said, Y chromosome analysis can be applied only to the males in the population, of course.

Mirroring mitochondrial analysis once again, Y chromosome analysis can be used to track human migration and founding populations. For example, an unusual Y chromosome polymorphism that was found among skeletal remains in modern Lebanon and Syria (the site of ancient Phoenicia) is also found among men living in seaports throughout the Mediterranean, including along the Atlantic coast of Portugal. This is consistent with the records of the Phoenicians as being great sailors in the ancient world.

Since surnames are also passed paternally among many western civilizations, there are ongoing genealogical projects to correlate British and Irish surnames (and other populations as well) with particular Y chromosome haplogroups. Many very common Irish and British surnames, such as Smith, Jones, Kelly, Murphy, and Brown, do not show a correspondence with a particular Y chromosome, indicating—not unexpectedly—that many men with the same last name are not genetically related to one another. Conversely, other surnames do have a shared Y chromosome haplogroup. Most individuals with the surname of Titchmarsh, Werrett, Herrick, or Attenborough have the same Y chromosome haplogroup as others with the same last name, suggesting that they all shared a male common ancestor many generations previously. The Y chromosome is a useful genealogical tool, since the spellings of names change much more often than the Y chromosome does, and DNA identifies genetic relationships better than names do.

Mitochondrial and Y chromosome DNA analysis for tracking migration and genealogies are so widely used that individual and national stories abound. Among the best known of these are the children fathered by Thomas Jefferson with his slave Sally Hemings and the Kohanim Y chromosome of Jewish priests. One historically interesting, if somewhat tragic, example of the use of mitochondrial DNA and Y chromosome analysis comes from villages in the Antioquia province in north-western Colombia. Genetic testing of the current inhabitants of these villages finds that more than 90% of the maternal lineage is derived from a Native American mitochondrial haplogroup, not surprisingly. However, more than 90% of the paternal lineage is a Y haplogroup that is European and found primarily in regions of Spain, rather than among Native Americans. Similar results have been found in other South American locations, consistent with a genetic heritage of a Native American mother and a Spanish conquistador father.

effects of limited migration due to a combination of geographical and cultural or linguistic barriers are seen with people of isolated groups throughout the world, such as the Basques, the Finns, and many island populations such as the Sardinians and Icelanders. Reduced migration is often also associated with small populations; we discuss some specific examples that combine these two factors.

We all have images and preconceptions about people from different groups. Some of these are well founded in genetics and are generally benign when it comes to cultivating positive human relationships. For example, Scandinavians do in fact tend to be taller and have lighter skin and hair than Italians or Greeks. Other preconceptions or prejudices that we have about some human groups do not have clear-cut genetic foundations or may have no genetic basis at all. It is naïve to believe that humans are all alike genetically or to think that any of us could escape these preconceptions. Humans are genetically diverse, and one of the primary explanations for our diversity relates to the subpopulations that arose from restricted migration. The interplay between these subpopulations and other evolutionary factors will be explored more fully below, and, as we will see, many alleles that contribute to genetic diversity are not in fact under selection but are the result of more random events.

KEY POINT Migration can allow two previously separated populations to intermix, or it can split one larger population into two distinct subpopulations. Barriers between populations may be physical or, in human populations, cultural or linguistic.

Genetic drift

An H–W equilibrium makes the assumption that the population is extremely large. However, populations have a finite size, and often the effective breeding population is fairly small. Thus, any subsequent generation is a statistical sample of the genetic variation that was present in the previous generation. As with any other type of sampling, this effect is especially pronounced if the population is small. Because allele frequencies change from generation to generation in random ways due to population size, a change in allele frequencies arising from small populations is known as genetic drift. Genetic drift is fundamentally equivalent to sampling any group of limited size and is illustrated in Figure 16.12.

Simulations of population growth show that the effects of genetic drift can be surprisingly large. We used one particular model in which the initial allele frequency was set to 0.5 and the effective population size was set to 100 individuals; this is smaller than the estimated size of most founding human populations but is not entirely unrealistic. The population was allowed to "evolve" under these conditions for 100 generations. The results for the first five simulations using this model are shown in Figure 16.13(a). For populations 1 and 5, the allele became fixed, that is, its frequency equals 1.0 and it is the only allele at this locus in the population. This can occur solely by chance, as in these two populations. In population 1, this occurred after about 25 generations, while for population 5, fixation occurred after 80 generations. On the other hand, in population 2, the frequency of the allele dropped to about 0.15 after about 90 generations.

The final allele frequencies for 300 simulations are graphed in Figure 16.13(b). In about 7% of the simulated populations, the allele became fixed; in 9% of the populations, the allele became extinct. The parameters of the simulation were not changed between the runs—the differing outcomes are purely the result of chance in the sampling from one generation to the next. While these particular results are merely illustrative, the effects of genetic drift cannot be underestimated. Populations that started out being similar to one another became quite different purely because of the chance effects arising from the limited effective population size.

Genetic drift, particularly in combination with limited migration, has had a very significant impact on human genetic diversity. The combined effects of genetic drift

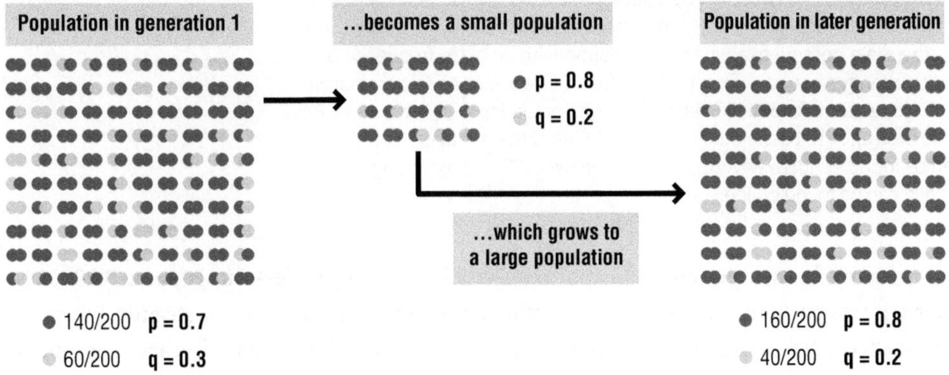

Figure 16.12 Finite population sizes change allele frequencies by genetic drift. Genetic drift refers to random fluctuations in allele frequencies because populations have a finite size. In this figure, the population goes through a bottleneck due to random events. The surviving small population has different allele frequencies from the original population due to random sampling. As the surviving population expands to form a larger population, the allele frequencies will be different from the original population.

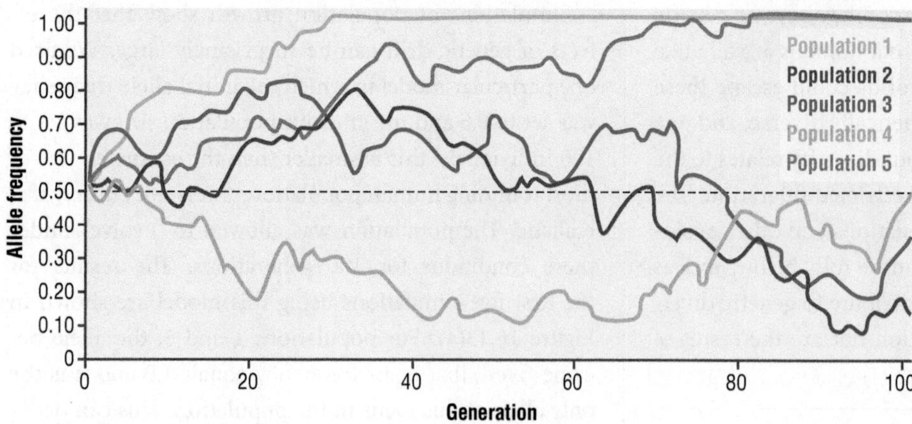

(a) Simulations of genetic drift

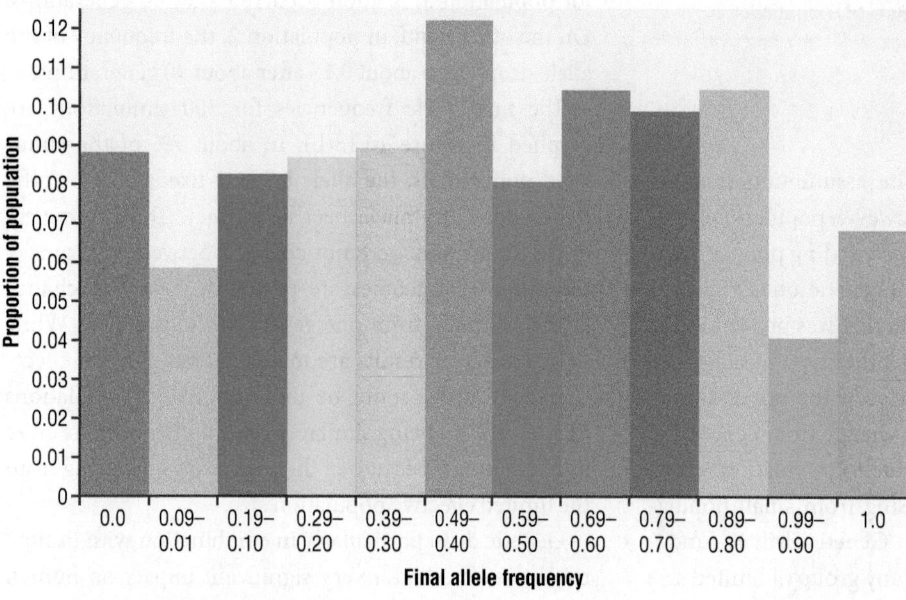

(b) Bar graph of 300 drift simulations

Figure 16.13 Simulating genetic drift. The effects of genetic drift can be simulated by various models. In the model shown here, the initial allele frequency was 0.5; the population size was set to 100, and changes in allele frequencies were simulated for 100 generations. (a) shows the first five simulations done under these conditions. In populations 1 and 5, the allele became fixed in the population with an allele frequency of 1, whereas in the other three, the allele frequency reached different values after 100 generations. (b) summarizes the allele frequency reached after 100 generations for 300 different simulated populations. For instance, in 7% of the populations, the allele became fixed after 100 generations, whereas in 9% of the populations, the allele became extinct. This simulation can be found at http://www.radford.edu/~rsheehy/Gen_flash/popgen/, but many other simulations for genetic drift can be found online.

and limited migration have apparently been especially significant in natural populations in two particular ways. To understand the impact of these effects, let's summarize some information about human population history.

The United Nations predicted a world population of about 8 billion people (8×10^9) in 2015. When agriculture was developed roughly 12,000 to 15,000 years ago, the world population was estimated to have been approximately 15 million people—that is, about 0.00002% of what it is now. The population grew steadily, but relatively slowly, for many thousands of years. For much of that time, this growth occurred in isolated populations in which the effective population size was perhaps 2000 people or fewer. Thus, these populations were distinct from one another; mutations arose in one that did not arise in another; the populations showed limited migration, and they demonstrated a large impact of genetic drift.

These small populations then became connected by the technological innovations that arose during the Industrial Revolution, and the rate of growth accelerated, so that, by 1800, there were 1 billion people on earth. Thus, while it took approximately 15,000 years to reach a population of 1 billion, it has only taken 200 years (or fewer than ten generations) to grow to more than 8 billion. Alleles that were present in those earlier isolated populations have been found as common polymorphisms within groups with similar geographical origins. Genomes therefore provide a record of the previous small population groupings with alleles that have not yet been fully "mixed" across the larger more mobile, global population. As we noted in Chapter 10, more than 90% of the polymorphisms found in a particular person are also found in others descended from that same continent or geographical location 500 or 1000 years ago; this is strong evidence for the effects of genetic drift in humans.

KEY POINT Genetic drift is the change in the frequency of an allele in a population due to random sampling and/or small population size.

Founder effect

The pattern of small isolated populations experiencing rapid growth is the prototype for the founder effect, one extreme version of genetic drift. You were introduced to the concept of the founder effect in Chapter 10, and several different examples were presented. For example, Iceland was founded by a small group of Vikings (perhaps a few thousand) in the ninth and tenth centuries. There was essentially no immigration to Iceland until after the Second World War, so the population was very isolated for about ten centuries. During this time, its population grew from a few thousand to about 270,000 people. Alleles that were present by chance in those original settlers are now common polymorphisms in Iceland but remain rare in other populations; similarly, some polymorphisms common in the rest of Europe are absent from the Icelandic population.

For example, a particular allele of the gene *BRCA2* is found only in Iceland, and nearly all cases of cancers due to this *BRCA2* mutation in Iceland are found on the same haplotype, which has been traced to a man named Einar the Cleric. Although this and other examples focus on single-gene traits, this same population is also helpful in understanding the genetic basis for some complex traits because of the "common disease, common variant" hypothesis developed in Chapter 10. Using the Icelandic database, candidate genes have also been identified for stroke, Alzheimer's disease, schizophrenia, and longevity, among many other complex traits.

While Iceland is one example in which the founder effect is seen, it is far from being the only example. The more than 6 million French Canadian Quebecois who live in Quebec are derived from a group of about 8500 people who moved from particular regions in France in several waves of migration beginning in 1608 and continuing for 150 years. They were isolated from the surrounding (British) population by language and religion, and, because of parish records of births, baptisms, weddings, and deaths, they are among the best-documented populations in the world.

More than 20 distinct single-gene traits reflect the history of French Canadians. In some cases, genetic diseases are much more common in the Quebecois population than in other populations. For example, hereditary tyrosinemia (due to mutations in the *FAH* gene) is an autosomal recessive disorder of tyrosine catabolism. The disease is uncommon, with an estimated worldwide frequency of eight to ten cases per million births, but there have been more than 500 cases in the Charlevoix Saguenay Lac St Jean region of eastern Quebec where one in 22 people is a carrier. Most of these cases are due to one particular mutation that affects a splice site in intron 12 of the gene, and more than 96% of these are on the same haplotype, suggesting a common origin. This mutation has only been found among the French Canadians, and genealogical records identified a cluster of common ancestors originating from north-west France.

In addition to some genetic diseases that are more common, there are alleles for other diseases that are quite different in the French Canadians from other populations in the world. For example, the neurological disorder Tay–Sachs disease is more common among Ashkenazi Jews than among the population at large, with approximately one in 30 people being a carrier. Nearly all of these cases are due to a 4-base pair insertion—and thus a frameshift mutation—in exon 11 of the *HexA* gene. French Canadians show a similar frequency of Tay–Sachs disease, with as many as one in 20 people being a carrier in some regions. In this case, however, the mutation is a 7.5 kb deletion of the *HexA* gene. This deletion has not been found among Jewish patients with Tay–Sachs disease, and only a single case of Tay–Sachs disease (in which one ancestor may have been Jewish) was due to the 4-base pair insertion.

Every population in the world that has been examined shows evidence of a founder effect for certain traits and haplotypes. In fact, the genome-wide association studies and the "common disease, common variant" hypothesis described in Chapter 10 arise directly as a consequence of these founder effects. While examples can be found in any population group, the Finns have developed a Finnish heritage database of about 40 genetic diseases and genetic conditions that, while not unique to Finland, are more common among the Finns than in the rest of Europe. We discuss the genetics of Finns further in Online Box 16.1.

WEBLINK: You can find Online Box 16.1 at **www.oup.com/uk/meneely**

KEY POINT The founder effect, which is genetic drift coupled with low migration, occurs when a small group establishes a new separate population. The genetic traits in this small group will predominate in subsequent generations, although they may not be common in the species as a whole.

Population bottlenecks

The second extreme form of genetic drift is that represented by population bottlenecks. Suppose that a large fraction of the population is wiped out, without regard to the genotype, by a flood or a fire, for example. As illustrated in Figure 16.12, because many individuals in the previous generation died, the next generation could have a very different genetic structure simply by chance. This is probably what has happened with many endangered species, or in species that were once endangered but have now grown in number. In each case, only a few individuals survived, and the alleles that they represent do not have the genetic diversity of the larger population that existed previously.

For example, cheetahs show almost no polymorphisms or heterozygosity and are so closely related that cheetahs in captivity can accept skin and tissue transplants from any other individual cheetah. Another example is the northern elephant seal found along the Pacific Coast in the United States, which was extensively hunted for fur in the nineteenth century. By 1900, there were fewer than 30 northern elephant seals. From that small group, the population has grown to hundreds of thousands of seals, all of which are homozygous at all loci examined so far. This extreme reduction in genetic diversity has made these, and other populations and species, much more vulnerable to pathogens and has increased the rate of sterility.

For humans, population bottlenecks have happened repeatedly, although we might not always have a historical record. We do, however, have historical records that capture the impact of the plague in Europe. The Black Death (bubonic plague due to *Yersinia pestis*) began in Venice in January of 1348; within 2 years, as much as two-thirds of the population of some cities in Europe had died. Over the next 200 years, outbreaks of the plague occurred almost annually, killing as much as 40% of the European population. Thus, the expansion to the modern population in Europe has occurred primarily during the past 600 years, and genetic diversity that was already reduced by the founder effect was further reduced by population bottlenecks. As we discussed in Box 4.4, however, those genotypes who survived the Black Death may not have been a random sample of the genotypes present in Europe at the time, so this may be an example that combines a population bottleneck and selection.

In addition to the impact of the founder effect, Iceland has experienced three severe bottlenecks. In the early fifteenth century, the Black Death killed more than 60% of the population. In 1707, a smallpox outbreak killed a third of the population again, while a volcanic eruption in 1783 released sulfur dioxide that killed another 25% of the population. Thus, the Icelandic population is extremely homogeneous, not only because of the founder effect but also because of ensuing population bottlenecks that further reduced the genetic variation.

KEY POINT Population bottlenecks occur when a large proportion of a population is killed by catastrophic events. The population that arises from the remaining individuals has reduced genetic diversity.

16.5 Selection and differential reproductive ability

The last important assumption made about a population at H–W equilibrium is that all gametes have an equal probability of contributing to the next generation, and each individual contributes the same number of gametes. Clearly, this is not a realistic assumption—individuals have differential reproductive ability no matter what

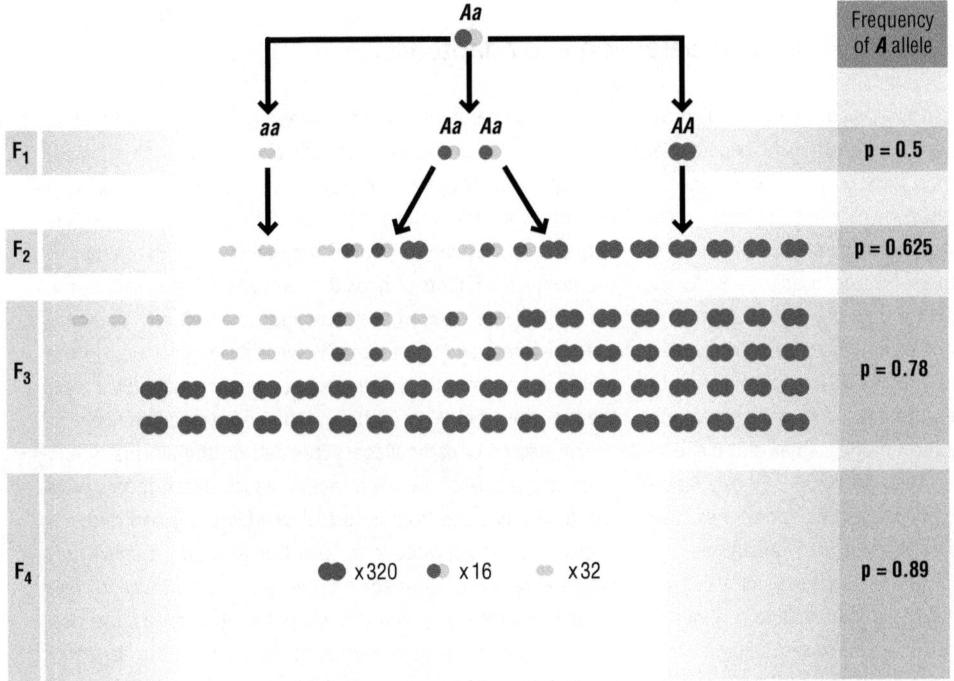

Figure 16.14 Selection changes allele frequencies due to differential reproductive ability of different genotypes and phenotypes. In the diagram and Figure 16.15, reproduction occurs by self-fertilization under different conditions of selection; the relative fitness of each genotype is reflected in the size of the symbol, as well as in the number of offspring. This figure shows directional selection for the ***A*** allele and the ***AA*** genotype. The ***Aa*** heterozygote at the top gives the four offspring shown in the F_1 generation, each of which reproduces by self-fertilization but with different numbers of offspring. The ***aa*** homozygote is the least fit and gives only two offspring. The ***AA*** homozygote is the fittest and gives six offspring. As shown in the summary on the right, the frequency of the ***A*** allele increases rapidly and will quickly become fixed.

species is being considered. Differential reproductive ability is the basis of natural selection; one common form of selection is summarized in Figure 16.14. Selection is self-evident; not all individuals in a population leave the same number of offspring, and some of the explanation for that differential fitness is genetic or heritable. We discussed this principle previously in Chapters 3 and 4. Selection will drive a change in allele frequencies and is one of the primary forces at work in a population.

But how does selection drive this change in allele frequencies? Because of natural selection, the genotypes in every generation that promote survival and reproduction in a particular environment—that is, the fittest genotypes—contribute more gametes to the next generation. In this way, the alleles and genotypes that enhance survival and reproduction are favored and will increase in frequency from one generation to the next. As a result, the population becomes better able to survive and reproduce in that environment. A trait that shows this effect is known as adaptive. Adaptive changes among finches in the shape of the beak were discussed in Chapter 1.

The fittest genotype depends, in large part, upon the effects of the environment. Thus, we commonly find that what is selected for, or advantageous in, one environment may be selected against, or disadvantageous in, another environment. In addition, the environment changes over time. The genome provides a record of selective forces in the past, but a changing environment means that these might not be the same forces acting today.

In addition to Darwin's finches, another well-known example of adaptation and selection is seen with the Biston moth in Europe during the Industrial Revolution; this example is recounted in Box 16.4.

KEY POINT Differential reproduction and selection change the frequency of alleles in a population. Genotypes that promote survival and reproduction in a given environment—those that are the fittest—contribute more gametes to the next generation.

We need to stress that selection does not produce the optimal genotype, only the one that is more fit than any of the current alternatives, a point we have made many times previously. It is sometimes imagined that natural

BOX 16.4 *An Historical Perspective* Selection and adaptation

Natural selection results in a reproduction advantage for the most favorable genotype in a particular environment at a particular time. With consistent selection in favor of certain genotypes, a trait becomes adapted—that is, certain polymorphisms confer such an advantage that they become common. This selective advantage may be small (a fitness increase of only 1–2%, for instance), which can make the advantageous alleles difficult to identify conclusively. With genome sequences for many species and populations now available to us, however, evidence for adaptation and selection can be objectively detected by genome comparisons, even if the fitness advantage is small or the effects of the gene had not previously been suspected. Since adaptation often occurs slowly, these genome-based approaches have yielded many clear examples of selection in many organisms.

Here is a famous example that pre-dates the availability of genome sequences in which selection, adaptation, and changes in the environment could be observed and recorded within historical time. The European peppered or Biston moth (*Biston betularia*), shown in Figure A, tends to sit on the white–gray bark of birch trees. One hundred and seventy-five years ago, the most common color for this moth was whitish gray, with a few individuals being a darker gray color. This color polymorphism is controlled by a single gene, and the gene is polymorphic, with both color alleles present in the population at observable frequencies.

European naturalists in the nineteenth century noted that, as industrial pollution increased and the air became smokier in England and Germany, the dark gray form of the moth increased in frequency, and the white form declined. In other words, the allele frequencies changed in favor of the darker allele. By examining the stomach contents of birds that preyed on Biston moths, the reason became clear. In the most industrialized environments, the most frequently eaten moths were the white ones. In the more rural environments, the most frequently eaten moths were the dark gray ones. What lay behind this observed difference?

The degree of camouflage depended on the air quality and color of the bark of the birch tree on which Biston moths were found. In areas free from industrial pollution, the bark of birch trees was its natural white–gray color. Consequently, the dark gray moths were conspicuous against the light-colored bark and were subject to increased predation, while the white moths blended into their surroundings and escaped. By contrast, the dark gray moths were better camouflaged when the air was smoky (and the bark was therefore darkened by soot) and escaped predation. However, the white moths in these environments found themselves to be conspicuous and were subject to heavy predation. More recently, as environmental regulations have helped clean up the air, the white variety has become more fit again, and the dark gray variety is decreasing in frequency.

Figure A Adaptation in the Biston moth. The Biston moth (*Biston betularia*), native to England, has two color polymorphisms—a dark gray phenotype on the left and a light gray-white phenotype on the right.

Source: Reproduced from Majerus, M.E.N. (2009). Industrial Melanism in the Peppered Moth, *Biston betularia:* An Excellent Teaching Example of Darwinian Evolution in Action. *Evo Edu Outreach* 2:63–74. With permission of Springer

selection is somehow "perfecting" a species, so that what exists is either ideal or is on track for becoming the ideal. This is a fairly common misconception that has persisted since the days of Darwin. The reality is rather different. Selection favors the fittest genotype among the possible alternatives, but it could be that none of those genotypes is "ideal" in some abstract sense. The commonly used phrase "survival of the fittest," which Darwin did not use himself until the later editions of his books, can be quite misleading in this regard. First, since fitness refers to reproduction, rather than simply survival, it should be "reproduction of the fittest." Second, because the fittest are not some ideal genotype, it should be "reproduction of the fittest among the alternative possibilities that are present in that generation in that environment at that time." But sometimes a pithy phrase wins out over the more accurate description.

Note also that fitness in an evolutionary sense is defined as the relative reproductive ability of a given genotype, and not by survival or healthiness or success in some cultural sense. Many genotypes are much less fit than the average for the population and die out. Genotypes that are fitter than average will have a reproductive advantage and will become more common over time. The rate at which allele frequencies (and thus genotype frequencies) change depends on the amount of this fitness difference. Even for newly arisen genotypes with a reproductive advantage, the increase in fitness may be fairly small; a fitness advantage of 0.02 or 0.05 is quite large for human populations. Our example in Figure 16.14 has a reproductive advantage of 0.07, and thus the change in allele frequency is quite rapid.

Usually, fitness is observed for the genotype of the organism as a whole, and not simply the genotype at one locus, but sometimes the genotype at a single locus will play a major role, which makes them easier to discuss. There are different types of selection at work in a population, some of which are discussed below. No one doubts the impact of natural selection, but it has only been with the development of genome sequencing projects that empirical and unbiased measures of selection are more readily obtained on a wider scale.

Depending on the alleles and the genotypes that are the fittest and the agents that are responsible for the reproductive differences, natural selection can affect allele frequencies and genetic diversity in a population in different ways. We summarize some of these in Figures 16.14 and 16.15, but these should be regarded as an overview. The actual effects of selection on allele and genotype frequencies can be more subtle than these simple diagrams suggest.

Directional selection

The type of selection that is probably the most familiar is termed **directional selection**; when most people think of "survival of the fittest," it is directional selection that they are thinking of. This situation is one in which there is selection in favor of a certain allele or against another allele, as shown in Figure 16.14. Let's assume that the *A* allele confers a reproductive advantage over the *a* allele in a particular environment. Over time—and assuming

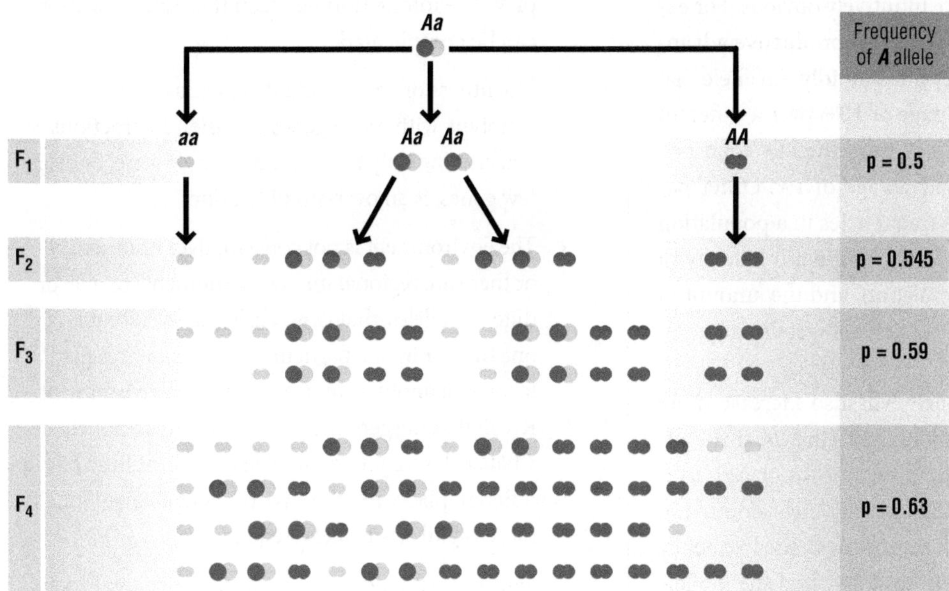

Figure 16.15 Balancing selection occurs when the heterozygote is the fittest. As in Figure 16.14, the individuals reproduce by self-fertilization, with the relative fitness of each genotype shown not only by the number of offspring but also in the sizes of each symbol. The original ***Aa*** heterozygote produces the four offspring shown in the F_1 generation. Each of these reproduces by self-fertilization but with different fitness. The ***aa*** homozygote produces a single offspring, while the ***AA*** homozygote produces two offspring. The frequency of the ***A*** allele increases, but the rate of increase is more gradual than in Figure 16.14, since the ***a*** allele is maintained by the fitness of the heterozygote. The ***A*** allele will not become fixed, and the frequencies of the two alleles will stabilize.

that the environment does not change—the frequency of the ***A*** allele will increase, while the frequency of the ***a*** allele decreases. Most of the individuals will then be ***AA***, and other genotypes will be less common. This is, in effect, how wild-type genotypes occur, because the ***A*** allele produces a functional activity and confers a selective advantage over any of the non-functional alleles. This does not mean that ***A*** is the best possible allele for the locus, but only that it confers a reproductive advantage over the ***a*** allele in that environment at that time.

Directional selection of this sort is sometimes the only one that people think about. It is sometimes called **purifying selection**, because it supposedly purifies the population of other less fit alleles, except as they arise by mutation and are maintained by drift and migration. However, the word "purifying" tends to reinforce the misconception that selection is creating a perfect specimen, so we prefer the term directional selection. In humans, directional selection helps to explain why most genetic diseases are rare.

When considering alleles under directional selection, we predict that the favorable allele will eventually become fixed and be the only allele at this locus in the population. Notice in Figure 16.14 that the frequency of the favorable ***A*** allele increases rapidly. The amount of time that it will take for an allele to become fixed will depend on a few parameters, most of which are intuitively obvious. For example, alleles that confer the largest reproductive advantage will increase in frequency most rapidly. An allele that confers a reproductive advantage of 10% over a different allele will increase in frequency more rapidly than one that confers a reproductive advantage of 1%. Other parameters that affect the allele frequencies in a population experiencing selection include the original frequency of the alleles, the size of the population, and the amount of time that has elapsed since the advantageous mutation appeared.

The frequency of the ***A*** allele will also increase more rapidly if ***A*** is incompletely dominant (that is, if the fitness of ***AA*** is greater than the fitness of ***Aa*** or ***aa***), as is the case in Figure 16.14. If ***A*** is completely dominant, such that ***AA*** and ***Aa*** have the same fitness, the frequency of the ***A*** allele will increase more slowly and the ***a*** allele may persist in the population in heterozygotes. Many recessive disease mutations with a large negative **selection differential** persist in the human population because the heterozygotes (carriers) have no selective disadvantage compared to the homozygotes without the mutation, that is, the wild-type allele is a simple dominant.

KEY POINT Directional selection indicates that selection for a particular allele is occurring. In a stable environment, this allele will increase in frequency over time and may become fixed.

Why does genetic diversity exist?

Since directional selection predicts that all individuals will become homozygotes for the favorable allele, why then do we find any genetic diversity at all? Why aren't individuals genetically identical, with all of them being homozygous for the fittest allele? Several explanations are usually advanced, and all of these are likely to be true for at least some genes and populations:

- The favorable allele is becoming fixed, but not enough time has yet elapsed for this to happen fully. This could be a newly arisen mutation that is more favorable than any previous allele, or it might be an allele that confers a very small selective advantage. Fixation also takes longer for an allele that is completely dominant, as noted above. No allele becomes fixed in a single generation, or even in a few generations; fixation takes time.
- The other factors that affect allele frequencies—most significantly, migration and drift—are also in play. The interaction between these and selection can be complicated.
- The fitness of an individual depends on the overall genotype with many genes and their interactions, so considering only the effect of a single gene, or even a few genes, is an over-simplification.
- The environment is not constant over time, and/or there are regional microenvironments that favor different alleles, that is, an allele that is favorable at one time or in one particular environment is not favored at another time or in another environment. Recall that the genome has no capacity to anticipate what alleles might be useful at a different time or at a different place, and the environment changes much more rapidly than allele frequencies.
- Selection is not directional at the particular locus being considered. While directional selection is the type most of us think of first, it is not the only type of selection that occurs and that affects allele frequencies. Examples of two other types of selection are discussed below, and some other types are considered in Box 16.5.

BOX 16.5 *Going Deeper* Types of selection

Most discussions of natural selection focus on **directional selection** and **balancing selection** or heterozygote advantage, and a few will also include **frequency-dependent selection**. While these are the most familiar examples, other types of selection have also been proposed or uncovered in particular circumstances. We briefly mention some of these here.

Disruptive selection occurs when the heterozygote is less fit than either homozygote. In a sense, it is the opposite of balancing selection in which the heterozygote is the fittest genotype. The most familiar examples of disruptive selection involve populations whose genomes have undergone chromosome rearrangements such as translocations or inversions. Individuals who are homozygous for the rearranged chromosome have nearly the same fertility as individuals who are homozygous for the normal or unrearranged chromosome. However, in a heterozygote, the rearranged and normal chromosome tend to pair or segregate poorly or produce gametes with substantial deletions, which results in lower fertility and fitness. Disruptive selection is sometimes considered as part of the process of speciation, discussed in Box 16.2.

Kin selection posits that an individual that is part of a social group, such as insects like ants or bees, may act "altruistically" as drones, so that their genetically related kin survive to reproduce, even if they do not. This behavior enhances the probability that their alleles will be passed to the next generation, even if they themselves do not reproduce. A highly controversial, and somewhat related, concept is group selection in which traits that are favorable to the species as a whole will increase or be maintained in a population, even if they are not necessarily advantageous to an individual with the trait or the allele. Although often proposed, group selection has been difficult to prove, and many evolutionary biologists discount the possibility that it occurs. However, the co-evolution of different species within communities as revealed by metagenomics and microbiomes, as will be described in Chapter 17, provides some support for a form of group selection.

Meiotic drive or gametic selection occurs when one gamete or haploid product becomes more frequent than predicted. Thus, unlike other types of selection that work on individuals or kindreds, gametic selection directly affects the allele. During meiosis I, the homologous chromosomes segregate to different poles on the spindle; during meiosis II, the sister chromatids separate to different poles. In nearly all discussions of meiosis, including ours in Chapter 6, the reasonable assumption is made that all of the meiotic products are found in equal frequency and have an equal chance of survival. This assumption underlies the familiar ratios of Mendelian genetics, such that half the gametes from a heterozygote are ***A*** and half are ***a***. A few examples are known in which this is not true, and the homologue with one of the alleles is significantly over-represented among the gametes; this is referred to as meiotic drive, segregation distortion, or gametic selection. These terms are based on the outcome of meiosis and may not occur by similar mechanisms.

Sexual selection arises when a trait confers an advantage to one sex, such that it becomes preferred among the competition for mates. Among birds, for example, the colorful and elaborate plumage common to males may be an example of sexual selection. Females prefer more colorful males as a mate, so alleles that confer color in males increase in frequency.

However, the same characteristics that make a mate more attractive can also confer a selective disadvantage from other agents; more colorful individuals are easier for predators to spot. The greatly exaggerated antlers of extinct Irish elk have been cited as one example of runaway sexual selection. The very large antlers may initially have been more attractive to females, but they became so large that the skeleton of the elk could not sustain the added weight, resulting in its death and eventual extinction.

Balancing selection with heterozygote advantage

Although we have just described directional selection in terms of one allele, it is important to recognize that what is being selected is the phenotypic outcome of the **genotype**. With directional selection, the phenotype conferred by the ***AA*** genotype is more advantageous than that arising from the ***aa*** genotype and increases in frequency until the ***A*** allele is fixed. As noted, this increase in frequency of the ***A*** allele will be even more rapid if ***AA*** has a greater fitness than ***Aa***.

However, the situation changes when the phenotype arising from a heterozygous genotype is favored, as shown in Figure 16.15. This situation of **heterozygote advantage** is an example of stabilizing or **balancing selection**. Heterozygotes will inevitably produce ***AA*** and ***aa*** individuals among their offspring. Under balancing selection, both of these genotypes result in phenotypes that are less fit than ***Aa***. As a result, both the ***A*** and ***a*** alleles and all three genotypes will be present in the population, and no allele or genotype becomes fixed. In other words, even when the allele frequencies are not themselves changing, the locus will be polymorphic with more than one common allele.

The exact frequencies of the ***A*** and ***a*** allele under balancing selection depend on a number of different parameters, most importantly the relative fitness of the three genotypes and the frequencies of the alleles in the prior generation. However, the important concept evolutionarily is that the frequencies of the different alleles are held in balance with one another.

KEY POINT A situation in which the heterozygote has greater fitness than the homozygotes is an example of stabilizing or balancing selection, and both alleles at this locus are maintained in the population.

Because many loci are polymorphic in human populations and we observe substantial phenotypic and genetic diversity, heterozygote advantage has been invoked as the explanation for many human traits. Sickle-cell anemia was recognized decades ago as an example of balancing selection and served as a classic illustration of this process. Sickle-cell anemia is common in many African—and thus African-American—populations, as well as in some regions around the Mediterranean Sea. This is also the region where falciparum malaria is common, as shown by the map in Figure 16.16(a).

Sickle-cell anemia arises from a particular recessive mutation in the β-globin gene. The "sickle-cell allele" is designated ***HbS***. In the polypeptide produced by the ***HbS*** allele, a glutamic acid found at position 6 in the β-globin protein is replaced by a valine, a mutation abbreviated Glu6Val, as shown in Figure 16.16(b). Until medical intervention was possible, the life expectancy for children homozygous for ***HbS*** was less than 5 years. Nonetheless, in many parts of Africa and in African-Americans, the ***HbS*** allele is quite common; in some regions of sub-Saharan West Africa, the frequency of the ***HbS*** allele is as high as 0.185, and, among African-Americans, it is estimated to be 0.05 or higher. How then did a mutation with such serious consequences become common in those populations?

The answer lies in heterozygote advantage. While the ***HbS*/*HbS*** homozygote cannot survive and reproduce without treatment, the ***HbS*** heterozygote is less likely to die of falciparum malaria. The malaria-causing protozoan *Plasmodium falciparum* can infect all humans but is less likely to cause cerebral malaria and other complications when it reproduces within sickled red blood cells. As a result, in malaria-endemic areas, heterozygotes have a higher fitness than homozygotes for the standard allele, since homozygotes for the standard allele are more sensitive to malaria. On the other hand, ***HbS*/*HbS*** is strongly selected against because the individuals have sickle-cell anemia. Thus, the locus is polymorphic for both alleles in regions where malaria is common.

Conversely, in environments where malaria is absent, there has been no selective advantage for the ***HbS*** allele in heterozygotes; although the same mutation almost certainly arose, it was lost. At least five different haplotypes have the same Glu6Val mutation, indicating that the same mutation has arisen independently multiple times in different populations and that the founder effect has not played a significant role before the present day. We learn more about sickle-cell anemia and the heterozygote advantage in Box 16.6.

The case for sickle-cell anemia and malaria resistance is so persuasive that many other examples of heterozygote advantage or balancing selection involving pathogen resistance have been proposed to explain the frequency of other disease-causing alleles in human populations. Some of these effects have recently been demonstrated based on genomic data, many years after they were originally proposed. For example, as noted in Chapter 10, the unusual Δ-F508 mutation in the *CFTR* gene is polymorphic in European populations and is responsible for as many as 75% of the cases of cystic fibrosis. One explanation for the frequency of the Δ-F508 allele is a founder effect, which is supported by the common haplotype that many affected individuals share. However, the high frequency of this allele is probably also due to heterozygote advantage during the time when Europe was being settled. The Δ-F508 mutation confers resistance to waterborne pathogens such as those that cause dehydrating diarrhea and cholera. Thus, heterozygotes for the allele were relatively resistant to cholera and had a selective advantage over both homozygotes in environments with contaminated water sources.

Another example in humans may connect an important modern resistance with a historical resistance to pathogens. As summarized in Figure 16.17(a), the *CCR5* gene encodes a non-essential chemokine receptor found on the surface of several different cell types in the immune system, including T cells. This is one of the receptors that has been co-opted by human immunodeficiency virus (HIV) for its entry into T cells. An allele with a 32-nucleotide deletion in the *CCR5* gene, known as *CCR5-Δ32*, is polymorphic among many European populations, with an allele frequency of 0.05 to 0.14, depending on the precise population, as shown in Figure 16.17(b). *CCR5-Δ32* individuals are generally healthy, although homozygotes have also been reported to have some reduced T-cell functions. Thus, there may be only a slight fitness disadvantage to the allele. However, homozygous *CCR5-Δ32* individuals are resistant to AIDS, since HIV cannot enter their T cells.

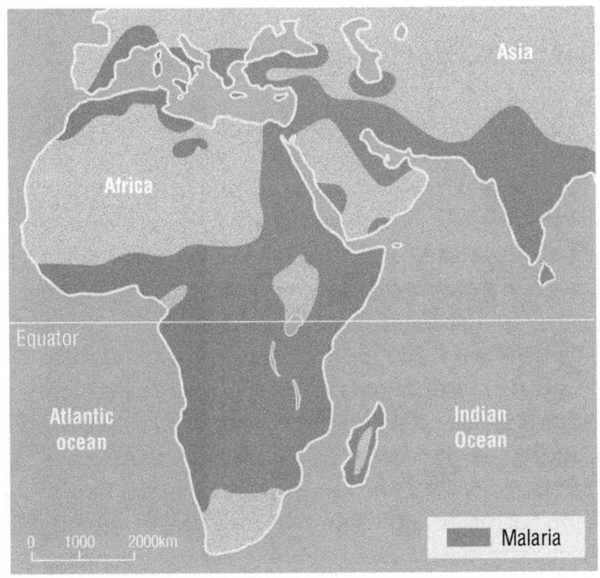

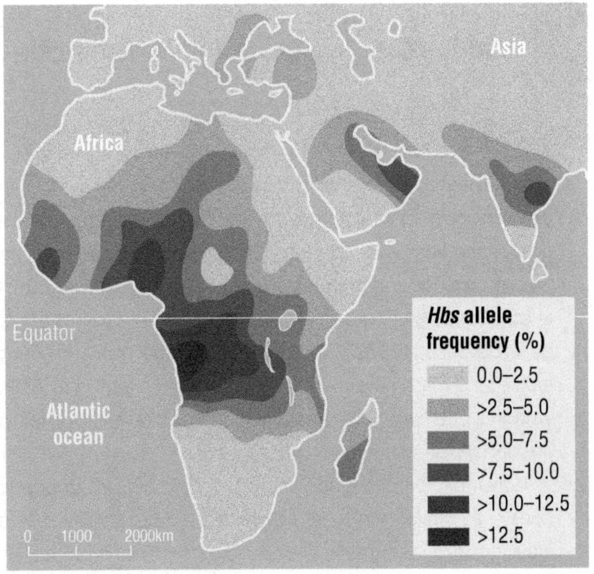

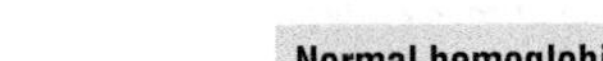
(a)

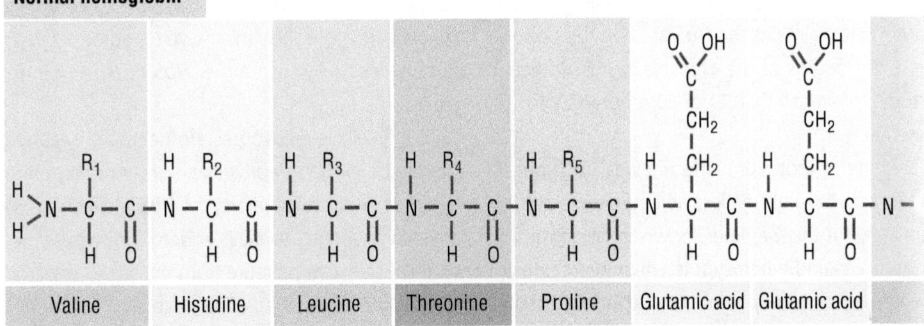

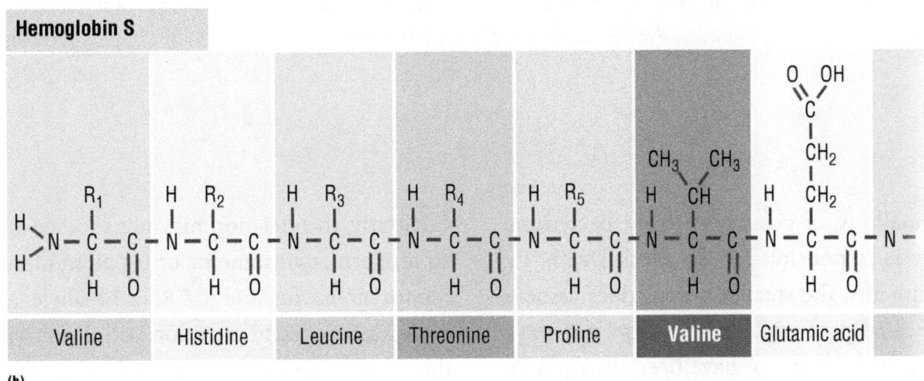

(b)

Figure 16.16 Sickle-cell anemia and malaria are an example of heterozygote advantage and balancing selection. (a) shows the prevalence of malaria and the ***HbS*** allele. Note that the ***HbS*** allele is the most common in locations where malaria is also common. ***HbA*/*HbS*** heterozygotes are resistant to malaria because the protozoan pathogen grows poorly in sickled blood cells. Thus, the ***HbS*** allele persists at high frequency, although ***HbS*/*HbS*** homozygotes do not reproduce. (b) The first six amino acids of β-globin encoded by the normal ***HbA*** allele and the ***HbS*** allele; note that glutamic acid in ***HbA*** has been replaced by valine in ***HbS***.

Source: Part (a) Encyclopædia Britannica, Inc.

HIV is a modern pathogen that has only circulated in human populations in the last century. There has not been enough time for selection to result in such a high allele frequency, and therefore HIV itself cannot have been the selective agent for *CCR5-Δ32*. Instead, it has been postulated that this deletion also confers resistance to the virus

BOX 16.6 *A Human Angle* The sickle-cell trait heterozygote advantage

According to eastern Nigerian mythology, an *Ogbanje* is a child who dies within a few months or years of birth and returns to a mythical land, only to re-enter the womb of his or her mother and be born again. Chinua Achebe's novel *Things Fall Apart* describes an *Ogbanje* who was reborn nine times to an Igbo woman, and whose comings and goings devastated her mother. The Yoruba, another population group within the borders of present-day Nigeria, but separated from the Igbos by the Niger river and considerable land mass, have a similar ancient myth referring to "Abiku," who, in a poem by Nobel Laureate Wole Soyinka, is "calling for the first and repeated time."

While the Yorubas, Igbos, and other West African subpopulations with similar myths are not connected genetically, they all are subject to strong selective pressure from falciparum malaria. The *Plasmodium falciparum* parasite, which causes malaria, and its *Anopheles gambiae* vector are most concentrated across Africa's equatorial belt. The high frequency of malaria in this area and the balancing selection that maintains the ***HbS*** allele in the population may underlie the origins of the *Ogbanje* myths. A recent socio-medical study confirmed that 70 of 80 reputed *Ogbanjes* had sickle-cell anemia.

How exactly are heterozygotes protected? It was first thought that sickled red blood cells might be less able to be infected by *P. falciparum* parasites, but this was later shown to be untrue. Indeed, ***HbA*** homozygotes and heterozygote children are equally likely to be infected with *P. falciparum*. The advantage stems from what happens next. A significant proportion of young children who are infected with *P. falciparum* will suffer a syndrome known as severe malaria, which is caused by the parasite making its host's red blood cells knobbed and sticky, so that they clump together. The knobbed cells stick to each other and to the lining of blood vessels in the brain, causing a type of stroke that is very often fatal. They also stick to vessels in the placenta, placing pregnant women and their unborn children at risk of death. The cell aggregation allows the parasite to sequester large clumps of red blood cells and increase its opportunities to replicate.

The parasite is unable to produce as many knobs in sickled cells, and the knob protein that is produced is aberrantly distributed on the surface of the infected sickled cells. (This knobbing defect is also seen in other hemoglobinopathies such as ***HbC*** and α-thalassemia, which also protect from severe malaria.) The end result is that heterozygotes are less likely than ***HbA*** homozygotes to progress to, or die of, severe malaria, while ***HbS*** homozygotes generally do not reach adulthood without medical intervention. Thus, in places where *P. falciparum* is endemic, and in the absence of treatments for malaria or sickle-cell anemia, heterozygotes are more likely to survive to reproductive age and reproduce successfully than either ***HbS*** or ***HbA*** homozygotes—they are fitter.

Sickle-cell heterozygotes are generally healthy but are more likely to be oxygen-deprived at high altitudes or during extremely vigorous exercise. Therefore, the ***HbS*** allele only provides a fitness advantage where malaria is endemic. However, the result of centuries of selective pressure from malaria in equatorial Africa, and a subsequent founder effect, explains why African-Americans, whose ancestors were forcibly brought from malaria-endemic Africa to the now malaria-free North America, continue to carry the ***HbS*** allele at a high frequency.

responsible for smallpox, or possibly even the bacterium *Y. pestis*, which was responsible for the Black Death. In laboratory experiments, the smallpox virus does associate with the CCR5 receptor, while *Y. pestis* does not, suggesting that the deletion may not have been involved in plague resistance. It is also possible that the receptor was used for entry into host cells by some other as yet unknown pathogen that was present in Europe thousands of years ago, as has happened with HIV in the last century.

The *CCR5-Δ32* mutation may be an example of a genetic variation that sat "unused" in the genome for centuries, only to re-emerge under a new selective pressure, rather like one of the sweaters in Chapter 3 that has suddenly come back in style. The deletion allele is not found in populations in Asia or Africa, so the original selective agent must not have been found in these locations. Curiously, this deletion may not provide an advantage in all modern environments or for all modern pathogens—homozygotes for the *CCR5-Δ32* allele are apparently more susceptible to infection with West Nile virus. Thus, the same allele can confer a selective advantage or disadvantage, depending on the environment and the local pathogens to which its host organism is exposed.

KEY POINT Heterozygote advantage, an example of balancing selection, is believed to be responsible for the maintenance of some alleles that cause genetic diseases when homozygous, since the heterozygous condition confers resistance to particular pathogens.

Another familiar polymorphic locus—the ABO blood locus—has also been widely postulated to be important for pathogen resistance; the allele frequencies may thus

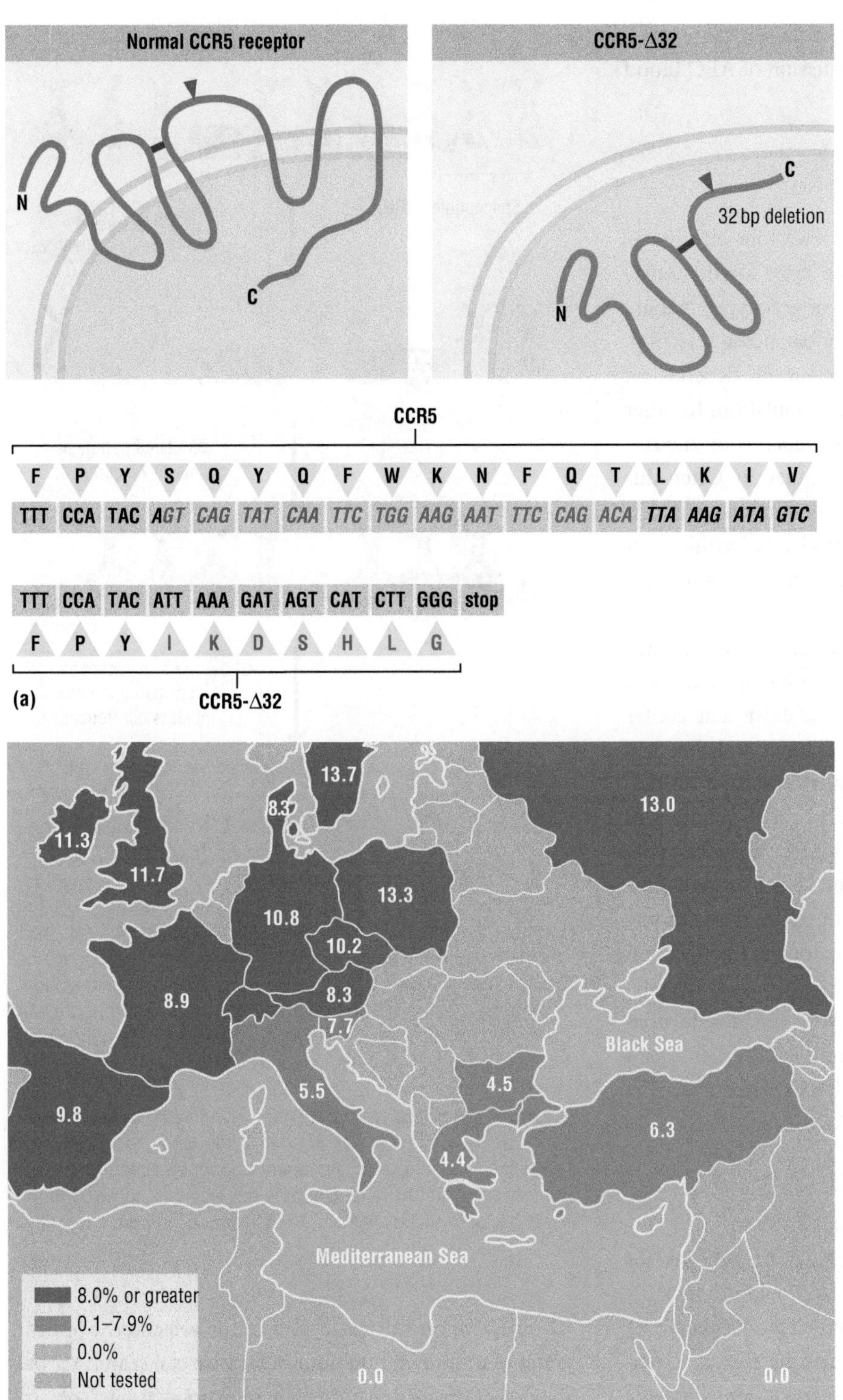

(b) Frequencies of the *CCR5-Δ32* allele

Figure 16.17 The *CCR5-Δ32* allele. (a) The chemokine receptor encoded by the normal *CCR5* gene crosses the cell membrane seven times, as shown, with a cross-link between transmembrane regions 3 and 4, indicated by a purple line. The blue arrow indicates the site of the 32-bp deletion found in the *CCR5-Δ32* allele. The deletion results in a frameshift mutation and a premature stop codon, as shown in the right-hand image, and with the sequences below the figure. This mutated version of the protein is not inserted into the membrane and cannot act as a receptor. (b) A map of Europe and the Mediterranean region showing the allele frequency of *CCR5-Δ32* in different countries.

Source: Part (b) reproduced from Eric de Silva, and Michael P.H. Stumpf (2004). HIV and the CCR5-Δ32 resistance allele. *FEMS Microbiology Letters*, Vol 241, Iss. 1, by permission of Oxford University Press.

be affected by balancing selection. Some of these associations were elaborated in our discussion of ABO blood types in Box 8.1.

Frequency-dependent selection

For directional selection and heterozygote advantage, the favorable genotype remains the same for that environment, regardless of other alleles or genotypes that are present. By contrast, another type of balancing selection, known as **frequency-dependent selection**, depends on the other individuals present in the population. In other words, the most favorable genotype depends on the frequency of other genotypes in the population. Often this is seen when pathogens and predators co-evolve and is a feature of the complex commensal relationships described in Chapter 17. One type of frequency-dependent selection is summarized in Figure 16.18.

For example, a pathogen has typically evolved to take advantage of some aspect of the host's biology such as a particular receptor. Mutations in the host that confer resistance to the pathogen will be favored and thus will increase in frequency, particularly if they have no deleterious effects on their own. A subsequent mutation in the pathogen may then allow it to infect hosts carrying these new, more frequent genotypes, so the original host genotype, or yet another allele in the host, may become the favorable one. The cycle of mutations in the host that confer resistance and mutations in the pathogen that circumvent this resistance continues, with the most favored genotypes and alleles depending on which other alleles are present. In any case, multiple alleles are maintained in the population.

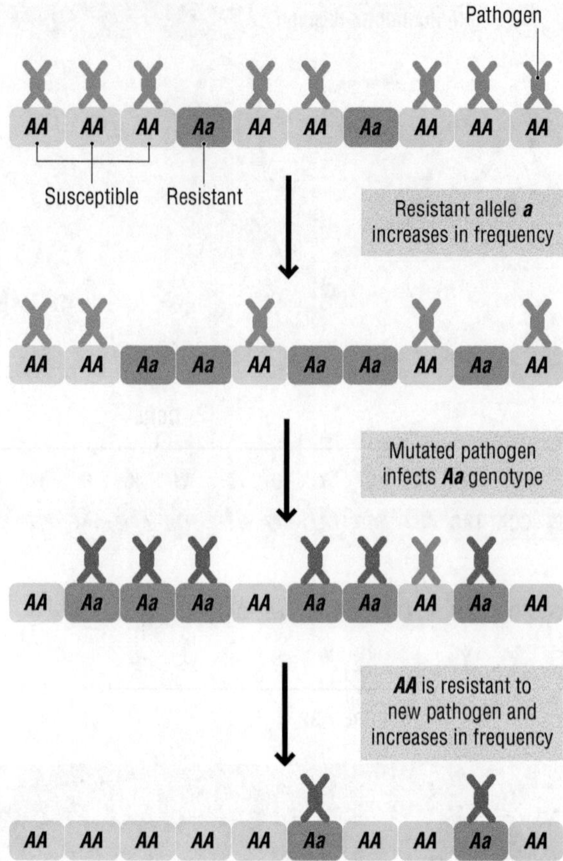

Figure 16.18 Frequency-dependent selection. While directional selection and heterozygote advantage are the most frequently considered types of natural selection, selection can work in other ways. Frequency-dependent selection often depends on the activities of a pathogen or a predator, or on other interactions among organisms in the environment. In this example, the ***AA*** genotype is susceptible to the pathogen, while the ***Aa*** genotype is resistant. The ***a*** allele, and thus the ***Aa*** genotype, will increase in frequency initially. However, a mutated pathogen arises that can infect the ***Aa*** genotype but not the ***AA*** genotype. Thus, the ***Aa*** genotype will decrease as the ***A*** allele increases in frequency, again due to its resistance to the mutated pathogen. Frequency-dependent selection can select for the rare allele, as in this example, or the common allele, but both alleles are maintained in the population.

Detecting selection

How can we tell if selection is occurring or has occurred for a particular locus? For many years, we could not look directly at genotypes or measure reproductive advantage over time, so selection was measured indirectly, if at all. Often the evidence for selection was an inference, applying logic that went something like: "It makes sense that this trait in this environment (such as fair skin in areas of low sunlight) should confer an advantage." Much of this evidence was accurate and informative, so these inferences should not be dismissed, although they sometimes led naturalists and geneticists to false, and even highly biased, conclusions.

Information from genome sequencing projects from different species and different populations has now given us objective and direct evidence for selection, even for genes of unknown function. In fact, we can search for the regions of the genome in which selection has been occurring without first knowing what genes in those regions actually do. One of the primary ways of doing this is to use the **Jukes–Cantor model**, named for the authors of the paper that first proposed using the relative frequencies of synonymous and non-synonymous base substitutions to analyze the evidence of selection in the genome.

In order to understand the Jukes–Cantor method of measuring selection, we need to return to the concept of neutral mutation, introduced in Chapter 4. A neutral mutation is one that neither increases nor decreases reproductive success. There is much debate about how frequently neutral mutations occur or how they can be detected, but a few measures are widely accepted. One of the best of these is the synonymous substitution rate, abbreviated dS. You will recall from Chapter 13 in our discussion of the genetic code that the third base in a codon can often vary without altering the amino acid that is specified by that codon. Thus, a mutation in the third base of a codon that does not change the amino acid is almost certainly a neutral mutation. The frequency of third-base substitutions that do not change the amino acid can be used as a good estimate of the dS.

For example, if we want to ask what genes have been selected for since the divergence of humans and chimpanzees, we begin by examining the coding sequences of comparable genes in the two species and determining how often the third base has changed without changing the amino acid it encodes. There will be many thousands of such sites. For example, consider a protein that is found in both chimpanzees and humans (which would be more than 98% of all proteins in either species). In our hypothetical example, suppose that both chimps and humans have a lysine at a particular position in this protein. But in the DNA sequence, chimps have AAG for the lysine codon, while humans have AAA. This third-base change is counted as a synonymous substitution, since the amino acid at this location in the protein remains the same, even though the DNA is altered.

By summing up all of the synonymous substitutions that have occurred, we have an estimate of how much random change has occurred overall between the two species; this estimated rate of synonymous change is designated dS.

The next step in the Jukes–Cantor model is to identify the non-synonymous substitutions—those that have changed an amino acid, for example. Returning to our hypothetical comparison of chimps and humans, the amino acid at a particular position in a hypothetical protein might be a glutamine (encoded by the DNA sequence CAA) in chimps but might be leucine (encoded by the DNA sequence CTA) in humans. This is an example of a non-synonymous change. Non-synonymous changes like this can be tallied in individual genes or for entire regions of the genome, and this rate of non-synonymous substitutions is abbreviated dN. Figure 16.19(a) illustrates synonymous and non-synonymous substitutions in two similar polypeptides.

The Jukes–Cantor model then uses the ratio of non-synonymous to synonymous substitutions (that is, dN/dS) to provide a direct measure of the impact of selection. If dN/dS is 1, then non-synonymous substitutions are found at the same frequency as synonymous ones, and selection at the locus or region under examination is effectively neutral. If dN/dS is less than 1, non-synonymous changes are much less common than synonymous changes and are not persisting in the population. We can infer that the non-synonymous substitutions are being selected against. In reality, nearly all loci in a genome will have dN/dS of much less than 1. This makes sense because most mutations are deleterious and selected against.

If we look at the dN/dS ratio across entire genomes of two species of subpopulations of a species, there are some loci with a value of dN/dS that is greater than 1. This means that non-synonymous substitutions have occurred and persisted much more frequently than expected, when compared to the rate of synonymous substitutions. A dN/dS value greater than 1 therefore implies that the non-synonymous changes must be conferring an advantage that has been selected for. Figure 16.19(b) illustrates how the ratio of dN/dS is used in a hypothetical region of the genome to detect a signature of selection. It is estimated that as much as 10% of the human genome shows such signatures of recent selection, typically conferring a reproductive advantage of 1–2%; analysis of additional populations and comparisons with ancient populations may result in a higher estimate of the genome that has been under selection.

KEY POINT Comparisons of the ratio of the rate of synonymous substitutions (dS) and non-synonymous substitutions (dN) in the DNA sequences of two populations via the Jukes–Cantor method give a direct measure of the effect of selection in a particular part of the genome. Most loci will have a dN/dS value of less than 1. If dN/dS is greater than 1, this indicates positive selection for a new allele or haplotype at this locus.

It is important to realize that the Jukes–Cantor and other methods provide an objective measure of selection without any preconceived ideas about what should be the favorable genotype. In the section below, we will discuss the two strongest signatures of selection found so far in the human genome. One of these examples agrees with our ideas about what should be the favorable genotype, while the other led to a new insight about the evolution of some human populations and gene function.

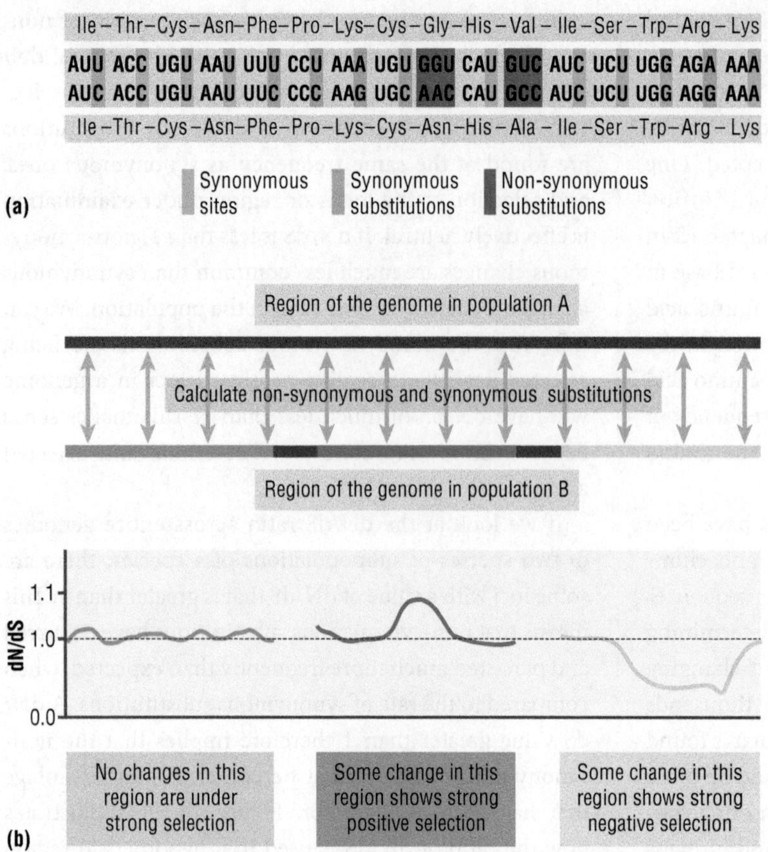

Figure 16.19 The Jukes–Cantor method uses synonymous and non-synonymous substitutions to find signatures of selection. The availability of genome sequences from different species and different populations has allowed objective analysis for evidence of selection in the genome. The Jukes–Cantor method uses the frequencies of synonymous and non-synonymous base substitutions to measure selection. (a) Two aligned sequences, with the identities of the encoded amino acid residues and substitutions noted. These sequences illustrate three concepts. Synonymous sites, shown in blue for third-base sites in these codons, are sites in the genome that have remained the same between the two genomes; this provides an overall standard for calculating the rate of substitutions. Synonymous substitutions, shown in green, are changes in the third base of a codon that do not change the amino acid sequence. This provides the frequency of total substitutions, abbreviated dS. Non-synonymous substitutions, shown in pink, are locations where the amino acid has been changed; the frequency of non-synonymous substitutions is abbreviated dN. (b) illustrates how these variables are used in a hypothetical region of the genome to detect a signature of selection. The ratio of dN/dS—the Jukes–Cantor ratio—is calculated for different regions of the genome from two different populations. If dN/dS is approximately 1, the rate of non-synonymous and synonymous changes are about the same, so there is no evidence for selection for any change in that region. In those few regions where dN/dS is greater than 1, the rate of non-synonymous changes is greater than expected from the rate of synonymous changes, so some change in this region must confer a selective advantage and is under positive selection. About 10% of the human genome shows such signatures of selection. If dN/dS is very low, the rate of synonymous changes is much greater than the rate of non-synonymous changes, so changes in this region are infrequent and are being selected against. Most protein coding regions will be like this.

How much of the human genome is under selection?

Before discussing these two specific examples, let's return to the general question about the effects of selection in modern human populations. It has been estimated by some geneticists that as much as 10% of our genes show evidence of selection when different human populations are compared. Others have put the number lower than that, and it is probably not possible to have an exact reconstruction of all of the selection forces our species has encountered. Still, it is clear that human genetic diversity shows evidence for selection at many loci. It may not be simple in all cases to decide what agent provides (or once provided) the selective force or when that selective force was felt. For most rare genetic diseases, there has never been a selective advantage in any environment, even in the heterozygote, so the trait remains rare. Selection in these cases has always been in favor of the "wild-type" or functional alleles. In our examples of balancing selection and heterozygote advantage, the selective agent was a pathogen. Our history has been dominated by infectious diseases whose impact has been blunted by modern antibiotics and treatment methods, at least in the areas of the world with good health care. Yet we retain the polymorphisms that conferred resistance long after their advantage has been lost.

Many aspects of lifestyles and culture, including availability of food, types of habitat, type and amount of physical activities, and climate, all have provided selection for or against some trait in some environment for some population at some time. Even when the selective agent is removed, the human population retains the polymorphisms. Among the traits for which we see such polymorphisms are salt-dependent hypertension, alcohol detoxification, starch digestion, sugar and fat metabolism, perspiration, and lactose utilization.

We may or may not even be able to recognize what the selective forces were, but genome analysis provides us with tools to look for their effects. The earliest comparisons of the genomes of different human populations, completed around 2005, found numerous regions of the genome in which dN/dS is greater than 1 in some populations, telling us that positive selection has occurred, that is, a non-synonymous substitution in this region of the genome has resulted in an advantageous change and has risen to a high frequency in that population but not in another population. Many of these effects were detected despite being rather small. However, the two strongest signatures for selection were found, by chance, on chromosome 2, albeit for different populations and different haplotypes. In each case, the adaptive mutation has been identified, but the reasons for the adaptation are quite different.

Lactase persistence

One signature for positive selection was found among European populations for a region on chromosome 2 that includes the gene encoding the enzyme lactase, the major enzyme that breaks down lactose (milk sugar) into the simpler sugars galactose and glucose, illustrated in Figure 16.20; the same signature for selection was not found among Han Chinese populations, for example. The advantageous mutation is not in the coding region of the lactase gene itself, which shows almost no sequence variation in the coding region, but rather in a region upstream of the start of the lactase gene, which plays a role in the regulation of lactase gene transcription.

In fact, this is an example in which our intuition and prior inferences about adaptation agree with the sequence data. It has been recognized that many people in some populations, including Europeans, continue to produce lactase throughout their lives and thus can digest milk into adulthood. This is known as lactase persistence. The frequency of this trait among the original populations is shown in Figure 16.21. Lactase persistence is clearly a derived condition, since other mammals and most human populations shut off transcription of the lactase gene sometime after weaning and cannot continue to digest milk into adulthood. We often refer to this condition as lactose intolerance, but, in fact, this is the ancestral condition, so the "mutation" results in lactase persistence.

As seen in Figure 16.21, lactase persistence is found in numerous populations throughout the world, with a similar signature for positive selection on chromosome 2 in all of these populations. This signature always affects a region upstream of the lactase gene responsible for the transcription of the gene. A few of the mutations and their population of origin are diagrammed in Figure 16.22. All of the mutations from a given population were also found on the same haplotype, that is, each of these mutations shows a different founder effect.

It was recognized more than 50 years ago that all of these lactase persistence populations have a cultural heritage of pastoralism and raised cattle, sheep, goats, or camels. The ability to digest milk results in a lifelong source of nutrition, which is expected to be advantageous among pastoralist populations where milk was common. Thus, whenever a mutation allowing continued transcription of the lactase gene arose by chance in one of these populations, the mutation had a selective advantage and became more frequent. Whenever the same mutation arose in a non-pastoralist population, there would be no selective advantage, so the mutation did not increase in frequency.

The results for the signature of selection for lactase persistence are striking. Among some historically pastoralist populations, more than 90% of the people are lactase-persistent. By contrast, fewer than 5% of the

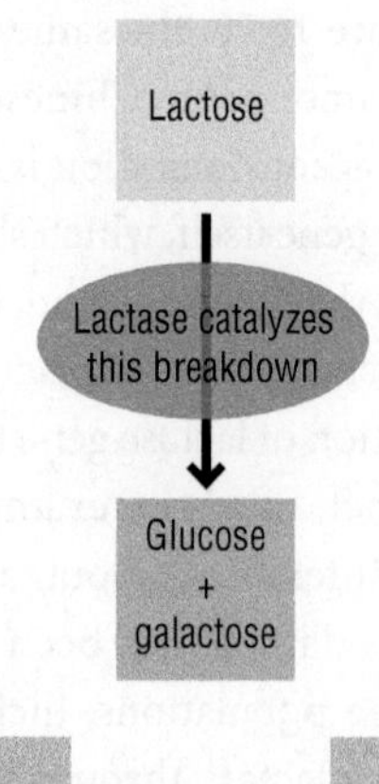

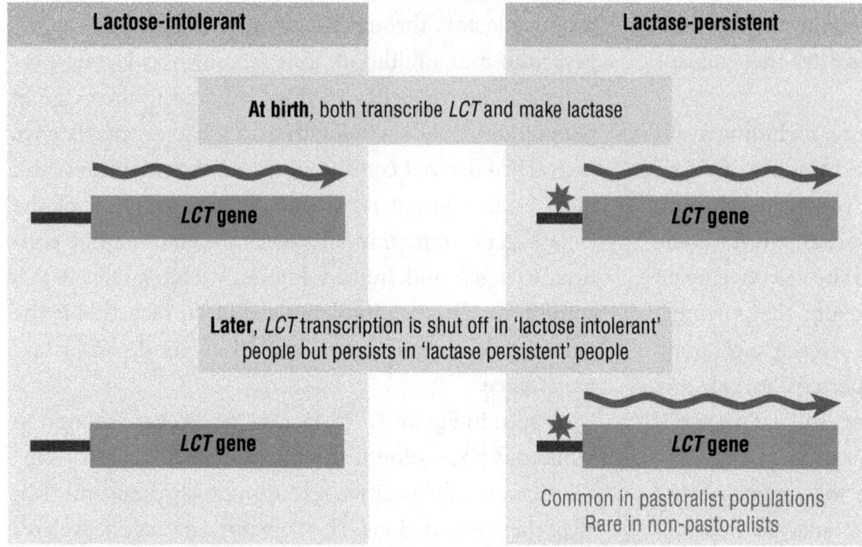

Figure 16.20 Lactase persistence. One of the strongest signatures for positive selection in the human genome surrounds the gene encoding the enzyme lactase. Lactase breaks down lactose, the common sugar in milk, into its simpler components, galactose and glucose. At birth, mammals transcribe the lactase gene. However, at some time after weaning, the transcription of lactase ceases in most people (and in most species of mammals). In some individuals, the lactase gene continues to be transcribed, which allows them to digest milk well after weaning and into adulthood. This derived condition of lactase persistence is particularly common in western cultures, so the ancestral condition in which lactase is shut off is often termed "lactose intolerance."

people in some non-pastoralist populations are lactase-persistent. In the pastoralist populations, lactase persistence has been calculated to have conferred a selective advantage of between 2% and 15%, with 5% being the best estimate of the increased fitness arising from this substitution.

It is also possible to estimate the time at which the lactase persistence phenotype arose. These estimates take into account the strength of selection, the frequency of the haplotype with the advantageous mutation, and the length of the haplotype. Among European and most pastoralist populations, lactase persistence arose approximately 8000–10,000 years ago, or shortly after the time when animal domestication became widespread, about 10,000 years ago.

KEY POINT Lactase persistence is an example of positive selection in European populations. It is found in populations that have a history of raising milk-producing animals and conferred a selective advantage by enabling individuals to digest milk products throughout their lives.

EDAR

A very long haplotype on chromosome 2 is found at a frequency of nearly 90% among Han Chinese populations but is almost completely absent elsewhere in the world. In fact, until the genomic analysis of populations included data from the Han Chinese, this region was not recognized as being polymorphic. This suggests that other regions in our genomes may be found to be under selection

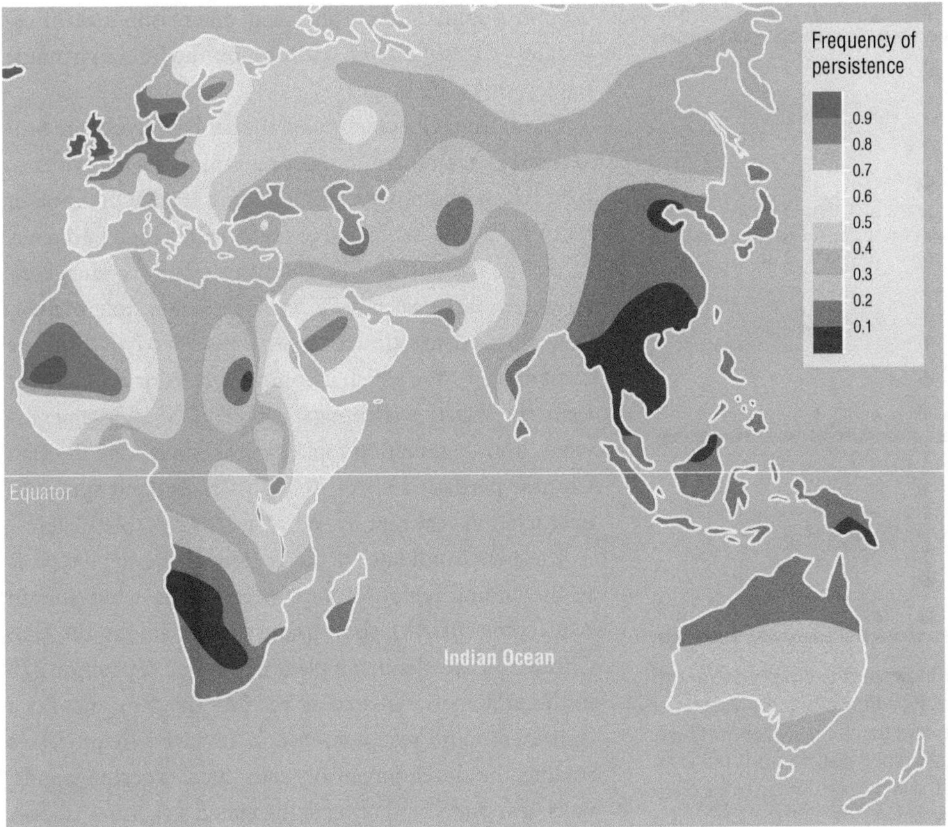

Figure 16.21 The distribution of lactase persistence. The frequency of lactase persistence among the indigenous peoples of each region is shown. The highest frequencies of lactase persistence are found in Europe, western Africa, and the Middle East where herding and pastoral lifestyles were common. Lactase persistence is very rare in regions among non-herding populations such as in South-east Asia, South Africa, and Australia.

Source: Itan et al. A worldwide correlation between lactase persistence phenotype and genotypes. *BMC Evolutionary Biology* (2010), 10:36 doi:10.1186/1471-2148-10-36. This article is published under license to BioMed Central Ltd.

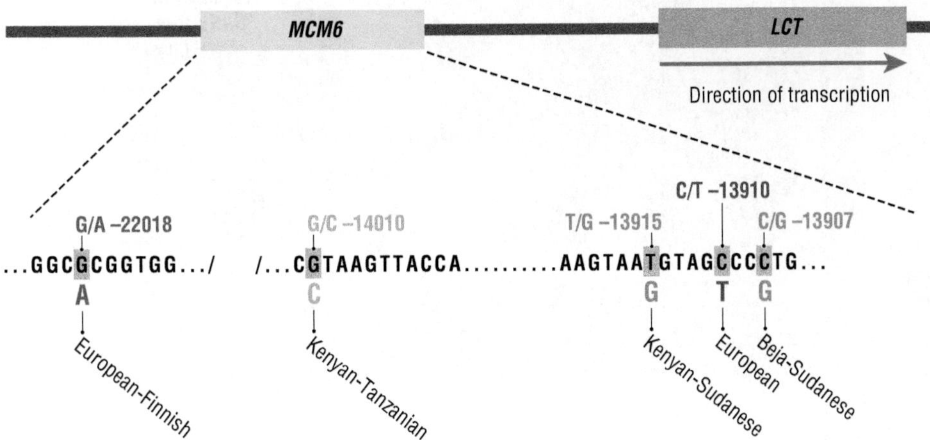

Figure 16.22 Single-nucleotide polymorphisms (SNPs) associated with lactase persistence. While lactase persistence is common in some parts of the world, the polymorphism and haplotype that gave rise to persistence is different in different regions, consistent with independent founder effects and strong positive selection in each population. The ancestral, non-persistent sequence is shown in black, with different SNPs shown. For example, non-Finnish Europeans who are lactase-persistent share a common haplotype with a T, rather than a C, at position −13,910. Note that these SNPs lie within the introns of the *MCM6* gene, which has a function unrelated to lactase persistence.

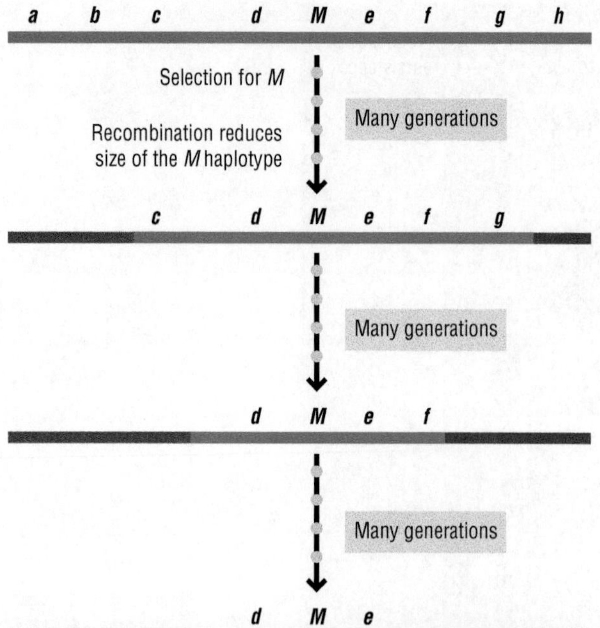

Figure 16.23 Haplotype length indicates the time and strength of selection. The length of a haplotype provides one indication of the time at which strong positive selection began, as shown here. The favorable ***M*** allele arose on the haplotype at the top in red. Over time, with continued selection for the ***M*** trait, the length of the original haplotype in the founder breaks down by recombination. Thus, a long haplotype indicates recent strong positive selection.

as more populations are analyzed, one reason that the estimates of the effects of selection on the human genome have such a wide range.

Recall from Chapters 9 and 10 that haplotypes are broken up by recombination, so long haplotypes are evidence for a recent origin, as shown in Figure 16.23. On the other hand, for a long haplotype to reach such a high frequency in one population and be absent in others, it must confer a very large selective advantage in that one population. Theoretical estimates are that the selective advantage must exceed 10%, which is comparable to, or even greater than, the selective advantage conferred by lactase persistence. Thus, a mutation must have arisen among the Han Chinese perhaps 15,000–30,000 years ago and spread to near fixation very rapidly. What is this favorable allele?

The best candidate in the region of the haplotype is in the coding region of a gene known as ectodysplasin A receptor (*EDAR*); the variant version among the Han Chinese has an alanine in place of a valine at position 370 in the protein, designated as V370A. The gene encodes a transmembrane receptor protein involved in processes such as the development of teeth, hair, exocrine glands, nails, tear ducts, and other skin-related structures, shown in Figure 16.24(a). The frequency of the EDAR V370A

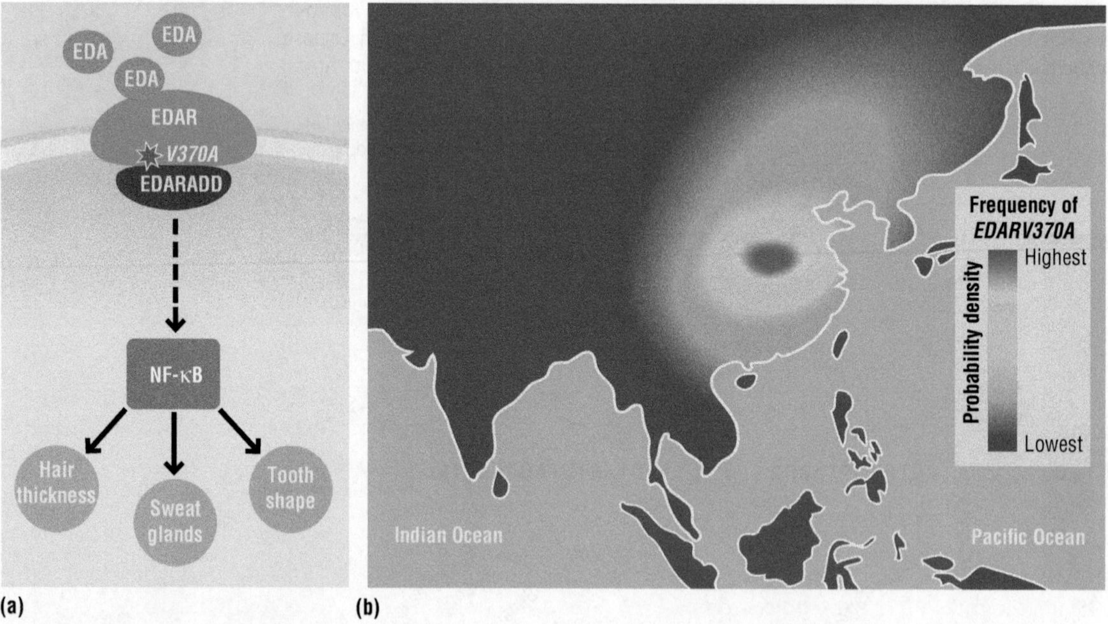

Figure 16.24 Selection at EDAR. The *EDAR* gene is involved in a number of biological processes involving the ectoderm, including the thickness and shape of hair shafts, tooth morphology, and the number and distribution of sweat glands. (a) The ligand EDA (ectodysplasin A) binds to its receptor EDAR, which, in turn, signals to the downstream transcription factor NF-κB. The signaling process involves the interaction of EDAR with its adaptor protein EDARADD. A T/C base change in *EDAR* results in the amino acid substitution of alanine (A) for valine (V) at position 370 in a region of EDAR that interacts with EDARADD. The precise mechanism by which this substitution affects the downstream signaling is not known. (b) The distribution of the ***EDARV370A*** allele in Asia. The *EDARV370A* substitution is exceptionally common in China, but essentially unknown outside of this region. This substitution has a very strong positive selection among the Chinese, possibly due to the effect on sweat glands that allows cooling in hot, humid environments.

Source: Part (b) reproduced from Kamberov et al. Modeling recent human evolution in mice by expression of a selected EDAR variant. *Cell.* (2013) Feb 14; 152(4):691-702. doi: 10.1016/j.cell.2013.01.016.

variant is shown in Figure 16.24(b); note the extremely high frequency in Chinese populations but its absence from the surrounding populations.

EDAR has pleiotropic effects, and the V370A variant was known to have numerous effects in humans, including increasing hair thickness and altering tooth morphology. Unlike the results with lactase persistence, however, it was not at all intuitively clear how any of these phenotypes might have provided such a strong adaptive advantage. Thus, a different type of experiment was done by producing the same mutation in the *EDAR* gene of mice and observing its effects on other phenotypes. When mice with the equivalent ***EDARV370A*** variant were created in the laboratory, mammary gland morphology was affected and, most notably, an increased number of sweat glands were seen.

Having seen the effects on sweat gland density in mice, the investigators then examined sweat gland density in humans. Consistent with the results in mice, people with the V370A variant have significantly more sweat glands than those who are V370. An increased number of sweat glands may allow mammals in hot, humid climates to sustain physical activity for longer periods of time—in effect, to cool the body by perspiration. Climatological data indicate that China was very warm and humid at about the time when this allele is estimated to have arisen, so this may be the source of the adaptive advantage. However, since *EDAR* is pleiotropic, this explanation remains likely but unproven.

16.6 Population genetics of bacteria and other asexually reproducing organisms

Although much of this chapter has focused on human populations, the basic principles apply to other populations as well, including bacteria. Bacterial populations experience selection, mutation, recombination, migration, genetic drift, and bottlenecks. However, the genetic structure of bacterial populations differs from much of what has been discussed so far because they reproduce by binary fission, in which a single cell gives rise directly to two daughter cells. A population that reproduces asexually is therefore derived from a single ancestor and is said to be **clonal**. New mutations that arise in bacteria will therefore be passed directly to daughter cells but can only be shuffled between genomes of different clonal lineages by horizontal gene transfer, if at all. Therefore, when advantageous mutations arise within different lineages of an asexual population, there will be multiple competing genotypes in the population. In sexual reproduction, by contrast, two parents contribute alleles to each offspring at every generation, so mutations can be combined together into an individual genome through this form of reproduction.

It has long been recognized that gene flow and population structure differ between asexually and sexually reproducing populations; the initial ideas were proposed in an early paper by Hermann J. Muller. The graphical representation of the population structure that Muller used to illustrate this point has proven to be a useful tool and is described in Communicating Genetics 16.2.

Analyzing the structure of genotypes within clonal populations is therefore the focus of bacterial population genetics. Bacteria are among the best-studied population genetics models among asexual organisms, but these same properties apply to viruses and yeasts and other unicellular eukaryotes that also reproduce asexually.

Experimental evolution in bacterial populations

As with all populations, the parameters used to assess population-level genetics are difficult to compute for natural populations of bacteria. Moreover, the environmental conditions that act on evolving genomes in nature cannot be controlled. Darwin himself suggested that observations were helpful to evolutionary biology but that experiments, if they could be performed, would provide a better approach for dissecting the principles of evolution. Thankfully, bacteria are extremely amenable to experiment. Their generation times are, for the most part, rapid, and populations are often very large. Bacteria therefore represent good model systems for evolutionary biology research and are among the favored models for hypothesis testing in experimental evolution.

Fitness is the relative propensity of an individual to leave progeny. For organisms that reproduce asexually by binary fission, fitness most often manifests as a reduced generation time or faster growth rate, which can be measured at the population level. In experiments, fitness is best assessed by growing two competing strains in the same medium and determining which strain ultimately

COMMUNICATING GENETICS 16.2 Muller plots

To understand the dynamics of allele frequencies in a population, it can helpful to depict the information visually. A method often used to depict bacterial populations is the Muller plot, named after the geneticist Hermann J. Muller.

In these plots, time is represented along one axis, while the frequency of each genotype is shown on the other axis; time is often calibrated in terms of generations. Figure A has an example with the time along the X-axis and genotype frequencies on the Y-axis. As the population divides from one generation to the next, new genotypes may arise by mutation; each new genotype is assigned a specific color or shading. At first, the new genotype will only be present in the one individual in which it arose, as indicated at the points of the colored shapes in Figure A. However, if the genotype survives, so that the mutation is transmitted to subsequent generations, the proportion of individuals with that genotype will increase over time.

In Figure A, the genotype of most of the original population is in white. A new genotype, represented in orange, arises and increases in frequency over time. Individuals with the genotype depicted in orange outcompete those with the original genotype (the white area on the plot) until the orange genotype is the only one present in this population. Subsequently other genotypes arise, represented in blue and yellow, and the proportion of individuals in the population with these genotypes increases over successive generations until they have outcompeted individuals with the orange genotype. Now all individuals in the population possess either the blue or the yellow genotype, with a larger proportion possessing the blue genotype.

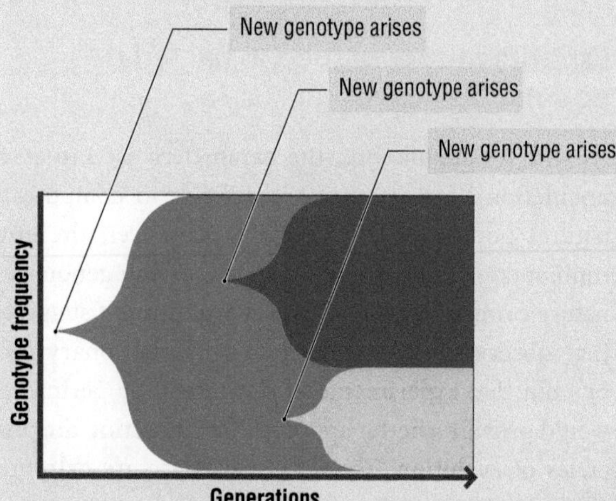

Figure A Muller plots. Muller plots are used to show the changing frequencies of genotypes (on the Y-axis) over time as generations (on the X-axis) and are especially useful for asexual organisms such as bacteria. In this plot, the initial genotype is shown in white, and the newly arisen mutations are shown in different colors. The size of the color regions is more significant than their positions on the Y-axis. Note that a new mutation (in orange) arises and becomes fixed. New mutations (in blue and then in yellow) arise subsequently and outcompete the orange phenotype.

Source: Adapted from: Jeffrey E. Barrick & Richard E. Lenski. Genome dynamics during experimental evolution. *Nature Reviews Genetics* 14, 827–839 (2013).

Figure A represents just one hypothetical scenario. The dynamics of the change in allele frequencies over time will depend on the relative fitness of the specific genotypes in a particular environment; as discussed in Section 16.6, growth rate is often used as a measure of fitness in asexually reproducing populations. Furthermore, since bacterial populations divide by binary fission, any newly arisen mutation can be passed on only to direct descendants of the original bacterium in which the mutation occurred. The colored areas on a Muller plot therefore represent the clonal descendants of the first bacterial cell of that specific genotype. Scenarios in which the genotype can only be passed to direct descendants also assume that mechanisms of horizontal gene transfer are not occurring. In reality, the rate at which beneficial mutations occur and the extent of horizontal gene transfer are likely to vary for different bacteria in natural populations occupying different environments.

Muller used plots like these to model how beneficial mutations would spread in populations of organisms that reproduce by sexual reproduction, compared to those that reproduce asexually. We have reproduced slightly modified versions of his graphical representation in Figure B, with the asexual population on the left and a sexual population on the right.

In both populations, beneficial mutations arise randomly. In asexually reproducing populations, new mutations can only be passed to the direct descendants or lineage. When new mutations arise in different individuals, the direct descendants carrying different beneficial mutations compete with one another, and the lineage with a mutation with a greater fitness advantage will eventually predominate. Note that the top three lineages in the figure on the left have mutations that have been maintained but have not conferred enough of a fitness advantage to outcompete the lineage at the bottom, which increases in frequency. Note also that the lineage at the bottom acquires additional mutations as it divides, some of which are also beneficial so that the combination of two beneficial mutations confers a further growth advantage.

In the absence of horizontal gene transfer, new combinations of alleles or mutations in asexual lineages arise only from the process of sequential mutation, since no genetic exchange or recombination occurs. In contrast, each individual in a sexually reproducing population has two parents, so there is the possibility of multiple beneficial mutations being combined in one offspring, as seen in the diagram on the right.

The differences between asexually and sexually reproducing populations in the spread of beneficial mutations will be

COMMUNICATING GENETICS 16.2 Continued

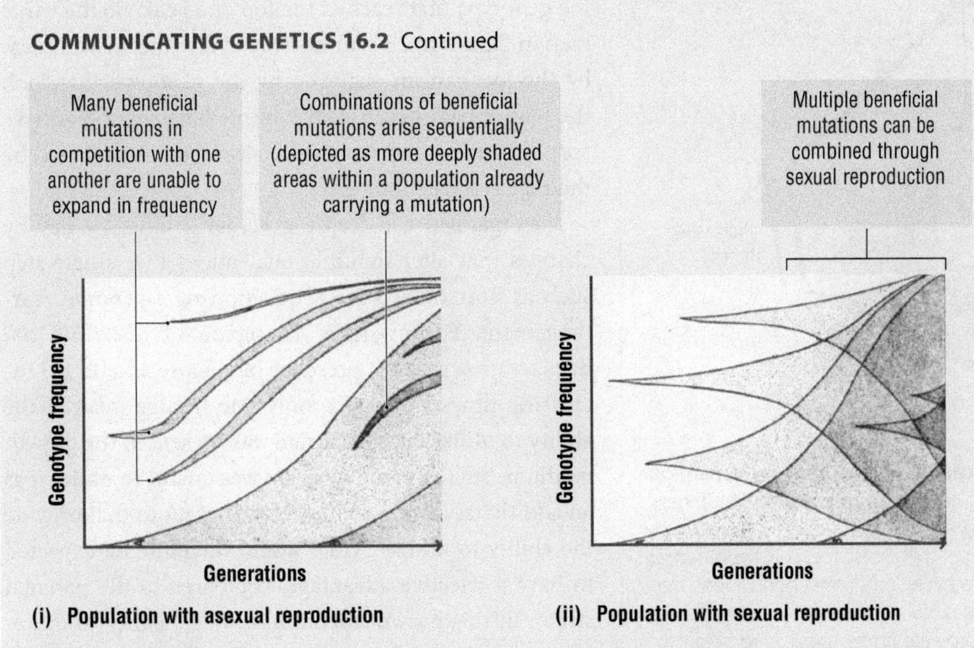
Many beneficial mutations in competition with one another are unable to expand in frequency
Combinations of beneficial mutations arise sequentially (depicted as more deeply shaded areas within a population already carrying a mutation)
Multiple beneficial mutations can be combined through sexual reproduction
Genotype frequency
Generations
Genotype frequency
Generations
(i) Population with asexual reproduction
(ii) Population with sexual reproduction

Figure B Muller's original plots of asexual and sexual populations. These show how new genetic combinations arise in each case. In the asexual population (i), new allelic combinations occur only through sequential mutations. In the sexual population (ii), new allele combinations occur through matings and recombination, as well as sequential mutations.

Source: Adapted from: Muller H.J. (1932). Some genetic aspects of sex. *Am Nat* 66: 118–138.

less dramatic if these mutations arise very rarely. In that case, it is more likely that one beneficial mutation will have expanded through the entire asexual population before the next mutation arises. This increases the likelihood that the asexually reproducing individuals would obtain multiple beneficial mutations sequentially.

represents a higher proportion of the population. The two competing strains are seeded into the same environment in equal densities, and the number of progeny from each genotype is counted after a set period of time. To allow for differential counting, at least one of the genotypes in question must be marked with a gene that produces an easily distinguishable phenotype. In experimental evolution, it is possible to compare different genotypes derived from a clonal population with their ancestors, since starting individuals and subcultures taken from different time points during population growth can be sampled and frozen. Thus, it is possible to determine whether an evolved genotype is better adapted to the environment, that is, whether it has become more or less fit under a given set of growth conditions.

KEY POINT Bacterial populations are used to study evolution experimentally, since the generation time is short and samples from many time points can be saved and compared.

The relative fitness for individuals in a population can be described with a **fitness landscape**. In such an approach, fitness can be represented as a three-dimensional map in which the height of the peak indicates higher fitness, as depicted in Figure 16.25. Such maps are notoriously complicated to interpret in detail, but an intuitive perspective can be gained by imaging this as any other topological landscape and thinking about the possible paths that would allow a cell or an organism to progress from one region of the landscape to another. Maps of a fitness landscape show that there can be multiple paths to increased fitness. Such a landscape can be used for any population, whether it reproduces sexually or asexually. In principle, a population or subpopulation can be at any point in this landscape. In practice, however, organisms that reproduce by binary fission and have little or no recombination evolve linearly and would need to progress across the landscape in a stepwise fashion—typically involving sequential point mutations.

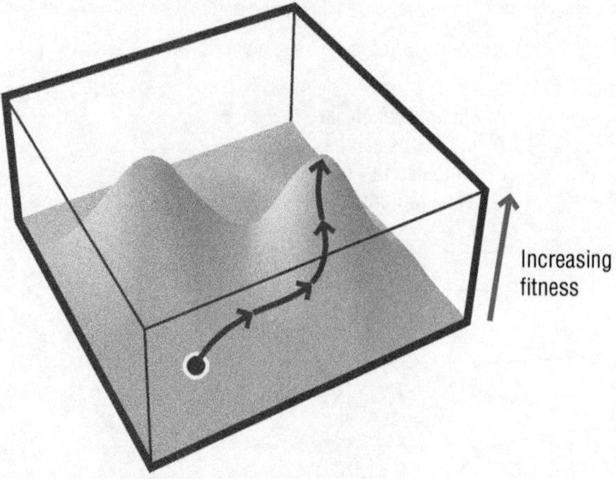

Figure 16.25 Fitness landscapes and evolutionary dynamics. Fitness is plotted as a topological map, with peaks of high fitness and valleys of lower fitness. Changes in the population dynamics can be imagined as paths across this landscape. This hypothetical adaptive landscape shows the evolution of two populations that originated from one ancestral genotype. These populations evolved independently and reached separate fitness peaks, the height of which indicates a high level of fitness.

Source: Reproduced from Elena SF, Lenski RE. Evolution experiments with microorganisms: the dynamics and genetic bases of adaptation. *Nat Rev Genet.* (2003) Jun; 4(6):457-69. Review. PubMed PMID: 12776215.

Selection acts on existing diversity, eliminating less fit genotypes and therefore making it impossible for a given genotype to tread on some parts of the landscape. In other words, the entire fitness landscape is not accessible to all genotypes in a species, and given populations can become "trapped" at peaks that are separated by valleys that would require several mutations to traverse. For example, the genotype that reached the top of a peak via the white path in Figure 16.25 is unlikely to evolve increased fitness by the mechanism achieved by the genotype that took the purple path because all intermediate genotypes have reduced fitness and would therefore be outcompeted by their ancestral genotype.

An organism can leapfrog across a fitness valley by changes that alter multiple nucleotides in a single step such as horizontal gene acquisition or a genome rearrangement. For example, 12 separate *Escherichia coli* lineages grown in the presence of glucose all evolved increasing fitness; however, only one lineage evolved the ability to utilize citrate. Citrate was present in the growth medium, but the parental strain was unable to utilize it as an additional carbon source. New mutations that provide the ability to utilize citrate would therefore be expected to have a selective advantage, compared to the parental strain, in this growth medium. A weak ability to utilize citrate (symbolized Cit$^+$) appeared in the lineage at about 31,000 generations; multiple subsequent mutations arose that increased the ability to utilize citrate.

The new Cit$^+$ phenotype arose from several sequential point mutations, each of which would have conferred a slight growth advantage, followed by a duplication event that conferred an even greater fitness advantage; the genomic basis for this citrate utilization phenotype is shown in Figure 16.26. The key component of this "novel" advantageous phenotype was that the genomes in the founding populations of bacteria had a citrate transporter gene that was not transcribed. In the one lineage that evolved the ability

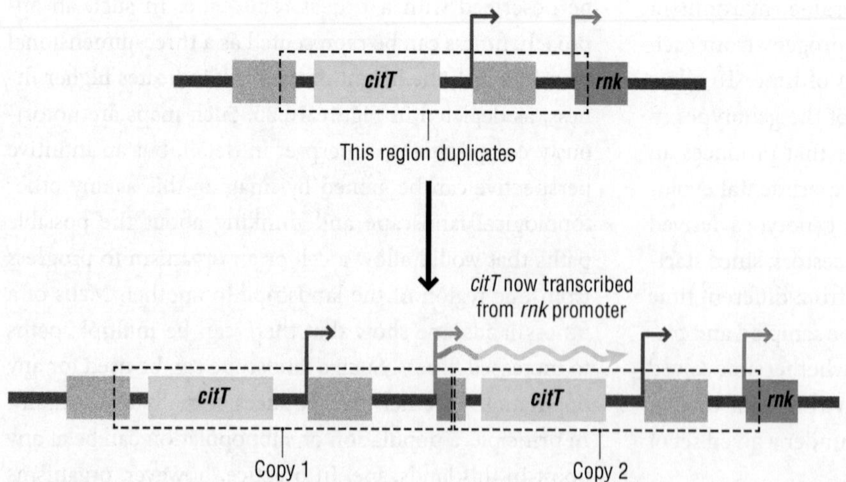

Figure 16.26 An evolutionary origin of citrate utilization in *E. coli*. In one long-term evolution experiment, fitness increased rapidly because the lineages acquired mutations that allowed them to utilize the citrate present in minimal growth medium as a carbon source. The underlying genomic change allowing citrate utilization is shown here. The original genome had a citrate transport coding region that was not transcribed. Mutations that duplicated this coding region and placed it next to an active promoter allowed citrate transport and utilization, which conferred a growth advantage in minimal medium.

to utilize citrate, a duplication in the genome occurred, which placed an active promoter upstream of the citrate transporter gene, thus allowing that lineage to transcribe the transporter gene and utilize citrate. The duplication is shown by the tandem copies of the genes in Figure 16.26.

Population genetics studies in *E. coli* and other bacteria, as well as in asexual unicellular eukaryotes, are a well-established field of study. Many of the features of sexually reproducing populations are also observed in these studies and are often more readily observed in asexual populations. For example, the effects of mutation are seen with **mutator** strains in which parts of the DNA repair machinery have been altered, resulting in a higher than normal rate of mutation. While individual cells may die, a bacterial population may be able to sustain a higher rate of mutation, which gives insights into the effects of mutations in all populations. Likewise, the effects of horizontal gene transfer are also readily observed. Horizontal gene transfer in bacteria is conceptually similar to a migrant being introduced into the population, although the type of genetic exchange is obviously quite different.

The long-term experimental evolution studies we have just described are based on the concepts of the founder effect and genetic drift, and directional selection is observed from the growth rates. A form of frequency-dependent selection occurs from the interactions among the microbes in a community, discussed in more depth in Chapter 17. These similarities between the results seen in asexual bacterial populations and sexual eukaryotic populations may not always occur by similar underlying molecular mechanisms, but the principles of population genetics are not so different.

16.7 Summary: the genetic structure of populations

To wrap up this chapter, we return to some reflections about human population genetics. As a species, humans have spent a relatively large part of our existence living in small isolated groups in far-flung microenvironments with distinct selective pressures. Over time, mutations arose, and the effects of limited migration, genetic drift, and selection turned some of them into polymorphisms in certain populations; in other populations, under other conditions, the mutation was lost.

Our current genetic diversity shows the impact of selection in those microenvironments, often in ways that no longer seem relevant to us. Because we have antibiotics to treat many bacterial infections and the impact of other diseases has been greatly reduced by improved sanitation, we may have forgotten the effect these selective forces had on our species. For many of us, we have a wide variety of food available to us at a temperature-controlled grocery store, so the effects of climate and nutrition also seem lost. We travel the world and communicate and reproduce with people from different cultures that were once isolated from us. But we did not evolve under these conditions. Our genomes remember those conditions, even if we have forgotten them, and preserve the relics of these past selective pressures.

In considering this genomic record of our history, we can call it experimentation by our ancestors as Delbrück did, or tinkering as Jacob did, or we can use another metaphor such as calling it a genetic scrapbook as we did here. All of these metaphors work, because they share a common explanation. Nothing in biology, including human genetic diversity, makes sense, except in the light of evolution. We might not recognize the events or situations, but the genome records their impact. With our genome analysis and knowledge of the effects of the evolutionary forces on populations and species, we can infer how the past is still with us.

CHAPTER CAPSULE

- Population genetics studies genetic variation in populations and the transmission or flow of alleles independent of the actual parents. It also measures and evaluates changes in the frequency of these alleles within the population. These can be measured by comparing allele or haplotype frequencies. Changes in gene flow over time allow the population to evolve.

- If five conditions are met, allele and genotype frequencies will remain stable through generations over time, and the population is said to be in Hardy–Weinberg equilibrium.
- The five conditions that must be met to maintain a Hardy–Weinberg equilibrium are:
  - Mating within the population occurs at random
  - No mutation occurs
  - There is no migration between populations
  - The population is extremely large
  - There is no selection acting on the population.

  Thus, if populations are different from one another, or if they change over time, some combination of these five factors has been at work.
- Non-random or assortative mating changes genotype frequencies but has no effect on allele frequencies.
- Mutations introduce new genetic variants into the population, which are then subject to natural selection and/or genetic drift.
- Migration can allow two previously separated populations to intermix or can split one larger population into two distinct subpopulations. Barriers between populations may be physical or, in human populations, cultural or linguistic.
- Genetic drift is the change in the frequency of an allele in a population due to random sampling and has a particular impact on small populations. Two specialized examples of genetic drift are the founder effect and population bottlenecks:
  - The founder effect occurs when a small group establishes a new separate population. The genetic traits in this small group will appear at high frequency in subsequent generations, although they may not be common in the species as a whole.
  - Population bottlenecks occur when a significant proportion of a population is lost due to catastrophic events or disease. There is reduced genetic diversity in the population that arises from the remaining individuals.
- Differential reproduction and selection change the frequency of alleles in a population. Genotypes that promote survival and reproduction in a given environment—those that are the fittest—contribute more gametes to the next generation. The most familiar type of selection is directional selection, in which the phenotype arising from a particular allele confers a reproductive advantage.
- A situation in which the heterozygote has greater fitness than the homozygotes is an example of balancing selection; both alleles are maintained in the population.

Balancing selection is believed to be responsible for the maintenance of some alleles that cause genetic diseases when homozygous, since the heterozygous condition confers resistance to particular pathogens.

- Comparisons of the ratio of the rate of synonymous substitutions (dS) and non-synonymous substitutions (dN) in the DNA sequences of two populations is one way to obtain a direct and unbiased measure of the effect of selection in the genome. If the ratio of dN/dS is 1, the rate of synonymous and non-synonymous substitutions is equal, and it is likely that no gene in this region is under selection. If dN/dS is much less than 1, this implies that non-synonymous changes are deleterious and are being selected against. If dN/dS is greater than 1, positive selection has occurred for a new allele in the region.
- Bacterial populations are used to study evolution experimentally, since the generation time is short and samples from many time points can be saved and compared.
- The genome is a record of the effects of natural selection and other evolutionary factors on the population.

STUDY QUESTIONS

Concepts and Definitions

16.1 Define "allele frequency." Why is the allele frequency a useful measure of the genetic structure of the population?

16.2 Define an effective population. What are some situations in which the effective population size differs from a simple census or count of the population?

16.3 What is a polymorphic locus? Explain how a locus might be polymorphic in some populations, but not in others.

16.4 Define positive and negative assortative mating, and speculate (from your observations of human behavior and human populations) about an example of each in humans. What may have been the likely effect of this assortative mating on human populations?

16.5 What is adaptation? What are some examples of adaptation? What might be the evidence that a particular trait is adaptive?

16.6 Explain directional selection and balancing selection, and compare and contrast their effects on a population.

Beyond the Concepts

16.7 Explain how these various concepts of a gene are related to one another:

a. The gene as a unit of inheritance

b. The gene as a locus on a chromosome

c. The gene as a DNA sequence

d. The gene as a description of the genetic structure of a population.

What are some of the properties of a gene that are more important for one of these gene concepts than for others?

16.8 The effects of assortative mating, mutation, migration, drift, and selection on populations can each be described by mathematical equations. While we have not introduced these equations, the parameters that are part of the equations are not too difficult to infer. For each effect or force listed in (a) to (d), what parameters will play a role, and thus be part of the equation, and what role will each one play?

Example: The effect of mutation on a population is a function of the rate of mutation and the size of the population. Mutation plays a larger role if the rate is high or if the population is small.

a. Assortative mating

b. Migration

c. Drift

d. Directional or balancing selection

16.9 The Jukes–Cantor model for detecting signatures of selection uses a parameter based on dN/dS.

a. Define dN and dS.

b. Why is it difficult to know precisely the amount of the human genome (or any genome) that is under selection?

c. What are the advantages of using dN/dS (or another similar measure) to determine the regions of the genome that have been under selection in a species?

d. What might be some of the challenges and limitations of using dN/dS when comparing two species?

e. dN and dS cannot be used for genes that encode functional non-coding RNAs, which are now recognized as a significant component of metazoan genomes. What might be used to determine sites that are neutral and sites that are functional and might be under selection for these regions of the genome?

f. In general terms, how do the length of the haplotype and dN/dS provide information about the strength and duration of selection?

16.10 While we can use the human genome to infer the effects of evolutionary forces on human populations in the past, it is also informative to think about what effects they may be exerting on us now. For example, modern human society has become far more mobile and inter-connected than in the past, and technological changes have had impacts on many aspects of our lives. Using the concepts from this chapter, speculate about how the genetic structure of future human populations might change. What are some of the (genetic or cultural) forces that could mitigate against these changes?

16.11 Natural populations of almost every species are genetically diverse. Why have natural populations not become fixed for the most favorable alleles at each locus? What type of evidence would be helpful in deciding among the possible explanations for genetic diversity in a natural population?

Applying the Concepts

16.12 The MN antigen system has two common co-dominant alleles, called L^M and L^N, encoded by a gene on chromosome 4. Like the ABO system, the MN blood type can be readily scored by an agglutination assay. In a certain village of 350 individuals, 168 were MM, 119 were MN, and 63 were NN.

a. What are the allele frequencies for the L^M and L^N alleles?

b. Is this village at Hardy–Weinberg equilibrium with respect to this blood type? Briefly explain your answer.

c. If the village is not at equilibrium, what do you think is the most likely explanation?

d. If no other evolutionary forces are at work in this village for this trait, what are the expected allele frequencies and genotype frequencies in the next generation?

e. Nearly all of the Aleut of the Arctic region are MM, while other Arctic populations, and most other populations throughout the world, have MM, MN, and NN individuals. Conversely, among the Indigenous Australians (or Aborigines), NN is much more common than MN, and MM is relatively infrequent. What might account for these varying frequencies of the L^M and L^N alleles among the different populations? How might you test your explanation?

16.13 In a particular population at equilibrium for a trait, 1% of the individuals are homozygous recessive ***a/a***.

a. Is the ***a*** allele a polymorphism in this population?

b. If two phenotypically wild-type parents produce an offspring, what is the probability that the offspring will be ***a/a***?

16.14 In a particular flowering plant, varieties are known with red flowers, pink flowers, and white flowers. In this plant, this is controlled by a single gene with two alleles. Two pink flowering plants are crossed to each other. The seeds are collected and planted, with the following results in the next generation.

Pink flowers: 17

Red flowers: 8

White flowers: 9

a. What are the genotypes of the pink, red, and white flowering plants?

b. In a large field, 104 plants were pink, 63 plants were red, and 33 were white.

i. What is the frequency of each of the alleles in this population?

ii. Is this field at Hardy–Weinberg equilibrium with respect to this trait? Use a χ^2 test to support your answer.

16.15 Red–green color blindness in humans is an X-linked recessive trait. In a particular large population, 9% of the men are red–green color-blind.

a. What percentage of the women in this population are expected to be carriers for red–green color blindness?

b. The red–green color blindness locus (more specifically, the X-linked opsin genes) are polymorphic in nearly all human populations, although red–green color blindness confers no selective advantage and may even have been slightly disadvantageous. Of the four forces that affect allele frequencies, which of these is the most likely to account for the high incidence of red–green color blindness? It may be helpful to review Box 9.2 for some insight.

c. What evidence would be a helpful test for your hypothesis in (b)?

16.16 The ABO locus in humans has three alleles, in which I^A and I^B are co-dominant and ***i*** is recessive to each. Suppose that, in a particular large population at equilibrium, the frequency of the I^A allele is 0.35 and the frequency of the I^B allele is 0.15.

a. What is the expected frequency of people with type AB blood in this population?

b. What is the expected frequency of people with type A blood?

Challenging

c. A blood bank collects 5000 pints of blood during a blood drive and finds that 1850 were type A, 635 were type B, 180 were type AB, and 2335 were type O. Calculate the frequencies of the I^A, I^B, and i alleles in this population.

16.17 Communicating Genetics 16.1 presents haplotypes from the Barbary macaques in Gibraltar. There are various hypotheses about the origins of these colonies. For example, some believe they may be the only survivors of an ancient population of European Barbary macaques; others believe they were imported to Gibraltar from Morocco or from Algeria, and some believe that some of the macaques that founded the current Gibraltar colony came from Morocco and others came from Algeria. Use the haplotype network shown in Figure A to decide which hypothesis seems to be the most likely, and explain why.

16.18 Easter Island (Rapa Nui) is one of the most remote inhabited islands in the world, located in the Pacific Ocean more than 3500 km west of Chile, with a current population of about 5800 residents. It is believed to have been settled by migrants from other Polynesian islands about 1000 years ago. The population is estimated to have grown to approximately 15,000 people during the seventeenth century, but, by the time Europeans arrived in 1722, the population had fallen to 2500; this decreased further to 111 people in 1877 due to disease and slave traders. While Easter Island is best known for its large stone statues or moai, the population has also been studied by geneticists.

a. From this brief history of the settlement of Rapa Nui, discuss the roles that the five genetic and evolutionary forces likely played on the nature of the genomes of the natives of Rapa Nui.

b. An analysis of the genomes of 27 native islanders (published in *Curr Biol* 2014, **24**: 2518–25) found that approximately 16% of their genomes were derived from Europeans, about 8% from Native Americans, and about 76% from Polynesians. Explain how these estimates were made.

Challenging

c. The authors hypothesized that the Native American DNA became part of the Rapa Nui genomes earlier than European DNA did. What type of evidence would be used to determine the time at which "foreign" DNA came to be part of the Rapa Nui genomes?

d. It is not known if the people of Rapa Nui traveled back and forth to South America or if South Americans traveled to, and settled on, Rapa Nui, an idea popularized by the Norwegian Thor Heyerdahl following his 1947 Kon-Tiki expedition by raft to the island. What kinds of genetic evidence might be helpful to resolve this question? (This is an open and debated question, with evidence, both genetic and otherwise, supporting both types of migration. An Internet search will find many helpful and accessible resources for the interested reader.)

16.19 Solomon Islands are located approximately 1800 km north-east of Australia. The native people of Solomon Islands have a complex genetic heritage, consistent with being located on a significant historical sea trade route. While the skin pigmentation of Solomon islanders is among the darkest outside of African populations, about 5–10% of the people have blond hair; blond hair is only common among native Europeans and is rare elsewhere in the world.

a. Postulate the genetic and evolutionary forces that led to blond hair being common among the people of Solomon Islands and Europe, but rare elsewhere in the world. (We will return to the explanation for the Solomon Islands in (e), after more information has been introduced.)

Looking back

b. A genome-wide association study (GWAS) analysis (of the type described in Chapter 10) of blond hair among the people of Solomon Islands (published in *Science* 2012, **336**: 554) identified a polymorphism in the *TYRP1* gene, a gene that is known among humans and other mammals to be involved in pigmentation. Outline briefly how this GWAS was done.

While most GWASs involve thousands of individuals, these studies identified the gene using only 43 blond-haired people and 42 dark-haired people. Why were so few individuals needed for this study?

Looking back

c. The study identified a C to T transition in exon 2 of the gene that resulted in a change from arginine to cysteine in the amino acid sequence, known as the ***R93C*** allele. Explain all of the main points in the preceding sentence in detail in order to demonstrate the relationships among the different concepts of the gene.

d. By direct analysis of the *TYRP1* gene using microarrays, the frequency of the ***R93C*** allele is 0.26; a comparison of the allele frequency with the hair color indicates that ***R93C*** is recessive to the ***R93*** allele found elsewhere in the world. Does the population of Solomon Islands appear to exhibit random mating for blond hair?

e. The ***R93C*** allele is absent from all other populations in the world, including among blond Europeans. Postulate why 5–10% of the people in Solomon Islands have blond hair.

CHAPTER 17

Metagenomes: Genome Analysis of Communities

IN A NUTSHELL

Genome analysis can be extended beyond individuals or single populations to study entire communities. Metagenomics involves the analysis of genome information extracted directly from organisms living in their native environments to address questions of biodiversity and community structure. The Five Great Ideas of Biology are integral in the study of metagenomes, which provides a powerful lens through which to view the otherwise invisible world of many microbial communities. Such a metagenomic approach can reveal remarkable degrees of interdependence and evidence that organisms have co-evolved for millions of years.

17.1 What is a metagenome?

We conclude this book with a topic that unifies the Five Great Ideas introduced in the Prologue and that requires an understanding of all of them. This is the new and growing field of metagenomics that emerged with current methods of genome sequencing and analyses.

A **metagenome** consists of all the genomes of the organisms that make up an environmental sample; metagenomics uses all or parts of the genomes obtained from that sample to provide us with an understanding of a community of organisms within a particular ecosystem. Metagenomics is essentially genome-informed ecology. It uses the theory and methods from genomics, evolution, genetics, cell biology, molecular biology, and the systems approaches of biochemistry and ecology to enable scientists to track multiple species and their relationships in time and space, much as biologists have tracked individuals or populations of cheetahs, whales, rodents, insects, or palm trees for decades and studied the impact they have on their environments. As we will point out in this chapter, metagenomics involves the interplay of all five of the Great Ideas, and many more connections are likely to be found by subsequent experimentation.

Much of metagenomics is very new, and its principles (and even some of its vocabulary) are not yet well established; while we will focus on a few examples to illustrate the integrated ideas, metagenomics is among the most rapidly changing fields in all of biology. As we wrote in the Prologue and Chapter 1, it is an exciting time to study biology.

Since most species on earth are microbial, and microbes are the dominant species in virtually all niches, ranging from surface water to soil, metagenomic studies largely focus on bacteria, archaea, microbial eukaryotes, and viruses. Many extreme environments, such as hot springs, are inhabited solely by microbes, and other niches—for example, the guts of invertebrates—are too small to contain anything much larger than single-celled organisms. Metagenomics provides us with a way to identify the individual microbes present in a community, which may otherwise be difficult—or impossible—to observe directly. Microbial species are also particularly amenable to metagenomic studies, as their genomes are relatively simple to sequence and annotate.

The living communities and ecosystems described by metagenomics include some of the same environments that are most familiar to you and that have been commonly studied such as the sea and soil. A microbial ecosystem can sometimes be referred to as a **microbiome**. We discuss the multiple interpretations of the word "microbiome" in Communicating Genetics 17.1. Microbiomes can be vast and complex systems, comprising many species that interact in multiple ways. The collection of microbial species and the microbial ecosystem associated with each individual metazoan constitutes that metazoan's own microbiome.

COMMUNICATING GENETICS 17.1 Is "microbiome" a "biome" word or an "omics" word?

What's in a word? Sometimes we recognize different meanings within words, depending on our own backgrounds and training. This certainly applies to the word "microbiome." Ecologists will immediately recognize the word "biome" contained within this word and thus might consider a microbiome as a "**microbial biome**." A biome is a community of organisms that occupy some particular habitat and is often classified by the major types of plants and animals found within it. Although the exact classification and scale of what is considered a biome varies, the term is broad and can encompass many ecosystems. An ecosystem, in turn, specifically includes both an interacting community of organisms (the biotic component) and the environmental factors with which they interact (the abiotic component). The biotic and abiotic components of the ecosystem are connected through the systems of nutrient cycling and energy flow that take place within the ecosystem. In defining an ecosystem then, there is an emphasis on the interactions between the biotic and abiotic components.

It is clear that the word "microbiome" is sometimes used in a way that captures more of an ecological view of the importance of the constituents of a habitat, the interactions among them, and their interactions with their environment. Although the use of "microbiome" usually emphasizes the biotic component, much like the use of the word "biome," the sense of community and importance of interactions among the constituents is often implied as well. Thus, "microbiome" can be viewed as an all-encompassing term that captures the sense of a collection of microorganisms together as an interconnected system—a full ecological community of microorganisms. This makes it distinct from a simple list of the particular microorganisms present in a location, which could more appropriately be referred to simply as the microbiota.

The difficulties in providing a precise definition were inherent in the first use of the "microbiome" by Joshua Lederberg; he used it "to signify the ecological community of commensal, symbiotic, and pathogenic microorganisms that literally share our body space and have been all but ignored as determinants of health and disease." This definition seems to not only consider a microbiome as a catalog of the microbiota but also refer to an ecological consideration of the interactions within such a community. As a practical matter, it is far easier to provide a catalog of the species that are present than to sort out their interactions. Thus, we will favor the use of the term "microbiome" to capture more of the system-level description, but the two meanings are obviously related and sometimes interchangeable.

In contrast to an ecological view of the word, those scientists with a background more in the world of genomics might recognize the "ome" in the word "microbiome" and see the word more as a description of the collective "microbial genome" associated with a particular environment (such as the human gut, for example). Indeed, some scientists and journal articles reference Joshua Lederberg as having used the term microbiome in this way. However, in many ways, the term "**metagenome**" more unambiguously captures the concept of a collective of genomes from a community of organisms. Such data can be gathered through metagenomic approaches that are designed to survey and sequence many different pieces of DNA from all the organisms present in a sample. Because these approaches do not require the microorganisms to be cultured in the laboratory, we know some microbial species solely through their DNA sequences and have very little information about their interactions with other species in their community. When organisms are defined solely through their genomic information, it can again blur the lines between metagenomes, microbiota, and microbiomes.

COMMUNICATING GENETICS 17.1 Continued

Wherever possible in this book, we attempt to refer to the collective genomes of a microbial community as a metagenome, rather than a microbiome, reserving the word microbiome to describe the collective ecological community of microorganisms. However, reflecting the complex, and sometimes ambiguous, nature of the term microbiome, you will inevitably find the word used in more than one way as you read here and elsewhere. The clearer part of the word microbiome then is the "micro" part, although maybe not—as there is nothing small about a microbiome, however it is defined!

FIND OUT MORE

Hooper, L. and Gordon, J. (2001) Commensal host–bacterial relationships in the gut. *Science* **292**: 1115–18

Lederberg, J. and McCray, A. T. (2001) 'Ome Sweet'Omics—a genealogical treasury of words. *Scientist* **15**: 8

Despite comprising separate species, hosts and their microbiome are not independent of one another. Some phenotypes of the host organism are even determined by the genotype(s) of other organisms with which they are closely associated and will be altered if the microbiome changes or is depleted. Together, the host and its microbiome can be referred to as a **holobiont** or **metaorganism**. Within the holobiont, the host metazoan, as well as microorganisms residing in or on it, may depend on each other for survival.

KEY POINT A metagenome is the sum total of all the genomes present in an ecological niche. The collection of microbial species and microbial ecosystems associated with a living organism is referred to as that organism's microbiome. The holobiont refers to the host and its associated microbes.

Thinking about microbiomes enlarges our perspective on living systems. We are well aware that we are the host to many different microbes, some of them pathogens but most of them not. But we do not typically think of microbes as permanent residents of our bodies, let alone as being essential for normal functions. We freely, and somewhat romantically, acknowledge that "no man is an island" and that we are all part of a greater ecosystem of plants and animals, land, sky, and sea. But we may not always think of our own bodies as being ecosystems or how we live in intimate functional relationships with billions of microbes, who depend on us as we depend on them. Metagenomics has given us insights into which individual microbes these are.

Deconstructing metagenomes

Great Idea: the cell is the fundamental unit of life

All microbiomes can be broken down to their individual component organisms, of which many or all may be unicellular. Thus, just as the cell is the fundamental unit of life, it can also be seen as the fundamental unit of the ecosystem.

One way to count, sort, and characterize microscopic cells within a liquid milieu is to use a **flow cytometer**, a cell sorter and analyzer that can "count" cells as they pass by a laser. Different cell types fluoresce naturally or can be tagged with fluorescent labels, allowing cell mixtures to be sorted and counted. Flow cytometers are most commonly used to sort the different cell types found in blood, but they have dozens of other applications in biology.

For example, in the 1980s, investigators on board a ship used a flow cytometer to sort the cells in their ocean water samples. They detected a cell population that showed unusual natural fluorescence, which then allowed them to identify a very abundant microorganism that had been overlooked for decades. The genus *Prochlorococcus* comprises unusually small bacteria that contain a red-fluorescing pigment and chlorophyll. *Prochlorococcus* cells are small enough to be difficult to visualize by light microscopy and don't grow on typical laboratory media, which had made them difficult to detect.

Prochlorococcus are photosynthetic cyanobacteria with a high surface-to-volume ratio, which facilitates nitrogen and phosphorus acquisition in the nutritionally dilute marine environment. Their discovery provided a key missing connection in our understanding of marine ecology, answering the question: where do marine organisms get their carbon? A wide variety of single-celled photosynthetic eukaryotes live in the ocean, but their numbers are far too small to feed all other forms of life. Breaking down the marine ecosystem into its component cells revealed that bacteria from the genus *Prochlorococcus*, and to a lesser extent the related bacterium *Synechococcus*, were the missing pieces of one of the most important ecological puzzles on earth.

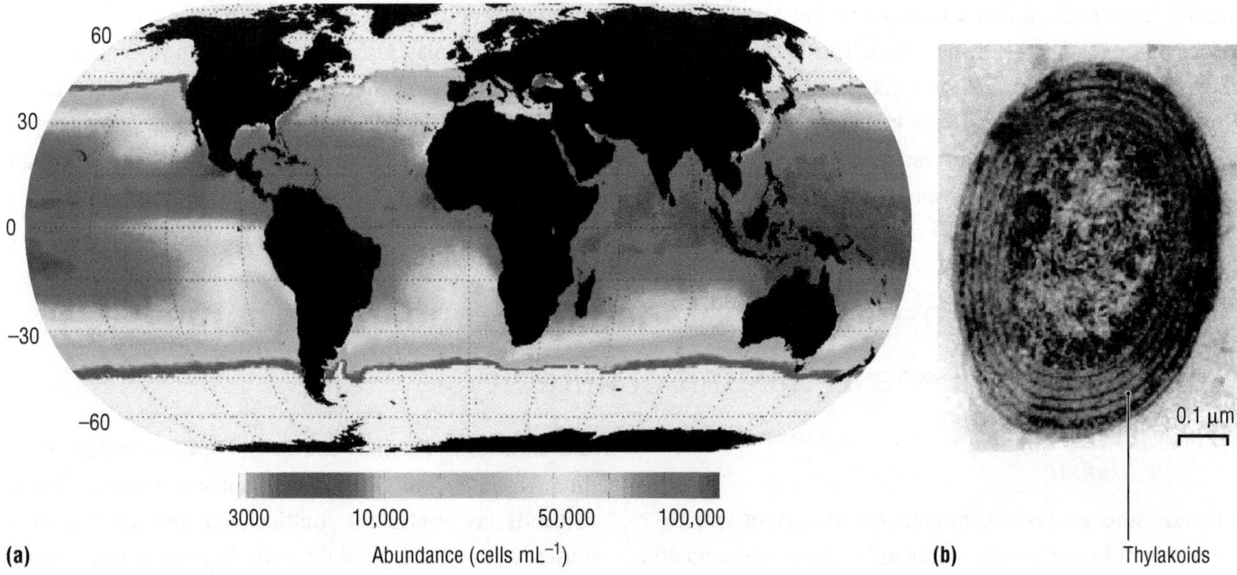

Figure 17.1 The "invisible forest." Half of the photosynthetic production on earth is accomplished by indispensable, but invisible, marine bacteria. (a) Heat map depicting the present global distribution (mean annual abundance at the sea surface) of the cyanobacterium *Prochlorococcus,* an important marine primary producer. (b) Electron micrograph of a longitudinal section of a *Prochlorococcus* strain. The electron-dense structures at the cell periphery are photosynthetic thylakoid membrances. Scale bar, 0.1 μm.

Source: Flombaum P et al. *PNAS* 2013;110:9824-9829. And Partensky F, Hess WR, Vaulot D. *Prochlorococcus*, a marine photosynthetic prokaryote of global significance. *Microbiol Mol Biol Rev.* 1999 Mar; 63(1):106-27.

We now know that up to a quarter of the photosynthetic capacity within the world's oceans comes from tiny bacteria in this genus, which are less than 0.8 μm long and have a cross-sectional diameter of less than 0.6 μm. A map of their global distribution and an electron micrograph of a *Prochlorococcus* cell are shown in Figure 17.1.

In addition to other pigments, *Prochlorococcus* cells carry an unusual version of chlorophyll, known as divinyl chlorophyll or chlorophyll b, the less abundant form of chlorophyll found in green land plants. Until the flow cytometer experiment, their existence was unknown, despite the fact that most marine environments contain as many as 10^4–10^6 *Prochlorococcus* cells per milliliter. In fact, this cyanobacterial genus is the most common taxon in ocean water between the latitudes 40 N and 40 S, from the surface to about 200 m below sea level, and is the most abundant microbial auxotroph. One of the authors of the original paper describing this genus has subsequently referred to these marine organisms as an "invisible forest."

Present-day ecological investigations of the ocean are performed by extracting total DNA and sequencing genomes in the sample. This type of metagenomic investigation has corroborated and deepened our understanding of *Prochlorococcus*'s ecology. Had the organisms not been identified by flow cytometry in the 1980s, they would inevitably have been recorded during the first metagenomic studies of the current millennium, which revealed that the Sargasso Sea contained not just an abundance of *Prochlorococcus* but a "conglomerate," that is, many different lineages of organisms within this genus. *Prochlorococcus* are well suited for metagenomics, as well as cellular microbiology research, since they have small genomes and considerable variation among strains due to mutations and horizontal gene acquisition, in addition to their abundance and natural unusual fluorescence.

The *Prochlorococcus* example serves to illustrate some of the properties of microbiomes and the challenges associated with studying them.

First, many of the organisms within a microbiome are difficult to culture in the laboratory and can only be found as living specimens in their natural environments.

Second, it is possible to detect microorganisms more easily now because we have more sensitive and sophisticated tools. Microbiomes and metagenomes have been in existence for millennia—certainly far in advance of humans. Just as the invention of the microscope allowed the visualization of single-celled organisms and organelles inside cells, the development of cell sorters, computers, and genomic methods has allowed us to analyze microbiomes and their metagenomes.

Third, interrelationships among organisms can result in ecosystem-level outputs, as exemplified by the amount of carbon fixed by *Prochlorococcus* through the process of photosynthesis. Roughly half of the *Prochlorococcus* cells arising from each nocturnal binary fission are eaten by larger microorganisms in the ocean. This is the first step in passing the carbon they have fixed by photosynthesis through an increasingly networked food web that sustains life in the ocean. Some varieties of *Prochlorococcus* thrive at low nitrogen or phosphorus levels and are major carbon fixers in ocean niches that have a low abundance of these molecules. Others, which can photosynthesize in low light levels, are more predominant in darker waters. Thus, different *Prochlorococcus* genotypes occupying specific habitats, so-called **ecotypes**, are distributed through the ocean in relation to ecologically important traits.

17.2 The composition of metagenomes and microbiomes

Great Idea: the gene as the unit of heredity

Identifying and classifying the species in an ecological habitat

Like any other ecosystem, the complete analysis of a microbiome would require us to count all of the species and all of their complex interactions. But this level of analysis is not realistic. How could scientists count all the *Prochlorococcus* cells in the ocean, not to mention the thousands of other bacterial species? While it may be difficult to isolate all species and cell types in an ecosystem as individual organisms and then identify them, genome-level techniques have made it much easier to detect, analyze, and quantify their DNA.

Metagenomic studies seek to determine which genomes are present in specific environments and how these genomes interact and function. Thus, one aspect of metagenomics is the cataloging of the microbial populations in a niche. Such a catalog should describe richness and species composition, while also serving as the foundation for studies that might seek to evaluate other features such as subspecies types and rates of change in the composition and biochemical capabilities within the community. Arising from this approach is a question of reproducibility and resilience. How similar are the microbiomes of similar niches, for example, the gut microbiomes of different organisms of the same species or oceans in different hemispheres? Do microbiomes change over time or with perturbation, and if they do, how and when do they recover? A second aspect, which we will revisit in Section 17.4, involves building a picture of physical, biochemical, and nutritional interactions that allow components of a metagenome to function as an integral community.

Ecologists are interested in which species are present in a given habitat and their relative abundances, as well as how they are organized within the habitat and any interactions that occur among them. Ecologists have historically cataloged living communities by counting and classifying the individuals and species within them directly or by culturing the organisms present. These methods are useful for multicellular species, but they underestimate microbial diversity because many species cannot be cultured and others have morphologies too similar to be distinguished microscopically. The primary method employed in metagenomics is DNA sequencing, which greatly augments traditional means of enumeration and identification.

Culture-dependent analysis of microbial communities

Traditionally, microbial communities were characterized by culturing samples and then identifying their members by standard microbiological methods. In this type of analysis, a sample is spread onto agar in a Petri dish, so that single cells are separated on the plate. Each cell multiplies to produce a colony of identical individuals. These colonies can then be aseptically subcultured onto fresh medium to yield separate **pure cultures** that can be identified by biochemical testing, staining and microscopy, and genetic analyses. Culture-dependent analysis has been standard practice for microbiological studies for more than 150 years; our understanding of the molecular biology and genetics of *Escherichia coli* and other model microorganisms, which provides the foundation for much of modern molecular biology, relied on an ability to grow and analyze pure cultures in the laboratory.

The analysis of microbial communities by culturing individual species has the advantage of creating stocks of organisms that can be evaluated further. On the other hand, it almost always underestimates the number of species present, since only a fraction of microorganisms can be cultured at all, and an even smaller fraction can be grown using common culture media. Any species with unusual growth requirements is missed by these methods. Our current state of knowledge suggests that

TOOL BOX 17.1 Culture barriers in bacteria

Why are so many organisms "unculturable," that is, what makes them difficult to grow in the laboratory? In the first place, one needs to know something about the nutritional and environmental requirements of a strain to culture it. Macro- and micronutrients must be provided in a Petri dish; the oxygen tension must be correct, and any other factors required for growth must be present. Culturing bacteria in the laboratory is a good way to look for "knowns" but much less efficient for finding "unknowns". Many unculturable organisms might ultimately be able to be cultured, but we need to understand their biochemistry first in order to develop protocols to grow them.

Second, some organisms live in symbiotic consortia in nature and so cannot be grown isolated in pure culture. Since pure culture methods are the mainstay of microbiological practice, it becomes impossible to grow them in the laboratory by conventional means. Some investigators have had success growing consortia of organisms. However, it is often necessary to have some sense of which species are required for co-survival in order to set up a multi-species culture.

In spite of all the pitfalls and challenges associated with culture, microbiologists will continue to attempt to culture new species. It is easier to test hypotheses in simple laboratory cultures than to do so *in situ* in ecosystems. Moreover, sequence data obtained during metagenomic studies are more easily ordered when reference genomes from known cultured organisms are available for comparison. Even though most of the published data on the marine cyanobacterium *Prochlorococcus* have been obtained using culture-independent methods, our understanding of this organism and its ecology is dependent on experiments performed on a handful of cultured strains, and on their genome sequences. But *Prochlorococcus* still illustrates the importance of *in situ* metagenomic research—it is only by studying the ecology of this organism in the ocean that seawater-based media have been devised to grow it in the laboratory.

It is important to emphasize that, although culture-independent methods tend to capture more species than do culture-based methods, some species are missed by culture-independent techniques as well. Methods that amplify and then clone 16S rRNA genes and use these clone libraries as the basis for a species catalog will miss species that have 16S rRNA genes that are not as close a match to the oligonucleotide primers used as other species in the community.

One such study compared such culture-independent and culture-dependent methods to catalog bacteria residing on apple leaves and found only partial overlap between the two methods. Twelve operational taxonomic units (OTUs) were found by both methods. The culture-independent method identified 21 OTUs that were not found in the culture-dependent method, while the culture-dependent method detected 11 OTUs of the order *Actinomycetales* that were missed by the culture-independent methods. Whole-genome shotgun sequencing of metagenomes (rather than targeted amplification of specific genes by the polymerase chain reaction (PCR)) further increases the species that can be identified, as illustrated by the example in Box 17.3 on ancient metagenomes. In addition, whole-genome sequencing does not require any prior knowledge of the community being studied, which is an advantage. In reality, the power of whole metagenome sequencing improves when some of the species identified have been previously characterized, since this will increase the number of DNA fragments that can be assigned taxonomically.

unculturable organisms represent at least half, and in some cases over 99%, of the microbial species in a given environment. Some of the barriers to culturing different species are discussed in Tool Box 17.1.

KEY POINT Detecting all the cell types and species in an ecosystem by culture or microscopy is desirable but challenging and often impossible. Identifying genomes using DNA sequencing techniques is a powerful surrogate for characterizing microbial communities.

Culture-independent analysis of microbial communities

The use of genome-based methods requires only the ability to isolate and analyze DNA from an organism, rather than growing the organism itself. Thus, **culture-independent** analyses of metagenomes avoid some of the challenges encountered with culturing microbes and can provide much more information on the diversity of species within a community. Metagenomic studies typically sequence a representative gene or genes from all organisms present; occasionally, they will use the entire complement of DNA present in a sample. Sometimes, and increasingly often, such studies may also examine **transcriptomes**, the entire set of RNA molecules within the sample, to determine which genes are being expressed. As with other genomic studies, the sequencing itself is usually quite rapid; the sorting and understanding of the large amount of data collected in these mega-projects is the rate-limiting step in our understanding of these communities.

For studies that use a single representative gene for species identification, the gene that encodes the RNA component of the small subunit of the ribosome (rRNA) is researchers' preferred target for PCR amplification. (PCR is described in Tool Box 4.1.) Once amplified, the rRNA gene can be sequenced and compared to known sequences in existing databases in order to identify microbes in a community.

WEBLINK: VIDEO 17.1 takes you through the process of identifying a species by its 16S rRNA sequence. Find it at **www.oup.com/uk/meneely**

The genes for 16S rRNA (for bacteria) or 18S rRNA (for eukaryotes) have many properties that make them ideal for this purpose.

1. They are universal, as all cellular organisms have ribosomes. Thus, the gene for the RNA component of the small subunit of the ribosome can be used to create phylogenies across the entire tree of life.
2. Because each newly derived 16S rRNA gene sequence 2 can be compared to every other sequence deposited in public databases, the identification of new species is relatively simple. Indeed, since methods for generating and analyzing 16S rRNA sequences were optimized in the late 1990s, the rate at which new bacterial species are reported has increased exponentially, as illustrated in Figure 17.2.
3. rRNA genes have long been known to correlate well with phylogenies based on other methods; therefore, the data obtained from them is phylogenetically and taxonomically meaningful.
4. Relatively short sequences are needed. The small subunit rRNA genes consist of conserved regions—typically those required for structure and function—interspersed with highly variable ones that are not under selection. The structure of 16S rRNA with the variable and conserved regions indicated is shown in Figure 17.3(a). Thus, it is possible to use primers for PCR that are complementary to conserved regions to amplify and sequence one or more variable regions. These regions are shown in Figure 17.3(b). Since the primers correspond to conserved sequences, the same primers can work for many different species.
5. rRNA genes represent vertical transmission or descent from common ancestors. rRNA genes are less likely to be transmitted horizontally than most other loci, and thus horizontal gene transfer is less likely to complicate identification.
6. rRNA genes usually occur in a single copy in the genomes of bacteria, although eukaryotes often have multiple copies. Because the copy number is defined and stable in bacteria, the 16S sequence can be used to quantify the presence of different microbial species within a habitat.

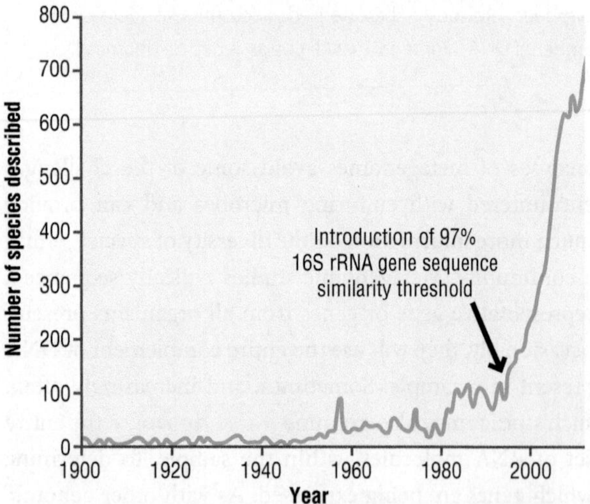

Figure 17.2 Sequencing of the 16S rRNA gene has greatly increased the number of species of bacteria and archaea that have been classified. Graph showing the number of species identified each year since 1900. Notice the sharp increase coincident with the use of 16S rRNA sequencing to identify species. A cut-off of 97% identity or more in the 16S rRNA sequence is often used to assign individuals to the same species.

Source: Reproduced from Chun, J. and Rainey, F.A. Integrating genomics into the taxonomy and systematics of the Bacteria and Archaea. *Int J Syst Evol Microbiol.* 2014 Feb;64(Pt 2):316-24.

In spite of all these advantages, small subunit rRNA gene profiling has some pitfalls that can underestimate diversity. The so-called "universal" 16S rRNA and 18S rRNA primers amplify the gene in most strains; however, an organism with a different sequence in the regions complementary to the primers will be missed. In other cases, the universality of the 16S rRNA genes is a disadvantage, since closely related taxa—for example, different *Prochlorococcus* ecotypes—may have highly similar sequences and indistinguishable 16S rRNA sequences. Both phylogenetic and ecological distinctions exist between low-light and high-light-adapted *Prochlorococcus*. These must be distinguished by analyzing other more highly variable regions of the genome.

KEY POINT The sequences of 16S rRNA genes generally work well for the identification of bacterial species and are the sequences most commonly used to identify microbes in metagenomes. However, it becomes important to look at other gene sequences in certain circumstances, particularly for very closely related species and strains.

The degree of species diversity found in a sample depends on the measurable differences in genomes,

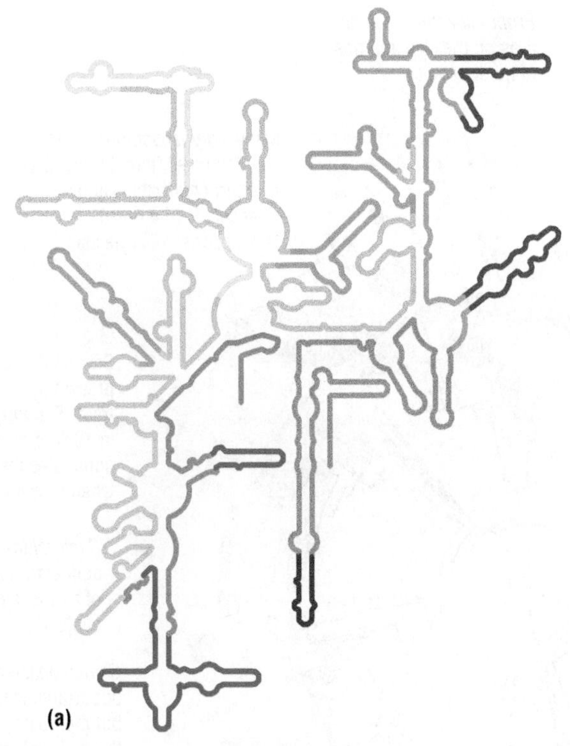

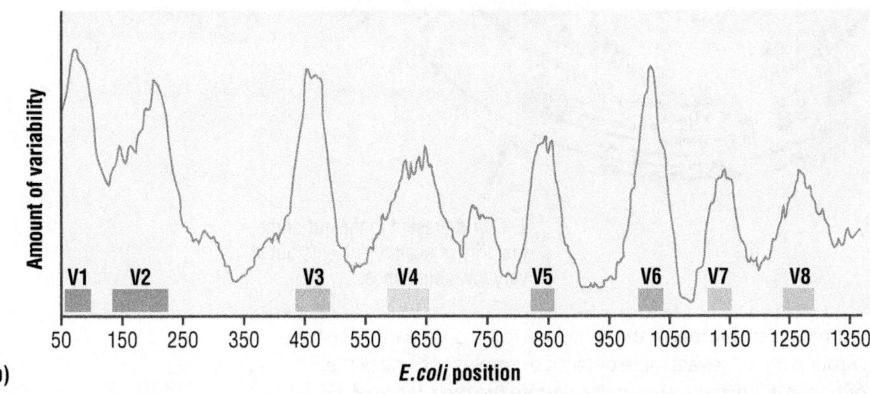

Figure 17.3 Structure and sequence conservation of the 16S rRNA. (a) The secondary structure of 16S rRNA from *Escherichia coli*, with the nine variable regions from structurally distinct regions highlighted in darker shading (b) Variable regions within the 16S rRNA gene illustrated, using the *E. coli* gene as a reference. Peaks indicate sequences that vary among species, while troughs reveal sequences that are highly conserved across species. This analysis reveals the hypervariable regions of the 16S rRNA gene, the first eight of which are shown (V1–V8) that are most taxonomically informative.

Source: Part (a) adapted from *Nature Reviews Microbiology* 12, 635–645 (2014) doi:10.1038/nrmicro3330.

as well as the way in which the investigators decide to categorize or bin the organisms that they find. Binning microbial species together into a grouping based on DNA sequence alone is tricky because the distinctions between strains with sequence polymorphisms are not always clear, and little else is known about these organisms. Investigators must decide the amount of DNA sequence variation that they will allow within one taxonomic category and how much variation is indicative of a different species. After binning, the resulting groupings are known as **operational taxonomic units (OTUs)**.

An OTU, sometimes called a **phylotype**, can be thought of as a species that has been defined by DNA sequences. The OTU data can then be shown in the form of phylogenetic trees, as was discussed in Section 4.7. One such tree showing OTUs identified from the human microbiome is shown in Figure 17.4. Overall, OTUs approximate species, but it is best to view such grouping as an organizational approach that allows computational cataloging of

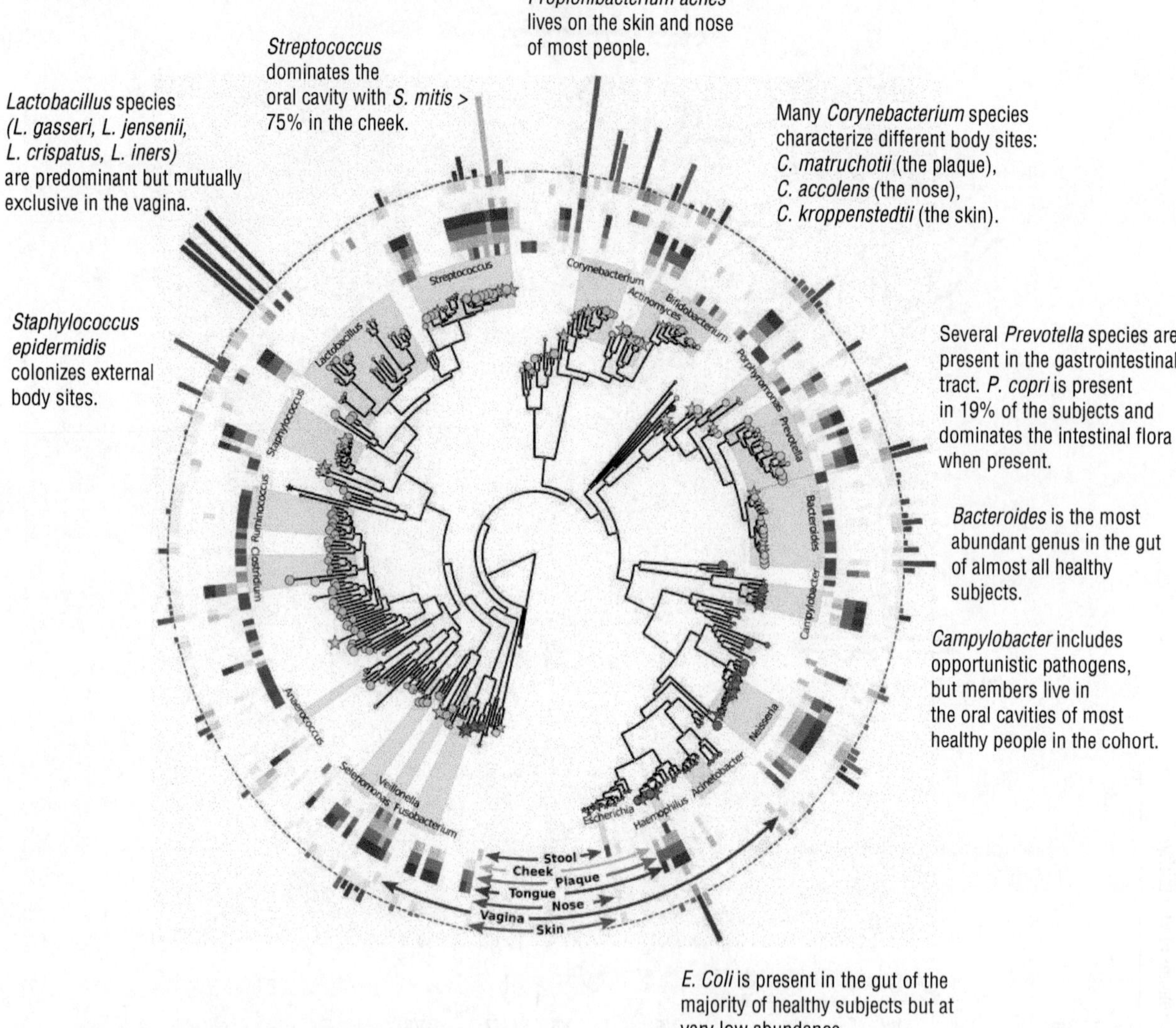

Propionibacterium acnes lives on the skin and nose of most people.
Streptococcus dominates the oral cavity with *S. mitis* > 75% in the cheek.
Lactobacillus species (*L. gasseri, L. jensenii, L. crispatus, L. iners)* are predominant but mutually exclusive in the vagina.
Many *Corynebacterium* species characterize different body sites: *C. matruchotii* (the plaque), *C. accolens* (the nose), *C. kroppenstedtii* (the skin).
Staphylococcus epidermidis colonizes external body sites.
Several *Prevotella* species are present in the gastrointestinal tract. *P. copri* is present in 19% of the subjects and dominates the intestinal flora when present.
Bacteroides is the most abundant genus in the gut of almost all healthy subjects.
Campylobacter includes opportunistic pathogens, but members live in the oral cavities of most healthy people in the cohort.
E. Coli is present in the gut of the majority of healthy subjects but at very low abundance.
Streptococcus
Corynebacterium
Actinomyces
Bifidobacterium
Porphyromonas
Prevotella
Bacteroides
Campylobacter
Neisseria
Acinetobacter
Haemophilus
Escherichia
Fusobacterium
Selenomonas
Veillonella
Anaerococcus
Clostridium
Ruminococcus
Staphylococcus
Lactobacillus
Stool
Cheek
Plaque
Tongue
Nose
Vagina
Skin

Figure 17.4 A map of diversity in the human microbiome. At the center is a radial phylogeny of species found in the Human Microbiome Project. Around the edges are more detailed examples of some of the species found. The intensities of colors around the edge of the phylogeny indicate the prevalence of the various species from the seven different body sites. The bars around the outer rim of the figure indicate the abundance of specific species, colored by the site in which they are most abundant.

Source: Adapted from Morgan, X. C. et al. (2013). Biodiversity and functional genomics in the human microbiome. *Trends in Genetics*, Vol. 29. Iss. 1. With permission from Elsevier.

metagenomic data, rather than rigid species definition. In fact, in light of the extensive amount of horizontal gene transfer that occurs among bacteria, the familiar concept of a species may not be a useful way to think about bacteria. We explore these and other issues in Box 17.1.

KEY POINT An important step in metagenome analysis is to determine to what extent sequence variants in a sample represent different individuals from the same taxonomic group, rather than different taxa. Gene sequences that are determined to be sufficiently similar are then grouped into the same operational taxonomic unit (OTU), which can be thought of as a species defined by DNA sequences only.

Sampling metagenomes

There are so many and such diverse bacterial populations on earth that it is impossible to enumerate and subtype all of them. Therefore, scientists study small samples and attempt to extrapolate their findings. As with any small sample, we must always question whether the sample is representative of the whole. There are many rare species, so it is highly unlikely that any sampling study will capture all the diversity that exists.

Sampling bacterial populations is analogous to other sampling situations, including mutant screens like those described in Chapter 15, particularly in Toolbox 15.1. Let's consider sampling a human

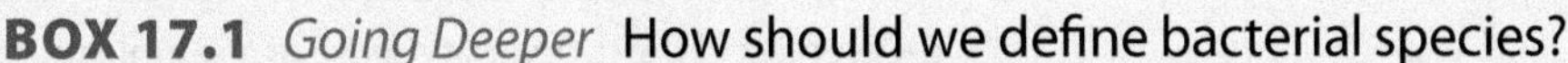

BOX 17.1 *Going Deeper* How should we define bacterial species?

A widely accepted definition of a species is based on the ability of sexually reproducing eukaryotes to mate and produce fertile progeny. Bacteria and archaea do not mate in this way, so we cannot use this as the basis for defining their species. Moreover, since most bacteria and archaea exchange DNA widely across taxonomic lines through mechanisms of horizontal gene transfer, the species concept is even more difficult to extrapolate from eukaryotes.

Here are some of the ways that bacteriologists therefore think about the species concept:

1. Species should be defined based on morphological features and biochemical capability. This is the archetypical way that species were defined. It was satisfactory for many species, although some got misclassified.
2. DNA–DNA hybridization. If genomic DNA from one bacterium is hybridized to the DNA of another, and a certain critical percentage of their genomes can hybridize to become double-stranded, then the two bacteria belong to the same species. This is a very robust way to distinguish bacterial species conceptually, but defining the cut-off is difficult because some bacteria exchange DNA more than others.
3. Organisms that do not vary more than a given percentage in the sequence of one or more core conserved genes are considered to belong to the same species. This is generally the method used to define operational taxonomic units (OTUs) in metagenomic studies.
4. Organisms that exchange DNA more with each other than with other organisms constitute a species.
5. Bacteria do not comprise species in the classical way that eukaryotes do. The borders between bacterial species are too "fuzzy" to be defined.

population to understand its composition and diversity as an example.

A census of the people at a shopping mall at a particular point in time will not provide a complete inventory of all types of people who live in the area, although it might be a useful starting point. A sample at the mall might over-estimate the frequencies of certain subgroups (such as individuals of a particular age), while at the same time missing any subgroups that never shop at the mall. Similarly, a sample for metagenomics drawn from a particular environment will over-estimate the frequency of some types of bacteria, while missing the existence of others. But metagenomic studies face an additional complication, since only part of the sample is analyzed; the data from one or a few genes represent just a **sample of the sample**.

The crux of this discussion is that every sampling method includes some biases. With human populations at the mall, we are aware of these sampling biases and can try to correct for them; with environmental sampling of bacteria, we may not be as aware of the biases that arise from sampling. Thus, while there are methods to correct for biases (such as taking and comparing multiple samples), it may be difficult to know the best ways to make these corrections in a metagenomic study.

Saturating metagenomic evaluations

One of the most important parameters to know about a sample is how well it describes the composition of the larger population. Suppose that our sample of bacteria identifies ten different OTUs. Are these the only OTUs that are found in this environment, so additional sampling would be pointless, or are there more OTUs to be found? If there are more, how many more are present, and how many more samples should be taken? In the genetics screens discussed in Box 15.2, this was described as **saturation** analysis—that is, recognizing when all of the genes that could reasonably be found by a particular mutant screen had been found, so that further sampling is not productive. Sampling for metagenomics faces the same questions and often describes them by **rarefaction curves**, as shown in Figure 17.5.

A rarefaction curve plots the number of specimens (or clones or DNA sequences) sampled on the X-axis with the number of different OTUs identified on the Y-axis. The curve shows how the number of OTUs changes as more specimens are analyzed and assigned to an OTU. If, after 100 specimens have been analyzed, only ten OTUs have been found and most of those have been found multiple times, it is reasonable to conclude our sample represents most of the OTUs that are present. On the other hand, if 100 specimens are analyzed and define 75 different OTUs (of which only a few are found multiple times), it is reasonable to conclude that further sampling is needed because many more OTUs remain to be found. To express this idea in another way, the relationship between the number of specimens sampled and the number

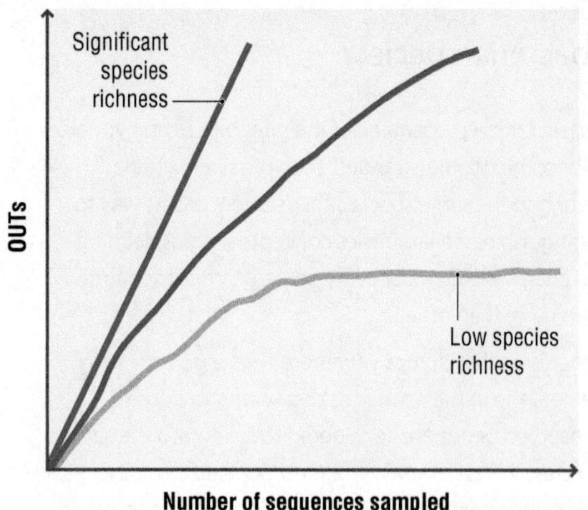

Figure 17.5 Rarefaction curves can be helpful in estimating species richness. The cumulative number of different OTUs identified is plotted on the Y-axis as a function of the number of samples/clones sampled (indicated along the X-axis). When species richness is low (green), then, after the first rounds, repeated sampling will not identify many more new species, and the curve will plateau. However, when there is more significant species richness (red), then each new sample will include new species identifications, and the curve will continue to climb. Situations between these extremes will show a curve that falls between the two shapes (blue).

Source: Wooley, J.C., Godzik, A. and Friedberg, I. (2010). A Primer on Metagenomics. *PLoS Comput Biol* 6(2): e1000667.

of OTUs identified shows the richness or taxonomic diversity of the population.

Comparing metagenomes

Very often, metagenomic studies involve the comparison of the composition of one sample to another to address a specific question or to identify a trend. In order to make a useful comparison, the data from each sample has to be analyzed and clustered into different groups based on some criteria; this allows us to determine the amount of diversity. Diversity can be defined in many ways, which tell us different things about a microbial community. Diversity **within** a sample is known as **α-diversity**, while diversity **between** different samples is known as **β-diversity**. In our shopping mall example, α-diversity is found among the shoppers in the mall at that particular time, while β-diversity compares those shoppers to the shoppers at another mall. These comparisons can inform us not only about the sample population itself—the shoppers—but also about the different environments from which the samples were drawn—the malls.

The number of new OTUs that appear in subsequent samplings from the same population can give some insight into changes in taxonomic richness or α-diversity. The simplest way to make comparisons between metagenomic samples is usually to measure and score α-diversity. If the α-diversity score of a sample increases as the weather gets warmer, for example, we could infer that the community in question is more diverse in summer than in winter. However, the comparison of α-diversity scores would fail to determine whether the actual species composition changed over time.

For metagenomic studies (and many other types of experiments where data are clustered), the statistical method to do this clustering is called **principal component analysis (PCA)**. The idea behind PCA is to group data based on different criteria in order to determine what criteria are diagnostic for the group (that is, the principal components) and which ones are less important. The criteria themselves are often not known when the studies begin, and it may not be important to identify them; the important aspect is that organisms that share these components cluster together.

The statistical methods for PCA are beyond the scope of this book, but we can return to our analysis of the clientele of a mall to illustrate how it works. Every individual who shops at the mall is a data point, and each data point has many components—gender, age, height, weight, appearance, dress, and so on. PCA helps to show that some of these aspects are more informative than others in describing the population at the mall, and thus of the larger population living in the region since the mall is a sample. For example, with shoppers at a mall, the principal components to describe the data points (that is, the shoppers) might be age and gender; the fraction of the shoppers wearing a green shirt is probably not as helpful in describing the population at the mall, or the larger population living in the area, so "green shirts" is not likely to cluster with any other component.

PCA identifies the principal components that describe the data points and how much those components contribute; most significantly, PCA provides quantitative values for the diversity. These values can be plotted on a graph, and the data points compared. In other words, if one axis is gender and another is age, a 38-year-old woman lies closer on the graph to a 41-year-old woman than to a 62-year-old man. This quantitative analysis allows the samples to be grouped and then compared.

Figure 17.6(a) shows data from the ongoing Human Microbiome Project. We can see that bacterial diversity in the samples clusters into distinct groupings that relate to where in the body the samples were taken. In this case, for example, the bacterial diversity in samples from the gut, represented by black dots, is distinctly different

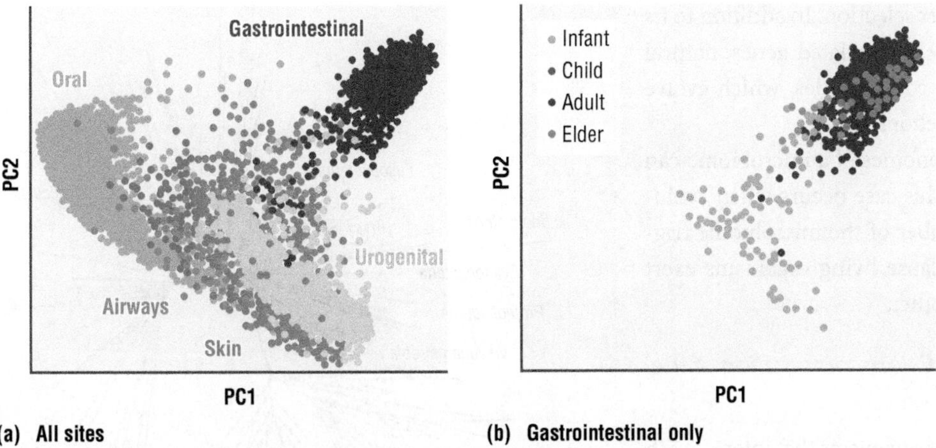

Figure 17.6 Microbial community structure in different habitats from the human body. Data from the Human Microbiome Project have been combined with additional data sets, and two coordinates from a principal component analysis (PC1 and PC2) have then been plotted to provide a visual display of the diversity found in the various communities. In (a), a comparison is made between microbial communities from different sites in the human body: airways (blue), skin (orange), urogenital (yellow), gastrointestinal (black), and oral (green). Bacteria from the different sites show different patterns of diversity. In (b), the data are only from gastrointestinal samples, and compares samples from infants (green), children (blue), adults (black), and elders (orange). The microbial communities in infants show a distinctly different pattern of diversity from that of adults and more similar to the skin and urogenital samples in (a) than to samples from the older individuals in (b).

Adapted from: Koren, O., et al. 2013. A guide to enterotypes across the human body: meta-analysis of microbial community structures in human microbiome datasets. *PLoS Computational Biology*. http://dx.doi.org/10.1371/journal.pcbi.1002863

from the diversity found in samples from the mouth (green dots) or other locations in the body. Interestingly, when data from stool samples alone (a read-out of the gut microbiome) are analyzed, the analysis clearly shows that there is a marked difference in the bacterial diversity in infants, compared to older individuals, as depicted in Figure 17.6(b).

The data generated from metagenomic studies being sequence-based are eminently suited to being analyzed using these quantitative and phylogenetic methodologies. Once we have a sense of the composition of a community living in one condition, it is then possible to use metagenomic studies to determine how this changes in time and space by making comparisons between or among samples. One might want to investigate the species found in an orchard, for example, to determine what happens as the seasons change, when its caretakers use different chemicals as pesticides, or when birds associated with the orchard have migrated south. There are ongoing studies doing just these kinds of analyses. These studies produce data that show patterns of change associated with different conditions. Importantly, such studies provide a framework for studying the evolution of life within an ecosystem.

KEY POINT Metagenomic diversity can be quantified. Both quantitative and qualitative comparisons can be made among metagenomic samples.

17.3 Selection at the level of the holobiont

Great Idea: evolution by natural selection

Having introduced microbiomes and some of the techniques for analyzing them, we now can dis-cuss a few studies of evolution at the level of the microbiome. In previous chapters, we have focused on the gene and the organism under selection. With microbiomes associated with a host, attention shifts to the **holobiont**, that is, the ecosystem comprising the host and its associated microbial community. Two important concepts about natural selection apply to such microbiomes.

First, the holobiont is under selection. In addition to its effects on organisms and their associated genes, natural selection also acts on living communities, which evolve along with their composite genomes.

Second, the component genomes of a microbiome can co-evolve. Co-evolution in this case occurs when evolutionary changes in one member of the microbiome triggers changes in another because living organisms exert selective pressure on one another.

The holobiont is under selection: composition of the mammalian gut microbiome

Living communities consist of organisms that interact with each other and their environment; the key composition of these communities is determined by their genomes, and these genomes continue to evolve. Just as Darwin deduced the principles of evolution from comparing finches in the Galapagos Islands, we can learn about selective forces and their consequences from comparing metagenomes.

The microbiome of the human gut illustrates some of the complex interactions between the host and microbe that can influence how natural selection might act on that holobiont as a whole. The number of microbial cells associated with a normal human body is thought to be greater than the number of human cells that make up the body by at least 3:1. The human intestine is one of the most diverse microbial habitats in the human body and is inhabited by several hundred microbial species. Among other functions, the gut microbiome produces hydrolase and lyase enzymes, which are not encoded in the human genome but which allow us to extract energy from complex carbohydrates. Therefore, one of the most fundamental requirements for living organisms—the ability to utilize nutrient sources in food—depends not only on our own genome but also on the genomes of the microbes that we have acquired.

Each human gut contains hundreds to thousands of different bacterial species. Most of these species are rod-shaped bacteria. *E. coli*, the earliest gut organism to be identified and the one that most people are familiar with, is present in almost all human gut microbiomes but in relatively low numbers; its importance in the natural physiology of the mammalian gut has probably been somewhat over-estimated because it was relatively easy to isolate, culture, and study. By contrast, obligate anaerobes, such as *Firmicutes* and *Bacteroidetes*, which are killed by oxygen and are therefore difficult to culture, are the most common bacterial phyla present. The rest of the gut microbiome consists of a large number of divisions, most notably *Actinobacillus*, *Proteobacteria* (including *E. coli*), and *Fusobacterium*.

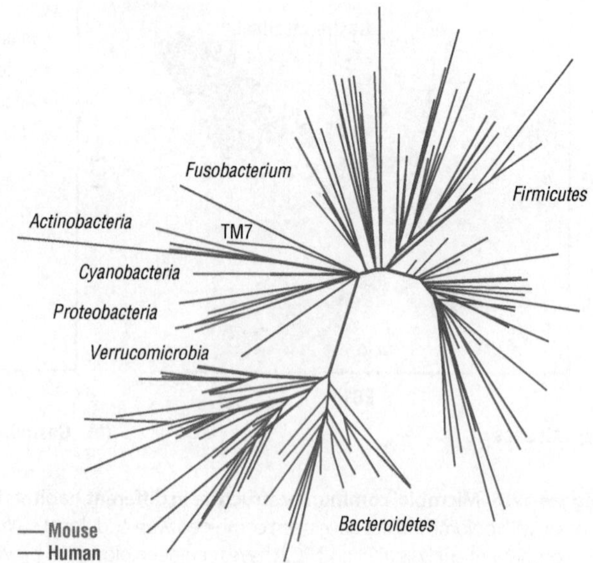

Figure 17.7 Phylogenetic tree of microbial species in mouse (red) and human (blue) cecae/colons. This tree has been constructed using 16S rRNA sequences. Although the specific OTUs identified in mice versus humans differ, the types and distributions across the various microbial divisions show a striking similarity and consist mostly of *Firmicutes* and *Bacteroidetes*. The bar represents 15% sequence divergence.

Source: Ley R.E., et al. 2005. Obesity alters gut microbial ecology. *Proceedings of the National Academy of Sciences*. 102 (31): 11070–11075. © National Academy of Sciences, USA.

The relative proportions of phyla vary with age, diet, and health status, as well as the genotype of the human host.

Firmicutes and *Bacteroidetes*, the bacteria that dominate the microbiome of the human intestine, also dominate the microbiome of the mouse intestine, although the exact species of bacteria are not the same between humans and mice. A phylogenetic tree constructed using 16S rRNA sequences from these two species provides a comparison of these microbiomes, as shown in Figure 17.7. Because the overall microbiomes of human and mouse intestines are similar, we can extrapolate certain findings from gut microbiome studies of mice to humans.

The holobiont is under selection: microbiome-modulated mammalian phenotypes

One of the first examples of how a microbiome and host interact to influence the overall holobiont's phenotype involved the study of obesity. Obesity is the result of caloric intake that exceeds energy expense; however, eating and exercise are not the only determinants of the body mass index. Obese humans generally have a greater proportion of *Firmicutes* in their intestinal flora, in contrast to non-obese humans, who have relatively more *Bacteroidetes*.

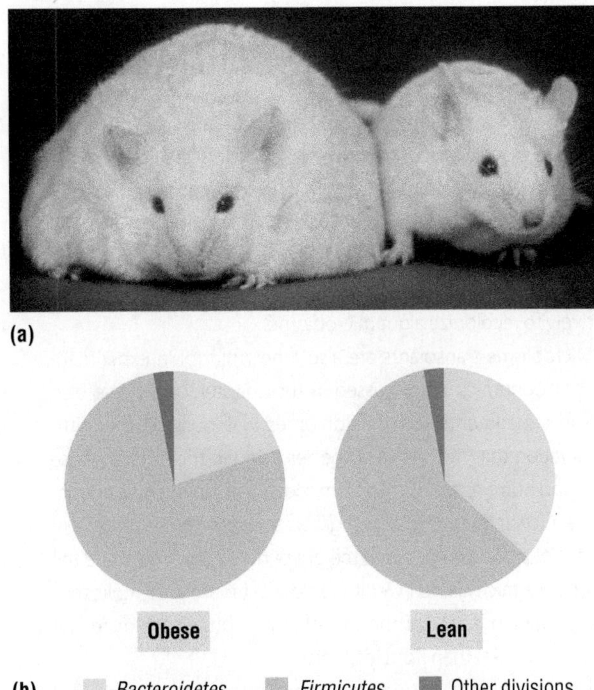

Figure 17.8 Microbiome composition can vary with the genetic composition of the host and can contribute to phenotypes. (a) Mice homozygous for the ***ob*** allele of the leptin gene are obese (left), compared to wild-type mice (right). (b) The relative proportion of *Firmicutes* and *Bacteroidetes* in obese and wild-type mice. Notice that the obese mice have a higher proportion of *Firmicutes* (green in the pie chart) and a lower proportion of *Bacteroidetes* (blue in the pie chart) in their gut microbiomes, compared to wild-type mice fed the same diet.

Source: Fraune S., Bosch T.C. 2010. Why bacteria matter in animal development and evolution. *Bioessays*. 32 (7): 571–580.

This is also true in mice, for which some genetic causes of obesity are known. Mouse mutants that are genetically obese (***ob/ob***) are pictured in Figure 17.8(a). As with humans, the obese (***ob/ob***) mice have a higher ratio of *Firmicutes* to *Bacteroidetes* than do the genetically non-obese mice, even when genetically obese and non-obese mice are fed the same diet; this is shown in Figure 17.8(b). Thus, there is a consistent correlation between obesity and the dominance of *Firmicutes* in the microbiome.

The important causative effect of the microbiome was shown by transplantation experiments. The intestinal microbiomes from ***ob/ob*** mice were transplanted or transfused into ***ob***/+ mice, which are not phenotypically obese. The ***ob***/+ mice began to gain weight, indicating that the dominance of *Firmicutes* in the microbiome results in weight gain, even in mice that are not genetically predisposed to obesity.

This experiment illustrates an important point about the complicated relationships between hosts and their microbiomes. The microbiome can result in phenotypes that are not encoded in the host's genome, and the composition of the microbiome is dependent, at least in part, on the genotype of the host. You are what you eat, but "you" includes your microbiome, since much of "you" is what your microbiome digests.

Microbiomes can be altered by environmental factors, so the genetics of the host are not the only (and maybe not even the primary) determinant. Obese people who consume a low-calorie diet see a shift in their microbiomes towards more *Bacteroidetes* and fewer *Firmicutes*, as they lose weight. Furthermore, individuals who receive antibiotics lose a fraction of their microbiome. In most individuals, the microbiome reconstitutes after antimicrobial therapy. In a significant and unfortunate minority, however, depletion of the intestinal microbiome allows overgrowth of organisms like *Salmonella* and *Clostridium difficile*, organisms that cause antibiotic-associated diarrhea, which can be protracted and even deadly. The best therapy for antibiotic-associated diarrhea has turned out to be a fecal transplant, which reconstitutes the intestinal flora, as described in Box 17.2.

KEY POINT Natural selection can occur at the level of the holobiont. Microbiome composition and richness is determined by both genetic and environmental factors.

Co-evolution and inter-organism interactions

As discussed in the previous section, the genomes of a holobiont's composite organisms will undergo independent genetic variation but will often experience evolution and selection as part of the community. Major evolutionary events (for example, population bottlenecks) can be synchronous if the entire holobiont is affected at once. As they may be experiencing similar selective forces, organisms that are intimately associated with one another **co-evolve**.

Following millions of years of close habitation, interdependence evolves among the organisms in the microbiome and their hosts. Such **symbiotic** interactions can affect nutrient supply, protection from parasites and predators, and growth or reproductive fitness, among other factors. Evolutionary change in a symbiont can trigger changes in the host that, in turn, influence the evolution of both species; this is an example of co-evolution. In other words, symbioses provide selective advantages but are themselves under selection.

Co-evolution occurs in other interactions as well—including antagonistic ones. Organisms that experience synchronous selective pressure and/or are co-evolving

BOX 17.2 *A Human Angle* Transplanting microbiomes

Babies acquire their first gut microorganisms from their mothers, as they pass through the birth canal, and even more as they feed. Thus, the microbiome is not inherited genetically, but some parts of it are acquired from a parent. However, most of our gut metagenome and the metagenomes of other habitats—collectively known as the human microbiome—is acquired from the environment or from other humans. Therefore, in a sense, we largely acquire our microbiomes horizontally. The OTUs of the gut microbiome vary substantially from one person to another. Intriguingly, the metabolic capacities of individuals consuming similar diets is remarkably similar, even when the precise composition of bacteria mediating these functions is different. This suggests that viewing our microbiome as our extracellular genome, rather than simply as a collection of different microorganisms, could be a helpful approach for understanding our own biology.

Since much of our microbiome is acquired during our lifetime, microbiomes can also be transplanted from one individual to another. If a recipient's microbiome is removed or depleted, it can be replaced with the microbiome of a donor. This typically establishes a richness and composition that is very similar to that of the microbiome of the donor. Microbiome transplants have been used to restore the microbiomes of individuals who suffer from infectious diarrhea because the infection has resulted in a loss of the protective effect of their healthy microbiomes. Ultimately, metagenomic analysis of microbiomes should enable the development of microbial therapeutics that can be used to seed new microbiomes, but, for now, fecal transplants can be used effectively to recolonize a gut microbiome.

Microbiome transplants are also a helpful tool in experimental metagenomics. As discussed in the chapter, one of the best known experiments with microbiomes was built on the initial observation that the bacterial species living in the guts of obese mice and humans are distinct from those living in the guts of non-obese mice and humans. Investigators transplanted bacteria from the gut of an obese mouse to the gut of non-obese mice, and the non-obese mice gained weight rapidly. Ultimately, the collective energy-mining ability of the microbiome from obese mice was shown to be less than from lean mice.

Similar studies have been performed by transplanting the gut microbiomes from healthy children and malnourished children into germ-free mice and then monitoring their nutritional statuses. These studies have shown that intractable, and often fatal, malnutrition can also be a disease of the microbiome.

have phylogenies that show similar branching patterns, the so-called congruent phylogenies, as they diverge together from their closest other relatives.

One example of co-evolution revealed by metagenomics involves insects of the louse family known as psyllids. Psyllids feed on plant sap and depend on bacterial endosymbionts to supply essential nutrients that they cannot get from feeding and cannot synthesize themselves. Their body cavities contain a bacteriome, a specialized organelle containing one or more bacterial endosymbionts. Organisms residing within the bacteriome are transmitted vertically, that is, from mother to progeny. When a phylogeny of psyllid species was constructed based on the nuclear gene *wingless*, the resulting tree was congruent with that constructed for the primary endosymbionts of the same psyllids based on the sequences of the 16S and 23S rRNA genes, as shown in Figure 17.9. Similar findings have been made with the endosymbionts of aphids and fruit flies. In other words, the symbiosis between the insects and the bacteria has been enduring and stable enough that both the insects and the bacteria form phylogenetic trees showing similar patterns of relationships and evolution.

Mutualistic symbiosis

Although co-evolution is commonly seen in symbiotic relationships, it does not require endosymbiosis (or even symbiosis, as we shall see). Ants belonging to the attine lineage have cultured fungi for over 50 million years by sowing and tending them in gardens fertilized with plant or insect waste. Queens carry these fungi on their bodies and initiate fungal gardens when they set up new colonies. The fungi are the principal food source for the colony and the only food for larvae and queens. The fungi reproduce asexually and depend on worker ants to propagate them and protect them from the competition of free-living fungi. Because both the ant and the fungus derive benefits from the partnership and neither can survive without it, this is known as an **obligate mutualistic symbiosis**.

Organisms living in a mutualistic symbiosis are part of the same ecosystem and provide mutually beneficial biochemistry or physiology. Mutualisms are much more common than once thought, and metagenomic research will likely reveal more such partnerships. Additionally, metagenomics is demonstrating that mutualisms rarely

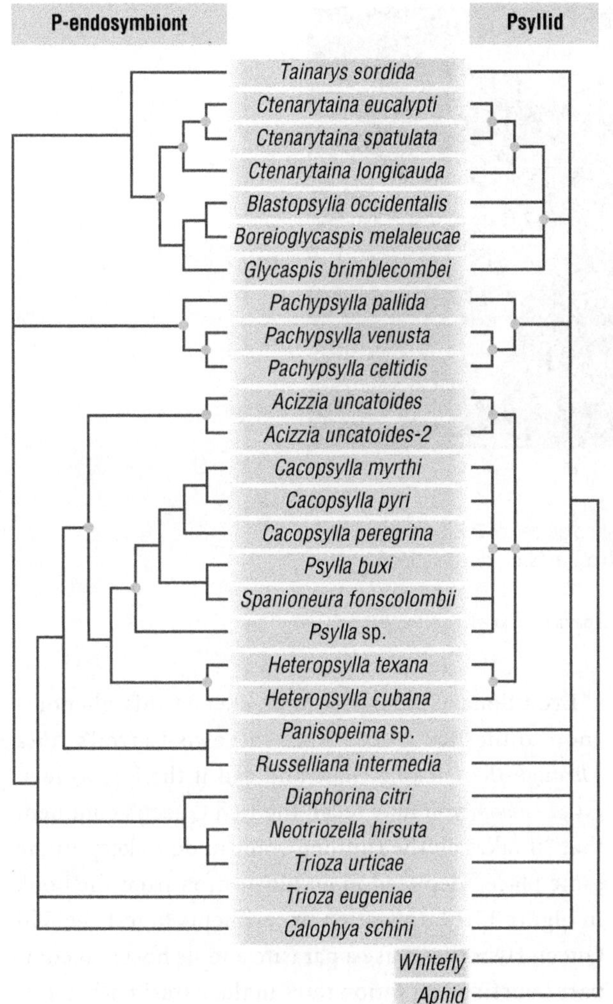

Figure 17.9 Co-evolution of insect hosts and their bacterial endosymbionts has resulted in congruent phylogenies. Comparisons of phylogenetic trees derived from 16S–23S rDNA of psyllid primary endosymbionts (listed as P-endosymbiont on the left) and from a nuclear host gene (*wingless*) from the psyllids themselves (right).

Source: Thao M.L., et al. 2000. Cospeciation of psyllids and their primary prokaryotic endosymbionts. *Applied and Environmental Microbiology* 66 (7): 2898–2905.

consist of just two species. Instead, they are complex networks among multiple organisms and have varying degrees of robustness.

Let's take a look at some of the other contributors to the ant–fungus symbiosis. The fungi cultivated by attine ants are nearly monocultures. Agricultural monocultures are susceptible to parasites—think of the specialized agricultural plant lines that humans cultivate and how a single parasite or pathogen can devastate a crop. Similarly, as shown in Figure 17.10, other fungi belonging to the genus *Escovopsis* can parasitize the ant fungal gardens. Worker ants bring in plant and insect material to fertilize the gardens, so it is very easy for *Escovopsis* to be imported into a garden. However, *Escovopsis* parasitism rarely overtakes the gardens that are actively being tended, as the ants also carry *Actinobacteria* that produce an antimicrobial chemical. The bacteria form dense clumps on the ants' cuticles, as illustrated in Figure 17.11(a). The antimicrobial chemical produced by the bacteria is lethal to *Escovopsis,* as shown in Figure 17.11(b), but is not lethal to the fungal cultivar. Both the antimicrobial-producing bacteria and the cultivar fungi are transported by the queen during her nuptial flight.

The four-way relationship among the ant species, its fungal cultivar, the parasitic *Escovopsis*, and the antimicrobial-producing *Actinobacteria* has continued for millions of years. Evidence for the evolutionary effect of this mutualism is seen in the genomes of these species. The ant, cultivar, and bacterial strains whose genomes are transmitted vertically through this mutualism have diverged from the ancestors in which the relationship emerged. There are now 13 known genera of attine ants, all of which cultivate fungal gardens. In principle, *Escovopsis* could become so virulent as to make such cultivation impossible, it could evolve resistance to the antimicrobial chemical, or the ant–bacterial interaction could break down. However, any of these events would have led to a collapse of the mutualism and would have placed the three co-dependent species and the parasite (which would then lack a host) at a selective disadvantage. Thus, there has been selective pressure for the co-evolution of these four species over time.

Host–parasite interactions

Let's now take a closer look at the evolution of the parasite in this community. Several ant species, all of which are descended from a common ancestor, currently cultivate fungi. Their fungi, also sharing a common ancestor, are transmitted through colonies vertically because queens carry them to their new colonies. However, the queens do not carry parasitic *Escovopsis* species because of the antifungal effects of the *Actinobacteria* they carry. Instead, the parasites are transmitted from one garden to another horizontally.

Based on the psyllid–endosymbiont example, you are probably not surprised that the phylogenies of the cultivated fungi and their attine gardeners are congruent. Fungal cultivars are passed to new colonies from ancestral ones by the founding queen, and, like the ant genera, extant fungal cultivars are monophyletic. However, it is more surprising that *Escovopsis*, which is transmitted horizontally between gardens, also has a phylogeny that is

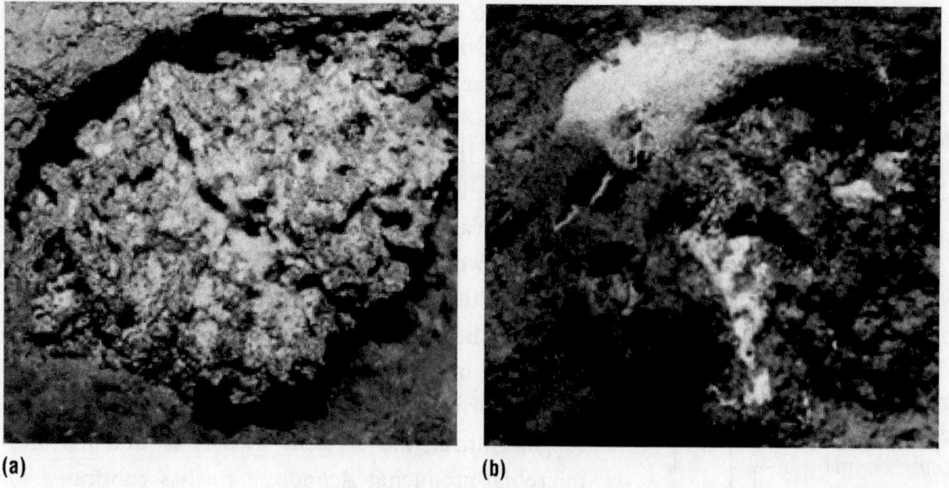

Figure 17.10 Attine ants cultivate gardens of fungus. (a) A healthy cultivated garden of *Trachymyrmex* fungus. (b) A *Trachymyrmex* fungal garden that has been devastated by an overgrowth of the *Escovopsis* parasitic fungus (white patches).

Source: Currie, C.R., Mueller U.G., Malloch, D. 1999. The agricultural pathology of ant fungus gardens. *Proceedings of the National Academy of Sciences*. 96 (14): 7998–8002. © National Academy of Sciences, USA.

not associated with geographical location and is also congruent with the attine ants and with the cultivated fungi, as shown in Figure 17.12.

What explains the congruence of the *Escovopsis* phylogeny with that of the ants or the cultivated fungi? The answer is that parasites and their hosts or targets also coevolve. If the host evolves resistance and the parasite does not evolve a new way to live on its altered host, the parasite will be unable to survive. Thus, hosts and their parasites are in what has been referred to as an evolutionary "arms race," in which, like mutualists, they must maintain coupled evolutionary histories in order to maintain their partnership.

Evolutionary biologists have likened this phenomenon to the Red Queen Race in Lewis Carroll's *Alice Through the Looking Glass* and call it the **Red Queen Hypothesis**; you may recall the Red Queen's comment that "it takes all the running you can do to keep in the same place," depicted in an illustration from the book in Figure 17.13. Evolution experiments to test the Red Queen Hypothesis use a parasite and its host and compare beneficial mutation rates in the parasite when it is propagated in its evolving host to mutation rates when repeatedly propagated in the ancestor of its host (so that evolution of the host was, in effect, kept constant).

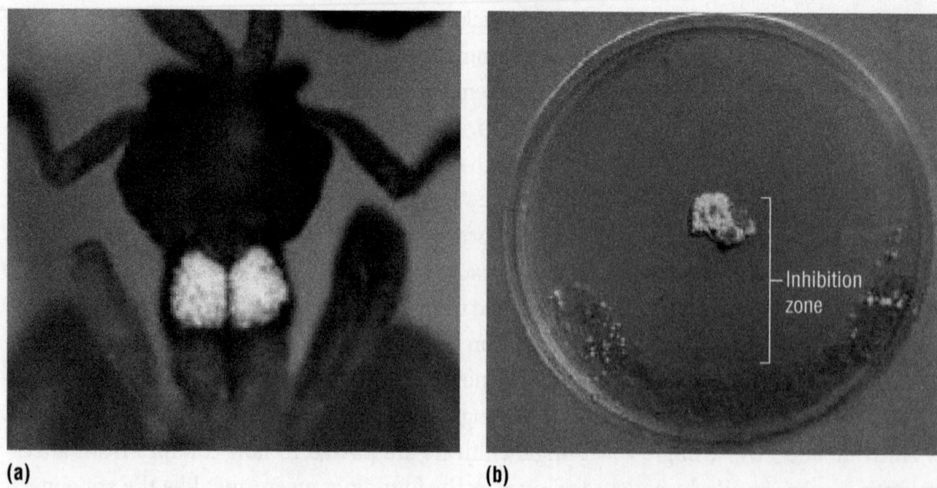

Figure 17.11 Fungus-growing ants use antimicrobial-producing bacteria to control fungal parasites in their gardens. (a) A patch of *Actinobacteria* (arrow) on an area of the cuticle on the underside of an attine ant. (b) A laboratory assay to show the effects of the *Actinobacteria* on the parasitic fungus *Escovopsis*. Note the significant clear zone of inhibited fungal growth that restricts the growth to the edge of the plate.

From: Currie C.R, Scott J.A., Summerbell R.C., Malloch D. 1999. Fungus-growing ants use antibiotic-producing bacteria to control garden parasites. *Nature* 398: 701–704.

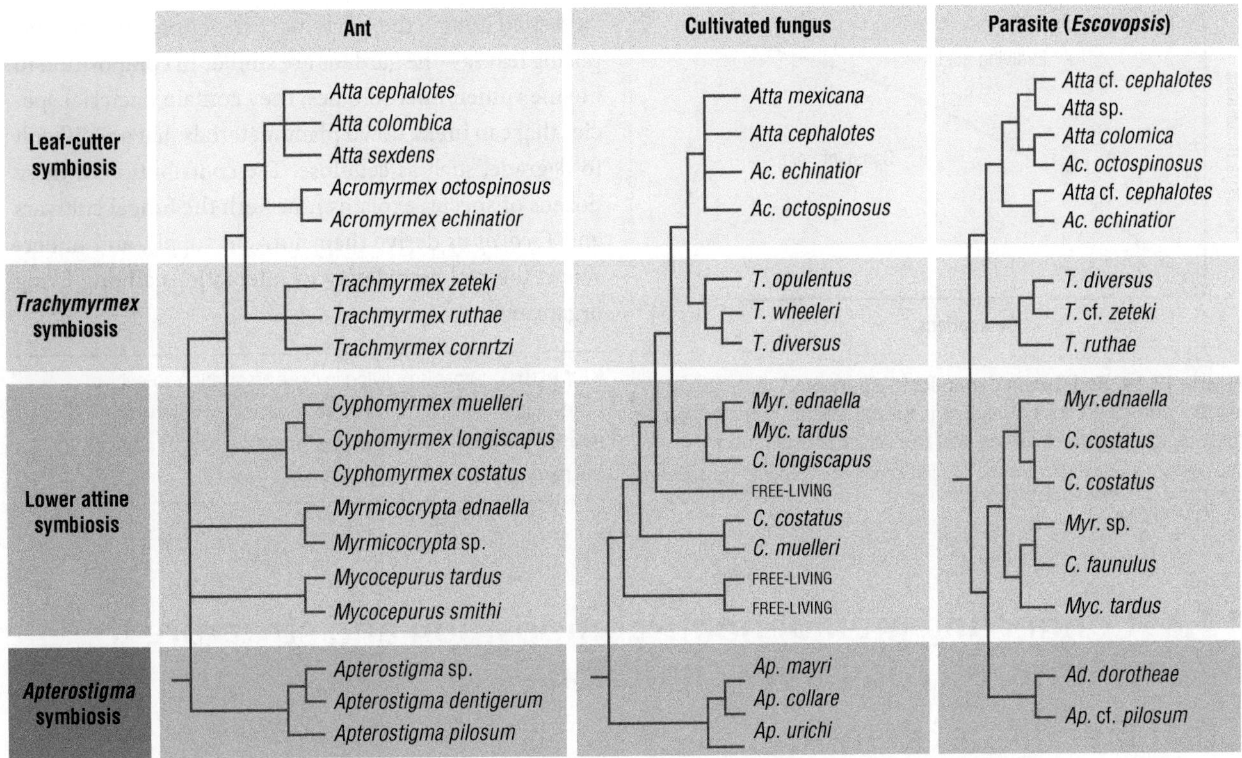

Figure 17.12 Phylogenies of attine ants, their cultivated fungi, and the associated parasitic fungi (*Escovopsis*) showing all three phylogenies with similar branching patterns. The trees are congruent with one another, despite the horizontal transmission of the parasite.

Source: Reproduced from Cameron R. Currie, et al. (2003). Ancient Tripartite Coevolution in the Attine Ant-Microbe Symbiosis. *Science* 299, 386.

The rate of mutation of the parasite is greater when the host is evolving than when it remains constant, as depicted in the graph in Figure 17.14. The Red Queen Hypothesis is one of many hypotheses that explain real-world host–parasite relationships.

Mutualisms are often multi-component and therefore effectively studied with metagenomics

The evolutionary tale of the ant, its fungal cultivar, the cultivar's parasite, and the antimicrobial-producing *Actinobacteria* would be relatively simple if those were the only characters in the mutualism story. But they are not. The *Actinobacteria* reside on the cuticle of the ant where nutrient availability could be unreliable. A vertically transmitted black yeast, which co-localizes with the *Actinobacteria* on the ant cuticle, is a good candidate for supplying the bacterium's nutrients.

While it is possible to keep adding to our knowledge of this consortium one species at a time, metagenomic approaches allow us to identify many more organisms associated with the ant and its various acquaintances in a single experiment. Metagenome analysis of an ant colony, for example, has revealed that—in addition to

"Well, in *our* country," said Alice, still panting a little, "you'd generally get to somewhere else—if you ran very fast for a long time as we've been doing."

"A slow sort of country!" said the Queen. "Now, *here*, you see, it takes all the running *you* can do, to keep in the same place.[12] If you want to get somewhere else, you must run at least twice as fast as that!"

Figure 17.13 The Red Queen Hypothesis. Excerpt from Lewis Carroll's *Through the Looking-Glass and What Alice Found There*, Chapter 2: "The Garden of Live Flowers," depicting The Red Queen's race after which the evolutionary hypothesis "The Red Queen Hypothesis" is named.

From Carol, L. *Through the Looking-Glass, and What Alice Found There* (1871). Illustration by John Tenniel.

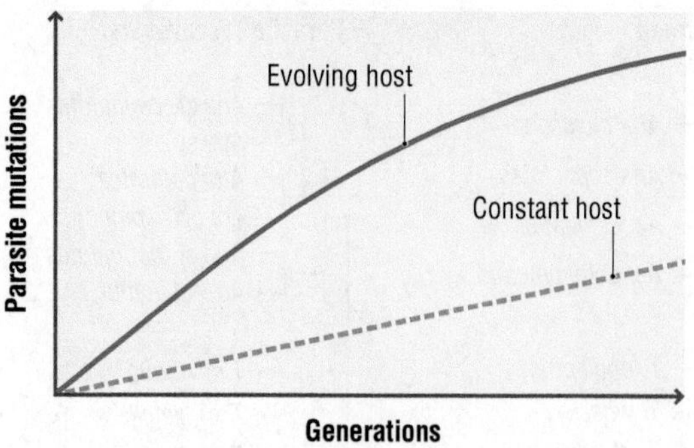

Figure 17.14 Red Queen dynamics in a parasite. Graph showing that non-deleterious mutations in the parasite are detected at a more rapid rate in the evolving host than in the constant host.

Source: Reproduced from Barrick, J.E. and Lenski, R.E. Genome dynamics during experimental evolution. *Nat Rev Genet.* 2013 Dec;14(12):827-39. doi: 10.1038/nrg3564.

cultivated fungi, the parasitic *Escovopsis*, and decomposing leaves—the gardens are similar in composition to bovine rumen microbiomes; they contain bacterial species that can break down plant materials that are difficult to degrade, such as cellulose. The contribution of these dozens of species explains how both the fungal cultivars and *Escovopsis* derive their nutrient supply, and underscores the real complexity of interactions among living organisms.

KEY POINT Organisms living in close association co-evolve. Co-evolution is a common feature in microbiomes, and the hallmarks of co-evolution can also be seen in the genomes of the participants.

17.4 Connecting metagenome composition and structure to metabolism and community function

Great Idea: life is chemistry

One of the Great Ideas introduced in the Prologue is that living organisms are composed of molecules. Consequently, life depends on chemistry. Cells are largely composed of oxygen, nitrogen, hydrogen, phosphorus, and sulfur-containing organic compounds, with many other elements present at small (but often essential) levels. Enzyme-catalyzed anabolic reactions build compounds containing these elements from simpler chemical precursors, while catabolic reactions break them down to provide energy or building blocks for the synthesis of new molecules.

An organism must be able to conduct all the catabolic and anabolic reactions it requires for life, but the genes encoding the necessary enzymes need not be encoded in the organism's own genome. A few, some, or even many could be provided by the microbiome. Thus, not every organism needs to be self-sufficient, but any habitat that supports life will have a strategy for obtaining and cycling carbon, nitrogen, and other elements essential for life. For example, plant sap is a food rich in carbohydrates but lacking many essential amino acids. As noted in Section 17.3, the endosymbionts of psyllids, aphids, and other sap-feeding organisms synthesize essential amino acids for their hosts. In return, they receive a carbohydrate-rich diet that the insects extract from the host plant's phloem. For larger organisms, symbiotic relationships of this nature often exist, but missing nutrients are typically provided by a more extensive consortium of organisms in the microbiome. When metabolism is viewed at the consortial level, rather than in one organism at a time, the complete complement of reactions that occur in very different niches are remarkably similar, as shown in Figure 17.15.

Whole metagenomic shotgun sequencing

16S rRNA profiling reveals which organisms are present but not what they can do. We can gain insights into all possible biochemical reactions occurring within a community, and therefore its metabolic capacity, by carrying out shotgun sequencing of the metagenome. (Shotgun sequencing of genomes is discussed in Tool Box 3.3.) Shotgun sequencing of ocean metagenomes has identified which digestive reactions are contributed by gut microorganisms. The advantages of shotgun sequencings-based metagenomics apply to both living and extinct microbial species. The approach has therefore provided some important insights in the field of bioarcheology, enabling us to explore the microbial composition of ancient microbiomes, as discussed in Box 17.3.

DNA sequences obtained from the direct sequencing of the entire metagenome are assembled into as few overlapping stretches of sequence as possible. These stretches of DNA, assembled from overlapping

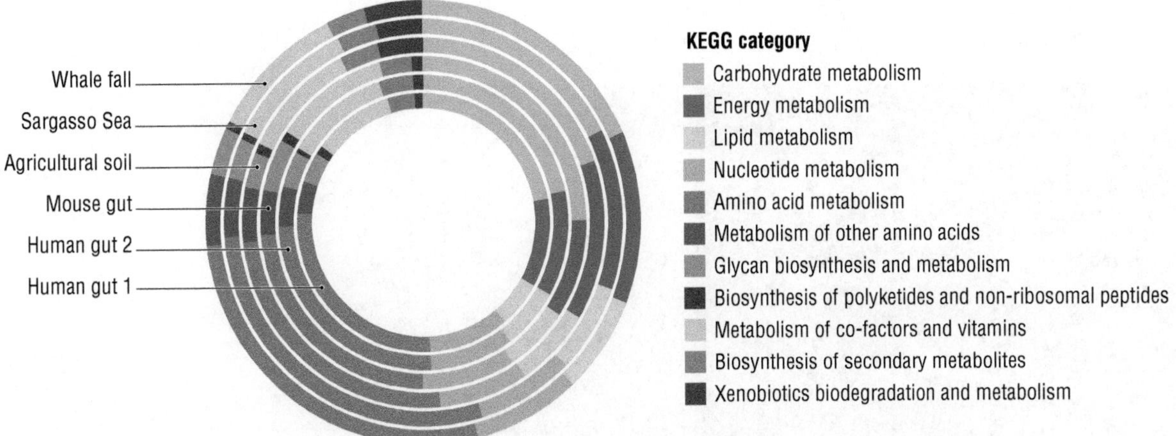

Figure 17.15 Abundance of genes involved in different metabolic categories in six different microbiomes. Three "environmental" microbiomes—whale "cemeteries" at the bottom of the ocean ("whale falls"), soil, and the Sargasso Sea—have proportions of key metabolic genes similar to human and mouse intestinal microbiomes. The categories have been predefined by KEGG, the Kyoto Encyclopedia for Genes and Genomes database, available at **www.genome.jp/kegg/kegg1a.html**.

Source: Adapted from Turnbaugh P.J., et al. 2007. The human microbiome project. *Nature* 449: 804–810.

shorter fragments, are referred to as contigs, as described in Tool Box 3.3. Because of the sheer number and extent of genomes and the short sequence reads produced by the newer sequencing techniques, the collections of contigs that are assembled are unlikely to fully cover entire genomes. However, significant portions of genomes can be assembled, and contigs that are 100 kb or larger are of a size sufficient for us to be able to extract substantial information about the bacterial species and the biochemical pathways their genomes encode. Investigators can use the assembled metagenomic sequence to determine which metabolic pathways are represented and their relative abundance.

BOX 17.3 *A Human Angle* Ancient metagenomes

Metagenomic analyses of ancient specimens tell us much about the social history, as well as the evolutionary paths, of present-day organisms. Although many bioarcheological studies have targeted specific organisms using PCR-based amplification followed by sequencing of specific genes or regions of the genome, the direct sequencing approach to metagenomes allows investigators to detect organisms that they would not have predicted were in the sample.

It is very difficult to identify organisms in fossilized or mummified archeological specimens, and the DNA contained within them is often fragmented. But massively parallel sequencing methods are well suited for short, denatured DNA fragments, and sequencing many-fold can improve coverage, and therefore the confidence in identifying each nucleotide base. The tools and methods of metagenomics have therefore seen wide application in bioarcheology.

A case in point is a medieval skeleton recovered from Geridu, close to Sardinia in Italy, which contained calcified nodules, a common feature of tuberculosis arising from infections by *Mycobacterium tuberculosis*. Part of the skeleton and the calcified nodules are shown in Figure A(i). The investigators used whole-genome sequencing to analyze DNA recovered from one of the nodules and assembled the genome of a *Brucella* strain, in addition to the fourteenth-century human DNA. *Brucella* is a less common cause of infected nodules, and very little is known about the epidemiology of this infectious agent before modern times.

After omitting single-nucleotide polymorphisms (SNPs) likely due to the types of DNA damage common in ancient specimens and those for which sequence coverage was insufficient, the investigators compared the medieval sequence to that of modern *Brucella* strains from Italy; this produced the tree shown in Figure A(ii). As can be seen in the tree, the ancient Geridu strain shares a common ancestor with the progenitor of modern-day strains from Italy. The Italian strains fall within a specific western Mediterranean phylogenetic cluster that is distinct from other *Brucella* strains, so the fact that the ancient genome clusters with them suggests that this western Mediterranean lineage is a long-standing one.

BOX 17.3 Continued

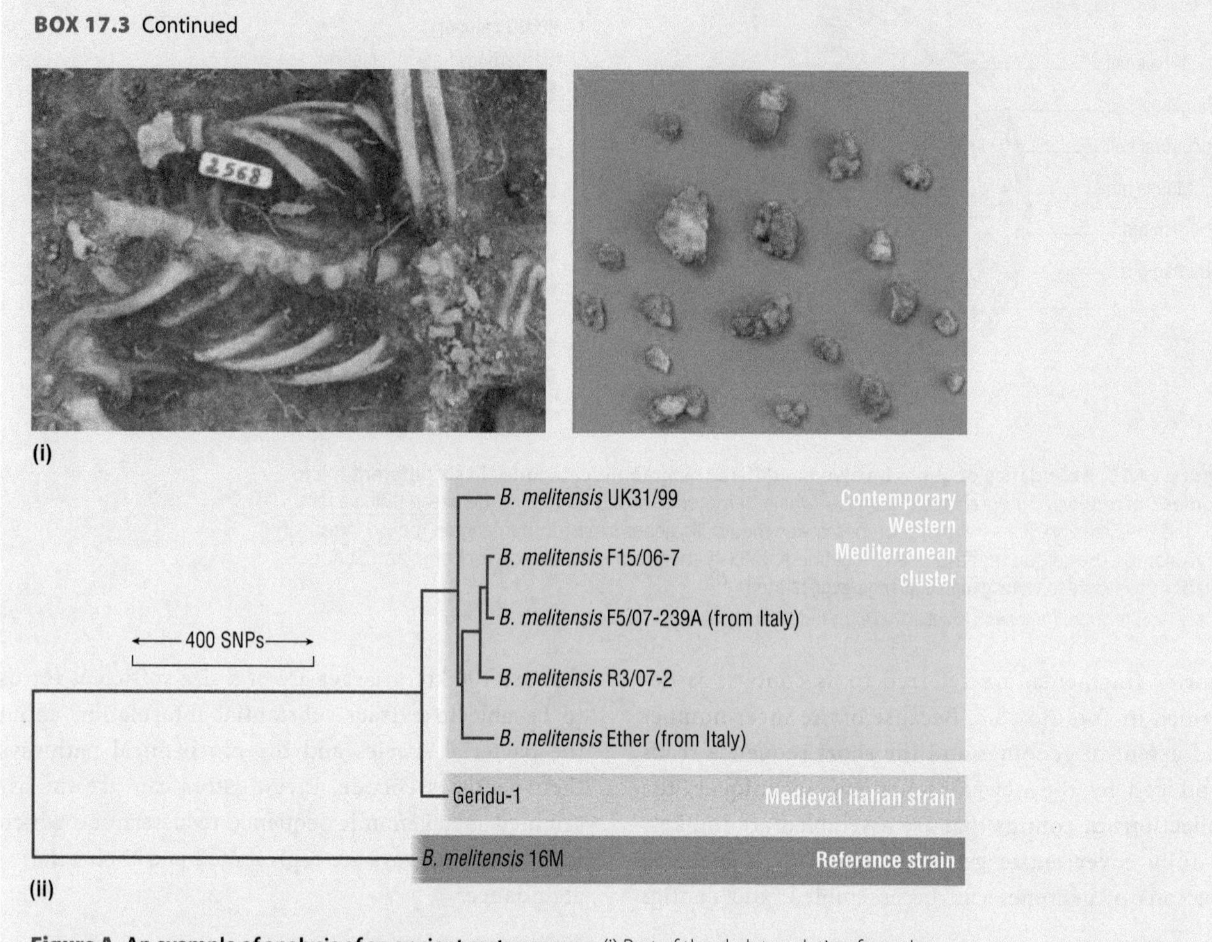

Figure A An example of analysis of an ancient metagenome. (i) Part of the skeleton dating from the fourteenth century, found in Geridu, Italy, with calcified nodules shown in the image on the right. (ii) Phylogenetic tree created with genomic data, showing the relationship between the Geridu *Brucella* strain and other known strains. Each strain has a letter and/or number code designation.

Source: Reproduced from Kay, G.L. et al. Recovery of a medieval *Brucella melitensis* genome using shotgun metagenomics. *MBio.* 2014 Jul 15;5(4):e01337-14.

Assigning sequences to taxa from whole metagenome projects can be challenging and sometimes impossible. Other ways to analyze and use the data include studying metagenome function, rather than composition. (This is how the data in Figure 17.15 were actually derived.) In the Sargasso Sea project, investigators looked at the variety of photosynthetic genes, whereas functional studies of ant gardens described in Section 17.3 revealed that their functional capacity is similar to that of herbivore rumen microbiomes. So-called "functional metagenomics" can provide important information about a microbiome. One only needs to listen to an orchestra to appreciate that the output of a system may be more important to most outsiders than who specifically constitutes it.

The consortium of scientists working on the Human Microbiome Project has used 16S rRNA sequences to determine the identity of organisms present in over 200 individuals. By analyzing metagenomic sequences from the same samples, it has been possible to determine the biochemical pathways encoded by the microbiome. Although the specific microbial OTUs in each microenvironment of the body vary widely among different individuals, the metabolic functions they provide are remarkably similar, as can be seen in Figure 17.16.

To describe one example, carbohydrate metabolism and vitamin co-factor biosynthesis are enriched in stool samples compared to other sites. Roughly, a third of the functional genes identified in the metagenomic sequence from each sample were devoted to these

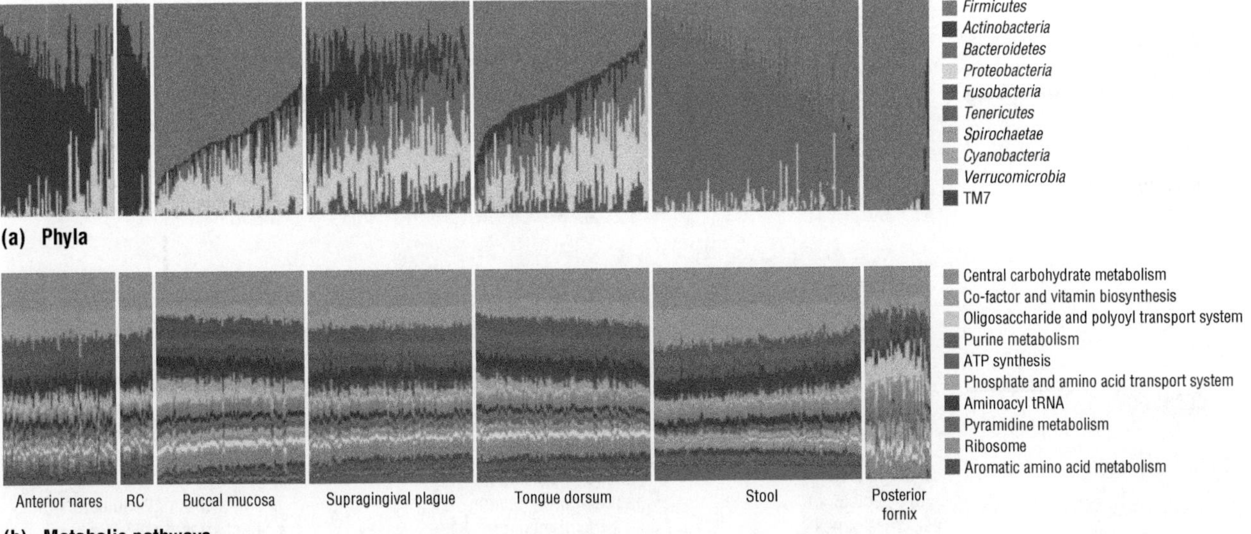

Figure 17.16 The particular microbial taxa carried by different individuals in different body areas varies, while metabolic pathways remain more stable within a healthy population. (a) Vertical lines represent the relative abundance of microbial OTUs (colored by phylum) across the 200 individuals analyzed (X-axis) for the seven body areas studied (RC is the retroauricular crease, behind the ear). Note that the precise microbial composition of the community varies from one individual to another and from one body area to another. (b) If we look instead at the metabolic pathways represented in each sample, there is much more uniformity between samples taken from different individuals for the same body area. There are also more marked similarities between the metabolic pathways represented in microbiomes from different body areas than was seen in (a).

Source: The Human Microbiome Project Consortium. 2012. Structure, function and diversity of the healthy human microbiome. *Nature*. 486: 207–214.

functions, even though the specific ratio of *Firmicuites* to *Bacteroidetes* varied from 95:5 to 5:95 across different samples. These findings suggest that the metabolic function of the microbiome is more important than the particular species that comprise it; it is the biochemical capability of the microbiome that is selected for. While specific bacterial species could more efficiently perform one function or another, the overall microbiomes of healthy individuals are selected to ensure a healthy and productive balance.

If the primary interest in a metagenomic search is to uncover the genes required for a specific metabolic process, experiments can be designed to specifically select those genes from the metagenome without the representation bias that can sometimes occur when researchers look for specific known sequences from an uncharacterized sample.

One case in point is bacterial resistance to antibiotics, a common and growing problem. Antibiotic resistance genes are commonly encountered in bacterial isolates from patients, but scientists and public health experts want to know how common these genes are in the microbiomes of healthy people or in other specific environments. This has prompted a series of studies in which metagenomic DNA is extracted, cut into fragments, and ligated into plasmid vectors. As outlined in Figure 17.17, the resulting libraries are transformed into laboratory strains of *E. coli*, which are then plated onto media containing antibiotics. Since only clones carrying resistance genes will allow the *E. coli* strains to grow in the presence of the antibiotic, resistance genes will be captured and can then be sequenced. Studies of this nature have characterized microbiome and metagenome "resistomes." Studies of specimens from healthy individuals and from soil have also recovered homologs of genes that encode resistance to antibiotics found in clinical isolates, providing some clues about the origin of antibiotic resistance genes and how they evolve. Studies using similar highly selective approaches have been used to clone genes encoding other functions—for example, the ability to use hydrocarbons as nutrients.

KEY POINT Whole metagenome shotgun sequencing and targeted functional screens of metagenomic DNA can uncover metabolic capacities of living communities.

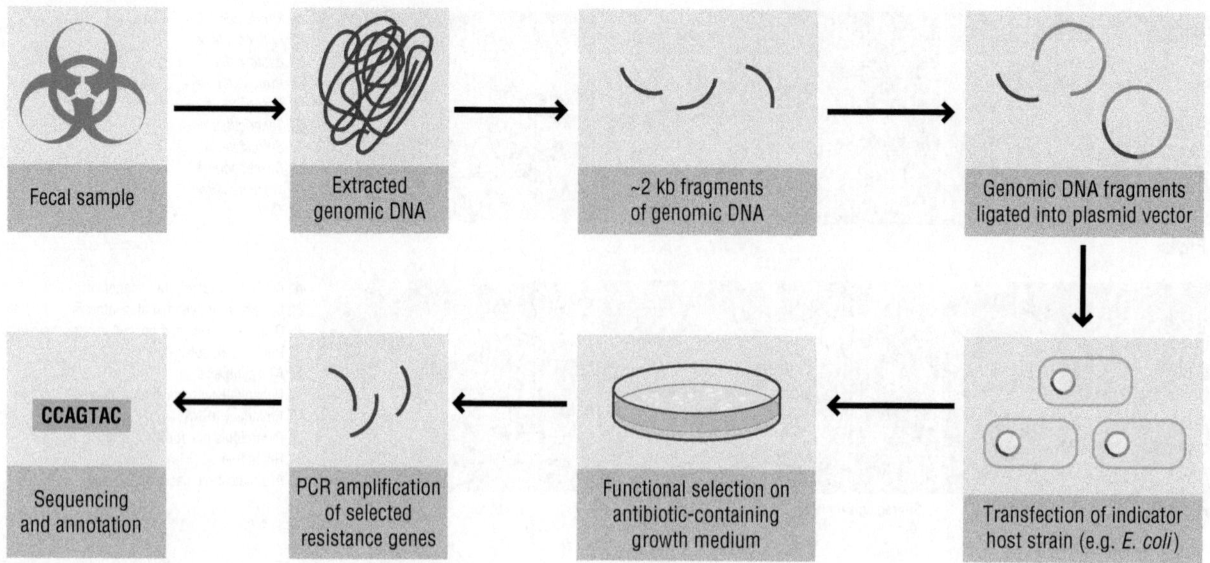

Figure 17.17 Metagenomes can be mined for genes encoding a function of interest. In this case, the objective is to identify antibiotic resistance genes in a stool sample. Genomic DNA is extracted from the fecal specimen of interest. To capture and identify known and unknown genes, the DNA is sheared into fragments larger than the average bacterial resistance gene (about 1 kb). The fragments are cloned into plasmid vectors, creating a metagenomic library. The library is transformed into an *E. coli* host strain, and transformants are selected on plates containing the antibiotic of interest. Successful capture and transformation of a resistance gene will result in colonies on the plates. The captured DNA can be amplified using primers specific to either end of the plasmid vector and then sequenced.

Source: Reproduced from Sommer, M.O. and Dantas, G. Antibiotics and the resistant microbiome. *Curr Opin Microbiol.* 2011 Oct;14(5):556-63. doi: 10.1016/j.mib.2011.07.005.

17.5 Metagenomes in action

Great Idea: biology as an organized system

It has become clear that our bodies might not develop properly or maintain good health if our own genome was the only source of genetic information. For example, if the microbiome of an embryonic zebrafish is removed by antimicrobial treatment, the zebrafish gut fails to mature. We therefore need to extend our knowledge of metagenomes beyond the "core" host genome to obtain a more complete picture of how organisms function and evolve. But knowing a gene is present in a metagenome is not the same as knowing it is used. Just as with studying "core" genomes, we can glean important insights into the function of a metagenome by exploring and understanding not only which genes are **present** but also which genes are **expressed**.

Let's return to the example we introduced at the beginning of the chapter. How does the abundant photosynthetic organism *Prochlorococcus* interact with other organisms in its community? To address this question, researchers extracted metagenomic RNA, rather than DNA, from ocean samples, converted it into cDNA, and then sequenced it. Previously, this chapter has described findings that have come from studying metagenomic DNA. In this case, however, the investigators were studying the transcription pattern of this ocean metagenome, not its sequence content. This is analogous to identifying the genes in a single genome and determining when those genes are expressed in the organism.

The transcriptional profile of a metagenome, compiled using RNA-seq, a technique described in Tool Box 3.2, is referred to as community RNA-seq. Using metagenomic DNA sequences and completed genomes as a scaffold, the researchers were able to attribute most of the transcripts

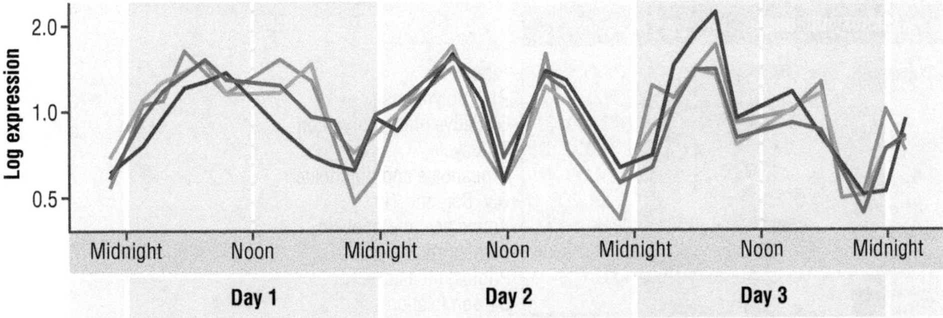

Figure 17.18 Transcriptional oscillations in *Prochlorococcus* species. RNA was extracted from ocean samples of *Prochlorococcus* collected at set time points over 2 days, and expression was measured by community RNA-seq. The normalized relative expression (Y-axis) of four photosystem I subunits (each represented by a different colored line) oscillates periodically in the course of each day. Some degree of this daily oscillation was seen in nearly half of the *Prochlorococcus* genome over the sampled time period.

Source: Adapted from Ottesen E.A., et al. 2014. Ocean microbes. Multispecies diel transcriptional oscillations in open ocean heterotrophic bacterial assemblages. *Science* 345 (6193): 207–212.

they identified to specific OTUs and then also to specific genes in those organisms. In this way, the researchers obtained valuable information about gene expression across the metagenome.

What additional insights did knowledge of the gene expression profiles in this ocean environment provide? *Prochlorococcus* is photosynthetic and, like other photosynthetic marine organisms, undergoes cycles of gene expression over the course of a 24-hour period. As shown in Figure 17.18, these daily, or diel, oscillations in gene expression from *Prochlorococcus* could be seen when the transcriptomes of samples taken from a particular geographical point were measured. The data show that the transcription of certain genes is turned on and off at specific times of the day. As might be expected, this could present an advantage for *Prochlorococcus*, which needs to express genes involved in photosynthesis during daylight hours and to conserve energy by not transcribing these genes in the dark. Almost half of the transcripts in *Prochlorococcus* showed diurnal periodicity.

What was surprising is that transcriptional cycling was also seen in transcripts from other genera in the ocean ecosystem, whether or not they are photosynthetic. In addition, the oscillations of transcriptional networks in non-photosynthetic bacteria shared the same diel periodicity as the *Prochlorococcus* bacteria with which they reside, as shown in Figure 17.19. Why is the gene expression of non-photosynthetic organisms cycling in synchrony with that of photosynthetic species? The most likely explanation is that the gene expression of heterotrophic organisms is regulated by nutrient availability, which, in turn, depends on when the photosynthetic organisms are expressing genes for enzymes used in photosynthesis. This is a clear example of how regulation within a few genomes in a community can orchestrate gene regulation in the metagenome. Given the large populations of these tiny marine organisms, coordinated biochemical reactions occur in the ocean on a much larger scale and have an impact that is far greater than was once thought.

Now that we have a sense of how *Prochlorococcus* might contribute to both the composition and richness of marine ecosystems and to gene regulation in its metagenome, let's take a final step back to recall the known factors contributing to the distribution of *Prochlorococcus* across the oceans. Knowledge of the genetics, metabolism, genomics, and ecology of *Prochlorococcus* has been used by researchers to build a hypothetical model of how ecotypes of this organism are distributed in the ocean. This, in turn, provides a map of nutrient availability, since *Prochlorococcus* lies at the foundation of the world's most extensive food chain. It is likely that similar, as well as additional, factors act on the other species in this community simultaneously, so that the nature and function of the metagenome may be best studied as a whole, as we unearth the evolution, genes, and genomes of its composite parts.

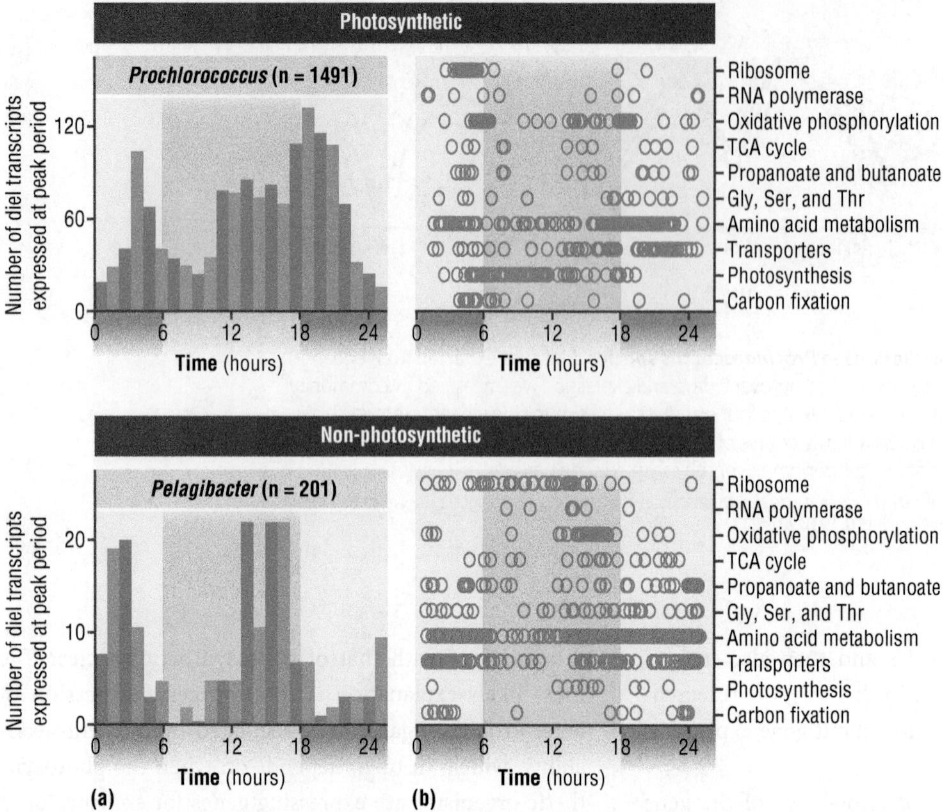

Figure 17.19 Temporal coordination of transcriptional profiles from co-residing marine bacteria. (a) The histograms show the number of periodically expressed RNA transcripts and the time of their peak expression during a 24-hour period. The bar at the bottom of the figure represents daylight and dark periods of the day. Both the photosynthetic *Prochlorococcus* and the non-photosynthetic *Pelagibacter* cluster clones show periodicity for some transcripts. (b) Transcripts sorted by functional pathways. Peak expression is marked in grey circles for all transcripts that could be assigned to a functional group, and orange circles denote transcripts showing significant periodicity. Photosynthesis pathway transcripts are few in the non-photosynthetic bacteria, but, in the photosynthetic *Prochlorococcus*, these transcripts are predominately expressed in daylight hours, highlighted in yellow. Note that, for some other groups of transcripts, such as those involved in oxidative phosphorylation and glycine, serine, and threonine metabolism, the periodicity seen in *Pelagibacter* is similar to that seen in *Prochlorococcus*.

Source: Adapted from Ottesen E.A., et al. 2014. Ocean microbes. Multispecies diel transcriptional oscillations in open ocean heterotrophic bacterial assemblages. Science 345 (6193): 207–212.

17.6 Summary: metagenomes, evolution, genomes, and genetics

It can be difficult to identify all the species in an ecosystem and to categorize their myriad interactions. Metagenomic studies can help identify the members of a community, particularly microbial species, even when culturing these species is difficult. Metagenomic studies employ methods, such as PCR to amplify genes, or direct sequencing of the DNA in a sample followed by the assembly and ordering the data. The sequences of key genes, such as those encoding the small subunit of the rRNA, or more extensive genomic information can then be used to identify the organisms that must have been present in the sample.

Other genes, and sometimes their transcription profiles, can be studied in order to identify the potential biochemical functions of members of the holobiont and how these may be interdependent. Metagenomic studies have therefore been a powerful tool for identifying and profiling the organisms in a community. Evidence suggests that entire microbiomes or holobionts co-evolve; thus natural selection is often acting within an interwoven system of such interactions, and an understanding of these relationships is required to really understand the community as a whole.

Writing without any knowledge of bacteria, genes, or genomes, Darwin said, "There is grandeur in this view of life," the perspective that integrates all of biology through the lens of evolution. Molecules and biochemistry, genes, cells, and the systems arising from their interactions are all Great Ideas, but they are not separate ideas; they are the players in the grand narrative of evolution by natural selection. Darwin began with an analysis of finches and other easily visible species. In Chapter 1, we saw how his analysis of visible characteristics also appeared in the genomes of the finches. We have concluded this book with the same principles but with species that are "visible" through the DNA sequences of their genomes.

CHAPTER CAPSULE

- A metagenome is the total of all genomes occupying a specific ecological niche.
- The community of microbial species associated with a living organism is referred to as that organism's microbiome.
- The term holobiont refers to the host organism and its associated microbes.
- Identifying genomes using culture-independent techniques is a powerful technique for characterizing microbial communities.
- Small subunit ribosomal RNA genes, such as 16S rRNA, are the sequences most commonly used to identify microbes comprising a metagenome.
- Small subunit genes that are similar in sequence are said to belong to the same operational taxonomic unit, or OTU, sometimes called a phylotype. This taxonomic grouping is defined by DNA sequence.
- Metagenomic diversity can be quantified. Diversity within a sample is known as α-diversity, while diversity between different samples is known as β-diversity.
- Microbiome composition and richness are determined by both genetic and environmental factors.
- Closely associated organisms within a community co-evolve. Thus, co-evolution is a common feature in microbiomes.
- Natural selection can occur at the level of the holobiont.
- In addition to identifying species within a community, whole metagenome sequencing can uncover functional capacities of this community.

STUDY QUESTIONS

Concepts and Definitions

17.1 How are the terms microbiome and metagenome related to one another?

17.2 What is an OTU, and how is this related to the term "species"?

17.3 Define "species richness," and discuss how it is important in metagenomics.

17.4 Define the terms "holobiont" and "co-evolution," and provide an example where both of these terms are involved.

17.5 Define α- and β-diversity, and discuss how each one is used in comparing samples from metagenomes.

17.6 What is a rarefaction curve, and how is it important in the sampling of metagenomes?

Beyond the Concepts

17.7 The analysis of metagenomes has only begun very recently. Much of the vocabulary, standard experimental approaches, and common body of knowledge that characterize more established fields have not yet been refined. However, analysis of ecosystems is a very well-established field of biology.

a. What are some features that make the analysis of metagenomes similar to or different from the well-established analysis of ecological communities?

b. What are some of the technological developments that have occurred to make metagenomics analysis possible?

17.8 Distinguish between culture-dependent and culture-independent analysis of microbiomes. How is each one done? What are the strengths and limitations of each approach?

17.9 What are some of the challenges with defining bacterial species? What are some methods that are currently used to assign bacteria to species or to OTUs?

17.10 Figure 17.16 shows data from two different types of metagenome analysis from different locations in the human body.

a. How is quantifying the identity of microbes present in a microbiome different from quantifying the metabolic capacity of a microbiome?

b. How do these relate to one another in different sites of the human body?

17.11 A fundamental concept in metagenomics is that the holobiont is under selection or is subject to other evolutionary pressures, that is, it is more than a collection of organisms living in the same habitat.

a. Discuss why this is such an important aspect of metagenomics.

b. What are some of the types of experimental evidence that are used to show this is occurring?

c. Give two examples that support the concept that the microbiome may provide essential functions.

17.12 Would you expect the microbiomes of the different finch species on the Galapagos Islands to have similar or disparate intestinal microbiomes? Explain your answer.

Applying the Concepts

17.13 Describe how sampling a metagenome uses similar principles to saturation analysis introduced for mutant screens in Chapter 15. What are some of the assumptions that are needed for this analysis, and how do they affect the results?

17.14 Suppose that you have been placed in charge of a project to carry out a metagenomics study.

a. Describe a biological community that you would be interested in analyzing by metagenomics.

b. Outline how you might carry out such an analysis.

c. What would be some of the specific research questions that you would design your analysis to address?

17.15 Many culture-independent methods use the sequences of 16S rRNA for the comparison.

a. Explain why this is an appropriate method to compare and classify bacterial OTUs.

b. What are some of the limitations of using 16S rRNA?

Challenging **c.** Suppose that you were interested in a similar analysis of a complex community, but you are only interested in the eukaryotic organisms, and specifically not in the bacteria or archaea that might be present. Suggest some genes whose sequences might be appropriate for such an approach.

17.16 Before rapid sequencing was possible, other methods were used to survey microbial diversity in samples. Torsvik *et al.* (*Appl Environ Microbiol* 1990, **56:** 782–787) extracted total DNA from soil, heated it to separate the double strands of the DNA in all of the microbial organisms into single strands, and then monitored the rate of re-association into double-stranded DNA. They found that the rate of re-association was quite slow. Their data were cited as evidence of the extreme microbial diversity in soil.

a. What is the rationale behind presuming that slow reannealing is equivalent to high diversity?

b. What would be the barriers to precise estimation of diversity from such data?

Challenging
Looking back **c.** While the rate of re-association overall was slow, some DNA fragments from the sample re-associated much more rapidly than the rest. Why would some fragments reanneal more rapidly than most of the metagenome, and what might be examples of some DNA elements that form double-stranded hybrids rapidly?

17.17 Figure Q17.1 shows the numbers of operational taxonomic units identified by 16S rRNA sequencing of DNA from five different locations, as part of the Human Microbiome

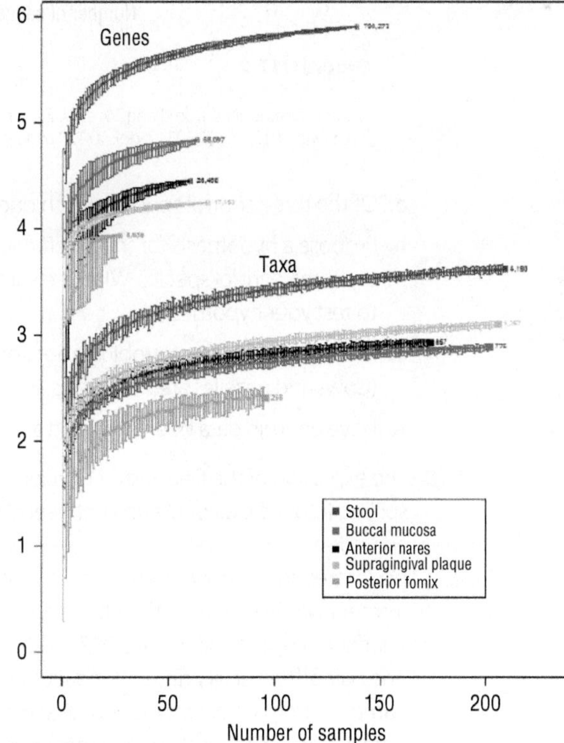

Figure Q17.1

Source: Reprinted by permission from Macmillan Publishers Ltd: Human Microbiome Project Consortium. Structure, function and diversity of the healthy human microbiome. *Nature.* (2012) Jun 13; Vol. 486, Iss. 7402.

Project, labeled "Taxa." It also shows the number of genes identified from assembled metagenomic sequences, labeled "Genes."

a. Describe what these data are showing.

b. Which location is the most diverse, and which is the least diverse?

c. The data are plotted as a rarefaction curve. In your opinion, were enough sequences assessed to make conclusions about the composition of the different locations?

17.18 Scientists compared the fecal microbiota from the giant panda, which feeds almost exclusively on bamboo, to that of several other mammals; the results are shown in Figure Q17.2. In the figure, the number of observed OTUs is plotted against the number of sequenced clones from each sample.

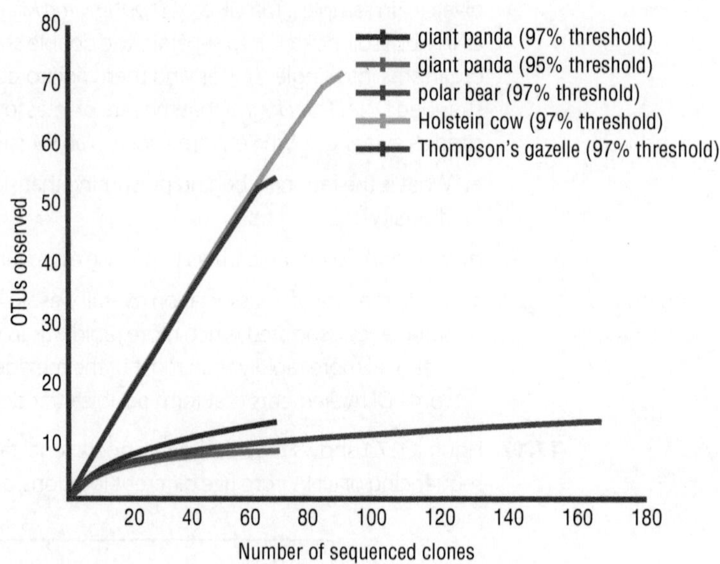

Figure Q17.2

Source: Reproduced from Fang, W. et al. Evidence for lignin oxidation by the giant panda fecal microbiome. *PLoS One*. 2012; 7(11):e50312. doi: 10.1371/journal.pone.0050312.

a. Of the three control species, which one has the microbiome with the least diversity?

b. Propose a hypothesis for the differences in the diversities among the microbiomes of the three control species. What are some additional mammals that might be analyzed to test your hypothesis?

c. Is the giant panda's microbiota more or less diverse than those of the other herbivores? (Cows and gazelles are herbivores, while polar bears are carnivores.)

d. Have enough data been collected to answer (c) conclusively?

17.19 The explosion of the Deepwater Horizon oil rig resulted in almost 800 million liters of oil spilling into the Gulf of Mexico between April and June 2010. Much of the oil spilled from an underwater plume that was only detected weeks after the spill began. Researchers performed metagenomic analyses of seawater samples collected close to (proximal) and further away from (distal) the plume and compared these samples to a control sample unaffected by the spill. Figure Q17.3A shows the results of 16S rRNA gene profiling using one pair of "universal primers" for bacteria and archaea. The data in Figure Q17.3B were compiled from whole metagenome sequences and whole transcriptome sequences. Data are presented as the major microbial subgroups represented in the sample. Reads corresponding to less abundant subgroups were combined into a single bin, labeled "other."

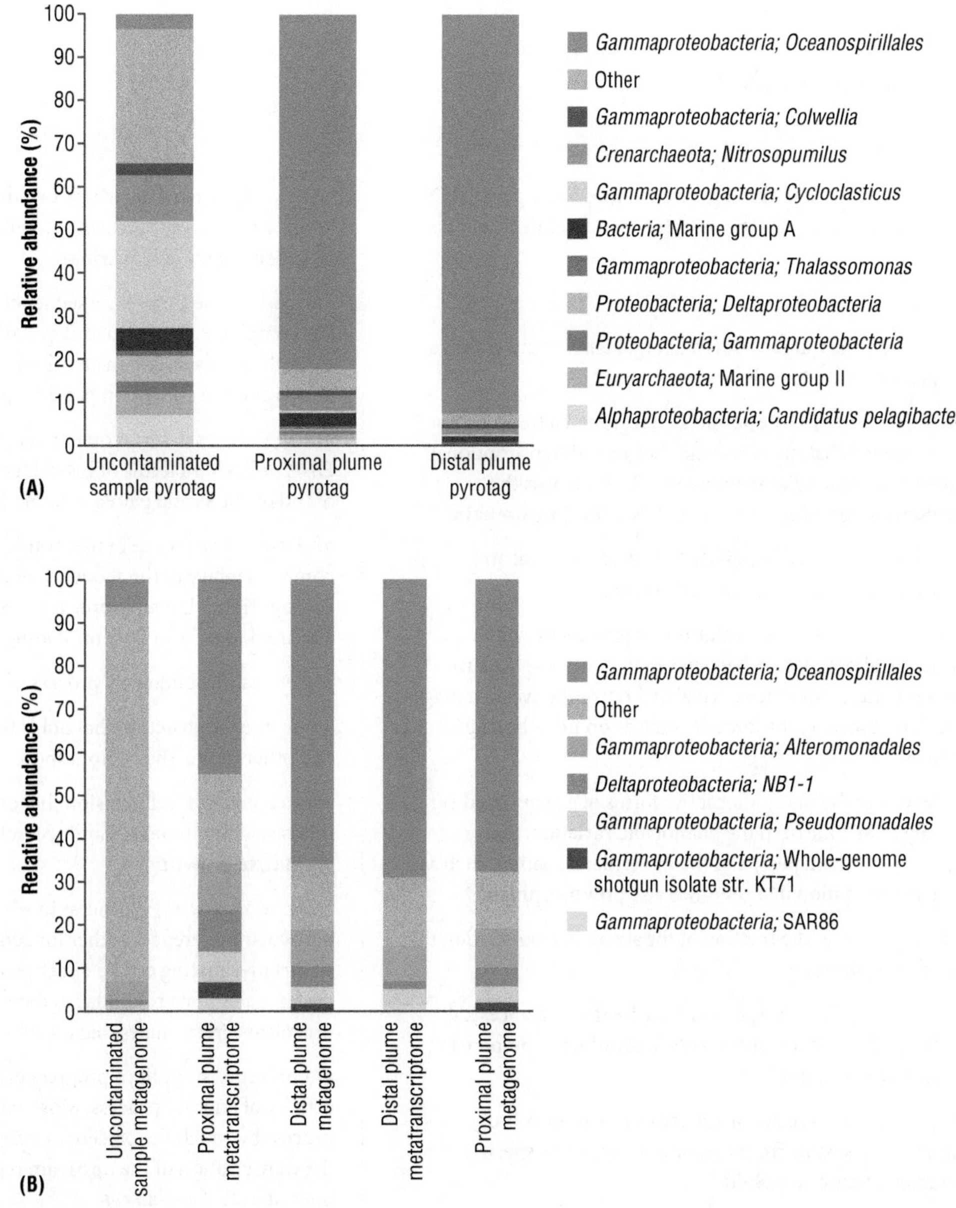

Figure Q17.3

Source: Reproduced from Mason, O.U. et al. Metagenome, metatranscriptome and single-cell sequencing reveal microbial response to Deepwater Horizon oil spill. *ISME J.* 2012 Sep; 6(9):1715-27. doi: 10.1038/ismej.2012.59.

a. Provide an explanation for why the 16S rRNA data and the whole metagenome data predict different relative abundances of different subgroups in the same samples.

b. The oil spill appears to have provided a selective advantage to one microbial subgroup. Which one was it?

c. The subgroup labeled "*Gammaproteobacteria*; whole genome shotgun isolate str. KT71" (in dark blue in Panel B) is similarly represented in whole metagenome reads proximal and distal to the plume. What explains the greater abundance for whole transcriptome reads in the proximal sample for this group?

d. If you were to compose a tweet (with a maximum of 140 characters) about the impact of the spill on microbial populations, what would your tweet read?

GLOSSARY

454 sequencing a technology used in **massively parallel DNA sequencing,** named for the machine made by Life Sciences.

A

α-diversity the range of different organisms found within a sample. Compare to **β-diversity**.

activation domain the amino acid region of a transcription factor protein that interacts with the general transcriptional proteins or the Mediator complex. Also known as the **interaction domain**. Compare to **DNA-binding domain**.

adaptation a heritable trait that increases survival and reproduction in a particular environment.

adaptive radiation an evolutionary process by which an ancestral population becomes rapidly diverse because different alleles provide survival and reproductive advantages in different local environments such as on neighboring islands or lakes.

allele one of the many alternative forms of a gene. Used very broadly to include both the phenotypic variation, such as round or wrinkled peas, and the molecular variation, such as a sequence alteration in a gene. See also **polymorphism**.

allele frequency the fraction of the alleles in a population that are of a specific type.

allopatric speciation species that arise from an ancestral species in different environments or locations. Compare to **sympatric speciation**.

allopolyploid a species or variety in which the sets of chromosomes come from two or more different species. Compare to **autopolyploid**.

alternative splicing a general term to describe processes by which more than one mRNA is made from precursor RNA because different sequences in the RNA are used as the splice donor and acceptor sites.

aminoacyl synthetase a family of enzymes that attach a specific amino acid to the 3′ end of a tRNA molecule based on the anticodon sequence of the tRNA.

anabolism a metabolic process that results in the assembly of a more complex macromolecule from simpler constituent parts. Compare to **catabolism**.

ancestral characteristic a characteristic that is found in the founding population or species from which other populations or species arise. Compare to **derived characteristic**.

aneuploidy a cell or organism that has either too many or too few copies of a particular chromosome.

anneal the process by which two different or separated single-stranded nucleic acids form a double-stranded molecule by complementary base pairing.

anticodon the three-base sequence in the tRNA molecule that has complementarity (in **anti-parallel** orientation) with the three bases of a codon in an mRNA. Base pairing between the codon and the anticodon provide specificity to translation.

anti-parallel a description to indicate that the two strands of a DNA or RNA molecule or hybrid have opposite polarity, with the 5′ end of one strand paired with the 3′ end of the other strand.

antisense a sequence of nucleotide bases that is complementary to the sequence of the mRNA or to the DNA coding strand. Usually applied to sequences made *in vitro*, but also used to refer to the non-coding strand.

apoptosis the biological process of programmed cell death.

ascus the sac structure that holds the haploid spores of yeast and other fungi. The haploid spores are ascospores.

asexual growth cell division that occurs by binary fission or mitosis, without meiosis and recombination. Also known as **vegetative growth**.

assortative mating a process in which organisms mate with particular preferences, either for genotypes or phenotypes. Assortative mating can be either positive, in which organisms prefer mates that are similar to themselves, or negative, in which organisms prefer mates that are different from themselves.

attenuation stopping a progress or reducing the strength or activity of another process. Most commonly used to refer to the process by which the presence or absence of tryptophan affects the transcription of the upstream region in the tryptophan operon in *Escherichia coli*.

autonomous element a transposable element that encodes all of the functions necessary for its own movement. Autonomous elements can also catalyze the movements of non-autonomous elements of the same type in the same genome.

autopolyploid a species or variety with more than two sets of chromosomes, which come from the same species. Compare to **allopolyploid**.

auto-regulation a process by which a gene regulates its own expression, often because the protein encoded by its mRNA binds to the regulatory region of the gene to affect subsequent transcription of the gene.

autosome any of the eukaryotic chromosomes in an organism other than the sex chromosomes.

auxotroph a mutant that cannot synthesize some essential nutrient for growth. Auxotrophic mutants can only grow if the nutrient is provided in the growth medium. Compare to **prototroph**.

B

β-diversity the range of different organisms found when different samples are compared. Compare to **α-diversity**.

bacteriophage or phage any of the many types of virus that infect bacteria.

balanced heterozygote an organism or a strain that remains as a heterozygote for several linked alleles or polymorphisms over many generations.

balancer chromosome a genetically altered chromosome that cannot recombine with its normal homologous chromosome. Any recessive mutations on the normal homologous chromosome will be inherited together if they are maintained as a heterozygote with a balancer chromosome.

balancing selection a form of natural selection in which an intermediate phenotype (and thus an intermediate genotype) is the most favoured. With balancing selection, more than one allele will be maintained in the population.

BLAST (Basic Local Alignment Search Tool) a suite of computational tools that identifies similar regions in nucleotide or amino acid sequences. Particular programs in the suite are names for the sequences they compare; thus, BLASTP is used for polypeptides and BLASTN is used for nucleic acids.

branch in a phylogenetic tree, the lines of descent connecting different species or individuals.

branch migration the process by which the Holliday junction formed during recombination between DNA molecules moves along the length of the molecules.

C

capsid the protein coat encasing a virus.

carrier an informal term for what is commonly a heterozygote, that is, an individual who does not show the phenotype for a trait but who may pass on the trait to the next generation.

catabolism a metabolic process that results in the breakdown of a complex macromolecule into simpler constituent parts. Compare to **anabolism**.

catabolite activator protein or CAP a protein in *Escherichia coli* that responds to the level of glucose in the growth medium, which ensures that the bacteria use glucose as a carbon source before using other sugars.

catabolite repression the process in *Escherichia coli* by which CAP shuts off or represses the utilization of other sugars, such as lactose, when glucose is also present in the growth medium.

centromere the region of the chromosome where sister chromatids are connected to one another and to which microtubules attach during mitosis and meiosis. In a **karyotype**, the centromere is usually seen as a constriction along the length of a chromosome. The proteins that form the attachment region for microtubules are properly referred to as the kinetochore, but the terms are often used to refer to similar structures. The kinetochore proteins and the corresponding DNA region comprise the centromere.

checkpoint one of several different genetically controlled mechanisms by which the mitotic or meiotic cell cycle is arrested until some process is completed. The checkpoints are typically named for the process that needs to be completed such as the DNA repair checkpoint or the spindle attachment checkpoint.

chiasma the physical connections between homologous chromosomes observed microscopically during crossing over during meiosis I. The plural is chiasmata.

chromatin the complex of DNA with associated proteins and RNA molecules that comprise the structure of the eukaryotic chromosome.

chromatin immunoprecipitation or ChIP a procedure in which an antibody against a transcription factor protein or other chromosomal protein is used to identify the binding sites on the chromosome for that protein across the genome.

chromosomal protein the general term for a protein associated with DNA in chromatin.

chromosome literally a colored body, the structures inside the cell made up of DNA and associated proteins that are responsible for the transmission and expression of genes, and thus contain most of the genome. While the term was originally applied only to eukaryotes, it is now also used for bacteria, although the structure of a bacterial chromosome is quite different from that of a eukaryotic chromosome.

cis-regulatory module or CRM the region that is responsible for the specific pattern of transcription for a eukaryotic gene. The CRM is usually located upstream of the gene and consists of one or several sets of **enhancer sequences**.

clade a group of organisms consisting of a common ancestor and all of the organisms descended from it.

cladogram a type of phylogenetic tree that shows the inferred relationships among all of the organisms descended from a common ancestor.

clonal a population that arises from a single cell that divides vegetatively or mitotically.

coding region the portion of a nucleotide base sequence of a gene that encodes the information that will be translated into the amino acid sequence of a polypeptide.

coding strand the DNA strand that has the same base sequence and polarity as the corresponding mRNA sequence, albeit with thymine, rather than uracil, and with deoxyribonucleotides, rather than ribonucleotides.

co-dominance a situation in which the phenotype of each of the two alleles is observed in the heterozygote.

codon three consecutive bases in an mRNA that are read together to encode a single amino acid in a polypeptide.

coefficient of coincidence defined as the ratio of observed number of double crossovers to the expected number, as a result of interference.

co-evolution the evolutionary changes that occur in two or more species as a result of the selective pressures that each species exerts on the other.

cohesin a complex of four polypeptide subunits that holds sister chromatids together during mitosis and meiosis.

cohesion any process by which two objects stick together, but usually applied in genetics to the processes that hold sister chromatids together during mitosis and meiosis.

"common disease, common variant" hypothesis the hypothesis that individuals who are affected by a genetic disorder that is common within a particular population will also have the same molecular variant that gives rise to the disorder. The prevalence of Δ-F508 as a causative allele for cystic fibrosis in Europeans is an example.

competent cell a cell, particularly in bacteria, that can take up DNA from the environment via transformation. Some species are naturally competent, while others can be induced to become competent by laboratory procedures.

complementary the sequence that forms stable base pairs in double-stranded nucleic acids. The base adenine is complementary to the bases thymine (in DNA) or uracil (in RNA), while the base guanine is complementary to the base cytosine.

complementary DNA or cDNA a DNA strand made *in vitro* using reverse transcriptase, so that it has the complementary sequence to part of an RNA molecule. The first strand, which is the one that is complementary to the RNA sequence, is usually then used to make a second strand, so that the cDNA molecule frequently consists of double-stranded DNA.

complementation test a genetic test to determine if two recessive mutations are alleles of the same gene. Mutations that are alleles of the same gene fail to complement.

complex traits traits that show some tendency to be inherited, but whose inheritance pattern cannot be described by the behavior of one or a few genes. Complex trait phenotypes are affected by variation in both genes and the environment. Complex traits are also called multifactorial traits because many genes are involved, or quantitative traits because many of the phenotypes can be measured numerically.

compound heterozygote a heterozygote in which both alleles of a gene are mutant, but the mutant alleles do not have the same base change, usually used in human genetics. Also known as **heteroallelic.**

conditional mutation a mutation that exhibits a mutant phenotype only under certain environmental conditions such as a slightly elevated growth temperature or particular nutritional conditions. Temperature-sensitive mutations are an example of conditional mutations.

conjugation the direct transfer of DNA from one bacteria to another via a process requiring cell-to-cell contact that requires an extracellular structure known as a conjugative **pilus**. Conjugation can be mediated by the **F plasmid**.

consanguinity a general term for many types of genetic relatedness.

constitutive expression a condition in which a gene is always expressed, particularly used for a gene whose expression is usually subject to regulation so that it is not always expressed.

contig the overlapping fragments or clones of DNA that are produced during genome assembly in a sequencing project, an ellipsis of "contiguous fragments."

copy number variation or CNV a type of sequence variation or polymorphism in a genome in which a particular block with a defined sequence is present a variable number of times in different individuals. Thus, the variation is in the number of the sequence blocks, rather than in the base sequence itself. The most conservative definition of CNV requires the sequence block to be at least 1000 bases in length.

core promoter or minimal promoter the region upstream of a gene that is needed for transcription initiation by RNA polymerase. Many eukaryotic core promoters have a TATA box about 25 nucleotides (or −25) upstream of the start of transcription and other necessary regulation sequences. The core is also called the minimal promoter, or sometimes simply the **promoter**, although the term "promoter" is also used more generally to refer to the core promoter and the ***cis*-regulatory module.**

co-regulated expression genes with coordinated expression, usually because they are under the control of the same regulatory factors.

co-repressor a molecule that interacts with the repressor protein and enables the repressor to bind to DNA and inhibit the transcription of a gene or genes. In bacteria, a co-repressor is often a product of an anabolic operon; for instance, tryptophan interacts with the *trp* repressor to shut off the transcription of the *trp* operon.

cosmid an artificial plasmid made *in vitro* using the *cos* sequence involved in packaging the genome of phage λ. Cosmids can contain approximately 40 kb phage λ.

CRISPR or clustered interspersed short palindromic repeat originally applied to a feature of many bacterial genomes, which were found to contain arrays of short repeated sequences in their genomes with spacer sequences between them. The spacer sequences are derived from the genomic fragments of viruses that previously infected the bacteria. CRISPR is now used to refer to a widely applicable technique for editing genomes, in which a specific sequence is used as the spacer to target the CRISPR-associated, or Cas, protein to particular sites at which the editing occurs.

crossing over the breakage and reunion process, especially during meiosis, by which homologous chromosomes exchange DNA sequences.

culture-independent a method to analyze the microorganisms in a sample that does not depend on the growth and division of the cells in the laboratory. Microorganisms were traditionally studied by culture-dependent methods, which required that they grow and divide under laboratory conditions.

cytological map a record of the organization of the chromosomes in a species based on an observed cellular structure such as a banding pattern.

D

deep sequencing also known as **massively parallel sequencing**, referring to the fact that the same DNA region is sequenced many times.

denaturation a general term for the process by which a biological molecule loses the structure that gives rise to its function. For proteins, denaturation (by heat, for example) causes the polypeptide to lose the three-dimensional shape needed to carry out its function. For nucleic acids, denaturation causes the base pairs that hold the two strands together to come apart, such that the strands separate from one another.

dendrogram a diagram, such as a family or a phylogenetic tree, that shows the relationships of individuals to one another based on their similarities.

deoxyribonucleic acid or DNA the nucleic acid that encodes the information responsible for genetic characteristics. DNA is usually represented as a double helical structure, with two separate strands that are joined together. The backbone of each strand is made of phosphate linkages, while the connections between the strands are made by the base pairing of specific deoxyribonucleotides of one strand with the other. Genetic information is encoded in the sequence of the nucleotide bases.

de-repression the expression of a gene or set of genes that are normally not transcribed under a particular condition, by the removal of a repressor protein, for example.

derived characteristic a characteristic not found in the ancestral founding population or species but found in some of its descendant populations or species. Compare with **ancestral characteristic**.

diauxic growth curve the diagram depicting the doubling of a bacterial culture when its medium has two different sugars. The curve has two distinct phases indicating the times when each sugar is being utilized as a carbon source.

dictyate arrest the process during prophase I of oogenesis in vertebrates when the oocyte stops before completing the rest of meiosis I. In humans, the dictyate arrest can last for several decades.

dideoxy sequencing an older, but still commonly used, method for determining the base sequence of DNA based on the incorporation of labeled, chain-terminating dideoxy nucleotides. Dideoxy sequencing is also referred to as Sanger sequencing (since it was developed by Frederick Sanger and Alan Coulson). In dideoxy sequencing, unlike **massively parallel sequencing**, each DNA template is sequenced once, and the sequencing reaction can be done either manually or by a sequencing machine.

diploid a eukaryotic organism or cell with two copies of every chromosome. Compare to **haploid** and **polyploid**.

directional selection selection that consistently increases or decreases the frequency of a particular allele or genotype. Also known as **purifying selection** since it removes unfavorable alleles from the gene pool.

disomy the condition in which an aneuploid cell that is haploid but has two copies of one chromosome.

DNA-binding domain the amino acid region of a **transcription factor** protein that interacts directly with the DNA sequence and thus determines the locations in the genome where the protein will bind to DNA. Compare to **activation domain**.

DNA polymerase a general term for one of the polypeptide complexes that are responsible for DNA replication and synthesis. Many different DNA polymerases exist.

DNA sequencing a technique or process by which the order of bases on a DNA molecule is determined *in vitro*. See also **dideoxy sequencing** and **massively parallel sequencing**.

domain a portion of a protein that folds into a particular structure and is responsible for a particular function such as the region that binds to DNA or that interacts with another protein. Compare to **motif**. Although domain and motif are often used interchangeably to refer to the same region of the polypeptide, "domain" properly is defined by the structure and its function, while "motif" is properly defined by the sequence of amino acids even when the structure and function are not known.

dosage compensation one of several distinct processes by which animals with different sex chromosomes regulate the level of X-linked transcription, so that both sexes make similar amounts of X-linked gene product. In mammals, dosage compensation occurs by **X chromosome inactivation**.

doubling time the length of time required for a cell to divide during **vegetative growth**, either by binary fission in bacteria or by mitosis in eukaryotic cells.

downstream region the region that lies to the 3′ side of a gene, as depicted on the coding strand, and thus is found after the translational stop codon. Compare to **upstream region**.

driver mutation in cancer cells, one of the mutations that caused the normal cell to become cancerous. Compare to **passenger mutation**.

E

ecotype a genetically distinct population that occupies a particular location and is adapted to these environmental conditions.

effective population a group of organisms occupying a location at a particular time that are capable of reproduction or genetic exchange. The effective population is usually smaller than the total number of individuals, since not all individuals in a population are capable of reproduction.

effector genes in a transcriptional regulatory network, the genes that carry out specific functions in the cell or encode cell-specific proteins. These lie at the lowest level of a eukaryotic transcriptional hierarchy and usually do not include transcription factors that regulate the expression of other genes.

endosymbiosis the process by which a single-celled organism, such as bacteria or archaea, became an inhabitant in the cytoplasm of a eukaryotic cell, often providing essential functions. Mitochondria and chloroplasts each arose as endosymbionts.

enhancer a sequence in the regulatory region of a gene that confers the specificity of transcription for that gene by activating transcription. Enhancer sequences are the binding sites of **transcription factor** proteins, and a gene may have many or few enhancer sequences. Compare to ***cis*-regulatory module**.

epigenetic change a heritable change in the phenotype of an organism that occurs without an underlying change in the DNA sequence. Epigenetic changes usually arise from stable modifications in chromatin. Because there is no change in the DNA sequence, epigenetic changes are often erased in the F_1 generation.

epistasis a genetic interaction between two genes in which the mutant phenotype of one gene masks the phenotype of the other gene. The term is sometimes used broadly to refer to any interaction between genes that affects phenotypes, rather than the more limited use employed here.

euchromatin a cytologically detected region of the chromosome that is less densely packed and thus stains more lightly with DNA dyes. Most protein-coding regions are found in euchromatin. Chromatin that stains more intensely is referred to as **heterochromatin**.

eukaryote an organism in which a nuclear membrane separates the genome from the cytoplasm.

euploidy having complete sets of chromosomes. Compare to **aneuploidy**.

evolutionarily accepted mutation a substitution that is found in the amino acid sequences of orthologous proteins from related organisms. Since the proteins in each species is functional, that particular amino acid change at that location must be compatible with the normal function of the protein, although the proteins may not function identically.

evolutionarily conserved a sequence or a structure that is highly similar in different organisms, indicating that the common ancestor to these organisms had a related sequence or structure. Evolutionary conservation is one indication that changes in the sequence or structure are disadvantageous.

execution point the term used in the analysis of the yeast cell cycle for when a cell carries out a particular cellular event.

exome sequencing the process of isolating and determining the base sequence of all of the **exons** in the genome of an organism.

exon a region in a eukaryotic precursor mRNA molecule that will be found in the final mRNA and includes the segments translated into an amino acid sequence of the polypeptide. Compare to **intron**.

expression profile the complete catalog of which genes are being transcribed in a particular tissue, stage, or environmental conditions. Also referred to as a **transcription profile**. An expression profile is usually obtained from a **microarray** or by **RNA-seq**.

expressivity the phenomenon whereby individuals with the same mutant genotype have somewhat different mutant phenotypes. Expressivity is expressed qualitatively, so that "variable expressivity" means that all of the individuals have a mutant phenotype but the mutant phenotype is not exactly the same.

extrachromosomal DNA any DNA in a cell that is not found as part of the primary chromosome.

F

F factor see **F plasmid**.

F plasmid a circular extrachromosomal DNA molecule in *E. coli* that encodes the functions necessary for **conjugation**. F originally stood for fertility.

feature one of the defined "spots" on a **microarray** slide with many thousands of copies of a particular oligonucleotide sequence. The nucleic acid in the sample is hybridized to the features on the microarray slide.

fitness the relative reproductive success of an individual genotype or phenotype in a population in a particular environment. Fitness reflects the relative contribution of the individual to the gene pool of the subsequent generation.

fitness landscape a plot showing the relative fitness of a population over space or over time, similar in concept to a topological map. The fitness landscape helps to indicate the direction that a population might evolve and directions that are unlikely to occur.

fixation an allele whose frequency is 1.0, so that no alternative alleles are present in the population, except for those few that arise by mutation.

flow cytometer a machine that sorts and counts cells in a sample based on fluorescent labels or other properties such as size or DNA content.

f-met-tRNA the tRNA molecule in bacteria that recognizes the initiator AUG methionine codon in an mRNA. As compared to the regular AUG methionine tRNA, f-met-tRNA has been post-transcriptionally modified by the addition of a formyl group.

founder effect a type of genetic drift that arises because a large population arose from a much smaller original population.

frameshift mutation a mutation that changes the translational reading frame of a protein-coding region, typically by the insertion or deletion of bases, the number of which is not evenly divisible by three.

frequency-dependent selection a form of natural selection in which the fitness of a genotype depends on its frequency in the population, often seen in predator–prey or pathogen–host relationships.

G

G band a cytologically visible band that arises when eukaryotic chromosomes at prophase or metaphase are stained with the DNA dye Giemsa.

gamete a reproductive cell, such as a sperm, pollen, or ovum, which has a haploid chromosome number.

gap gene one of the categories of genes found to govern segmentation in *Drosophila* embryos, so named because mutations in these genes result in embryos with large gaps in their segmentation pattern. The gap genes are generally the ones that act first to specify broad regions of the embryo. Compare to **pair-rule gene** and **segment polarity gene**.

gene conversion a non-reciprocal exchange of a base sequence between non-sister chromatids, arising during the process of recombination.

gene family or protein family Genes that are highly related in DNA sequence and that encode proteins with highly similar amino acid sequences. The proteins have similar, or even identical, functions.

gene pool the total number of alleles and genotypes in a population.

gene regulation the general term for any mechanisms by which the expression of a gene is controlled. Regulation often occurs at the initiation of transcription but can also occur during translation or, in eukaryotes, during splicing or by other processes such as those involving microRNAs.

genetic code the correspondence between particular codons and particular amino acids, which allows the specificity of information to be translated between mRNAs and polypeptides.

genetic diversity the total amount of genetic differences found among individuals in a population of a species.

genetic drift random changes in the frequency of an allele from one generation to the next that arise because the gametes passed on to the next generation in small population sizes are only a sampling of the gametes present in the previous generation.

genetic map a record of the locations of the genes on the chromosomes of a species. Distances between the genes reflect the frequency of recombination that occurs between the genes. Compare to **physical map**.

genetic screen an experimental strategy that finds individuals with mutant phenotypes for some particular biological process. Also known as a **mutant screen**.

genome reduction a process by which a genome becomes smaller and loses DNA sequences during evolution.

genome the entire DNA content and sequence of a species.

genome sequencing see **DNA sequencing**.

genome-wide association study or GWAS a method to identify the genes that contribute to a phenotype by searching the genomes of a large population of individuals for polymorphisms that are consistently associated with the phenotype because of linkage.

genome-wide screen a genetic screen in which genome sequence information is used to design specific molecular tools for the analysis of all or nearly all of the identified genes, without regard to what their mutant phenotype might be.

genomic island a block of horizontally acquired DNA found in a chromosome.

genotype the underlying genetic information of an organism. Ultimately, the genotype is the DNA sequence of the genes in the individual, but it is often represented by alleles or polymorphisms at specific loci.

genotype frequency the fraction of the genotypes in a population that are of a specific type.

growth curve a graph that depicts the cellular divisions or **doubling time** of an organism such as bacteria grown in particular culture medium.

H

haploid a eukaryotic organism or cell with one copy of every chromosome. Compare to **diploid** and **polyploid**.

haplotype polymorphisms in a region of the genome that continue to be inherited together as a unit over several or many generations because they are linked to each other.

Hardy–Weinberg equilibrium the balance in the allele frequencies in a population over time. A Hardy–Weinberg equilibrium is found when mutation, migration, and selection are absent in a large population in which individuals mate at random with respect to their genotypes. At equilibrium, allele frequencies and genotype frequencies do not change from one generation to the next.

hemizygous a diploid in which only one copy of an allele is present at a locus, rather than two. XY males are hemizygous for X-linked alleles, and individuals in which one copy of a gene has been deleted are hemizygous for the remaining allele. Since there is no corresponding allele, a hemizygous allele exerts its effect on the phenotype without regard to dominance.

heritability the fraction or percentage of the variance in a phenotype that can be attributed to genetic differences among the individuals. For example, the heritability of height in human populations is about 0.7, indicating that, within a particular population, approximately 70% of the variation in height is due to genetic differences. Two different terms, calculated by somewhat different methods, are included under the heading heritability. **Broad-sense heritability** includes all types of genotypic variance and is easier to compute, while **narrow-sense heritability** includes only the additional genetic variance and is more helpful in considering the amount the population can be changed by selective breeding.

heteroallelic a heterozygote in which both alleles are mutant, but the mutant alleles do not have the same base change. In human genetics, the equivalent term **compound heterozygote** is more commonly used. An individual who is heteroallelic or a compound heterozygote may not have the same phenotype as a true homozygote in which the two alleles are the same molecularly.

heterochromatin a cytologically detected region of the chromosome that is more densely packed and thus stains more intensely with DNA dyes. Constitutive heterochromatin stains more intensely at all stages of the cell cycle and in all cells, whereas facultative heterochromatin stains more intensely at only certain times in the cell cycle or in only some cells. Chromatin that stains less intensely is referred to as **euchromatin**.

heterogametic the sex that produces gametes with two different haploid sets such as the XY male that make Y-bearing sperm and X-bearing sperm. Compare to **homogametic**.

heterozygote advantage a type of natural or artificial selection that occurs when the most favored genotype is the heterozygote.

heterozygous an individual that has two different alleles for a gene, in particular one functional or wild-type allele and one mutant or altered allele.

high frequency of recombination or HFR a strain of *E. coli* in which the **F plasmid** has integrated into the chromosome. During **conjugation**, part of the F plasmid and some of the bacterial chromosome are transferred to the recipient cell.

histone one of the highly conserved and positively charged proteins that are found tightly associated with DNA in the eukaryotic chromosome. The four core histone proteins in the nucleosome are known as H2A, H2B, H3, and H4, while the linker and somewhat more variable histone is H1.

histone code a hypothesis that many of the functional elements of a genome, such as transcription start sites, origins of replication, and so on, depend on the particular constellation of histone modifications in the chromatin at that location.

holandric a gene found on the Y chromosome and thus found only in males. The term is somewhat archaic.

holobiont the host and its **microbiome** together. Also known as a **metaorganism**.

homeotic gene genes named for a mutant phenotype in which one normal body part is replaced by a different normal body part. For example, mutations in the *Drosophila* homeotic gene *Antennapedia* result in the formation of forelegs where antennae normally form. Homeotic phenotypes arise from mutations that disrupt the developmental pattern of the embryo.

homogametic the sex that produces gametes with the same haploid set such as the XX female that makes only X-bearing ova. Compare to **heterogametic**.

homologous chromosomes the corresponding members of a set of chromosomes in a diploid organism with the same genes and highly similar DNA content, although possibly with different alleles. One chromosome of each pair of homologs is inherited from each parent.

homology the descent of two sequences or structures from a common ancestor. Homology is a qualitative term describing the evolutionary history and, strictly speaking, cannot be expressed quantitatively. Homology is often inferred from sequence similarity, either for the DNA sequences or the amino acid sequences, which can be expressed quantitatively.

homozygous an individual in which the allele on each homolog is the same.

horizontal or lateral gene transfer the transfer of genetic information from one cell or organism to another cell or organism within the same generation.

hybridization used for nucleic acid sequences in which a fragment of DNA or RNA with a particular base sequence forms a double-stranded molecule with a fragment with the complementary base sequence.

I

identical by descent individuals who share an allele or haplotype or DNA sequence because they are descended from a common ancestor.

Illumina sequencing a technology used in **massively parallel DNA sequencing**, named for the machine made by the Illumina company.

inbreeding the mating of two individuals who are genetically related.

incomplete dominance a situation when neither allele in the heterozygote is fully dominant to the other, so that an intermediate phenotype is observed.

indel a contraction of "insertion–deletion," indicating that the genes or genomes of two individuals differ by an insertion or deletion of the sequence at a particular site. "Indel" is used, rather than attempting to distinguish whether the difference is an insertion into one genome or a deletion from the other genome.

inducer a molecule that binds to the **repressor** protein in an operon, so that the repressor can no longer bind to the **operator** sequence and block the transcription of the operon.

inducible expression a gene or set of genes whose transcription is capable of being activated by an environmental or nutritional condition.

informational suppressor a **suppressor mutation** that acts in one of the processes involved in the information flow in the cell such as translation or RNA splicing. Although the term can be used more broadly, it commonly refers to altered tRNA molecules that are capable of recognizing a stop codon, such that translation continues, rather than terminates. See also **nonsense suppressor**.

initiator element a sequence within the core promoter region of a eukaryotic gene that is necessary for the binding and activity of RNA polymerase and the general transcription factors.

integrase a protein that catalyzes **site-specific recombination** between two DNA loci or molecules, thus allowing the integration of one into another at a specific location. A well-studied integrase is that encoded by bacteriophage λ, which mediates its integration into the *Escherichia coli* chromosome.

interaction domain see **activation domain**.

interference the phenomenon by which the presence of one crossover between two homologous chromosomes reduces the frequency of another crossover occurring nearby.

intron a section of a eukaryotic precursor mRNA molecule that will be removed or spliced out to make the final mRNA when the exons are joined together. Originally derived from "intervening sequence" since these sequences are found in the gene or the pre-mRNA between the exons. Compare to **exon**.

J

Jukes–Cantor model a method to detect selection between two or more genomes by comparing the rates of synonymous and non-synonymous substitutions.

K

karyotype the chromosome content of an individual, particularly when the chromosomes have been stained for cytological visibility.

kinetochore the structure on the chromosome through which the microtubules of the spindle attach to the chromosome at the centromere.

L

lac repressor the protein that regulates the expression of the lactose operon in *E. coli* in response to the presence or absence of lactose in the growth medium.

lagging strand during DNA replication, the strand that is synthesized more slowly and in discontinuous Okazaki fragments because it is using the strand that reads 5′ to 3′ at the replication fork as its template. Compare to **leading strand**.

landmark events easily recognized cellular events such as the formation of a mitotic spindle or the formation of a bud in a yeast cell.

Law of Independent Assortment Mendel's "law" that genes located on different chromosomes are inherited independently of each other. See also the **Law of Segregation**.

Law of Segregation Mendel's "law" that the two alleles in a diploid organism separate from each other in the production of gametes. See also the **Law of Independent Assortment**.

lateral gene transfer see **horizontal gene transfer**.

leading strand during DNA replication, the strand that is synthesized more quickly and continuously because it is using the strand that reads 3′ to 5′ at the replication fork as its template. Compare to **lagging strand**.

ligase an essential enzyme during DNA replication that connects the ends of two DNA molecules.

linkage disequilibrium or LD a departure from the phenotypic ratios expected under a Hardy–Weinberg equilibrium because the alleles of the genes are genetically linked. Linkage is the cause of the disequilibrium in frequency, rather than the consequence.

linkage group a set of genes that are inherited together.

locus a term used to refer to the location of a gene on the chromosome, whether from a physical map or a genetic map. Because each gene occupies a characteristic location on the chromosome, locus is often used as a general term for a gene.

long non-coding RNA or long ncRNA a diverse category of transcripts that are longer than 200 bp, transcribed by RNA polymerase II, and capped and processed similar to mRNA transcripts, but which have no sustained open reading frame that could encode a protein. The functions of most long ncRNAs are not known, although they are widely believed to have roles in the regulation of gene expression.

loss-of-function a mutant allele that lacks the normal function of a gene.

lysogeny the ability of certain bacteriophage, such as λ, to integrate their genome into the chromosome of the bacterial host as a **prophage** and to be replicated along with the chromosome without lysing the bacterial host. Phages that are capable of lysogeny are referred to as lysogenic phages.

M

Manhattan plot the nickname for the type of plot used in genome-wide association studies, which shows the relative relationship between the occurrence of particular haplotypes and the presence of the complex-trait phenotype.

massively parallel sequencing the inclusive term for several different high-throughput technologies used to determine the sequence of bases in a DNA molecule or genome. In massively parallel sequencing, also referred to as deep sequencing, high-throughput sequencing, or next-generation sequencing (NGS), each DNA template is sequenced many times by a sequencing machine, in contrast to **dideoxy sequencing** in which a template is sequenced once, originally by hand but now by machines. Many different techniques have been developed by different companies, but familiar examples of next-generation sequencing methods include **454 sequencing** developed by Life Science Technology and Solexa or **Illumina sequencing** developed by Illumina.

mean in statistics, the arithmetic average of a series of n numbers, that is, the sum of the numbers divided by n.

meiosis the nuclear division process in the gonad, by which the diploid number of chromosomes is reduced to the haploid number found in gametes. Meiosis consists of two divisions—meiosis I in which homologous chromosomes separate and the ploidy is reduced, and meiosis II in which sister chromatids separate.

melting temperature for double-stranded nucleic acids, such as DNA, the temperature at which half of the molecules in solution are double-stranded and half are single-stranded.

metagenome a collective of genomes from a community of organisms in an environmental sample.

metaorganism also known as **holobiont**. The host and its microbiome together.

microarray a hybridization-based experimental procedure to examine the transcription of many genes simultaneously. A one-channel microarray labels all of the RNA found in a sample and is particularly used to determine the levels of transcription of different genes. A two-channel microarray uses different labels in different samples and is particularly used to compare transcriptional differences between them. The results from a microarray produce a **transcription profile** or an **expression profile**.

microbial biome the microbes that inhabit a particular environment or ecological niche and that are adapted to it.

microbiome the microbes that inhabit a particular ecosystem.

microRNA or miRNA an RNA product of approximately 22 nucleotides that regulates the expression of another gene by making a double-stranded hybrid with a complementary sequence on the target gene's mRNA. The double-stranded RNA hybrid either targets the mRNA for degradation or blocks its translation.

migration in population genetics, the movement of alleles or gene flow that occurs from one population to another.

minor allele frequency or MAF the frequency within a population of the second most common allele for a gene.

mismatch repair the mechanism that recognizes and replaces mispaired nucleotide base pairs that arise during DNA replication or recombination.

missense mutation a mutation in which one amino acid in the polypeptide sequence is replaced by another amino acid.

microtubule a hollow polymer comprising alternating subunits of α- and β-tubulin that is involved in cell shape, intracellular transport, and cell division as components of the spindle.

mobilizable DNA a genetic element that does not include all the genes necessary to mediate its own transfer from one cell to another but can move between cells if one or more factors is encoded elsewhere in the donor DNA.

model organism a commonly used and well-studied research organism from which inferences can be made about other less studied organisms.

molecular cloning a broad term to refer to the ability to isolate a DNA sequence, such as a gene, and propagate it to a high copy number *in vitro* to allow further analysis.

monogenetic trait a trait that is affected by a single gene. Compare to **polygenic** or **complex trait**.

monophyletic in phylogenetics, a group of organisms that includes an ancestral organism and all of its descendants. Also known as a **clade**. Compare to **polyphyletic**.

monosomy an aneuploid diploid cell that has a single copy of one chromosome, rather than two copies.

morphogen in developmental biology, a diffusible molecule whose concentration regulates the developmental fates or outcomes of a group of cells.

mosaicism having cells of more than one genotype within the same organism.

motif an amino acid sequence that recurs in recognizable patterns when polypeptide sequences are compared to each other. Compare to **domain**.

messenger RNA or mRNA the RNA transcripts within a cell that are translated into an amino acid sequence.

multifactorial trait see **complex trait**

multiple cloning site or MCS see **poly-linker**.

mutagen an agent that induces mutations.

mutant screen a method that finds individuals with mutant phenotypes for some particular biological process. Also known as a **genetic screen**.

mutation a change in the genotype that results in a heritable change in a phenotype.

mutator a phenotype in which the organism has a high frequency of mutations. Often, mutator strains have very active transposable elements. Mutations that affect DNA repair also result in a mutator phenotype.

N

natural selection the fundamental biological principle that allele frequencies change from one generation to the next because one allele or genotype has a greater reproductive ability in a particular environment than another. Selection takes many forms. In **directional selection**, one allele is favored and increases in frequency, while others decrease in frequency. If directional selection continues long enough under the same environmental conditions, the favorable allele will become **fixed**, in that no other allele will be found. Directional selection is also called **purifying selection**, since other alleles are eventually eliminated from the population. **Balancing selection** occurs when the most favorable genotype or phenotype is intermediate. Under balancing selection, more genetic diversity is maintained. **Heterozygote advantage** is one type of balancing selection. **Frequency-dependent selection** arises when the presence of other alleles and genotypes in the population affect which genotype or phenotype is the most favorable.

network biology the field of biology that attempts to characterize a biological system based on the interactions among its parts.

neutral mutation a mutation that has no apparent effect on the phenotype or overall fitness of an organism. Many changes in the third position of a codon do not change the amino acid and are considered neutral mutations.

next-generation sequencing or NGS see **massively parallel sequencing**.

node in phylogenetic trees, the point at which two branches diverge from one another.

non-autonomous element a **transposable element** that cannot move independently but can move if an autonomous element of the same type is present in the genome. Non-autonomous elements, which are generally more common than autonomous elements, usually have mutations that affect the function of transposase.

non-coding RNA or ncRNA a general term for an RNA molecule that does not encode a polypeptide as its functional product. Distinctions are made between **microRNAs** (miRNAs), for which the mature product is 22 nucleotides long, and **long ncRNAs** for which the mature product is more than 200 bases long, and other types of non-coding RNA such as **ribosomal RNA**. The roles of ncRNAs are diverse.

non-disjunction the process by which a pair of homologous chromosomes fails to separate from each other normally during one of the two meiotic divisions, resulting in gametes with too many or too few chromosomes.

nonsense mutation a mutation in which one codon in the mRNA is changed to become a stop codon, thus terminating the translation of the polypeptide prematurely.

nonsense suppressor a mutation in a tRNA molecule that makes it able to recognize a nonsense mutation or a stop codon and insert an amino acid, thus allowing translation to continue. One common type of **informational suppressor**.

non-synonymous substitution a change in the nucleotide sequence that also changes the amino acid sequence.

nucleoid the region in a bacterial cell that contains the chromosome, although it is not physically separated from other cellular components as the eukaryotic nucleus is.

nucleosome the stable core component of eukaryotic chromosomes, comprising a particle of two molecules each of the four core histones H2A, H2B, H3, and H4, and approximately 145 base pairs of DNA wrapped around it.

nucleus the intracellular organelle in a eukaryotic cell that contains the chromosomes, and thus most of the DNA. The nucleus is set off from other cellular components by the nuclear membrane, a lipid bilayer.

nullisomy an aneuploid cell that has no copies of a particular chromosome.

obligate mutualistic symbiosis a type of symbiotic interaction in which the species are dependent on one another, and both benefit.

oligonucleotide a short stretch of DNA or RNA, usually made *in vitro* with a defined base sequence. Informally referred to as an "oligo" and used in many types of experiments that rely on base pairing.

oncogene a gene that is capable of producing a cancerous cell when mutated. The oncogene, also known as the **proto-oncogene** is the wild-type version with normal cellular function, while the mutant version produces cancer.

operational taxonomic unit or OTU used instead of the term "species" when groups of organisms have been defined by their highly similar DNA sequence.

operator a short sequence of bases within, or immediately adjacent to, the promoter in a bacterial **operon** that serves as the recognition site for regulatory proteins such as **repressors** and **inducers**.

operon a set of adjacent genes that are transcribed together as one mRNA and whose transcription arises from a common promoter upstream of the first gene. Operons are common in bacteria and archaea, in which the genes transcribed together are functionally related.

origin of transfer the region of the **F plasmid** or the **HFR** that is the first to be transferred during **conjugation**.

ortholog an evolutionarily and functionally equivalent gene or protein in another species. Equivalence is usually recognized by highly similar nucleotide or amino acid sequences. The human β-globin gene and the mouse β-globin gene are orthologs of one another. Orthologs represent genes that were present in the common ancestor of the species being compared. Compare with **paralog**.

P

pair-rule gene in *Drosophila* development, genes that affect the patterning of pairs of embryonic segments. The expression of pair-rule genes is often regulated by the **gap genes**, and pair-rule genes often regulate the expression of **segment polarity genes**. Pair-rule genes were identified by mutations in which alternating embryonic segments were affected.

paleopolyploidy a doubling or other ploidy increase in a genome that occurred during the evolutionary history of a particular taxonomic group.

pan-genome the core genome of a bacterial species plus all of the DNA sequences that are known to have been acquired by horizontal gene transfer in any isolate of the species.

paralog a member of a **gene family or protein family** within the same species. The human β-globin gene and the human γ-globin genes are paralogs of each other. Paralogs are derived from a common ancestral gene, often by a gene duplication event. Compare with **ortholog**.

paraphyletic in phylogenetics, a group of organisms that share a common characteristic phenotype but excludes organisms derived from the common ancestor that lack the characteristic.

parental ditype among *Ascomycete* fungi, such as *Neurospora* and *Saccharomyces*, an **ascus** that contains two genotypes of **ascospores** or haploid products after meiosis, the same two genotypes as were found in the haploid parents. Asci with two genotypes that are different from the haploid parents are referred to as non-parental ditypes, while asci with spores of four genotypes are **tetratypes**.

passenger mutation in cancer cells, one of the mutations that arose after the cell became cancerous and as a consequence of that, although some passenger mutations are still necessary for the continued division of the cancer cell. Compare to **driver mutation**.

PCR see **polymerase chain reaction**

penetrance the percentage of genotypic mutant individuals that exhibit the associated mutant phenotype. Penetrance ranges from zero to 100%.

permissive temperature the temperature at which a temperature-sensitive or cold-sensitive mutation exhibits a wild-type phenotype. Compare to **restrictive temperature**.

phage see **bacteriophage**.

phage conversion expression of genes beneficial to a bacterial cell that are encoded within a phage genome it carries.

phenocopy a non-heritable mutant phenotype induced by an environmental agent, particularly one that is similar to a phenotype arising from a heritable mutation in a gene.

phenotype any characteristic or quality of a cell or an organism that can be measured using any assay. Phenotypes can have genetic and environmental contributions.

phosphodiester the type of chemical bond that is formed between the 3′ OH on one sugar (ribose or deoxyribose) and the 5′ phosphate on the adjacent one, in a nucleic acid.

photoreactivation the process in bacteria by which UV damage to DNA is repaired. The repair enzymes are activated in response to light.

phylogenetic tree a diagram depicting the relationships among organisms based on similarities and differences in certain characteristics, which allows inferences about their evolutionary histories.

phylogeny the evolution of a group of genetically related organisms.

phylogram a type of phylogenetic tree in which the length of the branches indicates the amount of divergence between the individuals depicted.

phylotype used instead of the term "species" when organisms (especially bacteria) have been defined by highly similar DNA sequence to a reference sample. Compare to **operational taxonomic unit**, which may not include similarity to a reference sample.

physical map a record of the locations of the genes in a species based on the base pair sequence of the genome. Distances on a physical map are measured in base pairs. Compare to **genetic map**.

pilus, conjugative the connection between two bacteria involved in conjugation. The plural is pili.

plasmid a circular DNA molecule that is replicated and propagated independently of the main chromosomes in bacteria and some microbial eukaryotes. Plasmids often contain genes that are necessary for growth under some environmental conditions, but dispensable for growth under other conditions such as antibiotic resistance, or that encode important functions that are not needed under all conditions such as the **F plasmid** involved in **conjugation**.

pleiotropy multiple different phenotypic effects that occur together as a result of a single mutation. The separate phenotypic effects are called phenes, although that term is generally out of use.

ploidy the number of sets of chromosomes in a cell or an individual.

polar body the smaller of the two cells produced by meiosis I in oogenesis, which may complete meiosis II or may degenerate, depending on the species. In humans, pre-natal diagnosis of some genetic conditions can be done by removing the polar body and testing its DNA sequence, a process known as **polar body biopsy**.

polar body biopsy see **polar body**

polar mutation a **nonsense mutation** in an upstream gene in an **operon** that also results in a mutant phenotype for the downstream gene, since translation cannot reinitiate.

polycistronic transcript an mRNA that encodes more than one polypeptide with different amino acid sequences such as those found for an operon. "Cistron" is an archaic term for a gene, so polycistronic transcripts encode more than one gene.

polygenic trait see **complex trait**.

poly-linker in a cloning vector, such as a plasmid, a relatively short nucleotide sequence that has been engineered *in vitro* to contain the recognition sequences for many different **restriction enzymes**, which allows the insertion of DNA fragments with different sequences at that location. Also known as a **multiple cloning site or MCS**.

polymerase chain reaction or PCR a technique to amplify defined regions of DNA by using sequence-specific primers and a thermostable DNA polymerase. The chain reaction occurs because the products of the preceding reaction cycle become the templates for the next reaction cycle. The region of the DNA sequence located between the primers is amplified hundreds of thousands of times, so a very small amount of DNA in the original can yield a much larger amount of PCR product for further analysis and experimentation.

polymorphic locus a locus in which the second most common allele has a frequency of 0.01 or greater, so that the most common allele has a frequency of less than 0.99.

polymorphism genetic variation at a locus. The terms "**allele**," "**mutation**," and "**polymorphism**" have similar meanings but different connotations, with polymorphism being the most inclusive term. "Polymorphism" usually refers to a naturally occurring variation in the DNA sequence which may not result in a detectable phenotypic variation other than the sequence itself. "Mutation" usually implies a genetic change that has been experimentally induced, that occurs rarely, or that substantially alters the phenotype. "Allele" often implies a detectable phenotype difference, which may arise from a natural or experimentally induced variation.

polyphyletic in phylogenetics, a group of organisms that share a common characteristic or phenotype, but the characteristic has been derived independently among species in the group, rather than from the common ancestor.

polyploid species or individuals with more than two full sets of chromosomes.

positional cloning an experimental strategy that uses the location of a gene on the genetic map to identify the DNA sequence of the gene.

position-specific weight matrix in genomics, a scoring system in which specific locations in an alignment are given greater weight or emphasis than other positions, which can be used to identify the positions within a **motif** that are most highly conserved during its evolution. A position-specific matrix is an alternative to a consensus sequence.

post-transcriptional gene silencing the original term used in plants for what has become known more generally as **RNA interference**.

postzygotic see **reproductive isolation**.

pre-initiation complex the large complex of proteins that assembles on the core promoter in eukaryotes, locally denatures the DNA, and positions RNA polymerase II for transcriptional initiation. At least six general transcription factors and RNA polymerase II comprise the pre-initiation complex, but additional ancillary proteins are also often necessary or important.

prezygotic see **reproductive isolation**.

primase an enzyme that is responsible for the synthesis of the first 10–20 nucleotides of an RNA primer molecule at a replication fork. The RNA primer is then used to extend transcription by a DNA polymerase, and the RNA primer made by primase is eventually replaced. Primase is related in sequence to RNA polymerases.

primer a short sequence that is used to initiate replication *in vivo*, or DNA synthesis or transcription *in vivo*.

principal component analysis or PCA a statistical method to cluster data based on different criteria.

proband the first member of a family affected by a genetic disorder to come to the attention of medical professionals or geneticists, and becomes the initial person in the study or pedigree.

probe a nucleotide fragment, typically short and synthesized to include a specific sequence, that is used to find a complementary sequence among a more complex library of DNA, cDNA, or RNA sequences.

product rule in probability, the probability that two independent events each occur as the product of their independent probabilities.

prokaryote an organism with no nuclear membrane to separate the genome from the cytoplasm. These are the true bacteria and the archaea.

promoter the DNA region needed for the transcription of a gene. Often used as **core promoter** or **minimal promoter** to indicate that it is the smallest region necessary for RNA polymerase to bind and transcribe a gene. The term is often used more informally to include not only the core promoter, but also the ***cis*-regulatory module** or the **enhancers**.

proofreading the process by which some types of DNA polymerase remove mismatched bases immediately after incorporation during DNA replication, allowing them to be replaced with the proper base. DNA polymerases that lack proofreading functions are known as error-prone polymerases.

prophage a **lysogenic** phage that has become incorporated into the bacterial chromosome.

proto-oncogene see **oncogene**.

prototroph a strain that can grow in the absence of an added nutrient because it has the wild-type function to synthesize that nutrient. The opposite of prototroph is **auxotroph**.

provirus a viral genome that has become integrated into the chromosome of its host. See also **prophage**.

pseudoautosomal region the region of the Y chromosome that can pair and exchange with the X chromosome during meiosis. Since sequences in this region are present on the Y chromosome, as well as the X chromosome, their inheritance appears to be similar to autosomal sequences despite their location on the Y chromosome.

pseudogene a member of a gene family that is non-functional because of mutation.

pure culture a culture of microbes or bacteria that consists of genetically identical or nearly identical cells.

purifying selection a type of **natural selection** that decreases the frequency of a particular allele or genotype from a population. Also known as **directional selection.**

purine an organic molecule consisting of two rings, which makes up part of the nucleotide bases that comprise the structure of nucleic acids. The naturally occurring purines in DNA and RNA are adenine and guanine.

pyrimidine an organic molecule with a single ring, which makes up part of the nucleotide bases that comprise the structure of nucleic acids. In DNA, the pyrimidines are cytosine and thymine, whereas in RNA, the pyrimidines are cytosine and uracil.

Q

quantitative trait see **complex trait.**

quantitative trait locus or QTLs one of several or many genes that contribute to the phenotype of a complex or quantitative trait.

R

rarefaction curve a plot of the number of different species that have been found on the Y-axis, compared to the number of samples on the X-axis. The curve's shape indicates the extent to which continued sampling is likely to identify additional species.

receptor tyrosine kinase a type of protein found in many intracellular signal transduction pathways. Receptor tyrosine kinases can be either monomers or dimers. The protein spans the cell membrane and interacts with polypeptides or small-molecule ligands on its extracellular domain. The binding of the ligand results in the phosphorylation of a tyrosine residue on the intracellular portion of the protein, as well as the phosphorylation of downstream target molecules. The phosphorylation triggers a signaling cascade, which results in the passage of the signal from the extracellular ligand to nuclear transcription factors to alter gene expression in response.

recombination any of several processes by which the genetic information in one generation is combined in new combinations in the succeeding generations. The processes include fertilization, independent assortment, horizontal gene transfer, and crossing over during meiosis, but recombination is often used to refer to crossing over.

recombination map see **genetic map.**

Red Queen Hypothesis an evolutionary hypothesis that, among **co-evolving** species, changes in one species exert selective pressures on the other species to change as well. Named for the Red Queen in *Lewis Carroll's Through the Looking Glass* who has to run faster to stay in place.

redundancy **robustness** that arises from duplicate nodes with the same function in a network.

regulatory region the entire DNA sequence necessary for the gene to be transcribed in the proper tissue at the proper time and in the proper amount. The regulatory region includes all of the enhancer and repressor sequences. For most genes, the regulatory region is found upstream of the core promoter, but many exceptions are known in which the regulatory region includes sequences downstream of the gene or internal to the coding sequence, such as intronic sequences.

regulon the complete set of genes controlled by a specific activator or repressor.

renaturation the re-formation of double-stranded nucleic acids from the separated single strands in solution.

replication the natural process by which DNA is copied to make a second nearly identical copy, by DNA polymerases and associated proteins. The related term DNA synthesis is applied more broadly to refer to both *in vivo* and *in vitro* processes, of which DNA replication is one.

replication fork the structure that forms by localized denaturation of the double helix at the origins of DNA replication, allowing replication to occur for both strands simultaneously. Because the origin is a denatured location, which is visible by electron microscopy, this is also referred to as a replication bubble.

replication machine the complex of proteins that carry out DNA replication, including DNA polymerase and the associated other proteins.

reporter gene a molecular construct that has the coding region of some gene whose product is easily assayed fused to the control region of the gene of interest. Among the proteins commonly used as reporters are β-galactoside and green fluorescence protein (GFP). Reporter genes can be either transcriptional or translational. A transcriptional reporter gene replaces most or all of the coding region of the gene of interest with the coding region of the reporter gene. A translational reporter gene places the coding region of the reporter gene in frame with the normal protein, so that a fusion protein is made between the normal protein and the reporter protein.

repressed a gene whose transcription has been shut off in response to some environmental or nutritional condition.

repressible a gene or set of genes whose transcription is capable of being repressed or shut off by an environmental or nutritional condition.

repressor a protein that turns off the transcription of genes in an operon. The *lac* repressor is a particularly well-studied example.

reproductive isolation the inability of populations to exchange genetic information with one another to produce viable and fertile offspring. Many different mechanisms of reproductive isolation are known. Mechanisms that prevent mating or fertilization are referred to as prezygotic, while the ones that prevent the embryo from mating into a reproductive adult are referred to as postzygotic.

restriction endonuclease a type of enzyme that cuts double-stranded DNA molecules at a specific sequence, each restriction endonuclease having its own recognition sequence and site for making the cut. Most restriction endonucleases are encoded by bacteria, with the first three letters of their name coming from the bacterium from which they are isolated, e.g. *Eco*RI from *Escherichia coli*.

restriction fragment length polymorphism or RFLP a type of genetic sequence variation in a genome that is recognized because the differences between individuals can be detected using **restriction endonucleases**.

restrictive temperature the temperature at which a **temperature-sensitive** or cold-sensitive mutation exhibits a mutant phenotype. Compare to **permissive temperature**.

retrovirus a virus with an RNA genome that encodes **reverse transcriptase** among its products, which converts the RNA genome into a DNA sequence.

reverse transcriptase an enzyme that can use an RNA template to make a complementary DNA sequence. Reverse transcriptases are encoded in the genome of **retroviruses** and are widely used *in vitro* to make **cDNA**.

ribosome the intracellular organelle at which translation occurs. The ribosome has two subunits, each of which has numerous characteristic proteins and several characteristic RNA molecules, called ribosomal proteins and **ribosomal RNA**.

ribosomal RNA or rRNA any of the RNA molecules that comprise structural and functional components of the ribosome and are involved in translation. Bacteria have three ribosomal RNA molecules, while eukaryotes have four.

RNA interference or RNAi a widely used experimental technique to reduce the expression of a gene in eukaryotes, using double-stranded RNA corresponding to part of the coding region of the gene. RNAi produces the equivalent of a mutant phenotype without altering the DNA sequence for the gene, using the same cellular machinery as **microRNA** molecules.

RNA sequencing or RNA-seq a high-throughput technique to determine the sequence and abundance of all of the RNA molecules in a cell or tissue. Although the method is called RNA-seq, the RNA is first converted to **cDNA**, and the cDNA is sequenced by one of the **massively parallel sequencing** methods.

robustness the ability of a system or network to maintain its function and overall structure when random parts are deleted.

root in phylogenetics, an individual or species that is distantly related to all of the individuals or species whose evolutionary history is being displayed by the phylogenetic tree. The distance from the root species provides a reference point to anchor the rest of the tree.

S

saturation a situation in which all of the genes that could be identified by a particular screen have been found.

secondary oocyte the larger of the two cells produced by meiosis I in oogenesis, which will go on to complete meiosis II to generate the ovum.

segment polarity gene in *Drosophila* development, genes that affect the patterning of structures within each embryonic segments. The expression of segment polarity genes is often regulated by the **pair-rule genes**. Segment polarity genes were identified by mutations in which the orientation or formation of each segment was altered, but the overall patterning of segmentation was not affected.

selectable marker a wild-type or mutant gene that is present in the genotype of some laboratory isolates to allow them to grow. For example, if the gene for kanamycin resistance is present on a plasmid and kanamycin is included in the growth medium, the kanamycin resistance gene is the selectable marker.

selection as applied to a **genetic screen**, a screen in which only certain genotypes can grow. For example, a growth medium containing the antibiotic kanamycin can be used to select for mutations or strains that are kanamycin-resistant.

selection differential the difference between the fitness of one particular phenotype or genotype and the average fitness of the population.

self-transmissible a conjugative plasmid, like the **F plasmid**, that encodes the genes that allow it to be passed to a recipient cell, with no additional genes necessary.

self-transmissible DNA a genetic element that includes all the genes necessary to mediate its own transfer from one cell to another.

semi-conservative a description of DNA replication by which the double-stranded molecule replicates to produce two daughter molecules, each of which has one parental or original template strand and one newly synthesized strand.

sequencing read the outcome of a sequencing reaction, referring primarily to the length of the sequence that is produced. **Massively parallel sequencing** methods result in shorter sequencing reads than **dideoxy sequencing**.

sex chromosomes the chromosomes that differ between the two sexes such as the X chromosome and the Y chromosome in mammals. The other chromosomes that do not differ between the sexes are called the **autosomes**.

sexual dimorphism the visible differences between the phenotypes of the two sexes in a species.

shotgun sequencing a strategy used in genomic sequencing and assembly in which genomic fragments are sequenced at random, without prior assembly of the genomic fragments. The sequences are then aligned computationally against one another to determine regions of overlap.

single-nucleotide polymorphism or SNP genetic variation that arises from a change in a single nucleotide at a locus. SNPs (pronounced "snips") are very common in natural populations and yield some of the most useful markers for association mapping and **genome-wide association studies.**

sister chromatids the two chromatids of a eukaryotic chromosome that arise immediately after DNA replication and that separate from one another at mitosis or meiosis II. The DNA molecules in sister chromatids, which are the products of the most recent round of DNA replication, are nearly identical to one another.

site-specific recombination crossing over between two different DNA molecules in which the double-stranded break that initiates recombination occurs at a specific nucleotide sequence in the two molecules. In addition to the specific sequences in the two molecules, the recombination requires a specific enzyme that recognizes and cuts at these sites.

SOS response a global response by cells to DNA damage by which the cell cycle is arrested and DNA repair is initiated.

speciation any of the processes by which new species arise from an ancestral species.

spliceosome the complex of proteins and small RNA molecules that carry out the splicing and alternative splicing of pre-mRNA molecules to produce mRNA in eukaryotes.

sporulation in fungi, the process by which haploid ascospores form from diploids in ascomycete fungi such as budding yeast and *Neurospora*. Sporulation in fungi is essentially the same as meiosis, although the haploid products can divide mitotically. The term is also used for some types of bacteria that can convert to a dormant and resistant cell-type.

sRNA one of the small RNA molecules that arises from the repeat sequences in a **CRISPR** array that help to guide Cas proteins to specific sequences.

standard deviation in statistics, the square root of the variance, and thus a measure of the differences between the individual data points and the mean of the population.

strand invasion the process during DNA recombination by which a single strand from one DNA molecule becomes the template for the synthesis of part of the other DNA molecule, thereby producing a recombinant.

supercoil a structure that arises from the winding of a closed circular DNA molecule such as a plasmid or a bacterial chromosome. Supercoils present challenges for DNA replication and are typically removed by the actions of topoisomerases.

suppression A genetic interaction between mutant alleles of two genes in which the mutation in the second gene, known as the **suppressor mutation**, overcomes or reverses the mutant phenotype of the first gene.

suppressor mutation a mutation in a second gene that overcomes the phenotypic effects of another mutation. As a result of a suppressor mutation, the phenotype of the double mutant more closely resembles wild-type than does the phenotype of the single mutation. Compare with **synthetic enhancer** mutation.

symbiotic interactions or symbiosis a general term for the interaction between members of two different species, typically used when both species derive benefits from the interactions.

sympatric speciation species that arise from an ancestral species within the same environments. Compare to **allopatric speciation.**

syncytium a group of nuclei that are found together in a common cytoplasm, usually because nuclear division has occurred without subsequent cell division or because separate cells have fused after formation.

synonymous substitution a change in the nucleotide sequence that does not change the amino acid sequence of the polypeptide that is encoded.

synteny a general term for genes that are located on the same chromosome. Genes may be syntenic but sufficiently far apart on the chromosome that they segregate independently due to recombination and thus are not linked in their inheritance. Synteny is important in comparing the genomes of closely related species because many linkage relationships have been preserved in evolution. A group of genes located together in two different species comprise a **syntenic block.**

synthetic enhancement A genetic interaction between mutant alleles of two genes in which the mutation in the second gene, known as the synthetic enhancer mutation, exacerbates or intensifies the mutant phenotype of the first gene. One common form is synthetic lethality, in which an individual that is mutant for either one of the two genes is viable, but a mutant in both genes is inviable.

synthetic enhancer mutation a mutation in a second gene that makes the phenotype of another mutation more severely mutant. In particular, each mutation alone has a wild-type phenotype, but the double mutant has a mutant phenotype. Compare with **suppressor mutation.**

T

telomerase the enzyme with both an RNA and a protein component that is responsible for DNA replication at the ends of linear eukaryotic chromosomes.

telomere the specialized structure at the ends of eukaryotic chromosomes. Telomeres, which comprise many copies of a short DNA sequence, prevent the fusion between the ends of chromosomes and are involved in the replication at the end of the chromosome.

temperate phage a type of phage that enters a bacterial cell and replicates within the cell but does not lyse the cell upon exit. An example is phage M13.

temperature-sensitive mutation a mutation that has a mutant phenotype at a slightly elevated temperature but a wild-type phenotype at a lower growth temperature. The temperature that exhibits the wild-type phenotype is known as the **permissive temperature.** The temperature that exhibits the mutant phenotype is known as the **restrictive temperature.**

template the strand of DNA that is used to make a copy of the other strand, based on complementary base pairing. Each strand serves as a template during DNA replication, while one strand is the template for transcription; the other strand is the **coding strand**.

test cross a mating to an individual or a strain that is homozygous recessive for all alleles being investigated. Since the homozygous recessive parent has no dominant alleles, all of the alleles in the other parent are revealed by the phenotypes of the offspring.

tetrad the four haploid products of a single meiotic division. In yeast and other fungi, these four haploid spores are found together in the **ascus**. An analogous use refers to the four chromatids present when homologous chromosomes are paired during prophase I of meiosis.

tetratype an ascus that has ascospores of four different genotypes.

thymidine (or pyrimidine) dimer the covalent linkage of adjacent thymidine residues on the same DNA strand. Thymidine dimers are induced by ultraviolet light and repaired in bacteria by **photoreactivation**.

topoisomerase a class of enzyme that unwinds **supercoils** in DNA molecules. There are several different classes of topoisomerases, which unwind supercoiled DNA by different mechanisms.

transcription the process by which an RNA molecule is made from a DNA template.

transcription factor a protein that binds to **enhancer** or repressor sequences to regulate transcriptional initiation of a gene in eukaryotes. The term "transcription factor" is used to refer to the proteins that bind to specific DNA sequences to regulate certain genes at specific times, while the term "general transcription factor" is used for the proteins that bind in the promoter region and are required for the transcription of all genes regulated by that RNA polymerase.

transcription profile the catalog of the transcriptions made by an organism in a particular tissue, stage, and environment. Since the expression of most genes is regulated at transcription, transcription profiles are also known as **expression profiles**. Compare with **transcriptome**.

transcriptional start site or TSS in a eukaryotic gene, the region upstream of a gene where transcription begins.

transcriptome the complete catalog of all of the transcripts that are made by an organism, regardless of the tissue, stage, or environmental conditions. Compare with **transcription profile.**

transduction the horizontal transfer of DNA from a host cell to a recipient cell, mediated by a virus. Transduction is particularly used to describe virus-mediated transfer in bacteria. **Generalized transduction** occurs when any portion of the bacterial genome can be transferred, whereas **specialized transduction** occur when only certain portions of the bacteria genome can be transferred.

transfection the process of experimentally introducing DNA into eukaryotic cells, often by using a physical method such as applying an electric current (electroporation) or a chemical method such as applying calcium phosphate.

transfer RNA or tRNA a non-coding RNA molecule of approximately 120 nucleotides that is responsible for recognizing the codon of mRNA and inserting a specific amino acid during translation.

transformation the process by which bacterial cells take up and incorporate DNA from their environment, an example of horizontal gene transfer. Transformation occurs naturally in some bacteria, but it is often used as a laboratory procedure to introduce plasmids or other DNA into bacteria. For this to occur, the laboratory species has to be made competent, usually by a chemical method.

transition a base substitution in which one purine replaces the other purine or one pyrimidine replaces the other pyrimidine. Compare with **transversion**.

translation the process by which the amino acid sequence of a polypeptide is synthesized using the base sequence of an mRNA.

transposable element a discrete DNA sequence element that can move from one location to another in the genome, found in both prokaryotes and eukaryotes. In bacteria, transposable elements include insertion sequences, which encode the transposable enzyme responsible for their movement, and transposons, larger elements that have insertion sequences as their ends and also encode other genes, particularly for antibiotic resistance. In eukaryotes, class I transposable elements or retrotransposons move via an RNA intermediate and are derived from **retroviruses**. Class II elements encode **transposase** and move via a DNA intermediate.

transposase a class of enzyme encoded within an insertion sequence or a class II **transposable element** in eukaryotes that catalyzes the excision and insertion of that transposable element. The transposase from one type of transposable element will not affect other types of transposable elements but will affect other elements of some type, regardless of their location in the genome.

transversion a base substitution in which a purine replaces a pyrimidine. Compare with **transition**.

trisomy an aneuploid diploid cell that has three copies of a chromosome, rather than two copies.

true breeding a variety or a strain that yields offspring that are phenotypically and genetically like itself. Most true-breeding varieties are homozygotes.

tumor suppressor a type of gene that suppresses the development of tumors in living cells. Tumor suppressor genes are often the ones whose normal functions are involved in the cell cycle and DNA replication and repair checkpoints. Mutations in tumor suppressor genes that result in a loss of gene function thus result in mis-regulated cell division or replication and repair, leading to the formation of tumors.

U

unequal crossing over a type of recombination that occurs between homologous chromosomes when the regions of the chromosomes are similar enough in sequence composition that local misalignment of the homologues can occur. It is referred to as unequal because the homologous chromosomes arising from recombination in such a misaligned region will have deletions and duplications with respect to each other.

upstream region the portion of the DNA sequence of a gene that is located to the 5′ side, that is, before the ATG translational start codon and the transcriptional start site if the sequence is viewed with the 5′ end on the left. Compare to **downstream region.**

UTR or untranslated region a sequence that is present on the mRNA but is not translated into part of the amino acid sequence. The 5′ UTR is located upstream of the AUG start codon, while the 3′ UTR is located downstream of the stop codon.

V

variance in statistics, a measure of the differences between individual data points and the mean of the population.

vegetative growth also known as **asexual growth.** Cell division that occurs without meiosis and recombination such as by binary fission or mitosis.

vertical gene transmission the passage of genetic information from one generation to the next, either in cells or in organisms.

W

wild-type for laboratory organisms, the standard isolate or strain to which all others are compared.

wobble applied to the base pair interactions between the codons of mRNA and the anticodons of tRNA. The third or 3′ position in the codon (and thus the most 5′ position in the anticodon) forms the least stable base pair and is usually less important in the specification of an amino acid than the other two positions.

X

X chromosome inactivation the process by which **dosage compensation** occurs in mammals. Female embryos transcriptionally inactivate one of their two X chromosomes in each cell, so that males and females make equivalent amounts of products from X-linked genes. Either X chromosome can be inactivated, but it remains inactivated during subsequent mitotic divisions.

X-linked or sex-linked genes located on the X chromosome. Because many genes map to the X chromosome and relatively few genes map to the Y chromosome, the terms sex-linked and X-linked are used interchangeably.

INDEX

E

F

G

H

T

U

V

W

X

Y

Z